Lecture Notes in Computer Science 4123

Commenced Publication in 1973
Founding and Former Series Editors:
Gerhard Goos, Juris Hartmanis, and Jan van Leeuwen

Editorial Board

Rudolf Ahlswede
Lars Bäumer   Ning Cai
Harout Aydinian   Vladimir Blinovsky
Christian Deppe   Haik Mashurian (Eds.)

General Theory
of Information Transfer
and Combinatorics

 Springer

Volume Editors

Edited by:
Rudolf Ahlswede
Universität Bielefeld, Fakultät für Mathematik
Universitätsstr. 25, 33615 Bielefeld, Germany
E-mail: ahlswede@math.uni-bielefeld.de

Assisted by:
Lars Bäumer
Ning Cai
Universität Bielefeld, Fakultät für Mathematik
Universitätsstr. 25, 33615 Bielefeld, Germany
E-mail: {baeumer,cai}@math.uni-bielefeld.de

In cooperation with:
Harout Aydinian
Vladimir Blinovsky *
Christian Deppe
Haik Mashurian
Universität Bielefeld, Fakultät für Mathematik
Universitätsstr. 25, 33615 Bielefeld, Germany
E-mail: {ayd,cdeppe,hmashur}@math.uni-bielefeld.de

* Russian Academy of Sciences
Institute of Problems of Information Transmission
Bol'shoi Karetnyi per. 19, 101447 Moscow, Russia
E-mail: blinov@postman.ru

Library of Congress Control Number: 2006937883

CR Subject Classification (1998): F.2, G.2-3, C.2, G.1.6, E.3-5

LNCS Sublibrary: SL 1 – Theoretical Computer Science and General Issues

ISSN   0302-9743
ISBN-10   3-540-46244-9 Springer Berlin Heidelberg New York
ISBN-13   978-3-540-46244-6 Springer Berlin Heidelberg New York

Springer is a part of Springer Science+Business Media

springer.com

© Springer-Verlag Berlin Heidelberg 2006
Printed in Germany

Typesetting: Camera-ready by author, data conversion by Scientific Publishing Services, Chennai, India
Printed on acid-free paper   SPIN: 11889342   06/3142   5 4 3 2 1 0

Preface

The Center for Interdisciplinary Research (ZiF) of the University of Bielefeld hosted a research group under the title "General Theory of Information Transfer and Combinatorics," abbreviated as GTIT-C, from October 1, 2001 to September 30, 2004. As head of the research group the editor shaped the group's scientific directions and its personal composition.

He followed ideas, problems and results which had occupied him during the past decade and which seem to extend the frontiers of information theory in several directions. The main contributions concern information transfer by channels. There are also new questions and some answers in new models of source coding. While many of the investigations are in an explorative state, there are also hard cores of mathematical theories. In particular, a unified theory of information transfer was presented, which naturally incorporates Shannon's Theory of Information Transmission and the Theory of Identification in the presence of noise as extremal cases. It provides several novel coding theorems. On the source coding side the concept of identification entropy is introduced. Finally, beyond information theory new concepts of solutions for probabilistic algorithms arose.

In addition to this book there will be a special issue of *Discrete Applied Mathematics* "General Theory of Information Transfer and Combinatorics" in three parts, which covers primarily work with a stronger emphasis on the second component, combinatorics. It begins with an updated version of "General Theory of Information Transfer" in order to make the theory known to a broader audience and continues with other new directions such as bioinformatics, search, sorting and ordering, cryptology and number theory, and networks with many new suggestions for connections.

It includes in a special volume works and abstracts of lectures devoted to the great Levon Khachatrian at the memorial held for him during the Opening Conference, November 4-9, 2002.

In a preparatory year, October 1, 2001 – September 30, 2002, guided by the general concepts and ideas indicated and described in greater detail in the present introduction, researchers and research institutions were approached worldwide in order to find out which possible participants might be and which more concrete projects could be realized in the main research year, October 1, 2002 to August 31, 2003.

Central events in this phase were two weekly preparatory meetings in February: General Theory of Information Transfer, abbreviated as GTIT, and Information in Natural Sciences, Social Sciences, Humanities and Engineering. Abstracts of the lectures can be found at
http://www.math.uni-bielefeld.de/ahlswede/zif.

The main goals were to test the applicability of the GTIT, particularly identification, and to strive for new information phenomena in the sciences, which can be modelled mathematically. Readers are strongly advised to read the Introduction for guidance.

Our special thanks go to the members of the administration of the "Zentrum für interdisziplinäre Forschung" (ZiF) in Bielefeld for a very pleasant cooperation and, in particular, to Gertrude Lübbe-Wolf, who as acting director authorized and generously supported this project, and to Ibke Wachsmuth, who continued her policy. Dr. Roggenhöfer, who was always responsive to new ideas and wishes is also thanked for his assistance.

June 2006   Rudolf Ahlswede

Table of Contents

I Probabilistic Models

V Information Measures – Error Concepts – Performance Criteria

VI   Search – Sorting – Ordering – Planning

VII   Language Evolution – Pattern Discovery – Reconstructions

VIII Network Coding

IX Combinatorial Models

Coverings

Partitions

Isoperimetry

Addresses of Contributers to the Project as well as Titles and Abstracts of their lectures delivered at two Preparatory Meetings, the Opening Conference, the Final Meeting and Seminars the Reader can find at http://www.math.uni-bielefeld.de/ahlswede/zif

More than restoring
strings of symbols transmitted
means transfer today

Rudolf Ahlswede

Introduction

The fundamental problem of communication is that of reproducing at one point either exactly or approximately a message selected at another point.

Claude E. Shannon

What is information?

$\boxed{Cn \text{ bits}}$ in Shannon's fundamental theorem

or

$\boxed{\log Cn \text{ bits}}$ in our Theory of Identification ?

Among the directions of research in GTIT-C, which could be established, we present in Chapters I-IX contributions of participants, which took shape in written form. The papers were thoroughly refereed. For the ease of reference they are numbered and in addition labelled by the letter B, which hints at this book.

There are other lists of publications we refer to. They are labelled by A,C, and D, where

R. Ahlswede et al. (Eds.): Information Transfer and Combinatorics, LNCS 4123, pp. 1–44, 2006.
© Springer-Verlag Berlin Heidelberg 2006

A indicates the editors present list of publications at the end of the book

C is used for papers mentioned in the preface, in particular in these comments

D gives the titles of the papers in the related Special Issue in Discrete Applied Mathematics (3 volumes)

It must be emphasized that there were several lectures and discussions with many ideas, which influenced work, which appears elsewhere. Also, several subjects with noticeable informational components are not ready for a theoretical and in particular mathematical frame.

For instance in **Bio-Chemistry** there are many advanced experiments concerning "signalling", where some kind of intensity is measured, but essentially nothing is known about transmission rates. Still we learnt a lot from several very stimulating lectures of Raimund Apfelbach about the "state of the art". Moreover, we are particularly thankful for his engagement in helping to shape a group for **Animal Communication** beyond his own work related to smelling.

In many lectures, especially also by young German emigrants to Great Britain, we were excited by an extremely rich and very active field in recent years. There are fascinating ways of encoding information (well-known a long time for bees and more recently for ants etc). Mathematical models are more in a beginning state. Our hope that in the kingdom of animals nature has chosen other forms of information transfer, like in identification, could not be verified. It seems that knowledge of these forms is a necessary prerequisite for setting up suitable experiments.

At least we know where one may start. Several abstracts from the April 2004 Final Meeting the reader can find at http://www.math.uni-bielefeld.de/ahlswede/zif. We also mention two instructive books [C10], [C49]. With advanced technology many new experiments have become possible here.

Similar observations can be made about **Psychology**. However, there have been signals encouraging careful reading, intuitive perception or enjoyment of G. Dueck's book [C25] with its classification of people into "right", "true" or "natural" and corresponding different key roles identification seems to play for them.

A last statement in this context we want to make about **Philosophical Ideas** concerning *information concepts and measures*, in particular about K.F. von Weizsäcker's monumental *attempt* [C100] to base all of Physics on an information concept – central even for time and space. We found that the newly discovered concept of *identification entropy* [B33], which has an unquestionable operational meaning, *is not covered by that approach and therefore challenges it!* Philosophers are called again to share a world with mathematicians and vice versa.

We hope to have made understandable why we did not receive contributions to Bio-Chemistry, Animal Communication, Psychology and Philosophy for this book.

There is, however, for instance a contribution to **Language Evolution** [B46], in which a **Conjecture of Nowak**, the man who has been running a five year project at the Institute for Advanced Study in Princeton on language evolution, was fairly generally established.

There are also contributions to **Genomic and Chemical Structure Theory**.

In **Engineering** a group of practical people working on the construction of alarm systems incorporating incentives from the theory of identification started to take shape, when sadly, the leading person, Sandor Csibi, unexpectedly passed away and there was no one who could take his role in combining the necessary engineering and mathematical know-how. We are very grateful to Sandor for devoting his last work, which he had pursuit with great enthusiasm, to this book.

In the course of the main research year (October 1, 2002 – August 31, 2003) more than 100 scientists from various disciplines were guests at the Center for Interdisciplinary Research and took part in the research group's activities. About 45 of them could be won for a longer stay (more than 1 month). Furthermore, permanent opportunities for conversations, work on joint papers and especially the research group's seminars held twice a week with all guests gave a frame for stimulating transfer of knowledge and ideas through lectures and discussions. The themes of the seminar talks were assembled into blocks along main directions of research, normally two in parallel.

The year ended with the "Wunschkonzert" symposium in August 2003, where speakers were invited with a prescribed title for their lecture, and two weeks as the beginning of intense efforts to solve problems, which came up and which were continued by the local Bielefeld group in the last year of the project and joined with the efforts of others at the final conference in April 2004, where every morning experts from typically one of seven groups we had established gave lectures. For the rest of the days (and nights) there were three rooms and a forest available for problem sessions and hikes. All invited speakers provided abstracts. Additional talks came from attendance, who came to us "in the last minute" and managed to present their work in problem sessions.

We now outline the general concepts and ideas leading to the project, driven by great expectations.

The Pre-Socratic Philosopher Protagoras is reported to have said (in our free translation) "Nothing exists. If something would exist, it could not be understood. If it could be understood, it could not be communicated" ([C22]).

These are perhaps the most basic three forms of nihilism. Kant was primarily concerned with the second and Wittgenstein with the third form.

Communication and thus also information has become a prime issue in the past century (For insightful description and assessments we recommend E.C. Cherry [C13] and J.R. Pierce [C74]).

The main credit for the advancements goes to the electrical engineers. The highlight was Shannon's Statistical Theory of Communication. Quite remarkably, Shannon insisted that his theory was intended to serve engineering purposes and was somewhat skeptical about attempts to apply it in others areas or at least asks for great care ([C85]). Still, we take the position that he was too modest. This conforms with C. Cherry [C12], from whom we quote some enlightening statements.

In the Appendix of [C12] one finds the following definition or explication of the concept "communication":

"The establishment of a social unit from individuals, by the shared usage of language or signs. The sharing of common sets of rules, for various goal–seeking activities. (There are many *shades of opinions*.)" We like to add that this has by now taken more concrete forms: Embodied Communication. Again in [C12] on page 41 we read:

"Perhaps the most important technical development which has assisted in the birth of communication theory is that of telegraphy. With its introduction the speed of transmission of "intelligence" arose. When its economic value was fully realized, the problems of compressing signals exercised many minds, leading eventually to the concept of "quantity of information" and to theories of times and speed of signalling" and on page 43:

"Hartley went further and defined information as the successive selection of signs or words from a given list, rejecting all "meaning" as a more subjective factor (it is the signs we transmit, or physical signs; we do not transmit their "meaning"). He showed that a message of N signs chosen from an "alphabet" or code book of S signs has S^N possibilities and that the "quantity of information" is most reasonably defined as the logarithm, that is, $H = N \log S$."

This concept of information is closely related to the idea of selection or discrimination and therefore sometimes called selective–information. It is also at the very basis of Shannon's statistical theory of communication.

This theory has by now been developed into a sophisticated mathematical discipline with many branches and facets. Sometimes more concrete engineering problems led to or gave the incentive to new directions of research and in other cases new discoveries were made by exploring inherent properties of the mathematical structures. Some of our views on the state of this theory, to which we also shall refer as the "Shannon Island", are expressed in [A35] and [A36].

The price for every good theory is simplification and its permanent challenge is reality.

"We live in a world vibrating with information" and in most cases we don't know how the information is processed or even what it is at the semantic and pragmatic levels. How does our brain deal with information? It is still worthwhile to read von Neumann's ideas about this [C98].

Cherry writes in [C12]:

"It is remarkable that human communication works at all, for so much seems to be against it; yet it does. The fact that it does depends principally upon the vast store of habits which one of us possess, the *imprints of all our past experiences*. With this, we can hear snatches of speech, the vague gestures and grimaces, and from this shreds of evidence we are able to make a continual series of inferences, guesses, with extra ordinary effectiveness."

We shall come to the issue of "prior knowledge" later and just mention that some aspects are accessible to a rigorous mathematical treatment.

There are various stimuli concerning the concepts of communication and information from the sciences, for instance from quantum theory in physics, the theory of learning in psychology, theories in linguistics, etc.

These hints give an idea of the size of the ocean around the Shannon Island.

We don't have the intention to drown in this ocean. However, since the ocean is large there ought to be some other islands. In fact there are.

Among those, which are fairly close to the Shannon Island we can see for instance

1.) Mathematical Statistics
2.) Communication Networks
3.) Computer Storage and Distributive Computing
4.) Memory Cells

Since those islands are close there is hope that they can be connected by dams.

A first attempt to explore connections between multi–user source coding and hypothesis testing was made in [A51] (IEEE Best paper award 1989). For interesting ideas about relations between multiple–access channels and communication networks see Gallager [C38]. A multitude of challenges to information theory comes from computer science. A proper frame for storage in memory cells is our abstract coding theory [A35],[A36].

However, a real step beyond the founding father Shannon was made with our creation of a theory of identification in the presence of noise. Randomization in the encoding was a key idea! The mathematical formulations were carried out together with G. Dueck (IEEE Best paper award 1991) and continued by many others (Anantharam, Bassalygo, Bäumer, Burnashev, Cai, Csibi, Csiszár, Han, Kleinewächter, Löber, Merhav, Narayan, Shamai, Steinberg, van der Meulen, Venkatesan, Verboven, Verdu, Wei, Winter, Yang, Yeong, Zhang, ...).

To fix first ideas, **Transmission (classical) concerns the question** "How many messages can we transmit over a noisy channel?" One tries to give an answer to the question "What is the actual message from $\mathcal{M} = \{1,\ldots,M\}$?"

On the other hand **in Identification it is asked** "How many possible messages can the receiver of a noisy channel identify?" One tries to give an answer to the question "Is the actual message i?"

Here i can be any member of the set of possible messages $\mathcal{N} = \{1,2,\ldots,N\}$.

Certain error probabilities are again permitted. **From the theory of transmission one cannot derive answers for these questions in the theory of identification!**

Actually, this theory initiated other research areas like **Common Randomness, Authentication in Cryptology, Alarm Systems**. It also led to the discovery of new methods which become fruitful also for the classical theory of transmission, for instance in studies of robustness like arbitrarily varying channels, optimal coding procedures in case of complete feedback, novel approximation problems for output statistics and generation of common randomness, the key issue in Cryptology.

Moreover our work on identification has led us to reconsider the basic assumptions of Shannon's Theory. It deals with "messages", which are elements of a *prescribed set of objects*, known to the communicators. The receiver wants to know the true message. This basic model occurring in all engineering work on communication channels and networks addresses a very special communication situation. More generally they are characterized by

(I)   The questions of the receivers concerning the given "ensemble", to be answered by the sender(s)
(II) The prior knowledge of the receivers
(III) The senders prior knowledge.

It seems that the whole body of present day Information Theory will undergo serious revisions and some dramatic expansions. **[A208] extends the frontiers of information theory in several directions.**

The main contributions concern information transfer by channels. There are also new questions and some answers in new models of source coding. While many of our investigations are in an explorative state, there are also hard cores of mathematical theories. In particular we present a unified theory of information transfer, which naturally incorporates Shannon's theory of information transmission and the theory of identification in the presence of noise as extremal cases. It provides several novel coding theorems. On the source coding side it strives for a concept of identification entropy.

Finally we mention as the perhaps most promising direction the study of **probabilistic algorithms with identification** as *concept of solution*. (For example: for any i, is there a root of a polynomial in interval i or not?)

The algorithm should be fast and have small error probabilities. *Every algorithmic problem* can be thus considered. This goes far beyond information theory. Of course, like in general information transfer also here a more general set of questions can be considered. As usual in Complexity Theory one may try to classify problems. What rich treasures do we have in the much wider areas of information transfer?!

The main goal of the research project was a further development of the GTIT and an exploration of phenomena in the sciences, particular for instance in psychology and in animal communication, where this theory is applied or is challenged with improvements of the model. This apparently requires a *highly interdisciplinary community*. By past experiences (see [A159]) Combinatorics is the mathematical discipline which provides most mathematical methods for formal solutions and, conversely, gets enriched the most by the problems we are dealing with.

This fact naturally gave the *duality in the structure of the project*, which is also expressed in the title.

We comment now chapterwise on the contributions to this book.

I

More than half a century ago in 1948 C. E. Shannon published his well-known paper [C83]. This laid the foundation of an important field now known as Information Theory. Two fundamental problems in communication were treated: source coding, also known as noiseless coding or data compression and (noisy) channel coding. Correspondingly two fundamental theorems were presented known as Shannon's Source- and Channel Coding Theorems. (In fact there are two Source Coding Theorems, lossy and lossless, in the well-known paper.)

By now Information Theory has been developed to a mature science. Many refinements and generalizations of concepts and models have been introduced. For quite a long time, until the late 80's, a majority of researchers in the area was mainly interested in **information transmission**. That is, the receivers in the communication have to **reproduce the messages** with certain fidelity levels. However the transmission is not the only goal of communication for human beings. Consider the following example: a man was injured in an accident on a highway. The people whose relatives were driving on the highway only want to know whether the poor man is their relative or not. If not, they do not care who he/she is. Based on this a new concept, identification, was introduced by R. Ahlswede and analysed together with G. Dueck [A59] for which a receiver only has to answer whether the message sent by the sender is the specified one or not, in any of all possible those "yes-no" questions. **The sender's encoding must be suitable at the same time for these questions.**

Still he provides less than what would be needed for the reproduction of a message. This relaxation dramatically speeds up the communication from exponentially fast to double exponentially fast or first order rates to second order rates. So far Identification has become a very active direction of research in Information Theory.

It was observed in [A60] that in Identification the second order rate is essentially determined by the first order rate of a random experiment set up by the communicators and whose outcome is known to both, sender and receiver, with high probability. In other words instead of the requirement for the receiver to recover the message sent by the sender with high probability it is required for the communicators to know the value of the same random variable with high probability. Thus a new concept, different from both transmission and identification, but with interesting connections to them was introduced. It is now called **common randomness**. A systematic presentation can be found in [A79], [A131]. Many interesting and important results and applications of common randomness have been obtained so far. **When we speak of GTIT today we mean it to include at its core the theory of information transmission, common randomness, identification and its generalizations and applications, but it goes far beyond it even outside communication theory when we think about probabilistic algorithms with identification (or more general tasks) as concepts of solution!**

Actually, the origin of the concepts **common randomness and common randomness capacity** took a fascinating path. Immediately after [A59] the results of [A60] were discovered – the papers appeared face by face in the same volume. An output process $Y_1, \ldots, Y_n$ produced by a DMC from an input process

$X_1, \ldots, X_n$ is not only known to the receiver of the channel W, but also to its sender, if there is a noiseless (passive) feedback channel. This common knowledge of the random process was used in [A60] for the randomization in a randomized identification procedure, which devotes a blocklength n to creating $Y_1, \ldots, Y_n$ and does then the identification in blocklength $\sqrt{n}$ (also called $\sqrt{n}$-trick). The size of the identification code obtained is of order $e^{e^{H(Y)n}}$! Making a best choice of X one gets the second order rate $C_F = \max_X H(Y)$, and the identification works if Shannon's transmission capacity $C_{Sh} = \max_X (H(Y) - H(Y|X))$ is positive.

Now the second idea was to wonder whether there is also or can be constructed also a random experiment (or process) in the original case of no feedback in [A59], where the second order identification capacity equals C_{Sh}. Well, just choose a channel λ-code $\{(u_i, D_i) : 1 \leq i \leq exp\{(C_{Sh} - \delta)n\}\}$ and define X^n as the RV taking codewords as values with equal probabilities.

Thus of course the sender knows X^n, but the receiver knows it almost, namely with an error probability not exceeding λ, if he uses the decoding sets D_i. This slight deviation from exact knowledge was not essential, the described experiment in conjunction with the $\sqrt{n}$-trick gave a second proof of the direct part of the coding theorem in [A59].

This discovery was followed up by R. Ahlswede and B. Verboven, a student from Leuven at that time, and led to solutions of identification problems for multi-way channels with noiseless feedback in [A71]. The paper contains a novel method by R. Ahlswede to prove weak converses by exploiting Schur concavity of the entropy function. In addition it has two new features, firstly it settles a rather rich class of channel models unheard of in multi-user theory for transmission, where it can be said - "cum grano salis" - that after struggles of more than 30 years the frontiers could not be moved very far beyond [A12], secondly the identification schemes are all **constructive** modulo the production of rich random experiments. This richness is measured by what was called **Mystery Numbers** or **Regions of k-tuples of Mystery Numbers** in [A71].

The constructions are based on Freivald's Lemma for hashing. As byproduct it gives also a constructive scheme for deterministic channels because they automatically have feedback. Shortly thereafter another construction was given for these special channels by Verdu and Wei [C96].

Mystery numbers have been subsequently called by R. Ahlswede in his lectures and papers, in particular also in [A208] **common randomness capacities**, a further enlightening development concerned what he formulated as a **PRINCIPLE::**

Second order identification capacity equals (first order) common randomness capacity.

After [A59], [A60], and [A71] a lot spoke for it and it became a driving dream leading to many results like [A100], coauthored by Z. Zhang, where the remarkable fact, that a wire-tapper cannot reduce identification capacity, if he cannot prohibit identification for 2 alternatives, and otherwise the identification capacity equals zero, was discovered and proved by arguments, which are by no means simple.

The same paper also started the investigation of identification in the presence of noisy (passive) feedback channels. This is discussed in [D1].

Continuing the line of children of the principle there are [A107] and striking work on the AVC with Ning Cai [A130] and arbitrarily varying MAC [A139], [A140], and above all for the maximal error concept for the AVC with complete feedback a determination of the capacity formula, which has a **trichotomy.**

Let's recall that the Ahlswede-dichotomy was for average error and no feedback [A29].

What was called "correlation in random codes", originally introduced in the pioneering paper [C8], can now be **understood as common randomness.**

Also its elimination in [A29] is an early version of what now Computer Scientists call **derandomization.**

Finally, we report on the removal of another heavy stone. Having understood how correlation in random codes, a form of common randomness, helps the communicators for AVC a next question is how a Slepian/Wolf type correlated source (U^n, V^n) [C89] helps the identification for a DMC W, when the sender knows U^n and the receiver knows V^n. Well, the principle says that it should be equivalent to asking how much common randomness can the communicators extract from (U^n, V^n), if they are assisted by the DMC W with capacity $C_{Sh}(W)$.

Now just notice that the case $C_{Sh}(W) = 0$ leads to the problem of finding what I. Csiszar asked for, and according to [C109] also D. Slepian, and named **Common Information.** It was determined by P. Gács and J. Körner [C36]. As expressed in their title the question was to know how this common information relates to Shannon's **mutual information,** in particular whether they are equal.

As we know the quantities are far apart, and under natural conditions, $C_{GK}(U, V)$ equals zero and it only depends on the positions of positivity of the joint distribution P_{UV}.

This got A. Wyner started, who believed that the quantity $C_W(U, V)$ he introduced was the right notion of common information. For one thing it does depend on the actual values of P_{XY}. On the other hand it satisfies $C_W(U, V) \geq I(U \wedge V)$ and is therefore rather big. The authors of [B38] gave a critical analysis about the problems at hand.

By the forgoing it is clear that the common randomness capacity of R. Ahlswede and V. Balakirsky, say $C_{AB}^W(U, V)$, equals $C_{GK}(U, V)$, if $C_{Sh}(w) = 0$. However, if $C_{Sh}(w) > 0$

$C_{AB}^W(U, V)$ nicely depends on the actual value of P_{UV}. Furthermore, $C_{GK}(U, V)$, which was always considered to be somewhat outside Information Theory proper, turns out be a common randomness capacity. The proof of the characterization of $C_{AB}^W(U, V)$ is a natural extension of the one in case $C_{Sh}(w) = 0$ given in [B38].

More importantly we feel that the analysis and discussion in [B38] are still of interest today, therefore we include that old (unpublished) manuscript here (as an Appendix).

"Der Mohr hat seine Schuldigkeit getan, der Mohr kann gehen"

The reader may consult [B17] to get help in deciding whether this is a quote from the "Merchant of Venice" and whether it is due to Shakespeare, Marlow (see also [C79]) or somebody else.

Here independently of the answer to this question the quote expresses our "gratefulness to the principle" for the inspirations we received from it.

After many lectures, in which we expressed firm believe in it we were challenged at the World Congress of the Bernoulli Society in Vienna in 1996.

At a slightly forced one night stay at a railway station in Serbia, we took a pencil ... (see the counter example in [A208]). There are two theories: theory of identification and theory of common randomness (see also [B10]). However, often they fall together as was shown by Y. Steinberg [C90] under reasonable conditions.

The first systematic investigation of common randomness started in [A79] and was continued after ideas had matured with [A131], in particular, with a revival of another old friend: balanced coloring for hypergraphs, which led also to a sharper converse for Wyner's wire-tap channel [C108].

Very remarkable work has been done since then by Csiszar and Narayan, and we are particular intrigued by the work of Venkatesan and Anantharam [C94], [C95].

In conclusion of the subject, we mention that common randomness and entanglement go into the center of Quantum Information Theory. But there according to [B24] already for simple channels identification and common randomness can be far apart.

The exploration of new concepts, ideas and models does not end at the the discovery of identification. It actually was a starting point for them. In [A208] more general communication systems were introduced and studied. Let $\mathcal{M}$ be a finite set of messages observed by a sender (or in general more than one sender) and Π be a set of partitions of $\mathcal{M}$. Then there are $|\Pi|$ receivers in a new communication system introduced in [A208], all access the output of the communication system, for example a noisy channel. Each receiver's responsible for a partition $\pi = \{M_{\pi,i}\}_i$ in Π in the sense that he has to find the subset $M_{\pi,i}$ in π, which contains the message observed by the sender. Let us first fix the communication system as a noisy channel and as the goal of the communicators the maximization of the size of the message sets. Then in the case that Π contains only partitions $\{\{m\} : m \in \mathcal{M}\}$ the problem is the transmission problem in Shannon's sense, coding for noisy channels. When $\Pi = \{\{\{m\}, \mathcal{M} \setminus \{m\}\}, m \in \mathcal{M}\}$ the problem becomes identification via noisy channels. Thus this model covers both transmission and identification.

Several interesting special problems are studied. For K-identification, that is Π consists of all partitions of $\mathcal{M}$ into a K-subset of $\mathcal{M}$ and its complement, lower and upper bounds of the optimal (second order) rates are found. An interesting relation between it and a well known combinatorial problem, the superimposed codes also called "r-cover-free families", is observed.

Several naturally defined sets Π of partitions for which the rates of transfer may not be larger than rates of transmissions, are presented.

The roles of feedback are discussed as well.

A family of problems on information transfer is introduced by the assumption that the receivers have certain prior knowledge. In particular it is called "K-separation" when the receiver knows initially that the message sent by the sender is in a K-subset of messages. It is shown in [A208] that K-identifiability and K-separability are closely related.

When one takes an information source as the communication system and uses variable-length codes to minimize the waiting time for the solution in which subset in $\pi \in \Pi$ the message is, several new problems of information transfer arise. Among them is an analogon to identification via a noisy channel, **identification for sources**, obtained by taking $\Pi = \{\{\{m\}, \mathcal{M} \setminus \{m\}\}, m \in \mathcal{M}\}$. Several contributions to identification theory are made in [A208] as well. They are identification capacity regions for multiple access channels and broadcast channels and the capacity region for simultaneous transmission and identification in the presence of noiseless feedback. Determining the identification capacity region for broadcast channels is surprising as determining the transmission capacity region for broadcast channels is among the well known hardest open problems in Multi-User Information Theory. Still many more problems, (for example, identification after group testing, binning via channels ...), solutions, and their relation to different types of problems in Information Theory, Statistics, Computer Science, and Combinatorics can be found in [A208] updated in [D1]. They can be recommended to the readers as a rich source of ideas, concepts, and proof techniques.

In [B1] more generalizations of identification for sources, namely generalized identification and generalized identification with decoder are posed. A probabilistic tool to study the bounds for the optimal waiting times of those models is introduced. Several results for identification for sources are obtained, including the limit of waiting times (in the worst case) for the code obtained by a nearly equal partition as the size of the alphabet of the uniformly distributed source goes to infinity, general bounds on the identification waiting times (in the worst case) and the limit of the average waiting times for block codes. Identification for sources is a new direction in the GTIT. In [B33] an interesting new quantity, identification entropy, is introduced and its operational properties are discussed. **It takes the role analogue to classical entropy in Shannon's Noiseless Coding Theorem.**

We have explained the role of common randomness for identification (The Principle!).

In the absence of feedback, one possibility to achieve the maximal possible rate of such a common random experiment is that the sender performs a uniform random experiment and transmits the result to the receiver using an ordinary transmission code. If noiseless feedback is available, the sender sends letters in such a way, that the entropy of the channel output (which he gets to know by the feedback channel) is maximized, where he can either use a deterministic or randomized input strategy, depending on the kind of code he may use. This interpretation proved to be the right one also for other kind of channels like the multiple access channel (see [A71]).

Thus the question arises if this equality is valid in general.

The answer is negative. In [A208] Ahlswede gives an example of a non-stationary memoryless channel with double exponentially growing input alphabet with identification capacity 1 and common randomness capacity 0. The structure of this channel has some similarities to the structure of ID-codes used in most of the achievability proofs for ID-coding theorems, thus it can be viewed as a channel with "built–in" ID-encoder.

In [B2] Kleinewächter presents a counterexample for the other direction. For given real numbers C_{ID} and C_{CR} with $0 < C_{ID} < C_{CR}$, constructed is a discrete channel with memory and noiseless passive feedback with identification capacity C_{ID} and common randomness capacity C_{CR}. This channel is constructed in such a way that it can be used in two ways. In one respect, the channel is good for the generation of common randomness, in the other it is suitable for identification.

It is quite reasonable to consider channels with memory. One may think for example of a system where data is transmitted by different voltage levels at high frequency. Because of the electrical capacity of the system it can be difficult to switch from a low voltage level to a high one and vice versa. There are also certain types of magnetic recording devices which have problems with long sequences of the same letter. These examples for instance lead to the notion of run length limited codes. A third example are systems requiring the use of binary codewords which have approximately the same number of zeroes and ones. This limitation arises if the system can only transmit an unbiased alternating current, therefore these codes are called DC-free.

In [A100] Ahlswede and Zhang gave bounds on the maximal rate for ID-codes for discrete memoryless channels with noisy feedback. Channels without feedback can in this model be described by a feedback channel that maps all input letters with probability 1 on one special feedback letter. The case with noiseless feedback is described by a feedback channel that has positive probability for a pair (y, z) of output respectively feedback letters only if $y = z$. In the cases of either no feedback or noiseless feedback, the upper and lower bound coincide and therefore this can be viewed as a **generalization and unification** of the results of [A59], [A60] and [C45], where the identification capacities for those channels were determined. Unfortunately in general there is a gap between the upper and lower bound.

Also in [B2] the upper bound on the size of *deterministic* codes given in [A100] for the channels where the *main channel is noiseless* is analyzed. The known bound states that the second order rate of deterministic ID codes is independently bounded by the maximal mutual information that channel input and feedback give about the channel output and the maximal entropy of the feedback. The improved bound is obtained by showing that these two quantities cannot be maximized independently, instead one has to choose an *input distribution that is good for both*, transmission and the generation of randomness. For the channels considered the new upper bound equals the known lower bound, therefore the deterministic identification capacity for these channels is now known.

In [B3] we see that the concept of identification is introduced in prediction theory. A framework of universal prediction of individual sequences is worked out in [C34]. There the authors used finite state machines to predict the outcoming letters of individual sequences according to the past parts of the sequences. As a main criterion the frequency of prediction errors is applied. Based on the work [C34] a universal predictor is developed and Markov predictors and Markov predictability of sequences are studied in [B3]. As a more general criterion, general loss functions are considered. After that the author of [B3] introduced identification into the prediction of sequences. That is, instead of predicting the outcoming letters in the sequences, the finite state machine has to answer an "easier" question, whether the outcoming letter is a specified one. Two identifiabilities, identifiability and strong identifiability, are defined according to the freedom of the users to use the finite state machines. Then a universal identification scheme is presented. The relations between predictability and identifiability are studied. For Markov machines it is shown that identifiability and strong identifiability asymptotically coincide. It was shown in [A59] that randomizations dramatically enlarge the sizes of message sets from exponential to double exponential in the blocklength. However it turns out here that randomization does not help at all in the asymptotic sense.

Next we turn to an application of identification in watermarking. Watermarking is a way to hide a secret message from a finite set, the watermark, in a data set, the covertext which is often used in protection of copyright. Watermarking may be viewed as a problem of transfer in a communication system in the sense that an attacker uses noisy channels to attack the watermark and a decoder tries to recover it. Watermarking identification codes were introduced by Y. Steinberg and N. Merhav [C91]. There the decoder identifies whether the watermark hidden in the covertext is the specific one instead of recovering it. This enables the size of watermarks to enlarge from exponential to double exponential. In the models in [C91] the attacker attacks the watermark with a single channel and the decoder knows either completely the covertext or nothing about the covertext. Since the assumption that the attack channel is single, meaning that the encoder (or the information hider) and the decoder know enough about the attack channel, the models are not robust. In [B4] robust models are studied. That is, the attacker is assumed to use an unknown channel from a set of attack channels to attack the watermark. More general assumptions on the knowledge of the decoder about the covertext are considered in two models, the component-wise key and block key, sent by the encoder to the decoder. Lower bounds on identification capacities for these models are obtained. Additionally to obtain the lower bound the authors in [B4] introduce two coding problems for common randomness and obtain their capacities for single channels. In the case of compound channels, lower and upper bounds are obtained for both models.

The successive refinement of information under error exponent or reliability constraints is discussed in [B5]. It was studied independently by Ahlswede [A50,A52] and by Koshelev, and then by Equitz and Cover. In the communication system the output of a discrete memoryless source is encoded and then sent to two users. The first user has to reproduce the output of the source in a Δ_1 level with respect to a distortion measure. The second user, who receives the message received by the first user and an additional message, has to reproduce the output of the source with a more accurate distortion level $\Delta_2 \leq \Delta_1$. An additional criterion, two levels of exponents of errors, was introduced in the model in [C46]. In [B5] the author refines the proof on the successive refinability condition in [C46] and re-establishes a result by E. Tuncel and K.Rose [C93], a single-letter characterization of the achievable rate region for successive refinement.

The coding problem for deterministic binary adder channels is considered in [B6]. The paper consists of two parts. In the first part the permanent user activity model and in the second part the partial user activity model are dealt with.

The E-capacity region for the multiple access channel is the object of the work [B7]. The multiple access channel is a multi-user channel with two senders (or more) and one receiver, whose capacity region was determined by R. Ahlswede ([A12]). Here by E-capacity region we mean the region of achievable rates under the constraint of the error probability exponent. The multiple access channels with states that are studied in [B7] are specified by a set of ordinary multiple access channels labeled by elements (which are called states) in a set $\mathcal{S}$ of states and a probability distribution Q on it. The channel takes values of channels labeled by $s_t \in \mathcal{S}$ at each time moment t independently. By combination of the cases where the two encoders, one encoder or/and the decoder know the states the author divides the problem into five models. Inner and outer bounds are derived for the capacity regions in the different models.

To realize identification via noisy channels it is necessary to find ways of constructing identification codes with the ability to correct errors as the proof of the achievability in [A59] is an existence proof. Constructive identification codes were first studied in [A71] as a special case of a feedback scheme, in which the channel is noiseless and therefore feedback automatically given. Then they were studied in [C96]. Notice that to achieve the capacity of identification codes, the length of codes are required to be sufficiently long whereas in practice the lengths of codes are bounded by technical constraints. So the constructive identification codes with length constraint are practically important. In this volume the paper [B8] is on this subject. Based on [C96] and their previous work [C18], the authors analyze a family of identification codes with length constraint and apply it to a simple kind of noiseless multiple access channel.

Codes with identifiable parent property were introduced by H.D.L. Hollmann, J.H. van Lint, J.P. Lennartz, and L.M.G.M. Tolhuizen [C51] for protection of copyright. Let a^n and b^n be two words of the same length from a finite alphabet. Then a descendant of them is a word of the same length such that each component coincides with the corresponding component of either a^n or b^n, who are called its parents. Identifiable parent property of a code means that one can discover at least one parent from the descendant of any pair of codewords in the code. R. Ahlswede and N. Cai observed in [B9] its relation with coding for multiple access channels. Its probabilistic version is coding for a multiple access channel such that two senders have the same codebook and the receiver has to decode the message at least from one of the two senders. This leads them to a coding problem for the multiple access channel, where the two senders are allowed to use different codebooks and again the receiver only needs to decode the message sent by anyone of the two senders. The capacity region is determined and the result shows that an optimal strategy for the receiver is to always decode the message from a fixed sender. The result has a simple consequence for the interference channel with one deterministic component which seems to be new.

II

Problems on GTIT, particularly, transmission, identification and common randomness, via a wire-tap channel with secure feedback are studied in the work [B10]. Recall that wire-tap channels were introduced by A. D. Wyner [C108] and were generalized by I. Csiszár and J. Körner [C19]. Its identification capacity was determined by R. Ahlswede and Z. Zhang in [A100]. In the article here secure feedback is introduced to wire-tap channels. Here by secure feedback we mean that the feedback is noiseless and that the wire-tapper has no knowledge about the content of the feedback except via his own output. Lower and upper bounds to the transmission capacity are derived. The two bounds are shown to coincide for two families of degraded wire-tap channels, including Wyner's original version of the wire-tap channel. The identification and common randomness capacities for the channels are completely determined. Also **here again identification capacity is much bigger than common randomness capacity**, because the common randomness used for the (secured) identification needs not to be secured!

In the work of Z. Zhang [B11] the scheme of encrypting the data X^n by using the key set K^n and function $f : K^n \times X^n \to Y^n$ is considered. Under given distribution of X^n the value of the conditional entropy $\frac{1}{n} H(K^n | Y^n)$ which is offered as the measure of the secrecy of the system is investigated. In several natural cases an expression for this measure (which is called 'key equivocation rate') in terms of sizes of alphabets and distributions of X^n is derived.

The secrecy system with ALIB encipherers was investigated in [A42] and is adapted in [B12] to satisfy the model of identification via channels. The smallest key rate of the ALIB encipherers needed for the requirement of security is analyzed.

Several papers in this section are devoted to the investigation of the problem of generating pseudorandom sequences and their statistical properties. This is an important task for cryptology since for instance these pseudorandom sequences can serve as a source of key bits needed for encryption. The first problem in [B13] is to find a proper test for sequences of elements from a finite alphabet to be pseudorandom. For a binary sequence with alphabet $\{-1, +1\}$ the authors choose the criterium of a small modul of the sum of subsets of the elements of the sequence maximized over the choice of the subset of positions of these elements and small correlation measure which is the absolute value of the sum of products of elements from some set of subsets of positions maximized over the special choice of the positions of these elements.

Then these measures are extended in [B14] to a nonbinary alphabet and as one generalization the frequency criterium is chosen, i.e. the deviation of the number of given patterns on given positions from the expected value, maximized over the choice of the positions. Relations are proved which show the equivalence (in some sense) of these different tests of pseudorandomness in the binary case. Also proved is that the number of the sequences with large measures of pseudorandomness is exponentially small in comparison with the number of all sequences.

Algorithms were introduced in [B15] for constructing pseudorandom sequences. These constructions can find applications in cryptology and simulations ([B16]). Also considered was the notion of f−complexity of the set of n−tuples which is the maximal number t s.t. arbitrary t positions have an arbitrary pattern in some n−tuple from this set. It was introduced in [A171].

More explanations are given by the excellent introductions also in earlier work, where C. Mauduit and A. Sarkozy explain their approach to cryptology. Roughly speaking their philosophy is that less can be more: instead of going after complex problems whose high complexity till now cannot be proved and therefore always there can be a bad end of a dream, they suggest to work with number theoretical functions of likely not highest complexity, but for which some degree of complexity can be proved.

Now about the work [B17] which approaches the study of the text authorship identification. Eventhough this problem is rather delicate the attempt to study it here in a particular case of the identification of Shakespeare text can be considered as an example for applying different methods of word choices and stylistic analysis to establish the author of the text. The reader can agree or disagree with the conclusions of this work, but in any case, it is interesting research of the authorship identification.

III

Since the end of the last century, quantum information and computation have become a more and more important and attractive research area for physicists, computer scientists, information theorists, and cryptographers.

Although the rules of quantum mechanics **are fairly simple**, even several experts find them counter-intuitive. The earliest antecedents of quantum information and computation may be found in the long-standing desire of physicists to better understand quantum mechanics. Generations of physicists have wrestled with quantum mechanics in an effort to make its predictions more palatable. One goal of research in quantum information and computation is to develop tools which sharpen our intuition about quantum mechanics and make its predictions more transparent to human minds.

In the recent years, the historic connection between information and physics has been revitalized as the methods in the theory of information and computation have been extended to treat the transmission and processing of intact quantum states, and interaction of such quantum information with classical information. As this book is devoted to the GTIT, we consider here also research on quantum information. The Theory of Quantum Mechanics shows that quantum information has a very different behavior than classical information. For example data can be copied freely, but can only be transmitted forward in time, to a receiver in the sender's forward light cone, whereas quantum entanglement cannot be copied but can connect any two points in space-time. Those differences not only enlarge the powers of information transfer and cryptography, but also provide a wide range for research on Theory of Information and Computation.

The problem of transferring quantum information from one quantum system into another quantum system is linked to the well-known work [C31] by A. Einstein, B. Podolski, and N. Rosen. They posed a far reaching question, but doubted that the answer can be given by Quantum Theory. As they pointed out later, it asserts the possibility to create simultaneously and at different places exactly the same random events. This phenomenon is often called "EPR effect", or for short EPR. So far the discussions on EPR have become a very attractive topic in the study of Quantum Information Theory. One can find a huge number of papers on or related to this topic. A beautiful work [B19] in this direction is included in this volume, where several nice and basic contributions are made. There are some anti-linear maps governing EPR tasks in the case that no reference bases are distinguished; some results on imperfect quantum teleportation and composition rules; quantum teleportation with distributed measurements; and some remarks on EPR with mixed state, triggered by a Lüders [C67] measurement. We expect that they contribute to the richness of the EPR study and related topics.

So far many important concepts in Classical Information Theory have been extended to Quantum Information Theory. Among them is the noisy channel, which is one of the most important objects in Information Theory. Classically a noisy channel is specified by a stochastic matrix in discrete communication systems, or more general a collection of conditional probability distributions on the output space indexed by the input symbols. The physical interpretation is that due to the classical noise, random errors may change input symbols into outputs of the channels. Such models have been extended into a quantum version, where the role of stochastic matrices is replaced by a quantum operation, that is, a linear trace preserving or more general trace-non-increasing, completely positive mapping. The physical interpretation is that a measurement on the environment, whose result is unknown by the communicators, is performed after a unitary transformation has been applied to the inputs coupled to the environment. In contrast to classical noisy channels, people have known very little about quantum noisy channels. A problem to study quantum channels is their characterization. A quantum channel of rank two is characterized by the rank and the determinant, or by the determinant alone under the assumption of trace preservation, up to unitary transformations. In the nice paper [B20] based on some identities for determinants of quantum channels of rank two, concurrences are calculated or estimated.

We recommend to our readers the excellent paper [B21], which surveys **several basic concepts** of **universality in quantum information processing** and deals with **various universal sets of quantum primitives** as well with their optimal use. To exploit the potential of the nature for quantum information processing appears to be very difficult. So it is very important for quantum information processing to explore what kind of quantum primitives form sets of primitives that are universal in a reasonable sense and that are easy to implement with various technologies for design of quantum registers. Finding rudimentary quantum information processing primitives, which perform efficient quantum information processing, and methods for their optimal use is fundamentally important. They help us to understand in depth the laws and limitations of quantum information processing and communication and also quantum mechanics itself. The search for sets of elementary, or even very rudimentary, but powerful, quantum computational primitives and their optimal use is one of the major tasks of quantum information processing research and has brought a variety of deep and surprising results. In this work, different types of universalities of sets of quantum gates and basic concepts used in their definitions are stated. Then based on them various results are presented. A rather comprehensive list of references in the area is provided. All this is quite transparent and helpful for people who want to get into this area.

One of the most exciting things in the area of computation brought by quantum mechanics perhaps is that quantum computation has more power than classical computation. Many excellent examples can be found in the literature. One of the well-known algorithms among the examples is Grover's Algorithm [C40], [C41]. The problem is to search a target element in an unstructured search universe of N elements using an oracle. For a classical algorithm the computation complexity is of order $\frac{N}{2}$ queries. It was shown in [C4] that no quantum algorithm can solve the search problem in fewer than $O(\sqrt{N})$ queries. Grover's algorithm solves the problem in approximately $\frac{\pi}{4}\sqrt{N}$ queries. So it is optimal in this sense. The problems have been generalized and discussed by different authors (for example see [C14] and its list of references). An alternative proof to the optimality of Grover's algorithm is presented by E. Arikan in [B22]. The proof is elegant and an information theoretic method is used in the analysis. Information theoretic analysis has been applied in classical computation and often is powerful. Here its power is shown in the analysis of quantum computation. It is reasonable to expect that it will play an important role in the Theory of Quantum Computation in the future.

As we have mentioned, many powerful methods and techniques in Classical Information Theory have been extended to Quantum Information Theory. A well-known example is in [C82], where the concept of typical sequences, an essential tool in Classical Information Theory is extended to typical subspaces. Those extensions not only provide a rich resource of ideas, tools and techniques in the study of Quantum Information Theory but also help us to better understand the physical meaning and structure of quantum information. It is considered by us as an important direction in the GTIT. The quantum channels closest to classical channels perhaps are the **classical quantum channels** introduced by A. Holevo [C50], which are among the few kinds of quantum channels, for which the **additivity of capacities** is known. Perhaps due to their similarity to classical communication systems classical quantum channels are among the quantum communication systems, for which the ideas and methods in Classical Information Theory often can be well extended. Several examples can be found in A. Winter's dissertation [C104]. In this volume there are two works of such extensions in Multi-user Information Theory. The first paper [B23] is on classical quantum multiple access channels. The coding and weak converse theorem for classical multiple access channels of [A12] was extended to classical quantum multiple access channels in [C105]. But the extension of the strong converse theorem has been open for several years. One reason, but likely not the only one, for it is that so far an analogue to the Blowing Up Lemma [A23] has not been discovered. The **Wringing Technique**, a powerful technique of [A44], could be extended in [B23] to the quantum case and then gave the desired strong converse. We expect that the work not only helps us to better **understand quantum multiple access channels**, but also brings new ideas and techniques into the whole area of quantum multi-user information theory. Readers can find more extensions of results in Multi-user Information Theory from classical to quantum in [C103]. The broadcast channel is a well known multi-user channel introduced in [C16] and in general its capacity region is still unknown. What we know is about its capacity region in two special cases, degraded broadcast channels ([C6], [C37], [A21])) and asymmetric broadcast channels ([C97], [C56]). Their extensions to a classical quantum version are obtained in [C103]. Some results, which extend the protocols for two users sharing different terminals of a correlated source to generate a private key in [A79], are obtained in [C103] for the quantum case as well. These extensions, however, are still incomplete. On the other hand an investigation of the difficulty may help understanding the difference between classical and quantum information systems. The readers who would like to study Quantum Multi-user Information Theory may find interesting open problems there as well.

Identification via channels was analyzed in [A59] in Classical Information Theory and today it has been developed as an important subject. Its extension to Quantum Information Theory began with the dissertation [C66]. Actually two models of identification arise in quantum which could be proved to be different! A strong converse theorem for classical quantum channels was obtained in [A161]. Once again this strong converse theorem shows the similarity of classical and classical quantum channels although more powerful inequalities and in particular a lemma of [B37] are needed in the proof. For both channels the identification capacities (in second order) are equal to the transmission capacities (in first order). However in general the situation is different. It was discovered in [C106] for the **ideal qubit channel**, one of the simplest quantum channels, that the **identification capacity is two, twice its transmission capacity.** The result shows the essential difference between classical and quantum channels. Classically such gaps occur in more artificial cases (see [A208]) or in cryptographic systems like wire-tap channels (see [A100] and [B4]). The paper [B24] continues the work along this line, where several interesting and nice results are presented. At first two alternative proofs for the capacity of the ideal qubit channel being equal to 2 are given. Then the author applies the results to prove the capacity formulas for quantum channels with two forms of feedback: passive classical feedback for quantum-classical channels, and coherent feedback for general quantum channels. The results are considered as generalization of Ahlswede/Dueck's earlier results in [A59] and [A60]. Due to the No-Cloning Theorem there is no direct way to extend the concept of feedback to a quantum channel. Actually for quantum channels feedback is a problematic issue. We hope that the discussion on feedback in the work will be helpful for readers concerned about ways to introduce feedback in transmission over quantum communication systems as well.

In the work of Nathanson [B25] the natural arithmetical structure of the so-called quantum integers is established.

IV

In [B26] the author discusses the foundations of Statistics focusing on the **fiducial argument**. This concept was introduced by Fisher in 1930, but unfortunately was misinterpreted by Fisher himself in his later work and by other statisticians. The author considers the fiducial argument as a first attempt to bridge the gap between two directions in Statistics: the Neyman-Pearson Theory and the Bayesian Theory. When interpreted properly, it leads to a unifying Theory of Statistics.

[B27] is an interesting discussion on the appropriateness of different asymptotic tools for sequential discrimination problems.

[B28] continues the work [A121] by R. Ahlswede, E. Yang and Z. Zhang, in which a new model was introduced and analyzed. "Identification" here has a new meaning. One motivation for its study was to estimate a joint distribution P_{XY} of a pair X, Y of random variables, by the observer of Y when he gets informed at a prescribed rate from the observer of X. Rather precise estimates of the error probability are given. **A new method described in the Inherently-Typical Subset Lemma was introduced.** It solves new cases of shadow problems.

A broad class of statistical problems arises in the framework of hypothesis testing in the spirit of identification for different kinds of sources, with complete or partial side information or without it. [B29] is devoted to the investigation of a hypothesis testing problem for arbitrarily varying sources with complete side information. [B30] considers the more difficult but more promising problem of hypothesis identification.

A very basic inequality, known as the Ahlswede-Daykin inequality and called Four Function Theorem by some authors, which is more general and also sharper than known correlation inequalities in Statistical Physics, Probability Theory, Combinatorics and Number Theory (see the preface and survey by Fishburn and Shepp in [C1]) is extended elegantly to function spaces. That is, the inequality of the same type holds for a Borel measure on $R^{[0,1]}$. We expect that it will have wide applications.

There are some earlier results stating upper bounds on the rate of convergence in the Central Limit Theorem. In [B32] the author proposes a new method for establishing a lower bound for the information divergence (or relative entropy), which is then used to determine the rate of convergence in the information-theoretic Central Limit Theorem.

V

Shannon (1948) has shown that a source $(\mathcal{U}, P, U)$ with output U satisfying Prob $(U = u) = P_u$, can be encoded in a prefix code $\mathcal{C} = \{c_u : u \in \mathcal{U}\} \subset \{0, q-1\}^*$ such that for the entropy

$$H(P) = \sum_{u \in \mathcal{U}} -p_u \log p_u \leq \sum p_u \|c_u\| \leq H(P) + 1,$$

where $\|c_u\|$ is the length of c_u.

In [B33] a prefix code $\mathcal{C}$ is used for another purpose, namely noiseless identification, that is every user who wants to know whether a u ($u \in \mathcal{U}$) of his interest is the actual source output or not can consider the RV C with $C = c_u = (c_{u_1}, \ldots, c_{u\|c_u\|})$ and check whether $C = (C_1, C_2, \ldots)$ coincides with c_u in the first, second etc. letter and stop when the first different letter occurs or when $C = c_u$. Let $L_\mathcal{C}(P, u)$ be the expected number of checkings, if code $\mathcal{C}$ is used.

Discovered is an identification entropy, namely the function

$$H_{I,q}(P) = \frac{q}{q-1}\left(1 - \sum_{u \in \mathcal{U}} P_u^2\right).$$

We prove that $L_C(P, P) = \sum_{u \in \mathcal{U}} P_u \, L_C(P, u) \geq H_{I,q}(P)$ and thus also that

$$L(P) = \min_{C} \max_{u \in \mathcal{U}} L_C(P, u) \geq H_{I,q}(P)$$

and related upper bounds, which demonstrate the operational significance of identification entropy in noiseless source coding similar as Boltzmann/Shannon entropy does in noiseless data compression.

It has been brought to our attention that in Statistical Physics an entropy $S_\alpha(P) = f(\alpha)(1 - \sum_{u \in \mathcal{U}} P_u^\alpha)$ has been used in Equilibrium Theory for more pathological cases, where Boltzmann's $H(P)$ fails.

Attempts to find operational justifications in Coding Theory have failed.

It is important here that $S_\alpha(P)$ (in particular also $S_2(P)$), which is to be compared with $H_{I,q}(P)$, does not have the parameter q, the size of the alphabet for coding. The factor $\frac{q}{q-1}$ equals the sum of the geometric series $1 + \frac{1}{q} + \frac{1}{q^2} + \dots$, which also has an operational meaning for identification. $H(P)$ also has a q in its formula, it is the basis of the log- function for which Shannon's result holds!

We emphasize, that storing the outcome of a source as a leaf in a prefix code **constitutes a data structure which is very practical.** Let for instance the c_u specify the person u out of a group $\mathcal{U}$ of persons, who has to do a certain service. Then every person traces along the tree to find out whether he/she has to go on service. We know that its expected reading time is always < 3 no matter how big $|\mathcal{U}|$ is. This goes so fast, because the persons care in this model only about themselves. If they don't have service, then they don't care in this model who has.

Finding out the latter takes time $\sim H(P)$ and goes to infinity as $H(P)$ does. Notice that $H_{I,q} \leq \frac{q}{q-1} \leq 2$.

The paper [B34] describes the concept of weakly chaotic (0-entropy) dynamical systems and how new ideas are needed to characterize them. Compression algorithms play a key role in this theory.

The difference between coarse optimality of a compression algorithm and the new concept, asymptotic optimality, are described. For instance the well-known Ziv/Lempel 77 and 78 compression algorithms are not asymptotically optimal. At the moment it is not clear whether the definition of "Asymptotic optimality" exactly captures what one is interested in regarding weakly chaotic systems. The main result states the asymptotic optimality of a compression algorithm similar to the **Kolmogorov Frequency Coding Algorithm.**

This compression algorithm is not of practical use because of its computational time cost. Presently no "fast" asymptotically optimal compression algorithm is known.

As entropy measure use is made of the **empirical entropy** $\hat{H}_l$ of a given string s as a **sequence** of numbers giving statistical measures of the average information content of the digits of the string.

Other conceptual instruments have been developed to understand physical phenomena related to weakly chaotic dynamical systems: self organized criticality, anomalous diffusion processes, transition to turbulence, formation of complex structures and others.

Among them are works using the generalized entropies $f(\alpha)(1 - \sum p_i^\alpha)$ mentioned above (lately called Tsallis entropies), many of these works are – according to the authors – " heuristic or experimental (mainly computer simulations) and few rigorous definitions and results can be found. Conversely there are rigorous negative results ([C92], [C9]) about the use of generalized entropies for the construction of invariants for 0-entropy measure preserving systems."

The book has no contributions to Algorithmic Information Theory ([C65]) or its applications. We mention work of Biochemists ([C27], [C28], [C29], [C30], and [C63] with several references) who think about information as a kind of program complexity in the context of selforganisation of matter.

Write efficient memories, or for short WEM, is a model for storing and updating repeatedly on a medium. It was introduced by R. Ahlswede and Z. Zhang in [A63]. Given a finite alphabet and a cost function of the pairs of letters in the alphabet a WEM code is a collection of subsets of codewords in the alphabet and a subset stands for a message. When a user is going to update the content of the memory to a new message, he changes the current codeword written on the medium to a codeword in the subset standing for the new message such that the cost he pays for the change is minimum in respect to the cost function. The goal is to maximize the set of messages under a constraint on the update cost. According to the knowledge of the encoder and decoder about the current codeword, the model is divided into four sub-models. In the original model, there is no assumption for the updating error. In [B35] R. Ahlswede and M. Pinsker introduced defect errors and localized errors to WEM. For binary alphabet and Hamming cost function, the capacities of the following WEM codes are obtained: in the case that only the encoder knows the current codeword the codes correcting defect errors and codes correcting localized errors and in the case that both encoder and decoder know the current codeword the code correcting localized errors.

Codes for k-ary trees are important in Computer Science and Data Compression. A stochastic generation of a k-ary tree was considered in [B36]. Starting from the root, extend k branches and append k children with probability p, or terminate with probability $1 - p$. This process gives the ideal codeword length if $0 \leq p \leq 1/k$. The expectation and variance of the length of ideal codes are determined. Also, the probability of obtaining infinite trees is established.

Perhaps many researchers in the area of Information Theory have not recognized that the definitions of several basic quantities, including that of the capacity of a noisy channel, the most important quantity in channel coding, are problematical in the precise sense. Probably it is because their research interests mainly focus on stationary memoryless (or more general ergodic) communication systems and for those systems the problems of non-proper definitions do not likely show up. R. Ahlswede noticed that information theory suffers from a lack of precision in terminology and it is necessary to clarify this. [B37] in this volume, first made available as a preprint of the SFB-343 in the year 2000 (see [A196]), opens an important discussion on this issue. The paper begins with basic concepts, namely, channels and codes and their performance parameters blocklength, error probability and code size, and moves on to concepts like capacities, converses,.... Then the author sketches the problems arising there. The readers may be very surprised about the difference it makes for systems when stationarity and/or component-wise independence are removed and the damage caused by misleading or less precise definitions in the history of research in Information Theory. Ways to get out of this dilemma are proposed with new concepts. Finally the author draws the reader's attention to important combinatorial methods in probabilistic coding theory and especially, shows its power in the proof of the strong converse for the identification coding theorem.

The following comments to
"On concepts of performance parameters for channels" are written
solely by R. Ahlswede

This contribution has appeared in the preprint series of the SFB 343 at Bielefeld University in the year 2000 [A195].

Its declared purpose was to open a discussion about basic concepts in Information Theory in order to gain more clarity.

This had become necessary after the work [A59] for identification gave also new interpretation to classical theory of transmission leading in particular to "A general formula for channel capacity" by S. Verdu and T.S. Han, which however, does not cover the results of [A2]. Moreover, discussions at meetings bore no fruits. Finally, quite amazingly the name information spectrum reminded us of the name code spectrum in [A47].

Also, some responses have send us through hell: per aspera ad asctra.

We give now a brief reaction to some comments and criticism we received.

1. Strong objections were made against statement α) after (10.3), which implies the claim that the inequality

$$\overline{\lambda} \geq F(\log M - \Theta) - e^{-\Theta}, \ \Theta > 0 \tag{a1}$$

is essentially not new. Particularly it has been **asserted that this inequality is not comparable to Shannon's**

$$\overline{\lambda} \geq \frac{1}{2} F(\log \frac{M}{2}) \tag{a2}$$

in (3.3) of Shannon's Theorem, because it is stronger.

Therefore we have to justify our statement by a proof. Indeed, just notice that Shannon worked with the constant $\frac{1}{2}$ for simplicity in the same way as one usually extracts from a code with average error probability $\overline{\lambda}$ a subcode of size $\frac{M}{2}$ with maximal probability of error not exceeding $2\overline{\lambda}$. However, again by the pigeon-hole principle for any $\beta \in (0, \frac{1}{2})$ there are $M\beta$ codewords with individual error probabilities $\leq \frac{\overline{\lambda}}{1-\beta}$. (This argument was used in [A1], [A2], [A5]).

Now just replace in Shannon's proof $\frac{1}{2}$ by β to get the inequality

$$\overline{\lambda} \geq (1 - \beta)F(\log M\beta). \tag{a3}$$

Equating now $F(\log M - \Theta)$ with $F(\log M\beta)$ we get $\beta = e^{-\Theta}$ and it suffices to show that

$$(1 - e^{-\Theta})F(\log M - \Theta) \geq F(\log M - \Theta) - e^{-\Theta}.$$

Indeed $e^{-\Theta} \geq e^{-\Theta}F(\log M - \Theta)$, because F is a distribution function. So (a3) is even slightly stronger than (a1).

Q.E.D.

For beginners we carry out the details.

Introduce $W^*(u|y) = \frac{\tilde{P}(u,y)}{Q(y)}$ and notice that $I(u,y) \leq \log M\beta$ is equivalent with $\frac{W^*(u|y)}{P(u)} \leq M\beta$ or, since $P(u) = M^{-1}$,

$$W^*(u|y) \leq \beta. \tag{b1}$$

Now concentrate attention on those pairs (u, y) for which (b1) holds.

Consider the bipartite graph $G = (\mathcal{U}, \mathcal{Y}, \mathcal{E})$ with vertix sets $\mathcal{U} = \{u_1, \ldots, u_M\}$, $\mathcal{Y}$, and edge set $\mathcal{E} = \{(u, y) : W^*(u|y) \leq \beta\}$

Clearly,

$$\tilde{P}(\mathcal{E}) = F(\log M\beta) \tag{b2}$$

We partition now $\mathcal{Y}$ into

$$\mathcal{Y}_+ = \{y \in \mathcal{Y} : \text{ exists } u \text{ with } W^*(u|y) > \beta\}, \ \mathcal{Y}_- = \mathcal{Y} \backslash \mathcal{Y}_+ \tag{b3}$$

and correspondingly we partition $\mathcal{E}$ into

$$\mathcal{E}_+ = \{(u, y) \in \mathcal{E} : y \in \mathcal{Y}_+\}, \ \mathcal{E}_- = \mathcal{E} \backslash \mathcal{E}_+. \tag{b4}$$

Clearly $\tilde{P}(\mathcal{E}) = \tilde{P}(\mathcal{E}_+) + P(\mathcal{E}_-)$. For $y \in \mathcal{Y}_+$ ML-decoding chooses a u with $W^*(u|y) > \beta$, but (u, y) is not in $\mathcal{E}$ and not in $\mathcal{E}_+$. Therefore all $(u', y) \in \mathcal{E}_+$ contribute to the error probability. The total contribution is $\tilde{P}(\mathcal{E}_+)$.

The contribution of the edges in $\mathcal{E}_-$ to the error probability is, if f is the ML-decoder,

$$\sum_{y \in \mathcal{Y}_-} Q(y)(1 - W^*(f(y)|y)) \geq \tilde{P}(\mathcal{E}_-)(1 - \beta)$$

and hence

$$\overline{\lambda} \geq \tilde{P}(\mathcal{E}_+) + \tilde{P}(\mathcal{E}_-)(1 - \beta)$$

(even slightly stronger than (a3)), written explicitly

$$\overline{\lambda} \geq \tilde{P}(\{(u, y) : \log \frac{W(y|u)}{Q(y)} \leq \log M - \Theta\}) - e^{\Theta} \tilde{P}(\{(u, y) :$$

$$for \ all \ u' \in \mathcal{U} \ \log \frac{W(y|u')}{Q(y)} \leq \log M - \Theta\}).$$

For those who do not accept it as Shannon's result it would be only consequential to name it then the Shannon/Ahlswede inequality.

2. We also have some good news. In Section 5 we argued that the optimistic capacity concept seems absurd and we provided convincing examples.

Investigations [B12] in Cryptography, made us aware that this concept, a dual to the pessimistic capacity, finds a natural place here, because **one wants to protect also against enemies having fortunate time for themself in using their wire-tapping channel!**

3. Finally, being concerned about performance criteria, we should not forget that in Information Theory, similarly as in Statistics, asymptotic theory gives a first coarse understanding, but never should be the last word.

In particular with the availability of a lot of computing power not only small, but even medium size samples call for algorithmic procedures with suitable parameters. This was the message from J. Ziv in his contribution [C112].

4. S. Dodunekov in [C23] explained that for linear codes with parameters blocklength n, dimension k and minimal distance d, fixing two parameters and optimizing the third of the quantities **(a)** $N_q(k, d)$, **(b)** $K_q(n, d)$, **(c)** $D_q(n, k)$ **the first one gives the most accurate description.**

5. It has been pointed out that there are different concepts of capacity, but that usually in a paper it is clearly explained which is used and a definition is given. It can be felt from those reactions that Doob's famous (and infamous with respect to Shannon and his work) criticism [C24] about the way in which Information Theory proceeds from Coding Theorem n to Coding Theorem $n+1$ keeps people alert.

That mostly researchers know which concepts they are using and that they sometimes even give definitions, that is still not enough.

For instance we have been reminded that our statement "there seems to be no definition of the weak converse in the book [C20]" is wrong and that a look at the index leads to the problem session, where a definition is given. This is correct.

However, it is not stated there on page 112 that this definition of a converse, referred to by us as "weak weak converse, ... or ...", is not the definition, which was established in Information Theory at least since the book [C107] by J. Wolfowitz came out in 1961 and was used at least by most of the Mathematicians and Statisticians working in the field.

Unfortunately this happened eventhough since 1974 we had many discussions and joint work with J. Körner and I. Csiszar, a great part of which entered the book and influenced the shape of the later parts of the book. It was also given to us for reading prior to publication and we have to apologize for being such a poor reader. Otherwise we would have noticed what we only noticed in 1998.

It is clear now why [A2] and [A3] are not cited in the book, because they don't fit into the frame.

This frame became the orientation for people starting to learn Information Theory via typical sequences. Shannon's stochastic inequalities ([C86]) perhaps were not taught anymore.

6. We know that the weak capacity has the additivity property for parallel channels. We draw attention to the fact that **Shannon (and also later Lovasz) conjectured** this property to hold also for his zero-error capacity (which was disproved by Haemers [C43]). Apparently, Shannon liked to have this property!

We all do, often naively, **time-sharing**, which is justified, if there is an additivity property!

We like to add that without thinking in terms of time-sharing we never would have discovered and proved our characterization of the (weak) capacity region for the MAC in [A12] (with a different proof in [A18]).

So our message is "Shannon also seems to think that additivity is an important property" and not "Shannon made a wrong conjecture".

The additivity property for quantum channels is of great interest to the community of Quantum Informationtheorists. This led M. Horodecki to his question quoted in [C52]. The answer is positive for degraded channels, but not in general!

7. Once the situation is understood it is time to improve it. We suggest below a unified description of capacity concepts with conventions for their notations.

In every science it is occasionally necessary to agree on some standards - a permanent fight against the second law of thermodynamics. We all know how important the settings of such standards have been in Physics, Chemistry, Biology etc. Every advice to the standards proposed here is welcome.

We start here with the case corresponding to B) under 4. above and restrict ourself to one-way channels.

We consider a general channel $\mathcal{K}$ with time structure, which is defined in (1.2) of [B37]. Recall that for a positive integer n and a non-negative real number λ

$N(n, \lambda)$ denotes for $\mathcal{K}$ the maximal cardinality of a code of block-length n with an error probability not exceeding λ.

Often λ depends on n and we write $\lambda(n)$. The sequence $\{\lambda(n)\}$ is typically of one of the following forms

(i) $\lambda(n) = 0$ for all $n \in \mathbb{N}$
(ii) $\lambda(n) = \lambda$, $0 < \lambda \leq 1$, for all $n \in \mathbb{N}$
(iii) $\lambda(n) = n^{-\alpha}$, $0 < \alpha$, for all $n \in \mathbb{N}$
(iv) $\lambda(n) = e^{-En}$, $0 < E$, for all $n \in \mathbb{N}$

We speak about zero, constant, polynomial, and exponential error probabilities.

With the sequence $\{N(n, \lambda(n))\}$ we associate a very basic and convenient performance parameter for the channel $\mathcal{K}$ the

rate-error function $R : \Lambda \to \mathbb{R}_+$,

where Λ is the space of all non-negative real-valued sequences and for every $\{\lambda(n)\} \in \Lambda$.

$R(\{\lambda(n)\})$ is the largest real number with $\frac{1}{n} \log N(n, \lambda(n)) \geq R(\{\lambda(n)\}) - \delta$ for every $\delta > 0$ and all large n.

How does it relate to capacities? In the four cases described we get for $R(\{\lambda(n)\})$ the values

(i') $C(0)$, that is, Shannon's zero error capacity
(ii') $C(\lambda)$, that is, the λ-capacity introduced in [A5]
(iii') $C(\alpha)$, that is, a novel α-capacity
(iv') $C(E)$, that is, the E-capacity introduced by Evgueni Haroutunian in [C47].

A word about notation is necessary. The functions $C(0)$, $C(\lambda)$, $C(\alpha)$, and $C(E)$ are distinguished only by their arguments, these will always appear explicitly. All our results have to be interpreted with this understanding.

This convention was made already in [A5] where not only the maximal error probability λ but also the average error probability $\overline{\lambda}$, the maximal error probability λ_R for randomized encoding, and the average error probability $\overline{\lambda_R}$ for randomized encoding were considered. For example, one of our theorems in [A5] says that

$$C(\lambda_R) = C(\overline{\lambda}) = C(\overline{\lambda_R}).$$

under certain conditions where $\lambda_R = \overline{\lambda} = \overline{\lambda_R}$. Taken literally this is a trivial statement. In the light of our notation it means that these functions coincide for certain values of the argument. This notation result is no confusion or ambiguity, and has the advantage of suggestiveness new and typographical simplicity.

An important point about the introduced rate-error function and the capacities is their existence for every channel $\mathcal{K}$.

The same is the case for the (ordinary) capacity

$$C = \inf_{0 < \lambda \leq 1} C(\lambda).$$

Our rate-error function may be called pessimistic and it has an optimistic twin $\overline{R}(\{\lambda(n)\})$, the largest real number with

$$\overline{\lim}_{n \to \infty} \frac{1}{n} \log N(n, \lambda(n)) \geq \overline{R}(\{\lambda(n)\}).$$

Correspondingly we get the optimistic capacities $\overline{C}(0)$, $\overline{C}(\lambda)$, $\overline{C}(\alpha)$, $\overline{C}(E)$, and $\overline{C}$. Of course for a DMC $\overline{C}(0) = C(0)$ etc.

They are relevant, for example, if a wire-tapper chooses the times of an attack.

Again all these quantities exist. Moreover, the error criteria $\overline{\lambda}$, λ_R, $\overline{\lambda}_R$ lead to analoga of the capacities in (iii)-(iv'), namely, $C(\overline{\alpha})$, $C(\alpha_R)$, $C(\overline{\alpha}_R)$, $C(\overline{E})$, $C(E_R)$, and $C(\overline{E}_R)$ and similarly for $\overline{C}$. In [A5] compound channels, a certain class of channels, were considered. The concepts are even more relevant for the more sophisticated AVC, for example.

8. Well-known is Shannon's **rate-distortion function** in source coding. It is amazing that our preceding analogue for channel coding was not introduced. However, it must be said that Shannon introduced his function "informationally" (and so does Toby Berger in [C7]) and not "operational" as we did . In channel coding he gave two definitions for the channel capacity, an informational and an operational one. This is very well discussed in the explanation of Aaron Wyner, which we cite in [B37]. Unfortunately, some well-known text books like [C17] or [C110] give the informational one. But $\{\max_{P^n} I(W^n | P^n)\}$ describes the operational capacity C only in special cases like the DMC and is too large for instance for averaged channels (memory!). Here lies one of the main roots for conceptional confusions about channel capacity!

9. As we have explained earlier a nice property to have is additivity of a capacity for parallel channels. This is the case for the ordinary capacity C, if $C = \overline{C}$ and this is exactly in the case where the **weak converse** holds. We also say in this case that the **weak capacity** exists (see [A3]). **So this quantity does not exist automatically.** This is sometimes overlooked and even more true for the strong capacity introduced in [A3], when the strong converse holds.

This must be kept in mind when we compare results. Generality is of course easier to obtain for capacities with weaker properties than for those with stronger properties.

In proving upper bounds like the weak converse it is helpful to prove first bounds, which can be obtained easier, like polynomial or exponential (see also soft converse in [A59]) weak converses discussed in [A208] and [D1].

10. We just indicate that in the spirit of [B37] one should introduce also

rate sequence-error functions

- saying in particular much more about non-stationary channels than rate-error functions,

classes of rate sequence-error functions

- catching tighter descriptions suggested in Section 11 of [B37].

Of course associated with these performance criteria are capacity concepts.

11. The discussion should be continued and should some time include other performance criteria like the analoga to (a) and (c) above. Also analoga of (a),(b),(c) for combinatorial channel models and also criteria for sources are to be classified.
It should be specific about distinctions stemming from

multi-way channels – feedback situations – non-block codes – delay – synchronization.

Beyond capacities for Shannon's transmission there are those for

identification (second order) – common randomness – general information transfer (first and second order).

In combinatorial channel models to be distinguished are the various error concepts:

failure of error detection or wrong detection – defects – insertions – deletions – localized errors – unidirectional errors – etc.

The case of feedback just brings us to search (see chapter VI) with the recently studied lies with cost constraints etc.
Also J. Körner's models of a combinatorial universe with information aspects, starting with ambiguous alphabets and going along trifferent paths, definitely must be included (Sperner Capacity).
Recent work by G. Katona and K. Tichler in search deserves immediate attention. There a test is a partition of the search space $\mathcal{X}$ into 3 sets $(\mathcal{Y}, \mathcal{N}, \mathcal{A})$. If the object x searched for satisfies $x \in \mathcal{Y}$ the answer is Yes, if it satisfies $x \in \mathcal{N}$ the answer is No, it is for $x \in \mathcal{A}$ arbitrary Yes or No.

Also to be classified are the performance criteria in the very important work on codes introduced by Kautz/Singleton [C55] and studied by Lindström, Dyachkov, Erdös and many others (see survey [D3]).
The results found an application in the probabilistic model of K-identification in [D1].
Finally, analogous performance criteria are to be defined in Statistics in particular in the interplay between Multi-user Source Coding Theory and Hypothesis Testing or Estimation starting with work [A51], [A66] with I. Csiszar and M. Burnashev and continued by many others (see survey [C44] by T.S. Han and S.I. Amari).
For comments on [B38] see Chapter I.

VI

Enrichments for the project are gained from relations between coding for channels with feedback and search problems (c.f. [A32], [A55]).

For example error-correcting codes with feedback, which were introduced by Berlekamp [C5] are equivalent to the following search problem. A search space $\mathcal{M} = \{1, \ldots, M\}$ is given and we want to find one (say defective) element. In every step we perform a test by choosing a subset of $\mathcal{M}$. When working correctly the test produce a "Yes", if the defective element is in the subset and otherwise it produces a "No". The main problem is that the tests not always give the correct answer. In our model we assume that the number of incorrect answers is restricted.

This search model is often described equivalently as "Renyi-Berlekamp-Ulam-Game". The models readily extends to q alternatives for answers $Q = \{0, 1, \ldots, q-1\}$. **The new idea, which was developed during the ZiF-period, is to consider error cost constraints.** That means, there is a function $\Gamma : Q \times Q \to \mathbb{N}$. The function Γ is meant to weigh the answers. Whenever an answers t to a question (a test), whose answer is s, is given the answer has weight $\Gamma(t, s)$. It is allowed to give false answers with total weight up to e.

In [B39] the authors assume some symmetry of Γ and weights 0 and 1. They provide a lower bound on the number of questions needed to solve the problem and prove that in infinitely many cases this bound is attained by (optimal) search strategies. Moreover they prove that, in the remaining cases, at most one question more than the lower bound is always sufficient to successfully find the unknown element. All strategies also enjoy the property that among all possible adaptive strategies they use the minimum amount of adaptiveness during the search process. **In [D10] the general weighted case is solved without any assumption on Γ.**

A coding scheme for delayed feedback, which shows that in this case the capacities of all memoryless channels with non-delayed feedback can be achieved, is given in [B40]. A characterization of the zero-error capacity of a DMC and the average-error capacity of an AVC, when the delay time increases linearly with the length of the codes, is also obtained.

In [B41] the Kraft inequality for d-DBS codes is sharpened, based on the work of Ambains-Bloch-Schweizer, who introduced these codes.

A generalization of the well-studied group testing problem is introduced in [B42] and an algorithm is given. Group testing is of interest in chemical and biological testing, DNA mapping and also in several computer science applications with conflict resolutions for packages in random-multiple access transportations with and without feedback (see survey by Gallager in [C38]).

A new suffix sorting algorithm to sort all suffixes of a string $x^n \in \{0, \ldots, k-1\}^n$ lexicographically is developed in [B43]. It computes the suffix sorting in $O(n)$ space and $O(n^2)$ time in the worst case. It has also the property that it sorts the suffixes lexicographically correctly according to the prefixes of length $\log_k \lceil \frac{n}{2} \rceil$ in the worst case in linear time.

In [B44] the comparison model and the linear model of monotonicity testing is considered. There are some general bounds on the comparison model and an analysis of the complexity, for example the monotonicity checking complexity of Boolean functions, is determined. A geometric interpretation of monotonicity checking is considered and a method to establish lower bounds in the linear model using this interpretation is developed.

The paper [B45] surveys two recent classes of randomized motion planning methods. The first are the so-called probabilistic roadmap methods which build a graph mapping of the free configuration space and which are suitable for answering multiple queries. The second class of methods explore the configuration space by growing trees from some designated configuration.

VII

In [B46] readers find an interesting application of Information Theory in the study of Language Evolution. The model was originally introduced by M.A. Nowak and D.C. Krakauer [C71], where the fitness of a language is introduced. For this model they showed if signals can be mistaken for each other, then the performance of such systems is limited. The performance cannot be increased over a fixed threshold by adding more and more signals. Nevertheless the concatenation of signals or phonemes to words increases significantly the fitness of the language. The fitness of such a signalling-system depends on the number of signals and on the probabilities to transmit individual signals correctly. R. Ahlswede, E. Arikan, L. Bäumer and C. Deppe investigated optimal configurations of signals in different metric spaces. In [B46] they **prove for all metrics with a positive semidefinite associated matrix a conjecture by Nowak** including all important metrics studied by different authors in this direction. The conjecture holds for all ultra-metric spaces. Especially the authors analyze the Hamming space. In this space the direct consequence of the theorem is that the fitness of the whole space equals the maximal fitness and the fitness of Hamming codes asymptotically achieves this maximum. These theoretical models of fitness of a language enable the investigations of traditional information theoretical problems in this context, in particular, for feedback problems, transmission problems for multi-way channels etc. It is shown that feedback increases the fitness of a language.

For a novel the frequency of its words can be ordered $f_1 \geq f_2 \geq f_3 \geq \ldots$. The most frequent has rank 1, the second most frequent has rank 2 and so on. **The stenographer J.B. Estoup [C33] observed that in a French text** $r \cdot f_r$ **is approximately constant.** This hyperbolic rank-frequency relationship was confirmed by very careful studies by Zipf [C111] giving him the harvest: **Zipf's Law.** Zipf argued that this vocabulary balance might be the result of two opposing forces, the tendency of the speaker to reduce the vocabulary (least effort going towards unification) and the auditors wish to associate meaning to speech (driving towards diversification).

Zipf did not offer a mathematical model, but viewed the situation as a two-person game, however without a definition of the pay-off function involved.

In idealization the f_i are proportional to $\frac{1}{i}$ and going to infinitely many letters $(f_1, f_2, \ldots)$ becomes proportional to a harmonic sequence. Accepting this idealization one comes to study the function $f : \mathbb{N} \to \mathbb{R}_+$ (Shannon used Zipf's Law to establish the entropy of English [C84]).

1961 B. Mandelbrot argued that a purely random mechanism will generate a text obeying Zipf's Law and Schroeder [C80] put this in the form "a monkey hitting typewriter keys *at random* will also produce a "language" obeying Zipf's Law".

But actually Mandelbrot, in the same paper, followed Zipf's game theoretic reflections by considering gametheoretic elements via coding of words. P. Harremoes and F. Topsoe in [B47] just accept the law **no matter how it comes about**, and try to describe it using the class of probability distributions, which are ordered $P_1 \geq P_2 \ldots$ and which are characterized by $P_i \geq i^{-a}$ for some $a > 1$ and infinitely many i and they call them *hyperbolic*.

Every distribution with infinite entropy is hyperbolic. Their interest is in the set of hyperbolic distributions with *finite entropy as candidates giving stability to a language*.

They considered a code-length-zero-sum 2 person game with a set $\mathcal{P}$ of probability distributions on $\mathbb{N}$ as set of strategies for Player I and the set $\kappa(\mathbb{N})$ of (idealized) *codes* $K : \mathbb{N} \to [0; \infty]$, for which $\sum_1^\infty \exp(-K(i)) = 1$, as set of strategies for Player II, and $< K, P >$ as a cost function for Player II. The game turns out to be in equilibrium if and only if $\sup_{P \in \mathcal{P}} H(P) = \sup_{P \in co(\mathcal{P})} H(P) < \infty$ and then $\sup H(P)$ is its value.

Furthermore, for any sequence $(P_n)_{n \geq 1}$ from $\mathcal{P}$ with $H(P_n) \to \sup_{P \in \mathcal{P}} H(P)$ there exists a P^*, a $\sup_{P \in \mathcal{P}} H(P)$ attractor, such that $P_n \to P^*$ (in total variation). Cases with **entropy loss**, $H(P^*) < \sup_{P \in \mathcal{P}} H(P)$ are possible. *This is where the hyperbolic distributions come in: they are exactly the attractors with entropy loss.* Turned positively, according to the authors, they are the guarantors of stability of a language providing the language the *potential to enrich itself (increase entropy) to higher and higher expressive powers without changing its basic structure.*

In [B48] the **notion of a motif**, which is a string of solid and wild characters, is investigated for data compression purposes. A major difficulty in using motifs is explained by the fact that their number can grow exponentially in the length of the sequence to be compressed. This is overcome by considering irredundant motifs. The authors present data compression techniques based on the notion of irredundant motifs and show that in several cases they provide an improvement in the rate of compression as compared with previous popular compression methods.

A new family of codes, the similarity codes, is discussed in [B49]. Such a code is specified by a similarity function depending on two vectors and fulfills the requirement that the similarity between any two distinct codewords is not greater than a threshold d. One can reformulate this condition in terms of semi-distances.

The similarity codes turn then into codes with minimal distance d. Motivated by biological applications, the authors consider the important subclass of LCS-codes by taking the length of the longest common subsequence of two strings as a measure for their similarity. These codes correspond to Levenshtein's insertion–deletion codes. Reverse-complement (RC-LCS) codes are those LCS-codes which are closed with respect to the operation of taking reverse-complements of the codewords. If, in addition, each codeword and its reverse-complement are distinct, the code is called a dRC-LCS-code. The authors give a lower bound for the rates of the discussed three types (LCS, RC-LCS and dRC-LCS) of codes.

One of the main tasks of **chemical graph theory** is the description of the chemical structure. As the topology of molecules determines a large number of their properties, a simple approach to achieve this task would be to have some measures (numbers) reflecting the main features of a topological structure. Such measures are called topological indices. A lot of such indices have been suggested during the last 50 years. One major drawback of them is, however, that they cannot discriminate well between isomers, often giving the same index value for different isomers. Another, relatively new approach is to use information measures (indices) for the characterization problem. In [B50] several information indices are reviewed. Then the author presents some numerical results of discriminating tests of indices on structural isomers and demonstrates the correlating ability of information indices on several classes of organic and organometallic compounds.

In [B51] a special case of the following graph–theoretical problem is investigated. Given two natural numbers d and k, find the largest possible number $n(d, k)$ of vertices in a graph with diameter k and maximum vertex degree d. In the case when $k = 2$ and $d = 6$, it is known that $n \geq 32$. The author gives an algorithm, which constructs all non-isomorphic largest graphs of diameter 2 and maximum vertex degree 6, at the same time showing that $n(6, 2) = 32$.

VIII

Combinatorial
Extremal
Problems

Information

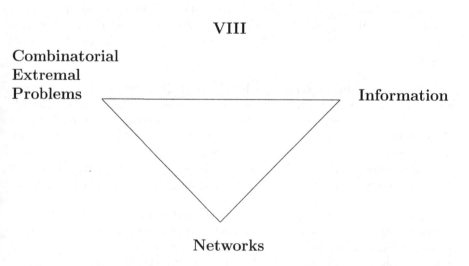

Networks

The founder of Information Theory Claude E. Shannon, who set the standards for efficient transmission of channels with noise by introducing the idea of coding - at a time where another giant John von Neumann was still fighting unreliability of systems by repetitions -, also wrote together with Peter Elias and Amiel Feinstein a basic paper on networks [C32] containing the - seemingly independently of L.R. Ford and D.R. Fulkerson [C35] discovered - Min Cut - Max Flow Theorem, saying that for flows of physical commodities like electric currents or water, satisfying Kirchhoff's laws, the maximal flow equals the minimal cut.

With the stormy development of Computer Science there is an ever increasing demand for designing and optimizing Information Flows over networks - for instance in the Internet.

Data, that is strings of symbols, are to be send from sources $s_1, \ldots, s_n$ to their destinations, sets of node sinks $D_1, \ldots, D_n$.

Computer scientist quickly realized that it is beneficial to copy incoming strings at processors sitting at nodes of the network and to forward copies to adjacent nodes. This task is called Multi-Casting.

However, quite surprisingly **they did not consider coding**, which means here to produce not only copies, but, more generally, new output strings as deterministic functions of incoming strings.

In [A155] a **Min-Max-Theorem was discovered and proved for Information Flows.**

Its statement can be simply explained. For one source only, that is $n = 1$, in the notation above, and $D_1 = \{d_{11}, d_{12}, \ldots, d_{1t}\}$ let F_{1j} denote the Max-Flow value, which can go for any commodity like water in case of Ford/Fulkerson from s_i to d_{1i}. The same water cannot go to several sinks. However, the amount of $\min_{1 \leq j \leq t} F_{1j}$ bits can go **simultaneously** to $d_{11}, d_{12}, \ldots$ and d_{1t}. Obviously, this is best possible. It has been referred to as ACLY-Min-Max-Theorem (**It also could be called Shannon's Missed Theorem**). To the individual F_{1j} Ford/Fulkerson's Min-Cut-Max-Flow Theorem applies.

It is very important that in the starting model there is no noise and it is amazing for how long Computer Scientists did the inferior Multicasting allowing only copies.

Network Flows with more than one source are much harder to analyze and lead to a wealth of old and new Combinatorial Extremal problems. **This is one of the most striking examples of an interplay between Information Transfer and Combinatorics.**

Two workshops in the ZiF-project GTIT-C were devoted to this.

Even nicely characterized classes of **error correcting codes** come up as being isomorphic to a complete set of solutions of flow problems **without errors!**

Also our characterization of optimal **Anticodes** obtained with the late Levon Khachatrian [A132] arises in such a role!

On the classical side for instance orthogonal **Latin Squares** - on which Euler went so wrong - arise.

The Min-Max-Theorem has been made practically more feasible by a polynomial algorithm by Peter Sanders, Sebastian Egner and Ludo Tolhuizen [C78] as well as by his competitors (or groups of competitors) in other parts of the world, leading to the joint publications [C53].

With NetCod 2005 - the first workshop on Network Coding Theory and Applications, April 7, 2005, Riva, Italy the **New Subject Network Coding** was put to start.

In preparation is a special issue of the Transaction on Information Theory and the Transactions on Networking dedicated to Networking and Information Theory.

Research into network coding is growing fast, and Microsoft, IBM and other companies have research teams who are researching this new field.

A few American universities (Princeton, MIT, Caltec and Berkeley) have also established research groups in network coding.

The holy grail in network coding is to plan and organize (in an automated fashion) network flow (that is to allowed to utilize network coding) in a feasible manner. Most current research does not yet address this difficult problem.

There may be a great challenge not only coming to **Combinatorics** but also to **Algebraic Geometry** and its present foundations (see [C58] and also [B52]).

An Introduction to the area of Network Coding is given in the book [C110]. For a discussion of some recent developments we refer to [B53].

In order to get outside opinions we have included under [B52] the Network Coding Side of a leading expert, Ralf Koetter.

The case $|D_i \cap D_j| = \varnothing$ for $i \neq j$ and $|D_i| = 1$ for $i = 1, \ldots, n$, that is, **each source sends its message to its sink** has an obvious symmetry and appeal. Soren Riis established the equivalence of this flow problem to a **guessing game**, which is **cooperative**. We include - as the only exception in this book - a draft, under [B53], serving two purposes: an essential widening of the scope of the book by inclusion of Game Theory and the additional stimulation for discussions Soren Riis asked for. An improved paper including the reactions will be published elsewhere.

Concerning Game Theory we like to add that there is another role of non-cooperative Game Theory in Computer Science recently emerged in the area of "selfish routing", where starting with work of Koutsoupias and Papadimitriou [C61] behaviour is modelled as being guided by Nash equilibria. This has entered Network Theory for instance in classical traffic models and **likely will enter also information flows in networks**!

IX

Basic covering problems are investigated in [B54]: the problem of thinnest coverings of spheres and ellipsoids with balls and ellipsoids in Hamming and Euclidean spaces. New bounds in terms of the *epsilon*-entropy of Hamming balls and spheres are established. The derived upper and lower bounds are optimal up to an additive logarithmic term on the dimension.

In [B55] the following covering problem for product spaces is investigated. Given a finite set $\mathcal{X}$ and a covering set system $\mathcal{E}$ for $\mathcal{X}$, let $\mathcal{X}^n$ and $\mathcal{E}^n$ be the Cartesian products of $\mathcal{X}$ and $\mathcal{E}$. The problem is to determine or estimate the minimal number needed for covering $\mathcal{X}^n$ by elements of $\mathcal{E}^n$. Upper and lower bounds are obtained for the minimal covering number. The main result of the paper was obtained by the author in 1971 independently of, but somewhat later than, the work of McEliece and Posner [C69] motivated by ideas of data compression and remained therefore unpublished.

However, the methods used in this paper turned out to be useful for investigation of more general covering problems in hypergraphs (see [A35], [A36]). Also the approach is purely combinatorial, no entropies and no games are involved and can be understood by readers with no background in Information Theory. Connections to other combinatorial problems like Shannon's zero-error capacity problem, which is exactly equivalent to the corresponding packing covering, are discussed as well. This justifies the present inclusion.

A perfect code is an important partition in the theory of error-correcting codes. The topics of [B56] are binary 1-perfect codes. In order to study them, the authors introduce a θ-centered function and testing sets of subsets of binary sequences for a given family of functions from the set of binary sequences to the set of real numbers. With them the authors proved the interesting result that for $r \leq \frac{n-1}{2}$ and in a binary 1-perfect code, the codewords of Hamming weight smaller than r are uniquely determined by the set of all codewords of Hamming weight equal to r. It is expected that the result is helpful to better understand the structure of binary 1-perfect codes.

The following extremal problem is considered in [B57]. Given a natural number n, partition a rectangle R into rectangles in such a way that any line parallel to a side of R intersects at most n of the smaller rectangles. What is the maximal number $f(n)$ of rectangles in such a partition? A simple construction shows that $f(n) \geq 3 \cdot 2^{n-1} - 2$ holds. By a stepwise refinement of the methods using finally also harmonic analysis, the authors come close to this lower bound and prove the upper bound $f(n) \leq 2n \cdot 2^n$. It is conjectured that the lower bound is tight.

Several new combinatorial extremal problems are raised and studied in [B58]. These problems are motivated by problems from other fields such as Information Theory, Complexity Theory and Probability Theory. One of the open problems stated here is a generalization of the isoperimetric problem for binary Hamming spaces. For a set of points V of a given size the problem is to minimize the number of points, not in V, that have at least k elements at distance one from V. For $k = 1$, we have the isoperimetric problem solved by Harper. Also an interesting observation is that guessing, which was introduced by Jim Massey [C68], is a very special case of the model of (what has been called in [A32], [A55]) **inspections**.

The edge–isoperimetric problem EIP for graphs have been studied by many authors. For a simple connected graph $G = (V, E)$ and a vertex subset M of a given size, the problem is to minimize the size of the edge boundary of M over all $M \subset V$. A related problem is to maximize the number of edges in M. Moreover, these two problems are equivalent for regular graphs. The problem was solved for the Hamming graph by Harper (1967). However, the problem is widely open for the Johnson graph $J(n, k)$. The vertex set of $J(n, k)$ is the set of $(0, 1)$-sequences of length n and weight k, and two vertices are adjacent iff the Hamming distance between them is two. Ahlswede and Katona [A31] solved the EIP for the case $k = 2$. For $k \geq 3$ the problem is still open. Unfortunately, the strong isoperimetric inequalities obtained, for general graphs, via eigenvalue techniques, do not give the desired result for the Johnson graph. In [B59] a new approach based on Young diagrams is used, to give another short proof of the result of Ahlswede and Katona, that quasi-star or quasi-ball are always optimal. There is no improvement on the second result of these authors, which specifies except for relatively few edge numbers who of the two configurations wins. Concerning the cases $k \geq 3$ there was a conjecture of Kleitman disproved for $k = 3$ in [A144]. The SFB-preprint [B60] has more results, cited sometimes in the literature, and is know better available.

In [B61] from 1981 as strengthening of the famous Erdős–Ko–Rado Theorem is presented. Surprisingly, this result was obtained for the solution of a geometric problem raised by M. Burnashev, in connection with a coding problem for Gaussian channels. Let $\mathcal{B}$ be a family of l-sets of an n-element set. Suppose that $\mathcal{B}$ satisfies the following "Triangle Property": $A \cap B \neq \varnothing$, $B \cap C \neq \varnothing \Rightarrow A \cap C \neq \varnothing$ for all $A, B, C \in \mathcal{B}$. In particular, the intersection property required in the EKR Theorem implies the Triangle Property. It is proved that for every family $\mathcal{B}$ satisfying the Triangle Property, with $n \geq 2l$ and $l \geq 3$, we have $|\mathcal{B}| \leq \binom{n-1}{l-1}$ and this bound is best possible. Intended inclusion in a book by G. Katona did not realize, because the book still does not exist.

The author of [B62] gives an excellent survey about algorithms for multileaf collimators. A multileaf collimator is used to modulate a radiation field in the treatment of cancer. The author considers several aspects of the construction of optimal treatment plans. The algorithms are presented in a very understandable way. The various definitions of several papers are unified. In the last part of the paper the author discusses advantages and disadvantages of the existing algorithms.

A communication network is modelled as an acyclic directed graph $G = (V, E)$ with some distinguished vertices called inputs and other distinguished vertices called outputs. The remaining vertices are called links. There are two parameters of particular interest in comparing networks: the size and the depth. The size (the number of edges) in some approximate sense corresponds to the cost of the network. The depth (the length of the longest path from an input to an output of the network) corresponds to the delay of the transmission in the network. Therefore in designing communication networks it is desirable to achieve smaller size and smaller depth. An (n, N, d)-connector or rearrangeable network is a network with n inputs, N outputs and depth d, in which for any injective mapping of input vertices into output vertices there exist n vertex–disjoint paths joining each input to its corresponding output. The problem of designing optimal connectors goes back to works of Shannon, Slepian et al. ([C87], [C88], [C15], [C3]) started in the 50's. In [B63] asymmetric connectors (connectors with $n \ll N$) of depth two are considered. A simple combinatorial construction of sparse connectors is given, which is based on the Kruskal/Katona Theorem for shadows of families of k-element subsets. Fault–tolerance of the constructed connectors is also considered. The results are in general and also in most special cases the presently best.

This ends the commentary. The reader is advised to look at titles of lectures and their abstracts in this book and the associated special issue of DAM, because a "lot of information is coded there".

Not visible are the oral communications with individual impacts. An example here stands for many. The Statistician W. Müller drew in a talk, filled with citations of original expositions by K.F. Gauss written in Latin and attempts of accurate translations, in particular attention to the following passage in a letter of Gauss to the Astronomer and Mathematician Bessel dated February 28, 1839 (see [C73]):

> Dass ich übrigens die in der Theoria Motus Corporum Coelestium ange-
> wandte Metaphysik für die Methode der kleinsten Quadrate späterhin
> habe fallen lassen, ist vorzugsweise auch aus einem Grunde geschehen,
> den ich selbst öffentlich nicht erwähnt habe. Ich muss es nämlich in
> alle Wege für weniger wichtig halten, denjenigen Werth einer unbekan-
> nten Grösse auszumitteln, dessen Wahrscheinlichkeit die grösste ist, die
> ja doch immer nur unendlich klein bleibt, als vielmehr denjenigen, an
> welchen sich haltend man das am wenigsten nachtheilige Spiel hat; . . .

The **astute reader notices an anticipation of A. Wald's Decision Theory**, based on zero-sum games between the Statistician and Nature, essentially a hundred years earlier. Well, this was even almost a hundred years before J. von Neumann discovered his Min-Max Theorem and then developed Game

Theory together with O. Morgenstern (see [C99]). Once it became clear that old books also should be read, a more detailed study of the collection of letters [C73] brought another very interesting news: between the position of accepting Kant's opinion concerning the role of the "äußere Anschauung" and the "innere Anschauung" for making **true** statements **a priori** possible and the position to reject them (which has been popular in modern science), Gauss rejects one (geometry is subject to experience), but accepts the other (giving truth to arithmetic statements).

On another occasion Gauss said that having read the Critique of Pure Reason six times (!) he started understanding.

References

[C1] I. Althöfer, N. Cai, G. Dueck, L.H. Khachatrian, M. Pinsker, A. Sárközy, I. Wegener, and Z. Zhang (editors), Numbers, Information and Complexity, Special volume in honour of R. Ahlswede on occasion of his 60th birthday, Kluwer Acad. Publ., Boston, Dordrecht, London, 2000.

[C2] E.H. Armstrong, A method of reducing disturbances in radio signating by a system of frequency modulation, Proc. Inst. Radio Eng., Vol. 24, 689-740, 1936.

[C3] V.E. Beneš, Optimal rearrangeable multistage connecting networks, Bell System Tech. J. 43, 1641–1656, 1964.

[C4] C.H. Bennett, E. Bernstein, G. Brassard, and U.V. Vazirani, Strength and weakness of quantum computing, SIAM J. on Computing, Vol. 26 no. 5, 1510-1523, Oct., 1997.

[C5] E.R. Berlekamp, Block coding with noiseless feedback, Phd-thesis, MIT, Cambridge, MA, 1964.

[C6] P.P. Bergmans, Random coding theorem for broadcast channels with degraded components, IEEE Trans. Inform. Theory, Vol. 19, No. 2, 197–207, 1973

[C7] T. Berger, Rate Distortion Theory: A Mathematical Basis for Data Compression, Prentice-Hall, Englewood Cliffs, 1971.

[C8] D. Blackwell, L. Breiman, and A.J. Thomasian, The capacity of a certain channel classes under random coding, Ann. Math. Statist. 31, 558–567, 1960.

[C9] F. Blume, Possible rates of entropy convergence, Ergodic Theory and Dynam. Systems, 17, No. 1, 45–70, 1997.

[C10] J.W. Bradbury and S.L. Vehrencamp, Principles of Animal Communication, Sinauer Assoc., Inc., Publishers Sunderland, Mass., USA, 1998.

[C11] R. Carnap and Y.B. Hillel, An Outless of a Theory of Semantic Information, in Language and Information, Massachusetts 1964.

[C12] E.C. Cherry, On Human Communication, a Review, a Survey and a Criticism, MIT–Press 1957, 1966.

[C13] E.C. Cherry, A history of information theory, Proc. Inst. Elec. Eng. (London), Vol. 98, pt 3, 383-393, 1951.

[C14] A.M. Childs and J. M. Eisenberg, Quantum algorithms for subset finding, e–print, quant-ph/0311038, 2003.

[C15] C. Clos, A study of non–blocking switching networks, Bell System Tech. J. 32, 406–424, 1953.

[C16] T.M. Cover, Broadcast channels, IEEE Trans. Inform. Theory, Vol. 18, 2-14, 1972.

[C17] T.M. Cover and J. Thomas, Elements of Information Theory, Wiley, 1991.

[C18] S. Csibi and E. C. van der Meulen, Error probabilities for identification coding and least length single sequence hopping, Numbers, Information and Complexity, special volume in honour of R. Ahlswede on occasion of his 60th birthday, edited by Ingo Althöfer, Ning Cai, Gunter Dueck, Levon Khachatrian, Mark S. Pinsker, Andras Sárközy, Ingo Wegener and Zhen Zhang, Kluwer Academic Publishers, Boston, Dordrecht, London, 221-238, 2000.

[C19] I. Csiszár and Körner, Broadcast channel with confidential message, IEEE Trans. Inform. Theory, Vol. 24, 339-348. 1978.

[C20] I. Csiszár and Körner, Information Theory: Coding Theorems for Discrete Memoryless Systems, Probability and Mathematical Statistics, Academic Press, New York-London, 1981.

[C21] L.D. Davisson, Universal noiseless coding, IEEE Trans. Inform. Theory, Vol IT-19, No. 6, 1973.

[C22] H. Diels and G. Plamböck, Die Fragmente der Vorsokratiker, Rowohlt, 1957.

[C23] S.M. Dodunekov, Optimization problems in coding theory, survey presented at a Workshop on Combinatorial Search, in Budapest, 23.-26. April, 2005.

[C24] J.L. Doob, Review of "A mathematical theory of communication", Mathematical Reviews 10, 133, 1949.

[C25] G. Dueck, Omnisophie: über richtige, wahre und natürliche Menschen, Springer-Verlag, Berlin Heidelberg, 2003.

[C26] H. Dudley, The vocoder, Bell. Lab. Rec., Vol. 18, 122-126, 1939.

[C27] M. Eigen, Self-organization of matter and the evolution of biological macro molecules, Naturwissenschaften 58, 465, 1971.

[C28] M. Eigen, Wie entsteht Information? Ber. Bunsenges. Phys. Chem. 80, 1060, 1976.

[C29] M. Eigen, Macromolecular evolution: dynamical ordering in sequence space, Ber. Bunsenges. Phys. Chem. 89, 658, 1985.

[C30] M. Eigen, Sprache und Lernen auf molekularer Ebene, in "Der Mensch und seine Sprache", A. Peise und K. Mohler (Eds), Berlin 1979.

[C31] A. Einstein, B. Podolsky, and N. Rosen, Can quantum-mechanical description of physical reality be considered complete? Phys. Rev., 47, 777-780, 1935.

[C32] P. Elias, A. Feinstein, and C.E. Shannon, A note on the maximum flow through a network, IEEE Trans. Inform. Theory, 11, 1956.

[C33] J.B. Estoup, Gammes Sténographique, Paris, 1916.

[C34] M. Feder, N. Merhav, and M. Gutman, Universal prediction of individual sequences, IEEE Trans. Inform. Theory, Vol. 38, 1258-1270, 1992.

[C35] L.R. Ford and D.R. Fulkerson, Flows in Networks, Princeton University Press, Princeton, N.J. 1962

[C36] P. Gács and J. Körner, Common information is far less than mutual information, Problems of Control and Information Theory/Problemy Upravlenija i Teorii Informacii 2, No. 2, 149–162, 1973.

[C37] R.G. Gallager, Traffic capacity and coding for certain broadcast channels (Russian), Problemy Peredači Informacii 10, No. 3, 3–14, 1974.

[C38] R.G. Gallager, A perspective on multi–access channels, IEEE Trans. Inf. Theory, Vol. 31, No. 2, M1985.

[C39] E.N. Gilbert, How good is Morsecode, Inform. Contr., Vol. 14, 559-565, 1969.

[C40] L.K. Grover, A fast quantum mechanical algorithm for database search, Proceedings, 28th Annual ACM Symp. on Theory of Computing (STOC), 212-219, 1996.

[C41] L.K. Grover, Quantum mechanics helps in searching for a needle in a haystack, Phys. Rev. Letters, 78, 2, 325-328, 1997.

[C42] G.B. Guercerio, K. Gödel, Spektrum der Wissenschaft, (Deutsche Ausgabe Scientific American), Biographie 1/2002.

[C43] W. Haemers, On some problems of Lovasz concerning the Shannon capacity of a graph, IEEE Trans. Inform. Theory, Vol. 25, No. 2, 231–232, 1979.

[C44] T.S. Han and S.I. Amari, Statistical inference under multiterminal data compression, information theory: 1948–1998, IEEE Trans. Inform. Theory 44, No. 6, 2300–2324, 1998.

[C45] T.S. Han and S. Verdu, New results in the theory of identification via channels, IEEE Trans. Inform. Theory, Vol. 38, No. 1, 14–25, 1993.

[C46] E.A. Haroutunian and A. N. Harutyunyan, Successive refinement of information with reliability criterion, Proc. IEEE Int Symp. Inform. Theory, Sorrento, Italy, June, 205, 2000.

[C47] E.A. Haroutunian, Upper estimate of transmission rate for memoryless channel with countable number of output signals under given error probability exponent, (in Russian), 3rd All-Union Conf. on Theory of Information Transmission and Coding, Uzhgorod, Publication house of Uzbek Academy of Sciences, Tashkent, 83-86, 1967.

[C48] R.V.L. Hartley, Transmission of information, Bell. Syst. Tech. J., Vol. 7, 535-563, 1928.

[C49] M.D. Hauser, The Evolution of Communication, MIT Press, Cambridge, Mass., London, England, 1997.

[C50] A.S. Holevo, Problems in the mathematical theory of quantum communication channels, Rep. Math. Phys., 12, 2, 273-278, 1977.

[C51] H.D.L. Hollmann, J.H. van Lint, J.P. Lennartz, and L.M.G.M. Tolhuizen, On codes with the identifiable parent property, J. Combin. Theory Ser. A, Vol. 82, No. 2, 121–133, 1998.

[C52] M. Horodecki, Is the classical broadcast channel additive? Oral Communication, Cambridge England, Dec. 2004.

[C53] S. Jaggi, P. Sanders, P. A. Chou, M. Effros, S. Egner, K. Jain, and L. Tolhuizen, Polynomial time algorithms for multicast network code construction, IEEE Trans. on Inform. Theory, Vol. 51, No. 6, 1973- 1982, 2005.

[C54] G. Katona and K. Tichler, When the lie depends on the target, presented at a Workshop on Combinatorial Search, in Budapest, April 23-26, 2005.

[C55] W. Kautz and R. Singleton, Nonrandom binary superimposed codes, IEEE Trans. Inform. Theory, Vol. 10, 363–377, 1964.

[C56] J. Körner and K. Marton, General broadcast channels with degraded message sets, IEEE Trans. Inform. Theory, Vol. 23, 60-64, 1977.

[C57] J. Körner, Some methods in multi-user communication – a tutorial survey. In: Information Theory, New Trends and Open Problems (ed. G. Longo) CISM Courses and Lectures No 219, Springer Verlag, Wien, 173-224, 1975.

[C58] R. Koetter and M. Medard, An algebraic approach to network coding, Transactions on Networking 11, No. 5, 782-795, 2003.

[C59] A.N. Kolmogorov, Logical basis for information theory and probability theory, IEEE Trans. Inform. Theory, Vol. 14, 663, 1968.

[C60] A.N. Kolmogorov, Three approaches to the quantitative definition of information, Internat. J. Comput. Math. 2, 157–168, 1968.

[C61] E. Koutsoupias and C. Papadimitriou, Worst-case equilibria, Proceedings of the 16th Symposium on Theoretical Aspects of Computer Science (STACS), 404-413, 1999.

[C62] K. Küpfmüller, Über Einschwingvorgänge in Wellen Filter, Elek. Nachrichten-
       tech., Vol. 1, 141-152, 1924.
[C63] B.O. Küppers, Der semantische Aspekt von Information und seine evolutions-
       biologische Bedeutung, Nova Acta Leopoldina, Bd. 72, Nr. 294, 195-219, 1996.
[C64] J.R. Larson, Notes on the theory of modulation, Proc. Inst. Radio Eng., Vol.
       10, 57-69, 1922.
[C65] M. Li and P. Vitanyi, An Introduction to Kolmogorov Complexity and Its
       Applications, Second Edition, Springer Verlag, 1997
[C66] P. Löber, Quantum channels and simultaneous ID coding, Dissertation, Uni-
       versität Bielefeld, Germany, 1999.
[C67] G. Lüders, Über die Zustandsänderung durch den Meßprozeß, Ann. d. Physik
       8, 322-328, 1951.
[C68] J.L. Massey, Guessing and entropy, Proc. IEEE Int. Symp. on Info. Th, page
       204, 1994.
[C69] R. McEliece and E.C. Posner, Hiding and covering in a compact metric space,
       Ann. Statist. 1, 729–739, 1973.
[C70] J. Monod, Zufall und Notwendigkeit, München, 1971.
[C71] M.A. Nowak and D.C. Krakauer, The evolution of language, PNAS 96, 14,
       8028-8033, 1999.
[C72] H. Nyquist, Certain factors affecting telegraph speed, Bell. Syst. Tech. J., Vol.
       3, 324-352, 1924.
[C73] G. Olms, Briefwechsel: Carl Friedrich Gauss und Friedrich Wilhelm Bessel,
       Hildesheim, 1975.
[C74] J.R. Pierce, The early days of Information Theory, IEEE Trans. Inform. Theory,
       Vol. I-19, No 1, 3-8, 1973.
[C75] M. Polanyi, Life's irreducible structure, Science 160, 1308, 1968.
[C76] V.A. Ratner, Molekulargenetische Steuerungssysteme, Stuttgart 1977.
[C77] A. Renyi, Probability Theory, Amsterdam 1970.
[C78] P. Sanders, S. Egner, and L. Tolhuizen, Polynomial Time Algorithms for Net-
       work Information Flow, Proceedings of the fifteenth annual ACM symposium
       on Parallel algorithms and architectures, San Diego, California, USA, 286 - 294,
       2003.
[C79] F. Schiller, Die Verschwörung des Fiesco zu Genua, 1783.
[C80] M. Schroeder, Fractals, Chaos, Power Laws, New York, W.H. Freeman, 1991.
[C81] E. Schrödinger, What is Life, Cambridge, 1944.
[C82] B. Schumacher, Quantum coding, Phys. Rev. A 51, 2738-2747, 1995.
[C83] C.E. Shannon, A mathematical theory of communication, Bell Syst. Techn. J.
       27, 339–425, 623–656, 1948.
[C84] C.E. Shannon, Predicition and entropy of printed English, Bell Sys. Tech. J 3,
       50-64, 1950.
[C85] C.E. Shannon, The Bandwagon, Institute of Radio Engineers, Transactions on
       Information Theory, Vol. 2, 1956.
[C86] C.E. Shannon, Certain results in coding theory for noisy channels, Inform. and
       Control 1, 6–25, 1957.
[C87] C.E. Shannon, Memory requirements in a telephone exchange, Bell System
       Tech. J. 29, 343–349, 1950.
[C88] D. Slepian, Two theorems on a particular crossbar switching network, unpub-
       lished manuscript, 1952.
[C89] D. Slepian and J. Wolf, Noiseless coding of correlated information sources,
       IEEE Trans. Inform. Theory, Vol. 19, 471–480, 1973.

[C90] Y. Steinberg, New converses in the theory of identification via channels, IEEE Trans. Inform. Theory 44, No. 3, 984–998, 1998.

[C91] Y. Steinberg and N. Merhav, Identification in presence of side information with application to watermarking, IEEE Trans. Inform. Theory, Vol. 47, 1410-1422, 2001.

[C92] F. Takens and E. Verbitski, Generalized entropies: Renyi and correlation integral approach, Nonlinearity, 11, No. 4, 771–782, 1998.

[C93] E. Tuncel and K. Rose Error Exponents in scalable source coding, IEEE Trans. Inform. Theory, Vol.49, 289-296, 2003.

[C94] S. Venkatesan and V. Anantharam, The common randomness capacity of a pair of independent discrete memoryless channels, IEEE Trans. Inform. Theory, Vol. 44, No. 1, 215–224, 1998.

[C95] S. Venkatesan and V. Anantharam, The common randomness capacity of a network of discrete memoryless channels, IEEE Trans. Inform. Theory, Vol. 46, No. 2, 367–387, 2000.

[C96] S. Verdù and V. Wei, Explicit construction of constant-weight codes for identification via channels, IEEE Trans. Inform. Theory, Vol. 39, 30-36, 1993.

[C97] E.C. van der Meulen, Random coding theorems for the general discrete memoryless broadcast channel, IEEE Trans. Inform. Theory, Vol. 21, 180–190, 1975.

[C98] J. von Neumann, The Computer and the Brain, Yale University Press, 1958.

[C99] J. von Neumann and O. Morgenstern, Theory of Games and Economic Behavior, Princeton University Press, Princeton, New Jersey, 1944.

[C100] C.F. von Weizsäcker, Die Einheit der Natur, München, 1971.

[C101] N. Wiener, The theory of communication, Phys. Today, Vol. 3, 31-32, Sept. 1950.

[C102] N. Wiener, Extrapolation, Interpolation, and Smoothing of Stationary Time Series, Technology Press of the Massachusetts Institute of Technology, Cambridge, Wiley, New York, Chapman & Hall, London, 1949

[C103] R. Wilmink, Quantum broadcast channels and cryptographic applications for separable states, Dissertation, Universität Bielefeld, Germany, 69 pages, 2003.

[C104] A. Winter, Coding theorems of quantum information theory, Dissertation, Universität Bielefeld, Germany, 1999.

[C105] A. Winter, The capacity region of the quantum multiple access channel, e-print, quant-ph/9807019, 1998, and IEEE Trans. Inform. Theory, Vol. 47, No. 7, 3059–3065, 2001.

[C106] A. Winter, Quantum and classical message identification via quantum channels, e-print, quant-ph/0401060, 2004.

[C107] J. Wolfowitz, Coding Theorems of Information Theory, Ergebnisse der Mathematik und ihrer Grenzgebiete, Heft 31, Springer-Verlag, Berlin-Göttingen-Heidelberg, Prentice-Hall, Englewood Cliffs, 1961; 3rd. edition, 1978.

[C108] A.D. Wyner, The wire-tap channel, Bell Sys. Tech. J. V. 54, 1355-1387, 1975.

[C109] A.D. Wyner and J. Ziv, A theorem on the entropy of certain binary sequences and applications II, IEEE Trans. Information Theory, Vol. 19, 772–777, 1973.

[C110] R.W. Yeung, A First Course in Information Theory, Information Technology: Transmission, Processing and Storage, Kluwer Academic/Plenum Publishers, New York, 2002.

[C111] G.K. Zipf, Human Behavior and the Principle of Least Effort, Addison-Wesley, Cambridge, 1949.

[C112] J. Ziv, Back from Infinity: a Constrained Resources Approach to Information Theory, IEEE Information Theory Society Newsletter, Vol. 48, No. 1, 1998.

Rudolf Ahlswede – From 60 to 66

On September 15th, 2004 Rudolf Ahlswede celebrated his 66th birthday and we describe here developments in his work in the last six years. For an account on the first 60 years we refer to the preface of the book "Numbers, Information and Complexity", special volume in honour of R. Ahlswede on occasion of his 60th birthday, edited by Ingo Althöfer, Ning Cai, Gunter Dueck, Levon Khachatrian, Mark S. Pinsker, Andras Sárközy, Ingo Wegener and Zhen Zhang, Kluwer Academic Publishers, Boston, Dordrecht, London, 2000. From there we just cite the following paragraph describing Ahlswede's approach to Information Theory:

"Ahlswede's path to Information Theory, where he has been world-wide a leader for several decades, is probably unique, because it went without any engineering background through Philosophy: Between knowing and not knowing there are several degrees of knowledge with probability, which can even quantitatively be measured – unheard of in classical Philosophy.

This abstract approach paired with a drive and sense for basic principles enabled him to see new land where the overwhelming majority of information theorists tends to be caught by technical details. Perhaps the most striking example is his creation of the Theory of Identification."

During the period to be described here came on January 30, 2002 as a shock the message of the sudden and unexpected passing of his collaborator and friend Levon Khachatrian. His untimely death disrupted a very fruitful cooperation. We refer the reader to the memorial text on Levon Khachatrian in this volume for an appreciation of person and work.

From 1975 to 2003 Ahlswede was full professor in Bielefeld. At the *Sonderforschungsbereich "Diskrete Strukturen in der Mathematik"* he was heading two research projects "Combinatorics on Sequence Spaces" and "Models with Information Exchange" from 1989 to 2000. Since fall 2003 he is emeritus. Neither the end of the Sonderforschungsbereich nor his new status of emeritus let him to slow down the pace of his research activities. Outwardly this can be seen by the 6 research projects, which he is currently heading or participating in with his group:

1. "General Theory of Information Transfer and Combinatorics", German Science Foundation (DFG), (2001-2005).
2. "Entanglement and Information", German Science Foundation (DFG), (2001-2005).
3. "Interactive Communication, Diagnosis and Prediction in Networks", German Science Foundation (DFG), (2001-2005).
4. "General Theory of Information Transfer and Combinatorics", Center for Interdisciplinary Research (ZiF), 2001-2004.
5. "Efficient Source Coding and Related Problems", INTAS, 2001-2004.
6. "Combinatorial Structure of Intractable Problems", RTN, EU, 2002-2006.

R. Ahlswede et al. (Eds.): Information Transfer and Combinatorics, LNCS 4123, pp. 45–48, 2006.

Ahlswede's generalization of Shannon's theory of transmission and the theory of identification in one broad unifying theory of "General Information Transfer" led to the research project at the ZiF. In this interdisciplinary venture information theorists and researchers from fields, where information plays an essential role, have come together to exchange ideas, to develop the theory further and find applications in other areas.

We mention here a sample of three results, which were outcomes of these efforts, one could choose others. For the first time a functional of a probability distribution could be determined, which deserves the name Identification Entropy since it serves the same role in the identification for sources as does the Boltzmann entropy in Shannon's classical data compression (see the article "Identification entropy" [B33]).

The second result is the proof of a conjecture of Nowak in the theory of evolution of human languages with coauthors Erdal Arikan, Lars Bäumer and Christian Deppe. The result shows the effects of word composition from single phonemes on the fitness of the individuals and the used language, respectively (see the article "Information theoretic models in language evolution" [B46]).

Finally we mention a third result, about which we quote from R. Kötter's homepage

> Like many fundamental concepts, network coding is based on a simple basic idea which was first stated in its beautiful simplicity in the the the seminal paper by R. Ahlswede, N. Cai, S.-Y. R. Li, and R. W. Yeung, "Network Information Flow", (IEEE Transactions on Information Theory, IT-46, pp. 1204-1216, 2000).

It was published in the year 2000 before the ZiF-project started.

For Ahlswede's contributions to combinatorial extremal problems the reader finds a rich resource in the article "Advances on Extremal Problems in Number Theory and Combinatorics", Proceedings of the European Congress of Mathematicians, Barcelona, Birkhäuser 2000 and also his work found entrance in the recent books "Global Methods for Combinatorial Isoperimetric Problems" by Harper, "Sperner Theory" by Engel and "Extremal Combinatorics" by Jukna.

On Ahlswede's agenda of research problems to be investigated was Quantum Information Theory for a very long time clearly before the great activity in this area. Since 1997 up to the present day this has been implemented with the coauthors Vladimir Blinovsky, Ning Cai, Peter Löber and Andreas Winter.

Progress was made for multi-user channels and the theory of identification, which is adaptable to the quantum theoretical setting.

In particular the concept of Common Randomness, which originated while studying identification in the presence of feedback, is closely linked with entanglement in quantum mechanics, the central property exploited in quantum cryptography in world-wide efforts.

As a student Ahlswede asked one of his teachers Kurt Reidemeister (for some time member of the Vienna School) for advice about how to combine in the

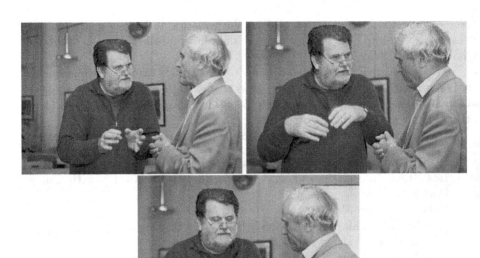

Embodied Communication

Final Conference of the ZiF Research Group, April 2004

studies interests in Philosophy and in Mathematics and received the answer: "Do it like I did it. First make it to an Ordinarius in Mathematics and then you have time for Philosophy."

In the instructive biography on Gödel [C42] one finds mentioned the principle **salva veritate** due to Leibniz: if two objects are identical and one has a property then also the other has that property. Identification asks for a converse. **Now there is time to think about this.**

Ahlswede has received many prizes and honors of which we mention here:

- 1988 IEEE Information Theory Society Best Paper Award (with Imre Csiszar)
- 1990 IEEE Information Theory Society Best Paper Award (with Gunter Dueck)
- 1996 Paul Erdös Monetary Award (with Levon Khachatrian)
- 1998/99 Humboldt – Japan Society Senior Scientist Award
- 2001 Russian Academy of Sciences, Honorary Doctor
- 2004 Membership in the European Academy of Sciences
- 2005 Announcement in Adelaide on September 8th: Shannon award for 2006

However, more important for him than the recognition of contemporaries is his belief that his work may survive some milder storms of history.

Information Theory and Some Friendly Neighbors – Ein Wunschkonzert
August 11 – 15, 2003

Harout Aydinian	Extremal problems under dimension constraints
Vladimir Blinovsky	On correlation inequalities
János Körner	Combinatorics and information theory
Faina Solov'eva	Automorphism group of perfect codes
Torleiv Klove	Even-weight codes for error correction
Andras Sarkózy	My work in Bielefeld
Julien Cassaigne	On combinatorics of words
Christian Mauduit	Construction of pseudorandom finite sequences
Katalin Gyarmati	On a family of pseudorandom binary sequences
Henk Hollmann	I.P.P. - codes
James L. Massey	Information theory after Shannon
Katalin Marton	Measure concentration problems (blowing-up)
Ulrich Tamm	Communication complexity
Peter Sanders	Polynomial time algorithms for network information Flow
Christian Bey	The complete intersection theorem and related results
Uwe Leck	Extremal problems in partially ordered sets
A. Jan Van Zanten	On the construction of snake-in-the-box codes
Andrew Barron	The role of information in the central limit problem
Ning Cai	Isoperimetric, diametric and intersection theorems
Lloyd Demetrius	Quantum statistics, entropy and the evolution of life span
Stefan Artmann	Life=information+computers? A philosophical look at artificial life
Holger Lyre	Is syntax without semantics possible?
Günter Kochendörfer	Signals, meaning, and information in the brain
Alon Orlitsky	Communication complexity and relatives
D. Werner Müller	Statistics and information
Frank Hampel	The foundations of statistics revisited, with a bridge to information theory
Peter Harremoes	A new approach to majorization
Evgueni Haroutunian	On logarithmically asymptotically optimal testing of hypotheses and identification
Anand Srivastav	Construction of sparse connectors
Kingo Kobayashi	Multi user information theory

R. Ahlswede et al. (Eds.): Information Transfer and Combinatorics, LNCS 4123, pp. 49–50, 2006.
© Springer-Verlag Berlin Heidelberg 2006

Ashot Haroutunian	The rate-reliability-distortion function: Main properties
Gabor Wiener	Approximate search
Andreas Winter	Quantum strong-converses and some consequences
Erdal Arikan	Search in quantum systems
Hiroshi Nagaoka	Some remarks on the quantum information geometry from estimation theoretical viewpoints
Imre Csiszar	On common randomeness in information theory
Venkat Anantharam	Identification via channels and common randomness
Yossef Steinberg	On identification for watermarking systems
Rudolf Ahlswede	Schlagerparade

I

Identification for Sources

R. Ahlswede, B. Balkenhol, and C. Kleinewächter

1 Introduction

1.1 Pioneering Model

The classical transmission problem deals with the question how many possible messages can we transmit over a noisy channel? Transmission means there is an answer to the question "What is the actual message?" In the identification problem we deal with the question how many possible messages the receiver of a noisy channel can identify? Identification means there is an answer to the question "Is the actual message u?" Here u can be any member of the set of possible messages.

Allowing randomized encoding the optimal code size grows double exponentially in the blocklength and somewhat surprisingly the second order capacity equals Shannon's first order transmission capacity (see [3]).

Thus Shannon's Channel Coding Theorem for Transmission is paralleled by a Channel Coding Theorem for Identification. It seems natural to look for such a parallel for sources, in particular for noiseless coding. This was suggested by Ahlswede in [4].

Let $(\mathcal{U}, P)$ be a source, where $\mathcal{U} = \{1, 2, \ldots, N\}$, $P = (P_1, \ldots, P_N)$, and let $\mathcal{C} = \{c_1, \ldots, c_N\}$ be a binary prefix code (PC) for this source with $\|c_u\|$ as length of c_u. Introduce the RV U with $\text{Prob}(U = u) = p_u$ for $u = 1, 2, \ldots, N$ and the RV C with $C = c_u = (c_{u_1}, c_{u_2}, \ldots, c_{u\|c_u\|})$ if $U = u$.

We use the PC for noiseless identification, that is user u wants to know whether the source output equals u, that is, whether C equals c_u or not. He iteratively checks whether $C = (C_1, C_2, \ldots)$ coincides with c_u in the first, second, etc. letter and stops when the first different letter occurs or when $C = c_u$. What is the expected number $L_{\mathcal{C}}(P, u)$ of checkings?

In order to calculate this quantity we introduce for the binary tree $T_{\mathcal{C}}$, whose leaves are the codewords $c_1, \ldots, c_N$, the sets of leaves $\mathcal{C}_{ik} (1 \leq i \leq N; 1 \leq k)$, where $\mathcal{C}_{ik} = \{c \in \mathcal{C} : c \text{ coincides with } c_i \text{ exactly until the } k\text{'th letter of } c_i\}$. If C takes a value in $\mathcal{C}_{uk}, 0 \leq k \leq \|c_u\| - 1$, the answers are k times "Yes" and 1 time "No". For $C = c_u$ the answers are $\|c_u\|$ times "Yes". Thus

$$L_{\mathcal{C}}(P, u) = \sum_{k=0}^{\|c_u\|-1} P(C \in \mathcal{C}_{uk})(k+1) + \|c_u\| P_u. \text{ [1]}$$

---

[1] Probability distributions and codes depend on N, but are mostly written without an index N.

R. Ahlswede et al. (Eds.): Information Transfer and Combinatorics, LNCS 4123, pp. 51–61, 2006.
© Springer-Verlag Berlin Heidelberg 2006

For code $\mathcal{C}$ $L_{\mathcal{C}}(P) = \max\limits_{1 \leq u \leq N} L_{\mathcal{C}}(P, u)$ is the expected number of checkings in the worst case and $L(P) = \min\limits_{\mathcal{C}} L_{\mathcal{C}}(P)$ is this number for a best code.

Analogously, if $\tilde{\mathcal{C}}$ is a randomized coding, $L_{\tilde{\mathcal{C}}}(P, u)$, $L_{\tilde{\mathcal{C}}}(P)$ and $\tilde{L}(P)$ were also introduced in [4].

What are the properties of $L(P)$ and $\tilde{L}(P)$? In analogy to the role of entropy $H(P)$ in Shannon's Noiseless Source Coding Theorem they can be viewed as approximations to a kind of "identification entropy" functional H_I.

Their investigation is left to future research. We quickly report now two simpler pioneering questions and partial answers from [4]. They shed some light on the idea that in contrast to classical entropy H, which takes values between 0 and ∞, the right functional H_I shall have 2 as maximal value.

Let us start with $P_N = \left(\frac{1}{N}, \ldots, \frac{1}{N}\right)$ and set $f(N) = L(P_N)$.

1. What is $\sup\limits_{N} f(N)$ or $\lim\limits_{N \to \infty} f(N)$?

   Starting with an identification code for $N = 2^{k-1}$ a new one for 2^k users is constructed by adding for half of all users a 1 as prefix to the codewords and a 0 for the other half. Obviously we are getting an identification code with twice as many codewords in this way. Now user u has to read the first bit. With probability $\frac{1}{2}$ he then stops and with probability $\frac{1}{2}$ he needs only an expected number of $f(2^{k-1})$ many further checkings. Now an optimal identification code is at least as good as the constructed one and we get the recursion

   $$f(2^k) \leq 1 + \frac{1}{2}f(2^{k-1}), f(2) = 1$$

   and therefore

   $$f(2^k) \leq 2 - 2^{-(k-1)}.$$

   On the other hand it can be verified that $f(9) = 1 + \frac{10}{9} > 2$ and more generally $f(2^k + 1) > 2$.

2. Is $\tilde{L}(P) \leq 2$?

   This is the case under the stronger assumption that encoder and decoder have access to a random experiment with unlimited capacity of common randomness (see [5]).

   For $P = (P_1, \ldots, P_N)$, $N \leq 2^n$ write $P^{(n)} = (P_1, \ldots, P_N, 0, \ldots, 0)$ with 2^n components. Use a binary regular tree of depth n with leaves $1, 2, \ldots, 2^n$ represented in binary expansions.

   The common random experiment with 2^n outcomes can be used to use 2^n cyclic permutations of $1, 2, \ldots, 2^n$ for 2^n deterministic codes. For each u we get equally often 0 and 1 in its representation and an expected word length $\leq 2 - \frac{1}{2^{n-1}} \leq 2$. The error probability is 0.

Remark 1. Note that the **same** tree $T_{\mathcal{C}}$ can be used by **all** users in order to answer their question ("Is it me or not?").

1.2 Further Models and Definitions

The model of identification for sources described can be extended (as for channels in the spirit of [4]) to *generalized identification* (GI) as follows.

There is now a set of users $\mathcal{V}$ (not necessarily equal to $\mathcal{U}$), where user $v \in \mathcal{V}$ has a set $\mathcal{U}_v \subset \mathcal{U}$ of source outputs of his interest, that is, he wants to know whether the source output u is in $\mathcal{U}_v$ or not.

Furthermore we speak of *generalized identification with decoding* (GID), if user v not only finds out whether the output is in $\mathcal{U}_v$, but also identifies it if it is in $\mathcal{U}_v$.

Obviously the two models coincide if $|\mathcal{U}_v| = 1$ for $v \in \mathcal{V}$. Also, they specialize to the original model in **1.1**, if $\mathcal{V} = \mathcal{U}$ and $\mathcal{U}_v = \{v\}$ for $v \in \mathcal{U}$.

For our analysis we use the following definition. We denote by $D(x)$ the set of all proper prefixes of $x \in \{0,1\}^*$, i.e.

$$D(x) \triangleq \{y \in \{0,1\}^* : y \text{ is prefix of } x \text{ and } \|y\| < \|x\|\}. \tag{1.1}$$

e stands for the empty word in $\{0,1\}^*$. For a set $A \subset \{0,1\}^*$ we extend this notion to

$$D(A) \triangleq \bigcup_{x \in A} D(x). \tag{1.2}$$

$\{0,1\}^*$ can be viewed as a binary, regular infinite tree with root e. A code $\mathcal{C}$ corresponds to the subtree $T_{\mathcal{C}}$ with root e and leaves $c_1, \ldots, c_N$.

In the sequel we use a specific example of a code for illustrations of concepts and ideas.

Example 1. Let $\mathcal{C}$ be the set of all words of length 3. Notice that $D(010) = \{e, 0, 01\}$ and $D(\{001, 010\}) = \{e, 0, 00, 01\}$.

The set $\mathcal{C}_v = \{c_u : u \in \mathcal{U}_v\}$ is a code for user v. For GID its codewords have to be uniquely decodable by user v in order to identify the source output. For this he uses the set of stop sequences

$$\mathcal{S}_v = \{y_1 \ldots y_k : y_1 \ldots y_{k-1} \in D(\mathcal{C}_v) \text{ and } y_1 \ldots y_k \notin D(\mathcal{C}_v)\}. \tag{1.3}$$

By definition of D $\mathcal{C}_v$ is contained in $\mathcal{S}_v$. We can also write

$$\mathcal{S}_v = \{xy : x \in \{0,1\}^*, y \in \{0,1\} \text{ with } x \in D(\mathcal{C}_v) \text{ and } xy \notin D(\mathcal{C}_v)\}. \tag{1.4}$$

(For $k = 1$ $y_1 \ldots y_{k-1}$ describes the empty word e or the root of the code tree which is element of each set $D(\mathcal{C}_v)$.)

Example 2. For the code of Example 1 we have for $\mathcal{C}_v = \{010\}$ $\mathcal{S}_v = \{1, 00, 011, 010\}$ and we have for $\mathcal{C}_v = \{001, 010\}$ $\mathcal{S}_v = \{1, 000, 001, 010, 011\}$.

With the families of sets of stop sequences $\mathcal{S}_v$ we derive first in Section 2 general lower bounds on the number of checkings for both models. In Section 3 we consider a uniform source and show that $\lim\limits_{N \to \infty} f(N) = 2$. Then, in Section 4, we derive bounds on the maximal individual (average) identification length, which is introduced in Section 2 C.

Finally, in Section 5, we introduce an *average identification* length for the case $\mathcal{V} = \mathcal{U}, \mathcal{U}_v = \{v\}$ for $v \in \mathcal{V}$ and derive asymptotic results.

2   A Probabilistic Tool for Generalized Identification

General supposition. We consider here prefix codes $\mathcal{C}$, which satisfy the Kraft inequality with equality, that is,

$$\sum_{u \in \mathcal{U}} 2^{-\|c_u\|} = 1. \tag{2.1}$$

We call them saturated, because they cannot be enlarged.

A. GID

For all $x \in \{0,1\}^*$ let $q_{\mathcal{C}}(P,x) = \begin{cases} 0, & \text{if } x \notin D(\mathcal{C}) \cup \mathcal{C} \\ P_u, & \text{if } x = c_u \\ q_{\mathcal{C}}(P,x0) + q_{\mathcal{C}}(P,x1), & \text{if } x \in D(\mathcal{C}). \end{cases}$

The general supposition implies that for any set of stopping sequences $\mathcal{S}_v$ we have $\mathcal{S}_v \subset D(\mathcal{C}) \cup \mathcal{C}$ and the probability for user v to stop in $x \in \mathcal{S}_v$ equals $q_{\mathcal{C}}(P,x)$. After stopping in x user v has read $\|x\|$ many bits. Therefore the average identification length of user v is

$$L_{\mathcal{C}}(P,v) = \sum_{x \in \mathcal{S}_v} q_{\mathcal{C}}(P,x)\|x\|. \tag{2.2}$$

By definition of $q_{\mathcal{C}}$ we get

$$L_{\mathcal{C}}(P,v) = \sum_{x \in D(\mathcal{C}_v)} q_{\mathcal{C}}(P,x). \tag{2.3}$$

By construction $\mathcal{S}_v$ forms a prefix code. Each codeword has to be uniquely decoded by user v. Furthermore the probabilities $q_{\mathcal{C}}(P,x)$, $x \in \mathcal{S}_v$, define a probability distribution on $\mathcal{S}_v$ by

$$P_{\mathcal{C},v}(x) \triangleq q_{\mathcal{C}}(P,x) \text{ for all } x \in \mathcal{S}_v. \tag{2.4}$$

By the Noiseless Coding Theorem $L_{\mathcal{C}}(P,v)$ can be lower bounded by the entropy $H(P_{\mathcal{C},v})$. More directly, using the grouping axiom we get

$$H(P_{\mathcal{C},v}) = \sum_{x \in D(\mathcal{C}_v)} q_{\mathcal{C}}(P,x) h\left(\frac{q_{\mathcal{C}}(P,x1)}{q_{\mathcal{C}}(P,x)}\right), \tag{2.5}$$

where h is the binary entropy function, and thus

$$L_{\mathcal{C}}(P,v) - H(P_{\mathcal{C},v}) = \sum_{x \in D(\mathcal{C}_v)} q_{\mathcal{C}}(P,x)\left(1 - h\left(\frac{q_{\mathcal{C}}(P,x1)}{q_{\mathcal{C}}(P,x)}\right)\right). \tag{2.6}$$

Suppose $P_u > 0$ for all $1 \leq u \leq N$, then

$$q_{\mathcal{C}}(P,x) > 0 \text{ and with } \left(\frac{q_{\mathcal{C}}(P,x1)}{q_{\mathcal{C}}(P,x)}\right) \leq 1 \text{ for all } x \in D(\mathcal{C})$$

it follows under the general supposition (2.1) for every user $v \in \mathcal{V}$ the average identification length satisfies

Theorem 1

$$L_{\mathcal{C}}(P, v) \geq H(P_{\mathcal{C},v}) \text{ with "=" iff } \frac{q_{\mathcal{C}}(P, x1)}{q_{\mathcal{C}}(P, x)} = \frac{1}{2} \text{ for all } x \in D(\mathcal{C}_v). \qquad (2.7)$$

Since P is fixed we write now $L_{\mathcal{C}}(v)$ for $L_{\mathcal{C}}(P, v)$.

B. GI

Suppose we have a node x and a user v with the properties

(a) all codewords having x as prefix are all elements of $\mathcal{C}_v$ or (b) they are all not in $\mathcal{C}_v$.

In this case user v can stop in x and decide whether v occurred or not. By construction of the stop sequences $\mathcal{S}_v$ in (1.3) only case (a) can occur. Therefore we have to start the following algorithm to generate modified sets $\mathcal{S}_v$.

1. If $\mathcal{C}_v$ contains two codewords different only in the last position, say $x_1 \ldots x_k 0$ and $x_1 \ldots x_k 1$ then
   (a) remove these two codewords from $\mathcal{C}_v$ and insert $x_1 \ldots x_k$. This new codeword has the probability $q_{\mathcal{C}}(P, x_1 \ldots x_k)$.
   (b) repeat step 1. Else continue with 2.
2. With the modified sets $\mathcal{C}_v$ construct the sets $\mathcal{S}_v$ as defined in (1.3).

The definition of $L_{\mathcal{C}}(P, v)$, $P_{\mathcal{C},v}$ and $H(P_{\mathcal{C},v})$ are as in (2.2), (2.4) and (2.5). Also the formulas (2.6) and (2.7) hold.

Example 3. Let $\mathcal{C}_v = \{000, 001, 010\}$. After step 1 of the algorithm we get $\mathcal{C}_v = \{00, 010\}$. With step 2 we define $D(\mathcal{C}_v) = \{\varnothing, 0, 01\}$ and $\mathcal{S}_v = \{1, 00, 010, 011\}$.

C. Maximal individual (expected) identification length $L(P)$

For a given probability distribution P and a given code $\mathcal{C}$ user v has uniquely to decode the codewords in $\mathcal{C}_v$.

Using (2.7) we can lower bound $L(P)$ as follows:

(i) Take the set of pairs $\mathcal{M} = \{(\mathcal{C}_v, v) : L(P) = L_{\mathcal{C}}(P, v)\}$.
(ii) Define

$$H_{\max}(P) = \max_{(\mathcal{C}_v, v) \in \mathcal{M}} H(P_{\mathcal{C},v}).$$

Then

$$L(P) \geq H_{\max}(P).$$

Remark 2. Note that

1.

$$\sum_{x \in D(\mathcal{C})} q_{\mathcal{C}}(P, x) = \sum_{u=1}^{N} P_u \|c_u\|.$$

2. Using the grouping axiom it holds

$$\sum_{x \in D(\mathcal{C})} q_{\mathcal{C}}(P, x) h \left(\frac{q_{\mathcal{C}}(P, x1)}{q_{\mathcal{C}}(P, x)} \right) = H(P)$$

for all codes $\mathcal{C}$.

3. If for each code $\mathcal{C}$ there exists a set $\mathcal{C}_v$ (in case B after modification) such that $D(\mathcal{C}_v) = D(\mathcal{C})$, then $L(P) = \sum_{u=1}^{N} P_u \|c_u\|$ where the code $\mathcal{C}$ is the Huffman–code for the probability distribution P.

Example 4. Suppose that $|\mathcal{V}| = \binom{N}{K}$, $K \geq \frac{N}{2}$, and $\{\mathcal{U}_v : v \in \mathcal{V}\} = \binom{[N]}{K}$.

1. In case A there exists for each code $\mathcal{C}$ a set $\mathcal{C}_v$ such that $D(\mathcal{C}_v) = D(\mathcal{C})$.
2. In case B with $K = \frac{N}{2}$ there exists for each code $\mathcal{C}$ a set $\mathcal{C}_v$ such that $D(\mathcal{C}_v) = D(\mathcal{C})$.
3. In case B if $K = N$ and thus $\mathcal{V} = \{v_1\}, \mathcal{U}_{v_1} = [N]$, then after modifying $\mathcal{C}_{v_1}$ the set $D(\mathcal{C}_{v_1})$ contains only the root of the tree which means the user v_1 has to read nothing from the received codeword (because he knows already the answer).

Remark 3. Example 4 is motivated by K–identification for channels!

3   The Uniform Distribution

Now we return to the original model of 1.1 with $\mathcal{V} = \mathcal{U}$ and $\mathcal{C}_v = \{c_v\}$ for each $v \in \mathcal{V}$. Let $P = (\frac{1}{N}, \ldots, \frac{1}{N})$. We construct a prefix code $\mathcal{C}$ in the following way. In each node (starting at the root) we split the number of remaining codewords in proportion as close as possible to $(\frac{1}{2}, \frac{1}{2})$.

1. Suppose $N = 2^k$. By construction our code $\mathcal{C}$ contains all binary sequences of length k. It follows that

$$q_{\mathcal{C}}(P, x) = \frac{1}{N} \frac{N}{2^{\|x\|}} = 2^{-\|x\|} \tag{3.1}$$

and by (2.3)

$$L_{\mathcal{C}}(P) = \sum_{x \in D(\mathcal{C}_v)} q_{\mathcal{C}}(P, x) = \sum_{i=0}^{k-1} 2^{-i} = 2 - 2^{-k+1} = 2 - \frac{2}{N}. \tag{3.2}$$

2. Suppose $2^{k-1} < N < 2^k$. By construction the remaining code contains only the codeword lengths $k - 1$ and k.

   By (2.3) we add the weights $(q_{\mathcal{C}}(P, x))$ of all nodes of a path from the root to a codeword (leave). Therefore in the worst case, N is odd and we have to add the larger weight.

At the root we split $(\frac{N-1}{2}, \frac{N-1}{2} + 1)$. Now we split again the larger one and in the worst case this number is again odd. It follows in general that

$$q_{\mathcal{C}}(P, x) \le \frac{1}{N} \left(\frac{N-1}{2^{\|x\|}} + 1 \right). \tag{3.3}$$

Therefore

$$L_{\mathcal{C}}(P) \le \sum_{i=0}^{k-1} \frac{1}{N} \left(\frac{N-1}{2^i} + 1 \right) = \sum_{i=0}^{k-1} 2^{-i} - \frac{1}{N} \sum_{i=0}^{k-1} 2^{-i} + \frac{1}{N} \sum_{i=0}^{k-1} 1$$

$$= 2 - \frac{1}{N} - \frac{2}{N} + \frac{2}{N^2} + \frac{k}{N} = 2 + \frac{k-3}{N} + \frac{2}{N^2}. \tag{3.4}$$

With $k = \lceil log_2(N) \rceil$ it follows

Theorem 2. *For* $P = \left(\frac{1}{N}, \dots, \frac{1}{N} \right)$

$$\lim_{N \to \infty} L_{\mathcal{C}}(P) = 2. \tag{3.5}$$

4 Bounds on $L(P)$ for General $P = (P_1, \dots, P_N)$

A. An upper bound

We will now give an inductive construction for identification codes to derive an upper bound on $L(P)$. Let $P = (P_1, \dots, P_N)$ be the probability distribution. W.l.o.g. we can assume that $P_i \ge P_j$ for all $i < j$. For $N = 2$ of course we assign 0 and 1 as codewords. Now let $N > 2$. We have to consider two cases:

1. $P_1 \ge 1/2$. In this case we assign 0 as codeword to message 1. We set $P_i'' = \frac{P_i}{\sum_{j=2}^{N} P_j}$ for $i = 2, \dots, N$. By induction we can construct a code for the probability distribution $P'' = (P_2'', \dots, P_N'')$ and messages 2 to N get the corresponding codewords for P'' but prefixed with a 1.

2. $P_1 < 1/2$. Choose ℓ such that $\delta_\ell = |\frac{1}{2} - \sum_{i=1}^{\ell} P_i|$ is minimal. Set $P_i' = \frac{P_i}{\sum_{j=1}^{\ell} P_j}$ for $i = 1, \dots, \ell$ and $P_i'' = \frac{P_i}{\sum_{j=\ell+1}^{N} P_j}$ for $i = \ell+1, \dots, N$. Analogous to the first case we construct codes for the distributions $P' = (P_1', \dots, P_\ell')$ (called the *left side*) and $P'' = (P_{\ell+1}'', \dots, P_N'')$ (called the *right side*). We get the code for P by prefixing the codewords from the left side with 0 and the codewords from the right side with 1.

Trivially this procedure yields a prefix code.

Theorem 3. *Let* $N \in \mathbb{N}$ *and let* $P = (P_1, \dots, P_N)$. *The previous construction yields a prefix code with* $L(P) \le 3$.

Proof. The case $N = 2$ is trivial. Now let $N \ge 3$.

Case 1. $P_1 \ge 1/2$: In this case we have $L(P) \le 1 + \max \left\{ P_1, L(P'') \sum_{i=2}^{N} P_i \right\}$, where $L(P'')$ denotes the corresponding maximal identification length for probability distribution P''. If the maximum is assumed for P_1 we have $L(P) \leq 2$, otherwise we get by induction $L(P) < 1 + 3 \cdot 1/2 < 3$.

Case 2. $P_1 < 1/2$ for $i = 1, \ldots, N$: In this case we have

$$L(P) \leq 1 + \max \left\{ L(P') \cdot \sum_{i=1}^{\ell} P_i, \quad L(P'') \cdot \sum_{i=\ell+1}^{N} P_i \right\}.$$

Choose ℓ' such that $\sum_{i=1}^{\ell'} P_i \leq 1/2 < \sum_{i=1}^{\ell'+1} P_i$. Obviously either $\ell = \ell'$ or $\ell = \ell' + 1$.

Subcase: $\ell = \ell'$. Suppose the maximum is assumed on the left side. Then without changing the maximal identification length we can construct a new probability distribution $P''' = (P_1''', \ldots, P_{\ell+1}''')$ by $P_1''' = \sum_{i=\ell+1}^{N} P_i$ and $P_i''' = P_{i-1}$ for $2 \leq i \leq \ell + 1$. Since $P_1''' \geq 1/2$ we are back in case 1. If the maximum is assumed on the right side then let $P_1''' = \sum_{i=1}^{\ell} P_i$ and $P_i''' = P_{i+\ell-1}$ for all $2 \leq i \leq n-\ell+1$. Notice that in this case $P_1''' \geq 1/3$ (because $P_1''' \geq 1/2 - P_2'''/2 \geq 1/2 - P_1'''/2$). Thus by induction $L(P''') \leq 1 + 3 \cdot 2/3 \leq 3$.

Subcase: $\ell = \ell' + 1$. If the maximum is on the right side we set $P_1''' = \sum_{i=1}^{\ell} P_i \geq 1/2$, $P_i''' = P_{i+\ell-1}$ for $2 \leq i \leq n - \ell + 1$ and we are again back in case 1. Now suppose the maximum is taken on the left side. Since $\sum_{i=1}^{\ell} P_i - 1/2 \leq 1/2 - \sum_{i=1}^{\ell'} P_i$ it follows that $\delta_\ell \leq P_\ell/2$. Because $P_{\ell'} \leq (2\ell')^{-1}$ we have $\delta_\ell \leq (4\ell')^{-1} = (4(\ell-1))^{-1}$. Also note that $\ell \geq 2$. The case $\ell = 2$ is again trivial. Now let $\ell > 2$. Then $L(P) < 3 \cdot (1/2 + \frac{1}{4(\ell-1)}) \leq 3 \cdot (1/2 + 1/8) < 3$.

5   An Average Identification Length

We consider here the case where not only the source outputs but also the users occur at random. Thus in addition to the source $(\mathcal{U}, P)$ and RV U, we are given $(\mathcal{V}, Q)$, $\mathcal{V} \equiv \mathcal{U}$, with RV V independent of U and defined by $\mathrm{Prob}\,(V = v) = Q_v$ for $v \in \mathcal{V}$. The source encoder knows the value u of U, but not that of V, which chooses the user v with probability Q_v. Again let $\mathcal{C} = \{c_1, \ldots, c_N\}$ be a binary prefix code and let $L_{\mathcal{C}}(P, u)$ be the expected number of checkings on code $\mathcal{C}$ for user u. Instead of $L_{\mathcal{C}}(P) = \max_{u \in \mathcal{U}} L_{\mathcal{C}}(P, u)$, the maximal number of expected checkings for a user, we consider now the average number of expected checkings

$$L_{\mathcal{C}}(P, Q) = \sum_{v \in \mathcal{V}} Q_v L_{\mathcal{C}}(P, v) \tag{5.1}$$

and the average number of expected checkings for a best code

$$L(P, Q) = \min_{\mathcal{C}} L_{\mathcal{C}}(P, Q). \tag{5.2}$$

(The models GI and GID can also be considered.)

We also call $L(P, Q)$ the average identification length. $L_{\mathcal{C}}(P, Q)$ can be calculated by the formula

$$L_{\mathcal{C}}(P, Q) = \sum_{x \in D(\mathcal{C})} q_{\mathcal{C}}(Q, x) q_{\mathcal{C}}(P, x). \tag{5.3}$$

In the same way as (5.3) we get the conditional entropy

$$H_{\mathcal{C}}(P\|Q) = \sum_{x \in D(\mathcal{C})} q_{\mathcal{C}}(Q,x) q_{\mathcal{C}}(P,x) h\left(\frac{q_{\mathcal{C}}(P,x1)}{q_{\mathcal{C}}(P,x)}\right). \tag{5.4}$$

5.1 Q Is the Uniform Distribution on $\mathcal{V} = \mathcal{U}$

We begin with $|\mathcal{U}| = N = 2^k$, choose $\mathcal{C} = \{0,1\}^k$ and note that

$$\sum_{\substack{x \in D(\mathcal{C}) \\ \|x\|=i}} q_{\mathcal{C}}(P,x) = 1 \text{ for all } 0 \le i \le k. \tag{5.5}$$

By (3.1) for all $x \in \{0,1\}^*$ with $\|x\| \le k$

$$q_{\mathcal{C}}(Q,x) = 2^{-\|x\|} \tag{5.6}$$

and thus by (5.3) and then by (5.5)

$$L_{\mathcal{C}}(P,Q) = \sum_{i=0}^{k-1} \sum_{\substack{x \in D(\mathcal{C}) \\ \|x\|=i}} 2^{-i} q_{\mathcal{C}}(P,x) \tag{5.7}$$

$$= \sum_{i=0}^{k-1} 2^{-i} = 2 - 2^{-k+1} = 2 - \frac{2}{N}. \tag{5.8}$$

We continue with the case $2^{k-1} < N < 2^k$ and construct the code $\mathcal{C}$ again as in Section 3. By (3.3)

$$q_{\mathcal{C}}(Q,x) \le \frac{1}{N}\left(\frac{N-1}{2^{\|x\|}} + 1\right). \tag{5.9}$$

Therefore

$$L_{\mathcal{C}}(P,Q) = \sum_{x \in D(\mathcal{C})} q_{\mathcal{C}}(Q,x) q_{\mathcal{C}}(P,x) \le \frac{1}{N} \sum_{x \in D(\mathcal{C})} (\frac{N-1}{2^{\|x\|}} + 1) q_{\mathcal{C}}(P,x)$$

$$= \frac{1}{N} \sum_{i=0}^{k-1} (\frac{N-1}{2^i} + 1) \sum_{\substack{x \in D(\mathcal{C}) \\ \|x\|=i}} q_{\mathcal{C}}(P,x) \le \frac{1}{N} \sum_{i=0}^{k-1} (\frac{N-1}{2^i} + 1) \cdot 1$$

$$= 2 + \frac{k-3}{N} + \frac{2}{N^2} \text{ (see (3.4))}. \tag{5.10}$$

With $k = \lceil log_2(N) \rceil$ it follows that

Theorem 4. *Let $N \in \mathbb{N}$ and $P = (P_1, \ldots, P_N)$, then for $Q = \left(\frac{1}{N}, \ldots, \frac{1}{N}\right)$*

$$\lim_{N \to \infty} L_{\mathcal{C}}(P,Q) = 2. \tag{5.11}$$

Example 4 with average identification length for a uniform Q^*

We get now

$$L_C(P, Q^*) = \sum_{x \in D(C)} \frac{|\{v : x \in D(C_v)\}|}{|\mathcal{V}|} q_C(P, x) \tag{5.12}$$

and for the entropy in (5.4)

$$H_C(P\|Q^*) = \sum_{x \in D(C)} \frac{|\{v : x \in D(C_v)\}|}{|\mathcal{V}|} q_C(P, x) h\left(\frac{q_C(P, x1)}{q_C(P, x)}\right). \tag{5.13}$$

Furthermore let $\mathcal{C}_0$ be the set of all codes C with $L_C(P, Q^*) = L(P, Q^*)$. We define

$$H(P\|Q^*) = \max_{C \in \mathcal{C}_0} H_C(P\|Q^*). \tag{5.14}$$

Then

$$L(P, Q^*) \geq H(P\|Q^*). \tag{5.15}$$

Case $N = 2^n$: We choose $C = \{0, 1\}^n$ and calculate $\frac{|\{v : x \in D(C_v)\}|}{|\mathcal{V}|}$. Notice that for any $x \in D(C)$ we have $2^{n - \|x\|}$ many codewords with x as prefix.

Order this set. There are $\binom{N-1}{K-1}$ $(K-1)$–element subsets of C containing the first codeword in this set. Now we take the second codeword and $K - 1$ others, but not the first. In this case we get $\binom{N-2}{K-1}$ further sets and so on.

Therefore $|\{v : x \in D(C_v)\}| = \sum_{j=1}^{2^{n-\|x\|}} \binom{2^n - j}{K-1}$ and (5.14) yields

$$L_C(P, Q^*) = \frac{1}{\binom{N}{K}} \sum_{x \in D(C)} \sum_{j=1}^{2^{n-\|x\|}} \binom{2^n - j}{K - 1} q_C(P, x)$$

$$= \frac{1}{\binom{2^n}{K}} \sum_{i=0}^{n-1} \left(\sum_{j=1}^{2^{n-i}} \binom{2^n - j}{K - 1}\right) \left(\sum_{\substack{x \in D(C) \\ \|x\|=i}} q_C(P, x)\right)$$

$$= \frac{1}{\binom{2^n}{K}} \sum_{i=0}^{n-1} \left(\sum_{j=1}^{2^{n-i}} \binom{2^n - j}{K - 1}\right) \quad \text{(by (5.5))}. \tag{5.17}$$

Lets abbreviate this quantity as $g(n, K)$. Its asymptotic behavior remains to be analyzed.

Exact values are

$$g(n, 1) = 2 - \frac{2}{2^n}, \quad g(n, 2) = \frac{2}{3} \frac{5 \cdot 2^{-n} - 9 + 4 \cdot 2^n}{2^n - 1}$$

$$g(n, 3) = -\frac{2}{7} \frac{49 \cdot 2^n - 70 + 32 \cdot 2^{-n} - 11 \cdot 4^n}{(2^n - 1)(2^n - 2)}, \quad g(n, 4) = \frac{4}{105} \frac{-2220 + 908 \cdot 2^{-n} - 705 \cdot 4^n + 1925 \cdot 2^n + 92 \cdot 8^n}{(2^n - 1)(2^n - 2)(2^n - 3)}$$

We calculated the limits $(n \to \infty)$

K	1	2	3	4	5	6	7	8	9
$\displaystyle\lim_{n\to\infty} g(n,K)$	2	$\frac{8}{3}$	$\frac{22}{7}$	$\frac{368}{105}$	$\frac{2470}{651}$	$\frac{7880}{1953}$	$\frac{150266}{35433}$	$\frac{13315424}{3011805}$	$\frac{2350261538}{513010785}$

This indicates that $\displaystyle\sup_{K}\lim_{n\to\infty} g(n,K) = \infty$.

References

1. C.E. Shannon, A mathematical theory of communication, Bell Syst. Techn. J. 27, 379–423, 623–656, 1948.

2. D.A. Huffman, A method for the construction of minimum redundancy codes, Proc. IRE 40, 1098–1101, 1952.

3. R. Ahlswede and G. Dueck, Identification via channels, IEEE Trans. Inf. Theory, Vol. 35, No. 1, 15–29, 1989.

4. R. Ahlswede, General theory of information transfer, Preprint 97–118, SFB 343 "Diskrete Strukturen in der Mathematik", Universität Bielefeld, 1997; General theory of information transfer:updated, General Theory of Information Transfer and Combinatorics, a Special Issue of Discrete Applied Mathematics, to appear.

5. R. Ahlswede and I. Csiszár, Common randomness in Information Theory and Cryptography, Part II: CR capacity, IEEE Trans. Inf. Theory, Vol. 44, No. 1, 55–62, 1998.

6. C.C. Campbell, Definition of entropy by means of a coding problem, Z. Wahrscheinlichkeitstheorie u. verw. Geb., 113–119, 1966.

On Identification

C. Kleinewächter

1 Introduction

In Shannon's classical model of transmitting a message over a noisy channel we have the following situation:

There are two persons called *sender* and *receiver*. Sender and receiver can communicate via a *channel*. In the simplest case the sender just puts some *input letters* into the channel and the receiver gets some *output letters*. Usually the channel is *noisy*, i.e. the channel output is a random variable whose distribution is governed by the input letters. This model can be extended in several ways: Channels with *passive feedback* for example give the output letters back to the sender. Multiuser channels like *multiple access channels* or *broadcast channels* (which will not be considered in this paper) have several senders or receivers which want to communicate simultaneously. Common to all these models of *transmission* is the task that sender and receiver have to perform: Both have a common *message set* M and the sender is given a *message* $i \in M$. He has to *encode* the message (i.e. transform it into a sequence of input letters for the channel) in such a way, that the receiver can *decode* the sequence of output letters so that he can decide with a small probability of error what the message i was. The procedures for encoding and decoding are called a *code* for the channel and the number of times the channel is used to transmit one message is called the *blocklength* of the code.

While this model of transmission is probably the most obvious model, there are also other ones. Maybe the receiver does not need to know what the exact message is, he is satisfied to get only partial information. Depending on the type of questions the receiver has about the messages, transmission codes can be rather inefficient. If one uses special codes suited to answer only these questions, then there might be more efficient codes with respect to the number of possible messages for given blocklength.

One of these models is identification. In the theory of identification, the receiver has a message of special interest to him, and he only wants to know whether the sender was given this special message or not. This special message is unknown to the sender (otherwise this problem would be trivial, the sender just has to encode one of two alternatives). As an application consider the following situation: Sender and receiver both have a large amount of text (like a book). The receiver now wants to know if both texts are equal (otherwise the receiver may want to order a copy of the sender's test). The trivial solution using a transmission scheme would be that the sender simply encodes the whole text, sends it over the channel, the receiver decodes the text and compares it to his text, but of course this is very inefficient. Usually the number of messages that one can transmit grows exponentially in the blocklength, so this method would

R. Ahlswede et al. (Eds.): Information Transfer and Combinatorics, LNCS 4123, pp. 62–83, 2006.

require codes with blocklengths proportional to the length of the text. The ratio between the logarithm of the number of messages and the blocklength is called the *rate* of the code and the maximal rate that is asymptotically achievable is called the capacity of the channel.

In [1] and [2] Ahlswede and Dueck showed that the maximal sizes of an identification code (*ID-code*) for a stationary discrete memoryless channel without and with noiseless feedback grow double exponentially fast in the blocklength and they determined the second order rate for these channels, if we allow an arbitrary small but positive probability of error that the answer we get is false. (For channels without feedback this holds only for randomized codes.) The rate of these codes is therefore defined as the ratio of the iterated logarithm of the number of messages and the blocklength.

Applied to our previous example this means that we need only codes of blocklength proportional to the logarithm of the length of the text for identification.

If we deal with ID-codes, we have to consider two kinds of errors, namely if message i is given to the sender the receiver, who is interested in message i, can decide that his message was not sent (missed identification) and a receiver who is interested in $j \neq i$ can decide that j was sent (false identification). These two errors will turn out to be of different nature. While missed identifications are caused by errors in the transmission, false identifications are also inherent to the codes.

In the case of a channel W without feedback the second order identification capacity is the classical Shannon capacity

$$C = \max_P I(P, W)$$

Here P is a distribution on the input letters and $I(P, W)$ is the *mutual information* of two random variables with joint distribution $P_{XY}(x, y) = P(x)W(y|x)$, that means these random variables behave like channel input respectively output.

The upper bound for channels without feedback was proved under the assumption that the allowed error probability vanishes exponentially fast. In [4] Han and Verdú proved the upper bound for constant error probability smaller than $1/2$. It is obvious, that if we allow error probabilities larger than $1/2$, then trivial codes, where the receiver simply guesses, achieve infinite rates. There are also other conditions on the error probabilities to avoid these trivialities, and mostly they lead to the same capacity, for example, if the sum of the maximal probability of accepting a false message plus the maximal probability of falsely rejecting a message should be less than 1. For the proof of Theorem 1 in Section 3 we show that an even slightly weaker condition suffices, namely that for each message the sum of error probabilities should be less than 1.

For a channel W with noiseless feedback (i.e. the sender immediately gets each letter that is received) the capacity is

$$\max_{x \in \mathcal{X}} H(W(\cdot|x))$$

for deterministic and

$$\max_P H(PW)$$

for randomized codes, if there are two different rows in W, otherwise they are 0. Identification codes for channels with feedback will be called *IDF-codes*.

The proof of the direct parts in [2] works in two stages: In the first stage, sender and receiver communicate over the channel with the goal to generate two random variables which are equal with high probability and have large entropy. These random variables are used to select a function which maps messages to a rather small set. In the second step (which contributes only negligible to the blocklength) the sender evaluates this function at the message and transmits it to the receiver. The receiver evaluates the function at the message he is interested in and compares the result with the value he got from the sender. He will vote for a positive identification if and only if they are equal.

This leads to the following interpretation of these results: If the identification capacity is positive, then it equals the common randomness capacity as defined in [8], i.e. the maximal entropy of an almost uniform random experiment which can be set up by communicating over the channel and is known with high probability to both sender and receiver. In the absence of feedback, one possibility to achieve the maximal possible rate of such a common random experiment is that the sender performs a uniform random experiment and transmits the result to the receiver using an ordinary transmission code. If noiseless feedback is available, the sender sends letters in such a way, that the entropy of the channel output (which he knows from the feedback channel) is maximized, where he can either use a deterministic or randomized input strategy, depending on the kind of code he may use. This interpretation proved to be the right one also for other kinds of channels like the multiple access channel (see [3]).

Thus the question arises if this equality is valid in general.

The answer is negative. In [7] Ahlswede gives an example of a non stationary memoryless channel with double exponentially growing input alphabet with identification capacity 1 and common randomness capacity 0. The structure of this channel has some similarities to the structure of ID-codes used in most of the achievability proofs for ID-coding theorems, thus it can be viewed as a channel with "built-in" ID-encoder.

In Section 3 we give a counterexample for the other direction. For given real numbers C_{ID} and C_{CR} with $0 < C_{ID} < C_{CR}$, we will explicitly construct a discrete channel with memory and noiseless passive feedback with identification capacity C_{ID} and common randomness capacity C_{CR}. This channel is constructed in such a way that it can be used in two ways. In one respect, the channel is good for the generation of common randomness, in the other it is suitable for identification.

It is quite reasonable to consider channels with memory. One may think for example of a system where data are transmitted by different voltage levels at high frequency. Because of the electrical capacity of the system it can be difficult to switch from a low voltage level to a high one and vice versa. There are also certain types of magnetic recording devices have problems with long sequences of the same letter. These examples for instance lead to the notion of run length limited codes. A third example are systems requiring the use of binary codewords which have approximately the same number of zeroes and ones. This limitation arises if the system can only transmit an unbiased alternating current, therefore these codes are called DC-free.

In [6] Ahlswede and Zhang gave bounds on the maximal rate for ID-codes for discrete memoryless channels with noisy feedback. Channels without feedback can in this model be described by a feedback channel that maps all input letters with probability 1 on one special feedback letter. The case with noiseless feedback is described by a feedback channel that has positive probability for a pair (y, z) of output respectively feedback letters only if $y = z$. In the cases of either no feedback or noiseless feedback, the upper and lower bound coincide and therefore this can be viewed as a generalization and unification of the results of [1], [2] and [4], where the identification capacities for those channels were determined. Unfortunately in general there is a gap between the upper and lower bound.

In Section 4 we improve the upper bound on the size of *deterministic* codes given in [6] for the channels where the *main channel is noiseless*. The known bound states that the second order rate of deterministic ID codes is independently bounded by the maximal mutual information that channel input and feedback give about the channel output and the maximal entropy of the feedback. We improve this bound by showing that these two quantities cannot be maximized independently, instead one has to choose an *input distribution that is good for both*, transmission and the generation of randomness. We will show for the channels considered the new upper bound equals the known lower bound, therefore the deterministic identification capacity for these channels is now known.

2   Auxiliary Results

In this section we will introduce some basic definitions and well know facts. All logarithms in this paper are assumed to be binary and we define $0 \log 0$ to be 0.

Definition 1. *For a finite or countable infinite set S we will denote by $\mathcal{P}(S)$ the set of all probability distributions on S.*

Definition 2 (Entropy). *Let P be a probability distribution on a finite set $\mathcal{X}$. The entropy of P is defined as*

$$H(P) \triangleq - \sum_{x \in \mathcal{X}} P(x) \log P(x).$$

Also if X is a random variable with distribution P, we define the entropy of X by

$$H(X) \triangleq H(P).$$

Entropy is a concave function, because the function $f(t) = -t \log t$ is concave. Furthermore, it is *Schur-concave*:

Lemma 1 (Schur-concavity of entropy). *Let $n \in \mathbb{N}$. Let P, Q be two probability distributions on $\{1, \ldots, n\}$ which are non increasing. If P and Q satisfy*

$$\sum_{i=1}^{k} P(i) \le \sum_{i=1}^{k} Q(i), \tag{1}$$

for all $k \in 1, \ldots, n$ then $H(P) \ge H(Q)$.

Entropy can be viewed as a measure of uncertainty of a random variable. For example, if one has to guess the value of a random variable X using yes-/no-questions about its value, an optimal strategy takes on the average approximately $H(X)$ questions. (They differ by at most 1 question.) To catch the dependency between random variables, we need the concept of conditional entropy:

Definition 3 (Conditional Entropy). *Let X, Y be random variables on finite sets $\mathcal{X}, \mathcal{Y}$ with distributions P resp. Q. $Q(\cdot|x)$ is the distribution of Y under the condition $X = x$. The conditional entropy of Y given X is defined by*

$$H(Y|X) \triangleq -\sum_{x \in \mathcal{X}} P(x) \sum_{y \in \mathcal{Y}} Q(y|x) \log Q(y|x).$$

The conditional entropy $H(Y|X)$ is the uncertainty about Y after having observed X, so the difference between $H(Y)$ and $H(Y|X)$ is the decrease in uncertainty about Y by observing X or loosely speaking the amount of information we "learn" about Y through observation of X. This quantity is called *mutual information*.

Definition 4 (Mutual information). *Let X and Y be random variables on finite sets $\mathcal{X}$ and $\mathcal{Y}$, respectively. Then we define the mutual information between X and Y by*

$$I(X \wedge Y) \triangleq H(Y) - H(Y|X).$$

The following lemma gives an upperbound on the difference of entropies of two distributions as a function of the l_1-distance (and the cardinality of $\mathcal{X}$, which is usually fixed):

Lemma 2. *Let P and Q be two distributions on a set $\mathcal{X}$ with*

$$\sum_{x \in \mathcal{X}} |P(x) - Q(x)| \le \delta \le \frac{1}{2}. \tag{2}$$

Then

$$|H(P) - H(Q)| \le -\delta \log \frac{\delta}{|\mathcal{X}|}. \tag{3}$$

Proof. Set $f(t) = -t \log t$. f is concave in $[0, 1]$. Thus for $0 \le \tau \le \frac{1}{2}, 0 \le t \le 1-\tau$ we have

$$f(t) - f(t + \tau) \le f(0) - f(\tau) \tag{4}$$

and

$$f(t + \tau) - f(t) \le f(1 - \tau) - f(1) \tag{5}$$

which gives

$$|f(t) - f(t + \tau)| \leq \max\{f(\tau), f(1 - \tau)\} = f(\tau). \tag{6}$$

Thus

$$|H(P) - H(Q)| \leq \sum_{x \in \mathcal{X}} |f(P(x) - f(Q(x)))|$$

$$\leq \sum_{x \in \mathcal{X}} f(|P(x) - Q(x)|)$$

$$\leq \delta \left(\sum_{x \in \mathcal{X}} f\left(\frac{|P(x) - Q(x)|}{\delta} \right) - \log \delta \right)$$

$$\leq \delta(\log |\mathcal{X}| - \log \delta)$$

where the last step follows from Jensen's inequality. $\square$

Another important application of entropy is counting sequences that have a given empirical distribution:

Definition 5 (*n*-type). *Let $n \in \mathbb{N}$. A probability distribution P on a finite set $\mathcal{X}$ is called n-type, if*

$$P(x) \in \left\{ \frac{0}{n}, \frac{1}{n}, \dots, \frac{n}{n} \right\} \text{ for all } x \in \mathcal{X}.$$

The set of all sequences in $\mathcal{X}^n$ that have empirical distribution P will be denoted by T_P. Vice versa if a sequence x^n is given, we will denote its empirical distribution by P_x.

n-types can be viewed as empirical distribution of sequences of length n. For fixed *n*-type P their number can be estimated as follows:

Lemma 3. *Let $n \in \mathbb{N}$. For an n-type P*

$$\frac{2^{nH(P)}}{(n+1)^{|\mathcal{X}|}} \leq |T_P| \leq 2^{nH(P)}. \tag{7}$$

A proof can be found for example in [10].

The following definition expands the notion of types to pairs of sequences:

Definition 6. *Let $\mathcal{X}$, $\mathcal{Z}$ be finite sets, let $W : \mathcal{X} \to \mathcal{Z}$ be a stochastic matrix, i.e. a matrix $W = (W(z|x))_{x \in \mathcal{X}, z \in \mathcal{Z}}$. Let $x^n = (x_1, \dots, x_n) \in \mathcal{X}^n$. Then we say that $z^n = (z_1, \dots, z_n) \in \mathcal{Z}^n$ has conditional type W given x^n if for all pairs $(x, z) \in \mathcal{X} \times \mathcal{Z}$*

$$<x, z|x^n, z^n> = <x|x^n> W(z|x) \tag{8}$$

where $<x|x^n> = |\{i|x_i = x\}|$ is the number of occurences of letter x in x^n and $< x, z|x^n, z^n > = |\{i|x_i = x \land z_i = z\}|$ is the number of occurences of the pair (x, z) in the sequence

$$((x_1, z_1), \dots, (x_n, z_n)).$$

The set of these sequences is denoted by $T_W(x^n)$.

As a corollary from Lemma (3) we get

Corollary 1. *Let $n \in \mathbb{N}$, let $\mathcal{X}, \mathcal{Z}$ be finite sets. For $x^n \in \mathcal{X}^n$ and a stochastic matrix $W : \mathcal{X} \to \mathcal{Z}$ (i.e. a matrix $W = (W(z|x))_{x \in \mathcal{X}, z \in \mathcal{Z}}$ with $\sum_{z \in \mathcal{Z}} W(z|x) = 1$ for all $x \in \mathcal{X}$) such that $T_V(X^n) \neq \varnothing$ we have*

$$\frac{2^{nH(Z|X)}}{(n+1)^{|\mathcal{X}||\mathcal{Y}|}} \leq |T_W(x^n)| \leq 2^{nH(Z|X)}, \tag{9}$$

where X, Z are random variables with joint distribution $P_{XZ}(x, z) = P(x)W(z|x)$.

3   Common Randomness Capacity Can Exceed Identification Capacity

In this section we will construct a channel where the *identification capacity is positive but smaller than the common randomness capacity*.

We will extend the definitions of [2] from memoryless channels with feedback to channels with memory. These channels will not be considered in general, but only a very specific subclass of them.

Definition 7 (Discrete channel with infinite memory). *Let $\mathcal{X}$ and $\mathcal{Y}$ be finite sets. Let*

$$\mathcal{W} = \{W | W : \mathcal{X} \to \mathcal{Y}\}$$

be the set of all stochastic matrices with input alphabet $\mathcal{X}$ and output alphabet $\mathcal{Y}$. A discrete channel V with infinite memory (here abbreviated as DC) with respect to the input letters is a series $V = (V_i)_{i \in \mathbb{N}}$ of maps, where $V_i : \mathcal{X}^{i-1} \to \mathcal{W}$. V_1 is understood to be an element of $\mathcal{W}$. Given an input sequence $x^n = (x_1, \ldots, x_n)$ of length n the transmission probability for output sequence $y^n = (y_1, \ldots, y_n)$ is given by

$$V^n(y^n|x^n) = \prod_{i=1}^{n} V_i(x_1, \ldots, x_{i-1})(y_i|x_i). \tag{10}$$

In our example the input alphabet $\mathcal{X}$ will be the set $\{1, 2\}$ and the channel will in fact "remember" only the first input letter.

We consider DCs with immediate noiseless feedback, i.e. after each transmitted letter the sender gets the output letter on the receivers side and his further encodings may depend on all previous output letters.

Formally, if we have N messages $1, \ldots, N$, then the encoding of message $j \in \{1, \ldots N\}$ is a vector $f_j = [f_j^1, \ldots f_j^n]$, where $f_j^i : \mathcal{Y}^{i-1} \to \mathcal{X}$. f_j^1 is understood to be an element of $\mathcal{X}$. The set of all functions of this kind is denoted by $\mathcal{F}_n$.

For a given DC V and an encoding f the probability $V^n(y^n|f)$ of receiving an output sequence $y^n = (y_1, \ldots, y_n)$ is given by

$$V^n(y^n|f) = V_1(y_1|f^1) \cdot V_2(f^1)(y_1|f^2(y_1)) \cdots \qquad (11)$$
$$V_n(f^1, f^2(y_1), \ldots, f^{n-1}(y_1, \ldots, y_{n-2}))(y_n|f^1(y_1, \ldots, y_{n-1})).$$

Definition 8 (Deterministic identification code for a DC). *Let* n, $N \in \mathbb{N}$, $0 \le \lambda \le 1$. *A deterministic* (n, N, λ) *identification code for a DC* V *with feedback is a family of pairs*

$$\{(f_j, \mathcal{D}_j)\}_{j \in \{1, \ldots, N\}}, \mathcal{D}_j \subset \mathcal{Y}^n \qquad (12)$$

with the following properties

$$V^n(\mathcal{D}_i|f_i) > 1 - \lambda \text{ for all } i \in \{1, \ldots, M\} \qquad (13)$$

$$V^n(\mathcal{D}_j|f_i) < \lambda \text{ for all } i, j \in \{1, \ldots, M\}, i \ne j \qquad (14)$$

Definition 9 (Randomized identification code for a DC). *Let* n, $N \in \mathbb{N}$, $0 \le \lambda \le 1$. *A randomized* (n, N, λ) *identification code for a DC* V *with feedback is a family of pairs*

$$\{(Q(\cdot|j), \mathcal{D}_j)\}_{j \in \{1, \ldots, N\}}, \quad Q(\cdot|j) \in \mathcal{P}(\mathcal{F}_n), \quad \mathcal{D}_j \subset \mathcal{Y}^n \qquad (15)$$

with the following properties

$$\sum_{f \in \mathcal{F}_n} Q(f|i) W^n(\mathcal{D}_i|f) > 1 - \lambda \text{ for all } i \in \{1, \ldots, M\} \qquad (16)$$

$$\sum_{f \in \mathcal{F}_n} Q(f|i) W^n(\mathcal{D}_j|f) < \lambda \text{ for all } i, j \in \{1, \ldots, M\}, i \ne j \qquad (17)$$

The definition of the common randomness capacity for a specific model is affected by the resources which are given to the sender and the receiver and this dependency lies in the definition of a permissible pair of random variables. Generally speaking, a pair of random variables is permissible (for length n), if it can be realized by a communication over the channel with blocklength n.

For our model it is the following (according to the notation and definitions of [8]):

A pair (K, L) of random variables is permissible, if K and L are functions of the channel output Y^n. If we allow randomization in the encoding, K may also depend on a random variable M which is generated in advance by the sender.

Definition 10 (Common randomness capacity). *The common randomness capacity* C_{CR} *is the maximal number* ν *such, that for a constant* $c > 0$ *and for all* $\epsilon > 0$, $\delta > 0$ *and for all* n *sufficiently large there exists a permissible pair* (K, L) *of random variables for length* n *on a set* $\mathcal{K}$ *with* $|\mathcal{K}| < e^{cn}$ *with*

$$Pr\{K \ne L\} < \epsilon$$

and

$$\frac{H(K)}{n} > \nu - \delta.$$

The upper bound on the cardinality $|\mathcal{K}|$ of the ground set guarantees that the small amount of probability where K and L may differ can give only small contribution to the entropy. It also ensures that $H(L) \approx H(K)$. Obviously, for a channel with noiseless passive feedback the common randomness capacity equals the maximal output entropy.

Theorem 1. *Let α, β be real numbers with $0 < \alpha < \beta$. There exists a DC V with common randomness capacity β and identification capacity α, that is if we denote by $N_f(n, \lambda)$ the maximal number N for which a deterministic (n, N, λ)-code for the DC V exists, then*

$$\lim_{n \to \infty} \frac{\log \log N_f(n, \lambda)}{n} = \alpha. \tag{18}$$

Analogously if we denote by $N_F(n, \lambda)$ the maximal number N for which a randomized (n, N, λ)-code exists, then

$$\lim_{n \to \infty} \frac{\log \log N_F(n, \lambda)}{n} = \alpha. \tag{19}$$

For the proof of Theorem 1 we give a construction for the channel. We show that from an (n, N, λ) identification code for the discrete memoryless channel with feedback induced by a certain matrix W_B we can construct an (n, N, λ) identification code for our DC V. Conversely we will show that a deterministic (n, N, λ) identification code for V induces a deterministic $(n, N - 1, \lambda)$ identification code for the DMC induced by W_B. For randomized codes, we will show that a code for the DC V leads to a code for the DMC W_B with a number of messages that is smaller by at most a constant factor that depends on the error probability λ.

Set $\mathcal{X} = \{1, 2\}$, $\mathcal{Y} = \{1, \ldots, \lceil 2^\beta \rceil\}$. Let P_0 be the probability distribution on $\mathcal{Y}$ with $P_0(1) = 1$, fix any two probability distributions P_1, P_2 on $\mathcal{Y}$ with

$$H(P_1) = \alpha \tag{20}$$

$$H(P_2) = \beta \tag{21}$$

$$P_1(i) \geq P_1(i + 1) \text{ for all } i = 1, \ldots |\mathcal{Y}| - 1. \tag{22}$$

Now define three stochastic matrices $W_A, W_B, W_C : \mathcal{X} \to \mathcal{P}(\mathcal{Y})$ by

$$W_A(\cdot|x) = P_0 \text{ for all } x \in \mathcal{X} \tag{23}$$

$$W_B(\cdot|x) = P_{x-1} \tag{24}$$

$$W_C(\cdot|x) = P_2 \text{ for all } x \in \mathcal{X}. \tag{25}$$

Finally we will define our DC V: Set $V_0 \triangleq W_A$ and

$$W_i(x_1, \ldots, x_{i-1}) = \begin{cases} W_B & \text{if } x_1 = 1 \\ W_C & \text{otherwise.} \end{cases} \tag{26}$$

By this construction the sender decides with his first letter whether he wants to use the channel for the maximal possible generation of common randomness (and giving up the possibility to transmit messages), or whether he also wants a channel with positive transmission capacity.

For this channel and an encoding $f_j = [f_j^1, \ldots, f_j^n]$ Equation (11) yields

$$V^n(y^n|f_j) = \begin{cases} P_0(y_1) \cdot W_B(y_2|f_j^2(y_1)) \cdot \ldots \cdot W_B(y_n|f_j^n(y_1, \ldots, y_{n-1})) & \text{if } f_j^1 = 1 \\ P_0(y_1) \cdot P_2(y_2) \cdot \ldots \cdot P_2(y_n) & \text{otherwise.} \end{cases}$$
(27)

It is obvious that this channel has a common randomness capacity of β, since the maximal output entropy is achieved, if the sender sends a 2 as first letter and arbitrary letters afterwards, which gives $(n-1) \cdot \beta$ bits of common randomness for blocks of length n.

The main tools in the calculation of the identification capacities of this channels will be Theorem 1 and Theorem 2 from [2] which give the capacities for memoryless channels:

Theorem 2. *Let W_i be a stochastic matrix. Let $0 < \lambda < 1/2$ be given. If there are two different rows in W_i, then for the maximal sizes $N_{f,W_i}(n, \lambda)$ of deterministic identification codes for W_i with error probability smaller than λ*

$$\liminf_{n \to \infty} n^{-1} \log\log N_{f,W_i}(n, \lambda) = \max_{x \in \mathcal{X}} H(W_i(\cdot|x)). \tag{28}$$

For the maximal sizes $N_{F,W_i}(n, \lambda)$ of randomized identification codes for W_i with error probability smaller than λ we have

$$\liminf_{n \to \infty} n^{-1} \log\log N_{F,W_i}(n, \lambda) = \max_{P \in \mathcal{P}(\mathcal{X})} H(PW_i). \tag{29}$$

Lemma 4. *Let $\lambda < 1/2$, $n \in I\!N$, let $\{(f_j, \mathcal{D}_j)\}_{j \in \{1, \ldots, N\}}$ be an (n, N, λ)-code for the DC defined before, then there exists at most one index j with $f_j^1 = 2$.*

Proof

Suppose we have $j \neq j'$ with $f_j^1 \neq 1$ and $f_{j'}^1 \neq 1$. Then by Equations (13), (14) and (27)

$$1/2 < 1 - \lambda < W^n(\mathcal{D}_j|f_j) = W^n(\mathcal{D}_j|f_{j'}) < \lambda < 1/2. \tag{30}$$

$\square$

Lemma 5. *Let $\lambda < 1/2$, $n \in I\!N$. Then*

$$N_f(n, \lambda) \leq N_{f,W_B}(n-1, \lambda) + 1.$$

Proof

Let $\{(f_j, \mathcal{D}_j)\}_{j \in \{1, \ldots, N\}}$ be a deterministic (n, N, λ) ID-code for the DC V. By Lemma (4) we can assume without loss of generality that $f_j^1 = 1$ for all $j \in \{1, \ldots, N-1\}$. Define a code $\{(f_j', \mathcal{D}_j')\}_{j \in \{1, \ldots, N-1\}}$ of length $n-1$ with $N-1$ codewords by the following rules

$$f_j'^i(y_1, \ldots, y_{i-1}) = f_j^{i+1}(1, y_1, \ldots, y_{i-1}) \text{ for all } i \tag{31}$$

$$\mathcal{D}_j' = \{(y_1, \ldots, y_{n-1}) \in \mathcal{Y}^{n-1} | (1, y_1, \ldots, y_{n-1}) \in \mathcal{D}_j\} \tag{32}$$

By Equation (27)

$$
\begin{aligned}
V^n((1, y_1, \ldots, y_{n-1})|f_j) &= W_B(y_1|f_j^2(1)) \cdot \ldots \cdot W_B(y_{n-1}|f_j^n(1, y_1, \ldots, y_{n-2})) \\
&= W_B(y_1|f_j'^1) \cdot \ldots \cdot W_B(y_{n-1}|f_j'^{n-1}(y_1, \ldots, y_{n-2})) \\
&= W_B^n(y^{n-1}|f'^j)
\end{aligned}
$$

and

$$V^n((0, y_1, \ldots, y_{n-1})|f_j) = 0. \tag{33}$$

Thus

$$W_B^n(\mathcal{D}_i'|f'^j) = V^n(\mathcal{D}_i|f^j) \text{ for all } i, j \in \{1, \ldots, N-1\} \tag{34}$$

i.e. $\{(f_j', \mathcal{D}_j')\}_{j \in \{1, \ldots, N-1\}}$ is an $(n-1, N-1, \lambda)$ ID-code for the DMC defined by matrix W_B. $\qquad\square$

Lemma 6. *Let* $\lambda < 1/2$, $n \in \mathbb{N}$. *Then*

$$N_f(n+1, \lambda) \geq N_{f, W_B}(n, \lambda).$$

Proof

Let $\{(f_j, \mathcal{D}_j)\}_{j \in \{1, \ldots, N\}}$ be an (n, N, λ) ID-code for the DMC W_B.
Define a code $\{(f_j', \mathcal{D}_j')\}_{j \in \{1, \ldots, N\}}$ of length $n+1$ with N codewords by

$$f_j'^1 = 1 \tag{35}$$

$$f_j'^i(y_1, \ldots, y_{i-1}) = f_j^{i-1}(y_1, \ldots, y_{i-2}) \text{ for all } i \in \{2, \ldots, n+1\} \tag{36}$$

$$\mathcal{D}_j' = \{(1, y_1, \ldots, y_n) \in \mathcal{Y}^{n+1} | (y_1, \ldots, y_n) \in \mathcal{D}_j\} \tag{37}$$

Therefore

$$V^{n+1}(\mathcal{D}_i'|f'^j) = V^n(\mathcal{D}_i|f^j) \text{ for all } i, j \in \{1, \ldots, N\} \tag{38}$$

i.e. $\{(f_j', \mathcal{D}_j')\}_{j \in \{1, \ldots, N\}}$ is an $(n+1, N, \lambda)$ ID-code for the DC W. $\qquad\square$

Now we will return to the proof of Theorem 1. By Theorem (2), Lemma 5 and Lemma 6

$$
\begin{aligned}
&\varlimsup_{n \to \infty} \frac{\log \log N_f(n, \lambda)}{n} \leq \varlimsup_{n \to \infty} \frac{\log \log N_{f, W_B}(n-1, \lambda) + 1}{n} \leq \\
&\max_{x \in \mathcal{X}} H(W_B(\cdot|x)) = \alpha
\end{aligned}
$$

and

$$
\begin{aligned}
&\liminf_{n \to \infty} \frac{\log \log N_f(n, \lambda)}{n} \geq \liminf_{n \to \infty} \frac{\log \log N_{f, W_B}(n-1, \lambda)}{n} \geq \\
&\max_{x \in \mathcal{X}} H(W_B(\cdot|x)) = \alpha.
\end{aligned}
$$

Thus

$$\lim_{n \to \infty} \frac{\log \log N_f(n, \lambda)}{n} = \alpha, \tag{39}$$

proving Theorem 1 for deterministic codes.

Since a deterministic code can also be viewed as a randomized code, this also gives the direct part of Theorem 1 for randomized codes. For the proof of the converse we need some further observations.

The second statement in Theorem (2) needs the following modification. The proof is rather lengthy, so instead of copying it we will only point out the modifications needed to adapt it for our purposes. An investigation of the proof of [2, Theorem 2 b] shows that the converse still holds if the bounds for the error probabilities are slightly weakened, namely replace Inequalities (4) and (5) in [2] by

$$\sum_{g \in \mathcal{F}_n} Q_F(g|i) \left(W^n(\mathcal{D}_i^C|g) + W^n(\mathcal{D}_j|g) \right) \leq 2 \cdot \lambda \tag{40}$$

for all $i, j \in \{1, \ldots, N\}$, $i \neq j$. The second order rate of a series IDF-code satisfying (40) with $\lambda < 1/2$ is upperbounded by $\max_{P \in \mathcal{P}(\mathcal{X})} H(P \cdot W)$.

To verify this, choose $\nu > 0$ with $2 \cdot \lambda + \nu < 1$ and define $\mathcal{E}_i^*$ as in [2, p. 35]. Equation (40) implies that the sets $\mathcal{D}_i \cap \mathcal{E}_i^*$ are pairwise different: Suppose we have $i \neq j$ with $\mathcal{D}_i \cap \mathcal{E}_i^* = \mathcal{D}_j \cap \mathcal{E}_j^*$. This would lead to the contradiction

$$\begin{aligned}
1 &= \sum_{g \in \mathcal{F}_n} Q_F(g|i) \left(W^n((\mathcal{D}_i \cap \mathcal{E}_i^*)^C|g) + W^n((\mathcal{D}_i \cap \mathcal{E}_i^*)|g) \right) \\
&= \sum_{g \in \mathcal{F}_n} Q_F(g|i) \left(W^n((\mathcal{D}_i \cap \mathcal{E}_i^*)^C|g) + W^n((\mathcal{D}_j \cap \mathcal{E}_j^*)|g) \right) \\
&\leq \sum_{g \in \mathcal{F}_n} Q_F(g|i) \left(W^n((\mathcal{D}_i)^C|g) + W^n((\mathcal{E}_i^*)^C|g) + W^n((\mathcal{D}_j)|g) \right) \\
&< 2\lambda + \nu < 1.
\end{aligned}$$

The rest of the proof can be applied as it is.

Lemma 7. *Let $\lambda < 1/2$, $n \in \mathbb{N}$. Then*

$$N_F(n, \lambda) \leq N'_{F, W_B}(n - 1, \lambda) \left\lceil \frac{1}{1 - 2 \cdot \lambda} \right\rceil.$$

where $N'_{F, W_B}(n - 1, \lambda)$ is the maximal size of a randomized identification code with the modified error probability defined by Equation (40) being less than λ.

Proof

Let $\{(Q(\cdot|j), \mathcal{D}_j)\}_{j \in \{1, \ldots, N\}}$ be a randomized (n, N, λ) IDF-code for the DC V. Let for all i $p_i = W^n(\mathcal{D}_i|g')$ and $q_i = \sum_{g \in \mathcal{F}_n|g^1 = 2} Q(g|i)$, where g' is any encoding function with $g'^1 = 2$. There exists a subset $I \subset \{1, \ldots, N\}$ of cardinality $|I| \geq \frac{N}{\lceil \frac{1}{1-2 \cdot \lambda} \rceil}$ with $|p_i - p_j| \leq 1 - 2 \cdot \lambda$ for all $i, j \in I$. This leads to the inequality

$$1 - p_i + p_j \geq 2 \cdot \lambda \tag{41}$$

and thus

$$\sum_{G \in \mathcal{F}_N : g^1 = 2} \left(V^n(\mathcal{D}_i^C) + V^n(\mathcal{D}_j) \right) \geq 2 \cdot \lambda \cdot q_i \tag{42}$$

and

$$\sum_{G \in \mathcal{F}_N : g^1 = 1} \left(V^n(\mathcal{D}_i^C) + V^n(\mathcal{D}_j) \right) < 2 \cdot \lambda \cdot (1 - q_i) \text{ for all } i, j \in I. \tag{43}$$

(Especially we have $q_i < 1$ for all $i \in I$.) Analogue to Lemma 5, we define the code $\{(Q'(\cdot|j), \mathcal{D}'_j)\}_{j \in I}$ of blocklength $n - 1$ with $|I|$ codewords by

$$Q'(g'|i) = \frac{Q(g|i)}{1 - q_i} \tag{44}$$

where $g' \in \mathcal{F}^{n-1}$ and $g \in \mathcal{F}^n$ are encoding function with $g^1 = 1$ and

$$g'^i(y_1, \dots, y_{i-1}) = g^{i+1}(1, y_1, \dots, y_{i-1}) \tag{45}$$

$$\mathcal{D}'_j = \{(y_1, \dots, y_{n-1}) \in \mathcal{Y}^{n-1} | (1, y_1, \dots, y_{n-1}) \in \mathcal{D}_j \} \tag{46}$$

Since

$$\sum_{G \in \mathcal{F}_N : g^1 = 1} \frac{Q(\cdot|i)}{1 - q_i} \left(V^n(\mathcal{D}_i^C) + V^n(\mathcal{D}_j) \right) < 2\lambda \tag{47}$$

this code satisfies the modified error condition of Equation (40) for the DMC W_B.[1]   □

Lemma 8. *Let W_B, P_1 and α be defined as before. Then the maximal output entropy for the DMC W_B is*

$$\max_{P \in \mathcal{P}(X)} H(PW_B) = H(P_1) = \alpha.$$

---

[1] Actually one may expect that also randomized identification codes for the DC V could only have one more message than the corresponding codes for the DMC W_B (as we could show for deterministic codes), but at least the conversion of codes we have used is not sufficient to show this. Here is an an informal recipe how to construct counterexamples (a formal construction is left out because this is neither relevant for the proof of our theorem nor very instructive): Suppose we have an IDF code for the DMC W_B with the property that all distributions of input letters are equal and the decoding sets consist only of sequences that are "typical" (there will be a formal definition in the next section). Then we can build a code for the DC V with twice as many code words as follows: The first half will be the original encoding functions and decoding sets both prefixed by "1" as in the previous lemma. For the other half, choose a suitable distribution on the original encoding function and any function that sends "2" as the first letter in such a way, that the errors of second kind with respect to decoding sets of the first half are little smaller than λ. For the decoding sets take the sets of the first half and remove some sequences to get the error of second kind for encoding functions of the first half are small enough. Then add some output sequences that are "typical" for encodings that start with a "2" to get the probability for errors of first kind right.

Proof. Consider a convex combination

$$P_\rho = (1 - \rho)P_0 + \rho P_1, \quad \rho \in [0, 1].$$

P_ρ is a non increasing probability distribution and

$$\sum_{i=1}^{k} P_\rho(i) = (1 - \rho) + \rho \sum_{i=1}^{k} P_1(i) \geq \sum_{i=1}^{k} P_1(i)$$

for all $k \in \mathcal{Y}$, i.e. P_ρ majorizes P_1, thus by Schur-concavity of entropy $H(P_\rho) < H(P_1)$. $\qquad\square$

Using our modified Version [2, Theorem 2b] we get therefore the desired result for randomized identification codes for our specific channel:

$$\varlimsup_{n \to \infty} \frac{\log \log N_F(n, \lambda)}{n} \leq \varlimsup_{n \to \infty} \frac{\log \log N'_{F,W_B}(n - 1, \lambda) \left\lceil \frac{1}{1 - 2 \cdot \lambda} \right\rceil}{n} \leq$$
$$\max_{P \in \mathcal{P}(X)} H(PW_B) = \alpha$$

and

$$\liminf_{n \to \infty} \frac{\log \log N_F(n, \lambda)}{n} \geq \liminf_{n \to \infty} \frac{\log \log N_f(n, \lambda)}{n} \geq$$
$$\max_{x \in \mathcal{X}} H(W_B(\cdot | x)) = \alpha.$$

4 The Deterministic Identification Capacity for Noiseless Channels with Noisy Feedback

In this section we improve the upper bound for deterministic ID-codes from [6] for the case of noiseless channels with noisy feedback. We show that our improved bound coincides with the lower bound of [6] for this type of channels.

Definition 11. *A channel with noisy feedback is a quadruple*

$$\{\mathcal{X}, W, \mathcal{Y}, \mathcal{Z}\}. \tag{48}$$

Here $\mathcal{X}$ is the input alphabet, $\mathcal{Y}$ and $\mathcal{Z}$ are the output alphabets for the receiver respectively the feedback and

$$W = (W(y, z | x))_{x \in \mathcal{X}, y \in \mathcal{Y}, z \in \mathcal{Z}} \tag{49}$$

is a stochastic matrix. To avoid trivialities, all alphabets are assumed to contain at least two elements. We want to consider finite alphabets, so we can also assume that all alphabets are the set of natural numbers less or equal to the given cardinality.

The channel is assumed to be memoryless, thus the transmission probabilities for sequences $x^n = (x_1, \ldots, x_n)$, $y^n = (y_1, \ldots, y_n)$ and $z^n = (z_1, \ldots, z_n)$ of length n are given by

$$W^n(y^n, z^n | x^n) = \prod_{t=1}^{n} W(y_t, z_t | x_t) \tag{50}$$

Now an encoding of message $j \in \{1, \ldots N\}$ is a vector-valued function $f_j = [f_j^1, \ldots f_j^n]$, where $f_j^i : \mathcal{Z}^{i-1} \to \mathcal{X}$. f_j^1 is an element of $\mathcal{X}$. If the message is j and the sender has received $z_1, \ldots, z_{j-1}$ as feedback sequence, he sends $f_j^i(z_1, \ldots, z_{j-1})$ as the ith letter. The set of all functions of this kind is again denoted by $\mathcal{F}_n$.

An encoding function f determines the joint distribution of the output random variables

$$(Y_1, \ldots, Y_n)$$

and the feedback random variables $(Z_1, \ldots, Z_n)$ by the following formula:

$$P(Y^n = y^n, Z^n = z^n | f) = W^n(y^n, z^n | f) = \prod_{t=1}^{n} W\left(y_t, z_t | f^t(z_1, \ldots, z_{t-1})\right) \quad (51)$$

Definition 12. *A deterministic (n, N, λ) IDF code for W is a family of pairs*

$$\{(f_i, \mathcal{D}_i) : i \in \{1, \ldots, N\}\} \quad (52)$$

with

$$f_i \in \mathcal{F}_n, \quad \mathcal{D}_i \subset \mathcal{Y}^n \quad (53)$$

and

$$W^n(\mathcal{D}_i | f_i) \geq 1 - \lambda, \quad W^n(\mathcal{D}_j | f_i) < \lambda \text{ for all } i, j \in \{1, \ldots, N\}, i \neq j \quad (54)$$

Let us denote by $N_f^*(n, \lambda)$ the maximal N such that an (n, N, λ) code for W exists.

The following result of Ahlswede and Zhang in [6] gives bounds on the growth of $N_f^*(n, \lambda)$.

Theorem AZ. *If the transmission capacity C of W is positive, then we have for all $0 < \lambda < 1/2$*

$$\liminf_{n \to \infty} n^{-1} \log \log N_f^*(n, \lambda) \geq \max I(Z \wedge U | X), \quad (55)$$

where the maximum is taken over all random variables (X, Y, Z, U) with joint distribution

$$P_{XYZU}(x, y, z, u) = p(x) W(y, z | x) q(u | x, z)$$

under the constraint

$$I(U \wedge Z | XY) \leq I(X \wedge Y). \quad (56)$$

Furthermore

$$\overline{\lim_{n \to \infty}} \, n^{-1} \log \log N_f^*(n, \lambda) \leq \min\{\max I(XZ \wedge Y), \max H(Z | X)\}, \quad (57)$$

where the maximum is taken over all random variables (X, Y, Z) with joint distribution

$$P_{XYZ}(x, y, z) = p(x) W(y, z | x).$$

We want to consider channels where the channel from the sender to the receiver is noiseless, i.e. $\mathcal{X} = \mathcal{Y}$ and $W(y, z | x) = 0$ for all $y \neq x$. We will in this case identify W with the matrix with entries $W(z | x) = W(x, z | x)$.

Theorem 3. *If W is a noiseless channel with noisy feedback, then for all $\lambda \in (0, 1/2)$*

$$\lim_{n \to \infty} n^{-1} \log \log N_f^*(n, \lambda) = \max\{\min\{H(X), H(Z|X)\}\} \qquad (58)$$

where the maximum is taken over all random variables X, Z with joint distribution of the form

$$P_{XZ}(x, z) = p(x) W(z|x) \qquad (59)$$

For $z^n \in \mathcal{Z}^n$ and $f \in \mathcal{F}^n$ let us denote

$$f(z^n) = \left(f^1, f^2(z_1), \ldots, f^n(z_1, \ldots, z_{n-1})\right)$$

Since for a noiseless channel input and output random variables are equal with probability 1, for given $z^n \in \mathcal{Z}^n$ $y^n = f(z^n)$ is by Equation (51) the unique sequence with $P_{Y^n Z^n}(y^n, z^n) > 0$ and thus the output random variable Y^n has distribution

$$P_{Y^n}(y^n) = \sum_{z^n \in f^{-1}(y^n)} P_{Z^n}(z^n) = \sum_{z^n \in f^{-1}(y^n)} \prod_{t=1}^{n} W\left(z_t | f(z_1, \ldots, z_{t-1})\right) \qquad (60)$$

Here $f^{-1}(y^n)$ is the inverse image of y^n under the encoding function f, i.e.

$$f^{-1}(y^n) = \{z^n \in \mathcal{Z}^n | f(z^n) = y^n\}. \qquad (61)$$

Nevertheless, we want to distinguish between the input and output random variables to stress whether we are considering feedback sequences produced by certain input sequences or output sequences as a function of feedback sequences.

The following lemma states that the output distributions generated by the encoding functions of an IDF-code must differ significantly on the decoding sets. Thus we can get an upper bound on the size of the code by *upperbounding the number of significantly different output distributions that can be generated.*

Lemma 9. *Let $\{(f_i, \mathcal{D}_i) : i \in \{1, \ldots, N\}\}$ be an (n, N, λ) IDF-code for W with $\lambda \in (0, 1/2)$. Set $\delta = 1/4 - \lambda/2$. Then*

$$|W(\mathcal{D}_i|f_i) - W(\mathcal{D}_i|f_j)| > \delta \text{ for all } i, j \in \{1, \ldots, N\}, i \neq j. \qquad (62)$$

Proof. Assume Equation (62) is false for i, j. Then

$$\frac{1}{2} < 1 - \lambda \leq W(\mathcal{D}_i|f_i) \leq W(\mathcal{D}_i|f_j) + \delta < \lambda + \delta < \frac{1}{2} \ . \qquad (63)$$

$\square$

Now for each given $X^n = x^n$ we have to split $\mathcal{Z}^n$ into one set of negligible probability and one set which contains only sequences of almost the same probability. The second set will be the $W - \delta$-*typical sequences under the condition x^n*.

Definition 13 ($W - \delta$-**typical sequences**). *Let $W : \mathcal{X} \to \mathcal{Z}$ be a stochastic matrix, let $x^n \in \mathcal{X}^n$. $z^n \in \mathcal{Z}^n$ is called $W - \delta$-typical under the condition x^n if for all $x \in \mathcal{X}, z \in \mathcal{Z}$*

$$\frac{1}{n} |<x,z|x^n, z^n> - <x|x^n> \cdot W(z|x)| \leq \delta \tag{64}$$

and additionally $< x, z|x^n, z^n > = 0$ if $W(z|x) = 0$. The set of all $W - \delta$-typical sequences under the condition x^n is denoted by $T_{W,\delta}(x^n)$. If z^n is not typical, we call it atypical.

In the sequel we will for sequences of length n choose $\delta_n = n^{-1/3}$ and refer to $W - \delta_n$-typical sequences simply as W-typical sequences and denote them by $T_W(x^n)$. The following lemma is the so–called *joint asymptotic equipartition property (joint AEP)* for our model. It is an analogue to the law of large numbers in probability theory.

Lemma 10. *For fixed $\mathcal{X}$ and $\mathcal{Z}$, for every stochastic matrix $W : \mathcal{X} \to \mathcal{Z}$ there exists a nullsequence $(\epsilon_n)_{n \in \mathbb{N}}$, such that for every distribution p on $\mathcal{X}$ and for all $x^n \in T_p$*

$$W^n(T_W(x^n)|x^n) \geq 1 - \epsilon_n \tag{65}$$

and

$$|T_W(x^n)| \leq 2^{n(H(Z|X)+\epsilon_n)} \tag{66}$$

Furthermore if $z^n \in T_W(x^n)$, then

$$\left|\frac{1}{n}\log W(z^n|x^n) + H(Z|X)\right| \leq \epsilon_n \tag{67}$$

where X, Z are random variables with joint distribution

$$P_{XZ}(x,z) = p(x)W(z|x). \tag{68}$$

Proof. These results are folklore in information theory, we supply some justification and references for the benefit of the non-expert reader. The first inequality is the second assertion in [10, Lemma 2.12], the second is a rewriting of the upper bound given by the second inequality in [10, Lemma 2.13]:

For $x^n = (x_1, \ldots, x_n)$ let $Z_1, \ldots, Z_n$ be independent random variables where Z_i has distribution $P_{Z_i} = W(\cdot|x_i)$. Then for all pairs $(x, z) \in \mathcal{X} \times \mathcal{Z}$ the random variable $<x, z|x^n, Z^n>$ has binomial distribution with expected value $< x|x^n > W(z|x)$. The variance is

$$<x|x^n> W(z|x)\,(1 - W(z|x)) \leq \frac{<x|x^n>}{4} \leq \frac{n}{4}. \tag{69}$$

We apply Chebyshev's inequality and get

$$P\left(n^{-1}|<x,z|x^n, z^n> - <x|x^n> \cdot W(z|x)| > \delta\right) \leq \frac{1}{4k\delta^2}. \tag{70}$$

Thus we get from the union bound that

$$W^n(T_W(x^n)|x^n) \geq 1 - \frac{|\mathcal{X}||\mathcal{Z}|}{4n\delta^2}. \tag{71}$$

The second inequality is a rather straightforward consequence of Lemma (2) and Corollary (1). For the third inequality notice that

$$W(z^n|x^n) = \prod_{(x,z)\in\mathcal{X}\times\mathcal{Z}:W(z|x)>0} W(z|x)^{<x,z|x^n,z^n>} \tag{72}$$

Thus

$$-\log W(z^n|x^n) = - \sum_{(x,z)\in\mathcal{X}\times\mathcal{Z}:W(z|x)>0} <x,z|x^n,z^n> \log W(z|x)$$

$$\geq - \sum_{(x,z)\in\mathcal{X}\times\mathcal{Z}:W(z|x)>0} (np(x)W(z|x) + \delta_n) \log W(z|x)$$

$$= nH(Z|X) - n\delta_n \sum_{(x,z)\in\mathcal{X}\times\mathcal{Z}:W(z|x)>0} \log W(z|x). \tag{73}$$

Analogously we have

$$-\log W(z^n|x^n) \leq nH(Z|X) + n\delta_n \sum_{(x,z)\in\mathcal{X}\times\mathcal{Z}:W(z|x)>0} \log W(z|x). \tag{74}$$

Since $\sum_{(x,z)\in\mathcal{X}\times\mathcal{Z}:W(z|x)>0} \log W(z|x)$ is fixed, this proves the third inequality for suitable ϵ_n. □

To approximate the output probability for a given sequence y^n by Equation (65) we can restrict ourselves to sequences z^n which are jointly typical under the condition $x^n = y^n$.

Since the probability of a decoding set is determined by the sizes of intersections of the inverse images of the decoding set with the different types of feedback sequences, we can then estimate the number of different encoding functions by the combination of sizes of intersections.

Lemma 11. *For every $0 < \lambda < \lambda' < 1/2$, n sufficiently large depending on λ and λ' and every (n, N, λ)-IDF-code $\{(f_i, \mathcal{D}_i) : i \in \{1,\ldots,N\}\}$ there exists an $(n+1, N, \lambda')$-IDF-code*
$\{(f_i', \mathcal{D}_i') : i \in \{1,\ldots,N\}\}$ *with*

$$f'(z^{n+1}) \in \bigcup_{i=1}^{N} \mathcal{D}_i' \Rightarrow z^n \in T_W(f(z^n)) \tag{75}$$

where $z^{n+1} = (z^n, 1)$.

Proof. For $i \in \{1,\ldots,n\}$ and $f_i = (f_i^1,\ldots,f_i^n)$ let $f_i' = (f_i^1,\ldots,f_i^{n+1})$ where

$$f_i^{n+1}(z^n) = \begin{cases} 1 \text{ if } z^n \in T_W(f(z^n)) \\ 2 \text{ otherwise} \end{cases} \tag{76}$$

and $\mathcal{D}_i' = \{(y_1,\ldots,y_n,1) : (y_1,\ldots,y_n) \in \mathcal{D}_i\}$. Obviously this code satisfies Implication (75) and Inequality (65) guarantees that the error probabilities are smaller than $\lambda + \epsilon_n < \lambda'$ for n sufficiently large. □

From now, we want to consider only IDF-codes respectively encoding functions that satisfy (75). We will denote the maximal size of such a code by $N'_f(n, \lambda)$. We introduce a new possibility for errors of first kind by declaring an error, if the feedback sequence is not jointly typical with the first n letters of the input sequence, but by the previous lemma, there is no loss in rate and the decrease in error performance is not significant. By considering only *typical* sequences, the sets also boil down to the required sizes. The next lemma shows that the number of feedback types that "typically" occur is at most polynomial in n.

Lemma 12. *Let P_X be a distribution on $\mathcal{X}$. The number of types of feedback sequences z^n for which there is an input sequence x^n of type p such that z^n is $W - \delta$-typical under the condition x^n is upperbounded by $(2n\delta + 1)^{|\mathcal{Z}|}$.*

Proof. The number of occurrences of each letter can take one of at most $(2n\delta+1)$ values.   □

We will now consider output sequences of fixed n-type p on $\mathcal{X}$ and upperbound the number of different probabilities of decoding sets we can generate on this set under the assumption that we produce an output sequence that is outside any decoding set, if the feedback sequence is not jointly typical with the input sequence:

Lemma 13. *Let p be an n-type on $\mathcal{X}$. Let $B(p, n)$ denote the number of different values that can occur as the output probability of a decoding set $\mathcal{D}$ with $\mathcal{D} \subset T_p$ under an encoding function that maps atypical feedback sequences outside $\mathcal{D}$. Then*

$$B(p,n) \leq \left(\frac{2^{nH(X)} + 2^{n(H(Z|X)+\epsilon_n)}}{2^{nH(X)}} \right)^{(2n\delta+1)^{|\mathcal{Z}|}} \tag{77}$$

where (X, Z) is a pair of random variables with joint distribution $P_{XZ}(x, z) = p(x)W(z|x)$.

Proof. There are at most $2^{nH(X)}$ output sequences of type p and for each of this sequences there are at most $2^{n(H(Z|X)+\epsilon_n)}$ feedback sequences that are jointly typical with that given sequence and have the same joint type. So if we condition on the event that the feedback sequence is of that joint type, the probability of that output sequence is l-typical, $l < 2^{n(H(Z|X)+\epsilon_n)}$, because it depends only on the number of feedback sequences that are mapped on x^n. Since the number of joint types that are $W - \delta$-jointly typical is less than $(2n\delta + 1)$ this proves the lemma, because the number of l-types on a set $\mathcal{U}$ is nondecreasing in both l and the cardinality of $\mathcal{U}$.   □

To get a bound on the number of different decoding functions, we have to consider all possible input types, but this number is only polynomial in n, so we have to upperbound

$$n^{-1} \log \log B(p, n)^{(n+1)^{|\mathcal{X}|}} :$$

To simplify notation, a non-integer argument for factorials or bounds of sums and products is understood to be rounded down.

Lemma 14.

$$\varlimsup_{n\to\infty} n^{-1} \log\log B(p,n)^{(n+1)^{|\mathcal{X}|}} \le H(X) \tag{78}$$

$$\varlimsup_{n\to\infty} n^{-1} \log\log B(p,n)^{(n+1)^{|\mathcal{X}|}} \le H(Z|X) \tag{79}$$

Proof

$$B(p,n) \le \left(\frac{\left(2^{nH(X)} + 2^{n(H(Z|X)+\epsilon_n)}\right)!}{\left(2^{nH(X)}\right)!\left(2^{n(H(Z|X)+\epsilon_n)}\right)!} \right)^{(2n\delta+1)^{|\mathcal{Z}|}}$$

$$\le \left(\frac{\prod_{i=1}^{2^{n(H(Z|X)+\epsilon_n)}} \left(2^{nH(X)} + i\right)}{\left(2^{n(H(Z|X)+\epsilon_n)}\right)!} \right)^{(2n\delta+1)^{|\mathcal{Z}|}}$$

$$= \left(\prod_{i=1}^{2^{n(H(Z|X)+\epsilon_n)}} \frac{2^{nH(X)} + i}{i} \right)^{(2n\delta+1)^{|\mathcal{Z}|}}$$

$$= \left(\prod_{i=1}^{2^{n(H(Z|X)+\epsilon_n)}} \left(1 + \frac{2^{nH(X)}}{i}\right) \right)^{(2n\delta+1)^{|\mathcal{Z}|}}$$

Now let us use an input distribution p that maximizes $\min(H(X), H(Z|X))$.

$$\varlimsup_{n\to\infty} \frac{1}{n} \log\log B(p,n) \le \varlimsup_{n\to\infty} \frac{1}{n} \log\log \left(\prod_{i=1}^{2^{n(H(Z|X)+\epsilon_n)}} \left(1 + \frac{2^{nH(X)}}{i}\right) \right)^{(2n\delta+1)^{|\mathcal{Z}|}}$$

$$= \varlimsup_{n\to\infty} \frac{1}{n} \log \sum_{i=1}^{2^{n(H(Z|X)+\epsilon_n)}} \log\left(1 + \frac{2^{nH(X)}}{i}\right)$$

$$\le \varlimsup_{n\to\infty} \frac{1}{n} \log\left(2^{n(H(Z|X)+\epsilon_n)} nH(X)\right)$$

$$= H(Z|X)$$

The other inequality can be proven analogously, since the expression on the right hand side of Equation 77 is symmetrical in $H(X)$ and $H(Z|X) + \epsilon_n$.

So for each type p the second order rate of a code that uses only input words of this type is bounded from above by the minimum of $H(X)$ and $H(Z|X)$. Because there are only polynomial in n many input types, this gives the desired result, namely let $B(n) = \max\{B(p,n) : p \text{ is n-type}\}$ and let $0 < \lambda < \lambda' < 1/2$. Then

$$\varlimsup_{n\to\infty} \frac{1}{n} \log\log N_f^*(n,\lambda) = \varlimsup_{n\to\infty} \frac{1}{n} \log\log N_f'(n,\lambda')$$

$$\le \varlimsup_{n\to\infty} \frac{1}{n} \log\log B(n)^{(n+1)^{|\mathcal{Z}|}}$$

$$\le \max_X \min\{H(X), H(Z|X)\}$$

To complete the proof of Theorem 3, we show that this coincides with the lower bound of Theorem AZ. We consider noiseless main channels, thus Inequality (56) reduces to

$$I(U \wedge Z|X) \leq H(X) \tag{80}$$

Let U be a random variable on $\mathcal{Z}$. Let

$$q'(u|x,z) = \begin{cases} 1; \text{ if } u = z \\ 0; \text{ otherwise} \end{cases} \tag{81}$$

and

$$q''(u|x,z) = \begin{cases} 1; \text{ if } u = 1 \\ 0; \text{ otherwise} \end{cases} \tag{82}$$

If for the input distribution which maximizes $\min\{H(X), H(Z|X)\}$ (if there are several, then take any of them) it holds that $H(X) \geq H(Z|X)$, then we set $q = q'$ and get $Z = U$ with probability 1. Therefore $I(U \wedge Z|X) = H(Z|X)$. Otherwise choose a convex combination $q = \alpha q' + (1 - \alpha)q''$ such that

$$I(U \wedge Z|X) = H(X). \tag{83}$$

On the boundary point for parameter 1 of the convex combination we have again $U = Z$, which gives $I(U \wedge Z|X) = H(Z|X) > H(X)$ by assumption, parameter 0 gives $I(U \wedge Z|X) = 0$. Thus by continuity of mutual information and mean value theorem such an α exists.

In both cases we get from Theorem AZ that

$$\liminf_{n \to \infty} n^{-1} \log \log N_f^*(n, \lambda) \geq \max \min\{H(X), H(Z|X)\} \tag{84}$$

which concludes the proof of Theorem 3.

Remarks

1. The upper bound of Theorem AZ applied to our case states that

$$\varlimsup_{n \to \infty} \log \log N_f^*(n, \lambda) \leq \min\{\max H(X), \max H(Z|X)\} \tag{85}$$

To see that *Theorem 3 really improves this bound*, consider the following channel: Let $\mathcal{X} = \mathcal{Y} = \mathcal{Z} = \{1, 2\}$. Let

$$W(z|x) = \begin{cases} 1/2; \text{ if } x = 1 \\ 1; \quad \text{ if } x = 2 \text{ and } z = 1 \\ 0; \quad \text{ if } x = 2 \text{ and } z = 2 \end{cases} \tag{86}$$

If one uses $p = (\beta, 1 - \beta)$ as input distribution, then $H(X) = h(\beta)$ and $H(Z|X) = \beta$ where $h(\beta) = -\beta \log \beta - (1 - \beta) \log(1 - \beta)$ is the binary entropy. So Inequality (85) gives

$$\varlimsup_{n \to \infty} n^{-1} \log \log N_f^*(n, \lambda) \leq 1, \tag{87}$$

while Theorem 3 gives

$$\varlimsup_{n \to \infty} n^{-1} \log \log N_f^*(n, \lambda) = \max_{\beta \in [0,1]} \min\{\beta, h(\beta)\} \approx 0.7729. \tag{88}$$

2. Our theorem can be viewed as a result about identification codes with a restriction on the input distributions that are allowed. Since the sender is not allowed to randomize in the encoding, the only possibility to generate randomness are the feedback letters. The sender now has to use input distributions that are simultaneously good for transmission and for the generation of randomness via the feedback channel. We conjecture that this is also true for feedback channels with noise:

Conjecture 1. *Let W be a channel with noisy feedback. Then for all $0 < \lambda < 1/2$*

$$\varlimsup_{n \to \infty} n^{-1} \log \log N_f^*(n, \lambda) \leq \max \min\{I(XZ \wedge Y), H(Z|X)\} \quad (89)$$

where the maximum is taken over all random variables (X, Y, Z) with joint distribution $P_{XYZ}(x, y, z) = p(x)W(y, z|x)$.

However, note that this conjecture in general does not coincide with the lower bound of Theorem AZ (although it does in more cases than the old bound did), so even if it was settled, determination of the capacity would still be an open problem .

References

1. R. Ahlswede and G. Dueck, Identification via channels, IEEE Trans. Inform. Theory, Vol. 35, 15–29, 1989.
2. R. Ahlswede and G. Dueck, Identification in the presence of feedback—a discovery of new capacity formulas, IEEE Trans. Inform. Theory, Vol. 35, 30–36, 1989.
3. R. Ahlswede and B. Verboven, On identification via multiway channels with feedback, IEEE Trans. Inform. Theory, vol. 37, 1519–1526, 1991.
4. T.S. Han and S. Verdú, New results in the theory of identification via channels, IEEE Trans. Inform. Theory, Vol. 38, 14–25, 1992.
5. T.S. Han and S. Verdú, Approximation theory of output statistics, IEEE Trans. Inform. Theory, Vol. 39, 752–772, 1993.
6. R. Ahlswede and Z. Zhang, New directions in the theory of identification, SFB 343 Diskrete Strukturen in der Mathematik, Bielefeld, Preprint 94-010, 1994.
7. R. Ahlswede, General theory of information transfer, Preprint 97–118, SFB 343 "Diskrete Strukturen in der Mathematik", Universität Bielefeld, 1997; General theory of information transfer:updated, General Theory of Information Transfer and Combinatorics, a Special Issue of Discrete Applied Mathematics, to appear.
8. R. Ahlswede and I. Csiszár, Common randomness in information theory and cryptograpy II: CR capacity IEEE Trans. Inform. Theory, Vol. 44, 225–240, 1998.
9. T.M. Cover and J.A. Thomas, Elements of information theory, Wiley Ser. Telecom., New York, 1991.
10. I. Cziszar and J. Körner, Information theory: Coding theorems for discrete memoryless systems, Academic Press, New York, 1981.

Identification and Prediction

L. Bäumer

1 Introduction

In this work the concept of identification is applied in the theory of prediction. This approach was suggested to us by our advisor Professor R. Ahlswede. This and other directions of research can be found also in [2]. Well known is Shannon's theory of transmission of messages over a noisy channel ([15]). Using the framework of Shannon's channel model a new concept of information transfer - called identification - was introduced by Ahlswede and Dueck in [1].

In the classical transmission model a sender wants to inform a receiver about a message by sending codewords over a channel. The channel may induce some errors and the goal is to have a large number of possible messages such that with sufficiently high probability the receiver should be able to decide which message had been sent. In identification via channels the receiver is no longer interested in what the actual message is, rather he is concerned about one particular message and only wants to know whether this message has occurred or not. However the sender does not know which message the receiver is interested in. Alternatively one can also think of several receivers, one for each message. Each receiver is interested whether his message has occurred or not. This modification of the problem actually leads to a general solution concept in mathematics. Whenever there is a problem in which the question has to be answered "What is the solution?" one can also formulate the corresponding identification problem by asking the questions "Is the solution equal to ...? ". We are going to apply this solution concept of identification to prediction problems.

In a typical prediction problem a person who has made observations $x_1, \ldots, x_t$ at time t has to answer the question "What is x_{t+1} ?". The starting point of the analysis here is to modify this problem by considering for every possible x a person that asks "Is $x_{t+1} = x$?".

In the formulation of the prediction problem it has to be specified how the data $x_1, x_2, \ldots$ is generated. Basically there are two different cases. In the probabilistic setting the sequence is generated by a random process. We will be mainly concerned with the deterministic setting where the sequence is thought to be arbitrary. This is the framework of the award winning paper by Feder, Merhav and Gutman ([8]). In this setting one wishes to deal with all sequences simultaneously. At first glance it may be surprising that if the sequence is arbitrary that the past can be helpful in predicting the future as they are not necessarily related and some care in defining the desired goals is necessary. The prediction scheme one is looking for shall use the past whenever it is helpful.

Information theorists have been concerned about prediction from the very beginning. Two ideas of Shannon shall be noted. In [16] he estimated the entropy of a language by giving persons who speak this language some text with gaps

R. Ahlswede et al. (Eds.): Information Transfer and Combinatorics, LNCS 4123, pp. 84–106, 2006.
© Springer-Verlag Berlin Heidelberg 2006

and asking them to make predictions about how to fill the gaps. In this way the persons use their enormous (unconscious) knowledge of the language and it is possible to get good estimates. In [17], inspired by Hagelbarger, he designed a *mind reading machine*. This machine is developed to play the game of matching pennies against human opponents. So it tries to predict human decisions between two alternatives at every time instant. The success of this machine is explained by the fact that "untrained" human opponents are not able to draw completely random bits. In our terminology the mind reading machine is a finite-state machine with eight states. The predictor presented in Chapter 2.1 is in this way a better mind reading machine as it outperforms for any sequence the best finite-state predictor, for that particular sequence. The price for this, apart from the complexity of the scheme, is the amount of information memorized from the past. In fact this predictor has infinite memory.

The thesis is organized as follows. In Chapter 2 we introduce the finite-state predictability of an individual sequence. This is the minimal asymptotic relative frequency of prediction errors made by the best finite-state predictor for that sequence. A predictor that achieves this performance simultaneously for all sequences in the long run (this will be called a universal predictor) is developed in Section 2.1. Section 2.2 deals with the generalization of the problem to general loss functions. In Chapter 3 we begin to work out the new approach of identification in prediction problems. We define the finite-state identifiability of a sequence. Actually we distinguish here two quantities the strong identifiability and the identifiability which differ in the way how restrictive the definitions are done. Then we show that the universal predictor that attains the finite-state predictability can also be used to derive a universal identification scheme (Section 3.1). Furthermore we compare the new notion of identifiability of a sequence with the predictability and derive relations between these quantities (Section 3.2). The analysis of a special class of finite-state machines, the Markov machines, enables us to show that asymptotically strong identifiability and identifiability coincide (Section 3.3). Motivated by the identification theory for channels where the consideration of randomized codes brought a big advantage we analyze the effects of randomized finite-state machines for identification. In Section 3.4 we show that asymptotically randomization does not increase the performance here.

2   Finite-State Predictability

We assume that there is a finite number of possibilities for the observations made at each time instant. Therefore we work throughout the thesis with a finite alphabet

$$\mathcal{X} = \{0, \ldots, M-1\}$$

of size $M \geq 2$. The set of all words of length n is denoted by $\mathcal{X}^n$. Words of length n are denoted as

$$x^n = (x_1, \ldots, x_n) \in \mathcal{X}^n.$$

The set of all infinite sequences of letters from $\mathcal{X}$ is denoted by $\mathcal{X}^\infty$ and a typical element of $\mathcal{X}^\infty$ will be denoted by $x^\infty \in \mathcal{X}^\infty$.

A deterministic predictor with infinite memory is a family $(b_t)_{t \geq 1}$ of functions $b_t : \mathcal{X}^{t-1} \to \mathcal{X}$. If x^{t-1} has been observed at time t so far then $b_t(x^{t-1})$ is the predicted letter. The performance criterion for a predictor of this form is the asymptotic relative frequency of prediction errors:

$$\frac{1}{n} \sum_{t=1}^{n} d(x_t, b_t(x^{t-1})),$$

where $d(x, y) = 0$ if $x = y$ and $d(x, y) = 1$ if $x \neq y$ (d is the Hamming distance).

If the sequence is thought to be an arbitrary individual sequence some care in defining the universal prediction problem has to be employed. Let $\mathcal{B} \triangleq \{(b_t)_{t \geq 1} : b_t : \mathcal{X}^{t-1} \to \mathcal{X}\}$ be the class of all deterministic predictors. Observe the following two facts.

1. For every individual sequence $x_1, x_2, \ldots$ there is one predictor $(b_t)_{t \geq 1} \in \mathcal{B}$ which makes no errors at all for that sequence, $b_t(x^{t-1}) = x_t$ for all $t \in \mathbb{N}$.
2. For every predictor $(b_t)_{t \geq 1} \in \mathcal{B}$ there is a sequence $\bar{x}_1, \bar{x}_2, \ldots$ for which this predictor makes errors at all time instants. Such a sequence is defined inductively by $\bar{x}_t \triangleq \bar{x}$ with $\bar{x} \neq b_t(\bar{x}^{t-1})$ for all $t \in \mathbb{N}$.

Therefore the search for a universal predictor that for all sequences is nearly as good as the best predictor from $\mathcal{B}$ for that particular sequence cannot be successful. To avoid these trivialities we will restrict the class $\mathcal{B}$ to some class $\mathcal{B}' \subset \mathcal{B}$ in a reasonable way and then try to achieve the performance of the best predictor from $\mathcal{B}'$. This class $\mathcal{B}'$ will be denoted as comparison class. But notice that, because of 2., every predictor from $\mathcal{B}$ is very bad for some sequences. Therefore we cannot hope to find a universal predictor in $\mathcal{B}$. This difficulty is avoided by allowing the predictors to be randomized.

Let us now describe how we restrict the class $\mathcal{B}$. The comparison class $\mathcal{B}'$ that we use will be the class of all *finite-state predictors*.

Definition 1. *A finite-state predictor is a triple* $(\mathcal{S}, g, f)$ *consisting of*

$$\mathcal{S} = \{1, \ldots, S\} \text{ a finite set of states,}$$

$$g : \mathcal{S} \times \mathcal{X} \to \mathcal{S} \text{ a next-state function,}$$

$$f : \mathcal{S} \to \mathcal{X} \text{ a prediction rule.}$$

An finite-state predictor works as follows. At time t it predicts the value of x_{t+1} depending on its current state s_t by

$$\hat{x}_{t+1} = f(s_t).$$

Then x_{t+1} is revealed and the machine changes its state to

$$s_{t+1} = g(s_t, x_{t+1})$$

according to the next-state function.

The specific labels of the states do not matter, therefore we assume without loss of generality that at the beginning the machine is always in state 1, i.e., $s_0 = 1$.

In this way, if g and x^n are given, a sequence $s_0, s_1, \ldots, s_{n-1}$ of states is generated. For this we use the following abbreviations.

Definition 2. *If x^n and a next-state function g are given and $s_0, s_1, \ldots, s_{n-1}$ is the generated state sequence then let*

$$\langle x^n|s, x\rangle \triangleq |\{t : s_t = s, x_{t+1} = x\}|,$$
$$\langle x^n|x\rangle \triangleq |\{t : x_t = x\}|,$$
$$\langle x^n|s\rangle \triangleq |\{t : s_t = s\}|.$$

The symbols for these counts do not indicate the dependence on the specific next-state function g but it should always be clear from the context which g is meant.

We can also allow probabilistic prediction rules f, i.e., we select $\hat{x}_{t+1}$ randomly with respect to a conditional probability distribution, given s_t. There are always optimal deterministic prediction rules meaning that if the next-state function g and the initial state s_0 are fixed then for given x^n a prediction rule that minimizes the relative frequency of prediction errors of the finite-state predictor is deterministic and given by

$$f(s) = \hat{x}, \text{ where } \hat{x} \text{ maximizes } \langle x^n|s, x\rangle \text{ over all } x \in \mathcal{X}. \tag{1}$$

This optimal rule for fixed g depends on the whole sequence x^n and in general cannot be determined while the data are observed or, as we shall call it, in a sequential way. The best prediction rule may depend on the whole sequence but anyway for each sequence there is a best finite-state predictor and although it cannot be determined sequentially it will serve us as a comparison for our sequential predictors.

Applying the optimal rule, as described in (1), to the sequence x^n yields a fraction of prediction errors equal to

$$\pi_S(x^n, g) \triangleq \frac{1}{n} \sum_{s=1}^{S} \left[\langle x^n|s\rangle - \max_{x \in \mathcal{X}}\{\langle x^n|s, x\rangle\} \right].$$

Definition 3. *The S-state-predictability of x^n is given by*

$$\pi_S(x^n) \triangleq \min_{g \in \mathcal{G}_S} \pi_S(x^n, g),$$

where $\mathcal{G}_S$ is the set of all $S^{|\mathcal{X}| \cdot S}$ next-state functions.

Definition 4. *The asymptotic S-state predictability of x^∞ is given by*

$$\pi_S(x^\infty) \triangleq \overline{\lim_{n \to \infty}} \pi_S(x^n).$$

Example 1. *Consider the sequence* $x^\infty = 01010101\ldots$
Then clearly $\pi_1(x^\infty) = \frac{1}{2}$ *and* $\pi_2(x^\infty) = 0$.

Definition 5. *The* finite-state predictability *of* x^∞ *is given by*

$$\pi(x^\infty) \triangleq \lim_{S \to \infty} \pi_S(x^\infty).$$

The limit in Definition 5 always exists because $\pi_S(x^\infty)$ is monotonically non-increasing in S.

2.1 A Universal Predictor

In this section, based on the results of Feder, Merhav and Gutman ([8]), we present a slightly generalized predictor that attains the finite-state predictability for all binary sequences. The first main step is to develop a predictor that attains the 1-state predictability universally, i.e., the predictor has to compete for each sequence with the best constant predictor. Our predictor works as follows: At time t it predicts

$$\hat{x}_{t+1} \triangleq \begin{cases} 0, & \text{with probability } \phi_t\left(\frac{\langle x^t|0\rangle + \gamma}{t + 2\gamma}\right) \\ 1, & \text{with probability } \phi_t\left(\frac{\langle x^t|1\rangle + \gamma}{t + 2\gamma}\right) \end{cases}$$

where $\gamma > 0$ is a constant and

$$\phi_t(\alpha) \triangleq \begin{cases} 0, & 0 \le \alpha < \frac{1}{2} - \epsilon_t \\ \frac{1}{2\epsilon_t}(\alpha - \frac{1}{2}) + \frac{1}{2}, & \frac{1}{2} - \epsilon_t \le \alpha \le \frac{1}{2} + \epsilon_t \\ 1, & \frac{1}{2} + \epsilon_t < \alpha \le 1 \end{cases}$$

and $(\epsilon_t)_{t \ge 0}$ is a sequence of parameters with $\epsilon_t > 0$ that will be specified later.

Let $\hat{\pi}(x^n)$ be the expected fraction of errors made by this predictor on the sequence x^n.

The following theorem shows that $\hat{\pi}(x^n)$ approaches $\pi_1(x^n)$ universally for all sequences.

Theorem 1. *Let* $\gamma > 0$. *For any sequence* $x^n \in \{0,1\}^n$ *and for* $\epsilon_t = \frac{1}{2\sqrt{t+2\gamma}}$ *it holds*

$$\hat{\pi}(x^n) \le \pi_1(x^n) + \delta_1(n, \gamma),$$

$$\text{where } \delta_1(n, \gamma) = O(\tfrac{1}{\sqrt{n}}). \tag{2}$$

Furthermore, for any sequence $x^n \in \{0,1\}^n$ *and for constant* $\epsilon_t = \epsilon$, $0 < \epsilon < \frac{1}{2}$, *it holds*

$$\hat{\pi}(x^n) \le \pi_1(x^n) + \frac{\epsilon}{1 - 2\epsilon} + \nu(n, \epsilon),$$

$$\text{where } \nu(n, \epsilon) = O(\tfrac{\log n}{n}). \tag{3}$$

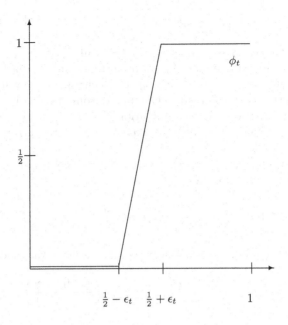

Fig. 1. The function ϕ_t

Remark 1.

1. *A natural choice of $\phi = \phi_t$ could have been*

$$\phi(\alpha) = \begin{cases} 0, & \alpha < \frac{1}{2} \\ \frac{1}{2}, & \alpha = \frac{1}{2} \\ 1, & \alpha > \frac{1}{2}. \end{cases}$$

This means we do majority voting and only if the number of ones and zeros is equal we flip a fair coin. But this is problematic for some sequences, e.g., $x^n = 0101\ldots0101$. $\pi_1(x^n) = \frac{1}{2}$ but the predictor would make 75% errors. The reason for this gap lies in the fact that $\frac{\langle x^t|0\rangle + \gamma}{t + 2\gamma}$ converges from above to $\frac{1}{2}$ which is a discontinuity point of ϕ. Thus it is crucial to make ϕ continuous.

2. *It was shown in [7] that the convergence rate of $O(\frac{1}{\sqrt{n}})$ is best possible.*

3. *As mentioned before it is essential that the universal predictor is randomized. There is no deterministic universal predictor.*

Proof of Theorem 1: Observe that $\pi_1(x^n) = \frac{1}{n}\min\{\langle x^n|0\rangle, \langle x^n|1\rangle\}$ depends only on the type of the sequence x^n, that is on the total number of 0's and 1's in the sequence. Let us show first that among all sequences of the same type the one for which our predictor performs worst is

$$\tilde{x}^n \triangleq \overbrace{0101\ldots01}^{2\langle x^n|1\rangle} \overbrace{00\ldots00}^{\langle x^n|0\rangle - \langle x^n|1\rangle} \tag{4}$$

where we assume without loss of generality that $\langle x^n|0\rangle \geq \langle x^n|1\rangle$.

For a sequence of some given type consider the sequence of absolute differences $C_t \triangleq |\langle x^t|0\rangle - \langle x^t|1\rangle|$. Then $C_0 = 0$ and $C_n = \langle x^n|0\rangle - \langle x^n|1\rangle$. We can think of these C_t as states in a state diagram. Let us call a pattern $(C_t = k, C_{t+1} = k+1, C_{t+2} = k)$ (for some integer k) an *upward loop* and similarly a *downward loop* as $(C_t = k, C_{t+1} = k-1, C_{t+2} = k)$. If we change an upward loop into a downward loop this corresponds to changing at some point of the sequence a 01 into a 10 or vice versa. So this operation does not change the type of the sequence but as we shall show next the expected number of errors made by our predictor is increased.

Assume first that $\langle x^t|0\rangle > \langle x^t|1\rangle$. Denote the expected number of errors incurred along an upward loop by

$$\alpha \triangleq 1 - \phi_t\left(\frac{\langle x^t|0\rangle + \gamma}{t + 2\gamma}\right) + \phi_{t+1}\left(\frac{\langle x^t|0\rangle + \gamma + 1}{t + 2\gamma + 1}\right)$$

and the expected number of errors incurred along a downward loop by

$$\beta \triangleq \phi_t\left(\frac{\langle x^t|0\rangle + \gamma}{t + 2\gamma}\right) + 1 - \phi_{t+1}\left(\frac{\langle x^t|0\rangle + \gamma}{t + 2\gamma + 1}\right).$$

Now we consider the difference

$$\alpha - \beta = \phi_{t+1}\left(\frac{\langle x^t|0\rangle + \gamma + 1}{t + 2\gamma + 1}\right) + \phi_{t+1}\left(\frac{\langle x^t|0\rangle + \gamma}{t + 2\gamma + 1}\right) - 2\phi_t\left(\frac{\langle x^t|0\rangle + \gamma}{t + 2\gamma}\right).$$

For the arguments in the equation above the following relations hold

$$\frac{\langle x^t|0\rangle + \gamma + 1}{t + 2\gamma + 1} > \frac{\langle x^t|0\rangle + \gamma}{t + 2\gamma} > \frac{\langle x^t|0\rangle + \gamma}{t + 2\gamma + 1} \geq \frac{1}{2}.$$

Now we distinguish two cases.

Case 1: $\frac{\langle x^t|0\rangle + \gamma}{t+2\gamma} \geq \frac{1}{2} + \epsilon_t$

Then $\phi_t\left(\frac{\langle x^t|0\rangle + \gamma}{t+2\gamma}\right) = 1$ and therefore $\alpha - \beta \leq 0$.

Case 2: $\frac{1}{2} \leq \frac{\langle x^t|0\rangle + \gamma}{t+2\gamma} < \frac{1}{2} + \epsilon_t$

Then using for the first two terms of the difference $\alpha - \beta$ a continuation of the sloping part of ϕ_t as an upper bound we get

$$\alpha - \beta \leq \frac{1}{2\epsilon_{t+1}}\left(\frac{\langle x^t|0\rangle + \gamma + 1}{t + 2\gamma + 1} - \frac{1}{2} + \frac{\langle x^t|0\rangle + \gamma}{t + 2\gamma + 1} - \frac{1}{2}\right) - \frac{2}{2\epsilon_t}\left(\frac{\langle x^t|0\rangle + \gamma}{t + 2\gamma} - \frac{1}{2}\right)$$

$$= \frac{1}{2\epsilon_{t+1}}\left(\frac{2\langle x^t|0\rangle - t}{t + 2\gamma + 1}\right) - \frac{1}{2\epsilon_t}\left(\frac{2\langle x^t|0\rangle - t}{t + 2\gamma}\right).$$

Therefore $\alpha - \beta \leq 0$ if

$$\epsilon_t(t + 2\gamma) \leq \epsilon_{t+1}(t + 2\gamma + 1).$$

So the function w given by $w(t) = \epsilon_t(t + 2\gamma)$ should be monotonically nondecreasing in t. This means that ϵ_t chosen to be constant or $\epsilon_t = \frac{1}{2\sqrt{t+2\gamma}}$ as in the theorem is possible. The case when $\langle x^t|0\rangle < \langle x^t|1\rangle$ is completely analogous and if $\langle x^t|0\rangle = \langle x^t|1\rangle$, then $\alpha - \beta = 0$.

So we have shown that if we are given a sequence of some type and we replace an upward loop by a downward loop we get a sequence of the same type for which the predictor makes a bigger expected number of errors. If we now iterate this process we will finally end up with the sequence of (4).

The expected number of errors the predictor makes on the sequence $\tilde{x}^n$ of (4) is therefore a uniform upper bound on $\hat{\pi}_1(x^n)$. Let $l_t \triangleq 1 - \phi_t$ then

$$n\hat{\pi}(\tilde{x}^n) = \sum_{k=1}^{\langle x^n|1\rangle} l_{2k-2}\left(\frac{k}{2k}\right) + \underbrace{\sum_{k=1}^{\langle x^n|1\rangle} l_{2k-1}\left(\frac{k-1+\gamma}{2k-1+2\gamma}\right)}_{\triangleq A}$$

$$+ \underbrace{\sum_{k=1}^{\langle x^n|0\rangle-\langle x^n|1\rangle} l_{k+2\langle x^n|1\rangle-1}\left(\frac{\langle x^n|1\rangle+k-1+\gamma}{2\langle x^n|1\rangle-1+k+2\gamma}\right)}_{\triangleq B} = \frac{\langle x^n|1\rangle}{2} + A + B.$$

Let us consider first the case when ϵ is fixed ($l_t = l$ for all t). In order to upperbound A observe that from the definition of l follows that

$$l\left(\frac{k-1+\gamma}{2k-1+2\gamma}\right) \le \frac{1}{2} + \frac{1}{2\epsilon}\left(\frac{1}{2} - \frac{k-1+\gamma}{2k-1+2\gamma}\right)$$

$$= \frac{1}{2} + \frac{1}{4\epsilon} \cdot \frac{1}{2k-1+2\gamma}.$$

Therefore

$$A \le \frac{\langle x^n|1\rangle}{2} + \frac{1}{4\epsilon}\sum_{k=1}^{\langle x^n|1\rangle} \frac{1}{2k-1+2\gamma}$$

$$\le \frac{\langle x^n|1\rangle}{2} + \frac{1}{4\epsilon}\int_1^{\langle x^n|1\rangle} \frac{du}{2u-1+2\gamma} + \frac{1}{4\epsilon} \cdot \frac{1}{2\gamma+1}$$

$$= \frac{\langle x^n|1\rangle}{2} + \frac{1}{8\epsilon}\ln\left(2\langle x^n|1\rangle - 1 + 2\gamma\right) - \frac{1}{8\epsilon}\ln\left(2\gamma+1\right) + \frac{1}{4\epsilon}\frac{1}{2\gamma+1}$$

$$\le \frac{\langle x^n|1\rangle}{2} + \frac{1}{8\epsilon}\ln\left(2n-1+2\gamma\right) + \frac{1}{4\epsilon}\frac{1}{2\gamma+1},$$

where we used the fact that $2\langle x^n|1\rangle \le n$ in the last inequality.

Now we consider the sum B. For the argument of l it is true that it is always larger than $\frac{1}{2}$ and that $\frac{\langle x^n|1\rangle+k-1+\gamma}{2\langle x^n|1\rangle-1+k+2\gamma} \ge \frac{1}{2} + \epsilon$ if

$$k \ge \frac{1 + 4\epsilon\langle x^n|1\rangle - 2\epsilon + 4\epsilon\gamma}{1 - 2\epsilon} \triangleq K.$$

For these k's l is zero and otherwise we can upperbound it by $\frac{1}{2}$. Therefore

$$B \leq \sum_{k=1}^{\lfloor K \rfloor} \frac{1}{2} \leq \frac{2\epsilon \langle x^n | 1 \rangle}{1 - 2\epsilon} + \frac{1}{2} \cdot \frac{1 - 2\epsilon + 4\epsilon\gamma}{1 - 2\epsilon} \leq \frac{n\epsilon}{1 - 2\epsilon} + \frac{1}{2} \cdot \frac{1 - 2\epsilon + 4\epsilon\gamma}{1 - 2\epsilon}.$$

If we combine the estimates for A and B we get

$$\hat{\pi}(x^n) \leq \underbrace{\frac{\langle x^n | 1 \rangle}{n}}_{\pi_1(x^n)} + \frac{\epsilon}{1 - 2\epsilon} + \frac{\ln(2n - 1 + 2\gamma)}{8\epsilon n} + \frac{1}{n} \left(\frac{1}{4\epsilon(2\gamma + 1)} + \frac{1 - 2\epsilon + 4\epsilon\gamma}{2 - 2\epsilon} \right),$$

which is the result claimed in (3). Now let us consider the case when ϵ is variable. We start by estimating the sum A. Since

$$l_{2k-1} \left(\frac{k - 1 + \gamma}{2k - 1 + 2\gamma} \right) \leq \frac{1}{2} + \frac{1}{2\epsilon_{2k-1}} \left(\frac{1}{2} - \frac{k - 1 + \gamma}{2k - 1 + 2\gamma} \right) = \frac{1}{2} + \frac{1}{2} \frac{1}{\sqrt{2k - 1 + 2\gamma}},$$

we get

$$A \leq \frac{\langle x^n | 1 \rangle}{2} + \frac{1}{2} \sum_{k=1}^{\langle x^n | 1 \rangle} \frac{1}{\sqrt{2k - 1 + 2\gamma}}$$

$$\leq \frac{\langle x^n | 1 \rangle}{2} + \frac{1}{2} \int_1^{\langle x^n | 1 \rangle} \frac{du}{\sqrt{2u - 1 + 2\gamma}} + \frac{1}{2\sqrt{2\gamma + 1}}$$

$$= \frac{\langle x^n | 1 \rangle}{2} + \frac{1}{2} \sqrt{2\langle x^n | 1 \rangle - 1 + 2\gamma} + \frac{1}{2} \left(\frac{1}{\sqrt{2\gamma + 1}} - \sqrt{2\gamma + 1} \right)$$

$$\leq \frac{\langle x^n | 1 \rangle}{2} + \frac{1}{2} \sqrt{n - 1 + 2\gamma} + \frac{1}{2} \left(\frac{1}{\sqrt{2\gamma + 1}} - \sqrt{2\gamma + 1} \right).$$

In order to estimate B observe that the nonzero components must satisfy

$$\frac{\langle x^n | 1 \rangle + k - 1 + \gamma}{2\langle x^n | 1 \rangle - 1 + k + 2\gamma} \leq \frac{1}{2} + \frac{1}{2\sqrt{2\langle x^n | 1 \rangle - 1 + k + 2\gamma}}.$$

The largest k satisfying this condition denoted as K can be upperbounded by

$$K \leq \frac{3}{2} + \sqrt{\frac{1}{4} + 2\langle x^n | 1 \rangle + 2\gamma}.$$

Since all non-zero terms of B are less than $\frac{1}{2}$ we get

$$B \leq \frac{K}{2} \leq \frac{3}{4} + \frac{1}{2} \sqrt{\frac{1}{4} + 2\langle x^n | 1 \rangle + 2\gamma} \leq \frac{3}{4} + \frac{1}{2} \sqrt{\frac{1}{4} + n + 2\gamma}.$$

Combining the estimates for A and B we derive that

$$\hat{\pi}(x^n) \leq \underbrace{\frac{\langle x^n|1\rangle}{n}}_{\pi_1(x^n)} + \frac{1}{2n}\left(\sqrt{n-1+2\gamma} + \sqrt{\frac{1}{4}+n+2\gamma}\right) + \frac{C_\gamma}{n},$$

where $C_\gamma \triangleq \frac{1}{2}\left(\frac{1}{\sqrt{2\gamma+1}} - \sqrt{2\gamma+1}\right) + \frac{3}{4}$.

This is the desired result of (2) and thus the proof of the theorem is complete.

$\square$

Next we deal with the problem how to achieve universally the performance $\pi_S(x^n, g)$ for a given next-state function g with a sequential predictor.

For each state $s \in \mathcal{S}$ the optimal prediction rule $\hat{x}_{t+1} = f(s)$ is fixed and thus we can extend Theorem 1 by considering S sequential predictors of the previously described form. For simplicity we choose $\gamma = 1$. Specifically let

$$\hat{p}_t(x|s) \triangleq \frac{\langle x^t|s, x\rangle + 1}{\langle x^t|s\rangle + 2} \qquad x \in \{0,1\}, s \in \mathcal{S}$$

and consider the predictor

$$\hat{x}_{t+1} = \begin{cases} 0, & \text{with probability } \phi_t(\hat{p}_t(0|s_t)) \\ 1, & \text{with probability } \phi_t(\hat{p}_t(1|s_t)), \end{cases}$$

where ϕ is as before with $\epsilon = \epsilon_{\langle x^t|s_t\rangle}$.

Now we can apply Theorem 1 to each subsequence of x^n which corresponds to a state $s \in \mathcal{S}$ and get

$$\hat{\pi}(x^n, g) \leq \frac{1}{n}\sum_{s=1}^{S}\min\{\langle x^n|s, 0\rangle, \langle x^n|s, 1\rangle\} + \langle x^n|s\rangle\,\delta_1(\langle x^n|s\rangle)$$

$$= \pi_S(x^n, g) + \sum_{s=1}^{S}\frac{\langle x^n|s\rangle}{n}\,\delta_1(\langle x^n|s\rangle)$$

$$\leq \pi_S(x^n, g) + \underbrace{\frac{S}{n}\sqrt{\frac{n}{S}+1} + \frac{S}{2n}}_{\delta_S(n)}. \tag{5}$$

Observe that, as there are less samples in each state, the convergence rate slows down (from $O(\frac{1}{\sqrt{n}})$ to $O(\sqrt{\frac{S}{n}})$).

The next problem we deal with is how to achieve sequentially the S-state predictability for fixed S.

Definition 6. *A refinement of a finite-state machine with next-state function g and S states is a finite-state machine with $\tilde{S} \geq S$ states and next-state function $\tilde{g}$ such that there exists a function $h : \tilde{S} \to S$ with the property that at each time instant $s_t = h(\tilde{s}_t)$ where s_t and $\tilde{s}_t$ are the states at time t generated by g, $\tilde{g}$ and any $x^n \in \mathcal{X}^n$.*

The next lemma shows that a refinement of a finite-state machine can only increase the performance of the finite-state predictor.

Lemma 1. *If the finite-state machine corresponding to $\tilde{g}$ is a refinement of the finite-state machine corresponding to g then for all $x^n \in \{0,1\}^n$ it holds*

$$\pi_S(x^n, g) \geq \pi_{\tilde{S}}(x^n, \tilde{g}).$$

Proof

$$\pi_S(x^n, g) = \frac{1}{n} \sum_{s=1}^{S} \min\{\langle x^n | s, 0\rangle, \langle x^n | s, 1\rangle\}$$

$$= \frac{1}{n} \sum_{s=1}^{S} \min\{ \sum_{\tilde{s}:h(\tilde{s})=s} \langle x^n | \tilde{s}, 0\rangle, \sum_{\tilde{s}:h(\tilde{s})=s} \langle x^n | \tilde{s}, 1\rangle\}$$

$$\geq \frac{1}{n} \sum_{s=1}^{S} \sum_{\tilde{s}:h(\tilde{s})=s} \min\{\langle x^n | \tilde{s}, 0\rangle, \langle x^n | \tilde{s}, 1\rangle\}$$

$$= \frac{1}{n} \sum_{\tilde{s}=1}^{\tilde{S}} \min\{\langle x^n | \tilde{s}, 0\rangle, \langle x^n | \tilde{s}, 1\rangle\} = \pi_{\tilde{S}}(x^n, \tilde{g}).$$

$\square$

Consider now a refinement $\tilde{g}$ of all S^{2S} possible S-state machines. The state $\tilde{s}_t$ of $\tilde{g}$ is the vector $(s_t^1, \ldots, s_t^M)$, where $s_t^i, i = 1, \ldots, S^{2S}$, is the state at time t of the i-th S-state machine g_i. From Lemma 1 it follows that for all g

$$\pi_{\tilde{S}}(x^n, \tilde{g}) \leq \pi_S(x^n, g)$$

and therefore also

$$\pi_{\tilde{S}}(x^n, \tilde{g}) \leq \pi_S(x^n).$$

Thus the sequential scheme based on $\tilde{g}$ asymptotically universally attains $\pi_S(x^n)$.

The disadvantages of this scheme are obviously that it is very complex, furthermore it attains the predictability only for a fixed given value of S. The rate of convergence also is not best possible.

In order to develop a predictor that universally attains the finite-state predictability and overcomes the disadvantages mentioned above we introduce Markov predictors and the Markov predictability of a sequence and show that it is equal to the finite-state predictability of the sequence. This enables us to design the desired prediction scheme.

Definition 7. *A Markov-Predictor of order $k \geq 1$ is a finite-state predictor with 2^k possible states where*

$$s_t = (x_{t-k+1}, \ldots, x_t).$$

The initial state $(x_{-k+1}, \ldots, x_0)$ does not affect the asymptotic performance of the Markov predictor. Therefore the choice of s_0 is irrelevant for our purposes. For instance it can be chosen to give the smallest possible value in (6) below (in

[8] for technical reasons the cyclic convention $x_{-i} = x_{n-i}$ for $i \in \{0, \ldots, k-1\}$ was used).

Then the k-th order Markov predictability of the finite sequence x^n is given by

$$\mu_k(x^n) \triangleq \frac{1}{n} \sum_{x^k \in \{0,1\}^k} \min\{\langle x^n | x^k, 0 \rangle, \langle x^n | x^k, 1 \rangle\}. \tag{6}$$

The asymptotic k-th order Markov predictability of the infinite sequence x^∞ is given by

$$\mu_k(x^\infty) \triangleq \overline{\lim_{n \to \infty}} \, \mu_k(x^n).$$

Finally the Markov predictability of x^∞ is given by

$$\mu(x^\infty) \triangleq \lim_{k \to \infty} \mu_k(x^\infty).$$

As the class of finite-state machines contains as a subclass the class of Markov machines it follows

$$\mu(x^\infty) \geq \pi(x^\infty).$$

The following theorem from [8, Theorem 2] establishes a converse inequality from which follows that Markov predictability and finite-state predictability are equivalent.

Theorem 2. *For all integers $k, S \geq 1$ and any finite sequence $x^n \in \{0,1\}^n$ it holds*

$$\mu_k(x^n) \leq \pi_S(x^n) + \sqrt{\frac{\ln S}{2(k+1)}}. \tag{7}$$

Remark 2. *The inequality of the theorem is meaningful only if the second term on the right hand side is small, i.e., if k is big compared to $\ln S$. Thus the theorem shows that no matter how clever a finite-state machine is chosen for a given sequence, if k is big enough the Markov predictor of the corresponding order will be almost as good.*

Now if in (7) we take the limit supremum as $n \to \infty$, then the limit $k \to \infty$ and finally the limit $S \to \infty$ we end up with $\mu(x^\infty) \leq \pi(x^\infty)$ which implies

$$\mu(x^\infty) = \pi(x^\infty).$$

Now it is clear how we can derive a sequential universal prediction scheme that attains $\mu(x^\infty)$ and thus $\pi(x^\infty)$.

We know that for fixed k we can achieve the k-th order Markov predictability by the predictor

$$\hat{x}_{t+1} = \begin{cases} 0, \text{ with probability } \phi_t(\hat{p}_t(0|x_{t-k+1}, \ldots, x_t)) \\ 1 \text{ with probability } \phi_t(\hat{p}_t(1|x_{t-k+1}, \ldots, x_t)), \end{cases} \tag{8}$$

where for $x \in \{0,1\}$

$$\hat{p}_t(x|x_{t-k+1}, \ldots, x_t) = \frac{\langle x^n | (x_{t-k+1}, \ldots, x_t), x \rangle + 1}{\langle x^n | (x_{t-k+1}, \ldots, x_t) \rangle + 2}.$$

To attain $\mu(x^\infty)$ the order k must grow the more data are available. There are two conflicting goals.

- Increasing the order fast in order to attain the Markov predictability as soon as possible.
- Increasing the order slowly in order to ensure that there are enough counts for each state.

It turns out that the order k is not allowed to grow faster than $O(\log t)$ in order to satisfy both requirements.

Let us denote by $\hat{\mu}_k(x^n)$ the expected fraction of errors made by the predictor (8) on the sequence x^n.

Then we know that

$$\hat{\mu}_k(x^n) \leq \mu_k(x^n) + \delta_{2^k}(n),$$

with δ_{2^k} as defined in (5) and $\delta_{2^k}(n) = O(\sqrt{\frac{2^k}{n}})$.

Divide the observed data into non-overlapping segments

$$x^\infty = x^{(1)}, x^{(2)}, \ldots$$

and apply the k-th order sequential predictor (8) to the k-th segment $x^{(k)}$. Let the length n_k of the k-th segment be at least $\alpha_k 2^k$, where $(\alpha_k)_k$ is a monotonically increasing sequence such that $\lim_{k\to\infty} \alpha_k = \infty$. Then

$$\hat{\mu}_k(x^{(k)}) \leq \mu_k(x^{(k)}) + \delta_{2^k}(n_k)$$

$$\leq \mu_k(x^{(k)}) + \frac{\sqrt{\alpha_k + 1}}{\alpha_k} + \frac{1}{2\alpha_k}$$

$$= \mu_k(x^{(k)}) + \xi(k),$$

where $\xi(k) = O(\frac{1}{\sqrt{\alpha_k}})$.

On an arbitrary long finite sequence x^n, where $n = \sum_{k=1}^{k_n} n_k$ and k_n denotes the number of segments in x^n, the above predictor achieves an average fraction of errors denoted by $\hat{\mu}(x^n)$ which satisfies

$$\hat{\mu}(x^n) = \sum_{k=1}^{k_n} \frac{n_k}{n} \hat{\mu}_k(x^{(k)}) \leq \sum_{k=1}^{k_n} \frac{n_k}{n} \mu_k(x^{(k)}) + \sum_{k=1}^{k_n} \frac{n_k}{n} \xi(k).$$

Now for any fixed $k' < k_n$ we obtain

$$\hat{\mu}(x^n) \leq \sum_{k=1}^{k'-1} \frac{n_k}{n} \mu_k(x^{(k)}) + \sum_{k=k'}^{k_n} \frac{n_k}{n} \mu_k(x^{(k)}) + \sum_{k=1}^{k_n} \frac{n_k}{n} \xi(k)$$

$$\leq \frac{1}{2} \sum_{k=1}^{k'-1} \frac{n_k}{n} + \sum_{k=1}^{k_n} \frac{n_k}{n} \mu_{k'}(x^{(k)}) + \sum_{k=1}^{k_n} \frac{n_k}{n} \xi(k).$$

From Lemma 1 it follows that

$$\sum_{k=1}^{k_n} \frac{n_k}{n} \mu_{k'}(x^{(k)}) \le \mu_{k'}(x^n).$$

Since $\xi(k)$ is monotonically decreasing and the lengths of the segments are monotonically increasing it follows that

$$\sum_{k=1}^{k_n} \frac{n_k}{n} \xi(k) \le \frac{1}{k_n} \sum_{k=1}^{k_n} \xi(k) \triangleq \bar{\xi}(k_n),$$

where by the Cesaro theorem $\lim_{n\to\infty} \bar{\xi}(k_n) = 0$.

Theorem 3. *For all sequences* $x^\infty \in \mathcal{X}^\infty$

$$\hat{\mu}(x^\infty) = \overline{\lim_{n\to\infty}} \, \hat{\mu}(x^n) = \mu(x^\infty) = \pi(x^\infty).$$

In summary we have shown that a sequential Markov predictor whose order is increased from k to $k+1$ after observing at least $n_k = \alpha_k 2^k$ data samples asymptotically achieves the performance of any finite-state predictor.

2.2 General Loss Functions

In this section we present a more general formulation of the prediction problem treated so far and give some references to related work.

It is possible to generalize our problem in the following way. Given is a finite set $\mathcal{B}$ of so called *strategies* and a *loss function* $l : \mathcal{B} \times \mathcal{X} \to \mathbb{R}$. At time t after having observed $x_1, \ldots, x_t$ one has to decide for a strategy, that is, select an element $b_{t+1} \in \mathcal{B}$. Then x_{t+1} is revealed and a loss of $l(b_{t+1}, x_{t+1})$ is incurred. Again the time average $\frac{1}{n} \sum_{t=1}^{n} l(b_t, x_t)$ is tried to be kept small and again it can be defined how good this can be done for a sequence by a finite-state machine.

Examples

1. If we set $\mathcal{B} = \mathcal{X}$ and l to be the Hamming distance then we are back to our original prediction problem.
2. If $\mathcal{B} = (0,1]$, $\mathcal{X} = \{0,1\}$ and $l(b,0) = -\log b$ and $l(b,1) = -\log(1-b)$ then we have the lossless coding problem. Here b_{t+1} has the interpretation of the estimated probability of the next letter to be a zero. The time average $\frac{1}{n} \sum_{t=1}^{n} l(b_t, x_t)$ then is the normalized length of a codeword of the sequential Shannon encoder based on the current letter probabilities from the data observed so far. This length can be attained using arithmetic coding techniques.
3. $\mathcal{B} = (0,1]$, $\mathcal{X} = \{0,1\}$. A sequential gambling problem can be formulated in this framework in the following way. At round t the player has to divide his capital. The share wagered on the next outcome is then doubled, i.e., if S_t is the player's capital after round t then $S_{t+1} = 2b_{t+1}S_t$ if $x_{t+1} = 0$ and $S_{t+1} = 2(1 - b_{t+1})S_t$ if $x_{t+1} = 1$. If l is as in 2., then the exponential growth rate of the player's capital $\frac{\log S_n}{n}$ is the time average of $1 - l(b_t, x_t)$.

4. There are also continuous alphabet applications. For instance prediction under the mean squared error criterion, i.e., $l(b, x) = (x - b)^2$.

General loss functions in the probabilistic setting were studied in [3]. There it was shown that if the data $x_1, x_2, \ldots$ are generated by a stationary ergodic source which is known and $\mathcal{B}$ consists of any measurable functions of the past $(x_1, x_2, \ldots, x_t)$ then the best strategy in order to minimize the *expected* time average loss is the one that attains the minimal conditional expectation of $l(b_{t+1}, x_{t+1})$ given the past. Furthermore, it was shown that this minimal loss is achievable almost surely under certain regularity conditions on the loss function even if the source is unknown a priori.

In the deterministic setting general loss functions were studied in [10]. Older work was devoted to another, in a way slightly more general problem, the so called *sequential compound decision problem* which was initiated by Robbins ([14]) and this was further studied by various authors ([4],[5],[13],[12]). In our language the problem is restricted to the case $S = 1$, i.e., the comparison class is only that of all constant predictors or strategies. It is more general because the observer has access only to *noisy* versions of the data $x_1, x_2, \ldots, x_t$.

3   Finite-State Identifiability

Now consider for every $x \in \mathcal{X}$ a person x who at time t has to answer the question "Is $x_{t+1} = x$?" .

We start by defining how good a sequence can be identified using a finite-state machine.

Definition 8. *A finite-state identification scheme is a triple $(\mathcal{S}, g, f)$ consisting of*

$$\mathcal{S} \text{ a set of } S \text{ states,}$$

$$g : \mathcal{S} \times \mathcal{X} \to \mathcal{S} \text{ a next-state function,}$$

$$f = (f_0, \ldots, f_{|\mathcal{X}|-1}) : \mathcal{S} \to \{0, 1\}^{|\mathcal{X}|} \text{ a decision rule.}$$

As before we can assume without loss of generality that the initial state is always 1, i.e., $s_0 = 1$.

The interpretation is that $f_x(s_t) = 1$ means that person x predicts that $x_{t+1} = x$ and $f_x(s_t) = 0$ means that person x predicts that $x_{t+1} \neq x$. Applied to some sequence x^n the fraction of errors person x makes is then given by

$$\eta_S(f, g, x^n, x) \triangleq \frac{1}{n} \sum_{t=1}^{n} (1 - f_x(s_{t-1})) \delta_{x_t, x} + f_x(s_{t-1})(1 - \delta_{x_t, x}),$$

where δ is the Kronecker symbol.

For a fixed next-state function g an optimal decision rule f is given by

$$f_x(s) = \begin{cases} 1, & \langle x^n | s, x \rangle > \langle x^n | s \rangle - \langle x^n | s, x \rangle \\ 0, & \langle x^n | s, x \rangle \leq \langle x^n | s \rangle - \langle x^n | s, x \rangle \end{cases}$$

for all $x \in \mathcal{X}$ and $s \in \mathcal{S}$.

If we apply this optimal f to the sequence x^n the fraction of errors person x makes is given by

$$\eta_S(g, x^n, x) \triangleq \frac{1}{n} \sum_{s=1}^{S} \min\{\langle x^n | s, x \rangle, \langle x^n | s \rangle - \langle x^n | s, x \rangle\}.$$

We can now define an average error criterion and a maximal error criterion. Furthermore we can distinguish the case where each person can use its own finite-state machine on the sequence (1. and 2. of Definition 9) and the more restrictive case where the persons have to use one finite-state machine (3. and 4. of Definition 9).

Definition 9. *1. The maximal S-state identifiability of the sequence x^n is given by*

$$\eta_S(x^n) \triangleq \max_{x \in \mathcal{X}} \min_g \eta_S(g, x^n, x).$$

2. The average S-state identifiability of the sequence x^n is given by

$$\bar{\eta}_S(x^n) \triangleq \frac{1}{|\mathcal{X}|} \sum_{x \in \mathcal{X}} \min_g \eta_S(g, x^n, x).$$

3. The strong maximal S-state identifiability of the sequence x^n is given by

$$\eta'_S(x^n) \triangleq \min_g \max_{x \in \mathcal{X}} \eta_S(g, x^n, x).$$

4. The strong average S-state identifiability of the sequence x^n is given by

$$\bar{\eta}'_S(x^n) \triangleq \min_g \frac{1}{|\mathcal{X}|} \sum_{x \in \mathcal{X}} \eta_S(g, x^n, x).$$

Definition 10. *The asymptotic maximal S-state identifiability of the sequence x^∞ is given by*

$$\eta_S(x^\infty) \triangleq \overline{\lim_{n \to \infty}} \, \eta_S(x^n).$$

The corresponding values of the asymptotic S-state identifiability are defined analogously in the other cases of Definition 9.

Definition 11. *The maximal finite-state identifiability of the sequence x^∞ is given by*

$$\eta(x^\infty) \triangleq \lim_{S \to \infty} \eta_S(x^\infty).$$

The corresponding values of the finite-state identifiability are defined analogously in the other cases of Definition 9.

The following relations follow easily from the definitions.

Lemma 2. *For all sequences $x^n \in \mathcal{X}^n$*

$$\eta'_S(x^n) \geq \eta_S(x^n) \geq \bar{\eta}_S(x^n), \tag{9}$$

$$\eta'_S(x^n) \geq \bar{\eta}'_S(x^n) \geq \bar{\eta}_S(x^n). \tag{10}$$

Remark 3. *In the binary case, $\mathcal{X} = \{0,1\}$, we have*

$$\eta_S(g, x^n, x) = \frac{1}{n} \sum_{s=1}^{S} \min\{\langle x^n | s, x \rangle, \langle x^n | s, 1 - x \rangle\} \tag{11}$$

$$= \eta_S(g, x^n, 1 - x) = \pi_S(g, x^n) \tag{12}$$

and this implies

$$\eta_S(x^n) = \bar{\eta}_S(x^n) = \eta'_S(x^n) = \bar{\eta}'_S(x^n) = \pi_S(x^n). \tag{13}$$

Thus, in the binary case identification of sequences gives no advantage over prediction.

3.1   A Universal Identification Scheme

Definition 12. *For a sequence $x^n \in \mathcal{X}^n$ and a letter $x \in \mathcal{X}$ let $\mathbf{1}_x x^n \in \{0,1\}^n$ be the sequence with*

$$(\mathbf{1}_x x^n)_t = \begin{cases} 1, & \text{if } x_t = x \\ 0, & \text{if } x_t \neq x. \end{cases}$$

Then it holds

$$\eta_1(g, x^n, x) = \frac{1}{n} \min\{\langle x^n | x \rangle, n - \langle x^n | x \rangle\} = \pi_1(\mathbf{1}_x x^n), \tag{14}$$

where the argument g of η_1 is the only possible constant next-state function as $S = 1$.

This suggests the following sequential identification scheme. At time t person x applies the predictor analyzed in Theorem 1 to the sequence $\mathbf{1}_x x^t$. If we denote by $\hat{\eta}_1(x^n, x)$ the expected fraction of identification errors person x makes using this scheme for the sequence x^n then Theorem 1 implies that

$$|\eta_1(g, x^n, x) - \hat{\eta}_1(x^n, x)| = O(\frac{1}{\sqrt{n}}) \text{ for all } x \in \mathcal{X}. \tag{15}$$

This means we know how to achieve sequentially the 1-state identifiability universally for all sequences. When $S = 1$ we can actually derive a formula for η_1 in terms of π_1.

Theorem 4. *For all sequences $x^n \in \mathcal{X}^n$ it holds*

$$\eta_1(x^n) = \min\{\pi_1(x^n), 1 - \pi_1(x^n)\}.$$

Proof. Note that $\eta_1(x^n) = \frac{1}{n} \max_x \min\{\langle x^n | x \rangle, n - \langle x^n | x \rangle\}$ and $\pi_1(x^n) = 1 - \max_x \frac{\langle x^n | x \rangle}{n}$.

Case 1: There exists $\bar{x} \in \mathcal{X}$ with $\langle x^n|\bar{x}\rangle \geq \frac{n}{2}$.
Then $\pi_1(x^n) = 1 - \frac{\langle x^n|\bar{x}\rangle}{n} \leq \frac{1}{2}$ and

$$\eta_1(x^n) = \frac{1}{n}\max\{n - \langle x^n|\bar{x}\rangle, \max_{x \neq \bar{x}}\langle x^n|x\rangle\}$$

$$= \frac{1}{n}(n - \langle x^n|\bar{x}\rangle) = \pi_1(x^n) \leq 1 - \pi_1(x^n),$$

where we used that $n - \langle x^n|\bar{x}\rangle = \sum_{x \neq \bar{x}}\langle x^n|x\rangle$.
Case 2: For all $x \in \mathcal{X}$ $\langle x^n|x\rangle < \frac{n}{2}$.
In this case $\pi_1(x^n) > \frac{1}{2}$ and

$$\eta_1(x^n) = \frac{1}{n}\max_x\langle x^n|x\rangle = 1 - \pi_1(x^n) < \pi_1(x^n).$$

$\square$

If $S > 1$ then η_S is not a function of π_S any longer. Nevertheless it is possible to determine some relations between these quantities and this will be done in the next section.

3.2   Relations Between Predictability and Identifiability

Theorem 5. *For all $S \geq 1$, for all sequences $x^n \in \mathcal{X}^n$ and all next-state functions g*

$$\pi_S(x^n, g) \geq \max_{x \in \mathcal{X}} \eta_S(g, x^n, x)$$

and

$$\pi_S(x^n) \geq \eta'_S(x^n).$$

Proof. Let f be the optimal prediction rule for g and x^n. Consider the following decision rule $\tilde{f} : \mathcal{S} \to \{0,1\}^{|\mathcal{X}|}$ with

$$\tilde{f}_x(s) = \begin{cases} 1, \text{ if } f(s) = x \\ 0, \text{ if } f(s) \neq x \end{cases}$$

for all $x \in \mathcal{X}$ and $s \in \mathcal{S}$.

Now observe that if there is no prediction error at some time instant then also no identification error occurs for all persons. As $\tilde{f}$ is not necessarily optimal the first inequality is proved. Let $\tilde{g}$ be a next-state function such that $\pi_S(x^n, \tilde{g}) = \min_g \pi_S(g, x^n)$. Then it holds

$$\pi_S(x^n) = \pi_S(\tilde{g}, x^n) \geq \max_{x \in \mathcal{X}} \eta_S(\tilde{g}, x^n, x) \geq \min_g \max_{x \in \mathcal{X}} \eta_S(g, x^n, x) = \eta'_S(x^n),$$

which is the second inequality.

$\square$

Note that η' is the biggest of all η-quantities.

A converse inequality is obtained by the following theorem.

Theorem 6. *For all $S \geq 1$ and for all sequences $x^n \in \mathcal{X}^n$*

$$\frac{1}{|\mathcal{X}|}\pi_{S^{|\mathcal{X}|}}(x^n) \leq \bar{\eta}_S(x^n).$$

Proof. Let $g_0, \ldots, g_{|\mathcal{X}|-1}$ be the optimal next state functions for person $0, \ldots, |\mathcal{X}|-1$, respectively. Let $f_0, \ldots, f_{|\mathcal{X}|-1}$ be the corresponding optimal decision rules. Let $\tilde{S} = S^{|\mathcal{X}|}$ and choose

$$\tilde{g} : \tilde{S} \times \mathcal{X} \to \tilde{S}$$

such that

$$\tilde{g}(s_0, \ldots, s_{|\mathcal{X}|-1}, x) = (g_0(s_0, x), \ldots, g_{|\mathcal{X}|-1}(s_{|\mathcal{X}|-1}, x))$$

and consider the following prediction rule $\tilde{f} : \tilde{S} \to \mathcal{X}$

$$\tilde{f}(s_0, \ldots, s_{|\mathcal{X}|-1}) = \begin{cases} x, & \text{if } f_x(s_x) = 1, \text{ if } x \text{ is not unique,} \\ & \text{choose arbitrarily any of these,} \\ \text{arbitrary, if } f_x(s_x) = 0 \text{ for all } x \in \mathcal{X}. \end{cases}$$

Then

$$\pi_{S^{|\mathcal{X}|}}(x^n) \leq \pi_{S^{|\mathcal{X}|}}(\tilde{g}, \tilde{f}, x^n) \leq \sum_{x \in \mathcal{X}} \min_g \eta_S(g, x^n, x) = |\mathcal{X}| \, \bar{\eta}_S(x^n).$$

$\square$

Note that $\bar{\eta}$ is the smallest of all η-quantities.

Theorem 7. *For all $S \geq 1$ and for all sequences $x^n \in \mathcal{X}^n$*

$$\bar{\eta}'_S(x^n) \leq \frac{2}{|X|}\pi_S(x^n).$$

Proof: For given $S \geq 1$ and $x^n \in \mathcal{X}^n$ let g and f be the optimal next-state function and prediction rule, respectively. Then we can define the following identification rule $\tilde{f} : S \to \{0,1\}^{|\mathcal{X}|}$ with

$$\tilde{f}_x(s) = \begin{cases} 1, \text{ if } f(s) = x \\ 0, \text{ if } f(s) \neq x \end{cases}$$

for all $x \in \mathcal{X}$ and $s \in \mathcal{S}$.

Now observe that if at some time instant there is no prediction error induced by the finite-state predictor given by g and f then there will be also no identification error induced by g and $\tilde{f}$. But if g and f produce a prediction error then there will be exactly two persons making an identification error if we use g and $\tilde{f}$. Therefore

$$2\pi_S(x^n) = 2\pi_S(g, f, x^n) = \sum_{x \in \mathcal{X}} \eta_S(g, \tilde{f}, x^n, x)$$

$$\geq \min_g \sum_{x \in \mathcal{X}} \eta_S(g, x^n, x) = |\mathcal{X}| \, \bar{\eta}'_S(x^n).$$

$\square$

Corollary 1. *For all sequences $x^\infty \in \mathcal{X}^\infty$*

$$\frac{1}{|\mathcal{X}|}\pi(x^\infty) \le \bar{\eta}(x^\infty) \le \frac{2}{|\mathcal{X}|}\pi(x^\infty).$$

Proof. Combining Theorem 6 and 7 and taking the $\overline{\lim}_{n\to\infty}$ and the $\lim_{S\to\infty}$ gives the desired result.

$\square$

Corollary 1 characterizes the average identifiability of any sequence in terms of the predictability of that sequence where upper and lower bound differ by a factor of 2.

3.3   Markov Machines for Identification

Similar to Definition 7 in Section 2.1 we now examine a special class of finite-state machines the class of Markov machines.

Definition 13. *For any $k \ge 1, x^n \in \mathcal{X}^n, x \in \mathcal{X}$ denote by*

$$\mu_k^I(x^n, x) \triangleq \frac{1}{n} \sum_{x^k \in \mathcal{X}^k} \min\{\langle x^n | x^k, x\rangle, \langle x^n | x^k\rangle - \langle x^n | x^k, x\rangle\}$$

the Markov identifiability of order k of the sequence x^n with respect to x. Furthermore let

$$\mu_k^I(x^\infty, x) \triangleq \overline{\lim_{n\to\infty}} \mu_k^I(x^n, x),$$

$$\mu^I(x^\infty, x) \triangleq \lim_{k\to\infty} \mu_k^I(x^\infty, x),$$

$$\mu^I(x^\infty) \triangleq \max_{x\in\mathcal{X}} \mu^I(x^\infty, x),$$

$$\bar{\mu}^I(x^\infty) \triangleq \frac{1}{|\mathcal{X}|} \sum_{x\in\mathcal{X}} \mu^I(x^\infty, x).$$

The result of [10, Theorem 2] which was derived for general loss functions and which is similar to Theorem 2 leads in our case to the following proposition.

Proposition 1. *For all $k \ge 1, S \ge 1$ and all sequences $x^n \in \mathcal{X}^n$*

$$\mu_k^I(x^n, x) \le \min_g \eta_S(g, x^n, x) + \sqrt{\frac{2\ln S}{k+1}}.$$

Theorem 8. *For all sequences $x^\infty \in \mathcal{X}^\infty$ it holds that*

$$\eta(x^\infty) = \eta'(x^\infty),$$

$$\bar{\eta}(x^\infty) = \bar{\eta}'(x^\infty).$$

Proof. Taking in Proposition 1 the limit supremum $n \to \infty$, the limit $k \to \infty$ and the limit $S \to \infty$ it follows that $\mu^I(x^\infty) \leq \eta(x^\infty)$. Therefore

$$\eta'(x^\infty) \geq \eta(x^\infty) \geq \mu^I(x^\infty) = \lim_{k\to\infty} \overline{\lim_{n\to\infty}} \underbrace{\max_{x\in\mathcal{X}} \mu^I_k(x^n, x)}_{\geq \eta'_{|\mathcal{X}|^k}(x^n)}$$

$$\geq \lim_{k\to\infty} \overline{\lim_{n\to\infty}} \, \eta'_{|\mathcal{X}|^k}(x^n) = \eta'(x^\infty).$$

□

Remark 4. *Only the asymptotic values for $S \to \infty$ of η, η' and $\bar\eta, \bar\eta'$ coincide. The values of η_S and η'_S do differ in general.*

If we compare the definitions of η and η' we see that the difference is the order of min *and* max. *Therefore Theorem 8 can be interpreted that asymptotically we have here a Minimax-Theorem.*

3.4   Effects of Randomization

In the theory of identification via channels one discovery was that randomized codes are tremendously superior compared with non-randomized codes whereas in the classical transmission model it doesn't affect the capacity (of the discrete memoryless channel).

In this section we consider randomized finite-state machines, i.e., we replace the next-state function $g : \mathcal{S} \times \mathcal{X} \to \mathcal{S}$ by a family

$$\mathcal{G} = \{G(\cdot|s,x) : s \in \mathcal{S}, x \in \mathcal{X}\} \cup G_0$$

of conditional probability distributions $G(\cdot|s,x)$ on $\mathcal{S}$ and an initial probability distribution G_0 on $\mathcal{S}$. The interpretation is that the initial state is chosen according to G_0 and then at each following time instant, if the machine is in state s and letter x occurs, the machine changes its state to s' with probability $G(s'|s,x)$. We consider randomized decision rules f where $f = (f_0, \ldots, f_{|\mathcal{X}|-1}) : \mathcal{S} \to [0,1]^{|\mathcal{X}|}$ with the interpretation that $f_x(s)$ is the probability that person x decides that the next symbol will be equal to x if the machine is in state s. Without loss of generality we can again restrict ourselves to deterministic decision rules, i.e., $f_x(s) = 0$ or 1 for all x and s. In order to see this, suppose we are given $\mathcal{G}$ and x^n. Then let for $t = 0, \ldots, n-1$ S_t be the random variable for the state at time t. The joint distribution of $S_0, \ldots, S_{n-1}$ is uniquely determined by $\mathcal{G}$ and x^n. Then the expected fraction of errors person x will make is given by

$$\eta^R_S(\mathcal{G}, f, x^n, x) \triangleq \frac{1}{n}\sum_{t=1}^{n}\sum_{s\in\mathcal{S}} P_{S_{t-1}}(s)(f_x(s)(1 - \delta_{x,x_t}) + (1 - f_x(s))\delta_{x,x_t})$$

$$= \frac{1}{n}\sum_{s\in\mathcal{S}}(f_x(s)\sum_{t=1}^{n} P_{S_{t-1}}(s)(1 - \delta_{x,x_t}) + (1 - f_x(s))\sum_{t=1}^{n} P_{S_{t-1}}(s)\delta_{x,x_t})$$

from which we see that $f_x(s) = 0$ or 1 is always an optimal choice resulting in an expected fraction of errors equal to

$$\eta_S^R(\mathcal{G}, x^n, x) \triangleq \frac{1}{n} \sum_{s \in \mathcal{S}} \min\{\sum_{t=1}^n P_{S_{t-1}}(s)(1 - \delta_{x,x_t}), \sum_{t=1}^n P_{S_{t-1}}(s)\delta_{x,x_t}\}.$$

Definition 14. *1.* $\eta_S^R(x^n, x) \triangleq \inf_{\mathcal{G}} \eta_S^R(\mathcal{G}, x^n, x)$,
2. $\eta_S^R(x^n) \triangleq \max_{x \in \mathcal{X}} \eta_S^R(x^n, x)$,
3. $\eta_S'^R(x^n) \triangleq \inf_{\mathcal{G}} \max_{x \in \mathcal{X}} \eta_S^R(\mathcal{G}, x^n, x)$,
4. $\bar{\eta}_S^R(x^n) \triangleq \frac{1}{|\mathcal{X}|} \sum_{x \in \mathcal{X}} \eta_S^R(x^n, x)$,
5. $\bar{\eta}_S'^R(x^n) \triangleq \inf_{\mathcal{G}} \frac{1}{|\mathcal{X}|} \sum_{x \in \mathcal{X}} \eta_S^R(\mathcal{G}, x^n, x)$.

The asymptotic quantities, $\eta_S^R(x^\infty)$, $\eta^R(x^\infty)$ etc., are defined analogously to Definitions 10 and 11.

Theorem 9. *For all sequences* $x^\infty \in \mathcal{X}^\infty$

$$\eta(x^\infty) = \eta^R(x^\infty) = \eta'^R(x^\infty).$$

Proof. From [10, Theorem 4] we can derive that

$$\mu_k^I(x^n, x) \leq \eta_S^R(x^n, x) + \sqrt{\frac{2 \ln S}{k + 1}}.$$

Taking the limit supremum as $n \to \infty$ and the limit as $k \to \infty$ and finally the limit $S \to \infty$ we obtain that $\mu^I(x^\infty, x) \leq \eta^R(x^\infty, x)$ and therefore

$$\mu^I(x^\infty) \leq \eta^R(x^\infty).$$

Together with Theorem 8 it follows

$$\eta'(x^\infty) = \eta(x^\infty) = \mu^I(x^\infty) \leq \eta^R(x^\infty) \leq \eta'^R(x^\infty) \leq \eta'(x^\infty).$$

$\square$

Theorem 9 shows that asymptotically randomization does not help here. The reason for this observation lies in the fact that deterministic Markov machines outperform asymptotically, as the number of states increases, any randomized finite-state machine.

References

1. R. Ahlswede and G. Dueck, Identification via channels, IEEE Trans. Inform. Theory, Vol. 35, No. 1, 15-29, 1989.
2. R. Ahlswede, General theory of information transfer, Preprint 97–118, SFB 343 "Diskrete Strukturen in der Mathematik", Universität Bielefeld, 1997; General theory of information transfer:updated, General Theory of Information Transfer and Combinatorics, a Special Issue of Discrete Applied Mathematics, to appear.

3. P.H. Algoet, The strong law of large numbers for sequential decisions under uncertainty, IEEE Trans. Inform. Theory, Vol. 40, No. 3, 609-633, 1994.

4. D. Blackwell, An analog to the minimax theorem for vector payoffs, Pac. J. Math., Vol. 6, 1-8, 1956.

5. D. Blackwell, Controlled random walks, Proc. Int. Congr. Mathematicians, 1954, Vol. III, Amsterdam, North Holland, 336-338, 1956.

6. T.M. Cover and A. Shenhar, Compound Bayes predictors for sequences with apparent Markov structure, IEEE Trans. Syst. Man Cybern., Vol. SMC-7, 421-424, 1977.

7. T.M. Cover, Behavior of sequential predictors of binary sequences, Proc. 4th Prague Conf. Inform. Theory, Statistical Decision Functions, Random Processes, 1965, Prague: Publishing House of the Czechoslovak Academy of Sciences, Prague, 263-272, 1967.

8. M. Feder, N. Merhav, and M. Gutman, Universal prediction of individual sequences, IEEE Trans. Inform. Theory, Vol. 38, No. 4, 1258-1270, 1992.

9. M. Feder, N. Merhav, and M. Gutman, Some properties of sequential predictors for binary Markov sources, IEEE Trans. Inform. Theory, Vol. 39, No. 3, 887-892, 1993.

10. M. Feder and N. Merhav, Universal schemes for sequential decision from individual data sequences, IEEE Trans. Inform. Theory, Vol. 39, No. 4, 1280-1292, 1993.

11. M. Feder and N. Merhav, Relations between entropy and error probability, IEEE Trans. Inform. Theory, Vol. 40, No. 1, 259-266, 1994.

12. J.F. Hannan and H. Robbins, Asymptotic solutions of the compound decision problem for two completely specified distributions, Ann. Math. Statist., Vol. 26, 37-51, 1957.

13. J.F. Hannan, Approximation to Bayes risk in repeated plays, Contributions to the Theory of Games, Vol. III, Annals of Mathematics Studies, Princeton, NJ, No. 39, 97-139, 1957.

14. H. Robbins, Asymptotically subminimax solutions of compound statistical decision problems, in Proc. 2nd Berkeley Symp. Math. Stat. Probab., 131-148, 1951.

15. C.E. Shannon, A mathematical theory of communication, Bell System Tech. J., Vol. 27, 379-423, 623-656, 1948.

16. C.E. Shannon, Prediction and entropy of printed English, Bell Sys. Tech. J., Vol. 30, 5-64, 1951.

17. C.E. Shannon, The mind reading machine, Bell Laboratories Memorandum, 1953, in *Shannon's Collected Papers*, A.D. Wyner and N.J.A. Sloane Eds., IEEE Press, 688-689, 1993.

Watermarking Identification Codes with Related Topics on Common Randomness

R. Ahlswede and N. Cai*

Abstract. Watermarking identification codes were introduced by Y. Steinberg and N. Merhav. In their model they assumed that

(1) the attacker uses a single channel to attack the watermark and both, the information hider and the decoder, know the attack channel;
(2) the decoder either completely he knows the covertext or knows nothing about it.

Then instead of the first assumption they suggested to study more robust models and instead of the second assumption they suggested to consider the case where the information hider is allowed to send a secret key to the decoder according to the covertext.

In response to the first suggestion in this paper we assume that the attacker chooses an unknown (for both information hider and decoder) channel from a set of channels or a compound channel, to attack the watermark. In response to the second suggestion we present two models. In the first model according to the output sequence of covertext the information hider generates side information componentwise as the secret key. In the second model the only constraint to the key space is an upper bound for its rate.

We present lower bounds for the identification capacities in the above models, which include the Steinberg and Merhav results on lower bounds. To obtain our lower bounds we introduce the corresponding models of common randomness. For the models with a single channel, we obtain the capacities of common randomness. For the models with a compound channel, we have lower and upper bounds and the differences of lower and upper bounds are due to the exchange and different orders of the max–min operations.

Keywords: Watermarking, identification, compound channel, common randomness.

1 Introduction

Watermarking technique is a way to embed secret information into a given message, say image, that cannot be removed nor deciphered without access to a secret key.

It can be used to protect copy right. Watermarking is now a major activity in audio, image, and video processing and standardization efforts for JPEG–2000, MPEG–4 and Digital Video Disks are underway.

---

* This paper is supported in part by INTAS 00–738.

R. Ahlswede et al. (Eds.): Information Transfer and Combinatorics, LNCS 4123, pp. 107–153, 2006.
© Springer-Verlag Berlin Heidelberg 2006

One way to analyze watermarking problems is to regard them as communication systems e.g., [16], [20], [27], [28], [29], [30], and [32]. In these systems the messages, which are called covertext, are generated by an information source. An information hider, whom we often call encoder because of his role in the system, has full access to the information source of covertexts and the set of secret messages. These secret messages are independent of the covertext, they are uniformly generated from the set, and will be called watermark. The role of the information hider, or encoder, is to embed the watermark in the covertext. When the embedding changes the covertext, it disturbs the message. To guarantee the quality of the watermarked message, we certainly would like not too much distortion. That is, for a given distortion measure, the distortion between the original covertext and the watermarked message in average may not exceed a given constant. An attacker wants to remove the watermark from the watermarked message without distorting the message too much i.e., the distortion between the covertext and the message corrupted by the attacker is not too large with respect to a certain distortion measure. Finally a decoder tries to recover the watermark from the corrupted message correctly with high probability. As the attacker is allowed to use a random strategy, we assume that the attacker uses a noisy channel to attack the watermark. Depending on the models the attacker may choose various channels and the encoder and decoder share different resources (e.g., secret key, side information, etc.).

Among huge contributions on watermarking we here briefly review two of them. In [28] P. Moulin and J.A. O'Sullivan obtained the capacity for the watermarking codes under the assumptions that the covertexts are generated from a memoryless source, the distortions are sum–type and the attack channels are compound channels whose states are known to the decoder but unknown to the encoder. The strategies of encoder–decoder and attacker are discussed as well.

Identification codes for noisy channels were introduced by R. Ahlswede and G. Dueck for the situation in which the receiver needs to identify whether the coming message equals a specified one. If not, then they don't care what it is [11]. It turned out that this weaker requirement dramatically increased the sizes of messages sets which could be handled: double exponential grown in the block lengths of codes. Identification is much faster than transmission!

Y. Steinberg and N. Merhav notice that in most cases people check watermarks in order to identify them (e.g. copyright) rather than recognize them and so they introduced identification codes to watermarking models [32]. In their models the attack channels are single memoryless channels. That means the attacker's random strategy is known by information hider (encoder) and decoder. They notice that the assumption is not robust and so suggested to study more robust models. As to the resources shared by encoders and decoders they consider two cases, the decoder either completely know the covertext or he knows nothing about it. (In all cases the attacker must not know the covertext because otherwise there would be no safe watermarking.)

By considering common randomness between encoder and decoder, they obtained lower bounds to the capacities of watermarking identification in both cases and the upper bounds easily followed from a theorem in [31]. The lower and upper bounds are tight in the former case but not in the latter case. As Y. Steinberg and N. Merhav only studied two extremal cases, they suggested to consider the more general case, that the decoder may obtain partial information, about the covertext, say key, from the encoder via a secure noiseless channel. The exponent of error probability was discussed as well.

In the present paper we deal with these two problems. But before turning to our result, we draw readers' attention to common randomness, which – as noticed in [12] – plays a central role in identification problems. It does so also in [32] and here R. Ahlswede and G. Dueck discovered in [12] that common randomness shared by encoder and decoder can be used to construct identification codes and therefore the rate of common randomness (in the sense of first order of logarithm) is not larger than the rate of identification codes (in the sense of the second order of logarithm). In general the capacities of common randomness shared by the encoder and the decoder may be smaller than the capacities of identification. Examples for discrete channels and Gaussian channels were presented in [5] and [17] respectively. Notice that the sizes of the input alphabets of the former channel is growing super exponentially as the length of codes and the sizes of the input alphabets of the latter is infinity. In fact it is seen from [31] that for any channel, whose input alphabet is exponentially increasing in the case that strong converse holds, the rates of common randomness and identification codes are the same.

The topic of common randomness has been become more and more popular e.g., [6], [9], [10], [23], [26], [33], [34], etc. Common randomness may be applied to cryptography, (e.g., [9], [18], [23], [26]), identification (e.g., [5], [11], [12], [10], [15], [18]), and arbitrarily varying channels (e.g., [1], [2], [8], [10]). For the first two applications the rates are important and the distributions of common randomnesses are required nearly uniformly. For cryptography certain secure conditions additionally needed. For the last application one has to face in the difficulty made by the jammer and find a smart way to generate the common randomness.

Now let us return to the two suggestions by Steinberg and Merhav. For the first suggestion we assume in our models, attackers are allowed to choose a channel arbitrarily from a set of memoryless channels to attack watermarks and neither encoders nor decoders know the attack channels. This is known as compound channel in Information Theory.

The assumption makes our models slightly more robust than that in [28] since in [28] the decoders are supposed to know the attack channels.

For the second suggestion we set up two models. In out first model we assume the encoder generates a random variable at time t according to component at time t of the output sequence of covertext source and certain probability and sends it to decoder via a secure channel. In this case the "key" actually is a side information of covertext shared by encoder and decoder. We obtain the first and the second models in [32] if we choose the side information equal to covertext almost surely and independent of covertext respectively. So our first model contain both models in [32]. In our second model the encoder is allowed to generate a key according to the covertext (but independently on watermark) in arbitrary way and sends the key to decoder through a secure channel with rate R_K. Obviously in our second model the key can be generated in a more general way than in our first model. For all combinations of above assumptions, we obtain lower bounds to the identification capacity, which contains both lower bounds in [32] as special cases.

To obtain our lower bounds to identification capacities, for each combination, we introduce a corresponding model of common randomness and obtain lower and upper bound to its capacity. For the single channel the two bound is closed for compound channel the gap between two bounds is up to the order of max–min. In addition, we show a lower bound to common randomness in [32] in fact is tight, which supports a conjecture in [32].

We must point out that our assumption of compound attack channels is still far from the most robust and practical assumption although according to our knowledge, it is most robust and practical existing assumption in this area. Actually the attacker has much more choices.

- He does not necessarily use a memoryless channel and stead he can chooses a channel with finite memory.
- The attacker may change the states time by time i.e., he may use an arbitrarily varying channel.
- The attacker knows output of the channel; even at time t, he know the output at time $t' > t$, since all outputs in fact are chosen by himself/herself. So the attacker may use this information to choose attack channel. This clearly makes the attack much more efficient.

So there is still a long way for us to achieve the most practical results and it provide a wide space for future research.

The rest part of the paper is organized as follows. In the next section we present the notation used in the paper. Our models and results are stated in Section 3 and Section 4 respectively. The direct parts of coding theorems of common randomness are proven in Section 5 and their converse parts are proven in Section 6. In Section 7 we briefly review the observation in [12] on the relation of identification and common randomness and therefore the lower bounds to the identification capacities from capacities of common randomness. Finally the converse theorem for a model in [32] is proven in Section 8.

2   The Notation

Our notation in this paper is fairly standard. log and ln stand for the logarithms with bases 2 and e respectively and a^z is often written as $exp_a[z]$. The random variables will be denoted by capital letters L, U, V, X, Y, Z etc. and their domains are often denoted by the correspondent script letters $\mathcal{L}, \mathcal{U}, \mathcal{V}, \mathcal{X}, \mathcal{Y}, \mathcal{Z}$ etc. But in some special cases it may be exceptional. When we denote a set by a script letter (for example, $\mathcal{X}$), its element is often denoted by the corresponding lower letter (for example x). $\mathcal{X}^n$ is the nth Cartesian power of the set $\mathcal{X}$ and $x^n = (x_1, x_2, \ldots x_n)$ is the sequence of length n. $Pr\{\mathcal{E}\}$ is the probability of that the event $\mathcal{E}$ occurs and $\mathbf{E}[\cdot]$ is the operator of expectation. P_X, P_{XY}, $P_{Z|X}$ etc. will stand for the distribution of random variable X , the joint distribution of the random variables (X, Y), the conditional distribution of random variable Z under the condition that X is given respectively. When we write a probability distribution as P^n, we mean that it is a product distribution of P and similarly a discrete memoryless channel of length n with stochastic matrix W is written as W^n.

Throughout this paper T_U^n, T_{UV}^n, $T_{U|VL}^n(v^n l^n)$ etc. will denote the sets of typical, joint typical, and conditional typical sequences and the corresponding sets of δ- typical, joint typical, and conditional typical sequences are written as $T_U^n(\delta)$, $T_{UV}^n(\delta)$, $T_{U|VL}^n(v^n l^n, \delta)$ etc.. We always understand these sets are not empty when we use the notation. When we introduce a set of typical sequences (for example, say T_Z^n), it is understood that the correspondent random variable(s) (i.e., Z in the example) with the (joint) type as distribution (P_Z) is introduced at the same time. For a subset $\mathcal{A}$ of sequences of length n we write $\mathcal{A}_U = \mathcal{A} \cap T_U^n$ and analogously $\mathcal{A}_{UV}$, $\mathcal{A}_{U|VL}(v^n l^n)$, $\mathcal{A}_U(\delta)$, $\mathcal{A}_{UV}(\delta)$, $\mathcal{A}_{U|VL}(v^n l^n, \delta)$ etc.

$|T_U^n|$ and the common values of $|T_{U|L}^n(l^n)|$, $l^n \in T_L^n$ some times are written as t_U, $t_{U|L}$ etc. respectively (the length n of the sequences are understood by the context). Analogously $t_U(\delta)$, $t_{Y|X}(\delta)$ etc, also are used.

3 The Models

Watermarking Identification Codes
In this subsection, we state our models for the simpler case that the attacker choose a single channel to attack the watermark and both the encoder (information hider) and the decoder know the attack channel. In the next subsection, we introduce the corresponding models of common randomness. In the last subsection of the section, we assume the attack chooses a channel unknown by both encoder and decoder from a set of channels and replace the single channel by a compound channel.

Let $\mathcal{V}$ be a finite set, and V be a random variable taking values in $\mathcal{V}$. Then the covertext is assumed to be generated by an memoryless information source $\{V^n\}_{n=1}^{\infty}$ with generic V. The watermark is uniformly chosen from a finite set $\{1, 2, \ldots, M\}$ independently on the context. The encoder is fully accessed the covertext and source of watermark and encodes the outputs of covertext v^n and of watermark m jointly to a sequence $x^n (= x^n(v^n, m))$ with the same length of sequence of covertext. The attack use a single discrete memoryless channel W to attack the watermarked sequence x^n i.e., to change x^n to y^n with probability $W^n(y^n|x^n) = \prod_{t=1}^{n} W(y_t|x_t)$. Usually for practical reason people assume that v^n, x^n, and y^n are chosen from the same finite alphabet, but for convenience of notation we assume they are from finite alphabets $\mathcal{V}$, $\mathcal{X}$, and $\mathcal{Y}$ respectively. The encoding mapping in general disturbs the covertext. To measure the distortion, we introduce a sum type distortion measure, watermarking distortion measure (WD–measure) ρ, such that for all $v^n = (v_1, \ldots, v_n) \in \mathcal{V}^n$, $x^n = (x_1, \ldots, x_n) \in \mathcal{X}^n$,

$$\rho(v^n, x^n) = \sum_{t=1}^{n} \rho(v_t, x_t), \tag{1}$$

where for all $v \in \mathcal{V}$, $x \in \mathcal{X}$ $0 \leq \rho(v, x) \leq \Delta$, for a positive constant Δ.

By definition, there should be certain distortion constraint to the output of attack channel. But now we are given a memoryless attack channel and we may omit the constraint simply by assume that the attack channel satisfies the constraint automatically. This clearly does not loss generality. Next we have to set up the key–resources shared by encoder and decoder, according to which we distinguish our watermarking identification codes into watermarking identification codes with side information (WIDCSI) and watermarking identification codes with secure key (WIDCK) as follows.

Watermarking identification codes with side information (WIDCSI)
In the first case, we assume that the encoder can generate "a component of a key", $L_t = l_t$ at the time t according to the current output of covertext $V_t = v_t$ and a given conditional distribution $P_{L|V}(\cdot|v)$. That is, the sender generates a sequence $L^n = (L_1, L_2, \ldots, L_n) = l^n = (l_1, l_2, \ldots, l_n)$ with probability $P_{L|V}^n(l^n|v^n)$ if the source outputs a sequence v^n of covertext and then sends it to the decoder. The latter try to recover the watermark from the invalidated message by the attacker with the help of the side information $L^n = l^n$. In this case the key-resource is actually governed by the conditional distribution $P_{L|V}$ or equivalently the joint probability distribution P_{VL}. So it can be understood as a pure side information at both sides of encoder and decoder instead of a " secure key". That is, if $\{V^n\}_{n=1}^{\infty}$ is a memoryless covertext with generic V, and $\{L^n\}_{n=1}^{\infty}$ is a side information observed by both encoder and decoder, then $\{(V^n, L^n)\}$ is a correlated memoryless source with generic (V, L). Thus the decoder can learn some thing about the covertext from the side information whereas the attacker knows nothing about it. A WIDCSI code becomes a "watermarking identification code with side information at transmitter and receiver" in [32] when V and L have the same alphabet and equal to each other almost surely and it becomes a "watermarking identification code with side information at the transmitter only" in [32] if V and L are independent.So the two codes defined in [32] is really the extreme cases of WIDCI codes.

Watermarking identification codes with secure key (WIDCK)
In this case we assume the encoder may generate a key $K_n = K_n(v^n)$ according to the whole output sequence $V^n = v^n$ of the random covertext V^n in an arbitrary way and send it to the decoder through a secure (noiseless) channel so that the attacker has absolutely has no knowledge about the covertext (except its distribution) nor the key. Since for given output v^n of the covertext the encoder may generate the K_n randomly, a WIDCSI code is a special WIDCK code. We

shall see that in general the latter is more powerful. Notice that a deterministic key function of output of covertext is a special random key. Finally of course the size of the key must be constraint. We require it exponentially increasing with the length of the code and its rate upper bounded by the key rate R_K. When the key rate is larger than the covertext entropy $H(V)$ the encoder certainly may inform the receiver the output of covertext. However "the rest part" of the key may serve as a common randomness between the communicators which increases the identification capacity (see [12], [10], and [32]).

Thus an $(n, R, \lambda_1, \lambda_2, D_1)$ WIDCSI code is a system $\{Q_m, \mathcal{D}_m(l^n) : l^n \in \mathcal{L}^n, m \in \mathcal{M}\}$ for $\mathcal{M} = \{1, 2, \ldots, M\}$ satisfying the following conditions.

- $Q_m, m = 1, 2, \ldots, M$ are stochastic matrices $Q_m : \mathcal{V}^n \times \mathcal{L}^n \longrightarrow \mathcal{X}^n$ such that for $m = 1, 2, \ldots, M$,

$$\sum_{v^n \in \mathcal{V}^n, l^n \in \mathcal{L}^n} P_{VL}^n(v^n, l^n) \sum_{x^n \in \mathcal{X}^n} Q_m(x^n | v^n, l^n) \rho(v^n, x^n) \leq D_1, \qquad (2)$$

where P_{VL} is the joint distribution of the generic (V, L).
- For all $l^n \in \mathcal{L}^n, m \in \mathcal{M}, \mathcal{D}_m(l^n) \subset \mathcal{Y}^n$ and for all $m \in \mathcal{M}$,

$$\sum_{v^n \in \mathcal{V}^n, l^n \in \mathcal{L}^n} P_{VL}^n(v^n, l^n) \sum_{x^n \in \mathcal{X}^n} Q_m(x^n | v^n, l^n) W^n(\mathcal{D}_m(l^n) | x^n) > 1 - \lambda_1, \quad (3)$$

and for all $m, m' \in \mathcal{M}, m \neq m'$,

$$\sum_{v^n \in \mathcal{V}^n, l^n \in \mathcal{L}^n} P_{VL}^n(v^n, l^n) \sum_{x^n \in \mathcal{X}^n} Q_m(x^n | v^n, l^n) W^n(\mathcal{D}_{m'}(l^n) | x^n) < \lambda_2. \quad (4)$$

λ_1 and λ_2 is called the errors of the first and the second kinds of the code
- The rate of the code is

$$R = \log \log M. \qquad (5)$$

Watermarking identification codes with secure key (WIDCK)
Next we define WIDCK code. Let $\{V^n\}_{n=1}^\infty$ be a memoryless covertext with generic V and alphabet $\mathcal{V}$, the attack channel W be memoryless, and WD-measure ρ be as (1). Then an $(n, R, R_K, \lambda_1, \lambda_2, D_1)$ WIDCK code is a system $\{Q_m^*, \mathcal{D}_m^*(k_n), W_{K_n} : m \in \mathcal{M}, k_n \in \mathcal{K}_n\}$ for $\mathcal{M} = \{1, 2, \ldots, M\}$ satisfying the following conditions.

- $\mathcal{K}_n$ is a finite set, which will be called the key book, with

$$\frac{1}{n} \log |\mathcal{K}_n| \leq R_K. \qquad (6)$$

R_K will be called key rate.
- W_{K_n} is a stochastic matrix, $W_{K_n} : \mathcal{V}^n \longrightarrow \mathcal{K}_n$. The output random variable will be denoted by K_n when the random covertext V^n is input to the channel W_{K_n} i.e., the pair of random variables (V^n, K_n) have joint distribution $P_{V^n K^n}(v^n, k^n) = P_V^n(v^n) W_{K_n}(k_n | v^n), v^n \in \mathcal{V} \ k_n \in \mathcal{K}_n$. In particular K_n

may be a deterministic function of output of covertext and in this case we write $K(\cdot)$ as a function defined on $\mathcal{V}^n$. Note that the choice of K_n does NOT depend on the message $m \in \mathcal{M}$ since the key should independent of the protected message.

– $Q_m^*, m = 1, 2, \ldots, M$ are stochastic matrices from $\mathcal{V}^n \times \mathcal{K}_n$ to $\mathcal{X}^n$, (the alphabet of the input of the attack channel), such that

$$\sum_{v^n \in \mathcal{V}^n} P_V^n(v^n) \sum_{k_n \in \mathcal{K}_n} W_{K_n}(k_n|v^n) \sum_{x^n \in \mathcal{X}^n} Q_m^*(x^n|v^n, k_n)\rho(v^n, x^n) \leq D_1. \quad (7)$$

– For all $k_n \in \mathcal{K}_n, m \in \mathcal{M}, \mathcal{D}_m(k_n) \subset \mathcal{Y}^n$ and for all $m \in \mathcal{M}$, the error of first kind

$$\sum_{v^n \in \mathcal{V}^n} P_V^n(v^n) \sum_{k_n \in \mathcal{K}_n} W_{K_n}(k_n|v^n) \sum_{x^n \in \mathcal{X}^n} Q_m^*(x^n|v^n, k_n)W^n(\mathcal{D}_m(k_n)|x^n) > 1 - \lambda_1,$$

$$(8)$$

and for all $m, m' \in \mathcal{M} \; m \neq m'$,

$$\sum_{v^n \in \mathcal{V}^n} P_V^n(v^n) \sum_{k_n \in \mathcal{K}_n} W_{K_n}(k_n|v^n) \sum_{x^n \in \mathcal{X}^n} Q_m^*(x^n|v^n, k_n)W^n(\mathcal{D}_{m'}(k_n)|x^n) < \lambda_2.$$

$$(9)$$

– Finally the rate of the code is defined in (5).

The capacities of the codes of the two types are defined in the standard way and denoted by $C_{WIDSI}((V, L), W, D_1)$ and $C_{WIDK}(V, W, R_K, D_1)$ respectively, where (V, L) and V are the generic of memoryless correlated source and source respectively, W is an attack memoryless channel, R_K is the key rate, and D_1 is the distortion criterion.

The Common Randomness

We speak of the common randomness between two (or among more than two) persons who share certain common resources, which may be correlated sources and/or (noisy or noiseless) channels. The common randomness between these two persons is just two random variables with common domain, which converges each other respect to probability. According to the resources different models are established.

For the purpose to build watermarking identification codes we need the following two kinds of common randomness. In the following two models of common randomness, the correlated source $\{(V^n, L^n)\}_{n=1}^\infty$ corresponds to the source of covertext and side information and the memoryless channel W corresponds the attack channel in the models of watermarking identification. The K_n in the Model II corresponds the key in the model of WIDCK.

Model I: Two-source with a constraint noisy channel

Let $\{(V^n, L^n)\}_{n=1}^\infty$ be a correlated memoryless source with two components, alphabets $\mathcal{V}$ and $\mathcal{L}$, and generic (V, L). Assume that there are two persons, say sender (or encoder) and receiver (or decoder). The sender may observe the whole output of the source (V^n, L^n) whereas only the output of the component L^n is observable for the receiver. To establish common randomness the sender may send message through memoryless channels W with input and output alphabets $\mathcal{X}$ and $\mathcal{Y}$ under certain constraint condition (specified below). The receiver is not allowed to send any message to the sender. The sender first chooses a channel code with set of codewords $\mathcal{U} \subset \mathcal{X}^n$ with the same length n as output sequence of the source and generates a random variable M, his/her "private randomness" taking values uniformly in a finite set $\mathcal{M}$ (, which is exponentially increasing as the length n of the source sequences increases) and independent of (V^n, L^n) of the output of the source. Assume a (sum type) distortion measure ρ in (1) and a criterion of distortion D_1 are given. According to the output $(V^n, L^n) = (v^n, l^n)$ of the source and the output of his/her private randomness $M = m$ the sender chooses a codeword $x_m(v^n, l^n) \in \mathcal{U} (\subset \mathcal{X}^n)$ such that the average of the distortion between the codeword and the component $V^n = v^n$ of the correlated source may not exceed D_1. Namely,

$$\frac{1}{n} \sum_{m \in \mathcal{M}} P_M(m) \sum_{v^n \in \mathcal{V}^n} \sum_{l^n \in \mathcal{L}^n} P_{VL}(v^n, l^n) \rho(x_m(v^n, l^n), v^n) \leq D_1. \tag{10}$$

The receiver receives an output sequence $y^n \in \mathcal{Y}^n$ with the probability

$$W^n(y^n | x_m(v^n, l^n))$$

if the sender input the codeword $x_m(v^n, l^n)$ to the channel. We also allow to choose $x_m(v, l^n)$ as a random input sequence instead of deterministic one (it is more convenient in the proof). Finally for a finite set $\mathcal{A}$ which typically increases exponentially when the length n of the source increases, i. e., for a constant κ

$$\frac{1}{n} \log |\mathcal{A}| \leq \kappa, \tag{11}$$

the sender creates a random variable F with range $\mathcal{A}$, according to the outputs of (V^n, L^n) and M, through a function

$$F : \mathcal{V}^n \times \mathcal{L}^n \times \mathcal{M} \longrightarrow \mathcal{A} \tag{12}$$

and the receiver creates a random variable G according to the output of the channel W^n and the output of the component L^n of the source, through a function

$$G : \mathcal{L}^n \times \mathcal{Y}^n \longrightarrow \mathcal{A}. \tag{13}$$

After the terminology in [10] we called the pair of random variables (F, G) generated in the above way permissible and say that a permissible pair (F, G) represents λ-common randomness if

$$Pr\{F \neq G\} < \lambda. \tag{14}$$

Typically λ should be an arbitrarily small but positive real number when length n of source sequences is arbitrarily large. It is not hard to see that under the conditions (11) and (14) by Fano inequality, the entropy rates $\frac{1}{n}H(F)$ and $\frac{1}{n}H(G)$ are arbitrarily close if λ in (14) is arbitrarily small. This was observed in [10]. Thus we can choose any one from the pair of entropy rates, say $\frac{1}{n}H(F)$ as the rate of common randomness.

A pair of real numbers (r, D_1) is called achievable for common randomness if for arbitrary positive real numbers ϵ, λ, μ and sufficiently large n (depending on ϵ, λ and μ) there exists a λ-common randomness satisfying (10) – (14), such that

$$\frac{1}{n}H(F) > r - \epsilon \tag{15}$$

and

$$\sum_{a \in \mathcal{A}} \mid Pr\{F = a\} - \frac{1}{|\mathcal{A}|} \mid < \mu. \tag{16}$$

The last condition says that the common randomness is required to be nearly uniform and we call it nearly uniform condition. We set it for reducing the errors of second kind of identification codes. The set of achievable pairs is called common randomness capacity region. For fixed D_1 the common randomness capacity (CR-capacity) is $C_{CRI}((V, L), W, D_1) = \max\{r : (r, D_1) \text{ is achievable}\}$.

Notice that there is no limit to the amount of sender's private randomness in the present model and the next model, Model II. However because of the limit of the capacity of the channel the " extra" private randomness is useless.

We remark here that this model is different from the model (i) in [10] in three points. First, the channel connect the sender and receiver is noiseless with constraint that rate $\leq R$ in the model (i) of [10] whereas in general it is noisy in current model. More importantly, because of the requirement of distortion the source not only plays a role of " side information" but also a role of "constrainer". That is, to fight for reducing the distortion the sender has to choose codewords properly. This makes the transformation more difficult. To see that let us consider an extremal case that the component L^n of the source is a constant. In this case the source makes no difference at all in the model (i) of [10] and therefore the common randomness capacity is trivially equal to capacity of the channel. But in this case for the present model the source makes difference i.e., because of it the sender may not choose the codewords freely and therefore the common randomness is reduced. To obtain the CR-capacity region for this model is also absolutely non-trivial. Finally in this model the sender and receiver observe the output $(V^n, L^n) = (v^n, l^n)$ and $L^n = l^n$ respectively. The common randomness before the transmission, is equal to $H(L^n) = I(V^n, L^n; L^n)$ the mutual information between the two observations. So it seems to be not surprising our characterization in Theorem 4.1 is quite different from that in Theorem 4.1 of [10] and it cannot obtain simply by substituting rate of noiseless channel by the capacity of the noisy channel.

Model II: Two-source with a constraint noisy channel and a noiseless channel

It is clear that our goal to study the common randomness of the model I is for the construction of WIDCSI-codes. Next to study WIDCK codes we introduced the Model II of common randomness. Actually our model is a little more general than that we really need. That is, we add "the side information". But for this we need to do almost no more work. Thus to define the Model II we only add a noiseless channel between the sender and receiver based on th Model I.

Namely we assume that the correlated source $\{(V^n, L^n)\}_{n=1}^{\infty}$, the noisy channel W, the distortion constraint (10), and the sender's private randomness M are still available. Additionally the sender may send a message k_n from a set of message $\mathcal{K}_n$ with rate $\frac{1}{n} \log |\mathcal{K}| \leq R_K$ to the receiver via noiseless channel. Again R_K is called key rate. Of course k_n is necessarily to be a function of the outputs of the source and sender's private randomness i.e., $k_n = k_n(v^n, m)$ for $v^n \in \mathcal{V}^n$, $m \in \mathcal{M}$. More generally the sender may use random strategies i.e., treats k_n is output of a channel W_K with input (v^n, m). To define the common randomness for this model we change (13) to

$$G : \mathcal{K}_n \times \mathcal{L}^n \times \mathcal{Y}^n \longrightarrow \mathcal{A}. \tag{17}$$

and keep the conditions (10), (11), (12), (14), (15), and (16) unchanged (but now the definition of function G has been changed due to the changing).

Analogously, one can define CR-capacity $C_{CRII}((V, L), W, R_K, D_1)$ for memoryless correlated source with generic (V, L), memoryless channel W, key rate R_K and the distortion criterion D_1 of this model.

The Models for Compound Channels

In this subsection we assume that the attacker employ a (stationary) memoryless channel from a family of channels satisfying attack distortion criterion to attack the watermark. Neither the sender nor receiver knows the which channel the attacker uses. These channels are known as compound channels in Information Theory. This assumption is slightly more robust and practical than that in [28] where the decoder has to know the attack channel in order to decode. In fact, according to our knowledge it is most robust assumption in this direction.

A compound channel is just a family of memoryless channels $\mathcal{W} = \{W(\cdot|\cdot, s) : s \in \mathcal{S}\}$ with common input and output alphabet $\mathcal{X}$ and $\mathcal{Y}$ respectively. $\mathcal{S}$ is a index set which is called state set and its members are called states. An output sequence $y^n \in \mathcal{Y}^n$ is output with the probability

$$W^n(y^n|x^n, s) = \prod_{t=1}^{n} W(y_t|x_t, s)$$

when the channel is governed by the state s and $x^n \in \mathcal{X}^n$ is input.

Underlie assumption for the attacker to use a compound channel to attack a watermarking transmission or identification code is that the attacker knows the input distribution P_n generated by the code. He then may employ such a compound channel that for all $s \in \mathcal{S}$

$$\frac{1}{n} \sum_{x^n \in \mathcal{X}^n} P_n(x^n) \sum_{y^n \in \mathcal{Y}^n} W^n(y^n|x^n, s)\rho'(x^n, y^n) \leq D_2,$$

where ρ' is a sum type distortion measure, attack distortion measure (AD-measure), may or may not be identify to WD-measure ρ and D_2 is the attack distortion criterion. In particular when the codewords are generated by an i. i. d. input distributions so that the input distribution generated by the code is an i. i. d. distribution

$$P^n(x^n) = \prod_{i=1}^{n} P(x_t)$$

a compound channel such that for all $s \in \mathcal{S}$

$$\sum_{x \in \mathcal{X}} P(x) \sum_{y \in \mathcal{Y}^n} W(y|x, s)\rho'(x, y) \le D_2$$

may be used. We always assume that all compound channels under the consideration satisfy the condition of distortion and do not worry it at all.

To adjust the models in the last two subsections to the compound channels the following modifications are necessary.

For *WIDCSI code* for compound channels: replace (3) and (4) by for all $l^n \in \mathcal{L}^n, m \in \mathcal{M}, \mathcal{D}_m(l^n) \subset \mathcal{Y}^n$ such that for all $m \in \mathcal{M}$, and $s \in \mathcal{S}$,

$$\sum_{v^n \in \mathcal{V}^n, l^n \in \mathcal{L}^n} P_{VL}^n(v^n, l^n) \sum_{x^n \in \mathcal{X}^n} Q_m(x^n|v^n, l^n)W^n(\mathcal{D}_m(l^n)|x^n, s) > 1 - \lambda_1, \quad (18)$$

and for all $m, m' \in \mathcal{M}$ $m \ne m'$, and $s \in \mathcal{S}$

$$\sum_{v^n \in \mathcal{V}^n, \ell^n \in \mathcal{L}^n} P_{VL}^n(v^n, \ell^n) \sum_{x^n \in \mathcal{X}^n} Q_m(x^n|v^n, \ell^n)W^n(\mathcal{D}_{m'}(l^n)|x^n, s) < \lambda_2 \quad (19)$$

respectively.

For *WIDCK* for compound channels: replace (8) and (9) by for all $k_n \in \mathcal{L}_n, m \in \mathcal{M}, \mathcal{D}_m(k_n) \subset \mathcal{Y}^n$ such that for all $m \in \mathcal{M}$, and $s \in \mathcal{S}$,

$$\sum_{v^n \in \mathcal{V}^n} P_V^n(v^n) \sum_{k_n \in \mathcal{K}_n} W_{K_n}(k_n|v^n) \sum_{x^n \in \mathcal{X}^n} Q_m^*(x^n|v^n, k_n)W^n(\mathcal{D}_m(k_n)|x^n, s) > 1 - \lambda_1,$$

$$(20)$$

and for all $m, m' \in \mathcal{M}$ $m \ne m'$, and $s \in \mathcal{S}$,

$$\sum_{v^n \in \mathcal{V}^n} P_V^n(v^n) \sum_{k_n \in \mathcal{K}_n} W_{K_n}(k_n|v^n) \sum_{x^n \in \mathcal{X}^n} Q_m^*(x^n|v^n, k^n)W^n(\mathcal{D}_{m'}(k_n)|x^n, s) < \lambda_2.$$

$$(21)$$

Here the fact that Q_m, Q_m^*, $\mathcal{D}_m(l^n)$ and $\mathcal{D}_m(k_n)$ are independent of the states governing the channels reflects the requirement that neither encoder nor decoder knows the states and that (18) – (21) hold for all $s \in \mathcal{S}$ is because the worst case to the encoder and decoder is considered.

For the *Common randomness in the models I and II:* for compound channels, replace (14) by, whenever any state s governs the channel,

$$Pr\{F \ne G|s\} < \lambda. \quad (22)$$

Again the functions F, G, codewords are independent of the states because the states are unknown for both encoder and the decoder.

Analogously, for compound channel $\mathcal{W}$ the corresponding capacities of watermarking identification codes and common randomness are denoted by $C_{WIDSI}((V, L), \mathcal{W}, D_1)$, $C_{WIDK}(V, \mathcal{W}, R_K, D_1)$, $C_{CRI}((V, L), \mathcal{W}, D_1)$ and $C_{CRII}((V, L), \mathcal{W}, R_K, D_1)$.

4   The Results

The Results on Common randomness

For given a correlated memoryless source $\{(V^n, L^n)\}_{n=1}^{\infty}$ whose generic has joint distribution P_{VL}, a memoryless channel W and distortion criterion D_1, let $\mathcal{Q}((V, L), W, D_1)$ be the set of random variable (V, L, U, X, Y) with domain $\mathcal{V} \times \mathcal{L} \times \mathcal{U} \times \mathcal{X} \times \mathcal{Y}$ and the following properties, where $\mathcal{U}$ is a finite set with cardinality $|\mathcal{U}| \leq |\mathcal{V}||\mathcal{L}||\mathcal{X}|$ and $\mathcal{X}$ and $\mathcal{Y}$ are input and output alphabets of the channel W respectively.

For all $v \in \mathcal{V}$, $l \in \mathcal{L}$, $u \in \mathcal{U}$, $x \in \mathcal{X}$, and $y \in \mathcal{Y}$

$$
\begin{aligned}
Pr\{(V, L, U, X, Y) = (v, l, u, x, y)\} \\
= P_{VLUXY}(v, l, u, x, y) \\
= P_{VL}(v, l) P_{UX|VL}(u, x|v, l) W(y|x).
\end{aligned}
\tag{23}
$$

For the given distortion measure ρ

$$
\mathbf{E}\rho(V, X) \leq D_1.
\tag{24}
$$

$$
I(U; V, L) \leq I(U; L, Y).
\tag{25}
$$

Then we have the coding theorem of common randomness in the model I for single channel W.

Theorem 4.1

$$
C_{CRI}((V, L), W, D_1) = \max_{(V, L, U, X, Y) \in \mathcal{Q}((V, L), W, D_1)} [I(U; L, Y) + H(L|U)].
\tag{26}
$$

For a given correlated source with generic (V, L) a channel W and positive real numbers R_K and D_1, we denote by $\mathcal{Q}^*((V, L), W, R_K, D_1)$ the set of random variables (V, L, U, X, Y) with domain as above and such that (23), (24) and

$$
I(U; V, L) \leq I(U; L, Y) + R_K
\tag{27}
$$

hold. Then

Theorem 4.2

$$
C_{CRII}((V, L), W, R_K, D_1) = \max_{(V, L, U, X, Y) \in \mathcal{Q}^*((V, L), W, R_K, D_1)} [I(U; L, Y) + H(L|U)] + R_K.
\tag{28}
$$

To state the coding theorem for compound channels we need new notation. For random variables (V, L, U, X) with alphabet $\mathcal{V} \times \mathcal{L} \times \mathcal{U} \times \mathcal{X}$ as above and the channel with input and output alphabets $\mathcal{X}$ and $\mathcal{Y}$ respectively, denote by $Y(W)$ the random variable such that the joint distribution $P_{LVUXY(W)} = P_{LVUX}W$ (consequently, $LVU \leftrightarrow X \leftrightarrow Y$ form a Markov chain). For a compound channel $\mathcal{W}$ with set of stares $\mathcal{S}$ and a state $s \in \mathcal{S}$ we also write $Y(W(\cdot|\cdot, s)) = Y(s)$. With the notation we write

$$I(U; L, Y(\mathcal{W})) = \inf_{s \in \mathcal{S}} I(U; L, Y(s))$$

and

$$I(U; Y(\mathcal{W})|L) = \inf_{s \in \mathcal{S}} I(U; Y(s)|L).$$

Sometimes just for the convenience, we also write $Y(s)$ as $\tilde{Y}(s)$ when we substitute P_{LVUX} by $P_{\tilde{L}\tilde{V}\tilde{U}\tilde{X}}$ and similarly $\tilde{Y}(\mathcal{W})$. Then

$$I(U; L, Y(\mathcal{W})) = I(U; L) + I(U; Y(\mathcal{W})|L). \tag{29}$$

Now for a compound channel we define $\mathcal{Q}_1((V, L), \mathcal{W}, D_1)$ as the set of random variables (V, L, U, X) such that its marginal distribution for the first two components is equal to the distribution P_{VL} and (24) and

$$I(U; V, L) \leq I(U; L, Y(\mathcal{W})) \tag{30}$$

hold. Analogously to set $\mathcal{Q}^*((V, L), W, R_K, D_1)$ we define $\mathcal{Q}_1^*((V, L), \mathcal{W}, R_K, D_1)$ the set of random variables (V, L, U, X) such that its marginal distribution for the first two components is equal to the distribution P_{VL} and (24) and

$$I(U; V, L) \leq I(U; L, Y(\mathcal{W})) + R_K. \tag{31}$$

hold. Then

Theorem 4.3

$$\sup_{(V,L,U,X)\in\mathcal{Q}_1((V,L),\mathcal{W},D_1)} [I(U; L, Y(\mathcal{W})) + H(L|U)] \leq C_{CRI}((V, L), \mathcal{W}, D_1)$$

$$\leq \inf_{W\in\mathcal{W}} \max_{(V,L,U,X,Y)\in\mathcal{Q}((V,L),W,D_1)} [I(U; L, Y) + H(L|U)]. \tag{32}$$

Theorem 4.4

$$\sup_{(V,L,U,X)\in\mathcal{Q}_1^*((V,L),\mathcal{W},R_K,D_1)} [I(U; L, Y(\mathcal{W})) + H(L|U)] + R_K$$

$$\leq C_{CRII}((V, L), \mathcal{W}, R_K, D_1)$$

$$\leq \inf_{W\in\mathcal{W}} \max_{(V,L,U,X,Y)\in\mathcal{Q}^*((V,L),W,R_K,D_1)} [I(U; L, Y) + H(L|U)] + R_K. \tag{33}$$

Notice the gaps of lower and upper bounds in both Theorems 4.3 and 4.4 are due to the orders of inf–sup.

The Results on Watermarking Identification Codes

We shall use the same notation as in the above part. Moreover for above sets $\mathcal{V}, \mathcal{X}$ and $\mathcal{Y}$ and a finite set $\mathcal{U}$ with cardinality bounded by $|\mathcal{V}||\mathcal{X}|$, a memoryless source with generic V, a memoryless cannel W, and compound channel $\mathcal{W}$, we define the following sets. Let $\mathcal{Q}^{**}(V, W, R_K, D_1)$ be the set of random variables (V, U, X, Y) with domain $\mathcal{V} \times \mathcal{U} \times \mathcal{X} \times \mathcal{Y}$ such that for all $v \in \mathcal{V}$, $u \in \mathcal{U}$, $x \in \mathcal{X}$, and $y \in \mathcal{Y}$

$$P_{VUXY}(v, u, x, y) = P_V(v)P_{UX|V}(u, x|v)W(y|x), \tag{34}$$

$$I(U; V) \leq I(U; Y) + R_K, \tag{35}$$

and (24) hold. Let $\mathcal{Q}_1^{**}(V, \mathcal{W}, R_K, D_1)$ be set of random variables (V, U, X) with domain $\mathcal{V} \times \mathcal{U} \times \mathcal{X}$ such that for all $v \in \mathcal{V}, u \in \mathcal{U}$ and $x \in \mathcal{X}$,

$$P_{VUX}(v, u, x) = P_V(v)P_{UX|V}(u.x|v), \tag{36}$$

$$I(U; V) \leq I(U; Y(\mathcal{W})) + R_K, \tag{37}$$

and (24) hold, where $I(U; Y(\mathcal{W})) = \inf_{W \in \mathcal{W}} I(U; Y(W))$. In particular, when the second component L^n of the correlated source $\{(V^n, L^n)\}_{n=1}^\infty$ is a constant, $\mathcal{Q}^*((V, L), W, R_K, D_1)$ and $\mathcal{Q}_1^*((V, L), \mathcal{W}, R_K, D_1)$ become $\mathcal{Q}^{**}(V, W, R_K, D_1)$ and $\mathcal{Q}_1^{**}(V, \mathcal{W}, R_K, D_1)$ respectively.

Theorem 4.5

$$C_{WIDSI}((V, L), W, D_1) \geq \max_{(V,L,U,X,Y) \in \mathcal{Q}((V,L),W,D_1)} [I(U; L, Y) + H(L|U)]. \tag{38}$$

Theorem 4.6

$$C_{WIDK}(V, W, R_K, D_1) \geq \max_{(V,U,X,Y) \in \mathcal{Q}^{**}(V,W,R_K,D_1)} I(U; Y) + R_K. \tag{39}$$

Theorem 4.7

$$C_{WIDSI}((V, L), \mathcal{W}, D_1) \geq \sup_{(V,L,U,X) \in \mathcal{Q}_1((V,L),\mathcal{W},D_1)} [I(U; L, Y(\mathcal{W})) + H(L|U)]. \tag{40}$$

Theorem 4.8

$$C_{WIDK}(V, W, R_K) \geq \sup_{(V,U,X) \in \mathcal{Q}_1^{**}(V,\mathcal{W},R_K,D_1)} I(U; Y(\mathcal{W})) + R_K. \tag{41}$$

Note that in Theorems 4.6 and 4.8 one may add side information L^n, the second component of the correlated source and then one can obtain the corresponding lower bound almost does not change the proofs.

A result on Watermarking Transmission Code with a common Experiment Introduced by Steinberg-Merhav
To construct watermarking identification code Y. Steinberg and N. Merhav in [32] introduced a code, which they call watermarking transmission code with common experiment, distortion measure ρ, and covertext P_V. They obtained there an inner bound to the its capacity region, which is sufficient for achieving their goal. We shall show their bound is tight and therefore actually the capacity region. Their definition and result on it and our proof will be presented it the last section.

5   The Direct Theorems for Common Randomness

In this section we prove the direct parts of Theorems 4.1 – 4.4. Since a DMC can be regarded as a special compound channel with a single member (i.e., $|\mathcal{S}| = 1$), we only have to show the direct parts of Theorems 4.3 and 4.4. To this end we need the following three lemmas for n–type $P_{\tilde{V}\tilde{L}\tilde{U}}$ over the product set $\mathcal{V} \times \mathcal{L} \times \mathcal{U}$ of finite sets $\mathcal{V}, \mathcal{L}$ and $\mathcal{U}$.

Lemma 5.1 (Uniformly covering). *For* $\ell^n \in T_{\tilde{L}}^n$, *let* $U_i(\ell^n)$ $i = 1, 2, \ldots, \lfloor 2^{n\alpha} \rfloor$ *be a sequence of independent random variables with uniform distribution over* $T_{\tilde{U}|\tilde{L}}^n(\ell^n)$ *and for any* $v^n \in T_{\tilde{V}|\tilde{L}}^n(\ell^n)$ *let* $\hat{\mathcal{U}}_{\tilde{U}|\tilde{V}\tilde{L}}^n(v^n\ell^n)$ *be the random set* $\{U_i(\ell^n) : i = 1, 2, \ldots, \lfloor 2^{n\alpha} \rfloor\} \cap T_{\tilde{U}|\tilde{V}\tilde{L}}^n(v^n\ell^n)$. *Then for all* $\varepsilon \in (0, 1]$

$$Pr\left\{\left||\hat{\mathcal{U}}_{\tilde{U}|\tilde{V}\tilde{L}}(v^n\ell^n)| - \lfloor 2^{n\alpha} \rfloor \frac{|T_{\tilde{U}|\tilde{V}\tilde{L}}^n(v^n\ell^n)|}{|T_{\tilde{U}|\tilde{L}}^n(\ell^n)|}\right| \geq \lfloor 2^{n\alpha} \rfloor \frac{|T_{\tilde{U}|\tilde{V}\tilde{L}}^n(v^n\ell^n)|}{|T_{\tilde{U}|\tilde{L}}^n(\ell^n)|}\varepsilon\right\} < 4 \cdot 2^{-\frac{\varepsilon^2}{4}2^{n\eta}}$$

(42)

for sufficiently large n if

$$\lfloor 2^{n\alpha} \rfloor > 2^{n\eta} \frac{|T_{\tilde{U}|\tilde{L}}^n(\ell^n)|}{|T_{\tilde{U}|\tilde{V}\tilde{L}}^n(v^n\ell^n)|}$$

Proof: Let

$$Z_i(v^n, \ell^n) = \begin{cases} 1 & \text{if } U_i(\ell^n) \in T_{\tilde{U}|\tilde{V}\tilde{L}}^n(v^n\ell^n), \\ 0 & \text{else,} \end{cases}$$

(43)

and $q = \frac{|T_{\tilde{U}|\tilde{V}\tilde{L}}^n(v^n\ell^n)|}{|T_{\tilde{U}|\tilde{L}}^n(\ell^n)|}$. Then $|\hat{\mathcal{U}}_{\tilde{U}|\tilde{V}\tilde{L}}^n(v^n\ell^n)| = \sum_{i=1}^{\lfloor 2^{n\alpha} \rfloor} Z_i(v^n\ell^n)$ and for $i = 1, 2, \ldots, \lfloor 2^{n\alpha} \rfloor$

$$Pr\{Z_i(v^n\ell^n) = z\} = \begin{cases} q & \text{if } z = 1 \\ 1 - q & \text{if } z = 0 \end{cases}$$

(44)

by the definitions of $U_i(\ell^n)$ and $Z_i(v^n, \ell^n)$.

Then by Chernov's bound, we have that

$$Pr\left\{\sum_{i=1}^{\lfloor 2^{n\alpha}\rfloor} Z_i(v^n\ell^n) \geq \lfloor 2^{n\alpha}\rfloor q(1+\varepsilon)\right\}$$

$$\leq e^{-\frac{\varepsilon}{2}\lfloor 2^{n\alpha}\rfloor q(1+\varepsilon)}\ E\ e^{\frac{\varepsilon}{2}\sum_{i=1}^{\lfloor 2^{n\alpha}\rfloor} Z_i(v^n,\ell^n)}$$

$$= e^{-\frac{\varepsilon}{2}\lfloor 2^{n\alpha}\rfloor q(1+\varepsilon)}\prod_{i=1}^{\lfloor 2^{n\alpha}\rfloor} E\ e^{\frac{\varepsilon}{2}Z_i(v^n,\ell^n)}$$

$$= e^{-\frac{\varepsilon}{2}\lfloor 2^{n\alpha}\rfloor q(1+\varepsilon)}[1+(e^{\frac{\varepsilon}{2}}-1)q]^{\lfloor 2^{n\alpha}\rfloor}$$

$$\leq e^{-\frac{\varepsilon}{2}\lfloor 2^{n\alpha}\rfloor q(1+\varepsilon)}\left[1+\left(\frac{\varepsilon}{2}+\left(\frac{\varepsilon}{2}\right)^2\right)q\right]^{\lfloor 2^{n\alpha}\rfloor}$$

$$\leq \exp_e\left\{-\frac{\varepsilon}{2}\lfloor 2^{n\alpha}\rfloor q(1+\varepsilon)+\frac{\varepsilon}{2}\lfloor 2^{n\alpha}\rfloor q\left(1+\frac{\varepsilon}{2}\right)\right\}$$

$$= e^{-\frac{\varepsilon^2}{4}\lfloor 2^{n\alpha}\rfloor q} < 2e^{-\frac{\varepsilon^2}{4}2^{n\eta}} \qquad (45)$$

if $\lfloor 2^{n\alpha}\rfloor > 2^{n\eta}q^{-1}$.

Here the first inequality follows from Chernov's bound; the second equality holds by (44); the second inequality holds because $e^{\frac{\varepsilon}{2}} < 1+\frac{\varepsilon}{2}+\left(\frac{\varepsilon}{2}\right)^2$ a by the assumption that $\varepsilon < 1$, $e^{\frac{\varepsilon}{2}} < e^{\frac{1}{2}} < 2$; and the third inequality follows from the well known inequality $1+x < e^x$. Similarly one can obtain

$$Pr\left\{\sum_{i=1}^{\lfloor 2^{n\alpha}\rfloor} Z_i(v^n\ell^n) \leq \lfloor 2^{n\alpha}\rfloor q(1-\varepsilon)\right\} < 2e^{-\frac{\varepsilon^2}{4}2^{n\eta}} \qquad (46)$$

if $\lfloor 2^{n\alpha}\rfloor > 2^{n\eta}q^{-1}$.

Finally we obtain the lemma by combining (45) and (46).

Lemma 5.2 (Packing). *Let $P_{\tilde{L}\tilde{U}}$ be an n-type, let $U_i(\ell^n)$, $i=1,2,\ldots,\lfloor 2^{n\alpha}\rfloor$ be a sequence of independent random variables uniformly distributed on $T^n_{\tilde{U}|\tilde{L}}(\ell^n)$ for an $\ell^n \in T^n_{\tilde{L}}$, and let $\mathcal{Y}$ be a finite set. Then for all n-types $P_{\tilde{L}\tilde{U}\tilde{Y}}$ and $P_{\tilde{L}\tilde{U}\overline{Y}}$ with common marginal distributions $P_{\tilde{L}\tilde{U}}$ and $P_{\overline{Y}} = P_{\tilde{Y}}$, all i, $\gamma > 0$ and sufficiently large n,*

$$Pr\left\{\frac{1}{\lfloor 2^{n\alpha}\rfloor}\sum_{i=1}^{\lfloor 2^{n\alpha}\rfloor}\left|T^n_{\tilde{Y}|\tilde{L}\tilde{U}}(\ell^n U_i(\ell^n))\cap\left[\bigcup_{j\neq i}T^n_{\tilde{Y}|\tilde{L}\tilde{U}}(\ell^n U_j(\ell^n))\right]\right| \geq t_{\tilde{Y}|\tilde{L}\tilde{U}}2^{-\frac{n}{2}\gamma}\right\} < 2^{-\frac{n}{2}\gamma} \qquad (47)$$

if $\lfloor 2^{n\alpha}\rfloor \leq \frac{t_{\tilde{U}|\tilde{L}}}{t_{\tilde{U}|\tilde{L}\overline{Y}}}2^{-n\gamma}$.

Here $t_{\tilde{Y}|\tilde{L}\tilde{U}}$, $t_{\tilde{U}|\tilde{L}}$, and $t_{\tilde{U}|\tilde{L}\overline{Y}}$ are the common values of $|T^n_{\tilde{Y}|\tilde{L}\tilde{U}}(\ell^n u^n)|$ for $(\ell^n, u^n) \in T^n_{\tilde{L}\tilde{U}}$, $|T^n_{\tilde{U}|\tilde{L}}(\ell^n)|$ for $\ell^n \in T^n_{\tilde{L}}$, and $|T^n_{\tilde{U}|\tilde{L}\overline{Y}}(\ell^n y^n)|$ for $(\ell^n, y^n) \in T^n_{\tilde{L}\overline{Y}}$, respectively.

Proof: For $i = 1, 2, \ldots, \lfloor 2^{n\alpha} \rfloor$, $y^n \in T^n_{\overline{Y}} = T^n_{\tilde{Y}}$, let

$$\hat{Z}_i(y^n) = \begin{cases} 1 & \text{if } y^n \in \bigcup_{j \neq i} T^n_{\overline{Y}|\tilde{L}\tilde{U}}\left(\ell^n U_j(\ell^n)\right) \\ 0 & \text{else} \end{cases} \tag{48}$$

and for all $u^n \in T^n_{\tilde{U}|\tilde{L}}(\ell^n)$

$$S_i(u^n) = \left| T^n_{\tilde{Y}|\tilde{L}\tilde{U}}(\ell^n u^n) \cap \left[\bigcup_{j \neq i} T^n_{\overline{Y}|\tilde{L}\tilde{U}}\left(\ell^n U_j(\ell^n)\right) \right] \right|. \tag{49}$$

Then

$$S_i(u^n) = \sum_{y^n \in T_{\tilde{Y}|\tilde{L}\tilde{U}}(\ell^n u^n)} \hat{Z}_j(y^n) \tag{50}$$

and

$$E\,\hat{Z}_i(y^n) = Pr\left\{ y^n \in \bigcup_{j \neq i} T^n_{\overline{Y}|\tilde{L}\tilde{U}}\left(\ell^n U_j(\ell^n)\right) \right\} \leq \sum_{j \neq i} Pr\{ y^n \in T^n_{\overline{Y}|\tilde{L}\tilde{U}}\left(\ell^n U_j(\ell^n)\right) \}$$

$$= \sum_{j \neq i} Pr\{ U_j(\ell^n) \in T^n_{\tilde{U}|\tilde{L}\overline{Y}}(\ell^n y^n) \} = (2^{\lfloor n\alpha \rfloor} - 1) \frac{t_{\tilde{U}|\tilde{L}\overline{Y}}}{t_{\tilde{U}|\tilde{L}}} < 2^{-n\gamma} \tag{51}$$

if $\lfloor 2^{n\alpha} \rfloor \leq \frac{t_{\tilde{U}|\tilde{L}}}{t_{\tilde{U}|\tilde{L}\overline{Y}}} 2^{-n\gamma}$.

Hence by (50) and (51) we have that $E\,S_i(u^n) \leq t_{\tilde{Y}|\tilde{L}\tilde{U}} 2^{-n\gamma}$ and i.e., $E\big[S_i(U_i(\ell^n))|U_i(\ell^n)\big] < t_{\tilde{Y}|\tilde{L}\tilde{U}} 2^{-n\gamma}$ (a.s.), so

$$E\,S_i\big(U_i(\ell^n)\big) = E\{ E\big[S_i(U_i(\ell^n))|U_i(\ell^n)\big] \} < t_{\tilde{Y}|\tilde{L}\tilde{U}} 2^{-n\gamma}. \tag{52}$$

Thus by Markov's inequality we have that

$$Pr\left\{ \frac{1}{\lfloor 2^{n\alpha} \rfloor} \sum_{i=1}^{\lfloor 2^{n\alpha} \rfloor} S_i\big(U_i(\ell^n)\big) \geq t_{\tilde{Y}|\tilde{L}\tilde{U}} 2^{-\frac{n}{2}\gamma} \right\} < 2^{-\frac{n}{2}\gamma},$$

i.e., (47).

Lemma 5.3 (Multi–Packing). *Under the conditions of the previous lemma, let* $U_{i,k}(\ell^n)$, $i = 1, 2, \ldots, \lfloor 2^{n\beta_1} \rfloor$, $k = 1, 2, \ldots, \lfloor 2^{n\beta_2} \rfloor$, *be a sequence of independent random variables uniformly distributed on* $T^n_{\tilde{U}|\tilde{L}}(\ell^n)$ *for a given* $\ell^n \in T^n_{\tilde{L}}$. *Then for all n–types* $P_{\tilde{L}\tilde{U}\tilde{Y}}$ *and* $P_{\tilde{L}\tilde{U}\overline{Y}}$ *in the previous lemma*

$$Pr\left\{ \frac{1}{\lfloor 2^{n\beta_2} \rfloor} \sum_{k=1}^{\lfloor 2^{n\beta_2} \rfloor} \frac{1}{\lfloor 2^{n\beta_1} \rfloor} \sum_{i=1}^{\lfloor 2^{n\beta_1} \rfloor} \left| T^n_{\tilde{Y}|\tilde{L}\tilde{U}}\left(\ell^n U_{i,k}(\ell^n)\right) \cap \left[\bigcup_{j \neq i} T^n_{\overline{Y}|\tilde{L}\tilde{U}}\left(\ell^n U_{j,k}(\ell^n)\right) \right] \right| \geq t_{\tilde{Y}|\tilde{L}\tilde{U}} 2^{-n\eta} \right\}$$
$$< 2^{-\frac{n}{2}\gamma} \tag{53}$$

if $\lfloor 2^{n\alpha} \rfloor \leq \frac{t_{\tilde{U}|\tilde{L}}}{t_{\tilde{U}|\tilde{L}\overline{Y}}} 2^{-n\gamma}$.

Proof: For $u^n \in T^n_{\tilde{U}|\tilde{L}}(\ell^n)$, let

$$S_{i,k}(u^n) = \left| T^n_{\tilde{Y}|\tilde{L}\tilde{U}}(\ell^n u^n) \cap \left[\bigcup_{j \neq i} T^n_{\tilde{Y}|\tilde{L}\tilde{U}}(\ell^n U_{j,k}(\ell^n)) \right] \right|.$$

Then we have shown in the proof to the previous lemma (c.f. (52))

$$E\, S_{i,k}(U_i(\ell^n)) < t_{\tilde{Y}|\tilde{L}\tilde{U}} 2^{-n\gamma}.$$

Thus (53) follows from Markov's inequality.

Now let us turn to the direct part of Theorem 4.3.

Lemma 5.4 (The Direct Part of Theorem 4.3). *For a compound channel* W,

$$C_{CRI}((V,L), W, D_1) \geq \sup_{(V,L,U,X) \in Q_1((V,L), W, D_1)} \left[I(U; L, Y(W)) + H(L|U) \right].$$
$$\tag{54}$$

Proof: We have to show for a given correlated memoryless source with generic (V, L), a compound channel W, $(V, L, U, X) \in Q_1((V, L), W, D_1)$ and sufficiently large n, the existence of the functions, F, G and $x_m(v^n, \ell^n)$ satisfying (10) – (13), (22), (15) and (16) with the rate arbitrarily close to $I(U; L, Y(W)) + H(L|U)$. Obviously the set of achievable rates of the common randomness is bounded and closed (i.e., compact). So without loss of generality, by uniform continuity of information quantities, we can assume that $E\rho(V, X) < D_1$, and $I(U; V, L) < I(U; L, Y(W))$. Because $I(U; V, L) = I(U; L) + I(U; V|L)$ and $I(U; L, Y(W)) = I(U; L) + I(U; Y(W)|L)$, there exists a sufficiently small but positive constant ξ, such that

$$I(U; Y(W)|L) - I(U; V|L) > \xi. \tag{55}$$

Without loss of generality, we also assume P_U is an n–type to simplify the notation. Then for arbitrary $\varepsilon_1 > 0$, by uniform continuity of information quantities, we can find $\delta_1, \delta_2 > 0$ with the following properties.

(a) For all $\ell^n \in T^n_L(\delta_1)$ with type $P_{\ell^n} = P_{\tilde{L}}$, there exists a $\delta' > 0$, such that $(v^n, \ell^n) \in T^n_{VL}(\delta'_2)$ yields that $T^n_{\hat{V}|\tilde{L}}(\ell^n) \subset T^n_{\hat{V}|\tilde{L}}(\ell^n, \delta_2)$, where $P_{\hat{V}\tilde{L}}$ is the joint type of (v^n, ℓ^n) and $P_{\hat{V}\tilde{L}} = P_{\tilde{L}} P_{V|L}$.

We call a pair (v^n, ℓ^n) of sequences with $\ell^n \in T^n_L(\delta_1)$, $(v^n, \ell^n) \in T^n_{VL}(\delta_2)$, (δ_1, δ_2)–typical and denote the set of (δ_1, δ_2)–typical sequences by $T^n(\delta_1, \delta_2)$.

Then we may require $\delta_2 \to 0$ as $\delta_1 \to 0$. Moreover (e.g., see [35]), there exist positive $\zeta_1 = \zeta_1(\delta_1)$, $\zeta_2 = \zeta_2(\delta_1, \delta_2)$, and $\zeta = \zeta(\delta_1, \delta_2)$ such that

$$P^n_L(T^n_L(\delta_1)) > 1 - 2^{-n\zeta_1} \tag{56}$$

$$P^n_{V|L}\{v^n : (v^n, \ell^n) \in T^n(\delta_1, \delta_2)|\ell^n\} > 1 - 2^{-n\zeta_2} \tag{57}$$

for all $\ell^n \in T^n_L(\delta_1)$ and

$$P^n_{VL}(T^n(\delta_1, \delta_2)) > 1 - 2^{n\zeta}. \tag{58}$$

(b) For all $\ell^n \in T_L^n(\delta_1)$ with type $P_{\ell^n} = P_{\tilde{L}}$ (say), one can find a joint type of sequences in $L^n \times \mathcal{U}^n$, say $P_{\tilde{L}\tilde{U}}$, with marginal distributions $P_{\tilde{L}}$ and P_U, sufficiently close to P_{LU}, (which will be specified below). We say that $P_{\tilde{L}\tilde{U}}$ is generated by the type $P_{\tilde{L}}$ of ℓ^n.

(c) For all $(v^n, \ell^n) \in T^n(\delta_1, \delta_2)$ with joint type $P_{v^n\ell^n} = P_{\tilde{V}\tilde{L}}$ (say), one can find a joint type $P_{\tilde{V}\tilde{L}\tilde{U}}$ of sequences in $\mathcal{V}^n \times \mathcal{L}^n \times \mathcal{U}^n$ with marginal distributions $P_{\tilde{V}\tilde{L}}$ and $P_{\tilde{L}\tilde{U}}$ and sufficiently close to P_{VLU} (which will be specified below), where $P_{\tilde{L}\tilde{U}}$ is the type generated by $P_{\tilde{L}}$. We say $P_{\tilde{V}\tilde{L}\tilde{U}}$ is generated by the joint type $P_{\tilde{V}\tilde{L}}$ of (v^n, ℓ^n).

(d) For all (δ_1, δ_2)–typical sequences (v^n, ℓ^n) with joint type $P_{\tilde{V}\tilde{L}}$ (say) and the joint type $P_{\tilde{V}\tilde{L}\tilde{U}}$ generated by $P_{\tilde{V}\tilde{L}}$, we let $(\tilde{V}, \tilde{L}, \tilde{U}, \tilde{X})$ be random variables with joint distribution $P_{\tilde{V}\tilde{L}\tilde{U}\tilde{X}}$ such that for all $v \in \mathcal{V}$, $\ell \in \mathcal{L}$, $u \in \mathcal{U}$ and $x \in \mathcal{X}$

$$P_{\tilde{V}\tilde{L}\tilde{U}\tilde{X}}(v, \ell, u, x) = P_{\tilde{V}\tilde{L}\tilde{U}}(v, \ell, u) P_{X|VLU}(x|v, \ell, u), \qquad (59)$$

and let $(\tilde{V}, \tilde{L}, \tilde{U}, \tilde{X}, \tilde{Y}(W))$ be random variables with joint distribution $P_{\tilde{V}\tilde{L}\tilde{U}\tilde{X}\tilde{Y}(W)}$ such that for all $v \in \mathcal{V}$, $\ell \in \mathcal{L}$, $u \in \mathcal{U}$, $x \in \mathcal{X}$, and $y \in \mathcal{Y}$

$$P_{\tilde{V}\tilde{L}\tilde{U}\tilde{X}\tilde{Y}(W)}(v, \ell, u, x, y) = P_{\tilde{V}\tilde{L}\tilde{U}\tilde{X}}(v, \ell, u, x) W(x|y), \qquad (60)$$

for any $W \in \mathcal{W}$ and $P_{\tilde{V}\tilde{L}\tilde{U}\tilde{X}}$ in (59). Then the following inequalities hold

$$E\rho(\tilde{V}, \tilde{X}) < D_1, \qquad (61)$$

$$|H(\tilde{L}) - H(L)| < \varepsilon_1, \qquad (62)$$

$$|I(\tilde{U}; \tilde{V}|\tilde{L}) - I(U; V|L)| < \varepsilon_1, \qquad (63)$$

and

$$|I(\tilde{U}; \tilde{Y}(W)|\tilde{L}) - I(U; Y(W)|L)| < \varepsilon_1, \qquad (64)$$

where $I(\tilde{U}; \tilde{Y}(W)|\tilde{L}) = \inf_{W \in \mathcal{W}} I(\tilde{U}; \tilde{Y}(W)|\tilde{L})$.

For arbitrarily small fixed ε_2 with $0 < \varepsilon_2 < \frac{1}{2}\xi$, for ξ in (55), we choose ε_1 (and consequently, δ_1, δ_2) so small that $\varepsilon_1 < \frac{1}{2}\varepsilon_2$ and an α such that

$$I(U; Y(W)|L) - \frac{\xi}{2} < \alpha < I(U; Y(W)|L) - \varepsilon_2 \qquad (65)$$

and $M = 2^{n\alpha}$ (say) is an integer. Notice that by (65) we may choose α arbitrarily close to $I(U; Y(W)|L) - \varepsilon_2$ and therefore arbitrarily close to $I(U; Y(W)|L)$ by choosing ε_2 arbitrarily small. Then by (55), (63) and (65) we have that

$$\alpha > I(U; V|L) + \frac{\xi}{2} > I(\tilde{U}; \tilde{V}|\tilde{L}) + \frac{\xi}{2} - \varepsilon_1 > I(\tilde{U}; \tilde{V}|\tilde{L}) + \frac{\xi}{4}, \qquad (66)$$

where the last inequality holds by our choice $\varepsilon_1 < \frac{1}{2}\varepsilon_2 < \frac{1}{4}\xi$, and by (64) and (65) we have

$$\alpha < I(\tilde{U}; \tilde{Y}(W)|\tilde{L}) + \varepsilon_1 - \varepsilon_2 < I(\tilde{U}; \tilde{Y}(W)|\tilde{L}) - \frac{\varepsilon_2}{2}. \qquad (67)$$

Denote by $t_{\tilde{U}|\tilde{L}}$ and $t_{\tilde{U}|\tilde{V}\tilde{L}}$ the common values of $|T^n_{\tilde{U}|\tilde{L}}(\ell^n)|$, $\ell^n \in T^n_{\tilde{L}}$ and $|T^n_{\tilde{U}|\tilde{V}\tilde{L}}(v^n, \ell^n)|$, $(v^n, \ell^n) \in T^n_{\tilde{V}\tilde{L}}$, respectively.

Then it is well known that $\frac{1}{n} \log \frac{t_{\tilde{U}|\tilde{L}}}{t_{\tilde{U}|\tilde{V}\tilde{L}}}$ arbitrarily close to $I(\tilde{U}; \tilde{V}|\tilde{L})$.

This means under our assumption that $\frac{1}{2}\varepsilon_2 < \frac{1}{4}\xi$, (66) implies that for all types $P_{\tilde{V}\tilde{L}\tilde{U}}$ generated by the joint types $P_{\tilde{V}\tilde{L}}$ of (δ_1, δ_2)–typical sequences

$$2^{\frac{n}{3}\varepsilon_2} \frac{t_{\tilde{U}|\tilde{L}}}{t_{\tilde{U}|\tilde{V}\tilde{L}}} < 2^{n\alpha} = M. \tag{68}$$

Next we let $\mathcal{Q}_{\mathcal{W}}(\ell^n u^n, \tau)$ be the set of conditional type $P_{\overline{Y}|\tilde{L}\tilde{U}}$, for a pair (ℓ^n, u^n) of sequences such that there exists a $W \in \mathcal{W}$ with $T^n_{\overline{Y}|\tilde{L}\tilde{U}}(\ell^n u^n) \subset T^n_{\tilde{Y}(W)|\tilde{L}\tilde{U}}(\ell^n u^n, \tau)$, where $P_{\tilde{L}\tilde{U}}$ is the type of (ℓ^n, u^n) and $P_{\tilde{L}\tilde{U}\tilde{Y}(W)}$ is the marginal distribution of the distribution in (60). Then

$$\bigcup_{P_{\overline{Y}|\tilde{L}\tilde{U}} \in \mathcal{Q}_{\mathcal{W}}(\ell^n u^n, \tau)} T^n_{\overline{Y}|\tilde{L}\tilde{U}}(\ell^n u^n, \tau) = \bigcup_{W \in \mathcal{W}} T^n_{\tilde{Y}(W)|\tilde{L}\tilde{U}}(\ell^n u^n, \tau), \tag{69}$$

and

$$|\mathcal{Q}_{\mathcal{W}}(\ell^n u^n, \tau)| < (n+1)^{|\mathcal{L}||\mathcal{U}||\mathcal{Y}|}. \tag{70}$$

Again for the common values $t_{\tilde{U}|\tilde{L}}$ of $|T^n_{\tilde{U}|\tilde{L}}(\ell^n)|$, $\ell^n \in T^n_{\tilde{L}}$, $t_{\tilde{U}|\tilde{L}\overline{Y}}$ of $|T^n_{\tilde{U}|\tilde{L}\overline{Y}}(\ell^n y^n)|$, $(\ell^n, y^n) \in T^n_{\tilde{L}\overline{Y}}$, $\lim_{n \to \infty} \frac{1}{n} \log \frac{t_{\tilde{U}|\tilde{L}}}{t_{\tilde{U}|\tilde{L}\overline{Y}}} = I(\tilde{U}; \overline{Y}|\tilde{L})$.

Thus, (67) yields that for all $P_{\tilde{V}\tilde{L}\tilde{U}}$ generated by the joint type of (δ_1, δ_2)–typical sequences, $(\ell^n, u^n) \in T^n_{\tilde{L}\tilde{U}}$, and $P_{\overline{Y}|\tilde{L}\tilde{U}} \in \mathcal{Q}_{\mathcal{W}}(\ell^n u^n, \tau)$,

$$M = 2^{n\alpha} < 2^{-\frac{n}{4}\varepsilon_2} \frac{t_{\tilde{U}|\tilde{L}}}{t_{\tilde{U}|\tilde{L}\overline{Y}}}, \tag{71}$$

if we choose τ so small (depending on ε_2) that for all $P_{\overline{Y}|\tilde{L}\tilde{U}} \in \mathcal{Q}_{\mathcal{W}}(\ell^n u^n, \tau)$

$$I(\tilde{U}; \overline{Y}|\tilde{L}) > I(\tilde{U}; Y(\mathcal{W})|\tilde{L}) - \frac{1}{8}\varepsilon_2$$

(recalling that by its definition $I(\tilde{U}; \tilde{Y}(\mathcal{W})|\tilde{L}) = \inf_{W \in \mathcal{W}} I(\tilde{U}; \tilde{Y}(W)|\tilde{L})$).

Now we are ready to present our coding scheme at rate α, which may arbitrarily close to $I(U; Y(\mathcal{W})|L)$.

Coding Scheme

1) Choosing Codebooks:

For all $\ell^n \in T^n_L(\delta_1)$ with type $P_{\tilde{L}}$, $P_{\tilde{L}\tilde{U}}$ generated by $P_{\tilde{L}}$ (cf. condition (b) above), we apply Lemma 5.1 with $\eta = \frac{\varepsilon_2}{3}$ and Lemma 5.2 with $\gamma = \frac{\varepsilon_2}{4}$ to random choice. Then since the numbers of sequences v^n, ℓ^n and the number of n–joint types are increasing exponentially and polynomially respectively, for all $\ell^n \in T^n_L(\delta_1)$ with type $P_{\tilde{L}\tilde{U}}$ generated by $P_{\tilde{L}}$, by (68), (71) we can find a subset $\mathcal{U}(\ell^n) \subset T^n_{\tilde{U}|\tilde{L}}(\ell^n)$ with the following property if n is sufficiently large.

If $(v^n, \ell^n) \in T^n(\delta_1, \delta_2)$ and has joint type $P_{\tilde{V}\tilde{L}}$ and $P_{\tilde{V}\tilde{L}\tilde{U}}$ is generated by $P_{\tilde{V}\tilde{L}}$ (cf. condition (c) above), then

$$\left| |\mathcal{U}_{\tilde{U}|\tilde{V}\tilde{L}}(v^n\ell^n)| - M\frac{t_{\tilde{U}|\tilde{V}\tilde{L}}}{t_{\tilde{U}|\tilde{L}}} \right| < M\frac{t_{\tilde{U}|\tilde{V}\tilde{L}}}{t_{\tilde{U}|\tilde{L}}}\varepsilon \tag{72}$$

for any $\varepsilon > 0$ (with $\varepsilon \to 0$ as $n \to \infty$), where

$$\mathcal{U}_{\tilde{U}|\tilde{V}\tilde{L}}(v^n\ell^n) \triangleq \mathcal{U}(\ell^n) \cap T^n_{\tilde{U}|\tilde{V}\tilde{L}}(v^n\ell^n). \tag{73}$$

For any $P_{\tilde{V}\tilde{L}\tilde{U}}$ generated by a joint type of (δ_1, δ_2)–typical sequence, (v^n, ℓ^n), and joint type $P_{\tilde{L}\tilde{U}\tilde{Y}}$ with marginal distribution $P_{\tilde{L}\tilde{U}}$ and any $P_{\tilde{Y}|\tilde{L}\tilde{U}} \in \mathcal{Q}_W(\ell^n u^n_{m'}(\ell^n), \tau)$ (notice that $\mathcal{Q}_W(\ell^n u^n, \tau)$ depends on $(\ell^n u^n)$ only through their joint type $P_{\ell^n u^n}$!)

$$M^{-1} \sum_{m=1}^{M} \left| T^n_{\tilde{Y}|\tilde{L}\tilde{U}}\left(\ell^n \tilde{u}^n_m(\ell^n)\right) \cap \left[\bigcup_{m' \neq m} T^n_{\tilde{Y}|\tilde{L}\tilde{U}}\left(\ell^n \tilde{u}^n_{m'}(\ell^n)\right) \right] \right| < 2^{-\frac{n}{8}\varepsilon_2} t_{\tilde{Y}|\tilde{L}\tilde{U}} \tag{74}$$

if we label the members of $\mathcal{U}(\ell^n)$ as $\tilde{u}^n_1(\ell^n), \tilde{u}^n_2(\ell^n), \dots, \tilde{u}^n_M(\ell^n)$. Consequently by (70) and the fact that $(\ell^n, u^n), (\ell'^n, u'^n)$ have the same type $\mathcal{Q}_W(\ell^n u^n) = \mathcal{Q}_W(\ell'^n u'^n)$,

$$M^{-1} \sum_{m=1}^{M} \left| T^n_{\tilde{Y}|\tilde{L}\tilde{U}}\left(\ell^n, \tilde{u}^n_m(\ell^n)\right) \cap \left[\bigcup_{m' \neq m} \bigcup_{P_{\tilde{Y}|\tilde{L}\tilde{U}} \in \mathcal{Q}_W(\ell^n u^n_{m'}(\ell^n))} T^n_{\tilde{Y}|\tilde{L}\tilde{U}}\left(\ell^n u^n_{m'}(\ell^n)\right) \right] \right|$$
$$< 2^{-\frac{n}{9}\varepsilon_2} t_{\tilde{Y}|\tilde{L}\tilde{U}}. \tag{75}$$

We call the subset $\mathcal{U}(\ell^n)$ the codebook for ℓ^n and its members $\tilde{u}^n_m(\ell^n)$, for $m = 1, 2, \dots, M$ codewords.

2) Choosing Input Sequence to Send through the Channel:
The sender chooses an input sequence $x^n \in \mathcal{X}^n$ according to the output (v^n, ℓ^n) of the correlated source observed by him and his private randomness as follows.

— In the case that outcome of the source is a (δ_1, δ_2)–typical sequence (v^n, ℓ^n) with joint type $P_{\tilde{V}\tilde{L}}$, the sender chooses a codeword in $\mathcal{U}_{\tilde{U}|\tilde{V}\tilde{L}}(v^n, \ell^n)$ in (73) randomly uniformly (by using his private randomness), say

$$\tilde{u}_m(\ell^n) \in \mathcal{U}_{\tilde{U}|\tilde{V}\tilde{L}}(v^n, \ell^n) \subset \mathcal{U}(\ell^n). \tag{76}$$

Then the sender chooses an input sequence $x^n \in \mathcal{X}^n$ with probability

$$P_{X|VLU}\left(x^n|v^n, \ell^n, \tilde{u}^n_m(\ell^n)\right) \tag{77}$$

by using the chosen $\tilde{u}^n_m(\ell^n)$ and his private randomness and sends it through the channel.

— In the other case i.e., a non–(δ_1, δ_2)–typical sequence is output, the sender chooses an arbitrarily fixed sequence, say x_e^n, and sends it through the channel.

— The codewords randomly chosen here and the random input of the channel generated here will be denoted by U'^n and X'^n in the part of analysis below.

3) Choosing the Common domain $\mathcal{A}$ of Functions F and G:
Let

$$J = \lfloor 2^{n(H(L) - 2\varepsilon_1)} \rfloor \tag{78}$$

and let e be an abstract symbol (which stands for that "an error occurs"). Then we define

$$\mathcal{A} = \left\{ \{1, 2, \ldots, M\} \times \{1, 2, \ldots, J\} \right\} \cup \{e\}. \tag{79}$$

4) Defining the Functions F and G:
To define functions F and G we first partition each $T_L^n \subset T_L^n(\delta_1)$ into J subsets with nearly equal size i.e., each subset has cardinality $\left\lfloor \frac{|T_L^n|}{J} \right\rfloor$ or $\left\lceil \frac{|T_L^n|}{J} \right\rceil$. Then we take the union of the jth subsets in the partitions over all $T_L^n \subset T_L^n(\delta_1)$ and obtain a subset $\mathcal{L}_j$ of $T_L^n(\delta_1)$. That is for $j = 1, 2, \ldots, J$

$$|\mathcal{L}_j \cap T_L^n| = \left\lfloor \frac{|T_L^n|}{J} \right\rfloor \text{ or } \left\lceil \frac{|T_L^n|}{J} \right\rceil. \tag{80}$$

4.1) Defining Function F:
The sender observes the output of the source and decides on the value of function F.

— In the case that the source outputs a (δ_1, δ_2)–typical sequence (v^n, ℓ^n), F takes value (m, j) if $\ell^n \in \mathcal{L}_j$, according to sender's private randomness $\tilde{u}_m(\ell^n)$ in (76) is chosen in the step 2) of the coding scheme.

— In the other case $F = e$.

4.2) Defining Function G:
The receiver observes the output ℓ^n of the component L^n (side information) of the correlated source and output of the channel y^n to decide on the value of function G. We use the abbreviation

$$\mathcal{Y}_m(\ell^n) = \bigcup_{P_{\overline{Y}|L\tilde{U}} \in \mathcal{Q}_W(\ell^n \tilde{u}_m^n(\ell^n), \tau)} T_{\overline{Y}|L\tilde{U}}^n \left(\ell^n \tilde{u}_m^n(\ell_n), \tau \right).$$

— In the case that $\ell^n \in T_L^n(\delta_1)$ and that there exists an $m \in \{1, 2, \ldots, M\}$ such that $y^n \in \mathcal{Y}_m(\ell^n) \setminus \left\{ \bigcup_{m' \neq m} \mathcal{Y}_m(\ell^n) \right\}$ G takes value (m, j) if $\ell^n \in \mathcal{L}_j$. Notice that this m must be unique if it exists.

— In the other case $G = e$.

Analysis

1) Distortion Criterion:

First we recall our assumption that the watermarking distortion measure ρ is bounded i.e.

$$0 \leq \rho \leq \Delta. \tag{81}$$

Then by (58)

$$\frac{1}{n} Pr\big((V^n, L^n) \notin T^n_{VL}(\delta_2)\big) E\big[\rho(V'^n, X'^n)|(V^n, L^n \notin T^n_{VL}(\delta_2)\big] < 2^{-n\xi}\Delta. \tag{82}$$

On the other hand, under the condition that

$$(V^n, L^n) \in T^n_{\tilde{V}\tilde{L}} \subset T^n_{VL}(\delta_2),$$

by definition $(V^n, L^n, U'^n) \in T^n_{\tilde{V}\tilde{L}\tilde{U}}$ with probability one for the joint type $P_{\tilde{V}\tilde{L}\tilde{U}}$ generated by $P_{\tilde{V}\tilde{L}}$.

So, by (60), (61) and the definition of (U'^n, X'^n) we have that

$$\frac{1}{n} E\big[\rho(V'^n, X'^n)|(V^n, L^n) \in T^n_{\tilde{L}\tilde{V}}\big]$$
$$= \sum_{(v,\ell,u)\in\mathcal{V}\times\mathcal{L}\times\mathcal{U}} P_{\tilde{V}\tilde{L}\tilde{U}}(v, \ell, u) \sum_x P_{X|VLU}(x|v, \ell, u)\rho(v, x)$$
$$= E\rho(\tilde{V}, \tilde{X}) < D_1. \tag{83}$$

Thus it follows from (82) and (83) that

$$\frac{1}{n} E\rho(V^n, X'^n) = Pr\big((V^n, L^n) \notin T^n_{VL}(\delta_2)\big) E\big[\rho(V^n, X'^n)|(V^n, L^n) \notin T^n_{VL}(\delta_2)\big]$$
$$+ \sum_{T^n_{\tilde{V}\tilde{L}}\subset T^n_{VL}(\delta_2)} Pr\big((V^n, L^n) \in T^n_{\tilde{V}\tilde{L}}\big) E\big[\rho(V^n, X'^n)|(V^n, L^n) \in T^n_{\tilde{V}\tilde{L}}\big]$$
$$< D_1, \tag{84}$$

for sufficiently large n.

2) The Condition of Nearly Uniformity

By the definition of function F in the step 4.1) of the coding scheme,
$Pr\{F = e\} \leq Pr\{(V^n, L^n) \notin T^n_{VL}(\delta_2)\} = 1 - P^n_{VL}(T^n_{VL}(\delta_2)\}$, and hence by (58),

$$|Pr\{F = e\} - |\mathcal{A}|^{-1}| \leq \max\{2^{-n\zeta}, |\mathcal{A}|^{-1}\} \longrightarrow 0 \ (n \to \infty). \tag{85}$$

Next fix an $\ell^n \in T^n_L(\delta_1)$ with type $P_{\tilde{L}}$ (say), let $P_{\tilde{L}\tilde{U}}$ be the joint type generated by $P_{\tilde{L}}$, and let $\mathcal{Q}(\tilde{L}\tilde{U})$ be the set of joint types $P_{\tilde{V}\tilde{L}\tilde{U}}$ with marginal distribution $P_{\tilde{L}\tilde{U}}$ and generated by the joint type of some (δ_1, δ_2)–typical sequence. Then $Pr\{U'^n = u^n|L^n = \ell^n\} > 0$, only if $u^n \in \mathcal{U}(\ell^n) = \{\tilde{u}^n_m(\ell^n) : m = 1, 2, \ldots, M\}$.

Moreover, for a (δ_1, δ_2)–typical sequence (v^n, ℓ^n) with joint type $P_{\tilde{V}\tilde{L}}$, $\tilde{u}_m^n(\ell^n)$ $\in \mathcal{U}(\ell^n)$, by the coding scheme

$$Pr\{V^n = v^n, U'^n = u_m^n(\ell^n)|L = \ell^n\}$$
$$= \begin{cases} P_{V|L}^n(V^n = v^n|\ell^n)|\mathcal{U}_{\tilde{U}|\tilde{V}\tilde{L}}(v^n\ell^n)|^{-1} & \text{if } u_m^n(\ell^n) \in \mathcal{U}_{\tilde{U}|\tilde{V}\tilde{L}}(v^n\ell^n) \\ 0 & \text{else.} \end{cases} \tag{86}$$

Recalling (73), then we have that for all $\ell^n \in T_{\tilde{L}}^n \subset T_L^n(\delta_1)$, $\tilde{u}_m^n(\ell^n) \in \mathcal{U}(\ell^n)$

$$Pr\{U'^n = \tilde{u}_m^n(\ell^n)|L = \ell^n\}$$
$$= \sum_{P_{\tilde{V}\tilde{L}\tilde{U}} \in \mathcal{Q}(\tilde{L}\tilde{U})} \sum_{v^n \in T_{\tilde{V}|\tilde{L}\tilde{U}}^n(\ell^n\tilde{u}_m^n(\ell^n))} P_{V|L}^n(v^n|\ell^n)|\mathcal{U}_{\tilde{U}|\tilde{V}\tilde{L}}(v^n\ell^n)|^{-1}. \tag{87}$$

By (72) we have that

$$[M(1+\varepsilon)]^{-1}\frac{|T_{\tilde{U}|\tilde{L}}^n(\ell^n)|}{|T_{\tilde{U}|\tilde{V}\tilde{L}}^n(v^n\ell^n)|} < |\mathcal{U}_{\tilde{U}|\tilde{V}\tilde{L}}(v^n\ell^n)|^{-1} < [M(1-\varepsilon)]^{-1}\frac{|T_{\tilde{U}|\tilde{L}}^n(\ell^n)|}{|T_{\tilde{U}|\tilde{V}\tilde{L}}^n(v^n\ell^n)|}. \tag{88}$$

On the other hand,

$$\sum_{P_{\tilde{V}\tilde{L}\tilde{U}} \in \mathcal{Q}(\tilde{L}\tilde{U})} \sum_{v^n \in T_{\tilde{V}|\tilde{L}\tilde{U}}^n(\ell^n\tilde{u}_m^n(\ell^n))} P_{V|L}^n(v^n|\ell^n)\frac{|T_{\tilde{U}|\tilde{L}}^n(\ell^n)|}{|T_{\tilde{U}|\tilde{V}\tilde{L}}^n(v^n\ell^n)|}$$
$$= \sum_{P_{\tilde{V}\tilde{L}\tilde{U}} \in \mathcal{Q}(\tilde{L}\tilde{U})} \sum_{v^n \in T_{\tilde{V}|\tilde{L}\tilde{U}}^n(\ell^n\tilde{u}_m^n(\ell^n))} P_{V^n|L}^n(T_{\tilde{V}|\tilde{L}}^n(\ell^n)|\ell^n)\frac{|T_{\tilde{U}|\tilde{L}}^n(\ell^n)|}{|T_{\tilde{V}|\tilde{L}}^n(\ell^n)||T_{\tilde{U}|\tilde{V}\tilde{L}}^n(v^n, \ell^n)|}$$
$$= \sum_{P_{\tilde{V}\tilde{L}\tilde{U}} \in \mathcal{Q}(\tilde{L}\tilde{U})} P_{V|L}^n(T_{\tilde{V}|\tilde{L}}^n(\ell^n)|\ell^n)$$
$$= Pr\{(V^n, \ell^n) \in T^n(\delta_1, \delta_2)|\ell^n\}, \tag{89}$$

where the first equality holds because the value of $P_{V|L}^n(v^n|\ell^n)$ for given ℓ^n depends on v^n through the conditional type; the second equality hold by the fact that $\frac{t_{\tilde{U}|\tilde{L}}}{t_{\tilde{V}|\tilde{L}}t_{\tilde{U}|\tilde{V}\tilde{L}}} = \frac{t_{\tilde{U}|\tilde{L}}}{t_{\tilde{V}\tilde{U}|L}} = \frac{1}{t_{\tilde{V}|\tilde{L}\tilde{U}}}$; and the last equality holds because $P_{\tilde{V}\tilde{L}\tilde{U}}$ is generated by $P_{\tilde{V}\tilde{L}}$ uniquely (see its definition in condition (c)).

Thus by combining (57), (87) – (89), we obtain for an $\eta > 0$ with $\eta \to 0$ as $n \to \infty$, $\varepsilon \to 0$

$$(1 - \eta)M^{-1} < Pr\{U'^n = \tilde{u}_m^n(\ell^n)|L = \ell^n\} < (1 + \eta)M^{-1}, \tag{90}$$

for $\ell^n \in T_{L(\delta_1)}^n$, $\tilde{u}_m^n(\ell^n) \in \mathcal{U}(\ell^n)$.

So for $m \in \{1, 2, \ldots, M\}$, $j \in \{1, 2, \ldots, J\}$,

$$
\begin{aligned}
Pr\{F = (m, j)\} &= Pr\{U'^n = \tilde{u}_m^n(L^n), L^n \in \mathcal{L}_j\} \\
&= \sum_{\ell^n \in \mathcal{L}_j} P_L^n(\ell^n) Pr\{U'^n = \tilde{u}_m^n(\ell^n) | L = \ell^n\} \\
&< (1 + \eta) M^{-1} P_L^n(\mathcal{L}_j). \quad (91)
\end{aligned}
$$

Since $|T_{\tilde{L}}^n| > 2^{n(H(\tilde{L}) + \frac{\varepsilon_1}{2})}$ for sufficiently large n, by (62) and (78), we have that $\frac{|T_{\tilde{L}}^n|}{J} > 2^{\frac{n}{2}\varepsilon_1}$ and hence by (80)

$$
|\mathcal{L}_j \cap T_{\tilde{L}}^n| \le \left\lceil \frac{|T_{\tilde{L}}^n|}{J} \right\rceil < \frac{|T_{\tilde{L}}^n|}{J} + 1 < \frac{|T_{\tilde{L}}^n|}{J} \left(1 + 2^{-\frac{n}{2}\varepsilon_1}\right).
$$

Because the value of $P_L^n(\ell^n)$ depends on ℓ^n through its type, this means that

$$
P_L^n(\mathcal{L}_j \cap T_{\tilde{L}}^n) < J^{-1} P_L^n(T_{\tilde{L}}^n) \left(1 + 2^{-\frac{n}{2}\varepsilon}\right)
$$

and consequently

$$
P_L^n(\mathcal{L}_k) < P_L^n\left(T_L^n(\delta_1)\right) J^{-1} \left(1 + 2^{-\frac{n}{2}\varepsilon_1}\right) \quad (92)
$$

which with (91) is followed by

$$
Pr\{F = (m, j)\} < M^{-1} J^{-1} (1 + \eta) \left(1 + 2^{-\frac{n}{2}\varepsilon_1}\right) P_L^n\left(T_L^n(\delta_1)\right). \quad (93)
$$

Similarly we have that

$$
Pr\{F = (m, j)\} > M^{-1} J^{-1} (1 - \eta) \left(1 - 2^{-\frac{n}{2}\varepsilon_1}\right) P_L^n\left(T_L^n(\delta_1)\right). \quad (94)
$$

Now (56), (93) and (94) together imply that for an $\eta' > 0$ with $\eta' \to 0$ as $n \to \infty$, $\eta \to 0$,

$$
\sum_{(m,j)} |Pr\{F = (m, j)\} - |\mathcal{A}|^{-1}| < \eta', \quad (95)
$$

which with (85) completes the proof of condition of nearly uniformity.

3) The Rate:

In (65) one can choose

$$
\alpha > I\left(U; Y(\mathcal{W}) | L\right) - \varepsilon' \text{ for any } \varepsilon' \text{ with } \varepsilon_2 < \varepsilon' < \frac{1}{2}\xi.
$$

Then by (58), (78), (79), (95), we know that for an $\eta'' > 0$ with $\eta'' \to 0$ as $n \to \infty$, $\eta' \to 0$

$$
\begin{aligned}
\frac{1}{n} H(F) &> \frac{1}{n} \log |\mathcal{A}| - \eta'' > I\left(U; Y(\mathcal{W}) | L\right) - \varepsilon' + H(L) - 2\varepsilon_1 - \eta' \\
&= I\left(U; Y(\mathcal{W}) | L\right) + I(U; L) + H(L|U) - \varepsilon' - 2\varepsilon_1 - \eta' \\
&= I\left(U; L, Y(\mathcal{W})\right) + H(L|U) - \varepsilon' - 2\varepsilon_1 - \eta',
\end{aligned}
$$

for sufficiently large n.

4) Estimation of Probability of Error:

In and only in the following three cases an error occurs.

Case 1

The source outputs a non-(δ_1, δ_2)-typical sequence whose probability is less than $2^{-n\zeta}$ by (58).

Now we assume that a (δ_1, δ_2)-typical sequence (v^n, ℓ^n) with joint type $P_{\tilde{V}\tilde{L}}$ is output. So the sender first chooses a $\tilde{u}_m^n(\ell^n) \in \mathcal{U}_{\tilde{U}|\tilde{V}\tilde{L}}(v^n, \ell^n)$, then an $x^n \in \mathcal{X}^n$ according to his private randomness and sends x^n through the channel. Consequently a $y^n \in \mathcal{Y}^n$ is output by the channel. Then in the following two cases an error occurs.

Case 2

A codeword $\tilde{u}_m(\ell^n) \in \mathcal{U}_{\tilde{U}|\tilde{V}\tilde{L}}(v^n\ell^n) \subset \mathcal{U}^n(\ell^n)$ is chosen and an output sequence

$$y^n \notin \mathcal{Y}_m(\ell^n) = \bigcup_{P_{\overline{Y}|\tilde{L}\tilde{U}} \in \mathcal{Q}_W(\ell^n, \tilde{u}_m(\ell^n))} T_{\overline{Y}|\tilde{L}\tilde{U}}^n\big(\ell^n \tilde{u}_m(\ell^n), \tau\big)$$

is output of the channel. Suppose now $W \in \mathcal{W}$ governs the channel. Then by (59), and (60) the probability that $y^n \in \mathcal{Y}^n$ is output of the channel under the condition that $(V^n, L^n) = (v^n, \ell^n) \in T^n(\delta_1, \delta_2)$ is output of the correlated source and $U'^n = \tilde{u}_m^n(\ell^n) \in \mathcal{U}_{\tilde{U}|\tilde{V}\tilde{L}}(v^n, \ell^n)$ is chosen is

$$Pr\{Y'^n = y^n | (V^n, L^n) = (v^n, \ell^n), U'^n = \tilde{u}_m^n(\ell^n)\}$$
$$= \sum_{x^n \in \mathcal{X}^n} P_{X|VLU}^n(x^n | v^n, \ell^n, \tilde{u}_m^n(\ell^n)) W^n(y^n | x^n)$$
$$= P_{\tilde{Y}(W)|\tilde{V}\tilde{L}\tilde{U}}^n\big(y^n | v^n, \ell^n, \tilde{u}_m(\ell^n)\big). \tag{96}$$

On the other hand

$$T_{\tilde{Y}(W)|\tilde{V}\tilde{L}\tilde{U}}^n\big(v^n \ell^n u_m^n(\ell^n), \tau\big) \subset T_{\tilde{Y}(W)|\tilde{L}\tilde{U}}^n\big(\ell^n u_m^n(\ell^n), \tau\big) \subset \mathcal{Y}_m.$$

So the probability that such an error occurs vanishes exponentially as n grows.

Case 3

A codeword $\tilde{u}_m^n(\ell^n)$ is chosen and a $y^n \in \mathcal{Y}_m \cap \left[\bigcup_{m' \neq m} \mathcal{Y}_{m'}\right]$ is output of the channel.

Now by (86), (88), (90), and simple calculation, we obtain that

$$[(1 - \eta)(1 - \varepsilon)]^{-1} P_{V|L}^n(v^n | \ell^n) \frac{t_{\tilde{U}|\tilde{L}}}{t_{\tilde{U}|\tilde{V}\tilde{L}}} < Pr\{V^n = v^n | L^n = \ell^n, U'^n = \tilde{u}_m^n(\ell^n)\}$$
$$< [(1 + \eta)(1 + \varepsilon)]^{-1} P_{V|L}^n(v^n | \ell^n) \frac{t_{\tilde{U}|\tilde{L}}}{t_{\tilde{U}|\tilde{V}\tilde{L}}} \tag{97}$$

for (δ_1, δ_2)-typical sequences (v^n, ℓ^n) with joint type $P_{\tilde{V}\tilde{L}}$ and $\tilde{u}_m(\ell^n) \in \mathcal{U}_{\tilde{U}|\tilde{V}\tilde{L}}(v^n, \ell^n)$, where $P_{\tilde{V}\tilde{L}\tilde{U}}$ is the type generated by $P_{\tilde{V}\tilde{L}}$.

Moreover, since $t_{\tilde{U}|\tilde{V}\tilde{L}} = \frac{t_{\tilde{V}\tilde{U}|\tilde{L}}}{t_{\tilde{V}|\tilde{L}}}$, $t_{\tilde{U}\tilde{V}|\tilde{L}} = t_{\tilde{U}|\tilde{L}}t_{\tilde{V}|\tilde{L}\tilde{U}}$, and since for given ℓ^n, the value of $P^n_{V|L}(v^n|\ell^n)$ depends on v^n through the conditional type,

$$P^n_{V|L}(v^n|\ell^n)\frac{t_{\tilde{U}|\tilde{L}}}{t_{\tilde{U}|\tilde{V}\tilde{L}}} = P^n_{V|L}(v^n|\ell^n)\frac{t_{\tilde{V}|\tilde{L}}}{t_{\tilde{V}|\tilde{L}\tilde{U}}} = P^n_{V|L}\big(T^n_{\tilde{V}|\tilde{L}}(\ell^n)|\ell^n\big)\frac{1}{t_{\tilde{V}|\tilde{L}\tilde{U}}}. \quad (98)$$

Further it is well known that for all

$$(v^n,\ell^n,u^n)\in T^n_{\tilde{V}\tilde{L}\tilde{U}},\lim_{n\to\infty}\frac{1}{n}\left(\log P^n_{\tilde{V}|\tilde{L}\tilde{U}}(v^n|\ell^n,u^n)-\log\frac{1}{t_{\tilde{V}|\tilde{L}\tilde{U}}}\right)=0.$$

So by (97) and (98), we have that

$$Pr\big\{V=v^n|L^n=\ell^n,U'^m=\tilde{u}^n_m(\ell^n)\big\}$$
$$< 2^{n\theta}P^n_{V|L}\big(T^n_{\tilde{V}|\tilde{L}}(\ell^n)|\ell^n\big)P^n_{\tilde{V}|\tilde{L}\tilde{U}}\big(v^n|\ell^n,\tilde{u}^n_m(\ell^n)\big)$$
$$\leq 2^{n\theta}P^n_{\tilde{V}|\tilde{L}\tilde{U}}\big(v^n|\ell^n,\tilde{u}^n_m(\ell^n)\big) \quad (99)$$

for (δ_1,δ_2)–typical sequences (v^n,ℓ^n) with type $P_{\tilde{V}\tilde{L}}$, $\tilde{u}^n_m\in\mathcal{U}_{\tilde{U}|\tilde{V}\tilde{L}}(\ell^n)\subset\mathcal{U}(\ell^n)$ and sufficiently large n, and a $\theta\to 0$ as $n\to\infty$.

We choose $\theta<\frac{1}{20}\varepsilon_2$.

Since $Pr\big\{(V^n,L^n)=(v^n,\ell^n),U'^m=u^n\big\}>0$ only if (v^n,ℓ^n) is (δ_1,δ_2) typical and $u^n\in\mathcal{U}_{\tilde{U}|\tilde{V}\tilde{L}}(v^n,\ell^n)$, by (96) and (99) we have that

$$Pr\big\{Y'^m=y^n|L^n=\ell^n,U'^m=\tilde{u}^n_m(\ell^n)\big\}$$
$$\leq\sum_{v^n\in\mathcal{V}^n}2^{n\theta}P^n_{\tilde{V}|\tilde{L}\tilde{U}}(v^n|\ell^n,u^n_m(\ell^n))\big\}P^n_{\tilde{Y}(W)|\tilde{V}\tilde{L}\tilde{U}}(y^n|v^n,\ell^n,u^n)$$
$$\leq 2^{n\theta}P^n_{\tilde{Y}(W)|\tilde{L}\tilde{U}}\big(y^n|\ell^n,u^n_m(\ell^n)\big) \quad (100)$$

for $\ell^n\in T^n_L(\delta_1)$, $\tilde{u}_m(\ell^n)\in\mathcal{U}(\ell^n)$ and $y^n\in\mathcal{Y}^n$ if $W\in\mathcal{W}$ governs the channel. Now we obtain an upper bound in terms of a product probability distribution

$$P^n_{\tilde{Y}(W)|\tilde{L}\tilde{U}}\big(y^n|\ell^n,u^n_m(\ell^n)\big)$$

whose value depends on y^n through the conditional type. Consequently by (75) and (100) we have that for all $\ell^n\in T^n_L(\delta_1)$, $\tilde{u}_m(\ell^n)\in\mathcal{U}(\ell^n)$ with joint type $P_{\tilde{L}\tilde{U}}$, $P_{\tilde{Y}|\tilde{L}\tilde{U}}\in\mathcal{Q}_W\big(\ell^n i,\tilde{u}_m(\ell^n),\tau\big)$

$$M^{-1}\sum_{m=1}^{M}Pr\left\{Y'^m\in T^n_{\tilde{Y}|\tilde{L}\tilde{U}}(\ell^n,u^n_m(\ell^n))\cap\left[\bigcup_{m'\neq m}\mathcal{Y}_{m'}(\ell^n)\right]|L^n=\ell^n,U'^m=\tilde{u}^n_m(\ell^n)\right\}$$
$$\leq 2^{n\theta}M^{-1}\sum_{m=1}^{M}P^n_{\tilde{Y}(W)|\tilde{L}\tilde{U}}\left\{T^n_{\tilde{Y}|\tilde{L}\tilde{U}}(\ell^n u^n_m(\ell^n))\cap\left[\bigcup_{m'\neq m}\mathcal{Y}_{m'}(\ell^n)\right]|\ell^n,u^n_m(\ell^n)\right\}$$
$$\leq 2^{n\theta}\cdot 2^{-\frac{n}{9}\varepsilon_2}P^n_{\tilde{Y}(W)|\tilde{L}\tilde{U}}\big\{T^n_{\tilde{Y}|\tilde{L}\tilde{U}}\big(\ell^n,u^n_m(\ell^n)\big)|\ell^n,\tilde{u}^n_m(\ell^n)\big\}$$
$$\leq 2^{-n\left(\frac{1}{9}\varepsilon_2-\theta\right)}<2^{-\frac{n}{20}\varepsilon_2},$$

$$(101)$$

where the last inequality holds by our choice $\theta < \frac{\varepsilon_2}{20}$. Recalling

$$\mathcal{Y}_m(\ell^n) = \bigcup_{P_{\tilde{Y}|\tilde{L}\tilde{U}} \in \mathcal{Q}_{\mathcal{W}}(\ell^n \tilde{u}_m(\ell^n), \tau)} T^n_{\tilde{Y}|\tilde{L}\tilde{U}}\big(\ell^n u^n_m(\ell^n)\big),$$

by the union bound and (101) we obtain that

$$M^{-1} \sum_{m=1}^{M} Pr\left\{ Y'^n \in \mathcal{Y}_m(\ell^n) \cap \left[\bigcup_{m' \neq m} \mathcal{Y}_m(\ell^n) \right] \bigg| L^n = \ell^n, U'^n = \tilde{u}_m(\ell^n) \right\}$$

$$< (n+1)^{|\mathcal{L}||\tilde{U}||\mathcal{Y}|} 2^{-\frac{n}{20}\varepsilon_2}$$

$$< 2^{-\frac{n}{21}\varepsilon_2} \tag{102}$$

for $\ell^n \in T^n_L(\delta_1)$, $\tilde{u}^n_m(\ell^n) \in \mathcal{U}(\ell^n)$ and sufficiently large n. Finally by (90) and (102) we obtain an upper bound to the probability that on error of this type occurs, under the condition $L^n = \ell^n \in T^n_L(\delta_1)$.

$$\sum_{m=1}^{M} Pr\{U'^n = \tilde{u}_m(\ell^n)|L^n = \ell^n\} Pr\left\{ Y'^n \in \mathcal{Y}_m(\ell^n) \cap \left[\bigcup_{m' \neq m} \mathcal{Y}_{m'}(\ell^n) \right] \bigg| L^n = \ell^n, U'^n = \tilde{u}_m(\ell^n) \right\}$$

$$< (1+\eta) \sum_{m=1}^{M} M^{-1} Pr\left\{ Y'^n \in \mathcal{Y}_m(\ell^n) \cap \left[\bigcup_{m' \neq m} \mathcal{Y}_{m'}(\ell^n) \right] \bigg| L^n = \ell^n, U'^n = \tilde{u}_m(\ell^n) \right\}$$

$$< (1+\eta) 2^{-\frac{n}{21}\varepsilon_2}, \tag{103}$$

which completes the proof because by definition
$$\sum_{m=1}^{M} Pr\{U'^n = \tilde{u}_m(\ell^n)|L^n = \ell^n\} = 1 \text{ for all } \ell^n \in T^n_L(\delta_1).$$

Remark: Our model of identification becomes that in [32] if L takes a constant value with probability one. So our proof of the lemma above provides a new proof of Theorem 4 in [32] (as special case) without using the Gelfand–Pinsker Theorem in [24].

Corollary 5.1 (Direct Part Theorem 4.1): *For all single channels W*

$$C_{CRI}\big((V,L), W, D_1\big) \geq \max_{(V,L,U,X,Y) \in \mathcal{Q}((V,L),W,D_1)} \big[I(U;L,Y) + H(L|U)\big].$$

Lemma 5.5 (Direct Part of Theorem 4.4): *For all compound channels $\mathcal{W}$*

$$C_{CRII}\big((V,L), \mathcal{W}, R_K, D_1\big)$$
$$\geq \sup_{(V,L,U,X) \in \mathcal{Q}_1^*((V,L),\mathcal{W},R_K,D_1)} \big[I\big(U;L,Y(\mathcal{W})\big) + H(L|U)\big] + R_K. \tag{104}$$

Proof: By the same reason as in the proof of the previous lemma, it is sufficient for us to show the availability of $I\big(U; L, Y(\mathcal{W})\big) + H(L|U) + R_K$ for (V, L, U, X) with $E\rho(V, X) < D_1$ and for some $\xi > 0$

$$I\big(U; Y(\mathcal{W})|L\big) + R_K - I(U; V|L) > \xi. \tag{105}$$

In the case $I\big(U; Y(\mathcal{W})|L\big) > I(U; V|L)$, by the previous lemma $I\big(U; LY(\mathcal{W})\big)$ $+ H(L|U)$ is achievable even if the noiseless channel is absent. So sender and receiver may generate $n\big(I(U; LY(\mathcal{W})) + H(L|U)\big)$ bits of common randomness and at the same time the sender sends R_K bits of his private randomness via the noiseless channel to the receiver to make additionally nR_K bits of common randomness. That is, the rate $I\big(U; L, Y(\mathcal{W})\big) + H(L|U) + R_K$ is achievable.

So, next we may assume that $I\big(U; Y(\mathcal{W})|L\big) \le I(U; V; |L)$. Moreover we can assume

$$I\big(U; Y(\mathcal{W})|L\big) > 0,$$

because otherwise $I\big(U; L, Y(\mathcal{W})\big) + H(L|U) + R_K = I(U; L) + H(L|U) + R_K = H(L) + R_K$ is achievable as follows. We partition $T_L^n(\delta_1)$ into $\mathcal{L}_j$, $j = 1, 2, \ldots, J$ as in the step 4) of the coding scheme in the proof of the previous lemma to get $n\big(H(L) - 2\varepsilon_1\big)$ bits of common randomness and get other nR_K bits of common randomness by using the noiseless channel. Thus it is sufficient for us to assume that

$$0 < I\big(U; Y(\mathcal{W})|L\big) \le I(U; V|L) < I\big(U; Y(\mathcal{W})|L\big) + R_K - \xi, \tag{106}$$

for a ξ with $0 < \xi < R_K$.

We shall use (δ_1, δ_2)–typical sequences, the joint types $P_{\tilde{L}\tilde{U}}$ and $P_{\tilde{V}\tilde{L}\tilde{U}}$ generated by the types $P_{\tilde{L}}$ and $P_{\tilde{U}\tilde{L}}$ respectively, and the random variables $(\tilde{V}, \tilde{L}, \tilde{U}, \tilde{X})$ and $(\tilde{V}, \tilde{L}, \tilde{U}, \tilde{X}, \tilde{Y}(\mathcal{W}))$ in (59) and (60) satisfying (61) – (64), which are defined in the conditions (a) – (d) in the proof of the previous lemma.

Instead of the choice α in (65) we now choose $\beta_1, \beta_2 > 0$ and $\beta_3 \ge 0$ for arbitrarily small but fixed ε_2 with $0 < \varepsilon_2 < \frac{1}{2}\xi$ such that

$$I\big(U; Y(\mathcal{W})|L\big) - \frac{3}{2}\varepsilon_2 < \beta_1 < I\big(U; Y(\mathcal{W})|L\big) - \varepsilon_2, \tag{107}$$

$$I(U; V|L) - I\big(U; Y(\mathcal{W})|L\big) + \xi \le \beta_2 \le R_K \tag{108}$$

and

$$0 \le \beta_3 = R_K - \beta_2. \tag{109}$$

Notice that the existence and positivity of β_2 are guaranteed by (106).

By adding both sides of the first inequalities in (107) and (108), we obtain that

$$\beta_1 + \beta_2 > I(U; V|L) + \left(\xi - \frac{3}{2}\varepsilon_2\right), \tag{110}$$

and by the first inequality in (107) and the equality in (109) we have that

$$\beta_1 + \beta_2 + \beta_3 > I(U; Y(\mathcal{W})|L) + R_K - \frac{3}{2}\varepsilon_2. \tag{111}$$

Let $\xi - \frac{3}{2}\varepsilon_2 = 2\eta$ and rewrite (110) as

$$\beta_1 + \beta_2 > I(U; V|L) + 2\eta. \tag{112}$$

Then $\eta > \frac{\xi}{8} > 0$ by our choice $\varepsilon_2 < \frac{1}{2}\xi$.

Next as in the proof to the previous lemma we fix an (arbitrary small) positive ε_2, η, choose ε_1 (and consequently δ_1, δ_2) sufficiently small so that $\varepsilon_1 < \min\left(\frac{1}{2}\varepsilon_2, \frac{1}{2}\eta\right)$. Then by (64) and the second inequality in (109) we have that

$$\beta_1 < I(\tilde{U}; \tilde{Y}(\mathcal{W})|\tilde{L}) - \frac{\varepsilon_2}{2}, \tag{113}$$

and by (65) and (112) we have that

$$\beta_1 + \beta_2 > I(\tilde{U}; \tilde{V}|\tilde{L}) + \frac{3}{2}\eta. \tag{114}$$

Without loss of generality we assume that $2^{n\beta_1}$, $2^{n\beta_2}$ and $2^{n\beta_3}$ are integers and denote by $M_1 = 2^{n\beta_1}$, $I = 2^{n\beta_2}$ and $K' = 2^{n\beta_3}$.

Then similarly as in the proof of the previous lemma, we have that for sufficiently large n, sufficiently small τ, all joint types $P_{\tilde{V}\tilde{L}\tilde{U}}$ generated by types of (δ_1, δ_2)–typical sequences and $\mathcal{Q}_{\mathcal{W}}(\ell^n u^n, \tau)$ in the proof of the previous lemma,

$$2^{n\eta} \frac{t_{\tilde{U}|\tilde{L}}}{t_{\tilde{U}|\tilde{V}\tilde{L}}} < M_1 I \tag{115}$$

and

$$M_1 < 2^{-\frac{n}{3}\varepsilon_2} \frac{t_{\tilde{U}|\tilde{L}}}{t_{\tilde{U}|\tilde{L}\tilde{Y}}}, \tag{116}$$

for all $P_{\tilde{Y}|\tilde{L}\tilde{U}} \in \mathcal{Q}_{\mathcal{W}}(\ell^n u^n, \tau)$.

Coding Scheme

1) Choosing the Codebook:
    We choose a codebook for all $\ell^n \in \mathcal{T}_L^n(\delta_1)$ in a similar way as in the step 1) of the coding scheme in the proof of the previous lemma. But we now use Lemma 5.1 for $\alpha = \beta_1 + \beta_2$ and Lemma 5.3 for $\gamma = \frac{\varepsilon_2}{3}$ instead of Lemmas 5.1 and 5.2. Thus by random choice we obtain subsets of $\mathcal{T}_U^n$ $\mathcal{U}^i(\ell^n) =$

$\{\tilde{u}_{m,i}^n(\ell^n) : m = 1, 2, \ldots, M_1\}$ for $i = 1, 2, \ldots, I$ for all $\ell^n \in T_L^n(\delta_1)$ such that for

$$\mathcal{U}^*(\ell^n) = \bigcup_{i=1}^{I} \mathcal{U}^i(\ell^n), \tag{117}$$

and $\mathcal{U}_{\tilde{U}|\tilde{V}\tilde{L}}^*(v^n\ell^n) = \mathcal{U}^*(\ell^n) \cap T_{\tilde{U}|\tilde{V}\tilde{L}}^n(v^n, \ell^n)$, where $P_{\tilde{V}\tilde{L}\tilde{U}}$ is the type generated by the joint type $P_{\tilde{V}\tilde{L}}$ of (δ_1, δ_2)–sequences (v^n, ℓ^n) as before, and with an abuse of notation the union in (117): counting it twice and labelling it as different elements $\tilde{u}_{m,i}^n(\ell^n)$ and $\tilde{u}_{m',i'}^n(\ell^n)$ if a codeword appears twice in it, the following holds.

$$\left| \mathcal{U}_{\tilde{U}|\tilde{V}\tilde{L}}^*(v^n\ell^n) - M_1 I \frac{t_{\tilde{U}|\tilde{V}\tilde{L}}}{t_{\tilde{U}|\tilde{V}\tilde{L}}} \right| < M_1 I \frac{t_{\tilde{U}|\tilde{V}\tilde{L}}}{t_{\tilde{U}|\tilde{V}\tilde{L}}} \varepsilon, \tag{118}$$

and for $\mathcal{Q}_\mathcal{W}(\ell^n v^n, \tau)$ in the proof of the previous lemmas and any conditional type $P_{\tilde{Y}|\tilde{L}\tilde{U}}$,

$$I^{-1} \sum_{i=1}^{I} M_1^{-1} \sum_{m=1}^{M_1} \left| T_{\tilde{Y}|\tilde{L}\tilde{U}}^n(\ell^n u_{m,i}^n(\ell^n)) \cap \left[\bigcup_{m' \neq m} \bigcup_{P_{\tilde{Y}|\tilde{L}\tilde{U}} \in \mathcal{Q}_\mathcal{W}} T_{\tilde{Y}|\tilde{L}\tilde{U}}^n(\ell^n u_{m',i}^n(\ell^n)) \right] \right|$$

$$< 2^{-\frac{n}{7}\varepsilon_2} t_{\tilde{Y}|\tilde{L}\tilde{V}} \tag{119}$$

here (118) and (119) are analogous to (72) and (75) respectively, and are shown in an analogous way.

2) Choosing Inputs of the Channels:
   In the current model, we have an additional noiseless channel with rate R_K except for the noisy channel which exists in the Model I. The sender chooses the inputs of the two channels as follows.
   2.1) Choosing the Input Sequence of the Noisy Channel:
        — In the case that the source outputs a (δ_1, δ_2)–typical sequence (v^n, ℓ^n) with joint type $P_{\tilde{V}\tilde{L}}$, by (118) for the type $P_{\tilde{V}\tilde{L}\tilde{U}}$ generated by $P_{\tilde{V}\tilde{L}}$, $\mathcal{U}_{\tilde{U}|\tilde{V}\tilde{L}}^*(v^n\ell^n) \neq \varnothing$. Then similarly to the Step 2) of the coding scheme in the proof of the previous lemma, the sender randomly and uniformly chooses a member of $\mathcal{U}_{\tilde{U}|\tilde{V}\tilde{L}}^*(v^n\ell^n)$, say $\tilde{u}_{m,i}^n(\ell^n)$, and according to the probability $P_{X|VLU}(x^n|v^n, \ell^n, \tilde{u}_{m,i}^n(\ell^n))$ chooses an input sequence x^n of the channel $\mathcal{W}$ and sends x^n through the channel.
        — In the case that the output of the source is non-(δ_1, δ_2)–typical, the sender sends an arbitrary fixed sequence x_e^n through the channel.
   2.2) Choosing the Input of the Noiseless Channel:
        — In the case that a (δ_1, δ_1)–typical sequence (v^n, ℓ^n) with joint type $P_{\tilde{V}\tilde{L}}$ is output of the correlated channel, the sender first spends $\log I = n\beta_2$ bits to send the index $i \in \{1, 2, \ldots, I\}$ to the receiver via the noiseless channel if a codeword $\tilde{u}_{m,i}^n(\ell^n) \in \mathcal{U}^i(\ell^n) \subset \mathcal{U}^*(\ell^n)$

is chosen in the substep 2.1) in the current coding scheme, then he randomly and uniformly chooses a $k' \in \{1, 2, \ldots, K'\}$ independent of the output of the source and sends it through the noiseless channel by using the rest of $nR_K - n\beta_2 = n\beta_3 = \log K'$ bits.

— In the case that a non–(δ_1, δ_2)–typical sequence is output, the sender sends a constant message through the noiseless channel.

3) Choosing the Common Range $\mathcal{A}$ of Functions F and G:

Let J be as in (78) and

$$\mathcal{A} = \left[\{1, 2, \ldots, M_1\} \times \{1, 2, \ldots, I\} \times \{1, 2, \ldots, K'\} \times \{1, 2, \ldots, J\} \right] \cup \{e\}. \tag{120}$$

4) Defining the Functions F and G:

Partition $T_L^n(\delta_1)$ into $\mathcal{L}_j$, $j = 1, 2, \ldots, J$ as in the step 4) of the coding scheme in the proof of the previous lemma and let $\mathcal{K}_n = \{1, 2, \ldots, I\} \times \{1, 2, \ldots, K'\}$.

4.1) Defining Function F:

The sender decides on the value of function F according to the output of the correlated source and his private randomness as follows.

— In the case that a (δ_1, δ_2)–typical sequence (v^n, ℓ^n) is output, F takes value (m, i, k', j) if $\ell^n \in \mathcal{L}_j$, $\tilde{u}_{m,i}^n(\ell^n) \in \mathcal{U}j(\ell^n) \cap \mathcal{U}_{\tilde{U}|\tilde{V}\tilde{L}}^*(v^n\ell^n)$ is chosen in step 2) of the current coding scheme, and k' is chosen for sending it via the noiseless channel in the last $n\beta_3$ bits (that means (i, k') is sent through the noiseless channel).

— In the other case $F = e$.

4.2) Defining Function G:

The receiver decides on the value of the function G according to the output $(i, k') \in \mathcal{K}_n$ of the noiseless channel, the output ℓ^n of the component L^n of the correlated source, and the output $y^n \in \mathcal{Y}^n$ of the noisy compound channel $\mathcal{W}$ as follows.

Let

$$\mathcal{Y}_{m.i}(\ell^n) = \bigcup_{P_{Y|\tilde{L}\tilde{U}} \in \mathcal{Q}_{\mathcal{W}}(\ell^n \tilde{u}_{m,i}^n(\ell^n), \tau)} T_{\tilde{Y}|\tilde{L}\tilde{U}}^n \left(\ell^n u_m^n, i(\ell^n) \right)$$

for $m = 1, 2, \ldots, M_1$, $i = 1, 2, \ldots, I$, and the type $P_{\tilde{L}\tilde{U}}$ generated by the type $P_{\tilde{L}}$ of $\ell^n \in \mathcal{L}_j \subset T_L^n(\delta_1)$.

— In the case that (i, k') is output of the noiseless channel, $\ell^n \in T_L^n(\delta_1)$ is output of the source, and there exists an $m \in \{1, 2, \ldots, M_1\}$ such that the output of the noisy compound channel $\mathcal{W}$,

$$y^n \in \mathcal{Y}_{m,i}(\ell^n) \setminus \left\{ \bigcup_{m' \neq m} \mathcal{Y}_{m',i}(\ell^n) \right\}, \, G \text{ takes value } (m, i, k', j) \text{ if } \ell^n \in$$

$\mathcal{L}_j$.

— In the other case $G = e$.

Analysis

1) – 3) Distortion Criterion, The Nearly Uniformity Condition, and the Rate.
One can verify the distortion criterion, the nearly uniformity condition and
the rate

$$\frac{1}{n} \log H(F) > \beta_1 + \beta_2 + \beta_3 + o(1) = I(U; Y(\mathcal{W})|L) + R_K + o(1)$$

(c.f. (111)), and obtain analogous inequalities

$$(1 - \eta)(M_1 I)^{-1} < Pr\{U'^n = u_{m,i}^n(\ell^n)|L = \ell^n\} < (1 + \eta)(M_1 I)^{-1} \quad (121)$$

to the inequalities in (90) for $\ell^n \in T_L^n(\delta_1)$, $u_{m,i}^n(\ell^n) \in \mathcal{U}^*(\ell^n)$ and random
variable U'^n chosen by the sender in step 2) of the coding scheme in the
same way as in parts 1) – 3) of the Analysis in the proof of the previous
lemma except that the roles of $\mathcal{U}(\ell^n)$ and (72) there are played by $\mathcal{U}^*(\ell^n) = \bigcup_{i=1}^{I} \mathcal{U}^i(\ell^n)$ and (118). Notice that in those parts of the proof of the previous
lemma (75) is not used, neither is (119) here correspondingly.

4) Estimation of Probability of Error:
By the same reason as in the proof of the previous lemma, the probabilities
of errors of the first two types, the error caused by that a non–(δ_1, δ_2)–
typical sequence is output and the error caused by that $\tilde{u}_{m,i}(\ell^n)$ is chosen
and $y^n \notin \mathcal{Y}_{m,i}(\ell^n)$ is output of the noisy compound channel exponentially
vanish as n grows.

Next by replacing $\mathcal{U}(\ell^n)$ and (75) by $\mathcal{U}^i(\ell^n)$ and (119), in the same way
as in the proof of the previous lemma we now obtain

$$(M_1 I)^{-1} \sum_{i=1}^{I} \sum_{m=1}^{M_1} Pr\left\{ Y'^n \in \mathcal{Y}_{m,i}(\ell^n) \cap \left[\bigcup_{m' \neq m} \mathcal{Y}_{m',i}(\ell^n) \right] |L^n = \ell^n, U'^n = u_{m,i}^n(\ell^n) \right\}$$

$$< 2^{-\frac{n}{21}\varepsilon_2}$$

instead of (102).

$$(122)$$

Finally analogously to in the way to obtain (103) in the proof of the
previous lemma from (90) and (102), we finish the proof by combining (121)
and (122).

Corollary 5.2 (Direct Part of Theorem 4.2): *For all single channels W*

$$C_{CRII}\big((V, L), W, R_K, D_1\big) \geq \max_{(V,L,U,X,Y)\in\mathcal{Q}^*((V,L),W,R_K,D_1)} \big(I(U; L, Y) + H(L|U)\big) + R_K.$$

6   The Converse Theorems for Common Randomness

To obtain single letter characterizations for the converse parts of coding theorems
for common randomness, we need a useful identity which appears in [22] (on
page 314).

Lemma 6.1. *(Csizár-Körner) Let (A^n, B^n) be an arbitrary pair of random sequences and let C be an arbitrary random variable. Then*

$$H(A^n|C) - H(B^n|C)$$
$$= \sum_{t=1}^{n}[H(A_t|A_{t+1}, A_{t+2}, \ldots, A_n, B^{t-1}, C) - H(B_t|A_{t+1}, A_{t+2}, \ldots, A_n, B^{t-1}, C)].$$
$$(123)$$

Proof
Let $(A_{t+1}, A_{t+2}, \ldots, A_n, B^t)$ to be understood as A^n and B^n when $t = 0$ and $t = n$, respectively. Then:

$$H(A^n|C) - (B^n|C)$$
$$= \sum_{t=0}^{n-1} H(A_{t+1}, A_{t+2}, \ldots, A_n, B^t|C) - \sum_{t=1}^{n} H(A_{t+1}, A_{t+2}, \ldots, A_n, B^t|C)$$
$$= \sum_{t=1}^{n} H(A_t, A_{t+1}, \ldots, A_n, B^{t-1}|C) - \sum_{t=1}^{n} H(A_{t+1}, A_{t+2}, \ldots, A_n, B^t|C)$$
$$= \sum_{t=1}^{n}[H(A_t, A_{t+1}, \ldots, A_n, B^{t-1}|C) - H(A_{t+1}, \ldots, A_n, B^{t-1}|C)]$$
$$- \sum_{t=1}^{n}[H(A_{t+1}, A_{t+2}, \ldots, A_n, B^t|C) - H(A_{t+1}, \ldots, A_n, B^{t-1}|C)]$$
$$= \sum_{t=1}^{n}[H(A_t|A_{t+1}, A_{t+2}, \ldots, A_n, B^{t-1}, C) - H(B_t|A_{t+1}, A_{t+2}, \ldots, A_n, B^{t-1}, C)].$$
$$(124)$$

Lemma 6.2 *(The converse part of Theorem 4.1)*
For single channel W,

$$C_{CRI}((V, L), W, D_1) \leq max_{(V,L,U,X,Y) \in \mathcal{Q}((V,L),W,D_1)}[I(U; LY) + H(L|U)].$$
$$(125)$$

Proof: Assume that for a source output of length n there are functions F and K such that for the channel W^n and the distortion measure (10) - (16) hold. Denote by X^n and Y^n the random input and output of the channel generated by the correlated source (V^n, L^n), sender's private randomness M, and the channel.
Then (10) be rewritten in terms of (V^n, X^n) as

$$\frac{1}{n}\mathbf{E}\rho(V^n, X^n) \leq D_1$$
$$(126)$$

Further by Fano inequality, (11) - (14), we have that

$$
\begin{aligned}
&H(F)\\
&\leq H(F) - H(F|G) + n\lambda \log \kappa + h(\lambda)\\
&= I(F; G) + n\lambda \log \kappa + h(\lambda)\\
&\leq I(F; L^n, Y^n) + n\lambda \log \kappa + h(\lambda)\\
&= I(F; Y^n|L^n) + I(F; L^n) + n\lambda \log \kappa + h(\lambda)\\
&\leq I(F; Y^n|L^n) + H(L^n) + n\lambda \log \kappa + h(\lambda)\\
&= I(F; Y^n|L^n) + \sum_{t=1}^{n} H(L_t) + n\lambda \log \kappa + h(\lambda)\\
&= \sum_{t=1}^{n} I(F; Y_t|L^n, Y^{t-1}) + \sum_{t=1}^{n} H(L_t) + n\lambda \log \kappa + h(\lambda),
\end{aligned}
\tag{127}
$$

where $h(z) = -z \log z - (1-z) \log(1-z)$ for $z \in [0,1]$ is the binary entropy. Here the first inequality follows from the Fano inequality, (11), (12) and (14); the second inequality holds by (13); and the third equality holds because the source is memoryless. Since $I(F; V^n, L^n) \leq H(F)$, the first four lines in (127) is followed by

$$
\begin{aligned}
0 &\leq I(F; L^n, Y^n) - I(F; V^n, L^n) + n\lambda \log \kappa + h(\lambda)\\
&\leq [I(F; Y^n|L^n) + I(F; L^n)] - [I(F; V^n|L^n) + I(F; L^n)] + n\lambda \log \kappa + h(\lambda)\\
&= I(F; Y^n|L^n) - I(F; V^n|L^n) + n\lambda \log \kappa + h(\lambda)\\
&= [H(Y^n|L^n) - H(Y^n|L^n, F)] - [H(V^n|L^n) - H(V^n|L^n, F)] + n\lambda \log \kappa + h(\lambda)\\
&= [H(Y^n|L^n) - H(V^n|L^n)] + [H(V^n|L^n, F) - H(Y^n|L^n, F)] + n\lambda \log \kappa + h(\lambda).
\end{aligned}
\tag{128}
$$

To obtain a single letter characterization we substitute A^n, B^n and C in (123) by V^n, Y^n and (L^n, F) respectively and so

$$
\begin{aligned}
&H(V^n|L^n F) - H(Y^n|L^n F)\\
&= \sum_{t=1}^{n} [H(V_t|V_{t+1}, V_{t+2}, \ldots, V_n, L^n, Y^{t-1}, F) - H(Y_t|V_{t+1}, V_{t+2}, \ldots, V_n, L^n, Y^{t-1}, F)].
\end{aligned}
\tag{129}
$$

Moreover because the source is memoryless, we have

$$
H(V^n|L^n) = \sum_{t=1}^{n} H(V_t|L_t).
\tag{130}
$$

We now substitute (128), (129); (130) and $H(Y^n|L^n) = \sum_{t=1}^{n} H(Y_t|L^n, Y^{t-1})$ into (127) and continue it;

$$0 \leq \sum_{t=1}^{n} [H(Y_t|L^n, Y^{t-1}) - H(V_t|L_t)] + \sum_{t=1}^{n} [H(V_t|V_{t+1}, V_{t+2}, \ldots, V_n, L^n, Y^{t-1}, F)$$
$$- H(Y_t|V_{t+1}, V_{t+2}, \ldots, V_n, L^n, Y^{t-1}, F)] + n\lambda \log \kappa + h(\lambda)$$
$$= \sum_{t=1}^{n} [H(Y_t|L^n, Y^{t-1}) - H(Y_t|V_{t+1}, V_{t+2}, \ldots, V_n, L^n, Y^{t-1}, F)]$$
$$- \sum_{t=1}^{n} [H(V_t|L_t) - H(V_t|V_{t+1}, V_{t+2}, \ldots, V_n, L^n, Y^{t-1}, F)] + n\lambda \log \kappa + h(\lambda)$$
$$= \sum_{t=1}^{n} I(Y_t; V_{t+1}, V_{t+2}, \ldots, V_n, F|L^n, Y^{t-1})$$
$$- \sum_{t=1}^{n} I(V_t; V_{t+1}, V_{t+2}, \ldots, V_n, L_1, L_2 \ldots, L_{t-1}, L_{t+1}, \ldots, L_n, Y^{t-1}, F|L_t)$$
$$+ n\lambda \log \kappa + h(\lambda)$$
$$\leq \sum_{t=1}^{n} [I(Y_t; V_{t+1}, V_{t+2}, \ldots, V_n L_1, L_2 \ldots, L_{t-1}, L_{t+1}, \ldots, L_n, Y^{t-1}, F|L_t)]$$
$$- \sum_{t=1}^{n} I(V_t; V_{t+1}, V_{t+2}, \ldots, V_n, L_1, L_2 \ldots, L_{t-1}, L_{t+1}, \ldots, L_n, Y^{t-1}, F|L_t)]$$
$$+ n\lambda \log \kappa + h(\lambda). \tag{131}$$

Let J be the random variable taking values in $\{1, 2, \ldots, n\}$ uniformly, and

$$U_J = (V_{J+1}, V_{J+2}, \ldots, V_n, L_1, L_2 \ldots, L_{J-1}, L_{J+1}, \ldots, L_n, Y^{J-1}, F). \tag{132}$$

Then J and (V_J, L_J) are independent i. e., $I(J; V_J, L_J) = 0$. Thus (131) is rewritten and continued in the following a few lines.

$$0 \leq nI(U_J; Y_J|L_J, J) - nI(U_J; V_J|L_J, J) + n\lambda \log \kappa + h(\lambda)$$
$$= n[I(U_J; L_J, Y_J|J) - I(U_J; L_J|J)] - [I(U_J; V_J, L_J|J) - I(U_J; L_J|J)]$$
$$+ n\lambda \log \kappa + h(\lambda)$$
$$= nI(U_J; L_J, Y_J|J) - nI(U_J; V_J, L_J|J) + n\lambda \log \kappa + h(\lambda)$$
$$\leq nI(U_J, J; L_J, Y_J) - n[I(U_J, J; V_J, L_J) - I(J; V_J, L_J)] + n\lambda \log \kappa + h(\lambda)$$
$$= nI(U_J, J; L_J, Y_J) - nI(U_J, J; V_J, L_J) + n\lambda \log \kappa + h(\lambda). \tag{133}$$

Next we denote by

$$(V'', L'', U'', X'', Y'') = (V_J, L_J, U_J J, X_J, Y_J) \tag{*}$$

for the uniformly distributed J and U_J in (132). Then, obviously (V'', L'') has the same probability distribution with the generic (V, L) of the correlated source, the conditional probability distribution $P_{Y''|X''} = W$, and $(V''L''U'', X'', Y'')$ forms a Markov Chain. Namely, the joint distribution of $(V'', L'', U'', X'', Y'')$ is $P_{V''L''U''X''Y''} = P_{VL}P_{U''X''|V''L''}W$. With the defined random variables, (126) is rewritten as

$$\mathbf{E}\rho(V'', X'') = \mathbf{E}[\mathbf{E}\rho(V'', X'')|J] = \mathbf{E}[\mathbf{E}\rho(V_J, X_J)|J] = \frac{1}{n}\mathbf{E}\rho(V^n, X^n) \le D_1.$$
$$(134)$$

Moreover, by substituting $(*)$ in (133) and then dividing both sides of resulting inequality by n, we obtain that

$$0 \le I(U''; L'', Y'') - I(U''; V'', L'') + o(1), \tag{135}$$

(as $\lambda \to 0$).

Because the set $\{P_{V,L,U,X,Y} : (V, L, U, X, Y) \in \mathcal{Q}((V, L), W, D_1)\}$ is a closed set, by (134) and (135) is is sufficient for us to complete the proof to show that

$$\frac{1}{n}H(F) \le I(U''; L'', Y'') + H(L''|U'') + o(1)$$

for $\lambda \to 0$. This is done by dividing both sides of (127) by n and continuing it by the following few lines.

$$\frac{1}{n}H(F)$$

$$\le \frac{1}{n}\sum_{t=1}^{n} I(F; Y_t|L^n, Y^{t-1}) + \frac{1}{n}\sum_{t=1}^{n} H(L_t) + \lambda \log \kappa + \frac{1}{n}h(\lambda),$$

$$\le \frac{1}{n}\sum_{t=1}^{n} I(V_{t+1}, V_{t+2}, \ldots, V_n, F; Y_t|L^n, Y^{t-1}) + \frac{1}{n}\sum_{t=1}^{n} H(L_t) + \lambda \log \kappa + \frac{1}{n}h(\lambda),$$

$$\le \frac{1}{n}\sum_{t=1}^{n} I(V_{t+1}, V_{t+2}, \ldots, V_n, L_1, L_2 \ldots, L_{t-1}, L_{t+1}, \ldots, L_n, Y^{t-1}, F; Y_t|L_t)$$

$$+ \frac{1}{n}\sum_{t=1}^{n} H(L_t) + \lambda \log \kappa + \frac{1}{n}h(\lambda)$$

$$= I(U_J; Y_J|L_J, J) + H(L_J|J) + \lambda \log \kappa + \frac{1}{n}h(\lambda)$$

$$\le I(U_J, J; Y_J|L_J) + H(L_J|J) + \lambda \log \kappa + \frac{1}{n}h(\lambda)$$

$$= I(U_J, J; Y_J|L_J) + H(L_J) + \lambda \log \kappa + \frac{1}{n}h(\lambda)$$

$$= I(U_J, J; Y_J|L_J) + I(U_J; L_J) + H(L_J|U_J) + \lambda \log \kappa + \frac{1}{n}h(\lambda)$$

$$\le I(U_J, J; Y_J|L_J) + I(U_J, J; L_J) + H(L_J|U_J) + \lambda \log \kappa + \frac{1}{n}h(\lambda)$$

$$= I(U_J, J; L_J, Y_J) + H(L_J|U_J) + \lambda \log \kappa + \frac{1}{n}h(\lambda)$$

$$= I(U''; L'', Y'') + H(L''|U'') + \lambda \log \kappa + \frac{1}{n}h(\lambda), \tag{136}$$

where the second equality holds because U_J is independent of J. Finally the upper bound to the size of $\mathcal{U}$ follows from the Support Lemma in [13] (as well on page 310 in the book [22]).

Lemma 6.3. *(The converse part of Theorem 4.2) For a single channel W,*

$$C_{CRI}((V,L),W,R_K,D_1) \leq \max_{(V,L,U,X,Y)\in\mathcal{Q}^*((V,L),W,R_K,D_1)}[I(U;L,Y)+H(L|U)]+R_K.$$

(137)

Proof: Let $\{(V^n,L^n)\}_{n=1}^{\infty}$ be a correlated source with generic (V,L), W be a noisy channel, and R_K and D_1 be the key rate and the distortion criterion in the Model II of common randomness respectively. Let F and G be functions satisfying (10) - (12), (17), and (14) - (16) in the Model II of common randomness (for output sequence of source of length n). Denote by X^n and K_n inputs of noisy channel W^n and the noiseless channel chosen by the sender according to the output of the correlated source and his/her private randomness. Then (126) holds and similarly to (127) by Fano inequality, we have that

$$
\begin{aligned}
&H(F)\\
&\leq I(F;G)+n\lambda\log\kappa+h(\lambda)\\
&\leq I(F;Y^n,L^n,K_n)+n\lambda\log\kappa+h(\lambda)\\
&= I(F;Y^n,L^n)+I(F;K_n|Y^n,L^n)+n\lambda\log\kappa+h(\lambda)\\
&= I(F;Y^n|L^n)+I(F;L^n)+I(F;K_n|Y^n,L^n)+n\lambda\log\kappa+h(\lambda)\\
&\leq I(F;Y^n|L^n)+H(L^n)+H(K_n|Y^n,L^n)+n\lambda\log\kappa+h(\lambda)\\
&\leq I(F;Y^n|L^n)+H(L^n)+H(K_n)+n\lambda\log\kappa+h(\lambda)\\
&\leq I(F;Y^n|L^n)+H(L^n)+nR_K+n\lambda\log\kappa+h(\lambda)\\
&= \sum_{t=1}^{n}I(F;Y_t|L^n,Y^{t-1})+\sum_{t=1}^{n}H(L_t)+nR_K+n\lambda\log\kappa+h(\lambda),
\end{aligned}
$$

(138)

where the second inequality holds by (17). Analogously to (128) we have

$$
\begin{aligned}
0 &\leq I(F;Y^n,L^n,K_n)-I(F;V^n,L^n)+n\lambda\log\kappa+h(\lambda)\\
&= I(F;Y^n,L^n)-I(F;V^n,L^n)+I(F;K_n|Y^n,L^n)+n\lambda\log\kappa+h(\lambda)\\
&\leq I(F;Y^n,L^n)-I(F;V^n,L^n)+H(K_n|Y^n,L^n)+n\lambda\log\kappa+h(\lambda)\\
&\leq I(F;Y^n,L^n)-I(F;V^n,L^n)+nR_K+n\lambda\log\kappa+h(\lambda).
\end{aligned}
$$

(139)

Note that we only used the basic properties of Shannon information measures, Lemma 6.1, and the assumption that the correlated source is memoryless in the estimation of $I(F;Y^n,L^n)-I(F;V^n,L^n)$ in the part of (128) - (131) and all these are available here. So we have the same estimation here i. e.,

$$I(F; Y^n, L^n) - I(F; V^n L^n)$$

$$\leq \sum_{t=1}^{n} I(Y_t; V_{t+1}, V_{t+2}, \ldots, V_n, L_1, L_2 \ldots, L_{t-1}, L_{t+1}, \ldots, L_n, Y^{t-1}, F | L_t)$$

$$- \sum_{t=1}^{n} I(V_t; V_{t+1}, V_{t+2}, \ldots, V_n, L_1, L_2 \ldots, L_{t-1}, L_{t+1}, \ldots, L_n, Y^{t-1}, F | L_t)$$

$$+ n\lambda \log \kappa + h(\lambda). \tag{140}$$

Let U_J and J be defined as in (132). Then (140) is rewritten as

$$I(F; Y^n, L^n) - I(F; V^n L^n) \leq nI(U_J, J; L_J, Y_J) - nI(U_J, J; V_J, L_J) + n\lambda \log \kappa + h(\lambda). \tag{141}$$

Let $(V'', L'', U'', X'', Y'')$ is defined as in the previous lemma.
Then (134) and $P_{V''L''U''X''Y''} = P_{VL} P_{U''X''|V''L''} W$ are certainly fulfilled. But now (139) - (141) lead us to

$$0 \leq I(U''; L'', Y'') - I(U''; V'', L'') + R_K + o(1). \tag{142}$$

In the same way as (136) we can show

$$\sum_{t=1}^{n} I(F; Y_t | L^n, Y^{t-1}) + \sum_{t=1}^{n} H(L_t) + nR_K + n\lambda \log \kappa + h(\lambda)$$

$$\leq nI(U''; L'', Y'') + nH(U'' | L'') + n\lambda \log \kappa + h(\lambda) \tag{143}$$

which with (138) yields

$$\frac{1}{n} H(F) \leq I(U''; L''Y'') + H(U'' | L'') + R_K + \lambda \log \kappa + \frac{1}{n} h(\lambda).$$

Again $|\mathcal{U}|$ is bounded by the Support Lemma. Thus our proof is finished.

Finally it immediately follows from Lemmas 6.2 and 6.3 that

Corollary 6.4. *For compound channel* $\mathcal{W}$,

1) (The converse part of Theorem 4.3:)

$$C_{CRI}((V, L), \mathcal{W}, D_1) \leq \inf_{W \in \mathcal{W}} \max_{(V,L,U,X,Y) \in \mathcal{Q}((V,L),\mathcal{W},D_1)} [I(U; L, Y) + H(L|U)] \tag{144}$$

and

2) (The converse part of Theorem 4.4:)

$$C_{CRII}((V, L), \mathcal{W}, R_K, D_1)$$

$$\leq \inf_{W \in \mathcal{W}} \max_{(V,L,U,X,Y) \in \mathcal{Q}^*((V,L),\mathcal{W},R_K,D_1)} [I(U; L, Y) + H(L|U)] + R_K. \tag{145}$$

7 Constructing Watermarking Identification Codes from Common Randomness

R. Ahlswede and G. Dueck found in [12] that a identification code with the same rate can be always obtained from the common randomness between a sender and and receiver under the condition

(*) *The sender can send a massage with arbitrarily small but positive rate (in the exponential sense).*

Thus under the condition (*) the capacity of identification is not smaller than that of common randomness. Note that the sets $\mathcal{Q}((V,L),W,D_1)$, $\mathcal{Q}^{**}(V,W,R_k,D_1)$, $\mathcal{Q}_1((V,L),W,D_1)$, and $\mathcal{Q}_1^{**}(V,W,R_k,D_1)$ are not empty implies the condition (*) in the Theorems 4.5, 4.6, 4.7, and 4.8 respectively. Consequently Theorems 4.5, 4.6, 4.7, and 4.8 follows from Theorems 4.1, 4.2, 4.3, and 4.4 respectively.

8 A Converse Theorem of a Watermarking Coding Theorem Due to Steinberg-Merhav

In order to construct identification codes in [32], Y. Steinberg and N. Merhav introduced the following code to build common randomness between sender and receiver and obtained an inner bound of the capacity region. This inner bound is sufficient for their goal. We shall show that it is as well tight. This would support their conjecture that the lower bound in their Theorem 4 ([32]) is tight although it does not imply it.

Let $\{V^n\}_{n=1}^{\infty}$ be a memoryless source with alphabet $\mathcal{V}$ and generic V and W be a noisy channel with input and output alphabets $\mathcal{X}$ and $\mathcal{Y}$ respectively. A pair of functions (f,g) is called an (n,M,J,δ,λ,D) watermarking transmission code with a common experiment, distortion measure ρ, distortion level D and covertext P_V if the followings are true.

— f is a function from $\mathcal{V}^n \times \{1,2,\ldots,M\}$ to $\{1,2,\ldots,J\} \times \mathcal{X}^n$.
— g is a function from $\mathcal{Y}^n$ to $\{1,2,\ldots,J\} \times \{1,2,\ldots,M\}$.

$$\frac{1}{M}\sum_{m=1}^{M}\sum_{v^n\in\mathcal{V}^n} P_V^n(v^n)W^n(\{y:g(y^n)=(f_J(v^n,m),m)\}|f_X(v^n,m)) \geq 1-\lambda,$$
(146)

where f_X and f_J are projections of f to $\mathcal{X}^n$ and $\{1,2,\ldots,J\}$ respectively.

$$\frac{1}{M}\sum_{m=1}^{M}\sum_{v^n\in\mathcal{V}^n} P_V^n(v^n)\rho(v^n,f_X(v^n,m)) \leq D.$$
(147)

For $m=1,2,\ldots,M$, there exists a subset $\mathcal{B}^{(m)} \subset \{1,2,\ldots,J\}$ of cardinality $|\mathcal{B}^{(m)}| \geq J2^{-n\delta}$ such that

$$J^{-1}2^{-n\delta} \leq P_V^n\{f_J(V^n,m)=j\} \leq J^{-1}2^{n\delta}$$
(148)

for all j and

$$\sum_{j \in \mathcal{B}(m)} P_V^n \{f_J(V^n, m) = j\} \geq 1 - \lambda. \tag{149}$$

g serves as a decoding function here. (148) and (149) play the same role as nearly uniform condition in construction of identification codes from common randomness. In fact one can find the nearly uniform condition (16) is stronger but for the purpose to construct identification codes the conditions (148) and (149) are strong enough.

A pair (R_1, R_2) is called achievable with distortion D if for all positive reals δ, λ, and ϵ there is an $(n, M, J, \delta, \lambda, D)$ code defined as above such that

$$\frac{1}{n} \log M > R_1 - \epsilon \tag{150}$$

and

$$\frac{1}{n} \log J > R_2 - \epsilon. \tag{151}$$

The set of achievable pair of rates is called capacity region and denoted by $\mathcal{R}$. Denote by $\mathcal{R}^{(*)}$ the subset of pair of real numbers such that there exist random variables (V, U, X, Y) taking values in $\mathcal{V} \times \mathcal{U} \times \mathcal{X} \times \mathcal{Y}$ such that $|\mathcal{U}| \leq |\mathcal{Y}| + |\mathcal{X}|$, for all

$v \in \mathcal{V}, u \in \mathcal{U}, x \in \mathcal{X}$ and $y \in \mathcal{Y}$,

$$P_{VUXY}(v, u, x, y) = P_V(v) P_{UX|V}(u, x|v) W(y|x),$$

$$\mathbf{E}\rho(V, X) \leq D,$$

$$0 \leq R_1 \leq I(U; Y) - I(U; V), \tag{152}$$

and

$$0 \leq R_2 \leq I(U; V). \tag{153}$$

It was shown in [32]

Theorem 8.1 *(Steinberg-Merhav)*

$$\mathcal{R}^* \subset \mathcal{R}. \tag{154}$$

We now show the opposite contained relation holds i. e.,

Theorem 8.2

$$\mathcal{R} \subset \mathcal{R}^*. \tag{155}$$

Proof: Let (f, g) be a pair of functions satisfying (146) - (151) for sufficiently large n (which is specified later) and Z_n be a random variable with uniform distribution over $\{1, 2, \dots, M\}$. Further let $f(V^n, Z_n) = (B_n, X^n)$, where B_n and X^n have ranges $\{1, 2, \dots, J\}$ and $\mathcal{X}^n$ respectively and Y^n be the random output of the channel W^n when X^n is input.

Then (148) and (149) are rewritten as

$$J^{-1} 2^{-n\delta} \leq P_{B_n|Z_n}(j|m) \leq J^{-1} 2^{n\delta} \tag{156}$$

for all $j \in \mathcal{B}^{(m)}$ and

$$P_{B_n|Z_n}(B_n \in \mathcal{B}^{(m)}|m) \geq 1 - \lambda \tag{157}$$

respectively. So,

$$
\begin{aligned}
H(B_n|Z_n) &= \sum_{m=1}^{M} P_{Z_n}(m) H(B_n|Z_n = m) \\
&\geq -\sum_{m=1}^{M} P_{Z_n}(m) \sum_{j \in \mathcal{B}^{(m)}} P_{B_n|Z_n}(j|m) \log P_{B_n|Z_n}(j|m) \\
&\geq -\sum_{m=1}^{M} P_{Z_n}(m) \sum_{j \in \mathcal{B}^{(m)}} P_{B_n|Z_n}(j|m) \log J^{-1} 2^{n\delta} \\
&= (\log J - n\delta) \sum_{m=1}^{M} P_{Z_n}(m) P_{B_n|Z_n}(B_n \in \mathcal{B}^{(m)}|m) \\
&\geq (\log J - n\delta)(1 - \lambda) \tag{158}
\end{aligned}
$$

where the second inequality holds by (156) and the last inequality follows from (157). Or equivalently

$$\frac{1}{n} \log J \leq \frac{\frac{1}{n} H(B_n|Z_n)}{1 - \lambda} + \delta. \tag{159}$$

Since $H(B_n) \leq \log J$, (159) implies that for a function θ such that $\theta(\delta, \lambda) \to 0$ as $\delta, \lambda \to 0$,

$$\frac{1}{n} \log J - \theta(\delta, \lambda) < \frac{1}{n} H(B_n|Z_n) \leq \frac{1}{n} H(B_n) \leq \frac{1}{n} \log J. \tag{160}$$

which says that B_n and Z_n are "nearly independent". Moreover because Z_n is independent of V^n, by Fano's inequality,

$$
\begin{aligned}
R_1 - \varepsilon < \frac{1}{n} \log M &= \frac{1}{n} H(Z_n) \\
&= \frac{1}{n} H(Z_n|V^n) \\
&\leq \frac{1}{n} H(B_n, Z_n|V^n) \\
&\leq \frac{1}{n} [H(B_n, Z_n|V^n) - H(B_n, Z_n|Y^n)] + \lambda \log JM + \frac{1}{n} h(\lambda) \\
&= \frac{1}{n} [I(B_n, Z_n; Y^n) - I(B_n, Z_n; V^n)] + \lambda \frac{1}{n} \log JM + \frac{1}{n} h(\lambda) \tag{161}
\end{aligned}
$$

where the second inequality follows from Fano's inequality. Since B_n is a function of V^n and Z_n, we have also

$$H(B_n, Z_n|V^n) \leq H(V^n, Z_n|V^n) = H(Z_n), \tag{162}$$

which and (160) are followed by

$$
\begin{aligned}
R_2 - \varepsilon < \frac{1}{n}\log J &< \frac{1}{n}H(B_n|Z_n) + \theta(\delta,\lambda) \\
&= \frac{1}{n}[H(B_n, Z_n) - H(Z_n)] + \theta(\delta,\lambda) \\
&\le \frac{1}{n}[H(B_n, Z_n) - H(B_n, Z_n|V^n)] + \theta(\delta,\lambda) \\
&= \frac{1}{n}I(B_n, Z_n; V^n) + \theta(\delta,\lambda).
\end{aligned}
\tag{163}
$$

So far we have had a non-single-letter characterization of the capacity region (161) and (163). In the rest part of the proof we shall reduce it to a single letter one.

First we substitute A^n, B^n, and C in (123) by V^n, Y^n, and (B_n, Z_n) respectively and obtain that

$$
\begin{aligned}
&H(V^n|B_n, Z_n) - H(Y^n|B_n, Z_n) \\
&= \sum_{t=1}^{n}[H(V_t|V_{t+1}, V_{t+2},\ldots, V_n, Y^{t-1}, B_n, Z_n) - H(Y_t|V_{t+1}, V_{t+2},\ldots, V_n, Y^{t-1}, B_n, Z_n)].
\end{aligned}
\tag{164}
$$

Next we note that $H(V^n) = \sum_{t=1}^{n} H(V_t)$ because the source is memoryless and $H(Y^n) = \sum_{t=1} H(Y_t|Y^{t-1})$. Therefore, we have

$$
\begin{aligned}
&I(B_n, Z_n; Y^n) - I(B_n, Z_n; V^n) \\
&= H(Y^n) - H(V^n) + [H(V^n|B_n, Z_n) - H(Y^n|B_n, Z_n)] \\
&= \sum_{t=1}^{n} H(Y_t|Y^{t-1}) - \sum_{t=1}^{n} H(V_t) + \sum_{t=1}^{n}[H(V_t|V_{t+1}, V_{t+2},\ldots, V_n, Y^{t-1}, B_n, Z_n) \\
&\quad - H(Y_t|V_{t+1}, V_{t+2},\ldots, V_n, Y^{t-1}, B_n, Z_n)] \\
&= \sum_{t=1}^{n}[H(Y_t|Y^{t-1}) - H(Y_t|V_{t+1}, V_{t+2},\ldots, V_n, Y^{t-1}, B_n, Z_n)] \\
&\quad - \sum_{t=1}^{n}[H(V_t) - H(V_t|V_{t+1}, V_{t+2},\ldots, V_n, Y^{t-1}, B_n, Z_n)] \\
&= \sum_{t=1}^{n} I(V_{t+1}, V_{t+2},\ldots, V_n, B_n, Z_n; Y_t|Y^{t-1}) \\
&\quad - \sum_{t=1}^{n} I(V_{t+1}, V_{t+2},\ldots, V_n, Y^{t-1}, B_n, Z_n; V_t) \\
&\le \sum_{t=1}^{n} I(V_{t+1}, V_{t+2},\ldots, V_n, Y^{t-1}, B_n, Z_n; Y_t) \\
&\quad - \sum_{t=1}^{n} I(V_{t+1}, V_{t+2},\ldots, V_n, Y^{t-1}, B_n, Z_n; V_t).
\end{aligned}
\tag{165}
$$

Moreover,

$$I(B_n, Z_n; V^n)$$

$$= \sum_{t=1}^{n} I(B_n, Z_n; V_t | V_{t+1}, V_{t+2}, \ldots, V_n)$$

$$\leq \sum_{t=1}^{n} I(V_{t+1}, V_{t+2}, \ldots, V_n, B_n, Z_n; V_t)$$

$$\leq \sum_{t=1}^{n} I(V_{t+1}, V_{t+2}, \ldots, V_n, Y^{t-1}, B_n, Z_n; V_t). \tag{166}$$

So we may let I be a random variable taking values in $\{1, 2, \ldots, n\}$ uniformly and

$$U' = (V_{I+1}, V_{I+2}, \ldots, V_n, Y^{I-1}, B_n, Z_n)$$

and conclude by (163), (164), (165), (166)

$$R_1 - \varepsilon \leq I(U'; Y_I | I) - I(U'; V_I | I) + \lambda \log JM + \frac{1}{n} h(\lambda)$$

$$\leq I(U', I; Y_J) - I(U', I; V_I) + I(I; V_I) + \lambda \log JM + \frac{1}{n} h(\lambda), \tag{167}$$

and

$$R_2 - \varepsilon \leq I(U'; V_I | I) \leq I(U', I; V_I) + \theta(\delta, \lambda). \tag{168}$$

Let $U = (U', I), V' = V_I, X = X_I$ and $Y = Y_I$. Then $P_{V'} = P_V$, $(V'U, X, Y)$ forms a Markov chain and (168) can be re-written as

$$R_2 \leq I(U; V') + \theta(\delta, \lambda),$$

and

$$EP(v', x') < D.$$

Further that $I(I; V_I) = 0$ (as the source is stationary) and (167) are followed by

$$R_1 \leq I(U; Y) - I(U; V') + \lambda \log JM + \frac{1}{n} h(\lambda) + \varepsilon.$$

Finally $|\mathcal{U}|$ is bounded by the support Lemma in the standard way.

References

1. R. Ahlswede, Channels with arbitrarily varying channel probability functions in the presence of noiseless feedback, Z. Wahrsch. verw. Gebiete 25, 239-252, 1973.
2. R. Ahlswede, Elimination of correlation in random codes for arbitrarily varying channels, Z. Wahrsch. verw. Gebiete 44, No. 2, 159-175, 1978.
3. R. Ahlswede, Coloring hypergraphs: a new approach to multi-user source coding I, J. Combin. Inform. System Sci. 4, No. 1, 76-115, 1979.

4. R. Ahlswede, Coloring hypergraphs: a new approach to multi-user source coding II, J. Combin. Inform. System Sci. 5, No. 3, 220-268, 1980.
5. R. Ahlswede, General theory of information transfer, Preprint 97–118, SFB 343 "Diskrete Strukturen in der Mathematik", Universität Bielefeld, 1997; General theory of information transfer:updated, General Theory of Information Transfer and Combinatorics, a Special Issue of Discrete Applied Mathematics, to appear.
6. R. Ahlswede and V. B. Balakirsky, Identification by means of a random process, (Russian), Problemy Peredachi Informatsii 32, No. 1, 144-160, 1996, translation in Problems Inform. Transmission 32, No. 1, 123-138, 1996.
7. R. Ahlswede and N. Cai, Information and control: matching channels, IEEE Trans. Inform. Theory, Vol. 44, No. 2, 542–563, 1998.
8. R. Ahlswede and N. Cai, The AVC with noiseless feedback and maximal error probability: a capacity formula with a trichotomy, Numbers, Information and Complexity, special volume in honour of R. Ahlswede on occasion of his 60th birthday, edited by Ingo Althöfer, Ning Cai, Gunter Dueck, Levon Khachatrian, Mark S. Pinsker, Andras Sárközy, Ingo Wegener and Zhen Zhang, Kluwer Academic Publishers, Boston, Dordrecht, London, 151-176, 2000.
9. R. Ahlswede and I. Csiszár, Common randomness in information theory and cryptography I, Secret sharing, IEEE Trans. Inform. Theory, Vol. 39, No. 4, 1121-1132, 1993.
10. R. Ahlswede and I. Csiszár, Common randomness in information theory and cryptograpy II, CR capacity, IEEE Trans. Inform. Theory, Vol. 44, No. 1, 225-240, 1998.
11. R. Ahlswede and G. Dueck, Identification via channels, IEEE Trans. Inform. Theory, Vol. 35, No. 1, 15-29, 1989.
12. R. Ahlswede and G. Dueck, Identification in the presence of feedback - a discovery of new capacity formulas, IEEE Trans. Inform. Theory, Vol. 35, No. 1, 30-36, 1989.
13. R. Ahlswede and J. Körner, Source coding with side information and a converse for degraded broadcast channels, IEEE Trans. Information Theory, Vol. 21, No. 6, 629-637, 1975.
14. C. Kleinewächter, On identification , this volume.
15. R. Ahlswede and Z. Zhang, New directions in the theory of identification via channels, IEEE Trans. Inform. Theory, Vol. 41, No. 4, 1040-1050, 1995.
16. M. Barni, F. Bartolini, A. De Rosa, and A. Piva, Capactiy of the watermark channel: how many bits can be hidden within a digital image?, Proc. SPIE, SPIE, Vol. 3657, 437-448, San Jose, CA, Jan, 1999.
17. M. Burnashav, On identification capacity of infinite alphabets or continuous time channel, IEEE Trans. Inf. Theory, Vol. 46, 2407–2414, 2000.
18. N. Cai and L. Y. Lam, On identification secret sharing schemes, Information and Computation, to appear.
19. T. M. Cover and J. A. Thomas, Elements of Information Theory, Wiley, 1991.
20. I. J. Cox, M. L. Miller, and A. Mckellips, Watermarking as communications with side information, Proc. of IEEE, Vol. 87, No. 7, 1127-1141, 1999.
21. I. Csiszár, Almost independence and secrecy capacity, Probl. Inform. Trans., Vol. 32, 40-47, 1996.
22. I. Csiszár and J. Körner, Information Theory: Coding Theorems for Discrete Memoryless Systems, Academic, 1981.
23. I. Csiszár and P. Narayan, Common randomness and secret key generation with a helper, IEEE Trans. Inform. Theory, Vol. 46, no 2, 344-366, 2000.
24. S.I. Gelfand and M.S. Pinsker, Coding for channels with random parameters, Problems of Control and Inform. Theory, Vol. 9, 19–31, 1980.

25. T. S. Han and S. Verdú, New results in the theory of identification via channels, IEEE Trans. Inform. Theory, Vol. 38, 14-25, 1992.
26. U. M. Maurer, Secret key argreenent by public discussion from common information, IEEE Trans. Inform. Theory, Vol. 39, 733-742, 1993.
27. N. Merhav, On random coding error exponents of watermarking codes, IEEE Trans. Inform. Theory, Vol. 46, No. 2, 420–430, 2000.
28. P. Moulin and J. A. O'Sulliovan, Information-theoretic analysis of information hiding, IEEE Trans. Inform. Theory, Vol. 49, No. 3, 563–593, 2003.
29. J. A. O'Sullivan, P. Moulin, and J. M. Ettinger, Information theoretic analysis of steganography, Proc. ISIT '98, 297, 1998.
30. S. D. Servetto, C. I. Podilchuk, and K. Ramchandran, Capacity issues in digital image watermarking, Proc. ICIP '98, 1998.
31. Y. Steinberg, New converses in the theory of identification via channels, IEEE Trans. Inform. Theory, Vol. 44, 984-998, 1998.
32. Y. Steinberg and N. Merhav, Indentification in the presence of side infromation with application to watermarking, IEEE Trans. Inform. Theory, Vol. 47, 1410–1422, 2001.
33. S. Venkatesan and V. Anantharam, The common randomness capacity of a pair of independent dicrete memoryless channels, IEEE Trans. Inform. Theory, Vol. 44, No. 1, 215-224, 1998.
34. S. Venkatesan and V. Anantharam, The common randomness capacity of network of discret mamoryless channels, IEEE Trans. Inform. Theory, Vol. 46, No. 2, 367-387, 2000.
35. R.W. Yeung, A First Course in Information Theory, Kluwer Academic, 2002.

Notes on Conditions for Successive Refinement of Information

A.N. Harutyunyan

Abstract. The successive refinement of information (or source divisibility) under error exponent or reliability constraint is discussed. By
rehabilitating Maroutian's result on multiple descriptions [13], we refine
our previous proof on successive refinability conditions reported in [7]
and restate the result by Tuncel and Rose [17]. In particular, it is noted that the successive refinement in "purely" reliability sense is always
possible.

Keywords: Successive refinement of information (source divisibility),
reliability (error exponent), rate-reliability-distortion function, hierarchical (scalable) source coding.

1 The Notes Subject

The concept of source *divisibility* was introduced by Koshelev [10]-[12] as a criterion of efficiency for source coding in hierarchical systems. The same concept
was independently defined by Equitz and Cover in [5] and named the *successive
refinement of information*. The authors became an essential mathematical input
from the results of [1], based on a wringing technique. The idea is to achieve the
rate-distortion limit at each level of successively more precise transmission. In
terms of those limits the problem statement has the following description. Assume that we transmit information to two users, the requirement of the first on
distortion is no larger than Δ_1 and demand of the second user is more accurate:
$\Delta_2 \leq \Delta_1$. It is well known from the rate-distortion theory that the value $R_P(\Delta_1)$
of the rate-distortion function for a source distributed according to a probabil-
ity law P is the minimal satisfactory transmission rate at the first destination.
Adding an information of a rate R' addressed to the second user the fidelity can
be made more precise providing no larger distortion than Δ_2. It is interesting to
know when it is possible to guarantee the equality $R_P(\Delta_1) + R' = R_P(\Delta_2)$. The
answer to this question is given in Koshelev's papers [10]-[12], and in [5] by Equitz
and Cover. Koshelev argued that the Markovity condition for the random variables (RV) characterizing the system is sufficient to achieve the rate-distortion
limit. Later on the same condition also as the necessity was established in [5],
where the authors exploited Ahlswede's result [1] on multiple descriptions without excess rate. Another proof of that result by means of characterization of the
rate-distortion region for the described hierarchical source coding situation and
an interpretation of the Markovity condition are given in Rimoldi's paper [15].

In [7], treating the above notion of successive refinement of information we
introduced an additional criterion to the quality of information reconstruction

R. Ahlswede et al. (Eds.): Information Transfer and Combinatorics, LNCS 4123, pp. 154–164, 2006.

– the *reliability*. The extension of the rate-distortion case to the rate-reliability-distortion one is the assumption that the messages of the source must be coded for transmission to receiver with a distortion not exceeding Δ_1 within error probability exponent (reliability) E_1, and then, using an auxiliary information, restored with a more precise distortion $\Delta_2 \le \Delta_1$ within error probability exponent E_2 (naturally $\ge E_1$). Let $R_P(E, \Delta)$ denotes the rate-reliability-distortion function introduced by Haroutunian and Mekoush [6] (it would be appropriate to refer also the paper [8] for details and a further analysis) which is actually the inverse of Marton's error exponent function [14]. And let R' be the rate of the additional encoding. Under the successive refinement of information (or divisibility of source) with reliability requirement, from (E_1, Δ_1) to (E_2, Δ_2), the condition $R_P(E_2, \Delta_2) = R_P(E_1, \Delta_1) + R'$ is meant.

The characterization of the rate-distortion region for this hierarchical transmission (also called the scalable source coding) has been independently obtained by Koshelev [10], Maroutian [13] (as a corollary from error exponents investigation for the same system), and Rimoldi [15]. Later on, the error exponents in the scalable source coding was studied in the paper [9] by Kanlis and Narayan.

In [17] Tuncel and Rose pointed out the difference of their result and the necessary and sufficient conditions for successive refinability under error exponent criterion previously derived by us and reported in [7]. In this paper, accepting the failure of our conditions which originates from [13], (where the author has attempted to find the attainable rate-reliability-distortion region for the hierarchical source coding), we restate the surprising result by Tuncel and Rose [17], amending Maroutian's result [13] and specializing it for both the cases $E_2 \ge E_1$ and vice versa.

2 The Communication System

Let $P^* = \{P^*(x), x \in \mathcal{X}\}$ be the probability distribution (PD) of messages x of the discrete memoryless source (DMS) X of finite alphabet $\mathcal{X}$. And let the reproduction alphabets of two receivers be the finite sets $\mathcal{X}^1$ and $\mathcal{X}^2$ accordingly, with the corresponding single-letter distortion measures

$$d_k : \mathcal{X} \times \mathcal{X}^k \to [0; \infty), \ k = 1, 2.$$

The distortions $d_k(\mathbf{x}, \mathbf{x^k})$ between a source N-length message $\mathbf{x}$ and its reproduced versions $\mathbf{x}^k$ are considered as averages of the per-letter distortions:

$$d_k(\mathbf{x}, \mathbf{x^k}) \triangleq \frac{1}{N} \sum_{n=1}^{N} d(x_n, x_n^k), \ k = 1, 2.$$

A code $(f, F) = (f_1, f_2, F_1, F_2)$ for the system (Fig. 1) consists of two encoders (as mappings of the source N-length messages space $\mathcal{X}^N$ into certain numerated finite sets $\{1, 2, ..., L_k(N)\}$):

$$f_k : \mathcal{X}^N \to \{1, 2, ..., L_k(N)\}, \ k = 1, 2,$$

and two decoders acting as converse mappings into the reproduction
N-dimensional spaces $\mathcal{X}^{1N}$ and $\mathcal{X}^{2N}$ in the following ways:

$$F_1 : \{1, 2, ..., L_1(N)\} \to \mathcal{X}^{1N},$$

$$F_2 : \{1, 2, ..., L_1(N)\} \times \{1, 2, ..., L_2(N)\} \to \mathcal{X}^{2N},$$

where in F_2 we deal with the Cartesian product of the two sets.

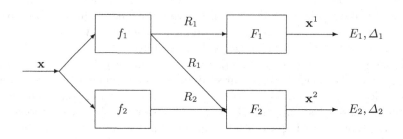

Fig. 1. The hierarchical communication system

Let the requirement of the first user on the averaged distortion be $\Delta_1 \geq 0$
and of the second one be $\Delta_2 \geq 0$. The sets definitions

$$\mathcal{A}_1 \stackrel{\triangle}{=} \{\mathbf{x} \in \mathcal{X}^N : F_1(f_1(\mathbf{x})) = \mathbf{x}^1, d_1(\mathbf{x}, \mathbf{x}^1) \leq \Delta_1\},$$

$$\mathcal{A}_2 \stackrel{\triangle}{=} \{\mathbf{x} \in \mathcal{X}^N : F_2(f_1(\mathbf{x}), f_2(\mathbf{x})) = \mathbf{x}^2, d_2(\mathbf{x}, \mathbf{x}^2) \leq \Delta_2\}.$$

will abbreviate the expressions for determination the probability of error (caused
by an applied code (f, F)) at the output of each decoder:

$$e_k(f, F, \Delta_k, N) \stackrel{\triangle}{=} 1 - P^*(\mathcal{A}_k), \; k = 1, 2,$$

where $P^*(\mathcal{A}_k)$ is the probability of the set $\mathcal{A}_k$.

We say that the nonnegative numbers (R_1, R_2) make an $(E_1, E_2, \Delta_1, \Delta_2)$-
achievable pair of coding rates if for every $\epsilon > 0$, $\delta > 0$, and sufficiently large N
there exists a code (f, F), such that

$$\frac{1}{N} \log L_k(N) \leq R_k + \epsilon,$$

$$e_k(f, F, \Delta_k, N) \leq \exp\{-N(E_k - \delta)\}, k = 1, 2.$$

Just these E_1 and E_2 are called reliabilities. Denote the set of $(E_1, E_2, \Delta_1, \Delta_2)$-
achievable rates for the system by $\mathcal{R}_{P^*}(E_1, E_2, \Delta_1, \Delta_2)$.

Let

$$P \stackrel{\triangle}{=} \{P(x), x \in \mathcal{X}\}$$

be a PD on $\mathcal{X}$ and

$$Q \triangleq \{Q(x^1, x^2 | x), \; x \in \mathcal{X}, \; x^k \in \mathcal{X}^k, \; k = 1, 2\}$$

be a conditional PD on $\mathcal{X}^1 \times \mathcal{X}^2$. In the sequel we shall use the divergence $D(P \parallel P^*)$ between PDs P and P^*

$$D(P \parallel P^*) \triangleq \sum_x P(x) \log \frac{P(x)}{P^*(x)},$$

and the related sets

$$\alpha(E_k, P^*) \triangleq \{P : D(P \parallel P^*) \leq E_k\}, \; k = 1, 2.$$

Let

$$I_{P,Q}(X \wedge X^1, X^2) \triangleq \sum_{x, x^1, x^2} P(x) Q(x^1, x^2 \mid x) \log \frac{Q(x^1, x^2 \mid x)}{\sum_x P(x) Q(x^1, x^2 \mid x)}$$

be the Shannon mutual information between RVs X and X^1, X^2 defined by PDs P and Q.

The separate informations $I_{P,Q}(X \wedge X^k)$, $k = 1, 2$, between the RVs X and X^k are defined similarly using the marginal distributions $Q(x^k \mid x)$:

$$\sum_{x^j, j \neq k} Q(x^j, x^k \mid x) = Q(x^k \mid x), \; j, k = 1, 2.$$

For a given quartet $(E_1, E_2, \Delta_1, \Delta_2)$ and a PD $P \in \alpha(E_1, P^*)$ let $\mathcal{Q}(P, E_1, E_2, \Delta_1, \Delta_2)$ be the set of those conditional PDs $Q_P(x_1, x_2 | x)$ that for the expectations on the distortions hold

$$\mathbf{E}_{P,Q_P} d_k(X, X^k) \triangleq \sum_{x, x^k} P(x) Q_P(x^k \mid x) d(x, x^k) \leq \Delta_k, \; k = 1, 2,$$

and only

$$\mathbf{E}_{P,Q_P} d_2(X, X^2) \leq \Delta_2$$

if P belongs to the subtraction of the sets $\alpha(E_2, P^*)$ and $\alpha(E_1, P^*)$, i.e.

$$P \in \alpha(E_2, P^*) \backslash \alpha(E_1, P^*),$$

and these all in case of $E_2 \geq E_1$.

And, respectively, in case of $E_1 \geq E_2$, under the notation $\mathcal{Q}(P, E_1, E_2, \Delta_1, \Delta_2)$ we mean the set of those conditional PDs Q_P for which

$$\mathbf{E}_{P,Q_P} d_k(X, X^k) \leq \Delta_k, \; k = 1, 2,$$

when $P \in \alpha(E_2, P^*)$, and only

$$\mathbf{E}_{P,Q_P} d_1(X, X^1) \leq \Delta_1$$

if $P \in \alpha(E_1, P^*) \backslash \alpha(E_2, P^*)$.

For $E_1 = E_2 = E$ we use the notation $\mathcal{Q}(P, E, \Delta_1, \Delta_2)$ instead of $\mathcal{Q}(P, E_1, E_2, \Delta_1, \Delta_2)$, but when $E \to 0$, $\alpha(E, P^*)$ consists only the PD P^*, hence the notation $\mathcal{Q}(P^*, \Delta_1, \Delta_2)$ will replace $\mathcal{Q}(P, E, \Delta_1, \Delta_2)$ one.

For DMS with the generic PD P^* denote by $R_{P^*}(E, \Delta)$ the rate-reliability-distortion function subject to the reliability $E > 0$ and distortion $\Delta \geq 0$, and by $R_{P^*}(\Delta)$ the rate-distortion one [16], see also the classical books [2]-[4]. Each of the pairs (E_1, Δ_1) and (E_2, Δ_2) in the considered hierarchical coding configuration of Fig. 1 determines the corresponding rate-reliability-distortion and the rate-distortion functions $R_{P^*}(E_k, \Delta_k)$ and $R_{P^*}(\Delta_k)$, $k = 1, 2$, respectively. It is known [6] that

$$R_{P^*}(E_k, \Delta_k) = \max_{P \in \alpha(E_k, P^*)} \min_{Q : \mathbf{E}_{P,Q} d_k(X, X^k) \leq \Delta_k} I_{P,Q}(X \wedge X^k), \qquad (1)$$

and, as a consequence when $E_k \to 0$,

$$R_{P^*}(\Delta_k) = \min_{Q : \mathbf{E}_{P,Q} d_k(X, X^k) \leq \Delta_k} I_{P,Q}(X \wedge X^k), \qquad (2)$$

therefore, equivalently for (1) in terms of (2)

$$R_{P^*}(E_k, \Delta_k) = \max_{P \in \alpha(E_k, P^*)} R_P(\Delta_k), \ k = 1, 2. \qquad (3)$$

Here it is relevant to point out also another consequence of (1). Let $R_{P^*}(E)$ denotes the special case of the rate-reliability-distortion function $R_{P^*}(E, \Delta)$ for source P^*, when $\Delta = 0$ and call it the *rate-reliability* function in source coding. Then the minimum asymptotic rate sufficient for the source lossless (zero-distortion) transmission under the reliability requirement, namely $R_{P^*}(E)$, immediately obtains from (1):

$$R_{P^*}(E) = \max_{P \in \alpha(E, P^*)} H_P(X), \qquad (4)$$

where $H_P(X)$ is the entropy of RV X distributed according to PD P:

$$H_P(X) \overset{\triangle}{=} - \sum_x P(x) \log P(x).$$

3   Achievable Rates Region

An attempt to find the entire $(E_1, E_2, \Delta_1, \Delta_2)$-achievable rates region

$$\mathcal{R}_{P^*}(E_1, E_2, \Delta_1, \Delta_2)$$

is made in Maroutian's paper [13]. However, our revision of his result stimulated by [17] brought us to a different region, although, in the particular case of $E_1, E_2 \to 0$, as a consequence, he has obtained the correct answer for the hierarchical source coding problem under fidelity criterion [11], known also due to

Rimoldi [15]. A careful treatment of the code construction strategy based on the type covering technique (by the way, used also by Tuncel and Rose in [17]) and the combinatorial method for the converse employed in [13], allow us to come to the results in Theorem 1 and Theorem 2, which together unify the refined result for the multiple descriptions [13] (or, in other terminology, for the hierarchical or scalable source coding).

Theorem 1. *For every* E_1, E_2 *with* $E_1 \geq E_2 > 0$, *and* $\Delta_1 \geq 0$, $\Delta_2 \geq 0$,

$$\mathcal{R}_{P^*}(E_1, E_2, \Delta_1, \Delta_2) = \bigcap_{P \in \alpha(E_2, P^*)} \bigcup_{Q_P \in \mathcal{Q}(P, E_1, E_2, \Delta_1, \Delta_2)} \{(R_1, R_2):$$

$$R_1 \geq \max \left(\max_{P \in \alpha(E_1, P^*) \setminus \alpha(E_2, P^*)} R_P(\Delta_1), I_{P, Q_P}(X \wedge X^1) \right), \tag{5}$$

$$R_1 + R_2 \geq I_{P, Q_P}(X \wedge X^1, X^2)\}.$$

or, equivalently,

$$\mathcal{R}_{P^*}(E_1, E_2, \Delta_1, \Delta_2) = \bigcap_{P \in \alpha(E_2, P^*)} \bigcup_{Q_P \in \mathcal{Q}(P, E_1, E_2, \Delta_1, \Delta_2)} \{(R_1, R_2):$$

$$R_1 \geq \max \left(R_{P^*}(E_1, \Delta_1), I_{P, Q_P}(X \wedge X^1) \right), \tag{6}$$

$$R_1 + R_2 \geq I_{P, Q_P}(X \wedge X^1, X^2)\}.$$

Note that the equivalence of (5) and (6) is due to the fact that $R_P(\Delta_1) \leq I_{P, Q_P}(X \wedge X^1)$ for each $P \in \alpha(E_2, P^*)$ and the representation (3).

Theorem 2. *In case of* $E_2 \geq E_1$,

$$\mathcal{R}_{P^*}(E_1, E_2, \Delta_1, \Delta_2) = \bigcap_{P \in \alpha(E_1, P^*)} \bigcup_{Q_P \in \mathcal{Q}(P, E_1, E_2, \Delta_1, \Delta_2)} \{(R_1, R_2):$$

$$R_1 \geq I_{P, Q_P}(X \wedge X^1), \tag{7}$$

$$R_1 + R_2 \geq \max \left(\max_{P \in \alpha(E_2, P^*) \setminus \alpha(E_1, P^*)} R_P(\Delta_2), I_{P, Q_P}(X \wedge X^1, X^2) \right)\}$$

or, equivalently,

$$\mathcal{R}_{P^*}(E_1, E_2, \Delta_1, \Delta_2) = \bigcap_{P \in \alpha(E_1, P^*)} \bigcup_{Q_P \in \mathcal{Q}(P, E_1, E_2, \Delta_1, \Delta_2)} \{(R_1, R_2):$$

$$R_1 \geq I_{P, Q_P}(X \wedge X^1), \tag{8}$$

$$R_1 + R_2 \geq \max \left(R_{P^*}(E_2, \Delta_2), I_{P, Q_P}(X \wedge X^1, X^2) \right)\}.$$

In this case, the equivalence of (7) and (8) is due to $R_P(\Delta_2) \leq I_{P,Q_P}(X \wedge X^1 X^2)$ for each $P \in \alpha(E_1, P^*)$ and the representation (3).

Now comparing these regions with the ones (specialized for the corresponding cases) derived in [17] one can conclude that the regions by Tuncel and Rose formulated in terms of the scalable rate-reliability-distortion function actually are alternative forms of $\mathcal{R}_{P^*}(E_1, E_2, \Delta_1, \Delta_2)$ characterized in Theorems 1 and 2.

Instead of proofs for the above theorems, we note that those statements are obtained as an outcome of a scrutiny of [13].

In case of equal requirements of the receivers on the reliability, i.e. $E_1 = E_2 = E$, we get from (5) and (7) a simpler region, denoted here by $\mathcal{R}_{P^*}(E, \Delta_1, \Delta_2)$.

Theorem 3. *For every $E > 0$ and $\Delta_1 \geq 0$, $\Delta_2 \geq 0$,*

$$\mathcal{R}_{P^*}(E, \Delta_1, \Delta_2) = \bigcap_{P \in \alpha(E,P^*)} \bigcup_{Q_P \in \mathcal{Q}(P,E,\Delta_1,\Delta_2)} \{(R_1, R_2):$$

$$R_1 \geq I_{P,Q_P}(X \wedge X^1), \tag{9}$$

$$R_1 + R_2 \geq I_{P,Q_P}(X \wedge X^1, X^2)\}.$$

Furthermore, with $E \to 0$ the definition of the set $\alpha(E, P^*)$ and (9) yield Koshelev's [10] result for the hierarchical source coding rate-distortion region, which was independently appeared then also in [13] and [15].

Theorem 4. *For every $\Delta_1 \geq 0$, $\Delta_2 \geq 0$, the rate-distortion region for the scalable source coding can be expressed as follows*

$$\mathcal{R}_{P^*}(\Delta_1, \Delta_2) = \bigcup_{Q_{P^*} \in \mathcal{Q}(P^*,\Delta_1,\Delta_2)} \{(R_1, R_2):$$

$$R_1 \geq I_{P^*,Q_{P^*}}(X \wedge X^1), \tag{10}$$

$$R_1 + R_2 \geq I_{P^*,Q_{P^*}}(X \wedge X^1 X^2)\}.$$

4   Successive Refinement with Reliability: The Conditions

As in [7], we define the notion of the successive refinability in terms of the rate-reliability-distortion function in the following way.

Definition. The DMS X with PD P^* is said to be successively refinable from (E_1, Δ_1) to (E_2, Δ_2) if the optimal rates pair

$$(R_{P^*}(E_1, \Delta_1), R_{P^*}(E_2, \Delta_2) - R_{P^*}(E_1, \Delta_1)), \tag{11}$$

is $(E_1, E_2, \Delta_1, \Delta_2)$-achievable, provided that $R_{P^*}(E_2, \Delta_2) \geq R_{P^*}(E_1, \Delta_1)$.

It is obvious that with $E \to 0$ we have the definition of the successive refinement in distortion sense [11], [5]. Another interesting special case is $\Delta_1 = \Delta_2 = 0$, then we deal with the successive refinement in "purely" reliability sense, namely the achievability of the optimal rates

$$(R_{P^*}(E_1), R_{P^*}(E_2) - R_{P^*}(E_1))$$

related to the corresponding rate-reliability functions for $E_2 \geq E_1$ (since only this condition ensures the inequality $R_{P^*}(E_2) \geq R_{P^*}(E_1)$).

Below we prove our refined conditions concerning the successive refinement of information in respect to the above definition, which are coincident with those obtained in [17]. The different conditions for two cases and their proofs employ the recharacterized results in Theorem 1 and 2 on multiple descriptions [13].

$E_1 \geq E_2$ **case:** For this situation, from (6) it follows that the rates pair (11) is achievable iff for each $P \in \alpha(E_2, P^*)$ there exists a $Q_P \in \mathcal{Q}(P, E_1, E_2, \Delta_1, \Delta_2)$ such that the inequalities

$$R_{P^*}(E_1, \Delta_1) \geq \max \left(R_{P^*}(E_1, \Delta_1), I_{P,Q_P}(X \wedge X^1) \right), \tag{12}$$

$$R_{P^*}(E_2, \Delta_2) \geq I_{P,Q_P}(X \wedge X^1, X^2) \tag{13}$$

hold simultaneously. These inequalities are satisfied for each $P \in \alpha(E_2, P^*)$ iff

$$R_{P^*}(E_1, \Delta_1) \geq I_{P,Q_P}(X \wedge X^1), \tag{14}$$

which is due to (12), and, meanwhile

$$R_{P^*}(E_2, \Delta_2) \geq I_{P,Q_P}(X \wedge X^1, X^2) \geq I_{P,Q_P}(X \wedge X^2) \geq R_{P^*}(\Delta_2) \tag{15}$$

for (13).

By the definition of the rate-reliability-distortion function (1) it follows that (14) and (15) hold for each $P \in \alpha(E_2, P^*)$ iff there exist a PD $\bar{P} \in \alpha(E_2, P^*)$ and a conditional PD $Q_{\bar{P}} \in \mathcal{Q}(\bar{P}, E_1, E_2, \Delta_1, \Delta_2)$, such that $X \to X^2 \to X^1$ forms a Markov chain in that order and at the same time

$$R_{P^*}(E_1, \Delta_1) \geq I_{\bar{P},Q_{\bar{P}}}(X \wedge X^1), \tag{16}$$

$$R_{P^*}(E_2, \Delta_2) = I_{\bar{P},Q_{\bar{P}}}(X \wedge X^2). \tag{17}$$

The conditions (16) and (17) are the same ones derived in [17] for the successive refinability under the reliability constraint in case of $E_1 \geq E_2$.

$E_2 \geq E_1$ **case:** Taking into account the corresponding rate-reliability-distortion region (8) it follows that $(R_{P^*}(E_1, \Delta_1), R_{P^*}(E_2, \Delta_2) - R_{P^*}(E_1, \Delta_1))$ is achievable iff for each $P \in \alpha(E_1, P^*)$ there exists a $Q_P \in \mathcal{Q}(P, E_1, E_2, \Delta_1, \Delta_2)$ such that

$$R_{P^*}(E_1, \Delta_1) \geq I_{P,Q_P}(X \wedge X^1) \tag{18}$$

and

$$R_{P^*}(E_2, \Delta_2) \geq \max \left(R_{P^*}(E_2, \Delta_2), I_{P,Q_P}(X \wedge X^1, X^2) \right). \tag{19}$$

For each $P \in \alpha(E_1, P^*)$, selecting $\bar{Q}_P$ as the conditional PD that minimizes the mutual information $I_{P,Q_P}(X \wedge X^1, X^2)$ among those which satisfy to (18) and (19), the optimal pair of rates $(R_{P^*}(E_1, \Delta_1), R_{P^*}(E_2, \Delta_2) - R_{P^*}(E_1, \Delta_1))$ will be achievable iff

$$R_{P^*}(E_1, \Delta_1) \geq I_{P,\bar{Q}_P}(X \wedge X^1) \tag{20}$$

and

$$R_{P^*}(E_2, \Delta_2) \geq \max\left(R_{P^*}(E_2, \Delta_2), I_{P,\bar{Q}_P}(X \wedge X^1, X^2)\right). \tag{21}$$

Since the inequalities have to be satisfied for each P from $\alpha(E_1, P^*)$, (20) and (21) are equivalent to

$$R_{P^*}(E_1, \Delta_1) \geq \max_{P \in \alpha(E_1, P^*)} I_{P,\bar{Q}_P}(X \wedge X^1) \tag{22}$$

and

$$R_{P^*}(E_2, \Delta_2) \geq \max_{P \in \alpha(E_1, P^*)} I_{P,\bar{Q}_P}(X \wedge X^1, X^2). \tag{23}$$

Then, recalling (1) again, the inequalities (22) and (23) in turn hold for each $P \in \alpha(E_1, P^*)$ iff

$$R_{P^*}(E_1, \Delta_1) = \max_{P \in \alpha(E_1, P^*)} I_{P,\bar{Q}_P}(X \wedge X^1) \tag{24}$$

and meantime

$$R_{P^*}(E_2, \Delta_2) \geq \max_{P \in \alpha(E_1, P^*)} I_{P,\bar{Q}_P}(X \wedge X^1, X^2). \tag{25}$$

Now, noting that the right-hand side of the last inequality does not depend on E_2 and the function $R_{P^*}(E_2, \Delta_2)$ is monotonically nondecreasing in E_2, we arrive to the conclusion that (25) will be satisfied for $\bar{Q}_P$ meeting (24) iff $\hat{E}_2 \geq E_2$, where

$$R_{P^*}(\hat{E}_2, \Delta_2) = \max_{P \in \alpha(E_1, P^*)} I_{P,\bar{Q}_P}(X \wedge X^1, X^2). \tag{26}$$

The derived here condition for $E_2 \geq E_1$ with specifications (24) and (26) was proved by Tuncel and Rose [17] in terms of the scalable rate-distortion function.

It must be noted also that the successive refinement in reliability sense in case of $E_2 \geq E_1$ is not possible if

$$\max_{P \in \alpha(E_1, P^*)} I_{P,\bar{Q}_P}(X \wedge X^1, X^2) > \max_P R_P(\Delta_2),$$

where the right-hand side expression is the value of the zero-error rate-distortion function (see [4]) for the second hierarchy (the maximum is taken over all PDs on $\mathcal{X}$), which can be obtained as the ultimate point of the rate-reliability-distortion function for extremely large reliabilities (i.e., when $E_2 \to \infty$ in (3)).

Finally note that we obtain the conditions for the successive refinability in distortion sense [11], [5], and [15], letting $E_1 = E_2 = E \to 0$, as it can be seen from (16) and (17).

Concluding Notes

In course of research [10]-[12], [5], and [15] we have known that the successive refinement of information in distortion sense is possible if and only if the Markovity condition for the source is fulfilled. As a correcting addendum to our previous work [7], we restated the surprising result of [17], at the same time revising Maroutian's region [13] for the hierarchical source coding problem under reliability criterion. In the more natural case $E_2 \geq E_1$, the restated necessary and sufficient condition asserts that the successive refinement is possible if and only if E_2 is larger than an indicated here threshold. Meanwhile it would be interesting to note (as a particular outcome of previous discussions) that the successive refinement in the "purely" reliability sense, i.e. when $\Delta_1 = \Delta_2 = 0$, is always possible for $E_2 \geq E_1$, since in that case Theorem 2 yields the achievability of the rates

$$R_1 = \max_{P \in \alpha(E_1, P^*)} H_P(X),$$

$$R_1 + R_2 = \max_{P \in \alpha(E_2, P^*)} H_P(X),$$

which are the corresponding values of the rate-reliability functions (4).

References

1. R. Ahlswede, The rate-distortion region for multiple descriptions without excess rate, IEEE Trans. Inform. Theory, Vol. 31, 721–726, November 1985.
2. T. Berger, Rate Distortion Theory: A Mathematical Basis for Data Compression. Englewood Cliffs, NJ, Prentice-Hall, 1971.
3. T. M. Cover and J. A. Thomas, Elements of Information Theory, Wiley, 1991.
4. I. Csiszár and J. Körner, Information Theory: Coding Theorems for Discrete Memoryless Systems, New York, Academic, 1981.
5. W.H.R. Equitz and T.M. Cover, Successive refinement of information, IEEE Trans. on Inform Theory, Vol. 37, No. 2, 269–275, 1991.
6. E. A. Haroutunian and B. Mekoush, Estimates of optimal rates of codes with given error probability exponent for certain sources, (in Russian), in Abstracts 6th Intern. Symposium on Inform. Theory, Vol. 1, Tashkent, U.S.S.R., 22–23, 1984.
7. E. A. Haroutunian and A. N. Harutyunyan, Successive refinement of information with reliability criterion, Proc. IEEE Int. Symp. Inform. Theory, Sorrento, Italy, 205, 2000.
8. E. A. Haroutunian, A. N. Harutyunyan, and A. R. Ghazaryan, On rate-reliability-distortion function for robust descriptions system, IEEE Trans. Inform. Theory, Vol. 46, No. 7, 2690–2697, 2000.
9. A. Kanlis and P. Narayan, Error exponents for successive refinement by partitioning, IEEE Trans. Inform. Theory, Vol. 42, 275–282, 1996.
10. V. N. Koshelev, Hierarchical coding of discrete sources, Problemy peredachi informatsii, Vol. 16, No. 3, 31–49, 1980.
11. V. N. Koshelev, An evaluation of the average distortion for discrete scheme of sequential approximation, Problemy peredachi informatsii, Vol. 17, No. 3, 20–30, 1981.

12. V. N. Koshelev, On divisibility of discrete sources with the single-letter-additive measure of distortion, Problemy peredachi informatsii, Vol. 30, No. 1, 31–50, 1994.
13. R. Sh. Maroutian, Achievable rates for multiple descriptions with given exponent and distortion levels (in Russian), Problemy peredachi informatsii, Vol. 26, No. 1, 83–89, 1990.
14. K. Marton, Error exponent for source coding with a fidelity criterion, IEEE Trans. Inform. Theory, Vol. 20, No. 2, 197–199, 1974.
15. B. Rimoldi, Successive refinement of information: Characterization of the achievable rates, IEEE Trans. on Inform Theory, Vol. 40, No. 1, 253–259, 1994.
16. C. E. Shannon, Coding theorems for a discrete source with a fidelity criterion. IRE National convention record, Part 4, 142–163, 1959.
17. E. Tuncel, K. Rose, Error exponents in scalable source coding, IEEE Trans. Inform. Theory, Vol. 49, No. 1, 289–296, 2003.

Coding for the Multiple-Access Adder Channel

B. Laczay*

Abstract. The coding problem for the multiple-access adder channel is considered, both for the case of permanent user activity and partial user activity. For permanent user activity, Khachatrian [10] has written an excellent survey for general, symmetric and non-symmetric rates. In this survey, we only deal with the special symmetric rate case, where all users have two codewords. The length of the shortest possible code is characterized, and amongst others, we present the code construction of Chang and Weldon [5]. We also deal with the case of signature coding (where we mean that one of the codewords for each user is the zero vector). As a code construction of this kind, we show Lindström's one [12].

We also consider partial user activity. For this case, the resulting upper and lower bounds on the length of the shortest possible code differs by a factor of two. There are some constructions for suboptimal codes, but we do not know about constructions with length approaching the upper bound. The signature code is similar to the B_m code examined by D'yachkov and Rykov [7]. It is interesting, that the upper and lower bounds for the length of B_m codes are the same as for signature codes.

1 The Problem

1.1 Multiple-Access Channel

From the viewpoint of information theory the **multiple-access channel** is a black-box operating in discrete time with a fixed number of inputs and one output. There are also extended models, with multiple outputs, the so called interference channels, but we do not deal with them now (c.f. [14,1]). We consider that one user "sits" at each input, so instead of inputs we usually refer to users. Let us denote the number of users with t. The input and output alphabets of the channel are denoted by I and O, respectively.

We will deal only with the case of memoryless channel, so to fully describe the channel, it is enough to give the channel transition probabilities $p_{y|x_1 x_2 \ldots x_t}$:

$$\mathbb{P}\big(Y = y \big| X_1 = x_1, X_2 = x_2, \ldots, X_t = x_t\big) = p_{y|x_1 x_2 \ldots x_t}$$
$$\forall (x_1, x_2, \ldots, x_t) \in I^t, \forall y \in O.$$

* This work was supported by the High Speed Networks Laboratory and the Center for Applied Mathematics and Computational Physics, BUTE. This work relates to Department of the Navy Grant N00014-04-1-4034 issued by the Office of Naval Research International Field Office. The United States Government has a royalty-free license throughout the world in all copyrightable material contained herein.

Here $X_1, X_2, \ldots, X_t$ denote the t inputs of the channel, while Y denotes the output.

Each user of the channel has a so called **component code**. A component code is a set of fixed codewords, one for each possible message of the user. We assume, that all these codewords of all users have a common length n. So the component code of the ith user can be written as

$$C_i = \left\{ \mathbf{x}_1^{(i)}, \mathbf{x}_2^{(i)}, \ldots, \mathbf{x}_{|C_i|}^{(i)} \right\} \subseteq I^n.$$

The **code** itself is the set of the component codes defined above:

$$\mathcal{C} = \{C_1, C_2, \ldots, C_t\}.$$

The **message** user i wants to send is denoted by the random variable $K_i \in \{1, 2, \ldots, |C_i|\}$. To send this message, the user transmits $\mathbf{x}_{K_i}^{(i)}$ through the channel. We will use a further restriction, that the codewords sent by the users are bit and block synchronized. This means, that at a given instant, all users are sending the same component of their codewords. Say, when user i is sending the mth component of his codeword ($[\mathbf{x}_{K_i}^{(i)}]_m$), then user j is also sending the mth component ($[\mathbf{x}_{K_j}^{(j)}]_m$). So we can treat the channel output as a vector of length n.

Since the channel is memoryless, we can write a simple formula for the distribution of the vectorial channel output, conditionally on that user i sends its k_i^{th} codeword:

$$\mathbb{P}\left(\mathbf{Y} = \mathbf{y} \middle| K_1 = k_1, K_2 = k_2, \ldots, K_t = k_t \right)$$
$$= \mathbb{P}\left(\mathbf{Y} = \mathbf{y} \middle| X_1 = \mathbf{x}_{k_1}^{(1)}, X_2 = \mathbf{x}_{k_2}^{(2)}, \ldots, X_t = \mathbf{x}_{k_t}^{(t)} \right)$$
$$= \prod_{m=1}^{n} \mathbb{P}\left(Y = [\mathbf{y}]_m \middle| X_1 = [\mathbf{x}_{k_1}^{(1)}]_m, X_2 = [\mathbf{x}_{k_2}^{(2)}]_m, \ldots X_t = [\mathbf{x}_{k_t}^{(t)}]_m \right).$$

To define the **error probability** of a given code $\mathcal{C}$, we must have a decoding function for each user:

$$d_i \colon O^n \to \{1, 2, \ldots, |C_i|\} \qquad \forall i \in [t].$$

(Here $[t]$ denotes $\{1, 2, \ldots, t\}$.) The aim of the decoding function d_i is to recover the message K_i of the ith user from the channel output vector ($\mathbf{Y}$). The error probability (P_e) of a given code is defined as the probability of making a mistake for at least one user, considering the optimal decoding functions:

$$P_e(\mathcal{C}) = \inf_{d_1, d_2, \ldots, d_t} \mathbb{P}\left(\{\exists i \in [t] \colon d_i(\mathbf{Y}) \neq K_i\} \right).$$

Here we consider, that the random variables $K_1, K_2, \ldots, K_t$ are independent and uniformly distributed over the set of possible messages ($\{1, 2, \ldots, |C_i|\}$):

$$\mathbb{P}(K_i = k) = \frac{1}{|C_i|} \qquad \forall i \in [t], \forall k \in \{1, 2, \ldots, |C_i|\}.$$

The **code rate** of a given code for user i is defined as

$$r_i(\mathcal{C}) = \frac{\log |C_i|}{n},$$

while the rate vector of a given code is formed by arranging these quantities into a vector:

$$\mathbf{r}(\mathcal{C}) = (r_1, r_2, \ldots, r_t).$$

The **rate region** of a channel is the set of rate vectors that can be reached by arbitrarily small error:

$$\mathcal{R} = \big\{ \mathbf{r} \colon \forall \varepsilon > 0 \colon \exists \mathcal{C} \colon P_e(\mathcal{C}) \leq \varepsilon \text{ and } \forall i \in [t] \colon r_i(\mathcal{C}) \geq [\mathbf{r}]_i \big\}.$$

Ahlswede [1] and van der Meulen [15] have determined the rate region for the case $t = 2$. Liao [11] has formulated the rate region for the general t user case.

1.2   Binary Adder Channel

The binary adder channel (Figure 1) is a special case of the multiple-access channel. Here the channel input alphabet is binary ($\mathcal{B} = \{0, 1\}$), while the output alphabet is the set of nonnegative integer numbers. The channel is deterministic: the output is the sum of the inputs, where the summation is the usual summation over $\mathbb{N}$ (and it is *not* the mod 2 summation). The channel transition probabilities can be written as

$$\mathbb{P}(Y = y | X_1 = x_1, X_2 = x_2, \ldots, X_t = x_t) = \begin{cases} 1 & \text{if } y = \sum_{i=1}^{t} x_i; \\ 0 & \text{otherwise.} \end{cases}$$

Using the vectorial form we can say, that the received vector is (almost sure) the sum of the sent codeword vectors:

$$\mathbf{Y} = \sum_{i=1}^{t} \mathbf{x}_{K_i}^{(i)}.$$

The rate region of this channel has been determined by Chang and Weldon [5]. For $t = 2$ it is shown in Figure 2. Ahlswede and Balakirsky [2] have given a code construction with high rates for the $t = 2$ case.

For a deterministic channel, like the adder channel, it is interesting to define the class of **uniquely decipherable** (u.d.) codes. Code $\mathcal{C}$ is a u.d. code for the adder channel, if the messages of the users can be recovered without error from the output of the channel ($P_e(\mathcal{C}) = 0$). I.e. if the users send different messages, then the received sum vector must be different. To formulate this, let us denote the message of the ith user with k_i. We call message-constellation the vector formed by the messages of the users:

$$\mathbf{k} = (k_1, k_2, \ldots, k_t) \in \bigotimes_{i=1}^{t} \{1, 2, \ldots, |C_i|\}.$$

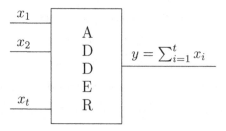

Fig. 1. The binary adder channel ($x_1, x_2, \ldots, x_t \in \mathcal{B}, y \in \mathbb{N}$)

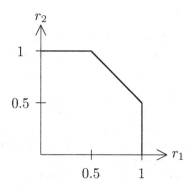

Fig. 2. The rate region of the binary adder channel for $t = 2$ users

For the adder channel, given the message-constellation **k**, the channel output vector is deterministic, and is denoted by $\mathbf{S(k)}$. In the case of the adder channel

$$\mathbf{S(k)} = \sum_{i=1}^{t} \mathbf{x}_{[\mathbf{k}]_i}^{(i)}.$$

Definition 1. *A multiple-access code is a uniquely decipherable (u.d.) code for the adder channel, if the received vector is unique considering all message-constellations:*

$$\mathbf{S(k)} = \mathbf{S(l)} \iff \mathbf{k} = \mathbf{l} \qquad \forall \mathbf{k}, \mathbf{l} \in \bigotimes_{i=1}^{t}\{1, 2, \ldots, |C_i|\}.$$

Khachatrian [10] has written an excellent survey on u.d. code constructions for the adder channel, with various rates.

In this article, we will deal only with a special class of u.d. codes with symmetric rates. If we take the sum rate

$$r_{\mathrm{sum}} = \sum_{i=1}^{t} r_i$$

of arbitrary u.d. codes for the adder channel, then from the results of Chang and Weldon [5] it follows, that the maximal achievable sum rate for large t is

$$r_{\text{sum}} \sim \frac{1}{2} \log t.$$

They have also given a u.d. code construction with two codewords per user with asymptotically the same rate. This means, that to maximize the sum rate, it is enough to consider codes with two codewords per user:

$$C_i = \left\{ \mathbf{x}_1^{(i)}, \mathbf{x}_2^{(i)} \right\} \quad \forall i \in [t].$$

In this case, the rates for all users are the same, namely

$$r_i = \frac{1}{n} \quad \forall i \in [t],$$

and because of this, we can even use the code length instead of the rate vector to compare codes.

As we have just mentioned, Chang and Weldon [5] and later Lindström [12] have given u.d. code constructions for the binary adder channel. Chang and Weldon's construction is, in fact, a statistical design that was given by Cantor and Mills [4]. Both constructions will be shown later in Sections 2.6 and 2.5.

We will deal also with the **signature coding** problem. Consider an alarming or signaling system: there are t stations, and some of them want to send an alarm signal: they send their codeword to the channel. The others, who do not want to signal an alarm, turn off their transmitter to conserve power. We can consider these stations as they are sending a zero vector to the channel.

The task of the receiver is to detect the alarming stations. This scenario is much like the general two codeword coding problem with one additional constraint: one of the codewords in all component codes should be the zero vector.

$$\mathcal{C} = \left\{ \left\{ \mathbf{0}, \mathbf{x}^{(1)} \right\}, \left\{ \mathbf{0}, \mathbf{x}^{(2)} \right\}, \ldots, \left\{ \mathbf{0}, \mathbf{x}^{(t)} \right\} \right\}.$$

We can define the class of **uniquely decipherable** (u.d.) signature codes, where the channel output must be different for all different set of active users. We will denote the channel output vector for a given set U of active users as $\mathbf{S}(U)$ ($U \subseteq [t]$). For the binary adder channel

$$\mathbf{S}(U) = \sum_{i \in U} \mathbf{x}^{(i)}.$$

Definition 2. *A signature code for the adder channel is uniquely decipherable (u.d.), if the received vector is unique considering all possible sets of active users:*

$$\mathbf{S}(U) = \mathbf{S}(V) \iff U = V \quad \forall U, V \subseteq [t].$$

1.3   User Models

Modeling the behavior of the users is as important as to model the channel itself. There are two fundamental user models, the permanent activity model and the partial activity model.

The **permanent activity** model is a rather simple one. Here the users considered to be active all the time, i.e. they always have some information to send. In the previous sections we considered this case.

In many real-life examples (think about a mobile phone network, for example) the users are not always active, they are just **partially active**, i.e. sometimes they have information to send, but most of the time they do not have. There are many possibilities to model a user population with this behavior. One interesting scenario is the m-out-of-t model: in this model we assume, that at most m users out of the t total ones are active at any instant, but this active set can vary arbitrarily with time as long as its size does not exceed m.

To make a good formalism for this partial activity case, we consider that the zero vector is added to each component code as a new codeword. The active users simply send the appropriate codeword belonging to their messages, while the inactive ones send the zero vector. This model corresponds to real life, where inactivity means simply turning off the transmitter.

This motivates the use of signature codes, introduced in the previous section. It those signature codes, for the active users there are only one usable codeword in their component codes. Thus we cannot transfer messages directly, but we can solve signaling and alarming tasks. For real information transfer, we will show a simple scenario a bit later.

The point in using m-out-of-t model is that, if m is significantly smaller than t, which is usually a realistic case, then a m-out-of-t code can be significantly shorter than a conventional t user multiple-access code.

For real information transmission with signature codes, one can use the following simple scenario. Consider a population of t users with m-out-of-t partial activity model. (Note, that the following construction works also for the permanent activity model.) We will create a u.d. multiple-access code with three codewords per user, one codeword is the all zero one, for signaling inactivity, while the other two are for the messages of the active users. Let us take a u.d. signature code C^* for $2t$ virtual users out of which at most m are active simultaneously (a code for the $2t$-out-of-m model):

$$C^* = \left\{ \left\{ \mathbf{0}, \mathbf{x}^{(1)} \right\}, \left\{ \mathbf{0}, \mathbf{x}^{(2)} \right\}, \ldots, \left\{ \mathbf{0}, \mathbf{x}^{(2t)} \right\} \right\}.$$

Create the component codes of the new multiple-access code by distributing two codewords from C^* for each of the t real users, and additionally, put the all zero codeword into the component codes:

$$C_i = \left\{ \mathbf{0}, \mathbf{x}^{(2i-1)}, \mathbf{x}^{(2i)} \right\} \qquad \forall i \in [t].$$

This way we have got a multiple-access code with two message codewords and one inactivity codeword for t users:

$$\mathcal{C} = \{C_1, C_2, \ldots, C_t\}.$$

Each user can transmit messages with the nonzero codewords, or signal inactivity with the zero one. If the number of simultaneously active users does not exceed m, then the code is u.d. Moreover, the partial activity of the users is exploited in the code, so the codeword length can be far below the length of a code for the permanent activity case.

2 Permanent Activity Case

In this section we survey some well known results regarding the u.d. coding problem of the multiple-access adder channel for the permanent activity case. We will present some bounds on the minimal length of u.d. codes, as well as some code constructions.

2.1 Equivalent Representations

There are many equivalent representations of the u.d. signature coding problem for the adder channel. In this section we show some of them.

Problem 1 (Coin Weighing Problem). *Given t coins, some of them are counterfeit, some are genuine, and we have to distinguish between them. The weight of the counterfeit coins are e.g. 9 grams, while the original ones weigh 10 grams. We have a scale, that weighs exactly. Give sets of coins, for which by measuring the weight of these sets, we can find the counterfeit coins. What is the minimal number of weighings required?*

We have to stress, that we consider here the problem, when the set of coins selected for the future weighings does not depend on the results of the previous weighings, so the sets to weigh is given before we start measuring at all. In terms of search theory this is called non-adaptive or parallel search. (There is another problem, when the next set to weigh can be selected according to the results of the previous weighings, but we do not deal with this so called adaptive search problem here.)

To show how this problem is related to the coding problem, first make some simplifications: let the weight of a counterfeit coin be one, and the weight of a genuine one be zero. Certainly this is still the same problem.

Now let us consider that each coin represents a user. Construct a codeword for each user: let the length of the codeword be equal to the number of weighings. Put 1 to the ith position if the coin associated with this user participates in the ith weighing. If the coin is not participating in that weighing, then put 0 there. If we give the zero vector as a common second codeword for all users, then we get a signature code.

Consider a given set of counterfeit coins, and consider that the users associated with the counterfeit coins send their non-zero codeword. (We can consider, that the other users send the zero vector.) The result of the weighings will be equal to the sum vector at the channel output. If we can determine the set of false coins from the output of the weighings, then we can determine the messages of the users from the sum vector, and vica versa. So the problems of planning the weighings and finding u.d. signature codes are equivalent.

Problem 2 (Combinatorial Detection Problem). *Given a finite set T construct a set $\mathcal{M}$ of subsets of it ($\mathcal{M} \subseteq \mathcal{P}(T)$, where $\mathcal{P}(T)$ is the power-set of T), in such a way that given an arbitrary $U \subseteq T$ we can determine U from the sizes of the intersections $|U \cap M|$ for all $M \in \mathcal{M}$. What is the minimal size of $\mathcal{M}$?*

To find the equivalence of this problem with the previous one is really easy. Here the sets in $\mathcal{M}$ denotes the sets of coins weighed together. So it follows, that this combinatorial detection problem is equivalent to the u.d. coding problem. But we will give a more formal proof of it using the matrix representation.

The **matrix representation** is simply another representation of the last combinatorial detection problem. If $T = [t]$, then we can represent one subset U of it with a binary column vector $\mathbf{u}$ of length t:

$$[t] \supseteq U \mapsto \mathbf{u} \in \mathcal{B}^t \text{ where } [\mathbf{u}]_i = \begin{cases} 1 & \text{if } i \in U; \\ 0 & \text{if } i \notin U \end{cases} \quad \forall i \in [t]. \tag{1}$$

The combinatorial detection problem can be formulated as to find column vectors $\mathbf{y}_1, \mathbf{y}_2, \ldots, \mathbf{y}_n \in \mathbf{B}^t$ in such a way, that if we know the values $\mathbf{y}_i^{\top}\mathbf{u}$ for all $i \in [n]$, then we can determine the vector $\mathbf{u}$. To complete the matrix representation, create a matrix from the row vectors $\mathbf{y}_i^{\top}$ by simply writing them below each other:

$$\mathbf{C} = \begin{pmatrix} \mathbf{y}_1^{\top} \\ \mathbf{y}_2^{\top} \\ \vdots \\ \mathbf{y}_n^{\top} \end{pmatrix}.$$

Now the problem is to find a matrix $\mathbf{C}$ for which

$$\mathbf{C}\mathbf{u} = \mathbf{C}\mathbf{v} \iff \mathbf{u} = \mathbf{v} \quad \forall \mathbf{u} \in \mathcal{B}^t, \forall \mathbf{v} \in \mathcal{B}^t, \tag{2}$$

or equivalently, introducing $\mathbf{w} = \mathbf{u} - \mathbf{v}$, the problem is to find a $\mathbf{C}$, for which

$$\mathbf{C}\mathbf{w} = \mathbf{0} \iff \mathbf{w} = \mathbf{0} \quad \forall \mathbf{w} \in \{-1, 0, 1\}^t. \tag{3}$$

Now we show the mapping between this matrix representation of the combinatorial detection problem and the u.d. coding problem of the multiple-access adder channel.

Let us write the above matrix $\mathbf{C}$ as column vectors $\mathbf{x}^{(1)}, \mathbf{x}^{(2)}, \ldots, \mathbf{x}^{(t)} \in \mathcal{B}^n$ written next to each other:

$$\mathbf{C} = \begin{pmatrix} \mathbf{x}^{(1)} & \mathbf{x}^{(2)} & \cdots & \mathbf{x}^{(t)} \end{pmatrix}.$$

Consider a multiple-access code, where each user has two codewords, one is the zero vector, the other is a column vector $\mathbf{x}^{(i)}$ from above:

$$\mathcal{C} = \{C_1, C_2, \ldots, C_t\},$$
$$C_i = \{\mathbf{0}, \mathbf{x}^{(i)}\} \quad \forall i \in [t].$$

It is easy to see, that this code is a u.d. signature code for the adder channel for t users. Let $T = [t]$ denote the population of the t users. At a given instant, each user sends that codeword from his component code, which corresponds to his message. Let k_i denote the message of the ith user, and let us map the messages and codewords in such a way, that "message 1" ($k_i = 1$) is associated with the $\mathbf{0}$ vector, and "message 2" ($k_i = 2$) is associated with the $\mathbf{x}^{(i)}$ vector. With these notations, the u.d. property in the case of the adder channel was the following (see Definition 1):

$$\sum_{i:[\mathbf{k}]_i=2} \mathbf{x}^{(i)} = \sum_{i:[\mathbf{l}]_i=2} \mathbf{x}^{(i)} \iff \mathbf{k} = \mathbf{l} \qquad \forall \mathbf{k}, \mathbf{l} \in \{1,2\}^t.$$

Let us denote the subset of users sending their nonzero codeword ($k_i = 2$) in constellation $\mathbf{k}$ with U and in constellation $\mathbf{l}$ with V ($U \subseteq T, V \subseteq T$). Let us represent these sets U and V with vectors $\mathbf{u}$ and $\mathbf{v}$ in the way given by our mapping (1). Now the uniquely decipherable property becomes

$$\mathbf{Cu} = \mathbf{Cv} \iff \mathbf{u} = \mathbf{v} \qquad \forall \mathbf{u} \in \mathcal{B}^t, \forall \mathbf{v} \in \mathcal{B}^t.$$

which is exactly the same formula as (2) which was given as the condition on $\mathbf{C}$.

2.2 Trivial Bounds on $N(t)$

First we present here a very simple statement about the minimal code length for the binary adder channel with two codewords per user.

Definition 3. *The **minimal code length** $N(t)$ is the length of the shortest possible u.d. code for a given number of users:*

$$N(t) = \min\{n \colon \exists \mathcal{C} \text{ u.d. code with length } n \text{ for } t \text{ users}\}.$$

Definition 4. *The **minimal signature code length** $N_s(t)$ is the length of the shortest possible u.d. signature code for a given number of users:*

$$N_s(t) = \min\{n \colon \exists \mathcal{C} \text{ u.d. signature code with length } n \text{ for } t \text{ users}\}.$$

Theorem 1.
$$\frac{t}{\log(t+1)} \leq N(t) \leq N_s(t) \leq t.$$

Proof.   For the lower bound, we simply use enumeration: there are 2^t possible message-constellations for t users ($\mathbf{k} \in \{1,2\}^t$). On the other side, the received vector has components in $\{0, 1, \ldots, t\}$, and in case of the shortest possible code it is of length $N(t)$. For the u.d. property, the number of possible received vectors cannot be smaller than the number of possible user subsets. Thus we have

$$(t+1)^{N(t)} \geq 2^t,$$

from which

$$N(t) \geq \frac{t}{\log(t+1)}.$$

(Here and from now, log represents the logarithm of base 2.)

The statement $N(t) \leq N_s(t)$ follows simply from the definition: all u.d. signature code is also a u.d. code, so the minimal u.d. code length cannot be greater than the minimal u.d. signature code length.

For the upper bound, we show a trivial u.d. signature code construction for t users with codeword length $n = t$: simply let the codewords of the ith user be $\mathbf{0}$ and $\mathbf{e}^{(i)}$ which is the ith unit vector. Using the matrix representation, we can say, that for t users let $\mathbf{C} = \mathbf{I}_t$ (which is the identity matrix of size $t \times t$). Since this matrix is invertible, this signature code certainly has the u.d. property. So $N_s(t) \leq t$. □

2.3 Upper Bound of Erdős and Rényi

Erdős and Rényi [8] has presented a nontrivial asymptotic upper bound on the minimal signature code length for the adder channel:

Theorem 2. *(Erdős–Rényi [8])*

$$\overline{\lim_{t \to \infty}} \frac{N_s(t) \log t}{t} \leq \log 9.$$

Proof.   The proof is based on random coding. We select a random signature code $\mathcal{C}$ of length $n + 1$ in such a way, that the first n components of the non-null codewords are i.i.d. uniformly distributed binary random variables, while the $(n+1)^{\text{th}}$ component is fixed to 1:

$$\mathbb{P}\left(\left[\mathbf{x}^{(j)}\right]_i = 0\right) = \mathbb{P}\left(\left[\mathbf{x}^{(j)}\right]_i = 1\right) = \frac{1}{2} \qquad \forall j \in [t], \forall i \in [n] \text{ (i.i.d.)},$$

$$\left[\mathbf{x}^{(j)}\right]_{n+1} = 1 \qquad\qquad\qquad\qquad \forall j \in [t].$$

We will work with the probability of the event

$$\text{Code } \mathcal{C} \text{ is not a u.d. signature code}, \qquad\qquad (*)$$

and we will give an upper bound for it, which will tend to 0 as $t \to \infty$.

We have introduced $\mathbf{S}(U)$ as the channel output when the set of active users is U. Simply we have that if $\mathcal{C}$ is not a u.d. signature code, then there are two different sets of active users, say U and V, for which $\mathbf{S}(U) = \mathbf{S}(V)$.

$$\mathbb{P}(\text{event } (*)) = \mathbb{P}\left(\bigcup_{U \neq V \subseteq [t]} \{\mathbf{S}(U) = \mathbf{S}(V)\}\right).$$

If there are two subsets U and V for which $\mathbf{S}(U) = \mathbf{S}(V)$, then there are also two disjoint subsets with the same property: $U \setminus V$ and $V \setminus U$ will suite. We know also, that $|U| = |V|$, since the $(n+1)^{\text{th}}$ component of the received vector is simply the number of active users, so if the received vectors are equal, then the sets of the active users must be of the same size:

$$\mathbb{P}\big(\text{event } (*)\big) = \mathbb{P}\left(\bigcup_{\substack{U \neq V \subseteq [t]: \\ |U| = |V|, U \cap V = \varnothing}} \{\mathbf{S}(U) = \mathbf{S}(V)\} \right).$$

We can use the so called union bound as an upper bound:

$$\mathbb{P}\big(\text{event } (*)\big) \leq \sum_{\substack{U \neq V \subseteq [t]: \\ |U| = |V|, \\ U \cap V = \varnothing}} \mathbb{P}\big(\mathbf{S}(U) = \mathbf{S}(V)\big)$$

$$= \sum_{k=1}^{\lfloor \frac{t}{2} \rfloor} \sum_{\substack{U, V \subseteq [t]: \\ |U| = |V| = k, \\ U \cap V = \varnothing}} \mathbb{P}\big(\mathbf{S}(U) = \mathbf{S}(V)\big), \tag{4}$$

since the common size of the active subsets is at most $\lfloor \frac{t}{2} \rfloor$.

The components of the codewords in $\mathcal{C}$ has a simple distribution, so it is easy to calculate $\mathbb{P}\big(\mathbf{S}(U) = \mathbf{S}(V)\big)$ for some fixed disjoint U and V of the same size k:

$$\mathbb{P}\big(\mathbf{S}(U) = \mathbf{S}(V)\big) = \mathbb{P}\left(\bigcap_{i=1}^{n} \bigcup_{\ell=0}^{k} \{[\mathbf{S}(U)]_i = \ell \text{ and } [\mathbf{S}(V)]_i = \ell\} \right)$$

$$= \left(\sum_{\ell=0}^{k} \binom{k}{\ell} 2^{-k} \binom{k}{\ell} 2^{-k} \right)^n,$$

since the components of $S(U)$ and $S(V)$ are independent, and for disjoint U and V, $S(U)$ and (V) has independent binomial distribution.

Let us introduce

$$Q(k) = \sum_{\ell=0}^{k} \binom{k}{\ell} 2^{-k} \binom{k}{\ell} 2^{-k}. \tag{5}$$

We have

$$Q(k) = \sum_{\ell=0}^{k} \binom{k}{\ell} \binom{k}{k-\ell} 2^{-2k} = \binom{2k}{k} 2^{-2k}.$$

For $1 < k$

$$\binom{2k}{k} 2^{-2k} \leq \frac{1}{\sqrt{\pi k}}.$$

(C.f. Gallager [9] Problem 5.8 pp. 530.) Thus

$$Q(k) \leq \frac{1}{\sqrt{\pi k}}. \tag{6}$$

Substituting this result into (4), we get

$$\mathbb{P}\big(\text{event } (*)\big) \leq \sum_{k=1}^{\lfloor \frac{t}{2}\rfloor} \sum_{\substack{U,V\subseteq[t]:\\ |U|=k,|V|=k,\\ U\cap V=\varnothing}} \left(\frac{1}{\sqrt{\pi k}}\right)^n$$

$$= \sum_{k=1}^{\lfloor \frac{t}{2}\rfloor} \binom{t}{k}\binom{t-k}{k}\left(\frac{1}{\sqrt{\pi k}}\right)^n. \tag{7}$$

Split the summation into two parts:

$$k = 1, 2, \ldots, \left\lfloor \frac{t}{2\ln^2 t}\right\rfloor,$$

and

$$k = \left\lfloor \frac{t}{2\ln^2 t}\right\rfloor + 1, \left\lfloor \frac{t}{2\ln^2 t}\right\rfloor + 2, \ldots, \left\lfloor \frac{t}{2}\right\rfloor.$$

(Here and from now, ln represents the natural logarithm.) For the first part, we use $\frac{1}{\sqrt{\pi k}} \leq \frac{1}{\sqrt{\pi}}$ and $\binom{t}{k}\binom{t-k}{k} \leq t^{2k}$:

$$\sum_{k=1}^{\lfloor \frac{t}{2\ln^2 t}\rfloor} \binom{t}{k}\binom{t-k}{k}\left(\frac{1}{\sqrt{\pi k}}\right)^n \leq \sum_{k=1}^{\lfloor \frac{t}{2\ln^2 t}\rfloor} t^{2k}\left(\frac{1}{\sqrt{\pi}}\right)^n$$

$$\leq \frac{(t^2)^{\frac{t}{2\ln^2 t}} - 1}{t^2 - 1}\left(\frac{1}{\sqrt{\pi}}\right)^n$$

$$\leq t^{\frac{t}{\ln^2 t}}\left(\frac{1}{\sqrt{\pi}}\right)^n$$

$$\leq \exp\left(\frac{t}{\ln t} - \frac{n}{2}\ln \pi\right),$$

which tends to zero as $t \to \infty$ if

$$\lim_{t\to\infty}\left(\frac{t}{\ln t} - \frac{n}{2}\ln \pi\right) = -\infty. \tag{8}$$

For the second part of the summation in (7), we use $\sum_{k=1}^{\lfloor \frac{t}{2}\rfloor}\binom{t}{k}\binom{t-k}{k} \leq 3^t$, which holds since selecting two subsets of $[t]$ is equivalent of partitioning it into three parts. So

$$\sum_{k=\lfloor \frac{t}{2\ln^2 t}\rfloor+1}^{\lfloor \frac{t}{2}\rfloor} \binom{t}{k}\binom{t-k}{k}\left(\frac{1}{\sqrt{\pi k}}\right)^n \le \sum_{k=1}^{\lfloor \frac{t}{2}\rfloor} \binom{t}{k}\binom{t-k}{k}\left(\frac{1}{\sqrt{\frac{\pi t}{2\ln^2 t}}}\right)^n$$

$$\le 3^t \left(\frac{1}{\sqrt{\frac{\pi t}{2\ln^2 t}}}\right)^n$$

$$= \exp\left(t\ln 3 - \frac{n}{2}\ln\frac{\pi t}{2\ln^2 t}\right),$$

which tends to zero as $t \to \infty$ if

$$\lim_{t\to\infty}\left(t\ln 3 - \frac{n}{2}\ln\frac{\pi t}{2\ln^2 t}\right) = -\infty. \tag{9}$$

Let us set

$$n = \left\lceil \frac{ct}{\ln t}\right\rceil.$$

In the first condition (8) this yields

$$\lim_{t\to\infty}\left(\frac{t}{\ln t} - \frac{n}{2}\ln\pi\right) \le \lim_{t\to\infty}\left(\frac{t}{\ln t} - \frac{ct\ln\pi}{2\ln t}\right)$$

$$= \left(1 - \frac{c\ln\pi}{2}\right)\lim_{t\to\infty}\frac{t}{\ln t},$$

which is $-\infty$ if $c > \frac{2}{\ln\pi} \approx 1.747$. In the second condition (9),

$$\lim_{t\to\infty}\left(t\ln 3 - \frac{n}{2}\ln\frac{\pi t}{2\ln^2 t}\right) \le \lim_{t\to\infty}\left(t\ln 3 - \frac{ct}{2\ln t}\ln\frac{\pi t}{2\ln^2 t}\right)$$

$$= \lim_{t\to\infty}\left(t\left(\ln 3 - \frac{c}{2}\frac{\ln\pi t - \ln\left(2\ln^2 t\right)}{\ln t}\right)\right)$$

$$= \left(\ln 3 - \frac{c}{2}\right)\lim_{t\to\infty} t,$$

which is $-\infty$ if $c > \ln 9 \approx 2.197$.

So we have shown, that for all $\varepsilon > 0$ and $n = \left\lceil \frac{(\varepsilon+\ln 9)t}{\ln t}\right\rceil$,

$$\lim_{t\to\infty}\mathbb{P}\big(\text{event }(*)\big) = 0.$$

This means, that for t large enough, the random code of length $n + 1$ we select is a u.d. signature code with positive probability. So there exists a u.d. signature code of this length:

$$N_s(t) \le \left\lceil\frac{(\varepsilon + \ln 9)t}{\ln t}\right\rceil + 1 \le \frac{(\varepsilon + \ln 9)t}{\ln t} + 2,$$

for t large enough, or equivalently

$$\varlimsup_{t\to\infty}\frac{N_s(t)\log t}{t} \le \log 9. \qquad \square$$

2.4   A Lower Bound

From our trivial lower bound (Theorem 1) and the upper bound of Erdős and Rényi (Theorem 2), for the minimal length of u.d. codes for the adder channel we have

$$1 \leq \liminf_{t\to\infty} N(t)\frac{\log t}{t} \leq \overline{\lim_{t\to\infty}} N_s(t)\frac{\log t}{t} \leq \log 9,$$

while the truth is that

$$\lim_{t\to\infty} N(t)\frac{\log t}{t} = \lim_{t\to\infty} N_s(t)\frac{\log t}{t} = 2.$$

Here we present an improved lower bound for the limes inferior. The upper bound for the limes superior will follow from the construction of Lindström (see Section 2.6).

Theorem 3. *(Chang–Weldon [5])*

$$\liminf_{t\to\infty} N(t)\frac{\log t}{t} \geq 2.$$

For the proof, we will need a lemma which is following from a bound of the discrete entropy (cf. e.g. [6] Theorem 9.7.1 pp. 235.):

Lemma 1. *Let X have an arbitrary distribution over the integers:*

$$\mathbb{P}(X = i) = p_i \qquad \forall i \in \mathbb{Z},$$

$$\sum_{i\in\mathbb{Z}} p_i = 1.$$

If X has variance $\mathbb{D}^2(X)$ and Shannon–entropy $\mathbb{H}(X) = -\sum_{i\in\mathbb{Z}} p_i \log p_i$, then

$$\mathbb{H}(X) \leq \frac{1}{2}\log\left(2\pi e\left(\mathbb{D}^2(X) + \frac{1}{12}\right)\right).$$

Lemma 2.

$$\mathrm{H}_{\mathrm{bin}}\left(n, \frac{1}{2}\right) \leq \frac{\log n}{2} + O(1),$$

where $\mathrm{H}_{\mathrm{bin}}(n, p)$ denotes the entropy of the binomial distribution with parameters (n, p).

Proof.   The variance of the binomial distribution with parameters $(n, \frac{1}{2})$ is $\frac{n}{4}$. Putting this into Lemma 1 we get

$$\mathrm{H}_{\mathrm{bin}}\left(n, \frac{1}{2}\right) \leq \frac{1}{2}\log\left(2\pi e\left(\frac{n}{4} + \frac{1}{12}\right)\right)$$

$$= \frac{1}{2}\log\left(\frac{\pi e}{2}\right) + \frac{1}{2}\log\left(n + \frac{1}{3}\right)$$

$$= \frac{1}{2}\log\left(\frac{\pi e}{2}\right) + \frac{1}{2}\left(\log n + O\left(\frac{1}{n}\right)\right)$$

$$= \frac{\log n}{2} + O(1). \qquad \square$$

Proof of Theorem 3. We will bound the same entropy in two different ways, and this will yield the bound. Consider an arbitrary u.d. code for the adder channel. Let us define the message-constellation $\mathbf{K}$ as a random vector variable with uniform distribution over all the possible constellations:

$$\mathbb{P}(\mathbf{K} = \mathbf{k}) = \frac{1}{2^t} \quad \forall \mathbf{k} \in \{1, 2\}^t$$

Since the code is u.d., the received vector $\mathbf{Y}$ must be different for all different $\mathbf{K}$, so

$$\mathbb{H}(\mathbf{Y}) = \mathbb{H}(\mathbf{K}) = t. \tag{10}$$

On the other hand, the entropy of $\mathbf{Y}$ can be upper bounded by the sum of the entropies of its components. Each component of the vector $\mathbf{Y}$ has a binomial distribution. To show this, considering only the jth bit of the codewords, split the user population into three groups. For the users in the first group both codewords has 0 in the jth position. For the users in the second group, one codeword is 0 and the other is 1 at the jth position, while for the third group both codewords are 1 at the jth position. If we denote the number of users in the first group with a_j, in the second group with b_j and in the third group with c_j the we can write

$$\mathbb{P}([\mathbf{Y}]_j = s) = \frac{2^{a_j} \binom{b_j}{s-c_j} 2^{c_j}}{2^{a_j+b_j+c_j}} = \binom{b_j}{s-c_j} 2^{-b_j} \quad \forall s \colon c_j \leq s \leq b_j + c_j,$$

where 2^{a_j} is the number of possible constellations for the users in the first group, $\binom{b_j}{s-c_j}$ is the number of possible constellations for users in the second group (select exactly $s - c_j$ users out of the b_i ones who have both zero and one at position j), and 2^{c_j} is the number of possible constellations for users in the third group. $2^{a_j+b_j+c_j}$ is the total number of possible constellations.

Now we can write

$$\mathbb{H}(\mathbf{Y}) \leq \sum_{j=1}^{n} \mathbb{H}([\mathbf{Y}]_j)$$

$$\leq \sum_{j=1}^{n} \mathrm{H}_{\text{bin}}\left(b_i, \frac{1}{2}\right),$$

and with Lemma 2

$$\mathbb{H}(\mathbf{Y}) \leq \sum_{j=1}^{n} \left(\frac{\log b_i}{2} + O(1)\right)$$

$$\leq n\left(\frac{\log t}{2} + O(1)\right). \tag{11}$$

Putting (10) and (11) together we get that

$$n \geq \frac{t}{\frac{\log t}{2} + O(1)},$$

so for the minimal n

$$N(t)\frac{\log t}{t} \geq \frac{2}{1 + O(\frac{1}{\log t})},$$

thus

$$\liminf_{t\to\infty} N(t)\frac{\log t}{t} \geq 2. \qquad \square$$

2.5   Chang and Weldon's Construction

In this section we show the u.d. code construction of Chang and Weldon [5] for the binary multiple-access adder channel. Their code construction is for $t_k = (k+2)2^{(k-1)}$ users, where k is an arbitrary natural number. Their code length is $n_k = 2^k$. If we put these together, we get

$$\lim_{k\to\infty} n_k\frac{\log t_k}{t_k} = 2,$$

from which

$$\varlimsup_{k\to\infty} N(t_k)\frac{\log t_k}{t_k} \leq 2.$$

First, we present a difference matrix representation of codes with two code-words per user. Given an arbitrary code

$$\mathcal{C} = \left\{ \{\mathbf{x}_1^{(1)}, \mathbf{x}_2^{(1)}\}, \{\mathbf{x}_1^{(2)}, \mathbf{x}_2^{(2)}\}, \ldots, \{\mathbf{x}_1^{(t)}, \mathbf{x}_2^{(t)}\} \right\},$$

we can create a so called difference matrix $\mathbf{D}$ by writing the differences of the two vectors in the component codes next to each other:

$$\mathbf{D} = \left(\mathbf{x}_2^{(1)} - \mathbf{x}_1^{(1)} \ \ \mathbf{x}_2^{(2)} - \mathbf{x}_1^{(2)} \ \cdots \ \mathbf{x}_2^{(t)} - \mathbf{x}_1^{(t)} \right)$$

It is easy to see, that given a difference matrix $\mathbf{D}$ (a matrix with all elements from $\{-1, 0, 1\}$) we can construct at least one code, for which this is the difference matrix. E.g. for all user $i \in [t]$ and for all code bit $j \in [n]$ let

$$[\mathbf{x}_1^{(i)}]_j = \begin{cases} 1 & \text{if } [\mathbf{D}]_{ij} = -1; \\ 0 & \text{if } [\mathbf{D}]_{ij} = 0; \\ 0 & \text{if } [\mathbf{D}]_{ij} = 1, \end{cases} \quad \text{and} \quad [\mathbf{x}_2^{(i)}]_j = \begin{cases} 0 & \text{if } [\mathbf{D}]_{ij} = -1; \\ 0 & \text{if } [\mathbf{D}]_{ij} = 0; \\ 1 & \text{if } [\mathbf{D}]_{ij} = 1. \end{cases}$$

Certainly we could also use 1 in both codewords if $[\mathbf{D}]_{ij} = 0$.

It is easy to show (with the same reasoning given in Section 2.1 at the matrix representation), that that the u.d. property can be expressed in the following property of matrix $\mathbf{D}$:

$$\forall \mathbf{w} \in \{-1, 0, 1\}^t : \mathbf{D}\mathbf{w} = \mathbf{0} \iff \mathbf{w} = \mathbf{0}. \qquad (12)$$

We have trivial u.d. difference matrices for $t = 1$ ($\mathbf{D}_0$) and for $t = 3$ ($\mathbf{D}_1$):

$$\mathbf{D}_0 = \begin{pmatrix} 1 \end{pmatrix}, \mathbf{D}_1 = \begin{pmatrix} 1 & 1 & 1 \\ 1 & -1 & 0 \end{pmatrix}.$$

Moreover, we can find a recursive construction for u.d. difference matrices $\mathbf{D}_k$ for $t = (k+2)2^{k-1}$ users. This is stated as the next theorem.

Theorem 4. *(Chang–Weldon [5]) The $\mathbf{D}_k$ matrices defined below are u.d. difference matrices for $t_k = (k+2)2^{k-1}$ users with code length $n_k = 2^k$:*

$$\mathbf{D}_0 = \begin{pmatrix} 1 \end{pmatrix}$$

$$\forall k \geq 1 : \mathbf{D}_k = \begin{pmatrix} \mathbf{D}_{k-1} & \mathbf{D}_{k-1} & \mathbf{I}_{2^{k-1}} \\ \mathbf{D}_{k-1} & -\mathbf{D}_{k-1} & \mathbf{0}_{2^{k-1}} \end{pmatrix}$$

Proof.   It is easy to see by induction, that $\mathbf{D}_k$ has $t_k = (k+2)2^{k-1}$ columns, since the two matrices $\mathbf{D}_{k-1}$ have two times $(k+1)2^{k-2}$ columns, and the additional identity and zero matrices has 2^{k-1} columns. The sum is really $t_k = (k+2)2^{k-1}$.

Now let us suppose, that for a given $\mathbf{w} \in \{-1, 0, 1\}^{t_k}$

$$\mathbf{D}_k\mathbf{w} = \mathbf{0}$$

We decompose vector $\mathbf{w}$ of length t_k into three vectors $\mathbf{w}_1, \mathbf{w}_2, \mathbf{w}_3$ of length $(k+1)2^{k-2}$, $(k+1)2^{k-2}$, and 2^{k-1} respectively:

$$\mathbf{w} = \begin{pmatrix} \mathbf{w}_1 \\ \mathbf{w}_2 \\ \mathbf{w}_3 \end{pmatrix}.$$

Now using the recursive definition of $\mathbf{D}_k$ we get

$$\begin{pmatrix} \mathbf{D}_{k-1} & \mathbf{D}_{k-1} & \mathbf{I}_{2^{k-1}} \\ \mathbf{D}_{k-1} & -\mathbf{D}_{k-1} & \mathbf{0}_{2^{k-1}} \end{pmatrix} \begin{pmatrix} \mathbf{w}_1 \\ \mathbf{w}_2 \\ \mathbf{w}_3 \end{pmatrix} = \begin{pmatrix} \mathbf{0} \\ \mathbf{0} \end{pmatrix},$$

which gives two equations:

$$\mathbf{D}_{k-1}\mathbf{w}_1 + \mathbf{D}_{k-1}\mathbf{w}_2 + \mathbf{w}_3 = \mathbf{0},$$
$$\mathbf{D}_{k-1}\mathbf{w}_1 - \mathbf{D}_{k-1}\mathbf{w}_2 = \mathbf{0}.$$

From this we have

$$2\mathbf{D}_{k-1}\mathbf{w}_1 = -\mathbf{w}_3$$

and since $\mathbf{w}_3$ has components in $\{-1, 0, 1\}$, it follows that $\mathbf{w}_3 = \mathbf{0}$, and therefore

$$\mathbf{D}_{k-1}\mathbf{w}_1 = \mathbf{0}$$
$$\mathbf{D}_{k-1}\mathbf{w}_2 = \mathbf{0}.$$

But $\mathbf{D}_{k-1}$ has property (12) by induction, so

$$\mathbf{w}_1 = \mathbf{0},$$
$$\mathbf{w}_2 = \mathbf{0}.$$

follows, and the proof is complete.   □

Chang and Weldon [5] have also given a decoding algorithm for their code. It is the following: let us suppose, that user i sends message k_i, so the received vector is

$$\mathbf{v} = \sum_{i=1}^{t} \mathbf{x}_{k_i}^{(i)}$$

If we introduce vector $\mathbf{u} \in \mathcal{B}^{t_k}$ which has 0 in the positions corresponding to users sending their "message 1" and 1 in the positions corresponding to users sending their "message 2"

$$[\mathbf{u}]_i = \begin{cases} 0 & \text{if } k_i = 1; \\ 1 & \text{if } k_i = 2; \end{cases} \quad \forall i \in [t_k].$$

then we can write, that

$$\mathbf{v} = \mathbf{D}_k \mathbf{u} + \sum_{i=1}^{t} \mathbf{x}_1^{(i)},$$

and therefore

$$\mathbf{D}_k \mathbf{u} = \mathbf{v} - \sum_{i=1}^{t} \mathbf{x}_1^{(i)}. \tag{13}$$

Now we make a decomposition:

$$\mathbf{v} - \sum_{i=1}^{t} \mathbf{x}_1^{(i)} = \begin{pmatrix} \mathbf{v}_1 \\ \mathbf{v}_2 \end{pmatrix} = \begin{pmatrix} \mathbf{D}_{k-1} & \mathbf{D}_{k-1} & \mathbf{I}_{2^{k-1}} \\ \mathbf{D}_{k-1} & -\mathbf{D}_{k-1} & \mathbf{0}_{2^{k-1}} \end{pmatrix} \begin{pmatrix} \mathbf{u}_1 \\ \mathbf{u}_2 \\ \mathbf{u}_3 \end{pmatrix},$$

where $\mathbf{v}_1$ and $\mathbf{v}_2$ has 2^{k-1} rows, $\mathbf{u}_1$ and $\mathbf{u}_2$ has $(k+1)2^{k-2}$ rows, and $\mathbf{u}_3$ has 2^{k-1} rows. This equation yields

$$\mathbf{v}_1 + \mathbf{v}_2 = 2\mathbf{D}_{k-1}\mathbf{u}_1 + \mathbf{u}_3.$$

Then

$$\mathbf{v}_1 + \mathbf{v}_2 \equiv \mathbf{u}_3 \mod 2,$$

and from this we get $\mathbf{u}_3 \in \mathcal{B}^t$. We have also

$$\mathbf{D}_{k-1}\mathbf{u}_1 = \frac{1}{2}(\mathbf{v}_1 + \mathbf{v}_2 - \mathbf{u}_3),$$

$$\mathbf{D}_{k-1}\mathbf{u}_2 = \frac{1}{2}(\mathbf{v}_1 - \mathbf{v}_2 - \mathbf{u}_3),$$

where the right hand side of the equations are known vectors. This is two instances of the same problem but with a smaller k like we started with in formula (13). In the $k = 0$ case we will have a trivial equation for $\mathbf{u}$ since $\mathbf{D}_0 = (1)$.

2.6 Lindström's Construction

Lindström presented a u.d. signature code construction for the multiple-access adder channel in [12]. His code is asymptotically optimal, since it sets an upper bound on the minimal u.d. signature code length which is equal to the lower bound in Theorem 3.

In the construction we will use $\mathcal{P}([t]) \to \mathbb{N}$ functions. Just to keep the original notation of Lindström, these functions will be noted as indexed variables like a_i rather than as functions like $a(i)$:

$$a : \mathcal{P}([t]) \to \mathbb{N} \qquad [t] \supseteq M \mapsto a_M \in \mathbb{N}.$$

In the followings we will give Lindström's u.d. signature code matrix $\mathbf{C}_s$ for $t = s2^{s-1}$ users where $s \in \mathbb{N}$. First let us fix an enumeration of the nonempty subsets of $[s]$:

$$M_1, M_2, \ldots, M_{2^s-1} \qquad (M_i \subseteq [s], M_i \neq \varnothing \quad \forall i \in [2^s - 1]).$$

Construction of matrix $\mathbf{A}_s$. We have $2^s - 1$ nonzero binary vectors of length s:

$$\mathbf{a}_1, \mathbf{a}_2, \ldots, \mathbf{a}_{2^s-1} \qquad (\mathbf{a}_i \in \mathcal{B}^s, \mathbf{a}_i \neq \mathbf{0} \quad \forall i \in [2^s - 1]).$$

Now introduce functions $a^{(j)} : \mathcal{P}([t]) \to \mathcal{B}$ based on these $\mathbf{a}_j$ vectors for all $j = 1, \ldots, 2^s - 1$ in the following way. For each $M \subseteq [t]$ let us represent M with a vector $\mathbf{m} \in \mathcal{B}^t$ where

$$[\mathbf{m}]_i = \begin{cases} 1 & \text{if } i \in M; \\ 0 & \text{if } i \notin M. \end{cases}$$

(This is the same way, as it was given at the matrix representation by (1).) Now let

$$a_M^{(j)} = \mathbf{m}^\top \mathbf{a}_j \mod 2 \qquad \forall M \subseteq [t]. \tag{14}$$

Let us define matrix $\mathbf{A}_s$ of size $2^s - 1 \times 2^s - 1$ in the following way:

$$[\mathbf{A}_s]_{ij} = a_{M_i}^{(j)}.$$

Construction of matrices $\mathbf{B}_s^{(k)}$. We have $2^s - s - 1$ subsets of $[s]$ with at least two elements:

$$L_1, L_2, \ldots, L_{2^s-s-1} \qquad (L_i \subseteq [s], |L_i| \geq 2 \quad \forall i \in [2^s - s - 1])$$

We will define a matrix $\mathbf{B}_s^{(k)}$ for each $k \in [2^2 - s - 1]$ in the following way. For each $j \in [|L_k| - 1]$ and for each nonempty subset M of L_k ($M \subseteq L_k, M \neq \varnothing$) introduce $b^{(k,j)} : \mathcal{P}(L_k) \setminus \{\varnothing\} \to \mathcal{B}$ such that the following condition holds:

$$\sum_{\substack{M \subseteq L_k \\ M \neq \varnothing}} b_M^{(k,j)} = 2^{j-1}.$$

This is possible for all k and j: e.g. order the nonempty subsets M of L_k arbitrarily, then for the first 2^{j-1} subsets M let $b_M^{(k,j)} = 1$, while for the other subsets M let $b_M^{(k,j)} = 0$. (This construction is valid for all $j \in [|L_k| - 1]$, since $2^{j-1} \leq 2^{|L_k|-2} < 2^{|L_k|} - 1$, which is the number of nonempty subsets of L_k.)
Now extend $b^{(k,j)}$ for all $M \subseteq [s]$:

$$\hat{b}^{(k,j)} : \mathcal{P}([t]) \to \mathcal{B} \quad \hat{b}_M^{(k,j)} = \begin{cases} 0 & \text{if } M = \varnothing; \\ b_M^{(k,j)} & \text{if } \varnothing \neq M \subseteq L_k; \\ b_{M \cap L_k}^{(k,j)} + |M \setminus L_k| \mod 2 & \text{if } M \not\subseteq L_k. \end{cases}$$

(15)

Finally define matrix $\mathbf{B}_s^{(k)}$ of size $2^s - 1 \times |L_k| - 1$ as

$$[\mathbf{B}_s^{(k)}]_{ij} = \hat{b}_{M_i}^{(k,j)}.$$

Construction of $\mathbf{C}_s$. Now we will construct matrix $\mathbf{C}_s$ of size $(2^s - 1) \times s2^{s-1}$, which will be the matrix of an u.d. code for the adder channel. Let us write the above defined $\mathbf{A}_s$ and $\mathbf{B}_s^{(k)}$ matrices next to each other:

$$\mathbf{C}_s = \left(\mathbf{A}_s \; \mathbf{B}_s^{(1)} \; \mathbf{B}_s^{(2)} \; \dots \; \mathbf{B}_s^{(2^s-s-1)} \right)$$

$\mathbf{C}_s$ has $2^s - 1$ rows, and it is easy to check that it has $s2^{s-1}$ columns:

$$2^s - 1 + \sum_{i=2}^{s} \binom{s}{i}(i-1) = \sum_{i=0}^{s} \binom{s}{i} i = s2^{s-1}.$$

Theorem 5. *The construction of Lindström gives u.d. signature codes for $t_s = s2^{s-1}$ users, where s is an arbitrary natural number. His code is of length $n_s = 2^s - 1$. Thus*

$$\lim_{s \to \infty} n_s \frac{\log t_s}{t_s} = 2.$$

For the minimal u.d. signature code length it follows, that

$$\varlimsup_{s \to \infty} N_s(t_s) \frac{\log t_s}{t_s} \leq 2.$$

From the values of t_s and n_s, the statement about the limes and the limes superior is trivial. It is important to note, that even if we had shown it for special values of t only, Lindström has given his construction for arbitrary t. This means, that he has proven, that

$$\varlimsup_{t \to \infty} N_s(t) \frac{\log t}{t} \leq 2.$$

What we need to proof in Theorem 5 is that this is a u.d. signature code. For this, we will start with some lemmas.

Lemma 3. *Choose an arbitrary* $\mathbf{a} \in \mathcal{B}^t$. *Let us define function "a"* $: \mathcal{P}([t]) \to \mathcal{B}$ *by construction (14). Then for every* $P \subseteq [t]$,

$$\sum_{M \subseteq P} a_M = \begin{cases} 0 & \text{if } \forall i \in P \colon [\mathbf{a}]_i = 0; \\ 2^{|P|-1} & \text{otherwise.} \end{cases}$$

Proof. If $[\mathbf{a}]_i = 0$ for all $i \in P$, then $a_M = 0$ for all $M \subseteq P$, and then the sum is trivially 0. If exists $i \in P$ with $[\mathbf{a}]_i = 1$, then for all M not containing i we have $a_{M \cup \{i\}} \equiv a_M + 1 \mod 2$, so one of a_M and $a_{M \cup \{i\}}$ is 1, the other is 0, thus $a_M + a_{M \cup \{i\}} = 1$. Then

$$\sum_{M \subseteq P} a_M = \sum_{M \subseteq P \setminus \{i\}} (a_M + a_{M \cup \{i\}}) = \sum_{M \subseteq P \setminus \{i\}} 1 = 2^{|P|-1}. \qquad \square$$

Lemma 4. *Choose an arbitrary* $\mathbf{a} \in \mathcal{B}^t$ *and* $\mathbf{b} \in \mathcal{B}^t$ *in such a way, that* $\mathbf{a} \neq \mathbf{0}$ *and* $\mathbf{b} \neq \mathbf{0}$. *Define* $a : \mathcal{P}([t]) \to \mathcal{B}$ *and* $b : \mathcal{P}([t]) \to \mathcal{B}$ *by construction (14). Then*

$$\sum_{M \subseteq [t]} a_M b_M = \begin{cases} 2^{t-1} & \text{if } \mathbf{a} = \mathbf{b}; \\ 2^{t-2} & \text{if } \mathbf{a} \neq \mathbf{b}. \end{cases}$$

Proof. If $\mathbf{a} = \mathbf{b}$ then $\forall M \subseteq [t] : a_M = b_M$ thus $a_M = b_M = 0$ or $a_M = b_M = 1$. In both cases $a_M b_M = a_M$ holds:

$$\sum_{M \subseteq [t]} a_M b_M = \sum_{M \subseteq [t]} a_M,$$

and then we continue with Lemma 3:

$$\sum_{M \subseteq [t]} a_M b_M = 2^{t-1}.$$

If $[\mathbf{a}]_i \neq [\mathbf{b}]_i$ for some i, then either $[\mathbf{a}]_i = 0$ and $[\mathbf{b}]_i = 1$ or $[\mathbf{a}]_i = 1$ and $[\mathbf{b}]_i = 0$. If $[\mathbf{a}]_i = 0$ and $[\mathbf{b}]_i = 1$ then for all M not containing i we have $a_M = a_{M \cup \{i\}}$ moreover one of b_M and $b_{M \cup \{i\}}$ is 0 while the other is 1, so

$$a_M b_M + a_{M \cup \{i\}} b_{M \cup \{i\}} = a_M b_M + a_M b_{M \cup \{i\}}$$
$$= a_M (b_M + b_{M \cup \{i\}})$$
$$= a_M.$$

If the case is the opposite ($[\mathbf{a}]_i = 1$ and $[\mathbf{b}]_i = 0$), then the result will be

$$a_M b_M + a_{M \cup \{i\}} b_{M \cup \{i\}} = b_M.$$

For both cases we can finish by Lemma 3:

$$\sum_{M \subseteq [t]} a_M b_M = \sum_{M \subseteq [t] \setminus \{i\}} \left(a_M b_M + a_{M \cup \{i\}} b_{M \cup \{i\}} \right)$$

$$= \begin{cases} \sum_{M \subseteq [t] \setminus \{i\}} a_M \\ \sum_{M \subseteq [t] \setminus \{i\}} b_M \end{cases}$$

$$= 2^{t-2} \qquad\qquad\qquad\qquad \square$$

Lemma 5. *Let a_M be defined arbitrarily for all nonempty subsets M of a given $L \subseteq [t]$, and let us extend this function over all subsets of $[t]$ by construction (15). Then for every $P \subseteq [t]$ for which $P \nsubseteq L$*

$$\sum_{M \subseteq P} a_M = 2^{|P|-1}.$$

Proof. First observe, that the third case of (15) is valid for all not empty $M \subseteq [t]$. Since $P \nsubseteq L$ there must exists an $i \in P$ for which $i \notin L$. From (15) we have for all $M \subseteq [t] \setminus \{i\}$, that

$$\begin{aligned}
a_{M \cup \{i\}} &\equiv a_{(M \cup \{i\}) \cap L} + \left| (M \cup \{i\}) \setminus L \right| & \mod 2 \\
&\equiv a_{M \cap L} + \left| (M \setminus L) \cup \{i\} \right| & \mod 2 \\
&\equiv a_{M \cap L} + \left| M \setminus L \right| + 1 & \mod 2 \\
&\equiv a_M + 1. & \mod 2
\end{aligned}$$

And from this for all M not containing i we have $a_M + a_{M \cup \{i\}} = 1$ (one term is 0 while the other is 1). Now with this i:

$$\sum_{M \subseteq P} a_M = \sum_{M \subseteq P \setminus \{i\}} \left(a_M + a_{M \cup \{i\}} \right)$$

$$= \sum_{M \subseteq P \setminus \{i\}} 1$$

$$= 2^{|P|-1}. \qquad\qquad\qquad \square$$

Now we present some important lemmas on the previously defined matrices $\mathbf{A}_s$ and $\mathbf{B}_s^{(k)}$. First define $\mathbf{d}_s^{(P)} \in \mathcal{B}^{2^s - 1}$ for all $P \subseteq [s]$ in the following way:

$$[\mathbf{d}_s^{(P)}]_i = \begin{cases} 1 & \text{if } M_i \subseteq P; \\ 0 & \text{if } M_i \nsubseteq P. \end{cases}$$

Lemma 6. *For the above defined vectors $\mathbf{d}_s^{(P)}$ and for matrix $\mathbf{A}_s$*

$$\mathbf{d}_s^{(P)\top} \mathbf{A}_s = \mathbf{0}^\top \quad \mod 2^{|P|-1}.$$

Proof. By the definition of $\mathbf{d}_s^{(P)}$ and $\mathbf{A}_s$ we have

$$
\begin{aligned}
\left[\mathbf{d}_s^{(P)\top}\mathbf{A}_s\right]_j &= \sum_{i:M_i\subseteq P}[\mathbf{A}_s]_{ij} \\
&= \sum_{i:M_i\subseteq P} a_{M_i}^{(j)} \\
&= \sum_{\substack{M\subseteq P \\ M\neq\varnothing}} a_M^{(j)} \\
&= \sum_{M\subseteq P} a_M^{(j)}.
\end{aligned}
$$

By Lemma 3 this is $0 \mod 2^{|P|-1}$. $\qquad\square$

Lemma 7. *For the above defined vectors $\mathbf{d}_s^{(P)}$ and for matrix $\mathbf{B}_s^{(k)}$*

$$
\mathbf{d}_s^{(P)\top}\mathbf{B}_s^{(k)} = \begin{cases} (2^{|P|-1}\ 2^{|P|-1}\ \cdots\ 2^{|P|-1}) & \text{if } P\not\subseteq L_k; \\ (\ 2^0\quad\ 2^1\quad \cdots\ 2^{|P|-2}) & \text{if } P = L_k. \end{cases}
$$

Proof. By the definition of $\mathbf{d}_s^{(P)}$ we have

$$
\begin{aligned}
\left[\mathbf{d}_s^{(P)\top}\mathbf{B}_s^{(k)}\right]_j &= \sum_{i:M_i\subseteq P}[\mathbf{B}_s^{(k)}]_{ij} \\
&= \sum_{i:M_i\subseteq P} \hat{b}_{M_i}^{(k,j)} \\
&= \sum_{\substack{M\subseteq P \\ M\neq\varnothing}} \hat{b}_M^{(k,j)}.
\end{aligned}
$$

For $P = L_k$ this is

$$
\sum_{\substack{M\subseteq L_k \\ M\neq\varnothing}} \hat{b}_M^{(k,j)} = \sum_{\substack{M\subseteq L_k \\ M\neq\varnothing}} b_M^{(k,j)} = 2^{j-1},
$$

given by the construction of $b^{(k,j)}$. For $P\not\subseteq L_k$ this is $2^{|P|-1}$ by Lemma 5. $\quad\square$

Lemma 8. $\mathbf{A}_s$ *is nonsingular.*

Proof. According to Lemma 4,

$$
\mathbf{A}_s{}^\top\mathbf{A}_s = \begin{pmatrix} 2^{s-1} & 2^{s-2} & \cdots & 2^{s-2} \\ 2^{s-2} & 2^{s-1} & \cdots & 2^{s-2} \\ \vdots & \vdots & \mathbb{D}^2ots & \vdots \\ 2^{s-2} & 2^{s-2} & \cdots & 2^{s-1} \end{pmatrix}.
$$

Thus

$$\det(\mathbf{A}_s)^2 = \det(\mathbf{A}_s{}^\top \mathbf{A}_s) = 2^{2+(s-2)2^s}.$$

So $\det(\mathbf{A}_s) \neq 0$.   □

Proof of Theorem 5. To prove that $\mathbf{C}_s$ is a u.d. code matrix, we need to show, that condition (3) given in section 2.1 holds:

$$\mathbf{C}_s \mathbf{w} = \mathbf{0} \iff \mathbf{w} = \mathbf{0} \qquad \forall \mathbf{w} \in \{-1, 0, 1\}^{s2^{s-1}}.$$

Let us partition $\mathbf{w}$ into $\mathbf{w}_0 \in \{-1, 0, 1\}^{2^s-1}$ and $\mathbf{w}_k \in \{-1, 0, 1\}^{|L_k|-1}$ for all $k \in [2^s - s - 1]$.

$$\mathbf{w} = \begin{pmatrix} \mathbf{w}_0 \\ \mathbf{w}_1 \\ \vdots \\ \mathbf{w}_{2^s-s-1} \end{pmatrix}.$$

Now the u.d. condition can be written as

$$\mathbf{A}_s \mathbf{w}_0 + \sum_{k=1}^{2^s-s-1} \mathbf{B}_s^{(k)} \mathbf{w}_k = \mathbf{0} \iff \mathbf{w} = \mathbf{0}. \qquad \forall \mathbf{w} \in \{-1, 0, 1\}^{s2^{s-1}}. \tag{16}$$

First note, that $\impliedby$ is trivial. To see $\implies$, we will distinguish two cases. If $\mathbf{w}_k = \mathbf{0}$ holds for all $k \geq 1$, then this is trivial, since $\mathbf{A}_s$ is nonsingular (Lemma 8). If there are some k for which $\mathbf{w}_k \neq \mathbf{0}$, then we select that k^*, which corresponds to the L_k of the largest size:

$$\mathbf{w}_{k^*} \neq \mathbf{0},$$
$$\mathbf{w}_k = \mathbf{0} \qquad \forall k \in [2^s - s - 1]: |L_k| > |L_{k^*}|.$$

If we multiply the left side of the equivalence in (16) by $\mathbf{d}_s^{(L_{k^*})\top}$, then we get

$$\mathbf{d}_s^{(L_{k^*})\top} \mathbf{A}_s \mathbf{w}_0 + \sum_{k=1}^{2^s-s-1} \mathbf{d}_s^{(L_{k^*})\top} \mathbf{B}_s^{(k)} \mathbf{w}_k = \mathbf{0},$$

and using Lemma 6 and Lemma 7

$$\left(2^0\ 2^1\ \cdots\ 2^{|L_{k^*}|-2}\right) \mathbf{w}_{k^*} \equiv \mathbf{0} \quad \mod 2^{|L_{k^*}|-1}$$

But since $\mathbf{w}_{k^*} \in \{-1, 0, 1\}^{|L_{k^*}|-1}$, this implies $\mathbf{w}_{k^*} = \mathbf{0}$, which is a contradiction with the choice of k^*.   □

3   Partial Activity m-out-of-t Model

In this section we present the signature coding results for the partial activity m-out-of-t model for the binary adder channel. Recall, that in this model there are t total users of the channel, out of which at most m are active at any instant. The inactive users send the zero vector from their component code, while the active ones send their other (non-zero) codeword. The received vector is the vectorial sum of the sent codewords, and from this we should recover the set of active users. We still use the simple notation $\mathbf{S}(U)$ for the received vector if the active users are those in set U:

$$\mathbf{S}(U) = \sum_{u \in U} \mathbf{x}^{(u)}.$$

Definition 5. *For the m-out-of-t model, the **minimal signature code length** $N(t, m)$ is the length of the shortest possible u.d. signature code for a given number t of total users and for a given maximal number m of simultaneously active ones:*

$$N(t, m) = \min\{n : \exists \mathcal{C} \text{ u.d. signature code with length } n$$
$$\text{for } t \text{ users out of which at most } m \text{ are active simultaneously}\}.$$

We show two bounds for $N(t, m)$ saying that for $1 \ll m \ll t$,

$$\frac{2m}{\log m} \log t \lesssim N(t, m) \lesssim \frac{4m}{\log m} \log t.$$

Based on Bose and Chowla's [3] work, Lindström [13] constructed a u.d. signature code with code length

$$n \sim m \log t.$$

This is the best known construction so far, thus there is no asymptotically optimal code construction for the m-out-of-t model.

3.1   Bounds for U.D. Signature Codes

The next theorem gives the asymptotic upper bound on $N(t, m)$. This follows from a similar theorem of D'yachkov and Rykov in [7], see the remark after Theorem 8 and 9. We present here a more simple proof.

Theorem 6. *(D'yachkov–Rykov [7]) For $N(t, m)$ we have that*

$$\varlimsup_{m \to \infty} \varlimsup_{t \to \infty} N(t, m) \frac{\log m}{m \log t} \leq 4.$$

Proof. Let the length of codewords be $n + 1$, let the last component of each non-zero codeword be fixed to 1, and the rest of the components be randomly chosen from $\mathcal{B} = \{0, 1\}$ with uniform distribution:

$$\mathbb{P}\left(\left[\mathbf{x}^{(u)}\right]_k = 0\right) = \mathbb{P}\left(\left[\mathbf{x}^{(u)}\right]_k = 1\right) = \frac{1}{2} \quad \forall u \in \mathcal{U}, \forall k : 1 \leq k \leq n \text{ (i.i.d.)},$$
$$\left[\mathbf{x}^{(u)}\right]_{n+1} = 1 \qquad\qquad\qquad \forall u \in \mathcal{U}.$$

We will give an upper bound on the probability of the event

$$\text{Code } \mathcal{C} = \left\{ \{\mathbf{0}, \mathbf{x}^{(1)}\}, \{\mathbf{0}, \mathbf{x}^{(2)}\}, \ldots, \{\mathbf{0}, \mathbf{x}^{(t)}\} \right\} \text{ is } not \text{ a u.d. signature code } (*)$$

and then show that for a given m and t, and for an n great enough this bound is less than 1. So the probability of randomly selecting a good code is definitely positive, and this means that there exists a good code for that n great enough, so we have an upper bound on the minimal code length.

A code is *not* a signature code, if and only if there are two different subsets U and V of $\mathcal{U}$ which contain at most m users, and the sum of the corresponding code vectors are the same:

$$\mathbb{P}\big(\text{event } (*)\big) = \mathbb{P}\left(\bigcup_{\substack{U \neq V \subseteq [t]: \\ |U| \leq m, |V| \leq m}} \{\mathbf{S}(U) = \mathbf{S}(V)\} \right).$$

If there are two subsets U and V which satisfy $\mathbf{S}(U) = \mathbf{S}(V)$, then there are also two disjoint subsets which satisfy it (e.g. $U \setminus V$ and $V \setminus U$). Moreover, the $(n+1)^{\text{th}}$ component is 1 in all codewords, so the last component of the received vector is the size of the active set. Thus, if $\mathbf{S}(U) = \mathbf{S}(V)$ then $|U| = |V|$, thus it is enough to take into account disjoint subsets of equal size:

$$\mathbb{P}\big(\text{event } (*)\big) = \mathbb{P}\left(\bigcup_{\substack{U \neq V \subseteq [t]: \\ |U| = |V| \leq m \\ U \cap V = \varnothing}} \{\mathbf{S}(U) = \mathbf{S}(V)\} \right).$$

Now we can calculate the upper bound with the so called union bounding:

$$\mathbb{P}\big(\text{event } (*)\big) = \sum_{k=1}^{m} \mathbb{P}\left(\bigcup_{\substack{U, V \subseteq [t]: \\ |U| = |V| = k \\ U \cap V = \varnothing}} \{\mathbf{S}(U) = \mathbf{S}(V)\} \right)$$

$$\leq \sum_{k=1}^{m} \sum_{\substack{U, V \subseteq [t]: \\ |U| = |V| = k \\ U \cap V = \varnothing}} \mathbb{P}\big(\mathbf{S}(U) = \mathbf{S}(V)\big).$$

Here $\mathbb{P}\big(S(U) = S(V)\big) = Q^n(k)$, where $Q(k)$ is defined by (5) and bounded by (6):

$$\mathbb{P}\big(\text{event } (*)\big) \le \sum_{k=1}^{m} \sum_{\substack{U,V\subseteq[t]: \\ |U|=|V|=k \\ U\cap V=\varnothing}} Q^n(k)$$

$$\le \sum_{k=1}^{m} \binom{t}{k}\binom{t-k}{k}\left(\frac{1}{\sqrt{\pi k}}\right)^n$$

$$\le \sum_{k=1}^{m} t^{2k}\left(\frac{1}{\sqrt{\pi k}}\right)^n$$

$$\le m \max_{k:\,1\le k\le m} t^{2k}\left(\frac{1}{\sqrt{\pi k}}\right)^n$$

$$= m\exp\left(\max_{k:\,1\le k\le m}\left(2k\log t - \frac{n}{2}\log\pi k\right)\right).$$

The exponent is convex in k, so the maximum is either at $k = 1$ or at $k = m$:

$$\mathbb{P}\big(\text{event } (*)\big) \le m\exp\left(\max\left\{2\log t - \frac{n}{2}\log\pi,\, 2m\log t - \frac{n}{2}\log\pi m\right\}\right).$$

If we want to ensure that a u.d. code exists, it is enough to show that the probability of randomly selecting a non-u.d. code is less than 1, namely $\mathbb{P}(\text{event } (*)) < 1$. This is surely satisfied if our upper bound tends to 0 as $t \to \infty$. For this we require

$$\lim_{t\to\infty} 2\log t - \frac{n}{2}\log\pi = -\infty, \tag{17}$$

and

$$\lim_{t\to\infty} 2m\log t - \frac{n}{2}\log\pi m = -\infty. \tag{18}$$

Let us set

$$n = \left\lceil \frac{4m}{\log m}\log t \right\rceil. \tag{19}$$

Then (17) holds for all $m \ge 2$:

$$\lim_{t\to\infty} 2\log t - \frac{n}{2}\log\pi \le 2\left(1 - \frac{m\log\pi}{\log m}\right)\lim_{t\to\infty}\log t = -\infty,$$

and (18) also holds:

$$\lim_{t\to\infty} 2m\log t - \frac{n}{2}\log\pi m \le 2m\left(1 - \frac{\log\pi m}{\log m}\right)\lim_{t\to\infty}\log t$$

$$= -2m\frac{\log\pi}{\log m}\lim_{t\to\infty}\log t$$

$$= -\infty.$$

So if we choose n as given in (19) then a u.d. signature code of length $n + 1$ will exist, so the length of the shortest possible u.d. signature code is bounded upper for large t:

$$N(t,m) \le \left\lceil \frac{4m}{\log m}\log t \right\rceil + 1.$$

It follows, that

$$\varlimsup_{t\to\infty} N(t,m)\frac{\log m}{m\log t} \leq 4.$$

$\square$

We show in the next theorem, that asymptotically for $1 \ll m \ll t$ we have that $N(t,m) \gtrsim \frac{2m}{\log m}\log t$. This is new, but closely relates to Theorem 9 of D'yachkov and Rykov.

Theorem 7. *For $N(t,m)$ we have that*

$$\liminf_{m\to\infty}\liminf_{t\to\infty} N(t,m)\frac{\log m}{m\log t} \geq 2.$$

Proof. Take an arbitrary u.d. signature code of length n for t users out of which at most m are active, and let U (the set of active users) be a discrete random variable with uniform distribution over the $\binom{t}{m}$ m-sized subsets of $[t]$:

$$\mathbb{P}(U = A) = \begin{cases} \binom{t}{m}^{-1} & \text{if } A \subseteq [t] \text{ and } |A| = m; \\ 0 & \text{otherwise.} \end{cases}$$

We will bound the entropy of $\mathbf{S}(U)$ in two different ways, to get an upper and a lower bound. Then by joining these bounds, we will get a lower bound on the code length.

First we set the lower bound on $\mathbb{H}(\mathbf{S}(U))$, which is the Shannon–entropy of the random variable $\mathbf{S}(U)$:

$$\mathbb{H}(\mathbf{S}(U)) = \mathbb{H}(U),$$

since for all different values of U the corresponding $\mathbf{S}(U)$ values are different for a u.d. signature code. But U has uniform distribution, so for it is entropy

$$\mathbb{H}(U) = \log\binom{t}{m},$$

and now we are ready to derive the lower bound on $\mathbb{H}(\mathbf{S}(U))$:

$$\begin{aligned} \mathbb{H}(\mathbf{S}(U)) &= \mathbb{H}(U) \\ &= \log\binom{t}{m} \\ &= \log\frac{t(t-1)\cdots(t-(m-1))}{m(m-1)\cdots 1} \\ &\geq m\log\frac{t}{m}. \end{aligned} \tag{20}$$

Now we will derive an upper bound, via bounding the entropy of the individual components of $\mathbf{S}(U)$. We can easily get the distribution of the ith component, if we introduce w_i, which is the number of codewords having 1 in their i^{th} component:

$$w_i = \left|\left\{\mathbf{x}^{(u)} : [\mathbf{x}^{(u)}]_i = 1\right\}\right|.$$

The number of all possible values of U is $\binom{t}{m}$. Moreover, the number of those U values for which $[\mathbf{S}(U)]_i = k$ can be enumerated. First we select k users out of the w_i ones those have 1 in the ith component of their codeword. Then we select $m - k$ more out of the $t - w_i$ ones those have 0 there. This is $\binom{w_i}{k}\binom{t-w_i}{m-k}$, if $\max\{0, m - (t - w_i)\} \le k \le \min\{w_i, m\}$. So the distribution of $[\mathbf{S}(U)]_i$ is hypergeometrical, with parameters $(m, w_i, t - w_i)$:

$$
\mathbb{P}([\mathbf{S}(U)]_i = k) = \begin{cases} \dfrac{\binom{w_i}{k}\binom{t-w_i}{m-k}}{\binom{t}{m}} & \text{if } \max\{0, m - (t - w_i)\} \le k \le \min\{w_i, m\}; \\ 0 & \text{otherwise.} \end{cases}
$$

If we introduce $H_{\text{hyp}}(m, a, b)$ which is the entropy of the hypergeometrical distribution with parameters (m, a, b), then we have that

$$
\mathbb{H}([\mathbf{S}(U)]_i) = H_{\text{hyp}}(m, w_i, t - w_i).
$$

Since the entropy of a vector can be bounded by the sum of the entropies of its components, we get the following upper bound on $\mathbb{H}(\mathbf{S}(U))$:

$$
\mathbb{H}(\mathbf{S}(U)) \le \sum_{i=1}^{n} \mathbb{H}([\mathbf{S}(U)]_i)
$$

$$
= \sum_{i=1}^{n} H_{\text{hyp}}(m, w_i, t - w_i)
$$

$$
\le n \max_{w} H_{\text{hyp}}(m, w, t - w),
$$

and using Lemma 1 we get

$$
\mathbb{H}(\mathbf{S}(U)) \le n \max_{w} \frac{1}{2} \log \left(2\pi e \left(\mathbb{D}^2_{\text{hyp}}(m, w, t - w) + \frac{1}{12} \right) \right),
$$

where $\mathbb{D}^2_{\text{hyp}}(m, w, t - w)$ denotes the variance of the hypergeometrical distribution with parameters $(m, w, t - w)$. Therefore

$$
\mathbb{H}(\mathbf{S}(U)) \le n \max_{w} \frac{1}{2} \log \left(2\pi e \left(m \frac{w}{t} \left(1 - \frac{w}{t} \right) \left(1 - \frac{m-1}{t-1} \right) + \frac{1}{12} \right) \right)
$$

$$
\le n \frac{1}{2} \log \left(\frac{1}{2} \pi e \left(m \left(1 - \frac{m-1}{t-1} \right) + \frac{1}{12} \right) \right). \tag{21}
$$

Combining (20) and (21) we get

$$
m \log \frac{t}{m} \le \mathbb{H}(\mathbf{S}(U)) \le n \frac{1}{2} \log \left(\frac{1}{2} \pi e \left(m \left(1 - \frac{m-1}{t-1} \right) + \frac{1}{12} \right) \right),
$$

which holds for all u.d. signature codes, including the shortest possible one. So

$$
N(t, m) \ge \frac{m \log \frac{t}{m}}{\frac{1}{2} \log \left(\frac{1}{2} \pi e \left(m \left(1 - \frac{m-1}{t-1} \right) + \frac{1}{12} \right) \right)},
$$

form which

$$
\liminf_{m \to \infty} \liminf_{t \to \infty} N(t, m) \frac{\log m}{m \log t} \ge 2. \qquad \square
$$

3.2 Bounds for B_m Codes

D'yachkov and Rykov [7] considered a special class of u.d. signature codes, namely the B_m codes.

Definition 6. *A B_m code is a set of t binary codewords of length n*

$$\mathcal{C} = \left\{ \mathbf{x}^{(1)}, \mathbf{x}^{(2)}, \ldots, \mathbf{x}^{(t)} \right\} \subseteq \mathcal{B}^n$$

which has the following property: all sums of exactly m (not definitely different) codewords are different.

It is obvious, that a B_m code $\mathcal{C}$ can be converted into a u.d. signature code $\mathcal{C}_s$ for the adder channel. What we have to do, is just to append a fixed 1 bit to the end of all the codewords in the B_m code:

$$\mathcal{C}_s = \left\{ \{\mathbf{0}, \mathbf{y}\} : \mathbf{y} \in \mathcal{B}^{n+1}, \exists \mathbf{x} \in \mathcal{C} : \forall j \in [n] : [\mathbf{y}]_j = [\mathbf{x}]_j \text{ and } [\mathbf{y}]_{n+1} = 1 \right\}.$$

The length of this signature code is $n + 1$. To see that this is really a u.d. signature code, we indirectly put up that there are two different subsets U and V of the users for which the sum vector is the same. If the size of U and V cannot differ, since then the $(n + 1)^{\text{th}}$ component of the sum vectors would also differ. So we can assume that $|U| = |V|$. Now take the following (exactly) m codewords: all the codewords in U plus $\mathbf{x}^{(1)}$ as many times as needed to get exactly m codewords ($m - |U|$ times). The sum vector of this multiset must be equal to the sum vector of all the codewords in V plus $\mathbf{x}^{(1)}$ $m - |V|$ times. But then we have found two multisets of codewords with m elements in the original B_m code $\mathcal{C}$, for which the sum vector is the same. This is a contradiction with the definition of the B_m codes.

D'yachkov and Rykov [7] have given upper and lower bounds on $N_B(t, m)$, which is the length of the shortest possible B_m code for t total users out of which at most m are active simultaneously. Their main result is that for $1 \ll m \ll t$,

$$\frac{2m}{\log m} \log t \lesssim N_B(t, m) \lesssim \frac{4m}{\log m} \log t. \tag{22}$$

More precisely, for the length of the shortest possible signature code, they have proven that

Theorem 8. *If $t \to \infty$ then for any fixed m*

$$N_B(t, m) \leq \frac{2m}{\log \dfrac{2^{2m}}{\binom{2m}{m}}} (1 + o(1)) \log t.$$

Theorem 9. *For any $m < t$*

$$N_B(t, m) \geq \frac{\log \dfrac{t^m}{m!}}{\mathrm{H}_{\mathrm{bin}}(m, \frac{1}{2})}.$$

From these theorems, formula (22) easily follows. Using the construction of signature codes from B_m codes, Theorem 6 also follows.

References

1. R. Ahlswede, Multi–way communication channels, Proceedings of the 2nd International Symposium on Information Theory, Hungarian Academy of Sciences, 23–52, 1971.
2. R. Ahlswede and V. B. Balakirsky, Construction of uniquely decodable codes for the two-user binary adder channel, IEEE Trans. Inform. Theory, Vol. 45, No. 1, 326-330, 1999.
3. R. C. Bose and S. Chowla, Theorems in the additive theory of numbers, Commentarii Mathematici Helvetici 37, 141–147, 1962.
4. D.G. Cantor and W.H. Mills, Determination of a subset from certain combinatorial properties, Canadian Journal of Mathematics 18, 42–48, 1966.
5. S. C. Chang and E. J. Weldon, Coding for T-user multiple-access channels, IEEE Trans. Inform. Theory, Vol. 25, No. 6, 684–691, 1979.
6. T. M. Cover and J. A. Thomas, Elements of Information Theory, John Wiley & Sons, 1991.
7. A. G. Dyachkov and V. V. Rykov, A coding model for a multiple-access adder channel, Problems of Information Transmission 17, 2, 26–38, 1981.
8. P. Erdős and A. Rényi, On two problems of information theory, Publications of the Mathematical Institute of the Hungarian Academy of Science 8, 229–243, 1963.
9. R. G. Gallager, Information Theory and Reliable Communication, John Wiley & Sons, 1968.
10. G. H. Khachatrian, A survey of coding methods for the adder channel, Numbers, Information, and Complexity (Festschrift for Rudolf Ahlswede), eds: I. Althöfer, N. Cai, G. Dueck, L. Khachatrian, M. Pinsker, A. Sárközy, I. Wegener, and Z. Zhang, Kluwer, 181-196, 2000.
11. H. Liao, A coding theorem for multiple access communications, Proceedings of the International Symposium on Information Theory, 1972.
12. B. Lindström, On a combinatory detection problem I, Publications of the Mathematical Institute of the Hungarian Academy of Science 9, 195–207, 1964.
13. B. Lindström, Determining subsets by unramified experiments, J. Srivastava, editor, A Survey of Statistical Designs and Linear Models, North Holland Publishing Company, 407–418, 1975.
14. C. E. Shannon, Two–way communication channels, Proceedings of the Fourth Berkeley Symposium on Mathematical Statistics and Probability, University of California Press, Vol. 1, 611–644, 1961.
15. E. C. van der Meulen, The discrete memoryless channel with two senders and one receiver, Proceedings of the 2nd International Symposium on Information Theory, Hungarian Academy of Sciences, 103–135, 1971.

Bounds of E-Capacity for Multiple-Access Channel with Random Parameter

M.E. Haroutunian[*]

Abstract. The discrete memoryless multiple-access channel with random parameter is investigated. Various situations, when the state of the channel is known or unknown on the encoders and decoder, are considered. Some bounds of E-capacity and capacity regions for average error probability are obtained.

1 Introduction

The discrete memoryless **multiple-access channel** (MAC) with two encoders and one decoder $W = \{W : \mathcal{X}_1 \times \mathcal{X}_2 \to \mathcal{Y}\}$ is defined by a matrix of transition probabilities

$$W = \{W(y|x_1, x_2), x_1 \in \mathcal{X}_1, x_2 \in \mathcal{X}_2, y \in \mathcal{Y}\},$$

where $\mathcal{X}_1$ and $\mathcal{X}_2$ are the finite alphabets of the first and the second inputs of the channel, respectively, and $\mathcal{Y}$ is the finite output alphabet.

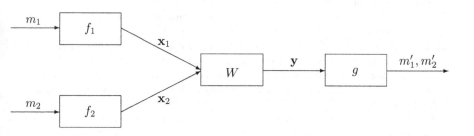

Fig. 1. Regular MAC

The **multiple-access channel** $W(Q)$ **with random parameter** is a family of discrete memoryless MAC $W_s : \mathcal{X}_1 \times \mathcal{X}_2 \to \mathcal{Y}$, where s is the channel state, varying independently in each moment with the same probability distribution $Q(s)$ on a finite set $\mathcal{S}$. In other words we have a set of conditional probabilities

$$W_s = \{W(y|x_1, x_2, s), x_1 \in \mathcal{X}_1, x_2 \in \mathcal{X}_2, y \in \mathcal{Y}\}, s \in \mathcal{S}.$$

If there is no distribution on the state sequence then the channel is called **arbitrarily varying MAC**.

[*] Work was partially supported by INTAS grant 00-738.

R. Ahlswede et al. (Eds.): Information Transfer and Combinatorics, LNCS 4123, pp. 196–217, 2006.
© Springer-Verlag Berlin Heidelberg 2006

Let $\mathcal{Q}(\mathcal{S})$ is the set of all probability distributions on $\mathcal{S}$.

The MAC with random parameter is memoryless and stationary, that is for N-length sequences

$$\mathbf{x}_1 = (x_{11}, x_{12}, \ldots, x_{1N}) \in \mathcal{X}_1^N, \quad \mathbf{x}_2 = (x_{21}, x_{22}, \ldots, x_{2N}) \in \mathcal{X}_2^N,$$

$$\mathbf{y} = (y_1, y_2, \ldots, y_N) \in \mathcal{Y}^N, \quad \mathbf{s} = (s_1, s_2, \ldots, s_N) \in \mathcal{S}^N,$$

the transition probabilities are given in the following way

$$W^N(\mathbf{y}|\mathbf{x}_1, \mathbf{x}_2, \mathbf{s}) = \prod_{n=1}^N W(y_n|x_{1n}, x_{2n}, s), \quad Q^N(\mathbf{s}) = \prod_{n=1}^N Q(s_n).$$

Here we investigate the so-called E-**capacity region**. The investigation of optimal rates of codes, ensuring when N increases the error probability exponential decrease with given exponent (reliability) E, is equivalent to studying of error exponents but sometimes is more expedient. E-capacity region is the generalization of the regular capacity region (reducing to the latter when $E \to 0$). In this paper some bounds of E-capacity region for the MAC with random parameter are obtained.

The MAC for the case of sending two independent messages (fig. 1) was considered by Ahlswede, who proved the simple characterization for the capacity region of this model [1], [2]. The E-capacity region was investigated in [3] and [4], where bounds for E-capacity regions for several models of discrete memoryless MAC were obtained. Various bounds of error probability exponents for MAC have been derived in [5], [6], [7], [8].

The oneway channel with random parameter was investigated by Gelfand and Pinsker [9]. They found the capacity of this channel in the situation, when the sequence $\mathbf{s}$ is known at the encoder, but unknown at the decoder. Some upper and lower bounds of E-capacity for channel with random parameter were obtained in [10], [11], [12].

Time varying MAC were considered in [13], [14], [15], [16], [17], [18]. Here we consider the MAC with random parameter, which can be considered in various situations, when the whole state sequence $\mathbf{s}$ is known or unknown at the encoders and at the decoder. The considered cases are not equally interesting from the mathematical point of view. The most interesting is the case, when the state sequence is known on one of the encoders and unknown on the other encoder and on the decoder.

2   Formulation of Results

Let $\mathcal{M}_1 = \{1, 2, \ldots, M_1\}$ and $\mathcal{M}_2 = \{1, 2, \ldots, M_2\}$ be the message sets of corresponding sources. The numbers

$$\frac{1}{N} \log M_i, \quad i = 1, 2,$$

are called transmission rates. We use the logarithmical and exponential functions to the base 2. Denote by $e(m_1, m_2, \mathbf{s})$ the probability of error transmission of the messages $m_1 \in \mathcal{M}_1$, $m_2 \in \mathcal{M}_2$ for given $\mathbf{s}$. We study the average error probability of the code:

$$\bar{e}(N, W(Q)) = \frac{1}{M_1 M_2} \sum_{m_1, m_2} \sum_{\mathbf{s} \in \mathcal{S}^N} Q^N(\mathbf{s}) e(m_1, m_2, \mathbf{s}). \tag{1}$$

Let E is a positive number called reliability. Nonnegative numbers R_1, R_2 are called E-achievable rates for MAC with random parameter $W(Q)$, if for any $\delta_i > 0$, $i = 1, 2$, for sufficiently large N there exists a code such that

$$\frac{1}{N} \log M_i \geq R_i - \delta_i, \ i = 1, 2, \tag{2}$$

and the average error probability satisfies the condition

$$\bar{e}(N, W(Q)) \leq \exp\{-NE\}. \tag{3}$$

The region of all E-achievable rates is called E-capacity region for average error probability and denoted $\overline{C}(E, W(Q))$. When $E \rightarrow 0$ we obtain the capacity region $\overline{C}(W(Q))$ of the channel $W(Q)$ for average probability of error.

For our notations of entropies, mutual informations, divergences as well as for the notions of types, conditional types and some basic relations we refer to [19], [20], [12]. In particular, we use the following notations: the set of all sequences $\mathbf{s}$ of type $Q = \{Q(s), s \in \mathcal{S}\}$ on $\mathcal{S}^N$ is denoted by $T_Q^N(S)$, the set of all sequences $(\mathbf{x}_1, \mathbf{x}_2)$ of conditional type $P = \{P(x_1, x_2|s), x_1 \in \mathcal{X}_1, x_2 \in \mathcal{X}_2, s \in \mathcal{S}\}$, for given vector $\mathbf{s} \in T_Q^N(S)$ is denoted by $T_{Q,P}^N(X_1 X_2|\mathbf{s})$, and so on. The following representation will be used in the proofs: for $\mathbf{s} \in T_Q^N(S)$, $(\mathbf{x}_1, \mathbf{x}_2) \in T_{Q,P}^N(X_1 X_2|\mathbf{s})$, $\mathbf{y} \in T_{Q,P,V}^N(Y|\mathbf{x}_1, \mathbf{x}_2, \mathbf{s})$

$$W^N(\mathbf{y}|\mathbf{x}_1, \mathbf{x}_2, \mathbf{s}) = \exp\{-N(H_{Q,P,V}(Y|X_1, X_2, S) + D(V\|W|Q, P))\}. \tag{4}$$

Note also that

$$D(Q' \circ P \circ V \| Q \circ P^* \circ W) = D(Q'\|Q) + D(P\|P^*|Q') + D(V\|W|Q', P). \tag{5}$$

1. Consider the case when s is known at the encoders and decoder
The code of length N for this model is a collection of mappings (f_1, f_2, g), where $f_1 : \mathcal{M}_1 \times \mathcal{S}^N \rightarrow \mathcal{X}_1^N$ and $f_2 : \mathcal{M}_2 \times \mathcal{S}^N \rightarrow \mathcal{X}_2^N$ are encodings and $g : \mathcal{Y}^N \times \mathcal{S}^N \rightarrow \mathcal{M}_1 \times \mathcal{M}_2$ is decoding. Denote

$$f_1(m_1, \mathbf{s}) = \mathbf{x}_1(m_1, \mathbf{s}), \ f_2(m_2, \mathbf{s}) = \mathbf{x}_2(m_2, \mathbf{s}),$$

$$g^{-1}(m_1, m_2, \mathbf{s}) = \{\mathbf{y} : g(\mathbf{y}) = (m_1, m_2, \mathbf{s})\},$$

then

$$e(m_1, m_2, \mathbf{s}) = W^N\{\mathcal{Y}^N - g^{-1}(m_1, m_2, \mathbf{s})|f_1(m_1, \mathbf{s}), f_2(m_2, \mathbf{s}), \mathbf{s}\}$$

is the error probability of the transmission of messages m_1 and m_2. Let random variables X_1, X_2, Y, S take values in alphabets $\mathcal{X}_1, \mathcal{X}_2, \mathcal{Y}, \mathcal{S}$ respectively with the following probability distributions :

$$Q = \{Q(s), s \in \mathcal{S}\},$$

$$P_i^* = \{P_i^*(x_i|s), x_i \in \mathcal{X}_i\},\ i = 1, 2,$$

$$P^* = \{P_1^*(x_1|s)P_2^*(x_2|s), x_1 \in \mathcal{X}_1, x_2 \in \mathcal{X}_2\},$$

$$P = \{P(x_1, x_2|s), x_1 \in \mathcal{X}_1, x_2 \in \mathcal{X}_2\},$$

with

$$\sum_{x_{3-i}} P(x_1, x_2|s) = P_i^*(x_i|s),\ i = 1, 2,$$

and joint probability distribution

$$Q \circ P \circ V = \{Q(s)P(x_1, x_2|s)V(y|x_1, x_2, s), s \in \mathcal{S},\ x_1 \in \mathcal{X}_1, x_2 \in \mathcal{X}_2, y \in \mathcal{Y}\},$$

where $V = \{V(y|x_1, x_2, s), s \in \mathcal{S}, x_1 \in \mathcal{X}_1, x_2 \in \mathcal{X}_2, y \in \mathcal{Y}\}$ is some conditional probability distribution.

The following region is called *random coding* bound:

$$\mathcal{R}_r(E, W(Q)) = co\{\bigcup_{P^*} \mathcal{R}_r(P^*, E, W(Q))\},$$

where $|a|^+ = \max(o, a)$, $co\{\mathcal{R}\}$ is the convex hull of the region $\mathcal{R}_r$ and $\mathcal{R}_r(P^*, E, W(Q)) = \{(R_1, R_2) :$

$$0 \le R_1 \le \min_{Q',P,V:D(Q'\circ P\circ V\|Q\circ P^*\circ W)\le E} |I_{Q',P,V}(X_1 \wedge X_2, Y|S)$$

$$+D(Q' \circ P \circ V\|Q \circ P^* \circ W) - E|^+,$$

$$0 \le R_2 \le \min_{Q',P,V:D(Q'\circ P\circ V\|Q\circ P^*\circ W)\le E} |I_{Q',P,V}(X_2 \wedge X_1, Y|S)$$

$$+D(Q' \circ P \circ V\|Q \circ P^* \circ W) - E|^+,$$

$$R_1+R_2 \le \min_{Q',P,V:D(Q'\circ P\circ V\|Q\circ P^*\circ W)\le E} |I_{Q',P,V}(X_1, X_2 \wedge Y|S)+I_{Q',P}(X_1 \wedge X_2|S)$$

$$+D(Q' \circ P \circ V\|Q \circ P^* \circ W) - E|^+\}.$$

The next region is called *sphere packing* bound:

$$R_{sp}(E, W(Q)) = co\bigcup_P R_{sp}(P, E, W(Q)),$$

where
$$R_{sp}(P, E, W(Q)) = \{(R_1, R_2) :$$

$$0 \leq R_1 \leq \min_{Q', V : D(Q' \circ P \circ V \| Q \circ P \circ W) \leq E} I_{Q', P, V}(X_1 \wedge Y | X_2, S),$$

$$0 \leq R_2 \leq \min_{Q', V : D(Q' \circ P \circ V \| Q \circ P \circ W) \leq E} I_{Q', P, V}(X_2 \wedge Y | X_1, S),$$

$$R_1 + R_2 \leq \min_{Q', V : \ D(Q' \circ P \circ V \| Q \circ P \circ W) \leq E} I_{Q', P, V}(X_1, X_2 \wedge Y | S)\}.$$

The following theorem takes place.

Theorem. For all $E > 0$, for MAC with random parameter the following inclusions are valid

$$\mathcal{R}_r(E, W(Q)) \subseteq \overline{C}(E, W(Q)) \subseteq \mathcal{R}_{sp}(E, W(Q)).$$

Corollary. When $E \to 0$, we obtain the inner and outer estimates for the channel capacity region, the expressions of which are similar but differ by the probability distributions P and P^*. The inner bound is:

$$\mathcal{R}_r(P^*, W(Q)) = \{(R_1, R_2) : \ 0 \leq R_i \leq I_{Q, P^*, W}(X_i \wedge Y | X_{3-i}, S), \ i = 1, 2,$$

$$R_1 + R_2 \leq I_{Q, P^*, W}(X_1, X_2 \wedge Y | S)\}.$$

2. The states are unknown at the encoders and decoder

For this model the mappings (f_1, f_2, g) are $f_1 : \mathcal{M}_1 \to \mathcal{X}_1^N$, $f_2 : \mathcal{M}_2 \to \mathcal{X}_2^N$ and $g : \mathcal{Y}^N \to \mathcal{M}_1 \times \mathcal{M}_2$. Then

$$f_1(m_1) = \mathbf{x}_1(m_1), \ f_2(m_2) = \mathbf{x}_2(m_2),$$

$$g^{-1}(m_1, m_2) = \{\mathbf{y} : g(\mathbf{y}) = (m_1, m_2)\},$$

and the error probability of the transmission of messages m_1 and m_2 is

$$e(m_1, m_2, \mathbf{s}) = W^N\{\mathcal{Y}^N - g^{-1}(m_1, m_2) | f_1(m_1), f_2(m_2), \mathbf{s}\}.$$

Consider the distributions

$$Q = \{Q(s), s \in \mathcal{S}\},$$

$$P_i^* = \{P_i^*(x_i), x_i \in \mathcal{X}_i\}, \ i = 1, 2,$$

$$P^* = \{P_1^*(x_1) P_2^*(x_2), x_1 \in \mathcal{X}_1, x_2 \in \mathcal{X}_2\},$$

$$P = \{P(x_1, x_2), x_1 \in \mathcal{X}_1, x_2 \in \mathcal{X}_2\},$$

$$V = \{V(y | x_1, x_2), x_1 \in \mathcal{X}_1, x_2 \in \mathcal{X}_2, y \in \mathcal{Y}\},$$

and

$$W^*(y | x_1, x_2) = \sum_{s \in \mathcal{S}} Q(s) W(y | x_1, x_2, s).$$

In this case the bounds in the theorem take the following form:
$$\mathcal{R}_r(P^*, E, W(Q)) = \{(R_1, R_2) :$$

$$0 \leq R_1 \leq \min_{P,V:D(P\circ V \| P^* \circ W^*) \leq E} |I_{P,V}(X_1 \wedge X_2, Y) + D(P \circ V \| P^* \circ W^*) - E|^+,$$

$$0 \leq R_2 \leq \min_{P,V:D(P\circ V \| P^* \circ W^*) \leq E} |I_{P,V}(X_2 \wedge X_1, Y) + D(P \circ V \| P^* \circ W^*) - E|^+,$$

$$R_1 + R_2 \leq \min_{P,V:D(P\circ V \| P^* \circ W^*) \leq E} |I_{P,V}(X_1, X_2 \wedge Y) + I_P(X_1 \wedge X_2) +$$

$$+ D(P \circ V \| P^* \circ W^*) - E|^+\},$$

and
$$R_{sp}(P, E, W(Q)) = \{(R_1, R_2) :$$

$$0 \leq R_1 \leq \min_{V:D(V\| W^* | P) \leq E} I_{P,V}(X_1 \wedge Y | X_2),$$

$$0 \leq R_2 \leq \min_{V:D(V\| W^* | P) \leq E} I_{P,V}(X_2 \wedge Y | X_1),$$

$$R_1 + R_2 \leq \min_{V:D(V\| W^* | P) \leq E} I_{P,V}(X_1, X_2 \wedge Y)\}.$$

Corollary. When $E \to 0$, we obtain the inner and outer estimates for the channel capacity region, the expressions of which as in the previous case are similar but differ by the probability distributions P and P^*. The inner bound is:

$$\mathcal{R}_r(P^*, W(Q)) = \{(R_1, R_2) : \ 0 \leq R_i \leq I_{P^*, W^*}(X_i \wedge Y | X_{3-i}), \ i = 1, 2,$$

$$R_1 + R_2 \leq I_{P^*, W^*}(X_1, X_2 \wedge Y)\}.$$

3. Now consider the case, when s is known on the decoder and unknown on the encoders
The code (f_1, f_2, g) in this case is a collection of mappings $f_1 : \mathcal{M}_1 \to \mathcal{X}_1^N$, $f_2 : \mathcal{M}_2 \to \mathcal{X}_2^N$ and $g : \mathcal{Y}^N \times \mathcal{S} \to \mathcal{M}_1 \times \mathcal{M}_2$. Then the error probability of the transmission of messages m_1 and m_2 will be

$$e(m_1, m_2, \mathbf{s}) = W^N \{ \mathcal{Y}^N - g^{-1}(m_1, m_2, \mathbf{s}) | f_1(m_1), f_2(m_2), \mathbf{s} \}. \tag{6}$$

For this model the following distributions are present in the formulation of the results
$$P_i^* = \{P_i^*(x_i), x_i \in \mathcal{X}_i\}, \ i = 1, 2,$$

$$P^* = \{P_1^*(x_1) P_2^*(x_2), x_1 \in \mathcal{X}_1, x_2 \in \mathcal{X}_2\},$$

$$P = \{P(x_1, x_2), x_1 \in \mathcal{X}_1, x_2 \in \mathcal{X}_2\},$$

$$Q' = \{Q'(s | x_1, x_2), s \in \mathcal{S}, x_1 \in \mathcal{X}_1, x_2 \in \mathcal{X}_2\},$$

$$V = \{V(y | x_1, x_2, s), s \in \mathcal{S}, x_1 \in \mathcal{X}_1, x_2 \in \mathcal{X}_2, y \in \mathcal{Y}\}.$$

Then the bounds in the theorem take the following form:
$\mathcal{R}_r(P^*, E, W(Q)) = \{(R_1, R_2) :$

$$0 \le R_1 \le \min_{Q', P, V : D(Q' \circ P \circ V \| Q \circ P^* \circ W) \le E} |I_{Q', P, V}(X_1 \wedge X_2, S, Y) +$$

$$+ D(Q' \circ P \circ V \| Q \circ P^* \circ W) - E|^+,$$

$$0 \le R_2 \le \min_{Q', P, V : D(Q' \circ P \circ V \| Q \circ P^* \circ W) \le E} |I_{Q', P, V}(X_2 \wedge X_1, S, Y) +$$

$$+ D(Q' \circ P \circ V \| Q \circ P^* \circ W) - E|^+,$$

$$R_1 + R_2 \le \min_{Q', P, V : D(Q' \circ P \circ V \| Q \circ P^* \circ W) \le E} |I_{Q', P, V}(X_1, X_2 \wedge S, Y) + I_P(X_1 \wedge X_2) +$$

$$+ D(Q' \circ P \circ V \| Q \circ P^* \circ W) - E|^+\},$$

and
$\mathcal{R}_{sp}(P, E, W(Q)) = \{(R_1, R_2) :$

$$0 \le R_1 \le \min_{Q', V : D(Q' \circ V \| Q \circ W | P) \le E} I_{Q', P, V}(X_1 \wedge Y, S | X_2), \tag{7}$$

$$0 \le R_2 \le \min_{Q', V : D(Q' \circ V \| Q \circ W | P) \le E} I_{Q', P, V}(X_2 \wedge Y, S | X_1), \tag{8}$$

$$R_1 + R_2 \le \min_{Q', V : D(Q' \circ V \| Q \circ W | P) \le E} I_{P, V}(X_1, X_2 \wedge Y, S)\}. \tag{9}$$

The statement of the corollary is the same with the inner bound of the channel capacity:

$$\mathcal{R}_r(P^*, W(Q)) = \{(R_1, R_2) : \ 0 \le R_i \le I_{Q, P^*, W}(X_i \wedge S, Y | X_{3-i}), \ i = 1, 2,$$

$$R_1 + R_2 \le I_{Q, P^*, W}(X_1, X_2 \wedge S, Y)\}.$$

4. The state of the channel is known on the encoders and unknown on the decoder

In this case $f_1 : \mathcal{M}_1 \times \mathcal{S}^N \to \mathcal{X}_1^N$ and $f_2 : \mathcal{M}_2 \times \mathcal{S}^N \to \mathcal{X}_2^N$ are encodings and $g : \mathcal{Y}^N \to \mathcal{M}_1 \times \mathcal{M}_2$ is decoding.

$$e(m_1, m_2, \mathbf{s}) = W^N \{\mathcal{Y}^N - g^{-1}(m_1, m_2) | f_1(m_1, \mathbf{s}), f_2(m_2, \mathbf{s}), \mathbf{s}\}$$

is the error probability of the transmission of messages m_1 and m_2. Let the auxiliary random variables U_1, U_2 take values correspondingly in some finite sets $\mathcal{U}_1, \mathcal{U}_2$. Then with

$$Q = \{Q(s), s \in \mathcal{S}\},$$

$$P_i^* = \{P_i^*(u_i, x_i | s), x_i \in \mathcal{X}_i, u_i \in \mathcal{U}_i\}, \ i = 1, 2,$$

$$P^* = \{P_1^*(u_1, x_1 | s) P_2^*(u_2, x_2 | s), x_1 \in \mathcal{X}_1, x_2 \in \mathcal{X}_2\},$$

$$P = \{P(u_1, u_2, x_1, x_2 | s), x_1 \in \mathcal{X}_1, x_2 \in \mathcal{X}_2\},$$

and

$$V = \{V(y | x_1, x_2, s), s \in \mathcal{S}, x_1 \in \mathcal{X}_1, x_2 \in \mathcal{X}_2, y \in \mathcal{Y}\}$$

the random coding bound in the theorem will be written in the following way:

$$\mathcal{R}_r(P^*, E, W(Q)) = \{(R_1, R_2) :$$

$$0 \le R_1 \le \min_{Q', P, V : D(Q' \circ P \circ V \| Q \circ P^* \circ W) \le E} |I_{Q', P, V}(U_1 \wedge U_2, Y) - I_{Q', P}(U_1 \wedge S) +$$

$$+ D(Q' \circ P \circ V \| Q \circ P^* \circ W) - E|^+,$$

$$0 \le R_2 \le \min_{Q', P, V : D(Q' \circ P \circ V \| Q \circ P^* \circ W) \le E} |I_{Q', P, V}(U_2 \wedge U_1, Y) - I_{Q', P}(U_2 \wedge S) +$$

$$+ D(Q' \circ P \circ V \| Q \circ P^* \circ W) - E|^+,$$

$$R_1 + R_2 \le \min_{Q', P, V : D(Q' \circ P \circ V \| Q \circ P^* \circ W) \le E} |I_{Q', P, V}(U_1, U_2 \wedge Y) - I_{Q', P}(U_1, U_2 \wedge S) +$$

$$+ I_{Q', P}(U_1 \wedge U_2) + D(Q' \circ P \circ V \| Q \circ P^* \circ W) - E|^+.$$

Corollary. When $E \to 0$, we obtain the inner estimate for the channel capacity region:

$$\mathcal{R}_r(P^*, W(Q)) = \{(R_1, R_2) : \ 0 \le R_i \le I_{Q, P^*, W}(U_i \wedge Y | U_{3-i}) - I_{Q, P^*}(U_1 \wedge S), \ i = 1, 2,$$

$$R_1 + R_2 \le I_{Q, P^*, W}(U_1, U_2 \wedge Y) - I_{Q, P^*}(U_1, U_2 \wedge S)\}.$$

This result was obtained by Pinsker [18].

5. The state is known on one of the encoders and unknown on the other encoder and on the decoder

For distinctness we shall assume, that the first encoder has an information about the state of the channel. Then the code will consist of the following mappings: $f_1 : \mathcal{M}_1 \times \mathcal{S}^N \to \mathcal{X}_1^N$ and $f_2 : \mathcal{M}_2 \to \mathcal{X}_2^N$ are encodings and $g : \mathcal{Y}^N \to \mathcal{M}_1 \times \mathcal{M}_2$ is decoding. The error probability of the transmission of messages m_1 and m_2 is

$$e(m_1, m_2, \mathbf{s}) = W^N \{\mathcal{Y}^N - g^{-1}(m_1, m_2) | f_1(m_1, \mathbf{s}), f_2(m_2), \mathbf{s}\}. \tag{10}$$

Let the auxiliary random variable U take values in some finite set $\mathcal{U}$. Then with

$$Q = \{Q(s), s \in \mathcal{S}\},$$

$$P_1^* = \{P_1^*(u, x_1 | s), x_1 \in \mathcal{X}_1, u \in \mathcal{U}\},$$

$$P_2^* = \{P_2^*(x_2), x_2 \in \mathcal{X}_2\},$$

$$P^* = \{P_1^*(u, x_1 | s) P_2^*(x_2), x_1 \in \mathcal{X}_1, x_2 \in \mathcal{X}_2\},$$

$$P = \{P(u, x_1, x_2 | s), x_1 \in \mathcal{X}_1, x_2 \in \mathcal{X}_2\},$$

and

$$V = \{V(y | x_1, x_2, s), s \in \mathcal{S}, x_1 \in \mathcal{X}_1, x_2 \in \mathcal{X}_2, y \in \mathcal{Y}\}$$

the random coding bound in the theorem will be:

$$\mathcal{R}_r(P^*, E, W(Q)) = \{(R_1, R_2) :$$

$$0 \leq R_1 \leq \min_{Q',P,V:D(Q' \circ P \circ V \| Q \circ P^* \circ W) \leq E} \left| I_{Q',P,V}(U \wedge Y, X_2) - I_{Q',P_1^*}(U \wedge S) + \right.$$

$$\left. + D(Q' \circ P \circ V \| Q \circ P^* \circ W) - E \right|^+, \tag{11}$$

$$0 \leq R_2 \leq \min_{Q',P,V:D(Q' \circ P \circ V \| Q \circ P^* \circ W) \leq E} \left| I_{Q',P,V}(X_2 \wedge Y, U) - I_{Q',P_1^*}(U \wedge S) + \right.$$

$$\left. + D(Q' \circ P \circ V \| Q \circ P^* \circ W) - E \right|^+, \tag{12}$$

$$R_1 + R_2 \leq \min_{Q',P,V:D(Q' \circ P \circ V \| Q \circ P^* \circ W) \leq E} \left| I_{Q',P,V}(U, X_2 \wedge Y) + I_{Q',P}(U \wedge X_2) - I_{Q',P_1^*}(U \wedge S) + \right.$$

$$\left. + D(Q' \circ P \circ V \| Q \circ P^* \circ W) - E \right|^+ \}. \tag{13}$$

Corollary. When $E \to 0$, we obtain the inner estimate for the channel capacity region:

$$\mathcal{R}_r(P^*, W(Q)) = \{(R_1, R_2) : \ 0 \leq R_1 \leq I_{Q,P^*,W}(U \wedge Y | X_2) - I_{Q,P_1^*}(U \wedge S),$$

$$0 \leq R_2 \leq I_{Q,P^*,W}(X_2 \wedge Y | U) - I_{Q,P_1^*}(U \wedge S),$$

$$R_1 + R_2 \leq I_{Q,P^*,W}(U, X_2 \wedge Y) - I_{Q,P_1^*}(U \wedge S)\}.$$

The proofs repeat principle steps, so we shall give each proof only for one case.

3   Proof of Outer Bound for the Case 3

Let $\delta > 0$ and a code (f_1, f_2, g) is given, with rates $R_i = (1/N) \log M_i$, $i = 1, 2$ and error probability

$$\bar{e}(N, W(Q)) \leq \exp\{-N(E - \delta)\}, \ E - \delta > 0.$$

According to (1) and (6) it means that

$$\frac{1}{M_1 M_2} \sum_{m_1, m_2} \sum_{\mathbf{s} \in \mathcal{S}^N} Q^N(\mathbf{s}) W^N \{\mathcal{Y}^N - g^{-1}(m_1, m_2 | \mathbf{s}) | f_1(m_1), f_2(m_2), \mathbf{s}\} \leq$$

$$\leq \exp\{-N(E - \delta)\}.$$

The left side of this inequality can only decrease if we take the sum by vectors $\mathbf{s}$ of some fixed type Q':

$$\sum_{\mathbf{s} \in T_{Q'}^N(S)} Q^N(\mathbf{s}) \sum_{(\mathbf{x}_1(m_1), \mathbf{x}_2(m_2)) \in f(\mathcal{M}_1, \mathcal{M}_2)} W^N \{ \mathcal{Y}^N - g^{-1}(m_1, m_2 | \mathbf{s}) | \mathbf{x}_1(m_1), \mathbf{x}_2(m_2), \mathbf{s} \} \le$$

$$\le M_1 M_2 \exp\{-N(E-\delta)\},$$

where $f(\mathcal{M}_1, \mathcal{M}_2)$ is the set of all codewords.
As

$$M_1 M_2 = \sum_P |f(\mathcal{M}_1, \mathcal{M}_2) \bigcap T_P^N(X_1, X_2)|$$

and the number of conditional types P does not exceed $(N+1)^{|\mathcal{X}_1||\mathcal{X}_2|}$, then there exists at least one conditional type P of the sequences $(\mathbf{x}_1, \mathbf{x}_2)$, such that

$$M_1 M_2 (N+1)^{-|\mathcal{X}_1||\mathcal{X}_2|} \le |f(\mathcal{M}_1, \mathcal{M}_2) \bigcap T_P^N(X_1, X_2)|. \tag{14}$$

Now for any type Q' and conditional type V we have

$$\sum_{(\mathbf{x}_1(m_1), \mathbf{x}_2(m_2)) \in T_P^N(X_1, X_2) \bigcap f(\mathcal{M}_1, \mathcal{M}_2)} \sum_{\mathbf{s} \in T_{Q', P}^N(S | \mathbf{x}_1(m_1), \mathbf{x}_2(m_2))} Q^N(\mathbf{s}) \cdot$$

$$\cdot W^N \{ T_{Q', P, V}^N(Y | \mathbf{x}_1(m_1), \mathbf{x}_2(m_2), \mathbf{s}) - -g^{-1}(m_1, m_2|\mathbf{s}) | \mathbf{x}_1(m_1), \mathbf{x}_2(m_2), \mathbf{s} \}$$

$$\le M_1 M_2 \exp\{-N(E-\delta)\}.$$

Taking into account that the conditional probability $Q^N(\mathbf{s}) W^N(\mathbf{y}|\mathbf{x}_1, \mathbf{x}_2, \mathbf{s})$ is constant for different $\mathbf{x}_1, \mathbf{x}_2 \in T_P^N(X_1, X_2)$, $\mathbf{s} \in T_{Q', P}^N(S|\mathbf{x}_1, \mathbf{x}_2)$, and $\mathbf{y} \in T_{Q', P, V}^N(Y|\mathbf{x}_1, \mathbf{x}_2, \mathbf{s})$, we can write

$$\sum_{(\mathbf{x}_1(m_1), \mathbf{x}_2(m_2)) \in T_P^N(X_1, X_2) \bigcap f(\mathcal{M}_1, \mathcal{M}_2)} \sum_{\mathbf{s} \in T_{Q', P}^N(S|\mathbf{x}_1(m_1), \mathbf{x}_2(m_2))}$$

$$\left\{ |T_{Q', P, V}^N(Y|\mathbf{x}_1(m_1), \mathbf{x}_2(m_2), \mathbf{s})| - \left| T_{Q', P, V}^N(Y|\mathbf{x}_1(m_1), \mathbf{x}_2(m_1), \mathbf{s}) \bigcap g^{-1}(m_1, m_2|\mathbf{s}) \right| \right\}$$

$$\le \frac{M_1 M_2 \exp\{-N(E-\delta)\}}{W^N(\mathbf{y}, \mathbf{s}|\mathbf{x}_1, \mathbf{x}_2)},$$

or according to (4)

$$\sum_{(\mathbf{x}_1(m_1), \mathbf{x}_2(m_2)) \in T_P^N(X_1, X_2) \bigcap f(\mathcal{M}_1, \mathcal{M}_2)} \sum_{\mathbf{s} \in T_{Q', P}^N(S|\mathbf{x}_1(m_1), \mathbf{x}_2(m_2))}$$

$$|T_{Q', P, V}^N(Y|\mathbf{x}_1(m_1), \mathbf{x}_2(m_2), \mathbf{s})| - \frac{M_1 M_2 \exp\{-N(E-\delta)\}}{\exp\{-N(H_{Q', P, V}(Y, S|X_1, X_2) + D(Q' \circ V \| Q \circ W | P))\}}$$

$$\leq \sum_{(\mathbf{x}_1(m_1),\mathbf{x}_2(m_2))\in T_P^N(X_1,X_2)} \sum_{\mathbf{s}\in T_{Q',P}^N(S|\mathbf{x}_1(m_1),\mathbf{x}_2(m_2))}$$

$$\left| T_{Q',P,V}^N(Y|\mathbf{x}_1(m_1),\mathbf{x}_2(m_1),\mathbf{s}) \bigcap g^{-1}(m_1,m_2|\mathbf{s}) \right|$$

$$= K.$$

The sets $g^{-1}(m_1,m_2|\mathbf{s})$ are disjoint for different $m_1 \in \mathcal{M}_1$, $m_2 \in \mathcal{M}_2$, so K can be upper bounded by diverse values:

$$K \leq \begin{cases} M_2 \exp\{NH_{Q',P,V}(Y,S|X_2)\}, \\ M_1 \exp\{NH_{Q',P,V}(Y,S|X_1)\}, \\ \exp\{NH_{Q',P,V}(Y,S)\}. \end{cases}$$

From (14) using each of last estimates we obtain correspondingly

$$M_1 M_2 (N+1)^{-|\mathcal{X}_1||\mathcal{X}_2|} \exp\{NH_{Q',P,V}(Y,S|X_1,X_2)\} -$$
$$-M_1 M_2 \exp\{N(H_{Q',P,V}(Y,S|X_1,X_2) + D(Q' \circ V\|Q \circ W|P) - E + \delta)\} \leq$$
$$\leq M_2 \exp\{NH_{Q',P,V}(Y,S|X_2)\},$$

$$M_1 M_2 (N+1)^{-|\mathcal{X}_1||\mathcal{X}_2|} \exp\{NH_{Q',P,V}(Y,S|X_1,X_2)\} -$$
$$-M_1 M_2 \exp\{N(H_{Q',P,V}(Y,S|X_1,X_2) + D(Q' \circ V\|Q \circ W|P) - E + \delta)\} \leq$$
$$\leq M_1 \exp\{NH_{Q',P,V}(Y,S|X_1)\},$$

$$M_1 M_2 (N+1)^{-|\mathcal{X}_1||\mathcal{X}_2|} \exp\{NH_{Q',P,V}(Y,S|X_1,X_2)\} -$$
$$-M_1 M_2 \exp\{N(H_{Q',P,V}(Y,S|X_1,X_2) + D(Q' \circ V\|Q \circ W|P) - E + \delta)\} \leq$$
$$\leq M_1 M_2 \exp\{NH_{Q',P,V}(Y,S)\}.$$

Now it is easy to obtain the following bounds:

$$M_1 \leq \frac{\exp\{NH_{Q',P,V}(Y,S|X_2) - H_{Q',P,V}(Y,S|X_1,X_2)\}}{(N+1)^{-|\mathcal{X}_1||\mathcal{X}_2|} - \exp\{N(D(Q' \circ V\|Q \circ W|P) - E + \delta)\}},$$

$$M_2 \leq \frac{\exp\{NH_{Q',P,V}(Y,S|X_1) - H_{Q',P,V}(Y,S|X_1,X_2)\}}{(N+1)^{-|\mathcal{X}_1||\mathcal{X}_2|} - \exp\{N(D(Q' \circ V\|Q \circ W|P) - E + \delta)\}},$$

$$M_1 M_2 \leq \frac{\exp\{NH_{Q',P,V}(Y,S) - H_{Q',P,V}(Y,S|X_1,X_2)\}}{(N+1)^{-|\mathcal{X}_1||\mathcal{X}_2|} - \exp\{N(D(Q' \circ V\|Q \circ W|P) - E + \delta)\}}.$$

The right sides of these inequalities can be minimized by the choice of types Q', V, meeting the condition

$$D(Q' \circ V\|Q \circ W|P) \leq E.$$

It is left to note that

$$H_{Q',P,V}(Y,S|X_2) - H_{Q',P,V}(Y,S|X_1,X_2) = I_{Q',P,V}(X_1 \wedge Y,S|X_2),$$

$$H_{Q',P,V}(Y,S|X_1) - H_{Q',P,V}(Y,S|X_1,X_2) = I_{Q',P,V}(X_2 \wedge Y,S|X_1),$$

$$H_{Q',P,V}(Y,S) - H_{Q',P,V}(Y,S|X_1,X_2) = I_{Q',P,V}(X_1,X_2 \wedge Y,S).$$

Hence (7), (8) and (9) are proved.

4   Proof of Inner Bound in the Case 5

The proofs of inner bounds are based on the random coding arguments. Here we bring the proof for case 5.

Let us fix positive integers N, M_1, type Q, conditional type P_1^*, $\delta > 0$. For brevity we shall denote $\mathbf{u}(m_1, \mathbf{s}), \mathbf{x}_1(m_1, \mathbf{s}) = \mathbf{ux}_1(m_1, \mathbf{s})$. Denote by $\mathcal{L}_{M_1}(Q, P_1^*)$ the family of all matrices

$$\mathbf{L}(Q, P_1^*) = \{\mathbf{ux}_1(m_1, \mathbf{s})\}_{m_1=\overline{1,M_1}}^{\mathbf{s}\in\mathcal{T}_Q^N(S)},$$

such that the rows $\mathbf{L_s}(Q, P_1^*) = (\mathbf{ux}_1(1, \mathbf{s}), \mathbf{ux}_1(2, \mathbf{s}), \ldots, \mathbf{ux}_1(M_1, \mathbf{s}))$ are not necessarily distinct vector pairs, majority of which are from $\mathcal{T}_{Q,P_1^*}^N(U, X_1|\mathbf{s})$.

Let us consider for any $m_1 \in \mathcal{M}_1$ and $\mathbf{s} \in \mathcal{T}_Q^N(S)$ the random event

$$\mathcal{A}_{Q,P_1^*}(m_1, \mathbf{s}) \overset{\triangle}{=} \{\mathbf{ux}_1(m_1, \mathbf{s}) \in \mathcal{T}_{Q,P_1^*}^N(U, X_1|\mathbf{s})\}.$$

Let us now consider the sets

$$\mathcal{S}(m_1, Q, P_1^*) \overset{\triangle}{=} \{\mathbf{s} \in \mathcal{T}_Q^N(S) : \mathcal{A}_{Q,P_1^*}(m_1, \mathbf{s})\}, \quad m_1 \in \mathcal{M}_1,$$

$$\mathcal{T}_Q^E(S) = \bigcup_{Q':D(Q'\|Q)\leq E} \mathcal{T}_{Q'}^N(S),$$

and the matrix

$$\mathbf{L}(Q, P_1^*, E) = \{\mathbf{ux}_1(m_1, \mathbf{s})\}_{m_1=\overline{1,M_1}}^{\mathbf{s}\in\mathcal{T}_Q^E(S)}.$$

We shall use the following modification of packing lemma [19].

Lemma 1. For each $E > 0$, $\delta \in (0, E)$ and any types P_1^*, P_2^* there exist M_2 vectors $\mathbf{x}_2(m_2)$ from $\mathcal{T}_{P_2^*}^N(X_2)$ and a matrix $\mathbf{L}(Q, P_1^*, E) = \{\mathbf{ux}_1(m_1, \mathbf{s})\}_{m_1=\overline{1,M_1}}^{\mathbf{s}\in\mathcal{T}_Q^E(S)}$, with

$$\frac{1}{N}\log M_1 \leq \min_{Q',P,V:D(Q'\circ P\circ V\|Q\circ P^*\circ W)\leq E} |I_{Q',P,V}(U \wedge Y, X_2) - I_{Q',P_1^*}(U \wedge S)+$$

$$+D(Q' \circ P \circ V\|Q \circ P^* \circ W) - E|^+, \tag{15}$$

$$\frac{1}{N}\log M_2 \leq \min_{Q',P,V:D(Q'\circ P\circ V\|Q\circ P^*\circ W)\leq E} |I_{Q',P,V}(X_2 \wedge Y, U) - I_{Q',P_1^*}(U \wedge S)+$$

$$+D(Q' \circ P \circ V\|Q \circ P^* \circ W) - E|^+, \tag{16}$$

$$\frac{1}{N}\log M_1 M_2 \leq \min_{Q',P,V:D(Q'\circ P\circ V\|Q\circ P^*\circ W)\leq E} |I_{Q',P,V}(U, X_2 \wedge Y) + I_{Q',P}(U \wedge X_2)-$$

$$-I_{Q',P_1^*}(U \wedge S) + D(Q' \circ P \circ V\|Q \circ P^* \circ W) - E|^+, \tag{17}$$

such that for each $Q' : D(Q'\|Q) \leq E$ and $\mathbf{s} \in T_{Q'}^N(S)$ the following inequality is true

$$\Pr\{\overline{\mathcal{A}}_{Q',P_1^*}(m_1,\mathbf{s})\} \leq \exp\{-\exp\{N\delta/4\}\}, \tag{18}$$

and for each $Q' : D(Q'\|Q) \leq E$ and $\mathbf{s} \in T_{Q'}^N(S)$, type $\hat{Q}$ such that $D(\hat{Q}\|Q) \leq E$, conditional types $P, \hat{P}, V, \hat{V}$, for sufficiently large N the following inequality holds

$$\frac{1}{M_1 M_2} \sum_{\mathbf{ux}_1(m_1,\mathbf{s}),\mathbf{x}_2(m_2)\in T_{Q',P}^N(U,X_1,X_2|\mathbf{s})} \left| T_{Q',P,V}^N(Y|\mathbf{ux}_1(m_1,\mathbf{s}),\mathbf{x}_2(m_2),\mathbf{s}) \bigcap \right.$$

$$\left. \bigcap_{(m_1',m_2')\neq(m_1,m_2)} \bigcup_{\mathbf{s}'\in\mathcal{S}(m_1,\hat{Q},P_1^*)} T_{\hat{Q},\hat{P},\hat{V}}^N(Y|\mathbf{ux}_1(m_1',\mathbf{s}'),\mathbf{x}_2(m_2'),\mathbf{s}') \right| \leq \tag{19}$$

$$\leq \left| T_{Q',P,V}^N(Y|\mathbf{ux}_1(m_1,\mathbf{s}),\mathbf{x}_2(m_2),\mathbf{s}) \right| \exp\{-N|E - D(\hat{Q}\circ\hat{P}\circ\hat{V}\|Q\circ P^*\circ W)|^+\}\times$$

$$\times \exp\{-N(D(P\|P^*|Q') - \delta)\}.$$

The proof of lemma 1 follows from lemma 2, which is proved in appendix.

Lemma 2. For each $E > 0$, $\delta \in (0,E)$ and any types Q, P_1^*, P_2^* there exist M_2 vectors $\mathbf{x}_2(m_2)$ from $T_{P_2^*}^N(X_2)$ and a matrix $\mathbf{L}(Q,P_1^*) = \{\mathbf{ux}_1(m_1,\mathbf{s})\}_{m_1=\overline{1,M_1}}^{\mathbf{s}\in T_Q^N(S)}$, with

$$\frac{1}{N}\log M_1 \leq \min_{P,V:D(P\circ V\|P^*\circ W|Q)\leq E} |I_{Q,P,V}(U\wedge Y,X_2) - I_{Q,P_1^*}(U\wedge S)+$$

$$+D(P\circ V\|P^*\circ W|Q) - E|^+, \tag{20}$$

$$\frac{1}{N}\log M_2 \leq \min_{P,V:D(P\circ V\|P^*\circ W|Q)\leq E} |I_{Q,P,V}(X_2\wedge Y,U) - I_{Q,P_1^*}(U\wedge S)+$$

$$+D(P\circ V\|P^*\circ W|Q) - E|^+, \tag{21}$$

$$\frac{1}{N}\log M_1 M_2 \min_{P,V:D(P\circ V\|P^*\circ W|Q)\leq E} |I_{Q,P,V}(U,X_2\wedge Y) + I_{Q,P}(U\wedge X_2) - I_{Q,P_1^*}(U\wedge S)+$$

$$+D(P\circ V\|P^*\circ W|Q) - E|^+, \tag{22}$$

such that for each $\mathbf{s} \in T_Q^N(S)$

$$\Pr\{\overline{\mathcal{A}}_{Q,P_1^*}(m_1,\mathbf{s})\} \leq \exp\{-\exp\{N\delta/4\}\}, \tag{23}$$

and for any $\mathbf{s} \in T_Q^N(S)$, types P, P', V, V', for sufficiently large N the following inequality holds

$$\frac{1}{M_1 M_2} \sum_{\mathbf{ux}_1(m_1,\mathbf{s}),\mathbf{x}_2(m_2)\in T_{\hat{Q},P}^N(U,X_1,X_2|\mathbf{s})} \left| T_{Q,P,V}^N(Y|\mathbf{ux}_1(m_1,\mathbf{s}),\mathbf{x}_2(m_2),\mathbf{s}) \bigcap \right.$$

$$\left. \bigcap_{(m_1',m_2')\neq(m_1,m_2)} \bigcup_{\mathbf{s}'\in\mathcal{S}(m_1,Q,P_1^*)} T_{Q,P',V'}^N(Y|\mathbf{ux}_1(m_1',\mathbf{s}'),\mathbf{x}_2(m_2'),\mathbf{s}') \right| \leq \tag{24}$$

$$\leq \left| T_{Q,P,V}^N(Y|\mathbf{u}\mathbf{x}_1(m_1,\mathbf{s}), \mathbf{x}_2(m_2), \mathbf{s}) \right| \exp\{-N|E - D(P' \circ V' \| P^* \circ W|Q)|^+\} \times$$
$$\times \exp\{-N(D(P\|P^*|Q) - \delta)\}.$$

Now we pass to the proof of the random coding bound for the case 5.

The existence of codewords $\mathbf{x}_2(m_2)$ and matrix $\mathbf{L}(Q, P_1^*, E) =$ $\{\mathbf{u}\mathbf{x}_1(m_1, \mathbf{s})\}_{m_1 = \overline{1,M_1}}^{\mathbf{s} \in T_Q^E(S)}$, satisfying (15), (16), (17), (18) and (19) is guaranteed by lemma 1. Note that for $Q' : D(Q'\|Q) > E$

$$Q^N(T_{Q'}^N(\mathcal{S})) \leq \exp\{-ND(Q'\|Q)\} < \exp\{-NE\}. \tag{25}$$

Let us apply the following decoding rule for decoder g: each $\mathbf{y}$ is decoded to such m_1, m_2 for which $\mathbf{y} \in T_{Q',P,V}^N(Y|\mathbf{u}\mathbf{x}_1(m_1,\mathbf{s}), \mathbf{x}_2(m_2), \mathbf{s})$ where Q', P, V are such that $D(Q' \circ P \circ V \| Q \circ P^* \circ W)$ is minimal.

The decoder g can make an error during the transmission of messages m_1, m_2, if $\overline{\mathcal{A}}_{Q',P_1^*}(m_1,\mathbf{s})$ takes place or if there exist types $\hat{Q} : D(\hat{Q}\|Q) \leq E$, $\hat{P}, \hat{V}$, some vector $\mathbf{s}' \in T_{\hat{Q}}(S)$, messages $(m_1', m_2') \neq (m_1, m_2)$, for which $\mathcal{A}_{\hat{Q},P_1^*}(m_1',\mathbf{s}')$ takes place, such that

$$\mathbf{y} \in T_{Q',P,V}^N(Y|\mathbf{u}\mathbf{x}_1(m_1,\mathbf{s}), \mathbf{x}_2(m_2), \mathbf{s}) \bigcap T_{\hat{Q},\hat{P},\hat{V}}^N(Y|\mathbf{u}\mathbf{x}_1(m_1',\mathbf{s}'), \mathbf{x}_2(m_2'), \mathbf{s}')$$

and

$$D(\hat{Q} \circ \hat{P} \circ \hat{V} \| Q \circ P^* \circ W) \leq D(Q' \circ P \circ V \| Q \circ P^* \circ W). \tag{26}$$

Denote

$$\mathcal{D} = \left\{ \hat{Q}, \hat{P}, \hat{V} : (26) \text{ is valid} \right\}.$$

Then from (18) and (25) average error probability (1) is upper bounded by the following way:

$$\overline{e}(N, W(Q)) \leq \frac{1}{M_1 M_2} \sum_{m_1,m_2} \sum_{Q':D(Q'\|Q)>E} \sum_{\mathbf{s} \in T_{Q'}^N(S)} Q^N(\mathbf{s})e(m_1, m_2, \mathbf{s})+$$

$$+\frac{1}{M_1 M_2} \sum_{m_1,m_2} \sum_{Q':D(Q'\|Q)\leq E} \sum_{\mathbf{s} \in T_{Q'}^N(S)} Q^N(\mathbf{s})e(m_1, m_2, \mathbf{s}) \leq$$

$$\leq (N+1)^{|S|} \exp\{-NE\}+$$

$$+\frac{1}{M_1 M_2} \sum_{m_2} \sum_{Q':D(Q'\|Q)\leq E} \sum_{\mathbf{s} \in T_{Q'}^N(S)} \sum_{m_1} Q^N(\mathbf{s}) \Pr\{\overline{\mathcal{A}}_{Q',P_1^*}(m_1,\mathbf{s})\}+$$

$$+\frac{1}{M_1 M_2} \sum_{Q':D(Q'\|Q)\leq E} \sum_{\mathbf{s} \in T_{Q'}^N(S)} Q^N(\mathbf{s}) \sum_{m_1:\mathcal{A}_{Q',P_1^*}(m_1,\mathbf{s})} \sum_P \sum_{\mathbf{x}_2(m_2) \in T_{Q',P}^N(X_2|\mathbf{u}\mathbf{x}_1(m_1,\mathbf{s}),\mathbf{s})}$$

$$e(m_1, m_2, \mathbf{s}) \leq$$

$$\leq \exp\{-N(E - \varepsilon)\} + \sum_{Q':D(Q'\|Q)\leq E} \sum_{\mathbf{s} \in T_{Q'}^N(S)} Q^N(\mathbf{s}) \exp\{-\exp\{N\delta/4\}\}+$$

$$+\frac{1}{M_1 M_2} \sum_{Q':D(Q'\|Q)\leq E} \sum_{\mathbf{s} \in T_{Q'}^N(S)} \sum_{m_1:\mathcal{A}_{Q',P_1^*}(m_1,\mathbf{s})} \sum_P \sum_{\mathbf{x}_2(m_2) \in T_{Q',P}^N(X_2|\mathbf{u}\mathbf{x}_1(m_1,\mathbf{s}),\mathbf{s})} Q^N(\mathbf{s}) \times$$

$$\times W^N \left\{ \bigcup_{\mathcal{D}} T^N_{Q',P,V}(Y|\mathbf{u}\mathbf{x}_1(m_1,\mathbf{s}),\mathbf{x}_2(m_2),\mathbf{s}) \bigcap \right.$$

$$\bigcap_{(m'_1,m'_2)\neq(m_1,m_2)} \bigcup_{\mathbf{s}'\in\mathcal{S}(m'_1,\hat{Q},P^*_1)} \left. T^N_{\hat{Q},\hat{P},\hat{V}}(Y|\mathbf{u}\mathbf{x}_1(m'_1,\mathbf{s}'),\mathbf{x}_2(m'_2),\mathbf{s}')|\mathbf{x}_1(m_1,\mathbf{s}),\mathbf{x}_2(m_2),\mathbf{s} \right\} \leq$$

$$\leq \exp\{-N(E-\varepsilon)\} + \exp\{-\exp\{N\delta/4\} + N\delta_1\}+$$

$$+\frac{1}{M_1 M_2} \sum_{Q':D(Q'\|Q)\leq E} \sum_{\mathbf{s}\in T^N_{Q'}(S)} \sum_{m_1:\mathcal{A}_{Q',P^*_1}(m_1,\mathbf{s})} \sum_{P} \sum_{\mathbf{x}_2(m_2)\in T^N_{Q',P}(X_2|\mathbf{u}\mathbf{x}_1(m_1,\mathbf{s})\mathbf{s})} \sum_{\mathcal{D}} Q^N(\mathbf{s})W^N(\mathbf{y}|\mathbf{x}_1,\mathbf{x}_2,\mathbf{s})\times$$

$$\left. \bigcap_{(m'_1,m'_2)\neq(m_1,m_2)} \bigcup_{\mathbf{s}'\in\mathcal{S}(m'_1,\hat{Q},P^*_1)} T^N_{\hat{Q},\hat{P},\hat{V}}(Y|\mathbf{u}\mathbf{x}_1(m'_1,\mathbf{s}'),\mathbf{x}_2(m'_2),\mathbf{s}') \right|.$$

According to (5), (19) and (26) we have

$$\bar{e}(N,W(Q)) \leq \exp\{-N(E-\varepsilon)\} + \exp\{-\exp\{N\delta/4\} + N\delta_1\}+$$

$$+\sum_{Q'}\sum_{P}\sum_{\mathcal{D}} \exp\{-N(Q'\|Q)\}\exp\{-N(H_{Q',P,V}(Y|X_1,X_2,S) + D(V\|W|Q',P))\}\times$$

$$\times \exp\{NH_{Q',P,V}(Y|X_1,X_2,S)\}\exp\{-N|E-D(\hat{Q}\circ\hat{P}\circ\hat{V}\|Q\circ P^*\circ W)|^+\}\times$$

$$\times \exp\{-N(D(P\|P^*|Q')-\delta)\} \leq \exp\{-N(E-\varepsilon_1)\}.$$

The inner bound for the case 5 is proved.

References

1. R. Ahlswede, Multi-way communication channels, 2nd Intern. Sympos. Inform. Theory. Tsahkadsor, Armenia, 1971, Budapest, Akad. Kiado, 23–52, 1973.
2. R. Ahlswede, The capacity region of a channel with two senders and two receivers, Annals Probability, Vol. 2. No. 2. 805–814, 1974.
3. E. A. Haroutunian, M. E. Haroutunian, and A. E. Avetissian, Multiple-access channel achievable rates region and reliability, Izvestiya Akademii Nauk Armenii, Matematika, Vol. 27, No. 5, 51–68, 1992.
4. M. E. Haroutunian, On E-capacity region of multiple-access channel, (in Russian) Izvestiya Akademii Nauk Armenii, Matematika, Vol. 38, No. 1, 3–22, 2003.
5. R. G. Gallager, A perspective on multiaccess channels, IEEE Trans. Inform. Theory, Vol. 31, No. 1, 124–142, 1985.
6. A. G. Dyachkov, Random constant composition codes for multiple access channels, Problems of Control and Inform. Theory, Vol. 13, No. 6, 357–369, 1984.
7. J. Pokorny and H.M. Wallmeier, Random coding bound and codes produced by permutations for the multiple-access channel, IEEE Trans. Inform. Theory, Vol. 31, 741–750, 1985.
8. Y. S. Liu and B. L. Hughes, A new universal coding bound for the multiple-access channel, IEEE Trans. Inform. Theory, Vol. 42, 376–386, 1996.
9. S. I. Gelfand and M. S. Pinsker, Coding for channel with random parameters, Problems of Control and Inform. Theory, Vol. 8, No. 1, 19–31, 1980.

10. E. A. Haroutunian and M. E. Haroutunian, E-capacity upper bound for channel with random parameter, Problems of Control and Information Theory, Vol. 17, No. 2, 99–105, 1988.

11. M. E. Haroutunian, Bounds of E-capacity for the channel with random parameter, Problemi Peredachi Informatsii, (in Russian), Vol. 27, No. 1, 14–23, 1991.

12. M. E. Haroutunian, New bounds for E-capacities of arbitrarily varying channel and channel with random parameter, Trans. IIAP NAS RA and YSU, Mathematical Problems of Computer sciences, Vol. 22, 44–59, 2001.

13. J. Jahn, Coding of arbitrarily varying multiuser channels, IEEE Trans. Inform. Theory, Vol. 27, No. 2, 212–226, 1981.

14. R. Ahlswede, Arbitrarily varying channels with states sequence known to the sender, IEEE Trans. Inform. Theory, Vol. 32, No. 5, 621–629, 1986.

15. R. Ahlswede and N. Cai, Arbitrarily varying multiple access channels, Part 1, IEEE Trans. Inform. Theory, Vol. 45, No. 2, 742–749, 1999.

16. R. Ahlswede and N. Cai, Arbitrarily varying multiple access channels, Part 2, IEEE Trans. Inform. Theory, Vol. 45, No. 2, 749–756, 1999.

17. A. Das and P. Narayan, Capacities of time-varying multiple-access channels with side information, IEEE Transactions on Information Theory, Vol. 48, No. 1, 4–25, 2002.

18. M. S. Pinsker, Multi-user channels, II Joint Swedish-Soviet Intern. workshop on Inform. Theory, Granna, Sweden, 160-165, 1985.

19. I. Csiszár and J. Körner, Information Theory. Coding Theorems for Discrete Memoryless Systems, Budapest, Akad. Kiado, 1981.

20. I. Csiszár, The method of types, IEEE Trans. Inform. Theory, Vol. 44, 2505-2523, 1998.

Appendix

Proof of the Lemma 2

Let P' and V' be such that $D(P' \circ V' \| P^* \circ W | Q) > E$, then we have

$$\exp\left\{ -N \left| E - D(P' \circ V' \| P^* \circ W | Q) \right|^+ \right\} = 1.$$

Since

$$\left| \mathcal{T}_{Q,P,V}^N(Y | \mathbf{u}\mathbf{x}_1(m_1, \mathbf{s}), \mathbf{x}_2(m_2), \mathbf{s}) \bigcap \bigcup_{(m_1', m_2') \neq (m_1, m_2)} \bigcup_{\mathbf{s}' \in \mathcal{S}(m_1', Q, P_1^*)} \mathcal{T}_{Q,P',V'}^N(Y | \mathbf{u}\mathbf{x}_1(m_1', \mathbf{s}'), \mathbf{x}_2(m_2'), \mathbf{s}') \right|$$

$$\leq \left| \mathcal{T}_{Q,P,V}^N(Y | \mathbf{u}\mathbf{x}_1(m_1, \mathbf{s}), \mathbf{x}_2(m_2), \mathbf{s}) \right|,$$

for the proof of (24) in this case it is enough to show that

$$\frac{\left| \mathcal{T}_{Q,P}^N(U, X_1, X_2 | \mathbf{s}) \bigcap f(\mathcal{M}_1, \mathcal{M}_2) \right|}{M_1 M_2} \leq \exp\left\{ -N(D(P \| P^* | Q) - \delta) \right\}.$$

To prove that (24) takes place for all P, V and P', V', it is enough to prove the inequality

$$\sum_{P,V} \sum_{P',V':D(P'\circ V'\|P^*\circ W|Q)\le E} \frac{1}{M_1 M_2} \sum_{\mathbf{ux}_1(m_1,\mathbf{s}),\mathbf{x}_2(m_2)\in\mathcal{T}^N_{Q,P}(U,X_1,X_2|\mathbf{s})} \left| \mathcal{T}^N_{Q,P,V}(Y|\mathbf{ux}_1(m_1,\mathbf{s}),\mathbf{x}_2(m_2),\mathbf{s}) \right.$$

$$\left. \bigcap_{(m'_1,m'_2)\ne(m_1,m_2)} \bigcup_{\mathbf{s}'\in\mathcal{S}(m'_1,Q,P^*_1)} \mathcal{T}^N_{Q,P',V'}(Y|\mathbf{ux}_1(m'_1,\mathbf{s}'),\mathbf{x}_2(m'_2),\mathbf{s}') \right| \times \quad (27)$$

$$\times \exp\{N(E - H_{Q,P,V}(Y|U,X_1,X_2,S) + D(P\|P^*|Q) - D(P'\circ V'\|P^*\circ W|Q) - \delta\} +$$

$$+ \sum_{P,V} \sum_{P',V':D(P'\circ V'\|P^*\circ W|Q)>E} \frac{|\mathcal{T}^N_{Q,P}(U,X_1,X_2|\mathbf{s})\bigcap f(\mathcal{M}_1,\mathcal{M}_2)|}{M_1 M_2} \times \exp\{N(D(P\|P^*|Q)-\delta\} \le 1.$$

Let us construct a random matrix $\widetilde{\mathbf{L}}(Q,P^*_1) = \{\mathbf{ux}_1(m_1,\mathbf{s})\}^{\mathbf{s}\in\mathcal{T}^N_Q(S)}_{m_1=\overline{1,M_1}}$, in the following way. We choose at random from $\mathcal{T}^N_{Q,P^*_1}(U)$ according to uniform distribution M_1 collections $\mathcal{J}(m_1)$ each of

$$J = \exp\{N(I_{Q,P^*_1}(U\wedge S) + \delta/2)\}$$

vectors $\mathbf{u}_j(m_1)$, $j = \overline{1,J}, m_1 = \overline{1,M_1}$.

For each $m_1 = \overline{1,M_1}$ and $\mathbf{s} \in \mathcal{T}^N_Q(S)$ we choose such a $\mathbf{u}_j(m_1)$ from $\mathcal{J}(m_1)$ that $\mathbf{u}_j(m_1) \in \mathcal{T}^N_{Q,P^*_1}(U|\mathbf{s})$. We denote this vector by $\mathbf{u}(m_1,\mathbf{s})$. If there is no such vector, let $\mathbf{u}(m_1,\mathbf{s}) = \mathbf{u}_J(m_1)$.

Next, for each m_1 and $\mathbf{s}$ we choose at random a vector $\mathbf{x}_1(m_1,\mathbf{s})$ from $\mathcal{T}^N_{Q,P^*_1}(X_1|\mathbf{u}(m_1,\mathbf{s}),\mathbf{s})$ if $\mathbf{u}(m_1,\mathbf{s}) \in \mathcal{T}^N_{Q,P^*_1}(U|\mathbf{s})$ and from $\mathcal{T}^N_{Q,P^*_1}(X_1|\mathbf{s})$ if $\mathbf{u}(m_1,\mathbf{s}) \notin \mathcal{T}^N_{Q,P^*_1}(U|\mathbf{s})$.

We choose also at random M_2 vectors $\mathbf{x}_2(m_2)$ from $\mathcal{T}^N_{P^*_2}(X_2)$.

First we shall show that for N large enough and any m_1 and $\mathbf{s}$ (23) takes place. Really,

$$\Pr\{\overline{\mathcal{A}}_{Q,P^*_1}(m_1,\mathbf{s})\} = \Pr\left\{\bigcap_{j=1}^J \mathbf{u}_j(m_1) \notin \mathcal{T}^N_{Q,P^*_1}(U|\mathbf{s})\right\} \le$$

$$\le \prod_{j=1}^J \left[1 - \Pr\left\{\mathbf{u}_j(m_1) \in \mathcal{T}^N_{Q,P^*_1}(U|\mathbf{s})\right\}\right] \le \left[1 - \frac{|\mathcal{T}^N_{Q,P^*_1}(U|\mathbf{s})|}{|\mathcal{T}^N_{Q,P^*_1}(U)|}\right]^J \le$$

$$\le \left[1 - \exp\{-N(I_{Q,P^*_1}(U\wedge S) + \delta/4)\}\right]^{\exp\{N(I_{Q,P^*_1}(U\wedge S)+\delta/2)\}}.$$

Using the inequality $(1 - t)^a \le \exp\{-at\}$, which is true for any a and $t \in (0,1)$, we can see that

$$\Pr\{\overline{\mathcal{A}}_{Q,P^*_1}(m_1,\mathbf{s})\} \le \exp\{-\exp\{N\delta/4\}\}.$$

To prove (27) it suffices to show that

$$\sum_{P,V} \sum_{P',V':D(P'\circ V'\|P^*\circ W|Q)\leq E} \mathbf{E}\left|T^N_{Q,P,V}(Y|\mathbf{u}\mathbf{x}_1(m_1,\mathbf{s}),\mathbf{x}_2(m_2),\mathbf{s})\bigcap\right.$$

$$\left.\bigcap_{(m_1',m_2')\neq(m_1,m_2)} \bigcup_{\mathbf{s}'\in\mathcal{S}(m_1',Q,P_1^*)} T^N_{Q,P',V'}(Y|\mathbf{u}\mathbf{x}_1(m_1',\mathbf{s}'),\mathbf{x}_2(m_2'),\mathbf{s}')\right|\times \quad (28)$$

$$\times\exp\{N(E-H_{Q,P,V}(Y|U,X_1,X_2,S)+D(P\|P^*|Q)-D(P'\circ V'\|P^*\circ W|Q)-\delta\}+$$

$$+\sum_{P,V}\sum_{P',V':D(P'\circ V'\|P^*\circ W|Q)>E}\mathbf{E}\frac{|T^N_{Q,P}(U,X_1,X_2|\mathbf{s})\bigcap f(\mathcal{M}_1,\mathcal{M}_2)|}{M_1M_2}\times\exp\{N(D(P\|P^*|Q)-\delta\}\leq 1.$$

To this end we estimate expectation

$$B=\mathbf{E}\left|T^N_{Q,P,V}(Y|\mathbf{u}\mathbf{x}_1(m_1,\mathbf{s}),\mathbf{x}_2(m_2),\mathbf{s})\bigcap\right.$$

$$\left.\bigcap_{(m_1',m_2')\neq(m_1,m_2)}\bigcup_{\mathbf{s}'\in\mathcal{S}(m_1',Q,P_1^*)} T^N_{Q,P',V'}(Y|\mathbf{u}\mathbf{x}_1(m_1',\mathbf{s}'),\mathbf{x}_2(m_2'),\mathbf{s}')\right|=$$

$$=\mathbf{E}\left|T^N_{Q,P,V}(Y|\mathbf{u}\mathbf{x}_1(m_1,\mathbf{s}),\mathbf{x}_2(m_2),\mathbf{s})\bigcap\bigcup_{m_1'\neq m_1}\bigcup_{\mathbf{s}'\in\mathcal{S}(m_1',Q,P_1^*)} T^N_{Q,P',V'}(Y|\mathbf{u}\mathbf{x}_1(m_1',\mathbf{s}'),\mathbf{x}_2(m_2),\mathbf{s}')\right|+$$

$$+\mathbf{E}\left|T^N_{Q,P,V}(Y|\mathbf{u}\mathbf{x}_1(m_1,\mathbf{s}),\mathbf{x}_2(m_2),\mathbf{s})\bigcap\bigcup_{m_2'\neq m_2}\bigcup_{\mathbf{s}'\in\mathcal{S}(m_1,Q,P_1^*)} T^N_{Q,P',V'}(Y|\mathbf{u}\mathbf{x}_1(m_1,\mathbf{s}'),\mathbf{x}_2(m_2'),\mathbf{s}')\right|+$$

$$+\mathbf{E}\left|T^N_{Q,P,V}(Y|\mathbf{u}\mathbf{x}_1(m_1,\mathbf{s}),\mathbf{x}_2(m_2),\mathbf{s})\bigcap\bigcup_{m_1'\neq m_1,m_2'\neq m_2}\bigcup_{\mathbf{s}'\in\mathcal{S}(m_1',Q,P_1^*)} T^N_{Q,P',V'}(Y|\mathbf{u}\mathbf{x}_1(m_1',\mathbf{s}'),\mathbf{x}_2(m_2'),\mathbf{s}')\right|.$$

The first summand can be estimated in the following way

$$\mathbf{E}\left|T^N_{Q,P,V}(Y|\mathbf{u}\mathbf{x}_1(m_1,\mathbf{s}),\mathbf{x}_2(m_2),\mathbf{s})\bigcap\bigcup_{m_1'\neq m_1}\bigcup_{\mathbf{s}'\in\mathcal{S}(m_1',Q,P_1^*)} T^N_{Q,P',V'}(Y|\mathbf{u}\mathbf{x}_1(m_1',\mathbf{s}'),\mathbf{x}_2(m_2),\mathbf{s}')\right|\leq$$

$$\leq\sum_{\mathbf{y}\in\mathcal{Y}^N}\Pr\{\mathbf{y}\in T^N_{Q,P,V}(Y|\mathbf{u}\mathbf{x}_1(m_1,\mathbf{s}),\mathbf{x}_2(m_2),\mathbf{s})\}\times$$

$$\times\sum_{m_1'\neq m_1}\Pr\{\mathbf{y}\in\bigcup_{\mathbf{s}'\in\mathcal{S}(m_1',Q,P_1^*)} T^N_{Q,P',V'}(Y|\mathbf{u}\mathbf{x}_1(m_1',\mathbf{s}'),\mathbf{x}_2(m_2),\mathbf{s}')\},$$

which follows from the independent choice of codewords. The first probability will be positive only when $\mathbf{y}\in T^N_{Q,P,V}(Y|\mathbf{x}_2(m_2),\mathbf{s})$. It can be estimated by the following way

$$\Pr\{\mathbf{y} \in \mathcal{T}_{Q,P,V}^{N}(Y|\mathbf{ux}_1(m_1,\mathbf{s}),\mathbf{x}_2(m_2),\mathbf{s})\} =$$

$$= \frac{\left|\mathcal{T}_{Q,P,V}^{N}(U,X_1|\mathbf{s},\mathbf{y},\mathbf{x}_2)\right|\left|\mathcal{T}_{Q,P}^{N}(X_2|\mathbf{s})\right|}{\left|\mathcal{T}_{Q,P_1^*}^{N}(U,X_1|\mathbf{s})\right|\left|\mathcal{T}_{P_2^*}^{N}(X_2)\right|} \leq$$

$$\leq (N+1)^{|\mathcal{U}||\mathcal{X}_1||\mathcal{X}_2||\mathcal{S}|}\exp\{-N(I_{Q,P,V}(U,X_1 \wedge Y,X_2|S) + I_{Q,P}(X_2 \wedge S))\}. \quad (29)$$

The second probability will be

$$\Pr\{\mathbf{y} \in \bigcup_{\mathbf{s}' \in \mathcal{S}(m_1',Q,P_1^*)} \mathcal{T}_{Q,P',V'}^{N}(Y|\mathbf{ux}_1(m_1',\mathbf{s}'),\mathbf{x}_2(m_2),\mathbf{s}')\} \leq$$

$$\leq \Pr\left\{\mathbf{y} \in \bigcup_{\mathbf{s}' \in \mathcal{S}(m_1',Q,P_1^*)} \mathcal{T}_{Q,P',V'}^{N}(Y|\mathbf{u}(m_1',\mathbf{s}'),\mathbf{x}_2(m_2),\mathbf{s}')\right\} \leq$$

$$\leq \Pr\left\{\mathbf{y} \in \bigcup_{\mathbf{u}_j(m_1') \in \mathcal{J}(m_1')} \bigcup_{\mathbf{s}' \in \mathcal{T}_{Q,P_1^*}^{N}(S|\mathbf{u}_j(m_1'))} \mathcal{T}_{Q,P',V'}^{N}(Y|\mathbf{u}_j(m_1'),\mathbf{x}_2(m_2),\mathbf{s}')\right\} \leq$$

$$\leq \sum_{\mathbf{u}_j(m_1') \in \mathcal{J}(m_1')} \Pr\left\{\mathbf{y} \in \mathcal{T}_{Q,P',V'}^{N}(Y|\mathbf{u}_j(m_1'),\mathbf{x}_2(m_2))\right\} \leq J\frac{|\mathcal{T}_{Q,P',V'}^{N}(U|\mathbf{y},\mathbf{x}_2(m_2))|}{|\mathcal{T}_{Q,P_1^*}^{N}(U)|} \leq$$

$$\leq (N+1)^{|\mathcal{U}|}\exp\left\{-N(I_{Q,P',V'}(U \wedge Y,X_2) - I_{Q,P_1^*}(U \wedge S) - \delta/2)\right\}. \quad (30)$$

Let us estimate the second expectation:

$$\mathbf{E}\left|\mathcal{T}_{Q,P,V}^{N}(Y|\mathbf{ux}_1(m_1,\mathbf{s}),\mathbf{x}_2(m_2),\mathbf{s}) \bigcap \bigcup_{m_2' \neq m_2} \bigcup_{\mathbf{s}' \in \mathcal{S}(m_1,Q,P_1^*)} \mathcal{T}_{Q,P',V'}^{N}(Y|\mathbf{ux}_1(m_1,\mathbf{s}'),\mathbf{x}_2(m_2'),\mathbf{s}')\right| \leq$$

$$\leq \sum_{\mathbf{y} \in \mathcal{Y}^N} \Pr\{\mathbf{y} \in \mathcal{T}_{Q,P,V}^{N}(Y|\mathbf{ux}_1(m_1,\mathbf{s}),\mathbf{x}_2(m_2),\mathbf{s})\} \times$$

$$\times \sum_{m_2' \neq m_2} \Pr\{\mathbf{y} \in \bigcup_{\mathbf{s}' \in \mathcal{S}(m_1,Q,P_1^*)} \mathcal{T}_{Q,P',V'}^{N}(Y|\mathbf{ux}_1(m_1,\mathbf{s}'),\mathbf{x}_2(m_2'),\mathbf{s}')\}.$$

The first probability will be positive only when $\mathbf{y} \in \mathcal{T}_{Q,P,V}^{N}(Y|\mathbf{ux}_1(m_1,\mathbf{s}),\mathbf{s})$. It is estimated by

$$\Pr\{\mathbf{y} \in \mathcal{T}_{Q,P,V}^{N}(Y|\mathbf{ux}_1(m_1,\mathbf{s}),\mathbf{x}_2(m_2),\mathbf{s})\} = \frac{\left|\mathcal{T}_{P,V}^{N}(X_2|\mathbf{y},\mathbf{ux}_1(m_1,\mathbf{s}),\mathbf{s})\right|}{\left|\mathcal{T}_{P_2^*}^{N}(X_2)\right|} \leq$$

$$\leq (N+1)^{|\mathcal{X}_2|}\exp\{-N(I_{Q,P,V}(X_2 \wedge Y,X_1,U,S))\}. \quad (31)$$

The second probability will be

$$\Pr\{\mathbf{y} \in \bigcup_{\mathbf{s}' \in \mathcal{S}(m_1, Q, P_1^*)} \mathcal{T}_{Q,P',V'}^N(Y|\mathbf{ux}_1(m_1, \mathbf{s}'), \mathbf{x}_2(m_2'), \mathbf{s}')\} \leq$$

$$\leq \sum_{\mathbf{u}_j(m_1) \in \mathcal{J}(m_1)} \Pr\left\{\mathbf{y} \in \mathcal{T}_{Q,P',V'}^N(Y|\mathbf{u}_j(m_1), \mathbf{x}_2(m_2'))\right\} \leq J \frac{|\mathcal{T}_{Q,P',V'}^N(X_2|\mathbf{y}, \mathbf{u}_j(m_1))|}{|\mathcal{T}_{P_2^*}^N(X_2)|} \leq$$

$$\leq (N+1)^{|\mathcal{X}_2|} \exp\left\{-N(I_{Q,P',V'}(X_2 \wedge U, Y) - I_{Q,P_1^*}(U \wedge S) - \delta/2)\right\}. \quad (32)$$

At last

$$\mathbf{E}\left|\mathcal{T}_{Q,P,V}^N(Y|\mathbf{ux}_1(m_1, \mathbf{s}), \mathbf{x}_2(m_2), \mathbf{s}) \bigcap \bigcup_{m_1' \neq m_1, m_2' \neq m_2} \bigcup_{\mathbf{s}' \in \mathcal{S}(m_1', Q, P_1^*)} \mathcal{T}_{Q,P',V'}^N(Y|\mathbf{ux}_1(m_1', \mathbf{s}'), \mathbf{x}_2(m_2'), \mathbf{s}')\right| \leq$$

$$\leq \sum_{\mathbf{y} \in \mathcal{Y}^N} \Pr\{\mathbf{y} \in \mathcal{T}_{Q,P,V}^N(Y|\mathbf{ux}_1(m_1, \mathbf{s}), \mathbf{x}_2(m_2), \mathbf{s})\} \times$$

$$\times \sum_{m_1' \neq m_1, m_2' \neq m_2} \Pr\{\mathbf{y} \in \bigcup_{\mathbf{s}' \in \mathcal{S}(m_1', Q, P_1^*)} \mathcal{T}_{Q,P',V'}^N(Y|\mathbf{ux}_1(m_1', \mathbf{s}'), \mathbf{x}_2(m_2'), \mathbf{s}')\},$$

which is positive only when $\mathbf{y} \in \mathcal{T}_{Q,P,V}^N(Y|\mathbf{s})$ and

$$\Pr\{\mathbf{y} \in \mathcal{T}_{Q,P,V}^N(Y|\mathbf{ux}_1(m_1, \mathbf{s}), \mathbf{x}_2(m_2), \mathbf{s})\} =$$

$$= \frac{|\mathcal{T}_{Q,P,V}^N(X_2|\mathbf{y}, \mathbf{s})||\mathcal{T}_{Q,P,V}^N(U, X_1|\mathbf{y}, \mathbf{x}_2(m_2), \mathbf{s})|}{|\mathcal{T}_{Q,P_1^*}^N(U, X_1|\mathbf{s})||\mathcal{T}_{P_2^*}^N(X_2)|} \leq$$

$$\leq (N+1)^{|\mathcal{U}||\mathcal{X}_1||\mathcal{X}_2||\mathcal{S}|} \exp\{-N(I_{Q,P,V}(X_2 \wedge Y, S) + I_{Q,P,V}(U, X_1 \wedge Y, X_2|S))\}. \quad (33)$$

The last probability to be estimated is

$$\Pr\{\mathbf{y} \in \bigcup_{\mathbf{s}' \in \mathcal{S}(m_1', Q, P_1^*)} \mathcal{T}_{Q,P',V'}^N(Y|\mathbf{ux}_1(m_1', \mathbf{s}'), \mathbf{x}_2(m_2'), \mathbf{s}')\} \leq$$

$$\leq \sum_{\mathbf{u}_j(m_1') \in \mathcal{J}(m_1')} \Pr\left\{\mathbf{y} \in \mathcal{T}_{Q,P',V'}^N(Y|\mathbf{u}_j(m_1'), \mathbf{x}_2(m_2'))\right\} \leq J \frac{|\mathcal{T}_{Q,P',V'}^N(U, X_2|\mathbf{y})|}{|\mathcal{T}_{Q,P_1^*}^N(U)||\mathcal{T}_{P_2^*}^N(X_2)|} \leq$$

$$\leq (N+1)^{|\mathcal{U}||\mathcal{X}_2|} \exp\left\{-N(I_{Q,P',V'}(U, X_2 \wedge Y) + I_{Q,P'}(U \wedge X_2) - I_{Q,P_1^*}(U \wedge S) - \delta/2)\right\}. \quad (34)$$

Now we can write

$$B \leq |\mathcal{T}_{Q,P,V}^N(Y|\mathbf{x}_2(m_2), \mathbf{s})| \Pr\{\mathbf{y} \in \mathcal{T}_{Q,P,V}^N(Y|\mathbf{ux}_1(m_1, \mathbf{s}), \mathbf{x}_2(m_2), \mathbf{s})\} \times$$

$$\times (M_1 - 1) \Pr\{\mathbf{y} \in \bigcup_{\mathbf{s}' \in \mathcal{S}(m_1', Q, P_1^*)} \mathcal{T}_{Q,P',V'}^N(Y|\mathbf{ux}_1(m_1', \mathbf{s}'), \mathbf{x}_2(m_2), \mathbf{s}')\} +$$

$$+ \left| \mathcal{T}^N_{Q,P,V}(Y|\mathbf{ux}_1(m_1,\mathbf{s}),\mathbf{s}) \right| \Pr\{\mathbf{y} \in \mathcal{T}^N_{Q,P,V}(Y|\mathbf{ux}_1(m_1,\mathbf{s}),\mathbf{x}_2(m_2),\mathbf{s})\} \times$$

$$\times (M_2 - 1) \Pr\{\mathbf{y} \in \bigcup_{\mathbf{s}' \in \mathcal{S}(m_1,Q,P_1^*)} \mathcal{T}^N_{Q,P',V'}(Y|\mathbf{ux}_1(m_1,\mathbf{s}'),\mathbf{x}_2(m_2'),\mathbf{s}')\} +$$

$$+ \left| \mathcal{T}^N_{Q,P,V}(Y|\mathbf{s}) \right| \Pr\{\mathbf{y} \in \mathcal{T}^N_{Q,P,V}(Y|\mathbf{ux}_1(m_1,\mathbf{s}),\mathbf{x}_2(m_2),\mathbf{s})\} \times$$

$$\times (M_1 M_2 - 1) \Pr\{\mathbf{y} \in \bigcup_{\mathbf{s}' \in \mathcal{S}(m_1',Q,P_1^*)} \mathcal{T}^N_{Q,P',V'}(Y|\mathbf{ux}_1(m_1',\mathbf{s}'),\mathbf{x}_2(m_2'),\mathbf{s}')\}.$$

According to (20), (21), (22) and (29)–(34) we have

$$B \leq (N+1)^{|\mathcal{U}|(|\mathcal{X}_1||\mathcal{X}_2||\mathcal{S}|+1)} \exp\{-N(I_{Q,P,V}(U,X_1 \wedge Y, X_2|S) + I_{Q,P}(X_2 \wedge S) - H_{Q,P,V}(Y|X_2,S))\} \times$$

$$\times \exp\left\{-N(I_{Q,P',V'}(U \wedge Y, X_2) - I_{Q,P_1^*}(U \wedge S) - \delta/2)\right\} \times$$

$$\times \exp\{N \min_{P,V:D(P\circ V\|P^*\circ W|Q)\leq E} |I_{Q,P,V}(U \wedge Y, X_2) - I_{Q,P_1^*}(U \wedge S) + D(P \circ V\|P^* \circ W|Q) - E|^+\} +$$

$$+ (N+1)^{2|\mathcal{X}_2|} \exp\{-N(I_{Q,P,V}(X_2 \wedge Y, X_1, U, S) - H_{Q,P,V}(Y|U,X_1,S))\} \times$$

$$\times \exp\left\{-N(I_{Q,P',V'}(X_2 \wedge U, Y) - I_{Q,P_1^*}(U \wedge S) - \delta/2)\right\} \times$$

$$\times \exp\{N \min_{P,V:D(P\circ V\|P^*\circ W|Q)\leq E} |I_{Q,P,V}(X_2 \wedge U, Y) - I_{Q,P_1^*}(U \wedge S) + D(P \circ V\|P^* \circ W|Q) - E|^+\} +$$

$$+ (N+1)^{|\mathcal{X}_2|(|\mathcal{X}_1||\mathcal{S}|+1)} \exp\{-N(I_{Q,P,V}(X_2 \wedge Y, S) + I_{Q,P,V}(U, X_1 \wedge Y, X_2|S) - H_{Q,P,V}(Y|S))\} \times$$

$$\times \exp\left\{-N(I_{Q,P',V'}(U, X_2 \wedge Y) + I_{Q,P'}(U \wedge X_2) - I_{Q,P_1^*}(U \wedge S) - \delta/2)\right\} \times$$

$$\times \exp\{N \min_{P,V:D(P\circ V\|P^*\circ W|Q)\leq E} |I_{Q,P,V}(U, X_2 \wedge Y) + I_{Q,P}(U \wedge X_2) - \tag{35}$$

$$-I_{Q,P_1^*}(U \wedge S) + D(P \circ V\|P^* \circ W|Q) - E|^+\}.$$

By using the following inequality $\min_x f(x) \leq f(x')$ from (35) we obtain

$$B \leq (N+1)^{|\mathcal{U}|(|\mathcal{X}_1||\mathcal{X}_2||\mathcal{S}|+1)} \exp\{-N(I_{Q,P,V}(U, X_1 \wedge Y, X_2|S) + I_{Q,P}(X_2 \wedge S) -$$

$$-H_{Q,P,V}(Y|X_2,S) - D(P' \circ V\|P^* \circ W|Q) + E - \delta/2)\} +$$

$$+ (N+1)^{2|\mathcal{X}_2|} \exp\{-N(I_{Q,P,V}(X_2 \wedge Y, X_1, U, S) -$$

$$-H_{Q,P,V}(Y|U,X_1,S) - D(P' \circ V\|P^* \circ W|Q) + E - \delta/2)\} +$$

$$+ (N+1)^{|\mathcal{X}_2|(|\mathcal{X}_1||\mathcal{S}|+1)} \exp\{-N(I_{Q,P,V}(X_2 \wedge Y, S) + I_{Q,P,V}(U, X_1 \wedge Y, X_2|S) -$$

$$-H_{Q,P,V}(Y|S) - D(P' \circ V\|P^* \circ W|Q) + E - \delta/2)\}. \tag{36}$$

Now notice that:

$$\mathbf{E}\frac{\left|T_{Q,P}^{N}(U,X_1,X_2|\mathbf{s})\bigcap f(\mathcal{M}_1,\mathcal{M}_2)\right|}{M_1 M_2} = \frac{\left|T_{Q,P}^{N}(U,X_1,X_2|\mathbf{s})\right|}{\left|T_{Q,P_1^*}^{N}(U,X_1|\mathbf{s})\right|\left|T_{P_2^*}^{N}(X_2)\right|} \leq$$

$$\leq (N+1)^{|\mathcal{U}||\mathcal{X}_1||\mathcal{S}|+|\mathcal{X}_2|}\exp\{-ND(P\|P^*|Q)\}, \tag{37}$$

and

$$D(P\|P^*|Q) = I_{Q,P}(U,X_1,S\wedge X_2).$$

Substitute (36) and (37) into (28) and note that:

$$I_{Q,P,V}(U,X_1\wedge X_2,Y|S) + I_{Q,P}(S\wedge X_2) - H_{Q,P,V}(Y|X_2,S)-$$
$$-I_{Q,P}(U,X_1,S\wedge X_2) + H_{Q,P,V}(Y|U,X_1,X_2,S) = 0,$$

$$I_{Q,P,V}(X_2\wedge U,X_1,Y,S) - H_{Q,P,V}(Y|U,X_1,S)-$$
$$-I_{Q,P}(U,X_1,S\wedge X_2) + H_{Q,P,V}(Y|U,X_1,X_2,S) = 0,$$

$$I_{Q,P,V}(U,X_1\wedge X_2,Y|S) + I_{Q,P,V}(Y,S\wedge X_2) - H_{Q,P,V}(Y|S)-$$
$$-I_{Q,P}(U,X_1,S\wedge X_2) + H_{Q,P,V}(Y|U,X_1,X_2,S) = 0.$$

It is easy to see that for N large enough (28) is true and hence lemma is proved.

Huge Size Codes for Identification
Via a Multiple Access Channel Under
a Word-Length Constraint

S. Csibi and E. von der Meulen*

In Memory of Sándor Csibi(1927-2003)

Abstract. It is well-known from pioneering papers published since 1989
that, for identification via a communication channel of given Shannon
capacity $\mathcal{C}$, the number $N(n)$ of messages which can be reliably identified using an identification code grows doubly exponentially with the
wordlength n, as $n \to \infty$. This paper provides contributions to the study
of identification plus transmission (IT) codes under a wordlength constraint $n \leq n_0$ for sending an identifier and message content over a
noiseless one-way channel. While the false identification probability no
longer vanishes under such constraint, it can be drastically suppressed
both by serial and parallel versions of an appropriately defined non-
algebraic forward error control (FEC). The main result of this paper
consists of exact and approximate expressions for the huge size $N(n)$,
and the corresponding second-order rate, of an asymptotically optimal
IT code sequence under a wordlength constraint, both with and without either of the two proposed FEC schemes. Also, upper bounds are
obtained on the false identification probability under such wordlength
constraint for both FEC versions, showing the drastic reduction possible
when applying FEC. Furthermore, it is outlined in this paper how the simultaneous transfer of identifiers and message contents by several active
users via a noiseless multiple access channel (MAC) could be handled,
using the concept of a least length single control sequence. Huge size identifier sets might be of design interest for certain prospective remote alarm

* Research of both authors was supported in part by the exchange program between
the Royal Belgian Academy of Sciences, Letters, and Fine Arts and the Hungarian Academy of Sciences; that of S. Csibi in part by the Hungarian National Research Foundation, Grant No. OTKA TP30651 (1999-2002), and by the Hungarian
Telecommunications Foundation (2002); that of E.C. van der Meulen in part by
INTAS Project 00-738 and by Project GOA/98/06 of Research Fund K.U. Leuven.
The material in this paper was presented in part by both authors at the Preparatory Meeting (February 18-23, 2002), and at the Opening Conference (November 3-9,
2002) of the 2002-2003 Research Year "General Theory of Information Transfer and
Combinatorics" held at the Center for Interdisciplinary Research (ZiF) of the Uni-
versity of Bielefeld. It was presented in part by the first author in a somewhat more
design-oriented way as an invited research seminar lecture at the Computer Science
and Telecommunications Department, Science School, University of Trento, Italy,
May 16, 2002. Sándor Csibi died on April 12, 2003. He was with the Department of
Telecommunications, Budapest University of Technology and Economics, Hungary.

R. Ahlswede et al. (Eds.): Information Transfer and Combinatorics, LNCS 4123, pp. 218–248, 2006.
© Springer-Verlag Berlin Heidelberg 2006

services, such as when occasional alarm services are to be conveyed via a simple MAC which is controlled by the same single common cyclically permutable sequence at all nodes involved. It is pointed out under which circumstances the use of huge IT codes might be worth further detailed investigation by designers interested in such prospective services.

1   Introduction

A widely used simple way of identification via a communication channel is to place the identifier (ID) into the heading of the codeword meant for transmitting the message content (MC). In this case, the identifier and message content are necessarily encoded jointly by the channel encoder, and both decoded by the channel decoder (cf. Fig. 1). For this situation and a binary communication channel the identifier set size is clearly $N(n) = 2^n$, for any $n \geq 1$, where n stands for the length of the binary representation of the identifier.

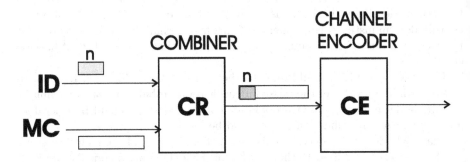

Fig. 1. The well-known simple way of handling an identifier

Identification via a communication channel under the constraint of joint channel coding and decoding might be well accomplished according to the scheme of Fig. 1, as long as there is no need at a given block length n for an identifier set size exceeding $N(n)$. However, the constraint of joint channel coding and decoding of both identifier and message content should be lifted if this is not the case. Obviously, more freedom is offered to the designer if also the identifier itself might be encoded, before combining it with the message block conveying the message content. This separate encoding is depicted in Fig. 2. Here two encoders are shown, a constant weight binary identifier encoder (CIE) and a channel encoder (CE). In this scheme, a single bit is retained (denoted by R_1 and "1" in Fig. 2) out of the M bits of the identification codeword assigned to the identifier, namely the one which depends on the message content to the conveyed. If no MC is to be sent, one out the M weights is drawn at random.

It is well-known that the scheme of Fig. 2 is not confined to a singly exponential growth, like the one of Fig.1, but might exhibit, by an appropriate choice of the parameters, an (at most) doubly exponential growth as $n \to \infty$ as shown by

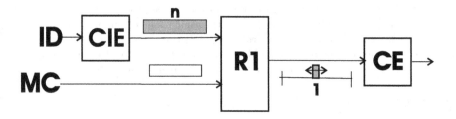

Fig. 2. More freedom for the designer

Ahlswede and Dueck [5]. Verdú and Wei [24] have shown how to explicitly construct an identification code the size of which achieves asymptotically a doubly exponential growth. It is those asymptotically good identification code designs which we have in mind in the present paper.

Notice that the joint channel coding and decoding is not the only constraint imposed in Fig. 1, but also the fact that distinct headings of length n are assigned to distinct identifiers. Thus the assignment is one-to-one and in this sense necessarily deterministic. A consequence of this fact is that the probability of false identification and, if the channel is noiseless, also that of missed identification equals zero.

The scheme of Fig. 2 could offer to the designer also additional freedom in this regard, if any value of the missed identification probability not exceeding λ_1 and of the false identification probability not exceeding λ_2 could be tolerated, and randomized identification codes could be allowed.

That such identification codes exist, with the size $N(n)$ of the identifier growing doubly exponentially with the length n of the identification code if a randomized code is selected in a suitable way, is the main content of the achievability part of the fundamental result of Ahlswede and Dueck [5].

Han and Verdú [14] initiated the study of conveying identifiers and message contents jointly via a communication channel of given Shannon capacity $\mathcal{C}$, and improved also the converse of the underlying theorem for identification from a soft version proved in [5] to a strong version. The first constructions of identification codes relying upon the complexity of random number generation are due to Ahlswede and Verboven [6], whereas the first explicit code constructions (which seem suitable also for real-time generation of identification codes) are due to Verdú and Wei [24].

In the present paper we will rely on these fundamental results concerning the sending of a single identifier (or any of a great amount of identifiers) over a communication channel. A brief account of this background is given in Section 2.

The third section of this paper contains new results obtained by the authors on the subject of identification plus transmission via a single, one-way channel under a wordlength constraint. In certain practical situations it may be required to constrain the wordlength n of an identification plus transmission (IT) code by a finite number n_0. Under such constraint the probabilities of missed and false identification do not necessarily fall below the preset thresholds λ_1 and λ_2.

In Section 3 we thoroughly analyze the effect this constraint has on the various parameters characterizing the IT code. We study hereby in particular a basic component of the IT code, which we have termed the IT-generating code. The Verdú-Wei code is such IT-generating (ITG) code. This ITG code is a binary constant-weight code, the rows of which correspond to the identifiers. The ones in each row correspond to messages. In identification plus transmission first a row is selected, and then a one is selected from this row. The position of this selected single bit is transmitted over the channel.

For simplicity we assume the transmission-channel to be noiseless, so that the probability of missed identification equals zero, and we only need to analyze the probability of false identification.

Following our previous paper [23], we establish an upper bound on the probability of false identification in terms of the parameters characterizing the Verdú-Wei code. By keeping two parameters of the Verdú-Wei code fixed, and letting the third one go to infinity, this upper bound will get as small as desired. However, when the constraint $n \leq n_0$ is imposed, this bound may still exceed λ_2.

Next, in Section 3, two kinds of forward error control (FEC) are introduced to further reduce drastically the probability of false identification under a wordlength constraint. FEC version 1 is a serial scheme and FEC version 2 a parallel scheme. In both FEC versions the position of the selected single bit is transmitted a fixed number times, either by consecutive transmissions or by parallel transmissions.

FEC version 1 was already to some extent introduced in [22] and [23], but here its properties are thoroughly investigated and the wordlength constraint is imposed. Both FEC versions result in a drop in size of the set of available identifiers, and in a reduction of the second-order rate of the ID code and of the probability of false identification. Our results are formulated in Theorems 3 and 4 below. A comparison of these two theorems shows that the new FEC version 2 provides essentially a much better performance than FEC version 1 in reducing the probability of false identification. The reduction in size of the identifier set is for both FEC versions rather drastic, but the remaining amount of available identifiers is still doubly exponential, i.e., huge.

A numerical example shows that for a moderate constraint n_0 and only a few (2 or 4) repetitions in the FEC schemes the probability of false identification in either version can be made very small.

In Section 4 a view is taken towards the application of the results obtained in Section 3 on identification plus transmission via a noiseless one-way channel under a wordlength constraint to multiple access communication. This addresses the problem of simultaneous identification plus transmission by several users over a common noiseless channel to a common receiver. This problem combines several problems of substantial importance and difficulty in information theory, i.e., that of multiaccess communication (as treated in [1], [7], [17], and [20]), that of least length control sequences (as investigated in [9], [10], [21], and [23]), and that of IT codes (as described in [24] and Section 3 of this paper). The imposition of a wordlength constraint might be particularly appropriate when several users are communicating simultaneously over a common multiple access channel. Obviously, it is a rather difficult task to solve the union of these three subproblems. In this paper we only outline the basic ideas of the subject of identification plus transmission via a multiple access channel under a wordlength constraint and discuss the underlying technique. Hereby we build forth on our earlier work [11].

In Section 5 we consider a class of important real life tasks where the use of huge identification codes, and our considered way of identification via a multiple access channel, are of much prospective interest. We mention in this regard already remote alarm services, which seem particularly appropriate for demonstrating circumstances where a plethora of identifiers makes practical sense.

In this paper we consider only conventional identification tasks. However, we are aware of other interesting kinds of networking tasks, which eventually might be formulated as so-called K-identification tasks, a notion introduced by Ahlswede [4]. A treatment of this topic is, however, outside the scope of this paper.

2   Recalling Some Well-Known Fundamental Results

In this section we describe in more detail some well-known fundamental relations that will serve as an immediate background of our subsequent investigations.

Ahlswede and Dueck [5] showed that there exist block codes for identification of length n the size N of which grows asymptotically as

$$N = N(n) = 2^{2^{nC}},$$

as $n \to \infty$, while the probabilities of missed and false identification are kept arbitrarily small. Here n stands for the length of a (binary) codeword of the identification code, and C for the Shannon capacity of the channel via which the codewords are to be conveyed. Thus the size N of the set of possible messages (identifiers) which can be sent for identification by the receiver exhibits asymptotically doubly exponential growth.

The above corresponds to the situation where the receiver is only interested in verifying whether a certain message is the transmitted message or not. Another way to formulate this is to say that the receiver is monitoring a certain message and wants to match that one to the message received from the sender. In this paper, as in [23], we will also use the term "identifier" to mean the actually transmitted message in the identification problem. An identifier match means the event that the message received from the sender is decoded into the message monitored by the receiver. An identifier match is thus a match at the receiver side between an arriving and a stored (monitored) identifier (typically a sequence of symbols). The fact that only one monitored identifier is matched at the receiver at any given time allows for a huge number of possible identifiers to be sent at the side of the source. This is in contrast with the classic Shannon formulation of the problem of reliable transmission of a message through a noisy channel, where the task of the receiver is to decide which one out of a large, but not huge, set of possible messages has been sent.

The task of identification is in this sense simpler than the task of decoding in the Shannon formulation.

We first recall the basic notions from [5]. A channel W with input alphabet A and output alphabet B is described by a sequence of conditional probability distributions $\{W^n : A^n \to B^n\}_{n=1}^{\infty}$.

An $(n, N, \lambda_1, \lambda_2)$ identification (ID) code is a collection

$$\{(Q_a, D_a), a = 1, \ldots, N\}$$

where Q_a is a probability distribution on A^n and $D_a \subset B^n$, such that the probability of missed identification is at most λ_1, for any $a = 1, \cdots, N$, and the probability of false identification is at most λ_2 for all $a \neq b$.

The rate of an $(n, N, \lambda_1, \lambda_2)$ ID code is defined as $\frac{1}{n} \log \log N$. We confine ourselves to logarithms of base 2, as in the present paper binary identification codes will be of our main interest.

R' is an achievable ID rate if for any $\gamma > 0$, $0 < \lambda_1 < 1$, $0 < \lambda_2 < 1$, and for all sufficiently large n, there exist $(n, N, \lambda_1, \lambda_2)$ ID codes with

$$\frac{1}{n} \log \log N > R' - \gamma.$$

The ID capacity $\mathcal{C}'$ of the channel is the maximum achievable ID rate.

For the ID capacity of the channel the following direct identification coding theorem holds:

Theorem 1. *[5] The ID capacity $\mathcal{C}'$ of any channel is greater than or equal to its Shannon capacity $\mathcal{C}$.*

Increasingly more general versions of the converse of Theorem 1 were proved consecutively in [5], [14], [15], and [4].

We next recall the notions introduced by Han and Verdú [14] for joint identification and transmission of identifier and message content over a channel W.

An $(n, N, M, \lambda_1, \lambda_2)$ identification plus transmission (IT) code consists of a mapping

$$f : \{1, \ldots, N\} \times \{1, \ldots, M\} \to A^n,$$

and a collection of subsets $D_{a,m}$ of B^n, $a \in \{1, \cdots, N\}, m \in \{1, \cdots, M\}$, such that suitably defined error probabilities (corresponding to the probabilities of missed and false identification) are at most λ_1 and λ_2, respectively. For the precise definition of an IT code see [14] or [24]. One can easily derive an $(n, N, \lambda_1, \lambda_2)$ ID code and N (n, M, λ_1) transmission codes from any given $(n, N, M, \lambda_1, \lambda_2)$ IT code. We shall not give these derivations here, but instead refer to [14] and [24]. The rate pair of an $(n, N, M, \lambda_1, \lambda_2)$ IT code is defined by

$$(R = \frac{1}{n} \log M, R' = \frac{1}{n} \log \log N).$$

Following [14], a pair (R, R') is said to be an achievable IT rate-pair, if for every $\gamma > 0$, $0 < \lambda_1 < 1$, $0 < \lambda_2 < 1$, and for all sufficiently large n, there exist $(n, N, M, \lambda_1, \lambda_2)$ IT codes such that

$$\frac{1}{n} \log M > R - \gamma,$$

$$\frac{1}{n} \log \log N > R' - \gamma.$$

Han and Verdú [14] proved both an achievability part and a converse part of the identification plus transmission coding theorem.

Theorem 2. *[14] For any finite-input channel $\mathcal{W}$ with Shannon capacity $\mathcal{C}$, the pair $(\mathcal{C}, \mathcal{C})$ is an achievable IT rate-pair, and conversely, every achievable IT rate-pair (R, R') must satisfy $\max(R, R') \le \mathcal{C}$.*

Theorem 2 holds in particular for a discrete memoryless channel (DMC), which will be of our interest in this paper.

The construction of IT codes in the proof of the achievability part of Theorem 2 can be viewed as the concatenation of a transmission code and a binary constant-weight code. This idea was further worked out in [24], where an explicit construction of optimal binary constant-weight codes for identification (and for identification plus transmission) was shown. The corresponding notions and results will be described next; they form important background material for understanding the results of Section 3.

Definition 1. *[24] An (S, N, M, K) binary constant-weight code C is a set of N binary S-tuples of Hamming weight M such that the pairwise overlap (maximum number of coincident 1's for any pair of codewords) does not exceed K.*

Any (S, N, M, K) binary constant-weight code can be described by an $N \times M$ matrix $s(a, m) \in \{1, \cdots, S\}$ such that, for every $a \in \{1, \cdots, N\}$, the row $(s(a, 1), \cdots, s(a, M))$ gives the locations of the M 1's in the ath codeword. In an IT-code, it is the position (location) of such one (a single bit) which will be selected for transmission over a channel $\mathcal{W}$.

Definition 2. *[24],[25] An (n, S, λ) transmission code $\overline{C}$ for channel $\{W^n : A^n \to B^n\}$ is a collection $\{(\phi(s), E_s) \in A^n \times \exp_2(B^n)\}, s = 1, \cdots, S\}$ such that the subsets E_s are nonoverlapping and*

$$W^n(E_s | \phi(s)) \ge 1 - \lambda.$$

In [24] the following three measures associated with a binary constant-weight code were defined:

(i) the weight factor $\beta := \frac{\log M}{\log S}$,

(ii) the second-order rate factor $\rho := \frac{\log \log N}{\log S}$,

(iii) the pairwise overlap fraction $\mu := \frac{K}{M}$.

Any (S, N, M, K) binary constant-weight code satisfies $\beta \le 1$ and $\rho \le 1$.

Proposition 1. *[24] Given an $(S, N, M, \mu M)$ binary constant-weight code $\{s(a, m), a = 1, \cdots, N; m = 1, \cdots, M\}, 0 \le \mu \le 1$, denoted by C, and an*

(n, S, λ) *transmission code* $\{(\phi(s), E_s), s = 1, \cdots, S\}$ *for channel* $\mathcal{W}$, *denoted by* $\overline{C}$, *the following is an* $(n, N, M, \lambda, \lambda + \mu)$ *IT-code* :

$$f(a, m) = \phi(s(a, m))$$

$$D_{a,m} = E_{s(a,m)}, \quad a = 1, \cdots, N, \quad m = 1, \cdots, M.$$

In [24] the following was observed. The rate-pair (R, R') of an IT code, constructed as in Proposition 1 from a transmission code $\overline{C}$ for a channel $\mathcal{W}$ with capacity $\mathcal{C}$ and a constant-weight code C, equals $(\beta\overline{R}, \rho\overline{R})$, where $\overline{R} = (\frac{1}{n})\log S$ is the rate of the transmission code, and β and ρ are the weight factor and second-order rate factor of the constant-weight code C. In order for (R, R') to approach the optimal rate-pair $(\mathcal{C}, \mathcal{C})$, as is theoretically achievable according to Theorem 2, one needs transmission codes whose rates $\overline{R}$ approach $\mathcal{C}$ and binary constant-weight codes C whose weight factor and second-order rate factor approach unity.

Definition 3. *[24] Consider a sequence* $\{C_i\}$ *of binary constant-weight codes* $C_i = (S_i, N_i, M_i, \mu_i M_i)$ *with weight factor* β_i, *second-order rate factor* ρ_i *and pairwise overlap fraction* μ_i.

The sequence of codes $\{C_i\}$ is said to be optimal for identification plus transmission if

$$\beta_i \to 1, \quad \rho_i \to 1, \quad \text{and} \quad \mu_i \to 0, \quad \text{as} \quad i \to \infty. \tag{1}$$

It may well happen that a sequence of binary constant-weight codes C_i satisfies condition (1), so that, when linked with a sequence of transmission codes $\overline{C}_i$ for a channel $\mathcal{W}$ (as in Proposition 1), the resulting IT-code sequence yields an optimal rate-pair (R, R') for identification plus transmission, but that the actual first-order rate $\frac{\log N_i}{S_i}$ of C_i tends to zero. We shall encounter such example shortly. This fact is immaterial, though, for joint identification and transmission with the combined scheme of a binary constant-weight code C and a transmission code $\overline{C}$ described in Proposition 1.

We next turn to the description of a sequence of codes $\{C_i\}$ which is optimal for identification plus transmission in the sense of Definition 3, as invented by Verdú and Wei [24]. Such codes are based on the concatenation of a pulse position modulation code with two Reed-Solomon codes. Concatenated codes were originally introduced by Forney [12] to solve a theoretical problem, but have turned out to be useful in a variety of applications [13].

Definition 4. *[24] Let* C_1 *and* C_2 *be codes with blocklength* n_i, *size* N_i, *and alphabet* $A_i, i = 1, 2$. *If* $N_1 = |A_2|$, *then the concatenated (or nested) code* $C_{12} = C_1 \circ C_2$ *with blocklength* $n_{12} = n_1 n_2$, *size* $N_{12} = N_2$ *and alphabet* $A_{12} = A_1$ *is constructed by using any one-to-one mapping* $h : A_2 \to C_1$:

$$C_{12} = \{(h(y_1), \cdots, h(y_{n_2})) : (y_1, \cdots, y_{n_2}) \in C_2\}.$$

Thus, to form the concatenation C_{12}, each letter of each codeword $(y_1, \cdots, y_{n_2})$ of C_2 is replaced by a different codeword of C_1. C_1 is called the inner code and C_2 the outer code.

One way to construct a binary constant-weight code is to concatenate a pulse position modulation (PPM) code C_1 with an outer code C_2 for which $|A_2| > 2$.

Definition 5. *[24] A [q] PPM code is a $(q, q, 1, 0)$ binary constant-weight code; it consists of all binary q-vectors of unit weight.*

If one concatenates a $[q]$ PPM code C_1 with an outer code C_2 with $|A_2| = q, n_2 = n'$, $N_2 = N$, and minimum distance $d_2 = d$, one obtains an $(n'q, N, n', n' - d)$ binary constant-weight code C_{12}. The parameters of C_{12} are thus $S_{12} = n'q, N_{12} = N, M_{12} = n'$, and $K_{12} = n' - d$.

Eventually we are interested in concatenating a $[q]$ PPM code C_1 with the concatenation of two Reed-Solomon codes C_2 and C_3.

Definition 6. *[24] Let q be a prime power (i.e. $q = p^m$), and denote the elements of the Galois field $GF(q)$ by $\{a_1, \cdots, a_q\}$. A $[q, k]$ Reed-Solomon (R-S) code $(k < q)$ is the set of q-vectors over $GF(q)$: $\{(p(a_1), \cdots, p(a_q)) : p(x)$ is a polynomial of degree $< k$ with coefficients from $GF(q)\}$.*

For a $[q, k]$ R-S code C_2, it holds that $n_2 = q, N_2 = q^k$, and $d_2 = q - k + 1$. The concatenation of a $[q]$ PPM code C_1 with a $[q, k]$ R-S code C_2 results in a $(q^2, q^k, q, k - 1)$ binary constant-weight code $C_{12} = C_1 \circ C_2$, i.e. $S_{12} = q^2$, $N_{12} = q^k$, $M_{12} = q$, and $K_{12} = k - 1$.

It was observed in [24] that the concatenation of a PPM code C_1 with a R-S code C_2 will not lead to binary constant-weight codes which are optimal for identification plus transmission in the sense of Definition 3. However, the concatenation of a PPM code C_1 with the concatenation $C_2 \circ C_3$ of two R-S codes will lead to such optimal codes. The following construction is due to Verdú and Wei [24].

Proposition 2. *[24] The $[q, k, t]$ three-layer concatenated code $C = C_{123} = C_1 \circ C_2 \circ C_3$, with $C_1 = [q]$ PPM, $C_2 = [q, k]$ Reed-Solomon and $C_3 = [q^k, q^t]$ Reed-Solomon, with $t < k < q = $ prime power, is a $(q^{k+2}, q^{kq^t}, q^{k+1}, kq^k + q^{t+1})$ binary constant-weight code.*

The parameters of C_{123} in Proposition 2 satisfy $S_{123} = q^{k+2}, N_{123} = q^{kq^t}, M_{123} = q^{k+1}$, and $K_{123} < kq^k + q^{t+1}$, and depend on the choice of q, k and t.

Consider now a sequence $\{C_i\}$ of binary constant-weight codes, with C_i being a $[q_i, k_i, t_i]$ three-layer concatenated code constructed as in Proposition 2. In order for this sequence to be optimal for identification plus transmission in the sense of Definition 3, Verdú and Wei [24] established the following conditions.

Proposition 3. *[24] Let C_i be a $[q_i, k_i, t_i]$ three-layer concatenated code as constructed in Proposition 2. The sequence of codes $\{C_i\}$ is optimal for identification plus transmission if (a) $t_i \to \infty$, (b) $\frac{t_i}{k_i} \to 1$, (c) $\frac{k_i}{q_i} \to 0$, and (d) $q_i^{t_i - k_i} \to 0$, as $i \to \infty$.*

The proof of Proposition 3 is based on the fact that, for the C_i considered, the parameters β_i, ρ_i, and μ_i satisfy the following:

$$\beta_i = \frac{k_i + 1}{k_i + 2}, \tag{2}$$

$$\rho_i = \frac{t_i}{k_i + 2} + \frac{\log k_i + \log \log q_i}{(k_i + 2) \log q_i}, \tag{3}$$

$$\mu_i \leq \frac{k_i}{q_i} + q_i^{t_i - k_i}. \tag{4}$$

Hence (1) holds if conditions (a) - (d) of Proposition 3 are satisfied.

It is observed in [24] that a simple sequence of parameters satisfying the conditions of Proposition 3 is given by $t_i = i, k_i = i + 1$, and q_i any increasing sequence of prime powers, and furthermore that the first-order rate $R_i = \frac{\log N_{123,i}}{S_i}$ tends to zero for any sequence $\{C_i\}$ satisfying condition (d) of Proposition 3.

As remarked before, the latter fact is of no importance for evaluating the performance of the joint identification and transmission scheme described in Proposition 1. In this joint identification and transmission scheme one aims for codes C_i which are optimal for identification, i.e. provide a high second-order rate, not necessarily a good first-order rate.

When considering the joint IT scheme of Proposition 1, consisting of a binary constant-weight code C linked with a transmission code $\overline{C}$, it is the combined scheme that forms the IT-code which allows joint identification and message content transmission. From it one can derive an $(n, N, \lambda_1, \lambda_2)$ ID code and N (n, M, λ_1) transmission codes ($\lambda_1 = \lambda, \lambda_2 = \lambda + \mu$). Although the joint scheme cannot be separated into these derived ID and transmission codes, it is clear that the binary constant-weight code C plays the major role in the identification part and $\overline{C}$ is the essential component for the transmission part of the joint IT-code. We shall refer to the binary constant-weight code C, underlying the scheme of Proposition 1, as an IT-generating (ITG) code, and denote it regularly by C_{ITG}. Similarly, binary constant-weight codes C_i, which form a sequence $\{C_i\}$ as in Definition 3, may be denoted by $C_{ITG,i}$. The particular binary constant-weight code described in Proposition 2 is such C_{ITG}, and conditions for a sequence $\{C_{ITG,i}\}$, when $C_{ITG,i}$ is a $[q_i, k_i, t_i]$ three-layer concatenated code, to be optimal for identification plus transmission are given by Proposition 3.

We further remark that, when linking a code C_{ITG} from Proposition 2 with a channel code (according to the procedure of Proposition 1), the position (one out of S) of the single bit (one out of M) needs to be transmitted over the channel. If the channel is binary, as will be our case, the binary representation of the position of the single bit to be transmitted is needed.

There are two properties of the binary constant-weight ITG code, constructed in Proposition 2 as a three-layer concatenated code, which are worthwhile mentioning. First, the size of this code ($N = q^{kq^t}$) exhibits a doubly exponential growth in q, thereby revealing the expected doubly exponential growth of the identifier set size. Second, for a sequence $\{C_i\}$ of $[q_i, k_i, t_i]$ three-layer concatenated codes which is optimal for identification plus transmission, it is a consequence of conditions (c) and (d) of Proposition 3 that the overlap fraction $\mu_i = \frac{K_i}{M_i} \to 0$ as $i \to \infty$. If the channel is noiseless this implies that the false identification probability will vanish. We will confine ourselves in this paper to a binary noiseless channel for transmitting the binary representation of the position of the single bit selected. For a noiseless channel the missed identification probability is zero.

In Fig. 3, the concatenated code of Proposition 2 is shown schematically. Whereas the component codes are ordered from right to left, the path of the message flow is from left to right.

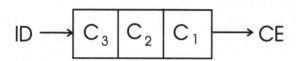

Fig. 3. From ID input to channel encoder

In practice, we are frequently interested in identification via a channel under a wordlength constraint $n \le n_0 < \infty$. Under such a constraint one should reckon with the fact that adopting the just described asymptotically optimal explicit construction might yield an unacceptably (too large) probability of false identification. Usually the false identification probability proves to be really critical in this respect. When the channel is noiseless this is the only quality-of-performance characteristic. A drastic suppression of the false identification probability with respect to an ITG code of blocklength S, the corresponding sequence of binary constant-weight codes of which is asymptotically suboptimal, might offer a solution to the problem. Two ways of doing so will be considered in the next section.

It has been pointed out in [24] that even if there is no simultaneous message transmission, the conveying of a single bit through the channel should be done by randomization. In this case one might choose the position of the bit to be transmitted uniformly from the M weights of $C = C_{ITG}$. We will propose an extension of such a design in the next section, not only for a reiterated (i.e. serial) but also for a simultaneous (i.e. parallel) transmission of more than one copy of the identifier, assuming an idealized model of scrambled messages, with the very same randomization also.

3   Elementary Forward Error Control Meant Especially for Huge Identification Codes

In the remote alarm services we have in mind, it is a rather common demand to constrain the wordlength n of a IT code by $n \le n_0 < \infty$.

While a wordlength n will be immediately related to the characteristics of the forward error correction (FEC) to be considered, it does not characterize the delay with respect to the instant the identification codeword has been initiated. Note that, when $n \leq n_0$, the error probabilities of missed and false identification may exceed the preset tolerance thresholds λ_1 and λ_2, respectively, since the basic ID and IT coding theorems (Theorems 1 and 2 above) are asymptotic results holding only for $n \to \infty$.

For a noiseless channel, the missed identification probability equals zero, and then the false identification probability becomes the only characterizer of the quality of performance. We confine ourselves in the current paper to such a channel (in particular a binary noiseless channel), just to focus on the most interesting features of our present subject of investigation.

By forward error correction (FEC) we mean a method of error control which concerns only the encoding part of the information transmission, not the decoding part.

In order to achieve an initial suppression of the false identification probability of the IT code based on the ITG code C_{ITG} of Proposition 2, we first put some restrictions on the parameters q, k, and t which define C_{ITG}. By the definition of a R-S code, we already know that these code parameters must satisfy $1 \leq t < k < q$. Now let $q = 2^m, m \geq 1$, and let t and k be fixed. Denote the corresponding three-layer concatenated code by $C_{ITG,m}$. For fixed t and k, and $m \to \infty$, the sequence $\{C_{ITG,m}\}$ is not optimal for identification plus transmission in the sense of Definition 1, as the four conditions of Proposition 3 are not all satisfied (in particular condition (a) is violated). However, this code sequence still has desirable properties, as we shall now demonstrate.

First, for the parameters β_m and ρ_m (defined as in (2) and (3) with $i = m$), we have

$$\lim_{m \to \infty} \beta_m = \beta_1 = \frac{k+1}{k+2} \tag{5}$$

$$\lim_{m \to \infty} \rho_m = \frac{t}{k+2}. \tag{6}$$

This implies that, if the rate $\overline{R} = \frac{1}{n} \log S$ of the transmission code in Proposition 1 approaches the capacity C of channel W, the rate-pair (R, R') of the IT-code formed by $C_{ITG,m}$ and this transmission code approaches the rate-pair

$$\left(\frac{k+1}{k+2} C, \frac{t}{k+2} C \right), \tag{7}$$

which is suboptimal with respect to the theoretically achievable rate-pair (C, C). By way of example, if $t = 1, k = 2$, (7) yields the pair $(\frac{3}{4}C, \frac{1}{4}C)$, and if $t = 2, k = 3$ then (7) yields $(\frac{4}{5}C, \frac{2}{5}C)$. Thus, when $C > 0$, the IT rate-pair (7) still provides a doubly exponentially large identifier set size and an exponentially large message set size.

For the identifier set size N, derived from the binary constant-weight code $C_{ITG,m}$ defined above (with t and k fixed, $1 \le t < k < q = 2^m$), we obtain the following expression:

$$N = N(m,t,k) = q^{kq^t} = 2^{mk2^{mt}} = 2^{2^{m[g(m,k)+t]}}, \tag{8}$$

where

$$g(m,k) := \frac{\log mk}{m} \to 0 \quad \text{as} \quad m \to \infty, \quad k \quad \text{fixed}. \tag{9}$$

Hence, for k fixed,

$$N = 2^{2^{m[t+o(1)]}} \simeq 2^{2^{mt}} \quad \text{as} \quad m \to \infty, \tag{10}$$

where $o(1) \to 0$ as $m \to \infty$. By the notation $f(m) \simeq g(m)$, as $m \to \infty$, we mean that $\log\log f(m) \sim \log\log g(m)$.

We next analyze the probability of false identification of the IT code corresponding to $C_{ITG,m}$ and establish an upper bound on it.

First, though, we review the rule used at the output of the channel for the identification of an arriving identifier, as discussed in [23], and this for an IT code based on any binary constant-weight code C_{ITG} as defined in Proposition 1.

We assume that all codewords $c_{IT} \in C_{ITG}$ are used for identification, so that to each identifier there is assigned one and only one codeword c_{IT}. Suppose that at the common output of the channel we want to decide whether or not identifier a has been sent (and if so, how to recover the associated message). Let a copy of the codeword $c_{IT}(a) \in C_{ITG}$, assigned to identifier a, be stored for this purpose at the output of the channel. Let codeword $c_{IT}(b)$ correspond to identifier b, and suppose identifier b and message m are sent by the sender. Then one of the $M = M_{ITG}$ ones of $c_{IT}(b)$, namely the single bit corresponding to message m $(1 \le m \le M)$ is selected, and its position s $(1 \le s \le S)$ is transmitted via a transmission code $\overline{C}$ over $\mathcal{W}$. Since $\mathcal{W}$ is assumed to be noiseless in our case, this amounts to mapping the position s into a binary sequence of length $n = \log S$ and transmitting this sequence over the binary noiseless channel $\mathcal{W}$ using n channel operations. After receiving the sequence corresponding to the position s at the channel output (or after decoding this sequence into s' if $\mathcal{W}$ is noisy, but here $s = s'$), the receiver declares identifier a if the codeword $c_{IT}(a)$ stored has a one at the location s (i.e., if the symbol one selected from $c_{IT}(b)$ covers any of the ones of $c_{IT}(a)$). This event is called an identifier match.

Furthermore, let (a', b') stand for any worst possible identifier pair, i.e., a pair such that the corresponding codewords $c_{IT}(a')$ and $c_{IT}(b')$ are the minimum Hamming distance apart. Define for an identifier pair (a, b), with $b \ne a$, the probability of false identification by

$$P(false, (a,b)) := P(\{b\ declared\}|\{a\ arrived\}).$$

Clearly, for identifier pair (a, b)

$$P(false, (a,b)) \le P(false, (a',b')),$$

where (a', b') is any worst possible identifier pair. Therefore we define

$$P(false) := P(false, (a', b')),$$

which takes the same value for any worst possible identifier pair. Thus $P(false)$ is the maximum probability of false identification, where the maximum is taken over all pairs of distinct codewords from C_{ITG}. If the channel W is noiseless then, as was shown in [23],

$$P(false) = \frac{K_{ITG}}{M_{ITG}}, \tag{11}$$

where $K_{ITG} = K$ and $M_{ITG} = M$ are the maximum number of coincident 1's for any pair of codewords (the worst possible correlation) and the Hamming weight, respectively, of the binary constant-weight code.

Now let us return to the binary constant-weight code $C_{ITG,m}$ discussed above, constructed as in Proposition 2, with $q = 2^m$ and t and k fixed. With $K_{ITG} = K_{123}$ and $M_{ITG} = M_{123}$, where K_{123} and M_{123} are given by the expressions immediately following Proposition 2, we obtain

$$P(false) < \frac{kq^k + q^{t+1}}{q^{k+1}} = \frac{k}{q}(1 + \frac{1}{kq^{k-t-1}}). \tag{12}$$

It follows that, if W is noiseless, $q = 2^m$, and t and k are fixed with $k - t > 1$, then

$$P(false) < \frac{k}{2^m}(1 + o(1)), \tag{13}$$

where $o(1) \to 0$ as $m \to \infty$. The upper bound in (13) provides indeed a considerable reduction of the false identification probability, as compared to when t and k are not fixed, but satisfy conditions (a) and (b) of Proposition 2, e.g., if $t = m$ and $k = m + 2$. By way of example, if $t = 1, k = 3$, and $m = 16$, then the upper bound in (12) yields

$$P(false) < 3 \cdot 2^{-16} + 2^{-32} < 2^{-14}, \tag{14}$$

whereas the choice $m = 16, t = 14$, and $k = 16$ substituted in (12) yields only

$$P(false) < 2^{-12} + 2^{-32}.$$

Notice that, if k and t are fixed, the choice $k - t = 1$ is also allowed. In this case the upper bound in (12) takes the form

$$P(false) < \frac{k + 1}{q}. \tag{15}$$

For $q = 2^m, m = 16, t = 1, k = 2$ (15) yields

$$P(false) < 3 \cdot 2^{-16},$$

which is actually smaller than the bound in (14).

The false identification probability can be further drastically suppressed by the usage of two kinds of non-algebraic forward error control, which will be introduced shortly. Before doing so, we elaborate on the threshold n_0 which was put forward as a constraint on the wordlength n of our IT code.

In the context of the IT code considered here, composed, as explained by Proposition 1 and Proposition 2, of a binary constant-weight code $C_{ITG} = (S, N, M, K)$ with $S = q^{k+2}, N = \underline{q}^{kq^t}, M = q^{k+1}, K < kq^k + q^{t+1}$, and $q = 2^m$, and an (n, S, λ) transmission code $\overline{C}$, there are various wordlengths to distinguish between.

First, there is the blocklength n of the overall IT code, which is also the blocklength of the transmission code $\overline{C}$ for sending the position of the selected bit over channel $\mathcal{W}$.

Secondly, there is the common length of each codeword of C_{ITG} which is given by

$$S = q^{k+2} = 2^{m(k+2)}, \tag{16}$$

and thirdly there is the parameter $m = \log q$ which represents the length of the binary representation of q. The relationship between these three parameters is given by (16) and

$$S = 2^n, \tag{17}$$

so that

$$n = \log S = m(k + 2). \tag{18}$$

Clearly, a threshold on n implies a threshold on S, and also on m, if k is given as in our case, and vice-versa.

Practical service objectives may impose a constraint S_0 on S, from which a tolerance threshold n_0 on n can be derived immediately as follows. If the threshold S_0 on S is specified, then define first the tolerance threshold imposed on m by

$$m_0 = \left\lfloor \frac{\log S_0}{k + 2} \right\rfloor, \tag{19}$$

and next the corresponding tolerance threshold on n by

$$n_0 = m_0(k + 2), \tag{20}$$

where $\lfloor x \rfloor$ stands for the largest integer less than or equal to x.

Lemma 1. *Let S_0 be a given threshold on the length S of a codeword from C_{ITG}, and let m_0 and n_0, defined by (19) and (20), be the corresponding thresholds on m and n, respectively. Define $q_0 = 2^{m_0}$. Then the wordlength S of a C_{ITG} code for which $q = q_0$, denoted by $S(q_0)$, satisfies*

$$S(q_0) \leq S_0. \tag{21}$$

Thus, the constraint $S \leq S_0$ is met for $S(q_0)$, where q_0 is derived from S_0.

Proof: $S(q_0) = 2^{m_0(k+2)} \leq 2^{\log S_0} = S_0$.

We are now ready to introduce two basic kinds of FEC, meant especially to drastically suppress the false identification probability of the above-mentioned IT codes. The price of using these is a drastic, though usually far from catastrophic, drop in the code size in one of the two approaches. The same drop in code size is still remarkable, but drastically less than in the previous case, for the other considered approach. We refer to these two approaches as FEC version 1 and FEC version 2. The first version corresponds to a serial scheme and the second one to a parallel scheme of FEC.

The reason for introducing these FEC schemes is that, under the constraint $S \leq S_0$, $m \leq m_0$, $n \leq n_0$, the probability of false identification might exceed λ_2 if no FEC is applied. Hence, these thresholds are assumed to be given and fixed in the ensuing discussion.

It turns out that even for FEC version 1 still appropriately huge code sizes might be produced at a word length $n \leq n_0 < \infty$ with respect to the code size of the original IT code. The code size drop for the second approach, FEC version 2, is much smaller, but this latter version is more complex than the former one.

If the thresholds were to diverge to infinity (i.e., if $n_0 \to \infty$), then a similar drop in code size would occur in both cases of FEC, which can be measured in terms of second-order rate. Hence these new identification codes are no more optimal, even asymptotically.

In the following discussion the IT code corresponding to the parameter m_0 plays an important role, as the FEC versions to be described are derived from it, and their performance will be measured relative to this original IT code. Therefore, we describe it here once more in detail.

If $m = m_0$, the C_{ITG} code from Proposition 2 is a binary constant-weight code (S, N, M, K) based on $q_0 = 2^{m_0}$, so that $S = S(q_0) = 2^{m_0(k+2)}$ and $N = N(q_0) = 2^{m_0 k 2^{m_0 t}}$, with $1 \leq t < k < 2^{m_0}$. We denote this C_{ITG} code by C_{ITG}^0. Associated with C_{ITG}^0 there is an $(n_0, S(q_0), \lambda)$ transmission code $\overline{C}_0$ for $\mathcal{W}$ with $n_0 = m_0(k + 2)$. The IT code formed by the juxtaposition of C_{ITG}^0 and $\overline{C}_0$ is denoted by C_{IT}^0. The maximum probability of false identification when using C_{IT}^0 is denoted by $P_0(false)$, and when $\mathcal{W}$ is noiseless, we have, according to (13)

$$P_0(false) < \frac{k}{2^{m_0}}(1 + o(1)). \tag{22}$$

For the code size of C_{IT}^0, denoted by N_0, we have from the above and (10)

$$N_0 = N(q_0) \simeq 2^{2^{m_0 t}} \quad \text{as } m_0 \to \infty \tag{23}$$

for t and k fixed, $1 \leq t < k < q_0$, and for the second-order rate of C_{IT}^0, denoted by R_0, we have, under the same assumptions,

$$R_0 = \frac{\log \log N_0}{n_0} \sim \frac{m_0 t}{n_0} = \frac{m_0 t}{m_0(k + 2)} = \frac{t}{k + 2} \tag{24}$$

in accordance with (7).

Description of FEC Version 1 of C_{IT}^0

Let the threshold m_0 for m be fixed, in function of threshold S_0 according to (19). Define the threshold for n by n_0 as in (20). Choose an integer $\nu \geq 1$, typically $\nu = 2$ or $\nu = 4$, in order to define a ν-fold repetition code for conveying the identifier towards the channel ν times. Define

$$m_\nu = \left\lfloor \frac{m_0}{\nu} \right\rfloor, \quad q_\nu = 2^{m_\nu}, \text{ and } n_\nu = (k+2)m_\nu. \tag{25}$$

Consider the $[q_\nu, k, t]$ three-layer concatenated code of Proposition 2, denoted by $C_{ITG,\nu}^1$, which is a $(S_\nu, N_\nu, M_\nu, K_\nu)$ binary constant-weight code with $S_\nu = q_\nu^{k+2}$, $N_\nu = q_\nu^{k(q_\nu)^t}$, and $M_\nu = q_\nu^{k+1}$. Recall that on the joint choice of q_ν, k and t the following constraint is imposed: $1 \leq t < k < q_\nu$. Whereas q_ν depends on the choice of ν, the parameters t and k are kept constant. Clearly for $\nu \geq 1$

$$S_\nu = 2^{(k+2)m_\nu} \leq 2^{(k+2)m_0} \leq S_0,$$

so that the length S_ν of each codeword of $C_{ITG,\nu}^1$ satisfies the constraint $S_\nu \leq S_0$. For $\nu = 1$, $C_{ITG,1}^1$ is the binary constant-weight code (S_1, N_1, M_1, K_1) based on $q_1 = 2^{m_1}$, t and k, with $m_1 = m_0$. Now consider the IT code $C_{IT,\nu}^1$ formed by the juxtaposition of $C_{ITG,\nu}^1$ and an $(n_\nu, S_\nu, \lambda_\nu)$ transmission code $\overline{C}_\nu^1$ for channel $\mathcal{W}$. Suppose the identifier b and message w are to be sent. In FEC version 1 this is done by applying the $C_{IT,\nu}^1$ code ν times in a consecutive order. This means that from the codeword $c_{IT,\nu}^1(b) \in C_{ITG,\nu}^1$, assigned to identifier b, the w-th "one" is selected, and its position, denoted by $s(b, w)$, is encoded into a codeword $\overline{c}_\nu^1(s(b, w))$ of length n_ν of $\overline{C}_\nu^1$. When $\mathcal{W}$ is noiseless and binary (as is our case) this amounts to mapping $s(b, w)$ into a binary sequence of length n_ν, which will be received errorfree at the output of channel $\mathcal{W}$, so that the position of the single bit of $c_{IT,\nu}^1(b)$ which was selected is known by the receiver. This procedure is repeated ν times. Thus the binary sequence of length n_ν, corresponding to the binary representation of the position $s(b, w)$ of the w-th bit of codeword $c_{IT,\nu}^1(b)$, is sent ν times over $\mathcal{W}$ with a total block length of $\nu \cdot n_\nu \leq n_0$. The combined sequence, consisting of ν times the same binary sequence of length n_ν, is nothing else than the codeword of a repetition code, for sending the same identifier ν times. If no message content w is to be transmitted at all, we select a single bit uniformly from the M weights of $c_{IT,\nu}^1(b)$ and transmit the position of this randomly selected bit; we do these uniform drawings independently over all ν successive codewords.

For decoding the identifier sent, we use the rule of an identifier match described above, but now for each of the ν consecutive blocks. A copy of just one of the N codewords of $C_{ITG,\nu}^1$ is stored at the output of the channel, $c_{IT,\nu}^1(a)$ say, if a is the identifier meant to be monitored at the output of the channel. We declare the stored identifier as the one having been sent, whenever the single bit received in all j codeword slots, $1 \leq j \leq \nu$, is covered within each codeword slot by one of the weights of the stored codeword $c_{IT,\nu}^1(a)$. Otherwise we declare a missed identification. If $\mathcal{W}$ is noiseless, and $b = a$, the receiver will identify the identifier which was sent correctly, and in each codeword slot the location of the single bit received will be the same. However, even if $\mathcal{W}$ is noiseless, an identifier match may also occur if another codeword $c^1_{IT,\nu}(b)$, $b \neq a$, different from the stored codeword $c^1_{IT,\nu}(a)$ has been sent. Such event corresponds to false identification. If the ν binary codewords $\overline{c}^1_{\nu,j}$ (conveying the positions of the single bits of the ν binary codewords $c^1_{IT,\nu,j}$) are decoded correctly, the entire message content might be recovered. This is certainly the case for any noiseless channel. For a noisy channel, error correction (or erasure) might need to be employed in order to decode the incoming codewords $\overline{c}^1_{\nu,j}$ correctly.

Let us next investigate what the drop in code size is of the IT code C^0_{IT} (with $\mathcal{W}$ noiseless), when it is replaced by a FEC version 1 with $\nu \geq 2$. We also investigate what the change in second order rate and of the probability of false identification is, when using FEC version 1, as compared to the original IT code C^0_{IT}, i.e. when using no FEC. We denote the code size of $C^1_{ITG,\nu}$ by $N_{1,\nu}$. This is the number of possible identifiers when using the IT code $C^1_{IT,\nu}$, consisting of the juxtaposition of $C^1_{ITG,\nu}$ with the transmission code $\overline{C}^1_\nu$. Since in FEC version 1 the same IT code $C^1_{IT,\nu}$ is repeated ν times, the number of possible identifiers (i.e. the code size) of a FEC version 1 of C^0_{IT} is also $N_{1,\nu}$. For the second order rate of this repetition code, denoted by $R_{1,\nu}$ we have

$$R_{1,\nu} = \frac{\log \log N_{1,\nu}}{n_0} \tag{26}$$

since the length of the total transmission over $\mathcal{W}$ is $n_0 = \nu n_\nu$.

We denote the maximum probability of false identification when using $C^1_{IT,\nu}$ only once by $P_{1,\nu,1}(false)$, and the same probability when using $C^1_{IT,\nu}$ ν times by $P_{1,\nu}(false)$.

Theorem 3. *Let m_0 be fixed, $q_0 = 2^{m_0}$, $\nu \geq 2$, $m_\nu = \lfloor \frac{m_0}{\nu} \rfloor$, $q_\nu = 2^{m_\nu}$, k and t fixed, $1 \leq t < k < q_\nu$. Then, for FEC version 1, the code size $N_{1,\nu}$, the second-order rate $R_{1,\nu}$, and the probability of false identification $P_{1,\nu}(false)$ one has:*

(i)
$$N_{1,\nu} = 2^{2^{m_\nu [g(m_\nu,k)+t]}},$$
$$\tag{27}$$

where $g(m_\nu, k) = \frac{\log m_\nu k}{m_\nu} \to 0$ as $m_\nu \to \infty$; hence

$$N_{1,\nu} \simeq 2^{2^{m_\nu t}} \quad \text{as } m_\nu \to \infty. \tag{28}$$

(ii)
$$R_{1,\nu} \sim \frac{t}{(k+2)\nu} \sim \frac{R_0}{\nu} \quad \text{as } m_\nu \to \infty, \tag{29}$$

(iii)
$$P_{1,\nu}(false) \leq P^{UB}_{1,\nu}, \tag{30}$$

with

$$P^{UB}_{1,\nu} \sim \left(P^{UB}_0\right)^\nu 2^{m_0(\nu-1)+\nu}, \tag{31}$$

and

$$P_0^{UB} = \frac{k}{2^{m_0}}. \tag{32}$$

Proof: (i) Since $N_{1,\nu} = 2^{m_\nu k 2^{m_\nu t}}$, (27) and (28) follow readily from (8) and (10). (ii) By (26) and (27)

$$R_{1,\nu} = \frac{m_\nu[g(m_\nu, k) + t]}{\nu n_\nu}.$$

Letting $m_\nu \to \infty$, we obtain

$$R_{1,\nu} \sim \frac{m_\nu t}{\nu n_\nu} = \frac{m_\nu t}{\nu m_\nu(k+2)} = \frac{t}{\nu(k+2)}. \tag{33}$$

Now compare (33) with (24). (iii) By independence

$$P_{1,\nu}(false) = (P_{1,\nu,1}(false))^\nu.$$

By (13)

$$P_{1,\nu,1}(false) < \frac{k}{2^{m_\nu}}(1 + o(1)) < \frac{k}{2^{m_0}}\left(2^{m_0\left(\frac{\nu-1}{\nu}\right)+1}\right)(1 + o(1)). \tag{34}$$

Hence

$$P_{1,\nu}(false) < \left(\frac{k}{2^{m_0}}\right)^\nu \left(2^{m_0(\nu-1)+\nu}\right)(1 + o(1)).$$

Denoting the righthand side of this inequality by $P_{1,\nu}^{UB}$, (30) and (31) follow from (32) and letting $m_\nu \to \infty$.

Description of FEC Version 2 of C_{IT}^0

Let the thresholds S_0, m_0, and n_0 be the same as in FEC version 1, and let $q_0 = 2^{m_0}$ as before. Choose again an integer $\nu \geq 1$, this time to define a ν-fold parallel code for conveying the identifier with ν codewords simultaneously over the channel. Let the parameters t and k again be fixed, $1 \leq t < k < q_0$.

Now consider the original C_{IT}^0 code, composed of C_{ITG}^0 and $\overline{C}_0$, defined prior to (22). C_{ITG}^0 is a binary constant-weight code (S, N, M, K) with $S = S(q_0)$ and $N = N(q_0)$ defined above. Next subdivide (partition) the $N(q_0)$ codewords, each of length $S(q_0)$, into ν subcodes of equal size

$$N_{2,\nu} = \left\lfloor \frac{N(q_0)}{\nu} \right\rfloor. \tag{35}$$

Denote these subcodes by $C_{ITG,j}^2$, $j = 1, \ldots, \nu$.

Observe that each $C_{ITG,j}^2$ is again a binary constant-weight code with $S = S(q_0)$, $N = N_{2,\nu}$, and the parameters M and K the same as for C_{ITG}^0. It is immaterial which codewords of C_{ITG}^0 are assigned to which subcode, and which codewords are dropped by the rounding off in (35), as only the code size $N_{2,\nu}$ matters. Suppose that the identifier b and message m are to be sent. In FEC version 2 this is done by encoding identifier b ν times, each time into a different codeword $c_{IT,j}^2(b)$, where each $c_{IT,j}^2(b)$ belongs to one and only one $C_{ITG,j}^2$, $j = 1, \ldots, \nu$. Next select from each codeword $c_{IT,j}^2(b)$, the w-th "one", and encode its position, denoted by $s_j(b, w)$ into a binary codeword $\bar{c}_j^2(s_j(b, w))$ of length n_0.

We assume the availability of ν parallel noiseless binary channels, denoted by $\mathcal{W}_j$, $j = 1, \ldots, \nu$. For each channel $\mathcal{W}_j$ there is an $(n_0, S(q_0), 0)$ transmission code $\overline{C}_j^2$, with $\bar{c}_j^2(s_j(b, w)) \in \overline{C}_j^2$. After n_0 channel operations the codeword $\bar{c}_j^2(s_j(b, w))$ is received errorfree at the output of $\mathcal{W}_j$. From it the position $s_j(b, w)$ of the single bit selected from $c_{IT,j}^2(b)$ can be deduced. Assume that one receiver observes the simultaneous outputs of all channels $\mathcal{W}_j$, i.e., he observes the vector $(\bar{c}_1^2(s_1(b, w)), \ldots, \bar{c}_\nu^2(s_\nu(b, w)))$, and thus knows the vector $(s_1(b, w), \ldots, s_\nu(b, w))$ of positions.

For decoding the identifier sent, we use again the rule of an identifier match, but now at the output of each channel $\mathcal{W}_j$. If identifier a is meant to be monitored by the common receiver, a copy of the corresponding codeword $c_{IT,j}^2(a) \in C_{ITG,j}^2$ is stored at the output of channel $\mathcal{W}_j$. The common receiver declares that identifier a was sent, whenever for each $j = 1, \ldots, \nu$, $c_{IT,j}^2(a)$ has a "one" on position $s_j(b, w)$. Otherwise he declares a missed identification. If $b = a$, the receiver will identify b correctly, since each $\mathcal{W}_j$ is noiseless. However, even if all $\mathcal{W}_j$'s are noiseless, an identifier match may occur at the output of a particular $\mathcal{W}_j$ if another codeword $c_{IT,j}^2(b)$, $b \neq a$, different from the stored codeword $c_{IT,j}^2(a)$, has been sent. If the latter event happens for all $j = 1, \ldots, \nu$, false identification occurs.

We now investigate what the change in code size is, if the original IT code C_{IT}^0 ($\mathcal{W}$ noiseless) is replaced by FEC version 2 with $\nu \geq 2$, and also what the possible change is of the second order rate and the probability of false identification. With $N_{2,\nu}$, defined in (35), being the common size of the subcodes $C_{ITG,j}^2$, $j = 1, \ldots, \nu$, the second order rate of FEC version 2, denoted by $R_{2,\nu}$, is given by

$$R_{2,\nu} = \frac{\log \log N_{2,\nu}}{n_0}. \tag{36}$$

We denote the maximum probability of false identification when using only one parallel channel (i.e. when $\nu = 1$, and then we are back in the situation of C_{ITG}^0) by $P_{2,1}(false)$, and the same probability when using FEC version 2 with $\nu \geq 2$ by $P_{2,\nu}(false)$.

Theorem 4. *Let $m_0 \geq 1$ be an integer, $q_0 = 2^{m_0}$, and let k and t be fixed, $1 \leq k < t < q_0$. Let $\nu \geq 2$ be a fixed integer. Then, for FEC version 2, the code size $N_{2,\nu}$, the second-order rate $R_{2,\nu}$, and the maximum probability of false identification $P_{2,\nu}(false)$ satisfy the following:*

(i)
$$N_{2,\nu} = \left\lfloor \frac{1}{\nu} 2^{2^{m_0[g(m_0,k)+t]}} \right\rfloor, \tag{37}$$

where $g(m_0, k) := \frac{\log m_0 k}{m_0} \to 0$ as $m_0 \to \infty$; hence

$$N_{2,\nu} \simeq \left\lfloor \frac{1}{\nu} 2^{2^{m_0 t}} \right\rfloor. \tag{38}$$

Furthermore

(ii)

$$R_{2,\nu} \sim \frac{t}{k+2} \sim R_0 \quad as \ m_0 \to \infty, \tag{39}$$

and

(iii)

$$P_{2,\nu}(false) \leq P_{2,\nu}^{UB}, \tag{40}$$

with

$$P_{2,\nu}^{UB} \sim \left(P_0^{UB} \right)^{\nu} = \left(\frac{k}{2^{m_0}} \right)^{\nu}. \tag{41}$$

Proof: (i) Since $N(q_0) = 2^{m_0 k 2^{m_0 t}}$, (37) and (38) follow from (8), (9), (10), and (35). (ii) It can be easily verified that for ν fixed, and m_0 greater than some constant $\delta(\nu)$,

$$2^{2^{m_0[g(m_0,k)+t]-1}} < N_{2,\nu} < 2^{2^{m_0[g(m_0,k)+t]}}, \tag{42}$$

hence

$$\lim_{m_0 \to \infty} \frac{\log \log N_{2,\nu}}{n_0} = \lim_{m_0 \to \infty} \frac{m_0[g(m_0,k)+t]}{m_0(k+2)} = \frac{t}{k+2}. \tag{43}$$

(iii) By the independence of the ν parallel channels,

$$P_{2,\nu}(false) = (P_{2,1}(false))^{\nu}. \tag{44}$$

By (13)

$$P_{2,1}(false) < \frac{k}{2^{m_0}}(1 + o(1)). \tag{45}$$

Letting

$$P_{2,\nu}^{UB} = \left(\frac{k}{2^{m_0}} \right)^{\nu}(1 + o(1)), \tag{46}$$

(40) and (41) follow from (44), (45), (46), and (32).

Remark 1: When comparing Theorem 3 and Theorem 4, we see that FEC Version 2 shows better performance than FEC Version 1. First, for $\nu \geq 2$, $N_{1,\nu}$ (given by (27) and (28)) is much smaller than $N_{2,\nu}$ (given by (37) and (38)). The drop in size from $N_{1,1} = N(q_0) = N_{2,1}$ to $N_{1,\nu}$ is quite drastic, but the reduction in size from $N_{2,1}$ to $N_{2,\nu}$ is much less significant. This difference becomes even clearer, when comparing the rates $R_{1,\nu}$ (approximately given by (29)) and $R_{2,\nu}$ (approximately given by (39)).

The approximate value of $R_{2,\nu}$, when m_0 is large, is ν times that of $R_{1,\nu}$, and, approximately, $R_{2,\nu} = R_{2,1} = R_0$, i.e., there is no rate loss in using FEC Version 2, provided m_0 is large. Finally, the approximate upper bound (31) to

$P_{1,\nu}(false)$ is a factor $2^{m_0(\nu-1)+\nu}$ larger than the corresponding upper bound (41) on $P_{2,\nu}(false)$.

Example: Assume $t = 1$, $k = 2$, $m_0 = 16$, so that $q_0 = 2^{16}$ and $n_0 = 64$.

Case (i). Let $\nu = 2$, so that $m_\nu = 8$.
For FEC Version 1 we obtain (cf. (27), (26), and (31)):

$$N_{1,\nu} = 2^{16 \cdot 2^8} = 2^{2^{12}}, \ R_{1,\nu} = \frac{12}{64} = \frac{3}{16}, \ P_{1,\nu}^{UB} \doteq \left(\frac{2}{2^{16}}\right)^2 \cdot 2^{18} = 2^{-12}.$$

The more precise bound (15) yields

$$P_{1,\nu}(false) < \left(\frac{3}{2^8}\right)^2 = \frac{9}{2^{16}} < 2^{-12}.$$

On the other hand, for FEC Version 2 we obtain (cf. (37), (36), and (41)):

$$N_{2,\nu} = \frac{1}{2} \cdot 2^{32 \cdot 2^{16}} = 2^{2^{21}-1} > 2^{2^{20}}, \ R_{2,\nu} > \frac{20}{64} = \frac{5}{16}, \ P_{2,\nu}^{UB} \doteq \left(\frac{2}{2^{16}}\right)^2 = 2^{-30}.$$

Applying (15) and (44) we get the bound

$$P_{2,\nu}(false) < \left(\frac{3}{2^{16}}\right)^2,$$

which is between 2^{-29} and 2^{-28}.

Case (ii). Let $\nu = 4$, so that $m_\nu = 4$.
For FEC Version 1 we then obtain by similar calculations:

$$N_{1,\nu} = 2^{8 \cdot 2^4} = 2^{2^7},$$

$$R_{1,\nu} = \frac{7}{64},$$

$$P_{1,\nu}^{UB} \doteq \left(\frac{2}{2^{16}}\right)^4 \cdot 2^{52} = 2^{-8}.$$

From (15) we get

$$P_{1,\nu}(false) < \left(\frac{3}{2^4}\right)^4 = \frac{81}{2^{16}},$$

which is between 2^{-10} and 2^{-9}.

For FEC Version 2 the corresponding results are:

$$N_{2,\nu} = \frac{1}{4} \cdot 2^{32 \cdot 2^{16}} = 2^{2^{21}-2} > 2^{2^{20}},$$

$$R_{2,\nu} > \frac{20}{64} = \frac{5}{16},$$

$$P_{2,\nu}^{UB} \doteq \left(\frac{2}{2^{16}}\right)^4 = 2^{-60}.$$

When applying the bound (15) we obtain

$$P_{2,\nu}(false) < \left(\frac{3}{2^{16}}\right)^4,$$

which is between 2^{-58} and 2^{-57}.

The example shows that for each ν, the performance of FEC Version 2 (as measured by the three quantities $N_{i,\nu}$, $R_{i,\nu}$, and $P_{i,\nu}^{UB}$, $i = 1, 2$) is far better than that of FEC Version 1. It also shows that the drop in size of $N_{1,\nu}$ (as compared to $N_{1,1}$) is much bigger than the drop in size of $N_{2,\nu}$. Furthermore, as ν increases, the performance of all three quantities $N_{1,\nu}$, $R_{1,\nu}$, and $P_{1,\nu}^{UB}$ gets worse, whereas in the case of FEC version 2 the values of $N_{2,\nu}$ and $P_{2,\nu}^{UB}$ only very slightly decrease, but $P_{2,\nu}^{UB}$ drastically decreases. We conclude that from the point of view of these three quantities FEC version 2 is far superior than FEC version 1, but from the viewpoint of implementation FEC version 2 is more complex than FEC version 1. Therefore, under certain circumstances, FEC version 1 might still be appealing.

Remark 2: In both Theorem 3 and Theorem 4 we allow the thresholds m_0 and m_ν to tend to infinity. We allowed this also in (23) and (24). Here we should like to clarify the meaning of this, as normally a threshold value should be finite and possibly not too large. We recall that m_0 is a threshold on m, and is related to the thresholds S_0 and n_0 by (19) and (20), so that as m_0 increases, also S_0 and n_0 increase. The asymptotic results of Theorems 1 and 2 only hold for $n \to \infty$. By imposing a constraint m_0 on m we restrict the value of n to $n \le n_0 = m_0(k+2)$. The asymptotic statements in Theorems 3 and 4 are to be interpreted as providing good approximations to $N_{i,\nu}$, $R_{i,\nu}$, and $P_{i,\nu}^{UB}$, $i = 1, 2$, for reasonably large, but finite values of m_0. One determining term in measuring how good the approximations are for $N_{i,\nu}$ and $R_{i,\nu}$ is the value of $g(m_\nu, k) = \frac{\log m_\nu k}{m_\nu}$. For $\nu = k = 2$, $t = 1$, and $m_0 = 256$, we find, by way of example, that $g(m_0, k) = \frac{9}{256} = 0.0352$ and $g(m_\nu, k) = \frac{8}{128} = 0.0625$. Hence for $m_0 \ge 256$, the approximations for $N_{i,\nu}$ and $R_{i,\nu}$ given in Theorems 3 and 4 are close to the true value. Also, for such value of m_0 the difference between $P_{i,\nu}^{UB}$, $i = 1, 2$, and its approximate value is negligible as can be seen from (12) and (13). Even for $m_2 = 100$, and thus $m_0 = 200$ and $n_0 = 800$, these approximations are close.

Remark 3: As discussed in Remark 1 and shown by the example, from the viewpoint of the quantities $N_{i,\nu}$, $R_{i,\nu}$, and $P_{i,\nu}^{UB}$, FEC version 2 is obviously better than FEC version 1, for the same value of ν, but FEC version 2 requires the availability of ν parallel channels and is therefore more complex in implementation. FEC version 1 requires only repeated transmissions. FEC version 1 can be viewed as a time-division approach and FEC version 2 as a frequency-division approach. Both TDM (time-division multiplexing) and FDM (frequency-division multiplexing) are techniques for splitting a big channel in many little channels. In TDM ν bit streams are multiplexed into one bit stream. This is done by sending the data in successive frames. In FDM the available bandwidth is split into ν equal parts. Thus, if the original channel has a usable bandwidth of W HZ each subchannel has W/ν HZ available. In FEC version 2 the ν codewords $\overline{c}_j^2(s_j(b, w))$, conveying the ν positions of the selected single bits, can be regarded as being bunched along the frequency axis, thereby retaining the original binary noiseless channel $\mathcal{W}$. In FEC version 1 the same bandwidth is used for $\nu = 1$ and $\nu \geq 2$. Recall that we only propose to use small values of ν, typically $\nu = 2$ or $\nu = 4$. For a discussion of bandwidth and the difference between TDM and FDM see [8]. For a discussion of the tradeoff between bandwidth and rate see [16]. We shall not go further into these questions here.

Another issue is the informationtheoretic interpretation of the introduction of extra, parallel channels in FEC version 2. Obviously, the capacity of a system of ν parallel channels (as allowed in FEC version 2), each of the same capacity $\mathcal{C}$, is equal to $\nu\mathcal{C}$. Thus, both the transmission capacity and the identification capacity of the IT code, when using ν parallel channels rather than one single channel, becomes significantly larger. However, in FEC version 2 we do not use the ν parallel channels to transmit at higher first-order or second-order rate, but rather to send the same identifier ν times, each time over a different component of the system of parallel channels, in order to reduce $P(false)$, as much as in FEC version 1 we send the same identifier ν times using ν successive frames. The system of ν parallel channels is only an informationtheoretic way of modeling what in a practical communication channel can be achieved with FDM.

Remark 4: In the description of FEC version 2 one could also assume the ν parallel channels $\mathcal{W}_j$ to be noisy, in which case the $(n_0, S(q_0), 0)$ transmission code $\overline{C}_j$ would need to be replaced by an $(n_0, S(q_0), \lambda)$ transmission code, and the decoding at the output of channel $\mathcal{W}_j$ would be subject to an error λ. However, if the transmission channel is noisy the probability of missed identification would also have to be controlled. In this paper we confine ourselves to noiseless channels, and focus on the suppression of $P(false)$, as only characterizer of the quality of performance, by FEC methods. The investigation of FEC, in combination with a wordlength constraint, when the underlying transmission channel is noisy is left as a problem for future study.

4 Such Identification Plus Transmission Codes Combined with Multiple Access

We now wish to combine the procedures discussed in Section 3, where the positions of ν single bits are conveyed to the receiver, with the principle of multiple access. We consider the model of multiple access communication, in which, out of a huge population of potential sources, only a relatively small number of users has a message to send at any given time over a common channel. This is e.g. the case if each of the many potential users only occasionally needs to send a message, such as an alarm, which occurs at this source rather infrequently. In this model, at any given time instant, a few sources are active sending a message (or ν consecutive messages) simultaneously over the common channel. This resembles the random access problem where at most M out of a total of T users can be active in the sense of sending a message over a common channel at any given time, as introduced in [20] and [7], and the M-active-out-of-T users collision channel without feedback, investigated in [17] and [1]. In that setup the size of the population of potential users is a finite number T. We however are interested in the situation where the size of the potential user population is vast, not precisely determined, and may grow in the future. Therefore, we like to resort to the ideas of identification theory, from which we know (cf. [5], [24]) that a huge number of identifiers can be made available.

Given an $(n, N, M, \lambda_1, \lambda_2)$ IT code, one of the N possible identifiers is assigned to each possible source from the outset. As discussed in Sections 1 and 2, this number N can be shown to grow doubly exponentially in the block length n, hence is not bounded as n tends to infinity.

It is well-known from [20], [7], [17], and [1] that, for a finite source population and *distinct* protocol sequences assigned to distinct sources, error-free performance for simultaneous transmission of messages over a common channel can be guaranteed, provided not too many sources are simultaneously active. However, for a source population of unknown large size no one-to-one assignment of distinct sequences to distinct sources is possible. But, as explained in [23], in this case a *single* common control sequence, assigned to each potential user in advance, can be used to control the access to the channel by a source which is activated following a demand. When using a single common control sequence, messages which are sent simultaneously may be successfully separated, but sources cannot anymore be identified, unless an IT code is used. We remark that the problem of the simultaneous transmission of messages over a common communication channel where the users sending these messages also need to be identified at the receiver's end, resembles much the problem raised (but no further studied) in [14] of the simultaneous transmission of a collection of messages (and their respective addresses) to their intended receivers.

We next recall some basic notions of asynchronous multiple access and the notion of a common single control (or hop) sequence.

The time axis is assumed to be partitioned into slots whose duration corresponds to the transmission time for one symbol. All users are assumed to know the slot boundaries but are otherwise unsynchronized. When a user transmits a symbol, he must transmit it exactly within a slot. This is also referred to as *asynchronous slotted multiple access*.

If in a particular slot, none of the users is sending a symbol, then the channel output in that slot is the silence symbol Λ. If exactly one user is sending a symbol in a particular slot, then the channel output in that slot is this symbol. If two or more users are sending symbols in a particular slot, then the channel output in that slot is the erasure symbol Δ. Hence the terminology of a multiple access *erasure* channel.

A user sends a message using a frame of $N^\star$ slots. A control sequence r_0 is a binary sequence $(r_{01}, \ldots, r_{0N^\star})$ of length $N^\star$ and weight $n^\star$. If source i wants to send a codeword $c_i^\star$ of length $n^\star$, he sends the symbols of $c_i^\star$ in consecutive order in those slots j for which $r_{0j} = 1$.

We are interested in conveying the positions of ν selected single bits of the original code C_{ITG}^0 when ν consecutive messages are sent, or of the ν selected bits of $C_{ITG,\nu}^1$ of FEC version 1, or of the selected single bits of $C_{ITG,j}^0$, $j = 1, \ldots, \nu$ of FEC version 2, through a noiseless multiple access erasure channel, when a common control sequence r_0 is used both at the simultaneously active senders and the receiver.

The positions to be sent are encoded by a multiple access code $\tilde{C}_0$ controlled by the single sequence s_0, common to all sources. The code $\tilde{C}_0$ is a binary constant-weight cyclically permutable code of weight $n^\star$ and length $N^\star$. Several constructions of binary constant-weight cyclically permutable codes are given in [1], where it is also shown how such codes provide a natural solution to the problem of constructing protocol-sequence sets for the M-active-out-of-T users collision channel without feedback.

In our situation of a common control sequence, the same kind of code $\tilde{C}_0$ is used as in [1]. In [10] and [23] it is shown how under certain conditions error-free performance can be achieved for multiple access controlled by a single hop sequence.

An important question is under which condition one might separate and decode without erasure a frame from a source arriving at instant t at the output end of the message register of the receiver with its front. In [23] the concepts of frame front, frame front coincidence, and frame match were introduced. It was shown in [10] (cf. Lemma 7 of [23]) that, in the considered setup of asynchronous multiple access with a common single control sequence r_0, if neither frame front coincidence nor overflow with respect to a certain threshold occurs at time instant t, then the just arriving frame from some source u can be separated at t and the frame decoded without error. Hence, codewords arriving from distinct codewords along successive frame matches can be separated most of the time.

After a codeword is successfully separated and decoded, an identifier match is carried out as described in Section 3. If the codeword is decoded successfully and the identification has been done correctly, the message block content, just transmitted by the position of a single symbol "one" of a codeword from C_{ITG}^0, is recovered together with the identifier in this case.

The idea to use for multiple access a common single control sequence for all network nodes involved was already proposed by Abramson in [2] for the Spread Aloha principle, prior to the papers by the first author on least length single sequence hopping [9,10]. For more on the Spread Aloha principle see [3], and for the use of single sequence hopping in deriving a direct sequence version of CDMA see [18].

The idea of the above multiple access codes combined with identification codes is to keep appropriately many silences between consecutive codewords when encoding the position of the bit of C_{ITG} to be sent through the channel. This scheme can also be applied for $C^1_{ITG,\nu}$ (FEC version 1) and for any of the encoder outputs of $C^2_{ITG,j}$, $j = 1, \ldots, \nu$ (assuming that the latter are being bunched e.g. along the frequency axis) as discussed in Remark 3. The use of such multiple access coding scheme for FEC version 2 (where ν consecutive frames are sent) has already been explained in [23].

It is important to note that, as in the above procedure we finally also decode the identifier anytime there is an identification match (either correctly or falsely), it does not matter at all that, for single sequence multiple access as such, just separation is possible. As a matter of fact, the present use of single sequence multiple access (combined with an IT code) is a school example of a case where mere separation by the control sequence makes sense.

We finally turn to the concept of *least length* single sequence multiple access, introduced in [10] and mentioned in [23]. The control sequence r_0 is said to be of least length N' if N' is the smallest value of N^* so that successful separation and decoding of an arriving frame is still possible. Least length control sequences are of course of practical importance as the time delay between the occurrence of a demand at a source and of the response to this at the decoded output is least in this case. But the problem of estimating the least possible sequence length in our case is also an interesting information-theoretic question. In the yet unpublished preprint [10], upper and lower bounds are obtained on the least possible sequence length N'.

The specific feature of the derivation of the lower bound in [10] is that, while an equation can be derived that is explicit with respect to the least sequence length N' for multiple access with distinct control sequences for the users, the corresponding equation for a single common control sequence includes N' only implicitly.

A lower bound on the solution of the underlying extremal set problem [19] in the case of distinct control sequences per user has been derived by Bassalygo and Pinsker [7]. However, an extension of the approach of [7] is necessary to solve the extremal set problem also for a single common control sequence. This extension is included in [10].

The upper bound on the solution of the underlying extremal set problem provided in [10] relies on a well-known result in [1].

5   Some Properties Worth Further Practical Investigations

The topic of identification via a channel is obviously of fundamental interest for understanding the properties, capabilities, and new design possibilities of communication systems meant to convey huge size identification codes in an affordable way via a multiple access channel.

The use of huge codes might make much sense, even if the size of the identifier set is far from being fully utilized in an actual application. However, the code construction at the place of the source should be appropriately simple, and should offer essential advantages with respect to well-known alternatives of both channel coding and identification.

Anyhow, when there is a plethora of identifiers available, these might be of design interest, even when the collection of available identifiers is limited to the more or less downsized identifier sets of FEC versions 1 and 2 derived from asymptotically optimal identification code sequences. Both of these versions might still make sense if the drop in the size of the original C_{IDG}^0 is large, provided of course the key features of the service offered this way are sufficiently appealing with respect to conventional practical solutions.

A typical task we have in mind is a remote alarm service embedded into a conventional cellular mobile wireless network, occupying only an irrelevant fraction of the total bandwidth of the conventional telephone and short message services of that network (see Fig. 4).

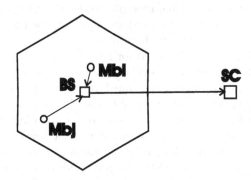

Fig. 4. Remote alarm service embedded into a cellular network

Assume that occasional alarms from any of a lot of sensor-transmitter pairs, spread over some geographical area, should be conveyed from base stations to some source center (SC) for evaluation and taking actions.

If the sensor-transmitter pairs are either mobiles (Mbi, Mbj) or might be moved (as the user is wandering), the actual positions of all sensor-transmitter pairs should rely upon either the Global Positioning System itself or some appropriate local positioning derived from this. The hexagon in Fig. 4 is the abstraction of the wireless cell of our actual interest. BS stands for the base station used, and SC for the remote service center taking actions each time an alarm is perceived from any of the mobiles within the considered wireless cell.

A vast amount of mobiles might be served coming from a very large service domain (SD) around BS, sojourning just within the wireless cell, provided all mobiles in SD can be distinguished by its single identifier (or by any of the identifiers within a large identifier set, assigned to it for a more secure design). We denote by RTS the total rate of the telecommunication services which is to be supported reliably by any of the wireless cells (not including the remote alarm utility). It is crucial for the economy of the telecommunication services that the total data rate of these should only be reduced by a negligible amount by the embedding of the otherwise independent remote alarm service. This is so if the total rate (RAS) of the remote alarm service is negligible with respect to RTS.

A broad range of tasks of practical interest might be considered this way, from less stringent to extreme requirements concerning complexity and cost at the place of the user.

A very human example of much interest within such emergency networks might be a prospective, really broadly accessible, remote alarm service for ailing or aged people or for those who are of much need to be monitored when traveling. However, an embedded remote alarm service in this case might make sense only if it is affordable by its public or by a private investor from the outset, even if the initial client population is small locally (within a certain geographical area) with respect to that finally expected. Obviously, the identifier set available should not prevent smooth further expansion, and even a flexible combination of identifiers placed conventionally into the heading of the message content (as in Fig. 1) and identifiers conveyed via a huge identifier code (as in Fig. 2) should not be excluded. The latter might offer the freedom to accept new clients by the remote alarm service seemingly without any limitation.

It is obviously relevant that the need for network control must be as small as possible. The use of the same single cyclically permutable sequence, assigned to all network nodes involved, might be of much interest in this respect.

In any case, most clients should be able to use the same single identifier over a geographical domain which is huge as compared to the diameter of a cell, and just a little fraction of the client population, being simultaneously present within the entire service domain, should log in for a codeword of the local identification code when entering the boundary of this domain. Obviously, the diameter of the said domain might be drastically increased, if a plethora of identifiers is available for the clients meant for identifying sensor-transmitter pairs.

There might be clients who want their identifier to be practically indiscoverable by a malevolent intruder. The service center might want to provide the same strict security, not necessarily steadily but just temporarily, anytime some of its clients seem to be endangered. In this case an appealing solution might be to assign to the clients in question not just a single identifier but a vast amount of identifiers. A distinct identifier should then be drawn randomly from this set, each time a true alarm or a signal for testing the connection is activated. In this case, an encrypted message should also be conveyed to the service center, including the identifier to be used next time at a subsequent occasion. Such kind of security tasks obviously need the hugest possible size of identification codes.

The client at the sensor-transmitter pair in question, or any other registered client not necessarily within the mobile cellular network, might want to get a positive acknowledgement about the perception of the alarm and eventually also about what actions have been taken at the service center. This option might be satisfied, even via a one-way remote alarm service, if all the registered parties in question have made their mobile numbers previously available to the service center.

These are essential features which should be considered both for each sensor-transmitter pair and for all products placed at the service center for the exclusive use of the client, in addition to the operational and amortization costs of the network and the service center itself.

It might be expected already from the basic properties and possibilities of huge size ID codes, derived from asymptotically optimal sequences of such codes as shown in this paper, that prospective public remote alarm services, relying either entirely or just in part on such huge codes, are worth further detailed scrutiny by competent practical designers. In particular this might be so if there is a plethora of identifiers available and all multiple access channels of possible interest are controlled by copies of the very same single sequence at all network nodes involved. Network control might be drastically simplified in this case.

Acknowledgment

The second author would like to express his gratitude for having had the opportunity to work with Sándor Csibi on this paper and for the many fruitful and illuminating research discussions he had with him over the years.

References

1. N.Q. A, L. Györfi, and J.L. Massey, Construction of binary constant-weight cyclic codes and cyclically permutable codes, IEEE Trans. Inform. Theory, Vol. 38, 940-949, 1992.
2. N. Abramson, Development of ALOHANET, IEEE Trans. Inform. Theory, Vol. 31, 119-123, 1985.
3. N. Abramson, Multiple access in wireless digital networks, Proc. IEEE, Vol. 82, 1360-1370, 1994.
4. R. Ahlswede, General theory of information transfer, Preprint 97–118, SFB 343 "Diskrete Strukturen in der Mathematik", Universität Bielefeld, 1997; General theory of information transfer:updated, General Theory of Information Transfer and Combinatorics, a Special Issue of Discrete Applied Mathematics, to appear.
5. R. Ahlswede and G. Dueck, Identification via channels, IEEE Trans Inform. Theory, Vol. 35, 15-29, 1989.
6. R. Ahlswede and B. Verboven, On identification via multi-way channels with feedback, IEEE Trans. Inform. Theory, Vol. 37, 1519-1526, 1991.
7. L.A. Bassalygo and M.S. Pinsker, Limited multiple-access to an asynchronous channel (in Russian), Problemy Peredachi Informatsii, Vol. 19, no. 4, 92-96, 1983.
8. D. Bertsekas and R. Gallager, Data Networks, New Jersey, Prentice-Hall, 1987.
9. S. Csibi, Two-sided bounds on the decoding error probability for structured hopping, single common sequence and a Poisson population, Proc. 1994 IEEE Internat. Symp. Inform. Theory, Trondheim, 290, 1994.
10. S. Csibi, On the decoding error probability of slotted asynchronous access and least length single sequence hopping, Preprint, Dept. Telecomm., Tech. Univ. Budapest, 1-54, 1997.

11. S. Csibi and E.C. van der Meulen, Properties of codes for identification via a multiple access channel under a word-length constraint, Preprint (A previous version of the present paper, distributed at the Preparatory Meeting, ZiF, Uni-Bielefeld, February 18-23, 2002, includes results achieved until that date only.), 1-8, 2002.
12. Jr.G.D. Forney, Concatenated Codes, Cambridge, MA, MIT Press, 1966.
13. Jr.G.D. Forney and G. Ungerboeck, Modulation and coding for linear Gaussian channels, IEEE Trans. Inform. Theory, Commemorative Issue 1948-1998, Vol. 44, 2384-2415, 1998.
14. T.S. Han and S. Verdú, New results in the theory of identification via channels, IEEE Trans. Inform. Theory, Vol. 38, 14-25, 1992.
15. T.S. Han and S. Verdú, Approximation theory of output statistics, IEEE Trans. Inform. Theory, Vol. 39, 752-772, 1993.
16. S. Lin and D.J. Costello, Error Control Coding, New Jersey, Prentice-Hall, 1983.
17. J.L. Massey and P. Mathys, The collision channel without feedback, IEEE Trans. Inform. Theory, Vol. 31, 192-204, 1985.
18. L. Pap, Performance analysis of DS unslotted packet radio networks with given auto- and crosscorrelation sidelobes, Proc. IEEE 3rd Internat. Symp. Spread Spectrum Techniques and Applications, Oulu, Finland, 343-345, 1994.
19. V. T.Sós, An additive problem in different structures, (a survey), Proc. 2nd Internat. Conf. Graph Theory, Combinatorics, Algorithms, and Applications, SIAM, 1991.
20. B.S. Tsybakov and N.B. Likhanov, Packet communication on a channel with feedback, Probl. Inform. Transm., Vol. 19, No. 2, 69-84, 1983.
21. E.C. van der Meulen and S. Csibi, Identification coding for least length single sequence hopping, Abstracts, 1996 IEEE Inform. Theory Workshop, Dan-Carmel, Haifa, 67, 1996.
22. E.C. van der Meulen and S. Csibi, Elementary forward error control for identification codes, Proc. 1997 IEEE Intern. Symp. Inform. Theory, Ulm, 160, 1997.
23. E.C. van der Meulen and S. Csibi, Error probabilities for identification coding and least length single sequence hopping, Numbers, Information and Complexity, special volume in honour of R. Ahlswede on occasion of his 60th birthday, edited by Ingo Althöfer, Ning Cai, Gunter Dueck, Levon Khachatrian, Mark S. Pinsker, Andras Sárközy, Ingo Wegener and Zhen Zhang, Kluwer Academic Publishers, Boston, Dordrecht, London, 221-238, 2000.
24. S. Verdú and V.K. Wei, Explicit constructions of constant-weight codes for identification via channels, IEEE Trans. Inform. Theory, Vol. 39, 30-36, 1993.
25. J. Wolfowitz, Coding Theorems of Information Theory, 3rd edition, New York, Springer, 1978.

Codes with the Identifiable Parent Property and the Multiple-Access Channel

R. Ahlswede and N. Cai

1 Introduction

We begin with

I. The identifiable parent property and some first results about it
If C is a q–ary code of length n and a^n and b^n are two codewords, then c^n is called a descendant of a^n and b^n if $c_t \in \{a_t, b_t\}$ for $t = 1, \ldots, n$. We are interested in codes C with the property that, given any descendant c^n, one can always identify at least one of the 'parent' codewords in C. We study bounds on $F(n, q)$, the maximal cardinality of a code C with this property, which we call the *identifiable parent property*. Such codes play a role in schemes that protect against piracy of software.

They have been introduced by Hollmann, van Lint, Linnartz and Tolhuizen [9]. We repeat first their concepts, basic examples and results.

Consider a code C of length n over an alphabet Q with $|Q| = q$ (i.e., $C \subset Q^n$). For any two words a^n, b^n in Q^n we define the *set of descendants* $D(a^n, b^n)$ by

$$D(a^n, b^n) := \{x^n \in Q^n | x_t \in \{a_t, b_t\}, t = 1, 2, \ldots, n\}. \tag{1.1}$$

Note that among the descendants of a^n and b^n we also find a^n and b^n themselves. For a code C we define the descendant code C^* by

$$C^* := \bigcup_{a^n \in C, b^n \in C} D(a^n, b^n). \tag{1.2}$$

For example, if C is the binary repetition code, then $C^* = F_2^n$. Similarly, if C is the ternary Hamming code of length 4, then $C^* = F_3^4$, since it is obvious that all words in a ball of radius 1 around a codeword are descendants of some pair containing that codeword.

If $c^n \in C^*$ is an element of $D(a^n, b^n)$, with $a^n \in C$, $b^n \in C$, then we call a^n and b^n *parents* of c^n. In general, an element of C^* has several pairs of parents. A trivial example are words of C themselves. We say that C has the "*identifiable parent property*" (IPP) if, for every descendant in C^*, at least one of the parents can be identified. In other words, for each $c^n \in C^*$ there is a codeword $\pi(c^n)$ in C such that each parent pair of c^n must contain $\pi(c)$.

Example: Consider the ternary Hamming code C of length 4, which has size 9. Since every pair of distinct codewords has distance 3, any descendant c^n in C^* has distance ≤ 1 to exactly one of the parents in a parent pair. There cannot be two codewords with distance 1 to c^n, so the unique codeword with distance ≤ 1

R. Ahlswede et al. (Eds.): Information Transfer and Combinatorics, LNCS 4123, pp. 249–257, 2006.
© Springer-Verlag Berlin Heidelberg 2006

to c^n is the identifiable parent. For the other parent there are then three choices if $c^n \notin \mathcal{C}$ (and of course eight choices if $c^n \in \mathcal{C}$).

We are interested in the *maximal size* of a code with the identifiable parent property. We define

$$F(n, q) := \max\{|\mathcal{C}| \mid \mathcal{C} \subseteq Q^n, \mathcal{C} \text{ has IPP}, |Q| = q\}.$$

Trivially, a code of cardinality 2 has IPP. If $q = 2$, a code of cardinality ≥ 3 does not have IPP. To see this, consider three binary words u_1, u_2, u_3. For $i = 1, 2, 3$, the i-th coordinate of c^n is determined by a majority vote over the corresponding coordinates of the three given words. Then c^n is clearly a descendant of any pair taken from the three words u_j. So from now on we assume $q \geq 3$.

As trivial cases we have $F(1, q) = q$, $F(2, q) = q$. (If x_t, $t = 1, 2$, is a symbol that occurs twice as t-th coordinate, then (x_1, x_2) has no identifiable parent.)

Theorem HLLT 1. $F(3, q) \leq 3q - 1$

For certain classes of codes, it is easy to see that IPP holds. We start with equidistant codes.

Theorem HLLT 2. *If $\mathcal{C}$ is an equidistant code of length n over an alphabet of size q and with distance d, then $\mathcal{C}$ has the identifiable parent property if d is odd or if d is even and $n < \frac{3}{2}d$.*

Theorem HLLT 3. *Let q be a prime power. If $q \geq n - 1$ then a (shortened, extended, or doubly extended) Reed–Solomon code over F_q with parameters $\left[n, \lceil \frac{n}{4} \rceil, n - \lceil \frac{n}{4} \rceil + 1\right]$ has IPP.*

Corollary. *If $q \geq n - 1$ and q is a prime power, then $F(n, q) \geq q^{\lceil \frac{n}{4} \rceil}$.*

Theorem HLLT 4. *We have $F(n, q) \leq 3q^{\lceil \frac{n}{3} \rceil}$.*

Theorem HLLT 5. *There is a constant c such that $F(n, q) \geq c \left(\frac{q}{4}\right)^{\frac{n}{3}}$.*

From the calculations it follows that we could take $c = 0.4$. For large q, Theorem 5 is better than the Corollary.

We expand here the model in the following direction.

II. Men and women model

Here we consider two sets of codewords $\mathcal{U}, \mathcal{V} \subset Q^n$ referred to as sets of men and of women. Naturally we define the descendant code $\mathcal{C}^*(\mathcal{U}, \mathcal{V})$ by

$$\mathcal{C}^*(\mathcal{U}, \mathcal{V}) = \bigcup_{u \in \mathcal{U}, v \in \mathcal{V}} D(u, v).$$

If $c^n \in \mathcal{C}^*(\mathcal{U}, \mathcal{V})$ is an element of $D(u, v)$, then we call u and v parents of c^n.

We say now that $(\mathcal{U}, \mathcal{V})$ has the identifiable parent property if for every descendant in $\mathcal{C}^*$ at least one of the parents can be identified. This means that for every $c^n \in \mathcal{C}^*$ there is a codeword $\pi(c^n)$ in $\mathcal{U} \cup \mathcal{V}$ such that each parent pair $\{u, v\}$ of c^n must contain $\pi(c^n)$.

III. Semicodes for MAC

The previous model suggests to look at the structure in terms of multiple–access channels (MAC) defined by a stochastic matrix $W : \mathcal{X} \times \mathcal{Y} \to \mathcal{Z}$. Then the IPP naturally leads to the new concept of semi codes with a new Coding Theorem determining the optimal rate $\bar{C}_{semi}$ for the average error concept (Theorem 1). The proof is by no means easy.

It has three basic ingredients: a wringing technique of [3], the blowing up method of [6] and the identity for entropies of [10] in the form of [7]. We analyze this model in Section 2. In Section 3 we mention directions of further research on identifiability.

2 Semicodes for the MAC

Let W be a stochastic matrix $\mathcal{X} \times \mathcal{Y} \to \mathcal{Z}$. We call a system $\left(\{u_i\}_{i=1}^{M_1}, \{v_j\}_{j=1}^{M_2}, \{\mathcal{E}_i\}_{i=1}^{M_1}, \{\mathcal{D}_j\}_{j=1}^{M_2}\right)$ an (n, M_1, M_2, λ)–semi–code of MAC W^n, if $u_i \in \mathcal{X}^n$ for $i = 1, \ldots, M_1$, $v_j \in \mathcal{Y}^n$ for $j = 1, \ldots, M_2$, $\mathcal{E}_i \cap \mathcal{E}_{i'} = \varnothing$ for $i \neq i'$, $\mathcal{D}_j \cap \mathcal{D}_{j'} = \varnothing$ for $j \neq j'$, $\mathcal{E}_i \cap \mathcal{D}_j = \varnothing$ for all i, j and

$$\frac{1}{M_1} \frac{1}{M_2} \sum_{i=1}^{M_1} \sum_{j=1}^{M_2} W^n(\mathcal{E}_i \cup \mathcal{D}_j | u_i, v_j) > 1 - \lambda. \tag{2.1}$$

Denote by $\bar{C}_{semi}(\lambda)$ the maximal real number such that, for all $\delta > 0$ and sufficiently large n there exists an (n, M_1, M_2, λ)–semi–code with $\frac{1}{n} \log M_1 M_2 > \bar{C}_{semi}(\lambda) - \delta$. We shall determine $\bar{C}_{semi}(\lambda)$ and show that it is independent of $\lambda \in (0, 1)$. The main issue is the (strong) converse theorem and our main idea is very similar to that in [3]. The following result (Lemma 4 of [3]) will play an important role.

Lemma A. *Let P and Q be probability distributions on $\mathcal{X}^n$ such that for a positive constant c*

$$P(x^n) \leq (1 + c)Q(x^n) \text{ for all } x^n \in \mathcal{X}, \tag{2.2}$$

then for any $0 < \gamma < c$, $0 \leq \varepsilon < 1$ there exist $t_1, \ldots, t_k \in \{1, \ldots, n\}$, where $0 \leq k \leq \frac{c}{\gamma}$ such that for some $\bar{x}_{t_1}, \ldots, \bar{x}_{t_k}$

$$P(x_t | \bar{x}_{t_1}, \ldots, \bar{x}_{t_k}) \leq \max((1 + \gamma)Q(x_t | \bar{x}_{t_1}, \ldots, \bar{x}_{t_k}), \varepsilon) \tag{2.3}$$

for all $x_t \in \mathcal{X}$ and all $t = 1, 2, \ldots, n$ and

$$P(\bar{x}_{t_1}, \ldots, \bar{x}_{t_k}) \geq \varepsilon^k. \tag{2.4}$$

To apply it, we modify its consequence (Corollary 2 in [3]) slightly

Corollary. *Let $\mathcal{U}_n \subset \mathcal{X}^n$ with $|\mathcal{U}_n| = M_1$, $\mathcal{V}_n \subset \mathcal{Y}^n$ with $|\mathcal{V}_n| = M_2$, $\mathcal{A} \subset \mathcal{U}_n \times \mathcal{V}_n$ with $|\mathcal{A}| \geq (1 - \lambda^*)M_1 M_2$ for some $\lambda^* \in (0, 1)$. Then for any $0 < \gamma < c \triangleq \frac{\lambda^*}{1 - \lambda^*}$,*

$0 \leq \varepsilon < 1$ *there exist* $t_1, \ldots, t_k \in \{1, \ldots, n\}$ *where* $k \leq \frac{\lambda^*}{\gamma(1-\lambda^*)}$ *and some*
$(\bar{x}_{t_1}, \bar{y}_{t_1}), \ldots, (\bar{x}_{t_k}, \bar{y}_{t_k})$ *such that for* $\bar{A} \triangleq \{(x^n, y^n) \in \mathcal{A} : x_{t_\ell} = \bar{x}_{t_\ell} \ y_{t_\ell} = \bar{y}_{t_\ell}, \text{ for}$
$\ell = 1, \ldots, k\}$

(a) $|\bar{A}| \geq \varepsilon^k |\mathcal{A}|$,
    and
(b) $\big((1+\gamma)Pr(\bar{X}_t = x)Pr(\bar{Y}_t = y) - \gamma - |\mathcal{X}||\mathcal{Y}|\varepsilon\big),$
    $\leq Pr(\bar{X}_t = x, \bar{Y}_t = y) \leq \max\big((1+\gamma)Pr(\bar{X}_t = x)Pr(\bar{Y} = y), \varepsilon\big)$
    for all $x \in \mathcal{X}$, $y \in \mathcal{Y}$, $1 \leq t \leq n$,
    where $(\bar{X}^n, \bar{Y}^n)$ *is a pair of RV's with uniform distribution on* $\bar{A}$.

Proof: The corollary is essentially the same as Corollary 2 of [3] and can be shown in the same way. But we give the proof because it is short.

Let P and Q be defined by $P(x^n, y^n) = \frac{1}{|\mathcal{A}|}$ if $(x^n, y^n) \in \mathcal{A}$ and $Q(x^n, y^n) = P_1(x^n)P_2(y^n)$ for $(x^n, y^n) \in \mathcal{X}^n \times \mathcal{Y}^n$, where P_1 and P_2 are marginal distributions of P, respectively. Then $P(x^n, y^n) \leq \frac{1}{1-\lambda^*}Q(x^n, y^n)$ and therefore one can apply Lemma A to $c = \frac{1}{1-\lambda^*} - 1 = \frac{\lambda^*}{1-\lambda^*}$ to obtain (a) and the second inequality of (b), which implies

$$Pr(\bar{X}_t = x, \bar{Y}_t = y) = 1 - \sum_{(x'.y') \neq (x,y)} Pr(\bar{X}_t = x', \bar{Y}_t = y')$$

$$\geq 1 - \sum_{(x',y') \neq (x,y)} \max\big((1+\gamma)Pr(\bar{X}_t = x')Pr(\bar{Y}_t = y'), \varepsilon\big)$$

$$\geq 1 - |\mathcal{X}||\mathcal{Y}|\varepsilon - (1+\gamma)(1 - Pr(\bar{X}_t = x)Pr(\bar{Y}_t = y))$$

$$= LHS \text{ of (b)}. \qquad \square$$

Another main tool here is the Blowing Up Lemma of [5]. Let d_H be Hamming distance and for all $B \subset \mathcal{Z}'^m$, $\Gamma^k B \triangleq \{z^n : \text{there is a } b^n \in B^n \text{ with } d_H(z^n, b^n) \leq k\}$, where $\mathcal{Z}'$ is a finite set. Then

Lemma AGK. (Blowing Up) *For any finite sets* $\mathcal{X}'$ *and* $\mathcal{Z}'$ *and sequence* $\{\varepsilon_n\}_{n=1}^\infty$ *with* $\varepsilon_n \to 0$, *there exist a sequence of positive integers* $\{\ell_n\}_{n=1}^\infty$ *with* $\ell_n/n \to 0$ *and a sequence* $\{\eta_n\}_{n=1}^\infty$ *with* $\eta \to 1$ *such that for every stochastic matrix* $V : \mathcal{X}' \to \mathcal{Z}'$ *and every* $n, x^n \in \mathcal{X}'^m$, $B \subset \mathcal{Z}'^m$
$W^n(B|x^n) \geq \exp\{-n\varepsilon_n\}$ *implies* $W^n(\Gamma^{\ell_n} B|x^n) \geq \eta_n$.

Remark: One can easily see that for a stochastic matrix $W : \mathcal{X} \times \mathcal{Y} \to \mathcal{Z}$ and any $y^n \in \mathcal{Y}^n$, the Blowing Up Lemma is still true for the channel $W^n(\cdot|\cdot, y^n) \triangleq \prod_{t=1}^n W(\cdot|\cdot, y_t)$. We shall actually employ this version of the Blowing Up Lemma.

Theorem 1. *For all* $\lambda \in (0,1)$,

$$\bar{C}_{semi}(\lambda) = \max_{X,Y} \max\{I(X \wedge Z) + H(Y), I(Y \wedge Z) + H(X)\}, \qquad (2.5)$$

where the first maximum is taken over all independent pairs of RV's (X, Y) with values in $\mathcal{X} \times \mathcal{Y}$, and Z is the corresponding output variable.

Proof

Converse: Let $\left(\{u_i\}_{i=1}^{M_1}, \{v_j\}_{j=1}^{M_2}, \{\mathcal{E}_i\}_{i=1}^{M_1}, \{\mathcal{D}_j\}_{j=1}^{M_2}\right)$ be an (n, M_1, M_2, λ)–semicode. Then (2.1) implies that

$$\frac{1}{M_1}\frac{1}{M_2}\sum_{i=1}^{M_1}\sum_{j=1}^{M_2} W^n(\mathcal{E}_i|u_i, v_j) > \frac{1-\lambda}{2} \tag{2.6}$$

or

$$\frac{1}{M_1}\frac{1}{M_2}\sum_{i=1}^{M}\sum_{j=1}^{M} W^n(\mathcal{D}_j|u_i, v_j) > \frac{1-\lambda}{2}, \tag{2.7}$$

must hold. W.l.o.g. assume (2.6) holds and therefore there is a subcode $\mathcal{A} \subset \{u_i : 1 \le i \le M_1\} \times \{v_j : 1 \le j \le M_2\}$ such that

$$|\mathcal{A}| > \frac{1-\lambda-2\mu}{2(1-\mu)}M_1M_2, \tag{2.8}$$

and for all $(u_i, v_j) \in \mathcal{A}$

$$W^n(\mathcal{E}_i|u_i, v_j) > \mu, \tag{2.9}$$

where μ is any positive constant less than $\frac{1-\lambda}{2}$.

We apply the Corollary to $\mathcal{A}$ with $\lambda^* \triangleq 1 - \frac{1-\lambda-2\mu}{2(1-\mu)} = \frac{1+\lambda}{2(1-\mu)}$, $\varepsilon = n^{-1}$ and $\gamma = n^{-\frac{1}{2}}$ and then get $t_1, \ldots, t_k$, $(\bar{x}_{t_1}, \bar{y}_{t_1}), \ldots, (\bar{x}_{t_k}, \bar{y}_{t_k})$, $\bar{\mathcal{A}}$ and $(\bar{X}^n, \bar{Y}^n)$ in the Corollary with

$$k \le \frac{\lambda^*}{\gamma(1-\lambda^*)} = \frac{1+\lambda}{1-\lambda-2\mu}n^{\frac{1}{2}} \tag{2.10}$$

and by (2.8) and (2.10)

$$|\bar{\mathcal{A}}| \ge \varepsilon^k|\mathcal{A}| \ge (1-\lambda^*)M_1M_2\varepsilon^k \ge \frac{1-\lambda-2\mu}{2(1-\mu)}M_1M_2\exp\left\{-\frac{1+\lambda}{1-\lambda-2\mu}n^{\frac{1}{2}}\log n\right\}. \tag{2.11}$$

Therefore

$$H(\bar{X}^n, \bar{Y}^n) = \log|\bar{\mathcal{A}}| \ge \log M_1M_2 + \log\frac{1-\lambda-2\mu}{2(1-\mu)} - \frac{1-\lambda}{1-\lambda-2\mu}n^{\frac{1}{2}}\log n. \tag{2.12}$$

Let (X_t, Y_t, Z_t) be the triple of RV's, for $t = 1, \ldots, n$, with distribution $Pr(X_t = x, Y_t = y, Z_t = z) = Pr(\bar{X}_t = x)Pr(\bar{Y}_t = y)W(z|x, y)$, and let $\bar{Z}^n$ be the output of the channel W^n for the input $(\bar{X}^n, \bar{Y}^n)$. Then by (b) of the corollary and the uniform continuity of information quantities,

$$\left|\left(I(X_t \wedge Z_t) + H(Y_t)\right) - \left(I(\bar{X}_t \wedge \bar{Z}_t) + H(\bar{Y}_t)\right)\right| < \alpha_n, \tag{2.13}$$

for all t and some sequence $(\alpha_n)_{n=1}^{\infty}$ with $\alpha_n \to 0$ as $n \to \infty$.

Recalling $\bar{\mathcal{A}} \subset \mathcal{A}$, we have (2.9) for all $(u_i, v_j) \in \bar{\mathcal{A}}$. Thus by applying the Blowing Up Lemma to $(u_i, v_j) \in \bar{\mathcal{A}}$, we obtain, for all $(u_i, v_j) \in \bar{\mathcal{A}}$

$$W^n(\Gamma^{\ell_n}\mathcal{E}_i | u_i, v_j) \geq \eta_n \text{ and } \eta_n \to 1, \frac{\ell_n}{n} \to 0 \text{ as } n \to \infty \qquad (2.14)$$

(c.f. the Remark after the Blowing Up Lemma).

Notice that $z^n \in \Gamma^{\ell_n}\mathcal{E}_i$ iff there is a $z'^n \in \mathcal{E}_i$ with $d_H(z^n, z'^n) \leq \ell_n$. We define "the decoding list" of z^n as $\mathcal{L}(z^n) \triangleq \{i : z^n \in \Gamma^{\ell_n}\mathcal{E}_i\}$. Then

$$|\mathcal{L}(z^n)| \leq \sum_{m=0}^{\ell_n} \binom{n}{m}(|\mathcal{Z}| - 1)^m \leq \exp\{n\beta_n\}, \text{ (say)} \qquad (2.15)$$

with $\beta_n \to 0$ as $n \to \infty$. Introduce a new RV J by setting $J = 0$ if $\bar{X}^n \in \mathcal{L}(\bar{Z}^n)$ and $J = 1$ else. Then

$$H(\bar{X}^n|\bar{Z}^n) = H(\bar{X}^n J|\bar{Z}^n) = H(\bar{X}^n|J\bar{Z}^n) + H(J|\bar{Z}^n) \leq H(\bar{X}^n|J\bar{Z}^n) + H(J)$$
$$\leq Pr(J = 0)H(\bar{X}^n|J = 0, \bar{Z}^n) + Pr(J = 1)H(X^n|J = 1) + \log 2$$
$$\leq Pr(J = 0)H(\bar{X}^n|J = 0, \bar{Z}^n) + (1 - \eta_n)n \log |\mathcal{X}| + \log 2 \text{ (by (2.14))}$$
$$\leq n\beta_n + (1 - \eta_n)n \log |\mathcal{X}| + \log 2 \text{ (by (2.15))}. \qquad (2.16)$$

Next we employ a technique of [10] which appears in 3.3 of [7]. Write for all $t \in \{1, 2, \ldots, n\}$

$$H(\bar{Y}_t|\bar{X}^n\bar{Y}^{t-1}\bar{Z}_{t+1}, \ldots, \bar{Z}_n) - H(\bar{Z}_t|\bar{X}^n\bar{Y}^{t-1}\bar{Z}_{t+1}, \ldots, \bar{Z}_n)$$
$$= H(\bar{Y}^t\bar{Z}_{t+1}, \ldots, \bar{Z}_n|\bar{X}^n) - H(\bar{Y}^{t-1}\bar{Z}_t, \ldots, \bar{Z}_n|\bar{X}^n), \qquad (2.17)$$

and obtain the following, by adding up both sides of (2.17) from 1 to n.

$$\sum_{t=1}^n \left(H(\bar{Y}_t|\bar{X}^n\bar{Y}^{t-1}\bar{Z}_{t+1}, \ldots, \bar{Z}_n) - H(\bar{Z}_t|\bar{X}^n\bar{Y}^{t-1}\bar{Z}_{t+1}, \ldots, \bar{Z}_n) \right)$$
$$= H(\bar{Y}^n|\bar{X}^n) - H(\bar{Z}^n|\bar{X}^n), \qquad (2.18)$$

In order to show

$$\sum_{t=1}^n (H(\bar{Y}_t|\bar{X}^n\bar{Y}^{t-1}\bar{Z}_{t+1}, \ldots, \bar{Z}^n) - H(\bar{Z}_t|\bar{X}^n\bar{Y}^{t-1}\bar{Z}_{t+1}, \ldots, \bar{Z}_n)) \leq \sum_{t=1}^n (H(\bar{Y}_t) - H(\bar{Z}_t|\bar{X}_t))$$

$$(2.19)$$

we have to prove for all t

$$I(\bar{Z}_t \wedge \bar{X}^{t-1}\bar{X}_{t+1}, \ldots, \bar{X}_n\bar{Y}^{t-1}\bar{Z}_{t+1}, \ldots, \bar{Z}_n|\bar{X}_t) \leq I(\bar{Y}_t \wedge \bar{X}^n\bar{Y}^{t-1}\bar{Z}_{t+1}, \ldots, \bar{Z}_n).$$

It is sufficient to show

$$I(\bar{Z}_t \wedge \bar{X}^{t-1}\bar{X}_{t+1}, \ldots, \bar{X}_n\bar{Y}^{t-1}\bar{Z}_{t+1}, \ldots, \bar{Z}_n|\bar{X}_t)$$
$$\leq I(\bar{Y}_t \wedge \bar{X}^{t-1}\bar{X}_{t+1}, \ldots, \bar{X}_n\bar{Y}^{t-1}\bar{Z}_{t+1}, \ldots, \bar{Z}_n|\bar{X}_t). \qquad (2.20)$$

Since $H(\bar{Z}_t | \bar{X}_t \bar{Y}_t) = H(\bar{Z}_t | \bar{X}_t \bar{Y}_t \bar{X}^{t-1} \bar{X}_{t+1}, \ldots, \bar{X}_n \bar{Y}^{t-1} \bar{Z}_{t+1}, \ldots, \bar{Z}_n)$,

$$I(\bar{Z}_t \wedge \bar{X}^{t-1} \bar{X}_{t+1}, \ldots, \bar{X}_n \bar{Y}^{t-1} \bar{Z}_{t+1}, \ldots, \bar{Z}_n | \bar{X}_t \bar{Y}_t) = 0. \qquad (2.21)$$

By adding (2.21) to LHS of (2.20), one obtains $I(\bar{Y}_t \bar{Z}_t \wedge \bar{X}^{t-1} \bar{X}_{t+1}, \ldots, \bar{X}_n \bar{Y}^{t-1} \bar{Z}_{t+1}, \ldots, \bar{Z}_n | \bar{X}_t)$, which implies (2.20) and therefore (2.19) holds.

Finally, (2.12), (2.13), (2.16), (2.18) and (2.19) together yield

$$\frac{1}{n} \log M_1 M_2 \leq \frac{1}{n} H(\bar{X}^n \bar{Y}^n) - \frac{1}{n} \log \frac{1 - \lambda - 2\mu}{2(1 - \mu)} + \frac{1 - \lambda}{1 - \lambda - 2\mu} n^{-\frac{1}{2}} \log n$$

$$\leq \frac{1}{n} \left(H(\bar{X}^n \bar{Y}^n) - H(\bar{X}^n | \bar{Z}^n) \right) + \beta_n + (1 - \eta_n) \log |\mathcal{X}|$$

$$+ \frac{1}{n} \log 2 - \frac{1}{n} \log \frac{1 - \lambda - 2\mu}{2(1 - \mu)} + \frac{1 - \lambda}{1 - \lambda - 2\mu} n^{-\frac{1}{2}} \log n$$

$$= \frac{1}{n} \left(I(\bar{X}^n \wedge \bar{Z}^n) + H(\bar{Y}^n | \bar{X}^n) \right) + \theta_n$$

$$= \frac{1}{n} \left(H(\bar{Z}^n) + H(\bar{Y}^n | \bar{X}^n) - H(\bar{Z}^n | \bar{X}^n) \right) + \theta_n$$

$$= \frac{1}{n} \left[H(\bar{Z}^n) + \sum_{t=1}^{n} \left(H(\bar{Y}_t | \bar{X}^n \bar{Y}^{t-1} \bar{Z}_{t+1}, \ldots, \bar{Z}_n) - H(\bar{Z}_t | \bar{X}^n \bar{Y}^{t-1} \bar{Z}_{t+1}, \ldots, \bar{Z}_n) \right) \right] + \theta_n$$

$$\leq \frac{1}{n} \sum_{t=1}^{n} \left(H(\bar{Z}_t) + H(\bar{Y}_t) - H(\bar{Z}_t | \bar{X}_t) \right) + \theta_n$$

$$= \frac{1}{n} \sum_{t=1}^{n} \left(I(\bar{X}_t \wedge \bar{Z}_t) + H(\bar{Y}_t) \right) + \theta_n$$

$$\leq \frac{1}{n} \sum_{t=1}^{n} \left(I(X_t \wedge Z_t) + H(Y_t) \right) + \alpha_n + \theta_n, \qquad (2.22)$$

where $\theta_n \triangleq \beta_n + (1 - \eta_n) \log |\mathcal{X}| + \frac{1}{n} \log 2 - \frac{1}{n} \log \frac{1-\lambda-2\mu}{2(1-\mu)} + \frac{1-\lambda}{1-\lambda-2\mu} n^{-\frac{1}{2}} \log n \to 0$ as $n \to \infty$.

Thus we conclude our proof of the converse part by setting (XYZ) as the triple achieving $\max_t \left(I(X_t \wedge Z_t) + H(Y_t) \right)$ and requiring $n \to \infty$ in (2.22).

Direct Part: The proof of the direct part can be done in the now standard way. It was actually first done in [1]. W.l.o.g. assume RHS of (2.5) is $I(X \wedge Z) + H(Y)$ and (X, Y, Z) is in the range of the maximum value. Then by letting $\{v_j : 1 \leq j \leq M_1\} = T_Y^n$, $\mathcal{D}_i = T_{Z|X,\delta}^n(u_i) \setminus \bigcup_{i' \neq i} T_{Z|X,\delta}^n(u_{i'})$ (δ is suitable) $\mathcal{E}_j = \varnothing$ and by independently randomly selecting u_i, $i = 1, 2, \ldots, \lfloor 2^{n(I(X \wedge Z) - \delta')} \rfloor$ on T_X^n one can get the desired code. We omit the details. $\square$

Remarks

1. Inspection of our results shows that we answered a basic question for the interference channel. We found the capacity region if one of the two channels is noiseless. Until now experts could not tell us whether this is known as a special case of complicated characterizations using several auxiliary RV's.

3   Further Results and Perspectives

IV. Screening design of experiments

Motivated by the original parent concept with no distinction between men and women we look at the special MAC with equal input alphabets $\mathcal{X} = \mathcal{Y} = Q$ and symmetric transmission probabilities

$$W(z|x,y) = W(z|y,x) \text{ for all } z \in \mathcal{Z} \text{ and } x,y \in Q$$

and at the situation where the codes $\mathcal{U}$ and $\mathcal{V}$ are equal. This communication situation came up for the first time in the theory of screening design of experiments (see the survey [11]), but now we look at the semicodes analogue to the above with $\mathcal{E}_i = \mathcal{D}_i$ for $i = 1, \ldots, M = M_1 = M_2$ and obtain the analogue to Theorem 1.

V. Semicodes for AVMAC

Next we tighten our models so that they give insight into the original problem. We replace the MAC by the AVMAC, the arbitrarily varying MAC, defined by a set of stochastic matrices $\mathcal{W} = \{w(\cdot|\cdot,\cdot,s) : s \in \mathcal{S}\}$ where $W(\cdot|\cdot,\cdot,s) : \mathcal{X}\times\mathcal{Y} \rightsquigarrow \mathcal{Z}$ and $s \in \mathcal{S}$.

We proved in [4] that its capacity region $\mathcal{R}(\mathcal{W})$ has the property:
$\mathcal{R}(\mathcal{W}) = \varnothing$ if and only if one of the following three conditions holds

(i) $\mathcal{W}$ is $(\mathcal{X},\mathcal{Y})$–symmetrizable, that is for a stochastic $\sigma : \mathcal{X} \times \mathcal{Y} \rightarrow \mathcal{S}$

$$\sum_s W(z|x,y,x)\sigma(s|x',y') = \sum_s W(z|x',y',s)\sigma(s|x,y)$$

for all $x,x' \in \mathcal{X}, y,y' \in \mathcal{Y}$ and $z \in \mathcal{Z}$.

(ii) $\mathcal{W}$ is $\mathcal{X}$–symmetrizable, that is for a stochastic $\sigma_1 : \mathcal{X} \rightarrow \mathcal{S}$

$$\sum_s W(z|x,y,s)\sigma_1(s|x') = \sum_s W(z|x',y,s)\sigma_1(s|x)$$

for all $x,x' \in \mathcal{X}, y \in \mathcal{U}$ and $z \in \mathcal{Z}$.

(iii) $\mathcal{W}$ is $\mathcal{Y}$–symmetrizable, that is for a stochastic $\sigma_2 : \mathcal{Y} \rightarrow \mathcal{S}$

$$\sum_s W(z|x,y,s)\sigma_2(s,y') = \sum_s W(z|x,y',s)\sigma_2(s|y)$$

for all $x \in \mathcal{X}, y,y' \in \mathcal{Y}$ and $z \in \mathcal{Z}$.

VI. Robust screening design of experiments

We can establish the analogue for the one code-set ($\mathcal{U} = \mathcal{V}$) situation and of course also the capacity formula.

VIII. For certain termites females can give birth to males without mating and to females after mating. This gives another structure of relatedness, which can be studied with respect to the identifiability property.

References

1. R. Ahlswede, Multi–way communication channels, Proceedings of 2nd International Symposium on Information Theory, Thakadsor, Armenian SSR, Sept. 1971, Akademiai Kiado, Budapest, 23–52, 1973.

2. R. Ahlswede, The capacity region of a channel with two senders and two receivers, Ann. Probability, Vol. 2, No. 5, 805–814, 1974.

3. R. Ahlswede, An elementary proof of the strong converse theorem for the multiple–access channel, J. Combinatorics, Information and System Sciences, Vol. 7, No. 3, 216–230, 1982.

4. R. Ahlswede and N. Cai, Arbitrarily varying multiple–access channels, Part I. Ericson's symmetrizability is adequate, Gubner's conjecture is true, IEEE Trans. Inf. Theory, Vol. 45, No. 2, 742–749, 1999.

5. R. Ahlswede and G. Dueck, Every bad code has a good subcode: a local converse to the coding theorem, Z. Wahrscheinlichkeitstheorie und verw. Geb., Vol. 34, 179–182, 1976.

6. R. Ahlswede, P. Gács, and J. Körner, Bounds on conditional probabilities with applications in multiuser communication, Z. Wahrscheinlichkeitstheorie und verw. Geb., Vol. 34, 157–177, 1976.

7. I. Csiszár and J. Körner, Information Theory: Coding Theorems for Discrete Memoryless Systems, Academic Press, 1981.

8. G. Dueck, The strong converse to the coding theorem for the multiple–access channel, J. Combinatorics, Information & System Science, 187–196, 1981.

9. H.D.L. Hollmann, J.H. van Lint, J.P. Linnartz, and L.M.G.M. Tolhuizen, On codes with the identifiable parent property, J. Combin. Theory Ser. A 82, No. 2, 121–133, 1998.

10. J. Körner and K. Marton, Images of a set via two channels and their role in multi–user comunication, IEEE Transactions on Information Theory, Vol. 23, 751–761, 1977.

11. J. Viemeister, Die Theorie Selektierender Versuche und Multiple–Access–Kanäle, Diplomarbeit, Universität Bielefeld, 1981.

II

Transmission, Identification and Common Randomness Capacities for Wire-Tape Channels with Secure Feedback from the Decoder

R. Ahlswede and N. Cai

Abstract. We analyze wire-tape channels with secure feedback from the legitimate receiver. We present a lower bound on the transmission capacity (Theorem 1), which we conjecture to be tight and which is proved to be tight (Corollary 1) for Wyner's original (degraded) wire-tape channel and also for the reversely degraded wire-tape channel for which the legitimate receiver gets a degraded version from the enemy (Corollary 2).

Somewhat surprisingly we completely determine the capacities of secure common randomness (Theorem 2) and secure identification (Theorem 3 and Corollary 3). Unlike for the DMC, these quantities are different here, because identification is linked to non-secure common randomness.

1 Introduction

The main results are mentioned in the abstract.

After having given standard concepts in Section 2 and known results and techniques for the wire-tape channel in Section 3, we state and prove Theorem 1 in Section 4. Our code construction relies upon a lemma for balanced coloring from [2], which has proved already useful for secrecy problems in [3].

The transmission capacities for the two kinds of degraded wire-tape channels are derived in Section 5. Particularly interesting is an example of a reversely degraded channel, where the channel $W_1' : \mathcal{X} \to \mathcal{Z}$ for the wiretapper is noiseless (for instance with binary alphabets) and the channel $W_2' : \mathcal{Z} \to \mathcal{Y}$ for the legal receiver is a noisy binary symmetric channel with crossover probability $p \in (0, 1/2)$. Here the wiretapper is in a better position than the legal user and therefore the capacity is zero, if there is no feedback. However, by our Corollary the capacity is positive, because the feedback serves as a secure key shared by sender and receiver.

In Section 6 a discussion based on the construction for transmission in Section 4 and known results and constructions for identification [8], [9], [15], and common randomness [9], [7], and all other references builds up the intuition for our solutions of the capacity problems for common randomness and identification in Section 7 and 8.

2 Notation and Definitions

Throughout the paper $\mathcal{U}$, $\mathcal{X}$, $\mathcal{Y}$ and $\mathcal{Z}$ are finite sets and their elements are written as corresponding lower letters e.g. u, x, y, and z. The letters U, X, Y,

R. Ahlswede et al. (Eds.): Information Transfer and Combinatorics, LNCS 4123, pp. 258–275, 2006.
© Springer-Verlag Berlin Heidelberg 2006

Z etc. will be used for random variables with values in the corresponding sets, $\mathcal{U}, \ldots T_X^n, T_{Y|X}^n(x^n), T_{XYZ}^n$, etc. are sets of n-typical, conditional typical and joint typical sequences, and sets of δ-typical, conditional typical and joint typical sequences are written as $T_{X,\delta}^n, T_{Y|X,\delta}^n(x^n), T_{XYZ,\delta}^n$, etc.

Then a (discrete memoryless) wire-tape channel is specified by a stochastic matrix $W : \mathcal{X} \to \mathcal{Y} \times \mathcal{Z}$, where $\mathcal{X}$ serves as input alphabet, $\mathcal{Y}$ as output alphabet of the legal receiver and $\mathcal{Z}$ as output alphabet of a wiretapper. The channel works as follows: the legal receiver receives an output sequence y^n and the wiretapper receives an output sequence z^n with probability

$$W^n(y^n z^n | x^n) = \prod_{t=1}^n W(y_t z_t | x_t).$$

In the case of transmission the sender's goal is to send to the receiver a message U uniformly distributed on an as large as possible set of messages with vanishing probability of error such that the wiretapper almost knows nothing about the message. Randomization at the sender side is allowed. The wiretapper, who knows the coding scheme, but not the message, tries to learn about the message as much as possible.

For given $\lambda, \mu > 0$, a (λ, μ)-code of length n with a set of messages $\mathcal{M}$ is a system $\{(Q_m : \mathcal{D}_m) : m \in \mathcal{M}\}$, where the Q_m's for $m \in \mathcal{M}$ are probability distributions on $\mathcal{X}^n$, and the $\mathcal{D}_m$'s are pairwise disjoint subsets of $\mathcal{Y}^n$, such that

$$|\mathcal{M}|^{-1} \sum_{m \in \mathcal{M}} \sum_{x^n \in \mathcal{X}^n} Q_m(x^n) \sum_{z^n \in \mathcal{Z}^n} W^n(\mathcal{D}_m, z^n | x^n) > 1 - \lambda, \qquad (2.1)$$

and

$$\frac{1}{n} I(U; Z^n) < \mu, \qquad (2.2)$$

if Z^n is the random output sequence generated by the message U through the channel. The transmission capacity of the wire-tape channel is the maximal non-negative number C_{wt} such that for $\mathcal{M}, \lambda, \mu, \varepsilon > 0$ and all sufficiently large length n, there exists a (λ, μ)-code with rate $\frac{1}{n} \log |\mathcal{M}| > C_{wt} - \varepsilon$. The security criterion (2.2) is strengthened in [11] to

$$I(U; Z) < \mu. \qquad (2.3)$$

In the current paper we assume the output y_t at time t is completely and immediately feedback to the sender via a secure noiseless channel such that the wiretapper has no knowledge about the feedback (except his own output z^n). Then for $\lambda, \mu > 0$, a (λ, μ)-code of length n for the wire-tape channel with secure feedback is a system $\{(Q, \mathcal{D}_m) : m \in \mathcal{M}\}$ where $\mathcal{D}_m, m \in \mathcal{M}$, are pairwise disjoint subsets of $\mathcal{Y}^n$ as before and Q is a stochastic matrix $Q : \mathcal{M} \times \mathcal{Y}^{n-1} \to \mathcal{X}^n$ with

$$Q(x^n | m, y^{n-1}) = \prod_{t=1}^n Q(x_t | m, y^{t-1})$$

for $x^n \in \mathcal{X}$, $y^{n-1} \in \mathcal{Y}^{n-1}$, and $m \in \mathcal{M}$, such that

$$|\mathcal{M}|^{-1} \sum_{m \in \mathcal{M}} \sum_{x^n \in \mathcal{X}} \sum_{z^n \in \mathcal{Z}^n} \sum_{y^n \in \mathcal{D}_m} Q(x^n|m, y^{n-1}) W^n(y^n, z^n|x^n) > 1 - \lambda \quad (2.4)$$

and (2.2) holds. The transmission capacity is defined analogously and denoted by C_{wtf}. In Theorem 1 in Section 4 we shall prove our (direct) coding theorem with the stronger security criterion (2.3).

3   Previous and Auxiliary Results

Our code construction is based on a coding lemma and a code for wire-tape channel without feedback. A balanced coloring lemma originally was introduced by R. Ahlswede [2] and we need its following variation.

Lemma 1. For all $\delta, \eta > 0$, sufficiently large n, all n-type P_{XY} and all $x^n \in T_X^n$, there exists a γ-coloring $c : T_{Y|X}^n(x^n) \to \{0, 1, 2, \ldots, \gamma - 1\}$ of $T_{Y|X}^n(x^n)$ such that for all joint n-type P_{XYZ} with marginal distribution P_{XY} and $\gamma^{-1}|T_{Y|XZ}^n(x^n, z^n)| > 2n\eta$, $x^n, z^n \in T_{XZ}^n$,

$$|c^{-1}(k)| \le \gamma^{-1}|T_{Y|XZ}^n(x^n, z^n)|(1 + \delta), \quad (3.1)$$

for $k = 0, 1, \ldots, \gamma - 1$, where c^{-1} is the inverse image of c.

Proof: Let us randomly and independently color $y^n \in T_{Y|X}^n(x^n)$ with γ colors and uniform distribution over $T_{Y|X}^n(x^n)$. Let for $k = 0, 1, \ldots, \gamma - 1$

$$S_k(y^n) = \begin{cases} 1 & \text{if } y^n \text{ is colored by } k \\ 0 & \text{else.} \end{cases} \quad (3.2)$$

Then for a joint type P_{XZY} and $z^n \in T_{Z|X}^n(x^n)$, by Chernoff bound,

$$Pr\left\{ \sum_{y^n \in T_{Y|XZ(x^n, z^n)}^n} S_k(y^n) > \gamma^{-1}|T_{Y|XZ}^n(x^n, y^n)|(1 + \delta) \right\}$$

$$\le e^{-\frac{\delta}{2}\gamma^{-1}|T_{Y|XZ}^n(x^n, z^n)|(1+\delta)} \prod_{y^n \in T_{Y|XZ}^n(x^n, z^n)} E e^{\frac{\delta}{2} S_k(y^n)}$$

$$= e^{-\frac{\delta}{2}\gamma^{-1}|T_{Y|XZ}^n(x^n, z^n)|(1+\delta)} \left[(1 - \gamma^{-1}) + \gamma^{-1} e^{\frac{\delta}{2}} \right]^{|T_{Y|XZ}^n(x^n, z^n)|}$$

$$= e^{-\frac{\delta}{2}\gamma^{-1}|T_{Y|XZ}^n(x^n, z^n)|(1+\delta)} \left[1 + (e^{\frac{\delta}{2}} - 1)\gamma^{-1} \right]^{|T_{Y|XZ}^n(x^n, z^n)|}$$

$$\le e^{-\frac{\delta}{2}\gamma^{-1}|T_{Y|XZ}^n(x^n, z^n)|(1+\delta)} \left[1 + \gamma^{-1}(\frac{\delta}{2} + \frac{\delta^2}{8}e) \right]^{|T_{Y|XZ}^n(x^n, z^n)|}$$

$$\leq \exp_e \left\{ -\frac{\delta}{2}\gamma^{-1}|T^n_{Y|XZ}(x^n, z^n)|(1+\delta) + \gamma^{-1}\left(\frac{\delta}{2} + \frac{\delta^2}{8}e\right)|T^n_{Y|XZ}(x^n, z^n)|\right\}$$

$$= \exp_e \left\{ -\frac{\delta}{2}\gamma^{-1}|T^n_{Y|XZ}(x^n, z^n)|\left(1-\frac{e}{4}\right)\delta\right\}$$

$$\leq e^{-\frac{e\delta^2}{24}\gamma^{-1}|T^n_{Y|XZ}(x^n,z^n))|}$$

$$\leq e^{-\frac{e\delta^2}{24}2^{n\eta}}, \tag{3.3}$$

if $\gamma^{-1}|T^n_{Y|XZ}(x^n, z^n)| > 2^{n\eta}$ and $\frac{\delta}{2} \leq 1$.
Here, to obtain the 2nd and 3rd inequalities, we use for $x \in [0, 1]$ the inequalities
$e^x \leq 1 + x + \frac{e}{2}x^2$ and $1 + x \leq e^x$ respectively.

(3.1) follows from (3.3) because the numbers of sequences z^n and n-types increase exponentially and polynomially respectively as the length increases.   □

To prove (the direct part of) the coding theorem for the wire-tape channel (without feedback) [11] Csiszár and Körner used a special code, Ahlswede's partition of typical input sequences into sets of code words, obtained by iterative maximal coding [1]. An easier proof appears in [2], part II, as consequence of the "link". We shall use its following improvement obtained with a Balanced Coloring Lemma of [2] and presented in [10].

For a given wire-tape channel such that for an input random variable X and its output random variables Y and Z for the legal user and wiretapper respectively

$$I(X;Y) - I(X;Z) > 0 \tag{3.4}$$

all $\lambda', \mu' > 0 \; 0 < \varepsilon' < I(X;Y) - I(X;Z)$ and sufficiently large n, there exists a set of codewords

$$\{u_{m,\ell} : m = 0, 1, 2, \ldots, M-1, \; \ell = 0, 1, 2, \ldots, L-1\}$$

in T^n_X having the following properties.

$$I(X;Y) - I(X;Z) - \varepsilon' < \frac{1}{n}\log M \leq I(X;Y) - I(X;Z) - \frac{\varepsilon'}{2} \tag{3.5}$$

$$I(X;Z) + \frac{\varepsilon'}{8} \leq \frac{1}{n}\log L < I(X;Z) + \frac{\varepsilon'}{4}. \tag{3.6}$$

For a set of properly chosen decoding sets $\{\mathcal{D}_{m,\ell}\}$,

$$\{(u_{m,\ell}, \mathcal{D}_{m,\ell}) : m = 0, 1, 2, \ldots, M-1, \ell = 0, 1, 2, \ldots, L-1\}$$

is a λ-code for the legal user.

Let $V, \tilde{Z}$ be random variables taking values in $\mathcal{M} \times \mathcal{Z}^n$, where $\mathcal{M} = \{0, 1, \ldots, M-1\}$, with probability for $(m, z^n) \in \mathcal{M} \times \mathcal{Z}^n$

$$Pr\{V, \tilde{Z}) = (m, z^n)) = \sum_{\ell=0}^{L-1} L^{-1}P^n_{Z|X}(z^n|u_{m,\ell}).$$

Then

$$I(V; \tilde{Z}) < \mu'. \tag{3.7}$$

4   The Coding Theorem for Transmission and Its Proof

Let $\mathcal{Q}$ be the set of quadruples of random variables (U, X, Y, Z) taking values in $\mathcal{U} \times \mathcal{X} \times \mathcal{Y} \times \mathcal{Z}$ for a finite set $\mathcal{U}$ with probability

$$Pr((U, X, Y, Z) = (u, x, y, z)) = P_{UX}(ux)W(yz|x) \tag{4.1}$$

for $(u, x, y, z) \in \mathcal{U} \times \mathcal{X} \times \mathcal{Y} \times \mathcal{Z}$.
    Then

Theorem 1. *The capacity of a wire-tape channel with feedback satisfies*

$$C_{wtf} \geq \max_{(U,X,Y,Z)\in\mathcal{Q}} \min[|I(U;Y) - I(U;Z)|^+ + H(Y|U, Z), I(U;Y)]. \tag{4.2}$$

Proof: For a $(U, X, Y, Z) \in \mathcal{Q}$, to show the achievability, one may introduce an auxiliary channel $P_{X|U}$ and construct a code for the channel

$$W'(y, z|u) = \sum_x P_{X|U}(x|u)W(y, z|x).$$

Then it is sufficient to show that $|I(X;Y) - I(X;Z)|^+ + H(Y|XZ)$ is achievable. Let us fix $\lambda, \mu, \varepsilon > 0$ and construct a (λ, μ)-code with rate

$$|I(X;Y) - I(X, Z)|^+ + H(Y|XZ) - \varepsilon. \tag{4.3}$$

To this end, let $\lambda', \mu', \varepsilon'$ be positive small real numbers specified later.
    Let $\mathcal{U} = \{u_{m,\ell} : m = 0, 1, 2, \ldots, M - 1, \ell = 0, 1, 2, \ldots, L - 1\}$ be the codebook if in the previous section for a sufficiently large n (3.4) holds i.e., $I(X;Y) - I(X;Z) > 0$.
    In the case that (3.4) does not hold we choose $M = 1$ and take a codebook of an arbitrary λ'-code for the legal user, with rate $I(X;Y) - \varepsilon' < R \triangleq \frac{1}{n} \log L \leq I(X;Y) - \frac{\varepsilon'}{2}$ as our codebook:

$$\mathcal{U} = \{u_{0,\ell} : \ell = 0, 1, 2, \ldots, L - 1\}.$$

Our code consists of N blocks of length n and sends a message $(U'_1, U'_2 U''_2, \ldots, U'_N U''_N)$ uniformly distributed on $\mathcal{M}' \times (\mathcal{M}' \times \mathcal{M}'')^{N-1}$, where

$$\mathcal{M}' = \{0, 1, 2, \ldots, M - 1\}, \mathcal{M}'' = \{0, 1, \ldots, L'' - 1\}, \tag{4.4}$$

and $L'' = \min\{L, 2^{n(H(Y|XZ) - \frac{3}{4})}\}$.
    In particular $M = 1$, $\mathcal{M}'$ is a dummy message set. Then the rate of the messages is

$$R^* = \frac{1}{n} \log M + \frac{1}{n} \log L'' - \frac{1}{nN} \log L'' \geq \frac{1}{n} \log M + \frac{1}{n} \log L'' - \frac{1}{N} \log |\mathcal{Y}|.$$

That is by (3.5), (3.6)

$$R^* \geq \begin{cases} I(X;Y)-I(X;Z)-\varepsilon'+\min\left[I(X;Z)+\frac{\varepsilon'}{8}, H(Y|XZ)-\frac{\varepsilon}{4}\right] - \frac{1}{N}\log|\mathcal{Y}| \\ \quad \text{if } I(X;Y)-I(X;Z(>0 \\ \min\left[I(X;Y)-\frac{\varepsilon'}{2}, H(Y|XZ)-\frac{\varepsilon}{4}\right] - \frac{1}{N}\log|\mathcal{Y}| \\ \quad \text{else.} \end{cases}$$

$$(4.5)$$

By choosing $\varepsilon' < \frac{\varepsilon}{2}$ and $N > 2\varepsilon^{-1}\log|\mathcal{Y}|$ in (4.5) we have

$$R^* > \min[|I(X;Y)-I(X;Z)|^+ + H(Y|XZ), I(X;Y)] - \varepsilon \qquad (4.6)$$

our desired rate.

In each block, we use a codebook

$$\mathcal{U} = \{u_{m,\ell} : m = 0,1,\dots,M-1, \ell = 0,1,2,\dots,L-1\}$$

defined as above. Suppose the sender wants to send a message $(m_1', m_2'm_2'',\dots, m_N'm_N'')$ to the receiver. Then our code consists of the following components.

1. In the first block the sender randomly chooses a $u_{m_1',\ell}$ from the codebook with uniform distribution on $\{u_{m_1'},j : j = 0,1,\dots,L-1\}$ and sends the codeword to the receiver. Then by choosing a proper decoder the receiver can decode $u_{m_1',\ell}$ and therefore m_1' correctly with probability $1 - \lambda'$.

2. From the first to the $N-1$st blocks, for all $u_{m,\ell} \in \mathcal{U}$, color all $T_{Y|\bar{X}}^n(u_{m,\ell}) \subset T_{Y|X,\delta_1}^n(u_{m,\ell})$ with L'' colors such that for a suitably small $\delta_2 > 0$ all n-joint type $P_{\bar{X}\bar{Y}Z}$ with $P_{\bar{X}} = P_X$ and

$$\sum_{yz} |P_{\bar{Y}\bar{Z}\bar{X}}(y,z|x) - P_{YZ|X}(yz|x)| < \delta_2. \qquad (4.7)$$

$T_{\bar{Y}|\bar{X}\bar{Z}}^n(u_{m,\ell}, z^n)$ is properly colored in the sense of Lemma 1.

3. For $j = 1,2,\dots,N-1$ after the sender receives output y^n of the jth block, he gives up if $y^n \notin T_{Y|X,\delta_1}^n(u(j))$, where $u(j)$ is the input sequence in $\mathcal{X}^n$ sent by the sender in the jth block. Then the probability for giving up at the jth block is exponentially small in n. In the case $y^n \in T_{Y|X,\delta_1}^n(u(j))$, y^n receives a coloring $c_{u(j)}(y^n) \in \{0,1,\dots,L''-1\}$ in the coloring for $T_{\bar{Y}|\bar{X}}^n(u(j))$, where $P_{\bar{X}\bar{Y}}$ is the joint type of $(u(j), Y^n)$.

   3.1. In the case $L \leq 2^{[H(Y|XZ)-\frac{3}{4}]}$ i.e., $L'' = L$, the sender sends $U_{m_{j+1}'m_{j+1}'' \oplus c_{m(j)}(y^n)} \triangleq u(j+1)$ in the codebook $\mathcal{U}$ in the $j+1$st block, where $\oplus$ is the addition modulo L''.

   3.2. In the case $L > 2^{n[H(Y|XZ)-\frac{3}{4}]}$, without loss of generality, we assume $L''|L$. Then the sender partitions $\{0,1,\dots,L-1\}$ into L'' segments of equal size. He randomly chooses an ℓ_{j+1}'' in the $m_{j+1}'' \oplus c_{u(j)}(y^n)$ segment with equal probability and sends $u_{m_{j+1}',\ell_{j+1}''}$ in the codebook in the $j+1$st block.

4. For $j = 1, 2, \ldots, N$ in the jth block the receiver decode separately by using a proper decoder and obtains a $\bar{u}(j)$ in the jth block. Thus $\bar{u}(j) = u(j)$ with probability λ' for a given j. Let $\lambda' < M^{-1}\lambda$, then $\bar{u}(j) = u(j)$ with probability larger than $1 - \lambda$ for all j. The receiver declares $m'_1 = \bar{m}'_1$ if $\bar{u}(1) = u_{\bar{m}'_1, \ell}$ for some ℓ. The receiver declares $m'_j m''_j = \bar{m}'_j \bar{m}''_j$ for $\bar{m}''_j = \ell_j \ominus c_{\bar{u}(j-1)}(y^n)$ if in the $j - 1$st block he receives y^n and $\bar{u}(j) = u_{\bar{m}'_j \ell_j}$ in the case $L'' = L$ and $\bar{u}(j) = u_{\bar{m}'_j, \ell'_j}$ for an ℓ'_j in the ℓ_jth segment in the case $L'' < L$, for $j = 2, 3, \ldots, N$. Obviously

$$(\bar{m}'_1, \bar{m}'_2 \bar{m}''_2, \ldots, \bar{m}'_N \bar{m}''_N) = (m'_1 m m'_2 m''_2, \ldots, m'_N m''_N)$$

if $\bar{u}(j) = u(j)$ for all j.

We have seen that the probability of error is smaller than λ and it is sufficient for us to verify the security criterion.

Denote by $\tilde{X}_j$, $\tilde{Y}_j$ and $\tilde{Z}_j$, the random input and outputs in the jth block generated by the code and the random message, $(U'_1, U'_2 U''_2, \ldots, U'_N U''_N)$ respectively, for $j = 1, 2, \ldots, N$. Notice here $\tilde{X}_j$, $\tilde{Y}_j$, and $\tilde{Z}_j$ are random sequences of length n. Let K_j be the coloring of the random output sequences of the legal receiver in the jth block. Write $U'^N = (U'_1, U'_2, \ldots, U'_N)$, $U''^N = (U''_1, U''_2, \ldots, U''_N)$ (where U'_1 is a dummy constant), $\tilde{X}^N = (\tilde{X}_1, \ldots, \tilde{X}_N)$, $\tilde{Y}^N = (\tilde{Y}_1, \ldots, \tilde{Y}_N)$ and $\tilde{Z}^N = (\tilde{Z}_1, \ldots, \tilde{Z}_N)$. Then we are concerned about an upper bound to $I(U'^N U''^N; \tilde{Z}^N)$.

At first we bound $I(U'^N; \tilde{Z}^N)$ with (3.7). Denote $\tilde{Z}^{\bar{j}} = (\tilde{Z}_1, \tilde{Z}_2, \ldots, \tilde{Z}_{j-1}, \tilde{Z}_{j+1}, \ldots, \tilde{Z}_N)$.

Then by symmetry, independent of $\tilde{Z}^{\bar{j}}$ and U'^{j-1}, given $U'_j = m$, the input of the channel in the jth block is uniformly distributed on the sub-codebook

$$\{u_{m,\ell} : \ell = 0, 1, \ldots, L - 1\}.$$

For $j = 1$ it immediately follows from the step 1 of the coding scheme. For $j > 1$, it is sufficient for us to show that $P_{U''_j \oplus K_{j-1} | U'^{j-1} \tilde{Z}^{\bar{j}}}$ is uniform. Indeed, for all ℓ, u'^{j-1}, and $z^{\bar{j}}$

$$Pr\{U''_j \oplus K_{j-1} = \ell | U'^{j-1} = u'^{j-1}, \tilde{Z}^{\bar{j}} = z^{\bar{j}}\}$$

$$= \sum_{m''=0}^{L''-1} L''^{-1} Pr\{K_{j-1} = \ell \ominus m'' | U^{j-1} = u'^{j-1}, Z^{\bar{j}} = Z^{\bar{j}}\} = L''^{-1}.$$

This means that for all j and $(V, \tilde{Z})$ in (3.7) we have

$$H(U'_j | U'^{j-1} \tilde{Z}^N) = H(U'_j | \tilde{Z}_j, U'^{j-1} Z^{\bar{j}}) = H(U | \tilde{Z})$$

and therefore by (3.7)

$$I(U'_j; U'^{j-1} \tilde{Z}^N) < \mu'$$

since U'_j and V have the same distribution.

Consequently

$$I(U'^N; Z^N) = \sum_{j=1}^{N} I(U_j; Z^N|U^{j-1}) \le \sum_{j=1}^{N} I(U_j; U'^{j-1}Z^N) \le N\mu'. \qquad (4.8)$$

Next we bound $I(U_j''; \tilde{Z}^N|U'^N U''^{j-1})$. At first we observe that by our coding scheme U_j'' is independent of $U'^N U''^{j-1}\tilde{Z}^i$ for all $i < j$ and therefore

$$I(U_j''; \tilde{Z}_i|U'^N U''^{j-1}\tilde{Z}^{i-1}) = 0, \text{ or}$$

$$I(U_j''; \tilde{Z}^N|U'^N U''^{j-1}) = \sum_{i=1}^{j-1} I(U_j''; \tilde{Z}_i|U'^N U''^{j-1}\tilde{Z}^{i-1})$$

$$+ I(U_j''; \tilde{Z}_j|U'^N U''^{j-1}\tilde{Z}^{j-1}) + I(U_j''; \tilde{Z}_{j+1}^N|U'^N U''^{j-1}\tilde{Z}^j)$$

$$= I(U_j''; \tilde{Z}_j|U'^N U''^{j-1}\tilde{Z}^{j-1}) + I(U_j''; \tilde{Z}_{j+1}^N|U'^N U''^{j-1}\tilde{Z}^j), \qquad (4.9)$$

where $\tilde{Z}_{j+1}^N = (\tilde{Z}_{j+1}, \dots, \tilde{Z}_N)$.

Moreover by our coding scheme under the condition given $U'^N U''^{j-1}\tilde{Z}^{j-1}$

$$U_j'' \Leftrightarrow U_j'' \oplus K_{j-1} \Leftrightarrow \tilde{Z}_j$$

form a Markov chain i.e., by the data processing inequality.

$$I(U_j''; \tilde{Z}_j|U'^N U''^{j-1}Z^{j-1}) \le I(U_j''; U_j'' \oplus K_{j-1}|U'^N U''^{j-1}Z^{j-1})$$
$$= I(U_j''; K_{j-1}|U'^N U''^{j-1}Z^{j-1}) \le I(U'^N U''^j Z^{j-1}; K_{j-1}). \qquad (4.10)$$

However, because $U'^N U''^j \tilde{Z}^{j-1} \Leftrightarrow \tilde{X}_{j-1}\tilde{Z}_{j-1} \Leftrightarrow K_{j-1}$ forms a Markov chain, (4.10) implies

$$I(U_j''; \tilde{Z}_j|U'^N U''^j Z^{j-1}) \le I(\tilde{X}_{j-1}\tilde{Z}_{j-1}; K_{j-1}). \qquad (4.11)$$

For $j - 1$

$$W_{j-1} = \begin{cases} 0 & \text{if } \tilde{Y}_{j-1} \in \mathcal{T}_{Y|X,\delta_1}^n(\tilde{X}_{j-1}) \\ 1 & \text{else,} \end{cases}$$

then recalling that the output of legal user is colored by Lemma 1 in the $j - 1$st block, by AEP we have

$$Pr\{K_{j-1} = k|\tilde{X}_{j-1} = x^n, \tilde{Z}_{j-1} = j^n W_{j-1} = 0\} \le L''^{-1}(1 + \delta).$$

Thus

$$H(K_{j-1}|\tilde{X}_{j-1}\tilde{Z}_{j-1}) \ge (1 - 2^{-n\theta})H(K_{j-1}|\tilde{X}_{j-1}\tilde{Z}_{j-1}W_{j-1} = 0)$$

$$\ge (1 - 2^{-n\theta})[\log L'' - \log(1 + \delta)],$$

for a $\theta > 0$ as $Pr(W_j = 0) > 1 - 2^{-n\theta}$. Thus for a $\mu'' > 0$ with $\mu'' \to 0$ as $\delta \to 0$,

$$I(\tilde{X}_{j-1}\tilde{Z}_{j-1}; K_{j-1}) = H(K_{j-1}) - \log L'' + \mu'' \le \mu'', \qquad (4.12)$$

for sufficiently large n. Similarly by the coding scheme under the condition given U'^N

$$U''^j Z^j \Leftrightarrow K_j \Leftrightarrow Z_{j+1}^N$$

forms a Markov chain and therefore

$$I(U_j''; Z_{j+1}^N | U''^N U''^{j-1}) \le I(U''^j Z^j; \tilde{Z}_{j+1}^N | U'^N) \le I(U''^j Z^j; K_j | U'^N) \le I(U'^N U''^j \tilde{Z}^j; K_j).$$
$$(4.13)$$

However, by the coding scheme $U'^N U''^j \tilde{Z}^j \Leftrightarrow \tilde{X}_j \tilde{Z}_j \Leftrightarrow K_j$ forms a Markov chain and so we can continue to bound (4.13) as

$$I(U_j''; Z_{j+1}^N | U'^N U''^{j-1} Z^j) \le I(\tilde{X}_j \tilde{Z}_j; K_j). \qquad (4.14)$$

By replacing $j - 1$ by j in (4.12) and applying it to (4.14) we have

$$I(U_j''; Z_{j+1}^N | U'^N U''^{j-1} Z^j) \le \mu''. \qquad (4.15)$$

Finally, we combine (4.8), (4.9), (4.10), (4.11), and (4.15), to obtain

$$I(U'^N U''^N; \tilde{Z}^N)$$

$$= I(U'^N; \tilde{Z}^N) + I(U''^N; \tilde{Z}^N | U'^N)$$

$$\le N\mu' + \sum_{j=2}^{N} I(U_j''; \tilde{Z}^N | U'^N U''^{j-1})$$

$$= N\mu' + \sum_{j=2}^{N} [I(U_j''; \tilde{Z}_j | U'^N U''^{j-1} \tilde{Z}^{j-1}) + I(U_j''; \tilde{Z}_{j+1}^N | U'^N U''^{j-1} \tilde{Z}_j)]$$

$$\le N\mu' + \sum_{j=2}^{N} [I(\tilde{X}_{j-1} \tilde{Z}_{j-1}; K_{j-1}) + I(U_j''; \tilde{Z}_{j+1}^N | U'^N U''^{j-1} \tilde{Z}_j)]$$

$$\le N\mu' + 2(N-1)\mu'' < \mu,$$

for sufficiently small μ' and μ''. This completes our proof.

5   Capacity of Two Special Families of Wire-Tape Channel

In this section we apply Theorem 1 to show the following upper bound of capacity, which is believed not to be tight in general, but is tight for wire-tape channels with certain Markovities.

Let Q' be the set of triples of random variables (X, Y, Z) with joint distribution $P_{XYZ}(x, y, z) = P_X(x)W(y, z|x)$ for $x \in \mathcal{X}$, $y \in \mathcal{Y}$, and $z \in \mathcal{Z}$. Then

Lemma 2. *For all wire-tape channels*

$$C_{wtf} \leq \max_{(X,Y,Z) \in Q'} \min[H(Y|Z), I(X;Y)]. \tag{5.1}$$

Proof: For a given (λ, μ)-code for the wire-tape channel, let X^n, Y^n, Z^n be the input and outputs generated by uniformly distributed messages U through the code. Then in the same way to show the converse coding theorem of a (two terminal) noisy channel with feedback, one obtains that

$$C_{wtf} \leq \frac{1}{n} \sum_{t=1}^{n} I(X_t; Y_t) + \varepsilon' \tag{5.2}$$

where $\varepsilon' \to 0$ as $\lambda \to 0$.

On the other hand, by the security condition and Fano's inequality we have

$$C_{wtf} = \frac{1}{n}H(U) \leq \frac{1}{n}H(U|Z^n) + \mu$$

$$\leq \frac{1}{n}H(U|Z^n) - \frac{1}{n}H(H|Y^n) + \lambda \log |\mathcal{X}| + \frac{1}{n}h(\lambda) + \mu$$

$$\leq \frac{1}{n}H(U|Z^n) - \frac{1}{n}H(U|Y^n Z^n) + \lambda \log |\mathcal{X}| + \frac{1}{n}h(\lambda) + \mu$$

$$= \frac{1}{n}I(U; Y^n|Z^n) + \varepsilon'' \leq \frac{1}{n}H(Y^n|Z^n) + \varepsilon''$$

$$= \frac{1}{n}\sum_{t=1}^{n}H(Y_t|Z^n Y^{t-1}) + \varepsilon'' \leq \frac{1}{n}\sum_{t=1}^{n}H(Y_t|Z_t) + \varepsilon'', \tag{5.3}$$

where $h(\lambda) = -\lambda \log \lambda - (1 - \lambda) \log(1 - \lambda)$ and $\varepsilon'' = \lambda \log |\mathcal{X}| + \frac{1}{n}h(\lambda) + \mu \to 0$ as $\lambda, \mu \to 0$.

Let $(UXYZ)$ be a quadruple of random variables with distribution

$$P_{UXYZ}(t, z, y, z) = \frac{1}{n}\sum_{t=1}^{n}P_{X_t Y_t Z_t}(x, y, z)$$

for $t \in \{1, 2, \ldots, n\}$, $x \in \mathcal{X}$, $y \in \mathcal{Y}$, $z \in \mathcal{Z}$. Then $(XYZ) \in Q'$ and by (5.2) and (5.3) for $\varepsilon = \max(\varepsilon', \varepsilon'')$

$$C_{wtf} \leq \min[H(Y|ZU), I(X;Y|U)] + \varepsilon \leq \min[H(Y|Z), I(X;Y)] + \varepsilon,$$

where $\varepsilon \to 0$ as $\lambda, \mu \to 0$. That is, (5.1).

Corollary 1. *For a wire-tape channel W such that there exist $W_1 : \mathcal{X} \to \mathcal{Y}$, and $W_2 : \mathcal{Y} \to \mathcal{Z}$ with*

$$W(y, z|x) = W_1(y|x)W_2(z|y), \tag{5.4}$$

for all $x \in \mathcal{X}$, $y \in \mathcal{Y}$, and $z \in \mathcal{Z}$ $C_{wtf} = \max_{(X,Y,Z) \in \mathcal{Q}'} \min[H(Y|Z), I(X;Y)]$.

Proof: By Markov condition (5.4), we have that for all $(X, Y, Z) \in \mathcal{Q}'$

$$I(X;Y) - I(X,Z) \geq 0 \tag{5.5}$$

and

$$I(X;Z|Y) = 0. \tag{5.6}$$

Thus

$$
\begin{aligned}
|I(X;Y) - I(X;Z)|^+ + H(Y|XZ) &= H(X|Z) - H(X|Y) + H(Y|XZ) \\
&= H(XY|Z) - H(X|Y) \\
&= H(Y|Z) + H(X|YZ) - H(X|Y) \\
&= H(Y|Z) + I(X;Z|Y) \\
&= H(Y|Z).
\end{aligned}
$$

Then corollary follows from Theorem 1 and Lemma 2.

Corollary 2. *For a wire-tape channel such that there exist* $W_1' : \mathcal{X} \to \mathcal{Z}$ *and* $W_2' : \mathcal{Z} \to \mathcal{Y}$ *with*

$$W(y, z|x) = W_1'(z|x) W_2'(y|z) \tag{5.7}$$

for $x \in \mathcal{X}$, $y \in \mathcal{Y}$, *and* $z \in \mathcal{Z}$

$$C_{wtf} = \max_{(X,Y,Z) \in \mathcal{Q}'} \min[H(Y|Z), I(X;Y)].$$

Proof: The Markov condition (5.7) implies that

$$I(X;Y) - I(X;Z) \leq 0 \tag{5.8}$$

and

$$H(Y|XZ) = H(Y|Z), \tag{5.9}$$

which yield

$$|I(X;Y) - I(X;Z)|^+ + H(Y|XZ) = H(Y|XZ) = H(Y|Z). \tag{5.10}$$

Thus the corollary follows from Theorem 1 and Lemma 2.

Example: An interesting example is a special channel for which W_1' is a noiseless channel and W_2' is a noisy channel in Corollary 2 e.g., W_1 is a noiseless binary channel, W_2'' is a binary symmetric channel with crossover probability $p \in \left(0, \frac{1}{2}\right)$. For this channel the wiretapper is in a better position than the legal user. So the capacity is zero without feedback. The feedback makes the capacity positive by our Corollary 2 as it serves as a secure key shared by sender and receiver.

6 Discussion: Transmission, Building Common Randomness and Identification

As goals of communications are considered transmission i.e., sending a given message from a set of messages, building common randomness i.e., to provide a random resource shared by users, and identification i.e., identifying whether an event of interest to a particular user occurs ([3], [4], [5], [13]).

Roughly saying in a given communication system, the capacity of transmission is upper bounded by the capacity of common randomness, since common randomness shared by a sender and receiver can be built by transmission whereas the capacity of identification is lower bounded by capacity of common randomness, if the former is positive, which is shown by a scheme in [5] to build identification codes by common randomness. That is,

$$\text{capacity of transmission} \leq \text{capacity of common randomness}$$
$$\leq \text{capacity of identification.} \tag{6.1}$$

However, in different communication systems equalities in (6.1) may or may not hold. In this section we illustrate the variety in two-terminal channels and wire-tape channels. More examples in more complicated communication systems can be found e.g. in [3], [12], [15].

First of all, obviously the first inequality in (6.1) is always an equality for a two terminal channel without feedback, because all information obtained by the receiver is from the transmission via the channel. Moreover, it has been shown in [4] that the second inequality is an equality and therefore the three quantities in (6.1) are actually the same if the channel is discrete memoryless. A channel with rapidly increasing alphabet (as the length of codes grows) for which the capacity of identification is strictly larger than capacity of common randomness was described in [6]. It was shown in [8] that under a certain condition the capacity of common randomness (which is equal to the capacity of transmission) for Gaussian channels is finite whereas the capacity of identification is infinite in the same communication system. We notice that Gaussian channels have continuous, or infinite alphabets. It is natural to expect that for a discrete channel whose input alphabet "reasonably" increases the last two quantities, or consequently the three quantities in (6.1) are equal. This was shown in [14] for all channels whose input alphabets exponentially increase as the lengths of codes linearly increase.

The situation of two terminal channels is different when feedback is present. In this case the capacity of identification, which is equal to the capacity of common randomness, is strictly larger than the capacity of transmission for simplest channels, namely discrete memoryless channels [5]. The reason is clear. On one hand, it is well known, feedback does not increase the capacity of transmission for discrete memoryless channels. On the other hand, the feedback provides a random resource, shared by sender and receiver, the random output, whose rate, roughly speaking, is input entropy. Obviously it increases common randomness between sender and receiver and therefore capacity of identification.

Next we turn to wire-tape channels without feedback. More precisely, we mean secure common randomness shared by sender and receiver, about which the wiretapper has (almost) no knowledge. By the same reason as for two terminal channels without feedback, the capacity of (secure) common randomness is not larger than the capacity of transmission over the wire-tape channel. In fact it is shown in [3], that it may not be larger than the capacity of transmission even in the case where a public forward channel with unbounded capacity is available to the sender and receiver. This intuitively is not surprising. R. Ahlswede and Z. Zhang observed in [7] that to keep the message to be identified in secret a secure common randomness with positive rate is sufficient and the *major part of common randomness between the legitimate communicator applied in the identification code in [5] can be publically sent.*

Based on this observation they show that the capacity of identification is strictly larger than the capacity of secure common randomness. A more detailed analysis in [9] shows that the amount of secure common randomness needed only depends on the probability of second error and security criterion and is independent of the rate of messages. For fixed criterion of error and security, a constant amount – or zero-rate – of secure common randomness is sufficient, if provided with sufficiently large public common randomness.

Let us return to our main topic wire-tape channels with secure feedback and investigate (6.1) in this communication system. We immediately find that the observation about wire-tape channels without feedback is still valid when feedback is present, because there is nothing in the observation which links to the existence of feedback. This means that the capacity of identification must be the capacity of "public" common randomness between sender and receiver i.e., the maximum rate of common randomness shared by the sender and the receiver, neglecting whether or how much the wiretapper knows about it once a positive amount of secure common randomness is provided. But now the public common randomness is the maximum output entropy for the channel $W_1 : \mathcal{X} \to \mathcal{Y}$ defined by

$$W_1(y|x) = \sum_{z \in \mathcal{Z}} W(y, z|x) \text{ for all } x \in \mathcal{X}, y \in \mathcal{Y}, \qquad (6.2)$$

or in other words $\max_{(X,Y,Z) \in \mathcal{Q}'} H(Y)$, for $\mathcal{Q}'$ as defined in Section 5. So we conclude that in this case the capacity of identification is either zero or $\max_{(X,Y,Z) \in \mathcal{Q}'} H(Y)$. The only problem left is to find suitable conditions for the positivity of the capacity. We shall discuss this later.

To see the relation of the first pair of quantities in (6.1), we take a look at our main result

Theorem 1. *The information theoretical meaning of mutual information in (4.2) is obvious. The capacity of transmission with security criterion can not exceed that without it. So we expect this term could be removed in the formula of capacity of common randomness. To investigate the remaining term in (4.2), let us recall our coding scheme in Section 4.*

From the first block to the second last block, the transmission in each block has two tasks, sending a secret message m'_j (in the jth block) with a rate $\sim |I(U;Y) - I(U;Z)|$; and generating a secure common randomness with a rate $\sim H(Y|UZ)$, which will be used as a private key to send message m''_{j+1} in the next block. This gives us a secure common randomness with rate $\sim H(Y|UZ)$. The reason for the fact that U occurs in the "condition" is that the key for the $j + 1$st block has to be independent of the message sent in the jth block. For secure common randomness itself this is not necessary. So we expect that the capacity of common randomness is $\max\limits_{(X,Y,Z)\in\mathcal{Q}'} H(Y|Z)$, which actually is shown in the next section.

But before this we have a remaining problem, namely the positivity of the capacity of identification, which should be discussed. First we notice that to have positive capacity of identification, the capacity of the channel W_1 in (6.2), where we do not count wiretapper's role, has to be positive. By counting wiretapper's role, we look for an input random variable X, the conditional entropy $H(Y|Z)$ for output random variable Y and Z has to be positive, because otherwise the wiretapper would know everything known by the legal receiver. *We shall show that the two necessary conditions together are sufficient for the positivity.*

7   The Secure Common Randomness Capacity in the Presence of Secure Feedback

Let $\mathcal{J}_n = \{0, 1, \ldots, J_n - 1\}$ be a finite set (whose size depends on n), $\lambda, \mu > 0$. An (n, J_n, λ, μ)-common randomness for the wire-tape channel with secure feedback is a pair of random variables (K_n, L_n) defined on the same domain $\mathcal{J}_n$ with the following properties.

There exists a random variable U taking value in a finite set $\mathcal{U}$ and three functions $\theta^n : \mathcal{U} \times \mathcal{Y}^{n-1} \to \mathcal{X}^n$, $\varphi : \mathcal{U} \times \mathcal{Y}^n \to \mathcal{J}_n$, and $\Psi : \mathcal{Y}^n \to \mathcal{J}_n$ such that for all $u \in \mathcal{U}$ and $y^{n-1} \in \mathcal{Y}^{n-1}$

$$\theta^n(u, y^{n-1}) = (\theta_1(u), \theta_2(u, y_1), \ldots, \theta_n(u, y^{n-1})), \tag{7.1}$$

$$K_n = \varphi(U, Y^n) \tag{7.2}$$

$$\text{and} \quad L_n = \Psi(Y^n), \tag{7.3}$$

where Y^n and Z^n are output random variables for the legal receiver and the wiretapper, respectively, generated by random variable U, encoding function θ^n, and the channel W.

I.e.

$$Pr((Y^n, Z^n) = (y^n, z^n)) = \sum_{u \in \mathcal{U}} Pr(U = u) W(y_1, z_1 | \theta_1(u)) \prod_{t=2}^n W(y_t, z_t | \theta_t(u, y^{t-1})).$$

$$\tag{7.4}$$

$$Pr(K_n \neq L_n) < \lambda, \tag{7.5}$$

$$\frac{1}{n}H(K_n|Z^n) > \frac{1}{n}\log J_n - \mu. \tag{7.6}$$

$\frac{1}{n}\log J_n$ is called rate of the code and the capacity of the (secure) common randomness, denoted by C_{wtf-cr}, is defined as the maximum achievable rate in the standard way.

Theorem 2

$$C_{wtf-cr} = \max_{(X,Y,Z)\in\mathcal{Q}'} H(Y|Z), \tag{7.7}$$

in particular, the RHS of (7.7) is achievable if (7.6) is replaced by a stronger condition

$$H(K_n|Z^n) > \log J_n - \mu. \tag{7.8}$$

Proof: The proofs to both, direct and converse parts, are straightforward. They immediately follow from the proofs for Theorem 1 and Lemma 2, respectively.

Let $(X', Y, Z) \in \mathcal{Q}'$ achieve the maximum at RHS (7.7). Apply Lemma 1 to color sets of typical remaining sequences $T_{Y'}^n \subset T_{Y,\delta}^n$ [1], then it follows from the proof of Theorem 1 (the part to show (4.11)) that for any fixed $\mu > 0$ and sufficiently large n

$$H(\tilde{K}|Z^n) > \log J_n - \mu,$$

where $\tilde{K}$ is the random J_n-coloring obtained from Lemma 1. Choose $K_n = L_n = \tilde{K}$, then the proof of the direct part is done. To show the converse part we apply Fano's inequality to (7.5). Then

$$\frac{1}{n}\log J_n \le \frac{1}{n}H(K_n|Z^n) + \mu$$

$$\le \frac{1}{n}H(K_n|Z^n) - \frac{1}{n}H(K_n|Y^n) + \mu + \frac{1}{n}\lambda\log J_n + \frac{1}{n}h(\lambda)$$

$$\le \frac{1}{n}H(K_n|Z^n) - \frac{1}{n}H(K_n|Y^n, Z^n) + \mu + \frac{1}{n}\lambda\log J_n + \frac{1}{n}h(\lambda)$$

$$\le \frac{1}{n}I(K_n; Y^n|Z^n) + \mu + \frac{1}{n}\lambda\log J_n + \frac{1}{n}h(\lambda)$$

$$\le \frac{1}{n}H(Y^n|Z^n) + \mu + \frac{1}{n}\lambda\log J_n + \frac{1}{n}h(\lambda).$$

Now the converse follows as in the proof for Lemma 2.

8   The Secure Identification Capacity in the Presence of Secure Feedback

In this section let us take a look at the coding theorem for identification codes. First we have to formally define the codes and capacity. An $(n, |\mathcal{M}|, \lambda_1, \lambda_2, \mu)$-

---

[1] More precisely, let $\mathcal{X}_0 = \{x_0\}$, $x^n = (x_0, x_0, \ldots, x_0)$, and (X, X', Y, Z) be random variables with joint distribution $Pr((X, X', Y, Z) = (x^n, x'^n, y^n, z^n)) = P_{X'YZ}(x'^n, y^n, z^n)$ for all x'^n, y^n, z^n and coloring for the "conditional" typical sequences $T_{Y|X}^n(x^n) = T_Y^n$.

identification code for a wire-tape channel with secure feedback is a system $\{Q, \mathcal{D}_m : m \in \mathcal{M}\}$ such that $Q : \mathcal{M} \times Y^{n-1} \to \mathcal{X}^n$ is a stochastic matrix with

$$Q(x^n|m, y^{n-1}) = Q_1(x_1|m) \prod_{t=2}^{n} Q_t(x_t|m, y^{t-1})$$

for $m \in \mathcal{M}$, $y^{n-1} \in \mathcal{Y}^{n-1}$, for all $m \in \mathcal{M}$

$$\sum_{x^n \in \mathcal{X}^n} \sum_{y^n \in \mathcal{D}_m} Q_n(x_1|m) \prod_{t=2}^{n} Q_t(x_t|m, y^{t-1}) W_1(y_t|x_t) > 1 - \lambda_1,$$

for $m, m' \in \mathcal{M}$ with $m \neq m'$

$$\sum_{x^n \in \mathcal{X}^n} \sum_{y^n \in \mathcal{D}'_m} Q_1(x_1|m) \prod_{t=2}^{n} Q_t(x_t|m, y^{t-1}) W_1(y_t|x_t) < \lambda_2,$$

and for all $m, m' \in \mathcal{M}$, $m \neq m'$ and $\mathcal{V} \subset Z^n$

$$\sum_{x^n \in \mathcal{X}^n} \sum_{y^n \in \mathcal{Y}^n} Q_1(x_1|m') \prod_{t=2}^{n} Q_t(x_t|m', y^{t-1}) W(y^n, \mathcal{V}|x^n)$$

$$+ \sum_{x^n \in \mathcal{X}^n} \sum_{y^n \in \mathcal{Y}^n} Q_1(x_1|m) \prod_{t=2}^{n} Q_t(x_t|m, y^{t-1}) W(y^n, \mathcal{V}^c|x^n) > 1 - \mu.$$

Then capacity of identification is defined in the standard way and denoted by C_{wtf-id}.

C_{wtf-id} is upper bounded by the RHS of (8.1), follows from the converse of the coding theorem of identification with feedback for channel W_1 [5]. In the case that II holds, one can construct a code achieving $H(Y)$ asymptotically from the code in [7] by replacing the ordinary code for W_1 by a uniform partition of output sequences for the legal receiver and a code for the wire-tape channel without feedback by a code for the same channel but with feedback.

Furthermore the two conditions in III

Theorem 3. *The following statements are equivalent.*

I $C_{wtf-id} = \max_{(X,Y,Z) \in \mathcal{Q}'} H(Y)$ (8.1)

II $C_{wtf} > 0$

III   There exists an $(X, Y, Z) \in \mathcal{Q}'$ such that

$$H(Y|Z) > 0$$

and the channel W_1 has positive capacity.

Proof: The converse of the coding theorem i.e., C_{wtf-id} is upper bounded by the right hand side of (8.1) follows from the converse of coding theorem of identification with feedback for channel W_1 [4]. In the case that II holds, one can construct a code achieving $H(Y)$ asymptotically from the code in [6] by replacing the ordinary code for W_1 by a uniform partition of output sequences for the legal receiver and a code for the wiretape channel without feedback by a code for the same channel but with feedback.

Furthermore the two conditions in III obviously are necessary for positivity of C_{wtf-id}. The only thing left to be proved is that III implies II. Let $(X_i, Y_i, Z_i) \in Q'$ for $i = 0, 1$ such that $H(Y_0|Z_0) > 0$ and $I(X_1, Y_1) > 0$. By Theorem 1, it is sufficient for us to find $(U, X, Y, Z) \in Q$ such that $I(U; Y) > 0$ and $H(Y|UZ) > 0$. Obviously we are done, if $I(X_0; Y_0) > 0$ or $H(Y_1|U_1, Z_1) > 0$. Otherwise we have to construct a quadruple of random variables $(U, X, Y, Z) \in Q$ from (X_0, Y_0, Z_0) and (X_1, Y_1, Z_1) such that $H(Y|UZ) > 0$ and $I(U; Y) > 0$. To this end, let $\mathcal{U} = \mathcal{X} \cup \{u_0\}$, (where u_0 is a special letter not in $\mathcal{X}$), and for all $u \in \mathcal{U}$, $x \in \mathcal{X}$, $y \in \mathcal{Y}$ and $z \in \mathcal{Z}$, let (U, X, Y, Z) be a quadruple of random variables such that

$$P_{UXYZ}(u, x, y, z) = \begin{cases} \frac{1}{2} P_{X_0 Y_0 Z_0}(x, y, z) & \text{if } u = u_0 \\ \frac{1}{2} P_{X_1 Y_1 Z_1}(x, y, z) & \text{if } u \in \mathcal{X} \text{ and } u = x \\ 0 & \text{otherwise.} \end{cases}$$

Then $(U, X, Y, Z) \in Q$, $P_{YZ|U}(y|u_0) = P_{Y_0 Z_0}(yz)$ for all $y \in \mathcal{Y}$ and $z \in \mathcal{Z}$. $P_0(u_0) = \frac{1}{2}$ and therefore

$$H(Y|UZ) = \sum_{u \in \mathcal{U}} P_U(u) H(Y|U = uZ) \geq \frac{1}{2} H(Y|U = u_0 Z) = \frac{1}{2} H(Y_0|Z_0) > 0.$$

On the other hand for

$$S = \begin{cases} 0 & \text{if } U = u_0 \\ 1 & \text{otherwise,} \end{cases}$$

for all $u \in \mathcal{X}, y \in \mathcal{Y}$

$$P_{UY|S}(u, y|S = 1) = P_{X_1 Y_1}(u, y)$$

and $P_s(1) = \frac{1}{2}$ and consequently

$$I(U; Y) = I(US; Y) \geq I(U; Y|S) \geq P_s(1) I(U; Y|S = 1) = \frac{1}{2} I(X_1; Y_1) > 0.$$

That is, (U, X, Y, Z) is as desired. We conclude with the

Corollary 3

$$C_{wtf-id} = \begin{cases} \max_{(X,Y,Z) \in Q'} H(Y|Z) \\ 0 \end{cases}$$

and $C_{wtf-id} = 0$ iff for all $(X, Y, Z) \in Q'$ $H(Y|Z) = 0$ or the capacity of W_1 is zero.

Proof: That for all $(X, Y, Z) \in \mathcal{Q}'$, $H(Y|Z) = 0$ implies that the wiretapper knows what the receiver receives with probability one no matter how the sender chooses the input and that the capacity of W_1 is zero means the sender may not change the output distributions at the terminal for the legal receiver. So in both cases $C_{wtf-id} = 0$. Thus the corollary follows from Theorem 3.

References

1. R. Ahlswede, Universal coding, Paper presented at the 7th Hawai International Conference on System Science, Jan. 1974, Published in [A21].
2. R. Ahlswede, Coloring hypergraphs: A new approach to multi-user source coding, Part I, J. Comb. Inform. Syst. Sci., Vol. 4, No. 1, 76-115, 1979; Part II, Vol. 5, No. 3, 220-268, 1980.
3. R. Ahlswede and I. Csiszár, Common randomness in information theory and cryptography, Part I: Secret sharing, IEEE Trans. Inf. Theory, Vol. 39, No. 4, 1121-1132, 1993; Part II: CR capacity, Vol. 44, No. 1, 55-62, 1998.
4. R. Ahlswede and G. Dueck, Identification via channels, IEEE Trans. Inform. Theory, Vol. 35, No. 1, 15-29, 1989.
5. R. Ahlswede and G. Dueck, Identification in the presence of feedback – a discovery of new capacity formulas, IEEE Trans. Inform. Theory, Vol. 35, No. 1, 30-39, 1989.
6. R. Ahlswede, General theory of information transfer, Preprint 97–118, SFB 343 "Diskrete Strukturen in der Mathematik", Universität Bielefeld, 1997; General theory of information transfer:updated, General Theory of Information Transfer and Combinatorics, a Special Issue of Discrete Applied Mathematics, to appear.
7. R. Ahlswede and Z. Zhang, New directions in the theory of identification via channels, IEEE Trans. Inform. Theory, Vol. 41, No. 4, 1040-1050, 1995.
8. M. Burnashev, On identification capacity of infinite alphabets or continuous time, IEEE Trans. Inform. Theory, Vol. 46, 2407-2414, 2000.
9. N. Cai and K.-Y. Lam, On identification secret sharing scheme, Inform. and Comp., 184, 298-310, 2002.
10. I. Csiszár, Almost independence and secrecy capacity, Probl. Inform. Trans., Vol. 32, 40-47, 1996.
11. I. Csiszár and J. Körner, Broadcast channel with confidential message, IEEE Trans. Inform. Theory, Vol. 24, 339-348, 1978.
12. I. Csiszár and P. Narayan, Common randomness and secret key generation with a helper, IEEE Trans. Inform. Theory, Vol. 46, No. 2, 344-366, 2000.
13. C.E. Shannon, A mathematical theory of communication, Bell. Sys. Tech. J., 27, 379-423, 1948.
14. Y. Steinberg, New converses in the theory of identification via channels, IEEE Trans. Inform. Theory, Vol. 44, 984-998, 1998.
15. S. Venkatesh and V. Anantharam, The common randomness capacity of a network of discrete memoryless channels, IEEE Trans. Inform. Theory, Vol. 46, 367-387, 2000.
16. A.D. Wyner, The wiretape channel, Bell. Sys. Tech. J., Vol. 54, 1355-1387, 1975.

A Simplified Method for Computing the Key Equivocation for Additive-Like Instantaneous Block Encipherers

Z. Zhang

Abstract. We study the problem of computing the key equivocation rate for secrecy systems with additive-like instantaneous block (ALIB) encipherers. In general it is difficult to compute the exact value of the key equivocation rate for a secrecy system $(f, \mathcal{C})$ with ALIB encipherer when the block length n becomes large. In this paper, we propose a simplified method for computing the key equivocation rate for two classes of secrecy systems with ALIB encipherers. 1) The function f is additive-like and the block encipherer C is the set of all n-length key words (sequences) of type P. 2) The function f is additive and the block encipherer C is a linear (n, m) code in the n-dimensional vector space $\mathrm{GF}(q)^n$. The method has a potential use for more classes of secrecy systems.

1 Introduction

The secrecy systems with additive-like instantaneous block (ALIB) encipherers was investigated by Ahlswede and Dueck[1]. The model of ALIB encipherers was given in detail there. We consider the same model in this paper. The only difference is that we consider the key equivocation criterion rather than error probability criterion. The key equivocation criterion was studied by Blom [2][3] for some other secrecy systems.

For readers' convenience, we briefly review the model of ALIB encipherers and the key observation of Ahlswede and Dueck[1]. Let $\mathcal{X}, \mathcal{K}, \mathcal{Y}$ be finite sets with

$$|\mathcal{X}| = |\mathcal{K}| = |\mathcal{Y}|$$

where the number of elements in a set $\mathcal{X}$ is denoted by $|\mathcal{X}|$. Without loss of generality, we assume that $\mathcal{X} = \mathcal{K} = \mathcal{Y}$. Let $(X_i)_{i=1}^{\infty}$ be a message source, where all the $X_i, i = 1, 2, \cdots$ are independent replicas of a random variable X with values in $\mathcal{X}$. The probability distribution of $X^n = (X_1, \cdots, X_n)$ is given by

$$\mathrm{Pr}(X^n = x^n) = \prod_{i=1}^{n} \mathrm{Pr}(X = x_i)$$

for all $x^n = (x_1, \cdots, x_n) \in \mathcal{X}^n$. Let $f : \mathcal{X} \times \mathcal{K} \to \mathcal{Y}$ be a function, where $f(x, \cdot)$ is bijective for each $x \in \mathcal{X}$ and $f(\cdot, k)$ is bijective for each $k \in \mathcal{K}$. The function $f^n : \mathcal{X}^n \times \mathcal{K}^n \to \mathcal{Y}^n$ denotes the n-fold product of f.

An (n, R) ALIB encipherer is a subset $\mathcal{C} \subset \mathcal{K}^n$ with $|\mathcal{C}| \leq 2^{nR}$. Given a pair $(f, \mathcal{C})$, we define a secrecy system which works as follows. A key word k^n is generated by a random key generator K^n according to the uniform distribution on

R. Ahlswede et al. (Eds.): Information Transfer and Combinatorics, LNCS 4123, pp. 276–284, 2006.

C. Using f^n and k^n, the sender encrypts the output x^n of the message source to the cryptogram $y^n = f^n(x^n, k^n)$ and sends it to the receiver over a noiseless channel. The receiver uses the same key word k^n and f^{-1} to decrypt the message $x^n = (f^{-1})^n(y^n, k^n)$, where the key word k^n is given to the receiver separately over a secure channel. The cryptanalyst intercepts the cryptogram y^n and attempts to decrypt x^n. Since the cryptanalyst does not know the actual key word k^n being used, he has to search for a correct key word by using his knowledge of the system. Suppose that the random key K^n and the source output X^n are mutual independent. Let $Y^n = f^n(X^n, K^n)$. Then the average uncertainty about the key when the cryptanalyst intercepts a cryptogram is the conditional entropy $H(K^n|Y^n)$. The quantity $H(K^n|Y^n)/n$ which is called key equivocation rate is used as a security criterion for the secrecy system (f, C). By the definition of the secrecy system, the joint probability distribution of X^n, K^n, Y^n is

$$\Pr(X^n = x^n, K^n = k^n, Y^n = y^n) = \Pr(X^n = x^n)\Pr(K^n = k^n)\delta(y^n, f^n(x^n, k^n))$$

where

$$\delta(y^n, f^n(x^n, k^n)) = \begin{cases} 1 & y^n = f^n(x^n, k^n) \\ 0 & \text{otherwise} \end{cases}$$

Then the conditional probability

$$\Pr(Y^n = y^n|K^n = k^n) = \sum_{x^n} \Pr(X^n = x^n)\delta(y^n, f^n(x^n, k^n)) \text{ for } k^n \in C$$

Define a discrete memoryless channel with transmission probability matrix $W = (W_{y|k}, k \in \mathcal{K}, y \in \mathcal{Y})$, where

$$W_{y|k} = \sum_x \Pr(X = x)\delta(y, f(x, k))$$

Then the transmission probabilities for n-words k^n, y^n are

$$\begin{aligned} W^n_{y^n|k^n} &= \prod_{i=1}^{n} W_{y_i|k_i} = \prod_{i=1}^{n} \sum_{x_i} \Pr(X_i = x_i)\delta(y_i, f(x_i, k_i)) \\ &= \sum_{x_1} \cdots \sum_{x_n} \prod_{i=1}^{n} \Pr(X_i = x_i)\delta(y_i, f(x_i, k_i)) \\ &= \sum_{x^n} \Pr(X^n = x^n)\delta(y^n, f^n(x^n, k^n)) \end{aligned}$$

The key observation of Ahlswede and Dueck [1] is that an (n, R) ALIB encipherer $C \subset \mathcal{K}^n$ can be regarded as an (n, R) code for the memoryless channel W. Furthermore, the random cryptogram Y^n is the output of the channel W^n when the input is the random key K^n.

The paper is organized as follows. In section 2, we derive some properties of the key equivocation for the secrecy system (f, C) which will be used in section 3. In section 3, we present the computational method of the key equivocation for the secrecy system (f, C) which is applied to two specific cases. Case 1): C is the set of all sequences of type P in $\mathcal{K}^n$. Case 2): f is additive and C is a linear subspace (linear code) of the n-dimensional vector space $GF(q)^n$ over the finite field $GF(q)$.

2 Properties of the Key Equivocation

Lemma 1. *For arbitrary secrecy system* $(f, \mathcal{C})$, *we have*

$$H(K^n | Y^n) = H(X^n | Y^n) \leq H(X^n)$$

Proof. By the properties of conditional entropies, we have

$$H(X^n, K^n | Y^n) = H(K^n | Y^n) + H(X^n | Y^n, K^n)$$
$$= H(X^n | Y^n) + H(K^n | X^n, Y^n)$$

The definition of the function f^n implies that any one of the random variables X^n, K^n, Y^n is a function of the remaining two others. Then $H(X^n | Y^n, K^n) = H(K^n | X^n, Y^n) = 0$.

The lemma is proved.

Lemma 2. *For arbitrary secrecy system* $(f, \mathcal{C})$, *we have*

$$H(K^n | Y^n) = H(X^n) + \log |\mathcal{C}| - H(Y^n)$$

Proof. By the properties of entropies, we have

$$H(K^n | Y^n) = H(K^n) + H(Y^n | K^n) - H(Y^n)$$

The definition of the function f^n and the independence of X^n and K^n imply the equality

$$H(Y^n | K^n) = H(X^n | K^n) = H(X^n)$$

Then $H(K^n | Y^n) = H(K^n) + H(X^n) - H(Y^n)$.

The lemma is proved by inserting $H(K^n) = \log |\mathcal{C}|$ into the last equality.

Lemma 3. *Let* $T : \mathcal{K}^n \to \mathcal{K}^n$ *be a 1 to 1 transformation.*
Denote $T\mathcal{C} = \{Tk^n; k^n \in \mathcal{C}\}$. *Suppose that* $(f, \mathcal{C})$ *satisfies the following conditions:*

1) $T\mathcal{C} = \mathcal{C}$;
2) $f^n(x^n, k^n) = T f^n(x^n, T^{-1} k^n)$ *or*
3) $f^n(x^n, k^n) = T f^n(T^{-1} x^n, T^{-1} k^n)$ *and* $\Pr(X^n = x^n) = \Pr(X^n = T^{-1} x^n)$ *for* $x^n \in \mathcal{X}^n, k^n \in \mathcal{C}$.
Then, we have

$$\Pr(Y^n = y^n) = \Pr(Y^n = Ty^n)$$

Proof. $\Pr(Y^n = y^n) = \sum_{k^n \in \mathcal{C}} |\mathcal{C}|^{-1} \sum_{x^n} \Pr(X^n = x^n) \delta(y^n, f^n(x^n, k^n))$
$\Pr(Y^n = Ty^n) = \sum_{k^n \in \mathcal{C}} |\mathcal{C}|^{-1} \sum_{x^n} \Pr(X^n = x^n) \delta(Ty^n, f^n(x^n, k^n))$
From conditions 2) and 1), we have

$\Pr(Y^n = Ty^n) = \sum_{T^{-1}k^n \in \mathcal{C}} |\mathcal{C}|^{-1} \sum_{x^n} \Pr(X^n = x^n) \delta(Ty^n, T f^n(x^n, T^{-1} k^n))$
$= \Pr(Y^n = y^n)$

The proof is similar for conditions 3) and 1).

3 Computation of the Key Equivocation

Theorem 1. *Let $T = \{T_1, T_2, \cdots, T_N\}$ be a transformation group. If for every $T \in T$, (f, C) satisfies the conditions of Lemma 3, then $\mathcal{Y}^n$ can be partitioned to equivalent classes which are denoted by $\mathcal{Y}_i, i = 1, 2, , M$. Furthermore, we have*

$$H(K^n | Y^n) = nH(X) + \log |C| + \sum_{i=1}^{M} |\mathcal{Y}_i| \Pr(Y^n = y_i^n) \log \Pr(Y^n = y_i^n)$$

where y_i^n is a fixed element of $\mathcal{Y}_i$.

Proof. y^n and y'^n are said to be equivalent if there exists a transformation $T \in T$ such that $y'^n = Ty^n$. Since T is a group, $TT' \in T$ if $T, T' \in T$. Hence, the above definition of equivalence of y^n and y'^n is an equivalent relation. Then $\mathcal{Y}^n$ can be partitioned to equivalent classes $\mathcal{Y}_i, i = 1, \cdots, M$. From Lemma 3, $\Pr(Y^n = y^n)$=constant for all $y^n \in \mathcal{Y}_i (1 \leq i \leq M)$. Therefore,

$$\begin{aligned}
H(Y^n) &= -\sum_{y^n} \Pr(Y^n = y^n) \log \Pr(Y^n = y^n) \\
&= -\sum_{i=1}^{M} \sum_{y_i^n \in \mathcal{Y}_i} \Pr(Y^n = y_i^n) \log \Pr(Y^n = y_i^n) \\
&= -\sum_{i=1}^{M} |\mathcal{Y}_i| \Pr(Y^n = y_i^n) \log \Pr(Y^n = y_i^n)
\end{aligned}$$

Thus, Theorem 1 is proved by Lemma 2 and $H(X^n) = nH(X)$.

We apply Theorem 1 to compute the key equivocation rate of the secrecy system (f, C) for two specific cases.

Case 1) The secrecy system (f, C) is defined as in section 1. $\mathcal{X} = \mathcal{Y} = \mathcal{K} = \{a_1, a_2, \cdots, a_r\}$. $C = T_P^n$ (set of all sequences of type P in $\mathcal{K}^n$) or $C = \bigcup_P T_P^n$ (union of T_P^n with different types of P), where P is a probability distribution on $\mathcal{K}$. Let $\tau : (1, 2, \cdots, n) \to (\tau(1), \tau(2), \cdots, \tau(n))$ be a permutation and SYM be the symmetric group of all permutations. For every $\tau \in$SYM, we define a transformation $T_\tau : \mathcal{K}^n \to \mathcal{K}^n$. by $T_\tau k^n = T_\tau(k_1, k_2, \cdots, k_n) = (k_{\tau(1)}, k_{\tau(2)}, \cdots, k_{\tau(n)})$.

It is easy to verify that $T = \{T_\tau; \tau \in$SYM $\}$ is a transformation group satisfying the condition of Theorem 1 (i.e. for every $T_\tau \in T$, (f, C) satisfies the conditions 3) and 1) of Lemma 3). Then we may use Theorem 1 to compute the key equivocation of the secrecy system (f, C). The equivalent classes $\mathcal{Y}_i, i = 1, 2, \cdots, M \leq (n + 1)^{r-1}$ are all different types of sequences in $\mathcal{Y}^n$. we have the following result.

Theorem 2. *For the secrecy system (f, C) given in the case 1), if $P = (P_1, \cdots, P_r)$ is a probability distribution on $\mathcal{K}$ and $C = T_P^n$, satisfying $\lim_{n \to \infty} P_s = \lim_{n \to \infty} \frac{N(a_s)}{n} = \Pi_s, 0 < \Pi_s < 1$ for $1 \leq s \leq r$.*
then

$$\lim_{n \to \infty} \frac{1}{n} H(K^n | Y^n) = H(X) + H(\Pi) - H(Q)$$

*where $N(a_s)$ is the number of a_s occurred in the sequence $k^n \in T_P^n$, $\Pi = (\Pi_1, \cdots, \Pi_r)$
and $Q = (Q_1, \cdots, Q_r)$ are probability distributions on $\mathcal{K}$ and $\mathcal{Y}$ respectively, with*

$$Q_t = \sum_{s=1}^{r} \Pi_s W_{a_t | a_s}, \quad 1 \le t \le r \tag{1}$$

$W = (W_{a_t | a_s})_{1 \le s, t \le r}$ *is the memoryless channel induced by the secrecy system*
$(f, \mathcal{C})$.

Proof. Since $\frac{1}{n} \log |T_P^n| = H(P) + o(1)$, by Theorem 1 and Lemma 2, it suffices
to prove

$$\lim_{n \to \infty} \frac{1}{n} H(Y^n) = H(Q)$$

Let $P(i), i = 1, 2, \cdots, M$ be all different types of sequences in $\mathcal{Y}^n$, and $y_i^n \in T_{P(i)}^n$.
Then, we have

$$\begin{aligned}
\frac{1}{n} H(Y^n) &= -\frac{1}{n} \sum_{i=1}^{M} |T_{P(i)}^n| \Pr(Y^n = y_i^n) \log \Pr(Y^n = y_i^n) \\
&= -\frac{1}{n} \sum_{i=1}^{M} \Pr(Y^n \in T_{P(i)}^n) \log \Pr(Y^n \in T_{P(i)}^n) \\
&\quad + \frac{1}{n} \sum_{i=1}^{M} \Pr(Y^n \in T_{P(i)}^n) \log |T_{P(i)}^n| \\
&= \sum_{i=1}^{M} \Pr(Y^n \in T_{P(i)}^n) H(P(i)) + o(1)
\end{aligned} \tag{2}$$

We intend to use Lemma 2.12 in the book of Csiśar and Körner[4]. The Lemma
is stated in our notation as follows.

Lemma 4 CK. *There exists a sequences $\varepsilon_n \to 0$ depending only on $|\mathcal{K}|$ and $|\mathcal{Y}|$
so that for every stochastic matrix $W : \mathcal{K} \to \mathcal{Y}$,*

$$W^n(T_{[W]_{\delta_n}}^n(k^n) | k^n) \ge 1 - \varepsilon_n \quad \text{for every } k^n \in \mathcal{K}^n,$$

*where $\delta_n \to 0$ and $\delta_n \sqrt{n} \to \infty$, as $n \to \infty$, $T_{[W]_\delta}^n(k^n)$ is the union of the sets
$T_V^n(k^n)$ for those conditional types V given k^n of sequences in $\mathcal{Y}^n$ which satisfy
$|V_{a_t | a_s} - W_{a_t | a_s}| \le \delta$ for every $a_s \in \mathcal{K}$ and $a_t \in \mathcal{Y}$, and $V_{a_t | a_s} = 0$ when
$W_{a_t | a_s} = 0$.*

*(A sequence $y^n \in \mathcal{Y}^n$ is said to have conditional type V given k^n, if $N(a_s, a_t) =
N(a_s) V_{a_t | a_s}$ for every $a_s \in \mathcal{K}$ and $a_t \in \mathcal{Y}$, where $N(a_s, a_t)$ is the number of
(a_s, a_t) occurred in the sequence (k^n, y^n), V is a stochastic matrix $V : \mathcal{K} \to \mathcal{Y}$.
The set of sequences $y^n \in \mathcal{Y}^n$ having conditional type V given k^n is denoted by
$T_V^n(k^n)$.)*

Now we use Lemma CK for the transmission probability matrix $W = (W_{y|k}, k \in
\mathcal{K}, y \in \mathcal{Y})$ and $k^n \in T_P^n$, We obtain

$$W^n\Big(\bigcup_V T_V^n(k^n) | k^n\Big) \ge 1 - \varepsilon_n \quad \text{for every } k^n \in T_P^n \tag{3}$$

where the union of the sets $T_V^n(k^n)$ is for those conditional types V given k^n of sequences in $\mathcal{Y}^n$ which satisfy $|V_{a_t|a_s} - W_{a_t|a_s}| \leq \delta_n$ for every $a_s \in \mathcal{K}$ and $a_t \in \mathcal{Y}$ and $V_{a_t|a_s} = 0$ whenever $W_{a_t|a_s} = 0$; δ_n and ε_n are given in Lemma CK.

Clearly, for any fixed conditional type V and $k^n \in T_P^n, T_V^n(k^n) \subset T_{Q(V)}^n$, where

$$Q(V) = (Q(V)_1, \cdots, Q(V)_r)$$

is a probability distribution on $\mathcal{Y}$ with $Q(V)_t = \sum_{s=1}^{r} P_s V_{a_t|a_s}$, $1 \leq t \leq r$. Let $Q(W) = (Q(W)_1, \cdots, Q(W)_r)$ be the probability distribution on $\mathcal{Y}$ defined by

$$Q(W)_t = \sum_{s=1}^{r} P_s W_{a_t|a_s}, \quad 1 \leq t \leq r.$$

Then from (3), we obtain

$$W^n(\bigcup_V T_{Q(V)}^n | k^n) \geq 1 - \varepsilon_n \quad \text{for every } k^n \in T_P^n, \tag{4}$$

where the union of the sets $T_{Q(V)}^n$ is for those conditional types V given k^n of sequences in $\mathcal{Y}^n$ which satisfy

$$\max_{1 \leq t \leq r} |Q(V)_t - Q(W)_t| \leq \delta_n.$$

Since for any two conditional types V given k^n and V' given k'^n, either $T_{Q(V)}^n = T_{Q(V')}^n$ or $T_{Q(V)}^n \cap T_{Q(V')}^n = \emptyset$ (empty set), hence only different $T_{Q(V)}^n$ need be contained in the union $\bigcup_V T_{Q(V)}^n$. Then from (4), we obtain

$$\sum_{i \in S_1} W^n(T_{P(i)}^n | k^n) \geq 1 - \varepsilon_n \text{ for every } k^n \in T_P^n \tag{5}$$

where

$$S_1 = \{i; \max_{1 \leq t \leq r} |P(i)_t - Q(W)_t| \leq \delta_n\}.$$

By the definition of Y^n and (5) we have

$$\sum_{i \in S_1} \Pr(Y^n \in T_{P(i)}^n) = \sum_{i \in S_1} \sum_{k^n \in T_P^n} |T_P^n|^{-1} W^n(T_{P(i)}^n | k^n)$$

$$= \sum_{k^n \in T_P^n} |T_P^n|^{-1} \sum_{i \in S_1} (T_{P(i)}^n | k^n) \geq 1 - \varepsilon_n.$$

Denote $S_2 = \{1, 2, \cdots, M\} - S_1$, then $\sum_{i \in S_2} \Pr(Y^n \in T_{P(i)}^n) < \varepsilon_n$. Since $P \to \Pi$, as $n \to \infty$, so $Q(W) \to Q$ as $n \to \infty$, further, by $\delta_n \to 0$, we obtain $H(P(i)) = H(Q) + o(1)$, for $i \in S_1$. On the other hand, by $\varepsilon_n \to 0$, $\sum_{i \in S_2} \Pr(Y^n \in P(i)) = o(1)$, Combining all results given above, we conclude from (2) that

$$\lim_{n \to \infty} \frac{1}{n} H(Y^n) = \lim_{n \to \infty} \sum_{i \in S_1} \Pr(Y^n \in T_{P(i)}^n) H(P(i) + o(1)$$
$$= H(Q).$$

Theorem 2 is proved.

For the binary additive case, i.e., $\mathcal{K} = \mathcal{X} = \mathcal{Y} = GF(2)$ (binary field), $f(x,k) = x + k, \mathcal{C} = T_P^n$, where $P = (P_0, P_1) = (n_0/n, n_1/n)$ is the frequency of the elements 0 and 1 occurred in the sequence $k^n = (k_1, k_2, \cdots, k_n)$. In this case, the channel $W = (W_{y|k}, k \in \mathcal{K}, y \in \mathcal{Y})$ defined in section 1 is a binary symmetric channel with crossover probability $\Pr(X = 1) = \varepsilon$.

Corollary 1. *For the binary additive secrecy system $(f, \mathcal{C})$ given above, if $\lim_{n\to\infty} n_1/n = r_1 (0 < r_1 < 1)$, then*

$$\lim_{n \to \infty} (1/n) H(K^n | Y^n) = h(\varepsilon) + h(r_1) - h(r_1(1 - \varepsilon) + \varepsilon(1 - r_1))$$

where $h(\varepsilon) = -\varepsilon \log \varepsilon - (1 - \varepsilon) \log(1 - \varepsilon)$ is the binary entropy function.

Remark. Though Corollary 1 is a special case of Theorem 2, we can prove it by a more elementary method. The key ideas are as follows. Similar to (2), $H(Y^n)/n$ can be expressed as $\frac{1}{n} H(Y^n) = \int_0^1 h(u) dF_n(u) + o(1)$, where $F_n(u)$ is the distribution function of $\frac{1}{n} W_H(Y^n)$, $W_H(Y^n)$ is the hamming weight of Y^n. Through simple calculation, we obtain

$$F_n(u) = \Pr((W_H(Y^n))/n) \le u) = \sum_{i; i \le nu} \Pr(Y^n \in T_{P(i)}^n)$$
$$= \sum_{i; i \le nu} \sum_{j=\max(0, i-n_0)}^{\min(i, n_1)} b(j, n_1, 1 - \varepsilon) b(i - j, n_0, \varepsilon)$$
$$= \sum_{m=0}^{\min(\lfloor nu \rfloor, n_0)} \sum_{j=0}^{\min(\lfloor nu \rfloor - m, n_1)} b(j, n_1, 1 - \varepsilon) b(m, n_0, \varepsilon)$$

where

$$b(j, n_1, (1 - \varepsilon)) = \binom{n_1}{j}(1 - \varepsilon)^j \varepsilon^{n_1 - j}$$

is the probability of the binomial distribution.

Using De Moivre-Laplace limit theorem, we obtain

$$\lim_{n \to \infty} F_n(u) = \begin{matrix} 0 & u < r_1(1 - \varepsilon) + (1 - r_1)\varepsilon \\ 1 & u > r_1(1 - \varepsilon) + (1 - r_1)\varepsilon \end{matrix}$$

Then, we have $\lim_{n\to\infty}(1/n)H(Y^n) = \lim_{n\to\infty} \int_0^1 h(u) dF_n(u) + o(1) = h(r_1(1 - \varepsilon) + (1 - r_1)\varepsilon)$.

Case 2) $\mathcal{K} = GF(q)$(finite field with q elements), $f(x,k) = x + k.\mathcal{C}$ is an m-dimensional linear subspace of $GF(q)^n$((n, m) linear code). For any $\alpha \in GF(q)^n$, we define a transformation $T_\alpha : \mathcal{K}^n \to \mathcal{K}^n$ by $T_\alpha k^n = k^n + \alpha$. It is easy to verify that $\mathcal{T} = \{T_\alpha; \alpha \in \mathcal{C}\}$ is a transformation group satisfying the condition of Theorem 1 (i.e., for every $T \in \mathcal{T}, (f, \mathcal{C})$ satisfies the conditions 2) and 1) of Lemma 3). Then, we may use Theorem 1 to compute the key equivocation of the secrecy system $(f, \mathcal{C})$. The equivalent classes $\mathcal{Y}_i, i = 1, 2, \cdots, M = q^{n-m}$, are all cosets of the linear code $\mathcal{C}$.

Denote $GF(q) = \{a_1, a_2, \cdots, a_q\}, P = (P_1, P_2, \cdots, P_q)$ is a probability distribution on $GF(q)$ with $P_t = \Pr(X = a_t), 1 \le t \le q$. Let $\mathcal{C} = \{k_1^n, k_2^n, \cdots, k_N^n\}$ $(N = q^m)$ be a linear (n, m) code, $\mathcal{Y}_i, i = 1, 2, \cdots, M = q^{n-m}$ be all cosets of $\mathcal{C}$, where $\mathcal{Y}_i = \mathcal{C} + y_i^n, y_i^n$ is the coset leader of $\mathcal{Y}_i (1 \le i \le M)$. Then

$H(X) = H(P), \frac{1}{n}\log|\mathcal{C}| = \frac{1}{n}\log q^m = \frac{m}{n}\log q$. Since when k^n runs through the code $\mathcal{C}, x^n = k^n + y_i^n$ runs through the coset $\mathcal{Y}_i$. Thus

$$\Pr(Y^n = y_i^n) = \sum_{k^n \in \mathcal{C}} q^{-m} W_{y_i^n|k^n}^n = \sum_{x^n \in \mathcal{Y}_i} q^{-m}\Pr(X^n = x^n)$$
$$= \sum_{j=1}^J q^{-m} T_{ij}\prod_{t=1}^q P_t^{nQ(j)_t}$$

where $Q(j) = (Q(j)_1, \cdots, Q(j)_q), j = 1, \cdots, J$ are all different types of sequences in $GF(q)^n, T_{ij}, j = 1, \cdots, J$ is the type distribution of the coset $\mathcal{Y}_i$, i.e., T_{ij} is the number of x^n in $\mathcal{Y}_i$ which are belong to $T_{Q(j)}^n$. Analogous to (2), we have

$$(1/n)H(Y^n) = -(1/n)\sum_{i=1}^M \Pr(Y^n \in \mathcal{Y}_i)\log\Pr(Y^n \in \mathcal{Y}_i) + (m/n)\log q.$$

Then, we obtain the following result.

Theorem 3. *For the secrecy system $(f, \mathcal{C})$ given in case 2), the key equivocation rate*

$$(1/n)H(K^n|Y^n) = H(P) + (1/n)\sum_{i=1}^M \Pr(Y^n \in \mathcal{Y}_i)\log\Pr(Y^n \in \mathcal{Y}_i)$$

where $\Pr(Y^n \in \mathcal{Y}_i) = \sum_{j=1}^J T_{ij}\prod_{t=1}^q P_t^{nQ(j)_t}$

An observation from Theorem 3 is that if the equivalent class entropy rate is smaller, then the key equivocation rate $(1/n)H(K^n|Y^n)$ is greater. We can improve the secrecy system $(f, \mathcal{C})$ by choosing $\mathcal{C}$ to be a linear cyclic (n, m) code. Let τ_j be a cyclic permutation, i.e. $\tau_j(1, \cdots, n) = (j, \cdots, n, 1, \cdots, j-1)(1 \le j \le n)$. Define T_{τ_j} by $T_{\tau_j}k^n = T_{\tau_j}(k_1, \cdots, k_n) = (k_j, \cdots, k_n, k_1, \cdots, k_{j-1})$. It is easy to check that $\mathcal{T} = \{T_\alpha, T_{\tau_j}, T_\alpha T_{\tau_j}; \alpha \in \mathcal{C}, 1 \le j \le n\}$ is a transformation group (notice that $T_{\tau_j}T_\alpha k^n = T_{\tau_j}(k^n + \alpha) = T_{\tau_j}k^n + T_{\tau_j}\alpha = T_\beta T_{\tau_j}k^n$, where $\beta = T_{\tau_j}\alpha$). Since T_α satisfies the conditions 2) and 1) of Lemma 3 and T_{τ_j} satisfies the conditions 3) and 1) of Lemma 3, hence Theorem 1 is also valid for the $(f, \mathcal{C})$ and $\mathcal{T}$ just given. The equivalent classes are union of some cosets of $\mathcal{C}$, i.e. $\bigcup_{j=1}^n (\mathcal{C} + T_{\tau_j}y_i^n)$.

Acknowledgement

This research work is supported by the ZiF research group "General Theory of Information Transfer and Combinatorics". I would like to express my gratitude to Professor R. Ahlswede for inviting me to participate in the ZiF research group and giving me his valuable ideas and suggestions for promoting my research work. I would like to thank Dr. Ning Cai also, by his ideas I have improved my original result in the binary case to the result for the general case in this paper.

References

1. R. Ahlswede and G. Dueck, Bad codes are good ciphers, Prob. Cont. Info. Theory 11, 337–351, 1982.
2. R.J. Blom, Bounds on key equivocation for simple substitution ciphers, IEEE Trans. Inform., Vol. 25, 8–18, 1979.
3. R.J. Blom, An upper bound on the key equivocation for pure ciphers, IEEE Trans. Inform., Vol. 30, 82–84, 1984.
4. I. Csiszár I. and J. Körner, Information Theory: Coding Theorem for Discrete Memoryless Systems, New York, Academic, 1981.

Secrecy Systems for Identification Via Channels with Additive-Like Instantaneous Block Encipherer

R. Ahlswede, N. Cai, and Z. Zhang*

Abstract. In this paper we propose a model of secrecy systems for identification via channels with ALIB encipherers and find the smallest asymptotic key rate of the ALIB encipherers needed for the requirement of security.

1 Introduction

Attention: This is the only paper in the collection which works with the **optimistic capacity**, which is the optimal rate achivable with arbitrary small error probability again and again as the blocklength goes to infinity.

The criticism of this concept made in [B34] has been supplemented by a new aspect:

in cryptology enemies strongest time in wire-taping must be taken into consideration!

The model of identification via channels was introduced by R. Ahlswede and G. Dueck [1] based on the following cases. The receivers of channels only are interested in whether a specified message was sent but not in which message was sent and the senders do not know in which message the receivers are interested. Sometimes the sender requires that the message sent can be identified only by legitimate receivers of the channel but not by any one else (e.g. wiretapper). For example, a company produces N kinds of products which are labelled by $j = 1, 2, \cdots, N$. The company wants to sell a kind of products only to the members of the company's association. For other customers it even does not want them to know what it is going to sell. In this case the company can use a secrecy system for identification via channels with additive-like instantaneous block (ALIB) encipherers, i.e. the sender encrypts the message (identification code) with a private key sending it via the channel and sends the same key only to the members of the company's association through a secure channel. The secrecy system with ALIB encipherers was investigated by R. Ahlswede and G. Dueck [2], but their model needs to be adapted to satisfy the requirement of identification via channels. In this paper we consider the model of secrecy systems for identification via channels with ALIB encipherers and investigate the

---

* Zhang's research was supported by the ZiF research group "General Theory of Information Transfer and Combinatorics".

R. Ahlswede et al. (Eds.): Information Transfer and Combinatorics, LNCS 4123, pp. 285–292, 2006.
© Springer-Verlag Berlin Heidelberg 2006

smallest asymptotic key rate of the ALIB encipherers needed for the requirement of security.

In Section 2, we review the necessary background of identification via channels. Our model is described in Section 3. Our result for symmetric channels is proved in Section 4.

2   Background

Let $\mathcal{X}, \mathcal{K}, \mathcal{Y}, \mathcal{Z}$ be finite sets. For simplicity, we assume that $\mathcal{X} = \mathcal{K} = \mathcal{Y} = \mathcal{Z} = GF(q)(q \geq 2)$. Let $W = \{W^n\}_{n=1}^{\infty}$ be a memoryless channel with transmission matrix $(w(z|x); x \in \mathcal{X}, z \in \mathcal{Z})$.

Definition 1. *A randomized* $(n, N_n, \mu_n, \lambda_n)$ *identification (Id) code for the channel* W^n *is a system* $\{(Q_i, D_i); 1 \leq i \leq N_n\}$, *where* Q_i *is a probability distribution (PD) of the random codeword* $X^n(i)$ *generated by a randomized encoder* $\varphi_n(i)$, *i.e.* $Q_i(x^n) = \Pr\{X^n(i) = x^n\}$, $x^n \in \mathcal{X}^n$, $D_i \subset \mathcal{Z}^n$ *is a decoding set.*

Denote by $Z^n(i)$ *the output of* W^n *when the input is* $X^n(i)$ *and* $Q_i W^n$ *the PD of* $Z^n(i)$. *Set* $\mu_n^{(i)} = Q_i W^n(D_i^c) = \Pr\{Z^n(i) \in \mathcal{Z}^n - D_i\}$ *and* $\lambda_n^{(j,i)} = Q_j W^n(D_i) = \Pr\{Z^n(j) \in D_i\}(j \neq i)$. $\mu_n = \max_{1 \leq i \leq N_n} \mu_n^{(i)}$ *and* $\lambda_n = \max_{1 \leq j, i \leq N_n, j \neq i} \lambda_n^{(j,i)}$ *are called the error probability of the first and second kind for the Id code, respectively,* $\frac{1}{n} \log \log N_n = r_n$ *is called the rate of the Id code.*

Definition 2. *Rate* R *is* (μ, λ)*–achievable if there exists a sequence of* $(n, N_n, \mu_n, \lambda_n)$ *Id codes for the channel* W^n $(1 \leq n < \infty)$ *satisfying the following conditions.*

1) $\overline{\lim_{n \to \infty}} \mu_n \leq \mu$, *2)* $\overline{\lim_{n \to \infty}} \lambda_n \leq \lambda$, *3)* $\liminf_{n \to \infty} r_n \geq R$.
The (μ, λ)*–Id capacity for the channel* W *is defined by* $D(\mu, \lambda|W) = \sup(R|R$ *is* (μ, λ)*–achievable).*

Theorem 1. *([1]) Let* $W = \{W^n\}_{n=1}^{\infty}$ *be an arbitrary channel. If there exists a number* ε *satisfying* $0 \leq \varepsilon \leq \mu$ *and* $0 \leq \varepsilon \leq \lambda$, *then it holds that* $D(\mu, \lambda|W) \geq C(\varepsilon|W)$, *where* $C(\varepsilon|W)$ *denotes the* ε*–channel capacity of the channel* W *which is defined as follows.*

Definition 3. *Rate* R *is* ε*–achievable if there exists a sequence of* (n, M_n, ε_n) *codes for the channel* $W^n (1 \leq n \leq \infty)$ *satisfying the following conditions.*
1) $\overline{\lim_{n \to \infty}} \varepsilon_n \leq \varepsilon$, *2)* $\liminf_{n \to \infty} \frac{1}{n} \log M_n \geq R$.
The ε*–channel capacity for the channel* W *is defined by*

$$C(\varepsilon|W) = \sup(R|R \text{ is } \varepsilon\text{–achievable}).$$

Theorem 1 is proved by using the following lemma.

Lemma 1. *([1]) Let* $\mathcal{M}$ *be an arbitrary finite set of size* $M = |\mathcal{M}|$. *Choose constants* τ *and* κ *satisfying* $0 < \tau \leq \frac{1}{3}$ *and* $0 < \kappa < 1$ *and* $\kappa \log(\frac{1}{\tau} - 1) \geq \log 2 + 1$, *where the natural logarithms are used. Define* $N = \lfloor e^{\tau M}/Me \rfloor$. *Then, there exist* N *subsets* $A_1, A_2, \cdots, A_N$ *of* $\mathcal{M}$ *satisfying* $|A_i| = \lfloor \tau M \rfloor$ $(1 \leq i \leq N)$ *and* $|A_i \cap A_j| < \kappa \lfloor \tau M \rfloor (i \neq j)$.

Using Lemma 1 the ID-code for proving Theorem 1 can be constructed as follows.

Let $\gamma > 0$ be an arbitrarily small constant and set $R = C(\varepsilon|W) - \gamma$. By Definition 3 R is ε–achievable as a rate of the transmission code. Therefore, there exists a sequence of (n, M_n, ε_n) codes for the channel $W^n (1 \leq n < \infty)$ satisfying the following conditions:

1) $\varlimsup_{n\to\infty} \varepsilon_n \leq \varepsilon$, 2) $\liminf_{n\to} \frac{1}{n} \log M_n \geq R$, where ε_n denotes the maximum decoding error probability of the code. Denote the (n, M_n, ε_n) code by $\mathcal{C}_n = \{c_1, c_2, \cdots, c_{M_n}\}$
($c_i \in \mathcal{X}^n$) and let E_i be the decoding region corresponding to $c_i (1 \leq i \leq M_n)$. Now we apply Lemma 1 by setting $\mathcal{M} = \{1, 2, \cdots, M_n\}$, $M = M_n$, $\tau = \tau_n = \frac{1}{(n+3)}$, $\kappa = \kappa_n = \frac{2}{\log(n+2)}$ and $N = N_n = \lfloor e^{\tau_n M_n}/M_n e \rfloor$. Since all conditions of Lemma 1 are satisfied, there exist N_n subsets $A_1, A_2, \cdots, A_{N_n}$ of $\mathcal{M}$ satisfying $|A_j| = \lfloor \tau_n M_n \rfloor (1 \leq j \leq N_n)$ and $| A_j \cap A_k | < \kappa_n \lfloor \tau_n M_n \rfloor (j \neq k)$. Define the subsets S_j $(1 \leq j \leq N_n)$ of $\mathcal{C}_n$ by $S_j = \bigcup_{i \in A_j} \{c_i\}$ and let Q_j denote the uniform distribution over S_j. Define $D_j = \bigcup_{i \in A_j} E_i$ as the decoding set corresponding to Q_j. It is shown that the constructed Id code $\{(Q_j, D_j); 1 \leq j \leq N_n\}$ can be used to prove Theorem 1.

Theorem 1 gives the direct theorem on the Id coding problem. We need the converse theorem also. Since the converse theorem is essentially related to the channel resolvability problem, we can introduce the channel resolvability instead.

Let $W = \{W^n\}_{n=1}^{\infty}$ be an arbitrary channel with input and output alphabets $\mathcal{X}$ and $\mathcal{Y}$ respectively. Let $Y = \{Y^n\}_{n=1}^{\infty}$ be the output from the channel W corresponding to a given input $X = \{X^n\}_{n=1}^{\infty}$. We transform the uniform random number U_{M_n} of size M_n into another input $\widetilde{X} = \{\widetilde{X}^n\}_{n=1}^{\infty}$. That is, $\widetilde{X}^n = f_n(U_{M_n})$, $f_n : \{1, 2, \cdots, M_n\} \to \mathcal{X}^n$.

Denote by $\widetilde{Y} = \{\widetilde{Y}^n\}_{n=1}^{\infty}$ the output from the channel W with an input $\widetilde{X}$. The problem of how we can choose the size M_n of the uniform random number U_{M_n} and the transform f_n such that the variational distance between $Y = \{Y^n\}_{n=1}^{\infty}$ and $\widetilde{Y} = \{\widetilde{Y}^n\}_{n=1}^{\infty}$ satisfies $\lim_{n\to\infty} d(Y^n, \widetilde{Y}^n) = 0$ is sometimes called the channel resolvability problem. In this problem, the criterion of approximation can be slightly generalized to $\varlimsup_{n\to\infty} d(Y^n, \widetilde{Y}^n) \leq \delta$, where δ is an arbitrary constant satisfying $0 \leq \delta < 2$.

Definition 4. *Rate R is δ–achievable for an input $X = \{X^n\}_{n=1}^{\infty}$ if there exists a sequence of transforms $\widetilde{X}^n = f_n(U_{M_n})(1 \leq n < \infty)$ satisfying*

$$\varlimsup_{n\to\infty} d(Y^n, \widetilde{Y}^n) \leq \delta \quad and \quad \varlimsup_{n\to\infty} \frac{1}{n} \log M_n \leq R,$$

where Y^n and $\widetilde{Y}^n$ denote the channel outputs corresponding to X^n and $\widetilde{X}^n$, respectively. The channel δ–resolvability for an input X is defined by

$$S_X(\delta|W) = \inf(R|R \ is \ \delta\text{–achievable for an input } X).$$

Theorem 2. *([3]) Let W be an arbitrary channel with time structure and X an arbitrary input variable. Then, it holds that $S_X(\delta|W) \leq \bar{I}(X;Y)$ for all $\delta \geq 0$, where Y denotes the channel output variable corresponding to X and $\bar{I}(X;Y)$ represents the sup-mutual information rate defined by*

$$\bar{I}(X;Y) = p - \varlimsup_{n \to \infty} \frac{1}{n} \log \frac{W^n(Y^n|X^n)}{P_{Y^n}(Y^n)}$$

$$= \inf \left(\alpha \Big| \lim_{n \to \infty} \Pr_{X^n Y^n} \left\{ \frac{1}{n} \log \frac{W^n(Y^n|X^n)}{P_{Y^n}(Y^n)} > \alpha \right\} = 0 \right). \tag{1}$$

3 Model

In this section we propose a model of the secrecy systems for identification via channels with ALIB encipherers. We keep the notations and assumptions given in Section 2 for reviewing the background of identification via channels.

Let $\{(Q_i, D_i) : 1 \leq i \leq N_n\}$ be the $(n, N_n, \mu_n, \lambda_n)$ Id code constructed as in the proof of Theorem 1 for the channel W. Recall that an (n, R) ALIB encipherer is a subset $C \subset \mathcal{K}^n$ with $|C| < e^{nR}$. Let $f : \mathcal{X} \times \mathcal{K} \to \mathcal{Y}$ be a function, where $f(x, \cdot)$ is bijective for each $x \in \mathcal{X}$ and $f(\cdot, k)$ is bijective for each $k \in \mathcal{K}$. $f^n : \mathcal{X}^n \times \mathcal{K}^n \to \mathcal{Y}^n$ denotes the n–fold product of f. Given a pair (f, C) we define a secrecy system which works as follows. If the sender wants to send a message $i(1 \leq i \leq N_n)$, he sends the random codeword $X^n(i)$ generated by the randomized encoder $\varphi_n(i)$. Before he transmits $X^n(i)$ he uses a random key generator K^n to generate k^n according to the uniform distribution on C. Then the sender encrypts $X^n(i)$ into the random cryptogram $Y^n(i) = f^n(X^n(i), K^n)$ and sends it to the receiver over the channel W^n. Suppose that $X^n(i)$ and K^n are mutually independent. The used key k^n is sent to the receiver over a secure channel. Denote by $\widetilde{Z}^n(i)$ the output of the channel W^n when the input is the cryptogram $Y^n(i)$. In general, the receiver cannot use the same key k^n to recover the received codeword $Z^n(i)$ from the received cryptogram $\widetilde{Z}^n(i)$ since the channel W^n is noisy. In order to solve this problem, we assume that $f(x, k) = x + k$, where $+$ operates in $GF(q)$. Then we have $Y^n(i) = X^n(i) + K^n$. Further, we need to assume that the channel W^n is memoryless with **symmetric transmission matrix**, more specifically, the output and input of the channel W^n have the following relation: $\widetilde{Z}^n(i) = Y^n(i) + E^n$, where $E^n = (E_1, E_2, \cdots, E_n)$ is a sequence of independent random variables with the same PD on $GF(q)$. Combining the two assumptions, we obtain $\widetilde{Z}^n(i) = X^n(i) + K^n + E^n = Z^n(i) + K^n$ or $Z^n(i) = \widetilde{Z}^n(i) - K^n$. Hence the receiver can get $Z^n(i)$ from $\widetilde{Z}^n(i)$ by using the same key k^n and decides that the message $i(1 \leq i \leq N_n)$ is sent if $Z^n(i) \in D_i$. Since the PD of $Z^n(i)$ is $Q_i W^n$ and $Q_i W^n(D_i^c) \leq \mu_n$, $Q_j W^n(D_i) \leq \lambda_n(j \neq i)$, the receiver can identify the message i with error probabilities of the first kind and second kind not greater than μ_n and λ_n, respectively. Another customer intercepts the channel output $\widetilde{Z}^n(i)$ and attempts to identify a message $j(1 \leq j \leq N_n)$ being sent. Since the customer does not know the actual key k^n being used, he has to use $\widetilde{Z}^n(i)$ and his knowledge of the system for deciding that the message j is sent. We need a security condition under which the customer can not decide for any fixed message $j(1 \leq j \leq N_n)$ being sent with small error probability. Such a condition was given by R. Ahlswede and Z. Zhang [4] for investigating the problem of identification via a wiretap channel. This condition is also suitable for our model. The condition is stated as follows.

Security Condition. For any pair of messages $(i,j)(1 \leq i \neq j \leq N_n)$ and $D \subset \mathcal{Z}^n$, it holds that $\widetilde{Q}_i W^n(D^c) + \widetilde{Q}_j W^n(D) > 1 - \delta_n$ and $\lim_{n \to \infty} \delta_n = 0$, where $\widetilde{Q}_i$ and $\widetilde{Q}_i W^n$ denote the PD of $Y^n(i)$ and $\widetilde{Z}^n(i)$ respectively.

From the identity $\widetilde{Q}_i W^n(D^c) + \widetilde{Q}_i W^n(D) = 1$ for any $i(1 \leq i \leq N_n)$ and any $D \subset \mathcal{Z}$ and the Security Condition, we obtain $\widetilde{Q}_j W^n(D) > 1 - \widetilde{Q}_i W^n(D^c) - \delta_n = \widetilde{Q}_i W^n(D) - \delta_n$ for any pair $(i,j)(1 \leq i \neq j \leq N_n)$. Therefore, the Security Condition means that $\widetilde{Q}_i W^n$ and $\widetilde{Q}_j W^n$ are almost the same for any pair (i,j) with $i \neq j$. Hence the customer can not decide on any fixed message $j(1 \leq j \leq N_n)$ being sent with small error probability.

We are interested in the following problem. What is the largest rate R of the ALIB encipherer C so that the distributions $\widetilde{Q}_i W^n(i = 1, 2, \cdots, N_n)$ satisfy the Security Condition.

4   Main Result

For the model of a secrecy system described in Section 3 we obtain the following main result.

Theorem 3. *1) Assume for the alphabets $\mathcal{X} = \mathcal{K} = \mathcal{Y} = \mathcal{Z} = GF(q)(q \geq 2)$ and that $W = \{W^n\}_{n=1}^{\infty}$ is a memoryless symmetric channel with the transmission matrix $(w(z|x) > 0; x \in \mathcal{X}, z \in \mathcal{Z})$.*
2) Assume that the function $f(x,k) = x + k$, where $+$ operates in the finite field $GF(q)$.
3) Suppose that the random key K^n has uniform distribution on the ALIB encipherer $C \subset \mathcal{K}^n$ and is mutually independent with each random codeword $X^n(i)(1 \leq i \leq N_n)$.
Then, the secrecy system for identification via the channel W with ALIB encipherers possesses the following properties.
1) The secrecy system can transmit N_n messages $i = 1, 2, \cdots, N_n$ with

$$\liminf_{n \to \infty} \frac{1}{n} \log \log N_n \geq \log q + \sum_{z \in \mathcal{Z}} w(z|x) \log w(z|x) - \gamma,$$

where $\gamma > 0$ is an arbitrarily small number and $x \in \mathcal{X}$ is fixed, the legitimate receiver can identify the message $i(1 \leq i \leq N_n)$ with arbitrarily small error probability.
2) The smallest asymptotic key rate R of the ALIB encipherer C is $R = -\sum_{z \in \mathcal{Z}} w(z|x) \log w(z|x)$ ($x \in \mathcal{X}$ is fixed) for the distributions $\widetilde{Q}_i W^n(i=1, 2, \cdots, N_n)$ satisfying the Security Condition. Hence, the other customer can not judge any fixed message $j(1 \leq j \leq N_n)$ being sent from $\widetilde{Q}_i W^n$ with small error probability.

Proof. 1) By assumption 1), the channel capacity of the channel W is $C(W) = C(0|W) = \log q + \sum_{z \in \mathcal{Z}} w(z|x) \log w(z|x)$. Using Theorem 1 with $\varepsilon = 0$, we obtain that the (μ, λ)–Id capacity of the channel W, $D(\mu, \lambda|W) \geq C(W)$ for $\mu \geq 0$, $\lambda \geq 0$. Hence, there exists a sequence of $(n, N_n, \mu_n, \lambda_n)$ Id codes for the channel $W^n(1 \leq n < \infty)$ satisfying the conditions: 1) $\lim_{n \to \infty} \mu_n = 0$; 2) $\lim_{n \to \infty} \lambda_n = 0$; 3) $\liminf_{n \to \infty} r_n \geq C(W) - \gamma$. Using the Id codes in the secrecy system, the property 1) holds.

2) By assumption 2), the random cryptogram $Y^n(i) = X^n(i) + K^n$, where the random key K^n has uniform distribution on an ALIB encipherer $C \subset \mathcal{K}^n$. R. Ahlswede and G. Dueck [2] have pointed out that $Y^n(i)$ and K^n can be regarded as the output and input of the channel denoted by $V = \{V^n\}_{n=1}^\infty$. In the case of identification, the channel V is a general channel rather than a memoryless channel. By assumption 3), the transmission probability of the channel V^n can be defined as $V^n_{y^n|k^n} = \sum_{x^n} Q_i(x^n)\delta(y^n, x^n + k^n)$, where

$$\delta(y^n, x^n + k^n) = \begin{cases} 1, & \text{if } y^n = x^n + k^n, \\ 0, & \text{otherwise.} \end{cases}$$

In order to prove property 2), we want to apply Theorem 2 for the general channel V. First, we consider the input U^n of the channel V^n which has uniform distribution on the ALIB encipherer $C = \mathcal{K}^n$. It is evident that the PD of the output $Y^n(i)$ corresponding to the input U^n is the uniform distribution on $\mathcal{Y}^n$, i.e. $\widetilde{Q}_i(y^n) = \Pr\{Y^n(i) = y^n\} = q^{-n}$ for any $y^n \in \mathcal{Y}$ and any $i(1 \leq i \leq N_n)$. By the assumption 1), it is also evident that the PD of the output $\widetilde{Z}^n(i)$ of the channel W^n corresponding to the input $Y^n(i)$ is the uniform distribution on $\mathcal{Z}^n$, i.e. $\widetilde{Q}_i W^n(z^n) = q^{-n}$ for any $z^n \in \mathcal{Z}^n$ and any $i(1 \leq i \leq N_n)$. Hence $\widetilde{Q}_i W^n(1 = 1, 2, \cdots, N_n)$ satisfy the Security Condition. But the key rate of $C = \mathcal{K}^n$ equals $\log q$, it can be reduced. Then, applying Theorem 2 for the input $U = \{U^n\}_{n=1}^\infty$ and $\delta = 0$, we obtain $S_U(0|V) \leq \overline{I}(U, Y(i))$, where $Y(i) = \{Y^n(i)\}_{n=1}^\infty$. We use formula (1) to compute $\overline{I}(U, Y(i))$. We have seen that $\Pr\{Y^n(i) = y^n\} = P_{Y^n(i)}(y^n) = q^{-n}$ for any $y^n \in \mathcal{Y}^n$ and $V^n_{y^n|k^n} = \sum_{x^n} Q_i(x^n)\delta(y^n, x^n + k^n) = \sum_{x^n \in S_i} |S_i|^{-1} \delta(y^n, x^n + k^n)$ for $k^n \in \mathcal{K}^n$, where $|S_i| = \tau_n M_n$, $\tau_n = \frac{1}{(n+3)}$, $\liminf_{n \to \infty} \frac{1}{n} \log M_n \geq C(W) - \gamma$. Then, the joint distribution of U^n and $Y^n(i)$

$$\Pr\{U^n = k^n, Y^n(i) = y^n\} = \begin{cases} q^{-n} |S_i|^{-1}, & \text{for } y^n \in S_i + k^n = \{x^n + k^n; x^n \in S_i\} \\ 0, & \text{otherwise.} \end{cases}$$

Hence,

$$\frac{1}{n} \log \frac{V^n(Y^n(i)|U^n)}{P_{Y^n(i)}(Y^n(i))} = \frac{1}{n} \log \frac{|S_i|^{-1}}{q^{-n}} = \log q - \frac{1}{n} \log |S_i|$$

$$= \log q - \frac{1}{n} \log M_n + \frac{1}{n} \log(n+3)$$

with probability one. Therefore, by formula (1):

$$\overline{I}(U;Y(i)) \leq \log q - C(W) + \gamma = -\sum_{z \in \mathcal{Z}} w(z|x) \log w(z|x) + \gamma.$$

Since γ is an arbitrarily small number, so $\overline{I}(U,Y(i)) = H(\{w(z|x); z \in \mathcal{Z}\})$, where $H(\cdot)$ is the entropy function. Then, we obtain $S_U(0|V) \leq H(\{w(z|x); z \in \mathcal{Z}\})$. By the definition 4, there exists a sequence of transforms $K^n = f_n(U_{M_n})(1 \leq n < \infty)$ satisfying $\lim_{n \to \infty} d(Y^n(i), \widetilde{Y}^n(i)) = 0$ and $\liminf_{n \to \infty} \frac{1}{n} \log M_n \leq H(\{w(z|x); z \in \mathcal{Z}\}) + \gamma$, where $Y^n(i)$ and $\widetilde{Y}^n(i)$ denote the outputs of channel V corresponding to the inputs U^n and K^n respectively.

In other words, there exists a sequence of (n, R) ALIB encipherers C with $R \leq H(\{w(z|x); z \in \mathcal{Z}\}) + \gamma$, such that if the random key K^n generates the key k^n according to the uniform distribution on C, then the random cryptogram $\widetilde{Y}^n(i) = X^n(i) + K^n$ satisfies $\lim_{n \to \infty} d(Y^n(i), \widetilde{Y}^n(i)) = 0$.

In the following, in order to avoid confusion, the PDs of $Y^n(i)$ and $\widetilde{Y}^n(i)$ are denoted by $Q_{Y^n(i)}$ and $\widetilde{Q}_i$, respectively, denote $\widetilde{Z}^n(i)$ the output of the channel W^n corresponding to the input $\widetilde{Y}^n(i)$. Now, we prove that the PD of $\widetilde{Z}^n(i)$, $\widetilde{Q}_i W^n(i = 1, 2, \cdots, N_n)$ satisfies the Security Condition. In fact, $Q_{Y^n(i)} W^n$ is the uniform distribution on $\mathcal{Z}^n$ and $Q_{Y^n(i)} W^n(D) + Q_{Y^n(i)} W^n(D^c) = 1$ for any $D \subset \mathcal{Z}^n$. On the other hand,

$$d(Q_{Y^n(i)} W^n, \widetilde{Q}_i W^n) = \sum_{z^n \in \mathcal{Z}^n} | Q_{Y^n(i)} W^n(z^n) - \widetilde{Q}_i W^n(z^n) |$$

$$\leq \sum_{z^n \in \mathcal{Z}^n} \sum_{y^n \in \mathcal{Y}^n} | Q_{Y^n(i)}(y^n) - \widetilde{Q}_i(y^n) | W^n_{z^n|y^n}$$

$$= d(Q_{Y^n(i)}, \widetilde{Q}_i).$$

Consequently, $\lim_{n \to \infty} d(Q_{Y^n(i)} W^n, \widetilde{Q}_i W^n) = 0$. Evidently, for any $i(1 \leq i \leq N_n)$,

$$| Q_{Y^n(i)} W^n(D^c) - \widetilde{Q}_i W^n(D^c) | \leq d(Q_{Y^n(i)} W^n, \widetilde{Q}_i W^n),$$

then,

$$\widetilde{Q}_i W^n(D^c) \geq Q_{Y^n(i)} W^n(D^c) - d(Q_{Y^n(i)} W^n, \widetilde{Q}_i W^n).$$

Similarly, for any $j(j \neq i)$,

$$Q_j W^n(D) \geq Q_{Y^n(j)} W^n(D) - d(Q_{Y^n(j)} W^n, \widetilde{Q}_j W^n).$$

Combine these two inequalities and set

$$\delta_n = 2[d(Q_{Y^n(i)} W^n, \widetilde{Q}_i W^n) + d(Q_{Y^n(j)} W^n, \widetilde{Q}_j W^n)].$$

We obtain $\widetilde{Q}_i W^n(D^c) + \widetilde{Q}_j W^n(D) > 1 - \delta_n$ and $\lim_{n \to \infty} \delta_n = 0$. Our proof is complete.

References

1. R. Ahlswede and G. Dueck, Identification via channels, IEEE Trans. Inform. Theory, Vol. 35, No. 1, 15–29, 1989.
2. R. Ahlswede and G. Dueck, Bad codes are good ciphers, Prob. Cont. Info. Theory 11, 337–351, 1982.
3. T.S. Han, Information-Spectrum Methods in Information Theory, Springer, Berlin, 2003.
4. R. Ahlswede and Z. Zhang, New directions in the theory of identification via channels, IEEE Trans. Inform. Theory, Vol. 41, No. 1, 14–25, 1995.
5. N. Cai and K.Y. Lam, On identification secret sharing schemes, Information and Computation, 184, 298–310, 2003.

Large Families of Pseudorandom Sequences of k Symbols and Their Complexity – Part I

R. Ahlswede, C. Mauduit, and A. Sárközy*

Dedicated to the memory of Levon Khachatrian

1 Introduction

In earlier papers we introduced the measures of pseudorandomness of finite binary sequences [13], introduced the notion of f–complexity of families of binary sequences, constructed large families of binary sequences with strong PR (= pseudorandom) properties [6], [12], and we showed that one of the earlier constructions can be modified to obtain families with high f–complexity [4]. In another paper [14] we extended the study of pseudorandomness from binary sequences to sequences on k symbols ("letters"). In [14] we also constructed *one* "good" pseudorandom sequence of a given length on k symbols. However, in the applications we need not only a few good sequences but large families of them, and in certain applications (cryptography) the complexity of the family of these sequences is more important than its size. In this paper our goal is to construct "many" "good" PR sequences on k symbols, to extend the notion of f–complexity to the k symbol case and to study this extended f–complexity concept.

2 A Special Case

First we will study the special case when k, the number of symbols (the "size of the alphabet") is a power of 2: $k = 2^r$. We will show that in this case any "good" PR binary sequence

$$E_N = (e_1, e_2, \ldots, e_N) \in \{-1, +1\}^N \tag{2.1}$$

defines a sequence on k symbols with "nearly as good" PR properties so that the constructions given in the binary case can be used in the $k = 2^r$ symbol case nearly as effectively.

First we have to recall several definitions from earlier papers. If E_N is a binary sequence of the form (2.1), then write

$$U(E_N; t, a, b) = \sum_{j=0}^{t-1} e_{a+jb}$$

---

* Research partially supported by the Hungarian National Foundation for Scientific Research, Grant No. T043623.

R. Ahlswede et al. (Eds.): Information Transfer and Combinatorics, LNCS 4123, pp. 293–307, 2006.
© Springer-Verlag Berlin Heidelberg 2006

and, for $D = (d_1, \ldots, d_\ell)$ with non–negative integers $d_1 < \cdots < d_\ell$

$$V(E_N, M, D) = \sum_{n=1}^{M} e_{n+d_1} e_{n+d_2} \cdots e_{n+d_\ell}.$$

Then the *well–distribution measure* of E_N is defined by

$$W(E_N) = \max_{a,b,t} |U(E_N, t, a, b)| = \max_{a,b,t} \left| \sum_{j=0}^{t-1} e_{a+jb} \right|,$$

where the maximum is taken over all $a, b, t \in \mathbb{N}$ and $1 \leq a \leq a + (t-1)b \leq N$, while the correlation measure of order ℓ of E_N is defined by

$$C_\ell(E_N) = \max_{M,D} |V(E_N, M, D)| = \max_{M,D} \left| \sum_{n=1}^{M} e_{N+d_1} e_{n+d_2} \cdots e_{n+d_\ell} \right|,$$

where the maximum is taken over all $D = (d_1, d_2, \ldots, d_\ell)$ and M such that $M + d_\ell \leq N$. Then the sequence E_N is considered as a "good" PR sequence if both these measures $W(E_N)$ and $C_\ell(E_N)$ (at least for small ℓ) are "small" in terms of N (in particular, both are $o(N)$ as $N \to \infty$). Indeed, it is shown in [5], [10] that for a "truly random" $E_N \in \{-1, +1\}$ both $W(E_N)$ and, for fixed ℓ, $C_\ell(E_N)$ are around $N^{1/2}$ with "near 1" probability.

In [13] a third measure was introduced, which will be needed here: the *combined* (well–distribution–correlation) *PR measure of order ℓ* is defined by

$$\begin{aligned} Q_\ell(E_N) &= \max_{a,b,t,D} \left| \sum_{j=0}^{t} e_{a+jb+d_1} e_{a+jb+d_2} \cdots e_{a+jb+d_\ell} \right| \\ &= \max_{a,b,t,D} |Z(a, b, t, D)| \end{aligned} \tag{2.2}$$

where

$$Z(a, b, t, D) = \sum_{j=0}^{t} e_{a+jb+d_1} e_{a+jb+d_2} \cdots e_{a+jb+d_\ell}$$

is defined for all $a, b, t, D = (d_1, d_2, \ldots, d_\ell)$ such that all the subscripts $a + jb + d_i$ belong to $\{1, 2, \ldots, N\}$ (and the maximum in (2.2) is taken over D's of dimension ℓ).

In [14] we extended these definitions to the case of k symbols. It is not at all clear how to do this extension and, indeed, in [14] we introduced two different ways of extension which are nearly equivalent. Here we will present only one of them which is more suitable for our purpose.

Let $k \in \mathbb{N}$, $k \geq 2$, and let $\mathcal{A} = \{a_1, a_2, \ldots, a_k\}$ be a finite set ("alphabet") of k symbols ("letters") and consider a sequence $E_N = (e_1, e_2, \ldots, e_N) \in \mathcal{A}^N$ of these symbols. Write

$$x(E_N, a, M, u, v) = |\{j : 0 \leq j \leq M - 1, e_{u+jv} = a\}|$$

and for $W = (a_{i_1}, \ldots, a_{i_\ell}) \in \mathcal{A}^\ell$ and $D = (d_1, \ldots, d_\ell)$ with non–negative integers $d_1 < \cdots < d_\ell$,

$$g(E_N, W, M, D) = |\{n : 1 \leq n \leq M, (e_{n+d_1}, \ldots, e_{n+d_\ell}) = W\}|.$$

Then the f–well–distribution ("f" for "frequency") measure of E_N is defined as

$$\delta(E_N) = \max_{a, M, u, v} \left| x(E_N, a, M, u, v) - \frac{M}{k} \right|$$

where the maximum is taken over all $a \in \mathcal{A}$ and u, v, M with $u + (M-1)v \leq N$, while the f–correlation measure of order ℓ of E_N is defined by

$$\gamma_\ell(E_N) = \max_{W, M, D} \left| g(E_N, W, M, D) - \frac{M}{k^\ell} \right|$$

where the maximum is taken over all $W \in \mathcal{A}^\ell$, and $D = (d_1, \ldots, d_\ell)$ and M such that $M + d_\ell \leq N$.

We showed in [14] that in the special case $k = 2$, $\mathcal{A} = \{-1, +1\}$ the f–measures $\delta(E_N)$, $\gamma_\ell(E_N)$ are between two constant multiples of the binary measures $W(E_N)$, resp. $C_\ell(E_N)$, so that, indeed, the f–measures can be considered as extensions of the binary measures.

Now let E_N be the binary sequence in (2.1), and to this binary sequence assign a sequence $\varphi(E_N)$ whose elements are the 2^n letters in the alphabet $\{-1, +1\}^r$, and whose length is $[N/r]$:

$$\varphi(E_N) = \big((e_1, \ldots, e_r), (e_{r+1}, \ldots, e_{2r}), \ldots, (e_{([N/r]-1)r+1}, \ldots, e_{[N/r]r})\big).$$

We will show that if E_N is a "good" PR binary sequence, then $\varphi(E_N)$ is also a "good" PR sequence on the $k = 2^r$ letters in the alphabet $\{-1, +1\}^r$. Indeed, this follows from the inequalities in the following theorem:

Theorem 1. *If E_N and $\varphi(E_N)$ are defined as above, then we have*

$$\delta\big(\varphi(E_N)\big) \leq \frac{1}{2^r} \sum_{s=1}^{r} \binom{r}{s} Q_s(E_N) \tag{2.3}$$

and, for $\ell \in \mathbb{N}$

$$\gamma_\ell(\varphi(E_N)) \le \frac{1}{2^{r\ell}} \sum_{s=1}^{r} \sum_{q=1}^{\ell} \binom{r}{s}\binom{\ell}{q} Q_{qs}(E_N). \tag{2.4}$$

Proof of Theorem 1. Clearly, for all $a = (\varepsilon_1, \ldots, \varepsilon_r) \in \{-1, +1\}^r$, M, u and v we have

$$x(\varphi(E_N), a, M, u, v)$$
$$= \left| \left\{ j : 0 \le j \le M - 1, (e_{(u+jv-1)r+1}, \ldots, e_{(u+jv)r}) = (\varepsilon_1, \ldots, \varepsilon_r) \right\} \right|$$
$$= \sum_{j=0}^{M-1} \prod_{i=1}^{r} \frac{e_{(u+jv-1)r+i}\varepsilon_i + 1}{2}$$
$$= \frac{M}{2^r} + \frac{1}{2^r} \sum_{s=1}^{r} \sum_{1 \le i_1 < \cdots < i_s \le r} \varepsilon_{i_1} \ldots \varepsilon_{i_s} \sum_{j=0}^{M-1} e_{(u+jv-1)r+i_1} \cdots e_{(u+jv-1)r+i_s}$$

whence

$$\left| x(\varphi(E_N), a, M, u, v) - \frac{M}{k} \right| = \left| x(\varphi(E_N), a, M, u, v) - \frac{M}{2^r} \right|$$
$$\le \frac{1}{2^r} \sum_{s=1}^{r} \sum_{1 \le i_1 < \cdots < i_s \le r} \left| \sum_{j=0}^{M-1} e_{(u-1)r+jvr+i_1} \cdots e_{(u-1)r+jvr+i_s} \right|$$
$$= \frac{1}{2^r} \sum_{s=1}^{r} \sum_{1 \le i_1 < \cdots < i_s \le r} \left| Z((u-1)r, vr, M-1, (i_1, \ldots, i_s)) \right|$$
$$\le \frac{1}{2^r} \sum_{s=1}^{r} \sum_{1 \le i_1 < \cdots < i_s \le r} Q_s(E_N) = \frac{1}{2^r} \sum_{s=1}^{r} \binom{r}{s} Q_s(E_N) \tag{2.5}$$

which proves (2.3).

Now let $\mathcal{A} = \{-1, +1\}^r$, $w = (a_{i_1}, \ldots, a_{i_\ell}) \in \mathcal{A}^\ell$, $a_{i_j} = (\varepsilon_1^{(j)}, \ldots, \varepsilon_r^{(j)})$ and $D = (d_1, \ldots, d_\ell)$. Then we have

$$g(\varphi(E_N), W, M, D)$$
$$= \left| \left\{ n : 1 \le n \le M, ((e_{(n+d_1-1)r+1}, \ldots, e_{(n+d_1)r}), \ldots, (e_{(n+d_\ell-1)r+1}, \ldots, e_{(n+d_\ell)r})) \right. \right.$$
$$= \left. \left. ((\varepsilon_1^{(1)}, \ldots, \varepsilon_r^{(1)}), \ldots, (\varepsilon_1^{(\ell)}, \ldots, \varepsilon_r^{(\ell)})) \right\} \right| = \sum_{n=1}^{M} \prod_{i=1}^{r} \prod_{j=1}^{\ell} \frac{e_{(n+d_j-1)+i}\varepsilon_i^{(j)} + 1}{2}$$
$$= \frac{M}{2^{r\ell}} + \frac{1}{2^{r\ell}} \sum_{s=1}^{r} \sum_{q=1}^{\ell} \sum_{\substack{1 \le i_1 < \cdots < i_s \le r \\ 1 \le j_1 < \cdots < j_q \le \ell}} \left(\prod_{\mu=1}^{s} \prod_{\nu=1}^{q} \varepsilon_{i_\mu}^{(j_\nu)} \right) \left(\sum_{n=1}^{M} \prod_{\mu=1}^{s} \prod_{\nu=1}^{q} e_{(n+d_{j_\nu}-1)r+i_\mu} \right)$$

so that, as in (2.5),

$$\left| g\big(\varphi(E_N), W, M, D\big) - \frac{M}{2^{r\ell}} \right| = \left| g\big(\varphi(E_N), W, M, D\big) - \frac{M}{k^\ell} \right|$$

$$\leq \frac{1}{2^{r\ell}} \sum_{s=1}^{r} \sum_{q=1}^{\ell} \sum_{\substack{1 \leq i_1 < \cdots < i_s \leq r \\ 1 \leq j_1 < \cdots < j_q \leq \ell}} \left| Z\big(0, r, M-1, (d_{j_1} r + i_1, d_{j_1} r + i_2, \ldots, d_{j_q} r + i_s)\big) \right|$$

$$\leq \frac{1}{2^{r\ell}} \sum_{s=1}^{r} \sum_{q=1}^{\ell} \sum_{\substack{1 \leq i_1 < \cdots < i_s \leq r \\ 1 \leq j_1 < \cdots < j_q \leq \ell}} Q_{qs}(E_N) = \frac{1}{2^{r\ell}} \sum_{s=1}^{r} \sum_{q=1}^{\ell} \binom{r}{s} \binom{\ell}{q} Q_{qs}(E_N)$$

whence (2.4) follows and this completes the proof of Theorem 1.

Finally, we will make some comments on the applicability of the construction described at the beginning of this section. First, we remark that in certain applications this simple construction can be used even in the case when k, the number of the given symbols, is not a power of 2; the price paid is a slight data expansion. E.g., consider the following problem in cryptography: assume that a plaintext is given which uses, say, $k = 80$ characters, and we want to encrypt it by using a PR sequence of letters taken from an alphabet of appropriate size as key. Then we consider the smallest power of 2 $\geq$ the number of characters: $2^7 > 80$ ($> 2^6$). Next to each of the characters we assign one of the 2^7 blocks of bits of length 7 taken from $\{0,1\}^7$, and we replace each character in the plaintext by the corresponding block from $\{0,1\}^7$, so that the plaintext is mapped into a sequence $a_1, a_2, \ldots, a_M$ whose elements belong to $\{0,1\}^7$. Now by using the algorithm described above with $r = 7$, we construct a PR sequence $b_1, b_2, \ldots, b_M$ of letters from the alphabet $\mathcal{A} = \{0,1\}^7$ (whose size is power of 2: $|\mathcal{A}| = 2^7$). Then we obtain the ciphertext $c_1, c_2, \ldots, c_M$ by taking $c_i \in \{0,1\}^7$ as the residue of $a_i + b_i$ modulo 2^7 (and to decipher $c_1, c_2, \ldots, c_M$, we subtract b_i from c_i modulo 2^7).

A further remark on the limits of the applicability of this method: this algorithm can be applied only if N is "much greater", than $k = 2^r$. Indeed, N must grow at least as fast as a large power of k, otherwise the inequalities in Theorem 1 become trivial or say very little.

3   A Construction in the General Case

We will construct a large family of sequences on k symbols with a given length which has good PR properties (for any $k \in \mathbb{N}$, $k \geq 2$). This construction will be the generalization of the construction given in [6] in the special case $k = 2$ (however, it is much more difficult to control the general case presented here).

We will need four definitions.

Definition 1. *A multiset is said to be a k–set if each element occurs with multiplicity less than k.*

(So that a 2–set is a set whose elements are distinct, each occurring only once; in this case we will also call the set "simple set".)

Definition 2. *If $k \in \mathbb{N}$, $k \geq 2$, $m \in \mathbb{N}$, $\mathcal{A}$ and $\mathcal{B}$ are multisets whose elements belong to $\mathbb{Z}_m$ [1] (= the ring of the residue classes modulo m) and $\mathcal{A}+\mathcal{B}$ represents every element of $\mathbb{Z}_m$ with multiplicity divisible by k, i.e., for all $c \in \mathbb{Z}_m$, the number of solutions of*

$$a + b = c, \; a \in \mathcal{A}, \; b \in \mathcal{B} \tag{3.1}$$

(the elements of $\mathcal{A}, \mathcal{B}$ counted with their multiplicity) is divisible by k (including the case when there are no solutions), then the sum $\mathcal{A} + \mathcal{B}$ is said to have property P_k.

Definition 3. *If $k, h, \ell, m \in \mathbb{N}$, $k \geq 2$ and $h, \ell \leq m$, then (h, ℓ, m) is said to be a k–admissible triple if there is no simple set $\mathcal{A} \subset \mathbb{Z}_m$ and k–set $\mathcal{B}$ with elements from $\mathbb{Z}_m$ such that $|\mathcal{A}| = h$, $|\mathcal{B}| = \ell$ (multiple elements counted with their multiplicity), and $\mathcal{A} + \mathcal{B}$ possesses property P_k.*

Definition 4. *If $k, h, \ell, m \in \mathbb{N}$, $k \geq 2$ and $h, \ell \leq m$, then (h, ℓ, m) is said to be a (k, k)–admissible triple if there are no k–sets $\mathcal{A}, \mathcal{B}$ with elements from $\mathbb{Z}_m$ such that $|\mathcal{A}| = h$, $|\mathcal{B}| = \ell$ (multiple elements counted with their multiplicity), and $\mathcal{A} + \mathcal{B}$ possesses property P_k.*

Note that in the special case $k = 2$ property P_2 is the property P introduced in [6], while both 2–admissibility and (2,2)–admissibility are the admissibility used there.

Theorem 2. *Assume that $k \in \mathbb{N}$, $k \geq 2$, p is a prime number, χ is a (multiplicative) character modulo p of order k (so that $k|(p - 1)$), $f(x) \in F_p[x]$ (F_p being the field of the residue classes modulo p) has degree $h(> 0)$, $f(x)$ has no multiple zero in $\bar{F}_p$ (= the algebraic closure of E_p), and define the sequence $E_p = \{e_1, \ldots, e_p\}$ on the k letter alphabet of the k-th (complex) roots of unity by*

$$e_n = \begin{cases} \chi(f(n)) & \text{for } (f(n), p) = 1 \\ +1 & \text{for } p \mid f(n). \end{cases}$$

Then

(i) we have

$$\delta(E_p) < 11hp^{1/2}\log p, \tag{3.2}$$

(ii) if $\ell \in \mathbb{N}$ is such that the triple (r, t, p) is k–admissible for all $1 \leq r \leq h$, $1 \leq t \leq \ell(k - 1)$, then

$$\gamma_\ell(E_p) < 10\ell hkp^{1/2}\log p. \tag{3.3}$$

---

[1] In classical notation this is $\mathbb{Z}/m\,\mathbb{Z}$ and $\mathbb{Z}_p$ stands for p–adic integers, but in this paper they don't occur and no confusion can happen.

Proof of Theorem 2. The proof of both (i) and (ii) will be based on

Lemma 1. *Assume that p is a prime number, χ is a non–principal character modulo p of order k, $f(x) \in F_p[x]$ has degree h and a factorization $f(x) = b(x - x_1)^{r_1} \dots (x - x_s)^{r_s}$ (where $x_i \neq x_j$ for $i \neq j$) in $\bar{F}_p$ with*

$$(k, r_1, \dots, r_s) = 1. \tag{3.4}$$

Let X, Y be real numbers with $0 < Y \le p$. Then

$$\left| \sum_{x < n \le X+Y} \chi\big(f(n)\big) \right| < 9sp^{1/2} \log p \le 9hp^{1/2} \log p. \tag{3.5}$$

Proof of Lemma 1. With h in the upper bound in (3.5), this is Theorem 2 in [13] where we derived it from A. Weil's theorem [17] (see also Lemma 1 and its proof in [6]). To see that (3.5) also holds in the slightly sharper form with the factor s in place of h, all we have to observe is that in the proof of Theorem 2 in [13], at a certain point (p. 374, line 6 from below) we bounded s by h from above; skipping this step we obtain (3.5) in the sharper form. (We are indebted to Igor Shparlinski for this observation.)

We will need Lemma 1 in the following slightly modified form:

Lemma 2. *The assertion of Lemma 1 also holds if assumption (3.4) is replaced by*

$$(k, r_1, \dots, r_s) < k \tag{3.6}$$

(i.e., there is an r_i with $k \nmid r_i$).

Note that this lemma is sharper than Lemma 3 in [14] since now $x_1, \dots, x_s \in F_p$ is not assumed.

Proof of Lemma 2. Write $\delta = (k, r_1, \dots, r_s)$ so that

$$\delta < k \tag{3.7}$$

by (3.6), and define the character χ_1 by $\chi_1 = \chi^\delta$; then by (3.7), χ_1 is a non–principal character. Write the polynomial $\varphi(x) = b^{-1}f(x) = (x - x_1)^{r_1} \dots (x - x_s)^{r_s} \in F_p[x]$ as the product of powers of distinct irreducible polynomials over F_p: $\varphi(x) = \big(\pi_1(x)\big)^{u_1} \dots \big(\pi_t(x)\big)^{u_t}$. Since irreducible polynomials cannot have multiple zeros, and distinct irreducible polynomials are coprime and thus cannot have a common zero, thus it follows that the exponents $u_1, \dots, u_t$ are amongst the exponents $r_1, \dots, r_s$ whence, by the definition of δ, we have $\delta \mid (u_1, \dots, u_t)$. Then writing $\psi(x) = \big(\pi(x)\big)^{u_1/\delta} \dots \big(\pi(x)\big)^{u_t/\delta}$, clearly we have $\psi(x) \in F_p[x]$ and

$$f(x) = b\varphi(x) = b\big(\psi(x)\big)^\delta.$$

It follows that

$$\left| \sum_{X < n \leq X+Y} \chi(f(n)) \right| = |\chi(b)| \left| \sum_{X < n \leq X+Y} (\chi(\psi(n)))^\delta \right| \leq \left| \sum_{X < n \leq X+Y} \chi_1(\psi(n)) \right|.$$
(3.8)

To estimate this sum, we will apply Lemma 1. Indeed, χ_1 is of order k/δ, and clearly $\psi(x)$ has the factorization $\psi(x) = (x - x_1)^{r_1/\delta} \ldots (x - x_s)^{r_s/\delta}$ in $\bar{F}_p$. Thus replacing χ and $f(x)$ in Lemma 1 by χ_1 and $\psi(x)$, condition (3.4) becomes

$$\left(\frac{k}{\delta}, \frac{r_1}{\delta}, \ldots, \frac{r_s}{\delta} \right) = 1$$

which holds trivially by the definition of δ. Thus, indeed, Lemma 1 can be applied to estimate the last sum in (3.8), and applying it, we obtain the desired upper bound.

(i) If a is a k–th root of unity, then writing

$$S(a, m) = \frac{1}{k} \sum_{t=1}^{k} (\bar{a}\chi(m))^t,$$
(3.9)

clearly we have

$$S(a, m) = \begin{cases} 1, & \text{if } \chi(m) = a \\ 0, & \text{if } \chi(m) \neq a. \end{cases}$$
(3.10)

If a is a k–th roof of unity, $u, v, M \in \mathbb{N}$ and

$$1 \leq u \leq u + (M - 1)v \leq p,$$
(3.11)

then we have

$$x(E_p, a, M, u, v) = \sum_{\substack{0 \leq j \leq M-1 \\ e_{u+jv} = a}} 1$$
(3.12)

where

$$\left| \sum_{\substack{0 \leq j \leq M-1 \\ e_{u+jv} = a}} 1 - \sum_{\substack{0 \leq j \leq M-1 \\ \chi(f(u+jv)) = a}} 1 \right| \leq \sum_{\substack{0 \leq j \leq M-1 \\ p | f(u+jv)}} 1.$$
(3.13)

By (3.9) and (3.10),

$$\sum_{\substack{0 \leq j \leq M-1 \\ \chi(f(u+jv))=a}} 1 = \sum_{j=0}^{M-1} S\big(a, f(u+jv)\big) = \sum_{j=0}^{M-1} \frac{1}{k} \sum_{t=1}^{k} \big(\bar{a}\chi\big(f(u+jv)\big)\big)^{t}$$

$$= \frac{1}{k} \sum_{\substack{0 \leq j \leq M-1 \\ (f(u+jv),p)=1}} 1 + \frac{1}{k} \sum_{t=1}^{k-1} \bar{a}^{t} \sum_{j=0}^{M-1} \chi^{t}\big(f(u+jv)\big)$$

$$= \frac{M}{k} - \frac{1}{k} \sum_{\substack{0 \leq j \leq M-1 \\ p|f(u+jv)}} 1 + \frac{1}{k} \sum_{t=1}^{k-1} \bar{a}^{t} \sum_{j=0}^{M-1} \chi^{t}\big(f(u+jv)\big)$$

whence

$$\left| \sum_{\substack{0 \leq j \leq M-1 \\ \chi(f(u+jv))=a}} 1 - \frac{M}{k} \right| \leq \frac{1}{k} \sum_{t=1}^{k-1} \left| \sum_{j=0}^{M-1} \chi^{t}\big(f(u+jv)\big) \right| + \frac{1}{k} \sum_{\substack{0 \leq j \leq M-1 \\ p|f(u+jv)}} 1. \quad (3.14)$$

Writing $g(x) = f(u + xv)$, it follows from (3.12), (3.13) and (3.14) that

$$\left| x(E_p, a, M, u, v) - \frac{M}{k} \right| \leq \frac{1}{k} \sum_{t=1}^{k-1} \left| \sum_{j=0}^{M-1} \chi^{t}\big(g(j)\big) \right| + 2 \sum_{\substack{0 \leq j \leq M-1 \\ p|g(j)}} 1. \quad (3.15)$$

The case $M = 1$ is trivial, thus we may assume that $M > 1$. Then by $v \geq 1$ and (3.11) we have $1 \leq v < p$ so that $(v, p) = 1$. It follows that the polynomials $f(x), g(x) \in F_p[x]$ have the same degree, and since $f(x)$ does not have multiple zeros, $g(x)$ does not have multiple zeros either. Moreover, $\chi_1 = \chi^t$ is also a character modulo p, and for $1 \leq t \leq k-1$ the character χ_1 is different from the principal character χ_0. Thus by Lemma 1 we have

$$\left| \sum_{j=0}^{M-1} \chi^{t}\big(g(j)\big) \right| = \left| \sum_{j=0}^{M-1} \chi_1\big(g(j)\big) \right| < 9hp^{1/2} \log p \text{ for } 1 \leq t \leq k-1. \quad (3.16)$$

Since f and g are of the same degree thus

$$\sum_{\substack{0 \leq j \leq M-1 \\ p|g(j)}} 1 \leq \sum_{\substack{0 \leq j < p \\ p|g(j)}} 1 \leq h. \quad (3.17)$$

It follows from (3.15), (3.16) and (3.17) that

$$\left| x(E_p, a, M, u, v) - \frac{M}{k} \right| \leq \frac{k-1}{k} \cdot 9hp^{1/2} \log p + 2h < 11hp^{1/2} \log p$$

which completes the proof of (3.2).

(ii) In order to prove (3.3), assume that $\ell \in \mathbb{N}$, $\ell \le N$, $b_1, \ldots, b_\ell$ are k-th roots of unity, $w = (b_1, \ldots, b_\ell)$, $D = (d_1, \ldots, d_\ell)$, $0 \le d_1 < \cdots < d_\ell$, $M \in \mathbb{N}$ and $M + d_\ell \le N$. Then

$$g(E_n, w, M, D) = \left|\left\{n : 1 \le n \le M, (e_{n+d_1}, \ldots, e_{n+d_\ell}) = w\right\}\right|$$
$$= \left|\left\{n : 1 \le n \le M, e_{n+d_1} = b_1, \ldots, e_{n+d_\ell} = b_\ell\right\}\right|. \quad (3.18)$$

Here we have

$$e_{n+d_1} = \chi\big(f(n + d_1)\big), \ldots, e_{n+d_\ell} = \chi\big(f(n + d_\ell)\big) \quad (3.19)$$

except for the values of n such that

$$f(n + d_i) \equiv 0 \ (\mathrm{mod}\ p) \text{ for some } 1 \le i \le \ell. \quad (3.20)$$

For fixed i, this congruence may have at most h solutions, and i may assume at most ℓ values. Thus the total number of solutions of (3.20) is $\le h\ell$. If n is not a solution of (3.20), then (3.19) holds, so that by (3.10), for all these n we have

$$\prod_{i=1}^{\ell} S\big(b_i, f(n + d_i)\big) = \begin{cases} 1 & \text{if } e_{n+d_1} = b_1, \ldots, e_{n+d_\ell} = b_\ell \\ 0 & \text{otherwise.} \end{cases} \quad (3.21)$$

For the exceptional values of n satisfying (3.20) (whose number is $\le h\ell$) again by (3.10) we have

$$\prod_{i=1}^{\ell} S\big(b_i, f(n + d_i)\big) = 0 \text{ or } 1. \quad (3.22)$$

It follows from (3.18), (3.21) and (3.22) that

$$\left| g(E_N, w, M, D) - \sum_{n=1}^{M} \prod_{i=1}^{\ell} S\big(b_i, f(n + d_i)\big) \right| \le h\ell \quad (3.23)$$

where we have

$$\sum_{n=1}^{M} \prod_{i=1}^{\ell} S\big(b_i, f(n + d_i)\big) = \sum_{n=1}^{M} \prod_{i=1}^{\ell} \frac{1}{k} \sum_{t_i=1}^{k} \big(\overline{b_i}\chi\big(f(n + d_i)\big)\big)^{t_i}$$

$$= \frac{1}{k^\ell} \sum_{t_1=1}^{k} \cdots \sum_{t_\ell=1}^{k} \overline{b_1^{t_1} \ldots b_\ell^{t_\ell}} \sum_{n=1}^{M} \chi\big((f(n + d_1))^{t_1} \ldots (f(n + d_\ell))^{t_\ell}\big)$$

$$= \frac{M}{k^\ell} + \frac{1}{k^\ell} \sum_{\substack{0 \le t_1, \ldots, t_\ell \le k-1 \\ (t_1, \ldots, t_\ell) \ne (0, \ldots, 0)}} \overline{b_1^{t_1} \ldots b_\ell^{t_\ell}} \sum_{n=1}^{M} \chi\big((f(n + d_1))^{t_1} \ldots (f(n + d_\ell))^{t_\ell}\big).$$

$$(3.24)$$

It follows from (3.23) and (3.24) that

$$\left| g(E_N, w, M, D) - \frac{M}{k^\ell} \right|$$

$$\leq \frac{1}{k^\ell} \sum_{\substack{0 \leq t_1, \ldots, t_\ell \leq k-1 \\ (t_1, \ldots, t_\ell) \neq (0, \ldots, 0)}} \left| \sum_{n=1}^{M} \chi\left(\left(f(n+d_1)\right)^{t_1} \ldots \left(f(n+d_\ell)\right)^{t_\ell} \right) \right| + h\ell. \quad (3.25)$$

Write $f(x) = Bf_1(x)$ where $B \in \mathbb{Z}_p$ and $f_1(x) \in \mathbb{Z}_p[x]$ is a unitary polynomial, and set $G(x) = f_1(x+d_1)^{t_1} \ldots f_1(x+d_\ell)^{t_\ell}$. Then the innermost sum in (3.25) can be rewritten in the following way:

$$\left| \sum_{n=1}^{M} \chi\left(\left(f(n+d_1)\right)^{t_1} \ldots \left(f(n+d_\ell)\right)^{t_\ell} \right) \right|$$

$$= \left| \chi(B^{t_1 + \cdots + t_\ell}) \right| \left| \sum_{n=1}^{M} \chi\left(G(n)\right) \right| \leq \left| \sum_{n=1}^{M} \chi\left(G(n)\right) \right|. \quad (3.26)$$

It suffices to show:

Lemma 3. *If k, f, h, ℓ are defined as in Theorem 2, then $G(x)$ has at least one zero (in $\bar{F}_p$) whose multiplicity is not divisible by k.*

Indeed, assuming that Lemma 3 has been proved, the proof of (3.3) can be completed in the following way: by Lemma 3, we may apply Lemma 2 with $G(x)$ in place of $f(x)$ (since then (3.6) holds by Lemma 3). The degree of $G(x)$ is clearly

$$ht_1 + \cdots + ht_\ell \leq \ell h(k-1) < \ell hk,$$

thus applying Lemma 2 we obtain

$$\left| \sum_{n=1}^{M} \chi\left(G(n)\right) \right| < 9\ell hkp^{1/2} \log p.$$

Each of the innermost sums in (3.25) can be estimated in this way. Thus it follows from (3.25) that

$$\left| g(E_N, w, M, D) - \frac{M}{k^\ell} \right| \leq \frac{1}{k^\ell} \sum_{\substack{0 \leq t_1, \ldots, t_\ell \leq k-1 \\ (t_1, \ldots, t_\ell) \neq (0, \ldots, 0)}} 9\ell hkp^{1/2} \log p + h\ell < 10\ell hkp^{1/2} \log p$$

for all w, M, D which proves (3.3). Thus it remains to prove the lemma:

Proof of Lemma 3: We will say that the polynomials $\varphi(x), \psi(x) \in F_p[x]$ are equivalent: $\varphi \sim \psi$ if there is an $a \in F_p$ such that $\psi(x) = \varphi(x + a)$. Clearly, this is an equivalence relation.

Write $f_1(x)$ as the product of irreducible polynomials over F_p. It follows from our assumption on $f(x)$ that these irreducible factors are distinct. Let us group these factors so that in each group the equivalent irreducible factors are collected. Consider a typical group $\varphi(x + a_1), \ldots, \varphi(x + a_r)$.

Then writing $G(x)$ as the product of irreducible polynomials over F_p, all the polynomials $\varphi(x + a_i + d_j)$ with $1 \le i \le r$, $1 \le j \le \ell$ occur amongst the factors, and for fixed i, j such a factor occurs t_j times. All these polynomials are equivalent, and no other irreducible factor belonging to this equivalence class will occur amongst the irreducible factors of $G(x)$.

Since irreducible polynomials have no multiple zeros and distinct irreducible polynomials cannot have a common zero, the conclusion of Lemma 3 fails, i.e., the multiplicity of each of the zeros of $G(x)$ is divisible by k, if and only if in each group, formed by equivalent irreducible factors $\varphi(x + a_i + d_j)$ of $G(x)$ each taken t_j times, every polynomial of form $\varphi(x + c)$ with $c \in F_p$ occurs with multiplicity divisible by k, i.e., the number of representation of c in the form $a_i + d_j$, counting this representation with multiplicity t_j, is divisible by k. In other words, if we write $\mathcal{A} = \{a_1, \ldots, a_r\}$ and $\mathcal{B}$ denotes the k–set whose elements are $d_1, \ldots, d_\ell$, each d_j taken with multiplicity $t_j \le k - 1$, for each group $\mathcal{A} + \mathcal{B}$ must possess property P_k. Now consider any of these groups (by $\deg f > 0$ there is at least one such group). Since $\mathcal{A} + \mathcal{B}$ possesses property P_k, $(|\mathcal{A}|, |\mathcal{B}|, p)$ is **not** a k–admissible triple. Here we clearly have

$$|\mathcal{A}| = r \le \deg f_1 = \deg f = h$$

and

$$|\mathcal{B}| = \sum_{j=1}^{\ell} t_j \le \ell(k - 1)$$

which contradicts our assumption on ℓ. Thus the conclusion of Lemma 3 cannot fail, and this completes the proof.

4   The Necessity of the k–Admissibility

Upper bound (3.3) in Theorem 2 is proved assuming certain k–admissibility. (The study of k–admissibility is a difficult problem to which we return in the next sections). Thus Theorem 2 could be applied more easily without this assumption, so that one might like to know whether this assumption is really necessary, or it can be dropped? Next we will show that, subject to certain mild conditions on the parameters involved, any negative example with a sum $\mathcal{A} + \mathcal{B}$ ($\mathcal{A}$ simple set, $\mathcal{B}$ k–set) having property P_k induces a construction of the type described in Theorem 2 with the property that conclusion (3.3) fails, i.e., certain correlation is large. (Sums $\mathcal{A} + \mathcal{B}$ of this type will be constructed later in Section 6.)

Assume that $k \in \mathbb{N}$, $k \ge 2$, p is a prime, $\mathcal{A} = \{a_1, \ldots, a_r\} \subset \{0, 1, \ldots, p - 1\}$, $\mathcal{B}$ is a k–set with elements from $\{0, 1, \ldots, p - 1\}$, $|\mathcal{A}| = r < p$, $|\mathcal{B}| = t < p$, the distinct elements of $\mathcal{B}$ are $d_1, \ldots, d_\ell$, their multiplicities are $t_1, \ldots, t_\ell$ ($1 \le t_i \le k - 1$), and $\mathcal{A} + \mathcal{B}$ has property P_k. Set $f(n) = (n + a_1) \ldots (n + a_r)$, and define the sequence $E_p = \{e_1, \ldots, e_p\}$ in the same way as in Theorem 2. Set $M = p - d_\ell$. We claim that assuming also

$$p \to \infty,$$

$$M = p - d_\ell \gg p \tag{4.1}$$

and

$$r\ell = o(p), \tag{4.2}$$

γ_ℓ cannot be "small":

$$|\gamma_\ell(E_p)| \neq o\left(\frac{p}{k^\ell}\right). \tag{4.3}$$

Consider the sum

$$S_M = \sum_{n=1}^{M} e_{n+d_1}^{t_1} \cdots e_{n+d_\ell}^{t_\ell}.$$

Here we have

$$e_{n+d_j}^{t_j} = \left(\chi\big(f(n + d_j)\big)\right)^{t_j} = \chi\left(\prod_{i=1}^{r}(n + a_i + d_j)^{t_j}\right)$$

except for n, j such that

$$n + a_i + d_j \equiv 0 \pmod{p} \text{ for some } 1 \leq i \leq r. \tag{4.4}$$

If n is such that there are no i (with $1 \leq i \leq r$), j satisfying (4.4), then we have

$$e_{n+d_1}^{t_1} \cdots e_{n+d_\ell}^{t_\ell} = \chi\left(\prod_{j=1}^{\ell}\prod_{i=1}^{r}(n + a_i + d_j)^{t_j}\right) = \chi\left(\prod_{c \in \mathcal{A} + \mathcal{B}}(n + c)\right). \tag{4.5}$$

Here every $c \in \mathcal{A} + \mathcal{B}$ is counted as many times as the number of solutions of

$$a + d = c, \ a \in \mathcal{A}, \ d \in \mathcal{B}$$

where the d's are counted with their multiplicity; for fixed $c \in \mathbb{Z}_p$, denote the number of solutions of this equation by $\varphi(c)$ (for $c \notin \mathcal{A} + \mathcal{B}$ we set $\varphi(c) = 0$). Then (4.5) can be rewritten as

$$e_{n+d_1}^{t_1} \cdots e_{n+d_\ell}^{t_\ell} = \prod_{\substack{c \in \mathbb{Z}_p \\ f(c) \neq 0}} \left(\chi(n + c)\right)^{\varphi(c)}.$$

Since $\mathcal{A} + \mathcal{B}$ possesses property P_k, $k \mid \varphi(c)$ for all $c \in \mathbb{Z}_p$, and we assumed that $n + c \neq 0$ if $\varphi(c) \neq 0$. Since χ is a character of order k, it follows that

$$e_{n+d_1}^{t_1} \cdots e_{n+d_\ell}^{t_\ell} = \prod_{\substack{c \in \mathbb{Z}_p \\ \varphi(c) \neq 0}} \left(\chi^k(n + c)\right)^{\varphi(c)/k} = \prod_{\substack{c \in \mathbb{Z}_p \\ \varphi(c) \neq 0}} 1 = 1$$

for every n for which (4.4) has no solution in i, j. In (4.4) the pair (i, j) can be chosen in $r\ell$ ways, and (i, j) determine n uniquely. Thus we have

$$|S_M - M| < r\ell. \tag{4.6}$$

On the other hand, assume that contrary to (4.3), we have

$$|\gamma_\ell(E_p)| = o\left(\frac{p}{k^\ell}\right)$$

so that, denoting the set of the k-th roots of unity by $\mathcal{A}$, for every ℓ–tuple $w = (\varepsilon_1, \ldots, \varepsilon_\ell) \in \mathcal{A}^\ell$ we have

$$g(E_p, w, M, D) = \frac{M}{k^\ell} + o\left(\frac{p}{k^\ell}\right).$$

It follows that

$$S_M = \sum_{(\varepsilon_1, \ldots, \varepsilon_\ell) \in \mathcal{A}^\ell} g\big(E_p, (\varepsilon_1, \ldots, \varepsilon_\ell), M, (d_1, \ldots, d_\ell)\big)\varepsilon_1^{t_1} \ldots \varepsilon_\ell^{t_\ell}$$

$$= \frac{M}{k^\ell} \sum_{(\varepsilon_1, \ldots, \varepsilon_\ell) \in \mathcal{A}^\ell} \varepsilon_1^{t_1} \ldots \varepsilon_\ell^{t_\ell} + o\left(\frac{p}{k^\ell} \sum_{(\varepsilon_1, \ldots, \varepsilon_\ell) \in \mathcal{A}^\ell} 1\right).$$

By $1 \le t_i \le k - 1$, the first sum is 0. Thus we have

$$S_M = o(p)$$

which contradicts (4.1), (4.2) and (4.6), and this completes the proof of our claim.

5   Concluding Remarks

We have just shown that the assumption on the k–admissibility in Theorem 2 cannot be dropped. Thus in order to be able to use the construction in Theorem 2, we need criteria for a triple (r, t, p) to be k–admissible. We will present sufficient criteria of this type in Part II. The complexity of the family that we have constructed will be also studied there. Finally we estimate the cardinality of a smallest family achieving a prescribed f–complexity by extending the result of [4] from binary to k–ary alphabets. Somewhat surprisingly we also improve the earlier results by establishing a uniformity property.

References

1. R. Ahlswede, Coloring hypergraphs: A new approach to multi–user source coding, Part I, J. Combinatorics, Information and System Sciences 4(1), 76–115, 1979; Part II, J. Combinatorics, Information and System Sciences 5, 3, 220–268, 1980.
2. R. Ahlswede, On concepts of performance parameters for channels, this volume.

3. R. Ahlswede and A. Winter, Strong converse for identification via quantum channels, IEEE Trans. on Inform., Vol. 48, No. 3, 569–579, 2002.

4. R. Ahlswede, L.H. Khachatrian, C. Mauduit, and A. Sárközy, A complexity measure for families of binary sequences, Periodica Math. Hungar., Vol. 46, No. 2, 107–118, 2003.

5. J. Cassaigne, C. Mauduit, and A. Sárközy, On finite pseudorandom binary sequences VII: The measures of pseudorandomness, Acta Arith. 103, 97–118, 2002.

6. L. Goubin, C. Mauduit, and A. Sárközy, Construction of large families of pseudorandom binary sequences, J. Number Theory 106, 56-69, 2004.

7. H. Halberstam and H.-E. Richert, Sieve Methods, Academic Press, London, 1974.

8. D.R. Heath–Brown, Artin's conjecture for primitive roots, Quat. J. Math. 37, 27–38, 1986.

9. C. Hooley, On Artin's conjecture, J. reine angew. Math. 225, 209–220, 1967.

10. Y. Kohayakawa, C. Mauduit, C.G. Moreira and V. Rödl, Measures of pseudorandomness for random sequences, Proceedings of WORDS'03, 159–169, TUCS Gen. Publ., 27, Turku Cent. Comput. Sci., Turku, 2003.

11. R. Lidl and H. Niederreiter, Introduction to Finite Fields and Their Applications, revised edition, Cambridge University Press, 1994.

12. C. Mauduit, J. Rivat, and A. Sárközy, Construction of pseudorandom binary sequences using additive characters, Monatshefte Math., 141, 197-208, 2004

13. C. Mauduit and A. Sárközy, On finite pseudorandom binary sequences, I. Measure of pseudorandomness, the Legendre symbol, Acta Arith. 82, 365–377, 1997.

14. C. Mauduit and A. Sárközy, On finite pseudorandom sequences of k symbols, Indag. Math. 13, 89–101, 2002.

15. A. Schinzel, Remarks on the paper "Sur certaines hypothèses concernant les nombres premiers", Acta Arith. 7, 1–8, 1961/1962.

16. A. Schinzel and W. Sierpiński, Sur certaines hypothèses concernant les nombres premiers, ibid. 4, 185–208, 1958; Corrigendum ibid. 5, 259, 1959.

17. A. Weil, Sur les courbes algébriques et les variétés qui s'en déduisent, Act. Sci. Ind. 1041, Hermann, Paris, 1948.

Large Families of Pseudorandom Sequences of k Symbols and Their Complexity – Part II

R. Ahlswede, C. Mauduit, and A. Sárközy

Dedicated to the memory of Levon Khachatrian

1 Introduction

We continue the investigation of Part I, keep its terminology, and also continue the numbering of sections, equations, theorems etc.
Consequently we start here with Section 6. As mentioned in Section 4 we present now criteria for a triple (r, t, p) to be k–admissible. Then we consider the f–complexity (extended now to k–ary alphabets) $\Gamma_k(\mathcal{F})$ of a family $\mathcal{F}$. It serves again as a performance parameter of key spaces in cryptography. We give a lower bound for the f–complexity for a family of the type constructed in Part I. In the last sections we explain what can be said about the theoretically best families $\mathcal{F}$ with respect to their f–complexity $\Gamma_k(\mathcal{F})$. We begin with straightforward extensions of the results of [4] for $k = 2$ to general k by using the same Covering Lemma as in [1].

But then we give an improvement (also of the earlier results) with respect to balancedness with the help of another old Covering Lemma from [1]. Finally this will again be improved by a more recent result on edge–coverings of hypergraphs from [2]. This has become a basic tool in Information Theory, for instance in the Theory of Identification. In the present context it gives families with a very strong balancedness property. A quantum theoretical analogue became a key tool for quantum channels [3]. It invites to investigate our cryptographical concepts in the quantum world.

2 Sufficient Criteria for k–Admissibility

We have shown in Part I that the assumption on the k–admissibility in Theorem 2 cannot be dropped. Thus in order to be able to use the construction in Theorem 2, we need criteria for a triple (r, t, p) to be k–admissible. We will prove three sufficient criteria of this type:

Theorem 3

(i) If $k, r, t \in \mathbb{N}$, $1 \leq t \leq k$, p is a prime and $r < p$, then the triple (r, t, p) is k–admissible.

(ii) If $k, r, t \in \mathbb{N}$, p is a prime and

$$(4t)^r < p, \tag{7.1}$$

then (r, t, p) is k–admissible.

R. Ahlswede et al. (Eds.): Information Transfer and Combinatorics, LNCS 4123, pp. 308–325, 2006.
© Springer-Verlag Berlin Heidelberg 2006

(iii) If $k \in \mathbb{N}$, $k \geq 2$, the prime factorization of k is $k = q_1^{\alpha_1} \ldots q_s^{\alpha_s}$ (where $q_1, \ldots, q_s$ are distinct primes and $\alpha_1, \ldots, \alpha_s \in \mathbb{N}$), and p is a prime such that each of $q_1, \ldots, q_s$ is a primitive root modulo p, then for every pair $r, t \in \mathbb{N}$ with $r, t < p$, the triple (r, t, p) is k-admissible.

Note that in the special case $k = 2$ this theorem gives Theorem 2 in [6].

Proof

(i) Assume that contrary to the assertion, there are $k, r, t \in \mathbb{N}$ and a prime p so that

$$1 \leq t \leq k, \tag{7.2}$$

$$r < p, \tag{7.3}$$

and the triple (r, t, p) is not k-admissible, i.e., there is an $\mathcal{A} \subset \mathbb{Z}_p$ and a k-set $\mathcal{B}$ whose elements belong to $\mathbb{Z}_p$ such that $|\mathcal{A}| = r$, $|\mathcal{B}| = t$ (multiple elements counted with their multiplicity) and the number of solutions of (3.1) is divisible by k for all $c \in \mathbb{Z}$.

Consider any $c \in \mathcal{A} + \mathcal{B}$ ($\mathcal{A}, \mathcal{B}$ are non–empty, thus $\mathcal{A} + \mathcal{B}$ is also non–empty). Since for this c (3.1) has at least one solution and the number of solutions is always divisible by k, thus (3.1) must have at least k solutions. On the other hand, clearly (3.1) may have at most $|\mathcal{B}| = t$ solutions so that we must have

$$|\mathcal{B}| = t \geq k. \tag{7.4}$$

It follows from (7.2) and (7.4) that

$$|\mathcal{B}| = t = k. \tag{7.5}$$

Since $\mathcal{B}$ is a k–set, the multiplicity of each element is $\leq k - 1$. Thus it follows from (7.5) that $\mathcal{B}$ must have at least two distinct elements: say, $b_o, b_o + d \in \mathcal{B}$, $d \neq 0$. Every element of $\mathcal{A} + b_o$ must have (at least) k representations in the form (3.1) whence, by (7.5), it follows easily that they also have a representation in the form $(a + b_o + d)$ with $a \in \mathcal{A}$ whence $\mathcal{A} + b_o = \mathcal{A} + b_o + rd$ for all $r \in \mathbb{N}$, thus $\mathcal{A} + b_o = \mathcal{A} + b_o + s$ for any $s \in \mathbb{Z}_p$, in particular for any $s \in \mathcal{A} + b_o$. Hence, $\mathcal{A} + b_o$ is an additive subgroup of $\mathbb{Z}_p$ thus $\mathcal{A} = \mathcal{A} + b_o = \mathbb{Z}_p$ which contradicts $|\mathcal{A}| = r$ and (7.3).

(ii) The proof is nearly the same as the proof of Theorem 2, (ii) in [6]. Thus will omit most of the details here, we will present only those critical steps where a slight modification is needed.

Assume that r, t, p satisfy (7.1), $\mathcal{A} \subset \mathbb{Z}_p$, $\mathcal{B}$ is a k–set whose elements belong to $\mathbb{Z}_p$, $|\mathcal{A}| = r$ and $|\mathcal{B}| = t$ (multiple elements counted with their multiplicity). It suffices to show that then there is a $c \in \mathbb{Z}_p$ for which the number of solutions of (3.1) (the b's counted with multiplicity) is greater than 0 and less than k. To show this, it suffices to prove that there are $m \in \mathbb{N}$, $c' \in \mathbb{Z}_p$ such that $(m, p) = 1$, and the number of solutions of

$$ma + mb = c', \ a \in \mathcal{A}, \ b \in \mathcal{B} \tag{7.6}$$

is greater than 0 and less than k. Again, the proof of this is based on Lemma 3 in [6]. We start out from this lemma, and we proceed in the same way as in [6]. In particular, we define $m, b_i, b_j, r_1, r_k, a_n, a_v$ in the same way. Then again, the numbers

$$mb_i + r_k = mb_j + ma_v$$

and

$$mb_j + r_1 = mb_j + ma_u$$

do not have any further representations in form (7.6). Since $\mathcal{B}$ is a k–set, the multiplicity of both b_i and b_j is less than k. Thus these numbers have more than 0 and less than k representations in form (7.6) (counting the b's with multiplicity) which completes the proof.

(iii) From a practical point of view this seems to be the most important of the three criteria. Namely, this criterion enables us to control even correlations of very high order provided that there are "many" primes p such that each of $q_1, \ldots, q_s$ is a primitive root modulo p. Partly because of the importance of this criterion, partly in order to help to understand the notion of k–admissibility and the related difficulties better, we will give a detailed discussion of this case in the next section. This discussion will lead not only to the proof of criterion (iii), but it will also provide negative examples. We will also show that, most probably, there are "many" primes p of the type described in (iii).

3 k–Good Primes: Negative Examples

Definition 5. *A number $m \in \mathbb{N}$ is said to be k–good if for any pair $r, t \in \mathbb{N}$ with $r < m$, $t < m$, the triple (r, t, m) is k–admissible. If for all $r < m$, $t < m$ the triple (r, t, m) is (k, k)–admissible, then m is said to be (k, k)–good.*

Theorem 4. *If $k \in \mathbb{N}$, $k \geq 2$, the prime factorization of k is $k = q_1^{\alpha_1} \ldots q_s^{\alpha_s}$ (where $q_1, \ldots, q_s$ are distinct primes and $\alpha_1, \ldots, \alpha_s \in \mathbb{N}$) and p is an odd prime such that each of $q_1, \ldots, q_s$ is a primitive root modulo p, then p is k–good.*

Proof of Theorem 4. We will need the following lemma:

Lemma 4. *If p is an odd prime and q is a prime which is a primitive root modulo p, then the polynomial $x^{p-1} + x^{p-2} + \cdots + x + 1$ is irreducible over F_q.*

Proof of Lemma 4. This is a trivial consequence of Theorem 2.47 in [11, p. 62].

We will prove the assertion of Theorem 4 by contradiction: assume that contrary to the statement of the theorem, there is a set $\mathcal{A} \subset \mathbb{Z}_p$ and a k–set $\mathcal{B}$ whose elements belong to $\mathbb{Z}_p$ so that

$$|\mathcal{A}| = r < p, \quad |\mathcal{B}| = t < p \tag{8.1}$$

and the sum $\mathcal{A} + \mathcal{B}$ has property P_k.

If $\mathcal{C}$ is a multiset whose elements belong to $\mathbb{Z}_p$, then let $Q_{\mathcal{C}}(x)$ denote the polynomial $\sum_{c \in \mathcal{C}} x^{s(c)}$ where $s(c)$ denotes the least non–negative element of the residue class c modulo p, and the elements c of $\mathcal{C}$ are to be taken with their multiplicity (so that if c occurs with multiplicity M in $\mathcal{C}$, then there is a term $M x^{s(c)}$ appearing in $Q_{\mathcal{C}}(x)$). Clearly we have $(x^p - 1) \mid x^u Q_{\mathcal{C}}(x) - Q_{\mathcal{C}+u}(x)$ (in $\mathbb{Z}[x]$), if $\mathcal{C}$ is a multiset of elements of $\mathbb{Z}_p$ and $u \in \mathbb{Z}_p$. It follows that $(x^p - 1) \mid (Q_{\mathcal{A}}(x)Q_{\mathcal{B}}(x) - Q_{\mathcal{A}+\mathcal{B}}(x))$:

$$Q_{\mathcal{A}}(x)Q_{\mathcal{B}}(x) = Q_{\mathcal{A}+\mathcal{B}}(x) + (x^p - 1)G(x) \text{ with } G(x) \in \mathbb{Z}[x]. \qquad (8.2)$$

Write $Q_{\mathcal{B}}(x) = \sum_{j=0}^{p-1} v_j x^j$ so that the v_j's are the multiplicities of the elements $j \in \mathbb{Z}_p$ in $\mathcal{B}$. It follows that $0 \leq v_j \leq k - 1$ for all $0 \leq j \leq p - 1$, and since

$$|\mathcal{B}| = \sum_{j=0}^{p-1} v_j > 0,$$

we have

$$(v_0, v_1, \ldots, v_{p-1}) \leq k - 1.$$

It follows that there is an i with $1 \leq i \leq s$, $q_i^{\alpha_i} \nmid (v_0, v_1, \ldots, v_{p-1})$. Write

$$q_i^{\beta} \| (v_0, v_1, \ldots, v_{p-1}) \qquad (8.3)$$

so that

$$0 \leq \beta < \alpha_i. \qquad (8.4)$$

Then every coefficient of $Q_{\mathcal{B}}(x)$ is divisible by q_i^{β}. Since $\mathcal{A} + \mathcal{B}$ has property P_k, the coefficients of $Q_{\mathcal{A}+\mathcal{B}}(x)$ are divisible by k and thus also by q_i^{β}. Thus by (8.2), every coefficient of $(x^p - 1)G(x)$ must be also divisible by q_i^{β}. Since the polynomial $x^p - 1$ is primitive (a polynomial $\in \mathbb{Z}[x]$ is said to be primitive if the greatest common divisor of its coefficients is 1), and by Gauss' lemma the product of primitive polynomials is also primitive, thus it follows that the coefficients of $G(x)$ are also divisible by q_i^{β}. Thus we may simplify (8.2) so that we divide the coefficients of $Q_{\mathcal{B}}(x)$, $Q_{\mathcal{A}+\mathcal{B}}(x)$ and $G(x)$ by q_i^{β}:

$$Q_{\mathcal{A}}(x) \left(\frac{1}{q_i^{\beta}} Q_{\mathcal{B}}(x) \right) = \left(\frac{1}{q_i^{\beta}} Q_{\mathcal{A}+\mathcal{B}}(x) \right) + (x^p - 1) \left(\frac{1}{q_i^{\beta}} G(x) \right). \qquad (8.5)$$

Since this equation holds over $\mathbb{Z}$, it also holds over $\mathbb{Z}_{q_i}$, i.e., in other words, we may consider (8.5) modulo q_i. The coefficients of $Q_{\mathcal{A}+\mathcal{B}}(x)$ are divisible by $q_i^{\alpha_i}$, thus by (8.4), the polynomial $\frac{1}{q_i^{\beta}} Q_{\mathcal{A}+\mathcal{B}}(x)$ is the zero polynomial. Since $(x^{p-1} + x^{p-2} + \cdots + 1) \mid (x^p - 1)$, thus it follows from (8.5) that

$$(x^{p-1} + x^{p-2} + \cdots + 1) \mid Q_{\mathcal{A}}(x) \left(\frac{1}{q_i^{\beta}} Q_{\mathcal{B}}(x) \right).$$

By Lemma 4 the polynomial $x^{p-1} + x^{p-2} + \cdots + 1$ is irreducible over F_{q_i}. Thus it follows that either

$$(x^{p-1} + x^{p-2} + \cdots + 1) \mid Q_{\mathcal{A}}(x) \qquad (8.6)$$

or

$$(x^{p-1} + x^{p-2} + \cdots + 1) \mid \left(\frac{1}{q_i^{\beta}} Q_{\mathcal{B}}(x) \right) ; \qquad (8.7)$$

note that by (8.3), the polynomial $\frac{1}{q_i^{\beta}} Q_{\mathcal{B}}(x)$ is not the 0 polynomial. Since by the definitions of $Q_{\mathcal{A}}(x)$ and $Q_{\mathcal{B}}(x)$ these polynomials are of degree at most $p - 1$, it would follow from (8.6) and (8.7) that $Q_{\mathcal{A}}(x)$, resp. $Q_{\mathcal{B}}(x)$, is a (non–zero) constant multiple of $x^{p-1} + x^{p-2} + \cdots + 1$, whence $|\mathcal{A}| \geq p$, resp. $|\mathcal{B}| \geq p$. This contradicts (8.1) which completes the proof of Theorem 4.

In Section 4 we mentioned that there are negative examples with sums $\mathcal{A} + \mathcal{B}$ having property P_k, i.e., examples for primes p which are not k–good. Now we will present examples of this type.

First we recall that in the special case $k = 2$ in [6] we proved that a prime p is 2–good if and only if 2 is a primitive root modulo p. There we presented several examples for sums $\mathcal{A} + \mathcal{B}$ possessing property P_2 (so that for the corresponding primes p, 2 is not a primitive root modulo p). Some of these examples follow:

Example 1. If $p = 7$, $\mathcal{A} = \{0, 1, 3\}$ and $\mathcal{B} = \{0, 1, 2, 4\}$, then $\mathcal{A} + \mathcal{B}$ possesses property P_2 so that the triples $(3, 4, 7)$ and $(4, 3, 7)$ are not 2–admissible.

Example 2. If $p = 17$, $\mathcal{A} = \{0, 3, 4, 5, 8\}$ and $\mathcal{B} = \{0, 3, 4, 5, 6, 9\}$, then $\mathcal{A} + \mathcal{B}$ has property P_2 so that $(5, 6, 17)$ and $(6, 5, 17)$ are not 2–admissible.

Example 3. If $p = 31$, $\mathcal{A} = \{0, 2, 5\}$ and $\mathcal{B} = \{0, 2, 4, 5, 6, 8, 9, 13, 14, 15, 16, 17, 20, 21, 23, 26\}$, then $\mathcal{A} + \mathcal{B}$ has property P_2, thus $(3, 16, 31)$ and $(16, 3, 31)$ are not 2–admissible.

One might like to present similar negative examples for other k (and p) values as well. To find examples of this type, one has to consider the proof of Theorem 4. We obtain that for fixed k and p, we have to look for non–trivial factorization of $x^p - 1$ over $\mathbb{Z}_k$ of the form

$$x^p - 1 = Q_1(x) Q_2(x) \qquad (8.8)$$

with

$$Q_1(x) = \sum_{a \in \mathcal{A}} x^a \text{ and } Q_2(x) = \sum_{d \in \mathcal{D}} t_d x^d.$$

(Here "non–trivial" means that both $Q_1(x)$ and $Q_2(x)$ have at least 2 terms.)

If we find a factorization of this form, then defining $\mathcal{B}$ so that it contains the elements $d \in \mathcal{D}$ each with multiplicity t_d, the sum $\mathcal{A} + \mathcal{B}$ possesses property P_k so that the triple $(|\mathcal{A}|, |\mathcal{B}|, p)$ is not k–admissible. The difficulty is that not only we have to find a non–trivial factorization of form (8.8), but also there is the additional restriction that all the coefficients of $Q_1(x)$ must be 0 or 1. This is the reason for that if k is a prime, then for $k > 2$ we can give only a sufficient condition for p being k–good. On the other hand, combining the proof of Theorem 4 and the argument above, we can prove that if k is a prime then a prime p is (k,k)–good if and only if k is a primitive root modulo p. (In [6] we proved this in the special case $k = 2$.)

Example 4. If $p = 13$, then we have

$$x^{13} - 1 = (1 + x + x^4 + x^6)(2 + x + 2x^2 + x^3 + 2x^5 + x^7)$$

over $\mathbb{Z}_3$. It follows that, writing $\mathcal{A} = \{0, 1, 4, 6\}$, $\mathcal{B} = \{0, 0, 1, 2, 2, 3, 5, 5, 7\}$, the sum $\mathcal{A} + \mathcal{B}$ possesses property P_3, so that $(4, 9, 13)$ is not 3–admissible, and thus $p = 13$ is not 3–good.

If we have a negative example for a certain $k \in \mathbb{N}$ and prime p, and $k \mid k'$, then one can use this example to construct negative examples for k' and p. E.g., starting out from Example 3, we obtain the following negative example for $k = 6$ and $p = 31$:

Example 5. If $p = 31$, $\mathcal{A} = \{0, 2, 4, 5, 6, 8, 9, 13, 14, 15, 16, 17, 20, 21, 23, 26\}$ and $\mathcal{B} = \{0, 0, 0, 2, 2, 2, 5, 5, 5\}$, then $\mathcal{A} + \mathcal{B}$ has property P_6, thus $(16, 9, 31)$ is not 6–admissible.

Finally, we will study the following question: is it true that for any $k \in \mathbb{N}$, $k \geq 2$ there are infinitely many k–good primes? Based on Theorem 4 and considering the work related to Artin's conjecture [8], [9] one would expect that the answer is affirmative, however, this is certainly beyond reach at the moment. On the other hand, we can prove that the affirmative answer would follow from Schinzel's Hypothesis H [15], [16] (see also [7, p. 21]) which generalizes the twin prime conjecture:

Hypothesis H. *If $k \in \mathbb{N}$, $F_1, \ldots, F_k$ are distinct irreducible polynomials in $\mathbb{Z}[x]$ (with positive leading coefficients) and the product polynomial $F = F_1 \ldots F_k$ has no fixed prime divisor, then there exist infinitely many integers n such that each $F_i(n)$ $(i = 1, \ldots, k)$ is a prime.*

Theorem 5. *If Hypothesis H is true, then for any primes $q_1 < \cdots < q_s$ there are infinitely many primes p so that each of $q_1, \ldots, q_s$ is a primitive root modulo p.*

Proof of Theorem 5. Let $r_1, \ldots, r_t$ be the odd primes amongst $q_1, \ldots, q_s$ (i.e., $\{r_1, \ldots, r_t\} = \{q_1, \ldots, q_s\} \setminus \{2\}$). For $i = 1, \ldots, t$, let u_i denote an arbitrary quadratic non–residue modulo r_i. Consider the linear congruence system

$$4x + 1 \equiv u_1 \pmod{r_1}$$
$$\vdots$$
$$4x + 1 \equiv u_t \pmod{r_t}.$$

Clearly, each of these linear congruences can be solved, and the moduli are coprime, thus this system has a unique solution modulo $r_1 \ldots r_t$. Let p_o be a positive element of this residue class so that

$$4p_o + 1 \equiv u_i \pmod{r_i} \ (\text{for } i = 1, \ldots, t). \tag{8.9}$$

Write

$$F_1(n) = p_o + nr_1 \ldots r_t$$

and

$$F_2(n) = 4F_1(n) + 1 = (4p_o + 1) + 4nr_1 \ldots r_t.$$

We will show that $F = F_1 F_2$ has no fixed prime divisor. $F_2(n)$ is always odd and $r_1 \ldots r_t$ is odd, thus $F_1(n)$ is odd infinitely often, whence $F_1(n)F_2(n)$ is also odd infinitely often. For $i = 1, 2, \ldots, t$, the number u_i is a quadratic non–residue modulo r_i, thus u_i cannot be congruent to 0 or 1 modulo r_i. By (8.9), it follows that

$$4F_1(n) \equiv 4p_o \equiv u_i - 1 \not\equiv 0 \ (\mathrm{mod}\ r_i)$$

and

$$F_2(n) \equiv 4p_o + 1 \equiv u_i \not\equiv 0 \ (\mathrm{mod}\ r_i)$$

so that $\big(r_i, F_1(n)F_2(n)\big) = 1$ for all i. Finally, if v is a prime different from each of $2, r_1, \ldots, r_t$, then

$$F_1(n)F_2(n) \equiv 0 \ (\mathrm{mod}\ v) \tag{8.10}$$

is a quadratic congruence which has at most 2 solutions modulo v. Since $v > 2$, there is at least one residue class modulo v which does not satisfy (8.10), so for all n from this residue class $v \nmid F_1(n)F_2(n)$.

Thus, indeed, $F_1 F_2$ has no fixed prime divisor, the polynomials $F_1, F_2 \in \mathbb{Z}[x]$ are linear and thus irreducible in $\mathbb{Z}[x]$, and their leading coefficients are positive, so that all the conditions in Hypothesis H hold. Since now this hypothesis is assumed to be true, there are infinitely many $n \in \mathbb{N}$ so that both

$$z = F_1(n) = p_o + nr_1 \ldots r_t \tag{8.11}$$

and

$$p = F_2(n) = 4z + 1 = (4p_o + 1) + 4nr_1 \ldots r_t \tag{8.12}$$

are primes. We will show that for such an n large enough, each of $2, r_1, \ldots, r_t$ is a primitive root modulo $p = p(n)$.

Since $p - 1 = 4z$ and z is a prime, all the positive divisors of $p - 1$ are $1, 2, 4, z, 2z$ and $4z$. Thus if $(g, p) = 1$ and g is not a primitive root modulo p, then we must have either

$$g^4 \equiv 1 \ (\mathrm{mod}\ p) \tag{8.13}$$

or

$$g^{\frac{p-1}{2}} \equiv 1 \ (\mathrm{mod}\ p). \tag{8.14}$$

Since now p is assumed to be large, (8.13) does not hold for $g = 2, r_1, \ldots, r_t$. Thus if one of these numbers is not a primitive root modulo p, then it must satisfy (8.14) whence, by Euler's lemma,

$$\left(\frac{g}{p}\right) = +1$$

(where $\left(\frac{g}{p}\right)$ denotes the Legendre symbol). Thus it suffices to show that

$$\left(\frac{g}{p}\right) = -1 \text{ for } g = 2, r_1, \ldots, r_t. \tag{8.15}$$

By (8.12) we have $p = 4z + 1$ where z is an odd prime, and thus p is of form $8k + 5$, whence (8.15) follows if $g = 2$. If $g = r_i$, $1 \le i \le t$, then by the quadratic reciprocity law we have

$$\left(\frac{r_i}{p}\right) = (-1)^{\frac{r_i-1}{2} \cdot \frac{p-1}{2}} \left(\frac{p}{r_i}\right). \tag{8.16}$$

By (8.12), $\frac{p-1}{2} = 2z$ is even and thus

$$(-1)^{\frac{r_i-1}{2} \cdot \frac{p-1}{2}} = +1. \tag{8.17}$$

Moreover, by (8.9) and (8.12) we have

$$p \equiv 4p_0 + 1 \equiv u_i \;(\text{mod } r_i)$$

whence, by the definition of u_i,

$$\left(\frac{p}{r_i}\right) = \left(\frac{u_i}{r_i}\right) = -1. \tag{8.18}$$

(8.15) with r_i in place of g follows from (8.16), (8.17) and (8.18), and this completes the proof of Theorem 5.

4   Extension of the Notion of f–Complexity and a Construction with High f–Complexity

In [4] we introduced the notion of f–complexity ("f" for family) of families of binary sequences. This notion can be generalized easily to families on k symbols:

Definition 6. *If $\mathcal{A}$ is a set of k symbols, $N, t \in \mathbb{N}$, $t < N$, $(\varepsilon_1, \ldots, \varepsilon_t) \in \mathcal{A}^t$, $i_1, \ldots, i_t$ are positive integers with $1 \le i_1 < \cdots < i_t \le N$, and we consider sequences $E_N = (e_1, \ldots, e_N) \in \mathcal{A}^N$ with*

$$e_{i_1} = \varepsilon_1, \ldots, e_{i_t} = \varepsilon_t, \tag{9.1}$$

then $(e_{i_1}, \ldots, e_{i_t}; \varepsilon_1, \ldots, \varepsilon_t)$ is said to be a specification of E_N of length t or a t–specification of E_N.

Definition 7. *The f–complexity of a family $\mathcal{F}$ of sequences $E_N \in \mathcal{A}^N$ on k symbols is defined as the greatest integer t so that for any t–specification (9.1) there is at least one $E_N \in \mathcal{F}$ which satisfies it. The f–complexity of $\mathcal{F}$ is denoted by $\Gamma_k(\mathcal{F})$. (If there is no $t \in \mathbb{N}$ with the property above, we set $\Gamma_k(\mathcal{F}) = 0$.)*

Note that the special case $k = 2$ of this definition is the notion of f–complexity of families of binary sequences introduced in [4].

One might like to show that the family constructed in Theorem 2, or at least a slightly modified version of it, is also of high f–complexity. Unfortunately, we have been able to prove only a partial result in this direction: we can handle only the case when k, the size of the alphabet, is a prime number (this, of course, includes the binary case). We will explain the difficulties arising in the case of composite k later. We hope to return to this case in a subsequent paper, and there we will present other constructions where the f–complexity can be handled also for composite k.

Theorem 6. *Assume that k, p are prime numbers, χ is a (multiplicative) character modulo p of order k (so that $k \mid p - 1$), $H \in \mathbb{N}$, $H < p$. Consider all the polynomials $f(x) \in F_p[x]$ with the properties that*

$$0 < \deg f(x) \leq H \tag{9.2}$$

and

$$\text{in } \bar{F}_p \text{ the multiplicity of each zero of } f(x) \text{ is less than } k. \tag{9.3}$$

For each of these polynomials $f(x)$, consider the sequence $E_p = E_p(f) = (e_1, \ldots, e_p)$ of k–th roots of unity defined as in Theorem 2:

$$e_n = \begin{cases} \chi\big(f(n)\big) & \text{for } \big(f(n), p\big) = 1 \\ +1 & \text{for } p \mid f(n). \end{cases}$$

Then we have

$$\delta(E_p) < 11 H p^{1/2} \log p. \tag{9.4}$$

Moreover, if $\ell \in \mathbb{N}$ and

(i) either

$$(4H)^\ell < p \tag{9.5}$$

(ii) or k is a primitive root modulo p and $\ell < p$,

then also

$$\gamma_\ell(E_p) < 10\ell H k p^{1/2} \log p \tag{9.6}$$

holds. Finally, we have

$$\Gamma_k(\mathcal{F}) \geq H. \tag{9.7}$$

Proof of Theorem 6. The proof is a combination and extension of Theorem 1 in [4] and Theorem 2 above, thus we will leave some details to the reader.

In order to prove (9.4), we argue in the same way as in the proof of (3.2) in the proof of Theorem 2. Again we set $g(x) = f(u + xv)$ and $\chi_1 = \chi^t$ with

$$1 \leq t \leq k - 1. \tag{9.8}$$

Then by (9.3) the multiplicity of the zeros of $g(x)$ is less than k, and since the order of χ is k and k is now a prime number, it follows from (9.8) that the character χ_1 is also of order k. Thus by Lemma 2, again (3.16) holds with H in place of h, and then we may complete the proof of (9.4) in the same way as the proof of (3.2) was completed.

Similarly, in order to prove (9.6), we argue as in the proof of (3.3) in the proof of Theorem 2. We define $B, f_1(x)$ and $G(x)$ as there: $f(x) = Bf_1(x)$, $f_1(x)$ is unitary,

$$G(x) = f_1(x + d_1)^{t_1} \ldots f_1(x + d_\ell)^{t_\ell} \tag{9.9}$$

with

$$0 \le t_1, \ldots, t_\ell \le k - 1, \quad (t_1, \ldots, t_\ell) \neq (0, \ldots, 0), \tag{9.10}$$

and again we get that (3.25) and (3.26) hold, and it suffices to show that the analogue of Lemma 3 holds.

Lemma 5. *If k, f, H, ℓ are defined as in Theorem 6, then $G(x)$ has at least one zero (in $\bar{F}_p$) whose multiplicity is not divisible by k.*

Indeed, assuming that Lemma 5 holds, the proof of (9.6) can be completed in the same way (with H in place of h) as the proof of (3.3) using Lemma 3. Thus it remains to prove Lemma 5.

Proof of Lemma 5. We argue as in the proof of Lemma 3, i.e., we consider the same equivalence relation as there, then we write $f_1(x)$ as the product of irreducible polynomials over F_p, and finally we group these factors so that in each group the equivalent irreducible factors are collected. However, there is a crucial difference with Lemma 3: while in Theorem 2 we assumed that $f(x)$ has no multiple zero, now this condition is relaxed to the weaker condition (9.3). It follows that now the irreducible factors may have an exponent not exceeding $k - 1$. So now a typical group of equivalent irreducible factors looks like $\varphi(x + a_1)^{s_1}, \ldots, \varphi(x + a_r)^{s_r}$ where

$$1 \le s_1, \ldots, s_r \le k - 1. \tag{9.11}$$

Then writing $G(x)$ in (9.9) as the product of irreducible polynomials over F_p, all the polynomials $\varphi(x + a_i + d_j)$ with $1 \le i \le r$, $1 \le j \le \ell$ occur amongst the factors, and for fixed i, j such a factor occurs with exponent exactly $s_i t_j$. Since now k is a prime, thus it follows from (9.10) and (9.11) that

$$\text{if } s_i t_j > 0 \text{ then } k \nmid s_i t_j. \tag{9.12}$$

The conclusion of Lemma 5 fails, i.e., the multiplicity of each of the zeros of $G(x)$ is divisible by k if and only if each of the factors $\varphi(x + a_i + d_j)$ occurs with an exponent divisible by k. This is so if and only if the following holds: if $\mathcal{A}$ denotes the k–set whose elements are $a_1, \ldots, a_r$, each a_i taken with multiplicity s_i, and $\mathcal{B}$ denotes the k–set whose elements are $d_1, \ldots, d_\ell$, each d_j taken with multiplicity t_j, then $\mathcal{A} + \mathcal{B}$ possesses property P_k. Take any of the groups formed by the equivalent irreducible factors (by (9.2) there is at least one such group), and consider the corresponding sum $\mathcal{A} + \mathcal{B}$ with property P_k. Then $(|\mathcal{A}|, |\mathcal{B}|, p)$ is **not** a (k, k)–admissible triple, and here we have

$$|\mathcal{A}| = \sum_{i=1}^{r} s_i \le \sum_{i=1}^{r} (k - 1) = r(k - 1) \le (\deg f_1)(k - 1) \le H(k - 1)$$

and

$$|\mathcal{B}| = \sum_{j=1}^{\ell} t_j \le \ell(k - 1).$$

It remains to show that assuming either (i) or (ii) (in Theorem 6), this is impossible.

(Observe that now we are studying (k, k)–admissibility instead of the k–admissibility occurring in the proof of Theorem 2; this is the price paid for relaxing the condition on the zeros of the polynomial $f(x)$ which is necessary for controlling the f–complexity. It is much more difficult to control (k, k)–admissibility than k–admissibility, since if we study (k, k)–admissibility then the set $\mathcal{A}$ in the sums $\mathcal{A} + \mathcal{B}$ considered also can be a multiset, thus we have more flexibility in constructing negative examples. Indeed, when k is composite, and both $\mathcal{A}$ and $\mathcal{B}$ can be k–sets, then it is easy to give negative examples of the type described in Example 5; this is why we cannot control the f–complexity for composite k.)

Assume first that (i) holds. Let $\bar{A}$ and $\bar{B}$ denote the set of the distinct elements of $\mathcal{A}$, resp. $\mathcal{B}$: $\bar{A} = \{a_1, \ldots, a_r\}$, $\bar{B} = \{d_1, \ldots, d_\ell\}$. Then by (9.2) and (9.5) we have

$$(4r)^\ell \le (4 \deg f_1)^\ell = (4 \deg f)^\ell \le (4H)^\ell < p$$

so that (9.1) in Theorem 3, (ii) holds with r and ℓ in place of t, resp. r. Thus the argument in the proof of Theorem 3, (ii) can be used with $k = 2$, and then we obtain that there is a $c \in \mathbb{Z}_p$ which has a unique representation in the form

$$d_j + a_i = c, \ d_j \in \bar{B}, a_i \in \bar{A}.$$

It follows that, considering also multiplicities,

$$a_i + d_j = c, \ a_i \in \mathcal{A}, d_\ell \in \mathcal{B}$$

has exactly $s_i t_j (> 0)$ solutions. By (9.12), this contradicts the assumption that $\mathcal{A} + \mathcal{B}$ has property P_k which completes the proof in this case.

Assume now that (ii) holds. Then we use the notations of the proof of Theorem 4, so that, by (9.11), $Q_{\mathcal{A}}(x) = \sum_{i=1}^{r} s_i x^{s(a_i)} \in F_k[x]$, by (9.10) $Q_{\mathcal{B}}(x) = \sum_{j=1}^{\ell} t_j x^{s(d_j)} \in F_k[x]$, and, since $\mathcal{A} + \mathcal{B}$ possesses property P_k, $Q_{\mathcal{A}+\mathcal{B}}(x) = 0$ in $F_k[x]$. Again, (8.2) holds, whence it follows that $x^{p-1} + x^{p-2} + \cdots + x + 1$ divides $Q_{\mathcal{A}}(x)Q_{\mathcal{B}}(x)$. Since it is now assumed that k is a primitive root modulo p, thus by Lemma 4 the polynomial $x^{p-1} + x^{p-2} + \cdots + x + 1$ is irreducible over

F_k. It follows that $x^{p-1} + x^{p-2} + \cdots + x + 1$ divides either $Q_A(x)$ or $Q_B(x)$, so that either $Q_A(x)$ or $Q_B(x)$ is a constant multiple of this polynomial, but this is impossible by $r \leq \deg f \leq H < p$ and $\ell < p$, and this completes the proof of (9.6). It remains to prove (9.7).

As in [4], we use

Lemma 6. *If T is a field and $g(x) \in T[x]$ is a non–zero polynomial, then it can be written in the form*

$$g(x) = \big(h(x)\big)^k g^*(x) \tag{9.13}$$

where the multiplicity of each zero of $g^*(x)$ (in $\bar{F}_p$) is less than k.

Proof of Lemma 6. The special case $k = 2$ of this lemma was stated and proved in [4] as Lemma 1, and the general case presented here can be proved in the same way, thus we leave the details to the reader.

To prove (9.7), we have to show that for any specification of length H:

$$e_{i_1} = \varepsilon_1, \ldots, e_{i_H} = \varepsilon_H \ (i_1 < \cdots < i_H), \tag{9.14}$$

there is a polynomial $f(x) \in F_p[x]$ which satisfies (9.2) and (9.3) so that $E_p = E_p(f) \in \mathcal{F}$, and this sequence $E_p = E_p(f)$ satisfies the specification (9.14).

By $H < p$, there is an integer i_{H+1} with $0 < i_{H+1} \leq p$, $i_{H+1} \notin \{i_1, \ldots, i_H\}$. Let ε_0 be a k–th root of unity with

$$\varepsilon_0 \neq 1, \tag{9.15}$$

and set

$$\varepsilon_{H+1} = \varepsilon_0 \varepsilon_1. \tag{9.16}$$

Denote the distinct k–th roots of unity by $\varphi_1, \ldots, \varphi_k$, let $v_1, \ldots, v_k$ be integers with

$$\chi(v_i) = \varphi_i \ (\text{for } i = 1, \ldots, k),$$

and define $y_1, \ldots, y_{H+1}$ by

$$y_i = v_z \text{ where } z = z(i) \text{ is defined by } \varphi_z = \varepsilon_i. \tag{9.17}$$

By the well–known interpolation theorem, there is a unique polynomial $g(x) \in F_p[x]$ with

$$\deg g(x) \leq H \tag{9.18}$$

and

$$g(i_j) = y_j \text{ for } j = 1, \ldots, H+1. \tag{9.19}$$

(This polynomial can be determined by using either Lagrange interpolation or Newton interpolation.) By Lemma 6 (with $T = F_p$), this polynomial $g(x)$ can be written in the form (9.13). Let

$$f(x) = g^*(x). \tag{9.20}$$

Then by Lemma 6, (9.3) holds. It follows from (9.13), (9.18) and (9.20) that

$$\deg f(x) = \deg g^*(x) \le \deg g(x) \le H. \tag{9.21}$$

By (9.17) and (9.19) we have

$$g(i_j) = y_j = v_{z(j)}$$

so that

$$\chi\big(g(i_j)\big) = \chi(v_{z(j)}) = \varphi_{z(j)} \ (\ne 0) \tag{9.22}$$

and thus

$$\big(g(i_j), p\big) = 1 \text{ for } j = 1, \ldots, H+1. \tag{9.23}$$

By (9.13), (9.17), (9.20), (9.22) and (9.23) we have

$$\chi\big(g(i_j)\big) = \chi\big((h(i_j))^k\big)\chi\big(g^*(i_j)\big) = \chi\big(f(i_j)\big) = \varphi_{z(j)} = \varepsilon_j \text{ for } j = 1, \ldots, H+1. \tag{9.24}$$

It follows from (9.15), (9.16) and (9.24) that

$$\chi\big(f(i_1)\big) \ne \chi\big(f(i_{H+1})\big)$$

and thus $f(x)$ is not constant, i.e.,

$$\deg f(x) > 0. \tag{9.25}$$

(9.2) follows from (9.21) and (9.25). Finally, it follows from (9.24) and the definition of $E_p(f)$ that $E_p(f)$ satisfies the specification (9.14) and this completes the proof of the theorem.

5 On the Cardinality of a Smallest Family Achieving a Prescribed f–Complexity and Multiplicity

We introduce first k–ary extensions of two quantities studied in [4].

Definition 8. *For positive integers* $j \le K \le N, M$ *and the alphabet* $\mathcal{A} = \{a_1, \ldots, a_k\}$ *set*

$$S(N, j, M, k) = \min\big\{|\mathcal{F}| : \mathcal{F} \subset \mathcal{A}^N, \forall(\varepsilon_1, \ldots, \varepsilon_j) \in \mathcal{A}^j \text{ and } 1 \le i_1 < \cdots < i_j \le N$$
$$\text{there are at least } M \text{ members } E_N = (e_1, \ldots, e_N) \text{ of } \mathcal{F}$$
$$\text{with } j\text{-specification } (e_{i_1}, \ldots, e_{i_j}; \varepsilon_1, \ldots, \varepsilon_j)\big\}. \tag{10.1}$$

We also say for the $\mathcal{F}$'s considered here that they *cover* every j–specification with multiplicity $\ge M$.

In particular for $M = 1$ and $j = K$ we get

$$S(N, K, k) \triangleq S(N, K, 1, k) = \min\big\{|\mathcal{F}| : \mathcal{F} \subset \mathcal{A}^N, \Gamma_k(\mathcal{F}) = K\big\}, \tag{10.2}$$

which counts how many sequences $E_N \in \mathcal{A}^N$ are needed to cover all K specifications, that is, to have f–complexity $\Gamma_k(\mathcal{F}) = K$.

Finding this number can be formulated as a covering problem for the hypergraph

$$\mathcal{H}H(N, K, k) = (\mathcal{V}(N, K, k), \mathcal{E}(N, k)),$$

where $\mathcal{E}(N, k) = \mathcal{A}^N$ is the edge set and the vertex set $\mathcal{V}(N, K, k)$ is defined as the set of K–specifications for $\mathcal{A}^N$ or, equivalently, as set of $(N-K)$–dimensional subcubes of $\mathcal{A}^N$ and thus

$$|\mathcal{V}(N, K, k)| = \binom{N}{K} k^K, |\mathcal{E}(N, k)| = k^N \tag{10.3}$$

$E_N \in \mathcal{E}(N, k)$ contains specification V if and only if $E_N \text{``} \in \text{''} V$. We derive now bounds on $S(N, K, k)$ and use (as in [4] for $k = 2$)

Lemma 7. (Covering Lemma 1 of [1]) *For any hypergraph* $(\mathcal{V}, \mathcal{E})$ *with*

$$\min_{v \in \mathcal{V}} \deg(v) \geq d \tag{10.4}$$

there exists a covering $\mathcal{C} \in \mathcal{E}$ *with*

$$|\mathcal{C}| \leq \left\lceil \frac{|\mathcal{E}|}{d} \log |\mathcal{V}| \right\rceil.$$

Theorem 7. *The cardinality* $S(N, K, k)$ *of a smallest family* $\mathcal{F} \subset \mathcal{A}^N$ *with*

f–*complexity* $\Gamma_k(\mathcal{F}) = K$ *satisfies*

$$k^K \leq S(N, K, k) \leq k^K \log \binom{N}{K} k^K \leq k^K K \log N \quad (K \geq k^3).$$

Proof: Application of Lemma 7 to our hypergraph $\mathcal{H}H(N, K, k)$ yields with $d = k^{N-K}$ a family $\mathcal{F}$ with $\Gamma_k(\mathcal{F}) \geq K$,

$$|\mathcal{F}| \leq \left\lceil \frac{k^N}{k^{N-K}} \log \binom{N}{K} k^K \right\rceil \leq k^K K \log N \quad (K \geq k^3)$$

and thus the upper bound for $S(N, K, k)$.

On the other hand one edge E_N covers exactly $\binom{N}{K}$ K–specifications and therefore by (10.3) necessarily as lower bound we have

$$S(N, K, k) \geq k^K.$$

We explained already in [4] that in order to make it difficult for an eavesdropper to identify a key $E_N \in \mathcal{F}$, when he has observed j positions, we must leave him many options. This can be achieved by constructing a family $\mathcal{F}$ of high f–complexity $\Gamma_k(\mathcal{F})$. Indeed for $j < \Gamma_k(\mathcal{F})$ the multiplicity $M_j(\mathcal{F})$, that is, the least multiplicity of every j–specification satisfies

$$M_j(\mathcal{F}) \geq k^{\Gamma_k(\mathcal{F})-j}, \tag{10.5}$$

because a j–specification can be extended to as many $\Gamma_k(f)$–specifications with the same support. Therefore

$$\min_{\mathcal{F}:\Gamma_k(\mathcal{F})\geq K} M_j(\mathcal{F}) \geq k^{K-j} \tag{10.6}$$

and thus

$$S(N, j, k^{K-j}, k) \leq S(N, K, k) \leq k^K K \log N \quad (K \geq k^3). \tag{10.7}$$

On the other hand, since $|\mathcal{V}(N, j, k)| = \binom{N}{j}k^j$ and an edge E_N covers exactly $\binom{N}{j}$ j–specifications, necessarily

$$S(N, j, k^{K-j}, k) \geq k^{K-j}\binom{N}{j}k^j\binom{N}{j}^{-1} = k^K. \tag{10.8}$$

Quite surprisingly, for $K \log N$ small relative to k^K the two bounds are very close to each other. The fact that $S(N, K, k)$ and therefore f–complexity contains almost complete information about the quantity $S(N, j, k^{K-j}, k)$ measuring multiplicity for the eavesdropper demonstrates the usefulness of our complexity measure. We summarize these findings.

Theorem 8. *The cardinality $S(N, j, k^{K-j}, k)$ of a smallest family $\mathcal{F} \subset \mathcal{A}^N$ which covers every j–specification with multiplicity $\geq k^{K-j}$ satisfies for all $j \leq K \leq N$*

$$k^K \leq S(N, j, k^{K-j}, k) \leq S(N, K, k) \leq k^K K \log N \quad (K \geq k^3).$$

6   Balanced Families with Prescribed f–Complexity

Definition 9. *A family $\mathcal{F} \subset \mathcal{A}^N$ with f–complexity $\Gamma_k(\mathcal{F}) = K$ is said to be c–balanced for some constant $c \in \mathbb{N}$, if no K–specification is covered by more than c sequences $E_N \in \mathcal{F}$.*

We improve now Theorem 7 by adding c–balancedness.

Theorem 9. *For $c = \log|\mathcal{V}(N, K, k)| = \log\binom{N}{K}k^K \leq K \log N$ $(K \geq k^3)$ the smallest c–balanced family $\mathcal{F} \subset \mathcal{A}^N$ with f–complexity $\Gamma_k(\mathcal{F}) = K$ has a cardinality meeting the bounds on $S(N, K, k)$ in Theorem 7.*

Proof: We replace Lemma 7 by a lemma on balanced coverings.

Definition 10. *A covering $\mathcal{C} \triangleq \{E_1, \ldots, E_L\}$ of a hypergraph $\mathcal{H}H = (\mathcal{V}, \mathcal{E})$ is called c–balanced for some constant $c \in \mathbb{N}$, if no vertex occurs in more than c edges of $\mathcal{C}$.*

Lemma 8. *(Covering Lemma 3 of [1, Part II]) A hypergraph $\mathcal{H}H = (\mathcal{V}, \mathcal{E})$ with maximal and minimal degrees $d_{\max} \triangleq \max\limits_{v \in \mathcal{V}} \deg(v)$ and $d_{\min} \triangleq \min\limits_{v \in \mathcal{V}} \deg(v) > 0$ has a c–balanced covering $\mathcal{C} = \{E_1, \ldots, E_L\}$ if*

(a) $L \geq \lceil |\mathcal{E}| d_{\min}^{-1} \cdot \log |\mathcal{V}| \rceil + 1$
(b) $c \leq L \leq c |\mathcal{E}| d_{\max}^{-1}$
(c) $\exp\left\{ -D\left(\lambda \| \frac{d_{\max}}{|\mathcal{E}|} \right) L + \log |\mathcal{V}| \right\} < \frac{1}{2}$ for $\lambda \triangleq \frac{c}{L}$

(Here D denotes the Kullback–Leibler divergence.)

Using Lemma 8 with $d_{\min} = d_{\max} = d = k^{N-K}$ and

$$c = \log |\mathcal{V}| = \log |\mathcal{B}(N, K, k)| = \log \binom{N}{K} k^K \leq K \log N \quad (K \geq k^3)$$

we get a c–balanced covering of said cardinality.

Remark: Using Theorem 9 also the bounds in Theorem 8 can be obtained in a c–balanced way with $c = K \log N$ by the previous reasoning.

Next we go for improvements of the balancedness property. It is known from probability theory that for large deviations the following inequality holds:

For a sequence $Z_1, Z_2, \ldots, Z_L$ of independent, identically distributed random variables with values in $[0, 1]$ and expectation $EZ_i = \mu$ for $0 < \varepsilon < 1$

$$Pr\left\{ \frac{1}{L} \sum_{i=1}^{L} Z_i \notin [(1-\varepsilon)\mu, (1+\varepsilon)\mu] \right\} \leq 2 \exp\left(-L \frac{\varepsilon^2 \mu}{2\ell n 2} \right).$$

This can be used to establish another balancedness property, which also gives a bound from below, but in exchange most, but not necessarily all, vertices satisfy it. This suggests to apply a more recent auxiliary result.

Lemma 9. *[2] Let $\mathcal{H}H = (\mathcal{V}, \mathcal{E})$ be an e–uniform hypergraph (all edges' cardinalities equal e) and P a probability distribution on $\mathcal{E}$. Consider a probability distribution Q on $\mathcal{V}$: $Q(v) \triangleq \sum\limits_{E \in \mathcal{E}} P(E) \frac{1}{e} 1_E(v)$.*

Fix $\varepsilon, \tau > 0$, and define the set of vertices $\mathcal{V}_0 = \left\{ v \in \mathcal{V} : Q(v) < \frac{\tau}{|\mathcal{V}|} \right\} \subset \mathcal{V}$, then there exist edges $E^{(1)}, \ldots, E^{(L)} \in \mathcal{E}$ such that for

$$\bar{Q}(v) \triangleq \frac{1}{L} \sum_{i=1}^{L} \frac{1}{e} 1_{E^{(i)}}(v)$$

(i) $Q(\mathcal{V}_0) \leq \tau$
(ii) $(1-\varepsilon)Q(v) \leq \bar{Q}(v) \leq (1+\varepsilon)Q(v)$ for all $v \in \mathcal{V} \setminus \mathcal{V}_0$
(iii) $L \leq \left\lceil \frac{|\mathcal{V}|}{e} \frac{2\ell n 2 \log(2|\mathcal{V}|)}{\varepsilon^2 \tau} \right\rceil$.

We apply this lemma now to the e–uniform hypergraph $\mathcal{H}H(N, K, k)$, whose edges have cardinality $e = \binom{N}{K}$. First notice that

$$L \leq \frac{\binom{N}{K}k^K}{\binom{N}{K}} \frac{3}{\varepsilon^2 \tau} \log \binom{N}{M} k^K = \frac{3}{\varepsilon^2 \tau} k^K \log \binom{N}{K} k^K \leq \frac{3}{\varepsilon^2 \tau} k^K K \log N (K \geq k^3).$$

Except for the constant $\frac{3}{\varepsilon^2 \tau}$ this is our previous bound.

Next choose as P the uniform PD on $\mathcal{E}(N, k)$. Then for all vertices $v \in \mathcal{V}(N, K, k)$

$$Q(v) = \sum_{E_N \in \mathcal{E}(N,k)} k^{-N} \binom{N}{K}^{-1} 1_{E_N}(v) = k^{-N} \binom{N}{K}^{-1} \deg(v)$$

$$= k^{-N} \binom{N}{K}^{-1} k^{N-K} = \frac{1}{\binom{N}{K}k^K} \tag{11.1}$$

and for $v \in \mathcal{V} \setminus \mathcal{V}_0$

$$(1 - \varepsilon) L e\, Q(v) \leq \sum_{i=1}^{L} 1_{e^{(i)}}(v) \leq (1 + \varepsilon) L e\, Q(v)$$

and for $\tau = 3/4$

$$(1 - \varepsilon) \frac{4}{\varepsilon^2} K \log N \leq \sum_{i=1}^{L} 1_{E^{(i)}}(v) \leq (1 + \varepsilon) \frac{4}{\varepsilon^2} K \log N. \tag{11.2}$$

This implies the uniformity property

$$\frac{1 - \varepsilon}{1 + \varepsilon} \leq \min_{v,v' \in \mathcal{V} \setminus \mathcal{V}_0} \left(\sum_{i=1}^{L} 1_{E^{(i)}}(v) \right) \left(\sum_{i=1}^{L} 1_{E^{(i)}}(v') \right)^{-1}$$

$$\leq \max_{v,v' \in \mathcal{V} \setminus \mathcal{V}_0} \left(\sum_{i=1}^{L} 1_{E^{(i)}}(v) \right) \left(\sum_{i=1}^{L} 1_{E^{(i)}}(v') \right)^{-1} \leq \frac{1 + \varepsilon}{1 - \varepsilon}. \tag{11.3}$$

By choosing τ small most vertices are in $\mathcal{V} \setminus \mathcal{V}_0$.

Now comes a **surprise**. Our hypergraph has strong symmetries and by (11.1) $Q(v)$ is independent of v. Therefore for $\tau = 3/4 < 1$ $\mathcal{V}_0 = \phi$ and (11.3) holds for all vertices. We have established

Theorem 10. *For every $\varepsilon \in (0, 1)$ there is a family $\mathcal{F} \subset \mathcal{A}^N$ with f–complexity $\Gamma_K(\mathcal{F}) = K, k^K \leq |\mathcal{F}| \leq \frac{4}{\varepsilon^2} k^K \log N (K \geq k^3)$ such that for every K–specification the number of sequences $E_N \in \mathcal{F}$ which cover this specification lies between $\frac{4(1-\varepsilon)}{\varepsilon^2} K \log N$ and $\frac{4(1+\varepsilon)}{\varepsilon^2} K \log N$.*

7   Conclusion

We have constructed large families of sequences of k symbols with strong pseudorandom properties. We have also introduced and studied the notion of f–complexity of families of sequences on k symbols, and we have shown that the f–complexity of the family constructed by us is large if k, the size of the alphabet is a prime number but we have not been able to control the case when k is composite. We have also shown what are essentially minimal cardinalities of families with prescribed complexity and which additional multiplicity properties they may have.

One might like to construct families of large complexity for composite k as well; we will return to this problem in a subsequent paper.

References

1. R. Ahlswede, Coloring hypergraphs: A new approach to multi–user source coding, Part I, J. Combinatorics, Information and System Sciences 4, 1, 76–115, 1979; Part II, J. Combinatorics, Information and System Sciences 5, 3, 220–268, 1980.
2. R. Ahlswede, On concepts of performance parameters for channels, this volume.
3. R. Ahlswede and A. Winter, Strong converse for identification via quantum channels, IEEE Trans. on Inform., Vol. 48, No. 3, 569–579, 2002.
4. R. Ahlswede, L.H. Khachatrian, C. Mauduit, and A. Sárközy, A complexity measure for families of binary sequences, Periodica Math. Hungar., Vol. 46, No. 2, 107–118, 2003.
5. J. Cassaigne, C. Mauduit, and A. Sárközy, On finite pseudorandom binary sequences VII: The measures of pseudorandomness, Acta Arith. 103, 97–118, 2002.
6. L. Goubin, C. Mauduit, and A. Sárközy, Construction of large families of pseudorandom binary sequences, J. Number Theory, 106, 56-69, 2004.
7. H. Halberstam and H.-E. Richert, Sieve Methods, Academic Press, London, 1974.
8. D.R. Heath–Brown, Artin's conjecture for primitive roots, Quat. J. Math. 37, 27–38, 1986.
9. C. Hooley, On Artin's conjecture, J. reine angew. Math. 225, 209–220, 1967.
10. Y. Kohayakawa, C. Mauduit, C.G. Moreira and V. Rödl, Measures of pseudorandomness for random sequences, Proceedings of WORDS'03, 159–169, TUCS Gen. Publ., 27, Turku Cent. Comput. Sci., Turku, 2003.
11. R. Lidl and H. Niederreiter, Introduction to Finite Fields and Their Applications, revised edition, Cambridge University Press, 1994.
12. C. Mauduit, J. Rivat, and A. Sárközy, Construction of pseudorandom binary sequences using additive characters, Monatshefte Math., 141, 197-208, 2004
13. C. Mauduit and A. Sárközy, On finite pseudorandom binary sequences, I. Measure of pseudorandomness, the Legendre symbol, Acta Arith. 82, 365–377, 1997.
14. C. Mauduit and A. Sárközy, On finite pseudorandom sequences of k symbols, Indag. Math. 13, 89–101, 2002.
15. A. Schinzel, Remarks on the paper "Sur certaines hypothèses concernant les nombres premiers", Acta Arith. 7, 1–8, 1961/1962.
16. A. Schinzel and W. Sierpiński, Sur certaines hypothèses concernant les nombres premiers, ibid. 4, 185–208, 1958; Corrigendum ibid. 5, 259, 1959.
17. A. Weil, Sur les courbes algébriques et les variétés qui s'en déduisent, Act. Sci. Ind. 1041, Hermann, Paris, 1948.

On a Fast Version of a Pseudorandom Generator

K. Gyarmati*

Abstract. In an earlier paper I constructed a large family of pseudo-random sequences by using the discrete logarithm. While the sequences in this construction have strong pseudorandom properties, they can be generated very slowly since no fast algorithm is known to compute ind n. The purpose of this paper is to modify this family slightly so that the members of the new family can be generated much faster, and they have almost as good pseudorandom properties as the sequences in the original family.

1 Introduction

In this work I will continue the work initiated in [5]. C. Mauduit and A. Sárközy [9, pp. 367-370] introduced the following measures of pseudorandomness:
For a finite binary sequence $E_N = \{e_1, e_2, \ldots, e_N\} \in \{-1, +1\}^N$ write

$$U(E_N, t, a, b) = \sum_{j=0}^{t-1} e_{a+jb}$$

and, for $D = (d_1, \ldots, d_k)$ with non-negative integers $d_1 < \cdots < d_k$,

$$V(E_N, M, D) = \sum_{n=1}^{M} e_{n+d_1} e_{n+d_2}, \ldots e_{n+d_k}.$$

Then the *well-distribution measure* of E_N is defined as

$$W(E_N) = \max_{a,b,t} |U(E_N(t,a,b)| = \max_{a,b,t} \left| \sum_{j=0}^{t-1} e_{a+jb} \right|,$$

where the maximum is taken over all $a, b, t \in \mathbb{N}$ and $1 \le a \le a + (t-1)b \le N$. The *correlation measure of order* k of E_N is defined as

$$C_k(E_N) = \max_{M,D} |V(E_N, M, D)| = \max_{M,D} \left| \sum_{n=1}^{M} e_{n+d_1} e_{n+d_2}, \ldots e_{n+d_k} \right|,$$

where the maximum is taken over all $D = (d_1, d_2, \ldots, d_k)$ and M with $M + d_k \le N$. In [6] I introduced a further measure: Let

$$H(E_N, a, b) = \sum_{j=0}^{[(b-a)/2]-1} e_{a+j} e_{b-j},$$

---

* Research partially supported by Hungarian Scientific Research Grants OTKA T043631 and T043623.

R. Ahlswede et al. (Eds.): Information Transfer and Combinatorics, LNCS 4123, pp. 326–342, 2006.

and then the *symmetry measure* of E_N is defined as

$$S(E_N) = \max_{1 \leq a < b \leq N} |H(E_N, a, b)| = \max_{1 \leq a < b \leq N} \left| \sum_{j=0}^{[(b-a)/2]-1} e_{a+j} e_{b-j} \right|.$$

A sequence E_N is considered as a "good" pseudorandom sequence if each of these measures $W(E_N)$, $C_k(E_N)$ (at least for small k) and $S(E_N)$ is "small" in terms of N (in particular all are $o(N)$ as $N \longrightarrow \infty$). Indeed, it was proved in [3, Theorem 1, 2] and in [6, Theorem 1, 2] that for a truly random sequence $E_N \subseteq \{-1, +1\}^N$ each of these measures is $\ll \sqrt{N \log N}$ and $\gg \sqrt{N}$.

Throughout the paper we will use the following notations: $\| x \|$ is the distance of x from the closest integer, $e(\alpha) = e^{2\pi i \alpha}$, $\overline{\mathbb{F}_p}$ is the algebraic closured of the field $\mathbb{F}_p$. Finally, if p is a prime, α and m are natural numbers we say that $p^{\alpha} \| m$ if $p^{\alpha} \mid m$ but $p^{\alpha+1} \nmid m$.

Numerous binary sequences have been tested for pseudorandomness by J. Cassaigne, S. Ferenczi, C. Mauduit, J. Rivat and A. Sárközy. The sequences with the strongest pseudorandom properties have been constructed in [4], [5], [9], [11] and [12]. As concerning the strength of the pseudorandom properties these constructions are nearly equally good. But in the construction given by A. Sárközy in [12] and extended by me in [5], the generation of the sequences in question is much more slowly than in the other constructions. Indeed Sárközy's construction is the following:

Let p be an odd prime, $N = p - 1$ and define $E_N = \{e_1, \ldots, e_N\} \subseteq \{-1, +1\}^N$ by

$$e_n = \begin{cases} +1 & \text{if } 1 \leq \text{ind } n \leq \frac{p-1}{2}, \\ -1 & \text{if } \frac{p+1}{2} \leq \text{ind } n \leq p - 1. \end{cases} \tag{1}$$

Here ind n denotes the index or discrete logarithm of n modulo p, defined as the unique integer with

$$g^{\text{ind } n} \equiv n \pmod{p}, \tag{2}$$

and $1 \leq \text{ind } n \leq p - 1$, where g is a fixed primitive root modulo p. In [5] I extended this construction to a large family of binary sequences with strong pseudorandom properties by replacing n by a polynomial $f(n)$ in (1) (in the same way as the Legendre symbol construction in [9] was extended in [4].)

Indeed in [5] I proved for the generalized sequence:

Theorem A. *For all $f \in \mathbb{F}_p[x]$ with $k = \deg f$ we have $W(E_{p-1}) \leq 38kp^{1/2}(\log p)^2$.*
Moreover if one of the following conditions holds:

a) *f is irreducible;*
b) *If f has the factorization $f = \varphi_1^{\alpha_1} \varphi_2^{\alpha_2} \ldots \varphi_u^{\alpha_u}$, where $\alpha_i \in \mathbb{N}$ and the φ_i's are irreducible over $\mathbb{F}_p$, then there exists a β such that exactly one or two φ_i's have the degree β;*
c) *$\ell = 2$;*
d) *$(4\ell)^k < p$ or $(4k)^{\ell} < p$.*

Then

$$C_\ell(E_{p-1}) < 10k\ell 4^\ell p^{1/2}(\log p)^{\ell+1}.$$

Finally, if $f(x) \not\equiv f(t-x)$ for all $t \in \mathbb{Z}_p$, then $S(E_{p-1}) < 88kp^{1/2}(\log p)^3$.

As we pointed out earlier these constructions are nearly as good as the others, but the problem is that it is slow to compute e_n since no fast algorithm is known to compute ind n. The Diffie-Hellman key-exchange system utilizes the difficulty of computing ind n.

In this paper my goal is to improve on the construction in Theorem A by replacing the sequence

$$e_n = \begin{cases} +1 \text{ if } 1 \leq \text{ind } f(n) \leq \frac{p-1}{2}, \\ -1 \text{ if } \frac{p+1}{2} \leq \text{ind } f(n) \leq p-1 \text{ or } p \mid f(n) \end{cases} \tag{3}$$

by a sequence which can be generated faster. I will show that this is possible at the price of giving slightly weaker upper bounds for the pseudorandom measures. Throughout this paper we will use the following:

Notation. *Let p be an odd prime, g be a primitive root modulo p. Define* ind n *by (2). Let $f \in \mathbb{F}_p[x]$ be a polynomial of degree $k \geq 1$, and $f = ch^d$ where $c \in \mathbb{F}_p$ and $h \in \mathbb{F}_p[x]$ is not a perfect power of a polynomial over $\mathbb{F}_p[x]$. Moreover let*

$$p - 1 = mh$$

with $m, h \in \mathbb{N}$, and let x be relative prime to m: $(x, m) = 1$.

The crucial idea of the construction is to reduce ind n modulo m:

Construction 1. *Let* ind$^*n$ *denote the following function: For all $1 \leq n \leq p-1$*

$$\text{ind } n \equiv x \cdot \text{ind}^*n \pmod{m}$$

*(*ind$^*n$ *exists since $(x, m) = 1$.) Define the sequence $E_{p-1} = \{e_1, \ldots, e_{p-1}\}$ by*

$$e_n = \begin{cases} +1 \text{ if } 1 \leq \text{ind}^* f(n) \leq \frac{m}{2}, \\ -1 \text{ if } \frac{m}{2} < \text{ind}^* f(n) \leq m \text{ or } p \mid f(n). \end{cases} \tag{4}$$

Note that this construction also generalizes the Legendre symbol construction described in [4] and [9]. Indeed in the special case $m = 2$, $x = 1$ the sequence e_n defined in (4) becomes

$$e_n = \begin{cases} +1 \text{ if } \left(\frac{f(n)}{p}\right) = -1, \\ -1 \text{ if } \left(\frac{f(n)}{p}\right) = 1 \text{ or } p \mid f(n). \end{cases}$$

(In the special case $m = p - 1$, $x = 1$ we obtain the original construction given in (3)).

We will show that the construction presented above has good pseudorandom properties, each of the measures $W(E_{p-1})$, $C_k(E_{p-1})$ is small under certain conditions on the polynomial f. In the case of the well-distribution measure we can control the situation completely.

Theorem 1. *If $m/(m,d)$ is even we have*

$$W(E_{p-1}) \le 36kp^{1/2}\log p \log(m+1).$$

While in the other case, when $m/(m,d)$ is odd we have:

$$W(E_{p-1}) = \frac{p-1}{m} + O(kp^{1/2}\log p \log(m+1)).$$

In the case of the correlation measures the situation is slightly more difficult. When the order of the correlation measure is odd we have:

Theorem 2. *If $f \in \mathbb{F}_p$, $k = \deg f$ and ℓ are odd integers while m is an even integer, then we have*

$$C_\ell(E_{p-1}) < 9k\ell 4^\ell p^{1/2}(\log p)^{\ell+1}.$$

Otherwise we need the same conditions on the polynomial f as in [5] in the original construction. If the degree of the polynomial is small depending on m, the same upper bound holds as in [5], while in the general case I will prove a slightly weaker result.

Theorem 3. *i) Suppose that m is even or m is odd with $2m \mid p-1$, and at least one of the following 4 conditions holds:*

a) *f is irreducible;*
b) *If f has the factorization $f = \varphi_1^{\alpha_1}\varphi_2^{\alpha_2}\ldots\varphi_u^{\alpha_u}$ where $\alpha_i \in \mathbb{N}$ and the φ_i's are irreducible over $\mathbb{F}_p$, then there exists a β such that exactly one or two φ_i's have the degree β;*
c) *$\ell = 2$;*
d) *$(4\ell)^k < p$ or $(4k)^\ell < p$.*

Then

$$C_\ell(E_{p-1}) < 9k\ell 4^\ell p^{1/2}(\log p)^{\ell+1} + \frac{\ell! k^{\ell(\ell+1)}}{m^\ell}p. \tag{5}$$

ii) Moreover if we also have $2^\beta \parallel m$ and $k = \deg f < 2^\beta$ then

$$C_\ell(E_{p-1}) < 9k\ell 4^\ell p^{1/2}(\log p)^{\ell+1}.$$

For fixed m by Heath-Brown's work on Linnik's theorem [7] the least prime number p with $m \mid p-1$ is less than $cm^{5.5}$. Thus the condition $\deg f < 2^\beta \parallel m \mid p-1$ is not too restrictive.

If $m^{2\ell} < p$ holds, then the first term majorizes the second term in (5), thus the upper bound becomes $O\left(p^{1/2}(\log p)^{\ell+1}\right)$ where the implied constant factor may depend on k and ℓ.

The study of the symmetry measure also considered in [5] would lead to further complications and I could control it only under the further assumption $\deg f \le 2^{\beta+2}$ where β is defined by $2^\beta \parallel m$. Thus, I do not go into the details of this here.

In applications one should balance between the strength of the upper bounds and the speed of the generation of the sequence depending on our priorities. We will show in section 3 that the sequence described in (4), in particular ind $^*f(n)$, can be computed faster than the original construction. Indeed, if the prime factors of m are smaller than $\log p$ then ind$^* f(n)$ can be computed by $O((\log p)^6)$ bit operations.

In [2] R. Ahlswede, L.H. Khachatrian, C. Mauduit and A. Sárközy introduced the notion of f-complexity of families of binary sequences as a measure of applicability of the constructions in cryptography.

Definition 1. *The complexity $C(\mathcal{F})$ of a family $\mathcal{F}$ of binary sequence $E_N \in \{-1,+1\}^N$ is defined as the greatest integer j so that for any $1 \leq i_1 < i_2 < \cdots < i_j \leq N$, and for $\varepsilon_1, \varepsilon_2, \ldots, \varepsilon_j$, we have at least one $E_N = \{e_1, \ldots, e_N\} \in \mathcal{F}$ for which*

$$e_{i_1} = \varepsilon_1, \ e_{i_2} = \varepsilon_2, \ldots, e_{i_j} = \varepsilon_j.$$

We will see that the f-complexity of the family constructed in (4) is high.

Theorem 4. *Consider all the polynomials $f \in \mathbb{F}_p[x]$ with*

$$0 < \deg f \leq K.$$

For each of these polynomials f, consider the binary sequence $E_{p-1} = E_{p-1}(f)$ defined by (4), and let $\mathcal{F}$ denote the family of all binary sequences obtained in this way. Then we have

$$C(\mathcal{F}) > K.$$

In [10] C. Mauduit and A. Sárközy proved an inequality involving the pseudorandom measures W and C_2. The following is a generalization of their inequality:

Theorem 5. *For all $E_N \in \{-1,+1\}^N$, $3\ell^2 \leq N$ we have*

$$W(E_N) \leq 3\ell N^{1-1/(2\ell)} \left(C_{2\ell}(E_N)\right)^{1/(2\ell)}.$$

Here the constant factor 3ℓ could be improved by using a more difficult argument, I will return to this in a subsequent paper.

In section 4 we will prove Theorem 5 and using Theorems 1,2 and 3 we will show that Construction 1 provides a natural example for that the inequality in Theorem 5 is the best possible apart from a constant factor. Moreover, Construction 1 gives us a sequence for which the correlation measures of small order are small while the well-distribution measure is possibly large.

2 Proofs

2.1 Proof of Theorem 1

First we note that the sequence defined in (4) by the polynomial $f = h^d$ and the modulus m, remains the same sequence if we replace in Construction 1 the polynomial $f = h^d$ by the polynomial $h^{d/(m,d)}$ and the modulus m by the modulus $m/(m,d)$. Thus in order to prove this theorem it is sufficient to study the case when $(m,d) = 1$.

The proof of the theorem is very similar to the proof of Theorem 1 in [6]. By the formula

$$\frac{1}{m} \sum_{\chi:\chi^m=1} \overline{\chi}^j(a)\chi(b) = \begin{cases} 1 \text{ if } m \mid \text{ind } a - \text{ind } b, \\ 0 \text{ if } m \nmid \text{ind } a - \text{ind } b, \end{cases}$$

we obtain

$$e_n = 2 \sum_{\substack{1 \le j \le m/2 \\ jx \equiv \text{ind } f(n) \pmod{m}}} 1 - 1 = \frac{2}{m} \sum_{1 \le j \le m/2} \sum_{\chi:\chi^m=1} \overline{\chi}(f(n))\chi(g^{jx}) - 1.$$

Thus

$$e_n = \frac{2}{m} \sum_{1 \le j \le m/2} \sum_{\chi \ne \chi_0:\chi^m=1} \overline{\chi}(f(n))\chi(g^{xj}) + \frac{(-1)^m - 1}{2m}. \tag{6}$$

Assume now that $1 \le a \le a + (t-1)b \le N$. Then we have

$$|U(E_{p-1}, t, a, b)| = \left| \frac{2}{m} \sum_{\chi \ne \chi_0:\chi^m=1} \left(\sum_{i=0}^{t-1} \overline{\chi}(f(a+ib)) \right) \left(\sum_{j=1}^{[m/2]} \chi^j(g^x) \right) \right.$$
$$\left. + \frac{((-1)^m - 1)t}{2m} \right|. \tag{7}$$

We will prove the following:

$$S \stackrel{\text{def}}{=} \left| \frac{1}{m} \sum_{\chi \ne \chi_0:\chi^m=1} \left(\sum_{i=0}^{t-1} \overline{\chi}(f(a+ib)) \right) \left(\sum_{j=1}^{[m/2]} \chi^j(g^x) \right) \right|$$
$$\le 18kp^{1/2}(\log p)^2. \tag{8}$$

If m is even we obtain the statement of Theorem 1 immediately from (7) and (8). If m is odd using the triangle inequality we get

$$|U(E_{p-1}, t, a, b)| = \frac{t}{m} + O(kp^{1/2}(\log p)^2)$$

which completes the proof of Theorem 1. Thus in order to prove Theorem 1, we have to verify (8).

We will use the following lemma:

Lemma 1. *Suppose that p is a prime, χ is a non-principal character modulo p of order z, $f \in \mathbb{F}_p[x]$ has s distinct roots in $\overline{F}_p$, and it is not a constant multiple of a z-th power of a polynomial over $\mathbb{F}_p$. Let y be a real number with $0 < y \le p$. Then for any $x \in \mathbb{R}$:*

$$\left| \sum_{x < n \le x+y} \chi(f(n)) \right| < 9sp^{1/2} \log p.$$

Proof of Lemma 1
This is a trivial consequence of Lemma 1 in [1]. Indeed, there this result is deduced from Weil theorem, see [13].

Consider $\sum_{i=0}^{t-1} \overline{\chi}(f(a+ib))$ in (7), and here, let the order of χ be z. Since $\chi^m = 1$ we have $z \mid m$. On the other hand $f = ch^d$ is not a constant multiple of a z-th power of a polynomial over $\mathbb{F}_p$, since $1 = (m,d) = (z,d)$ (because of $z \mid m$) and h is not a perfect power of any polynomial over $\mathbb{F}_p$.

Using Lemma 1 we have:

$$\left| \sum_{i=0}^{t-1} \overline{\chi}(f(a+ib)) \right| \leq 9kp^{1/2} \log p$$

and thus by (8)

$$S \leq \frac{9kp^{1/2} \log p}{m} \sum_{\chi \neq \chi_0 : \chi^m = 1} \left| \sum_{j=1}^{[m/2]} \chi^j(g^x) \right|.$$

Lemma 2

$$\sum_{\chi \neq \chi_0 : \chi^m = 1} \left| \sum_{j=1}^{[m/2]} \chi^j(g^x) \right| \leq \sum_{\chi \neq \chi_0 : \chi^m = 1} \frac{2}{|1 - \chi(g^x)|} < 2m \log(m+1).$$

Proof of Lemma 2. This is Lemma 3 in [5] with m in place of d, $m/2$ in place of $(p-1)/2$ and g^x in place of g, respectively, and it can be proved in the same way.

Using Lemma 2 we obtain

$$S < 18kp^{1/2} \log p \log(m+1)$$

which proves (8) and this completes the proof of Theorem 1.

2.2   Proof of Theorem 2 and 3

In this section we may suppose that m is even: In Theorem 2 m cannot be odd. If m is odd in Theorem 3, then considering $2m$ in place of m and f^2 in place of f in Construction 1 we generate the same sequence; however in this case we have $(2m, 2d) > 1$.

To prove Theorems 2 and 3, consider any $\mathcal{D} = \{d_1, d_2, \ldots, d_\ell\}$ with non-negative integers $d_1 < d_2 < \cdots < d_\ell$ and positive integers M with $M + d_\ell \leq p-1$. Then arguing as in [12, p. 382] with $f(n+d_j)$ in place of $n+d_j$, m in place of $p-1$, and g^x in place of g from (6) and since m is even we obtain:

$$|V(E_N, M, \mathcal{D})| \leq \frac{2^\ell}{m^\ell} \sum_{\substack{\chi_1 \neq \chi_0 \\ \chi_1^m = 1}} \cdots \sum_{\substack{\chi_\ell \neq \chi_0 \\ \chi_\ell^m = 1}} \left| \sum_{n=1}^{M} \chi_1(f(n+d_1)) \cdots \chi_\ell(f(n+d_\ell)) \right| \times \prod \left| \sum_{\ell_j=1}^{m/2} \overline{\chi}_j(g^{x\ell_j}) \right|.$$

$$(9)$$

Now let χ be a modulo p character of order m; for simplicity we will choose χ as the character uniquely defined by $\chi(g) = e\left(\frac{x^*}{m}\right)$ where $xx^* \equiv 1 \pmod{m}$. Then

$$\chi(g^x) = e\left(\frac{1}{m}\right). \tag{10}$$

Let $\chi_u = \chi^{\delta_u}$ for $u = 1, 2, \ldots, \ell$, whence by $\chi_1 \neq \chi_0, \ldots, \chi_\ell \neq \chi_0$, we may take

$$1 \leq \delta_u < m.$$

Thus in (9) we have

$$\left|\sum_{n=1}^{M} \chi_1(f(n+d_1))\ldots\chi_\ell(f(n+d_\ell))\right| = \left|\sum_{n=1}^{M} \chi^{\delta_1}(f(n+d_1))\ldots\chi^{\delta_\ell}(f(n+d_\ell))\right|$$

$$= \left|\sum_{n=1}^{M} \chi\left(f^{\delta_1}(n+d_1)\ldots f^{\delta_\ell}(n+d_\ell)\right)\right|.$$

If $f^{\delta_1}(n+d_1)\cdots f^{\delta_\ell}(n+d_\ell)$ is not a perfect m-th power, then this sum can be estimated by Lemma 1, whence

$$\left|\sum_{n=1}^{M} \chi(f^{\delta_1}(n+d_1)\cdots f^{\delta_\ell}(n+d_\ell))\right| \leq 9s\ell p^{1/2}\log p.$$

Therefore by (9) and the triangle-inequality we get:

$$|V(E_N, M, D)| \leq \frac{2^\ell}{m^\ell} \sum_{\substack{\chi_1 \neq \chi_0 \\ \chi_1^m = 1}} \cdots \sum_{\substack{\chi_\ell \neq \chi_0 \\ \chi_\ell^m = 1}} 9s\ell p^{1/2}\log p \left|\prod_{j=1}^{\ell}\left(\sum_{l_j=1}^{m/2} \chi^{\delta_j}(g^{x\ell_j})\right)\right|$$

$$+ \frac{2^\ell}{m^\ell} \sum_{\substack{1 \leq \delta_1, \ldots, \delta_\ell \leq m, \\ f^{\delta_1}(n+d_1)\cdots f^{\delta_\ell}(n+d_\ell) \text{ is} \\ \text{a perfect } m\text{-th power}}} (p-1) \left|\prod_{j=1}^{\ell}\left(\sum_{l_j=1}^{m/2} \chi^{\delta_j}(g^{x\ell_j})\right)\right|$$

$$= \sum\nolimits_1 + \sum\nolimits_2. \tag{11}$$

From Lemma 2 the same way as in [12, p.384] we have

$$\sum\nolimits_1 \leq 9k\ell p^{1/2}(\log p)^{\ell+1}. \tag{12}$$

It remains to estimate $\sum_2$. First we claim that in Theorem 2 and in Theorem 3 (ii) we have $\sum_2 = 0$.

Indeed in these cases I will show that if $f^{\delta_1}(n+d_1)\ldots f^{\delta_\ell}(n+d_\ell)$ is a perfect m-th power, then there exists a δ_i which is even. Then, if δ_i is even, by (10) and $m \nmid \delta_i$ $(1 \leq \delta_i \leq m-1)$ we have

$$\sum_{\ell_j=1}^{m/2} \chi^{\delta_i}(g^{x\ell_j}) = \sum_{\ell_j=1}^{m/2} e\left(\frac{\delta_i/2}{m/2}\ell_j\right) = 0,$$

which means that in $\sum_2$ the product is 0, whence $\sum_2 = 0$. From this, (11) and (12) Theorem 2 and 3 (ii) follows.

Let us see the proof of those cases for which there exists an even δ_i. In the case of Theorem 2 if $f^{\delta_1}(n+d_1)\cdots f^{\delta_\ell}(n+d_\ell)$ is a perfect m-th power, then m divides the degree of $f^{\delta_1}(n+d_1)\cdots f^{\delta_\ell}(n+d_\ell)$ which is $k(\delta_1 + \cdots + \delta_\ell)$. Contrary to our statement, suppose that all δ_i are odd. Then using that k and ℓ are also odd we get that $k(\delta_1 + \cdots + \delta_\ell)$ is odd, which contradicts $2 \mid m \mid k(\delta_1 + \cdots + \delta_\ell)$. In the case of Theorem 3 (ii) we will use the following lemma, which is Lemma 5 of [5] with m in place of $p - 1$.

Lemma 3. *Suppose that the conditions of Theorem 3 hold.*
Then if $1 \le \delta_1, \ldots, \delta_\ell \le m - 1$, and $f^{\delta_1}(n + d_1)\cdots f^{\delta_\ell}(n + d_\ell)$ is a perfect m-th power, then there is a δ_i ($1 \le i \le \ell$) and an integer $1 \le \alpha \le k$ such that $m \mid \alpha\delta_i$.

By Lemma 3 we have

$$m \mid \alpha\delta_i \text{ and } \frac{m}{(m,\alpha)} \mid \delta_i.$$

By the conditions of Theorem 3 we have $2^\beta \parallel m$ and $k < 2^\beta$. Thus $(m, \alpha) \le \alpha \le k < 2^\beta$. Therefore $2 \mid \frac{m}{(m,\alpha)}$, whence δ_i is even. This completes the proof of Theorem 2 and Theorem 3 (ii).

In order to prove Theorem 3 (i) we need a generalization of Lemma 3. This is the following:

Lemma 4. *Suppose that the conditions of Theorem 3 (i) hold. If $1 \le \delta_1, \ldots, \delta_\ell \le m - 1$ and $f^{\delta_1}(n + d_1)\cdots f^{\delta_\ell}(n + d_\ell)$ is a perfect m-th power, then there is a permutation $(\rho_1, \ldots, \rho_\ell)$ of $(\delta_1, \ldots, \delta_\ell)$ such that for all $1 \le i \le \ell$ there exists an α_i with $1 \le \alpha_i \le k^i$ and*

$$m \mid \alpha_i\rho_i.$$

We postpone the proof of Lemma 4.

Now, from this lemma we verify that $\sum_2 \le \frac{\ell! k^{\ell(\ell+1)}}{m^\ell} p$. Consider a fixed ℓ-tuple $(\delta_1, \ldots, \delta_\ell)$ for which $f^{\delta_1}(n+d_1)\ldots f^{\delta_\ell}(n+d_\ell)$ is a perfect m-th power. We will prove that

$$\prod_{j=1}^{\ell} \left| \sum_{\ell_j}^{m/2} \chi^{\delta_j}(g^{x\ell_j}) \right| \le \frac{k^{\ell(\ell+1)/2}}{2^\ell}. \tag{13}$$

Indeed, by Lemma 4 we have a permutation $(\rho_1, \ldots, \rho_\ell)$ of $(\delta_1, \ldots, \delta_\ell)$ such that for all $1 \le i \le \ell$ there exists an α_i with $1 \le \alpha_i \le k^i$ and $m \mid \alpha_i\rho_i$. By this, $0 < \alpha_i\rho_i < \alpha_i m$ and $\alpha_i \le k^i$ we get

$$m \le \alpha_i\rho_i \le (\alpha_i - 1)m, \ \frac{1}{\alpha_i} \le \frac{\rho_i}{m} \le 1 - \frac{1}{\alpha_i} \text{ and } \frac{1}{k^i} \le \frac{1}{\alpha_i} \le \left\| \frac{\rho_i}{m} \right\|.$$

By this, (10) and $|1 - e(\alpha)| \ge 4 \|\alpha\|$ we have

$$\left| \sum_{\ell_j=1}^{m/2} \chi^{\rho_j}(g^{x\ell_j}) \right| \le \frac{2}{|1 - \chi^{\rho_j}(g^x)|} = \frac{2}{|1 - e(\rho_j/m)|} \le \frac{1}{2\|\rho_j/m\|} \le \frac{k^j}{2}. \tag{14}$$

Taking the term-wise product in (14) for $j = 1, \ldots, \ell$ we obtain (13). Thus

$$\sum\nolimits_2 \leq p \frac{k^{\ell(\ell+1)/2}}{m^\ell} \sum_{\substack{1 \leq \delta_1, \ldots, \delta_\ell \leq m, \\ f^{\delta_1}(n+d_1) \cdots f^{\delta_\ell}(n+d_\ell) \text{ is} \\ \text{a perfect } m\text{-th power}}} 1. \tag{15}$$

Next we give an upper bound for

$$r \overset{\text{def}}{=} \sum_{\substack{1 \leq \delta_1, \ldots, \delta_\ell \leq m, \\ f^{\delta_1}(n+d_1) \cdots f^{\delta_\ell}(n+d_\ell) \text{ is} \\ \text{a perfect } m\text{-th power}}} 1. \tag{16}$$

The number of different permutations $(\rho_1, \ldots, \rho_\ell)$ of $(\delta_1, \ldots, \delta_\ell)$ is $\ell!$. Consider a fixed permutation $(\rho_1, \ldots, \rho_\ell)$. Then by Lemma 4 we have $m \mid \alpha_i \rho_i$ where $1 \leq \alpha_i \leq k^i$. Thus $\frac{m}{(m, \alpha_i)} \mid \rho_i$. Since $1 \leq \rho_i \leq m$ we have that ρ_i may assume $(m, \alpha_i) \leq \alpha_i \leq k^i$ values. Therefore

$$r \leq \ell! \prod_{i=1}^{\ell} k^i = \ell! k^{\ell(\ell+1)/2}. \tag{17}$$

By (15), (16) and (17) we have

$$\sum\nolimits_2 \leq \ell! \frac{k^{\ell(\ell+1)}}{m^\ell} p$$

which proves Theorem 3 (i). It remains to prove Lemma 4.

Proof of Lemma 4

We will need the following definition and lemma:

Definition 2. *Let $\mathcal{A}$ and $\mathcal{B}$ be multi-sets of the elements of $\mathbb{Z}_p$. If $\mathcal{A} + \mathcal{B}$ represents every element of $\mathbb{Z}_p$ with multiplicity divisible by m, i.e., for all $c \in \mathbb{Z}_p$, the number of solutions of*

$$a + b = c \quad a \in \mathcal{A}, \ b \in \mathcal{B}$$

(the a's and b's are counted with their multiplicities) is divisible by m, then the sum $\mathcal{A} + \mathcal{B}$ is said to have property P.

Lemma 5. *Let $\mathcal{A} = \{a_1, a_2, \ldots, a_r\}$, $\mathcal{D} = \{d_1, d_2, \ldots, d_\ell\} \subseteq \mathbb{Z}_p$. If one of the following two conditions holds*

(i) $\min\{r, \ell\} \leq 2$ and $\max\{r, \ell\} \leq p - 1$,
(ii) $(4\ell)^r \leq p$ or $(4r)^\ell \leq p$,
then there exist $c_1, \ldots, c_\ell \in \mathbb{Z}_p$ and a permutation $(q_1, \ldots, q_\ell)$ of $(d_1, \ldots, d_\ell)$ such that for all $1 \leq i \leq \ell$

$$a + d = c_i \quad a \in \mathcal{A}, \ d \in \mathcal{D}$$

has at least one solution, and the number of solutions is less than $i+1$. Moreover for all solution $a \in \mathcal{A}$, $d \in \mathcal{D}$ we have $d \in \{q_1, q_2 \ldots, q_i\}$, and $d = q_i$, $a = c_i - q_i$ is always a solution.

Proof of Lemma 5

We will prove Lemma 5 by induction on i. It was proved in [4, Theorem 2] that for all sets $\mathcal{A}$ and $\mathcal{D}$ with the conditions of Lemma 5, we have a $c \in \mathbb{Z}_p$ such that

$$a + d = c \quad a \in \mathcal{A}, \ d \in \mathcal{D}$$

has exactly one solution.

This proves Lemma 5 in the case $i = 1$. Suppose that Lemma 5 holds for $i = j$. Then we will prove that it also holds for $i = j + 1$. By the induction hypothesis we have $c_1, \ldots, c_j$ and a permutation $(q_1, \ldots, q_j)$ of $(d_1, \ldots, d_j)$ according to Lemma 5. Let $\mathcal{D}' = \mathcal{D} \setminus \{q_1, \ldots q_j\}$. Since Lemma 5 is true for $i = 1$ we have that there exists $c_{j+1} \in \mathbb{Z}_p$ such that

$$a + d = c_{j+1} \quad a \in \mathcal{A}, \ d \in \mathcal{D}'$$

has exactly one solution. Let this unique solution be $\alpha = \alpha_{i+1}$ and $d = q_{j+1}$. Then for the solution of

$$a + d = c_{j+1} \quad a \in \mathcal{A}, \ d \in \mathcal{D}$$

we have $d \in \{q_1, q_2, \ldots, q_{j+1}\}$ which completes the proof of Lemma 5.

Now we return to the proof of Lemma 4. The following equivalence relation was defined in [4] and also used in [5]: We will say that the polynomials $\varphi(x), \psi(x) \in \mathbb{F}_p[x]$ are equivalent, $\varphi \sim \psi$, if there is an $a \in \mathbb{F}_p$ such that $\psi(x) = \varphi(x + a)$. Clearly, this is an equivalence relation.

Write f as the product of irreducible polynomials over $\mathbb{F}_p$. Let us group these factors so that in each group the equivalent irreducible factors are collected. Consider a typical group $\varphi(x + a_1), \ldots, \varphi(x + a_r)$. Then f is of the form $f(x) = \varphi^{\alpha_1}(x + a_1) \ldots \varphi^{\alpha_r}(x + a_r)g(x_r)$ where $g(x)$ has no irreducible factors equivalent with any $\varphi(x + a_i)$ $(1 \leq i \leq r)$.

Let $h(n) = f^{\delta_1}(n + d_1) \cdots f^{\delta_\ell}(n + d_\ell)$ be a perfect m-th power where $1 \leq \delta_1, \ldots, \delta_\ell < m$. Then writing $h(x)$ as the product of irreducible polynomials over $\mathbb{F}_p$, all the polynomials $\varphi(x + a_i + d_j)$ with $1 \leq i \leq r$, $1 \leq j \leq \ell$ occur amongst the factors. All these polynomials are equivalent, and no other irreducible factor belonging to this equivalence class will occur amongst the irreducible factors of $h(x)$.

Since distinct irreducible polynomials cannot have a common zero, each of the zeros of h is of multiplicity divisible by m, if and only if in each group, formed by equivalent irreducible factors $\varphi(x + a_i + d_j)$ of $h(x)$, every polynomial of form $\varphi(x + c)$ occurs with multiplicity divisible by m. In other words writing $\mathcal{A} = \{a_1, \ldots, a_1, \ldots, a_r, \ldots, a_r\}$, $\mathcal{D} = \{d_1, \ldots, d_1, \ldots, d_\ell, \ldots, d_\ell\}$ where a_i has the multiplicity α_i in $\mathcal{A}$ (α_i is the exponent of $\varphi(x + a_i)$ in the factorization of $f(x)$) and d_i has the multiplicity δ_i in $\mathcal{D}$ (where $h(n) = f^{\delta_1}(n + d_1) \cdots f^{\delta_\ell}(n + d_\ell)$ is a perfect m-th power), then for each group $\mathcal{A} + \mathcal{D}$ must possess property P.

Let $\mathcal{A}'$ and $\mathcal{D}'$ be the simple set version of $\mathcal{A}$ and $\mathcal{D}$, more exactly, let $\mathcal{A}' = \{a_1, \ldots, a_r\}$ and $\mathcal{D}' = \{d_1, \ldots, d_\ell\}$. $\mathcal{A}'$ and $\mathcal{D}'$ satisfy the conditions of Lemma 5. So by Lemma 5 for the multi-sets $\mathcal{A}$ and $\mathcal{D}$ we have the following: There exist

$c_1, \ldots, c_\ell \in \mathbb{Z}_p$ and a permutation $(q_1, \ldots, q_\ell) = (d_{j_1}, \ldots, d_{j_\ell})$ of $(d_1, \ldots, d_\ell)$ such that if

$$a + d = c_i \quad a \in \mathcal{A}', \ d \in \mathcal{D}',$$

then we have

$$d \in \{q_1, \ldots, q_i\} = \{d_{j_1}, \ldots, d_{j_i}\}$$

and $d = q_i$, $a = c_i - q_i$ is a solution. Here $(j_1, \ldots, j_\ell)$ is a permutation of $(1, \ldots, \ell)$. Define ρ_i's by $\rho_i = \delta_{j_i}$ (so $(\rho_1, \ldots, \rho_\ell) = (\delta_{j_1}, \ldots, \delta_{j_\ell})$ is the same permutation of $(\delta_1, \ldots, \delta_\ell)$ as the permutation $(q_1, \ldots, q_\ell) = (d_{j_1}, \ldots, d_{j_\ell})$ of $(d_1, \ldots, d_\ell)$). Returning to the multi-set case, using these notation we get that the number of the solutions

$$a + d = c_i \quad a \in \mathcal{A}, \ d \in \mathcal{D}$$

is of the form

$$\epsilon_{i,1}\alpha_{i,1}\rho_1 + \epsilon_{i,2}\alpha_{i,2}\rho_2 + \cdots + \epsilon_{i,i}\alpha_{i,i}\rho_i$$

where $\epsilon_{i,j} \in \{0,1\}$, $\alpha_{i,j} \in \{\alpha_1, \ldots, \alpha_r\}$ for $1 \leq j \leq i$ and $\epsilon_{i,i} = 1$. (We study the number of the solutions by multiplicity since $\mathcal{A}$ and $\mathcal{D}$ are multi-sets).

Since $\mathcal{A} + \mathcal{D}$ posses property $\mathcal{P}$ we have that for all $1 \leq i \leq \ell$

$$m \mid \epsilon_{i,1}\alpha_{i,1}\rho_1 + \epsilon_{i,2}\alpha_{i,2}\rho_2 + \cdots + \epsilon_{i,i}\alpha_{i,i}\rho_i. \tag{18}$$

By induction on i we will prove that

$$m \mid \alpha_{1,1}\alpha_{2,2}, \ldots, \alpha_{i,i}\rho_i. \tag{19}$$

Indeed, for $i = 1$ by (18) and $\epsilon_{1,1} = 1$ we get $m \mid \alpha_{1,1}\rho_1$. We will prove that if (19) holds for $i \leq j - 1$, then it also holds for $i = j$.

By the induction hypothesis we have

$$m \mid \alpha_{1,1}\rho_1, \ m \mid \alpha_{1,1}\alpha_{2,2}\rho_2, \ \ldots, \ m \mid \alpha_{1,1}\alpha_{2,2} \ldots, \alpha_{j-1,j-1}\rho_{j-1}. \tag{20}$$

Multiplying (18) for $i = j$ by $\alpha_{1,1} \ldots \alpha_{j-1,j-1}$ we get:

$$m \mid \epsilon_{j,1}\alpha_{j,1}\alpha_{1,1} \ldots \alpha_{j-1,j-1}\rho_1 + \epsilon_{j,2}\alpha_{j,2}\alpha_{1,1} \ldots \alpha_{j-1,j-1}\rho_2 + \cdots$$
$$+ \epsilon_{j,j}\alpha_{j,j}\alpha_{1,1} \ldots \alpha_{j-1,j-1}\rho_i.$$

From this using (20) and $\epsilon_{j,j} = 1$ we get

$$m \mid \alpha_{1,1} \ldots \alpha_{j,j}\rho_j$$

which was to be proved.

$\alpha_{1,1}, \ldots, \alpha_{i,i} \in \{\alpha_1, \ldots, \alpha_r\}$ where α_i's are exponents of irreducible factors of f, thus $1 \leq \alpha_{i,i} \leq \deg f = k$. Therefore $\alpha_{1,1}\alpha_{2,2} \ldots \alpha_{i,i} \leq k^i$ and by (19) this completes the proof of Lemma 4.

2.3   Proof of Theorem 4

The proof is exactly the same as in [2, Theorem 1], the only difference is in the definitions of q and r: now we choose q, r as integers with $(q, p) = (r, p) = 1$ and $1 \leq \text{ind}^* q \leq \frac{m}{2}$, $\frac{m}{2} < \text{ind}^* r \leq m$.

3 Time Analysis

Construction 1 depends on the key g^x where g is a primitive root and $(x, m) = 1$. We only need g^x, it is not necessary to know the value of g or x. First we prove that it is easy to find a key g^x.

Suppose that the factorization of m is known: $m = p_1^{\alpha_1} \ldots p_r^{\alpha_r}$ where $p_1, \ldots, p_r$ are primes. The condition $(x, m) = 1$ is equivalent with that $y = g^x$ is not a perfect p_i-th power for any $1 \le i \le r$ in $\mathbb{F}_p$. In other words, using Fermat's theorem we have that

$$y^{(p-1)/p_i} \equiv 1 \pmod{p} \tag{21}$$

does not hold for all $1 \le i \le r$. By using the iterated squaring method to check (21), it takes $O\left((\log p)^3\right)$ bit operations (see e.g. in [8]).

We will choose a random $y \in \mathbb{Z}_p$, and by (21) we check that $y = g^x$ weather satisfies $(x, m) = 1$ or not. For a fix primitive root g, the number of x's with this property is $\varphi(m)\frac{p-1}{m} \gg \frac{p}{\log\log p}$. Thus after $c \log\log p$ attempts we will find a suitable key g^x with high probability.

Next we prove that $\mathrm{ind}^* n$ can be computed fast. Indeed, first we determine $\mathrm{ind}^* n$ modulo prime power divisor q^α of m by $O\left(\alpha q(\log p)^3\right)$ bit operations. If we know $\mathrm{ind}^* n$ modulo $p_i^{\alpha_i}$ for all $1 \le \alpha_i \le r$ where $m = p_1^{\alpha_1} \ldots p_r^{\alpha_r}$, then using the Chinese Remainder theorem we have determined the value $\mathrm{ind}^* n$ modulo m, which gives $\mathrm{ind}^* n$ because of $1 \le \mathrm{ind}^* n \le m$. Thus to compute $\mathrm{ind}^* n$ we use

$$O((\log m)^4 + (\log p)^3(\alpha_1 p_1 + \cdots + \alpha_r p_r))$$
$$\le O((\log m)^4 + (\log p)^3(\alpha_1 + \cdots + \alpha_r) \max_{1 \le i \le r} p_i)$$
$$\le O((\log p)^4 \max_{1 \le i \le r} p_i)$$

bit operations.

Let us see the proof of that $\mathrm{ind}^* n$ can be computed modulo prime power divisors q^α of m by $O(\alpha q(\log p)^3)$ bit operations. We will prove this by induction on α. When $\alpha = 0$ the statement is trivial. Suppose that we already know $\mathrm{ind}^* n$ modulo q^i:

$$\mathrm{ind}^* n \equiv s \pmod{q^i}.$$

From this we compute $\mathrm{ind}^* n$ modulo q^{i+1} by $O(q(\log p)^3)$ bit operations if $q^{i+1} \mid m$. In order to prove this statement we will use the following lemma, which is a trivial consequence of the properties of the primitive roots and Fermat's theorem.

Lemma 6. $q^\alpha \mid m$. *Then*

$$\mathrm{ind}^* n \equiv s \pmod{q^\alpha}$$

holds if and only if

$$n/g^{sx} \text{ is a perfect } q^\alpha\text{-th power modulo } p$$

which is equivalent with

$$(n/g^{sx})^{(p-1)/q^\alpha} \equiv 1 \pmod{p}. \tag{22}$$

By Lemma 6 we have that n/g^{sx} is a perfect q^i-th power. By Lemma 6, using (22), we check that which of the numbers

$$n/g^{sx},\ n/g^{(s+q^i)x},\ n/g^{(s+2q^i)x},\ldots,\ n/g^{s+(q-1)q^i x}$$

is a perfect q^{i+1}-th power. This takes $O\left(q(\log p)^3\right)$ bit operations. There is surely one which is a perfect q^{i+1}-th power, because s, $s+q^i,\ldots,\ s+(q-1)q^i$ run over the residue classes modulo q^{i+1} which are congruent to s modulo q^i. By Lemma 6, $n/g^{s+jp^i x}$ is a perfect p^{i+1}-th power if and only if $\text{ind}^* n \equiv s + jq^i x$ (mod q^{i+1}). This completes the proof of the statement.

4   An Extension of an Inequality of Mauduit and Sárközy

C. Mauduit and A. Sárközy [10] expressed the connection between the well-distribution measure and the correlation measure of order 2 in a quantitative form: For all $E_N \in \{-1,+1\}^N$

$$W(E_N) \leq 3\sqrt{NC_2(E_N)}. \tag{23}$$

They also gave a construction for which $W(E_N) \gg \sqrt{NC_2(E_N)}$. Their result shows that (23) is sharp apart from a constant factor. The following theorem generalizes (23) for the correlation measures of higher order:

Theorem 5. *For all* $E_N \in \{-1,+1\}^N$, $3\ell^2 \leq N$ *we have*

$$W(E_N) \leq 3\ell N^{1-1/(2\ell)}\left(C_{2\ell}(E_N)\right)^{1/(2\ell)}.$$

By Theorem 3 we get for $N = p - 1$:

$$C_\ell(E_N) \ll_\ell k^{\ell(\ell+1)} \frac{p}{m^\ell} \tag{24}$$

if $m < \frac{p^{1/(2\ell)}}{(\log p)^{1+1/\ell}}$. We will see that if ℓ is even, m is odd and small enough, then by Theorem 5 and Theorem 1 we have that the upper bound in (24) is sharp apart from a constant factor. Thus in case of even ℓ and odd m Construction 1 provides a natural example for a sequence whose correlation measures of small orders are small while the well-distribution measure is possibly large. Indeed, by Theorem 1 if $m < \frac{1}{2k}p^{1/2}/(\log p)^2$ we have

$$W(E_{p-1}) \gg \frac{p}{m}.$$

By Theorem 5 we fixed

$$\frac{p}{m} \ll W(E_N) \ll \ell p^{1-1/(2\ell)}\left(C_{2\ell}(E_N)\right)^{1/(2\ell)},$$

which implies

$$\frac{1}{\ell^{2\ell}}\frac{p}{m^{2\ell}} \ll C_{2\ell}(E_N).$$

Comparing this with (24) we get that Theorem 5 is sharp apart from a constant factor. While the construction of A. Sárközy and C. Mauduit [10] showing that (23) is sharp used probabilistic methods, Construction 1 is explicit.

Proof of Theorem 5

The proof is nearly the same as in [10], however we have to handle larger product of e_i's than in [10].

Let

$$W(E_N) = \sum_{j=0}^{t-1} e_{a+jb} = \sum_{\substack{a \leq i < m \\ i \equiv a \ (\text{mod } b)}} e_i$$

where $m = a + tb \leq N + b$. If $N < i \leq N + b$, let $e_i = 1$. Then

$$(W(E_N))^{2\ell} = \Big(\sum_{\substack{a \leq i < m \\ i \equiv a \ (\text{mod } b)}} e_i \Big)^{2\ell} \leq \sum_{h=0}^{b-1} \Big(\sum_{\substack{a \leq i < m \\ i \equiv h \ (\text{mod } b)}} e_i \Big)^{2\ell}$$

$$= \sum_{\substack{r \leq 2\ell, \ a \leq i_1 < i_2 < \cdots < i_r < m \\ i_1 \equiv i_2 \equiv \cdots \equiv i_r \ (\text{mod } b)}} X_r \cdot e_{i_1} e_{i_2} \ldots e_{i_r}$$

$$= \sum_{\substack{j \leq \ell, \ a \leq i_1 < i_2 < \cdots < i_{2j} < m \\ i_1 \equiv i_2 \equiv \cdots \equiv i_{2j} \ (\text{mod } b)}} X_{2j} \cdot e_{i_1} e_{i_2} \ldots e_{i_{2j}}. \tag{25}$$

Here $r \leq 2\ell$ because originally all the products are in the form of $e_1^{\alpha_1} \ldots e_{m-1}^{\alpha_{m-1}}$ (where $\alpha_1 + \cdots + \alpha_{m-1} = 2\ell$) but $e_i^{\alpha_i} = 1$ if α_i is even and $e_i^{\alpha_i} = e_i$ if α_i is odd. The sum $\alpha_1 + \cdots + \alpha_{m-1} = 2\ell$ is even, so the number of odd α_i's is even. Thus in (25) we may suppose that $r = 2j$ where $j \in \mathbb{N}$.

Let s denote the number of i's with $a \leq i < m$ and for which i belongs to a fixed residue class modulo b (here s is the number of the terms in $\sum_{\substack{a \leq i < m \\ i \equiv h \ (\text{mod } b)}} e_i$ for any h, s does not depend on h on the value of the fixed residue class). Using the multinomial theorem:

$$X_{2j} = \sum_{\substack{\alpha_1 + \cdots + \alpha_s = 2\ell \\ \alpha_1, \ldots, \alpha_{2j} \text{ are odd} \\ \alpha_{2j+1}, \ldots, \alpha_s \text{ are even}}} \frac{(2\ell)!}{\alpha_1! \ldots \alpha_s!} \leq \sum_{\substack{\alpha_1 + \cdots + \alpha_s = 2\ell \\ \alpha_1, \ldots, \alpha_{2j} \text{ are odd} \\ \alpha_{2j+1}, \ldots, \alpha_s \text{ are even}}} (2\ell)!.$$

For $1 \leq i \leq 2j$ let $\alpha_i = 2\beta_i - 1$ and for $2j + 1 \leq i \leq s$ let $\alpha_i = 2\beta_i - 2$. Then

$$X_{2j} \leq (2\ell)! \sum_{\substack{\beta_1 + \cdots + \beta_s = s + \ell - j \\ \forall i: \ \beta_i > 0}} 1 = (2\ell)! \binom{s + \ell - j - 1}{s - 1}$$

$$\leq \frac{(2\ell)!}{(\ell - j)!} (s + \ell - j - 1)^{\ell - j} \leq (2\ell)^{\ell+j} (s + \ell - j - 1)^{\ell - j}$$

$$\leq (2\ell)^{\ell+j} (N + \ell)^{\ell - j} = 2^{\ell+j} \ell^{\ell+j} (N + \ell)^{\ell - j}. \tag{26}$$

By (25) and the triangle-inequality we have

$$(W(E_N))^{2\ell} \leq \sum_{j=0}^{\ell} |X_{2j}| \sum_{\substack{1 \leq d_1 < d_2 < \cdots < d_{2j-1} < m-a \\ 0 \equiv d_1 \equiv d_2 \equiv \cdots \equiv d_{2j-1} \pmod{b}}} \left| \sum_{i=a}^{m-1-d_{2j-1}} e_i e_{i+d_1} \cdots e_{i+d_{2j-1}} \right|. \tag{27}$$

By the definition of the correlation measure we have:

$$\left| \sum_{i=a}^{m-1-d_{2j-1}} e_i e_{i+d_1} \cdots e_{i+d_{2j-1}} \right| \leq C_{2\ell}(E_N) + 1.$$

Thus from (26) and (27) we obtain

$$(W(E_N))^{2\ell} \leq \sum_{j=0}^{\ell} 2^{\ell+j} \ell^{\ell+j} (N+\ell)^{\ell-j} \sum_{\substack{1 \leq d_1 < d_2 < \cdots < d_{2j-1} < m-a \\ 0 \equiv d_1 \equiv d_2 \equiv \cdots \equiv d_{2j-1} \pmod{b}}} (C_{2j}(E_N) + 1)$$

$$= \sum_{j=0}^{\ell} 2^{\ell+j} \ell^{\ell+j} (N+\ell)^{\ell-j} N^{2j-1} (C_{2j}(E_N) + 1)$$

where by definition $C_0(E_N) = N$. Using that for $1 \leq j \leq \ell - 1$ $C_{2j}(E_N) \leq N$ we obtain

$$(W(E_N))^{2\ell} \leq \sum_{j=0}^{\ell-1} 2^{\ell+j} \ell^{\ell+j} (N+\ell)^{\ell+j} + 4^\ell \ell^{2\ell} N^{2\ell-1} (C_{2\ell}(E_N) + 1). \tag{28}$$

By $1 + x \leq e^x$ we have

$$\sum_{j=0}^{\ell-1} 2^{\ell+j} \ell^{\ell+j} (N+\ell)^{\ell+j} = 2^\ell \ell^\ell (N+\ell)^\ell \sum_{j=0}^{\ell-1} 2^j \ell^j (N+\ell)^j$$

$$= 2^\ell \ell^\ell (N+\ell)^\ell (1 + 2\ell(N+\ell))^{\ell-1}$$

$$= 2^{2\ell-1} \ell^{2\ell-1} N^{2\ell-1} \left(1 + \frac{\ell}{N}\right)^\ell \left(1 + \frac{2\ell^2+1}{2\ell N}\right)^{\ell-1}$$

$$\leq 2^{2\ell-1} \ell^{2\ell-1} N^{2\ell-1} e^{2\ell^2/N} \leq 4^\ell \ell^{2\ell-1} N^{2\ell-1}.$$

From this and (28) we obtain

$$(W(E_N))^{2\ell} \leq 4^\ell \ell^{2\ell} N^{2\ell-1} (C_{2\ell}(E_N) + 1 + \frac{1}{\ell}) \leq 9^\ell \ell^{2\ell} N^{2\ell-1} C_{2\ell}(E_N),$$

which proves Theorem 5.

I would like to thank to Professor András Sárközy for the valuable discussions and to the referee Christian Elsholtz for his careful reading and constructive comments.

References

1. R. Ahlswede, C. Mauduit, and A. Sárközy, Large families of pseudorandom sequences of k symbols and their complexity, Part I, Part II, this volume.
2. R. Ahlswede, L.H. Khachatrian, C. Mauduit, and A. Sárközy, A complexity measure for families of binary sequences, Periodica Math. Hungar. 46, 107-118, 2003.
3. J. Cassaigne, C. Mauduit, and A. Sárközy, On finite pseudorandom binary sequences VII: The measures of pseudorandomness, Acta Arith. 103, 97-118, 2002.
4. L. Goubin, C. Mauduit, and A. Sárközy, Construction of large families of pseudorandom binary sequences, J. Number Theory 106, No. 1, 56-69, 2004.
5. K. Gyarmati, On a family of pseudorandom binary sequences, Period. Math. Hungar. 49, No. 2, 45-63, 2004.
6. K. Gyarmati, On a pseudorandom property of binary sequences, Ramanujan J. 8, No. 3, 289-302, 2004.
7. D. R. Heath-Brown, Zero-free regions for Dirichlet L-functions and the least prime in an arithmetic progression, Proc. London Math. Soc. 64, 265-338, 1992.
8. N. Koblitz, A Course in Number Theory and Cryptography, Graduate Texts in Mathematics 114, Springer-Verlag, New-York, 1994.
9. C. Mauduit and A. Sárközy, On finite pseudorandom binary sequences I: Measures of pseudorandomness, the Legendre symbol, Acta Arith. 82, 365-377, 1997.
10. C. Mauduit and A. Sárközy, On the measures of pseudorandomness of binary sequences, Discrete Math. 271, 195-207, 2003.
11. C. Mauduit, J. Rivat, and A. Sárközy, Construction of pseudorandom binary sequences using additive characters, Monatshefte Math., 141, No. 3, 197-208, 2004.
12. A. Sárközy, A finite pseudorandom binary sequence, Studia Sci. Math. Hungar. 38, 377-384, 2001.
13. A. Weil, Sur les courbes algébriques et les variétés qui s'en déduisent, Act. Sci. Ind. 1041, Hermann, Paris, 1948.

On Pseudorandom Sequences and Their Application

J. Rivat* and András Sárközy*

Abstract. A large family of finite pseudorandom binary sequences is presented, and also tested "theoretically" for pseudorandomness. The optimal way of implementation is discussed and running time analysis is given. Numerical calculations are also presented.

1 Introduction

In the last century numerous papers have been written on pseudorandom (briefly, PR) sequences. In these papers a wide range of goals, approaches, tools is presented, even the concept of "pseudorandomness" is interpreted in different ways (depending mostly on the applications in mind). In the majority of the papers constructions of PR sequences are given and/or tested for pseudorandomness. In most papers PR sequences of real numbers taken from $[0, 1)$ are considered, much less is known on PR binary sequences, although PR sequences of this type are also needed in applications (simulation, cryptography). Thus when a PR binary sequence is needed, then typically one constructs a sequence by using either a random bit generator (which can be both hardware-based or software-based), or a mathematical principle. In the latter case, one describes a mathematical algorithm which maps certain parameters to a well-defined binary sequence; the values of these parameters are chosen randomly from a certain set (this is the *seed*). In either of the two cases, we *do not have a priori control about the PR quality* of the sequence to be constructed, thus when the construction is over, *one has to test the numerical sequence* obtained by using certain statistical tests (that a truly random sequence must pass).

Motivated by these facts, Mauduit and Sárközy initiated a comprehensive study of finite pseudorandom binary sequences focusing "... on construction and testing, more exactly, on apriori or, as Knuth [11] calls it, *theoretical* testing". They wrote "... our goal is not the search for new constructions superior to all previous ones; this would be too optimistic. Instead, we are aiming at constructions superior to the previous ones at least in certain special situations, besides we will gather new information on random-type properties of special binary sequences playing an important role in number theory and in other fields of mathematics". Since then more than 10 related papers have been written. We

* Research partially supported by Hungarian National Foundation for Scientific Research, Grant No T 029 759 and MKM fund FKFP-0139/1997. This paper was completed while the authors were visiting the Zentrum für interdisziplinäre Forchung, Universität Bielefeld, Germany.

R. Ahlswede et al. (Eds.): Information Transfer and Combinatorics, LNCS 4123, pp. 343–361, 2006.
© Springer-Verlag Berlin Heidelberg 2006

feel we have arrived to the point to utilize our theoretical conclusions in the applications as well; to see what are the most promising constructions and to look for the most effective and economical ways to adapt them to different fields in the applications. In this paper our goal is to make the first steps in this direction by focusing on the case when our only goal is to guarantee possibly "good" PR properties (in a possibly effective way).

(Note that usually binary sequences consisting of 0, 1 bits are considered. However, in our case it will be more convenient to study sequences consisting of −1 and +1; clearly, this difference is insignificant).

2   The Measures of Pseudorandomness

In particular, in [12] Mauduit and Sárközy proposed to use the following measures of pseudorandomness.

Consider a finite pseudorandom binary sequence

$$E_N = \{e_1, \ldots, e_N\} \in \{-1, +1\}^N. \tag{1}$$

Then the *well-distribution measure* of E_N is defined as

$$W(E_N) = \max_{a,b,t} \left| \sum_{j=0}^{t-1} e_{a+jb} \right| \tag{2}$$

where the maximum is taken over all $a, b, t \in \mathbb{N}$ such that $1 \le a \le a+(t-1)b \le N$, while the *correlation measure of order* k of E_N is defined as

$$C_k(E_N) = \max_{M,D} \left| \sum_{n=1}^{M} e_{n+d_1} e_{n+d_2} \cdots e_{n+d_k} \right| \tag{3}$$

where the maximum is taken over all $D = (d_1, \ldots, d_k)$ and M such that $0 \le d_1 < \cdots < d_k \le N - M$. Then the sequence is considered as a "good" pseudorandom sequence if both these measures $W(E_N)$ and $C_k(E_N)$ (at least for "small" k) are "small" in terms of N (in particular, both are $o(N)$ as $N \to \infty$). Indeed, it is shown in [5] that for a "truly random" $E_N \in \{-1, +1\}^N$, both $W(E_N)$ and for fixed k, $C_k(E_N)$ are around $N^{1/2}$ with "near 1" probability. Thus for "really good" PR sequence we expect the measures (2) and (3) to be not much greater than $N^{1/2}$. (In [5], other important properties of the two measures are studied as well).

We remark that quantities like (2) or (3) often occur in the literature, and even the word "correlation" (or autocorrelation) is often used in connection with expressions of form (3). However, in our case definitions (2) and (3) have two important characteristics:

(i) In both cases (2) and (3), we also take the maximum in terms of the length of the sum, in other words, "incomplete sums" are also considered.

(ii) When the word "correlation" is used, then typically sums of form (3) are considered with "small", often fixed d_1, d_2, ..., d_k and "large" M (typically $M \to +\infty$); in this case we may speak of "short range" correlation. In our case (3), "long range" correlation is also considered.

In [12] other possible measures of pseudorandomness are also mentioned (see also [7]), but we decided to restrict ourselves to these two measures (2) and (3).

3   The Q-Construction and Its Pseudorandomness

In [12] the Legendre symbol was also studied, and it was tested for pseudorandomness. More exactly, let p be an odd prime, write $N = p - 1$, and define the binary sequence E_N by

$$E_N = \{e_1, e_2, \ldots, e_N\}, \quad e_n = \left(\frac{n}{p}\right) \text{ for } n = 1, \ldots, N, \qquad (4)$$

(where $\left(\frac{n}{p}\right)$ denotes the Legendre symbol). Then it was shown in [12] that for the sequence (4) we have

$$W(E_N) \ll N^{1/2} \log N \qquad (5)$$

and

$$C_k(E_N) \ll k N^{1/2} \log N \qquad (6)$$

for all $k < N$ (where $\ll$ is Vinogradov's notation, i.e., $f(x) \ll g(x)$ means that $f(x) = O(g(x))$). (In [12] (5) and (6) are stated in a slightly different form).

[12] was followed by a series of papers in which numerous other sequences were constructed and tested for pseudorandomness. Still the Legendre symbol sequence (4) is the best PR sequence constructed, but recently Goubin, Mauduit and Sárközy [6] have extended construction (4) considerably (and this construction was also studied in [1]). This construction and its most important properties are described in the following theorem (proved in [6]):

Theorem 1. *If p is a prime number, $f(x) \in \mathbb{F}_p[x]$ ($\mathbb{F}_p$ being the field of the modulo p residue classes) has degree k (> 0), $f(x)$ has no multiple zero in $\overline{\mathbb{F}_p}$ ($=$ the algebraic closure of $\mathbb{F}_p$), and the binary sequence $E_p = \{e_1, \ldots, e_p\}$ is defined by*

$$e_n = \begin{cases} \left(\frac{f(n)}{p}\right) & \text{for } (f(n), p) = 1, \\ +1 & \text{for } p \mid f(n), \end{cases} \qquad (7)$$

then we have

$$W(E_p) < 10 k p^{1/2} \log p. \qquad (8)$$

Moreover, assume that also $\ell \in \mathbb{N}$, and one of the following assumptions holds:

(i) $\ell = 2$;
(ii) $\ell < p$, and 2 is a primitive root modulo p;
(iii) $(4k)^\ell < p$.

Then we also have

$$C_\ell(E_p) < 10k\ell p^{1/2} \log p. \tag{9}$$

The crucial tool in the proofs of (8) and (9) is an estimate for incomplete character sums of the form

$$\sum_{A < x < B} \chi(f(x))$$

where $\chi \neq \chi_0$ is a character modulo p and $f(x) \in \mathbb{F}_p[x]$; this estimate was deduced in [12] from a theorem of Weil [22] by using an inequality of Vinogradov. The $f(n) = n$ special case of this theorem corresponds to the basic Legendre symbol construction studied in (4), (5) and (6). If the degree k of the polynomial $f(n)$ grows, then the upper bounds in (8) and (9) get weaker. However, this slight loss is more than compensated by the fact that in this way we obtain a "large" family of "good" pseudorandom sequences. Indeed, it is shown in [1] that this family is not only large, but it is also of "rich", "complex" structure which can be very well utilized (e.g., in cryptography).

The construction described in Theorem 1 will play a role of basic importance in the remaining part of this paper. Thus in order to be able to refer to it in a short form, we will call it Q-sequence, Q-construction (Q for "quadratic", since the construction is based on the use of quadratic residues).

4   Our PR Measures and the Standard Statistical Tests

In [13] (which is an excellent monograph and we will often refer to it) the following definition is presented.

Definition 1. *"A pseudorandom bit generator (PRBG) is a deterministic algorithm which, given a truly random binary sequence of length k, outputs a binary sequence of length $l \gg k$" which "appears" to be random. The output of the PRBG is called a* pseudorandom bit sequence*".*

Moreover, referring to "ad hoc" techniques for PR bit generation, [13] writes: "In order to gain confidence that such generators are secure, they should be subjected to a variety of statistical tests designed to detect the specific characteristics expected of random sequences. A collection of such tests is given in 5.4. As the following example demonstrate, passing these statistical tests is a *necessary* but not *sufficient* condition for a generator to be secure". Next the *linear congruential generator* is presented, and the conclusion is: "While such generators are commonly used for simulation purposes and probabilistic algorithms, and pass the statistical tests of 5.4, they are predictable and hence entirely insecure for cryptographic purposes"...

We will show that our Q-sequence passes (or nearly passes) the statistical tests mentioned above (as a consequence of Theorem 1) so that it can be used very well for simulation purposes and probabilistic algorithms; this is one of our main goals here. Moreover, from cryptographic aspect the situation is not as negative as in the case of the linear congruential method: a limited cryptographic application of the construction is possible, and with some work it can be made more secure; we will return to this briefly in section 7.

The "five basic tests" of [13], 5.4 are (we adopt our notation, apart from that, we quote [13]):

(i) *Frequency test (monobit test).* (...) Let n_-, n_+ denote the number of -1's and $+1$'s in E_N, respectively. The statistic used is

$$X_1 = \frac{(n_- - n_+)^2}{N} \tag{10}$$

which approximately follows a χ^2 distribution with one degree of freedom if $N \geq 10$.

(ii) *Serial test (two bit test).* (...) Let n_-, n_+ denote the number of -1's and $+1$'s in E_N, respectively, and let n_{--}, n_{-+}, n_{+-}, n_{++} denote the number of occurrences of $(-1, -1)$, $(-1, +1)$, $(+1, -1)$, $(+1, +1)$ in E_N, respectively. (...) The statistic used is

$$X_2 = \frac{4}{N-1}(n_{--} + n_{-+} + n_{+-} + n_{++}) - \frac{2}{N}(n_-^2 + n_+^2) + 1 \tag{11}$$

which approximately follows a χ^2 distribution with two degrees of freedom if $N \geq 21$.

(iii) *Poker test.* Let m be a positive integer such that $\lfloor N/m \rfloor \geq 5 \cdot 2^m$ and let $k = \lfloor N/m \rfloor$. Divide the sequence E_N into k non-overlapping parts each of length m, and let n_i be the number of occurrences of the i-th type of sequence of length m, $1 \leq i \leq 2^m$. (...) The statistic used is

$$X_3 = \frac{2^m}{N}\left(\sum_{i=1}^{2^m} n_i^2\right) - k \tag{12}$$

which approximately follows a χ^2 distribution with $2^m - 1$ degrees of freedom. Note that the poker test is a generalization of the frequency test: setting $m = 1$ in the poker test yields the frequency test.

(iv) *Runs test.* (...) The expected number of runs of -1's (or $+1$'s) of length i in a random sequence of length N is $m_i = (N - i + 3)/2^{i+2}$. Let k be equal to the largest integer i for which $m_i \geq 5$. Let B_i, G_i be the number of runs of -1's (or $+1$'s) of length i in E_N for each i, $1 \leq i \leq k$. The statistic used is

$$X_4 = \sum_{i=1}^{k} \frac{(B_i - m_i)^2}{m_i} + \sum_{i=1}^{k} \frac{(G_i - m_i)^2}{m_i} \tag{13}$$

which approximately follows a χ^2 distribution with $2k - 2$ degrees of freedom.

(v) *Autocorrelation test.* Let d be a fixed integer, $1 \leq d \leq \lfloor N/2 \rfloor$. The number of bits in E_N not equal to their d-shifts is

$$A(d) = -\sum_{i=1}^{N-d} \frac{e_i e_{i+d} - 1}{2} = \frac{N-d}{2} - \frac{1}{2}\sum_{i=1}^{N-d} e_i e_{i+d}$$

The statistic used is

$$X_5 = 2 \left(A(d) - \frac{N-d}{2} \right) / (N-d)^{1/2} \qquad (14)$$

which approximately follows a $\mathcal{N}(0,1)$ distribution if $N - d \geq 10$. Since small values of $A(d)$ are as unexpected as large values of $A(d)$, a two-sided test should be used.

Now we will show that the statistics X_1, X_2, X_5 can be controlled very well (in fact, nearly optimally) while X_4 satisfactorily in terms of our PR measures W and C_ℓ. (We will return to the case of the statistic X_3 in a remark at the end of this section).

Theorem 2. *For all binary sequences E_N of form (1) we have*

(i)

$$X_1 \leq \frac{1}{N}(W(E_N))^2;$$

(ii)

$$X_2 \leq \frac{2}{N} \left((C_2(E_N)^2) + (W(E_N))^2 \right) + 21; \qquad (15)$$

(iii) *writing*

$$Y_i = \frac{(B_i - m_i)^2 + (G_i - m_i)^2}{m_i}$$

so that

$$X_4 = \sum_{i=1}^{k} Y_i \qquad (16)$$

we have

$$Y_i \leq \frac{2}{m_i} \left(3 + \frac{i+2}{2^{i+2}} W(E_N) + \frac{1}{2^{i+2}} \sum_{\ell=2}^{i+2} \binom{i+2}{\ell} C_\ell(E_N) \right)^2 \qquad (17)$$

for $i \leq W(E_N)$, and

$$Y_i = 2m_i \qquad (18)$$

for $i > W(E_N)$;

(iv)

$$X_5 \leq \frac{C_2(E_N)}{(N-d)^{1/2}}$$

Proof of Theorem 2

(i) Clearly we have

$$n_- = -\frac{1}{2} \sum_{i=1}^{N} (e_i - 1), \quad n_+ = \frac{1}{2} \sum_{i=1}^{N} (e_i + 1)$$

so that

$$X_1 = \frac{1}{N}(n_- - n_+)^2 = \frac{1}{N}\left(-\sum_{i=1}^{N} e_i\right)^2 \le \frac{1}{N}(W(E_N))^2.$$

(ii) If $|a| \le |b|$ we will write $a = \theta(b)$. Clearly we have

$$\frac{1}{N-1} = \frac{1}{N} + \frac{1}{(N-1)N} = \frac{1}{N} + \theta\left(\frac{2}{N^2}\right)$$

so that

$$\frac{4}{N-1}(n_{--}^2 + n_{-+}^2 + n_{+-}^2 + n_{++}^2) \tag{19}$$

$$= \frac{4}{N}(n_{--}^2 + n_{-+}^2 + n_{+-}^2 + n_{++}^2) + \theta\left(\frac{2}{N^2}\right)\theta\left(4N^2\right)$$

$$= \frac{4}{N}(n_{--}^2 + n_{-+}^2 + n_{+-}^2 + n_{++}^2) + \theta(8)$$

Moreover we have

$$n_- = n_{--} + n_{-+} + \theta(1)$$

whence

$$n_-^2 = (n_{--} + n_{-+})^2 + \theta(2(n_{--} + n_{-+})) + \theta(1) \tag{20}$$
$$= n_{--}^2 + 2n_{--}n_{-+} + n_{-+}^2 + \theta(3N)$$

and in the same way

$$n_+^2 = n_{+-}^2 + 2n_{+-}n_{++} + n_{++}^2 + \theta(3N). \tag{21}$$

It follows from (11), (19), (20) and (21) that

$$X_2 = \frac{2}{N}\left((n_{--} - n_{-+})^2 + (n_{+-} - n_{++})^2\right) + \theta(8) + \theta(12) + \theta(1) \tag{22}$$

Here we have

$$(n_{--} - n_{-+})^2 \tag{23}$$

$$= \left(\frac{1}{4}\sum_{i=1}^{N-1}(e_i - 1)(e_{i+1} - 1) + \frac{1}{4}\sum_{i=1}^{N-1}(e_i - 1)(e_{i+1} + 1)\right)^2$$

$$= \left(\frac{1}{2}\left(\sum_{i=1}^{N-1} e_i e_{i+1} - \sum_{i=1}^{N-1} e_{i+1}\right)\right)^2$$

$$\le \frac{1}{4}(C_2(E_N) + W(E_N))^2$$

$$\le \frac{1}{2}((C_2(E_N))^2 + (W(E_N))^2)$$

and in the same way

$$(n_{+-} - n_{++})^2 \le \frac{1}{2}\left((C_2(E_N))^2 + (W(E_N))^2\right). \tag{24}$$

(15) follows from (22), (23) and (24).

(iii) Clearly we have

$$B_i = \frac{(-1)^i}{2^{i+1}}(e_1 - 1)\cdots(e_i - 1)(e_{i+1} + 1)$$
$$+ \sum_{n=1}^{N-i-1} \frac{(-1)^i}{2^{i+2}}(e_n + 1)(e_{n+1} - 1)\cdots(e_{n+i} - 1)(e_{n+i+1} + 1)$$
$$+ \frac{(-1)^i}{2^{i+1}}(e_{N-i} + 1)(e_{N-i+1} - 1)\cdots(e_N - 1).$$

Here the absolute value of both the first and the last term is at most 1, so that their contribution is $\theta(2)$ (where again $\theta(\dots)$ is defined as previously). Taking the term-by-term product in the sum in the middle, the contribution of the products of the -1's and $+1$'s is

$$(N - i - 1)\frac{1}{2^{i+2}} = m_i + \theta(1),$$

and all the other terms can be collected in form of sums of type

$$\pm \frac{1}{2^{i+2}} \sum_n e_{n+j_1} e_{n+j_2} \cdots e_{n+j_\ell} \tag{25}$$

where n runs over consecutive integers so that the contribution of such a sum is

$$\theta(W(E_N)/2^{i+2}) \quad \text{for } \ell = 1$$

and

$$\theta(C_\ell(E_N)/2^{i+2}) \quad \text{for } \ell > 1,$$

and here $(j_1, \dots, j_\ell)$ runs over all ℓ-tuples with $0 < \ell \le i + 2$,

$$0 \le j_1 < j_2 < \cdots < j_\ell \le i + 1.$$

For fixed ℓ the number of these ℓ-tuples is $\binom{i+2}{\ell}$, so that altogether

$$B_i = m_i + \theta\left(3 + \frac{1}{2^{i+2}}\left((i + 2)W(E_N) + \sum_{\ell=2}^{i+2}\binom{i + 2}{\ell}C_\ell(E_N)\right)\right).$$

Since exactly the same estimate can be given for G_i, (17) follows (for all k).

If $i > W(E_N)$ then for all $1 \le n \le N - i + 1$ we have

$$\left|\sum_{j=n}^{n+i-1} e_j\right| \le W(E_N) < i,$$

thus both -1 and $+1$ occur amongst e_n, e_{n+1}, $\ldots$, e_{n+i-1} so that there is no run of length i, hence

$$B_i = G_i = 0$$

which proves (18).

(iv) Clearly we have

$$|X_5| = \left| \left(-\sum_{i=1}^{N-d} e_i e_{i+d} \right) / (N-d)^{1/2} \right| \leq \frac{C_2(E_N)}{(N-d)^{1/2}}.$$

and this completes the proof of the theorem.

Combining Theorems 1 and 2 we get

Corollary 1. *For the Q-sequence described in Theorem 1 we have*

$$X_1 \leq 100 \ k^2 (\log p)^2,$$
$$X_2 \leq 1000 \ k^2 (\log p)^2 + 21,$$
$$X_5 \leq \frac{20 \ k p^{1/2} \log p}{(p-d)^{1/2}}.$$

Remarks

(i) Specifying Theorem 2 to the case of the Q-sequence, we get a good estimate for the statistic X_4 only if k in (16) is much smaller than the one described in the definition of the runs test. Namely, the number of runs of length i can be estimated by using the bounds for the correlations of order $\leq i$, however, the estimates obtained in this way become too weak for large i.

(ii) Trying to estimate the statistic X_3 in the poker test, the difficulty is that we have to divide E_N into *non-overlapping* parts; it is for this reason that it is not enough to use correlations of different orders. However, in [12] we also introduced a third PR measure, the *combined* (well-distribution – correlation) *PR-measure of order k*, defined as

$$Q_k(E_N) = \max_{a,b,t,D} \left| \sum_{j=0}^{t} e_{a+jb+d_1} e_{a+jb+d_2} \cdots e_{a+jb+d_k} \right|;$$

and we also proved that for the special Q-sequence formed by the Legendre symbol (described in (4)) we have

$$Q_k(E_N) \ll k N^{1/2} \log N.$$

(Later we dropped the use of this measure Q_k. Namely, it provides better insight into the PR properties of the given sequence if we separate the estimates of the well-distribution measure and the correlation, besides the arguments and formulas become easier to follow.)

It could be shown similarly that for the general Q-sequence, described in Theorem 1, we have

$$Q_k(E_N) \ll k\ell N^{1/2} \log N.$$

Using this, we could also estimate the statistic X_3 reasonably well.

The *theoretical* testing performed above lead to the conclusion that for the Q-sequence described in Theorem 1, *uniformly for all choices of the seed* (the coefficients of $f(n)$), it is guaranteed that the value of three of the five basic statistics is either within the passing limit, or in the worst case just a little (by at most a factor $O(\log^2 N)$) greater than this limit, and reasonably good uniform upper bounds can be given for the values of the two other statistics as well. These facts can be considered as a *strong tendency towards pseudorandomness* which has two consequences of basic importance:

(i) In not very demanding applications we may accept the sequences made by the Q-construction *without any further numerical testing*.

(ii) Even if in certain special applications we need special numerical sequences which pass all the five basic tests, we may expect that substituting randomly chosen seeds successively, in a very *few tries* we arrive to a numerical sequence which *passes all the five tests*.

5　Trying to Eliminate the Logarithm Factor

We have seen that three times out of five cases, the theoretical uniform upper bound is worse than the limit for passing the test by an $O(\log^2 N) = O(\log^2 p)$ factor only. Can one eliminate this unwanted log factor, or at least a part of it ? Can one do this by choosing p (and/or $f(n)$) in the appropriate way ? The mathematically provable answer seems to be beyond reach at the moment; however, one can give a more or less convincing heuristics.

Specifying the Q-construction to the case (4) (the Legendre symbol), (8) in Theorem 1 gives

$$\max_{0<X<Y<p} \left| \sum_{X<n<Y} \left(\frac{n}{p}\right) \right| \leq W(E_p) < 10p^{1/2} \log p. \tag{26}$$

Can one improve on this, can one prove

$$\max_{0<X<Y<p} \left| \sum_{X<n<Y} \left(\frac{n}{p}\right) \right| = o(p^{1/2} \log p) \ ?$$

It seems hopeless to prove this without any unproved hypothesis. On the other hand, Montgomery and Vaughan [15] proved: "Suppose that GRH [the Generalized Riemann Hypothesis] is true. Then for any non-principal character χ modulo q and any x,

$$\sum_{n \leq x} \chi(n) \ll q^{1/2} \log \log q.$$

This estimate is essentially best possible, for Paley [18] has shown that there are infinitely many fundamental discriminants $D \equiv 1 \bmod 4$ for which

$$\max_x \left| \sum_{n \leq x} \left(\frac{D}{n} \right) \right| > \frac{1}{7} D^{1/2} \log \log D."$$

These facts may suggest that, perhaps, our upper estimates for X_1, X_2 and X_5 also hold with the unwanted $\log^2 p$, resp. $\log p$ factors reduced to $\log \log p$; however it seems hopeless to prove this without GRH, and it seems to be very difficult even under GRH. Moreover, even if this reduction to $\log \log p$ is possible, this is the best we can achieve *uniformly* in p.

On the other hand, still one may hope that even this unwanted $\log \log p$ factor can be eliminated for appropriate *values* of p. A result of Montgomery and Vaughan [16] seems to point in this direction: they proved that for a positive proportion of the primes p we have

$$\max_{0 \leq X < Y \leq p} \left| \sum_{X < n < Y} \left(\frac{n}{p} \right) \right| \ll p^{1/2}.$$

However, even if the logarithm factors in the upper bounds for X_1, X_2 and X_5 can be eliminated completely for certain special values of p, it will be very difficult to show this.

6   Choosing p

We have just proposed a restriction on the choice of p. However, there is an other, even more important requirement when we choose p. Namely, the upper estimate (9) in Theorem 1 is conditional: it holds under the condition that one of (i), (ii) and (iii) in Theorem 1 holds. Thus when we choose our parameters, we have to do this so that we should be able to use one of the three assumptions.

Clearly, from practical point of view (i) is the least useful condition, since it ensures the control of the correlation of order 2 only. (iii) is much more useful, and if we want completely unconditional construction and estimates, then this is the best one of the three. On the other hand, the inequality in (iii) poses quite strong restriction on the choice of k and ℓ, so that if we want to control high order correlations as well, then we must keep the degree of $f(n)$ quite small. Probably this inequality is very far from being best possible; it would be very desirable to improve on it.

However, out of the three, (ii) is far the most convenient condition to use, provided that there is a p at hand with the property that 2 is a primitive root modulo p. So what is known about the primes with this property? For $p > 2$, let $g(p)$ denote the least positive integer which is a primitive root modulo p. Murata [17] writes: "Numerical examples show that, in most cases, $g(p)$ are very small. Among the first 19862 odd primes up to 223051, $g(p) = 2$ happens for 7429 primes (37.4%) (...) And we can support this observation by a probabilistic argument $(\ldots)$ we can surmise that, for almost all prime p, $g(p)$ is not very far from $\frac{p-1}{\varphi(p-1)}$. The function $\frac{p-1}{\varphi(p-1)}$ fluctuates irregularly, but we can prove the asymptotic formula:

$$\pi(x)^{-1} \sum_{p \leq x} \frac{p-1}{\varphi(p-1)} = C + O\left(\frac{\log\log x}{\log x}\right),$$

$$C = \prod_p (1 + \frac{1}{(p-1)^2}) \approx 2.827."$$

Heath-Brown [8] writes: "In 1927 Artin conjectured that any integer k, other than -1 or a perfect square, is a primitive root for infinitely many primes. It was shown by Hooley [9], [10, Chapter 3] that the conjecture is true for k, providing that the Riemann Hypothesis holds for the Dedekind zeta function of each field $\mathbb{Q}(k^{1/q})$ (where q runs over primes). Indeed, under this assumption, Hooley proved that there are asymptotically $c\,\mathrm{Li}(x)$ primes $p \leq x$ for which k is a primitive root, where c is a certain constant depending on k." Moreover, he proved in [8]: "There are at most two (positive) integers for which Artin's conjecture fails." So that almost certainly there are infinitely many primes p such that 2 is a primitive root modulo p, and a positive proportion of the primes p is expected to have this property. Thus if we check consecutive primes p for 2 being a primitive root modulo p, we may expect to hit a prime with this property in bounded many tries. Moreover, as Murata's remark shows, one may speed up this search for a good prime by restricting ourselves to primes such that $\frac{p-1}{\varphi(p-1)}$ is "small", say < 4 (anyway, we need the factorization of $p-1$ to decide whether 2 is a primitive root modulo p or not); this requirement is in good accordance with the one formulated at the end of Section 5.

If there is a table of prime numbers at hand, then in the search for a good p we may also utilize the well-known and easy-to-prove elementary fact that if p is a prime of the form $4q+1$ where q is also a prime, then 2 is a primitive root modulo p. (If a stronger, quantitative version of Schinzel's well-known "Hypothesis H" [19], [20] is true, then there are $\gg x/\log^2 x$ primes of this form up to x). Note that a prime of this form also suits the requirement at the end of Section 5 ideally (certain heuristic considerations seem to indicate that, perhaps, character sums of type studied in Section 5 are easier to estimate from above if $\varphi(p-1)/(p-1)$ is possibly large).

7   Length of the Sequence Generated

So far we have studied the "complete" Q-sequence $E_p = \{e_1, \ldots, e_p\}$ described in (7). However, in modular constructions it is customary (in cryptographic applications, even necessary) also study much shorter, "truncated" sequences $E_M = \{e_1, e_2, \ldots, e_M\}$ with M much smaller than p. So the question is: how small can one make M so that E_M still possesses certain PR character?

If $M > cp$ (with $c > 0$), then clearly, our estimates above can be used equally well, the sequence preserves its PR nature basically intact. If M decreases to about $N^{1/2}$, the rate of the error term to the trivial estimate grows, and around $N^{1/2}$ our estimates become trivial. Of course, it is possible that the sequence still preserves its PR character, only we cannot prove it, only the techniques fail. So what is really going on at this point? Again, as in Section 5, nothing can be proved anymore, but at least a good heuristics can be given.

If there is a non trivial upper bound for $W(E_M)$ with, say, $M = \lfloor p^c \rfloor$, then this must also cover the simplest special case (4) (the Legendre symbol) so that we must have

$$\left| \sum_{n \leq M} \left(\frac{n}{p} \right) \right| < M$$

It follows that the least quadratic non-residue is less than $M \leq p^c$, but this is not known for $c < \frac{1}{4}e^{-1/2}$ (still Burgess' [4] $c = \frac{1}{4}e^{-1/2} + \varepsilon$ is the best known exponent). Thus we cannot expect any unconditional proof for $M < p^c$ with c small. Again, under GRH the situation is different: "on GRH one has

$$\sum_{n \leq x} \chi(n) \ll_\varepsilon x^{1/2} q^\varepsilon \text{ "}$$

[15] (where χ belongs to the modulus q) so that we may expect that our PR measures are "small" even for $M \sim p^\varepsilon$. Thus the truncated sequence almost certainly preserves its PR character down to p^ε, and perhaps even much further. However, for sure we cannot go below $\log p \log \log p$ since it is known [14] that the GRH implies that the least quadratic non residue mod p is infinitely often $\gg \log p \log \log p$.

Summarizing: M must be chosen from the interval

$$[\log p(\log p \log p)^{1+c}, p];$$

if our priority is pseudorandomness, then we have to choose M near the top end, say $M > cp$, while in, say, cryptographic applications, we may wish to choose M closer to the lower end, but then it is advised to check the numerical sequence constructed by using one or more of the basic statistical tests listed above.

8   Choosing the Seed (the Polynomial $f(n)$)

Let K be a positive integer to be fixed later; at this point it suffices to assume that, say, $3 \leq K < p^{1/2}$. Consider all the polynomials $h(x) \in \mathbb{F}_p[x]$ of the form

$$h(x) = a_K x^K + \sum_{i=0}^{T} a_i x^i \quad \text{with } T = \lfloor K/3 \rfloor \qquad (27)$$

(here we modify slightly the construction given in [6]) where a_K, a_T, a_0, a_1, ..., a_{T-1} are chosen in random way with

$$a_K, \ a_T \in \mathbb{F}_p \setminus \{0\}, \ a_0, \ a_1, \ldots, \ a_{T-1} \in \mathbb{F}_p. \qquad (28)$$

Each of these polynomials $h(x)$ can be written in the form

$$h(x) = (r(x))^2 h^\star(x) \tag{29}$$

where $r(x) \in \mathbb{F}_p[x]$, $h^\star(x) \in \mathbb{F}_p[x]$ and $h^\star$ has no multiple zero in $\overline{\mathbb{F}_p}$ (the algebraic closure of F_p). (See Lemma 1 in [1]). Then clearly $r(x) \mid (h(x), h'(x))$ whence

$$r(x) \mid (Kh(x) - xh'(x)) = (K - T)a_T x^T + \cdots$$

and thus

$$0 \leq \deg r(x) \leq T. \tag{30}$$

It follows from (29) and (30) that

$$\deg h^\star(x) = \deg h(x) - 2 \deg r(x) \tag{31}$$
$$\geq K - 2T = K - 2\lfloor K/3 \rfloor \geq \lfloor K/3 \rfloor > 0.$$

By (31) and the definition of $h^\star(x)$, the polynomial $h^\star(x)$ is of degree $\geq \lfloor K/3 \rfloor > 0$ (this is why we needed the gap between the x^T and x^K terms in (27)) and it has no multiple zero. Thus we may apply Theorem 1 with $h^\star(x)$ and $\deg h^\star(x)$ ($\leq \deg h(x) = K$) in place of f(x) and k, respectively. We obtain that for the Q-sequence $E_p = \{e_1, \ldots, e_p\}$ defined by

$$e_n = \begin{cases} \left(\frac{h^\star(n)}{p}\right) & \text{for } (h^\star(n), p) = 1 \\ +1 & \text{for } p \mid h^\star(n) \end{cases} \tag{32}$$

we have

$$W(E_p) < 10Kp^{1/2} \log p \tag{33}$$

and assuming that either $\ell < p$ and 2 is a primitive root mod p, or

$$4K^\ell < p,$$

we have

$$C_\ell(E_p) < 10Klp^{1/2} \log p. \tag{34}$$

(Note that (31) ensures not only $\deg h^\star(x) > 0$, but also that $h^\star(n)$ in (32) is of "not very small" degree.)

The analysis of the Q-sequences given in the previous sections applies to this family $\mathcal{F}$ of Q-sequences of form (32). *The size of the family $\mathcal{F}$ is huge:* there are more than $p^{K/3}$ polynomials $h(x)$ of form (27), and uniformly over all the random choices in (28), the polynomial $h(x)$ in (27) defines a "good" PR sequence by (32). It is possible that different polynomials $h(x)$ reduce to the same polynomial $h^\star(x)$; however, it was shown by Ahlswede, Khachatrian, Mauduit and Sárközy [1] that there are not only many different sequences in $\mathcal{F}$, but also the structure of $\mathcal{F}$ is "rich", "complex", which may pay very well in the applications.

It remains to study how to choose K from the interval $3 \leq K < p^{1/2}$. If K grows, then our bounds in (33) and (34) for $W(E_p)$ and $C_\ell(E_p)$ become weaker, and for $K \gg p^{1/2}$ they become trivial. Thus to guarantee good PR properties,

K must be much less than $P^{1/2}$, so that we may choose K from the interval, say,

$$0 < K < p^{1/4}. \tag{35}$$

Again, it is a matter of our priorities how to choose K from the interval (35). If our top priority is good PR properties then we choose K near the lower end, while if we choose K near the top end, then the PR properties may get weaker, still this choice pays in applications where our priority is to construct "large" families of PR sequences (like in cryptography).

9   A Further Construction

Recently the second author [21] studied the following binary sequence: let p be an odd prime, let g be a primitive root modulo p, and let $\operatorname{ind} a$ denote the (modulo p) index of a (to the base g) so that

$$g^{\operatorname{ind} a} \equiv a \bmod p,$$

and also assume

$$1 \leq \operatorname{ind} a \leq p - 1.$$

Then write $N = p - 1$, and define the sequence $E_N = \{e_1, \ldots, e_N\}$ by

$$e_n = \begin{cases} +1 & \text{if } 1 \leq \operatorname{ind} n \leq (p-1)/2, \\ -1 & \text{if } (p+1)/2 \leq \operatorname{ind} n \leq p - 1. \end{cases}$$

Sárközy [21] showed that this sequence has "good" PR properties: we have

$$W(E_N) < 20N^{1/2}\log^2 N$$

and, for all $\ell \in \mathbb{N}$, $\ell < p$,

$$C_\ell(E_N) < 27\ell 8^\ell N^{1/2}(\log N)^{\ell+1}.$$

Comparing these estimates with the ones in Theorem 1, there the bounds are just slightly better; the really important difference is that there a large family of "good" PR sequences is constructed.

We learned recently that the idea to exploit $\operatorname{ind} x < p/2$ or $> p/2$ was also used in the Blum-Micali (BM) generator [3], defined precisely in the next section. However, the two constructions are very different, and the Blum-Micali generator was proved to pass the next bit test, while the construction above ensures better control over the PR properties.

10   Running Time Analysis and Comparison with Other Sequences

The Blum-Micali (BM) algorithm [3] (see also [13], p. 189) defines a binary sequence $E_N = \{e_1, \ldots, e_N\} \in \{-1, +1\}^N$ as follows: let p an odd prime number, g a primitive root modulo p and x_0 an integer (the *seed*) such that $1 \leq x_0 < p$. Then for $n = 1, \ldots, N$ we compute

$$x_n = g^{x_{n-1}} \bmod p \quad \text{and} \quad e_n = \begin{cases} +1 & \text{if } 1 \leq x_{n-1} \leq (p-1)/2, \\ -1 & \text{if } (p+1)/2 \leq x_{n-1} \leq p-1. \end{cases}$$

Generating each pseudorandom bit using BM requires essentially one modular exponentiation (modulo p), so the whole sequence will cost at least $O(N \log^2 p)$. The Blum-Blum-Shub [2] (BBS) algorithm defines a sequence

$$\{b_1, \ldots, b_N\} \in \{0,1\}^N$$

as follows: let p and q be two distinct (random) prime numbers congruent to 3 modulo 4, and compute $n = pq$. Select a random integer s (the *seed*), $1 \leq s < n$ such that $(s, n) = 1$, and compute $x_0 = s^2 \bmod n$. Then for $i = 1, \ldots, N$:

$$x_i = x_{i-1}^2 \bmod n, \quad b_i = \text{least significant bit of } x_i.$$

A ± 1 sequence can be deduced immediately by writing $e_i = 2b_i - 1$.

Generating each pseudorandom bit using BBS requires essentially one modular squaring (modulo n), so the whole sequence will cost between $O(N \log n)$ and $O(N \log^2 n)$ depending on the squaring method used.

The construction of a Q-sequence (7) is very simple when we are going to take $N \equiv p$. In that case we just compute a table of all Legendre symbols in $O(p) = O(N)$ bit operations, and then it remains to compute all consecutive $f(n)$. If the polynomial f has a special form, this step can be optimized. Otherwise, the most straightforward method will lead to $O(NK)$ bit operations. Of course if we do not take K "small", the running time becomes greater than in case of the BM or BBS generators, but we have much better control over the PR properties. The case when we take N much "smaller" than p is just a little more complicated: we will have to compute each Legendre symbol instead of using a table. This will multiply the running time by $O(\log p)$.

11 Numerical Calculations

To illustrate our theoretical results by numerical data we have used the "Statistical Test Suite for random and pseudorandom number generators for cryptographic applications" (sts-1.4) from the National Institute of Standards and Technology (NIST) to submit our Q-sequences to several basic tests.

First, we have written a Q-sequences generator for "small" values of p, which is very fast and produces good pseudorandom sequences. To save space, here we present an example with a relatively small p and a polynomial of small degree. We take $p = 1000003$, and the pseudorandom polynomial

$$f(x) = x^{32} + 637854\, x^9 + 514861\, x^8 + 755545\, x^7 + 883229\, x^6$$
$$+ 237063\, x^5 + 741922\, x^4 + 631773\, x^3$$
$$+ 687734\, x^2 + 928348\, x + 283971.$$

To illustrate the behavior of the sequence e_n obtained in this manner, we have plotted in figure 1 the sum $\sum_{n \leq x} e_n$, compared to $\pm\sqrt{x}$. We used sts-1.4 to

Fig. 1. "Walk" of our specific Q-sequence, compared with $\pm\sqrt{x}$

analyze the quality of pseudorandomness of the sequence obtained. (Of course in the applications, p can be chosen much greater). Issuing the command `assess 50000` (which means that the size of each bit-stream is 50000 bits), we could generate 20 such bit-streams and the output of the program `sts-1.4` is reproduced in figure 2. Column `C1` up to `C10` correspond to the a frequency specific to the test. Then `P-VALUE` is the result of the application of a χ^2–test, and `PROPORTION` the proportion of sequences that pass the test. We have not limited the tests to those discussed in this paper. We have also included some other classical tests like cumulative sums (`Cusum`), binary matrix rank test (`Rank`), discrete Fourier transform (`FFT`), approximate entropy (`Apen`). The column `PROPORTION` shows that 100% of our Q-sequences passed all requested basic statistical tests.

C1	C2	C3	C4	C5	C6	C7	C8	C9	C10	P-VALUE	PROPORTION	STATISTICAL TEST
2	2	1	1	4	2	0	3	3	2	0.739918	1.0000	Frequency
4	1	2	2	1	0	0	4	3	3	0.350485	1.0000	Block-Frequency
2	1	2	3	3	3	0	2	1	3	0.834308	1.0000	Cusum
1	2	3	3	3	3	1	1	1	2	0.911413	1.0000	Cusum
6	2	2	1	0	2	2	2	0	3	0.162606	1.0000	Runs
2	3	2	3	1	4	3	2	0	0	0.534146	1.0000	Long-Run
1	0	2	0	2	4	0	3	4	4	0.162606	1.0000	Rank
0	1	5	3	0	1	2	4	2	2	0.213309	1.0000	FFT
0	3	4	3	1	1	3	3	0	2	0.437274	1.0000	Apen
2	6	1	1	3	0	1	1	1	4	0.090936	1.0000	Serial
6	2	2	1	0	3	1	2	0	3	0.122325	1.0000	Serial

Fig. 2. Statistical tests for $p = 1000003$

Secondly, we have implemented a version of our Q-sequences generator, using the package
`free-lip-1.1`,
written by Arjen K. Lenstra, which is included in `sts-1.4`. This version which can handle arbitrary large integers is of course much slower (minutes instead of seconds), both because of the increased size of p and the necessity to compute

each Legendre symbol individually. This is a price to pay for more security. Concerning the statistics, the results are similar.

12 Conclusions

A large family finite pseudorandom binary sequences is presented. These sequences possess "very good" PR properties. The construction uses the Legendre symbol modulo p, and polynomials $f(n)$ over $\mathbb{F}_p$. It involves a large number of parameters: p, the degree of $f(n)$ and, mostly, the (almost) random coefficients of $f(n)$. This fact ensures large flexibility in adapting the construction for different purposes. Several general principles are discussed how to choose our parameters optimally depending on our priorities (good PR qualities, short running time, long sequence, some sort of cryptographical security or a combination of these).

References

1. R. Ahlswede, L. Khachatrian, C. Mauduit, and A. Sárközy, A complexity measure for families of binary sequences, Periodica Mathematica Hungarica 46, 107–118, 2003.

2. L. Blum, M. Blum, and M. Shub, A simple unpredictable pseudorandom number generator, SIAM Journal on Computing 15, 364–383, 1986.

3. M. Blum and S. Micali, How to generate cryptographically strong sequences of pseudorandom bits, SIAM Journal on Computing 13, 850–864, 1984.

4. D.A. Burgess, The distribution of quadratic residues and non residues, Mathematika 4, 106–112, 1957.

5. J. Cassaigne, C. Mauduit, and A. Sárközy, On finite pseudorandom binary sequences VII: The measures of pseudorandomness, Acta Arithmetica 103, 97–118, 2002.

6. L. Goubin, C. Mauduit, and A. Sárközy, Construction of large families of pseudorandom binary sequences, Journal of Number Theory 106, 56–69, 2004.

7. K. Gyarmati, On a pseudorandom property of binary sequences, Ramanujan J. 8, No. 3, 289–302, 2004.

8. D.R. Heath-Brown, Artin's conjecture for primitive roots, Quarterly Journal of Math. 37, 27–38, 1986.

9. C. Hooley, On Artin's conjecture, J. reine angew. Math. 226, 209–220, 1967.

10. C. Hooley, Application of Sieve Methods to the Theory of Numbers, Cambridge Tracts in Mathematics, 1970.

11. D.E. Knuth, The Art of Computer Programming, Vol. 2, Addison Wesley, 2 ed., 1981.

12. C. Mauduit and A. Sárközy, On finite pseudorandom binary sequences I: Measure of pseudorandomness, the Legendre symbol, Acta Arithmetica 82, 365–377, 1997.

13. A. Menezes, P. van Oorschot, and S. Vanstone, Handbook of Applied Cryptography, CRC Press, Inc., 1997.

14. H.L. Montgomery, Topics in Multiplicative Number Theory, Springer Verlag, New-York, 1971.

15. H.L. Montgomery and R.C. Vaughan, Exponential sums with multiplicative coefficients, Inventiones Mathematicae 43, 69–82, 1977.

16. H.L. Montgomery and R.C. Vaughan, Mean values of character sums, Canadian Journal of Mathematics 31, 476–487, 1979.

17. L. Murata, On the magnitude of the least prime primitive root, Journal of Number Theory 37, 47–66, 1991.

18. R.E.A.C. Paley, A theorem on characters, Journal London Math. Soc. 7, 28–32, 1932.

19. A. Schinzel, Remarks on the paper "Sur certaines hypothses concernant les nombres premiers", Acta Arithmetica 7, 1–8, 1961/62.

20. A. Schinzel and W. Sierpiński, Sur certaines hypothses concernant les nombres premiers, Acta Arithmetica 4, 185–208, 1958. Corrigendum: ibid 5, 259, 1959.

21. A. Sárközy, A finite pseudorandom binary sequence, Studia Sci. Math. Hungar. 38, 377–384, 2001.

22. A. Weil, Sur les courbes algbriques et les varits qui s'en dduisent, Vol. 1041 of Act. Sci. Ind., Hermann, Paris, 1948.

Authorship Attribution of Texts: A Review

M.B. Malyutov

Abstract. We survey the authorship attribution of documents given some prior stylistic characteristics of the author's writing extracted from a corpus of known works, e.g., authentication of disputed documents or literary works. Although the pioneering paper based on word length histograms appeared at the very end of the nineteenth century, the resolution power of this and other stylometry approaches is yet to be studied both theoretically and on case studies such that additional information can assist finding the correct attribution.

We survey several theoretical approaches including ones approximating the apparently nearly optimal one based on Kolmogorov conditional complexity and some case studies: attributing Shakespeare canon and newly discovered works as well as allegedly M. Twain's newly-discovered works, Federalist papers binary (Madison vs. Hamilton) discrimination using Naive Bayes and other classifiers, and steganography presence testing. The latter topic is complemented by a sketch of an anagrams ambiguity study based on the Shannon cryptography theory.

Keywords: micro-style, macro-style analysis, anagrams presence testing and ambiguity.

1 Micro-style Analysis

1.1 Introduction

The importance of dactyloscopy (fingerprint) and DNA profiling in forensic and security applications is universally recognized after successful testing of their resolution power and standardization of analyzing tools. Much less popular so far is a similar approach to the attribution of disputed texts based on statistical study of patterns appearing in texts written by professional writers. The best tests and their power are yet to be estimated both theoretically and by intensive statistical examination of stylometric differences between existing canons. If this work will prove that conscious and unconscious style features of different professionals can be discriminated as well or nearly as well as fingerprints of different persons, stylometry will change its status from a *hobby* to a *forensic tool* of comparable importance to those mentioned above. One obstacle for implementing this program is the *evolution and enrichment of styles* during professional careers of writers. Thus *plots of style characters vs. time of production* seem more relevant tools than constant characters. Rates of change for characters may vary. Also, authors can work in several forms, for instance, prose and verse which may have different statistical properties. Therefore, an appropriate *preprocessing* must be applied to the texts analyzed to avoid heterogeneity of forms in, for example,

R. Ahlswede et al. (Eds.): Information Transfer and Combinatorics, LNCS 4123, pp. 362–380, 2006.

parts of a dramatic corps. Finally, a reliable stylometry analysis should take into account all available information about a disputed work, say time of its preparation, and thus teams of "classifiers" should consist of specialists in different fields, certainly including literary experts.

Especially appealing are those case studies where the stylometric evidence helps to identify an otherwise unexpected candidate for authorship or deny a popular candidate, if this attribution is confirmed later by credible evidence. One example of such success is the denial of Quintus Curtius Snodgrass articles' attribution to Mark Twain, later confirmed by credible documents, see section 1. A recently discovered play "Is he dead" was also attributed to Mark Twain. Why not study this play by tools of stylometry before claiming its attribution?

Much more dramatic is the famous Shakespeare controversy with the attribution result so far unavailable. Various stylometry and other tests point to the same person, although much more careful testing is needed. It would be extremely encouraging if credible evidence would prove one day the correctness of the stylometry results in this case study.

1.2   Survey of Micro-stylometry Tools

The pioneering stylometric study (Mendenhall, 1887, 1901) was based on histograms of word-length distribution of various authors. These papers showed significant difference of these histograms for different languages and also for different authors (Dickens vs. Thackeray) using the same language. The second paper describes the histograms for Shakespeare contemporaries commissioned and supported by A. Hemminway. This study demonstrated a significant difference of Shakespearean histogram from those of all but one contemporaries studied (including the Bacon's), and at the same time it called attention to the practical striking identity of Shakespearean and C. Marlowe's histograms (Marlowe allegedly perished two weeks before the first Shakespearean work was published). The identity was shown by a method close to the contemporary bootstrap. However, Williams (1975) raised some doubts about the validity of the Bacon-Shakespeare divergence of styles, pointing to the lack of homogeneity of the texts that were analyzed (Bacon used different literary forms, which in my opinion only strengthens discrepancy of their styles). This objection deserves careful statistical analysis; its cost is now minor (hours vs. months before) because of the availability of software and texts in electronic form. Stability of word-length distribution for a given author also deserves further statistical study.

Ever since T. Mendenhall's pioneering work, word-length histograms have become a powerful tool that has been used to attribute authorship in several case studies including an inconclusive one over a disputed poem (Moore vs. Livingstone) controversy, and a successful rejection of Quintus Curtius Snodgrass articles' attribution to M. Twain, as described in Brinegar, 1963.

The frequencies and histograms mentioned above characterize the stationary distribution of words or letters when an author has a large body (canon) of known work. Another popular attribution tool of this kind is a *Naive Bayes (NB) classifier of Mosteller and Wallace* (1964) developed during their long and very costly work over binary authorship attribution (Madison vs. Hamilton) of certain *Federalist papers* supported by federal funding.

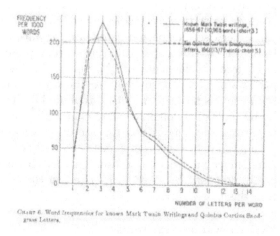

Fig. 1. Histograms of word length in Mark Twain and Quintus Curtius Snodgrass

After fitting appropriate parametric family of distributions (Poisson or negative binomial), they follow the Bayes rule for odds (*posterior odds is the product of prior odds times the likelihood ratio*)when multiplying the odds: Madison vs. Hamilton, by the sequence of likelihood ratios corresponding to the frequencies of a certain collection of relatively frequent function words, obtaining astronomical odds in favor of Madison.

This classifier presumes independence of function words usage, which is obviously **false**. This premise should be kept in mind when estimating significance of similar studies (see, for example, the attribution study of certain Shakespeare works as a byproduct of cardiac diagnosis software, well-advertised by the Boston Globe on August 5, 2003, or certain Moliere-Corneille controversy studies). The NB-attribution can often be confirmed by other stylometric tests, although the NB-likelihood ratios cannot be taken seriously. The NB-classifier is routinely used also for screening out bulk or junk e-mail messages, see Katirai, 1999, De Vel et all, 2001.

In contrast, Thisted and Efron, 1987, use the *new words usage distribution* in a newly discovered non-attributed anapest poem "Shall I die, shall I fly?", found in the Yale University library, 1985.

I will touch on only one detail in their application of a popular estimation method for the number of unseen biological species (first invented by Turing and Good for breaking the Enigma code), namely neglecting the enrichment of an author's language with time. Thus the distribution of new words in a disputed work *preceding* the canon of an author and that for a text *following* the canon, can be significantly different, for example if Marlowe or Shakespeare wrote the poem. Therefore, this particular application of the Turing-Good method seems inappropriate. Also, the comparative power of their inference in appropriate cases seems unknown.

More promising and popular now tools use *modeling of long canons as Markov chains of some order composed of English letters and auxiliary symbols*. Given a non-attributed text T and a collection of firmly attributed (to author k) canons T(k) of approximately the same length for training the Markov model of, say, order 1, with transition probabilities $P(k, i, j)$ between symbols i and j, k=1,..., M, the log likelihood of T being written by the k-th author is

$$\sum \log(p(k, i, j)) N(i, j) + \log \pi_k(x(1)),$$

where the sum is over all i and j, $N(i, j)$ is the frequency of i followed by j, π_k denotes the stationary probability of the k-th Markov chain, and $x(1)$ is the first symbol in T. Second order Markov chain modeling admits similar expressions for the likelihood. The author with *maximal likelihood* is chosen, which is practically equivalent to *minimizing the cross entropy of empirical and fitted Markov distributions* and to *minimizing the prediction error probability of a next symbol given the preceding text* (Rosenfeld, 1994, 1996, Zhao, 1999), see also Khmelev, 2000, who considers his work as an extension of the classical approach of A. Markov, 1913, 1916. Markov, 1916 applies Markov modeling to the authorship attribution, improving an earlier less satisfactory approach by Morozov, 1915. *The power of this inference can be approximated theoretically for large sizes of canons T(k) and T under rather natural conditions of asymptotic behavior of their sizes* (Kharin and Kostevich, personal communication). Some regularization of small transition frequencies is worthwhile.

In a canon apparently written jointly by several authors (say, the *King James English bible*), a Hidden Markov modeling is more appropriate.

Even better attribution performance in certain tests is shown in Kukushkina et al, 2001, by the now very popular *conditional complexity of compression minimizing* classifier discussed also by Cilibasi and Vitanyi, 2003, available from the web-site of the first author. There, the idea (approximating a more abstract *Kolmogorov conditional complexity* concept which may appear theoretically the best authorship classifier) is the following: every good compressor automatically adapts to patterns in the text which it is compressing, reading the text from its beginning (some compressors use various extensions of the Markov modeling described above, including those based on the variants of the Lempel-Ziv algorithm). Let us define *concatenated texts* $C(k) = T(k)T$ as texts starting with $T(k)$ and proceeding to T without stop, and corresponding compressed texts $T'(k)$ and $C'(k)$. Define the conditional compressing complexity (CCC) to be the difference between the lengths of compressed texts $|C'(k)| - |T'(k)|$ and choose the author with minimal CCC. Certainly, this definition depends on the compressor used. In the tests described in Kukushkina et al, 2001, the best attributing performance was shown to be that of the compressor rar.

A comparable performance is shown by some ad hoc classification methods such as Support Vector Machines, (see Bosh and Smith, 1998, Burges, 1998).

These methods are based on sets of characters chosen ad hoc and not unified between different applications which does not permit a valid comparison.

I skip also any discussion of methods based on grammar parsing since these methods are yet not fully automated. Also, their application for classifying very old texts, such as those written by Shakespearean contemporaries, seems doubtful.

2 Shakespeare Controversy

2.1 Introduction

Controversy concerning authorship of the works traditionally attributed to W. Shakespeare dates back several centuries. A **bibliography** of material relevant to the controversy that was compiled by J. Galland in 1947 is about **1500 pages** long (see *Friedmans*, 1957). A comparable work written today might well be at least four times as large. Resolving the controversy would certainly aid our understanding of what the author intended to convey in his works and thus would contribute to a better insight into the history of culture. Methodology developed during this investigation would also be useful in other applications, including the attribution of newly discovered non-attributed texts. The goal of this part of our rather personal overview is to stimulate further research by scholars with diverse areas of expertise in order to resolve the Shakespeare authorship mystery. My own contribution is minor and concerns the existence of certain steganography in the sonnets and plausibility of longer messages hidden there. I review in more detail the arguments in favor of only one alternative candidate, whom I personally regard as the most likely one.

If additional incentive to undertake this study is needed, note that the Calvin Hoffman prize, presently worth about one million British pounds, will be awarded to the person who resolves this controversy.

The orthodox side, consisting of those who believe the traditional figure to be the true author of these works or simply of those who find it appropriate to maintain this version, mostly keeps silent about arguments put forth against the authorship of W. Shaxpere (W.S.) from Stratford on Avon (*this one of several spellings of the name is used to distinguish the traditional figure from the as yet undecided author of the Shakespeare canon*). When not silent, the orthodox accuse the heretics of being lunatics or snobbish. A collection of their arguments can be found in Matus, 1994.

2.2 Documentary and Literary Arguments

Anti-Stratfordian snobbish lunatics (including to some extent M. Twain, S. Freud, Ch. Chaplin, Ch. Dickens, B. Disraeli, J. Galsworthy, V. Nabokov, W. Whitman, R. Emerson, J. Joyce, and H. James: *"divine William is the biggest and most successful fraud ever practiced"*) point out numerous documentary and literary reasons for rejecting or doubting the authorship of W.S.

One early survey of these grave doubts in *several hundred pages* was written by a US presidential hopeful *Donnelly*, 1888. Similar doubts were expressed in many subsequent books including recent ones, see *Mitchell*, 1996, and *Price*, 2001. Scarce documents related to W.S. revealed there allow the following scenario of his career.

His education in Stratford or any literary work there is not documented. A rather ambiguous record about his marriage is kept in the local church. Abandoning Stratford just after the birth of his twins and being warned of severe persecution over next stealing rabbits in the woods of his landlord, he apparently wandered for several years in constant fear of a severe punishment imposed on tramps in the Elizabethan time. Eventually, he was employed in valet horse parking at one of London theaters, later on he was apparently promoted to its security (since he is mentioned in several complaints over his part in assaults against alternative theaters: these were also centers of criminal activities such as gambling, prostitution, etc., there were frequent fights between them which forced the London mayor to transfer them out of City). Being a talented organizer, W.S. has later become an ambitious administrator, producer and shareholder of the theater occasionally performing secondary scenic roles,, and likely also an informer of the ESS (he seems to be the only theater functionary avoiding arraignment after the Essex revolt involving a performance of an allegedly W.S.' play!). W.S. has probably bought a respect of censors for popular plays to pass smoothly. He used to make around a *thousand pounds a year* for his apparently mostly undercover activity (compare this to *only a twice larger sum which was paid by Elizabeth to her prime minister W. Cecil!*). He argued fiercely with dramatists for changes in their plays to make them more popular, and he was not sensitive to authors' rights in publications which brought him pennies as compared to his other activities. Thus he apparently cared little if any plays were published under his name (if he could fluently read at all). His Last Will clearly shows that he did not keep any printed matters, without mentioning manuscripts. His death was not even noticed by contemporary poets.

There is evidence that W.S. lent money to dramatists for writing plays performed and published under his name and ruthlessly prosecuted those failing to give the money back in time. This is revealed by Mitchell and Price in their discussions of *Groatsworth of Wit* published in 1592 after the death of well-known dramatist R. Green, where apparently W.S. is called *Terence and Batillus* with the obvious meaning of appropriating somebody else's plays. In a recently found manuscript (see

`http://ist-socrates.berkeley.edu/~ahnelson/Roscius.html`)

written during W.S.'s retirement in Stratford prior to 1623 (First Folio) he was called *our humble Roscius* by a local educated Stratfordian author, meaning a famous Roman who profited from special laws allowing him to hawk or sell seats in the theater, and who was not known as an actor/playwright, merely as a businessman who profited on special favor[1].

---

[1] A possible visual pattern for W.S. is father Doolittle from "My fair lady", who also did little for creating the treasure of arts bearing his name.

As a Russian scholar, I knew several Russian *Terences* in Math Sciences who used their Communist party privileges to produce remarkable lists of publications "borrowed" from others, say persons *condemned as dissidents or enemies of the State*, who were meant to be forgotten in the Soviet Union, and for whom any reference to their work was strictly forbidden. It is not sufficiently remembered that Elizabethan England was an equally closed society with its ruthless censorship and persecution. Well-known, Oscar-winning scenarios written during the McCarthy era in the US by blacklisted authors under false names were milder similar stories.

This concise overview cannot touch on the *hundreds of grave very different questions* raised in the books mentioned above[2]. W.S.'s authorship (WSA) of a substantial part of Shakespeare is hardly compatible with *any* of them and *my subjective log likelihood of WSA* to answer *all of them does not exceed negative 40* (compare with *naive Bayes classifier* discussed before). In my experience as a statistical consultant in forensic cases (especially a disputed paternity) involving DNA profiling, a much milder mismatch would be sufficient for a court to reject paternity. Some scholars would prefer an explanation of the existing documents not to be based on miracles as holds for WSA. Forensic (in addition to literary) experts must play a decisive role in resolving the controversy as shown further.

The major issues for anti-Stratfordians to resolve are: whose works were published under the Shakespeare name, and why this disguise of authorship happened in the first place and then remained hidden for such a long time.

Francis Bacon became the first candidate for an alternate author, probably because his knowledge of vast areas of culture matched well with that shown in the Shakespeare works.

(a) Francis Bacon   (b) Ignatius Donnelly

Fig. 2.

---

[2] See also a vast recent collection in http://www2.localaccess.com/marlowe/

The pioneering stylometric study (Mendenhall, 1901) of Shakespeare contemporaries using histograms of their word-length distribution demonstrated the unlikelihood of Bacon's authorship of Shakespeare.

Century-long fruitless mining for cryptography in Shakespeare, allegedly installed there by F. Bacon, and multi-million expenditures for digging the ground in search of the documents proving that F. Bacon wrote Shakespearean works, are brilliantly analyzed in *Friedmans*, 1957. The father of American military cryptography William Friedman and his wife started their careers in cryptography assisting the deceptive (by their opinion) Bacon cryptography discovery in Shakespeare by E. Gallup (which was *officially endorsed by general Cartier, the head of the French military cryptography* those days!). This amusing book, full of historic examples, exercises and humor, should be read by everyone studying cryptography!

Up to now, one of most attractive alternative candidates has been *Edward de Vere, 17th earl of Oxford*. De Vere's life seems by many to be reflected in the sonnets and *Hamlet*. Both de Vere and F. Bacon headed branches of the *English Secret Service* (ESS). De Vere was paid an enormous sum annually by Queen Elizabeth allegedly for heading the *Theater Wing* of the ESS, which was designed in order to prepare plays and actors to serve the propaganda and intelligence collecting aims of the Queen's regime[3]. De Vere's active public support of the corrupt establishment of the official Anglican church in the dramatic Marprelate religious discussions confirms him as one of the principal Elizabethan propaganda chiefs.

(a) Mary Sidney Herbert, countess of Pembroke

(b) William Friedman

Fig. 3.

---

[3] See www.shakespeareauthorship.org/collaboration.htm referring to Holinshed's chronicles commissioned by W. Cecil, the head of Elizabethan Privy Council.

Other major candidates for Shakespeare authorship include R. Manners, 5th earl of Rutland, W. Stanley, 6th earl of Derby and several other members of an aristocratic Inner Circle surrounding the Queen and including F. Bacon, Edward de Vere and *Mary Sidney Herbert* (who ran a *literary academy* at her estate in Wiltshire for the University Wits) together with her sons. *Judging by the works that were firmly attributed with reasonable certainty to each of them*, none seems to have been a genius in poetry.

Some from this circle might have been able to produce plots and first versions of plays, but these attempts would need a master in order to be transformed into masterpieces. Some of these people may in fact have done the *editing* work on some of the Shakespeare works (Mary Sidney and her sons). One should also consider that the voluntary hiding of authorship on any of their parts seems unlikely. Due to the wide extent of the Inner Circle, authorship information would inevitably have become known to everyone. And yet, to the true author of the plays and poems there should have been *dramatic reasons to not claim the works universally recognized as "immortal"*. Note also that the author of the works mastered more than 30,000 English words (as estimated by Efron and Thiested (1975)) compared to about 3000 words used by an average poet. He had also mastered Greek, Latin and several contemporary European languages. In addition, he must have had a profound knowledge of classical literature, philosophy, mythology, geography, diplomacy, court life and legal systems, science, sport, marine terminology and so forth.

The role of paper *Mendenhall, 1901,* may be informally compared with that of a hunting dog. Due to the discovery contained in it, a *famous poet, translator and playwright Christopher Marlowe emerged as one of main candidates.* In an **unprecedented** petition by *Elizabethan Privy Council* Marlowe's important service on behalf of the ESS was acknowledged, and granting him Masters degree by Cambridge University was requested in spite of his frequent long absences (see *Nicholl, 1992.). His blank iambic pentameter, developed further in Shakespearean works, remained the principal style of English verse for several centuries.* In 29, ambitious Marlowe was among the most popular London dramatists during his allegedly last 5 years.

Arraigned into custody after T. Kyd's confessions under torture, and let out on bail by his ESS guarantors, Marlowe was allegedly killed by an ESS agent (in the presence of another one responsible for *smuggling agents to the continent*) at their *conventional departure house in Deptford*, owned by a close associate of Elizabeth, almost immediately *after a crucial evidence of Marlowe's heresy* was received by the court, implying an *imminent death sentence*. There is evidence of Marlowe's involvement in the Marprelate affair which made him a *personal enemy of the extremely powerful ruthless archbishop Whitgift of Canterbury*, who did everything possible to expose Marlowe for ages as a heretic and eliminate him (see the well-known *anathema* written by cleric T. Bird, O. Cromwell's teacher, in Nicholl, 1992).

A Marlowe's friend T. Penry, publisher of the Marprelate pamphlets, was hanged previous evening *two miles from Deptford* and *his body has never after been accounted for, in spite of many petitions by Penry's relatives.*

Then, two weeks after Marlowe's supposed demise, the manuscript of the poem *Venus and Adonis*, which had been anonymously submitted to a publisher some months before, was amended with a dedication to the earl of Southampton that listed for the *very first time* the name of W. Shakespeare as author (any link between the earl and W.S. seems unlikely, Marlowe was likely the earl's tutor in Cambridge).

There are numerous documentary and literary reasons to believe that Marlowe's death was faked by his ESS chiefs (who expected further outstanding service from him) in exchange for his obligation to live and work forever after under alternate names. These arguments are shown on the intelligent and informative web-site http://www2.prestel.co.uk/rey/ of a popular Shakespearean actor and former top manager of British Airlines, P. Farey. One of them is *obvious* : spending the whole last day in Deptford (apparently, awaiting the companion), Marlowe defied a strict regulation of daily reporting to the court, hence *he knew beforehand that he would never come back* under his name.

Farey also reviews extracts from the sonnets and other works of Shakespeare hinting at their authorship by Marlowe *after the Deptford affair.* He gives the results of various stylometric tests, showing that the *micro-styles of Marlowe and Shakespeare are either identical, or else the latter's style is a natural development and enrichment of the former.* The micro-style *fingerprint* would give strong evidence for Marlowe's authorship of Shakespearean work, if further comprehensive study confirms that their style patterns are within the natural evolutionary bounds while other contemporary writers deviate significantly in style[4].

Some scholars believe that the ingenious propaganda chiefs of the ESS partly inspired and paid for the production of C. Marlowe, and perhaps of some other politically unreliable dramatists, and directed this production using the Shakespeare pipeline to avoid problems with scrupulous censorship proceedings and also convert dissidents into a kind of unnamed slaves (an early prototype of Stalin labor camps for researchers).

During O. Cromwell puritan revolt in forties-fifties of 17th century all theaters were closed, many intelligence documents were either lost or burnt, and the revival of interest to the Shakespearean creative work came only in 18th century making the authorship attribution problematic.

2.3 Micro-style Analysis

The stylometric tables in the section Stylometrics and Parallelisms, Chapter *Deception in Deptford*, found on the Farey's website include convincing tables of word-length usage frequencies, including those made by T. Mendenhall, as

---

[4] Imagine James Bond let out on bail and reported killed by his colleague soon after a DNA test proved his unauthorized crime. Will you believe in his death if his DNA was repeatedly found later on his victims?

well as of *function words, feminine endings, run-on lines*, etc., in both Marlowe and Shakespeare as functions of presumable dates of writing corresponding texts.

Again, more careful statistical study of these and more powerful micro-style tests described in section 1 is desirable. Farey's plots clearly show the evolution of styles, which has not been taken into account (or even denied) in many previous studies. For example, some Russian linguists have claimed that the proportion of function words is constant inside the canon during the whole writer's life. This claim was used by them to reject Sholokhov's authorship of the first part of his Nobel prize-winning novel.

2.4   Macro-style Analysis

An interesting controversial comparative study of Shakespeare's and Marlowe's *macro-styles*[5] exists on the web-site of late Alfred Barkov

http://www.geocities.com/shakesp_marlowe/

Barkov's analysis of the *inner controversies* in Marlowe's and Shakespeare works including *Hamlet*, well-known for a long time, enables him to claim that the texts were intentionally used to encode the story in such a way that the authors' actual messages remain misunderstood by laymen while being understandable to advanced attentive readers. Barkov calls this style *menippea*, considering it similar to the *satira menippea*, a style found in many classical works and discussed by prominent Russian philosopher M.Bakhtin (1984). Menippeas often appear in closed societies, since authors tend to use Aesopian language to express their views. This language was very characteristic for Marlowe: he used his poetic genius to provoke Elizabethan enemies by his ambiguous statements to expose their views for subsequent reporting to the ESS (see *Nicholl*, 1992).

Barkov's analysis of the inner controversies in Hamlet is parallel to the independent analysis of other authors. For instance, the well-known contemporary novelist publishing under the nickname *B. Akunin*, presented recently his version of Hamlet in Russian (available in the Internet via the search inside the web-library www.lib.ru) with a point of view rather similar to that of Barkov, including the sinister decisive role played by *Horatio*.

2.5   Cryptpgraphy Mining

In November 2002, a Florida linguist, R. Ballantine, sent me her decipherment of *Marlowe's anagrams* in consecutive bi-lines (that is, pairs of lines) of most of Shakespeare and also of some other works, revealing the author's amazing life story as a *master English spy both in Britain and overseas* up to 1621. Her stunning overview with commentaries based also on her previous 20 years of documentary studies is almost 200 pages long. Her novels covering Marlowe's life

---

[5] Namely, a sophisticated architecture of their works and well-known ambiguity of many statements inside them.

Fig. 4. Roberta Ballantine

until the Deptford affair are more than thousand pages long. I was challenged to make a judgment about the validity of her findings, which stimulated my interest in the topic.

Irrespective of the authenticity of the historic information conveyed in her overview, the story is so compelling that it might become a *hit of the century* if supplied with dialogues and elaboration and published as a fiction novel by a master story teller (see several chapters of her unpublished novels and anagram examples on the web-site:

`http://www.geocities.com/chr_marlowe/`

Barkov claims that Ballantine's *deciphered anagram texts follow the menippea macro-style of Marlowe's works.* If established as true, this story will constitute a bridge between golden periods of poetry and theater in the South-Western Europe and Britain because in it C. Marlowe is revealed as a close friend of such leading late Renaissance figures as M. Cervantes and C. Monteverdi, as well as the main rival in love and theater of Lope de Vega.

It is almost unbelievable that the author of Shakespearean works could pursue additional goals while writing such magnificent poetry. However, caution is needed: Thompson and Padover, 1963, p. 253, claim that Greek authors of tragedies used to anagram their names and time of writing in the first lines of their tragedies (a kind of *water marking*), which Marlowe could well learn from his best teachers in the King's school, Canterbury and University of Cambridge; a similar tradition was shared by Armenian ancient writers as a protection against plagiarism of copyists, as described in Abramyan, 1974. Also, *first announcing discoveries by anagrams was very popular in those times (Galileo, Huygens, Kepler, Newton among other prominent authors)*; anagrams were certainly used by professional spies.

Attempting to establish cryptographic content in Shakespeare after the discouraging book *Friedmans*, 1957, is very ambitious. Moreover, serious doubts remain concerning the appropriateness of anagrams as a hidden communication (or *steganography*) tool, as will be discussed further.

It is natural to consider two stages in the analysis of the validity of deciphered anagrams. The first question to address is the *existence of anagrams* in the texts. This we have attempted to test statistically starting from our observation that all the anagrams deciphered in Shakespeare contain various forms of Marlowe's signature at the beginning.

R. Ballantine has considered bi-lines as suitable periods for anagramming case-insensitive letters. After deciphering an initial bi-line, she proceeds to the very next one, and so on, until the final signature. In a given play, the first bi-line that begins an anagramming is usually at the beginning of a dialogue, or after a special, but otherwise meaningless sign, a number of which appear in early editions of Shakespearean works.

Following Thompson and Padover, 1963, we mine for Marlowe's signature in the first bi-lines of sonnets, which makes for an easier test, since a disastrous multiplicity-of-decisions problem is avoided in this way. Besides, 154 sonnets, with only a small part of them deciphered so far, constitute a homogeneous sample of 14 lines (7 bi-lines) each (with a single exception). Hence we chose to focus on the sonnets for statistical testing of the presence of anagrams leaving aside almost all other Shakespearean works, which allegedly also contain anagrams.

An important requirement is a careful choice of an accurate published version which has varied over time. I was fortunate to find help from an expert in the field, Dr. D. Khmelev, University of Toronto, who was previously involved in a joint Shakespeare-Marlowe stylometry study with certain British linguists.

For a given bi-line b, let us introduce the event $M = \{b$ contains the set of case-insensitive letters M,A,R,L,O,W,E $\}$ (*event M is equivalent for this name to be a part of an anagram*) Using a specially written code, Khmelev showed (by my request):

Proposition 1. *The numbers of first, second, etc. bi-lines in the sonnets for which event M occurs are respectively 111, 112, 88, 98, 97, 101, 102 out of 154 sonnets.*

Our first corollary follows:

Proposition 2. *Let us test the null hypothesis of homogeneity: event M has the same probability for all consecutive bi-lines in sonnets versus the alternative that the first bi-line contains this set of letters more often than subsequent ones. It is also assumed that these events for all bi-lines are independent. Then the P-value of the null hypothesis (i.e. the probability of the frequency deviation to be as large or more under the null hypothesis) is less than four per cent.*

Proof. We apply a standard two-sample test for equality of probabilities based on the normalized difference between frequencies $f_i, i = 1, 2$, of containing the case-insensitive set of letters 'm', 'a', 'r', 'l', 'o', 'w', 'e' inside the first and all other bi-lines respectively which has approximately standard normal distribution for such a big sample; f_1 is near 72.1 per cent, f_2 is almost 65 per cent. Thus the approximate normalized difference of frequencies $(f_1 - f_2)/\sqrt{\bar{f}(1 - \bar{f})(1 + 1/6)}/154$ is around 1.78, where $\bar{f} := (f_1 + 6f_2)/7$, and *the normal approximation to the binomial probability of this or larger deviation (P-value) is near 3.75 per cent which is a rather unlikely event.*[6]

Apparently, this anomaly in homogeneity of bi-lines signals that the first bi-lines were specially designed to include this set of letters as part of an anagram signature. Note that signatures may vary over sonnets. Thus our estimate is an upper bound for the P-value of bi-lines homogeneity versus several variants of Marlowe's signature in the first bi-line.

Thus, the *existence of anagrams hidden by Marlowe in Shakespeare* looks rather likely.

Of course, *other explanations of this statistical anomaly* might also be possible. To deal with this possibility, I applied to a recognized expert in statistics on Shakespeare and on English verses in general who is with the University of Washington. Unfortunately, she turned out to be a Stratfordian, and so she chose not to reply at all.

A much more difficult task is to study the *authenticity (or uniqueness) of the anagrams deciphered by R. Ballantine.* This is due to a *notorious ambiguity of anagrams* which seems to be overlooked by those who have used anagrams to claim priority, see above. An amazing example of this ambiguity is shown on pp. 110-111, Friedmans, 1957, namely: **3100** different meaningful lines-anagrams in Latin exist for the salutation "Ave Maria, gratia plena, Dominus tecum". These are referred to a book published in 1711.

A *theory of anagram ambiguity* can be developed along the lines of the famous approach to cryptography given in C. Shannon's *Communication Theory of Secrecy Systems* written in 1946 and declassified in 1949. An *English text is modeled in it as a stationary ergodic sequence of letters* with its entropy per letter characterizing the uncertainty of predicting the next letter given a long preceding text. The binary entropy of English turns out to be around 1.1 (depending on the author and style), estimated as a result of long experimentation.

Shannon showed that this value of the entropy implies the existence of around $2^{1.1N}$ meaningful English texts of large length N. Due to the ergodicity of long texts, the frequencies of all letters in all typical long messages are about the same, and so all typical texts could be viewed as almost anagrams of each other. Thus, the *number of anagrams to a given text seems to grow with the same exponential rate as the number of English texts.* We can prove this plausible **conjecture** in a more artificial approximation of English text as an i.i.d. multinomial sequence of symbols. Let us first further simplify the setting for transparency:

---

[6] A more detailed study of numbers in Proposition 1 ignoring the multiplicity of hypotheses shows that the case insensitive set 'marlowe' is located anomalously often in *the first two bi-lines* of the sonnets (**the homogeneity P-value is around 0.2 per cent**). Another popular (according to Ballantine) signature 'Kit M.' turns out to be found unusually often (the homogeneity P-value is around 5 per cent) in the *last two bi-lines* concluding the sonnets.

Proposition 3. *Consider an i.i.d. three-nomial N-sequence of three letters A, B and C with rational probabilities $p(A) = L(A)/N, p(B) = L(B)/N$ such that $Np(A) := L(A)$ and $Np(B) := L(B)$ are integers. Our claim is: Number $N(A, B)$ of N-sequences with $L(A)$ letters A and $L(B)$ letters B satisfies:*

$$\log N(A, B)/N = H(A, B) = -[p(A) \log p(A) + \cdots + p(C) \log p(C)](1 + o(1)).$$

Proof. Follows immediately from the method of types (see e.g. Cover and Thomas, 1991). The fraction above is asymptotically the number of typical N-sequences as we stated above.

A generalization to a general multinomial case without the condition of all probabilities being multiples of $1/N$ is straightforward. A generalization to a model of stationary ergodic source can be formulated and proved using the techniques also developed in Cover and Thomas, 1991, say their *sandwich* argument, in proving the equipartition theorem.

Thus the number of meaningful English anagrams for n bi-lines is the n-th power of that for a single bi-line, if deciphering is independent for subsequent bi-lines, and also *exponential* in the length of text. This is a discouraging result for considering anagrams as a communication tool beyond other disadvantages of anagrams, namely excessive complexity of encoding and decoding. Moreover, the aim of putative anagrams that would become known to an addressee only after the long process of publication is unclear, unless an ESS editor would pass it directly to an addressee. Again a parallelism: many of *M. Bulgakov's menippeas* with hidden anti-Soviet content were prepared for publication by an active informer of the Stalin secret police!

There still remains hope that R. Ballantine's claim about the uniqueness of the anagrams she deciphered may prove correct due to the following reasons:

Every one of her deciphered anagrams starts with one of the variants of Marlowe's signature, which restricts the remaining space on the first bi-line, and makes the combination of remaining letters *atypical*, thereby narrowing the set of meaningful anagrams. Furthermore, the names and topics conveyed by Marlowe in the hidden text, may be familiar to his intended receiver (say, the earl of Southampton or M. Sidney Herbert with her sons), who might decipher the anagrams using a type of Bayesian inference, looking for familiar names and getting rid of possible anagrams that did not make sense for him/her. Existence of other keys unknown to us is also possible.

It should also be noted that the hidden sentence on the first bi-line is usually continued on the next bi-line (run-on line) giving the decipherer additional information as to how to start deciphering the next bi-line, and so forth. Surely, these arguments are rather shaky. Only a **costly experimentation in deciphering anagrams by specially prepared experts** can lead to sound results about the authenticity of anagrams deciphered from these texts. Various types of software are available to ease the deciphering of anagrams, although it is questionable if any of them is suitable for these archaic texts.

In summary, the anagram problem in Shakespeare remains unresolved, although I regard it as worthy of further study.
C. Shannon himself developed an important theory for breaking codes. His *Unicity theory* specifies the minimal length of encoded messages that admit a unique decoding of a hidden message by a codebreaker due to the redundancy of English. Unfortunately, *his main assumption of the key and message independence, crucial for his results about unicity in cryptography, is obviously not valid for anagrams*, which use special keys for each bi-line depending on the combination of letters in the bi-line.
Our statistical result on the special structure of the first bi-lines shows that the encoding (if it took place at all!) had to be iterative: if the poetic bi-line was not suitable for placing Marlowe's anagram-signature there, the line and hidden message were to be revised in order to make the enciphering possible. This is exactly a situation where *knowledge of an incredible number of English words, demonstrated by Shakespeare, could have been put to perfect use permitting flexibility in the choice of a relevant revised text!*

2.6 Hopes for Genetic Evidence

It turns out that the *critical argument against Marlowe's authorship of Shakespeare* is the inquest by Queen Elizabeth's personal coroner found in 1935 (made in violation of several instructions) stating that Marlowe was killed on May 30, 1593. The question of the validity of this inquest is discussed by *Farey* and *Nicholl,* 1992 in detail. If the inquest was faked and C. Marlowe's survival for several more years is proved, then his authorship of Shakespearean works becomes very likely: Marlowe could have written these masterpieces with abundant features to be ascribed to him, and he had more than enough reasons to hide under a fictitious name.
One long-shot way to prove Marlowe's survival is as follows. A mysterious posthumous mask is kept in Darmstadt, Germany, ascribed to Shakespeare by two reasons: The Encyclopaedia Britannica states that it matches perfectly the known portraits of the bard (which are likely actually versions of Marlowe's portraits as shown brilliantly, say, on the title page of the web-site of a recent award-winning documentary film *Much ado about something.* A second reason is the following: this mask was sold to its penultimate owner-collectioner together with a posthumous portrait of apparently the same dead man in laurels lying in his bed.
The mask contains 16 hairs that presumably belonged to the portrayed person. A specialist from the University of Oxford has claimed in a personal letter to me his ability to extract mitochondrial DNA from these hairs and match it with that from the bones of W.S. (or W.S.'s mother Mary Arden) and Marlowe's mother Kate or any of W.S.'s or Marlowe's siblings. As is well-known, mtDNA is inherited strictly from maternal side since sperm does not contain mitochondria. This study is in the planning stage, and serious legal, bureaucratic, financial and experimental obstacles must first be overcome before the study can proceed.

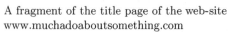

A fragment of the title page of the web-site
www.muchadoaboutsomething.com

The posthumous mask ascribed
to Shakespeare

Fig. 5.

3 Conclusion

The problem of Shakespeare authorship is old, and the documents are scarce.
Therefore, only a statistical approach, e.g., comparing the likelihoods of hypotheses based on the *fusion of all kinds of evidence*, seems feasible in trying to
resolve it.

An explosion in computing power, emergence and development of new methods of investigation and their fusion let me believe that in this framework the
Shakespeare controversy will eventually be resolved with sufficient conviction in
spite of the four-century long history of puzzles and conspiracies.

The methods that are now developing are promising and could also very well
apply in other similar problems of authorship attribution, some of which might
even have significant security applications.

Acknowledgements

This study was proposed to the very obliged author by R. Ahlswede at the beginning of the author's two-month stay with the program *Information transfer*
at ZIF, University of Bielefeld. The author thanks many colleagues (especially
R. Ballantine!) for their helpful remarks, E. Haroutunian for citing Abramyan.
I am extremely grateful to E. Gover, D. Massey and I. Malioutov for considerable improvement of my English and style and to D. Malioutov for TEX-nical
help.

References

1. A. Abramyan, The Armenian Cryptography (in Armenian), Yerevan University Press, 1974.
2. M. Bakhtin, Problemy Poetiki Dostoevskogo, English translation, University of Minnesota Press, 1984.
3. R. Bosch and J. Smith, Separating hyperplanes and the authorship of the disputed Federalist Papers, American Mathematical Monthly 105, 7, 601-608, 1998.
4. C. Brinegar, Mark Twain and the Quintus Curtis Snodgrass Letters: A statistical test of authorship, Journal of American Statistical Association 58, 301, 85-96, 1963.
5. R. Cilibasi and P. Vitanyi, Clustering by compression, CWI manuscript, 2003, submitted.
6. C. Burges, A Tutorial on support vector machines for pattern recognition, Data Mining and Knowledge Discovery 2, No. 2, 955-974, 1998.
7. T. Cover and J. Thomas, Elements of Information Theory, Wiley, N.Y., 1991.
8. I. Donnelly, The Great Cryptogram, 1, 1888, reprinted by Bell and Howell, Cleveland, 1969.
9. B. Efron and R. Thisted, Estimating the number of unseen species; How many words did Shakespeare know? Biometrika 63, 435-437, 1975.
10. R. Thisted and B. Efron, Did Shakespeare write a newly discovered poem? Biometrika 74, 445-455, 1987.
11. W. Friedman and E. Friedman, The Shakespearean Ciphers Exposed, Cambridge University Press, 1957.
12. H. Katirai, Filtering junk e-mail, 1999, see his web-site http://members.rogers.com/hoomank/
13. D. Khmelev and F.J. Tweedy, Using markov chains for identification of writers, Literary and Linguistic Computing 16, No. 4, 299-307, 2001.
14. O. Kukushkina, A. Polikarpov, and D. Khmelev, Text authorship attribution using letter and grammatical information, Problems of Information Transmission 37, 2, 172-184, 2001.
15. A. Markov, On application of statistical method, Comptes Rendus of Imper. Academy of Sciences, Ser. VI, X, 153, 1913; 239, 1916.
16. I. Matus, Shakespeare, in Fact, Continuum, N.Y., 1994.
17. T.A. Mendenhall, The characteristic curves of composition, Science 11, 237-249, 1887.
18. T.A. Mendenhall, A mechanical solution to a literary problem, Popular Science Monthly 60, 97-105, 1901.
19. J. Mitchell, Who Wrote Shakespeare, Thames and Hudson Ltd., London, 1996.
20. F. Mosteller and D. Wallace, Inference and Disputed Authorship, Addison-Wesley, Reading, 1964.
21. Ch. Nicholl. The Reckoning, second edition, Chicago University Press, 1992.
22. D. Price, Shakespeare's Unorthodox Biography, Greenwood Press, London, 2001.
23. R. Rosenfeld, A maximum entropy approach to adaptive statistical language, Modeling. Computer, Speech and Language 10, 187–228, 1996, a shortened version of the author's PhD thesis, Carnegie Mellon University, 1994.
24. J.W. Thompson and S.K. Padover, Secret Diplomacy; Espionage and Cryptography, F. Ungar Pub. Co., N.Y, 1500-1815, 1963.

25. O. De Vel, A. Anderson, M. Corney, and G. Mohay, Multi-Topic E-mail Authorship Attribution Forensics, Proc. Workshop on Data Mining for Security Applications, 8th ACM Conference on Computer Security (CCS'2001), 2001.
26. C. Williams, Word-length distribution in the works of Shakespeare and Bacon, Biometrika 62, 207-212, 1975.
27. J. Zhao, The impact of cross-entropy on language modeling, PhD thesis, Mississippi State University, 1999.
http://www.isip.msstate.edu/publications/courses/ece_7000_speech
/lectures/1999/lecture_06/paper/paper_v1.pdf

III
Raum-Zeit und Quantenphysik – Ein Geburtstagsständchen für Hans-Jürgen Treder

A. Uhlmann

Gefragt nach einem Vortrag für Hans-Jürgen Treders 75. Geburtstag, hatte ich kurz entschlossen geantwortet, etwas zum Verhältnis von Relativitätstheorie und Quantentheorie sagen zu wollen. Das war etwas leichtsinnig; denn nur ein paar Beobachtungen habe ich anzubieten, Lösungen nicht.

Die großen physikalischen Theorien des 20-ten Jahrhunderts, die wir zurecht bewundern, überdecken und beherrschen weite Teile der Physik. Doch nicht in allen arbeiten sie gleich gut. In manchen finden wir sie perfekt. In anderen meint man, ein leichtes Klappern der Maschinerie heraushören zu können. Es liegt eine Spannung zwischen Geometrie und dynamischen Größen wie Energie und Impuls. Wir finden sie in der Fortschreibung der Teilung in Kinematik und Dynamik, die wir von der Klassischen Mechanik kennen. Kinematik ist die geometrische Analyse der mechanischen Bewegung, Dynamik fragt nach ihrer Verursachung.

Es war ein 2000 Jahre während Geheimnis, dass die Euklidische Geometrie nicht die allein denkbare ist. Dann aber brauchte es keine hundert Jahre, um auch der Zeit die Starre des Absoluten zu nehmen. Und kaum hatte Albert Einstein in seiner Speziellen Relativitätstheorie Raum und Zeit zu einem universellen Block vereinigt und Minkowski ihn geometrisch analysiert, so zeigte er mit der Allgemeinen Relativitätstheorie dessen Veränderbarkeit und Bedingtheit.

Bald darauf entdeckten Werner Heisenberg und Erwin Schrödinger eine geschlossene, in sich stimmige Form der Quantenphysik. Max Born erkannte die zufällige Natur beobachtbarer Quantenprozesse. Danach ist die Antwort eines Quantensystem auf einen definierten Eingriff nur bedingt determiniert: Welche Reaktionen mit welcher Wahrscheinlichkeit erlaubt sind, ist vom Systemzustand abzulesen. Welche der erlaubten Reaktionen jedoch im Einzelprozess erfolgt, ist Zufall.

1 Raum-Zeit und Geometrie

Der Relativitätstheorie liegt die Mannigfaltigkeit der Raum-Zeit-Punkte, die auch "Welt-Punkte" genannt werden, zugrunde. Ihre metrischen Bindungen sind ihr vornehmster Forschungsgegenstand. In ihnen, im metrischen Tensor, ist die Geometrie von Raum und Zeit kodiert.

Paare von Weltpunkten können, zumindest in nicht zu großen Gebieten, in Klassen eingeteilt werden: Entweder A ist vor B, oder B ist vor A, oder keines von beiden findet statt.

R. Ahlswede et al. (Eds.): Information Transfer and Combinatorics, LNCS 4123, pp. 381–393, 2006.

A vor B zeigt die Möglichkeit an, eine Wirkung über eine Kausalkette von A nach B zu übertragen. Eine etwas allgemeinere, gegen die Vielschichtigkeit des Kausalitätsproblems weniger anfällige Definition, ist die: Von A nach B kann ein Signal gesendet werden, ist "Information" übertragbar. Umgekehrt kann, wenn A nach B kommt, also B vor A, in A ein Signal mit Ursprung in B empfangen werden.

Nicht weniger interessant sind die Fälle, in denen Signale weder von A nach B noch von B nach A gelangen können. Diese Paare nennt man entweder zueinander "raumartig" oder voneinander "kausal unabhängig". Denn Ereignisse, die an ihnen stattfinden oder stattfinden könnten, sind mit Notwendigkeit voneinander unabhängig. Andererseits nennt man ein solches Paar von Welt-Punkten aber auch raumartig, weil man ihm, wenigstens lokal, eine Entfernung zuordnen kann, die sich verhält, wie man es von einem ordentlichen Abstandsbegriff erwartet.

Die kausale Unabhängigkeit von Ereignissen, die in den unmittelbaren Umgebungen raumartiger Paare stattfinden, ist grundlegend für jedwede relativistische Quantenphysik: Simultane Messungen oder andere Eingriffe in räumlich voneinander getrennten Weltgebieten beeinflussen sich nicht in vorhersehbarer Weise. Sie verhalten sich wie unabhängige Zufallsgrößen. Die Schwierigkeiten mit nicht-vertauschbaren Observablen finden hier ihr vorübergehendes Ende.

Treten wir nun ein wenig in die Fußstapfen des Laplaceschen Dämons: Hätten wir in einem Raum-Zeit-Gebiet die Liste der Abstände aller raumartigen Paare von Weltpunkten, dann hätten wir auch den kompletten metrischen Tensor. Diesen Zweck würde auch jede gekürzte Liste erfüllen, die nur Paare berücksichtigt, die nicht weiter als eine vorgegebene Länge, beispielsweise 1 mm, auseinander liegen. Jede derartige Liste impliziert die Kenntnis der möglichen zeitlichen Abläufe. Jede Änderung der räumlichen Geometrie zieht unausweichlich die Veränderung zeitlicher Abläufe nach sich.

Analoge virtuelle Listen kann man aus Zeitmessungen gewonnen denken. Man hat Kurven zu betrachten, längs denen Signalfortschreibung möglich ist. Man nennt sie "vorwärts gerichtete" Weltkurven. An jedem Punkt einer derartigen Weltkurve zeigt die Tangente in die von diesem Weltpunkt aus erreichbare Zukunft.

Beschränken wir unsere Liste auf zeitartige Weltkurven und schließen lichtartige Weltkurven, die eine Sonderrolle beanspruchen, aus. Letztere sind Weltkurven, die durchfahren werden ohne Zeit zu verbrauchen. Auf ihnen steht die Zeit still. Sie gehören zu Grenzen, die raumartige und zeitartige Regimes trennen.

Es dauert seine Zeit, eine zeitartige Weltkurve zu durchlaufen. Diese Dauer ist eine Invariante der Kurve. Indem man alle zeitartigen Weltkurven, die A und B verbinden, zur Konkurrenz zulässt, entsteht ein neues Optimierungsproblem: Es muss nach einer Kurve gesucht werden, deren Verbrauch an Zeit, um von A nach B zu gelangen, von keiner anderen Kurve übertroffen wird. Besagte maximale Zeitdauer ist die "zeitliche Distanz", die B von A trennt.

Auch unsere Liste der zeitlichen Abstände der Paare von Raum-Zeit-Punkten eines Weltgebiets, die kausale Abhängigkeiten zulassen, reicht aus, um die Geometrie des Gebiets komplett zu bestimmen. Es langt auch jede gekürzte

Liste, die nur Paare von Weltpunkten enthält, deren zeitliche Distanz eine obere Schranke, zum Beispiel eine Millisekunde, nicht überschreitet.

Dass die Länge von Raumkurven minimiert, die Dauer von Zeitkurven aber maximiert werden muss, unterscheidet das Räumliche vom Zeitlichen.

2   Quantale Grundstrukturen

Auch in der Quantenphysik gibt es Gegenden von fundamentaler Bedeutung, die in markanter Weise nicht dynamisch sind.

Versuchen wir, dies an Hand der 2-Niveau-Systeme zu verstehen. Ihre klassische Version wäre ein System, das nur zwei verschiedene Zustände einnehmen kann, zum Beispiel ein Schalter, der nur die Stellungen EIN und AUS zulässt. Alle anderen Eigenschaften, (Aufbau, Zuverlässigkeit usw.), die eine konkrete Realisierung des Schalters betreffen, werden vernachlässigt. Als Abstraktum ist der klassische Schalter ein ziemlich triviales System.[1]

Ein quantales 2-Niveau-System ist wesentlich komplexer. Bei ihm werden sowohl die beobachtbaren physikalischen Größen als auch ihre Zustände mit Hilfe von 2x2-Matrizen (Operatoren) beschrieben. Zählen wir ein paar physikalisch interessante Beispiele auf:

- Die Polarisation des Photons.
- Der Spin des Elektrons.
- Die beiden Energie-Niveaus des Ammoniak Moleküls, die den Ammoniak Maser ermöglichen, und ihre Überlagerungen.
- Organische Farbstoffe, etwa das Fuchsin, die ihre Farbe einem 2-Niveau Subsystem verdanken.
- Physikalisch von ganz anderer Art ist ein 2-Niveau System, das den neutralen K-Mesonen zugeordnet wird, um die eigenartigen Verhältnisse zu beschreiben, die sie mit der schwachen und der starken Wechselwirkung eingehen.

Um in den genannten Beispielen von der abstrakten Grundstruktur zur Dynamik zu kommen, müssen eine oder mehrere Matrizen zu Observablen erklärt werden: Die Helizität beim Photon, Drehimpulse beim Spin des Elektrons, die Energie beim Ammoniak und beim Fuchsin, die Hyperladung bei den neutralen K-Mesonen.

Woran sehen wir, dass es sich bei einem 2-Niveau-System um einen Elektronen-Spin handelt? Offenbar dadurch, dass wir es als Teilsystem eines anderen, umfassenderen erkennen, welches eine genauere Beschreibung des Elektrons gestattet. Diese Identifizierung entsteht, indem wir die Einbettung des kleineren Systeme in ein oder mehrere größere Systeme analysieren: Der Spin des Elektrons verbleibt, wenn wir andere Freiheitsgrade, zum Beispiel den Impuls, ignorieren.

Analog bleibt nur die Polarisation des Photons übrig, wenn wir von Energie (Frequenz) und Ausbreitungsrichtung des Photons absehen.

---

[1] Allerdings nur, wenn er als einzelnes Objekt auftritt und nicht in Massen, wie bei einem Prozessor.

Weitere wichtige Kriterien für die physikalische Identifizierung von Quantensystemen folgen aus Erhaltungssätzen. Die Einstufung als Teilsystem ist keineswegs herabsetzend. Es ist auch nicht als bloße Approximation an etwas, das wir genauer kennen, zu verstehen. Im Gegenteil: Jedes Quantensystem ist Teilsystem von anderen Systemen, die zusätzliche Freiheitsgrade besitzen. Reduktion und Erweiterung von Quantensystemen sind für die "Quantenwelt" von grundsätzlicher Bedeutung. In dieser hierarchischen Struktur verbleibt jede Observable eines Systems als Observable in jedem größeren. Und umgekehrt bestimmt der Zustand eines Obersystems die Zustände aller seiner Teilsysteme. Letztere nennt man deshalb auch "reduziert" oder "marginal" bezüglich des Obersystems.

Kurz gesagt: Quantensysteme, ob endlich oder nicht, sind hierarchisch miteinander verknüpft. Von einem physikalischen System ist es wichtig zu wissen, von welchen anderen Systemen es ein Teilsystem ist und, falls das zutrifft, wie es in diesen untergebracht ist, wie es in ihnen eingebettet ist.

Ich vermerke noch, dass der Quantentheorie die Annahme eines "größten", allumfassenden Systems fremd ist. Aus sich heraus bietet sie keinerlei Handhabe für ein solches Konstrukt. Was sollte man auch von dem Versuch halten, alle in der Natur tatsächlich oder virtuell angelegten Freiheitsgrade zu umfassen und über alles, was in Zukunft entdeckt werden wird, vielleicht aber auch für immer verborgen bleibt, etwas aussagen zu wollen[2]?

Um zu erklären, welche der Regeln und Gesetze ich hier als "nicht-dynamisch" bezeichnen will, komme ich nochmals auf die beobachtbaren Größen zurück. Die Frage ist hier, was von der Identifizierung einer Matrix (oder eines Operators) mit einer konkreten beobachtbaren Größe abhängt, und was von dieser Identifizierung ganz unabhängig ist. Den unabhängigen, nicht-dynamischen Teil will ich für die Zwecke des Vortrages "quantale Grundstruktur" nennen.

Bereits unsere kleine Liste von 2-Niveau Systemen ist geeignet, sich der Frage zu nähern: Was hängt davon ab, ob wir eine Matrix mit der Beschreibung der Helizität des Photons, oder mit einer Spinkomponente des Elektrons, oder mit dem Energie-Operator identifizieren, und was nicht. Was bleibt, wenn wir die Besonderheiten der einzelnen 2-Niveau Systeme ignorieren? Sehr wenig, wird man zurecht sagen. Denn die Vielfalt der Kräfte und Wechselwirkungen, die Spezifika ihrer Realisierungen, gehen verloren.

Und doch verbleibt nicht nur ein dürrer Rest, sondern etwas sehr Wichtiges: Das Überlagerungsprinzip und die Gesetze, die die Zufälligkeit von Quantenprozessen kontrollieren. Es sind Regeln, die von allen quantalen Vorgängen befolgt werden müssen. Sie sind von jener Klarheit und Schärfe, die wir von elementaren logischen Operationen kennen.

Versichern wir uns zunächst an Hand eines Beispiels, wie die elementare Logik mit "Eigenschaften" umgeht, die entweder wahr oder falsch sind, zutreffen oder nicht zutreffen können: Eine natürliche Zahl ist entweder eine Primzahl, oder sie ist es nicht. Also bilden in der Menge aller Zahlen die Primzahlen eine gut

---

[2] Auch in der Kosmologie beschränkt man sich auf eine überschaubare Zahl von Begriffen und Parametern.

definierte Teilmenge. Es ist zwar tautologisch, zu sagen, eine Zahl sei prim, wenn sie zu dieser Teilmenge gehört. Aber durch diesen Trick entsteht eine Isomorphie zwischen Eigenschaften und Teilmengen einer Gesamtmenge: Die Eigenschaften werden durch eine Familie vom Teilmengen repräsentiert.

Besonders interessant wird es, wenn nicht alle, sondern nur gewisse Teilmengen als Eigenschaften zugelassen werden. Dann wird man verlangen, dass logische Operationen, zum Beispiel das UND, das ODER und das NEIN, in einer solchen Familie von Teilmengen, sprich Eigenschaften, ausgeführt werden können. Diesen Trick hat uns G. Boole verraten.

In der Quantenphysik gibt es das nämliche Problem und die Willkür bei seiner Lösung ist nicht allzu groß; denn das Überlagerungsprinzip ist zu berücksichtigen: Damit eine Menge von Zuständen eines quantalen Systems eine Eigenschaft im logischen Sinne sein kann, darf die Bildung von Überlagerungen nicht aus ihr herausführen. Eine quantale Eigenschaft muss also mit zwei oder mehreren Zuständen auch allen denkbaren Überlagerungen zukommen. Sie muss in diesem Sinne geschlossen sein.

Nach Birckhoff und von Neumann haben wir es mit einer notwendigen Bedingung zu tun. Nur für eine solche Menge kann eine Messapparatur existieren, die entscheiden kann, ob ein Zustand zu ihr gehört oder nicht.

Zwar muss nicht jede Menge von Zuständen, die die genannte Voraussetzung erfüllt, notwendigerweise eine Eigenschaft sein. Der Punkt ist, dass es nur Eigenschaften gibt und Messgeräte geben kann, die mit dem Überlagerungsprinzip im Einklang sind. Am Rande beinhaltet diese Behauptung eine sehr scharfe und weitreichende Verneinung der Existenz verborgener Parameter.

Ein, zwei Beispiele zur Erläuterung. Wählen wir in dem Raum, der einem Schrödingerschen Teilchen zugänglich ist, ein Teilgebiet G aus. Die Menge der Zustände, die bei einer Ortsmessung mit Notwendigkeit das Teilchen in G auffinden lässt, ist eine Eigenschaft. Es ist die Eigenschaft, in G lokalisiert zu sein. Trifft sie zu, so ist die Schrödingerfunktion nur in G von Null verschieden. Im zweiten Beispiel verlangen wir, dass die Messung der kinetischen Energie mit Notwendigkeit einen Wert ergibt, der nicht größer als ein vorgegebener Wert ist. Auch diese Vorschrift definiert eine Eigenschaft.

In diesen Formulierungen sollte die Wendung "mit Notwendigkeit" aufhorchen lassen. Man könnte an ihrer Stelle auch "mit Wahrscheinlichkeit Eins" sagen[3]. Sie führt uns auf eine weitere Besonderheit, die sie von der Klassischen Logik deutlich unterscheidet. Es ist die Rolle des Zufalls und der Wahrscheinlichkeit. Sie markiert eine Grenze zwischen Klassischer und Quantenphysik, vielleicht sogar die wichtigste.

Ein Schrödinger-Teilchen muss sich weder in einem Raumgebiet G aufhalten noch muss es zwingend außerhalb von G angetroffen werden. Es muss überhaupt nicht in irgendeinem Teilgebiet lokalisiert sein! Um dem damit verbundenen Dilemma zu entgehen, kommt der Zufall zu Hilfe:

Soll experimentell entschieden werden, ob das Schrödinger-Teilchen sich in G aufhält oder nicht, so ist es fast immer der pure Zufall, ob die Antwort

---

[3] Für tatsächliche Messungen ist der Unterschied irrelevant.

JA oder NEIN lautet. In welchem Zustand sich das Teilchen auch anfangs befunden hatte, nach dem Test ist es entweder in G oder nicht in G. Dabei ist die Wahrscheinlichkeit, es in G anzutreffen, durch den Zustand vor der Messung festgelegt. Jedoch ist es innerhalb dieses Rahmens ein zufälliges Ereignis, ob der Experimentator JA oder NEIN findet. Findet er im Einzelfall JA, so darf er sicher sein, dass sich das Teilchen nach seiner Messung tatsächlich in G aufhält: Der Test präpariert einen neuen Zustand. Sobald wir eine Information über den Systemzustand erzwingen, passt sich dieser dem hierzu erforderlichen Eingriff an! (Ich muss zugeben, dass ich ein einfaches aber etwas akademisches Beispiel gewählt habe, da durch den Rand von G der Raum scharf in verschiedene Teile getrennt wird. Das ist energetisch nur angenähert möglich.)

Wir sehen hier eine weiteres, für die Quantenphysik fundamentales nichtdynamisches Gesetz. Als Folge des Testens einer Quanten-Eigenschaft wird ein neuer Zustand hergestellt, der diese Eigenschaft entweder definitiv besitzt oder definitiv nicht besitzt. Welcher der beiden möglichen neuen Zustände hergestellt wird, ist nichts als Zufall. Bei häufiger Wiederholung der gleichen Messung im jeweils gleichen Zustand nähert man sich einer berechenbaren Wahrscheinlichkeitsverteilung. Dadurch und durch nichts anderes wird der Zufall reguliert. All das gilt unabhängig vom konkreten physikalischen Charakter des betrachteten Quantensystems und unabhängig von der Art der Wechselwirkungen. Im Rahmen der Quantenphysik ist es ein universelles Gesetz.

Nun muss ich noch auf eine weitere, bisher nicht erwähnte seltsame Eigenart quantaler Zustände zu sprechen kommen: Zustände sind generisch nicht lokalisiert. Jede Veränderung in einem begrenzten Raumgebiet zieht unmittelbar globale Veränderungen nach sich.

Insbesondere ist der Übergang von einem Zustand in einen neuen, der als Ergebnis des Testens einer Eigenschaft erfolgt, global und instantan[4]. Diese Behauptung kann man experimentell nur approximativ prüfen. Aber es gibt gute Experimente, die, bei vorsichtiger Abschätzung der Messfehler, für die Ausbreitung derartiger Zustandsänderungen vielfache Lichtgeschwindigkeiten messen. Und es gibt, wie schon gesagt, gute theoretische Gründe für die Behauptung, dass sich diese Änderungen prinzipiell nicht mit endlicher Geschwindigkeit ausbreiten können.

Die vermutlich eindrucksvollsten Experimente handeln von Messungen, die simultan in der Nähe zweier raumartig getrennter Weltpunkte ausgeführt werden. Der Zustand des zu messenden Systems kann so eingestellt werden, dass man aus der Messung am Weltpunkt A auf das Messergebnis im Weltpunkt B schließen kann. (Einstein-Podolski-Rosen-Effekt.) Einzelne Ereignisse, die derart streng korrelieren[5], ohne jedoch kausale Abhängigkeiten zu erlauben, müssen rein zufällig sein. Jegliches Abweichen von der Zufälligkeit erlaubt die Übertragung von Information mit beliebig hoher Geschwindigkeit. Daher sind die in einer Einstein- oder Minkowski-Welt gültigen Schranken für kausale Prozesse mit dem Bornschen Postulat über die Zufälligkeit quantaler Prozesse auf Gedeih und Verderb verbunden!

---

[4] In der älteren Literatur wird vom Kollaps der Wellenfunktion gesprochen.
[5] Es handelt sich nicht um statistische Korrelationen!

Diesen Aussagen, die in der Klassischen Physik kein Analogon besitzen, setzt die Theorie noch ein geometrisches Sahnehäubchen auf. In der Menge der Zustände eines Systems gibt es Metriken[6], die die Bornschen Wahrscheinlichkeiten für Quantenprozesse in Abstände konvertieren: Je näher der Zustand X am Zustand Y liegt, um so größer ist die Wahrscheinlichkeit, dass X in Y übergeht. Ist eine Zustandsänderung zeitlos, so verläuft sie auf einer kürzesten Verbindung im Raum der Zustände. Wenn aber die Veränderung des Zustandes an einen zeitlichen Vorgang geknüpft ist, wie wir es von der zeitabhängigen Schrödinger-Gleichung kennen, dann ist die entsprechende Kurve gekrümmt und keine Kürzeste.

Jeder mögliche Quantenprozess, der den Zustand X mit dem Zustand Y verbindet, wird im Raum aller Zustände als Kurve von X nach Y beschrieben. Die Länge dieser Kurve ist genau dann minimal, wenn der Prozess raum-zeitlich instantan ist, also keine Zeit verbraucht. Es ist die raum-zeitliche "Oberfläche", die nicht fähig ist, derartige Prozesse darzustellen. Sie wirkt eher wie ein Zensor, der nur Teile des Quantengeschehens zur Besichtigung frei gibt.

3 Spektralität

Eine der notwendigen Voraussetzungen für das Verständnis von Dynamik wird unter dem Namen "Spektralität" geführt. Ihr physikalischer Ursprung sind die Linien- und Energiespektren der Atome und Moleküle, ihr mathematischer ist in der Operatortheorie begründet.

Relativistisch gesehen fordert Spektralität, dass Massen und Energien niemals negativ sein dürfen.

Nicht-relativistisch spielen nur Energiedifferenzen eine Rolle. Hier darf die Energie nicht unbegrenzt ins Negative abwandern. Anderenfalls hätten wir einen sich nie füllenden energetischen Abgrund, eine fatale energetische Instabilität.

Die Spektralitätsforderung ist in ein Postulat gegossene Erfahrung. Wir müssen sie cum grano salis per Hand hinzufügen. Auf den ersten Blick scheint sie von allen anderen theoretischen Grundannahmen unabhängig zu sein. Jedoch: Wenngleich wir es mit einem plausiblen und anschaulichen Postulat zu tun haben, werden mit ihm allerlei Fallen aufgestellt.

Wir wissen, dass die Klassische Physik keine Handhabe für die energetische Stabilität von Coulombschen Systemen bietet. Zumindest für das nicht-relativistische Regime wird dieser Mangel durch die Quantentheorie beseitigt. Dank Dyson, Lenard, Lieb, Thirring und anderen wissen wir, dass Vielteilchensysteme mit dominierenden Coulomb-Kräften energetisch[7] stabil sein können: Die Bindungsenergie pro Teilchen bleibt beschränkt. Voraussetzung ist, dass das System insgesamt (so gut wie) neutral ist, dass keine nennenswerten Mengen an Antiteilchen vorhanden und mindestens die Hälfte der Teilchen Fermionen[8]

---

[6] Mathematisch von Fubini und Study bzw. von Bures entdeckt.

[7] ... und auch thermodynamisch.

[8] Fermionen brauchen Platz, Bosonen nicht notwendigerweise.

sind. Das ist so bei "gewöhnlichen" makroskopischen Dingen, einem Stück Eisen, Blumentöpfen, bei Hunden und Katzen, usw.

Wir müssen jedoch folgern, dass energetische Stabilität nicht nur von der Art der Wechselwirkung abhängt. Es ist auch wichtig, welche stabilen Teilchen massenhaft vorhanden sind und in welchem Verhältnis sie anzutreffen sind. Spektralität hängt nicht nur von den fundamentalen Wechselwirkungen ab! Offenbar können wir annehmen, dass die kosmologische Evolution dieses Problem für uns gelöst hat: Was an "normaler Materie" übrig geblieben ist, ist energetisch stabil. Das gilt aber nur, solange die Gravitation nicht dominant wird. Für sie ist die Bindungsenergie pro Teilchenmasse bereits in der Newtonschen Näherung nicht beschränkt.

Ich erinnere mich an einen Besuch bei Hans-Jürgen. Vor den Buchwänden stehend, griff ich die Kirchhoffsche Mechanik heraus, ein meisterhaft geschriebenes Lehrbuch. Ich war erstaunt, wie selbstverständlich Kirchhoff den Zusammenhang von Symmetrie und Erhaltung behandelt. Schon in der zweiten Hälfte des 19. Jahrhunderts hatten die Physiker erkannt, dass die allgemeinen Erhaltungssätze aus den Symmetrien von Raum und Zeit folgen. Jeder Erhaltungssatz der Mechanik, der nicht von der speziellen Natur des physikalischen Systems abhängt, muss einen Grund haben. Er kann nicht einfach vom Himmel gefallen sein. In Sonderheit erweist sich der Erhaltungssatz für die Energie als Konsequenz der Homogenität der Zeit.

In der Quantenphysik ist dieser Zusammenhang noch inniger. Die Symmetrien legen die korrespondierenden dynamischen Größen durch ihre Erzeugenden bis auf einen Faktor fest. Dieser Faktor ist das mit i multiplizierte (und nach Dirac normierte) Plancksche Wirkungsquantum.

Erinnern wir uns rückblickend an die abstrakten Quantensysteme, in denen das Überlagerungsprinzip und die quantalen Wahrscheinlichkeiten kodiert sind. In ihnen finden wir das Plancksche Wirkungsquantum nicht. Größen mit der physikalischen Dimension einer Wirkung sind in ihnen nicht vorhanden. Erst wenn wir Entfernung und Geschwindigkeit mit dynamischen Variablen wie Masse, Impuls und Energie verbinden wollen, bekommt es seine überragende Bedeutung. Sein Auftritt verbindet rein raum-zeitliche Konstrukte mit dynamischen.

Und was ist mit i, einer der beiden Wurzeln aus -1 ? Heisenberg und Schrödinger haben die in der oberen Hälfte der Gaußschen komplexen Zahlenebene liegende Wurzel in ihren Vertauschungsregeln und Bewegungsgleichungen benutzt. Denkt man sich die reelle Achse der Gaußschen Ebene als Zeitachse, so vergeht die Zeit, indem wir ihre obere Hälfte links liegen lassen, sie im mathematisch positiven Sinne umfahren.

So gesehen hat Spektralität eine physikalisch nicht leicht zu verstehende Konsequenz: Man kann die Zeit analytisch in die obere Halbebene fortsetzen. Zeitabhängige Erwartungswerte können daher oft als stetige Randwerte analytischer Funktionen dargestellt werden. Eine analytische Funktion wird als Ganzes bereits im Kleinen vollständig bestimmt. Jede Abänderung zieht unweigerlich ihre globale Veränderung nach sich. Auch diese simple Tatsache stützt die Behauptung über die nicht-lokale Natur von (generisch allen) Quantenprozessen.

Analytische Fortsetzung physikalisch relevanter Größen in sogenannte unphysikalische Bereiche wie komplexe Impulse, Energien, Zeitvariable usw. ist eine weit ausgebaute Methode der Quantenphysik. Die Streutheorie macht exzessiven Gebrauch von ihr. In der Relativistischen Quantentheorie wird sie herangezogen, um das Spin-Statistik Theorem zu beweisen, aber auch den Unruh-Effekt, die Minkowski-Variante der Hawkinschen Strahlung schwarzer Löcher und manches mehr.

Es ist eigenartig, dass für einige fundamentale Sätze der Physik keine Beweismethode gefunden worden ist, die die analytische Fortsetzung in anscheinend unphysikalische Wertebereiche vermeidet.

Spektralität ist eine sehr vernünftige Forderung, da sie als energetische Stabilität verstanden werden kann. Weiterführende Konsequenzen ergeben sich jedoch erst in Verbindung mit Symmetrien, die von zeitartigen Transformationen herrühren. Es ist das Dreieck a) dynamische Observable, b) Symmetrie und c) Erhaltungssatz, das immer wieder durchlaufen wird, um Postulaten wie der Spektralität Bedeutung zu verleihen. Wird dieses Dreieck an einer Stelle zerbrochen, kommt man schnell in Beweis- oder Definitionsnot, als wäre ein Kompass verloren gegangen.

Und das geschieht, wenn wir die Spezielle Relativitätstheorie verlassen und Quantentheorie auf allgemeineren Raum-Zeit-Geometrien zu treiben beabsichtigen. Der allgemein beliebte Langrange-Formalismus reicht dann selbst für freie Teilchen nicht zur korrekten Quantisierung aus. Es muss noch auf die richtige Massen- und Energieabhängigkeit geachtet werden. Im hochsymmetrischen Minkowskischen Fall erzielt man mit energetisch stationären Lösungen die gewünschte Spektralität. Stationarität aber wird als Invarianz gegen Zeitverschiebungen verstanden. Einige Autoren meinen deshalb, vermutlich etwas zu voreilig, dass man nur in einem asymptotischen Sinn von Teilchen sprechen dürfe, und das auch nur unter der Annahme, wir hätten es mit einer Welt zu tun, die für hinreichend große Entfernungen vom eigentlichen Geschehen sich immer genauer einer Minkowski Welt annähert.

4 Die Welt ohne Symmetrien

Was erwartet uns in einer Raum-Zeit ohne Symmetrien [9]? Auch hier gibt es Transformationen mit zeitartigen Trajektorien im Überfluss. Jedoch lassen sie die metrische Struktur nicht invariant. Daher haben wir keinen Grund, auf eine von ihnen zu zeigen und sie der Energie (oder dem Impuls) zuzuordnen (oder für sie einen Erhaltungssatz fordern).

Es entsteht eine Art Definitionsnot: Der Begriff des universellen Erhaltungssatzes, des allgemeinen Integrals der Bewegung, wird fragwürdig. Spezielle Integrale konkreter physikalischer Systeme darf man hingegen erwarten.

---

[9] Nur von Raum-Zeit-Symmetrien ist hier die Rede!

Wie schon gesagt, bemerkt man das angedeutete Problem schmerzlich bei dem Versuch, auch nur das einfachste freie Quantenfeld auf einer symmetrielosen Raum-Zeit zu etablieren. Trotz vieler Anstrengungen und schöner Teilerfolge muss die Aufgabe als ungelöst angesehen werden. Allerdings sind 3- und 4-dimensionale Probleme ohne Symmetrien mathematisch notorisch überaus schwierig. Doch muss wohl auch noch Einiges zur begrifflichen Analyse des Problems getan werden.

Würde in der heute üblichen Manier ein Wettbewerb um die raffinierteste Gleichung der Physik ausgeschrieben werden, so würde sicherlich die Einsteinsche Gleichung der Allgemeinen Relativitätstheorie als unangefochtener Sieger hervorgehen. Schon kurz nach ihrer Aufstellung zeigte H. Cartan ihre Einzigartigkeit unter sehr wenigen und sehr vernünftigen Annahmen. Nur zwei Konstanten bleiben frei: Die auch heute noch nicht voll verstandene Kosmologische Konstante und die Gravitationskonstante. Zum Besonderen gehört, dass der Energie-Impuls Tensor einem Ausdruck gleichgesetzt wird, der allein aus der Metrik heraus erklärt ist. Denken wir etwa an die eingangs diskutierte Liste räumlicher Abstände. Sie bestimmt nicht nur die gesamte Metrik sondern auch die Verteilung von Energie und Impuls, wenn auch in einer etwas delikaten Weise.

Ganz im Gegensatz dazu ist der Energie-Impuls Tensor für kein physikalisches System ohne Kenntnis des metrischen Tensors bildbar. Ohne Metrik ist er undefiniert.

Die weltpunktweise Erhaltung der Energie ist diesem Tensor als Existenzbedingung auferlegt worden. Infolge des allgemeinen Mangels an Symmetrie, kann hieraus kein Erhaltungssatz für Weltgebiete hergeleitet werden, wie klein oder groß sie auch seien.

Man kann vom Energie-Impuls Tensor aber Positivität von Energie und Ruhemasse verlangen: Spektralität ist keine Folge der Einsteinschen Gleichungen. Sie kann nur ad hoc gefordert werden. Aber man kann sie wenigstens formulieren.

Die Situation wird wesentlich besser, wenn weitere Annahmen erfüllt sind. So kann eine Art Gesamtenergie definiert werden, wenn asymptotisch flache "Zeitschlitze" existieren. Für die Schwarzschild-Lösung bekommt man so den "richtigen" Wert.

Ein Zeitschlitz ist eine Teilmenge der Raum-Zeit, deren Weltpunkte paarweise raumartig angeordnet sind und die nicht vergrößert werden kann, ohne dieses Charakteristikum zu zerstören. Ein Zeitschlitz ist ein Moment in der zeitlichen Entwicklung, ein Augenblick von idealer Kürze.

Über den Schwarzschild-Fall hinaus hat man starke Aussagen für fast-stationäre[10] Zeitschlitze beweisen können. In besagtem Fall ist die Energiedichte der Riemannschen skalaren Krümmung proportional. Von der gesamten Energie konnte man jetzt zeigen, dass sie größer ist als die Energiesumme der in ihr eventuell enthaltenen Schwarzen Löcher, eine 30 Jahre alte Vermutung von Penrose.

---

[10] Gemeint ist das Verschwinden der zweiten Fundamentalform sowie asymptotische Flachheit.

Ohne die vorausgesetzte Spektralität brechen die mathematisch sehr aufwendigen und intelligenten Beweise zusammen und die genannten und weitere schöne Ergebnisse sind nicht mehr richtig.

Der Energie-Impuls Tensor reflektiert im Idealfall alles, was dynamisch geschieht. Er wirkt dabei wie ein Filter, in den alles Dynamische eingebracht wird, und der nur einen genau ausgesuchten Anteil der dynamischen Komplexität hindurch lässt. Kennt man daher besagten Tensor, so wissen wir keineswegs welche Vielfalt an physikalischen Prozessen sich hinter ihm verbirgt.

Das Ausführen dieser Reduktion ist die schwierige Weise, die Einsteinschen Gleichungen zu lesen. Sie besteht in der Vorgabe eines physikalischen Systems, z.B. eines elektromagnetischen Feldes, und der Suche nach der Metrik. Die Metrik soll dann derart sein, dass der mit ihr gebildete Energie-Impuls Tensor den Einsteinschen Gleichungen genügt.

Eine einfache und naheliegende Hypothese meint, der Energie-Impuls Tensor sei als Erwartungswert zu verstehen und der metrische Tensor eine Zufallsgröße, wie man es von Quantensystemen gewohnt ist. Das ist vermutlich zu kurz gedacht, obwohl niemand weiß, wie es ist.

5   Die Metrik und der Zufall

Wie verträgt sich die Allgemeine Relativitätstheorie mit dem Zufall, wie kommt sie mit der Quantenphysik zurecht? Sie tut es nicht perfekt, aber erstaunlich gut.

Bei einer hypothetischen quantenphysikalischen Ausmessung des metrischen Tensors ist die zu erwartende Unschärfe extrem gering und ganz außerhalb aller Experimentierkunst. Denn wir haben nicht Entfernungen[11] und Impuls gleichzeitig zu messen, sondern Entfernungen und Geschwindigkeiten. Die metrischen Verhältnisse stellen sich somit als ein "fast klassisches" Gebilde dar.

Dieser Schluss wird erst brüchig, wenn es um sehr dichte Sterne oder um schwarze Löcher geht. Bei letzteren erlauben Symmetrien die willkürfreie Definition besonderer dynamischer Größen. Damit ist auch die Quantenphysik wieder in ihrem Element. Und nicht nur das, der vorhergesagte Quanteneffekt, die Hawkinsche Strahlung, hat einen besonders guten Beweisstatus in der Wightmanschen axiomatischen Quantenfeldtheorie. Er gehört zu den Glanzpunkten des Zusammenwirkens von Quanten- und Allgemeiner Relativitätstheorie.

Werfen wir nochmals einen Blick auf das Problem der Messungen an einem Quantensystem. Hierzu nehmen wir an, in einem Raumgebiet G werde eine Messung durchgeführt, die das gesamte Raumgebiet betrifft. Die Messung präpariert dann momentan und global einen neuen Zustand, indem in der Menge aller zu G raumartig gelegenen Weltpunkte eine neue Anfangbedingung für die weitere quantale Evolution gesetzt wird. Ist das Messergebnis bekannt, so hat man erfahren, worin die Zustandveränderung besteht.

Nach dieser Wiederholung des schon früher Gesagten, fragen wir uns nun, *wann* das Messergebnis bekannt sein kann.

---

[11] Messungen der Entfernungen werden meist als Ortsmessungen bezeichnet.

Das Messergebnis ist eine Information. Es unterliegt somit den Einsteinschen Kausalitätsforderungen. Wird, wie angenommen, ganz G zur Durchführung der Messung benötigt, so muss diese Information von allen in G liegenden Punkten abgerufen werden. Folglich kann das Ergebnis nur an Weltpunkten verfügbar sein, die kausal nach allen in G liegenden Weltpunkten kommen[12], die also in Bezug auf G zeitlich später sind. Ist r die lineare Ausdehnung von G und c die Lichtgeschwindigkeit, so müssen wir mindestens die Zeit r/c auf das Messergebnis warten. Erst danach wird uns die Veränderung bekannt, die zur Messzeit eingetreten ist. Vorher wird sie vor uns geheim gehalten!

Es ist auch dieser Trick, der das Zusammenspiel von Raum-Zeit Metrik mit quantalen Prozessen widerspruchsfrei gestattet. Wären die Messergebnisse unmittelbar bekannt und könnten wir nicht kompatible Messungen beliebig schnell hintereinander ausführen, so wäre Informationstransport mit Überlichtgeschwindigkeit möglich, wenn auch mit einem wesentlich anderen Mechanismus als dem bereits früher angesprochenen.

In der Tat gibt es in der Quanten-Informationstheorie eine Reihe von Protokollen, die zu akausalen Prozessen führen, wenn man annimmt, quantenphysikalisch inkompatible Messungen und Manipulationen könnten ohne Zeitverlust hintereinander ausgeführt werden. Ich halte eine solche Annahme für falsch.

Die eben kurz dargestellten Überlegungen sind auch allgemein relativistisch gut denkbar. Hierzu müssen wir lediglich annehmen, dass auch für die Metrik die Anfangsbedingungen (ein ganz klein wenig!) geändert werden und die Veränderung der Metrik mit der Verarbeitung der Information mitlaufen. Man kann sich so vorstellen, dass das Ergebnis der Messung in Übereinstimmung mit allen kausalen Forderungen bekannt wird – und nicht früher. Dabei ist die Veränderung der Metrik abrupt allenfalls auf allen Zeitschlitzen, die G enthalten. Auf ihnen kann infolge eines Eingriffs in ein Quantensystem eine Änderung des Zustands als ein sogenannter instantaner Quantensprung[13] erscheinen. Die damit verbundenen Veränderungen in Zustandsräumen der Quantensysteme verlaufen hingegen glatt, siehe das unter dem Stichwort "geometrisches Sahnehäubchen" Gesagte.

Ich hatte schon betont, dass zur Bildung des Energie-Impuls Tensors, wie schlechthin zu allen dynamischen Observablen, die Beihilfe des metrischen Tensors unabdingbar ist. Beeinflusst er auch den nicht-dynamischen Bereich, die quantalen Grundstrukturen? Diese Frage ist zu bejahen. Die Wahrscheinlichkeiten, die für die Änderung des Zustandes bei Eingriffen zuständig sind, hängen von der raum-zeitlichen Metrik ab. Dies zeigen schon die einfachsten Beispiele. Auch die (äquivalente) Frage, welche Operatoren als Observable zulässig sind, wird durch die Geometrie mit entschieden.

Damit rückt auch die Art und Weise, wie Quantensysteme ineinander verschachtelt sind, in den Zugriff der Metrik. Da bei einer Messung in G die Zustandsänderung in allen Zeitschlitzen, die G enthalten, vor sich gehen muss, können weitere Abhängigkeiten entstehen, die logisch sehr schwer zu kontrollieren sind, von ihrer konkreten Form ganz zu schweigen.

---

[12] Diese Weltpunkte liegen im Durchschnitt aller von G ausgehenden Zukunftskegel.

[13] Eine historisch korrekte, heute vor allem von Werbefachleuten benutzte Bezeichnung.

Der Zugriff der raum-zeitlichen Geometrie auf die quantale Grundstruktur ist im stationären Fall gut verstanden. Allgemein ist das aber nicht so. Einen großen Teil der Schwierigkeiten sieht man bereits bei dem Versuch, einfachen Quantensystemen zeitabhängige Randbedingungen aufzuerlegen. Eine systematische Untersuchung dieser Aufgabe ist mir nicht bekannt.

Hiermit schließe ich meine subjektiven und stückhaften Ausführungen. Manchmal hänge ich der Vorstellung nach, es gäbe eine Quantenwelt, die uns auf irgendeine Weise Raum und Zeit als Benutzeroberfläche zur Verfügung stellt.

Quantum Information Transfer from One System to Another One

A. Uhlmann

Abstract. The topics of the paper are: a) Anti-linear maps governing EPR tasks in the absence of distinguished reference bases. b) Imperfect quantum teleportation and its composition rule. c) Quantum teleportation by distributed measurements. d) Remarks on EPR with an arbitrary mixed state, and triggered by a Lüders measurement.

1 Introduction

The problem of transferring "quantum information" from one quantum system into another one has its roots in the 1935 paper [1] of A. Einstein, B. Podolski, and N. Rosen. These authors posed a far reaching question, but they doubt the answer given by quantum theory. The latter, as was pointed out by them, asserts the possibility to create simultaneously and at different places exactly the same random events. The phenomena is often called "EPR effect" or, simply, "EPR".

Early contributions to the EPR problem are due to Schrödinger, [2]. Since then a wealth of papers had appeared on the subject, see [3] and [4] for a résumé. Even to-day some authors consider it more a "paradox" than a physical "effect", because EPR touches the question, whether and how space and time can live with the very axioms of quantum physics, axioms which, possibly, are prior to space and time[1].

Quantum information theory considers EPR as a map or as a "channel", as an element of protocols transferring "quantum information" from one system to another one or supporting the transmission of classical information, [5], [6], [4]. One of our aims is to present a certain calculus for EPR and EPR-like processes or, more general, for processes triggered by measurements. We begin, therefore, with some selected fundamentals of quantum measurements.

The following treatment of EPR has its origin in the identification problem in comparing two or more quantum systems. It is by far not obvious how to identify two density operators, say ϱ^a and ϱ^b, belonging to two different Hilbert spaces, $\mathcal{H}_a$ and $\mathcal{H}_b$. Often one fixes two bases, $\{\phi_j^a\}$ and $\{\phi_j^b\}$, and defines ϱ^a and ϱ^b to be "equal" one to another if they have the same matrix representation with respect to the reference bases. The more, one needs a stable synchronization if several tasks have to be done in the course of time. It seems, therefore, worthwhile to postpone the selection of the reference bases as long as possible. If that can be done, it can be done using the s-maps, [12], of the EPR section. These maps are anti-linear. The anti-linearity in the EPR problem is usually masked by the reference bases: The bases provide conjugations which create, combined

---

[1] Sometimes it is helpful to think space-time a user interface above the quantum world.

R. Ahlswede et al. (Eds.): Information Transfer and Combinatorics, LNCS 4123, pp. 394–412, 2006.
© Springer-Verlag Berlin Heidelberg 2006

with the "natural" anti-linearity, the suggestion of an unrestricted linearity. An interesting, though quite different approach, [23], is founded on Ohya's idea of compound states [22]. Also in [24] there is a side remark on anti-linearity.

For pure states in quantum systems with finitely many degrees of freedom, there is a duality between pure states and maximal properties in the sense of von Neumann and Birkhoff. In the section "inverse" EPR we show by an example the meaning of the mentioned duality.

We proceed with an analysis of the beautiful quantum teleportation protocol of Bennett at al [18]. Here we prove a composition rule for imperfect (i. e. not faithful) quantum teleportation. Then we show its use in quantum teleportation with distributed measurements by an example with a 5-partite system. In the following EPR example based on a 4-partite Hilbert space one observes "entanglement swapping", see Zukowski et al [20], Bose et all [21].

There is a short section on polar decompositions of the s-maps, including a quite elementary link to operator representations.

Finally we show how to handle, again by some anti-linear maps, an EPR task in a bi-partite system if its state is mixed and if a measurement is performed in one of its subsystems by a projection operator of any rank.

Remarks on notation: In this paper the Hermitian adjoint of a map or of an operator A is denoted by A^*. The scalar product in Hilbert spaces is assumed linear in its second argument.

2   Preliminaries

The implementation independence in quantum information theory is guaranteed by the use of Hilbert spaces, states (density operators), and operations between and on them. It is *not* said, what they physically describe in more concrete terms, whether we are dealing with spins, polarizations, energy levels, particle numbers, or whatever you can imagine. Because of this, the elements of quantum information theory, to which the EPR-effect belong, are of rather abstract nature.

Let a physical system be is described by an Hilbert space $\mathcal{H}$. A quantum state of the system is then given by a density operator ω, a positive operator with finite trace, the latter normalized to be one. Thus every positive trace-class operator different from the zero operator uniquely defines a state. One only has to divide it by its trace.

Every vector $\psi \in \mathcal{H}$, $\psi \neq 0$, defines a vector state, the density operator of which is the projection operator, say P_ψ, onto the 1-dimensional subspace generated by ψ. It is common use to speak of "the state ψ" if the state can be described by P_ψ. The vector states of our system are called *pure* if the properties of linear independent vectors do not coincide.

The quantum version of Boolean Logics is due to Birckhoff and J. von Neumann, [7]. According to them a *property*, a quantum state can have, is a subspace of $\mathcal{H}$ or, equivalently, a projection operator onto that subspace. Not every subspace may be considered a property. The point is, that there are no other properties a quantum state can have. This well established postulate excludes some hidden parameter dreams.

Here we shall assume that every subspace defines a property, and that two different subspaces encode different properties.

Looking at these two concepts, states and properties, there is a certain "degeneracy". A vector can denote a state or a property. A (properly) minimal projection operator represents *either* a maximal property *or* the density operator of a pure state. What applies depends on the context. The existence of maximal properties is a special feature of physical systems with a finite number of degrees of freedom.

Let $\mathcal{H}_0$ be a subspace of $\mathcal{H}$ denoting a property. A state, given by a density operator ω, possesses property $\mathcal{H}_0$, if and only if its support is in $\mathcal{H}_0$. That is, ω must annihilate the orthogonal complement of $\mathcal{H}_0$. If $\omega = P_\psi$ is a vector state, this is equivalent to $\psi \in \mathcal{H}_0$.

Let P_0 denote the ortho-projection onto $\mathcal{H}_0$. A test, whether ω has property P_0 results in one bit of information: Either the answer is YES or it is NO.

i) The probability of outputting the answer YES is $p := \mathrm{Tr}\, P_0\omega$.

ii) If p is not zero, and if the answer is YES, then the test has *prepared* the new state $P_0\omega P_0$. Multiplying by p^{-1} gives its density operator.

As p is the probability of the change from ω to $P_0\omega P_0$, it is the *transition probability* between the two states in question.

An *executable measurement within* $\mathcal{H}$ must be a finite orthogonal decomposition of $\mathcal{H}$ into subspaces. The subspaces have to be properties. Denoting by P_j the orthogonal projections onto the subspaces, the requirement reads

$$\sum_{j=1}^m P_j = 1, \quad P_i P_k = 0 \text{ if } i \neq k. \tag{1}$$

Remark: The phrase "executable" asserts the possible existence of an apparatus doing the measurement. A general observable can be approximated (weakly) by such devises. Important physical quantities like energy, momentum, and position in Schrödinger theory represent examples of observables, which can be approximated by executable ones without being executable themselves.

Remark: In saying that the measurement is "in" $\mathcal{H}$ we exclude measurements in an upper-system containing the system in question as a sub-system. A larger system allows for properties not present in the smaller one. Measuring properties of a larger system is reflected in the smaller one by so-called POVMs, "positive operator valued measures".

To be a measurement, the device testing the properties P_j should output a definite signal a_j if it decides to prepare the state $P_j\omega P_j$. Well, $a_1, \ldots, a_m$ constitute the letters of an alphabet. The device randomly decides what letter to choose. The probability of a decision in favor of the letter a_j is $\mathrm{Tr}\, P_j\omega$ with ω the density operator of the system's state. Thus, the classical information per probing the properties (1) is

$$H(p_1, \ldots, p_n) = - \sum p_j \log_2 p_j, \quad p_j = \mathrm{Tr}\, P_j \omega.$$

A little more physics come into the game in assuming that the alphabet consists of m different complex numbers. Then the operator

$$A := \sum_1^n a_j P_j \tag{2}$$

is an *observable* for the measurement of the properties (1). Clearly, executable observables are normal operators, $AA^* = A^*A$, and their spectra are finite sets.

One observes that information theory is not interested in the nature of the alphabet that distinguishes the outcomes of a measurement. It suffices for its purposes to discriminate the outcomes and to know the state that is prepared. *Portability* is gained that way.

It is standard that two properties can be checked simultaneously if and only if their ortho-projections commute. Otherwise one gets in trouble with the probability interpretation. Two observables, A and B, can be measured (or approximated by such procedures) simultaneously provided they commute. Executing a set $A_1, \ldots, A_n$ of mutually commuting observables will be called a *distributed measurement*.

Non-relativistically a distributed measurement may consist of several measuring devices, sitting on different (possibly overlapping) places in space, but being triggered at the same time.

Relativistically, every measurement is done in a certain space-time or "world" region. A particular case of a distributed measurement consists of devices doing their jobs in disjunct, mutually space-like world regions: Quantum theory does not enforce restrictions for measurements (or "interventions" a la A. Peres) for space-like separated world regions. EPR and quantum teleportation make use of it in an ingenious way.

Thinking in terms of the evolution of states in the course of time, these tasks update the initial conditions of the evolution. The choice of the new Cauchy data is done randomly and governed by transition probabilities.

In Minkowski space the problem is somehow delicate. According to Hellwig and Kraus [8] it is consistent to let take place the state change at the boundary of the past of the region. The past of the world region is the union of all backward light-cones terminating in one of the world points of the region the measurement is done. Finkelstein [9] has argued that it is also possible to allow the change at the light-like future of the world region in question. We, [11], think it even consistent to assume a slightly stronger rule: *The state changes accompanied by a measurement in a space-time region takes place at the set of those points, which are neither in the past nor in the future of that region.* The assumed region of influence is bounded to the past a la Hellwig and Kraus and to the future according to Finkelstein. The remarkable experiments of Zbinden et al [10] agree with it.

3   EPR

Let us consider a bi-partite quantum system composed of two Hilbert spaces $\mathcal{H}_a$ and $\mathcal{H}_b$ and one of its vectors

$$\mathcal{H} := \mathcal{H}_a \otimes \mathcal{H}_b, \quad \psi \in \mathcal{H}. \tag{3}$$

In such a bi-partite system $\mathcal{H}_a$ characterizes a subsystem, the a-system, which is embedded in the system of the Hilbert space $\mathcal{H}$. The same is with the b-system.

We assume the state of the composed system is the vector state defined by ψ. We are interested in what is happening if a property is checked in the a-system. A *local subspace* of $\mathcal{H}$ is a direct product of two subspaces, one of $\mathcal{H}_a$, the other one of $\mathcal{H}_b$. A *local property* of $\mathcal{H}$ is, therefore, a projection operator of the form $P_a \otimes P_b$. P_a and P_b are projectors from the subsystems. Similarly one proceeds in multi-partite systems.

If P_a is a property of $\mathcal{H}_a$, the local property in the composed system that checks nothing in the b-system reads $P_a \otimes 1_b$. If so, and if the test of P_a outputs YES, the newly prepared state is again a vector state. The state change is

$$\psi \mapsto (P_a \otimes 1_b)\psi. \tag{4}$$

Is something to be seen in the b-system by such a change? Posing and answering the question is an essentially part of the EPR problem. In pointing out the intrinsic anti-linearity in the EPR problem we follow [12] and [13].

Let us consider maximal properties, i. e. rank one projection operators, of the a-system,

$$P_a = \frac{|\phi^a\rangle\langle\phi^a|}{\langle\phi^a, \phi^a\rangle}, \quad \phi^a \in \mathcal{H}_a. \tag{5}$$

Then the state prepared in (4) must be a product vector, the first factor being a multiple of ϕ^a. Therefore, given ψ, there *must* be a map from $\mathcal{H}_a$ into $\mathcal{H}_b$ associating to any given ϕ^a its partner in the product state. Let us denote this map by

$$\mathcal{H}_a \ni \phi^a \mapsto \mathbf{s}_\psi^{ba}\phi^a \in \mathcal{H}_b.$$

It is defined by

$$\left(|\phi^a\rangle\langle\phi^a| \otimes 1_b\right)\psi = \phi^a \otimes \mathbf{s}_\psi^{ba}\phi^a, \quad \forall \phi^a \in \mathcal{H}_a. \tag{6}$$

We see: *If in testing the property ϕ^a the answer is YES, the same is true with certainty if in the b-system one is asking for the property* $\mathbf{s}_\psi^{ba}\phi^a$.

It becomes clear by inspection of (6) that $\mathbf{s}_\psi^{ba}$ is an *anti-linear* map from $\mathcal{H}_a$ into $\mathcal{H}_b$ which depends linearly on ψ. We also may ask the same question starting from the b-system, resulting in an anti-linear map $\mathbf{s}_\psi^{ab}$ from $\mathcal{H}_b$ into $\mathcal{H}_a$,

$$\left(1_a \otimes |\phi^b\rangle\langle\phi^b|\right)\psi = \mathbf{s}_\psi^{ab}\phi^b \otimes \phi^b, \quad \forall \phi^b \in \mathcal{H}_b. \tag{7}$$

Let us go back to (6) and let us choose a vector ϕ^b in $\mathcal{H}_b$. Taking the scalar product (6) with $\phi^a \otimes \phi^b$, one easily finds

$$\langle\phi^a \otimes \phi^b, \psi\rangle = \langle\phi^b, \mathbf{s}_\psi^{ba}\phi^a\rangle.$$

By symmetry, or by using (7) appropriately, one finally arrives at the identity

$$\langle \phi^a \otimes \phi^b, \psi \rangle = \langle \phi^b, \mathbf{s}_\psi^{ba} \phi^a \rangle = \langle \phi^a, \mathbf{s}_\psi^{ab} \phi^b \rangle \tag{8}$$

which is valid for all $\phi^a \in \mathcal{H}_a$ and $\phi^b \in \mathcal{H}_b$. Obviously, taking into account their anti-linearity, the two maps between the Hilbert spaces of the subsystems are Hermitian adjoints one from another.

$$(\mathbf{s}_\psi^{ab})^* = \mathbf{s}_\psi^{ba}, \quad (\mathbf{s}_\psi^{ba})^* = \mathbf{s}_\psi^{ab}$$

Finally, by the linearity of the s-maps with respect to $\psi \in \mathcal{H}$, one arrives at the following recipe for their construction:

$$\psi = \sum a_{jk} \phi_j^a \otimes \phi_k^b \ \Rightarrow\ \mathbf{s}_\psi^{ba} \phi^a = \sum a_{jk} \langle \phi^a, \phi_j^a \rangle \phi_k^b \tag{9}$$

$$\Rightarrow\ \mathbf{s}_\psi^{ab} \phi^b = \sum a_{jk} \langle \phi^b, \phi_k^b \rangle \phi_j^b$$

The s-maps obey some simple rules if local operations are applied to them. The most obvious is

$$\varphi = (A \otimes B)\psi \ \leftrightarrow\ \mathbf{s}_\varphi^{ab} = A \mathbf{s}_\psi^{ab} B^* . \tag{10}$$

Let us now escape from the formalism to a short discussion. We assume, as starting point, the bi-partite system in a pure state $\psi \in \mathcal{H}$. We can assume that ψ and an arbitrarily chosen ϕ^a are unit vectors. P_a denotes the projection operator of the 1-dimensional subspace generated in $\mathcal{H}_a$ by ϕ^a.

What can be seen from ψ in the subsystems? This is encoded in the reduced states, in the density operators ϱ_ψ^a and ϱ_ψ^b respectively. In more general terms: The state of a subsystem is given by the expectation values of the operators accessible within the subsystem. All what an owner, say Bob, can learn within his subsystem $\mathcal{H}_b$ *without* resources from outside, he has to learn from ϱ^b. Any belief, he could learn anything else from its quantum system alone, is nothing than a reanimation of the hidden parameter story.

The reduced density operators can be calculated by partial traces. In the case at hand a definition for the b-system is

$$\langle \psi, (\mathbf{1}_a \otimes B)\psi \rangle = \operatorname{Tr} \varrho_\psi^b B, \quad \forall B \in \mathcal{B}(\mathcal{H}_b) .$$

The reduced density operators can also be expressed by the s-maps,

$$\varrho_\psi^a = \mathbf{s}_\psi^{ab} \mathbf{s}_\psi^{ba}, \quad \varrho_\psi^b = \mathbf{s}_\psi^{ba} \mathbf{s}_\psi^{ab} \tag{11}$$

The probability, p, for a successful test of ϕ^a is $\langle \phi^a, \varrho_\psi^a \phi^a \rangle$. The maximal possible probability appears if ϕ^a is an eigenvector to the largest eigenvalue of ϱ_ψ^a.

The square roots of the eigenvalues $p_i > 0$ of ϱ_ψ^a are the Schmidt-coefficients of the Schmidt decomposition of ψ and, according to (11), also the singular values of $\mathbf{s}_\psi^{ab}$.

Let $\{\phi_j^a\}$ be the vectors of a basis and P_a^j the ortho-projection onto the space generated by ϕ_j^a. Let us now ask what is going on if we test the properties P_a^j. We can use any operator

$$A = \sum a_j P_a^j$$

with mutually different numbers a_j. The probability p_j' of preparing ϕ_j^a is $\langle \phi_a^j, \varrho_\psi^a \phi_a^j \rangle$. It is well known, that the probability vector $\{p_j'\}$ is majorized by the set of eigenvalues $\{p_j\}$ of ϱ^a. Any probability vector, which is majorized by the vector of its eigenvalues, can be gained this way by the use of a suitable basis of $\mathcal{H}_a$. Consequently, in measuring A, one can produce a message with an entropy not less than the entropy of the eigenvalue distribution of ϱ_ψ^a. If and only if the chosen basis is an eigen-basis of ϱ_ψ^a, we get the minimally possible entropy.

Enhancing the entropy of Alice's side is not useful for Bob. Though his system will definitely be in the state $\phi_j^b = \mathbf{s}^{ba}\phi_j^a$ if on Alice's side the state ϕ_j^a is prepared, he cannot always make too much use of it. While Alice is preparing states which *must* be mutually orthogonal, and hence distinguishable, the vector states on Bob's side do not share this necessarily. Indeed, Bob's state are mutually orthogonal if and only if Alice had minimized the entropy, i. e. if she had chosen an eigen-basis of her density operator.

Let us repeat it from another perspective. Let Alice perform some measurements using the observable A. Assume that just before any measurement, the state of the bi-partite system is the vector state ψ. Then, whenever the device answers "a_j", the state of the a-system changes to ϕ_j^a. The state of the b-system becomes $\phi_j^b = \mathbf{s}^{ba}\phi_j^a$. Bob *knows* this state iff he knows ψ and which of the values a_j the measuring device has given to Alice. Now, if Alice uses an eigen-basis of the density operator ϱ_ψ^a then Bob himself is able to measure which state he get and, therefore, which a_j Alice has obtained. On the contrary, if Alice does not use a basis of eigenvectors, Bob's possible states are not orthogonal and he cannot distinguish exactly between them. Therefore, the gain in entropy in the a-system by using a measurement basis distinct from the eigenvector basis is compensated by a loss of Bob's possibility to distinguish between the states he gets.

One can prove the assertion by calculating

$$\langle \phi_j^b, \phi_k^b \rangle = \langle \mathbf{s}^{ba}\phi_j^a, \mathbf{s}^{ba}\phi_k^a \rangle = \langle \phi_k^a, \mathbf{s}^{ab}\mathbf{s}^{ba}\phi_j^a \rangle$$

or, by (11),

$$\langle \phi_j^b, \phi_k^b \rangle = \langle \phi_k^a, \varrho_\psi^a \phi_j^a \rangle . \tag{12}$$

If *all* von Neumann measurements of Alice are on equal footing, and Bob can *always* discover the state prepared by Alice within his system to any precision, the EPR settings is "perfect" or "tight". In the tight case the reduced density operator ϱ_ψ^a of Alice is equal to $(\dim \mathcal{H}_a)^{-1}\mathbf{1}_a$, i.e. to the unique tracial state of her system. This state is like "white quantum paper", there is no quantum information at all in it. The "more white" Alice's "quantum paper" ϱ_ψ^a is, the better EPR is working. That somewhat fabulous language can be made precise substituting "more mixed" or "less pure" for "more white".

A further remark should be added to our short and incomplete account of the EPR mechanism. It is a well known theorem that $\mathcal{H}_a \otimes \mathcal{H}_b$ is canonically isomorph to the space $\mathcal{L}^2(\mathcal{H}_a, \mathcal{H}_b^*)$ of Hilbert-Schmidt mappings form $\mathcal{H}_a$ into $\mathcal{H}_b^*$. On the other hand, $\mathcal{H}_b^*$ is canonically anti-linearly isomorphic to $\mathcal{H}_b$, a fact used by P. A. Dirac to establish his bra-ket correspondence $|.\rangle \leftrightarrow \langle.|$ Composing both maps one immediately see the isomorphism between $\mathcal{H}_a \otimes \mathcal{H}_b$ and the space of anti-linear Hilbert-Schmidt maps $\mathcal{L}^2(\mathcal{H}_a, \mathcal{H}_b)_{\text{anti}}$. The isomorphism is an isometry expressed by

$$\langle \varphi, \psi \rangle = \text{Tr } s_\psi^{ab} s_\varphi^{ba} = \text{Tr } s_\psi^{ba} s_\varphi^{ab} \tag{13}$$

with ψ and φ from $\mathcal{H} = \mathcal{H}_a \otimes \mathcal{H}_b$.

4   "Inverse" EPR

In the preceding section we have considered three vectors: ψ from the composite Hilbert space (3) and ϕ^a, ϕ^b from its constituents. In the EPR setting ψ is a given pure state which is to test whether it enjoys the local properties defined either by ϕ^a, by ϕ^b, or by both. In the "dual" or "inverse" EPR setting their roles are just reversed: ψ appears as a non-local property which is to check. $\phi^a \otimes \phi^b$ is the state to be tested for the property ψ. Because transition probabilities are symmetric in their arguments, one can enroll the EPR setting backwards. The trick has been clearly seen and used by C. Bennett, G. Brassard, C. Crepeau, R. Jozsa, A. Peres, and W. Wootters in their famous quantum teleportation paper [18], see also [25].

To demonstrate what is going on, let us consider a simple but instructive example. Here $\mathcal{H}$ is of dimension four, and its two factors 2-dimensional. Dirac's bra-ket notation is used, but anti-linear maps should be applied to kets only! In the example we choose the vectors

$$\psi = \frac{1}{\sqrt{2}}(|00\rangle + |11\rangle), \quad \phi^a = |x\rangle, \quad \phi^b = |0\rangle$$

with $x = 0, 1$. Alice is trying to send a bit-encoded message to Bob by choosing $|x\rangle$ accordingly one after the other. Bob's input is always $|0\rangle$. By doing so, they enforce the bi-partite system into the state

$$|x\,0\rangle = |x\rangle \otimes |0\rangle, \quad x = 0, 1$$

Then it is checked whether it has the property ψ. If $x = 1$, the measuring apparatus will necessarily answer the question with NO because the state is orthogonal to ψ. If, however, $x = 0$, the answer is YES with probability 0.5 and NO with the same probability. The input state $|00\rangle$ of the bi-partite system now has changed as follows:

$$\text{YES} \mapsto \psi, \quad \text{NO} \mapsto \psi' = \frac{1}{\sqrt{2}}(|00\rangle - |11\rangle).$$

Let now q be the probability of an input $x = 0$. Then the input ensemble is transformed by the measurement in the following way:

$$\{|00\rangle, |10\rangle; q, 1-q\} \rightarrow \{\psi, \psi', |10\rangle; \frac{q}{2}, \frac{q}{2}, 1-q\}$$

The classical information encoded in the input state is not lost. It could be regained by measuring the property

$$|\psi\rangle\langle\psi| + |\psi'\rangle\langle\psi'|,$$

a task which does not change the states involved. Now, next, Bob and Alice perform local measurements by testing the properties

$$P_b := |0\rangle\langle0|_b, \quad P_a := |0\rangle\langle0|_a.$$

If ψ or ψ' is the state of $\mathcal{H}$, the states of the local parts will be $(1/2)\mathbf{1}_a$ and $(1/2)\mathbf{1}_b$ respectively. The answer is either YES or NO with equal property $1/2$ as seen from

$$(|0\rangle\langle0|_a \otimes \mathbf{1}_b)\psi = \frac{1}{\sqrt{2}}|00\rangle = (\mathbf{1}_a \otimes |0\rangle\langle0|_b)\psi$$

and from the similar relation with ψ'. There is a strong correlation: Either both devices return YES or both say NO. Therefore, if the input of Alice is $|0\rangle$, the output is either YES for Alice as well as for Bob, or it is NO for both. If, however, $|1\rangle$ is the input of Alice, then $|10\rangle$ becomes the state of $\mathcal{H}$. It follows that Alice gets necessarily NO and Bob YES.

We see that Bob and Alice would have the full information of the message, Alice had encoded in her system, if both parties could communicate their measurement results – even if Alice has forgotten her original message. No information is lost, but it is non-locally distributed after testing the property ψ.

A particular interesting case is the transmission of information from Alice to Bob, who knows neither the result of testing the property ψ nor has he obtained any information from Alice. He knows, which property has been checked, but does not know the result.

Though there is no classical information transfer, Bob gets some information from Alice by testing in his system property P_b. Considering all intermediate state changes as done by a quantum black box, the process is stepwise described by

$$\{|0\rangle\langle0|_a, |1\rangle\langle1|_a\} \mapsto \{\frac{1}{2}\mathbf{1}_b, |0\rangle\langle0|_b\}$$

and can be represented as an application of the stochastic cp-map

$$T : \begin{pmatrix} \omega_{00} & \omega_{01} \\ \omega_{10} & \omega_{11} \end{pmatrix}_a \mapsto \frac{1}{2}\begin{pmatrix} \omega_{00} + 2\omega_{11} & 0 \\ 0 & \omega_{00} \end{pmatrix}_b.$$

A message encoded by Alice with probabilities q or $1-q$ per letters 0 or 1 carries an information $H(q, 1-q)$. The Holevo bound for the quantum message Bob obtains by measuring P_b can be calculated to be

$$H(1 - \frac{1}{2}q, \frac{1}{2}q) - q.$$

Its maximum is reached at $q = (2/5)$. In that case Bob receives approximately 0.322 bit per letter, while Alice has encoded her message with 0.962 bit per letter.

The channel is rotationally symmetric. From this and standard convexity reasoning one gets: The Holevo ("one-shot") capacity of the channel will be reached with a diagonal ensemble at $q = 2/5$. Therefore,

$$\chi^*(T) = H(\frac{4}{5}, \frac{1}{5}) - \frac{2}{5}$$

What we have just discussed above is a slight variation of protocols invented independently by Aharonov and Albert, [14], and by R. D. Sorkin [15]. The latter claimed it to be an example of a measurement "forbidden by Einstein causality". More recently Beckman et al [16], adding a remarkable collection of similar measurements, have extended and sharpened Sorkin's assertion. On the other hand, Vaidman [17] presented teleportation protocols of non-local measurements. We, B. Crell and me, [11], think the causality considerations of Sorkin and Beckmann et al not conclusive: While a measurement allows for instantaneous changes of states, the output of an apparatus includes classical information processing which has to go on in the world region the device is working. To detect the output of the signal can only be possible in the intersection of all future cones originating in world points of the measuring region. Bob can detect Alice's message not before his world lines have crossed all the future light cones originating from the world points at which the measuring process is going on. Hence, though the state change has taken place, Bob can be informed only after a time delay of the order "radius of the measuring device / velocity of light". Before that time has elapsed, the state change is hidden to Bob – as required by causality.

Generally, [11], the rule with which quantum theory outlines the defect of being not causal, is as follows. Let A and B be two non-commuting observables which we like to measure sequentially, say A before B. Let G_A and G_B denote the world region at which the measurements should take place. Then G_B *must be in the complete future of G_A*, that is G_B must be in the intersection of all forward cones originating in the world points of G_A. Similar it is with unitary moves if combined with measurements not commuting with them.

The return to the general case of inverse EPR with ψ an arbitrary vector of a bi-partite system with Hilbert space $\mathcal{H}$ is formally straightforward: Checking the property ψ if the system is in a product state $\phi_1^a \otimes \phi_1^b$ one comes across

$$|\psi\rangle\langle\psi| \, \phi_1^a \otimes \phi_1^b = \langle\psi, \phi_1^a \otimes \phi_1^b\rangle \, \psi.$$

If Alice and Bob can communicate, and they can check with which probability their states enjoy the property $\phi_2^a \otimes \phi_2^b$. The transition amplitude for an affirmative answer can be expressed, according to (8), by

$$\langle\psi, \phi_1^a \otimes \phi_1^b\rangle \langle\phi_2^a \otimes \phi_2^b, \psi\rangle = \langle\phi_1^a, \mathbf{s}_\psi^{ab}\phi_1^b\rangle^* \langle\phi_2^a, \mathbf{s}_\psi^{ab}\phi_2^b\rangle.$$

5 Imperfect Quantum Teleportation

Quantum teleportation has been invented by Bennett et al [18]. "Perfect" or faithful quantum teleportation starts within a product of three Hilbert spaces of equal finite dimension and with a maximal entangled vector in the last two. It is triggered by a von Neumann measurement in the first two spaces using a basis of maximally entangled vectors. The measurement randomly chooses one of several quantum channels. The information, which quantum channel has been activated, is carried by the classical channel. It serves to reconstruct, by a unitary move, the desired state at the destination.

All those possible "perfect" or "tight" schemes, together with their dense coding counterparts, have been reviewed by R. F. Werner [19].

Following [18] and analyzing their computations, one can decompose the chosen quantum channel into two parts, an inverse EPR and an EPR setting. As one can identify two particular s-map with them, one is tempted to use two general s-maps. In doing so one can treat a more general setup. But even in "perfect" circumstances the explicit use of the mentioned decomposition may be of some interest.

Let $\mathcal{H}$ be a tri-partite Hilbert space

$$\mathcal{H}_{abc} = \mathcal{H}_a \otimes \mathcal{H}_b \otimes \mathcal{H}_c . \tag{14}$$

There is no restriction on the dimensions of the factor spaces. The *input* is a vector $\phi^a \in \mathcal{H}_a$, possibly unknown, and a known vector φ^{bc}, the "ancilla", out of $\mathcal{H}_b \otimes \mathcal{H}_c$. The teleportation protocol is to start with the initial vector

$$\varphi^{abc} := \phi^a \otimes \varphi^{bc} \in \mathcal{H}_{abc} . \tag{15}$$

Now one performs a measurement on $\mathcal{H}_a \otimes \mathcal{H}_b$. Instead of a complete von Neumann measurement we ask just whether a property, given by a vector ψ^{ab}, is present or not. In doing "nothing" on the c-system, one is checking a local property of the abc-system. If the check runs affirmative, the vector state ψ^{ab} is prepared in $\mathcal{H}_{ab}$, inducing a state change in the larger abc-system:

$$(|\psi^{ab}\rangle\langle\psi^{ab}| \otimes \mathbf{1}^c)(\phi^a \otimes \varphi^{bc}) = \psi^{ab} \otimes \phi^c, \tag{16}$$

with a vector $\phi^c \in \mathcal{H}_c$ yet to be determined. Indeed,

$$\phi^a \mapsto \phi^c$$

represents the teleportation channel which is triggered by an affirmative check of the property defined by ψ^{ab}. Letting ϕ^a as a free variable, we introduce the *teleportation map* $\mathbf{t}^{ca}$ by

$$\mathbf{t}^{ca}\phi^a \equiv \mathbf{t}^{ca}_{\psi,\varphi}\phi^a = \phi^c, \quad \psi \equiv \psi^{ab}, \quad \varphi \equiv \varphi^{bc} \tag{17}$$

The teleportation map $\mathbf{t}^{ca}$ is governed by the *composition rule*, [12],

$$\mathbf{t}^{ca}_{\psi,\varphi} = \mathbf{s}^{cb}_{\varphi} \mathbf{s}^{ba}_{\psi} . \tag{18}$$

The s-maps being Hilbert-Schmidt, the t-maps must be of trace class and linear. Indeed, *every trace class map from $\mathcal{H}_a$ into $\mathcal{H}_c$ can be gained as a t-map, provided its rank does not exceed the dimension of $\mathcal{H}_b$.* Of course, this fact can be obtained also directly, without relying on the decomposition rule, [27,28,26,29,30,31], where also cases with a mixed ancilla have been studied.

Proof of (18).   Let us abbreviate the left hand side of (16) by ψ^{abc}. Choosing in $\mathcal{H}_b$ an ortho-normal basis $\{\phi_j^b\}$ gives the opportunity to write

$$\varphi^{bc} = \sum \phi_j^b \otimes \mathbf{s}_\varphi^{cb} \phi_j^b$$

and hence

$$\psi^{abc} = \psi^{ab} \otimes \sum_j \langle \psi^{ab}, \phi^a \otimes \phi_j^b \rangle \, \mathbf{s}_\varphi^{cb} \phi_j^b \, .$$

We choose in $\mathcal{H}_a$ an ortho-normal basis, $\{\phi_k^a\}$, to resolve the scalar product in the last equation:

$$\psi^{abc} = \psi^{ab} \otimes \sum_{jk} \langle \phi_k^a, \phi^a \rangle \, \langle \mathbf{s}_\psi^{ba} \phi_k^a, \phi_j^b \rangle \, \mathbf{s}_\varphi^{cb} \phi_j^b \, .$$

Using anti-linearity,

$$\psi^{abc} = \psi^{ab} \otimes \mathbf{s}_\varphi^{cb} \sum_k \langle \phi^a, \phi_k^a \rangle \sum_j \langle \phi_j^b, \mathbf{s}_\psi^{ba} \phi_k^a \rangle \, \phi_j^b$$

The summation over j results in $\mathbf{s}_\psi^{ba} \phi_k^a$. Next, again by anti-linearity, the sum over k comes down to

$$\mathbf{s}_\psi^{ba} \sum_k \langle \phi_k^a, \phi^a \rangle \, \phi_k^a = \mathbf{s}_\psi^{ba} \phi^a$$

and we get finally

$$\psi^{abc} = \psi^{ab} \otimes \mathbf{s}_\varphi^{cb} \, \mathbf{s}_\psi^{ba} \phi^a$$

and the composition rule is proved.

Distributed measurements

The next aim is to present an extension of the composition rule to multi-partite systems. In a multi-partite system one can distribute the measurements and the entanglement resources over some pairs of subsystems. With an odd number of subsystems we get *distributed teleportation,* with an even number something like *distributed EPR.*

At first let us see, as an example, distributed teleportation with five subsystems.

$$\mathcal{H} = \mathcal{H}_a \otimes \mathcal{H}_b \otimes \mathcal{H}_c \otimes \mathcal{H}_d \otimes \mathcal{H}_e \, . \tag{19}$$

The input is an unknown vector $\phi^a \in \mathcal{H}_a$, the ancillary vectors are selected from the bc- and the de-system,

$$\varphi^{bc} \in \mathcal{H}_{bc} = \mathcal{H}_b \otimes \mathcal{H}_c, \quad \varphi^{de} \in \mathcal{H}_{de} = \mathcal{H}_d \otimes \mathcal{H}_e, \tag{20}$$

and the vector of the total system we are starting with is

$$\varphi^{abcde} = \phi^a \otimes \varphi^{bc} \otimes \varphi^{de} . \tag{21}$$

The channel is triggered by measurements in the ab- and in the cd-system. Suppose these measurements are successful and they prepare the vector states

$$\psi^{ab} \in \mathcal{H}_{ab} = \mathcal{H}_a \otimes \mathcal{H}_b, \quad \psi^{cd} \in \mathcal{H}_{cd} = \mathcal{H}_c \otimes \mathcal{H}_d . \tag{22}$$

Then we get the relation

$$(|\psi^{ab}\rangle\langle\psi^{ab}| \otimes |\psi^{cd}\rangle\langle\psi^{cd}| \otimes \mathbf{1}^e)\varphi^{abcde} = \psi^{ab} \otimes \psi^{cd} \otimes \phi^e \tag{23}$$

and the vector ϕ^a is mapped onto ϕ^e. Introducing the s-maps corresponding to the vectors

$$\psi^{ab} \to \mathbf{s}^{ba}, \quad \varphi^{bc} \to \mathbf{s}^{cb}, \quad \psi^{cd} \to \mathbf{s}^{dc}, \dots,$$

the *factorization rule* becomes

$$\phi^e = \mathbf{t}^{ea}\phi^a, \quad \mathbf{t}^{ea} = \mathbf{s}^{ed}\,\mathbf{s}^{dc}\,\mathbf{s}^{cb}\,\mathbf{s}^{ba} . \tag{24}$$

Next we consider a setting with *four* Hilbert spaces, $\mathcal{H}_b$ to $\mathcal{H}_e$. The input state is

$$\varphi^{bcde} = \varphi^{bc} \otimes \varphi^{de}$$

and we perform a test to check whether the property ψ^{cd} is present or not. Let the answer be YES. Then the subsystems bc and de become disentangled. The cd system gets ψ^{cd} and, hence, the entanglement of this vector state. The previously unentangled systems $\mathcal{H}_b$ and $\mathcal{H}_e$ will now be entangled.

The newly prepared state is

$$\chi^{bcde} := (\mathbf{1}_b \otimes |\psi^{cd}\rangle\langle\psi^{cd}| \otimes \mathbf{1}_e)\,\varphi^{bcde} . \tag{25}$$

With

$$\psi^{cd} = \sum \lambda_j \phi_j^c \otimes \phi_j^d$$

we obtain

$$\chi^{bcde} = \sum \lambda_j \lambda_k [(\mathbf{1}_b \otimes |\phi_j^c\rangle\langle\phi_j^c|)\varphi^{bc}] \otimes [(|\phi_j^d\rangle\langle\phi_j^d| \otimes \mathbf{1}_e)\varphi^{de}] .$$

Let us denote just by $\mathbf{s}^{bc}$ and $\mathbf{s}^{de}$ the s-maps of φ^{bc} and φ^{de} respectively. They allow to rewrite χ^{bcde} as

$$\chi^{bcde} = \sum \lambda_j \lambda_k (\mathbf{s}^{bc}\phi_k^c \otimes \phi_j^c) \otimes (\phi_j^d \otimes \mathbf{s}^{ed}\phi_k^d)$$

which is equal to

$$\chi^{bcde} = \sum \lambda_k (\mathbf{s}^{bc}\phi_k^c) \otimes \psi^{cd} \otimes (\mathbf{s}^{ed}\phi_d^c) . \tag{26}$$

The Hilbert space $\mathcal{H}_c \otimes \mathcal{H}_d$ is decoupled from $\mathcal{H}_b$ and $\mathcal{H}_e$. The vector state of the latter can be characterized by a map from $\mathcal{H}_c \otimes \mathcal{H}_d$ into $\mathcal{H}_b \otimes \mathcal{H}_e$.

$$\varphi^{be} := (\mathbf{s}^{bc} \otimes \mathbf{s}^{ed}) \, \psi^{cd} \tag{27}$$

is indicating how the entanglement within the be-system is produced by entanglement swapping, and how the three vectors involved come together to achieve it.

Addendum: A rearrangement lemma.

The starting point is a collection of bi-partite spaces and vectors,

$$\psi_j \in \mathcal{H}_{ab}^j, \quad \mathcal{H}_{ab}^j = \mathcal{H}_a^j \otimes \mathcal{H}_b^j, \quad j = 1, \dots, m \tag{28}$$

from which we build

$$\mathcal{H}_{ab} = \mathcal{H}_{ab}^1 \otimes \cdots \otimes \mathcal{H}_{ab}^j, \quad \psi = \psi_1 \otimes \cdots \otimes \psi_m \,. \tag{29}$$

We abbreviate the s-maps accordingly,

$$\psi_j \leftrightarrow \mathbf{s}_j^{ab} \leftrightarrow \mathbf{s}_j^{ba} \tag{30}$$

We now change to the rearranged Hilbert space

$$\mathcal{H}_{AB} = \mathcal{H}_A \otimes \mathcal{H}_B = (\mathcal{H}_a^1 \otimes \dots \mathcal{H}_a^m) \otimes (\mathcal{H}_b^1 \otimes \dots \mathcal{H}_b^m) \,. \tag{31}$$

The Hilbert spaces (29) and (31) are unitarily equivalent in a canonical way:

$$V : \quad \mathcal{H}_{ab} \mapsto \mathcal{H}_{AB} \tag{32}$$

is defined to be the linear map satisfying

$$V (\phi_1^a \otimes \phi_1^b \otimes \cdots \otimes \phi_m^a \otimes \phi_m^b) = (\phi_1^a \otimes \cdots \otimes \phi_m^a) \otimes (\phi_1^b \otimes \cdots \otimes \phi_m^b) \tag{33}$$

This is a unitary map, $V^{-1} = V^*$.

Assume we need the s-maps of

$$\varphi := V \, \psi \tag{34}$$

with ψ given by (29). The rearrangement lemma we have in mind reads

$$\mathbf{s}_\varphi^{AB} = V (\mathbf{s}_1^{ab} \otimes \cdots \otimes \mathbf{s}_m^{ab}) \, V^{-1} \,. \tag{35}$$

The proof uses the fact that both sides are multi-linear in the vectors ψ_j. Therefore, it suffices to establish the assertion in the case, the ψ_j are product vectors. But then the proof consists of some lengthy but easy to handle identities.

6   Polar Decompositions

Let us come back to the s-maps. It is worthwhile to study their polar decompositions. As we already know (11) it is evident that we should have

$$\mathbf{s}_\psi^{ba} = (\varrho_\psi^b)^{1/2} \mathbf{j}_\psi^{ba} = \mathbf{j}_\psi^{ba} (\varrho_\psi^a)^{1/2}, \tag{36}$$

$$\mathbf{s}_\psi^{ab} = (\varrho_\psi^a)^{1/2}\mathbf{j}_\psi^{ab} = \mathbf{j}_\psi^{ab}(\varrho_\psi^b)^{1/2}.$$

The j-maps are anti-linear partial isometries with left (right) supports equal to the support of their left (right) positive factor. From Alice's point of view, who can know her reduced density operator but not the state from which it is reduced, $\mathbf{j}_\psi^{ab}$ is a non-commutative phase. It is in discussion whether and how relative phases of this kind can be detected experimentally.

One outcome of the polar decomposition is a unique labelling of purifications. If ϱ^a denotes a density operator on $\mathcal{H}_a$, then all its purifications can be gained by the chain

$$\varrho^a \mapsto \mathbf{j}^{ba}(\varrho^a)^{1/2} = \mathbf{s}_\psi^{ba} \mapsto \psi$$

where $\mathbf{j}^{ba}$ runs through all those anti-linear isometries from a to b whose right supports are equal to the support of ϱ^a.

The uniqueness of the polar decomposition and (11) yields

$$(\mathbf{j}_\psi^{ba})^* = \mathbf{j}_\psi^{ab}, \quad \varrho_\psi^b = \mathbf{j}_\psi^{ba}\,\varrho_\psi^a\,\mathbf{j}_\psi^{ab}. \tag{37}$$

Now we can relate the expectation values of the reduced density operators: Assume the bounded operators A and B on $\mathcal{H}_a$ and $\mathcal{H}_b$ are such that

$$B^*\,\mathbf{j}_\psi^{ba} = \mathbf{j}_\psi^{ba}A. \tag{38}$$

Then one gets, as a little exercise in anti-linearity,

$$\mathrm{Tr}\ \varrho_\psi^a A = \mathrm{Tr}\ \varrho_\psi^b B. \tag{39}$$

It is possible to express the condition (38) for the validity of (39) by an anti-linear operator J_ψ acting on $\mathcal{H}_a \otimes \mathcal{H}_b$. To this end we define J_ψ as the anti-linear extension of

$$J_\psi(\phi^a \otimes \phi^b) = \mathbf{j}_\psi^{ab}\phi^b \otimes \mathbf{j}_\psi^{ba}\phi^a. \tag{40}$$

With this definition it is to be seen that (38) is as strong as

$$J_\psi(A \otimes B) = (A \otimes B)^* J_\psi. \tag{41}$$

(40) is a crossed tensor product, $\tilde{\otimes}$. With every pair of maps, one from $\mathcal{H}_a$ to $\mathcal{H}_b$ and one in the opposite direction, and both either linear or anti-linear, one can build the crossed tensor product $\tilde{\otimes}$. An important example is (40), where the two factors are j-maps. We may formally write

$$J_\psi = \mathbf{j}_\psi^{ab}\,\tilde{\otimes}\,\mathbf{j}_\psi^{ba}$$

for the just defined anti-linear operator acting on $\mathcal{H}_{ab}$.

Now let the factors of $\mathcal{H}_{ab}$ be of equal dimension and ψ "completely entangled". In a more mathematical language ψ is called a cyclic and separating vector, a so-called GNS-vector[2] or a "GNS vacuum", for the representation

$$A \mapsto A \otimes 1_b$$

---

[2] GNS stands for I. M. Gelfand, M. A. Naimark, I. E. Segal.

of the algebra $\mathcal{B}(\mathcal{H}_a)$. In this context, J_ψ is an elementary example of Tomita-Takeski's *modular conjugation*. That ψ is completely entangled can be expressed also in terms of s-maps: $\mathbf{s}_\psi^{ab}$ must be invertible. (Its inverse, if it exists, must be unbounded for infinite dimensional Hilbert spaces.)

There are two further operators, particularly tied to the modular conjugation. The first is introduced by

$$(A \otimes 1_b)\,\psi = S_\psi(A^* \otimes 1_b)\psi\,. \tag{42}$$

S_ψ can also be gained by the help of the twisted cross product

$$S_\psi = (\mathbf{s}_\psi^{ba})^{-1}\,\tilde{\otimes}\,\mathbf{s}_\psi^{ba}\,. \tag{43}$$

It is standard to write the polar decomposition of the anti-linear S-operator

$$S_\psi = J_\psi \sqrt{\Delta_\psi}\,. \tag{44}$$

Δ_ψ is called the Tomita-Takesaki *modular operator*. The distinguished role of these and similar "modular objects" becomes apparent in the theory of general von Neumann algebras where they play an exposed and quite natural role. From them I borrowed the notations for the s- and the j-maps. In the elementary case we are dealing with, one has

$$\Delta_\psi = \varrho_\psi^a \otimes (\varrho_\psi^b)^{-1}\,.$$

See [32] for a physically motivated introduction. Further relations between the s- and j-maps and to modular objects can be found in [12] and [13].

7   From Vectors to State

With $\varrho \equiv \varrho^{ab}$ we may write similar to (6),

$$(|\phi^a\rangle\langle\phi^a| \otimes 1_b)\,\varrho^{ab}(|\phi^a\rangle\langle\phi^a| \otimes 1_b) = |\phi^a\rangle\langle\phi^a| \otimes \Phi_\varrho^{ba}(|\phi^a\rangle\langle\phi^a|), \quad \forall \phi^a \in \mathcal{H}_a \tag{45}$$

For every decomposition

$$\varrho^{ab} = \sum c_{jk}|\psi_j\rangle\langle\psi_k|, \quad \mathcal{H}_{ab} \ni \psi_j \leftrightarrow \mathbf{s}_j^{ab} \tag{46}$$

there is a representation

$$\Phi_\varrho^{ba}(|\phi^a\rangle\langle\phi^a|) = \sum c_{jk}\mathbf{s}_j^{ba}|\phi^a\rangle\langle\phi^a|)\mathbf{s}_k^{ab}\,. \tag{47}$$

Similarly one defines Φ_ϱ^{ab}. The maps are linear in ϱ^{ab} and can be defined for every trace class operator ϱ. Moreover, their domain of definition can be extended to the bounded operators of the subsystems: Let X and Y denote bounded operators on $\mathcal{H}_a$ and $\mathcal{H}_b$ respectively, then

$$X \mapsto \Phi_\varrho^{ba}(X), \quad Y \mapsto \Phi_\varrho^{ab}(Y) \tag{48}$$

are well defined and anti-linear in X or Y. The equation

$$\text{Tr } X\Phi_\varrho^{ab}(Y^*) = \text{Tr } Y\Phi_\varrho^{ba}(X^*) = \text{Tr } \varrho\,(X \otimes Y) \tag{49}$$

is valid. Proving them at first for finite linear combinations of rank one operators, one finds the maps (48) mapping the bounded operators of one subsystem into the trace class operators of the other one. Indeed, the finite version of (49) provides us with estimates like

$$\| \Phi_\varrho^{ba}(X^*) \|_1 \leq \| X \|_\infty \| \varrho \|_1 \,. \tag{50}$$

We now have a one-to one correspondence

$$\Phi_\varrho^{ab} \leftrightarrow \Phi_\varrho^{ba} \leftrightarrow \varrho \tag{51}$$

That we have a map from the bounded operators of $\mathcal{H}_a$ into the trace class operators of $\mathcal{H}_b$ is physically quite nice. It is an opportunity to reflect on testing a property P_a of $\mathcal{H}_a$ once more, but under the condition that $\varrho \equiv \varrho^{ab}$ is in any (normal) state. The rank of P_a is not necessarily finite. The rule of Lüders, [33], says that the prepared state is $\omega^a := P_a\varrho^a P_a$ if one finds the property P_a valid and ϱ^a is the reduced density matrix of ϱ in the a-system before the test. The EPR channel asks for ω^b, the density operator of the b-system after an affirmative checking of the property P_a. This density operator is given by a Φ-map:

If ϱ is the density operator of $\mathcal{H}_{ab}$ and if a local measurement establishes property P_a, then the state ω^b of the b-system is given by

$$\omega^b = \Phi_\varrho^{ba}(P_a)\,. \tag{52}$$

The proof is by looking at the effect in the bi-partite system resulting from a local measurement. Let ϕ_j^a be a basis of the support space of P_a. One obtains

$$(P_a \otimes 1_b)\,|\psi\rangle\langle\psi|\,(P_a \otimes 1_b) = \sum |\phi_j^a\rangle\langle\phi_k^b| \otimes \mathbf{s}_\psi^{ba}|\phi_k^a\rangle\langle\phi_j^a|\,\mathbf{s}_\psi^{ab}$$

and this is, up to normalization, the state prepared by the local measurement. Next we sandwich the equation between $1_a \otimes B$ and take the trace. At the left hand we get $\text{Tr } \omega^b B$. On the right we obtain $\Phi_\varrho^{ba}(P_a)$. Now we have seen from (49) that (52) is correct for pure states. By linearity and (50) we get the assertion.

It may be worthwhile to compare (49) with the now well known "duality" between super-operators T of $\mathcal{H}_a$ and operators on $\mathcal{H}_a \otimes \mathcal{H}_b$. Here the Hilbert spaces are of equal finite dimension. One selects a maximally entangled vector ψ and defines

$$\rho := (T \otimes \text{id}_b)(|\psi\rangle\langle\psi|) \tag{53}$$

to express the structure of T by that of ρ. This trick is due to A. Jamiolkowski, [34], and is now refined and much in use after the papers of B. Terhal [36] and of Horodecki et al [35]. Comparing (47) and (48), one can connect both approaches as follows:

From ϱ we get a map Φ_ϱ^{ab}. From a maximally entangled ψ we get an anti-linear map $\mathbf{s}_\psi^{ab}$, enabling the correspondence (53) to be expressed by

$$\varrho \leftrightarrow T, \quad T(X) = \mathbf{s}_\psi^{ab}\, \Phi_\varrho^{ba}(X)\, \mathbf{s}_\psi^{ba} \tag{54}$$

In a certain way, anti-linearity is the prize for eliminating the reference state ψ in Jamiolkowski's approach.

Acknowledgements

I thank Bernd Crell for valuable comments.

References

1. A. Einstein, B. Podolsky, and N. Rosen, Can quantum-mechanical description of physical reality be considered complete ? Phys.Rev. 47, 777-780, 1935.
2. E. Schrödinger, Die gegenwärtige Situation in der Quantenmechanik, Naturwissenschaften 35, 807–812, 823–828, 844–849, 1935.
3. A. Peres, Quantum Theory: Concepts and Methods, Kluwer Academic Publ., Dortrecht 1993
4. M. A. Nielsen and I. L. Chuang, Quantum Computation and Quantum Information, Cambridge University Press 2000
5. A. K. Ekert, Quantum cryptography based on Bell's theorem, Phys. Rev. Lett. 67, 661-663, 1991.
6. Ch. H. Bennett and S. J. Wiesner, Communication via one– and two–particle operators on Einstein– Podolski–Rosen states, Phys. Rev. Lett. 69, 2881-2884, 1992.
7. G. Birckhoff and J. von Neumann, The logic of quantum mechanics, Ann. of Math. 37, 823-843, 1936.
8. K.-E. Hellwig and K. Kraus, Formal Description of Measurements in Local Quantum Field Theory, Phys. Rev. D 1, 566-571, 1970.
9. J. Finkelstein, Property Attribution and the Projection Postulate in Relativistic Quantum Theory, Phys. Lett. A 278, 19-24, 2000.
10. B. Zbinden, J. Brendel, N. Gisin, and W. Tittel, Experimental test of non-local quantum correlations in relativistic configurations, Phys. Rev. Lett. 84, 4737-4740, 2000.
11. For the discussion of that point of view I am indebted to B. Crell, Leipzig.
12. A. Uhlmann, Quantum channels of the Einstein-Podolski-Rosen kind, In: (A. Borowiec, W. Cegla, B. Jancewicz, W. Karwowski eds.), Proceedings of the XII Max Born Symposium FINE DE SIECLE, Wroclaw 1998, Lecture notes in physics, Springer, Vol. 539, 93-105, 2000.
13. A. Uhlmann, Operators and maps affiliated to EPR channels, In: (H.-D. Doebner, S. T. Ali, M. Keyl, R. F. Werner eds.), Trends in Quantum Mechanics, World Scientific, 138-145, 2000.
14. Y. Aharonov and D. Z. Albert, Can we make sense out of the measurement process in relativistic quantum mechanics? Phys. Rev. D 24, 359-370, 1981.
15. R. D. Sorkin, Impossible measurements on quantum fields, In: (Bei-Lok Hu, T. A. Jacobson eds.,) Directions in General Relativity, V 2, Cambridge University Press, 293-305, 1993.

16. D. Beckman, D. Gottesman, M. N. Nielsen, and J. Preskill, Causal and localizable quantum operations, Phys. Rev. A 64, 2001.

17. L. Vaidman, Instantaneous measurement of non-local variables, Phys. Rev. Lett. 90, 2003.

18. C. Bennett, G. Brassard, C. Crepeau, R. Jozsa, A. Peres, and W. Wootters, Teleporting an unknown quantum state via dual classical and Einstein-Podolsky-Rosen channels, Phys. Rev. Lett. 70, 1895-1898, 1993.

19. R. F. Werner, All teleportation and dense coding schemes, in: (D. Bouwmeester, A. Ekert, A. Zeilinger, eds.) The Physics of Quantum Information, Springer Verlag, Berlin, Heidelberg, New York, 2000.

20. M. Zukowski, A. Zeilinger, M. A. Horne, and A. K. Ekert, Event-ready-detectors Bell experiment via entanglement swapping, Phys. Rev. Lett. 71, 4287, 1993.

21. S. Bose, V. Vedral, and P. L. Knight, Multiparticle schemes for entanglement swapping, Phys. Rev. A 57 822, 1998.

22. M. Ohya, Note on quantum probability, Nuovo Cim. 38, 402-406, 1983.

23. V. P. Belavkin and Ohya, Entanglement and compound states in quantum information theory, quant-ph/0004069.

24. D. I. Fivel, Remarkable Phase Oscillations Appearing in the Lattice Dynamics of Einstein-Podolsky-Rosen States, Phys. Rev. Lett. 74, 835-838, 1995.

25. G. Brassard, Teleportation as Quantum Computation, Physica D 120, 43-47, 1998.

26. S. Albeverio and Shao-M. Fei, Teleportation of general finite dimensional quantum systems. Phys. Lett. A 276, 8-11, 2000.

27. R. Horodecki, M. Horodecki and P. Horodecki, Teleportation, Bell's inequalities and inseparability, Phys. Lett. A 222, 21-25, 1996.

28. T. Mor and P. Horodecki, Teleportation via generalized measurements, and conclusive teleportation, e-print, quant-ph/9906039.

29. C. Trump, D. Bruß, and M. Lewenstein, Realistic teleportation with linear optical elements Phys. Lett. A, 279, 7-11, 2001.

30. K. Banaszek, Fidelity balance in quantum operations, Phys. Rev. Lett., 86, 1366-1369, 2001.

31. J. Rehacek, Z. Hradil, J. Fiurasek, and C. Bruckner, Designing optimal CP maps for quantum teleportation. Phys. Rev. A 64, 2001.

32. R. Haag, Local Quantum Physics, Springer Verlag, Berlin, Heidelberg, New York, 1993.

33. G. Lüders, Über die Zustandsänderung durch den Meßprozeß, Ann. d. Physik 8, 322-328, 1951.

34. A. Jamiolkowski, Linear transformations which preserve trace and positive semi-definiteness of operators, Rep. Math. Phys. 3, 275-278, 1972.

35. M. Horodecki, P. Horodecki, and R. Horodecki, General teleportation channel, singlet fraction and quasi-distillation, e-print, quant-ph/9807091, 1998.

36. B. M. Terhal, Positive maps for bound entangled states based on unextendible product bases, Linear Algebra Appl. 323, 61-73, 2000.

On Rank Two Channels

A. Uhlmann

Abstract. Based on some identities for the determinant of completely positive maps of rank two, concurrences are calculated or estimated from below.

1 Introduction

In the paper I present some identities which are useful in the study of rank two completely positive maps, including attempts to calculate concurrences. It complements my earlier papers [8] and [13].

Let us consider a map, Φ, from the algebra $\mathcal{M}_m$ of $m \times m$–matrices into another matrix algebra. Φ is of rank k if the rank of the matrix $\Phi(X)$ never exceeds k. Then one can reduce Φ to a map into a matrix algebra $\mathcal{M}_k$. If Φ is of rank two, then the trace and the determinant characterize $\Phi(X)$ up to unitary transformations. Thus, for trace preserving maps one essentially remains with $\det \Phi(X)$. As shown in the next section, there is a remarkable and, perhaps, not completely evident way to express that quantity.

The bridge to higher ranks is provided by the use of the second symmetric function, which seems, because of the identity

$$2 \det Z = (\operatorname{tr} Z)^2 - \operatorname{tr} Z^2, \quad Z \in \mathcal{M}_2 \tag{1}$$

quite natural, see Rungta et al [12]. These, and several other authors restrict themselves to trace preserving channels, resulting in $\operatorname{tr} Z = 1$, $Z = \Phi(X)$. A review, pointing to the main definitions and most applications is by Wootters [10]. Mintert et al [11] recently derived a lower bound for the concurrence. It seems to be equivalent, though expressed quite differently, with our estimate (44) in case of rank two.

To consider $det\Phi$ is most efficient for completely positive map of length two. The *length* of a cp-map Φ is the minimal number of Kraus operators, necessary to write down Φ as a Kraus representation. Now, if

$$\Phi(X) = \sum A_j X A_j^* \tag{2}$$

is any Kraus representation of Φ, then the linear space, generated by the Kraus operators A_j, depends on Φ only. The linear space will be called the *Kraus space* of Φ, and it is denoted by Kraus(Φ). Clearly, the dimension of the Kraus space is the length of Φ.

We devote a section to compute explicitly $\det \Phi$ for some channels of rank two and, with one exception, of length two, and the last section to concurrences.

For instance, in tracing out the 2-dimensional part, the partial trace of a $2 \times m$ quantum system is a channel of rank m and of length 2. In the example (see

R. Ahlswede et al. (Eds.): Information Transfer and Combinatorics, LNCS 4123, pp. 413–424, 2006.
© Springer-Verlag Berlin Heidelberg 2006

below) the partial trace is embedded in a one parameter family (7) of channels. Later on we shall see in the $2 \otimes 2$ case, how the whole family can be treated straightforwardly and similar to the way opened by Wootters, partly together with Hill, in their beautiful papers [4] and [6] which has their roots already in Bennett et al [5].

Example 1a: A prominent example of a trace-preserving cp-map of rank m and length two is the partial trace of a $2 \times m$ quantum system into its m-dimensional subsystem,

$$\text{tr}_2 : \quad \mathcal{M}_{2m} = \mathcal{M}_2 \otimes \mathcal{M}_m \mapsto \mathcal{M}_m. \tag{3}$$

Writing the matrices in block format,

$$\text{tr}_2 X \equiv \text{tr}_2 \begin{pmatrix} X_{00} & X_{01} \\ X_{10} & X_{11} \end{pmatrix} = X_{00} + X_{11}, \tag{4}$$

a valid Kraus representation reads

$$\text{tr}_2(X) = A_1 X A_1^* + A_2 X A_2^*, A_1 = \begin{pmatrix} \mathbf{1} & \mathbf{0} \end{pmatrix}, A_2 = \begin{pmatrix} \mathbf{0} & \mathbf{1} \end{pmatrix}, \tag{5}$$

with $\mathbf{0}$ and $\mathbf{1}$ the $(m \times m)$-null and -identity matrices. The Kraus space consists of $(2 \times 2m)$-matrices $\begin{pmatrix} a\mathbf{1} & b\mathbf{1} \end{pmatrix}$. Alternatively, the Kraus space can be generated space by

$$B_1 = \begin{pmatrix} \mathbf{1}_m & \mathbf{1}_m \end{pmatrix}, \quad B_2 = \begin{pmatrix} \mathbf{1}_m & -\mathbf{1}_m \end{pmatrix}, \tag{6}$$

and one can embed tr_2 within the trace preserving cp-maps

$$X \mapsto (1-p) B_1 X B_1^* + p B_2 X B_2^* = X_{00} + X_{11} + (1-2p)(X_{01} + X_{10}). \tag{7}$$

With $0 < p < 1$ one gets "phase-damped" partial traces. $\diamond$

2   The Determinant

What are the merits of the rank two property of a channel? As already mentioned, these trace-preserving cp–maps are governed by just one function on the input system, by $\det \Phi(X)$. Wootters, [6], has used this fact efficiently to calculate the 2×2 entanglement of formation. His proof is based on the so-called concurrence constructions, see next section. While there is a richness of variants in extending the original concept of concurrence for higher ranks, there seems to be a quite canonical one for rank two cp-maps.

In a 2-dimensional Hilbert space there is, up to a phase factor, an exceptional anti-unitary operator, the spin-flip θ_f. (The index "f" remembers Fermi and "fermion".) We choose a reference basis, $|0\rangle$, $|1\rangle$, to fix the phase factor according to

$$\theta_f(c_0|0\rangle + c_1|1\rangle) = c_1^*|0\rangle - c_0^*|1\rangle, \tag{8}$$

or, in a self-explaining way, by

$$\theta_f \begin{pmatrix} c_0 \\ c_1 \end{pmatrix} = \begin{pmatrix} 0 & 1 \\ -1 & 0 \end{pmatrix}_{\text{anti}} \begin{pmatrix} c_0 \\ c_1 \end{pmatrix} = \begin{pmatrix} c_1^* \\ -c_0^* \end{pmatrix}.$$

We need $\theta_f^* = \theta_f^{-1} = -\theta_f$ and the well known equation

$$\theta_f X^* \theta_f X = -(\det X)\, \mathbf{1}. \tag{9}$$

One remembers that the Hermitian adjoint ϑ^* of an anti-linear operator ϑ in any Hilbert space is defined by

$$\langle \psi, \vartheta^* \varphi \rangle = \langle \varphi, \vartheta \psi \rangle.$$

In particular, θ_f is skew Hermitian.

Applying (9) to a rank two cp-map (2) results in

$$(\det \Phi(X))\, \mathbf{1} = -\sum_{jk} \theta_f A_j X^* A_j^* \theta_f A_k X A_k^*$$

and, taking the trace,

$$\det \Phi(X) = -\frac{1}{2} \mathrm{tr} \sum_{jk} (A_k^* \theta_f A_j) X^* (A_j^* \theta_f A_k) X \tag{10}$$

Now we insert $X = |\psi\rangle\langle\varphi|$. Respecting the anti-linearity rules one obtains

$$\det \Phi(|\psi\rangle\langle\varphi|) = -\sum_{j<k} \langle \varphi, (A_k^* \theta_f A_j)\varphi \rangle \cdot \langle (A_j^* \theta_f A_k)\psi, \psi \rangle.$$

This bilinear expression we rewrite further. Consider

$$\langle \varphi, A_k^* \theta_f A_j \varphi \rangle = \langle A_k \varphi, \theta_f A_j \varphi \rangle = -\langle A_j \varphi, \theta_f A_k \varphi \rangle,$$

where $\theta_f^* = -\theta_f$ has been used. The last relation tells us that only the Hermitian part of the operator sandwiched by φ is important. This offers to define the Hermitian anti-linear operators

$$\vartheta_{jk} = \frac{1}{2} \big(A_j^* \theta_f A_k - A_k^* \theta_f A_j \big). \tag{11}$$

Inserting in the determinant expression and adsorbing the minus sign yields

$$\det \Phi(|\psi\rangle\langle\varphi|) = \sum_{j<k} \langle \varphi, \vartheta_{jk}\varphi \rangle \langle \vartheta_{jk}\psi, \psi \rangle = \sum_{j<k} \langle \psi, \vartheta_{jk}\psi \rangle^* \langle \varphi, \vartheta_{jk}\varphi \rangle \tag{12}$$

Before becoming more acquainted with ϑ_{jk} by examples, let us discuss some of their invariance properties. Taking care with the anti-linearity, one gets

$$\Big(\sum a_j A_j\Big)^* \theta_f \Big(\sum b_k A_k\Big) - \Big(\sum b_k B_k\Big)^* \theta_f \Big(\sum a_j B_j\Big) = \sum_{jk} a_j^* b_k^* \vartheta_{jk} \tag{13}$$

First conclusion

The linear space generated by the anti-linear operators θ_{jk} does not depend on the chosen Kraus operators.

Let us call this space the *derived Kraus space of* Φ, denoted by Kraus'(Φ). (Notice: The set of Hermitian anti-linear operators form a complex-linear space. Kraus'(Φ) is one of its subspaces.) In particular,

$$A, B \in \text{Kraus}(\Phi) \implies A^* \theta_f B - B^* \theta_f A \in \text{Kraus}'(\Phi), \tag{14}$$

and, consequently,

$$\text{If Kraus}(\Phi_1) = \text{Kraus}(\Phi_2), \text{ then Kraus}'(\Phi_1) = \text{Kraus}'(\Phi_2) \tag{15}$$

The following items are mutually equivalent for rank two cp-maps Φ.

- The vector $|in\rangle$ obeys $\Phi(|in\rangle\langle in|) = |out\rangle\langle out|$.
- With a unique $C \in \text{Kraus}(\Phi)$ it holds $A|in\rangle = (\text{tr } AC^*)|out\rangle$ for all $A \in \text{Kraus}(\Phi)$.
- For all $\vartheta \in \text{Kraus}'(\Phi)$ it holds $|in\rangle \perp \vartheta|in\rangle$.

The second item is valid for all cp-maps. It does not depend on the rank. From a Kraus representation of Φ with operators A_j one gets the numbers λ_j from $A_j|in\rangle = \lambda_j|out\rangle$. These relations define a linear form over Kraus(Φ) which can be uniquely written as indicated in the second item. Because item one can take place if and only if the determinant of $\Phi(|in\rangle\langle in|)$ vanishes, the third item is a simple consequence of (12).   ◇

Let us now consider the case of two different sets, $\{A_j\}$ and $\{\tilde{A}_j\}$, of Kraus operators belonging both to Φ. This aim is reached by

$$\tilde{A}_k = \sum_j u_{jk} A_j$$

if and only if the u_{jk} are the entries of a unitary matrix. The induced transformation of the operators (11) reads

$$\tilde{\vartheta}_{mn} = \sum_{jk} u_{jm} u_{kn} \vartheta_{jk}$$

We now see, by anti-linearity of the ϑ operators,

$$\sum \tilde{\vartheta}_{mn} X \tilde{\vartheta}_{mn} = \sum u_{jm} u_{kn} u_{rm}^* u_{sn}^* \vartheta_{jk} X \vartheta_{rs}.$$

By the unitarity condition it becomes evident that

$$\sum \tilde{\vartheta}_{mn} X \tilde{\vartheta}_{mn} = \sum \vartheta_{jk} X \vartheta_{jk}$$

holds. Thus, the anti-linear, completely positive map

$$\Phi'(X) := \sum_{j<k} \vartheta_{jk} X \vartheta_{jk} \tag{16}$$

is *uniquely* associated to Φ. Let us call Φ' the *(first) derivative of Φ*. If one needs linearity, $\Phi'(X^*)$ is offered, a completely co-positive map. As one can see from (12),

$$\det \Phi(|\psi\rangle\langle\varphi|) = \langle \varphi, \Phi'(|\varphi\rangle\langle\psi|)\psi\rangle \tag{17}$$

Another way to express the same is by Gram matrices G_φ with matrix entries $\langle \varphi, \vartheta_{jk}\varphi\rangle$,

$$\det \Phi(|\psi\rangle\langle\varphi|) = -\frac{1}{2}\operatorname{tr} G_\varphi G_\psi^* \tag{18}$$

There may be further useful quantities by replacing the trace by other algebraic invariant operations.

3   Examples

At first we continue with example 1a to show the automatic appearance of Wootters' conjugation, and to see what happens with the phase-damped partial trace of a 2×2–system. Next we look at a Kraus space of dimension three. The channels belonging to it describe certain "inverse EPR" tasks: Alice and Bob input pure states $|0x\rangle$, and a device tests "a la Lüders" whether the system is in a certain maximally entangled state or not. Then Alice is asking whether her state is $|0\rangle$ or $|1\rangle$. In the third collection of examples we treat 1-qubit cp-maps of length two. As in the first example there is, essentially, only one ϑ_{12}, denoted simply by ϑ.

Example 1b: Here we call attention to Example 1a, restricted, however, to $m = 2$. Then tr_2 is of rank and of length two. Applying the recipe (11) and using the operators B_j of (6), we start calculating

$$\vartheta = \frac{\sqrt{p(1-p)}}{2}\,(B_1^*\theta_f B_2 - B_2^*\theta_f B_1).$$

At first, we see

$$\begin{pmatrix} 1 & 0 \\ 0 & 1 \\ 1 & 0 \\ 0 & 1 \end{pmatrix} \begin{pmatrix} 0 & -1 \\ 1 & 0 \end{pmatrix}_{\text{anti}} \begin{pmatrix} 1 & 0 & -1 & 0 \\ 0 & 1 & 0 & -1 \end{pmatrix} = \begin{pmatrix} 0 & 1 & 0 & -1 \\ -1 & 0 & 1 & 0 \\ 0 & 1 & 0 & -1 \\ -1 & 0 & 1 & 0 \end{pmatrix}_{\text{anti}}$$

We have to take the Hermitian part. An anti-linear operator is Hermitian if every matrix representations is a symmetric matrix. We obtain, up a factor, Wootters' conjugation

$$\vartheta = \sqrt{p(1-p)}\begin{pmatrix} 0 & 0 & 0 & -1 \\ 0 & 0 & 1 & 0 \\ 0 & 1 & 0 & 0 \\ -1 & 0 & 0 & 0 \end{pmatrix}_{\text{anti}} = -\sqrt{p(1-p)}\theta_f \otimes \theta_f \tag{19}$$

We infer from the last equation: *The derived Kraus space of the phase-damped partial traces in 2×2-systems is generated by Wootters conjugation.* $\diamond$

Example 2:   Consider the 1-qubit-channels

$$\Phi_q\left(\begin{pmatrix} x_{00} & x_{01} \\ x_{10} & x_{11} \end{pmatrix}\right) = \begin{pmatrix} (1-q)x_{00} & 0 \\ 0 & x_{11} + qx_{00} \end{pmatrix}. \tag{20}$$

with $0 < q < 1$. We easily see

$$\det \Phi(X) = (1-q)x_{00}(x_{11} + qx_{00}).$$

The channels are entanglement breaking and of length three. The operators

$$A_1 = \begin{pmatrix} 0 & 0 \\ 0 & 1 \end{pmatrix}, \quad A_2 = \sqrt{1-q}\begin{pmatrix} 1 & 0 \\ 0 & 0 \end{pmatrix}, \quad A_3 = \sqrt{q}\begin{pmatrix} 0 & 0 \\ 1 & 0 \end{pmatrix} \tag{21}$$

can be used to Kraus represent the channels:

$$\Phi_q(X) = A_1 X A_1 + A_2 X A_2 + A_3 X A_3^*,$$

where the dependence on q of the A_j has not been made explicit. (A_1 and A_2 are Hermitian.) $|1\rangle\langle 1|$ is a fix-point of (20) All Φ_q belong to the same Kraus space which consists of all operators A satisfying $\langle 1|A|1\rangle = 0$. See also Verstraede and Verschelde, [3], (theorem5).

A straightforward calculation yields

$$\vartheta_{12} = -\frac{1}{2}\sqrt{1-q}\begin{pmatrix} 0 & 1 \\ 1 & 0 \end{pmatrix}_{anti}, \quad \vartheta_{23} = \sqrt{q(1-q)}\begin{pmatrix} 1 & 0 \\ 0 & 0 \end{pmatrix}_{anti}. \tag{22}$$

and $\vartheta_{13} = 0$. Therefore, the first derivative of Φ_q becomes

$$\Phi_q'(X^*) = \frac{1-q}{4}\begin{pmatrix} x_{11} + 4qx_{00} & x_{01} \\ x_{10} & x_{00} \end{pmatrix}, \tag{23}$$

and, after some elementary calculations, we get

$$\operatorname{tr} X\Phi'(X^*) = \det \Phi_q(X) - \frac{1-q}{2}\det X. \tag{24}$$

This also makes sense for $q = 0$, getting the identity map, and for $q = 1$, resulting in a degenerate length two channel. The deviation from being of length two is indicated by the commutator

$$\vartheta_{12}\vartheta_{23} - \vartheta_{23}\vartheta_{12} = \frac{1-q}{2}\sqrt{q}\begin{pmatrix} 0 & 1 \\ -1 & 0 \end{pmatrix}. \tag{25}$$

One may wonder whether it is useful to examine more generally the space of linear operators generated by the commutators of the operators ϑ_{jk}. However, I do not know the meaning of it. Is it an indication of a co-homology like sequence?   ◇

Example 3:   Let us now turn to completely positive 1-qubit–maps of length two. The reader may consult [8] and [13] for other proofs and aspects.

In the case at hand, we get a Kraus space generated by two operators, say A and B. For our next purpose we rewrite (11),

$$\vartheta = \frac{1}{2}\left(A^*\theta_f B - B^*\theta_f A\right) = \begin{pmatrix} \alpha_{00} & \alpha_{01} \\ \alpha_{10} & \alpha_{11} \end{pmatrix}_{\text{anti}}, \tag{26}$$

and obtain the following matrix entries:

$$\alpha_{00}^* = a_{00}b_{10} - a_{10}b_{00}, \quad \alpha_{11}^* = a_{01}b_{11} - a_{11}b_{01},$$

$$\alpha_{01}^* = \alpha_{10}^* = \frac{1}{2}(a_{00}b_{11} + a_{01}b_{10} - a_{10}b_{01} - a_{11}b_{00}). \tag{27}$$

There are a lot of possibilities in choosing A and B in order to obtain a pre-described ϑ. For instance, setting $B = 1$ in (27), one arrives at

$$B = 1 \; \Rightarrow \; \vartheta = \begin{pmatrix} -a_{10}^* & \frac{1}{2}(a_{00} - a_{11})^* \\ \frac{1}{2}(a_{00} - a_{11})^* & a_{01}^* \end{pmatrix}_{\text{anti}}.$$

Therefore, every anti-linear and Hermitian ϑ can be gained via (26) with a suitable A and with $B = 1$.

More general cases can be seen better after a unitary change of Φ. $\tilde{\Phi}$ is *unitarily equivalent* to Φ, if for all X

$$\tilde{\Phi}(X) = U_1\Phi(U_2 X U_2^*)U_1^*, \quad \tilde{\vartheta} = U_2^*\vartheta U_2$$

with a special unitary U_1 and a unitary U_2. (The the unitaries with $\det U = 1$ commute with θ_f.) As is known, see Ruskai et al [1] and the early paper of Gorini and Sudarshan [2], every 1-qubit–channel of length two is unitarily equivalent to a "normal form" with Kraus operators

$$A = \begin{pmatrix} a_{00} & 0 \\ 0 & a_{11} \end{pmatrix}, \quad B = \begin{pmatrix} 0 & b_{01} \\ b_{10} & 0 \end{pmatrix}, \tag{28}$$

and these Kraus operators imply

$$\vartheta = \begin{pmatrix} z_0^2 & 0 \\ 0 & -z_1^2 \end{pmatrix}_{\text{anti}}, \quad z_0^2 = (b_{10}a_{00})^*, \quad z_1^2 = (b_{01}a_{11})^* \tag{29}$$

The map Φ is called *non-degenerate* if $\det \vartheta^2 \neq 0$. Then $z_0 z_1 \neq 0$. There are two cases if Φ is degenerate. Either one of the numbers z_1, z_2 is zero, but the other one not. Or, both are zero. (An example is $a_{11} = b_{10} = 0$ but $a_{00}b_{01} \neq 0$.)

4   Concurrence

Concurrence, originally introduced with respect to partial traces, can be consistently defined for all channels, and even for all positive maps. For trace-preserving cp-map this fact can be understood by the the Stinespring dilatation theorem. If Φ is not of rank two, one replaces in the definitions below $\det \Phi$ according to

$$\det \Phi(X) \implies \frac{1}{2}\left((\operatorname{tr} X)^2 - \operatorname{tr} X^2\right), \tag{30}$$

which does not change anything if $\Phi(X)$ is 2×2. In some cases one can replace the condition of being rank two by demanding $\Phi(X)$ to possess not more than two different, but degenerated, eigenvalues. See [15].

After repeating, for convenience, the definition and some general knowledge, a more detailed treatment for rank two (and length two) cp-maps will be given, though not exhaustive.

Let Φ be a positive map of rank two. $C(\Phi; X)$, the Φ-concurrence, is defined for all positive operators X of the input space by the following properties:

(i) $C(\Phi; X)$ is homogeneous of degree one,

$$C(\Phi; \lambda X) = \lambda C(\Phi; X), \quad \lambda \geq 0.$$

(ii) $C(\Phi; X)$ is sub-additive,

$$C(\Phi; X + Y) \leq C(\Phi; X) + C(\Phi; Y)$$

(iii) $C(\Phi; X)$ is the largest function with properties (i) and (ii) above, satisfying for all vectors ψ of the input space

$$C(\Phi; |\psi\rangle\langle\psi|) = \sqrt{\det \Phi(|\psi\rangle\langle\psi|)} \tag{31}$$

Let us draw a conclusion. Let be Z_1 an operator on the input and Z_2 one on the output space. Then

$$\tilde{\Phi}(X) = Z_2 \Phi(Z_1 X Z_1^*) Z_2^* \Rightarrow C(\tilde{\Phi}; X) = |\det Z_2|^2 C(\Phi; Z_1 X Z_1^*). \tag{32}$$

Indeed, the concurrence of $\tilde{\Phi}$ as given by (32) fulfills (i) and (ii), and both functions coincide for positive operators of rank one. $\diamond$

There are other, equivalent possibilities to define C. It is not difficult to show that

$$C(\Phi; X) = \inf\left\{ \sum \sqrt{\det \Phi(|\psi_j\rangle\langle\psi_j|)}, \quad \sum |\psi_j\rangle\langle\psi_j| = X \right\}. \tag{33}$$

holds. Next, just because the square root of the determinant is a concave function in dimension two, a further valid representation is given by

$$C(\Phi; X) = \inf\left\{ \sum \sqrt{\det \Phi(X_j)}, \quad \sum X_j = X \right\}, \tag{34}$$

so that the $X_j \geq 0$ can be arbitrarily chosen up to the constraint of summing up to X. Notice, that a similar trick with the determinant (or the second symmetric function) in the definition of concurrence would fail because the determinant is not concave on the cone of positive operators.

For cp-maps of rank and length two more can be said about the variational problem involved in the definitions above. This is due to the fact that the derived Kraus space is 1-dimensional, as explained in the preceding section. The appropriate extension of Wootters procedure goes this way:

Step 1. For two positive operators, X and Y, of the input space we need

$$\{\lambda_1 \geq \lambda_2 \geq \ldots\} = \text{eigenvalues of } (X^{1/2}YX^{1/2})^{1/2} \tag{35}$$

to define

$$C(X,Y) := \max\{0, \lambda_1 - \sum_{j>1}\lambda_j\}. \tag{36}$$

Step 2. We replace Y by $\vartheta X\vartheta$,

$$C(\Phi; X) = C(X, \vartheta X\vartheta), \tag{37}$$

and we are done, [9].

To see a first use, let us return to the $2 \otimes 2$ case, Φ being a partial trace. It was shown, see example 1b, that Wootters' $\vartheta = -\theta_f \otimes \theta_f$ must be replaced by $\sqrt{p(1-p)}\vartheta$ for the phase-damped partial traces of example 1a. The relevant eigenvalues (35), which give (37) vie (36), have to be multiplied accordingly. Therefore, the concurrence of the phase-damped partial trace is Wootters' concurrence multiplied by the factor $\sqrt{p(1-p)}$.

A similar reasoning applies for all length two, rank two channels: All cp-maps with the same Kraus space induce, up to a numerical factor, the same concurrence. Many details can be seen for length two 1-qubit cp-maps by further discussing example 3 of the preceding section.

Example 3a: In dimension two there are only two eigenvalues, λ_1, λ_2, to be respected in (35). Therefore, the right hand side of (36) is equal to $\lambda_1 - \lambda_2$. However, combining

$$(\lambda_1 - \lambda_2)^2 = (\text{tr}\,\xi)^2 - 4\det\xi, \quad \xi = (X^{1/2}YX^{1/2})^{1/2}$$

with the identity

$$(\text{tr}\,\xi)^2 = \text{tr}\,\xi^2 + 2\det\xi,$$

yields

$$(\lambda_1 - \lambda_2)^2 = \text{tr}\,\xi^2 - 2\det\xi.$$

Finally, removing the auxiliary operator ξ, we obtain

$$C(X,Y)^2 = \text{tr}\,(XY) - 2\sqrt{\det(XY)}. \tag{38}$$

With the Kraus operators A, B of Φ, and with ϑ given by (26), the relation (38) provides us with

$$C(\Phi; X)^2 = \text{tr}\,(X\vartheta X\vartheta) - 2(\det X)(\det\vartheta^2)^{1/2}. \tag{39}$$

Let Φ be in the normal form (29) so that ϑ is diagonal with entries z_0^2 and $-z_1^2$ as in (29). Then we arrive at

$$\text{tr}\,X\vartheta X\vartheta = (z_0^* x_{00} z_0)^2 - (z_0^* x_{01} z_1)^2 - (z_0 x_{10} z_1^*)^2 + (z_1^* x_{11} z_1)^2,$$

$$(\det X)(\det\vartheta^2)^{1/2} = (z_0 z_0^* z_1 z_1^*)(x_{00}x_{11} - x_{01}x_{10}).$$

Combining these two expressions as dictated by (39) results in

$$C(\Phi; X)^2 = (z_0 z_0^* x_{00} - z_1 z_1^* x_{11})^2 - (z_0 z_1^* x_{10} - z_1 z_0^* x_{01})^2. \qquad (40)$$

The number within the second delimiter is purely imaginary and, therefore, C is the sum of two positive quadratic terms. This observation remains true if we allow for any Hermitian operator in (40).

The square of the concurrence (39) is a positive semi-definite quadratic form of maximal rank two on the real-linear space of Hermitian Operators. The concurrence is a Hilbert semi-norm.

There is a further curious observation: The concurrence of our 1-qubit cp-map in normal form is equal to the absolute value of the complex number

$$c(X) := z_0 z_0^* x_{00} - z_1 z_1^* x_{11} + z_0 z_1^* x_{10} - z_1 z_0^* x_{01}$$

Following Kossakowski [14], it is tempting to ask, whether $c(X)$ is to replaced by a Quaternion for positive, but not completely positive maps of rank two.

Given $X = X^*$, its squared concurrence is

$$C^2(\Phi; X)^2 = l_1^2(X) + l_2^2(X) \qquad (41)$$

with real

$$l_1(X) = z_0 z_0^* x_{00} - z_1 z_1^* x_{11}, \quad l_2(X) = i(z_0 z_1^* x_{10} - z_1 z_0^* x_{01}) \qquad (42)$$

The value of l_1, together with the trace of X, determine x_{00} and x_{11} uniquely. (We exclude the trivial case $z_1 = z_2 = 0$.) The value of l_2 now determines a line of constant squared concurrence crossing X. Along this line only the off-diagonal entries of X vary. Explicitly, along

$$y_{01} = z_0 z_1^* t + x_{01}, \quad y_{10} = z_0^* z_1 t + x_{10} \qquad (43)$$

we get $l_2(Y) = l_2(X)$, y_{jk} denote the matrix entries of Y. For positive X we know $C \geq 0$, and there is no ambiguity in taking the square root in (41). It is a particular property of every rank two, length two cp-map that its concurrence remains constant along a certain bundle of *parallel* lines.

In the degenerate case, $z_1 z_2 = 0$, it holds $l_2 = 0$ always: After fixing x_{00} and x_{11} we get planes of constant C^2. ◇

If Φ is not of length two, there are lines, crossing a given positive X, along which the concurrence is a linear function, but not necessarily a constant one. By this reason, and by the possibility of bifurcations, [7] general expressions similar to (37) seem to be unknown. However, an estimation from below is available. To this end let us look at

$$\left(\sum C(X, \vartheta_{jk} X \vartheta_{jk})^2 \right)^{1/2}$$

The terms within the sum can be seen as squared concurrences of length two channels. Therefore, every term is the square of a Hilbert semi-norm, and the

whole expression fulfills again the requirements (i) and (ii) in the definition of
Φ-concurrence at the beginning of the present section. Because of (12), and by
its very construction, the expression coincides for positive rank one operators
with $C(\Phi; X)$. But the latter is the largest function with these properties. This
proves the inequality

$$C(\Phi; X)^2 \geq \sum_{j>k} C(X, \vartheta_{jk} X \vartheta_{jk})^2, \quad X \geq 0 \tag{44}$$

Sometimes one can say more, as the further treatment of example 2 shall show.

Example 2a: Remembering (20)

$$\Phi_q\left(\begin{pmatrix} x_{00} & x_{01} \\ x_{10} & x_{11} \end{pmatrix}\right) = \begin{pmatrix} (1-q)x_{00} & 0 \\ 0 & x_{11} + qx_{00} \end{pmatrix}$$

and (22)

$$\vartheta_{12} = -\frac{1}{4}\sqrt{1-q}\begin{pmatrix} 0 & 1 \\ 1 & 0 \end{pmatrix}_{\text{anti}}, \quad \vartheta_{23} = \frac{1}{2}\sqrt{q(1-q)}\begin{pmatrix} 1 & 0 \\ 0 & 0 \end{pmatrix}_{\text{anti}},$$

we need to know

$$C(X, \vartheta_{12} X \vartheta_{12}), \quad C(X, \vartheta_{23} X \vartheta_{23}). \tag{45}$$

The first one belongs to the phase-damping 1-qubit channels. As it is not in
normal form, we compute it directly:

$$\operatorname{tr} X \vartheta_{12} X \vartheta_{12} = \frac{1-q}{8}(|x_{01}|^2 + x_{00} x_{11}), \quad \sqrt{\vartheta_{12}^2} = \frac{1-q}{16},$$

yielding

$$C(X, \vartheta_{12} X \vartheta_{12})^2 = (1-q)|x_{01}|^2/4. \tag{46}$$

For the other C we simply specify (40) and get

$$C(X, \vartheta_{23} X \vartheta_{23})^2 = q(1-q)x_{00}^2/4. \tag{47}$$

As a particular case of (44), we arrive at the inequality

$$C(\Phi_q; X) \geq \frac{1}{2}\sqrt{(1-q)}\sqrt{(qx_{00}^2 + |x_{01}|^2)} \tag{48}$$

for positive X. If $x_{01} = 0$, the right hand side of (48) becomes linear. Therefore,
by convexity of C, equality must hold, i.e.

$$C\left(\Phi_q; \begin{pmatrix} x_{00} & 0 \\ 0 & x_{11} \end{pmatrix}\right) = \frac{1}{2}\sqrt{q(1-q)}\, x_{00}.$$

References

1. M. B. Ruskai, S. Szarek, and E. Werner, An analysis of completely–positive trace–preserving maps on 2x2 matrices, Lin. Alg. Appl. 347, 159-187, 2002.
2. V. Gorini and E. C. G. Sudarshan, Extreme affine transformations, Commun. Math. Phys. 46, 43-52, 1976.
3. F. Verstraete and H. Verschelde, On one-qubit channels, quant-ph/0202124.
4. S. Hill and W. Wootters, Entanglement of a pair of quantum bits, Phys. Rev. Lett. 78, 5022-5025, 1997.
5. C. Bennett, D. Di Vincenzo, J. Smolin, and W. Wootters, Mixed state entanglement and quantum error correction, Phys. Rev. A 54, 3824-3851, 1996.
6. W. Wootters, Entanglement of formation of an arbitrary state of two qubits, Phys. Rev. Lett. 80, 2245, 1998.
7. F. Benatti, A. Narnhofer, and A. Uhlmann, Broken symmetries in the entanglement of formation, Int. J. Theor. Phys. 42, 983-999, 2003.
8. A. Uhlmann, On 1-qubit channels, J. Phys. A, Math. Gen. 34, 7074-7055, 2001. revised version: e-print, quant-ph/0011106, 2001.
9. A. Uhlmann, Fidelity and concurrence of conjugated states, Phys. Rev. A 62, 2000.
10. W. K. Wootters, Entanglement of formation and concurrence, Quantum Information and Computation 1, 27-47, 2002.
11. F. Mintert, M. Kuś, and A. Buchleitner, Concurrence of mixed bipartite quantum states of arbitrary dimensions, e-print, quant-ph/0403063.
12. R. Rungta, V. Buzek, C. M. Caves, M. Hillery, G. J.Milburn, and W. K. Wootters, Universal state inversion and concurrence in arbitrary dimensions, Phys. Rev. A 64, 042315, 2001.
13. A. Uhlmann, Concurrence and foliations induced by some 1-qubit channels, Int. J. Theor. Phys., 42, 983-999, 2003.
14. A. Kossakowski, Remarks on positive maps of finite-dimensional simple Jordan algebras, Rep. Math. Phys. 46 393-397, 2000.
15. S. Fei, J. Jost, X. Li-Jost, and G. Wang, Entanglement of formation for a class of quantum states, e-print, quant-ph/0304095.

Universal Sets of Quantum Information Processing Primitives and Their Optimal Use

J. Gruska*

Progress in science is often done by pessimists
Progress in technology is always done by optimists

Abstract. This paper considers several concepts of universality in quantum information processing and deals with various (sometimes surprising) universal sets of quantum primitives as well as with their optimal use.

1 Introduction

Nature offers enormous variety of ways – let us call them technologies – several, more or less powerful, quantum information processing primitives can be exhibited, implemented and used.

Since it appears to be very difficult to exploit the potential of nature for quantum information processing, it is of major importance for quantum information processing to explore which kind of quantum primitives form sets that are universal, in some relevant sense, and that are reasonably easy to implement with some available or potential technologies. Moreover, also from the point of view of understanding the laws and limitations of quantum information processing and communication as well as of quantum mechanics itself, the problems of finding rudimentary and yet powerful quantum information processing primitives, as well as methods for their optimal use, are of large experimental and also fundamental importance.

The search for such quantum computation universal primitives, and for their optimal use, is therefore one of the major tasks of the current quantum information processing research (both theoretical and experimental) that starts to attack the task of building processors for few, or not so few, qubits applications.

The search for sets of elementary, or even very rudimentary, but powerful, quantum computational primitives and for their optimal use, has brought a variety of deep and surprising results that are much encouraging for some of the main current challenges of the field – i.e. the design of powerful quantum information processing processors.

One of the outcomes that looked very surprising at first, has been the discovery of the high computational power of quantum projective measurements, and

---

* Paper was written partially during the author's stay with the Imai Quantum Computing Project, Tokyo in 2004 and their support, as well as support of the grant GACR 201/04/1153 is to be acknowledged.

R. Ahlswede et al. (Eds.): Information Transfer and Combinatorics, LNCS 4123, pp. 425–451, 2006.
© Springer-Verlag Berlin Heidelberg 2006

the understanding that a small set of measurements (observables) is sufficient to simulate any quantum (unitary) operation. It has then turned out that it is actually quite natural that quantum projective measurements are a proper way to realize quantum unitary operations (in a special but meaningful and interesting way). This in turn led to the design of two new abstract models for quantum computing: projective measurement based quantum circuits and quantum Turing machines.

A problem related to the search for universal sets of quantum computation primitives, which is also crucial for quantum computing, is how to synthesize any arbitrary unitary transformation using primitives of some given universal set. This problem is often decomposed into a couple of subproblems. One of them is the synthesis of arbitrary unitary transformations from one- and two-qubit unitary transformations, or measurements. Second problem is that of synthesis of arbitrary two-qubit and one-qubit transformations from a simpler set of (elementary, in some sense), unitary one- and two-qubit transformations. Another problem is that of the synthesis of one- and two-qubit transformations using primitives provided by a particular technology. For example, from RF-pulses in the case of NMR information processing. Yet another possibility is to simulate unitary operations through projective measurements.

Concerning primitives for quantum communication several surprising discoveries have been made. For example, an addition of a simple state as a given resource can transform a non-universal set of primitives to a universal set. Moreover, the high computational power of the nearest neighbor exchange interactions has been discovered.

It is also quite clear that optimization of quantum circuits will be of similar importance as in the case of classical circuits. This time, however, optimization seems to be a much more difficult task. In spite of that, several interesting and important results have already been obtained concerning optimization of special circuits – e.g. the circuits for two- and three-qubit unitary gates.

In this paper we assume a basic knowledge of quantum information processing concepts. For example, on the level of Gruska (1999, 2003).

2   Universality and Optimality in Classical Computing

In *classical computing*, the most widely used set of gates, i.e. AND-, OR- and NOT-gates, is universal and so is also the set consisting of AND- and NOT-gates, or even the set consisting of only a single NOR- or NAND-gate.

The problem of optimality for classical circuits with such sets of gates has been solved quite satisfactorily. Optimization methods for classical circuits are heavily used and have major applications.

In the case of *classical reversible computing*, both the Toffoli gate $T(x, y, z) = (x, y, (x \wedge y) \oplus z)$ and the Fredkin gate $F(x, y, z) = (x, \bar{x}y + xz, \bar{x}z + xy)$ are universal, if constant inputs are allowed, as well as additional (ancilla) "wires" with identity gates. Otherwise, a slightly more convenient universal set is that of gates T, CNOT and NOT and auxiliary ancilla wires. Actually, CNOT can be deleted because it can be realized by a circuit with two Toffoli gates and two NOT gates and one ancilla wire.

Perhaps the most useful definition of universality for sets of reversible gates reads as follows:

Definition 1. *A set of reversible gates $\mathcal{G}$ is universal if for every n and all permutations $\pi \in S_{2^n}$ there exists a constant a_n such that some circuit consisting of the gates from $\mathcal{G}$ computes π using at most a_n ancilla wires.*

Interesting enough, there is neither one- nor two-inputs classical universal reversible gate. However, as discussed later, there are universal two inputs quantum gates.

Reversible classical circuits started to be of importance recently for several reasons: (a) they are useful in some applications, like signal processing, communication, cryptography, where circuits should be information lossless; (b) In some technologies (see Shende et al. (2002)), the loss of information due to irreversibility implies energy loss and in nanotechnologies switching devices with gain are not easy to build; (c) reversible classical circuits are special cases of quantum circuits.

Synthesis of reversible circuits from reversible elements and their optimization are also interesting problems. The goal is to minimize either the number of ancilla wires (for minimizing auxiliary storage), or the total number of gates from some universal set, or, still, the depth of the circuits.

Toffoli (1982) provided a synthesizing algorithm, but it needed a lot of ancilla wires (up to $n - 3$ for n-input circuits using TOFFOLI, CNOT and NOT gates). However, Shende et al. (2002) showed that every even (odd) permutation[1] can be implemented without any (with one) ancilla using the gates TOFFOLI, CNOT and NOT.

An optimization problem for an important class of CNOT-circuits has been solved by Patel et al. (2002). The basic observation is that each circuit consisting of CNOT-gates realizes a so called xor-linear gate[2], and every xor-linear gate can be realized by a CNOT-circuit. They provide an algorithm that implements any xor-linear n qubit gate using $\mathcal{O}(n^2/lgn)$ CNOT-gates and they also show that this result is asymptotically optimal.

The optimization of $\{CNOT, NOT\}$ circuits has been solved by Iwama and Yamashita (2003). They found a complete set of transformation rules that can transform any $\{CNOT, NOT\}$-circuit into an optimal one. However, the time complexity of this optimization grows exponentially.

The above optimization results are of interest also for quantum computing, where CNOT gates play an important role while, at the same time, its physical implementation is highly nontrivial because it can create entangled outputs from non-entangled inputs.

---

[1] By Toffoli (1982), every reversible circuit with n inputs and no $n \times n$ gate implements an even permutation.

[2] An n qubit gate U is called xor-linear if $U(x \oplus y) = U(x) \oplus U(y)$ holds for every $x, y \in \{0,1\}^n$.

3   Basic Concepts and Types of Universality

¿From the physical implementation point of view, the most basic requirements and tools for quantum information processing are the following: (a) to use a scalable physical system with long decoherence time to implement well specified qubits; (b) to create a fiducial initial state, say the state $|0^{(n)}\rangle$; (c) to implement all the gates from a universal set (or all primitives from a universal set); (c) to perform the standard basis projective measurement (or some other projective measurements, for example the Bell measurement). Additional important requirements are the ability to perform two-qubit gates on any pair of qubits and to use maximal parallelism. A search for technologies that meet the above requirements is of up most importance.

¿From the quantum state evolution point of view, a set of primitives has to be used for implementing, or at least arbitrarily well approximating, any unitary transformation. Finally, from the quantum computation point of view, a set of primitives is needed for simulating a universal classical computer.

3.1   Basic Notations, Concepts and Definitions

Let us first introduce some of the very basic concepts that are used in the definitions of various notions of universality for quantum information processing.

For a given integer n, $U(n)$ $(SU(n))$ is used to denote the group of unitary operators of degree n (with determinants equal to 1).[3] Similarly, $O(n)$ $(SO(n))$ is used to denote the set of real, orthogonal, and therefore unitary, matrices of degree n (with determinants equal to 1).

By an n-qubit (operation) gate we understand a unitary gate (operation) over the 2^n-dimensional Hilbert space. A gate is called *real*, or an *rgate*, if the corresponding operator matrix in the standard basis contains only real numbers.

Let $\mathcal{G}$ be a set of quantum gates. A $\mathcal{G}$-circuit is a quantum circuit where all gates are from $\mathcal{G}$. A k-qubit $\mathcal{G}$-ancilla is a quantum state $C|x\rangle$, where C is a $\mathcal{G}$-circuit with k qubit inputs and outputs and $x \in \{0,1\}^k$.

The next important concepts are those of equivalence of two gates and of two sets of gates, as well as two concepts of approximability.

Definition 2. *Two n-qubit gates G_1 and G_2 are locally equivalent if there are n one-qubit gates $U_1, \ldots, U_n$ and n one qubit gates $V_1, \ldots, V_n$ such that $G_1 = (\bigotimes_{i=1}^{n} U_i) \otimes G_2 \otimes (\bigotimes_{i=1}^{n} V_i)$.*

That is, two gates are called locally equivalent if any one of them can be implemented using the second one and one-qubit gates (that is using local one qubit unitary operations only).

In the following, it is assumed that whenever a gate is available, it can be used also in the reverse way, with "output wires" as the input ones and vice versa.

---

[3] The global phase of a state has no impact on measurement. Therefore, from the computation point of view global phases can be ignored. As a consequence, any unitary operator in $U(n)$ can be represented in a normal form by a matrix from $SU(n)$.

Definition 3. *A set of gates $\mathcal{G}_1$ is said to be adequate for a set of gates $\mathcal{G}_2$ if every gate form $\mathcal{G}_2$ can be implemented by a $\mathcal{G}_1$-circuit. Sets of gates $\mathcal{G}_1$ and $\mathcal{G}_2$ are called* equivalent *if $\mathcal{G}_1$ is adequate for $\mathcal{G}_2$ and vice versa.*

Two types of approximability of one gate by another one are of importance. The most general one is approximability using an ancilla state; the second one, less general, is without an ancilla (Shi, 2002).

Definition 4. *An operator*

$$U : \mathcal{H}_{2^r} \to \mathcal{H}_{2^r}$$

is ε-approximated, for an $\varepsilon > 0$, by an operator

$$\bar{U} : \mathcal{H}_{2^n} \to \mathcal{H}_{2^n},$$

using an ancilla state $|\alpha\rangle \in \mathcal{H}_{2^{n-r}}$, if for any state $|\phi\rangle \in \mathcal{H}_{2^r}$,

$$||\bar{U}(|\phi\rangle \otimes |\alpha\rangle) - U(|\phi\rangle) \otimes |\alpha\rangle|| \leq \varepsilon ||\alpha||.$$

Definition 5. *An operator*

$$U : \mathcal{H}_{2^n} \to \mathcal{H}_{2^n}$$

is ε-wa-approximated (approximated "without an ancilla") by an operator

$$\bar{U} : \mathcal{H}_{2^n} \to \mathcal{H}_{2^n},$$

for an $\varepsilon > 0$, if for any state $|\phi\rangle \in \mathcal{H}_{2^n}$,

$$||\bar{U}(|\phi\rangle) - U(|\phi\rangle)\rangle|| \leq \varepsilon.$$

Types of Universality for Quantum Information Processing. Now we define several types of universality for sets of quantum gates.

Definition 6. *A set of gates $\mathcal{G}$ is called fully universal (f-universal) if every gate can be realized, up to a global phase factor, by a $\mathcal{G}$-circuit.*[4]

Since the number of quantum unitaries has cardinality of continuum, no finite set of gates can be f-universal. The requirement of f-universality is therefore practically too strong. The next two concepts of universality are therefore more appropriate from several points of view.

Definition 7. *A set of gates $\mathcal{G}$ is called densely universal (d-universal) if there exists an integer n_0 such that for any $n \geq n_0$, the subgroup generated by $\mathcal{G}$ is dense in $SU(2^n)$.*

---

[4] The claim that a gate G can be approximated (realized) up to a global phase factor means that for an appropriate real ϕ the gate $e^{\phi i}G$ is approximated (realized).

Definition 8. *A set of gates $\mathcal{G}$ is called universal if there is an integer n_0 such that any n-qubit unitary gate with $n \geq n_0$, can be, for any $\varepsilon > 0$, ε-approximated by a $\mathcal{G}$-circuit.*

From the classical computation point of view a still weaker concept of universality is sufficient.

Definition 9. *A set of real gates $\mathcal{G}$ is called computationally universal (c-universal) if there is an integer n_0 such that any n-qubit real unitary gate with $n \geq n_0$, can be, for any $\varepsilon > 0$, ε-approximated by a $\mathcal{G}$-circuit.*

Of importance is also the concept of *efficient universality*, see also Ahoronov (2003).

Definition 10. *A gate is efficiently approximable by a set of gates G if it can be ε-approximated, for any $\varepsilon > 0$, using $polylog(\frac{1}{\varepsilon})$ gates from G. A set of gates $\mathcal{G}$ is called efficiently universal (e-universal), if it can be used to ε-approximate with arbitrarily small $\varepsilon > 0$ any quantum circuit having n qubits and t one- and two-qubit gates, with only poly-logarithmic overhead in $(n, t, \frac{1}{\varepsilon})$.*

Efficient universality therefore does not mean that any n qubit unitary operation U can be implemented using a polynomial number (with respect to n and $\frac{1}{\varepsilon}$), for an arbitrary $\varepsilon > 0$, of gates from a given universal set of gates $\mathcal{G}$. This is not possible to achieve in general. Efficient universality only means that if U can be implemented using some number of t of one- and two-qubit gates, then the total number of gates needed from $\mathcal{G}$ to implement U is polynomial with respect to n, t and $\frac{1}{\varepsilon}$.

Another important concept of universality is that of *fault-tolerant universality*, or *ft-universality* in short. A set of gates $\mathcal{G}$ is ft-universal if it is universal and all gates of $\mathcal{G}$ have fault-tolerant implementations.

A set of gates $\mathcal{G}$ that is universal in some of the above senses is, in quantum computing jargon, often called a *gate library*. A gate is also often called universal in one of the above senses in case the set consisting of that gate (and it inverse) is universal provided some constant (in standard bases) inputs are also allowed.

Remark 1. *From a practical point of view there is still another notion of universality. It is related to the problem how reliable have to be our primitives - gates, channels - so that we can have really universal quantum information processing. The theory of fault-tolerant computation solved, in some sense, this problem. It has shown that there are thresholds - values δ with the property that if all primitives, gates and wires, have an error rate smaller than δ, then, using proper error-correction schemes, computations and communications can be done sufficiently well (stabilized) and we can speak, also from a practical point of view, about the existence of universal computations. Thresholds δ depend on the error correction schemes and their estimates vary from 10^{-6} to 10^{-4}. They are therefore very hard to achieve. A potential way out that may provide "beatable thresholds" could be to work with some combination of operations that are errorless, plus some operations that can be erroneous. See page 433 for an analysis of one such approach.*

4   Universal Sets of Complex Unitary Gates

We now consider various universal sets of gates. The following gates will play by that an important role:

- $\sigma_x = X, \sigma_y = Y, \sigma_z = Z, K = \sigma_z^{\frac{1}{2}}, T = \sigma_z^{\frac{1}{4}}$. where $\sigma_0 = I, \sigma_1 = \sigma_x, \sigma_2 = \sigma_y$ and $\sigma_3 = \sigma_z$ are Pauli operators;
- $CNOT = \Lambda_1(\sigma_x)$, DCNOT, as well as TOFFOLI = TOF = $\Lambda_2(\sigma_x)$, where $DCNOT(x, y) = (y, x \oplus y)$;
- HADAMARD = $H = \frac{1}{2}(\sigma_x + \sigma_z)$, SWAP and $\sqrt{\text{SWAP}}$;
- so called elementary rotation gates

$$R_\alpha(\theta) = \cos\frac{\theta}{2} I - i\sin\frac{\theta}{2}\sigma_\alpha, \quad \text{for } \alpha \in \{x, y, z\}.$$

Observe that $\Lambda_1(\sigma_x) = (H \otimes I)\Lambda_1(\sigma_z)(H \otimes I)$. Therefore, gates $\Lambda_1(\sigma_x)$ and $\Lambda_1(\sigma_z)$ are locally equivalent. Observe also that for any real α,

$$\sigma_z^\alpha = \begin{pmatrix} 1 & 0 \\ 0 & e^{i\pi\alpha} \end{pmatrix} = \Lambda_0(e^{i\pi\alpha}).$$

The first universal gate was discovered by Deutsch (1989). This 3-qubit gate is a generalization of the Toffoli gate and has the form

$$U_D = \begin{pmatrix} 1 & & \mathbf{0} & & \\ & 1\,0 & 0 & 0 \\ \mathbf{0} & 0\,1 & 0 & 0 \\ & 0\,0 & i\cos\theta & \sin\theta \\ & 0\,0 & \sin\theta & i\cos\theta \end{pmatrix},$$

where θ is an irrational multiple of π.

Deutsch's result has been improved to construct two-qubit universal gates. For example, Barenco (1995) showed the universality of the following two-qubit gate

$$U_B = \begin{pmatrix} 1\,0 & 0 & 0 \\ 0\,1 & 0 & 0 \\ 0\,0 & e^{i\alpha}\cos\theta & -ie^{i(\alpha-\phi)}\sin\theta \\ 0\,0 & -ie^{i(\alpha+\phi)}\sin\theta & e^{i\alpha}\cos\theta \end{pmatrix},$$

where α, θ, ϕ are irrational multiples of π.

Shortly afterwards, Barenco et al. (1995), Deutsch et al. (1995) and Lloyd (1995) showed that almost any 2-qubit gate forms, with its inverse, a universal set of gates.

A simple example of a universal two-qubit gate, due to Tamir (2004), is the gate

$$\begin{pmatrix} R_y(\alpha) & \mathbf{0} \\ \mathbf{0} & R_z(\beta) \end{pmatrix}$$

where α, β and π are linearly independent over the rationals.

The above results are nice and interesting, but the underlying universal gates are still too complex for implementation. Implementation of one-qubit gates is not a problem for most of quantum technologies, but this is not the case with two qubits gates. The first really satisfactory result has been due to Barenco et al. (1995):

Theorem 1. *The CNOT gate with all one-qubit gates form a universal set.*

The proof is in principle simple and its basic idea will be discussed and analyzed in Section 9.

Theorem 1 can be easily improved into the following, also very nice, form:

Theorem 2. *The CNOT gate with elementary rotation gates form a universal set of gates.*

This follows from the well known result that any one-qubit gate can be expressed, up to a phase factor, as a triple product of two of the elementary R_α-gates.

It is natural to ask how it is possible that such a simple two qubit gate as the CNOT-gate plays a so prominent role and whether in Theorem 1 the CNOT cannot be replaced by some other gate. This problem was solved by Brylinskis (2001), as a significant improvement of Theorem 1, using the concept of an *entangling gate*, which is a gate that can produce entangled states when applied to unentangled states.

Theorem 3. *Any entangling two-qubit gate with all one-qubit gates (or only all elementary rotation gates) form a universal set.*

Remark 2. *Any two qubit gate that is not a product of two one-qubit gates and is not locally equivalent to the SWAP gate is entangling. An entangling gate is called a perfect entangler if it can map a product state into a maximally entangled state. CNOT and $\sqrt{SWAP}$ gates are perfect entanglers.*

The CNOT gate is an important primitive in optics-based quantum information processing. In the case of superconductor- and spin-based quantum computing, basic role is rather played by the gates iSWAP and $\sqrt{SWAP}$. These gates lead to results similar to those with CNOT concerning the complexity of circuits designed with them and with one-qubit gates. In general, for different technologies, different two qubit gates or sets of gates e^{iHt}, for different t, and a Hamiltonian H, are considered as elementary, and the circuit design task is then to decompose unitaries in terms of these elementary gates and one-qubit gates.

Let us now call a two-qubit gate U locally universal (l-universal) if this gate and all one-qubit gates form an f-universal set of gates. Clearly, any gate that is locally equivalent to an entangling gate is locally universal.

Example 1. *An interesting example of a locally universal gate is the gate*

$$R = \frac{1}{\sqrt{2}} \begin{pmatrix} 1 & 0 & 0 & 1 \\ 0 & 1 & -1 & 0 \\ 0 & 1 & 1 & 0 \\ -1 & 0 & 0 & 1 \end{pmatrix}$$

that transforms the standard basis into the Bell basis. This gate is also a solution of so called Yang-Baxter equation

$$(R \otimes I_2)(I_2 \otimes R)(R \otimes I_2) = (I_2 \otimes R)(R \otimes I_2)(I_2 \otimes R),$$

see Kauffman and Lomonoco (2004), which is a natural structure for consider-ing the topology of braids, knots and links.[5] This relates quantum topology and quantum computing.

The following are finite, interesting and important universal sets of gates:

- SHOR=$\{TOF, H, \sigma_z^{\frac{1}{2}}\}$, see Shor (1996).
- KITAEV = $\{\Lambda_1(\sigma_z^{\frac{1}{2}}), H\}$, see Kitaev (1997).
- KLZ1 = $\{CNOT, \Lambda_1(\sigma_z^{\frac{1}{2}}), \sigma_z^{\frac{1}{2}}\}$, see Knill et al. (1998).
- BMPRV=$\{CNOT, H, \sigma_z^{\frac{1}{4}}\}$, see Boykin et al. (1999).

Since sets KITAEV and SHOR are equivalent and gates in SHOR can be simulated by KLZ1-circuits, the universality of the set KLZ1 follows.

4.1   A Thin Border Between Non-universality and Universality

A border is thin between universality and non-universality in the case of quan-tum primitives. Indeed, Gottesman-Knill theorem (see Gottesman (1998)) tells, that quantum circuits with operators in so called *Clifford set* (or with Clifford operators) can be simulated on classical computers in polynomial time. However, if the set of Clifford operators is "slightly enlarged", by one of special (mixed) states, we get already an f-universal set of quantum primitives. We deal now with this quite surprising result, due to Bravyi and Kitaev (2004).

Let us first note that n-qubit circuits with operators from the Clifford set $\{CNOT, H, K = \sigma_z^{\frac{1}{2}}\}$ generate so called Clifford group $\mathcal{C}_n$ of operators that contains n-qubits Pauli group $\mathcal{P}_n$. (Pauli operators can be generated by Clifford operators. Indeed, $\sigma_z = K^2$, $\sigma_x = H\sigma_z H$.)

It has been shown that if the operators of the Clifford set are extended by the ability to create the state $|0\rangle$, by the measurement of eigenvalues of Pauli operators on qubits, and by one of the states

1. $|H\rangle = \cos\frac{\pi}{8}|0\rangle + \sin\frac{\pi}{8}|1\rangle$;
2. $|G\rangle = \cos\beta|0\rangle + e^{i\frac{\pi}{4}}\sin\beta|1\rangle$, where $\cos(2\beta) = \frac{1}{\sqrt{3}}$;
3. or by a one qubit mixed state ρ that is close, with respect to fidelity, to the H-type or G-type states – states that can be obtained from the state $|H\rangle$ or $|G\rangle$ by an operator of the Clifford group; then a universal set of primitives is again obtained.

---

[5] Let us remind that a knot is an embedding of a circle, taken up to topological equivalence and a link is an embedding of a collection of circles.

Let us first illustrate that states $|H\rangle$ and $|G\rangle$ are not "fallen from the heavens". Indeed $|H\rangle$ is an eigenvector of the operator H and the state $|G\rangle$ is an eigenvector of the operator $G = e^{\frac{i\pi}{4}} KH$. (This operator is actually again a "nice operator". Indeed, $G\sigma_x G^\dagger = \sigma_z$, $G\sigma_z G^\dagger = \sigma_y$ and $G\sigma_y G^\dagger = \sigma_x$.)

We explain now how to show the case (1); the case (2) can be shown in a similar way. The case (3) can be shown by demonstrating that from such a state ρ one can obtain one of the states $|H\rangle$ or $|G\rangle$ by a distillation algorithm that uses the state $|0\rangle$, Clifford operators and measurements of eigenvalues of Pauli operators.

It is easy to verify that $HK|H\rangle = e^{\frac{i\pi}{8}}|A_{-\pi/4}\rangle$, where

$$|A_\theta\rangle = \frac{1}{\sqrt{2}}(|0\rangle + e^{i\theta}|1\rangle),$$

Claim. If we have sufficiently many copies of the state

$$|A_\theta\rangle = \frac{1}{\sqrt{2}}(|0\rangle + e^{i\theta}|1\rangle),$$

then we can implement the operator $\Lambda_0(e^{i\theta}) = \Lambda(e^{i\theta})$ using the Clifford set operations and Pauli operator eigenvalues measurements. Indeed, an application of this operator on a qubit $|\psi\rangle$ can be done by the circuit shown in Figure 1. This circuit applies randomly one of the operators $\Lambda_0(e^{\pm i\theta})$ and it is known which one, due to the classical outcomes of measurements. By repeating the process several times we get, sooner or later, that the operator $\Lambda_0(e^{i\theta})$ is applied, and for $\theta = \frac{\pi}{4}$ we get the operator T that enlarges the Clifford set to a universal set of operations.

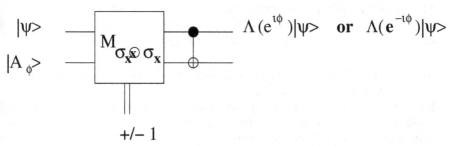

Fig. 1. Implementation of the operator $\Lambda_0(e^{\pm i\theta})$

5 Computationally Universal Sets of Gates

The concept of evolutionary universality can be seen as too strong from the computational point of view. Indeed, Bernstein and Vazirani (1993) have first shown that for having universal quantum computation it is sufficient to work with real amplitudes. The basic idea is simple. Since each complex number can be seen as a pair of real numbers, using one additional ancilla qubit quantum computation with complex amplitudes can be simulated by quantum computation with real amplitudes or by quantum circuits with real unitary gates. Moreover, it has been shown by Adleman et al. (1997) that the set of amplitudes that are really needed is very small, for example $A = \{0, \pm 3/5, \pm 4/5, \pm 1\}$, or $B = \{0, \pm 1/\sqrt{2}, \pm 1\}$, or $C = \{0, \pm \cos \theta, \pm \sin \theta, \pm 1\}$, for various θ.

All that led naturally to a new concept of universality, as already mentioned. A set of real gates $\mathcal{G}$ forms a **computationally universal set of gates**, if it can approximate with arbitrary precision any real unitary matrix. This holds, for example, if $\mathcal{G}$ generates a dense subgroup in the group of orthogonal matrices.

Rudolph and Grover (2002) have shown that a two-qubit real gate

$$G = \begin{pmatrix} 1 & 0 & 0 & 0 \\ 0 & 1 & 0 & 0 \\ 0 & 0 & \cos \phi & -\sin \phi \\ 0 & 0 & \sin \phi & \cos \phi \end{pmatrix},$$

with ϕ being an irrational multiple of π, is computationally universal.

The next task has been to find a small set of very simple real gates that would be computationally universal. The main candidates for such a set were for a long time Toffoli and Hadamard gates, but the proof of that was done only recently by Shi (2002), see also Aharonov (2003). This has been a surprising result since Hadamard and CNOT gates do not form a universal set of gates because computations with such gates can be efficiently simulated classically. Shi (2002) has also shown another surprising result. The CNOT gate and any one qubit gate that does not preserve the computational basis and is not the Hadamard gate form a computationally universal set of gates. Hence, it is not sufficient to add to the CNOT gate the Hadamard gate, but adding (almost) any other one-qubit gate is fine – this yields a computationally universal set of gates. This is a surprising result in spite of the fact that it follows easily from the results of Barenco et al. (1995) that the CNOT gate plus one one-qubit gate that rotates by an angle which is not a rational multiple of π form a computationally universal set of gates. Shi (2003) has actually also shown, in a very nontrivial way, that:

Theorem 4. – *The Toffoli gate and any one-qubit gate changing the computational basis form a computationally universal set of gates.*
– *The CNOT gate and any one-qubit gate such that its square does not preserve the computational basis form a universal set of gates.*

Shi's results are also interesting from several other points of view. Since the Toffoli gate is universal for classical reversible computing, Shi's result means that the full power of quantum computation is obtained by adding just the Hadamard gate.

Remark 3. *There are several ways to see what kind of power the Hadamard gate represents. On one side, the Hadamard gate is a simple form of the Fourier* *transform, so one can say that, in some sense, quantum Fourier transform is what distinguishes classical and quantum computing. On the other hand, the Hadamard gate can be seen as performing a random coin tossing, so one can say that it is just quantum random bit tossing that needs to be added to get quantum out of the classical computation.*

6   Fault-Tolerantly Universal Sets Gates

Informally, a fault-tolerantly universal set of gates is a universal set of gates such that all gates of the set can operate well also in a noisy environment. A more formal requirement is that there exists a quantum error correcting code such that all gates of the set can be performed on logical qubits without a need to decode them first and in such a way that the propagation of single-qubit errors to other qubits in the same codeword is excluded.

The above requirements impose strong restrictions on operations that can be applied and also on the type of error correcting codes that can be used. Because of that it has not been easy to design a universal set of fault-tolerant gates. One reason has been the fact that the first universal sets of gates contained at least one "irrational" gate, that is a gate performing, in some sense, a rotation by an irrational multiple of π: and for such gates a direct fault-tolerant implementation is impossible. On the other hand, some codes are not suitable to carry on their logical qubits a universal set of gates in a fault-tolerant way. For example, if Steane's $(7, 1, 3)$ code is used (Steane, 1995), one can implement a nice set of operations

$$CNOT, H, \sigma_z^{\frac{1}{2}},$$

but this set of operations is not universal, as discussed above.

The following sets of gates have been shown to be fault-tolerantly universal:

1. SHOR= $\{T, H, \sigma_z^{\frac{1}{2}}\}$, due to Shor (1996).
2. KITAEV = $\{\Lambda_1(\sigma_z^{\frac{1}{2}}), H\}$, due to Kitaev (1997).
3. BMPRV = $\{CNOT, H, \sigma_z^{\frac{1}{4}}\}$, due to Boykin et al. (1999).

Shor (1996) has shown fault-tolerance of the SHOR basis, which was quite non-trivial, especially concerning the Toffoli gate. Universality of the SHOR basis follows from the fact that it is equivalent to KITAEV basis, see Boykin et al. (1999), which was shown to be universal by Kitaev (1997).

It is not easy to show fault-tolerance of some fault-tolerant gates. One general method has been developed by Boykin et al. (1999) and can be used, for example, to show the fault-tolerance of such non-trivial gates as the Toffoli gate and $\sigma_z^{\frac{1}{4}}$.'

Remark 4. *The design of fault-tolerant gates is not the only way to fight decoherence. Another important way is to search for technologies that allow to implement some gates in a fault-tolerant way, due to special properties of some* *physical systems. For example, this could be done with sufficient control of any-ones - quasi-particles with unusual statistics. Special technologies could, in prin-ciple, realize some sets of operations exactly. However, it is not clear whether some universal set of operations can be realized exactly. It seems more realistic to consider important, but not fully universal sets of operations that can be realized exactly. It is then crucial to find out how many additional, but possibly faulty operations needs to be added to obtain a universal set of quantum computational primitives. A very specific version of this problem is discussed in Section 3.*

7   Encoded Universality of Heisenberg Exchange Interactions

Encoded universality refers to the capability to generate, or to approximate, all unitary matrices on a subspace of a Hilbert space created by some logical qubits.

This new concept of universality, that seems to be connected with the use of quantum error-correcting codes, is related also to the attempts to find universal information processing primitives that are yet simpler than unitary operations and physically even more rudimentary.

One such primitive is Heisenberg physical nearest neighbor exchange interaction. This interaction is not universal for quantum computation in general, but, surprisingly, it can be universal on properly encoded logical qubits.

An example is the following encoding of the standard basis states of qubits by a row of 8 qubits, where the first four qubits encode the first basis state $|0_L\rangle$, and the next four qubits encode the second basis state $|1\rangle$, see Hsieh et al. (2003):

$$|0_L\rangle = \frac{1}{2}(|01\rangle - |10\rangle) \otimes (|01\rangle - |10\rangle)$$

$$|1_L\rangle = \frac{1}{\sqrt{3}}(|11\rangle \otimes |00\rangle - (\frac{1}{\sqrt{2}}(|01\rangle + |10\rangle) \otimes (\frac{1}{\sqrt{2}}(|01\rangle + |10\rangle)) + |00\rangle \otimes |11\rangle).$$

The exchange of the first two (or the last two) qubits of each logical qubit realizes the operation $|0_L\rangle \rightarrow -|0_L\rangle$ and $|1_L\rangle \rightarrow |1_L\rangle$, therefore, up to a phase factor, it actually encodes the σ_z operation on logical qubits. On this basis one can show that the Hamiltonian for the $\sigma_z^{1/4}$ operation, realized by the nearest neighbors interaction is

$$e^{i\frac{\pi}{8}E_{1,2}},$$

where

$$E_{i,i+1} = \frac{1}{2}(\sigma_{x,i} \otimes \sigma_{x,i+1} + \sigma_{y,i} \otimes \sigma_{y,i+1} + \sigma_{z,i} \otimes \sigma_{z,i+1} + I \otimes I) = \text{SWAP-gate}$$

is the interaction between the ith and $(i + 1)$th qubit. With this notation, an exact encoded Hadamard gate can be obtained using as Hamiltonian $H = e^{it_1 E_{1,2}} e^{it_2 E_{2,3}} e^{it_1 E_{1,2}}$, where $t_1 = \frac{1}{2}\arcsin\sqrt{\frac{2}{3}}$ and $t_2 = \arccos\sqrt{\frac{1}{3}}$. To obtain an encoded realization of the CNOT gate, numerical methods have been used (with

27 parallel nearest neighbor exchange interactions, or 50 serial gates). As a consequence, a single two qubit exchange interaction forms a universal set (therefore no single qubit operations are needed)[6] with respect to encoded universality.

8   Efficiency of Universal Sets of Quantum Primitives

The need to minimize the impact of decoherence and to minimize the size of quantum circuits as well as their depth (computation time) rises as a natural and important issue with respect to the efficiency of different universal sets of quantum primitives.

The Solovay-Kitaev theorem, Kitaev et al. (2002), implies that for evolutionary and computational universality, and any fixed and sufficiently large k, the number of gates from a universal set that are required to approximate any unitary matrix on k qubits within ε, grows only in $polylog(\frac{1}{\varepsilon})$ steps. As a consequence, it is not costly to replace one universal basis by another one – it requires only poly-logarithmic overhead. (However, it is not clear how far this holds for other concepts of universality.) This implies that any gate from one finite universal set can be approximated with precision ε using $polylog(\frac{1}{\varepsilon})$ gates from other finite universal set of gates.

Of course, the above results are asymptotic and as such they have their limits.

9   Compilation of Unitaries

In principle, any general purpose compiler for unitaries will consist of decomposition and optimization methods that are technology independent and of methods that take into the account the specifics of a given technology. Since there is no clear candidate on a scalable and reliable technology yet, the emphasis in the research is so far on technology independent decomposition and optimization methods as discussed below.

Two very basic questions concerning the decomposition of n-qubit unitaries into one-and two-qubit gates are the following:

- What is the total number of one-and two-qubits gates needed to decompose an arbitrary n qubit unitary operation, for an arbitrary n?
- What is the total number of CNOT gates (or of some other entangling two qubit gates) needed to decompose an arbitrary n qubit unitary, for an arbitrary n?

Barenco et al. (1995) have shown that any n qubit gate can be realized by $\mathcal{O}(n^3 4^n)$ CNOT and one-qubit gates. This has been improved, step by step, to $\mathcal{O}(n^2 4^n)$, $\mathcal{O}(n4^n)$ and, finally, by Vartiainen et al. (2003) and by Möttönen et al. (2004) to $\mathcal{O}(4^n)$ gates. More exactly, to $4^n - 2^{n+1}$ the CNOT gates and 4^n

---

[6] Heisenberg interaction is not only very simple, it is also a strong interaction and therefore it should permit very fast implementation, in the GHZ range, as several implementation proposals have been suggesting, see Hsieh et al. (2003).

one-qubit gates. Since an n qubit unitary is represented by a matrix with 4^n elements the above result is clearly asymptotically tight. Concerning the CNOT gates only, the best known upper bound is $\mathcal{O}(4^n)$ due to Vartiainen et al. (2003) and the best lower bound, due to Shende et al. (2003), is $\lceil (4^n - 3n - 1)/4 \rceil$ (the same lower bound holds also for some other gates.)

The basic idea for decomposition is borrowed from the QR-decomposition in linear algebra using Given's rotation matrices $G_{i,j,k}$ that are "two-level matrices" which operate non-trivially only on the j-th and k-th basis vectors, and nullify the elements on the i-th column and k-th row. Then the overall decomposition of a unitary matrix U into a unit matrix has the form

$$\left(\prod_{i=2^n-1}^{1} \prod_{j=i+1}^{2^n} G_{i,j,j-1} \right) U = I.$$

Each two-level matrix can then be implemented using $\Lambda_{n-1}(V)$ and $\Lambda_{n-1}(\text{NOT})$ matrices, where V is a unitary 2×2 matrix and $\Lambda_k(V)$ denotes a matrix with k control bits that control the application of the matrix V. A $\Lambda_{n-1}(V)$ matrix can be implemented with $\mathcal{O}(n^2)$ one- and two-qubit gates. Moreover, $\mathcal{O}(n)$ of $\Lambda_{n-1}(\text{NOT})$ gates are needed between each two $\Lambda_{n-1}(V)$ gates and this leads to the total of $\mathcal{O}(n^3 4^n)$ gates. An improvement to $\mathcal{O}(4^n)$ has been achieved by using Gray-code ordering of the basis states.

An optimization method for quantum circuits, which is based on the existence of the above decomposition, and which concentrates on an optimization of the $\Lambda_{n-1}(\text{NOT})$-gates, is due to Aho and Svore (2003).

Fortunately, in some important cases, like the Quantum Fourier Transform, an n-qubit unitary can be realized by a circuit with a polynomial number (in n) of one- and two-qubit gates.

The decomposition into one- and two-qubit gates is not always a necessity. In some technologies, see Wang et al. (2000), $\Lambda_{n-1}(V)$ gates can be realized in a straightforward way. It may also be case, with some technologies, that some other qubit gates can be easily implemented. All that makes the problem of optimal decomposition of unitary matrices very complex.

10 Optimal Design of Quantum Circuits for Two- and Three-Qubit Gates

The problem of decomposing unitary gates into one and two qubit gates has been consider in general by Tucci (1999). However, his decomposition can be far from optimal for two-qubit unitaries. This case is important because it is very likely that circuits implemented in the near future will use only one type of two-qubit gate and that the number of gates used for a given task will be crucial, especially for experimentalists.

Another general and recursive decomposition method of any unitary matrix into one- and two-qubit unitary matrices, based on the Cartan decomposition of the Lie group $su(2^n)$, is due to Khaneja and Glaser (2000).

10.1 Optimal Universal Circuit Schemes for Two-Qubit Gates

There has been significant progress recently toward optimal (in some sense) realization of two-qubit gates. The main problems in this area can be formulated as follows:[7]

- Given an entangling two-qubit gate G, what is the smallest number of gates G and of one-qubit (elementary) gates of a circuit which would implement an arbitrary given two-qubit gate? (Find the best possible upper and lower bounds.)
- Given an entangling gate G, find the smallest possible (with respect to the number of gates G and one-qubit (elementary) gates), universal circuit for implementation of any two-qubit gate?
- Solve the above problems for special classes of entangling gates, or for specific entangling gates, as CNOT, or double CNOT (DCNOT), or $\sqrt{\text{SWAP}}$.
- If G is an entangling two-qubit gate and n_G is the minimal number of gates G needed to realize (with one-qubit gates) any two-qubit gate, then, for any $1 \le k \le n_G$, determine necessary and sufficient conditions for a two-qubit gate to be implementable by a circuit with k gates G (and one-qubit gates).

Solutions to these problems should contribute to the realization of few qubits processors because they provide estimations of the effort and of the overhead for experimentalists.

Consider first the case that a two-qubit Controlled-U gate G is given for a one-qubit operation U. Since

$$U = e^{i(n_x \sigma_x + n_y \sigma_y + n_z \sigma_z)}$$

the controlled-U operation U_c can be written as

$$U_c = (I \otimes e^{\frac{-i\gamma}{2}\sigma_z} U_1^\dagger) e^{\frac{i\gamma}{2}\sigma_z \otimes \sigma_z} (I \otimes U_1)$$

for some one-qubit unitary U_1 and $\gamma = \sqrt{n_x^2 + n_y^2 + n_z^2}$. Without loss of generality, we can therefore view any Controlled-U gate as having the form $U_c = e^{\frac{i\gamma}{2}\sigma_z \otimes \sigma_z}$. It has been shown by Zhang et al. (2003) that the upper bound for the number of such controlled gates is $\lceil \frac{3\pi}{2\gamma} \rceil$. Zhang et al. (2003) also provide a procedure for designing near optimal circuits for any two-qubit gate with U_c being the only two-qubit gate used. Since for any entangling gate G two applications of the gate are sufficient to design a Controlled-U gate for any one-qubit gate U, the above mentioned quantum circuit design procedure provides good solutions for entangling gates also.

---

[7] For most technologies, and compared with one-qubit gates, two-qubit gates are much more difficult to implement and much more costly (take longer time, require complicated manipulations and exhibit stronger decoherence). Therefore, concerning the complexity of one- and two-qubits circuits, only the number of two-qubit gates actually counts.

However, the key problem is how many CNOT and one-qubit gates are necessary and sufficient to implement any two-qubit gate. Moreover, since each one-qubit gate can be expressed as a composition of any two of the elementary rotation gates R_x, R_y and R_z, it is of interest, and actually of high practical importance, to determine the minimal number of (elementary) gates R_x, R_y, R_z and $CNOT$ gates needed to implement an arbitrary two-qubit gate.

Progress in the exploration of efficient realizations of two-qubit gates has been remarkable in the last few years. We discuss here only the best outcomes, so far, due to Vidal and Dawson (2003), Shende et al. (2003) and Vatan and Williams (2003). (References to earlier results can be found in their papers.) The main result is that 3 CNOT gates and 10 one-qubit and CNOT gates in total are sufficient to realize any two qubit gate, and that, in general, 3 CNOT gates and 9 gates in total are necessary. Moreover, each two-qubit gate can be realized using 3 CNOT gates, in a total of 18 gates from the set containing the CNOT gate and any two or the three gates from the set $\{R_x, R_y, R_z\}$ – with the exception of gates R_y and R_z, where 19 is the upper bound. The above result is optimal (see Shende et al., 2003)) if temporary storage is not allowed (because of cost). The universal two-qubit circuit scheme with three CNOT gates and 10 basic gates, or 18 gates from the set $\{CNOT, R_y, R_z\}$ is shown in Figure 2. Moreover, for gates from $SO(4)$ only 12 gates R_y, R_z are needed (see Vatan and Williams, 2003). For all of these gate libraries the optimal solution can be achieved by a single universal circuit.

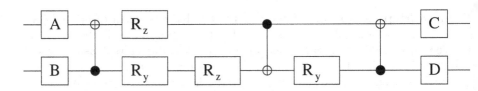

Fig. 2. A universal 2-qubit circuit

With the following criterion, due to Shende et al. (2003), it is possible to determine the number of CNOT gates needed to realize a two-qubit gate with the help of one-qubit operations:

Theorem 5. *Let*

$$E = \begin{pmatrix} 0 & 0 & 0 & -1 \\ 0 & 0 & -1 & 0 \\ 0 & -1 & 0 & 0 \\ 1 & 0 & 0 & 0 \end{pmatrix},$$

and for any matrix $U \in SU(4)$ let $\gamma(U) = UEUE$. Then

1. *U can be realized by a circuit with no CNOT gate if and only if $\gamma(U) = I$.*
2. *U can be realized by a circuit with one CNOT gate if and only if $\mathrm{Tr}(\gamma(U)) = 0$ and $\gamma(U)^2 = -I$.*
3. *U can be realized using two CNOT gates if and only if $\mathrm{Tr}(U))$ is real.*

Using this criterion one can show, for example, that the SWAP gate cannot be realized by a circuit with two CNOT gates and one-qubit gates.

The above results remain valid if the CNOT gate is replaced by any other maximally entangling gate, for example by the iSWAP- or $\sqrt{\text{SWAP}}$-gate. However, the situation is different (see Vatan and Williams (2004)), if not maximally entangling gates are used. In such a case it may happen that up to 6 two-qubit gates are needed.

Remark 5. *The above results are based on the following two important and interesting decompositions:*

- *Any two-qubit unitary matrix U has a unique decomposition*

$$U = (A_1 \otimes B_1)e^{i(\theta_x X \otimes X + \theta_y Y \otimes Y + \theta_z Z \otimes Z)}(A_2 \otimes B_2),$$

  where $\frac{\pi}{4} \geq \theta_x \geq \theta_y \geq |\theta_z|$.
- *In the magic basis any $U \in SO(4)$ is an element of $SU(2) \otimes SU(2)$ (see, for example, Vatan and Williams (2003)).*

Open problem 1. *(1) Are there two two-qubit gates G_1 and G_2 such that any two-qubit gate can be implemented by a circuit with one-qubit gates and at most two of the gates G_1 and G_2? (b) Design an algorithm that constructs, for any two-qubit entangling gate U a minimal universal circuit scheme, with respect to the number of U-gates, that uses U-gates as the only kind of two qubit gate.*

Going a Step Down. Closely related to the above problem of optimal implementation of two qubit gates using a fixed two qubit gate, is the following problem: what is the minimal time to realize a two-qubit unitary using a fixed two-qubit entangling Hamiltonian and (fast) one-qubit unitaries?

It has been shown by Childs et al. (2003) (their paper contains further references), that if $U = e^{iH_0}$ is a two qubit unitary and H a two-qubit entangling Hamiltonian, then the minimal time required to simulate U using H and fast one-qubit unitaries is the minimal t such that there exists a vector $\bar{m}$ of integers satisfying

$$\lambda(H_1) + \pi\bar{m} \prec \frac{\lambda(H + \tilde{H})}{2}t,$$

where $\lambda(A)$ denotes the vector of eigenvalues of a Hermitian matrix A and $\tilde{H} = (Y \otimes Y)H^T(Y \otimes Y)$ (and $\prec$ is the majorization of vectors relation).

10.2   Universal Circuit Schemes for 3-Qubits Gates

The optimal realization of 3-qubits gates using a fixed two-qubit gate and one-qubit gates seems much more complex, but at the same time much more important. Indeed, compared with two-qubit gates, 3-qubit gates have quite specific properties and exhibit different phenomena. For example, the classification of entanglement and non-locality in the case of 3-qubit states is much more complex and requires special investigations. It is also believed that the 3-qubit case allows to get a deeper insight into the difference between non-locality and entanglement.

A universal circuit scheme with 40 the CNOT gates and 98 one-qubit elementary gates, R_y and R_z, due to Vatan and Williams (2004), is, so far, the most efficient general way of implementing 3 qubit gates. This circuit scheme has the form

$$(T_1 \otimes O_1)N_1(T_2 \otimes O_2)M(T_3 \otimes O_3)N_2(T_4 \otimes O_4),$$

where O_i and T_i are one- and two-qubit gates and the two special 3 qubit gates N and M are defined as follows:

$$N(a,b,c) = e^{i(a\sigma_x\sigma_x+b\sigma_y\sigma_y+c\sigma_z\sigma_z)},$$

$$M(a,b,c) = e^{i(a\sigma_x\sigma_x+b\sigma_y\sigma_y+c\sigma_z\sigma_z+dII)},$$

These two gates have simple implementations using 10 and 11 CNOT gates, respectively.

The above universal 3-qubit circuit has also been obtained using the general, already mentioned. decomposition method of Khaneja and Glaser (2001) and therefore it is likely that a more efficient universal circuit can be found. However, this is still an open problem.

Given a two qubit gate G, there is a way to find out for any other two qubit gate U whether G and U are locally equivalent and therefore a circuit for U can be designed that uses G only once (and one-qubit gates). Indeed, Makhlin (2000) has showed a set of three real polynomial invariants for two qubits gates that fully characterize the entangling properties of gates in such a way that two two-qubit gates G and U are locally equivalent if and only if they have the same values for these invariants. In addition, given G, there exists a procedure for designing a circuit for U with only one instance of the gate G.

11   Projective Measurements as Universal Primitives

Surprisingly, projective measurements alone are sufficient to create universal sets of quantum primitives in the sense that they can be used to simulate universal sets of unitary primitives, in a special way. More precisely, the following results have been obtained:

- Raussendorf and Briegel (2000) have shown that one-qubit projective measurements applied to a special fixed *cluster state* form a universal set of quantum primitives.
- Nielsen (2001) has shown that 4-qubit measurements are sufficient to simulate all unitary operations.
- Leung (2001) has shown that (a) almost any maximally entangling 4 qubit measurement is universal; (b) 2-qubit measurements are sufficient to simulate all 2 qubit operations.

- Leung (2003) has shown that there is a finite set of four 2-qubit measurements that can realize all 2-qubit unitary operations, if four ancilla qubits are available.
- Perdrix (2004) has shown that a set of measurements consisting of one two-qubit and three one-qubit measurements forms a universal set of quantum measurements if a one-qubit ancilla as an additional resource is available. (These resources are clearly minimal.)

The above results imply, very surprisingly, and in contrary to the widespread belief, that unitary dynamics is not necessary for universal quantum computation and that projective measurements are sufficient. Actually, Bell measurements play an important role and the results obtained are actually based upon the information processing power of quantum teleportation.

Minimal Projective Measurement Resources. In this section the universality of a simple set of measurements will be illustrated. Two basic facts that will be used are the following: (a) The operations CNOT, H and $T = \sigma_z^{1/4}$, form a universal set of unitary operations; (b) teleportation and its simplified version, called state transfer, can be used to perform unitary operations indirectly.

The basic teleportation scheme is shown in Figure 3a,[8] where B stands for the Bell measurement, which transforms the initial state

$$|\phi\rangle|\text{EPR}\rangle = \frac{1}{2} \sum_{i=0}^{3} |\Phi_i\rangle \otimes (\sigma_i|\phi\rangle).$$

where $|\Phi_0\rangle = |\Phi^+\rangle$, $|\Phi_1\rangle = |\Psi^+\rangle$, $|\Phi_2\rangle = |\Psi^-\rangle$ and $|\Phi_3\rangle = |\Phi^-\rangle$ are the Bell states. into a state $|\Phi_i\rangle \otimes \sigma_i|\Phi_i\rangle$, while information about i is obtained in the classical world, so that the correction σ_i can be applied.

A modification of the basic teleportation scheme, see Figure 3b, allows to perform indirectly, or remotely, (and randomly) any unitary 1-qubit operation U. Indeed, if the Bell measurement is applied to the first two qubits of the initial state $(I \otimes U)|\text{EPR}\rangle$, then, with probability $\frac{1}{4}$, "Bob's particle" is in the state $U|\phi\rangle$ and with probability $\frac{3}{4}$ a correction $U\sigma_j U^\dagger$ needs to be performed, which can be done, recursively, again indirectly, until no more corrections are needed. This way one can expect on average 4 attempts until such a process succeeds.

A small modification of the above scheme is shown in Figure 3c. It is based on the fact that the states $(I \otimes U\sigma_j)|\text{EPR}\rangle$ form an orthonormal basis and therefore a single measurement of the first two qubits of the state $|\phi\rangle|\text{EPR}\rangle$, with respect to such a basis (denoted here as "B_U"), creates an appropriate state – and it does not matter into which of the above states the measurement collapses.

---

[8] In this figure, double-lined boxes denote measurements, simple lines denote qubits and bold lines denote measurement outcomes. Gates connected to a measurement box by a bold line are conditioned on the measurement outcomes. If this is the case, that is if the applied sequence of gates applied depends on results of measurements, we speak of *adaptive quantum computations*.

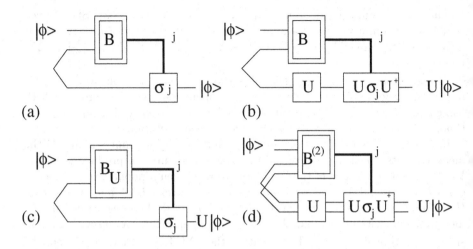

Fig. 3. Teleportation of quantum operations

Another variation of the above modified teleportation scheme, see Figure 3d, allows, in a similar way, to perform indirectly any 2-qubit unitary operation U. In this case the target state is

$$I_2 \otimes U(\frac{1}{2} \sum_{i,j=0}^{1} |ijij\rangle)$$

Since with the CNOT is the only two-qubit unitary operation we can have universal computation, the key point is to show that the target 4-qubit state

$$|\phi_{in}\rangle = I_2 \otimes \text{CNOT}(\sum_{i,j=0}^{1} |ijij\rangle = \frac{1}{2}(|0000\rangle + |0101\rangle + |1011\rangle + |1110\rangle)),$$

can be created using 2-qubit measurements only. This can be done, by Leung (2003), as follows:

1. With one- and two-qubit measurements create the state

$$|\phi_1\rangle = \frac{1}{2}(|0\rangle + |1\rangle) \otimes |0\rangle \otimes (|00\rangle + |11\rangle).$$

2. Apply to $|\phi_1\rangle$ a measurement with two projectors $P_+ = |\Phi_0\rangle\langle\Phi_0| + |\Phi_1\rangle\langle\Phi_1|$ and $P_- = |\Phi_2\rangle\langle\Phi_2| + |\Phi_3\rangle\langle\Phi_3|$. If the projector P_+ is chosen, the resulting state is

$$|\phi_2\rangle = \frac{1}{2\sqrt{2}}(|0\rangle + |1\rangle)(|000\rangle + |011\rangle + |101\rangle + |110\rangle).$$

3. Finally, measure the parity of the 1st and 3rd qubits, In case the outcome is even, the resulting state is the required target state $|\phi_{in}\rangle$.

The above results have been improved first by Leung (2003). She showed that if a four qubit ancilla is available then the following set of four two qubit measurements, defined by four observables $X \otimes X$, $Z \otimes Z$, $X \otimes Z$ and $\frac{1}{\sqrt{2}}(X+Y)$, is universal.

Leung's results, based on a generalized form of teleportation, were recently improved, to get minimal ancilla and measurement resources, by Perdrix (2004), using another version of teleportation, called *state transfer*, which captures only this part of teleportation which is necessary for computation.

In a similar way as above, for any one qubit unitary mappings U and V the measurements specified by observables shown in Figure 4a, produce the state $V\sigma U^{\dagger}|\phi\rangle$, where σ is one of the Pauli operators, chosen randomly. In the special cases $U = H$ and $V = I$, the output has the form $\sigma H|\phi\rangle$ and in the case $U = T$ and $V = H$ the output has the form $\sigma HT|\phi\rangle$. In these two cases only the measurements with observables X, Z, $\frac{1}{\sqrt{2}}(X+Y)$ and $X \otimes Z$ are used. Figure 4b shows how to realize, up to a Pauli matrix, the CNOT operation. Again, no new observable is used. Since the set of unitaries $\{CNOT, H, HT\}$ is universal, the above set of observables/measurements is universal, and they require only one ancilla qubit.

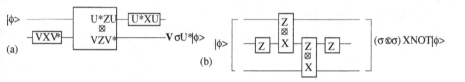

Fig. 4. Two schemes for providing universal state transfer

Measurements are basic operations also in a new measurement based model of quantum Turing machines, see Perdrix and Jorrand (2004). These Turing machines operate on tapes of qubits and their (classical) transition functions map

states × outcomes of measurements → states × observables × movements of heads

The minimal resources needed for universality are one-qubit ancilla and a similar set of observables as in the case of quantum measurements based circuits.

12   Universal Randomized Quantum Circuits

In classical computing, for every m there is a universal circuit C_m, with $m + 2^m$ inputs, such that if the first m inputs get a binary (input) string i and the remaining inputs a 2^m bit a specification p_f of a m-nary Boolean function $f : \{0,1\}^m \rightarrow \{0,1\}^m$, then the first m output bits contain $f(i)$, see Figure 5a.

A natural question is whether we can also have universal quantum circuits in the above sense. Since the number of $m \times m$ unitary operations is infinite for any

m, it seems to be intuitively clear that such universal circuits cannot exist. More exactly, it seems that given an m we cannot have a quantum circuit as shown in Figure 5b, that would realize a unitary mapping D_m, which would take as an input an m qubit state $|\phi\rangle$ and for any unitary operation U on m qubits a state $|s_U\rangle$, for encoding U, and would produce $U|\phi\rangle$ on the first m qubits. Actually, we can show that even a stronger claim holds (see Nielsen and Chuang, 1997).

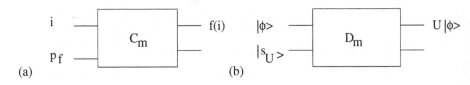

(a)   (b)

Fig. 5. Universal classical and quantum circuits – basic schemes

Indeed, let us assume that such a circuit D_m is universal up to a global phase at least for a set S of n distinct $m \times m$ unitary mappings. Hence, for two different unitary operations U_1 and U_2 from S the following should hold, for any m-qubit state $|\phi\rangle$:

$$D_m(|\phi\rangle|s_{U_1}\rangle) = |U_1|\phi\rangle\rangle|g_1\rangle$$

$$D_m(|\phi\rangle|s_{U_2}\rangle) = |U_2|\phi\rangle\rangle|g_2\rangle$$

Since unitary operations preserve scalar product we have

$$\langle s_{U_1}|s_{U_2}\rangle = \langle g_1|g_2\rangle\langle\phi|U_2^\dagger U_1|\phi\rangle.$$

Since the left side of the above equation does not depend on $|\phi\rangle$, if $\langle g_1|g_2\rangle \neq 0$, it has to hold $U_2^\dagger U_1 = \gamma I$ for some complex γ. However, this contradicts the assumption that unitary transformations from S are different up to a global phase factor. Consequently, it has to hold that $\langle g_1|g_2\rangle = 0$ and also $\langle s_{U_1}|s_{U_2}\rangle = 0$. Therefore, the program-states $|s_{U_1}\rangle$ and $|s_{U_2}\rangle$ have to be mutually orthogonal. Hence, the circuit D_m can be a universal circuit for a set S of n m-qubit unitary transformations only if it has at least $\log n$ qubits to encode the n unitary transformations from S.

On the other hand, quite surprisingly, we can have, for any m, a randomized universal quantum device, called a *universal programmable array* R_m, that can implement, with a known positive (but exponentially small) probability, any m-qubit gate.

The basic idea behind R_m is actually another simple modification of the teleportation idea, similar to that of Section 11. Let us illustrate the idea for $m = 1$. The universal quantum device R_1 is shown in Figure 6.

R_1 has three-qubit inputs and outputs. The first input qubit is in an arbitrary state $|q\rangle$ and, for any one-qubit unitary transformation U the other two qubits

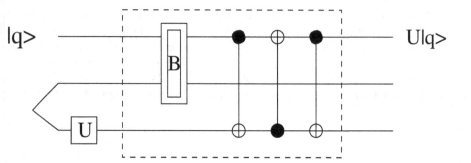

Fig. 6. The universal randomized quantum device R_1

are in the state $(I \otimes U)|\mathrm{EPR}\rangle$. It is now easy to verify that the total input state can be expressed as follows:

$$|q\rangle \otimes (I \otimes U)|\Phi_0\rangle = \frac{1}{2}\sum_{i=0}^{3}|\Phi_i\rangle\sigma_i U|q\rangle.$$

If the Bell measurement is now performed on the first two qubits, then with probability $\frac{1}{4}$ the last qubit will be in one of the states $\sigma_i U|q\rangle$, and it will be clear for which i this has happened. The last three CNOT gates realize the SWAP operation and make the first output to be in the state $U|q\rangle$.

A generalization to a universal circuit for unitary transformations for any m qubit states is quite straightforward: $2m$ ancilla qubits are needed and a unitary transformation U will be encoded by the state

$$|s_U\rangle = (I_m \otimes U)\bigotimes_{i=1}^{m}|\Phi_{i,m+i}^{+}\rangle,$$

where $|\Phi_{i,m+i}^{+}\rangle$ is the state $|\Phi^{+}\rangle$ shared between the ith and $(m+i)$th qubit. Bell measurements on the first $2m$ qubits provide $U|\phi\rangle$ on the last m qubits.

13   Interactions as Primitives

From the Hilbert space view of quantum mechanics, unitaries, entanglement and measurements are primitives. Going deeper into physics one sees interactions between quantum systems according to some Hamiltonians as the most fundamental and universal primitives and resources. An important problem of QIPC then becomes that of finding efficient ways to convert interactions as primitives and resources into other resources like unitaries, entanglement and communication channels for quantum state transfers, with or without the help of local (unitary) operations. This problem has been recently explored with interesting outcomes.

One such primitive is Heisenberg nearest neighbor exchange interaction. This interaction is not universal for quantum computation per se, but, surprisingly, it can become universal on properly encoded logical qubits as already discussed in Section 7.

Power of nearest neighbor interactions in a chain of spins with nearest neighbor XY coupling, that is with the Hamiltonian

$$H = \sum_{i,j} \frac{\omega_{ij}}{2} \left(\sigma_{x,i} \otimes \sigma_{x,i} + \sigma_{y,j} \otimes \sigma_{y,j} \right)$$

has been explored, for example, by Yung et al. (2003). In the case of three spins, interactions can realize perfect transfer of quantum state from the first to third spin, can create one ebit among them and can realize simultaneous communication of one bit in both directions. The creation of entanglement is possible also in case of longer chains of spins. In the case of quantum state transfers this can be done only approximately in general.

References

1. L.M. Adleman, J. DeMarrais, and M.-D.A. Huang, Quantum computability, SIAM Journal of Computing, 26, 5, 1524–1540, 1997.

2. D. Aharonov, A simple proof that Toffoli and Hadamard are quantum universal, quant-ph/0301040, 2003.

3. A.V. Aho and K.M. Svore, Computing quantum circuits using the palindrom transformation, quant-ph/0311008, 2003.

4. A. Barenco, A universal two-bit gate for quantum computation, Proceedings of Royal Society London A, 449, 679–683, 1995.

5. A. Barenco, C.H. Bennett, R. Cleve, D.P. DiVincenzo, N. Margolus, P.W. Shor, T. Sleator, J.A. Smolin, and H. Weinfurter, Elementary gates of quantum computation, Physical Review A, 52, 5, 3457–3467, 1995.

6. E. Bernstein and U. Vazirani, Quantum complexity theory, Proceedings of 25th ACM STOC, 11–20, 1993.

7. P.O. Boykin, T. Mor, M. Pulver, V. Roychowdhury, and F. Vatan, On universal and fault-tolerant quantum computing, quant-ph/9906054, 1999.

8. S. Bravyi and A. Kitaev, Universal quantum computation based on a magic state distillation, quant-ph/0403025, 2004.

9. M.J. Bremner, J.L. Dodd, M.A. Nielsen, and D. Bacon, Fungible dynamics: there are only two types of entangling multiple-qubit interactions, quant-ph/0307148, 2003.

10. J.-L. Brylinski and R. Brylinski, Universal quantum gates, quant-ph/0108062, 2001.

11. A.M. Childs, H.L. Haselgrove, and M.A. Nielsen, Lower bounds on the complexity of simulating quantum gates, quant-ph/0307190, 2003.

12. D. Deutsch, Quantum computational networks, Proceedings of Royal Society of London A, 425, 73–90, 1989.

13. D. Deutsch, A. Barenco, and A.K. Ekert, Universality in quantum computation, Proceedings of Royal Society London A, 449, 669–677, 1995.

14. D. Gottesman, The Heisenberg representation of quantum computers, quant-ph/9807006, 1998.

15. J. Gruska, Quantum Computing, McGraw-Hill, 1999-2004. See also additions and updatings of the book on http://www.mcgraw-hill.co.uk/gruska.

16. J. Gruska, Quantum entanglement as a new quantum information processing resource, New Generation Computing, 21, 279–295, 2003.
17. M. Hsieh, J. Kempe, S. Myrgren, and K. B. Whaley, An explicit universal gate-set for exchange-only quantum computation, quant-ph/0309002, 2003.
18. K. Iwama and S. Yamashita, Transformation rules for CNOT-based quantum circuits and their applications, New generation computing, 21, 4, 297–318, 2003.
19. L.H. Kauffman and S.J. Lomonaco, Braiding operators are universal quantum gates, quant-ph/0401090, 2004.
20. N. Khaneja and S.J. Glaser, Cartan decomposition of $SU(2^n)$, constructive controllability of spin systems and universal quantum computing, quant-ph/0010100, 2000.
21. A.Y. Kitaev, A.H. Shen, and M.N. Vyalyi, Classical and quantum computation, American Mathematical Society, 2002.
22. A. Kitaev, Quantum Computations: Algorithms and Error Correction, Russian Mathematical Surwey, 52, 1191–1249, 1997.
23. E.Knill, R.Laflamme and W.H. Zurek, Resilent quantum computation: error models and thresholds, Proceedings of the Royal Society of London, Series A, 454, 365–384, 1998.
24. E.Knill, R.Laflamme, and W.H. Zurek, Accuracy threshold for quantum computation, quant-ph/9611025, 1998.
25. D.W. Leung, Two-qubit projective measurements are universal for quantum computation, quant-ph/0111077, 2001.
26. D.W. Leung, Quantum computations by measurements, quant-ph/0310189, 2003.
27. D.W. Leung, I.L. Chuang, F. Yamahuchi, and Y. Yamamoto, Efficient implementation of selective recoupling in heteronuclear spin systems using Hadamard matrices, quant-ph/9904100, 1999.
28. S. Lloyd, Almost any quantum gate is universal, Physical Review Letters, 75, 346–349, 1995.
29. Yuriy Makhlin, Nonlocal properties of two-qubit gates and mixed states and optimization of quantum computation, quant-ph/0002045, 2000.
30. M. Möttönen, J.J. Vartiainen, V. Bergholm, and M.M. Salomaa, Universal quantum computation, quant-ph/0404089, 2004.
31. M.A. Nielsen, Universal quantum computation using only projective measurements, quantum memory and preparation of the 0 state, quant-ph/0108020, 2001.
32. M.A. Nielsen and I.L. Chuang, Programmable quantum gate arrays, quant-ph/9703032, 1997.
33. K.N. Patel, I.L. Markov, and J.P. Hayes, Efficient synthesis of linear reversible circuits, quant-ph/0302002, 2003.
34. S. Perdrix, State transfer instead of teleportation in measurement-based quantum computation, quant-ph/0402204, 2004.
35. S. Perdrix and P. Jorrand, Measurement-based quantum Turing machines and questions of universalities, quant-ph/0402156, 2004.
36. R. Raussendorf and H.J. Briegel, Quantum computing by measurements only, Phys. Rev. Lett, 86, 2004.
37. T. Rudolph and L. Grover, A 2 rebit gate universal for quantum computing, quant-ph/0210087, 2002.
38. V.V. Shende, S.S. Bullock, and I.L. Markov, Recognizing small-circuit structure in two-qubit operators and timing Hamiltonians to compute controlled-not gates, quant-ph/0308045, 2003.
39. V.V. Shende, I.L. Markov, and S.S. Bullock, Minimal universal two-qubit cnot-based circuits, quant-ph/0308033, 2003a.

40. V.V. Shende, A.K. Prasard, I.L. Markov, and J.P. Hayes, Synthesis of reversible logic circuits, quant-ph/0207001, 2002.
41. Y. Shi, Both Toffoli and controlled-not need little help to do universal computation, quant-ph/0205115, 2002.
42. P.W. Shor, Fault-tolerant quantum computation, Proceedings of 37th IEEE FOCS, 56–65, 1996.
43. A.M. Steane, Error correcting codes in quantum theory, Physics Review Letters, 77, 5, 793–797, 1996.
44. D. Stepanenko and N.E. Bonesteel, Universal quantum computation through control of spin-orbit coupling, quant-ph/0403023, 2004.
45. T. Toffoli, Reversible computing, Tech. Memo MIT/LCS/TM-151, MIT Lab for CS, 1980.
46. R.R. Tucci, A rudimentary quantum compiler, quant-ph/9902062, 1999.
47. J.J. Vartiainen, M. Möttönen, and M.M. Salomaa, Efficient decomposition of quantum gates, quant-ph/0312218, 2003.
48. F. Vatan and C. Williams, Optimal realization of a general two-qubit quantum gate, quant-ph/0308006, 2003.
49. F. Vatan and C. Williams, Realization of a general three-qubit quantum gate, quant-ph/0401178, 2004.
50. G.Vidal and C.M. Dawson, A universal quantum circuit for two-qubit transformations with three cnot gates, quant-ph/0307177, 2003.
51. X. Wang, A. Sorensen, and K. Molmer, Multi-bit gates for quantum computing, quant-ph/0012055, 2000.
52. M.-H. Yung, D.W. Leung, and S. Bose, An exact effective two-qubit gate in a chain of three qubits, quant-ph/0312105, 2003.
53. J. Zhang, J. Vala, S. Sastry, and B. Whaley, Optimal quantum circuit synthesis from controlled-u gates, quant-ph/0308167, 2003.

An Upper Bound on the Rate of Information Transfer by Grover's Oracle

E. Arikan

Abstract. Grover discovered a quantum algorithm for identifying a target element in an unstructured search universe of N items in approximately $\pi/4\sqrt{N}$ queries to a quantum oracle. For classical search using a classical oracle, the search complexity is of order $N/2$ queries since on average half of the items must be searched. In work preceding Grover's, Bennett et al. had shown that no quantum algorithm can solve the search problem in fewer than $O(\sqrt{N})$ queries. Thus, Grover's algorithm has optimal order of complexity. Here, we present an information-theoretic analysis of Grover's algorithm and show that the square-root speed-up by Grover's algorithm is the best possible by any algorithm using the same quantum oracle.

Keywords: Grover's algorithm, quantum search, entropy.

1 Introduction

Grover [1], [2] discovered a quantum algorithm for identifying a target element in an unstructured search universe of N items in approximately $\pi/4\sqrt{N}$ queries to a quantum oracle. For classical search using a classical oracle, the search complexity is clearly of order $N/2$ queries since on average half of the items must be searched. It has been proven that this square-root speed-up is the best attainable performance gain by any quantum algorithm. In work preceding Grover's, Bennett et al. [4] had shown that no quantum algorithm can solve the search problem in fewer than $O(\sqrt{N})$ queries. Following Grover's work, Boyer et al. [5] showed that Grover's algorithm is optimal asymptotically, and that square-root speed-up cannot be improved even if one allows, e.g., a 50% probability of error. Zalka [3] strengthened these results to show that Grover's algorithm is optimal exactly (not only asymptotically). In this correspondence we present an information-theoretic analysis of Grover's algorithm and show the optimality of Grover's algorithm from a different point of view.

2 A General Framework for Quantum Search

We consider the following general framework for quantum search algorithms. We let X denote the state of the target and Y the output of the search algorithm. We assume that X is uniformly distributed over the integers 0 through $N - 1$. Y is also a random variable distributed over the same set of integers. The event $Y = X$ signifies that the algorithm correctly identifies the target. The probability of error for the algorithm is defined as $P_e = P(Y \neq X)$.

R. Ahlswede et al. (Eds.): Information Transfer and Combinatorics, LNCS 4123, pp. 452–459, 2006.

The state of the target is given by the density matrix

$$\rho_T = \sum_{x=0}^{N-1} (1/N)|x\rangle\langle x|, \tag{1}$$

where $\{|x\rangle\}$ is an orthonormal set. We assume that this state is accessible to the search algorithm only through calls to an oracle whose exact specification will be given later. The algorithm output Y is obtained by a measurement performed on the state of the quantum computer at the end of the algorithm. We shall denote the state of the computer at time $k = 0, 1, \ldots$ by the density matrix $\rho_C(k)$. We assume that the computation begins at time 0 with the state of the computer given by an initial state $\rho_C(0)$ independent of the target state. The computer state evolves to a state of the form

$$\rho_C(k) = \sum_{x=0}^{N-1} (1/N)\rho_x(k) \tag{2}$$

at time k, under the control of the algorithm. Here, $\rho_x(k)$ is the state of the computer at time k, conditional on the target value being x. The joint state of the target and the computer at time k is given by

$$\rho_{TC}(k) = \sum_{x=0}^{N-1} (1/N)|x\rangle\langle x| \otimes \rho_x(k). \tag{3}$$

The target state (1) and the computer state (2) can be obtained as partial traces of this joint state.

We assume that the search algorithm consists of the application of a sequence of unitary operators on the joint state. Each operator takes one time unit to complete. The computation starts at time 0 and terminates at a predetermined time K, when a measurement is taken on $\rho_C(K)$ and Y is obtained. In accordance with these assumptions, we shall assume that the time index k is an integer in the range 0 to K, unless otherwise specified.

There are two types of unitary operators that may be applied to the joint state by a search algorithm: oracle and non-oracle. A non-oracle operator is of the form $I \otimes U$ and acts on the joint state as

$$\rho_{TC}(k+1) = (I \otimes U)\,\rho_{TC}(k)\,(I \otimes U)^\dagger = \sum_x (1/N)|x\rangle\langle x| \otimes U\rho_x(k)U^\dagger. \tag{4}$$

Under such an operation the computer state is transformed as

$$\rho_C(k+1) = U\rho_C(k)U^\dagger. \tag{5}$$

Thus, non-oracle operators act on the conditional states $\rho_x(k)$ uniformly; $\rho_x(k+1) = U\rho_x(k)U^\dagger$. Only oracle operators have the capability of acting on conditional states non-uniformly.

An oracle operator is of the form $\sum_x |x\rangle\langle x| \otimes O_x$ and takes the joint state $\rho_{TC}(k)$ to

$$\rho_{TC}(k+1) = \sum_x (1/N)|x\rangle\langle x| \otimes O_x \rho_x(k) O_x^\dagger. \tag{6}$$

The action on the computer state is

$$\rho_C(k+1) = \sum_x (1/N) O_x \rho_x(k) O_x^\dagger. \tag{7}$$

All operators, involving an oracle or not, preserve the entropy of the joint state $\rho_{TC}(k)$. The von Neumann entropy of the joint state remains fixed at $S[\rho_{TC}(k)] = \log N$ throughout the algorithm. Non-oracle operators preserve also the entropy of the computer state; the action (5) is reversible, hence $S[\rho_C(k+1)] = S[\rho_C(k)]$. Oracle action on the computer state (7), however, does not preserve entropy; $S[\rho_C(k+1)] \neq S[\rho_C(k)]$, in general.

Progress towards identifying the target is made only by oracle calls that have the capability of transferring information from the target state to the computer state. We illustrate this information transfer in the next section.

3 Grover's Algorithm

Grover's algorithm can be described within the above framework as follows. The initial state of the quantum computer is set to

$$\rho_C(0) = |s\rangle\langle s| \tag{8}$$

where

$$|s\rangle = \sum_{x=0}^{N-1} (1/\sqrt{N})|x\rangle. \tag{9}$$

Since the initial state is pure, the conditional states $\rho_x(k)$ will also be pure for all $k \geq 1$.

Grover's algorithm uses two operators: an oracle operator with

$$O_x = I - 2|x\rangle\langle x|, \tag{10}$$

and a non-oracle operator (called 'inversion about the mean') given by $I \otimes U_s$ where

$$U_s = 2|s\rangle\langle s| - I. \tag{11}$$

Both operators are Hermitian.

Grover's algorithm interlaces oracle calls with inversion-about-the-mean operations. So, it is convenient to combine these two operations in a single operation, called Grover iteration, by defining $G_x = U_s O_x$. The Grover iteration takes the joint state $\rho_{TC}(k)$ to

$$\rho_{TC}(k+1) = \sum_x (1/N)|x\rangle\langle x| \otimes G_x \rho_x(k) G_x^\dagger \tag{12}$$

In writing this, we assumed, for notational simplicity, that G_x takes one time unit to complete, although it consists of the succession of two unit-time operators.

Grover's algorithm consists of $K = (\pi/4)\sqrt{N}$ successive applications of Grover's iteration beginning with the initial state (8), followed by a measurement on $\rho_C(K)$ to obtain Y. The algorithm works because the operator G_x can be interpreted as a rotation of the x–s plane by an angle $\theta = \arccos(1-2/N) \approx 2/\sqrt{N}$ radians. So, in K iterations, the initial vector $|s\rangle$, which is almost orthogonal to $|x\rangle$, is brought into alignment with $|x\rangle$.

Grover's algorithm lends itself to exact calculation of the eigenvalues of $\rho_C(k)$, hence to computation of its entropy. The eigenvalues of $\rho_C(k)$ are

$$\lambda_1(k) = \cos^2(\theta k) \tag{13}$$

of multiplicity 1, and

$$\lambda_2(k) = \frac{\sin^2(\theta k)}{N - 1} \tag{14}$$

of multiplicity $N - 1$. The entropy of $\rho_C(k)$ is given by

$$S(\rho_C(k)) = -\lambda_1(k) \log \lambda_1(k) - (N - 1)\lambda_2(k) \log \lambda_2(k) \tag{15}$$

and is plotted in Fig. 1 for $N = 2^{20}$. (Throughout the paper, the unit of entropy is bits and log denotes base 2 logarithm.) The entropy $S(\rho_c(k))$ has period $\pi/\theta \approx (\pi/2)\sqrt{N}$.

Our main result is the following lower bound on time-complexity.

Proposition 1. *Any quantum search algorithm that uses the oracle calls $\{O_x\}$ as defined by (10) must call the oracle at least*

$$K \geq \left(\frac{1 - P_e}{2\pi} + \frac{1}{\pi \log N}\right) \sqrt{N} \tag{16}$$

times to achieve a probability of error P_e.

For the proof we first derive an information-theoretic inequality. For any quantum search algorithm of the type described in section 2, we have by Fano's inequality,

$$H(X|Y) \leq \mathcal{H}(P_e) + P_e \log(N - 1) \leq \mathcal{H}(P_e) + P_e \log(N), \tag{17}$$

where for any $0 \leq u \leq 1$

$$\mathcal{H}(u) = -\delta \log \delta - (1 - \delta) \log(1 - \delta). \tag{18}$$

On the other hand,

$$\begin{aligned}
H(X|Y) &= H(X) - I(X;Y) \\
&= \log N - I(X;Y) \\
&\geq \log N - S(\rho_C(K))
\end{aligned} \tag{19}$$

where in the last line we used Holevo's bound [6, p. 531].

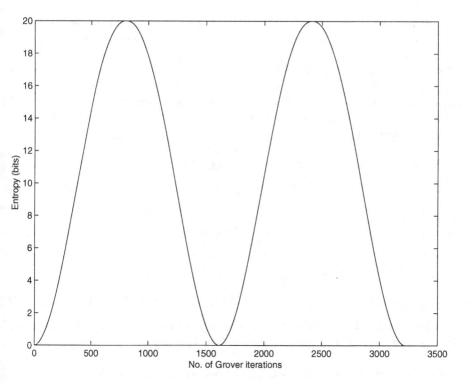

Fig. 1. Evolution of entropy in Grover's algorithm

Let μ_k be the largest eigenvalue (sup-norm) of $\rho_C(k)$. We observe that μ_k begins at time 0 with the value 1 and evolves to the final value μ_K at the termination of the algorithm. We have

$$S(\rho_C(K)) \leq -\mu_K \log \mu_K - (1-\mu_K) \log[(1-\mu_K)/(N-1)] \leq \mathcal{H}(\mu_K) + (1-\mu_K) \log N. \tag{20}$$

since the entropy is maximized, for a fixed μ_K, by setting the remaining $N-1$ eigenvalues equal to $(1 - \mu_K)/(N - 1)$. Combining (19) and (20),

$$\mu_K \log N \leq P_e \log N + \mathcal{H}(P_e) + \mathcal{H}(\mu_K) \leq P_e \log N + 2 \tag{21}$$

Now, let

$$\Delta = \sup\{|\mu_{k+1} - \mu_k| : k = 0, 1, \ldots, K-1\}. \tag{22}$$

This is the maximum change in the sup norm of $\rho_C(k)$ per algorithmic step. Clearly, $K \geq \frac{1-\mu_K}{\Delta}$. Using the inequality (21), we obtain

$$K \geq \frac{1 - P_e + 2/\log N}{\Delta}. \tag{23}$$

Thus, any upper bound on Δ yields a lower bound on K. The proof will be completed by proving

Lemma 1. $\Delta \leq 2\pi/\sqrt{N}$.

We know that operators that do not involve oracle calls do not change the eigenvalues, hence the sup norm, of $\rho_C(k)$. So, we should only be interested in bounding the perturbation of the eigenvalues of $\rho_C(k)$ as a result of an oracle call. We confine our analysis to the oracle operator (10) that the Grover algorithm uses.

For purposes of this analysis, we shall consider a continuous-time representation for the operator O_x so that we may break the action of O_x into infinitesimal time steps. So, we define the Hamiltonian

$$H_x = -\pi |x\rangle\langle x| \tag{24}$$

and an associated evolution operator $O_x(\tau) = e^{-i\tau H_x} = I + (e^{i\pi\tau} - 1)|x\rangle\langle x|$. The operator O_x is related to $O_x(\tau)$ by $O_x = O_x(1)$.

We extend the definition of conditional density to continuous time by

$$\rho_x(k_0 + \tau) = O_x(\tau)\rho_x(k_0)O_x(\tau)^\dagger \tag{25}$$

for $0 \leq \tau \leq 1$. The computer state in continuous-time is defined as

$$\rho_C(t) = \sum_x (1/N)\rho_x(t). \tag{26}$$

Let $\{\lambda_n(t), u_n(t)\}$, $n = 1, \ldots, N$, be the eigenvalues and associated normalized eigenvectors of $\rho_C(t)$. Thus,

$$\rho_C(t)|u_n(t)\rangle = \lambda_n(t)|u_n(t)\rangle, \quad \langle u_n(t)|\rho_C(t) = \lambda_n(t)\langle u_n(t)|, \quad \langle u_n(t)|u_m(t)\rangle = \delta_{n,m}. \tag{27}$$

Since $\rho_C(t)$ evolves continuously, so do $\lambda_n(t)$ and $u_n(t)$ for each n.

Now let $(\lambda(t), u(t))$ be any one of these eigenvalue-eigenvector pairs. By a general result from linear algebra (see, e.g., Theorem 6.9.8 of Stoer and Bulirsch [7, p. 389] and the discussion on p. 391 of the same book),

$$\frac{d\lambda(t)}{dt} = \langle u(t)|\frac{d\rho_C(t)}{dt}|u(t)\rangle. \tag{28}$$

To see this, we differentiate the two sides of the identity $\lambda(t) = \langle u(t)|\rho_C(t)|u(t)\rangle$, to obtain

$$\frac{d\lambda(t)}{dt} = \langle u'(t)|\rho_C(t)|u(t)\rangle + \langle u(t)|\frac{d\rho_C(t)}{dt}|u(t)\rangle + \langle u(t)|\rho_C(t)|u'(t)\rangle$$

$$= \langle u(t)|\frac{d\rho_C(t)}{dt}|u(t)\rangle + \lambda(t)[\langle u'(t)|u(t)\rangle + \langle u(t)|u'(t)\rangle]$$

$$= \langle u(t)|\frac{d\rho_C(t)}{dt}|u(t)\rangle + \lambda(t)\frac{d}{dt}\langle u(t)|u(t)\rangle$$

$$= \langle u(t)|\frac{d\rho_C(t)}{dt}|u(t)\rangle$$

where the last line follows since $\langle u(t)|u(t)\rangle \equiv 1$. Differentiating (26), we obtain

$$\frac{d\rho_C(t)}{dt} = \sum_x -(i/N)[H_x, \rho_x(t)] \tag{29}$$

where $[\cdot, \cdot]$ is the commutation operator. Substituting this into (28), we obtain

$$\left| \frac{d\lambda(t)}{dt} \right| = \left| \langle u(t)| - \frac{i}{N} \sum_x [H_x, \rho_x(t)] \, |u(t)\rangle \right|$$

$$\leq \frac{2}{N} \left| \sum_x \langle u(t)|H_x \rho_x(t)|u(t)\rangle \right|$$

$$\overset{(a)}{\leq} \frac{2}{N} \sqrt{\sum_x \langle u(t)|H_x^2|u(t)\rangle} \sqrt{\sum_x \langle u(t)|\rho_x^2(t)|u(t)\rangle}$$

$$\overset{(b)}{=} \frac{2}{N} \sqrt{\sum_x \pi^2 |\langle u(t)|x\rangle|^2} \sqrt{N\langle u(t)|\rho_C(t)|u(t)\rangle}$$

$$= \frac{2\pi}{\sqrt{N}} 1 \cdot \sqrt{\lambda(t)}$$

$$\leq \frac{2\pi}{\sqrt{N}}$$

where (a) is the Cauchy-Schwarz inequality, (b) is due to (i) $\rho_x^2(t) = \rho_x(t)$ as it is a pure state, and (ii) the definition (26). Thus,

$$|\lambda(k_0 + 1) - \lambda(k_0)| = \left| \int_{k_0}^{k_0+1} \frac{d\lambda(t)}{dt} dt \right| \leq 2\pi/\sqrt{N}. \tag{30}$$

Since this bound is true for any eigenvalue, the change in the sup norm of $\rho_C(t)$ is also bounded by $2\pi/\sqrt{N}$.

4   Discussion

The bound (16) captures the $\sqrt{N}$ complexity of Grover's search algorithm. As mentioned in the Introduction, lower-bounds on Grover's algorithm have been known before; and, in fact, the present bound is not as tight as some of these earlier ones. The significance of the present bound is that it is largely based on information-theoretic concepts. Also worth noting is that the probability of error P_e appears explicitly in (16), unlike other bounds known to us.

References

1. L.K. Grover, A fast quantum mechanical algorithm for database search, Proceedings, 28th Annual ACM Symposium on the Theory of Computing (STOC), 212-219, 1996.
2. L.K. Grover, Quantum mechanics helps in searching for a needle in a haystack, Phys. Rev. Letters, 78, 2, 325-328, 1997.
3. C. Zalka, Grover's quantum searching is optimal, Phys. Rev. A, 60, 2746, 1999.

4. C.H. Bennett, E. Bernstein, G. Brassard, and U.V. Vazirani, Strength and weaknesses of quantum computing, SIAM Journal on Computing, Vol. 26, No. 5, 1510-1523, 1997.
5. M. Boyer, G. Brassard, P. Hoeyer, and A. Tapp, Tight bounds on quantum computing, Proceedings 4th Workshop on Physics and Computation, 36-43, 1996; Fortsch. Phys. 46, 493-506, 1998.
6. M.A. Nielsen and I.L. Chuang, Quantum Computation and Quantum Information, Cambridge University Press, 2000.
7. J. Stoer and R. Bulirsch, Introduction to Numerical Analysis, Springer, NY, 1980.

A Strong Converse Theorem for Quantum Multiple Access Channels

R. Ahlswede and N. Cai

Abstract. With the wringing technique developed by the first author for classical multiple access channels we show that the strong converse theorem holds also for quantum multiple access channels, if classical messages are transmitted.

Keywords: classical quantum multiple access channel; strong converse; wringing technique; non–stationary memoryless classical quantum channel.

1 Introduction

The coding theorem for memoryless channels, the most fundamental theorem in Information Theory, was presented by C.E. Shannon with a sketch of its proof in his celebrated work [23] of 1948. The first formally satisfactory proof of the coding theorem appeared in 1954 in [12] by A. Feinstein, who attributed the part called weak converse of the theorem to R.M. Fano [11] (in 1952). The strong converse for memoryless channels was first proved by J. Wolfowicz [29] in 1957. A.S. Holevo [15] and B. Schumacher–M. Westmoreland [22] extended the coding theorem to quantum memoryless channels for transmission of classical messages. The theorem is known as HSW Theorem in Quantum Information Theory — named after the authors — and is one of most important results in Quantum Information Theory. With the *concept of typical subspaces* A. Winter [27] successfully extended Wolfowitz's method to a quantum version and proved the strong converse for quantum memoryless channels. He also presented an elegant new proof for the direct part of the coding theorem. At the same time a proof of the strong converse was given by T. Ogawa and H. Nagaoka [18] and their method of proof may be regarded as an extension of that by S. Arimoto [6]. The coding theorem, direct part and strong converse, for non–stationary memoryless classical quantum channels was proved by A. Winter [28] who extended Ahlswede's method of [1] to the quantum case.

A significant difference between Shannon's Coding Theorem and the HSW Coding Theorem is that in the terminology of classical Information Theory the former is a single–letter characterization but the latter is not. Hilbert spaces of unbounded dimensions are involved in the capacity formula in the HSW Theorem. In fact it is one of the most important and challenging problems in Quantum Information Theory to get a computable characterization. A family of quantum memoryless channels, known as classical quantum channels, whose HSW Theorem has single–letter characterization, was introduced by A.S. Holevo [14].

R. Ahlswede et al. (Eds.): Information Transfer and Combinatorics, LNCS 4123, pp. 460–485, 2006.

An observation by A.S. Holevo [16] shows that the HWS Theorem for general quantum channels can be easily derived from the HWS Theorem for a classical quantum channel by regarding those channels as a classical quantum channel with a "super alphabet". So it seems to be natural to consider classical quantum multi–user channel when we try to extend the Classical Multi–user Information Theory to Quantum before a single–letter form of the HWS Theorem is obtained.

C.E. Shannon, the founder of Information Theory, also started Multi–user Information Theory in [24]. The only multi–user channel whose single–letter capacity region is completely known is the multiple access channel (MAC). The Coding Theorem for MAC, the direct part and weak converse, was proved by R. Ahlswede [2]. Its strong converse theorem was shown by G. Dueck [9] (with the Ahlswede, Gacs, Körner Blowing Up Lemma and a wringing technique for mutual Information) and R. Ahlswede [3] (with a stronger wringing technique for probability alone). To the best of our knowledge the quantum model of the MAC was first studied by A.E. Allahverdyan and D.B. Saakian [5] and the Coding Theorem for the classical quantum MAC, the direct part and weak converse, was shown by A. Winter [26].

However, till now the strong converse resisted all efforts of proof. Already for the classical MAC the strong converse theorem could not be proved for many years, because average errors are genuinely used and even nowadays there is no way to do this via typical sequences and also not via Arimoto's method. In the quantum case it gets even worse, because we have no analogue of the Blowing Up Method for operators (instead for words). Consequently Dueck's approach fails also. Fortunately Ahlswede's wringing method for probabilities, invented for classification of the role of fundamental methods, works also for the quantum situation. However the Packing Lemma of [3] requires now a more demanding analytical proof than the original application of Chebyshev's inequality. We present necessary definitions and state formally our results in the next section and provide the plan to prove the main result in Section 3. In Section 4 we show a strong converse for quantum non–stationary memoryless channels (slightly stronger than that in [28]), which we need in the proof to the main theorem. Finally the proof of the main theorem is completed in Section 5.

2   Definitions and Results

Throughout the paper the script letters $\mathcal{X}, \mathcal{Y}, \mathcal{Z}, \ldots$ stand for finite sets and $\mathcal{X}^n$, $\mathcal{Y}^n, \mathcal{Z}^n, \ldots$ are their Cartesian powers. We denote their elements and the random variables taking values in them, by corresponding lower letters, $x, y, z, \ldots, x^n, y^n$, $z^n, \ldots$ and capital letters $X, X', Y, \tilde{Y}, Z, \ldots, Z^n, Y^n, \ldots$ respectively. The probability distribution of random variable X, the conditional distribution of random variables X given $Y, \ldots$ are written as $P_X, P_{X|Y}, \ldots$. When we speak of P_X, we mean the underlying random variable X is automatically defined and similarly for $P_{X|Y}$. As in the standard way (e.g., [17], $|i\rangle, |j\rangle, |\alpha\rangle, |\beta\rangle, \ldots$ (the "ket") stand for normalized column vectors in Hilbert spaces and their dual vectors are written as $\langle i|, \langle j|, \langle \alpha|, \langle \beta|, \ldots$ (the "bra"). The density operators or states are denoted by Greek letters e.g. $\rho, \sigma, \varphi, \dots$. We write von Neumann entropy of state ρ and the quantum relative entropy of ρ to σ as $S(\rho)$ and $D(\rho\|\sigma)$, respectively, i.e., $S(\rho) = -th(\rho \log \rho)$ and $D(\rho\|\sigma) = tr(\rho \log \rho) - tr(\rho \log \sigma)$. Here and throughout the paper log is the logarithm with base 2 whereas ℓn is the logarithm with base e. Also throughout the paper we assume all Hilbert spaces in the discussion are over the complex field and have finite dimensions.

One way to define classical quantum channels is to specify them by a set of states $\Phi = \{\varphi(x) : x \in \mathcal{X}\}$ in the same Hilbert space $\mathcal{H}H$, labelled by a finite set $\mathcal{X}$. We call Φ a classical quantum channel, or shortly a $c - q$ channel, with input alphabet $\mathcal{X}$ and output space $\mathcal{H}H$. $x \in \mathcal{X}$ and $\varphi(x)$ are called input letter and output state, respectively. We assume that the receiver, or decoder, of the channel receives the state $\varphi(x)$ from the output of the channel, if the sender, or encoder, of the channel inputs a letter $x \in \mathcal{X}$ to the channel. The Holevo quantity of a $c - q$ channel $\Phi = \{\varphi(x) : x \in \mathcal{X}\}$ with respect to the input distribution P is denoted by $\chi(P; \Phi)$

$$
\begin{aligned}
\chi(P; \Phi) &= S(\sigma) - \sum_{x \in \mathcal{X}} P(x) S(\varphi(x)) \\
&= \sum_{x \in \mathcal{X}} P(x) \big[-tr(\varphi(x) \log \sigma) + tr(\varphi(x) \log \varphi(x)) \big] \\
&= \sum_{x \in \mathcal{X}} P(x) D(\varphi(x)\|\sigma),
\end{aligned}
\tag{2.1}
$$

where $\sigma = \sum_{x \in \mathcal{X}} P(x)\varphi(x)$. (Perhaps it should be called ... mutual information.)

A non–stationary memoryless classical quantum channel, or shorter non–stationary memoryless $c - q$ channel, is specified by a sequence of $c - q$ channels $\{\Phi_n\}_{n=1}^{\infty}$ with common input alphabet $\mathcal{X}$ and output space $\mathcal{H}H$ such that an output state

$$
\varphi^{\otimes n}(x^n) = \varphi_1(x_1) \otimes \varphi_2(x_2) \otimes \cdots \otimes \varphi_n(x_n)
$$

is output from the channel if the sequence $x^n = (x_1, \dots, x_n) \in \mathcal{X}^n$ is input to the channel, where $\Phi_t = \{\varphi_t(x) : x \in \mathcal{X}\}$ for $t = 1, 2, \dots$, and $\otimes$ is the tensor product. An $(n, M, \bar{\lambda})$–code for the non–stationary memoryless $c - q$ channel, $(\mathcal{U}, \mathcal{D})$ consists of a subset $\mathcal{U} \subset \mathcal{X}^n$ and a measurement $\mathcal{D}$ on the outputspace $\mathcal{H}H^{\otimes n}$ described by $\{\mathcal{D}_{u^n} : u^n \in \mathcal{U}\}$, where $\mathcal{D}_{u^n}$ corresponds to input u^n, such that $|\mathcal{U}| = M$ and

$$
M^{-1} \sum_{u^n \in \mathcal{U}} tr\big[\varphi^{\otimes n}(u^n) \mathcal{D}_{u^n}\big] > 1 - \bar{\lambda}.
\tag{2.2}
$$

Here $\mathcal{U}$ is called codebook, members of $\mathcal{U}$ are called codewords, n is called the length of the code, or of the codewords, and $\bar{\lambda}$ is called the average probability of error. $\frac{1}{n} \log M$ is called rate of the code. An $(n, M, \bar{\lambda})$–code is used to transmit a set of classical messages of size M as follows. In the case the encoder has to send the mth message in the set, he sends the mth codeword, say $u^n(m) \in \mathcal{U}$ through the non–stationary memoryless $c - q$ channel. To decode the message, the decoder performs the measurement $\mathcal{D}$ to the outputstate and he decodes the output to the m'th message of the set if the outcome of measurement $\mathcal{D}$ is $u^n(m')$. Then the average fidelity of transmission is bounded from below by (2.2). A positive real number R is said to be $\bar{\lambda}$–achievable for a $c-q$ channel if for all $\varepsilon > 0$ there exists an $(n, M, \bar{\lambda})$–code for the channel with rate $\frac{1}{n} \log M > R - \varepsilon$ if n is sufficiently large. The maximum $\bar{\lambda}$–achievable rate for a channel is called its $\bar{\lambda}$–capacity, denoted by $C(\bar{\lambda})$. The capacity of the channel is defined as $C = \inf\limits_{\bar{\lambda} > 0} C(\bar{\lambda})$.

Instead of capacity for average probability of error, we have capacity for maximum probability of error, if we replace (2.2) by

$$tr\left[\varphi^{\otimes n}(u)\mathcal{D}_{u^n}\right] > 1 - \lambda \text{ for all } u^n \in \mathcal{U}. \tag{2.3}$$

However by the pigeon–hole principle we know that the coding problems for average probability and maximum probability of error are equivalent for two terminals channels.

We call a non–stationary memoryless $c - q$ channel specified by $\{\Phi_n\}_{n=1}^{\infty}$ stationary, or a stationary memoryless $c - q$ channel, if for all n $\Phi_n = \Phi_1$. The coding theorem for stationary memoryless $c - q$ channels is an important special case of the HSW Theorem. Coding theorem with strong converse for non–stationary memoryless $c - q$ channels were shown by A. Winter in [28] and his strong converse has the following form.

Theorem W. (Winter [28]) *Given a non–stationary memoryless $c - q$ channel $\{\Phi_n\}_{n=1}^{\infty}$, for all $0 < \lambda < 1$ and $\varepsilon > 0$, there is an $n_o = n_o(\lambda, \varepsilon)$ (independent of the channel) such that if there exists an (n, M, λ)–code for the channel with $n > n_o$, then*

$$\frac{1}{n} \log M \leq \frac{1}{n} \sum_{t=1}^{n} \max_{P_t} \mathcal{X}(P_t; \Phi_t) + \varepsilon.$$

To prove our main result, we need a slightly stronger version of the strong converse for a non–stationary memoryless $c - q$ channel.

Theorem 1. *Let $\{\Phi_n\}_{n=1}^{\infty}$ be a non–stationary memoryless $c - q$ channel such that for diagonalizations of $\varphi_n(x)$, $n = 1, 2, \ldots$, $x \in \mathcal{X}$,*

$$\varphi_n(x) = \sum_{j=1}^{d} W_n(j|x)|j_n(x) >< j_n(x)| \tag{2.4}$$

where $d = \dim \mathcal{H}H$ and $\mathcal{H}H$ is the output space,

$$\inf\limits_{n,x,j}{}^{+} W_n(j|x) = w > 0, \tag{2.5}$$

where the $\inf^{+}$ is infimum of $W_n(j|x)$ for $1, 2, \ldots,$, $j = 1, 2, \ldots, d$ and $x \in \mathcal{X}$ with $W_n(j|x) > 0$. Then there exists a function h defined on $(0, 1)$ and depending on w in (2.5) such that for all (n, M, λ)–codes for the channel $\{\Phi_n\}_{n=1}^{\infty}$ and sufficiently large n,

$$\frac{1}{n} \log M \le \sum_{t=1}^{n} \frac{1}{n} \chi(P_{X_t}; \Phi_t) + \frac{1}{\sqrt{n}} h(\lambda), \qquad (2.6)$$

where $X^n = (X_1, X_2, \ldots, X_n)$ is the random variable uniformly distributed over the codebook $\mathcal{U}$. In other words P_{X^n} is the empirical distribution over the codebook and P_{X_t} is the marginal distribution of its t–th component.

We shall explain the role of the theorem in the proof of the main result in the next section and we prove it in Section 4.

Remarks

1. As one can always use a finite set of $c - q$ channels to approach an infinite set of $c - q$ channels by a quantum version (see Section VII in [28]) of ε–net techniques in Classical Information Theory and the condition (2.5) automatically holds, in the case that Φ_n, $n = 1, 2, \ldots$ are taken from a finite set of $c - q$ channels, the condition (2.5) can be removed from Theorem 1. Consequently one may replace h by a function independent of w. This can be done in exactly the same way as in [28]. However the current form of the theorem is sufficient for our purposes.
2. By considering the previous remark, Theorem W follows from Theorem 1.

Now let us turn to our main object, (stationary) memoryless classical quantum multiple access channels, or for short $c - q$ MAC. In general, like for classical MAC, a $c - q$ MAC has more than one sender (or encoder), for simplicity in the current paper, we consider $c - q$ MAC with two encoders. Thus a $c - q$ MAC is specified by a set $\Phi = \{\varphi(x, y) : x \in \mathcal{X}, y \in \mathcal{Y}\}$ of states in common Hilbert space $\mathcal{H}H$ labelled by the elements in a Cartesian product $\mathcal{X} \times \mathcal{Y}$ of two finite sets $\mathcal{X}$ and $\mathcal{Y}$. Again $\mathcal{H}H$ is called output space and $\mathcal{X}$ and $\mathcal{Y}$ are called input alphabets. We address the two encoders as $\mathcal{X}$– and $\mathcal{Y}$–encoders. The decoder receives the state

$$\varphi^{\otimes n}(x^n, y^n) = \varphi(x_1, y_1) \otimes \varphi(x_2, y_2) \otimes \cdots \otimes \varphi(x_n, y_n) \qquad (2.7)$$

if the $\mathcal{X}$–encoder and the $\mathcal{Y}$–encoder send input sequences $x^n = (x_1, x_2, \ldots, x_n) \in \mathcal{X}^n$ and $y^n = (y_1, y_2, \ldots, y_n) \in \mathcal{Y}^n$, respectively, over the $c - q$ MAC. An $(n, M, N, \bar{\lambda})$–code $(\mathcal{U}, \mathcal{V}, \mathcal{D})$ for a $c - q$ MAC Φ consists of a subset $\mathcal{U} \subset \mathcal{X}^n$, a subset $\mathcal{V} \subset \mathcal{X}^n$, with cardinalities $|\mathcal{U}| = M$, $|\mathcal{V}| = N$ respectively and a measurement $\mathcal{D} = \{\mathcal{D}_{u^n, v^n} : u^n \in \mathcal{U}, v^n \in \mathcal{V}\}$, where $\mathcal{D}_{u^n, v^n}$ corresponds an outcoming (u^n, v^n), such that

$$M^{-1} N^{-1} \sum_{u^n \in \mathcal{U}} \sum_{v^n \in \mathcal{V}} tr\left[\varphi^{\otimes n}(u^n, v^n) \mathcal{D}_{u^n, v^n}\right] > 1 - \bar{\lambda}. \qquad (2.8)$$

We call $\mathcal{U}$ and $\mathcal{V}$ $\mathcal{X}$– and $\mathcal{Y}$–codebooks and their members codewords. $\bar{\lambda}$ and n are called average probability of error and length of code (or of codewords), respectively.

Define $R_1 = \frac{1}{n} \log |M|$ and $R_2 = \frac{1}{n} \log |N|$. (R_1, R_2) is called pair of rates of the code. A pair (R_1, R_2) of positive real numbers is called $\bar{\lambda}$–achievable if for all $\varepsilon > 0$ and sufficiently larger n there exists an $(n, M, N, \bar{\lambda})$–code for the $c - q$ MAC with rates $\frac{1}{n} M > \mathcal{R}_1 - \varepsilon$ and $\frac{1}{n} \log N > \mathcal{R}_2 - \varepsilon$. The set of achievable pairs of real numbers is called $\bar{\lambda}$–capacity region of the $c - q$ MAC and denoted by $\mathcal{R}(\bar{\lambda})$. Then the capacity region of the $c - q$ MAC Φ is defined as

$$\mathcal{R} = \bigcap_{1 > \bar{\lambda} > 0} \mathcal{R}(\bar{\lambda}). \tag{2.9}$$

As a special case of a $c - q$ MAC $\Phi = \{\varphi(x, y) : x \in \mathcal{X}, y \in \mathcal{Y}\}$ becomes a classical MAC, when $\varphi(x, y)$ $x \in \mathcal{X}$, $y \in \mathcal{Y}$ can be simultaneously diagonalized by the same basis. One may define the capacity region (λ–capacity region) for maximum probability of error in an analogous way by replacing (2.8) by

$$tr\left[\varphi(u^n, v^n) \mathcal{D}_{u^n, v^n}\right] > 1 - \lambda$$

for all $u^n \in \mathcal{U}$, $v^n \in \mathcal{V}$.

However, it turned out that unlike for two terminal channels capacity regions for maximum probability of error and average probability of error of the same $c - q$ MAC may be different even for the special case of a classical MAC (c.f. [8]).

Throughout the paper for $c - q$ MAC we only consider the average probability of error and to present the formulas of capacity region of a $c - q$ MAC for it we introduce the following notation.

For a given $c - q$ MAC Φ with input alphabets $\mathcal{X}$ and $\mathcal{Y}$, and a pair of random variables (X, Y) distributed on $\mathcal{X} \times \mathcal{Y}$ we write

$$\chi(X, Y; \Phi) = \chi(P_{XY}; \Phi), \tag{2.10}$$

where $\chi(\cdot; \cdot)$ is the Holevo quantity defined in (2.1),

$$\chi(X : \Phi|Y) = \sum_{y \in \mathcal{Y}} P_Y(y) \chi\left(P_{X|Y}(\cdot|y); \Phi(\cdot, y)\right)$$

$$= \sum_{y \in \mathcal{Y}} P_Y(y) \left[S(\sigma_{XY,2}(y)) - \sum_{x \in \mathcal{X}} P_{X|Y}(x|y) S\big(\varphi(x, y)\big) \right], \quad (2.11)$$

where $\Phi(\cdot, y) = \{\varphi(x, y) : x \in \mathcal{X}\}$ for all $y \in \mathcal{Y}$ is a $c - q$ channel with input alphabet $\mathcal{X}$ and $\sigma_{XY,2} = \sum_x P_{X|Y}(x, y) \varphi(x, y)$ for all $y \in \mathcal{Y}$, and

$$\chi(Y; \Phi|X) = \sum_{x \in \mathcal{X}} P_X(x) \chi\left(P_{Y|X}(\cdot|x); \Phi(x, \cdot)\right)$$

$$= \sum_{x \in \mathcal{X}} P_X(x) \left[S(\sigma_{XY,1}(x)) - \sum_{y \in \mathcal{Y}} P_{Y|X}(y|x) S\big(\varphi(x, y)\big) \right] \tag{2.12}$$

where $\Phi(x, \cdot) = \{\varphi(x, y) : y \in \mathcal{Y}\}$ for all x is a $c - q$ channel with input alphabet $\mathcal{Y}$ and $\sigma_{XY,1}(x) = \sum_y P_{Y|X}(y|x) \varphi(x, y)$.

Then it immediately follows from (2.1), (2.10) – (2.12) that

$$\chi(X, Y; \Phi) = \sum_{x \in \mathcal{X}} \sum_{y \in \mathcal{Y}} P_{XY}(x, y) D(\varphi(x, y) \| \sigma_{XY}), \qquad (2.13)$$

where $\sigma_{XY} = \sum_{x \in \mathcal{X}} \sum_{y \in \mathcal{Y}} P_{XY}(x, y) \varphi(x, y)$,

$$\chi(X; \Phi | Y) = \sum_{y \in \mathcal{Y}} P_Y(y) \sum_{x \in \mathcal{X}} P_{X|Y}(x|y) D(\varphi(x, y) \| \sigma_{XY,2}(y)), \qquad (2.14)$$

and

$$\chi(Y; \Phi | X) = \sum_{x \in \mathcal{X}} P_X(x) \sum_{y \in \mathcal{Y}} P_{Y|X}(y|x) D(\varphi(x, y) \| \sigma_{XY,1}(x)). \qquad (2.15)$$

For a pair of random variables (X, Y) with values in $\mathcal{X} \times \mathcal{Y}$ we let $\mathcal{R}(X, Y)$ be the set of pairs of positive real numbers (R_1, R_2) satisfying

$$R_1 + R_2 \leq \chi(X, Y; \Phi), \qquad (2.16)$$
$$R_1 \leq \chi(X; \Phi | Y), \qquad (2.17)$$

and

$$R_2 \leq \chi(Y; \Phi | X). \qquad (2.18)$$

Denote by $\partial(\mathcal{X}) \times \partial(\mathcal{Y})$ the set of pairs of independent random variables X taking values on $\mathcal{X}$ and Y taking values on $\mathcal{Y}$ (i.e. $P_{XY}(x, y) = P_X(x) P_Y(y)$) and

$$\mathcal{R}^* = \text{conv} \left[\bigcup_{(X,Y) \in \partial(\mathcal{X}) \times \partial(\mathcal{Y})} \mathcal{R}(X, Y) \right], \qquad (2.19)$$

where $\text{conv}(A)$ stand for the convex closure of the set A.

Then the Coding Theorem for the $c - q$ MAC of A. Winter [26] says that

$$\mathcal{R} = \mathcal{R}^*. \qquad (2.20)$$

The main contribution of this paper is the strong converse for all $c - q$ MAC.

Theorem 2. *Given a $c - q$ M Φ, there exists a function $\tilde{h}$ defined on $(0, 1)$ such that for all $(n, M, N, \bar{\lambda})$-codes with $\bar{\lambda} \in (0, 1)$, there exist random variables $\tilde{X}_n = (\tilde{X}_1, \ldots, \tilde{X}_n)$ and $\tilde{Y}^n = (\tilde{Y}_1, \ldots, \tilde{Y}_n)$ taking values in $\mathcal{X}^n$ and $\mathcal{Y}^n$, respectively, with*

$$\frac{1}{n} \log M + \frac{1}{n} \log N \leq \frac{1}{n} \sum_{t=1}^{n} \chi(\tilde{X}_t, \tilde{Y}_t; \Phi) + \frac{\log n}{\sqrt{n}} \tilde{h}(\bar{\lambda}), \qquad (2.21)$$

$$\frac{1}{n} \log M \leq \frac{1}{n} \sum_{t=1}^{n} \chi(\tilde{X}_t; \Phi | \tilde{Y}_t) + \frac{\log n}{\sqrt{n}} \tilde{h}(\bar{\lambda}), \qquad (2.22)$$

and

$$\frac{1}{n}\log N \le \frac{1}{n}\sum_{t=1}^{n}\chi(\tilde{Y}_t;\Phi|\tilde{X}_t) + \frac{\log n}{\sqrt{n}}\tilde{h}(\bar{\lambda}), \qquad (2.23)$$

and for all t, $\tilde{X}_t$ and $\tilde{Y}_t$ are independent.

But the dependence is not necessary. It is a consequence of the dependence of h on w in (2.5) in Theorem 1. So by Remark 1 and on inspection of the proof for Theorem 2, one may obtain an $\tilde{h}^*$ independent of the channel.

3 Outline of Ideas to Extend Ahlswede's Strong Converse Proof for Classical MAC to $c - q$ MAC

Roughly speaking, his proof is divided into three steps.

In the first step Ahlswede extended Augustin's strong converse theorem in [7], which says that for an arbitrary (n, M, λ) code for a (classical) non–stationary memoryless channel $\{W_n\}_{n=1}^{\infty}$

$$\log M \le \sum_{t=1}^{n} I(X_t; Y_t) + \frac{3}{1-\lambda}|\mathcal{X}|\sqrt{n},$$

where $X^n = (X_1, \ldots, X_n)$ is random variable uniformly distributed on the codebook. $Y^n = (X_1, \ldots, Y_n)$ is the output random variable for input X^n, and I is Shannon's mutual information.

By applying the strong converse in the first step to the classical MAC, one may obtain an outer bound for the achievable rate pair (R_1, R_2) such that

$$R_1 + R_1 \underset{\sim}{<} \frac{1}{n}\sum_{t=1}^{n} I(X_tY_t; Z_t)$$

$$R_1 \underset{\sim}{<} \frac{1}{n}\sum_{t=1}^{n} I(X_t; Z_t|Y_t)$$

and

$$R_2 \le \frac{1}{n}\sum_{t=1}^{n} I(Y_t; Z_t|X_t),$$

formally like the capacity region. But now (X_t, Y_t) may not be independent! So the outer bound is not tight.

The reason is due to application of the strong converse of non–stationary channels in step 1 requiring the maximum error criterion, whereas for MAC the capacity regions of maximum and average error criteria may be different [8]. By the Markov inequality one may obtain a subcode $\mathcal{A}$ from the original code for MAC. But the uniform distributions over $\mathcal{A}$ are not independent. To solve the problem R. Ahlswede discovered the following technique for the second step.

Lemma A. (Wringing Technique [3]) *Let P and Q be probability distributions over a Cartesian power Z^n of a finite set Z such that for a positive constant c*

$$P(z^n) \leq (1 + c)Q(z^n) \text{ for all } z^n \in \mathcal{Z}^n,$$

then for any $0 < \gamma < c$, $0 \leq \varepsilon < 1$ there exist $t_1, \ldots, t_k \in \{1, 2, \ldots, n\}$, where $0 \leq k \leq \frac{c}{\gamma}$ such that for some $\bar{z}_{t_1}, \ldots, \bar{z}_{t_k}$,

$$P(z_t | \bar{z}_{t_1}, \ldots, \bar{z}_{t_k}) \leq \max\left[(1 + \gamma)Q(z_t | \bar{z}_{t_1}, \ldots, \bar{z}_{t_k}), \varepsilon\right]$$

and

$$P(\bar{z}_{t_1}, \ldots, \bar{z}_{t_k}) \geq \varepsilon^k.$$

With the lemma Ahlswede obtained a further subcode $\mathcal{B}$ from the subcode $\mathcal{A}$ (without loosing too much rate) such that the uniform distribution over the codebook for it is nearly componentwise independent. Notice that the uniform distribution for $\mathcal{B}$ is not necessarily independent, but componentwise independence is sufficient for the purpose.

In the third step Ahlswede first combined the results in the first two steps to obtain an outer bound in terms of nearly componentwise independent input distributions and then by some calculation based on the continuity of Shannon information quantities showed the outer bound is arbitrarily close to the capacity region of the classical MAC.

We now plan to finish our proof for Theorem 2 paralleling these three steps. We first inspect the second step, the wringing technique, and find that we can take it into our proof almost without doing additional work, because the wringing technique is only applied in the input space and by definition our input space is classical. That means the only thing, which we need to do, is replace the codebook for the classical MAC by one for the $c - q$ MAC, and then consider the consequence at the output.

As by Fannes inequality [10] von Neumann entropy is continuous the extension to the quantum version in the third step is not so difficult. The only difficult part for the extension is the first part. Winter's strong converse (Theorem W) in [28] for non–stationary $c - q$ channels is in terms of optimal input distribution whereas the strong converse, which we need, is in terms of empirical distributions over codebooks. Hayashi and Nagaoka [13] extended a general capacity formula due to Verdu and Han [25] to classical quantum channels. But it turned out that they obtained a "ratewise" strong converse for stationary $c - q$ channels in terms of optimal input distributions by applying their formula.

So to prove our strong converse for $c - q$ MAC, Theorem 2, we have to show a strong converse for non–stationary $c - q$ channels in terms of the uniform distribution over the codebook i.e. our Theorem 1. In summary we state our plan as follows.

In the first step we prove Theorem 1 as an auxiliary result.

In the second step we apply the wringing technique to our codebook to obtain a subcode whose uniform distribution $P_{\tilde{X}^n \tilde{Y}}$ is nearly componentwise independent (i.e. $P_{\tilde{X}_t \tilde{Y}_t}(x, y)$ is arbitrarily close to $P_{\tilde{X}_t}(x)P_{\tilde{Y}_t}(y)$ for all $x \in \mathcal{X}$, $y \in \mathcal{Y}$).

Finally we finish our proof in the third step by calculation based on continuity of von Neumann entropy.

4   A Strong Converse for Classical Quantum Non–stationary Channels

We begin with our proof to Theorem 1 with a lemma which can be considered as a quantum analogue of Lemma 1 in [3].

Let $\mathcal{U}$ be a finite index set and let

$$\{\rho(u) : u \in \mathcal{U}\} \tag{4.1}$$

be a set of states labelled by indices in $\mathcal{U}$ in a given complex Hilbert space $\mathcal{H}H$ of finite dimension d. Further let σ be a state in $\mathcal{H}H$. For a given real number $r(u)$ we denote the projector of $\mathcal{H}H$ onto subspace

$$\rho(u) - 2^{r(u)}\sigma \geq 0 \tag{4.2}$$

by $\mathcal{P}^+\big(u, \sigma, r(u)\big)$. That is, $\mathcal{P}^+(u, \sigma, r)$ is the projector with

$$\mathcal{P}^+\big(u, \sigma, r(u)\big) = \sum_{\alpha_j(u) \geq 0} |e_j(u) >< e_j(u)| \tag{4.3}$$

if $\rho(u) - 2^{r(u)}\sigma$ is diagonalized as

$$\rho(u) - 2^{r(u)}\sigma = \sum_{j=0}^{d-1} \alpha_j(u)|e_j(u) >< e_j(u)|. \tag{4.4}$$

Then we have

Lemma 1. *Let $\mathcal{U}$, $\{\rho(u) : u \in \mathcal{U}\}$, σ, and $\mathcal{P}^+\big(u, \sigma, r(u)\big)$ be defined as above and let $\{\mathcal{D}_u : u \in \mathcal{U}\}$ be a measurement in $\mathcal{H}H$ (i.e., $0 \leq \mathcal{D}_u \leq I$ for all $u \in \mathcal{U}$, $\sum_{u \in \mathcal{D}_u} \mathcal{D}_u = I$, and $\mathcal{D}_u$ corresponds to the outcome "u", where I is the identity operator in $\mathcal{H}H$) such that for a positive real number δ*

$$tr\rho(u)\mathcal{D}_u - tr\rho(u)\mathcal{P}^+\big(u, \sigma, r(u)\big) \geq \delta \tag{4.5}$$

for all $u \in \mathcal{U}$. Then

$$|\mathcal{U}| \leq \delta^{-1}2^{|\mathcal{U}|^{-1}\sum_{u \in \mathcal{U}} r(u)}. \tag{4.6}$$

Proof: For all $u \in \mathcal{U}$, by (4.3), (4.4) and (4.5),

$$2^{r(u)}tr\sigma\mathcal{D}_u = tr\rho(u)\mathcal{D}_u - tr\big[\rho(u) - 2^{r(u)}\sigma\big]\mathcal{D}_u$$
$$\geq tr\rho(u)\mathcal{D}_u - tr\big[\rho(u) - 2^{r(u)}\sigma\big]\mathcal{P}^+\big(u, \sigma, r(u)\big)$$
$$\geq tr\rho(u)\mathcal{D}_u - tr\rho(u)\mathcal{P}^+\big(u, \sigma, r(u)\big) \geq \delta, \tag{4.7}$$

where the first inequality follows from (4.3), (4.4), and the fact that for any operator π with diagonlization $\pi = \sum_{i=0}^{d-1} p_i|e_i >< e_i|$, the projector $\mathcal{P}_\pi^+ = \sum_{p_i \geq 0} |e_i >< e_i|$ maximizes $tr(\pi\tau)$ among the operators τ with $0 \leq \tau \leq I$, and the last inequality holds by condition (4.5).

Next notice that $\delta 2^{-|\mathcal{U}|^{-1} \sum\limits_{u \in \mathcal{U}} r(u)} = \left[\prod\limits_{u \in \mathcal{U}} \delta 2^{-r(u)} \right]^{\frac{1}{|\mathcal{U}|}}$. Then (4.7) is followed by

$$\delta 2^{-|\mathcal{U}|^{-1} \sum\limits_{u \in \mathcal{U}} r(u)} \leq \left[\prod\limits_{u \in \mathcal{U}} tr(\sigma \mathcal{D}_u) \right]^{\frac{1}{|\mathcal{U}|}}. \tag{4.8}$$

However geometric means may not exceed arithmetic means and so we can continue to bound (4.8) by

$$\delta 2^{-|\mathcal{U}|^{-1} \sum\limits_{u \in \mathcal{U}} r(u)} \leq \frac{1}{|\mathcal{U}|} \sum\limits_{u \in \mathcal{U}} tr(\sigma \mathcal{D}_u) = |\mathcal{U}|^{-1} tr \left(\sigma \sum\limits_{u \in \mathcal{U}} \mathcal{D}_u \| ! \right) = |\mathcal{U}|^{-1} tr(\sigma I) = |\mathcal{U}|^{-1}.$$

That is (4.6).   □

To apply Lemma 1 in the proof of Theorem 1 we have to estimate $\left[tr\rho \mathcal{P}^+ (\rho, \sigma, r) \right]$ for projector

$$\mathcal{P}^+(\rho, \sigma, r) = \sum\limits_{\alpha_j \geq 0} |e_j >< e_j| \tag{4.9}$$

onto the subspace $\rho - 2^r \sigma \geq 0$, where σ and ρ are states and the diagonalization of $\rho - 2^r \sigma$ is

$$\rho - 2^r \sigma = \sum\limits_{j=0}^{d-1} \alpha_j |e_j >< e_j|. \tag{4.10}$$

For $j = 0, 1, 2, \ldots, d-1$, let $P(j) = \langle e_j | \rho | e_j \rangle$ and $Q(j) = \langle e_j | \sigma | e_j \rangle$. Then P and Q are probability distributions on $\{0, 1, \ldots, d-1\}$ since ρ and σ are states: Thus by (4.9) and (4.10) one may rewrite

$$tr\rho \mathcal{P}^+(\rho, \sigma, r) = \sum\limits_{j : P(j) - 2^r Q(j) \geq 0} P(j) = \sum\limits_{j : \log \frac{P(j)}{Q(j)} \geq r} P(j). \tag{4.11}$$

A natural way to estimate $\left[tr\rho \mathcal{P}^+(\rho, \sigma, r) \right]$ is applying Chebyshev's inequality to (4.11). But it will lead us to an estimation in terms of Shannon information quantities, which is not what we want to have.

Based on the works [20], [21] by Petz, T. Ogawa and H. Nagaoka provided an estimation in [19].

Lemma ON ([19]). *For all $s \in [0, 1)$*

$$\log \left[tr\rho \mathcal{P}^+(\rho, \sigma, r) \right] \leq 2^{-rs} tr(\rho^{1+s} \sigma^{-s}). \tag{4.12}$$

It can be shown that

$$\frac{1}{s} tr\rho^{1+s} \sigma^{-s} \to D(\rho \| \sigma) \text{ as } s \to 0, \tag{4.13}$$

which was used by M. Hayashi and H. Nagaoka to obtain a strong converse theorem for stationary memoryless $c-q$ channels in terms of optimal input distributions.

However the convergence in (4.13) is not sufficient for us to obtain a strong converse theorem for non–stationary channel and, roughly saying, we need that

$$\frac{1}{s}\log\left[tr(\rho_t(u)^{1+s}\sigma_t^{-s})\right] \to D\big(\rho_t(u)\|\sigma_t\big) \text{ as } s\to 0 \qquad (4.14)$$

uniformly in t for sets $\big\{\rho_t(u):u\in\mathcal{U}\big\}$ and σ_t, $t=1,2,\ldots,n$ as states. Precisely our estimation is done by the following lemma.

Lemma 2. *For a given positive real number w there exists a positive $a=a(w)$ such that for all states ρ whose minimum positive eigenvalue is not smaller than w, all states σ, and all $s\in\big[0,\frac{1}{2}\big)$*

$$D(\rho\|\sigma)s \le \log[tr\rho^{1+s}\sigma^{-s}] \le D(\rho\|\sigma)s + d^{\frac{1}{2}}\big[a+tr(\rho\sigma^{-1})\big]s^2, \qquad (4.15)$$

where d is the dimension of the Hilbert space $\mathcal{H}H$.

Proof: Let ρ and σ have diagonalizations

$$\rho = \sum_j W(j)|j\rangle\langle j| \qquad (4.16)$$

and

$$\sigma = \sum_y Q(y)|y\rangle\langle y|, \qquad (4.17)$$

respectively. We may assume that the support of ρ contains the support of σ, because otherwise all terms in (4.15) are infinity and we need to do nothing. So in the following we assume all summations run over positive eigenvalues of σ and do not worry about "zero denominators". Then by (4.15) and (4.16)

$$tr\rho^{1+s}\sigma^{-s} = \sum_j\sum_y tr\left[W(j)\left(\frac{W(j)}{Q(y)}\right)^s |j\rangle\langle j||y\rangle\langle y|\right]$$
$$= \sum_j\sum_y W(j)\left(\frac{W(j)}{Q(y)}\right)^s |\langle j|y\rangle|^2. \qquad (4.18)$$

Let $f(s) = tr(\rho^{1+s}\sigma^{-s})$ and $g(s) = \log f(s)$. Then by Taylor expansion we have for $s\in\big[0,\frac{1}{2}\big)$ the estimate

$$g(s) = g(0) + g'(0)s + g''(s_0)s^2 \qquad (4.19)$$

for some $s_0\in[0,s]$. Next we have to calculate the derivations of g. By simple calculation we obtain from (4.18)

$$f'(s) = \sum_j\sum_y W(j)|\langle j|y\rangle|^2\left(\frac{W(j)}{Q(y)}\right)^s \ell n\frac{W(j)}{Q(y)}, \qquad (4.20)$$

and

$$f''(s) = \sum_j \sum_y W(j)|\langle j|y\rangle|^2 \left(\frac{W(j)}{Q(y)}\right)^s \left(\ell n \frac{W(j)}{Q(y)}\right)^2. \qquad (4.21)$$

Next we substitute $f(s) = tr(\rho^{1+s}\sigma^{-s})$ and (4.20) into

$$g'(s) = \frac{f'(s)}{f(s)} \cdot \log e, \qquad (4.22)$$

and $f(s) = tr(\rho^{1+s}\sigma^{-s})$, (4.20) and (4.21) into

$$g''(s) = \left[\frac{f''(s)}{f(s)} - \frac{(f'(s))^2}{f^2(s)}\right] \log e, \qquad (4.23)$$

respectively.

Then we obtain

$$g'(s) = \frac{\sum_j \sum_y W(j)|\langle j|y\rangle|^2 \left(\frac{W(j)}{Q(y)}\right)^s \log \frac{W(j)}{Q(y)}}{tr(\rho^{1+s}\sigma^{-s})} \qquad (4.24)$$

and

$$g''(s) = \left[\frac{\sum_j \sum_y W(j)|\langle j|y\rangle|^2 \left(\frac{W(j)}{Q(y)}\right)^s \left(\ell n \frac{W(j)}{Q(y)}\right)^2}{tr(\rho^{1+s}\sigma^{-s})} \right.$$
$$\left. - \left(\frac{\sum_j \sum_y W(j)|\langle j|y\rangle|^2 \left(\frac{W(j)}{Q(y)}\right)^s \ell n \frac{W(j)}{Q(y)}}{tr(\rho^{1+s}\sigma^{-s})}\right)^2\right] \log e. \qquad (4.25)$$

We are ready to see from (4.25) that for all s

$$g''(s) \geq 0. \qquad (4.26)$$

Indeed by (4.18), $Q_s = \left\{Q_s(j,y) = \frac{W(j)|\langle j|y\rangle|^2\left(\frac{W(j)}{Q(y)}\right)^s}{tr(\rho^{1+s}\sigma^{-s})}\right\}_{(j,y)}$ is a probability distribution. So we may define a random variable $Z(s)$ taking value $\frac{W(j)}{Q(y)}$ with probability $Q_s(j,y)$ and rewrite (4.25) as $g''(s) = \text{Var}\left[\ell n Z(s)\right] \log e$. Moreover by (4.24), (4.16), and (4.17) we have that

$$g'(0) = \sum_j \sum_y W(j)|\langle j|y\rangle|^2 \log W(j) - \sum_j \sum_y W(j)|\langle j|y\rangle|^2 \log Q(y)$$
$$= tr(\rho \log \rho) - tr(\rho \log \sigma) = D(\rho\|\sigma), \qquad (4.27)$$

Which with (4.19) and (4.26) and $g(0) = 0$ yields the first inequality in (4.15).

To obtain the second inequality in (4.15), we have to choose a at the right hand side of (4.15) according to w. We observe that $\lim\limits_{x\to\infty} \frac{x^{\frac{1}{2}}(\ell n \ x^2)\log e}{x} = 0$ and so there exists an $A > 1$ such that for all $s \in \left[0, \frac{1}{2}\right]$, $x \in [A, \infty)$

$$x^s(\ell n \ x)^2 \log e \le x^{\frac{1}{2}}(\ell n \ x)^2 \log e \le x, \tag{4.28}$$

We choose

$$a = \max\left\{ \max_{x\in[w,1]}(\ell n \ x)^2 \log e, \ \max_{x\in[1,A]} x^{\frac{1}{2}}(\ell n \ x)^2 \log e\right\}$$

and then for all $s \in \left[0, \frac{1}{2}\right]$, $x \in [w, A]$,

$$x^s(\ell n \ x)^2 \log e \le a. \tag{4.29}$$

By (4.28) and (4.29) we upperbound

$$x^s(\ell n \ x)^2 \log e \le x + a \tag{4.30}$$

for $s \in \left[0, \frac{1}{2}\right]$ and $x \in [w, \infty)$. Notice that by our assumption that for all j $w \le W(j)$, for all j, y, $\frac{W(j)}{Q(y)} \in [w, \infty)$.

Thus we may apply the upper bound in (4.30) with $x = \frac{W(j)}{Q(y)}$ to (4.25). This gives us that for all $s \in \left[0, \frac{1}{2}\right]$

$$g''(s) \le \frac{1}{tr(\rho^{1+s}\sigma^{-s})} \sum_j \sum_y |\langle j|y\rangle|^2 \left(\frac{W(j)}{Q(y)}\right)^s \left(\ell n \frac{W(j)}{Q(y)}\right)^2 \log e$$

$$\le \frac{1}{tr(\rho^{1+s}\sigma^{-s})} \sum_j \sum_y |\langle j|y\rangle|^2 \left(\frac{W(j)}{Q(y)} + a\right) = \frac{tr(\rho\sigma^{-1}) + a}{tr(\rho^{1+s}\sigma^{-s})}. \tag{4.31}$$

Since for $s > 0$ $\sigma^{-s} \ge I$ and $x^{\frac{3}{2}}$ is convex $\cup$, for $s \in \left[0, \frac{1}{2}\right]$

$$tr(\rho^{1+s}\sigma^{-s}) \ge tr\rho^{1+s} = \sum_j W(j)^{1+s} \ge \sum_j W(j)^{\frac{3}{2}} \ge d\left[\sum_j \frac{1}{d} W(j)\right]^{\frac{3}{2}} = d^{-\frac{1}{2}}. \tag{4.32}$$

Finally the second inequality in (4.15) follows from (4.19), (4.27), (4.31), (4.32) and $g(0) = 0$.

$\square$

Proof of Theorem 1. To conclude the section we prove Theorem 1. Let $\{\Phi\}_{n=1}^{\infty}$ be a non–stationary $c - q$ and let $(\mathcal{U}, \mathcal{D})$ be an (n, M, λ)–code for it. Without loss of generality, we assume it is a code for the maximum probability of error because by Markov's inequality one always may obtain an $\left(n, \left\lfloor \frac{\lambda - \bar\lambda}{\lambda} M \right\rfloor, \lambda\right)$– code for maximum probability of error from an $(n, M, \bar\lambda)$–code with average probability of error for all $\lambda \in (\bar\lambda, 1)$. So, we have for all $u^n \in \mathcal{U}$ that (2.3) holds.

Let $X^n = (X_1, \ldots, X_n)$ be the sequence of random variables uniformly distributed on the codebook $\mathcal{U}$ and for $t = 1, 2, \ldots, n$

$$\sigma_t = E\varphi_t(X_t) = \sum_{x \in \mathcal{X}} P_{X_t}(x)\varphi_t(x) \tag{4.33}$$

where $\varphi_t(x) \in \Phi_t = \{\varphi_t(x) : x \in \mathcal{X}\}$. Let

$$\sigma^{\otimes n} = \sigma_1 \otimes \sigma_2 \otimes \cdots \otimes \sigma_n. \tag{4.34}$$

To prove the theorem, we should apply Lemma 1 to $\{\varphi^{\otimes n}(u^n) : u^n \in \mathcal{U}\}$ and $\sigma^{\otimes n}$, where

$$\varphi^{\otimes n}(u^n) = \varphi_1(u_1) \otimes \varphi_2(u_2) \otimes \cdots \otimes \varphi_1(u_n) \tag{4.35}$$

for $u^n = (u_1, \ldots, u_n)$. To this end we set for all $u^n \in (u_1, \ldots, u_n) \in \mathcal{U}$

$$r(u^n) = \sum_{t=1}^{n} \left\{ D\big(\varphi_t(u_t)\|\sigma_t\big) + \frac{d^{\frac{1}{2}}[a + tr(\varphi_t(u_t)\sigma_t^{-1})] + \log\frac{2}{1-\lambda}}{\sqrt{n}} \right\}, \tag{4.36}$$

where a is the constant in (4.15), Lemma 2 (defined in the proof). We have to verify (4.5) for $\delta = \frac{1-\lambda}{2}$, which will be done by applying Lemma ON and Lemma 2. By Lemma ON, we have that for all $u^n \in \mathcal{U}$ and $s \in [0, 1)$

$$tr\big[(\varphi^{\otimes n}(u^n))\mathcal{P}^+\big(u^n, \sigma^{\otimes n} r(u^n)\big)\big] \leq 2^{-r(u^n)s} tr\big([\varphi^{\otimes n}(u^n)]^{1+s}[\sigma^{\otimes n}]^{-s}\big)$$

$$= 2^{-r(u^n)s} \prod_{t=1}^{n} tr\big[\varphi_t^{1+s}(u_t)\sigma_t^{-s}\big]$$

$$= 2^{-r(u^n)s} 2^{\sum_{t=1}^{n} \log tr[\varphi_t^{1+s}(u_t)\sigma_t^{-s}]}, \tag{4.37}$$

where the first equality follows from (4.34) and (4.35). Next we bound

$$\log tr\big[\varphi_t^{1+s}(u_t)\sigma_t^{-s}\big]$$

by the second inequality in (4.15), Lemma 2 and notice that by condition (2.5) of Theorem 1, for all t, and all $u \in \mathcal{X}$, the minimum positive eigenvalue of $\varphi_t(u)$ is not smaller than w.

Then by Lemma 2, we have that for all $s \in [0, \frac{1}{2})$

$$\log\big[tr\varphi_t^{1+s}(u_t)\sigma_t^{-s}\big] \leq D(\varphi_t(u_t)\|\sigma_t)s + d^{\frac{1}{2}}\big[a + tr(\varphi_t(u_t)\sigma_t^{-1})\big]s^2, \tag{4.38}$$

By substitution of (4.38) into (4.37), we further bound $tr\big(\varphi^{\otimes n}(u^n)\mathcal{P}^+$ $\big(u^n, \sigma^{\otimes n}, r(u^n)\big)$ as follows:

$$tr\big(\varphi^{\otimes n}(u^n)\mathcal{P}^+\big(u^n, \sigma^{\otimes n}, r(u^n)\big) \leq$$

$$\exp_2\left\{-r(u^n)s + \sum_{t=1}^{n}[D\big(\varphi_t(u_t)\|\sigma_t\big)s + d^{\frac{1}{2}}\big[a + tr\big(\varphi_t(u_t)\sigma_t^{-1}\big]s^2\right\} \tag{4.39}$$

for all $s \in \big(0, \frac{1}{2}\big]$. We choose $s = \frac{1}{\sqrt{n}}$ in (4.39) and then substitute (4.36) into it.

Thus we obtain

$$tr\big(\varphi^{\otimes n}(u^n)\mathcal{P}^+\big(u^n, \sigma^{\otimes n}r(u^n)\big)\big)$$

$$\leq \exp_2\left\{\frac{-1}{\sqrt{n}}\sum_{t=1}^{n}\left[D\big(\varphi_t(u_t)\|\sigma_t\big) + \frac{d^{\frac{1}{2}}[a + tr(\varphi_t(u_t)\sigma_t^{-1})] + \log\frac{2}{1-\lambda}}{\sqrt{n}}\right]\right\}$$

$$+ \frac{1}{\sqrt{n}}\sum_{t=1}^{n}D\big(\varphi_t(u_t)\|\sigma_t\big) + \frac{1}{n}\sum_{t=1}^{n}d^{\frac{1}{2}}\big[a + tr\big(\varphi_t(u_t)\sigma_t^{-1}\big)\big]$$

$$= 2^{-\frac{1}{n}\sum_{t=1}^{n}\log\frac{2}{1-\lambda}} = 2^{-\log\frac{2}{1-\lambda}} = \frac{1-\lambda}{2}, \tag{4.40}$$

which with assumption (2.3) yields that (4.5) holds for $\varphi^n(u^n)$, σ^n, $r(u^n)$, $u^n \in \mathcal{U}$, and $\delta = \frac{1-\lambda}{2}$. That is, the conditions of Lemma 1 are satisfied, and thus (for $\{\rho(u) : u \in \mathcal{U}\} = \{\varphi^{\otimes n}(u^n) : u \in \mathcal{U}\}$, $\sigma = \sigma^{\otimes n}$, $u = u^n$) get

$$\log M \leq -M^{-1}\sum_{u^n \in \mathcal{U}}r(u^n) - \log\frac{1-\lambda}{2}. \tag{4.41}$$

Finally we recall the definition of random variables $X^n = (X_1, \ldots, X_n)$, the definition of $r(u^n)$ in (4.36) and treat $r(u^n)$, $D\big(\varphi_t(u)\|\sigma_t\big)$ and $tr(\varphi_t(u)\sigma_t^{-1})$ as functions of u^n and u. Then it follows from (4.41), (4.36), (4.33), and (2.1) that

$$\frac{1}{n}\log M \leq \frac{1}{n}\sum_{x^n \in \mathcal{X}^n}Pr(X^n = x^n)r(x^n) - \frac{1}{n}\log\frac{1-\lambda}{2} = \frac{1}{n}Er(X^n) - \frac{1}{n}\log\frac{1-\lambda}{2}$$

$$= \frac{1}{n}\sum_{t=1}^{n}E\big[D\big(\varphi_t(X_t)\|\sigma_t\big)\big] + n^{-\frac{3}{2}}\sum_{t=1}^{n}d^{\frac{1}{2}}\big[a + E\big[tr\big(\varphi_t(X_t)\sigma_t^{-1}\big)\big]\big]$$

$$- \log\frac{1-\lambda}{2}(n^{-1} + n^{-\frac{1}{2}})$$

$$= \frac{1}{n}\sum_{t=1}^{n}\sum_{x \in \mathcal{X}}P_{X_t}(x)D\big(\varphi_t(x)\|\sigma_t\big) + n^{-\frac{3}{2}}s^{\frac{1}{2}}\sum_{t=1}^{n}\big[a + tr\big[E\big(\varphi_t(X_t)\big)\sigma_t^{-1}\big]\big]$$

$$- \log\frac{1-\lambda}{2}(n^{-1} + n^{-\frac{1}{2}})$$

$$= \frac{1}{n}\sum_{t=1}^{n}\chi(P_{X_t}; \Phi_t) + n^{-\frac{3}{2}}d^{\frac{1}{2}}\sum_{t=1}^{n}\big[a + tr(\sigma_t \cdot \sigma_t^{-1})\big]$$

$$- \log\frac{1-\lambda}{2}(n^{-1} + n^{-\frac{1}{2}})$$

$$\leq \frac{1}{n}\sum_{t=1}^{n}\chi(P_{X_t}; \Phi_t) + n^{-\frac{1}{2}}d^{\frac{1}{2}}(a + d) - 2n^{-\frac{1}{2}}\log\frac{1-\lambda}{2}$$

$$= \sum_{t=1}^{n}\frac{1}{n}\chi(P_{X_t}; \Phi_t) + \frac{1}{\sqrt{n}}\left[d^{\frac{1}{2}}(a + d) - 2\log\frac{1-\lambda}{2}\right]. \tag{4.42}$$

Finally we complete our proof by choosing $h(\lambda) = d^{\frac{1}{2}}(a+d) - 2\log\frac{1-\lambda}{2}$ in (4.42). □

5   The Proof of the Main Result

In previous section we have shown Theorem 1, which completed the first step of the plan to prove Theorem 2. In this section we shall finish our proof according to the plan in Section 3. The second step is to apply wringing technique Lemma A and it directly follows from the Lemma the

Corollary A. *For given finite sets $\mathcal{X}$ and $\mathcal{Y}$, $\mathcal{U} \subset \mathcal{X}^n$ and $\mathcal{V} \in \mathcal{Y}^n$ and a subset $\mathcal{A} \subset \mathcal{U} \times \mathcal{V}$ with cardinality $|\mathcal{A}| \geq \beta|\mathcal{U}| \times |\mathcal{V}|$ for a $\beta \in (0,1)$, a $\gamma \in (0, \beta^{-1} - 1)$, and $\varepsilon > 0$ there exists $t_1, t_2, \ldots, t_k \in \{1, 2, \ldots, n\}$ and $(x_{t_1}, y_{t_1}), \ldots, (x_{t_k}, y_{t_k})$ for a $k \leq \frac{\beta^{-1}-1}{\gamma}$ such that the section of $\mathcal{A}$ at $(x_{t_1}, y_{t_1}), (x_{t_2}, y_{t_2}), \ldots, (x_{t_k}, y_{t_k})$,*

$$\mathcal{B} = \left\{(u^n, v^n) \in \mathcal{A} : (u_{t_i}, v_{t_i}) = (x_{t_i}, y_{t_i}), i = 1, 2, \ldots, k\right\}$$

has the following properties.

$$|\mathcal{B}| \geq \varepsilon^k |\mathcal{A}|, \tag{5.1}$$

and for the pair $(\bar{X}^n, \bar{Y}^n)$ of sequences of random variables uniformly distributed on B,

$$(1 + \gamma)Pr(\bar{X}_t = x)Pr(\bar{Y}_t = y) - \gamma - |\mathcal{X}||\mathcal{Y}|\varepsilon$$
$$\leq Pr(\bar{X}_t = x, \bar{Y}_t = y) \leq \max\left((1 + \gamma)Pr(\bar{X}_t = x)Pr(\bar{Y}_t = y), \varepsilon\right) \tag{5.2}$$

for all $x \in \mathcal{X}$, $y \in \mathcal{Y}$ and $t = 1, 2, \ldots, n$.

The corollary is actually Corollary 2 in [3]. (The only difference is that we now remove the assumption in [3] that $\mathcal{U} \times \mathcal{V}$ is a codebook for the MAC but this is obviously not an essential assumption.) (5.1) and the first inequality in (5.2) are a simple consequence of Lemma 1 with the choice of P as uniform distribution on $\mathcal{A}$ and Q as uniform distribution of $\mathcal{U} \times \mathcal{V}$. The second inequality in (5.2) follows the first inequality in (5.2) and the fact that $\sum_{(x,y)\in\mathcal{X}\times\mathcal{Y}} Pr(\bar{X}_t = x)Pr(\bar{Y}_t = y) = 1$.

So we omit the details. Readers can make them by themselves or read [3].

Now let Φ be a $c-q$ MAC. We have to find an $\tilde{h}$ such that for all $(n, M, N, \bar{\lambda})$-codes and the channel (2.23) – (2.25) hold. Suppose that we are given an $(n, M, N, \bar{\lambda})$-code $(\mathcal{U}, \mathcal{V}, \mathcal{D})$ for the $c-q$ MAC Φ. Let $\lambda = \frac{1+\bar{\lambda}}{2}$ and $\mathcal{A} = \left\{(u^n, v^n) : tr\left[\varphi(u^n, v^n)\mathcal{D}_{u^n, v^n}\right] > 1 - \lambda, u^n \in \mathcal{U}, v^n \in \mathcal{V}\right\}$. Then we obtain that

$$|\mathcal{A}| > \left(1 - \frac{\bar{\lambda}}{\lambda}\right)MN = \frac{1-\bar{\lambda}}{1+\bar{\lambda}}MN \tag{5.3}$$

by applying Shannon's well known approach to (2.8). By definition for all $(u^n, v^n) \in \mathcal{A}$

$$tr\left[\varphi(u^n, v^n)\mathcal{D}_{u^n, v^n}\right] > 1 - \lambda = \frac{1-\bar{\lambda}}{2}. \tag{5.4}$$

Let $\beta = \frac{1-\bar\lambda}{1+\bar\lambda}$ $\gamma = \frac{1}{\sqrt{n}}$, and $\varepsilon = \frac{1}{\sqrt{n}}$. Then corollary A is followed by that there exists a $\mathcal{B} \subset \mathcal{A}$ in Corollary A such that for a $k \leq \frac{2\bar\lambda}{1-\bar\lambda}\sqrt{n}$,

$$|\mathcal{B}| \geq n^{-\frac{\bar\lambda}{1-\bar\lambda}\sqrt{n}}|\mathcal{A}| \geq \frac{1-\bar\lambda}{1+\bar\lambda}n^{-\frac{\bar\lambda}{1-\bar\lambda}\sqrt{n}}MN, \qquad (5.5)$$

where the last inequality follows from (5.3), and (5.2) holds for the pair of sequences of random variables $(\bar{X}^n, \bar{Y}^n)$ uniformly distributed on $\mathcal{B}$. That is, for all $x \in \mathcal{X}$, $y \in \mathcal{Y}$, and $t = 1, 2, \dots, n$,

$$\frac{1}{\sqrt{n}}\big[Pr(\bar{X}_t = x)Pr(\bar{Y}_t = y) - 1 - |\mathcal{X}||\mathcal{Y}|\big]$$

$$\leq Pr(\bar{X}_t = x, \bar{Y}_t = y) - Pr(\bar{X}_t = x)Pr(\bar{Y}_t = y)$$

$$\leq \frac{1}{\sqrt{n}}\big[Pr(\bar{X}_t = x)Pr(\bar{Y}_t = y) + 1\big], \qquad (5.6)$$

where we use the assumption $\gamma = \varepsilon = \frac{1}{\sqrt{n}}$ and to obtain the last inequality we use the obvious inequality

$$\max\left\{\left(1 + \frac{1}{\sqrt{n}}\right)Pr(\bar{X}_t = x)Pr(\bar{Y}_t = y), \frac{1}{\sqrt{n}}\right\} \leq \left(1 + \frac{1}{\sqrt{n}}\right)Pr(\bar{X}_t = x)Pr(\bar{Y}_t = y) + \frac{1}{\sqrt{n}}.$$

We first treat our $c-q$ MAC channel as a stationary memoryless $c-q$ channel with input alphabet $\mathcal{X} \times \mathcal{Y}$ and it is obvious that $(\mathcal{B}, \mathcal{D})$ is an $(n, |\mathcal{B}|, \lambda)$–code for it, where $\lambda = \frac{1+\bar\lambda}{2}$.

So we may apply Theorem 1 to it and obtain

$$|\mathcal{B}| \leq \sum_{t=1}^{n}\frac{1}{n}\chi(P_{\bar{X}_t\bar{Y}_t}; \Phi) + \frac{1}{\sqrt{n}}h(\lambda) = \frac{1}{n}\sum_{t=1}^{n}\chi(\bar{X}_t, \bar{Y}_t; \Phi) + \frac{1}{\sqrt{n}}h(\lambda), \qquad (5.7)$$

which together with (5.5) implies

$$\frac{1}{n}\log M + \frac{1}{n}\log N \leq \frac{1}{n}\sum_{t=1}^{n}\chi(\bar{X}_t, \bar{Y}_t; \Phi) + \frac{1}{\sqrt{n}}h(\lambda) + \frac{1}{\sqrt{n}}\frac{\bar\lambda}{1-\bar\lambda}\log n + \frac{1}{n}\log\frac{1+\bar\lambda}{1-\bar\lambda}. \qquad (5.8)$$

Next we get for $y^n \in \mathcal{Y}^n$

$$\mathcal{B}(y^n) = \big\{(u^n, y^n) : (u^n, y^n) \in \mathcal{B}\big\}. \qquad (5.9)$$

Then by our definitions

$$\mathcal{B}(y^n) = \phi \text{ if } y^n \notin \mathcal{V}^n, \qquad (5.10)$$

$$Pr(\bar{Y}^n = y^n) = \frac{|\bar{\mathcal{B}}(y^n)|}{|\mathcal{B}|}, \qquad (5.11)$$

and $\{\mathcal{B}(v^n) : v^n \in \mathcal{V}, \mathcal{B}(v^n) \neq \phi\}$ is a partition of $\mathcal{B}$. We partition $\mathcal{V}$ into two parts according to $P_{\bar{Y}^n}$,

$$\mathcal{V}^+ = \left\{v^n \in \mathcal{V} : Pr(\bar{Y}^n = v^n) \geq (nN)^{-1}\right\} \tag{5.12}$$

and

$$\mathcal{V}^- = \left\{v^n \in \mathcal{V} : Pr(\bar{Y}^n = v^n) < (nN)^{-1}\right\}. \tag{5.13}$$

Then by (5.10) and (5.11)

$$1 = Pr(\bar{Y}^n \in \mathcal{V}) < Pr(\bar{Y}^n \in \mathcal{V}^+) + (nN)^{-1}|\mathcal{V}| = Pr(\bar{Y}^n \in \mathcal{V}^+) + \frac{1}{n},$$

or

$$Pr(\bar{Y}^n \in \mathcal{V}^+) > 1 - \frac{1}{n}. \tag{5.14}$$

Now we combine (5.11) and (5.12) with (5.5) and obtain that for all $v^n \in \mathcal{V}^+$

$$|\mathcal{B}(v^n)| \geq (nN)^{-1}|\mathcal{B}| \geq \frac{1-\bar{\lambda}}{1+\bar{\lambda}} n^{-\left(\frac{\bar{\lambda}}{1-\bar{\lambda}}\sqrt{n}+1\right)} M, \tag{5.15}$$

which together with (5.14) yields

$$\sum_{v^n \in \mathcal{V}^n} Pr(\bar{Y}^n = v^n) \log |\mathcal{B}(v^n)| \geq \sum_{v^n \in \mathcal{V}^+} Pr(\bar{Y}^n = v^n) \log |\mathcal{B}(v^n)|$$

$$\geq \left(1 - \frac{1}{n}\right)\left[\log M - \left(\frac{\bar{\lambda}}{1-\bar{\lambda}}\sqrt{n} + 1\right)\log n + \log\frac{1-\bar{\lambda}}{1+\bar{\lambda}}\right]$$

$$\geq \left(1 - \frac{1}{n}\right)\left[\log M - \sqrt{n}\log n\left(\frac{1}{1-\bar{\lambda}} + 1 + \log\frac{1+\bar{\lambda}}{1-\bar{\lambda}}\right)\right]$$

$$\geq \log M - \frac{1}{n}\log|\mathcal{X}^n| - \sqrt{n}\log n\left(\frac{1}{1-\bar{\lambda}} + 1 + \log\frac{1+\bar{\lambda}}{1-\bar{\lambda}}\right)$$

$$\geq \log M - \sqrt{n}\log n\left(\frac{2-\bar{\lambda}}{1-\bar{\lambda}} + \log\left(\frac{1+\bar{\lambda}}{1-\bar{\lambda}}|\mathcal{X}|\right)\right),$$

i.e.,

$$\frac{1}{n}\log M \leq \frac{1}{n}\sum_{v^n \in \mathcal{V}^n} Pr(\bar{Y}^n = v^n)\log|\mathcal{B}(v^n)| + \frac{\log n}{\sqrt{n}}\left(\frac{2}{1-\bar{\lambda}} + \log\left(\frac{1+\bar{\lambda}}{1-\bar{\lambda}}|\mathcal{X}|\right)\right). \tag{5.16}$$

Here we use the convention "$0\log 0 = 0$". For $v^n \in \mathcal{V}$, we let

$$\mathcal{U}_{\mathcal{B}}(v^n) = \left\{u^n \in \mathcal{U} : (u^n, v^n) \in \mathcal{B}(v^n)\right\}. \tag{5.17}$$

Then obviously by definition $(\mathcal{U}_\mathcal{B}(v^n), \mathcal{D})$ is an $(n, |\mathcal{B}(v^n)|, \lambda)$–code for non-stationary memoryless $c - q$ channel $\{\Phi_t(\cdot, v_t)\}_t$ where for $v^n = (v_1, v_2, \ldots, v_n)$, $\Phi = \{\varphi(x, y) = x \in \mathcal{X}, y \in \mathcal{Y}\}$, $\Phi_t(\cdot, v_t) = \{\varphi(x, v_t) : x \in \mathcal{X}\}$, and $P_{\bar{X}^n | \bar{Y}^n}(\cdot | v^n)$ is the uniform distribution on $\mathcal{B}(v^n)$. Denote by $w(x, y)$, the minimum positive eigenvalue of $\varphi(x, y)$ and by $w = \min\limits_{x,y} w(x, y)$. Then $w > 0$ since $|\mathcal{X}||\mathcal{Y}| < \infty$, and (2.5) holds. Consequently by Theorem 1 we have that

$$
\frac{1}{n} \log |\mathcal{B}(v^n)| \leq \sum_{t=1}^n \frac{1}{n} \left[S \left(\sum_{x \in \mathcal{X}} P_{\bar{X}_t | \bar{Y}_t^n}(x | v^n) \varphi(x, v^n) \right) \right.
$$
$$
\left. - \sum_{x \in \mathcal{X}} P_{\bar{X}_t | \bar{Y}_t^n}(x | v^n) S(\varphi(x, v_t)) \right] + \frac{h(\lambda)}{\sqrt{n}}. \tag{5.18}
$$

Notice that $\varphi(x_t, v_t)$ depends on $x^n = (x_1, \ldots, x_n)$ through x_t for fixed v. One may rewrite (5.18) as

$$
\frac{1}{n} \log |\mathcal{B}(v^n)| \leq \sum_{t=1}^n \frac{1}{n} \left[S \left(\sum_{x^n \in \mathcal{X}^n} P_{\bar{X}_t^n | \bar{Y}^n}(x^n | v^n) \varphi(x_t, v_t) \right) \right.
$$
$$
\left. - \sum_{x^n \in \mathcal{X}^n} P_{\bar{X}_t^n | \bar{Y}_t^n}(x^n | v^n) S(\varphi(x_t, w_t)) \right] + \frac{h(\lambda)}{\sqrt{n}}. \tag{5.19}
$$

Next by the concavity $(\cap)$ of von Neumann entropy and Jensens inequality we have

$$
\sum_{y^n \in \mathcal{Y}^n} P_{\bar{Y}^n}(y^n) S \left(\sum_{x^n \in \mathcal{X}^n} P_{\bar{X}^n | \bar{Y}^n}(x^n | y^n) \varphi(x_t, y_t) \right)
$$
$$
= \sum_{y_t \in \mathcal{Y}} P_{\bar{Y}_t}(y) \sum_{i \neq t} \sum_{y_i \in \mathcal{Y}} Pr(\tilde{Y}_i = y_i, i \neq t | \tilde{Y}_t = y_t) S \left(\sum_{x^n \in \mathcal{X}^n} P_{\bar{X}^n | \bar{Y}^n}(x^n, y^n) \varphi(x_t, y_t) \right)
$$
$$
\leq \sum_{y_t \in \mathcal{Y}} P_{\bar{Y}_t}(y_t) S
$$
$$
\left[\sum_{i \neq t} \sum_{y_i \in \mathcal{Y}} Pr(\bar{Y}_i - y_i, i \neq t | \bar{Y}_t = y_t) \sum_{x^n \in \mathcal{X}^n} Pr(\bar{X}^n = x^n | \bar{Y}^n = y^n) \varphi(x_t, y_t) \right]
$$
$$
= \sum_{y_t \in \mathcal{Y}} P_{\bar{Y}_t}(y_t) S \left(\sum_{x_t \in \mathcal{X}} P_{\bar{X}_t | \bar{Y}_t}(x_t | y_t) \varphi(x_t, y_t) \right) = \sum_{y \in \mathcal{Y}} P_{\bar{Y}_t}(y) S(\sigma_{\bar{X}_t \bar{Y}_t, 2}(y))
$$
$$
\tag{5.20}
$$

for $\sigma_{\bar{X}_t \bar{Y}_t, 2}(y) = \sum_{x \in \mathcal{X}} P_{\bar{X}_t | \bar{Y}_t}(x | y) \varphi(x, y)$.

This and the fact

$$\sum_{y^n \in \mathcal{Y}^n} P_{\bar{Y}^n}(y^n) \sum_{x^n \in \mathcal{X}} P_{\bar{X}^n|\bar{Y}^n}(x^n|y^n) S\big(\varphi(x_t, y_t)\big)$$

$$= \sum_{y \in \mathcal{Y}} P_{\bar{Y}_t}(y) \sum_{x \in \mathcal{X}} P_{\bar{X}_t|\bar{Y}_t}(x|y) S\big(\varphi(x_t, y_t)\big)$$

together with (2.11) imply that

$$\frac{1}{n} \sum_{t=1}^{n} P_{\bar{Y}^n}(y^n) \log |\mathcal{B}(y^n)|$$

$$\leq \frac{1}{n} \sum_{t=1}^{n} \left[\sum_{y \in \mathcal{Y}} P_{\bar{Y}_t}(y) \left(S\big(\sigma_{\bar{X}_t \bar{Y}_t, 2}(y)\big) - \sum_{x \in \mathcal{X}} P_{\bar{X}_t|\bar{Y}_t}(x|y) S\big(\varphi(x_t, y_t)\big) \right) \right] + \frac{h(\lambda)}{\sqrt{n}}$$

$$= \frac{1}{n} \sum_{t=1}^{n} \chi(\bar{X}_t; \Phi|\bar{Y}_t) + \frac{h(\lambda)}{\sqrt{n}}. \tag{5.21}$$

Recalling that $P(\bar{Y}^n = y^n) = 0$, if $y^n \notin \mathcal{V}^n$, by combining (5.16) with (5.21), we have that

$$\frac{1}{n} \log M \leq \frac{1}{n} \sum_{t=1}^{n} \chi(\bar{X}_t; \Phi|\bar{Y}_t) + \frac{\log n}{\sqrt{n}} \left(\frac{2}{1-\lambda} + \log \left(\frac{1+\bar{\lambda}}{1-\bar{\lambda}} |\mathcal{X}| \right) \right) + \frac{h(\lambda)}{\sqrt{n}}$$

$$\leq \frac{1}{n} \sum_{t=1}^{n} \chi(\bar{X}_t; \Phi|\bar{Y}_t) + \frac{\log n}{\sqrt{n}} \left(\frac{2}{1-\bar{\lambda}} + \log \left(\frac{1+\bar{\lambda}}{1-\bar{\lambda}} |\mathcal{X}| \right) + h\left(\frac{1+\bar{\lambda}}{2} \right) \right), \tag{5.22}$$

where in the last step we use our choice $\lambda = \frac{1+\bar{\lambda}}{2}$. By interchanging the roles of $\bar{X}^n$ and $\bar{Y}^n$, we obtain in the same way that

$$\frac{1}{n} \log N \leq \frac{1}{n} \sum_{t=1}^{n} \chi(\bar{Y}_t|\bar{X}_t) + \frac{\log n}{\sqrt{n}} \left(\frac{2}{1-\bar{\lambda}} + \log \left(\frac{1+\bar{\lambda}}{1-\bar{\lambda}} |\mathcal{Y}| \right) + h\left(\frac{1+\bar{\lambda}}{2} \right) \right). \tag{5.23}$$

So for our proof has not been done and we only finished the second step of our plan although (5.8), (5.22), and (5.23) have the same form as (2.23), (2.24), and (2.25), because $(\bar{X}_t, \bar{Y}_t)$ may not be independent. We have to replace $(\bar{X}_t, \bar{Y}_t)$ by a pair of random variables $(\tilde{X}_t, \tilde{Y}_t)$ with distribution

$$P_{\tilde{X}_t \tilde{Y}_t}(x, y) = P_{\bar{X}_t}(x) P_{\bar{X}_t}(y) \tag{5.24}$$

for $x \in \mathcal{X}, y \in \mathcal{Y}$, for $t = 1, 2, \ldots, n$. This is finished in the last step.

In the calculation we need two basic inequalities from Quantum Information Theory.

Strong Convexity of the Trace Distance (P. 407 [17])

Let $\{P_i\}$ and $\{Q_i\}$ be two probability distributions and let $\{\rho_i\}$ and $\{\sigma_i\}$ be two sets of states. Then

$$tr\left(\left\|\sum_i P_i\rho_i - \sum_i Q_i\sigma_i\right\|\right) \le \sum_i |P_i - Q_i| + \sum_i P_i tr|\rho_i - \sigma_i|. \qquad (5.25)$$

Fanne's Inequality (Continuity of von Neumann Entropy ([10] also P. 512 [17])
For two states ρ and σ

$$|S(\varphi) - S(\sigma)| \le \frac{1}{2}tr|\rho - \sigma| \log \frac{2d}{tr|\rho - \sigma|}. \qquad (5.26)$$

In the following let us denote by $PQ^+ = \{z \in \mathcal{Z} : P(z) \ge Q(z)\}$ for two probability distributions P and Q on the same set Z. Then the second inequality in (5.6) implies that

$$\sum_{x\in\mathcal{X}y\in\mathcal{Y}} |P_{\bar{X}_t\bar{Y}_t}(x,y) - P_{\bar{X}_t}(x)P_{\bar{Y}_t}(y)|$$
$$\le 2 \sum_{(x,y)\in P_{\bar{X}_t\bar{Y}_t}(P_{\bar{X}_t}P_{\bar{Y}_t})} +P_{\bar{X}_t\bar{Y}_t}(x,y) - P_{\bar{X}_t}(x)P_{\bar{Y}_t}(y)$$
$$\le \frac{2}{\sqrt{n}}\left(1 + |\mathcal{X}||\mathcal{Y}|\right). \qquad (5.27)$$

So by (5.25) and (5.27) we have

$$tr\left(\left\|\sum_{x\in\mathcal{X}y\in\mathcal{Y}} P_{\bar{X}_t\bar{Y}_t}(x,y)\varphi(x,y) - \sum_{x\in\mathcal{X}y\in\mathcal{Y}} P_{\bar{X}_t}(x)P_{\bar{Y}_t}(y)\varphi(x,y)\right\|\right)$$
$$\le \sum_{x\in\mathcal{X}y\in\mathcal{Y}} |P_{\bar{X}_t\bar{Y}_t}(x,y) - P_{\bar{X}_t}(x)P_{\bar{Y}_t}(y)|$$
$$\le \frac{2}{\sqrt{n}}\left(1 + |\mathcal{X}||\mathcal{Y}|\right) \qquad (5.29)$$

Moreover, by (5.27)

$$\left|\sum_{x,y} P_{\bar{X}_t\bar{Y}_t}(x,y)S(\varphi(x,y)) - \sum_{x,y} P_{\bar{X}_t}(x)P_{\bar{Y}_t}(y)S(\varphi(x,y))\right|$$
$$\le \sum_{x,y} |P_{\bar{X}_t\bar{Y}_t}(x,y) - P_{\bar{X}_t}(x)P_{\bar{Y}_t}(y)|S(\varphi(x,y))$$
$$\le \log d \sum_{x,y} |P_{\bar{X}_t\bar{Y}_t}(x,y) - P_{\bar{X}_t}(x)P_{\bar{Y}_t}(y)|$$
$$\le \frac{2}{\sqrt{n}}\left(1 + |\mathcal{X}||\mathcal{Y}|\right) \log d. \qquad (5.30)$$

Now (5.26) and (5.30) imply that

$$|\chi(\bar{X}_t, \bar{Y}_t; \Phi) - \chi(\tilde{X}_t, \tilde{Y}_t; \Phi)|$$

$$= \left| \left[S\left(\sum_{x,y} P_{\bar{X}_t \bar{Y}_t}(x,y)\varphi(x,y) \right) - \sum_{x,y} P_{\bar{X}_t \bar{Y}_t}(x,y)S(\varphi(x,y)) \right] \right.$$

$$\left. - \left[S\left(\sum_{x,y} P_{\bar{X}_t}(x)P_{\bar{Y}_t}(y)\varphi(x,y) \right) - \sum_{x,y} P_{\bar{X}_t}(x)P_{\bar{Y}_t}(y)S(\varphi(x,y)) \right] \right|$$

$$\leq \left| S\left(\sum_{x,y} P_{\bar{X}_t \bar{Y}_t}(x,y)\varphi(x,y) \right) - S\left(\sum_{x,y} P_{\bar{X}_t}(x)P_{\bar{Y}_t}(y)\varphi(x,y) \right) \right|$$

$$+ \left| \sum_{x,y} P_{\bar{X}_t \bar{Y}_t}(x,y)S(\varphi(x,y)) - \sum_{x,y} P_{\bar{X}_t}(x)P_{\bar{Y}_t}(y)S(\varphi(x,y)) \right|$$

$$\leq \frac{1}{\sqrt{n}}(1 + |\mathcal{X}||\mathcal{Y}|)\log\frac{\sqrt{n}d}{1 + |\mathcal{X}||\mathcal{Y}|} + \frac{2}{\sqrt{n}}(1 + |\mathcal{X}||\mathcal{Y}|)\log d$$

$$< \frac{1}{\sqrt{n}}(1 + |\mathcal{X}||\mathcal{Y}|)\log\frac{\sqrt{n}d^3}{1 + |\mathcal{X}||\mathcal{Y}|}$$

$$< \frac{1}{\sqrt{n}}(1 + |\mathcal{X}||\mathcal{Y}|)\log\sqrt{n}d^3. \tag{5.31}$$

Next let us turn to estimate the difference $|\chi(\bar{X}_t; \Phi|\bar{Y}_t) - \chi(\tilde{X}_t; \Phi|\tilde{Y}_t)|$. To this end we have to upper bound the difference $\left| \sum_{y \in \mathcal{Y}} P_{\bar{Y}_t}(y)S\left(\sigma_{\bar{X}_t \bar{Y}_t, 2}(y)\right) - \sum_{y \in \mathcal{Y}} P_{\bar{Y}_t}(y)S\left(\sigma_{\tilde{X}_t \tilde{Y}_t, 2}(y)\right) \right|$ for

$$\sigma_{\bar{X}_t \bar{Y}_t, 2}(y) = \sum_{x \in \mathcal{X}} P_{\bar{X}_t|\bar{Y}_t}(x|y)\varphi(x,y) \text{ and } \sigma_{\tilde{X}_t \tilde{Y}_t, 2}(y) = \sum_{x \in \mathcal{X}} P_{\bar{X}_t}(x)\varphi(x,y).$$

Since for all y $\sum_x P_{\bar{X}_t \bar{Y}_t}(x,y) = \sum_x P_{\bar{X}_t}(x)P_{\bar{Y}_t}(y) = P_{\bar{Y}_t}(y)$, if $P_{\bar{Y}_t}(y) \neq 0$, by (5.6)

$$P_{\bar{Y}_t}(y) \sum_{x \in \mathcal{X}} |P_{\bar{X}_t|\bar{Y}_t}(y|x) - P_{\bar{X}_t}(x)| = \sum_{x \in \mathcal{X}} |P_{\bar{X}_t \bar{Y}_t}(x,y) - P_{\bar{X}_t}(x)P_{\bar{Y}_t}(y)|$$

$$= 2\Sigma^+\left(P_{\bar{X}_t \bar{Y}_t}(x,y) - P_{\bar{X}_t}(x)P_{\bar{Y}_t}(y)\right)$$

$$\leq \frac{2}{\sqrt{n}}\left(P_{\bar{Y}_t}(y) + |\mathcal{X}|\right)$$

$$\leq \frac{2}{\sqrt{n}}(1 + |\mathcal{X}|), \tag{5.32}$$

where the sum Σ^+ is taken over all $x \in \mathcal{X}$ with $P_{\bar{X}_t|\bar{Y}_t}(x|y) \geq P_{\bar{X}_t}(x)$.

Thus by (5.26), (5.25), and (5.32), we have that

$$
\left| \sum_{y \in \mathcal{Y}} P_{\bar{Y}_t}(y) S(\sigma_{\bar{X}_t | \bar{Y}_t, 2}(y)) - \sum_{y \in \mathcal{Y}} P_{\bar{Y}_t}(y) S(\sigma_{\tilde{X}_t \tilde{Y}_t, 2}(y)) \right|
$$

$$
\leq \sum_{y \in \mathcal{Y}} P_{\bar{Y}_t}(y) |S(\sigma_{\bar{X}_t \bar{Y}_t, 2}(y)) - S(\sigma_{\tilde{X}_t \tilde{Y}_t, 2}(y))|
$$

$$
\leq \sum_{y \in \mathcal{Y}} P_{\bar{Y}_t}(y) \cdot \frac{1}{2} tr |\sigma_{\bar{X}_t \bar{Y}_t, 2}(y) - \sigma_{\tilde{X}_t \tilde{Y}_t, 2}(y)| \log \frac{2d}{tr |\sigma_{\bar{X}_t \bar{Y}_t, 2}(y) - \sigma_{\tilde{X}_t \tilde{Y}_t, 2}(y)|}
$$

$$
\leq \frac{1}{2} \sum_{y \in \mathcal{Y}} P_{\bar{X}_t}(y) tr |\sigma_{\bar{X}_t \bar{Y}_t, 2}(y) - \sigma_{\tilde{X}_t \tilde{Y}_t, 2}(y)| \log \frac{2d}{P_{\bar{X}_t}(y) tr |\sigma_{\bar{X}_t \bar{Y}_t, 2}(y) - \sigma_{\tilde{X}_t \tilde{Y}_t, 2}(y)|}
$$

$$
\leq \frac{1}{2} \sum_{y \in \mathcal{Y}} P_{\bar{Y}_t}(y) \sum_{x \in \mathcal{X}} |P_{\bar{X}_t | \bar{Y}_t}(x|y) - P_{\bar{X}_t}(x)| \log \frac{2d}{P_{\bar{Y}_t}(y) \sum_{x \in \mathcal{X}} |P_{\bar{X}_t | \bar{Y}_t}(x|y) - P_{\bar{X}_t}(x)| \varphi(x, y)}
$$

$$
\leq \frac{1}{2} \sum_{y \in \mathcal{Y}} \frac{2}{\sqrt{n}} (1 + |\mathcal{X}|) \log \frac{d\sqrt{n}}{1 + |\mathcal{X}|}
$$

$$
= \frac{1}{\sqrt{n}} (1 + |\mathcal{X}|) |\mathcal{Y}| \log \frac{d\sqrt{n}}{1 + |\mathcal{X}|}
$$

$$
\leq \frac{1}{\sqrt{n}} (1 + |\mathcal{X}|) |\mathcal{Y}| \log d\sqrt{n} \tag{5.33}
$$

where the second inequality holds by (5.26); the fourth inequality follows from (5.25) and the monotonicity of $z \log \frac{2d}{z}$ in the interval $\left[0, \frac{2d}{e}\right]$; and the fifth inequality follows from (5.32).

Considering that by (2.11)

$$
\chi(\bar{X}_t; \Phi|\bar{Y}_t) = \sum_{y \in \mathcal{Y}} P_{\bar{Y}_t} S(\sigma_{\bar{X}_t \bar{Y}_t, 2}(y)) - \sum_{x, y} P_{\bar{X}_t \bar{Y}_t}(x, y) S(\varphi(x, y))
$$

and $\chi(\tilde{X}_t; \Phi|\tilde{Y}_t) = \sum\limits_{y \in \mathcal{Y}} P_{\bar{Y}_t} S(\sigma_{\tilde{X}_t \tilde{Y}_t, 2}(y)) - \sum_{x, y} P_{\bar{X}_t}(x) P_{\bar{Y}_t}(y) S(\varphi(x, y))$, we add up (5.30) and (5.33),

$$
|\chi(\bar{X}_t; \Phi|\bar{Y}_t) - \chi(\tilde{X}_t; \Phi|\tilde{Y}_t)|
$$

$$
\leq \left| \sum_{y \in \mathcal{Y}} P_{\bar{Y}_t}(y) S(\sigma_{\bar{X}_t \bar{Y}_t, 2}(y)) - \sum_{y \in \mathcal{Y}} P_{\bar{Y}_t}(y) S(\sigma_{\tilde{X}_t \tilde{Y}_t, 2}(y)) \right|
$$

$$
+ \left| \sum_{x, y} P_{\bar{X}_t \bar{Y}_t}(x, y) S(\varphi(x, y)) - \sum_{x, y} P_{\bar{X}_t}(x) P_{\bar{Y}_t}(y) S(\varphi(x, y)) \right|
$$

$$
\leq \frac{1}{\sqrt{n}} (2 + |\mathcal{Y}| + 3|\mathcal{X}||\mathcal{Y}|) \log d + \frac{1}{2\sqrt{n}} (1 + |\mathcal{X}|) |\mathcal{Y}| \log n. \tag{5.34}
$$

Next, we exchange the roles of $\bar{X}_t$ and $\bar{Y}_t$ in (5.34), in the same way we obtain that

$$
|\chi(\bar{Y}_t; \Phi|\bar{X}_t) - \chi(\tilde{Y}_t; \Phi|\tilde{X}_t)| \leq \frac{1}{\sqrt{n}} (2 + |\mathcal{X}| + 3|\mathcal{X}||\mathcal{Y}|) \log d + \frac{1}{2\sqrt{n}} (1 + |\mathcal{Y}|) |\mathcal{X}| \log n.
$$

$$
\tag{5.35}
$$

Finally we set

$$\tilde{h}(\bar{\lambda}) = h\left(\frac{1+\bar{\lambda}}{2}\right) + \frac{2+\bar{\lambda}}{1-\bar{\lambda}} + \log\left(\frac{1+\bar{\lambda}}{1-\bar{\lambda}}|\mathcal{X}||\mathcal{Y}|\right) + (3+6|\mathcal{X}||\mathcal{Y}|)\log d$$

and combine (5.8) with (5.31), (5.22) with (5.34), and (5.23) with (5.35), respectively. Then (2.23) – (2.25) follow.   □

References

1. R. Ahlswede, Beiträge zur Shannonschen Informationstheorie im Falle nichtstationärer Kanäle, Z. Wahrsch. und verw. Gebiete, Vol. 10, 1–42, 1968.
2. R. Ahlswede, Multi–way communication channels, 2nd Intern. Symp. on Inform. Theory, Thakadsor, 1971, Publ. House of the Hungarian Acad. of Sci., 23–52, 1973.
3. R. Ahlswede, An elementary proof of the strong converse theorem for the multiple–access channel, J. Comb. Inform. Sys. Sci., Vol. 7, 216–230, 1982.
4. R. Ahlswede, Performance parameters for channels, this volume.
5. A.E. Allahverdyan and D.B. Saakian, Multi–access channels in quantum information theory, http://xxx.lanl.gov/abs/quant-ph/9712034V1, 1997.
6. S. Arimoto, On the converse to the coding theorem for discrete memoryless channels, IEEE Trans. Inform. Theory, Vol. 19, 375–393, 1963.
7. U. Augustin, Gedächtnisfreie Kanäle für diskrete Zeit, Z. Wahrscheinlichkeitstheorie u. verw. Gebiete, Vol. 6, 10–61, 1966.
8. G. Dueck, Maximal error capacity regions are smaller than average error capacity regions for multi–user channels, Problems of Control and Information Theory, Vol. 7, 11–19, 1978.
9. G. Dueck, The strong converse to the coding theorem for the multiple–access channel, J. Comb. Inform. Syst. Sci., Vol. 6, 187–196, 1981.
10. M. Fannes, A continuity property of the entropy density for spin lattice systems, Common. Math. Phys. 31, 291–294, 1973.
11. R.M. Fano, Class Notes for Transmission of Information, Course 6.574, MIT, Cambridge, Mass., 1952.
12. A. Feinstein, A new basic theorem of information theory, IRE trans., Inform. Theory, Vol. 4, 2–22, 1954.
13. M. Hayashi and H. Nagaoka, General formulas for capacity of classical–quantum channels, http//xxx.lanl.gov/abs/quant-ohy/0206186V1, 2002.
14. A.S. Holevo, Problems in the mathematical theory of quantum communication channels, Rep. Math. Phys. 12, 2, 273–278, 1977.
15. A.S. Holevo, The capacity of the quantum channel with general signal states, IEEE Trans. Inform Theory, Vol. 44, 269–273, 1998.
16. A.S. Holevo, Coding theorems for quantum channels, http://xxx.lanl.giv/abs/quant–ph/9809023V1, 1998.
17. M.A. Nielsen and I. Chuang, Quantum Computation and Quantum Information, Cambridge University Press, 2000.
18. T. Ogawa and H. Nagaoka, Strong converse to the quantum channel coding theorem, IEEE Trans. Inform. Theory, Vol. 45, 2486–2489, 1999.
19. T. Ogawa and H. Nagaoka, Strong converse and Stein's lemma in quantum hypothesis testing, IEEE Trans. Inform. Theory, Vol. 46, 2428–2433, 2000.

20. D. Petz, Quasientropies for states of a von Neumann algebra, Publ. RIMS, Kyoto Univ. Vol. 21, 787–800, 1985.
21. D. Petz, Quasientropies for finite quantum systems, Rep. Math. Phys., Vol. 23, 57–65, 1986.
22. B. Schumacher and M. Westmorelang, Sending classical information via noisy quantum channel, Phys. Rev. A, Vol. 56, 1, 131–138, 1997.
23. C.E. Shannon, A mathematical theory of communication, Bell Sys. Tech. Journal, Vol. 27, 379–423, 1948.
24. C.E. Shannon, Two-way communication channels, Proc. 4th Berkeley Symp. Math. Statist. and Prob., Unvi. of Calif. Press, Berkeley, Vol. 1, 611–644, 1961.
25. S. Verdú and T.S. Han, A general formula for channel capacity, IEEE Trans. Inform. Theory, Vol. 40, 1147–1157, 1994.
26. A. Winter, The capacity region of the quantum multiple access channel, http://xxx.lanl.gov/abs/quant-ph/9807019, 1998.
27. A. Winter, Coding theorem and strong converse for quantum channels, IEEE Trans. Inform. Theory, Vol. 45, 2481–2485, 1999.
28. A. Winter, Coding theorem and strong converse for nonstationary quantum channels, Preprint 99–033, SFB 343 "Diskrete Strukturen in der Mathematik", Universität Bielefeld, 1999.
29. J. Wolfowitz, The coding of message subject to chance errors, Illinois J. Math. 1, 591–606, 1957.
30. J. Wolfowitz, Coding Theorems of Information Theory, Springer–Verlag, Berlin, Heidelberg, New York, 3-rd edition, 1978.

Identification Via Quantum Channels in the Presence of Prior Correlation and Feedback

A. Winter

Abstract. Continuing our earlier work (quant-ph/0401060), we give two alternative proofs of the result that a noiseless qubit channel has identification capacity 2: the first is direct by a "maximal code with random extension" argument, the second is by showing that 1 bit of entanglement (which can be generated by transmitting 1 qubit) and negligible (quantum) communication has identification capacity 2. This generalizes a random hashing construction of Ahlswede and Dueck: that 1 shared random bit together with negligible communication has identification capacity 1.

We then apply these results to prove capacity formulas for various quantum feedback channels: passive classical feedback for quantum–classical channels, a feedback model for classical–quantum channels, and "coherent feedback" for general channels.

1   Introduction

While the theory of identification via noisy channels[4,5] has generated significant interest within the information theory community (the areas of, for instance, common randomness[3], channel resolvability[16] and watermarking[31] were either developed in response or were discovered to have close connections to identification), the analogous theory where one uses a *quantum channel* has received comparably little attention: the only works extant at the time of writing are Löber's[24] starting of the theory, a strong converse for discrete memoryless classical-quantum channels by Ahlswede and Winter[6], and a recent paper by the present author[33].

This situation may have arisen from a perception that such a theory would not be very different from the classical identification theory, as indeed classical message transmission via quantum channels, at a fundamental mathematical level, does not deviate much from its classical counterpart. In [20,28,27,32] coding theorems and converses are "just like" in Shannon's classical channel coding theory with Holevo information playing the role of Shannon's mutual information. (Though we have to acknowledge that it took quite a while before this was understood, and that there are tantalising differences in detail, e.g. additivity problems[30].)

In our recent work[33], however, a quite startling discovery was made: it was shown that — contrary to the impression the earlier papers[24,6] gave — the identification capacity of a (discrete memoryless, as always in this paper) quantum channel is in general *not equal to its transmission capacity*. Indeed, the

R. Ahlswede et al. (Eds.): Information Transfer and Combinatorics, LNCS 4123, pp. 486–504, 2006.

identification capacity of a noiseless qubit was found to be 2. This means that for quantum channels the rule that identification capacity equals common randomness capacity (see the discussion by Ahlswede[1] and Kleinewächter[22]) fails dramatically, even for the most ordinary channels!

In the present paper we find some new results for identification via quantum systems: after a review of the necessary definitions and known results (section 2) and a collection of statements about what we called "random channels" in our earlier paper[33], we first give a direct proof that a qubit has identification capacity 2, in section 4. (Our earlier proof[33] uses a reduction to *quantum identification*, which we avoid here.) Then, in section 5, we show the quantum analogue of Ahlswede and Dueck's result [5] that 1 bit of shared randomness plus negligible communication are sufficient to build an identification code of rate 1, namely, 1 bit of entanglement plus negligible (quantum) communication are sufficient to build an identification code of rate 2. In section 6 we briefly discuss the case of more general prior correlations between sender and receiver.

In section 7, we turn our attention to feedback channels: we first study quantum–classical channels with passive classical feedback, and prove a quantum generalization of the capacity formula of Ahlswede and Dueck[5]. Then, in section 8, we introduce a feedback model for general quantum channels which we call "coherent feedback", and prove a capacity formula for these channels as well which can be understood as a quantum analogue of the feedback identification capacity of Ahlswede and Dueck[5]. We also comment on a different feedback model for classical–quantum channels.

2 Review of Definitions and Known Facts

For a broader review of identification (and, for comparison, transmission) via quantum channels we refer the reader to the introductory sections of our earlier paper[33], to Löber's Ph.D. thesis[24], and to the classical identification papers by Ahlswede and Dueck[4,5]. Here we are content with repeating the bare definitions:

We are concerned with quantum systems, which are modeled as (finite) Hilbert spaces $\mathcal{H}$ (or rather the operator algebra $\mathcal{B}(\mathcal{H})$). *States* on these systems we identify with density operators ρ: positive semidefinite operators with trace 1.

A *quantum channel* is modeled in this context as a completely positive, trace preserving linear map $T : \mathcal{B}(\mathcal{H}_1) \longrightarrow \mathcal{B}(\mathcal{H}_2)$ between the operator algebras of Hilbert spaces $\mathcal{H}_1, \mathcal{H}_2$.

Definition 1 (Löber[24], Ahlswede and Winter[6]). *An* identification code *for the channel T with error probability λ_1 of first, and λ_2 of second kind is a set $\{(\rho_i, D_i) : i = 1, \ldots, N\}$ of states ρ_i on $\mathcal{H}_1$ and operators D_i on $\mathcal{H}_2$ with $0 \leq D_i \leq \mathbb{1}$, such that*

$$\forall i \quad \mathrm{Tr}\big(T(\rho_i)D_i\big) \geq 1 - \lambda_1, \forall i \neq j \quad \mathrm{Tr}\big(T(\rho_i)D_j\big) \leq \lambda_2.$$

For the identity channel $\mathrm{id}_{\mathcal{C}^d}$ of the algebra $\mathcal{B}(\mathcal{C}^d)$ of a d–dimensional system we also speak of an identification code on $\mathcal{C}^d$.

For the special case of memoryless channels $T^{\otimes n}$ (where T is implicitly fixed), we speak of an $(n, \lambda_1, \lambda_2)$–ID code, and denote the largest size N of such a code $N(n, \lambda_1, \lambda_2)$.

An identification code as above is called simultaneous if all the D_i are coexistent: this means that there exists a positive operator valued measure (POVM) $(E_k)_{k=1}^K$ and sets $\mathcal{D}_i \subset \{1, \ldots, K\}$ such that $D_i = \sum_{k \in \mathcal{D}_i} E_k$. The largest size of a simultaneous $(n, \lambda_1, \lambda_2)$–ID code is denoted $N_{\mathrm{sim}}(n, \lambda_1, \lambda_2)$.

Most of the current knowledge about these concepts is summarized in the two following theorems.

Theorem 1 (Löber[24], Ahlswede and Winter[6]). *Consider any channel T, with transmission capacity $C(T)$ (Holevo[20], Schumacher and Westmoreland [28]). Then, the simultaneous identification capacity of T,*

$$C_{\mathrm{sim-ID}}(T) := \inf_{\lambda_1, \lambda_2 > 0} \liminf_{n \to \infty} \frac{1}{n} \log \log N_{\mathrm{sim}}(n, \lambda_1, \lambda_2) \geq C(T).$$

(With log and exp in this paper understood to basis 2.)

For classical–quantum (cq) channels T (see Holevo[19]), even the strong converse for (non–simultaneous) identification holds:

$$C_{\mathrm{ID}}(T) = \lim_{n \to \infty} \frac{1}{n} \log \log N(n, \lambda_1, \lambda_2) = C(T),$$

whenever $\lambda_1, \lambda_2 > 0$ and $\lambda_1 + \lambda_2 < 1$. □

That the (non–simultaneous) identification capacity can be larger than the transmission capacity was shown only recently:

Theorem 2 (Winter[33]). *The identification capacity of the noiseless qubit channel, $\mathrm{id}_{\mathcal{C}^2}$, is $C_{\mathrm{ID}}(\mathrm{id}_{\mathcal{C}^2}) = 2$, and the strong converse holds.* □

The main objective of the following three sections is to give two new proofs of the achievability of 2 in this theorem.

3   Random Channels and Auxiliary Results

The main tool in the following results (as in our earlier paper[33]) are *random channels* and in fact *random states*[25,26,18]:

Definition 2. *For positive integers s, t, u with $s \leq tu$, the random channel $R_s^{t(u)}$ is a random variable taking values in quantum channels $\mathcal{B}(\mathcal{C}^s) \longrightarrow \mathcal{B}(\mathcal{C}^t)$ with the following distribution:*

There is a random isometry $V : \mathcal{C}^s \longrightarrow \mathcal{C}^t \otimes \mathcal{C}^u$, by which we mean a random variable taking values in isometries whose distribution is left–/right–invariant under multiplication by unitaries on $\mathcal{C}^t \otimes \mathcal{C}^u$/on $\mathcal{C}^s$, respectively, such that

$$R_s^{t(u)}(\rho) = \mathrm{Tr}_{\mathcal{C}^u}\left(V \rho V^*\right).$$

Note that the invariance demanded of the distribution of V determines it uniquely — one way to generate the distribution is to pick an arbitrary fixed isometry $V_0 : \mathcal{C}^s \longrightarrow \mathcal{C}^t \otimes \mathcal{C}^u$ and a random unitary U on $\mathcal{C}^t \otimes \mathcal{C}^u$ according to the Haar measure, and let $V = UV_0$.

Remark 1. Identifying $\mathcal{C}^{tu}$ with $\mathcal{C}^t \otimes \mathcal{C}^u$, we have $R_s^{t(u)} = \mathrm{Tr}_{\mathcal{C}^u} \circ R_s^{tu(1)}$. Note that $R_s^{t(1)}$ is a random isometry from $\mathcal{C}^s$ into $\mathcal{C}^t$ in the sense of our definition, and that the distribution of $R_s^{s(1)}$ is the Haar measure on the unitary group of $\mathcal{C}^s$.

Remark 2. The one–dimensional Hilbert space $\mathcal{C}$ is a trivial system: it has only one state, 1, and so the random channel $R_1^{t(u)}$ is equivalently described by the image state it assigns to 1, $R_1^{t(u)}(1)$. For $s = 1$ we shall thus identify the random channel $R_1^{t(u)}$ with the random state $R_1^{t(u)}(1)$ on $\mathcal{C}^t$. A different way of describing this state is that there exists a random (Haar distributed) unitary U and a pure state ψ_0 such that $R_1^{t(u)} = \mathrm{Tr}_{\mathcal{C}^u}\big(U\psi_0 U^*\big)$ — note that it has rank bounded by u. These are the objects we concentrate on in the following.

Lemma 1 (see Bennett et al.[8], [9] Winter[33]). Let ψ be a pure state, P a projector of rank (at most) r and let U be a random unitary, distributed according to the Haar measure. Then for $\epsilon > 0$,

$$\Pr\left\{\mathrm{Tr}(U\psi U^* P) \geq (1+\epsilon)\frac{r}{d}\right\} \leq \exp\left(-r\frac{\epsilon - \ln(1+\epsilon)}{\ln 2}\right).$$

For $0 < \epsilon \leq 1$, and $\mathrm{rank}\, P = r$,

$$\Pr\left\{\mathrm{Tr}(U\psi U^* P) \geq (1+\epsilon)\frac{r}{d}\right\} \leq \exp\left(-r\frac{\epsilon^2}{6\ln 2}\right),$$

$$\Pr\left\{\mathrm{Tr}(U\psi U^* P) \leq (1-\epsilon)\frac{r}{d}\right\} \leq \exp\left(-r\frac{\epsilon^2}{6\ln 2}\right).$$

$\square$

Lemma 2 (Bennett et al.[8], [9]). For $\epsilon > 0$, there exists in the set of pure states on $\mathcal{C}^d$ an ϵ-net $\mathcal{M}$ of cardinality $|\mathcal{M}| \leq \left(\frac{5}{\epsilon}\right)^{2d}$; i.e., $\forall \varphi$ pure $\exists \widehat{\varphi} \in \mathcal{M}$ $\|\varphi - \widehat{\varphi}\|_1 \leq \epsilon$. $\square$

With these lemmas, we can prove an important auxiliary result:

Lemma 3 (see Harrow et al.[18]). For $0 < \eta \leq 1$ and $t \leq u$, consider the random state $R_1^{t(u)}$ on $\mathcal{C}^t$. Then,

$$\Pr\left\{R_1^{t(u)} \notin \left[\frac{1-\eta}{t}\mathbb{1}; \frac{1+\eta}{t}\mathbb{1}\right]\right\} \leq 2\left(\frac{10t}{\eta}\right)^{2t}\exp\left(-u\frac{\eta^2}{24\ln 2}\right).$$

Proof. We begin with the observation that $R_1^{t(u)} \in [\alpha\mathbb{1}; \beta\mathbb{1}]$ if and only if for all pure states (rank one projectors) φ,

$$\mathrm{Tr}\big(R_1^{t(u)}\varphi\big) = \mathrm{Tr}\big(R_1^{tu(1)}(\varphi \otimes \mathbb{1}_u)\big) \begin{cases} \geq \alpha, \\ \leq \beta. \end{cases}$$

Due to the triangle inequality, we have to ensure this only for φ from an $\eta/2t$–net and with $\alpha = \left(1 - \frac{\eta}{2}\right)/t$, $\beta = \left(1 + \frac{\eta}{2}\right)/t$. Then the probability bound claimed above follows from lemmas 1 and 2, with the union bound. $\quad\square$

4   ID Capacity of a Qubit

Here we give a new, direct proof of theorem 2 — in fact, we prove the following proposition from which it follows directly.

Proposition 1. *For every $0 < \lambda < 1$, there exists on the quantum system $\mathcal{B}(\mathcal{C}^d)$ an ID code with*

$$N = \left\lceil \frac{1}{2}\exp\left(\left(\frac{\lambda}{3000}\frac{d}{\log d}\right)^2\right)\right\rceil$$

messages, with error probability of first kind equal to 0 and error probability of second kind bounded by λ.

Proof. We shall prove even a bit more: that such a code exists which is of the form $\{(\rho_i, D_i) : i = 1, \ldots, N\}$ with

$$D_i = \mathrm{supp}\,\rho_i, \quad \mathrm{rank}\,\rho_i = \delta := \alpha\frac{d}{\log d}, \quad \rho_i \leq \frac{1+\eta}{\delta}D_i. \qquad (1)$$

The constants $\alpha \leq \lambda/4$ and $\eta \leq 1/3$ will be fixed in the course of this proof. Let a *maximal* code $\mathcal{C}$ of this form be given. We shall show that if N is "not large", a random codestate as follows will give a larger code, contradicting maximality.

Let $R = R_1^{d(\delta)}$ (the random state in dimension d with δ–dimensional ancillary system, see definition 2), and $D := \mathrm{supp}\,R$. Then, according to the Schmidt decomposition and lemma 3,

$$\mathrm{Pr}\left\{R \notin \left[\frac{1-\eta}{\delta}D; \frac{1+\eta}{\delta}D\right]\right\} = \mathrm{Pr}\left\{R_1^{\delta(d)} \notin \left[\frac{1-\eta}{\delta}\mathbb{1}_\delta; \frac{1+\eta}{\delta}\mathbb{1}_\delta\right]\right\}$$

$$\leq 2\left(\frac{10\delta}{\eta}\right)^{2\delta}\exp\left(-d\frac{\eta^2}{24\ln 2}\right). \qquad (2)$$

This is $\leq 1/2$ if

$$d \geq \left(\frac{96\ln 2}{\eta^2}\log\frac{10}{\eta}\right)\delta\log\delta,$$

which we ensure by choosing $\alpha \leq \lambda\left(\frac{96\ln 2}{\eta^2}\log\frac{10}{\eta}\right)^{-1} \leq \lambda/4$.

In the event that $\frac{1-\eta}{\delta}D \leq R \leq \frac{1+\eta}{\delta}D$, we have on the one hand

$$\mathrm{Tr}(\rho_i D) \leq \mathrm{Tr}\left(\frac{1+\eta}{\delta}D_i\frac{\delta}{1-\eta}R\right) \leq 2\mathrm{Tr}(RD_i). \tag{3}$$

On the other hand, because of $R_1^{d(\delta)} = \mathrm{Tr}_{\mathcal{C}^\delta}R_1^{d\delta(1)}$, we can rewrite

$$\mathrm{Tr}(RD_i) = \mathrm{Tr}\big(R_1^{d\delta(1)}(D_i \otimes \mathbb{1}_\delta)\big),$$

hence by lemma 1

$$\mathrm{Pr}\{\mathrm{Tr}(RD_i) > \lambda/2\} \leq \exp(-\delta^2). \tag{4}$$

So, by the union bound, eqs. (3) and (4) yield

$$\mathrm{Pr}\Big\{\mathcal{C} \cup \{(R,D)\}\ \text{has error probability of}$$

$$\text{second kind larger than } \lambda \text{ or violates eq. (1)}\Big\} \leq \frac{1}{2} + N\exp(-\delta^2).$$

If this is less than 1, there must exist a pair (R, D) extending our code while preserving the error probabilities and the properties of eq. (1), which would contradict maximality. Hence,

$$N \geq \frac{1}{2}\exp\left(\delta^2\right),$$

and we are done, fixing $\eta = 1/3$ and $\alpha = \lambda/3000$.   □

The *proof of theorem 2* is now obtained by applying the above proposition to $d = 2^n$, the Hilbert space dimension of n qubits, and arbitrarily small λ. That the capacity is not larger than 2 is shown by a simple dimension counting argument[33], which we don't repeat here.   □

5   ID Capacity of an Ebit

Ahlswede and Dueck[5] have shown that the identification capacity of any system, as soon as it allows — even negligible — communication, is at least as large as its common randomness capacity: the maximum rate at which shared randomness can be generated. (We may add, that except for pathological examples expressly constructed for that purpose, in practically all classical systems for which these two capacities exist, they turn out to be equal[5,2,22,1].)

Their proof relies on a rather general construction, which we restate here, in a simplified version:

Proposition 2 (Ahlswede and Dueck[5]). *There exist, for $\lambda > 0$ and $N \geq 4^{1/\lambda}$, functions $f_i : \{1,\ldots,M\} \longrightarrow \{1,\ldots,N\}$ $(i = 1,\ldots,2^M)$ such that the distributions P_i on $\{1,\ldots,M\} \times \{1,\ldots,N\}$ defined by*

$$P_i(\mu,\nu) = \begin{cases} \frac{1}{M} & \text{if } \nu = f_i(\mu), \\ 0 & \text{otherwise.} \end{cases}$$

and the sets $\mathcal{D}_i = \mathrm{supp}\, P_i$ form an identification code with error probability of first kind 0 and error probability of second kind λ.

In other words, prior shared randomness in the form of uniformly distributed $\mu \in \{1, \ldots, M\}$ between sender and receiver, and transmission of $\nu \in \{1, \ldots, N\}$ allow identification of 2^M messages. □

(In the above form it follows from proposition 4 below: a perfect transmission code is at the same time always an identification code with both error probabilities 0.)

Thus, an alternative way to prove that a channel of capacity C allows identification at rate $\geq C$, is given by the following scheme: use the channel $n - O(1)$ times to generate $Cn - o(n)$ shared random bits and the remaining $O(1)$ times to transmit one out of $N = 2^{O(1)}$ messages; then apply the above construction with $M = 2^{Cn-o(n)}$. More generally, a rate R of common randomness and only negligible communication give identification codes of rate R.

The quantum analogue of perfect correlation (i.e., shared randomness) being pure entanglement, substituting quantum state transmission wherever classical information was conveyed, and in the light of the result that a qubit has identification capacity 2, the following question appears rather natural (and we have indeed raised it, in remark 14 of our earlier paper[33]): Does 1 bit of entanglement plus the ability to (even only negligibly) communicate result in an ID code of rate 2, asymptotically?

Proposition 3. *For $\lambda > 0$, $d \geq 2$ and $\Delta \geq \left(\frac{900}{\lambda^2} \log \frac{30d}{\lambda}\right) \log d$, there exist quantum channels $T_i : \mathcal{B}(\mathcal{C}^d) \longrightarrow \mathcal{B}(\mathcal{C}^\Delta)$ $(i = 1, \ldots, N' = \left\lceil \frac{1}{2} \exp(d^2) \right\rceil)$, such that the states $\rho_i = (\mathrm{id} \otimes T_i)\Phi_d$ (with state vector $|\Phi_d\rangle = \frac{1}{\sqrt{d}} \sum_{j=1}^d |j\rangle|j\rangle$), and the operators $D_i = \mathrm{supp}\, \rho_i$ form an identification code on $\mathcal{B}(\mathcal{C}^d \otimes \mathcal{C}^\Delta)$ with error probability of first kind 0 and error probability of second kind λ.*

In other words, sender and receiver, initially sharing the maximally entangled state Φ_d, can use transmission of a Δ-dimensional system to build an identification code with $\left\lceil \frac{1}{2} \exp(d^2) \right\rceil$ messages.

Proof. Let a maximal code $\mathcal{C}$ as described in the proposition be given, such that additionally

$$D_i = \mathrm{supp}\, \rho_i, \quad \mathrm{rank}\, \rho_i = d, \quad \rho_i \leq \frac{1+\lambda}{d} D_i. \tag{5}$$

Consider the random state $R = R_1^{d\Delta(d)}$ on $\mathcal{C}^{d\Delta} = \mathcal{C}^d \otimes \mathcal{C}^\Delta$, and $D := \mathrm{supp}\, R$. Now, by Schmidt decomposition and with lemma 1 (compare the proof of proposition 1), for $\eta := \lambda/3$

$$\Pr\left\{ R \notin \left[\frac{1-\eta}{d}D; \frac{1+\eta}{d}D\right] \right\} = \Pr\left\{ R_1^{d(\Delta d)} \notin \left[\frac{1-\eta}{d}\mathbb{1}_d; \frac{1+d}{\delta}\mathbb{1}_d\right] \right\}$$
$$\leq 2\left(\frac{10d}{\eta}\right)^{2d} \exp\left(-d\Delta\frac{\eta^2}{24\ln 2}\right). \tag{6}$$

The very same estimate gives

$$\Pr\left\{\mathrm{Tr}_{\mathcal{C}^{\triangle}}R \notin \left[\frac{1-\eta}{d}\mathbb{1}_d; \frac{1+\eta}{d}\mathbb{1}_d\right]\right\} = \Pr\left\{R_1^{d(\triangle d)} \notin \left[\frac{1-\eta}{d}\mathbb{1}_d; \frac{1+\eta}{d}\mathbb{1}_d\right]\right\}$$
$$\leq 2\left(\frac{10d}{\eta}\right)^{2d}\exp\left(-d\triangle\frac{\eta^2}{24\ln 2}\right). \tag{7}$$

By choosing $\triangle \geq \left(\frac{144\ln 2}{\eta^2}\log\frac{10}{\eta}\right)\log d$, as we indeed did, the sum of these two probabilities is at most $1/2$.

In the event that $\frac{1-\eta}{d}D \leq R \leq \frac{1+\eta}{d}D$, we argue similar to the proof of proposition 1 (compare eq. (3)):

$$\mathrm{Tr}(\rho_i D) \leq \mathrm{Tr}\left(\frac{1+\lambda}{d}D_i\frac{d}{1-\eta}R\right) \leq 3\mathrm{Tr}(RD_i). \tag{8}$$

On the other hand (compare eq. (4)),

$$\Pr\{\mathrm{Tr}(RD_i) > \lambda/3\} \leq \exp(-d^2), \tag{9}$$

by lemma 1 and using $\triangle^{-1} \leq \lambda/6$.

In the event that $\frac{1-\eta}{d}\mathbb{1} \leq \mathrm{Tr}_{\mathcal{C}^{\triangle}}R \leq \frac{1+\eta}{d}\mathbb{1}$, there exists an operator X on $\mathcal{C}^d$ with $\frac{1}{1+\eta}\mathbb{1} \leq X \leq \frac{1}{1-\eta}\mathbb{1}$, such that

$$R_0 := \sqrt{R}(X \otimes \mathbb{1})\sqrt{R} \quad \text{(which has the same support } D \text{ as } R\text{)}$$

satisfies $\mathrm{Tr}_{\mathcal{C}^{\triangle}}R_0 = \frac{1}{d}\mathbb{1}$. By the Jamiołkowski isomorphism[21] between quantum channels and states with maximally mixed reduction, this is equivalent to the existence of a quantum channel T_0 such that $R_0 = (\mathrm{id} \otimes T_0)\Phi_d$. Observe that $R_0 \leq \frac{1+\lambda}{d}D$ and $\mathrm{Tr}(R_0 D_i) \leq \frac{3}{2}\mathrm{Tr}(RD_i)$.

So, putting together the bounds of eqs. (6), (7), (8) and (9), we get, by the union bound,

$$\Pr\Big\{\mathcal{C} \cup \{(R_0, D)\} \text{ has error probability of}$$
$$\text{second kind larger than } \lambda \text{ or violates eq. (5)}\Big\} \leq \frac{1}{2} + N'\exp(-d^2).$$

If this is less than 1, there will exist a state $R_0 = (\mathrm{id} \otimes T_0)\Phi_d$ and an operator D enlarging the code and preserving the error probabilities as well as the properties in eq. (5), which contradicts maximality.

Hence, $N' \geq \frac{1}{2}\exp(d^2)$, and we are done. $\qquad\square$

This readily proves, answering the above question affirmatively:

Theorem 3. *The identification capacity of a system in which entanglement (EPR pairs) between sender and receiver is available at rate E, and which allows (even only negligible) communication, is at least $2E$. This is tight for the case that the available resources are only the entanglement and negligible communication.* $\qquad\square$

Remark 3. Just as the Ahlswede–Dueck construction of proposition 2 can be understood as an application of random hashing, we are tempted to present our above construction as a kind of "quantum hashing": indeed, the (small) quantum system transmitted contains, when held together with the other half of the prior shared entanglement, just enough of a signature of the functions/quantum channels used to distinguish them pairwise reliably.

6   General Prior Correlation

Proposition 2 quantifies the identification capacity of shared randomness, and proposition 3 does the same for shared (pure) entanglement. This of course raises the question what the identification capacity of other, more general, correlations is: i.e., we are asking for code constructions and bounds if (negligible) quantum communication and n copies of a bipartite state ω between sender and receiver are available.

For the special case that the correlation decomposes cleanly into entanglement and shared randomness,

$$\omega = \sum_{\mu} p_{\mu} \Psi_{\mu}^{AB} \otimes |\mu\rangle\langle\mu|^{A'} \otimes |\mu\rangle\langle\mu|^{B'},$$

with an arbitrary perfect classical correlation (between registers A' and B') distributed according to p and arbitrary pure entangled states Ψ_{μ}, we can easily give the answer (let the sender be in possession of AA', the receiver of BB'):

$$C_{\text{ID}} = H(p) + 2\sum_{\mu} p_{\mu} E(\Psi_{\mu}^{AB}); \tag{10}$$

here, $H(p)$ is the entropy of the classical perfect correlation p; $E(\Psi^{AB}) = S(\Psi^A)$ is the entropy of entanglement [7], with the reduced state $\Psi^A = \text{Tr}_B \Psi^{AB}$. The archievability is seen as follows: by entanglement and randomness concentration [7] this state yields shared randomness and entanglement at rates $R = H(p)$ and $E = \sum_{\mu} p_{\mu} E(\Psi_{\mu})$, respectively (without the need of communication — note that both users learn which entangled state they have by looking at the primed registers). Proposition 3 yields an identification code of rate $2E$, while proposition 4 below shows how to increase this rate by R.

That the expression is an upper bound is then easy to see, along the lines of the arguments given in our earlier paper for the capacity of a "hybrid quantum memory" [23,33].

Proposition 4 (Winter[33]). *Let $\{(\rho_i, D_i) : i = 1, \ldots, N\}$ be an identification code on the quantum system $\mathcal{H}$ with error probabilities λ_1, λ_2 of first and second kind, respectively, and let $\mathcal{H}_C$ be a classical system of dimension M (by this we mean a Hilbert space only allowed to be in a state from a distinguished orthonormal basis $\{|\mu\rangle\}_{\mu=1}^{M}$). Then, for every $\epsilon > 0$, there exists an identification code $\{(\sigma_f, \widetilde{D}_f) : f = 1, \ldots, N'\}$ on $\mathcal{H}_C \otimes \mathcal{H}$ with error probabilities $\lambda_1, \lambda_2 + \epsilon$ of first* *and second kind, respectively, and $N' \geq \left(\frac{1}{2}N^\epsilon\right)^M$. The f actually label functions (also denoted f)* $\{1, \ldots, M\} \longrightarrow \{1, \ldots, N\}$, *such that*

$$\sigma_f = \frac{1}{M} \sum_\mu |\mu\rangle\langle\mu| \otimes \rho_{f(k)}.$$

In other words, availability of shared randomness (μ on the classical system $\mathcal{H}_C$) with an identification code allows us to construct a larger identification code.   □

The general case seems to be much more complex, and we cannot offer an approximation to the solution here. So, we restrict ourselves to highlighting two questions for further investigation:

1. What is the identification capacity of a bipartite state ω, together with negligible communication? For noisy correlations, this may not be the right question altogether, as a look at work by Ahlswede and Balakirsky[2] shows: they have studied this problem for classical binary correlations with symmetric noise, and have found that — as in common randomness theory[3] — one ought to include a limited rate of communication and study the relation between this additional rate and the obtained identification rate. Hence, we should ask: what is the identification capacity of ω plus a rate of C bits of communication? An obvious thing to do in this scenario would be to use part of this rate to do entanglement distillation of which the communication cost is known in principle ([12], [13], [14]). This gives entanglement as well as shared randomness, so one can use the constructions above. It is not clear of course whether this is asymptotically optimal.

2. In the light of the code enlargement proposition 4, it would be most interesting to know if a stronger version of our proposition 3/theorem 3 holds: Does entanglement of rate E increase the rate of a given identification code by $2E$?

7   Identification in the Presence of Feedback: Quantum–Classical Channels

Feedback for quantum channels is a somewhat problematic issue, mainly because the output of the channel is a quantum state, of which there is in general no physically consistent way of giving a copy to the sender. In addition, it should not even be a "copy" for the general case that the channel outputs a mixed state (which corresponds to the distribution of the output), but a copy of the exact *symbol* the receiver obtained; so the feedback should establish correlation between sender and receiver, and in the quantum case this appears to involve further choices, e.g. of basis. The approach taken in the small literature on the issue of feedback in quantum channels (see Fujiwara and Nagaoka[15], Bowen[10], and Bowen and Nagarajan[11]) has largely been to look at active feedback, where the receiver decides what to give back to the sender, based on a partial evaluation of the received data.

We will begin our study by looking at a subclass of channels which do not lead into any of these conceptual problems: quantum–classical (qc) channels, i.e., destructive measurements, have a completely classical output anyway, so there is no problem in augmenting every use of the channel by instantaneous passive feedback.

Let a measurement POVM $(M_y)_{y \in \mathcal{Y}}$ be given; then its qc–channel is the map

$$T : \rho \longmapsto \sum_y \mathrm{Tr}(\rho M_y) |y\rangle\langle y|,$$

with an orthogonal basis $(|y\rangle)_y$ of an appropriate Hilbert space $\mathcal{F}$, say. We will denote this qc–channel as $T : \mathcal{B}(\mathcal{H}) \longrightarrow \mathcal{Y}$.

For a qc–channel T, a *(randomized) feedback strategy* F for block n is given by states $\rho_{t:y^{t-1}}$ on $\mathcal{H}_1$ for each $t = 1, \dots, n$ and $y^{t-1} \in \mathcal{Y}^{t-1}$: this is the state input to the channel in the tth timestep if the feedback from the previous rounds was $y^{t-1} = y_1 \dots y_{t-1}$. Clearly, this defines an output distribution Q on $\mathcal{Y}^n$ by iteration of the feedback loop:

$$Q(y^n) = \prod_{t=1}^n \mathrm{Tr}(\rho_{t:y^{t-1}} M_{y_t}). \tag{11}$$

Remark 4. We could imagine a more general protocol for the sender: an initial state σ_0 could be prepared on an ancillary system $\mathcal{H}_A$, and the feedback strategy is a collection Φ of completely positive, trace preserving maps

$$\varphi_t : \mathcal{B}(\mathcal{F}^{\otimes(t-1)} \otimes \mathcal{H}_A) \longrightarrow \mathcal{B}(\mathcal{H}_A \otimes \mathcal{H}),$$

where $\mathcal{F}$ is the quantum system representing the classical feedback by states from an orthogonal basis: this map creates the next channel input and a new state of the ancilla (potentially entangled) from the old ancilla state and the feedback.

This more general scheme allows for memory and even quantum correlations between successive uses of the channel, via the system $\mathcal{H}_A$. However, the scheme has, for each "feedback history" y^{t-1} up to time t, a certain state $\sigma_{t-1:y^{t-1}}$ on $\mathcal{H}_A$ (starting with σ_0), and consequently an input state $\rho_{t:y^{t-1}}$ on $\mathcal{H}$:

$$\rho_{t:y^{t-1}} = \mathrm{Tr}_{\mathcal{H}_A}\left(\varphi_t(|y^{t-1}\rangle\langle y^{t-1}| \otimes \sigma_{t-1:y^{t-1}})\right),$$

$$\sigma_{t:y^t} = \frac{1}{\mathrm{Tr}(\rho_{t:y^{t-1}} M_{y_t})} \mathrm{Tr}_{\mathcal{H}}\left(\varphi_t(|y^{t-1}\rangle\langle y^{t-1}| \otimes \sigma_{t-1:y^{t-1}})\right).$$

It is easy to check that the corresponding output distribution Q of this feedback strategy according to our definition (see eq. (11)) is the same as for the original, more general feedback scheme. So, we do not need to consider those to obtain ultimate generality.

An $(n, \lambda_1, \lambda_2)$-*feedback ID code* for the qc–channel T with passive feedback is now a set $\{(F_i, D_i) : i = 1, \ldots, N\}$ of feedback strategies F_i and of operators $0 \leq D_i \leq \mathbb{1}$, such that the output states $\omega_i = \sum_{y^n} Q_i(y^n) |y^n\rangle\langle y^n|$ with the operators D_i form an identification code with error probabilities λ_1 and λ_2 of first and second kind, respectively. Note that since the output is classical — i.e., the states are diagonal in the basis $(|y^n\rangle)$ —, we may without loss of generality assume that all $D_i = \sum_{y^n} D_i(y^n) |y^n\rangle\langle y^n|$, with certain $0 \leq D_i(y^n) \leq 1$.

Finally, let $N_F(n, \lambda_1, \lambda_2)$ be the maximal N such that there exists an $(n, \lambda_1, \lambda_2)$-feedback ID code with N messages. Note that due to the classical nature of the channel output codes are automatically simultaneous.

To determine the capacity, we invoke the following result:

Lemma 4 (Ahlswede and Dueck[5], Lemma 4). *Consider a qc–channel* $T : \mathcal{B}(\mathcal{H}) \to \mathcal{Y}$ *and any randomized feedback strategy F for block n. Then, for $\epsilon > 0$, there exists a set $\mathcal{E} \subset \mathcal{Y}^n$ of probability $Q(\mathcal{E}) \geq 1 - \epsilon$ and cardinality $|\mathcal{E}| \leq \exp\left(n \max_\rho H(T(\rho)) + \alpha\sqrt{n}\right)$, where $\alpha = |\mathcal{Y}|\epsilon^{-1/2}$.*

The *proof* of Ahlswede and Dueck[5] applies directly: a qc–channel with feedback is isomorphic to a classical feedback channel with an infinite input alphabet (the set of all states), but with finite output alphabet, which is the relevant fact. □

This is the essential tool to prove the following generalization of Ahlswede's and Dueck's capacity result[5]:

Theorem 4. *For a qc–channel T and $\lambda_1, \lambda_2 > 0$, $\lambda_1 + \lambda_2 < 1$,*

$$\lim_{n\to\infty} \frac{1}{n} \log\log N_F(n, \lambda_1, \lambda_2) = C_{\mathrm{ID}}^F(T) = \max_\rho H\left(T(\rho)\right),$$

unless the transmission capacity of T is 0, in which case $C_{\mathrm{ID}}^F(T) = 0$.

In other words, the capacity of a nontrivial qc–channel with feedback is its maximum output entropy and the strong converse holds.

Proof. Let's first get the exceptional case out of the way: $C(T)$ can only be 0 for a constant channel (i.e., one mapping every input to the same output). Clearly such a channel allows not only no transmission but also no identification.

The archievability is explained in the paper of Ahlswede and Dueck[5]: the sender uses $m = n - O(1)$ instances of the channel with the state ρ each, which maximizes the output entropy. Due to feedback they then share the outcomes of m i.i.d. random experiments, which they can concentrate into $nH(T(\rho)) - o(n)$ uniformly distributed bits. (This is a bit simpler than in the original paper[5]: they just cut up the space into type classes.) The remaining $O(1)$ uses of the channel (with an appropriate error correcting code) are then used to implement the identification code of proposition 2 based on the uniform shared randomness.

The strong converse is only a slight modification of the arguments of Ahlswede and Dueck[5], due to the fact that we allow probabilistic decoding procedures: first, for each message i in a given code, lemma 4 gives us a set $\mathcal{E}_i \subset \mathcal{Y}^n$ of cardinality $\leq K = \exp\left(n \max_\rho H(T(\rho)) + 3|\mathcal{Y}|\epsilon^{-1/2}\sqrt{n}\right)$, with probability

$1 - \epsilon/3$ under the feedback strategy F_i, where $\epsilon := 1 - \lambda_1 - \lambda_2 > 0$. Now let $c := \lceil \frac{3}{\epsilon} \rceil$, and define new decoding rules by letting

$$\widehat{D}_i(y^n) := \begin{cases} \frac{1}{c}\lfloor cD_i(y^n) \rfloor & \text{for } y^n \in \mathcal{E}_i, \\ 0 & \text{for } y^n \notin \mathcal{E}_i. \end{cases}$$

(I.e., round the density $D_i(y^n)$ down to the nearest multiple of $1/c$ within $\mathcal{E}_i$, and to 0 outside $\mathcal{E}_i$.) It is straightforward to check that in this way we obtain an $(n, \lambda_1 + \frac{2}{3}\epsilon, \lambda_2)$–feedback ID code.

The argument is concluded by observing that the new decoding densities are (i) all distinct (otherwise $\lambda_1 + \frac{2}{3}\epsilon + \lambda_2 \geq 1$), and (ii) all have support

$$\leq K = \exp\left(n \max_\rho H(T(\rho)) + 3|\mathcal{Y}|\epsilon^{-1/2}\sqrt{n} \right).$$

Hence

$$N \leq \binom{|\mathcal{Y}|^n}{K}(c+1)^K \leq \left[(c+1)|\mathcal{Y}|^n \right]^{2^{n \max_\rho H(T(\rho)) + O(\sqrt{n})}},$$

from which the claim follows. □

8   Identification in the Presence of Feedback: "Coherent Feedback Channels"

Inspired by the work of Harrow[17] we propose the following definition of "coherent feedback" as a substitute for full passive feedback: by Stinespring's theorem we can view the channel T as an isometry $U : \mathcal{H}_1 \longrightarrow \mathcal{H}_2 \otimes \mathcal{H}_3$, followed by the partial trace Tr_3 over $\mathcal{H}_3$: $T(\rho) = \mathrm{Tr}_3(U\rho U^*)$. *Coherent feedback* is now defined as *distributing*, on input ρ, the bipartite state $\Theta(\rho) := U\rho U^*$ among sender and receiver, who get $\mathcal{H}_3$ and $\mathcal{H}_2$, respectively.

A *coherent feedback strategy* Φ for block n consists of a system $\mathcal{H}_A$, initially in state σ_0, and quantum channels

$$\varphi_t : \mathcal{B}\big(\mathcal{H}_A \otimes \mathcal{H}_3^{\otimes(t-1)}\big) \longrightarrow \mathcal{B}\big(\mathcal{H}_A \otimes \mathcal{H}_3^{\otimes(t-1)} \otimes \mathcal{H}_1\big),$$

creating the tth round channel input from the memory in $\mathcal{H}_A$ and the previous coherent feedback $\mathcal{H}_3^{\otimes(t-1)}$. The output state on $\mathcal{H}_2^{\otimes n}$ after n rounds of coherent feedback channel alternating with the φ_t, is

$$\omega = \mathrm{Tr}_{\mathcal{H}_A \otimes \mathcal{H}_3^{\otimes n}}\left[(\Theta \circ \varphi_n \circ \Theta \circ \varphi_{n-1} \circ \cdots \circ \Theta \circ \varphi_1)\sigma_0 \right],$$

where implicitly each Θ is patched up by an identity on all systems different from $\mathcal{H}_1$, and each φ_t is patched up by an identity on $\mathcal{H}_2^{\otimes(t-1)}$.

Now, an $(n, \lambda_1, \lambda_2)$–*coherent feedback ID code* for the channel T with coherent feedback consists of N pairs (Φ_i, D_i) of coherent feedback strategies Φ_i (with output states ω_i) and operators $0 \le D_i \le \mathbb{1}$ on $\mathcal{H}_2^{\otimes n}$, such that the (ω_i, D_i) form an $(n, \lambda_1, \lambda_2)$–ID code on $\mathcal{H}_2^{\otimes n}$.

As usual, we introduce the maximum size N of an $(n, \lambda_1, \lambda_2)$–coherent feedback ID code, and denote it $N_{|F\rangle}(n, \lambda_1, \lambda_2)$. It is important to understand the difference to $N_F(n, \lambda_1, \lambda_2)$ at this point: for the qc–channel, the latter refers to codes making use of the classical feedback of the measurement result, but coherent feedback — even for qc–channels — creates entanglement between sender and receiver, which, as we have seen in section 5, allows for larger identification codes.

We begin by proving the analogue of lemma 4:

Lemma 5. *Consider a quantum channel $T : \mathcal{B}(\mathcal{H}_1) \to \mathcal{B}(\mathcal{H}_2)$ and any feedback strategy Φ on block n with output state ω on $\mathcal{H}_2^{\otimes n}$. Then, for $\epsilon > 0$, there exists a projector Π on $\mathcal{H}_2^{\otimes n}$ with probability $\mathrm{Tr}(\omega \Pi) \ge 1 - \epsilon$ and rank*

$$\mathrm{rank}\, \Pi \le \exp\left(n \max_{\rho} S(T(\rho)) + \alpha\sqrt{n} \right),$$

where $\alpha = (\dim \mathcal{H}_2)\epsilon^{-1/2}$.

Proof. The feedback strategy determines the output state ω on $\mathcal{H}_2^{\otimes n}$, and we choose complete von Neumann measurements on each of the n tensor factors: namely, the measurement M of an eigenbasis $(|m_y\rangle)_y$ of $\widetilde{\omega}$, the entropy–maximizing output state of T (which is unique, as easily follows from the strict concavity of S).

Defining the qc–channel $\widetilde{T} := M \circ T$ (i.e., the channel T followed by the measurement M), we are in the situation of lemma 4, with $\mathcal{Y} = \{1, \ldots, \dim \mathcal{H}_2\}$. Indeed, we can transform the given quantum feedback strategy into one based solely on the classical feedback of the measurement results, as explained in remark 4. Note that the additional quantum information available now at the sender due to the coherent feedback does not impair the validity of the argument of that remark: the important thing is that the classical feedback of the measurement results collapses the sender's state into one depending only on the message and the feedback.

By lemma 6 stated below, $\max_{\rho} H\big(\widetilde{T}(\rho)\big) = S(\widetilde{\omega})$, so lemma 4 gives us a set $\mathcal{E}$ of probability $Q(\mathcal{E}) \ge 1 - \epsilon$ and $|\mathcal{E}| \le \exp\big(nS(\widetilde{\omega}) + \alpha\sqrt{n}\big)$. The operator

$$\Pi := \sum_{y^n \in \mathcal{E}} |m_{y_1}\rangle\langle m_{y_1}| \otimes \cdots \otimes |m_{y_n}\rangle\langle m_{y_n}|$$

then clearly satisfies $\mathrm{Tr}(\omega \Pi) = Q(\mathcal{E}) \ge 1 - \epsilon$, and $\mathrm{rank}\, \Pi = |\mathcal{E}|$ is bounded as in lemma 4. □

Lemma 6. *Let $T : \mathcal{B}(\mathcal{C}^{d_1}) \longrightarrow \mathcal{B}(\mathcal{C}^{d_2})$ be a quantum channel and let $\widetilde{\rho}$ maximize $S(T(\rho))$ among all input states ρ. Denote $\widetilde{\omega} = T(\widetilde{\rho})$ (which is easily seen to be the unique entropy–maximizing output state of T), and choose a diagonalisation $\widetilde{\omega} = \sum_j \lambda_j |e_j\rangle\langle e_j|$. Then, for the channel $\widetilde{T}$ defined by*

$$\widetilde{T}(\rho) = \sum_j |e_j\rangle\langle e_j| T(\rho) |e_j\rangle\langle e_j|$$

(i.e., T followed by dephasing of the eigenbasis of $\widetilde{\omega}$),

$$\max_\rho S\big(\widetilde{T}(\rho)\big) = S(\widetilde{\omega}) = \max_\rho S\big(T(\rho)\big).$$

Proof. The inequality "$\geq$" is trivial because for input state $\widetilde{\rho}$, T and $\widetilde{T}$ have the same output state.

For the opposite inequality, let us first deal with the case that $\widetilde{\omega}$ is strictly positive (i.e., 0 is not an eigenvalue). The lemma is trivial if $\widetilde{\omega} = \frac{1}{d_2}\mathbb{1}$, so we assume $\widetilde{\omega} \neq \frac{1}{d_2}\mathbb{1}$ from now on. Observe that $\mathcal{N} := \{T(\rho) : \rho \text{ state on } \mathcal{C}^{d_1}\}$ is convex, as is the set $\mathcal{S} := \{\tau \text{ state on } \mathcal{C}^{d_2} : S(\tau) \geq S(\widetilde{\omega})\}$, and that $\mathcal{N} \cap \mathcal{S} = \{\widetilde{\omega}\}$. Since we assume that $\widetilde{\omega}$ is not maximally mixed, $\mathcal{S}$ is full–dimensional in the set of states, so the boundary $\partial S = \{\tau : S(\tau) = S(\widetilde{\omega})\}$ is a one–codimensional submanifold; from positivity of $\widetilde{\omega}$ (ensuring the existence of the derivative of S) it has a (unique) tangent plane H at this point:

$$H = \Big\{\xi \text{ state on } \mathcal{C}^{d_2} : \operatorname{Tr}\big[(\xi - \widetilde{\omega})\nabla S(\widetilde{\omega})\big] = 0\Big\}.$$

Thus, H is the unique hyperplane separating $\mathcal{S}$ from $\mathcal{N}$:

$$\mathcal{S} \subset H^+ = \Big\{\xi \text{ state on } \mathcal{C}^{d_2} : \operatorname{Tr}\big[(\xi - \widetilde{\omega})\nabla S(\widetilde{\omega})\big] \geq 0\Big\},$$
$$\mathcal{N} \subset H^- = \Big\{\xi \text{ state on } \mathcal{C}^{d_2} : \operatorname{Tr}\big[(\xi - \widetilde{\omega})\nabla S(\widetilde{\omega})\big] \leq 0\Big\}.$$

Now consider, for phase angles $\underline{\alpha} = (\alpha_1, \ldots, \alpha_{d_2})$, the unitary $U_{\underline{\alpha}} = \sum_j e^{i\alpha_j} |e_j\rangle\langle e_j|$, which clearly stabilizes $\mathcal{S}$ and leaves $\widetilde{\omega}$ invariant. Hence, also H and the two half-spaces H^+ and H^- are stabilized:

$$U_{\underline{\alpha}} H U_{\underline{\alpha}}^* = H, \qquad U_{\underline{\alpha}} H^+ U_{\underline{\alpha}}^* = H^+, \qquad U_{\underline{\alpha}} H^- U_{\underline{\alpha}}^* = H^-.$$

In particular, $U_{\underline{\alpha}} \mathcal{N} U_{\underline{\alpha}}^* \subset H^-$, implying the same for the convex hull of all these sets:

$$\operatorname{conv}\left\{\bigcup_{\underline{\alpha}} U_{\underline{\alpha}} \mathcal{N} U_{\underline{\alpha}}^*\right\} \subset H^-.$$

Since this convex hull includes (for $\tau \in \mathcal{N}$) the states

$$\sum_j |e_j\rangle\langle e_j| \tau |e_j\rangle\langle e_j| = \frac{1}{(2\pi)^{d_2}} \int d\underline{\alpha}\, U_{\underline{\alpha}} \tau U_{\underline{\alpha}}^*,$$

we conclude that for all ρ, $\widetilde{T}(\rho) \in H^-$, forcing $S\big(\widetilde{T}(\rho)\big) \leq S(\widetilde{\omega})$.

We are left with the case of a degenerate $\widetilde{\omega}$: there we consider perturbations $T_\epsilon = (1-\epsilon)T + \epsilon\frac{1}{d_2}\mathbb{1}$ of the channel, whose output entropy is maximized by the same input states as T, and the optimal output state is $\widetilde{\omega}_\epsilon = (1-\epsilon)\widetilde{\omega} + \epsilon\frac{1}{d_2}\mathbb{1}$. These are diagonal in any diagonalising basis for $\widetilde{\omega}$, so $\widetilde{T}_\epsilon = (1-\epsilon)\widetilde{T} + \epsilon\frac{1}{d_2}\mathbb{1}$.

Now our previous argument applies, and we get for all ρ,

$$S\big(\widetilde{T}_\epsilon(\rho)\big) \leq S(\widetilde{\omega}_\epsilon) \leq (1-\epsilon)S(\widetilde{\omega}) + \epsilon \log d_2 + H(\epsilon, 1-\epsilon).$$

On the other hand, by concavity, $S\big(\widetilde{T}_\epsilon(\rho)\big) \geq (1-\epsilon)S\big(\widetilde{T}(\rho)\big) + \epsilon \log d_2$. Together, these yield for all ρ,

$$S\big(\widetilde{T}(\rho)\big) \leq S(\widetilde{\omega}) + \frac{1}{1-\epsilon}H(\epsilon, 1-\epsilon),$$

and letting $\epsilon \to 0$ concludes the proof.   □

We are now in a position to prove

Theorem 5. *For a quantum channel T and $\lambda_1, \lambda_2 > 0$, $\lambda_1 + \lambda_2 < 1$,*

$$\lim_{n \to \infty} \frac{1}{n} \log \log N_{|F\rangle}(n, \lambda_1, \lambda_2) = C_{\mathrm{ID}}^{|F\rangle}(T) = 2\max_\rho S\big(T(\rho)\big),$$

unless the transmission capacity of T is 0, in which case $C_{\mathrm{ID}}^{|F\rangle}(T) = 0$.

In other words, the capacity of a nontrivial quantum channel with coherent feedback is twice its maximum output entropy and the strong converse holds.

Proof. The trivial channel is easiest, and the argument is just as in theorem 4. Note just one thing: a nontrivial channel with maximal quantum feedback will always allow entanglement generation (either because of the feedback or because it is noiseless), so — by teleportation — it will always allow quantum state transmission.

For archievability, the sender uses $m = n - O(\log n)$ instances of the channel to send one half of a purification Ψ_ρ of the output entropy maximizing state ρ each. This creates m copies of a pure state which has reduced state $T(\rho)$ at the receiver. After performing entanglement concentration[7], which yields $nS(T(\rho)) - o(n)$ EPR pairs, the remaining $O(\log n)$ instances of the channel are used (with an appropriate error correcting code and taking some of the entanglement for teleportation) to implement the construction of proposition 3, based on the maximal entanglement.

The converse is proved a bit differently than in theorem 4, where we counted the discretised decoders: now we have operators, and discretisation in Hilbert space is governed by slightly different rules. Instead, we do the following: given an identification code with feedback, form the uniform probabilistic mixture Φ of the feedback strategies Φ_i of messages i — formally, $\Phi = \frac{1}{N}\sum_i \Phi_i$. Its output state ω clearly is the uniform mixture of the output states ω_i corresponding to message i: $\omega = \frac{1}{N}\sum_i \omega_i$. With $\epsilon = 1 - \lambda_1 - \lambda_2$, lemma 5 gives us a projector Π of rank $K \leq \exp\big(n\max_\rho S(T(\rho)) + 48(\dim \mathcal{H}_2)^2 \epsilon \sqrt{n}\big)$ such that $\mathrm{Tr}(\omega\Pi) \geq 1 - \frac{1}{2}(\epsilon/24)^2$. Thus, for half of the messages (which we may assume to be $i = 1, \ldots, \lfloor N/2 \rfloor$), $\mathrm{Tr}(\omega_i\Pi) \geq 1 - (\epsilon/24)^2$.

Observe that the ω_i together with the decoding operators D_i form an identification code on $\mathcal{B}(\mathcal{H}_2^{\otimes n})$, with error probabilities of first and second kind λ_1

and λ_2, respectively. Now restrict all ω_i and D_i ($i \leq N/2$) to the supporting subspace of Π (which we identify with $\mathcal{C}^K$):

$$\widetilde{\omega}_i := \frac{1}{\mathrm{Tr}(\omega_i \Pi)} \Pi \omega_i \Pi, \qquad \widetilde{D}_i := \Pi D_i \Pi.$$

This is now an identification code on $\mathcal{B}(\mathcal{C}^K)$, with error probabilities of first and second kind bounded by $\lambda_1 + \frac{1}{3}\epsilon$ and $\lambda_2 + \frac{1}{3}\epsilon$, respectively, as a consequence of the gentle measurement lemma[32]: namely, $\frac{1}{2}\|\omega_i - \widetilde{\omega}_i\|_1 \leq \frac{1}{3}\epsilon$. So finally, we can invoke Proposition 11 of our earlier paper[33], which bounds the size of identification codes (this, by the way, is now the discretisation part of the argument):

$$\frac{N}{2} \leq \left(\frac{5}{1 - \lambda_1 - \epsilon/3 - \lambda_2 - \epsilon/3} \right)^{2K^2} = \left(\frac{15}{\epsilon} \right)^{2^n \max_\rho 2S(T(\rho)) + O(\sqrt{n})},$$

and we have the converse. □

Remark 5. For cq–channels $T : \mathcal{X} \longrightarrow \mathcal{B}(\mathcal{H})$ (a map assigning a state $T(x) = \rho_x$ to every element x from the finite set $\mathcal{X}$), we can even study yet another kind of feedback (let us call it *cq–feedback*): fix purifications Ψ_x of the ρ_x, on $\mathcal{H} \otimes \mathcal{H}$; then input of $x \in \mathcal{X}$ to the channel leads to *distribution* of Ψ_x between sender and receiver. In this way, the receiver still has the channel output state ρ_x, but is now entangled with the sender.

By the methods employed above we can easily see that in this model, the identification capacity is

$$C_{\mathrm{ID}}^{FF}(T) \geq \max_P \left\{ S\left(\sum_x P(x)\rho_x \right) + \sum_x P(x)S(\rho_x) \right\}.$$

Archievability is seen as follows: for a given P use a transmission code of rate $I(P;T) = S(\sum_x P(x)\rho_x) - \sum_x P(x)S(\rho_x)$ and with letter frequencies P in the codewords[20,28]. This is used to create shared randomness of the same rate, and the cq–feedback to obtain pure entangled states which are concentrated into EPR pairs[7] at rate $\sum_x P(x)E(\Psi_x) = \sum_x P(x)S(\rho_x)$; then we use eq. (10).

The (strong) converse seems to be provable by combining the approximation of output statistics result of Ahlswede and Winter[6] with a dimension counting argument as in our previous paper's [33] Proposition 11, but we won't follow on this question here.

Remark 6. Remarkably, the coherent feedback identification capacity $C_{\mathrm{ID}}^{|F\rangle}(T)$ of a channel is at present the only one we actually "know" in the sense that we have a universally valid formula which can be evaluated (it is single–letter); this is in marked contrast to what we can say about the plain (non–simultaneous) identification capacity, whose determination remains the greatest challenge of the theory.

Acknowledgements

Thanks to Noah Linden (advocatus diaboli) and to Tobias Osborne (doctor canonicus) for help with the proof of lemma 6.
The author was supported by the EU under European Commission project RESQ (contract IST-2001-37559).

References

1. R. Ahlswede, General theory of information transfer, General Theory of Information Transfer and Combinatorics, a Special issue of Discrete Mathematics, to appear.
2. R. Ahlswede and V. B. Balakirsky, Identification under random processes, Probl. Inf. Transm., 32, 1, 123–138, 1996.
3. R. Ahlswede and I. Csiszár, Common randomness in information theory and cryptography I: secret sharing, IEEE Trans. Inf. Theory, Vol. 39, No. 4, 1121–1132, 1993; Common Randomness in information theory and cryptography II: common randomness capacity, IEEE Trans. Inf. Theory, Vol. 44, no. 1, 225–240, 1998.
4. R. Ahlswede and G. Dueck, Identification via channels, IEEE Trans. Inf. Theory, Vol. 35, No. 1, 15–29, 1989.
5. R. Ahlswede and G. Dueck, Identification in the presence of feedback — a discovery of new capacity formulas, IEEE Trans. Inf. Theory, Vol. 35, No. 1, 30–36, 1989.
6. R. Ahlswede and A. Winter, Strong converse for identification via quantum channels, IEEE Trans. Inf. Theory, Vol. 48, No. 3, 569–579, 2002. Addendum, ibid., Vol. 49, No. 1, 346, 2003.
7. C.H. Bennett, H.J. Bernstein, S. Popescu, and B. Schumacher, Concentrating partial entanglement by local operations, Phys. Rev. A, 53(4), 2046–2052, 1996.
8. C.H. Bennett, P. Hayden, D.W. Leung, P.W. Shor, and A. Winter, Remote preparation of quantum states, e–print, quant-ph/0307100, 2003.
9. P. Hayden, D. W. Leung, P. W. Shor, and A. Winter, Randomizing quantum states: constructions and applications, e–print quant-ph/0307104, 2003.
10. G. Bowen, Quantum feedback channels, e–print, quant-ph/0209076, 2002.
11. G. Bowen and R. Nagarajan, On feedback and the classical capacity of a noisy quantum channel, e–print, quant-ph/0305176, 2003.
12. I. Devetak and A. Winter, Distillation of secret key and entanglement from quantum states, e–print, quant-ph/0306078, 2003.
13. I. Devetak and A. Winter, Relating quantum privacy and quantum coherence: an operational approach, e–print, quant-ph/0307053, 2003.
14. I. Devetak, A. W. Harrow and A. Winter, A family of quantum protocols, e–print, quant-ph/0308044, 2003.
15. A. Fujiwara and H. Nagaoka, Operational capacity and pseudoclassicality of a quantum channel, IEEE Trans. Inf. Theory, Vol. 44, No. 3, 1071–1086, 1998.
16. T. S. Han and S. Verdú, Approximation theory of output statistics, IEEE Trans. Inf. Theory, Vol. 39, No. 3, 752–772, 1993.
17. A. W. Harrow, Coherent communication of classical messages, Phys. Rev. Lett., 92, 9, 097902, 2004.
18. A. Harrow, P. Hayden, and D. W. Leung, Superdense coding of quantum states, e–print, quant-ph/0307221, 2003.

19. A.S. Holevo, Problems in the mathematical theory of quantum communication channels, Rep. Math. Phys., 12, 2, 273–278, 1977.
20. A. S. Holevo, The capacity of the quantum channel with general signal states, IEEE Trans. Inf. Theory, Vol. 44, No. 1, 269–273, 1998.
21. A. Jamiołkowski, Linear transformations which preserve trace and positive semi-definiteness of operators, Rep. Math. Phys., 3, 275–278, 1972.
22. C. Kleinewächter, On identification, this volume.
23. G. Kuperberg, The capacity of hybrid quantum memory, IEEE Trans. Inf. Theory, Vol. 49, No. 6, 1465–1473, 2003.
24. P. Löber, Quantum channels and simultaneous ID coding, Dissertation, Universität Bielefeld, Bielefeld, Germany, 1999. Available as e–print quant-ph/9907019.
25. E. Lubkin, Entropy of an n–system from its correlation with a k–reservoir, J. Math. Phys., 19, 1028–1031, 1978.
26. D. Page, Average entropy of a subsystem, Phys. Rev. Lett., 71, 9, 1291–1294, 1993.
27. T. Ogawa and H. Nagaoka, Strong converse to the quantum channel coding theorem, IEEE Trans. Inf. Theory, Vol. 45, No. 7, 2486–2489, 1999.
28. B. Schumacher and M.D. Westmoreland, Sending classical information via noisy quantum channels, Phys. Rev. A, 56, 1, 131–138, 1997.
29. C.E. Shannon, A mathematical theory of communication, Bell System Tech. J., 27, 379–423 and 623–656, 1948.
30. P.W. Shor, Equivalence of additivity questions in quantum information theory, Comm. Math. Phys. 246, No. 3, 453–472, 2004.
31. Y. Steinberg and N. Merhav, Identification in the presence of side information with applications to watermarking, IEEE Trans. Inf. Theory, Vol. 47, No. 4, 1410–1422, 2001.
32. A. Winter, Coding theorem and strong converse for quantum channels, IEEE Trans. Inf. Theory, Vol. 45, No. 7, 2481–2485, 1999.
33. A. Winter, Quantum and classical message identification via quantum channels, e–print, quant-ph/0401060, 2004.

Additive Number Theory and the Ring of Quantum Integers*

M.B. Nathanson**

In memoriam Levon Khachatrian

Abstract. Let m and n be positive integers. For the quantum integer $[n]_q = 1 + q + q^2 + \cdots + q^{n-1}$ there is a natural polynomial addition such that $[m]_q \oplus_q [n]_q = [m+n]_q$ and a natural polynomial multiplication such that $[m]_q \otimes_q [n]_q = [mn]_q$. These definitions are motivated by elementary decompositions of intervals of integers in combinatorics and additive number theory. This leads to the construction of the ring of quantum integers and the field of quantum rational numbers.

1 The Quantum Arithmetic Problem

For every positive integer n we have the *quantum integer*

$$[n]_q = 1 + q + q^2 + \cdots + q^{n-1}.$$

Then

$$\mathcal{F} = \{[n]_q\}_{n=1}^{\infty}$$

is a sequence of polynomials in the variable q. This sequence arises frequently in the study of q-series and of quantum groups (cf. Kassel [1, Chapter IV]). Adding and multiplying polynomials in the usual way, we observe that

$$[m]_q + [n]_q \neq [m+n]_q$$

and

$$[m]_q \cdot [n]_q \neq [mn]_q.$$

This suggests the problem of introducing new operations of addition and multiplication of the polynomials in a sequence so that addition and multiplication of quantum integers behave properly. We can state the problem more precisely as follows. Define "natural" operations of *quantum addition*, denoted $\oplus_q$, and *quantum multiplication*, denoted $\otimes_q$, on the polynomials in an arbitrary sequence $\mathcal{F} = \{f_n(q)\}_{n=1}^{\infty}$ of polynomials such that $f_m(q) \oplus_q f_n(q)$ and $f_m(q) \otimes_q f_n(q)$

---

* 2000 Mathematics Subject Classification: Primary 11B75, 11N80, 05A30, 16W35, 81R50. Secondary 11B13. Key words and phrases. Quantum integers, quantum addition and multiplication, polynomial functional equations, additive bases.

** This work was supported in part by grants from the NSA Mathematical Sciences Program and the PSC-CUNY Research Award Program.

R. Ahlswede et al. (Eds.): Information Transfer and Combinatorics, LNCS 4123, pp. 505–511, 2006.
© Springer-Verlag Berlin Heidelberg 2006

are polynomials, not necessarily in $\mathcal{F}$. We want to construct these operations so that, when applied to the polynomial sequence $\mathcal{F} = \{[n]_q\}_{n=1}^\infty$ of quantum integers, we have

$$[m]_q \oplus_q [n]_q = [m+n]_q \tag{1}$$

and

$$[m]_q \otimes_q [n]_q = [mn]_q \tag{2}$$

for all positive integers m and n. We would like these operations to determine the quantum integers uniquely.

2 Combinatorial Operations on Intervals of Integers

Let A and B be sets of integers, and let m be an integer. We define the *sumset*

$$A + B = \{a + b : a \in A \text{ and } b \in B\},$$

the *translation*

$$m + A = \{m + a : a \in A\},$$

and the *dilation*

$$m * A = \{ma : a \in A\}.$$

We write $A \oplus B = C$ if $A + B = C$ and every element of C has a unique representation as the sum of an element of A and an element of B.

Let $[n] = \{0, 1, 2, \ldots, n-1\}$ denote the set of the first $n-1$ nonnegative integers. Then

$$
\begin{aligned}
[m + n] &= \{0, 1, 2, \ldots, m+n-1\} \\
&= \{0, 1, 2, \ldots, m-1\} \cup \{m, m+1, m+2, \ldots, m+n-1\} \\
&= \{0, 1, 2, \ldots, m-1\} \cup m + \{0, 1, 2, \ldots, n-1\} \\
&= [m] \cup (m + [n]),
\end{aligned}
$$

and

$$[m] \cap (m + [n]) = \varnothing.$$

Moreover,

$$
\begin{aligned}
[mn] &= \{0, 1, 2, \ldots, mn-1\} \\
&= \{0, 1, 2, \ldots, m-1\} \oplus \{0, m, 2m, \ldots, m(n-1)\} \\
&= \{0, 1, 2, \ldots, m-1\} \oplus m * \{0, 1, 2, \ldots, n-1\} \\
&= [m] \oplus (m * [n]).
\end{aligned}
$$

If $m_1, \ldots, m_r$ are positive integers, then, by induction, we have the partition

$$[m_1 + m_2 + \cdots + m_r] = \bigcup_{j=1}^{r} \left(\sum_{i=1}^{j-1} m_i + [m_j] \right)$$

into pairwise disjoint sets, and the direct sum decomposition

$$[m_1 m_2 \cdots m_r] = \bigoplus_{j=1}^{r} \left(\prod_{i=1}^{j-1} m_i * [m_j] \right).$$

Associated to every set A of integers is the *generating function*

$$f_A(q) = \sum_{a \in A} q^a.$$

This is a formal Laurent series in the variable q. If A and B are finite sets of nonnegative integers and if m is a nonnegative integer, then $f_A(q)$ is a polynomial, and

$$f_{m+A} = q^m f_A(q)$$

and

$$f_{m*A}(q) = f_A(q^m).$$

If A and B are disjoint, then

$$f_{A \cup B}(q) = f_A(q) + f_B(q).$$

If $A + B = A \oplus B$, then

$$f_{A \oplus B}(q) = f_A(q) f_B(q).$$

The generating function of the interval $[n]_q$ is the quantum integer $[n]_q$. Since $[m] \cap (m + [n]) = \varnothing$, we have

$$
\begin{aligned}
[m + n]_q &= f_{[m+n]}(q) \\
&= f_{[m] \cup (m+[n])}(q) \\
&= f_{[m]}(q) + f_{m+[n]}(q) \\
&= f_{[m]}(q) + q^m f_{[n]}(q) \\
&= [m]_q + q^m [n_q].
\end{aligned}
$$

Similarly,

$$
\begin{aligned}
[mn]_q &= f_{[mn]}(q) \\
&= f_{[m] \oplus (m*[n])}(q) \\
&= f_{[m]}(q) f_{m*[n]}(q) \\
&= f_{[m]}(q) f_{[n]}(q^m) \\
&= [m]_q [n]_{q^m}.
\end{aligned}
$$

These identities suggest natural definitions of quantum addition and multiplication. If $\mathcal{F} = \{f_n(q)\}_{n=1}^{\infty}$ is a sequence of polynomials, we define

$$f_m(q) \oplus_q f_n(q) = f_m(q) + q^m f_n(q) \tag{3}$$

and

$$f_m(q) \otimes_q f_n(q) = f_m(q)f_n(q^m). \tag{4}$$

Then

$$[m]_q \oplus_q [n]_q = [m+n]_q$$

and

$$[m]_q \otimes_q [n]_q = [mn]_q.$$

More generally, if $\mathcal{F} = \{f_n(q)\}_{n=1}^{\infty}$ is any sequence of functions, not necessarily polynomials, then we can define quantum addition and multiplication by (3) and (4). We shall prove that the only nonzero sequence $\mathcal{F} = \{f_n(q)\}_{n=1}^{\infty}$ of functions such that

$$f_m(q) \oplus_q f_n(q) = f_{m+n}(q)$$

and

$$f_m(q) \otimes_q f_n(q) = f_{mn}(q)$$

is the sequence of quantum integers.

3   Uniqueness of Quantum Arithmetic

Let $\mathcal{F} = \{f_n(q)\}_{n=1}^{\infty}$ be a sequence of polynomials in the variable q that satisfies the addition and multiplication rules for quantum integers, that is, $\mathcal{F}$ satisfies the *additive functional equation*

$$f_{m+n}(q) = f_m(q) + q^m f_n(q) \tag{5}$$

and the *multiplicative functional equation*

$$f_{mn}(q) = f_m(q)f_n(q^m) \tag{6}$$

for all positive integers m and n. Nathanson [2] showed that there is a rich variety of sequences of polynomials that satisfy the multiplicative functional equation (6), but there is not yet a classification of all solutions of (6). There is, however, a very simple description of all solutions of the additive functional equation (5).

Theorem 1. *Let $\mathcal{F} = \{f_n(q)\}_{n=1}^{\infty}$ be a sequence of functions that satisfies the additive functional equation (5). Let $h(q) = f_1(q)$. Then*

$$f_n(q) = h(q)[n]_q \quad \text{for all } n \in \mathbb{N}. \tag{7}$$

Conversely, for any function $h(q)$ the sequence of functions $\mathcal{F} = \{f_n(q)\}_{n=1}^{\infty}$ defined by (7) is a solution of (5). In particular, if $h(q)$ is a polynomial in q, then $h(q)[n]_q$ is a polynomial in q for all positive integers n, and all polynomial solutions of (5) are of this form.

Proof. Suppose that $\mathcal{F} = \{f_n(q)\}_{n=1}^{\infty}$ is a solution of the additive functional equation (5). Define $h(q) = f_1(q)$. Since $[1]_q = 1$ we have

$$f_1(q) = h(q)[1]_q.$$

Let $n \geq 2$ and suppose that $f_{n-1}(q) = h(q)[n-1]_q$. From (5) we have

$$
\begin{aligned}
f_n(q) &= f_1(q) + q f_{n-1}(q) \\
&= h(q)[1]_q + q h(q)[n-1]_q \\
&= h(q)([1]_q + q[n-1]_q) \\
&= h(q)[n]_q.
\end{aligned}
$$

It follows by induction that $f_n(q) = h(q)[n]_q$ for all $n \in \mathbb{N}$.

Conversely, multiplying (5) by $h(q)$, we obtain

$$h(q)[m+n]_q = h(q)[m]_q + q^m h(q)[n]_q,$$

and so the sequence $\{h(q)[n]_q\}_{n=1}^{\infty}$ is a solution of the additive functional equation (5) for any function $h(q)$. This completes the proof.

We can now show that the sequence of quantum integers is the only nonzero simultaneous solution of the additive and multiplicative functional equations (5) and (6).

Theorem 2. *Let $\mathcal{F} = \{f_n(q)\}_{n=1}^{\infty}$ be a sequence of functions that satisfies both functional equations (5) and (6). Then either $f_n(q) = 0$ for all positive integers n, or $f_n(q) = [n]_q$ for all n.*

Proof. The multiplicative functional equation implies that $f_1(q) = f_1(q)^2$, and so $f_1(q) = 0$ or 1. Since $\mathcal{F} = \{f_n(q)\}_{n=1}^{\infty}$ also satisfies the additive functional equation, it follows from Theorem 1 that there exists a function $h(q)$ such that $f_n(q) = h(q)[n]_q$ for all positive integers n, and so $h(q) = 0$ or 1. It follows that either $f_n(q) = 0$ for all n or $f_n(q) = [n]_q$ for all n. This completes the proof.

4   The Ring of Quantum Integers

We can now construct the ring of quantum integers and the field of quantum rational numbers. We define the function

$$[x]_q = \frac{1 - q^x}{1 - q}$$

of two variables x and q. This is called the *quantum number* $[x]_q$. Then

$$[0]_q = 0,$$

and for every positive integer n we have

$$[n]_q = \frac{1 - q^n}{1 - q} = 1 + q + \cdots + q^{n-1},$$

which is the usual quantum integer. The negative quantum integers are

$$[-n]_q = \frac{1 - q^{-n}}{1 - q} = -\frac{1}{q^n}[n]_q = -\left(\frac{1}{q} + \frac{1}{q^2} + \cdots + \frac{1}{q^n}\right).$$

Then

$$\begin{aligned}
[x]_q \oplus_q [y]_q &= [x]_q + q^x [y]_q \\
&= \frac{1 - q^x}{1 - q} + q^x \frac{1 - q^y}{1 - q} \\
&= \frac{1 - q^{x+y}}{1 - q} \\
&= [x + y]_q
\end{aligned}$$

and

$$\begin{aligned}
[x]_q \otimes_q [y]_q &= [x]_q [y]_{q^x} \\
&= \frac{1 - q^x}{1 - q} \frac{1 - q^{xy}}{1 - q^x} \\
&= \frac{1 - q^{xy}}{1 - q} \\
&= [xy]_q.
\end{aligned}$$

The identities

$$[x]_q \oplus_q [y]_q = [x + y]_q \qquad \text{and} \qquad [x]_q \otimes_q [y]_q = [xy]_q \tag{8}$$

immediately imply that the set

$$[\mathbf{Z}]_q = \{[n]_q : n \in \mathbf{Z}\}$$

is a commutative ring with the operations of quantum addition $\oplus_q$ and quantum multiplication $\otimes_q$. The ring $[\mathbf{Z}]_q$ is called the *ring of quantum integers*. The map $n \mapsto [n]_q$ from $\mathbf{Z}$ to $[\mathbf{Z}]_q$ is a ring isomorphism.

For any rational number m/n, the quantum rational number $[m/n]_q$ is

$$[m/n]_q = \frac{1 - q^{m/n}}{1 - q} = \frac{\frac{1 - \left(q^{1/n}\right)^m}{1 - q^{1/n}}}{\frac{1 - \left(q^{1/n}\right)^n}{1 - q^{1/n}}} = \frac{[m]_{q^{1/n}}}{[n]_{q^{1/n}}}.$$

Identities (8) imply that addition and multiplication of quantum rational numbers are well-defined. We call

$$[\mathcal{Q}]_q = \{[m/n]_q : m/n \in \mathcal{Q}\}$$

the *field of quantum rational numbers*.

If we consider $[x]_q$ as a function of real variables x and q, then

$$\lim_{q \to 1} [x]_q = x$$

for every real number x.

We can generalize the results in this section as follows:

Theorem 3. *Consider the function*

$$[x]_q = \frac{1 - q^x}{1 - q}$$

in the variables x and q. For any ring R, not necessarily commutative, the set

$$[R]_q = \{[x]_q : x \in R\}$$

is a ring with addition defined by

$$[x]_q \oplus_q [y]_q = [x]_q + q^x [y]_q.$$

and multiplication by

$$[x]_q \otimes_q [y]_q = [x]_q [y]_{q^x}$$

The map from R to $[R]_q$ defined by $x \mapsto [x]_q$ is a ring isomorphism.

Proof. This is true for an arbitrary ring R because the two identities in (8) are formal.

References

1. C. Kassel, Quantum Groups, Graduate Texts in Mathematics, Vol. 155, Springer-Verlag, New York, 1995.
2. M. B. Nathanson, A functional equation arising from multiplication of quantum integers, J. Number Theory 103, No. 2, 214–233, 2003.

IV

The Proper Fiducial Argument

F. Hampel

Abstract. The paper describes the proper interpretation of the fiducial argument, as given by Fisher in (only) his first papers on the subject. It argues that far from being a quaint, little, isolated idea, this was the first attempt to build a bridge between aleatory probabilities (the only ones used by Neyman) and epistemic probabilities (the only ones used by Bayesians), by implicitly introducing, as a new type, frequentist epistemic probabilities. Some (partly rather unknown) reactions by other statisticians are discussed, and some rudiments of a new, unifying general theory of statistics are given which uses upper and lower probabilities and puts fiducial probability into a larger framework. Then Fisher's pertaining 1930 paper is being reread in the light of present understanding, followed by some short sections on the (legitimate) aposteriori interpretation of confidence intervals, and on fiducial probabilities as limits of lower probabilities.

Keywords: Fiducial argument, fiducial probability, R.A. Fisher, foundations of statistics, statistical inference, aleatory probabilities, epistemic probabilities, structure of epistemic probabilities, upper and lower probabilities, frequentist statistics, axiom of frequentist epistemic probability, Bayesian statistics, intersubjective statistics, bets, odds, fair bets, successful bets, confidence interval, aposteriori interpretation of confidence intervals, Neyman-Pearson statistics, Behrens-Fisher problem.

1 Introduction

At first glance, it may be surprising to find in a book with stress on information theory and related mathematics an article on a subtle concept in the foundations of probability theory and statistics. However, this concept of fiducial probabilities, as well as the seemingly unrelated one of upper and lower probabilities (which however allows to put the former concept into the proper perspective), may (and I believe eventually will) have profound influence on statistics and other areas of stochastics, including information theory. Apart from that, I find it about time for the fiducial argument to be clarified and put into a broader framework, roughly 3/4 of a century after its invention, and after many hot and confused discussions about it.

Related is also the fine distinction (made already by Bernoulli [4], but later suppressed) between probabilities that are actually known to us ("epistemic probabilities") and "probabilities" (a misnomer due to false speculations about the long unpublished Ars conjectandi [4], cf. Shafer [44], or Brönnimann [5])

R. Ahlswede et al. (Eds.): Information Transfer and Combinatorics, LNCS 4123, pp. 512–526, 2006.
© Springer-Verlag Berlin Heidelberg 2006

that are merely hypothetically assumed (essentially "aleatory probabilities", cf. Sec. 3). The fiducial argument is a bridge which sometimes allows to derive the former from the latter.

While probability theory assumes perfect knowledge of the underlying stochastic model, in applied statistics (but also often in other applications such as information theory) we have only partial knowledge. It seems more realistic in many situations to describe this partial knowledge of the underlying random mechanism by upper and lower probabilities (which were also already used in [4], cf. Sec. 2).

I arrived at the main topic of this paper via an apparent detour. (Proper) fiducial probabilities arose (somewhat surprisingly) as a special case in a side branch of my inference theory using upper and lower probabilities [26], and understanding of the general theory may help (and certainly has helped me) to understand the "mysterious" fiducial theory, and find an appropriate place for it in a larger framework. On the other hand, my experience both with the literature and with many oral discussions is that there still exists a lot of confusion, not only about the fiducial argument, but also about related concepts such as aleatory and epistemic probabilities, a frequentist interpretation of epistemic probabilities, and the difference between Fisher's and Neyman's interpretation of confidence intervals. Since all this is also part of my theory, an understanding of these (historical) concepts is obviously required for a full understanding of my theory. But what is mysterious to me is that more than 70 years after Fisher's [12] first (and correct) paper about the fiducial argument, there is still no clarity about it, and most descriptions of it (following Fisher's erroneous later work) are from half true to plainly wrong and nonsensical. Therefore I thought it perhaps worthwhile to try to explain the (proper) fiducial argument and its surroundings in more detail.

The fiducial argument was meant to be a new mode of inference, making superfluous the appeal to a (usually unknown) apriori distribution to be entered into Bayes' theorem. From 1930 to about 1960, it was one of the "hottest" topics of debate in statistics, with participation of top statisticians across the whole spectrum, from J.W. Tukey to A.N. Kolmogorov. As some mathematical contradictions (within the later, false interpretation by Fisher) could be derived, the whole debate fell into oblivion soon after Fisher's death (in 1962), and many young statisticians today have never even heard of a fiducial argument or probability.

To give briefly one of the simplest examples: Let (entirely within a frequentist framework) a random variable X have the distribution $N(\theta, 1)$ (normal with unknown location θ and known variance 1), where θ may be anywhere on the real line. Then for every fixed $c \in R^1$, and for every (true, unknown, fixed) parameter θ it is true that $P(X \leq \theta + c) = \Phi(c)$ (with Φ being the cumulative standard normal). Equivalently, $P(\theta \geq X - c) = \Phi(c)$. Now assume we have observed (a realization of the random variable) $X = x$ (e.g., $X = 3$). Then Fisher, using the "modus ponens", plugs in the observed x (this is the controversial "fiducial argument") and obtains $P(\theta \geq x - c) = \Phi(c)$ (e.g., with $c = 1$, $P(\theta \geq 3 - 1) = P(\theta \geq 2) = \Phi(1)$ (this is interpreted as a "fiducial probability" for θ, namely the probability that $\theta \geq 2$; letting c move from $-\infty$ to $+\infty$, one obtains the whole "fiducial probability distribution" of θ).

Now what is random? θ? But θ was a fixed unknown constant. Moreover, P is actually P_θ (this θ is often suppressed). Are there two different θ's in the same formula?

In 1930 [12] and 1933 [13], Fisher gave correct (though cumbersome, brief and incomplete, hence apparently misunderstood) interpretations of this "tempting" result. But starting in 1935 [14], he really believed he had changed the status of θ from that of a fixed unknown constant to that of a random variable on the parameter space with known distribution (cf. [16]). Apparently he needed this unfounded and false assumption in order to "solve" the Behrens-Fisher problem (the test for equality of means of two independent normal samples with unknown and possibly different variances, as opposed to the two-sample t-test). The "solution" was shown to be mathematically wrong; but Fisher was intuitively fully convinced of the importance of "fiducial inference", which he considered the jewel in the crown of the "ideas and nomenclature" for which he was responsible ([51], p. 370); and he vigorously defended his false interpretation up to his last statistics book [16]. Later on, most statisticians, unable to separate the good from the bad in Fisher's arguments, considered the whole fiducial argument Fisher's biggest blunder, or his one great failure (cf., e.g., [51], [10]), and the whole area fell into disrepute.

By contrast, I consider the (properly interpreted) fiducial argument the first (though highly incomplete) attempt to bridge the gap between a wholly aleatory Neyman-Pearson theory and a wholly epistemic Bayesian theory, either of which comprising only one-half of what statistics should be [27]; and Fisher does so by introducing (implicitly) frequentist(!) intersubjective epistemic probabilities (for a brief explanation of concepts, see Sec. 3). These concepts have strong implications for the everyday practical use of statistics, such as the aposteriori interpretation of confidence intervals (see Sec. 5). I thus agree with Fisher, not in his formal later interpretation of the fiducial argument (which is wrong), but about the importance of the basic idea (and the correctness of his first interpretation, which he later denied).

I can only speculate about the reasons why the fiducial argument was not clarified earlier. Some reasons might be:

1. Lack of distinction between aleatory and epistemic probabilities (cf. Sec. 3). I believe Fisher felt the distinction intuitively, but he never clearly formulated it. For the Neyman-Pearson school, there exist only aleatory probabilities (very strictly so!), and for (strict) Bayesians there exist only epistemic probabilities (perhaps apart from a few simple cases where Hacking's principle – "If an aleatory probability is known to be p, this p should be used as epistemic probability" – might be applicable), hence the two schools basically cannot even talk to each other (cf. also [27]).

2. Almost nobody seems to have checked on which probability space the (proper) fiducial probabilities can be defined! (Cf. the example above.) While the axiomatic foundation of probability spaces was done only a few years after Fisher's first fiducial paper [32], I find it surprising that apparently none of

the later mathematical statisticians (with only one exception [41] known to me) has asked this basic question.

3. There seems to be an implicit conviction that there can be no frequentist epistemic probabilities (apart, perhaps, again from simple uses of Hacking's principle). This leaves only "logical" and subjectivist Bayesian results for scientists who really want to learn from data (and not just obey behavioristic rules), both of which are unsatisfactory in principle for them.

4. Within frequentist statistics, it seems often impossible for the thinking of statisticians to get away from the deeply entrenched paradigm of independent repetitions of the SAME experiment, although apart from a few applications, such as quality control and simulation studies, they hardly ever exist in science. Most scientists do DIFFERENT independent experiments each time, and frequentist properties of statistical methods can and should be evaluated with regard to such sequences of experiments. (If this should be considered an enlargement of the formal definition of "frequentist", then I find it long overdue. This point certainly was clear already to Fisher, for example, and other early writers.)

5. It is very tempting to believe that something which formally looks like a probability distribution is actually a probability distribution, without regard to the restrictions and interpretations under which it was derived. I am talking, of course, about the general misinterpretation of the "fiducial probability distribution".

6. Perhaps a major reason is Fisher's highly intuitive and condensed style of writing, which requires "reading from within" (trying to "feel" what he meant and thought) rather than "reading from the outside" (superficially taking words literally). - In addition, probably few statisticians went "back to the roots", to Fisher's first papers on the topic; it is the custom in our scientific enterprise to try to be always at the forefront of research; and the forefront in this case was leading astray because of Fisher's later blunder. (Still, the fact that some kind of blunder became known to exist, might have motivated a few more statisticians to study the origins more carefully.)

This paper contains some reflections on various reactions to the fiducial argument by other writers (Sec. 2), and mainly (throughout the paper, and specifically in Sec. 4) a description of what I consider the proper fiducial argument, based on a new reading and interpretation of Fisher's first pertaining papers ([12], [13]) in relation to my more general approach [26]. It seems necessary to explain a few rudiments of my approach before Section 4 (Sec. 3). Some remarks on the dispute between Fisher and Neyman about the proper interpretation of confidence intervals are also added (Sec. 5), as well as a short section on fiducial probabilities as limiting and special cases of lower probabilities (Sec. 6).

The paper discusses only one-dimensional problems. The emphasis is on frequentist properties, as opposed to full conditionality and coherence; elsewhere I have shown that (symmetrical) optimal compromises between the two desiderata can be defined (including Bayes solutions closest to a frequentist interpretation, and vice versa), and that they can be numerically very close to each other ([23], [26]).

2   Some Reactions to the Fiducial Argument

The history of Fisher's work on fiducial probabilities is excellently described by Zabell [51]. Briefly, Fisher discovered and solved the argument 1930 [12] for the correlation parameter, and 1933 [13] he solved it for the variance in normal samples (with a critical discussion of Jeffreys' approach to this problem). In 1935 [14], Fisher "solved" the Behrens-Fisher problem by assuming that the "fiducial distribution" is an ordinary probability distribution of a random parameter, and from then on he had to defend himself not only against lack of understanding, but also against justified criticisms. He tried - in vain - to escape these criticisms by some spurious conditioning arguments, but he was unable to retract gracefully from an untenable position (after previously having criticized Bayesians for the same reason). He even accepted the Bayesian derivation (with an improper prior) for the Behrens-Fisher test given by Jeffreys, after having criticized Jeffreys and stressing the differences of their approaches in 1933 [13]. His last major authoritative (though partly false) claims are in 1955 [15] and in his book 1956 [16]. (It should be clear that by the proper fiducial argument I mean Fisher's argument of 1930 and 1933, and not anything building upon his later erroneous work.)

It may be noted that already quite early, Fisher spoke occasionally and briefly of "fiducial inequalities" (in situations with discrete variables), thus faintly foreshadowing the use of upper and lower probabilities in these cases. In his 1956 book [16] and earlier, he contrasted fiducial probabilities and likelihoods as the main inference tools for continuous and discrete data, resp.; it seems to me that likelihoods might here better be replaced by upper and lower probabilities (while maintaining a central auxiliary role in both situations, of course).

By restricting himself to proper probabilities, Fisher obtained a rather limited theory. This reminds me of Bayes [2] who in his scholium implicitly made the same restriction. Neither seems to have thought - or known - about upper and lower probabilities, although they had been introduced implicitly much earlier by James (Jacob) Bernoulli [4] in the short and fragmentary Part IV of his Ars conjectandi (cf. also Lambert [36], Shafer [44], and Brönnimann [5]).

Upper and lower probabilities in statistics were rediscovered by Koopman [34], Kyburg [35], C. A. B. Smith [45] and I. J. Good [18], cf. also Fine [11]. Special mathematical structures were thoroughly investigated by Choquet [6] and partly rediscovered (in the unpublished first version of [46]) by Strassen [46], for solving a problem in information theory; cf. also [30] and [31]. Huber ([28], [29]) discusses their use in statistics. Both Bayes and Fisher could have avoided the main limitations of their approaches by explicitly allowing upper and lower probabilities.

A number of statisticians, starting with Bartlett, tried to check the Behrens-Fisher test or to find conditions under which Fisher's new methods could be justified by other, objective arguments (cf., e.g., [1], [48], [49]). A number of other statisticians were more indirectly inspired by Fisher's work, trying to find something new in a similar direction. They include Fraser [17] trying to exploit group structures if they happen to be available, and especially Dempster ([8], [9]), whose work on different kinds of upper and lower probabilities led to the theory of belief functions by Shafer [43], Smets and others. Probably many if not most of the first, older generation of statisticians working on upper and lower probabilities (a new research area in statistics) have at one time or other thought hard about fiducial probabilities and were motivated by them for their own work.

Kolmogorov [33] discusses fiducial probabilities in a summarizing report on contemporary British statistics. In Footnote 12, he suggests the introduction of a new axiom: if all conditional probabilities of an event, given the parameters, exist and are equal, then the unconditional probability exists and equals this value. At first, I was puzzled by this remark. Later, I thought that maybe this can be interpreted to be the axiomatic introduction of a new kind of probability (an epistemic one, to use present wording) which does not depend on any (unknown) parameters. (We may call it the axiom of frequentist epistemic probability.) Viewed this way, it may make deep sense, although it is still highly incomplete (for example, the underlying probability space and the epistemic interpretation are not discussed). - Incidentally, Kolmogorov [33] quite naturally discusses sequences of different experiments (as opposed to repetitions of the same experiment); and he partly argues in favor of unconditional statements in applications, for practical reasons.

One of the greatest mysteries for me around the fiducial argument is why the extremely important 1957 paper by Pitman [41] seems to have gone virtually unnoticed in the discussion of the fiducial argument. (I owe the reference to Robert Staudte.) Pitman gives a lucid and deep mathematical description of the fiducial argument, what is right and what is wrong about it, and mathematical conditions under which it can be applied. He is not afraid of calling a mistake a mistake, such as Fisher's [16] claim that the parameter has now the status of a random variable (simple acknowledgment of this fact would have made most past discussions of "fiducial probabilities" superfluous). In the same sentence (p. 325) he asserts that nevertheless "... we are able to make statements about the unknown parameter with a calculable probability of being right" (this is precisely my viewpoint). Pitman also discusses the fiducial distribution of several parameters, while making it clear that Fisher's integrations for the "Behrens-Fisher solution" were simply not permissible. He does not go further into the conceptual interpretation, but any truly informed mathematical paper about the fiducial argument has to incorporate what he has to say. Nevertheless Pitman's paper is not cited in Cox and Hinkley [7], Walley [50], Zabell [51], Efron [10], to name a few prominent works discussing fiducial inference; and this, although Pitman is not entirely unknown in mathematical statistics, the journal in which the paper appeared is not totally obscure, and the paper was an invited review article on Fisher's most important book on foundations, and was written in Stanford, a place not totally provincial in statistics.

Cox and Hinkley [7] just give the usual "on the one hand ... on the other hand ..." type of discussion of the fiducial argument; and in exercise 7 of Ch. 7, p. 248 (accessible through the index), they try to discredit fiducial probabilities (without using the word here) by using an operation (curtailing the distribution) which was (correctly) explicitly forbidden in Fisher's early works (though not in his later, incorrect works).

Most statisticians, after making up their mind that the whole fiducial argument was just a big blunder, tried to forget it. C.R. Rao [42] may have been one of the last famous statisticians to include it in an ordinary applied statistics textbook, though with reservations. But although the argument in its later form had been proven wrong, some statisticians, such as J.W. Tukey, still thought "there may be something to it". I learned this attitude from him in the seventies, and in our last telephone conversation in July 2000, a few days before his death, he confirmed to me that this still was his opinion.

In his 1996 Fisher Lecture, Efron [10] gave (among other things) a reasonable sounding discussion of the desirability of something like fiducial inference in future statistics. But then (end of Section 8), out of the blue he suddenly speculates: "Maybe Fisher's biggest blunder will become a big hit in the 21st century!" I agree of course about the hit (though, slowly as statistics progresses, it may well be the 22nd century), but just by rereading his paper I can't find any reason or justification for his optimism (unless there is a - direct or indirect - connection with my talk on foundations at Stanford in the Berkeley-Stanford Colloquium in March 1994). In any case, it is gratifying to see (also in the discussion of the paper) that some statisticians might be willing to take a fresh look at the fiducial argument, recognizing its basic importance for statistics.

3   Some Rudiments of a New General Theory of Statistics

Before rereading Fisher's 1930 paper, it seems necessary to explain some of the concepts which are hidden in and behind Fisher's early work, and which should make this work more comprehensible.

The broadest view of my emerging theory is given in [20], some more solutions in [21], the probably most readable introduction in [24], and the most recent highly condensed survey in [26]. I first noticed the connection with fiducial probabilities in an abstract [19], and later in a chapter in [20].

Let me now try to explain my view of statistics.

I became more and more convinced that we have to distinguish between aleatory and epistemic probabilities. Aleatory probabilities (as in dice throwing) are supposed to be probabilities occurring objectively in Nature (in "random experiments"). (Of course, I am aware that the whole concept of an objectively existing Nature - as well as that of probabilities - can be criticized philosophically, but without it we could have no science, and we have been quite successful with it. Here I am trying to keep the discussion on a reasonably pragmatic level.) Usually, aleatory probabilities are unknown to us (except in simulation studies, or under Laplacian assumptions of symmetry), but we are trying to learn something about them. Aleatory probabilities are frequentist, that is, they obey the law of large numbers. This (besides the usual axioms) gives them an (approximate, but arbitrarily accurate) operational interpretation: In a long sequence of (usually different!) independent experiments, all with probability of "success" (different "successes"!) equal p, the observed fraction of "successes" will be close to p (with the usual precise mathematical formulation).

But many statisticians, and users of statistics, also want to learn something, and want to know some probabilities (and not only approximately, if it can be done). Probabilities which refer to our (personal or intersubjective, assumed or true!) state of knowledge are called epistemic. It would be nice if we could derive known epistemic probabilities from the unknown aleatory probabilities.

But the Neyman-Pearson theory considers only aleatory probabilities. Neyman ([38], [39]) explicitly excluded inductive inference, hence all learning processes and the epistemic probabilities which they could lead to. This is very satisfactory for pure mathematicians, because it keeps the basis of the theory mathematically and conceptually simple, and it may be tolerable for decision theorists; but it is frustrating for many applied statisticians and users of statistics who actually want to learn something from their data (also in a broader context), and not just "behave inductively" without any thinking being allowed.

On the other hand, all Bayesian theories (and we have to distinguish at least between Bayes - if we want to call him a Bayesian -, Laplace, Jeffreys, and the Neo-Bayesians, who may be further split up) use, at least in principle, only epistemic probabilities (except perhaps for the conditional distributions of the observations, given the parameters, which may be considered aleatory, but which are quickly transformed away). They start with epistemic prior distributions for the parameters, and they end with epistemic posterior distributions or some predictions or decisions derived from them. Bayesian probabilities may be subjective (as with the Neo-Bayesians) or "logical", "canonical" or "objective" (a very dubious term), as with the other Bayesian schools mentioned; these logical probabilities are intersubjective, that means, they are the same for scientists with the same data (and model) and the same background knowledge. But none of them has a frequentist interpretation (unless the prior chosen happens to be a true aleatory prior). The concept of a true, unknown aleatory probability distribution which governs the success of Bayesian claims and bets, is alien to (strict) Neo-Bayesian theory, and the self-assuring success of Bayesian "fair bets" results from them being evaluated by their own subjective priors. We shall see that Fisher, in effect, tried to introduce frequentist intersubjective epistemic probabilities (without using these qualifying terms).

It seems natural to describe epistemic probabilities by bets or odds (or odd ratios), as has been commonly done already centuries ago. Bayesian (pairs of) fair bets ("If I am willing to bet $p : q$ on A, I am also willing to bet $q : p$ on A^c") are two-sided bets and correspond 1:1 to ordinary probabilities ("$P(A) + P(A^c) = 1$"). But if we are not in a decision situation, but in an inference situation, we may also consider one-sided bets, expressing partial lack of knowledge about the true probability (up to total ignorance); in order to avoid "sure loss" with bets both on A and on A^c, we must have $P(A) + P(A^c) \leq 1$ and are thus led to (some form of) upper and lower probabilities. (The Bayesian claims for equality to 1 are circular and just not true, except for enforced decisions.)

If the expected gain of my one-sided bet, evaluated under the true (unknown) aleatory (not epistemic!) probability of A is nonnegative, I call my bet "successful". Obviously, I cannot bet successfully on A (except $0 : 1$) without any knowledge about its probability; but the amazing fact is that in general I can find nontrivial successful bets if I have independent past observations from the same parametric model.

The bets cannot, of course, in general be conditionally successful given any fixed past observation, and at the same time informative, because usually there may be totally misleading past observations; but they can be successful when averaged also over the distribution of the past observations. Their success can be operationally checked and empirically validated by considering long sequences of independent successful bets (from different experiments!); with bounded and sufficiently variable gains the average gain will most likely be $> -\epsilon$ for n large enough.

My theory is mainly for prediction, because I find prediction in general more important in practice than parameter estimation (cf. [22]), and it can be checked empirically. But the theory can also be done for random (!) parameter sets; and then, in some cases, it just gives the (proper) fiducial probabilities (as frequentist epistemic proper probabilities).

4   Rereading Fisher's 1930 Fiducial Paper

(It may be useful for the reader to get a copy of this paper, e.g. from [3].)

In the beginning of the paper, Fisher [12] attacks Bayes (incorrectly, but mildly) and the then Bayesians (to discuss this part would mean another section - there appear to be some parallels between the history of Bayes' argument and of the fiducial argument); and then he discusses likelihood (describing it, like the fiducial argument, more as an empirical discovery rather than an invention). Starting at the bottom of p. 532 (p. 433 in [3]), he discusses the fiducial argument with the example of the correlation coefficient, even giving a table for $n = 4$ which for every ρ gives the upper 95% value (now "confidence limit") for r. And this relationship "implies the perfectly objective fact that in 5% of samples r will exceed the 95% value corresponding to the actual value of ρ in the population from which it is drawn." (And he goes on to define ρ_r, the "fiducial 5% value of ρ" corresponding to a given observed r.) Thus the actual value of ρ (an unknown constant) may be anything (or have any prior distribution, for that matter): among all unselected(!) pairs (ρ, r) (typically from different experiments!), in 5% of all cases ρ will be smaller than ρ_r. The random event "$\rho < \rho_r$" (where the randomness is in ρ_r!) has an epistemic (known to us), frequentist probability of 5%. We can bet on its truth successfully $1 : 19$, and we can even bet on its complement successfully $19 : 1$; that means, we have not only a lower, but a proper probability for this event and a Bayesian (pair of) fair bet(s). The bets can be validated by taking any sequence of (different) independent experiments with arbitrary values of ρ, the (unselected!) observed values of r and the corresponding values of ρ_r; in exactly 5% of all cases (in the long run), ρ will be found to be less than ρ_r.

"If a value $r = .99$ were obtained from the sample, we should have a fiducial 5% ρ equal to about .765. The value of ρ can then only be less than .765 in the event that r has exceeded its 95% point, an event which is known to occur just once in 20 trials. In this sense ρ has a probability of just 1 in 20 of being less than .765." Here (if "in this sense" is interpreted correctly) Fisher is still correct, but dangerously close to his later mistake of considering the "fiducial distribution" (which he defines next) as an ordinary probability distribution. The event "$\rho < .765$" can correctly be included in a sequence of events of the form "$\rho < \rho_r$", all having epistemic probability $1/20$; but the other members of any such sequence ("almost surely") don't have $\rho_r = .765$ (with some dubiously "random" ρ's floating around), but rather any sequence of fixed, predetermined, arbitrary, haphazardly taken, even aleatorily random values of ρ, random values of r whose distributions depend on them, and the corresponding values of ρ_r determined by r. (By contrast, when we wanted to define a "probability distribution for ρ", .765 would have to be a fixed value in repetitions of the same experiment - there is nothing of that here.)

I think one of the reasons why we still have problems with fiducial probabilities, is that we lack an adequate formalism for frequentist epistemic probabilities. For a start, let me offer the following. Given an ordinary, aleatory parametric model, consider a class of random "claims" or "statements" $\{S\}$ depending on a random variable X on that model. In the simplest case, this is a single random statement (e.g., "$\theta \geq X - 1$"), or a complementary pair of such statements. We can call a random claim "assessable" if it has the same (aleatory) probability under all parameter values; this probability value is taken to define the epistemic (since it is known to us) probability P of the random claim (cf. also [33]). We then define, for each θ, a mapping $V = V_\theta$ (depending on θ) from the space of possible realizations of the random claims to the two-point space $\Omega = \Omega_S = \{$true, false$\}$ with $V(S) = $ "true" if S is true. Obviously, the probability of a randomly (via X) selected claim to be true is P, independently of θ (but the set of all claims which are true is different for each θ). Hence we obtain a fixed probability distribution on Ω (belonging to our random claim), independently of θ.

If we have several random claims on the same probability space, we can build the Cartesian product of the Ω_S's, and no matter what happens with the joint distributions, the marginal distributions of the single Ω_S's are still correct. Sometimes we may derive new assessable claims. In particular, if we have a sequence of independent claims (based on independent aleatory experiments), with the same probability of being true, we can apply the law of large numbers, and hence we can bet successfully (and even fairly) $P : (1 - P)$ on the truth of any one claim (whose realization is randomly selected by some X_S) and will come out even in the long run. (The problem of selection among different possible random claims or successful bets is still not quite solved in the general theory (cf. [26]): one idea has a deeper interpretation, and the other is mathematically more elegant.)

The "fiducial distribution of a parameter θ for a given statistic T" may be considered as a collection, or shorthand notation, for all (mutually compatible) successful claims or bets about θ derivable from T. (Note that Fisher writes: "for a given statistic T"! T will be different next time.) At most it might perhaps be called a "random distribution", depending on T. From it can be derived epistemic probabilities and successful bets on events of the form "$a < \theta < b$" (etc., by simple linear operations), but not, e.g., of the form "$\theta^2 < a$" or "$|\theta| < b$" (cf. [41]).

Replacing equal probabilities by an infimum of probabilities in the introduction of epistemic probabilities, most of the argument (except the fair bet aspect) can be done also with lower probabilities. There are some connections with belief function theory [43], but the interpretation is different.

It seems that Zabell ([51], e.g. p. 374), and probably many other readers, were confused by Fisher's wild switching between aleatory and epistemic probabilities. But at least in his first papers, Fisher was, by intuition, always right; and the point (Fisher's, subconsciously, and mine) is that we need both types of probability integrated in order to get a complete theory of statistics. One type is derived from the other in a perfectly objective way.

Returning to [12]: In his paper, Fisher finally compares fiducial and Bayes solutions. Curiously, he first argues by logic and not by insight that the two pertaining distributions "must differ not only numerically, but in their logical meaning" because the results will in general differ even though the Bayesian prior may be aleatorily true. But then he goes into details by considering a prior for which the posterior probability of $\rho < .765$, given $r = .99$, is not 5%, but 10%. He correctly argues that (with the Bayesian sampling) in 10% of all cases where r happens to be exactly $= .99$, ρ will be less than .765. "Whereas apart from any sampling for ρ [!], we know that if we take a number of samples of 4, from the same or from different populations [!], and for each calculate the fiducial 5% value for ρ, then in 5% of all cases the true value of ρ will be less than the value we have found. There is thus no contradiction between the two statements. The fiducial probability is more general and, I think, more useful in practice ..." [exclamation marks added]. The sequences of events considered by both arguments in a sequence of experiments are clearly very different.

Here Fisher claims that if an (aleatory, true) Bayesian prior happens to be known, both methods give valid though different answers (for different questions). Later (e.g., in [16]), he strongly insists that the fiducial argument must only be used if nothing is known about the parameter. There is something to both attitudes. Clearly, the fiducial argument is correct and leads to successful bets even if a Bayesian prior is known. (By the way, this is also true if an inefficient statistic is used for the fiducial argument, a point against which Fisher later argues in his Author's Note in [3], p. 428.) The problem is the efficiency, or the information, or the selection problem for successful bets alluded to above. According to present results, if an aleatory Bayesian method is available, it should be used - apart from questions of robustness or stability, the one big area Fisher refused to look at.

5 The Aposteriori Interpretation of Confidence Intervals

As all adherents of the Neyman-Pearson school know, a 95% confidence interval has a probability of 95% of covering the true unknown parameter apriori, before the data are in. After the data are in, the probability is 0 or 1, but we don't know which one. That is all the theory says. But as probably most of those of us know who tried to teach Neyman-Pearson statistics to critical, intelligent, unspoilt users of statistics, these scientists have a strong intuition that even "after the fact" there is, or should be, something with 95% probability; and they are very frustrated when they are told their intuition is entirely wrong. Some may become overly critical of statistics as a whole, while some others will humbly believe the "experts", like the werewolf in Christian Morgenstern's German poem [37] (cf. also Hampel [24]).

Now the explanation of the conflict is simple. Both sides are right, in a way. Since Neyman considers only aleatory probabilities, for him 0 or 1 are the only possibilities. But the scientist using statistics can bet 19 : 1 that the unknown fixed parameter is in the fixed (but randomly derived) confidence interval, and in a long sequence of such bets with different independent experiments (and different confidence intervals with the same level), she will be right in 95% of all cases (or at least 95%, if she uses conservative confidence intervals), so her bets are successful. This means, she correctly has a frequentist epistemic probability (or lower probability, for the conservative intervals) of 95% for the event or claim that the parameter is covered, in full accordance with her intuition.

By the way, she would be rather stupid - though not wrong - offering the same bet again and again in the case of many independent replications of the SAME experiment, because after a while she could have learned much more about the parameter - unless the information that it was the same experiment was withheld from her, or some such artificial device.

It should be noted that the aposteriori interpretation of confidence intervals (and thus the implicit fiducial argument and a subconscious switch between aleatory and epistemic probability) was probably centuries old (cf. the related Endnote 8 in [51]); certainly around 1900 interpretations like "the odds are 1 : 1 that the true mean is within ±1 probable error" were commonplace (cf., e.g., "Student's" writings [47], [40]; cf. also the remarks on Maskell in [51], p. 371). It is Neyman's merit that he clarified the purely aleatory argument; but by restricting himself to it, he cut himself off from one half of what statistics ought to be.

Incidentally, the above explanation can perhaps throw new light on the dispute between Neyman and Fisher around and about the invention of confidence intervals. At first, Fisher seemed to believe that Neyman's intervals are essentially the same as his fiducial intervals (apart from the point of uniqueness and efficiency related to sufficiency etc.). But a short time later, he "mysteriously" seemed to change his mind and claimed that the two methods were very different, after all, without giving reasons. My guess and interpretation is that Fisher, more or less consciously, always included the epistemic interpretation with his intervals and in the beginning naively thought that Neyman did the same (given the formal similarity of what they wrote, and the lack of formalism for the epistemic aspect), until he suddenly (or perhaps creepingly) discovered that Neyman's view was in fact much more primitive.

6 From Lower Probabilities to Fiducial Probabilities

As said before, with discrete models we have to use lower probabilities to describe successful bets. Moreover, even if ideally we have a model with a fiducial proper probability, in reality (e.g., with a digital computer) the data are always discretised. But when the discretisation gets finer and finer, the lower probabilities of an event and its complement converge to proper probabilities adding to one.

A simple example is the following [20]. Let X be uniformly distributed on $[\theta, \theta + 1]$ (θ real), and let (for every n) Y_n be X rounded upwards to the nearest multiple of $1/n$. Then for every c between 0 and 1, and all θ, $P_\theta(Y_n \le \theta + c) \le c$, and $P_\theta(Y_n > \theta + c) \le 1 - c + 1/n$, hence we can bet with epistemic lower probability $1 - c$ on $[\theta < y_n - c]$, and with odds $(c - 1/n) : (1 - c + 1/n)$ on $[\theta \ge y_n - c]$. The sum of the lower probabilities is $1 - 1/n \to 1$ as $n \to \infty$.

Thus, fiducial probabilities are just a limiting case of lower probabilities, though interesting in their own right because they allow fair (pairs of) bets (with a frequentist interpretation!). Hence they produce something similar to the Bayesian omelette, after all, without breaking the Bayesian eggs.

References

1. M.S. Bartlett, Complete simultaneous fiducial distributions, Ann. Math. Statist, 10, 129–138, 1939.
2. T. Bayes, An essay towards solving a problem in the doctrine of chances, Philos. Trans. Roy. Soc. London A 53, 370–418, 1763. Reprinted in Biometrika, 45, 293–315, 1958, and in Studies in the History of Statistics and Probability, Pearson, E. S., Kendall, M. G. (eds.), Griffin, London, 131–153, 1970.
3. J.H. Bennett, Collected Papers of R. A. Fisher, Vol.II (1925-31), The University of Adelaide, 1972.
4. J. Bernoulli, Ars Conjectandi, Thurnisiores, Basel, 1713. Reprinted in Werke von Jakob Bernoulli, Vol. 1, Birkhäuser Verlag, Basel, 1975.
5. D. Brönnimann, Die Entwicklung des Wahrscheinlichkeitsbegriffs von 1654 bis 1718, Diplomarbeit, Seminar für Statistik, Swiss Federal Institute of Technology (ETH) Zurich, 2001.
6. G. Choquet, Theory of capacities, Ann. Inst. Fourier, 5, 131–295, 1953/54.
7. D.R. Cox and D.V. Hinkley, Theoretical Statistics, Chapman and Hall, London, 1974.
8. A.P. Dempster, Upper and lower probabilities induced by a multivalued mapping, Ann. Math. Statist, 38, 325–339, 1967.
9. A.P. Dempster, A generalization of Bayesian inference, J. Roy. Statist. Soc. B, 30, 205–245, 1968.
10. B. Efron, R.A. Fisher in the 21st Century: Invited paper presented at the 1996 R.A. Fisher Lecture, Statistical Science, 13, 2, 95–122, 1998.
11. T.L. Fine, Theories of Probability, Academic, New York, 1973.

12. R.A. Fisher, Inverse probability, Proc. of the Cambridge Philosophical Society, 26, 528–535, 1930. Reprinted in Collected Papers of R. A. Fisher, ed. J. H. Bennett, Volume 2, 428–436, University of Adelaide, 1972.

13. R.A. Fisher, The concepts of inverse probability and fiducial probability referring to unknown parameters, Proc. Roy. Soc. London, Ser. A, 139, 343–348, 1933.

14. R.A. Fisher, The fiducial argument in statistical inference, Annals of Eugenics, 6, 391–398, 1935.

15. R.A. Fisher, Statistical methods and scientific induction, J. Roy. Statist. Soc. Ser. B, 17, 69–78, 1955.

16. R.A. Fisher, Statistical Methods and Scientific Inference, Oliver and Boyd, London, 1956; 2nd ed. 1959, reprinted 1967.

17. D.A.S. Fraser, The Structure of Inference, Series in Probability and Mathematical Statistics, Wiley, New York, 1968.

18. I.J. Good, Subjective probability as the measure of a non-measurable set, in E.Nagel, P.Suppes and A.Tarski (eds.), Logic, Methodology and Philosophy of Science, Stanford Univ. Press, Stanford, 319–329, 1962.

19. F. Hampel, Fair bets, successful bets, and the foundations of statistics. Abstract, Second World Congress of the Bernoulli Society, Uppsala, 1990.

20. F. Hampel, Some thoughts about the foundations of statistics, in New Directions in Statistical Data Analysis and Robustness, S.Morgenthaler, E.Ronchetti, and W.A. Stahel (eds.), Birkhäuser Verlag, Basel, 125–137, 1993.

21. F. Hampel, On the philosophical foundations of statistics: Bridges to Huber's work, and recent results, in Robust Statistics, Data Analysis, and Computer Intensive Methods; In Honor of Peter Huber's 60th Birthday, H.Rieder (ed.), No. 109 in Lecture Notes in Statistics, Springer-Verlag, New York, 185–196, 1996.

22. F. Hampel, What can the foundations discussion contribute to data analysis? And what may be some of the future directions in robust methods and data analysis?, J. Statist. Planning Infer., 57, 7–19, 1997.

23. F. Hampel, How different are frequentist and Bayes solutions near total ignorance?, Proc. 51th Session of the ISI, Contrib. Papers, Book 2, Istanbul, 25–26, 1997.

24. F. Hampel, On the foundations of statistics: A frequentist approach, in Estatística: a diversidade na unidade, M.S. de Miranda and I.Pereira, (eds.), Edições Salamandra, Lda., Lisboa, Portugal, 77–97, 1998.

25. F. Hampel, Is statistics too difficult?, Canad. J. Statist, 26, 3, 497–513, 1998.

26. F. Hampel, An outline of a unifying statistical theory, Proc. of the 2nd Internat. Symp. on Imprecise Probabilities and their Applications, Cornell University, 26-29 June 2001, G. de Cooman, T.L. Fine, and T.Seidenfeld (eds.), Shaper Publishing, Maastricht, 205-212, 2000.

27. F. Hampel, Why is statistics still not unified?, Proc. 53rd Session of the ISI, Contrib. Papers, Book 1, Seoul, 73–74, 2001.

28. P.J. Huber, The use of Choquet capacities in statistics, Bull. Internat. Statist. Inst. 45, 181–188, 1973.

29. P.J. Huber, Kapazitäten statt Wahrscheinlichkeiten? Gedanken zur Grundlegung der Statistik, Jber. Deutsch. Math. Verein., 78, 2, 81–92, 1976.

30. P.J. Huber, Robust Statistics, Wiley, New York, 1981.

31. P.J. Huber and V. Strassen, Minimax tests and the Neyman–Pearson lemma for capacities, Ann. Statist. 1, 2, 251–263, 1973. Corr: 2, 223–224.

32. A. Kolmogoroff, Grundbegriffe der Wahrscheinlichkeitsrechnung. Ergebnisse der Mathematik und ihrer Grenzgebiete, Springer, Berlin, 1933.

33. A.N. Kolmogorov, The estimation of the mean and precision of a finite sample of observations, in Russian, (Section 5: Fisher's fiducial limits and fiducial probability), Bull. Acad. Sci. U.S.S.R., Math. Ser., 6, 3–32, 1942.

34. B.O. Koopman, The bases of probability, Bulletin of the American Mathematical Society, 46, 763–774, 1940.

35. H.E. Kyburg, Probability and the Logic of Rational Belief, Wesleyan University Press, Middletown, 1961.

36. J.H. Lambert, Neues Organon, oder Gedanken über die Erforschung und Bezeichnung des Wahren und dessen Unterscheidung von Irrtum und Schein, Leipzig, 1764. Reprinted by Olms of Hildesheim as the first two volumes of LAMBERT's Philosophische Schriften, 1965.

37. C. Morgenstern, Gesammelte Werke, R. Piper & Co. Verlag, München, 1965. Selected translations in: M. Knight, Christian Morgenstern's Galgenlieder, University of California Press, Berkeley and Los Angeles, 1963.

38. J. Neyman, 'Inductive behavior' as a basic concept of philosophy of science, Rev. Internat. Statist. Inst., 25, 7–22, 1957.

39. J. Neyman, Foundations of behavioristic statistics, Foundations of Statistical Inference, V.P. Godambe, and D.A. Sprott (eds.), Holt, Rinehart and Winston of Canada, Toronto, 1971.

40. E.S. Pearson and J. Wishart (eds.), "Student's" Collected Papers, Cambridge University Press, 1958.

41. E.J.G. Pitman, Statistics and science, J. Amer. Statist. Assoc, 52, 322–330, 1957.

42. C.R. Rao, Linear Statistical Inference and Its Applications, Wiley, London, 1965, (2nd edition 1973).

43. G. Shafer, A Mathematical Theory of Evidence, Princeton University Press, Princeton, N. J., 1976.

44. G. Shafer, Non-Additive Probabilities in the Work of Bernoulli and Lambert, Archive for History of Exact Sciences, Springer-Verlag, Vol. 19, 4, 309–370, 1978.

45. C.A.B. Smith, Consistency in statistical inference and decision, J.Roy. Statist. Soc. B, 23, 1–37, 1961.

46. V. Strassen, Messfehler und Information, Z. Wahrscheinlichkeitstheorie verw. Geb., 2, 273–305, 1964.

47. "Student", The probable error of a mean, Biometrika, 6, 1–25, 1908.

48. J.W. Tukey, A smooth invertibility theorem, Ann. Math. Statist, 29, 581–584, 1958.

49. J.W. Tukey, Handouts for the Wald Lectures 1958, in The Collected Works of John W. Tukey, Volume VI: More Mathematical (1938 – 1984), C.L. Mallows (ed.), Wadsworth & Brooks/Cole Advanced Books & Software, Pacific Grove, CA. Ch.10, 119 – 148, 1990.

50. P. Walley, Statistical Reasoning with Imprecise Probabilities, Chapman and Hall, London, 1991.

51. S.L. Zabell, R.A. Fisher and the fiducial argument, Statistical Science, 7, 3, 369–387, 1992.

On Sequential Discrimination Between Close Markov Chains

M.B. Malyutov and D.M. Malyutov

Abstract. The appropriateness of the Wald-type logarithmic asymptotics for the mean length of sequential discrimination strategies between close alternatives has been already challenged in the well-known controversy over comparative performances of the asymptotically optimal Chernoff's discrimination strategies and ad hoc heuristic rules of Box and Hill in the seventies.

We continue this discussion by showing a poor performance of the Wald-type asymptotic bounds for the mean length of asymptotically optimal sequential discrimination strategies between the simplest types of Markov chains by simulation. We propose some weak remedies against this disaster and some alternative asymptotic tools.

Keywords: Sequential test, maximal error probability, mean length of strategies, Markov chain.

1 Introduction

One of the most popular results in Wald (1947) is his logarithmic asymptotic lower bound for the mean length of sequential discrimination strategies between simple hypotheses which turned out to be asymptotically attained by his Sequential Probability Ratio Test. This asymptotic approach was generalized and extended to numerous discrimination settings between composite hypotheses and change-point problems by Chernoff, Kiefer and Sacks, Lai, and many other authors including the first author. The accuracy of the Wald-type logarithmic asymptotics for the mean length of sequential discrimination strategies between close alternatives has been already challenged in the well-known controversy over comparative performances of the asymptotically optimal Chernoff's discrimination strategies and the ad hoc heuristic rules of Box and Hill (1967) in the seventies (see, e.g. Blot and Meeter, (1973)). Under error probability α interesting in practice, the residual term $o(\log |\alpha|)$ may well exceed the principal term of the Wald-type bound, see our discussion further.

We continue this discussion by showing a poor performance of the Wald-type asymptotic bounds for the mean length of asymptotically optimal sequential discrimination strategies between simplest types of Markov chains *by simulation*. We propose some weak remedies against this disaster (section 2) and some alternative asymptotic tools. We study the performance of several interrelated sequential discrimination strategies theoretically and by statistical simulation for two particular examples.

R. Ahlswede et al. (Eds.): Information Transfer and Combinatorics, LNCS 4123, pp. 527–534, 2006.
© Springer-Verlag Berlin Heidelberg 2006

In section 2 we outline an application of general asymptotically optimal sequential discrimination strategies proposed in Malyutov and Tsitovich (2001) (abbreviated further as MT-2001) for an illustrative example of testing the correlation significance for a first-order autoregression with small noise. Here, for a certain range of parameters we can show the attainability of our asymptotic lower bound *if we permit an initial transition period to the equilibrium equilibrium for the observed MC.* An extension of the results in section 2 to the discrimination between statistical hypotheses about a general conservative dynamical system perturbed by small noise is straightforward.

Two simplified versions of discrimination strategies between Markov Chains (MC) are introduced in section 3 and studied in section 4 by simulation for testing a regular binary random number generator versus a Markov chain with *transition probabilities very close to those in the null hypothesis.* We end up with the conclusions formulated in Section 5.

Our examples are related to the general setting in MT-2001 which we now introduce. Let X be a finite set with m_X elements, $\mathcal{P}$ be a Borel set of transition probability matrices for Markov chains on state space X satisfying the conditions formulated in the next paragraph. We denote by $p(x, y), x \in X, y \in X$, elements of the matrix $P \in \mathcal{P}$. Under the convention $0/0 := 1$ we assume that for some $C > 0$

$$\sup_{P,Q \in \mathcal{P}} \max_{x \in X, y \in X} \frac{p(x, y)}{q(x, y)} \leq C < \infty \tag{1}$$

and for every $P \in \mathcal{P}$ MC with transition probability matrix P is aperiodic and irreducible which implies the existence and uniqueness of the stationary distribution $\mu := \mu_P$ with $\mu_P(x) > 0$ for every $x \in X$. It follows from (1) that $p(x, y) = 0$ for any $P \in \mathcal{P}$ entails $q(x, y) = 0$ for all $Q \in \mathcal{P}$. Our statistical decisions are based on the log-likelihood probability ratios:

$$z(P, Q, x, y) := \ln p(x, y)/q(x, y).$$

$$I(x, P, Q) := \sum_{y \in X} p(x, y)z(P, Q, x, y)$$

is the Kullback-Leibler divergence (*cross-entropy*). The set $\mathcal{P}$ is partitioned into Borel sets $\mathcal{P}_0$, $\mathcal{P}_1$ and the indifference zone $\mathcal{P}_+ = \mathcal{P} \setminus (\mathcal{P}_1 \cup \mathcal{P}_0)$. We test $H_0 : P \in \mathcal{P}_0$ versus $H_1 : P \in \mathcal{P}_1$, every decision is good for $P \in \mathcal{P}_+$. Suppose that the divergence between the hypotheses is positive, i.e.

$$\min_{i=0,1} \inf_{P \in \mathcal{P}_i, Q \in \mathcal{P}_{1-i}} \max_{x \in X} I(x, P, Q) \geq \delta_0 > 0. \tag{2}$$

The probability law of $X_i, i = 0, 1, \ldots$, is denoted by $\mathbf{P}_P$ and the expectation is denoted by $\mathbf{E}_P$. In particular

$$I(x, P, Q) = \mathbf{E}_P(z(P, Q, X_0, X_1)|X_0 = x).$$

A *strategy* s consists of a stopping (Markov) time N and a measurable binary decision δ, $\delta = r$ means that H_r , $r = 0, 1$, is accepted. Introduce α-*strategies* s satisfying

$$\max_{r=0,1} \sup_{P \in \mathcal{P}_r} \mathbf{P}_P(\delta = 1 - r) \leq \alpha.$$

$\mathbf{E}_P^s N$ is the *mean length* (MEAL) of a strategy s. **Define:** $I(\mu, P, Q) := \sum_{x \in X} \mu(x) I(x, P, Q)$, where μ is a probability distribution on X, $I(P, Q) := I(\mu_P, P, Q)$ and $I(P, \mathcal{R}) := \inf_{Q \in \mathcal{R}} I(P, Q)$ for $\mathcal{R} \subset \mathcal{P}$; $A(P) := \mathcal{P}_{1-r}$ for $P \in \mathcal{P}_r$ as the alternative set to P (in $\mathcal{P}$). For $P \in \mathcal{P}_+$, if $I(P, \mathcal{P}_0) \leq I(P, \mathcal{P}_1)$, then $A(P) := \mathcal{P}_1$, otherwise, $A(P) := \mathcal{P}_0$. Finally, $k(P) = I(P, A(P))$. It follows from (2) that

$$k_0 := \inf_{P \in \mathcal{P}} k(P) > 0, P \in \mathcal{P}$$

since $\mu_P(x) > 0$ for all $x \in X$ and for any $P \in \mathcal{P}$. It is proved in MT-2001 that for every $P \in \mathcal{P}$ and α-strategy s

$$\mathbf{E}_P^s N \geq \rho(\alpha, P) + O\left(\sqrt{\rho(\alpha, P)}\right), \tag{3}$$

as $\alpha \to 0$, where $\rho(\alpha, P) = |\ln \alpha|/k(P)$ is the well-known principal term of the MEAL first appearing in Wald (1947), and the following α-strategy s^1 attaining equality in (3) is constructed depending on a parameter β, $\beta < \alpha$. Strategy s^1 consists of conditionally i.i.d. loops. Every loop contains two phases. Based on the first

$$N_1 = N_1(\alpha, \delta_0) \tag{4}$$

observations of a loop, we estimate the matrix P by the MLE $\hat{P} \in \mathcal{P}$ (or its orthogonal projection on $\mathcal{P}$ as a subset of $\mathbf{R}^{m_X^2}$). Let us enumerate measurements of the second phase anew and introduce $L_k(\hat{P}, Q) = \sum_{i=1}^k z(\hat{P}, Q, X_{i-1}, X_i)$. We stop observations of the loop at the first moment N_2 such that

$$\inf_{Q \in A(\hat{P})} L_{N_2}(\hat{P}, Q) > |\ln \beta| \tag{5}$$

or

$$N_2 > N_0 := 2k_0^{-1} |\ln \alpha|, \tag{6}$$

stop all experiments and accept the hypothesis H_r (i.e. $\delta = r$), if (5) holds and $1 - r$ is the index of the set $A(\hat{P})$. After event (6) we begin a new loop. Strategies s^1 and s^2 (see section 3) do not revise the estimate $\hat{P}$ during the second phase of a loop which we modify for strategy s^3 introduced in section 3. In both new strategies the rule (5) is replaced by comparing (with the prescribed level) only the *likelihood ratios with respect to the closest alternatives* to $\hat{P}$ which is numerically much easier to implement. Our simulation in section 4 shows that s^3 is much better than s^2. Note also that for attaining an asymptotic equality in (3) it is assumed in MT-2001 that $P(N_2 > N_0) \to 0$ as $\alpha \to 0$ making the probability of more than one loop negligible. This holds, if

$$EI(P, \hat{P})/\delta_0 \to 0 \text{ as } \alpha \to 0. \tag{7}$$

We study the situation of *close alternatives (δ_0 small)* in both our examples further which is dubious to deal with the conventional *asymptotic approach of* *large deviations* common in discrimination problems, see e.g. Chernoff (1972). Le Cam's theory of *contiguous alternatives might give a better approximation* which we plan to study in future. Namely, the misclassification probability under the hypotheses at distance of order $cn^{-1/2}$, where n is the sample size of the first stage, can be shown to be normal with parameter depending on c. Hence we will be able to choose parameters in such a way that the unfavorable outcome of the first loop would take place with probability less than α.

In sections 3 and 4 the condition (7) is impractical since $L = |\ln \alpha|$ cannot be very large. Our simulation shows that under the parameters studied only α-strategies with several times larger MEAL than $\rho(\alpha, P)$ seem to be attainable which appears around twice less than the sample size of the best static strategy. It is an open problem whether our strategies can be modified to require the MEAL equivalent to the lower bound proved so far.

Remark. The most promising revision of our strategies would be following: generalizing to the first stage of our strategies the recent methods of supervised discrimination *maximizing margin*. These methods use the estimation of the likelihood function only in the vicinity of the *margin* points crucial for discrimination getting rid of the idea to approximate the likelihood function globally (and plug in the estimated parameters there).

In the next section 2 we study ßmall noisecase where condition (7) can be achieved under mild conditions in theMC transition period, where the *signal-noise ratio exceeds considerably that for the stationary distribution*. Even simpler is to justify the condition (7) in non-stationary ßignal plus small noisemmodels which we do not consider here.

2   Testing Correlation in a First-Order Autoregression

Here we illustrate the previously exposed general results for an example of sequential discrimination between $i.i.d.N(0, \varepsilon^2)$ measurements versus a first order autoregression with small correlation. We view this as an example of conservative dynamical systems perturbed by small noise which can be treated similarly.

Consider a Markov chain $X_0, X_1, \ldots$ with joint distribution P_θ:

$$X_t = \theta X_{t-1} + \varepsilon e_t, t = 1, 2, \ldots,$$

where $|\theta| \leq \Theta < 1$ is an unknown correlation, the noise εe_t is $i.i.d.N(0, \varepsilon^2)$ and we can choose X_0 to be, say, 1 (or more generally is random with constant mean as $\varepsilon \to 0$).

We test $H_0 = \{\theta = 0\}$ versus the composite hypothesis $H_1 = \{|\theta| \geq d > 0\}$, $\{0 < |\theta| < d$ being an indifference zone. The marginal distribution of X_t is well-known to converge exponentially fast as $t \to \infty$ to $N(0, \varepsilon^2(1 - \theta^2))$ for every initial state. We study here the performance of strategy s^1 for small d, ε and α. The loglikelihood of P_θ versus $P_{\dot\theta}$ up to moment T is $Z_0 + \sum_1^{T+1} Z_t$, where

$$Z_t := [(X_t - \dot\theta X_{t-1})^2 - (X_t - \theta X_{t-1})^2]/(2\varepsilon^2).$$

First averaging Z_t given X_{t-1}, we get

$$I(x, \theta, \dot{\theta}) = (\theta - \dot{\theta})^2 x^2 / (2\varepsilon^2),$$

and then averaging over the stationary distribution, we get the stationary cross-entropy

$$I(\theta, \dot{\theta}) := \mathbf{E}_\theta(Z_t) = (\theta - \dot{\theta})^2 (1 - \theta^2).)$$

In particular, $I(0, \theta) = \theta^2, I(\theta, 0) = \theta^2(1 - \theta^2)$. It is straightforward from the above calculations that the Fisher information $J(x, \theta)$ of X_t given that $X_{t-1} = x$ is $x^2 / (2\varepsilon^2)$. Thus

$$E(\sum_0^T J(X_t)) = \varepsilon^{-2} \sum_{t=0}^T (\theta^{2t}) / 2$$

is not less asymptotically for large T than $1/[2(1 - D^2)\varepsilon^2]$ implying that the variance of the preliminary MLE $\hat{\theta}$ based on $\sqrt{\rho(\alpha, P_d)}$ observations is $1/\sqrt{\rho(\alpha, P_d)}$, if we assume that $\varepsilon^2(1 - D^2) = o(d^2/L)$.

This implies the attainment of the lower bound (3) by s^1 along the lines of MT-2001. The bound (3) holds, if the transition period to the stationary distribution is not sufficient to discriminate with error probability less than α, which is also straightforward to rephrase in terms of the model parameters.

3   Testing Random Number Generator vs. Markov Chain

Our basic hypothesis H_0 deals with a Bernoulli binary equally likely distributed (P_0) sequence of measurements $X_1, X_2, \ldots$. We test it versus an alternative hypothesis H_1 that the sequence observed is a stationary Markov chain with transition probabilities $P_r := (p_{ij}, i, j = 1, 2)$, where $r := (r_1, r_2), r_1 := p_{11} - 1/2, r_2 := p_{22} - 1/2$, such that for certain $d > 0$

$$I(P_0, P_r) := -\ln[16 p_{11}(1 - p_{11}) p_{22}(1 - p_{22})]/4 \geq d^2.$$

Note that $I(P_0, P_r) = ||r||_2^2(1 + o(1))$ as $||r||_2 \to 0$, where

$$||r||_2^2 := r_1^2 + r_2^2.$$

For very small d considered in our simulation, we can approximately view the set of alternatives as the exterior to the figure which is very close to a circle of radius d with center in $(1/2, 1/2)$. We skip a more lengthy general expression for $I(P_r, P_{\hat{r}})$ unused in our presentation of the simulation results. The local optimization method used in our program for finding the set of P_{r*} minimizing $I(P_{\hat{r}}, P_{\hat{r}})$ over $P_{\hat{r}}$ on the border of the alternative set to a preliminary estimate $P_{\hat{r}}$, always unexpectedly found a *unique* minimizing point $A(\hat{r})$ (which simplifies the strategy considerably). We are not aware if this is true in general situations. Now we introduce two sequential algorithms for the discrimination between H_0 and H_1. The only difference of the strategy s^2 from s^1 is that the rule (3) is replaced with

$$L_{N_2}(P_{\hat{r}}, P_{A_{\hat{r}}}) > |\ln \alpha|. \tag{8}$$

To simplify the strategy further, we abandon another parameter $\beta < \alpha$ in the definition of s^1 which apparently does not change the performance of s^3 considerably (see the corresponding figures further) whereas the evaluation of the recommended in MT-2001 update is clumsy.

Now, the strategy s^3 deviates from s^2 only in being more greedy: we continue to update recurrently the preliminary estimate for the true parameter r during the second phase in parallel to counting the likelihood ratios, and if a loop ended undecidedly, we plug the updated estimate for r into the likelihood ratio, find the closest alternative, and start the new second phase. Therefore, only second phases occur in loops after the first one.

3.1 Static Discrimination

Now we discuss the performance of the best static (non-sequential) strategy with maximal error probability α. Our large sample discrimination problem is obviously asymptotically equivalent to the discrimination of the zero mean hypothesis for the bivariate rotationally invariant Normal distribution versus the spherically invariant alternative dealt with in Example 5.17 of Cox and Hinkley (1974). It is shown there that the best critical region is the exterior to a circle of certain radius, and that the distribution under the alternative is the non-central Chi-Square with two degrees of freedom. Using the power diagrams of non-central Chi-Square in Sheffe (1958), we find that the radius providing the equality of maximal errors under the null hypothesis and the alternative is approximately $0.39d$. This finally determines the sample size such that the maximal error probabilities equal specified levels which we compare with the empirical MEALs of strategies s^3, s^2 found by simulation.

4 Simulation Results

We present a series of simulation results (using MATLAB) for s^2 and s^3 with various parameters of the model described in the preceding section. The code is available by request.

The table below summarizes the results of a few trials of our simulation. In the table, $N_1 = K_1\sqrt{L}/d^2$, where $L = |\ln(\alpha)|$, n is the number of times strategies were run under the same parameters, the empirical MEAL (EMEAL) is the average number of random numbers taken before the decision, the ENOL is the average number of loops over n runs. Note that all the parameters in simulations 2-4 are taken the same with n=100, and with $n = 1000$ in simulation 5 for estimating the variance of the performance parameters.

In the plots $d^2 = 0.001, p := p_{11} = p_{22}$, methods 1 and 2 represent respectively s^2 and s^3. Figures 3 and 6 plot respectively EMEAL and FE under various alternatives. Both are maximal, as natural, in the middle of the indifference zone. Other figures illustrate the performance of our strategies under H_0. The empirical MEAL, number of loops and empirical error rate are plotted versus changing values of various parameters of our strategies. The main news is that

Table 1. Please write your table caption here

	d^2	α	K_1	p	n	EMEAL	FE	ENOL
(1)	s^2	0.0002	0.01	2500	0.5015	200	263723.82	0.02	4.37
	s^3						93836.39	0.0	1.905
(2)	s^2	0.001	0.02	500	0.5	100	42434.89	0.09	4.27
	s^3						15553.15	0.03	1.89
(3)	s^2	0.001	0.02	500	0.5	100	49191.76	0.04	4.8
	s^3						15879.13	0.01	1.97
(4)	s^2	0.001	0.02	500	0.5	100	52232.04	0.04	4.97
	s^3						16412.41	0.03	1.98
(5)	s^2	0.001	0.02	500	0.5	1000	42934.085	0.057	4.24
	s^3						15729.03	0.01	1.90

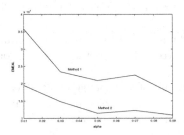

Fig. 1. Empirical meal (EMEAL) vs.Alpha

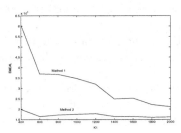

Fig. 2. Empirical meal (EMEAL) vs. K1

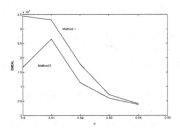

Fig. 3. Empirical meal (EMEAL) vs. P

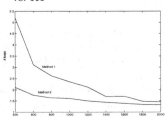

Fig. 4. Empirical number of loops (ENQL) vs. K1

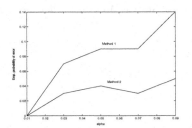

Fig. 5. Frequency of errors (FE) vs. Alpha

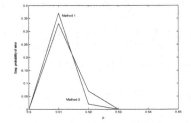

Fig. 6. Frequency of errors (FE) vs. P

the EMEAL exceeds the theoretical principal term approximately four times under best parameters of our strategies.

5 Conclusions

1. Conventional asymptotic expansions of the MEAL in terms of $\ln \alpha$, where α is the maximal error probability can be of dubious importance for discrimination between close hypotheses.
2. Our simulation shows that strategy s^3 which keeps updating the preliminary estimate of the true parameter is clearly preferable as compared to the theoretically justified strategy s^1.
3. Further work to find a valid expansion for the MEAL of discriminating between close hypotheses seems necessary incorporating Le Cam's contiguity techniques.
4. Although seemingly suboptimal, the strategy s^3 is clearly preferable to the best static discrimination strategies even for discrimination between close hypotheses.
5. Use of the MC transition periods for preliminary estimation of true parameters may be fruitful in discrimination between almost deterministic ergodic Markov chains.
6. Simplified versions of the best controlled and of the change-point detection strategies, also studied in MT-2001, as well as the development of s^1, and of strategies in Lai (1998) should also be examined by simulation.

References

1. W.J. Blot and D.A. Meeter, Sequential experimental design procedures, Journal of the American Statistical Association, 68, 343, 586-593, 1973.
2. G.E.P. Box and W.J. Hill Discrimination between mechanistic models, Technometrics, 9, 57-71, 1967.
3. H. Chernoff, Sequential Analysis and Optimal Design, SIAM, Philadelphia, Theoretical Statistics, Chapman and Hall, London, 1972.
4. T.L. Lai, Information bounds and quick detection of parameter changes in stochastic systems, IEEE Trans. Inform. Theory, Vol. 44, No. 7, 2917–2929, 1998.
5. M.B. Malyutov and I.I. Tsitovich, Second order optimal sequential discrimination between Markov chains, Mathematical Methods of Statist., 10, No. 4, Allerton Press, 446–464, 2001.
6. H. Sheffe, Analysis of Variance, Wiley, N.Y., 1958.
7. A. Wald, Sequential Analysis, Wiley, N.Y., 1947.

Estimating with Randomized Encoding the Joint Empirical Distribution in a Correlated Source

R. Ahlswede and Z. Zhang

1 Introduction

In order to put the present model and our results into the right perspectives we describe first key steps in multiuser source coding theory.

We are given a discrete memoryless double source (DMDS) with alphabets $\mathcal{X}$, $\mathcal{Y}$, and generic variables X, Y, i.e., a sequence of independent replicas (X_t, Y_t), $t = 1, 2, \ldots$ of the pair of random variables (X, Y) taking values in the finite sets $\mathcal{X}$ and $\mathcal{Y}$, respectively.

I. Slepian and Wolf considered the problem of encoding the source output blocks $X^n \triangleq X_1 \ldots X_n$ resp. $Y^n \triangleq Y_1 \ldots Y_n$ by two separate encoders in such a way that a common decoder could reproduce both blocks with small probability of error. They proved that such an encoding is possible with rates (R_1, R_2) if and only if

$$R_1 \geq H(X|Y), \quad R_2 \geq H(Y|X), \quad R_1 + R_2 \geq H(X,Y). \qquad (1.1)$$

II. It may happen, however, that what is actually required at the decoder is to answer a certain question concerning (X^n, Y^n). Such a question can of course be described by a function F of (X^n, Y^n). The authors of [5] are interested in those functions for which the number k_n of possible values of $F(X^n, Y^n)$ satisfies

$$\lim_{n \to \infty} \frac{1}{n} \log k_n = 0. \qquad (1.2)$$

This means that the questions asked have only "a few" possible answers. For example, X_t and Y_t may be the results of two different quality control tests performed on the ith item of a lot. Then for certain purposes, e.g., for determining the price of the lot, one may be interested only in the frequencies of the various possible pairs (x, y) among the results, their order, i.e., the knowledge of the individual pairs (X_t, Y_t), being irrelevant. In this case $k_n \leq (n + 1)^{|\mathcal{X}||\mathcal{Y}|}$, and (1.2) holds. A natural first question is whether or not it is always true in this case that, for large n, arbitrarily small encoding rates permit the decoder to determine $F(X^n, Y^n)$.

The authors of [5] also consider other choices of F and first obtain the following result. For every DMDS with

$$H(X|Y) > 0, \quad H(Y|X) > 0$$

there exists a binary question (function F with only two possible values) such that in order to answer this question (determine $F(X^n, Y^n)$) one needs encoding rates as specified in (1.1).

R. Ahlswede et al. (Eds.): Information Transfer and Combinatorics, LNCS 4123, pp. 535–546, 2006.
© Springer-Verlag Berlin Heidelberg 2006

As a matter of fact, almost all randomly selected functions F are of this kind. Since the reason for this unexpected phenomenon might be that randomly selected functions are very irregular, we next study more regular functions. A function F of special interest is the joint composition (joint type) of the two source blocks hinted at in the quality control example. In this respect our main result is that for determining the joint type of X^n and Y^n when Y^n is completely known at the decoder, X^n must be encoded with just as large a rate as if X^n were to be fully reproduced except for (exactly specified) singular cases. Actually, this analogous result is proved in [5] for a class of functions F which include, in addition to the joint type, the Hamming distance and — for alphabet size at least three — the parity of the Hamming distance.

As a consequence of these results one obtains that in the case of encoding both X^n and Y^n, the rates must satisfy

$$R_1 \geq H(X|Y), \ R_2 \geq H(Y|X), \tag{1.3}$$

in order that the joint type or the Hamming distance of X^n and Y^n can be determined by the decoder. In particular, it follows that for a DMDS with independent components (i.e., when X and Y are independent random variables (RV's)) nothing can be gained in rates, if instead of (X^n, Y^n) only the joint type or the Hamming distance of X^n and Y^n is to be determined by the decoder. For a DMDS with dependent components such a rate gain is possible, although it remains to be seen whether this always happens and to what extent. At present a complete solution to this problem is available only in the binary symmetric case. In fact, it readily follows from a result of Körner and Marton, that our necessary conditions (1.3) are also sufficient. Let us emphasize that their result concerns "componentwise" functions F

$$F(X^{n'}, Y^n) \triangleq \big(F_1(X_1, Y_1), F_1(X_2, Y_2), \ldots, F_1(X_n, Y_n)\big), \tag{1.4}$$

where F_1 is defined on $\mathcal{X} \times \mathcal{Y}$.

In the binary symmetric case (i.e. $Pr\{X = Y = 0\} = Pr\{X = Y = 1\}$, $Pr\{X = 0, Y = 1\} = Pr\{X = 1, Y = 0\}$), they proved for the particular F with $f_1(x, y) \triangleq x + y \pmod 2$ that (R_1, R_2) is an achievable rate pair for determining $F(X^n, Y^n)$ if and only if (1.3) holds. Now observe that the types of X^n and of Y^n can be encoded with arbitrarily small rates and that those two types and the mod 2 sum $F(X^n, Y^n)$ determine the Hamming distance and also the joint type of X^n, Y^n.

Notice that the problem of F–codes is outside the usual framework of rate–distortion theory except for "componentwise" functions F, cf. (1.4). Still, a complete description of the achievable F rate region, e.g., for $F(x, y) \triangleq P_{x,y}$, may be as hard a problem as to determine the achievable rate region for reproducing X^n, Y^n within a prescribed distortion measure. We draw attention to the fact that for the latter problem it is also the projection of the achievable rate region to the R_1–axis which could be determined (Wyner–Ziv, [10]).

III. The authors of [9] consider a new model: identification via compressed data. To put it in perspective, let us first review the traditional problems in traditional rate–distortion theory for sources. Consider the diagram shown in Fig 1,

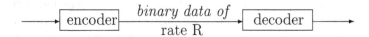

Fig. 1. Model for source coding

where $\{X_t\}_{t=1}^{\infty}$ is an independent and identically distributed (i.i.d.) source taking values in a finite alphabet $\mathcal{X}$. The encoder output is a binary sequence which appears at a rate of R bits per symbol. The decoder output is a sequence $\{\hat{X}_n\}_{n=1}^{\infty}$ which takes values in a finite reproduction alphabet $\mathcal{Y}$. In traditional source coding theory, the decoder is required to recover $\{X_t\}_{t=1}^{\infty}$ either completely or with some allowable distortion. That is, the output sequence $\{\hat{X}_t\}_{t=1}^{\infty}$ of the decoder must satisfy

$$\frac{1}{n}\sum_{i=1}^{n} E_\rho(X_t, \hat{X}_t) \le d \tag{1.5}$$

for sufficiently large n, where $\mathbf{E}$ denotes the expected value,

$$\rho : \mathcal{X} \times \mathcal{Y} \to [0, +\infty)$$

is a distortion measure, and d is the allowable distortion between the source sequence and the reproduction sequence. The problem is then to determine the infimum of the rate R such that the system shown in Fig. 1 can operate in such a way that (1.5) is satisfied. It is known from rate distortion theory that the infimum is given by the rate distortion function of the source $\{X_t\}_1^{\infty}$.

Let us now consider the system shown in Fig. 2,

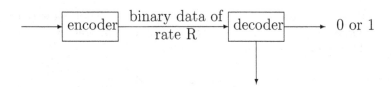

Fig. 2. Model for joint source coding and identification

where the sequence $\{Y_t\}_1^{\infty}$ is a sequence of i.i.d. random variables taking values from $\mathcal{Y}$. Knowing Y^n, the decoder is now required to be able to identify whether or not the source sequence X^n and the sequence Y^n have some prescribed relation F in such a way that two kinds of error probabilities, the probabilities for misacceptance (false identification) and the probabilities for misrejection, satisfy some prescribed conditions. In parallel with rate distortion theory, we consider in this paper the following relation F defined by:

$$n^{-1} \sum_{t=1}^{n} \rho(X_t, Y_t) \le d. \tag{1.6}$$

That is, the values X^n and Y^n are said to have relation F if (1.6) is satisfied. The problem we are interested in is to determine the infimum of the rate R such that the system shown in Fig. 2 can operate so that the error probability of misrejection, that is the decoder votes for 0 even though F holds, and the error probability of misacceptance, that is the decoder votes for 1 even though F does not hold, satisfy constraints on the error exponents α and β, say. So the goal of the decoder is to identify whether X^n is close to Y^n (in the sense of relation F) or not. The encoder is cooperative.

It must be remarked that in this model the minimum achievable rate is shown to always equal zero, if we only require that the two kinds of error probabilities go to zero as n goes to infinity. So the exponential decay of error probabilities makes the problem meaningful. The regions of pairs of exponents (α, β) are studied as functions of rate R and fidelity criterion d for general correlated sources. Complete characterizations are obtained, if X^n and Y^n are independent.

IV. Now we come to our new model of estimating the joint empirical distribution (joint type) not exactly like in [5], but within some accuracy only. This "computational" aspect was motivated by [9]. Furthermore the help of randomization was understood in [6] and [7].

We consider the following model. The encoder knows a word $x^n \in \mathcal{X}^n$ and the receiver knows a word $y^n \in \mathcal{Y}^n$. The encoder sends information of at most ℓ bits to the receiver, who uses these bits and his own observation $y^n \in \mathcal{Y}^n$ to estimate the joint type. The question is how accurate the estimate can be. It can be formalized as follows:

A randomized encoding is a pair $\mathcal{E} = \{\mathcal{M}, Q(\cdot|\cdot)\}$, where

$$\mathcal{M} = \{1, 2, \ldots, M\}, M = 2^k, \quad \text{and} \quad Q(\cdot|x^n) \in \mathcal{P}(\mathcal{M}), x^n \in \mathcal{X}^n. \tag{1.5}$$

Here and elsewhere $\mathcal{P}(\cdot)$ denotes the set of probability distributions (abbreviated as PD) or probability vectors of a set in brackets.

The decoder uses a decoding function

$$g : \mathcal{M} \times \mathcal{Y}^n \to \mathcal{P}(\mathcal{X} \times \mathcal{Y}). \tag{1.6}$$

Next we describe performance criteria for the code $\mathcal{C} = (\mathcal{E}, g)$. For any two PD's $P = (P_1, \ldots, P_s)$ and $Q = (Q_1, \ldots, Q_s)$ define the norms

$$\|P - Q\|_1 = \sum_{i=1}^{s} |P_i - Q_i|, \tag{1.7}$$

$$\|P - Q\|_2 = \sqrt{\sum_{i=1}^{s} |P_i - Q_i|^2}, \tag{1.8}$$

and the "individual errors" based on them for the code $\mathcal{C} = (\mathcal{E}, g)$

$$\bar{e}_{\mathcal{C}}^{(i)}(x^n, y^n) = \sum_{j \in \mathcal{M}} Q(j|x^n) \|g(j, y^n) - P_{x^n y^n}\|_i \quad (i = 1, 2). \tag{1.9}$$

This leads to two notions of maximal errors of the code $(\mathcal{C}, g)$

$$\bar{e}_{\mathcal{C}}^{(i)} = \max_{x^n, y^n} \bar{e}_{\mathcal{C}}^{(i)}(x^n, y^n); \ (i = 1, 2). \tag{1.10}$$

Finally, we get the best possible maximal errors (for parameters n and M)

$$\bar{e}^{(i)}(n, M) = \min_{\mathcal{C}: |\mathcal{M}| = M} \bar{e}_{\mathcal{C}}^{(i)}. \tag{1.11}$$

We mention two other kinds of criteria for the measurement of the estimation error.

Let J be a RV with distribution $\Pr(J = j) = Q(j|x^n)$ and use the RV's

$$\Delta_{x^n y^n}^{(i)}(J) = \|g(J, y^n) - P_{x^n y^n}\|_i \tag{1.12}$$

to define

$$e_{\mathcal{C}}^{(i)}(x^n, y^n, \delta) = \Pr\left(\Delta_{x^n y^n}^{(i)}(J) > \delta\right); \ i = 1, 2; \tag{1.13}$$

and

$$e_{\mathcal{C}}^{(i)}(x^n, y^n, \varepsilon) = \min\left\{\delta : \Pr\left(\Delta_{x^n y^n}^{(i)}(J) > \delta\right) < \varepsilon\right\}; \ i = 1, 2. \tag{1.14}$$

Actually, all these definitions lead to similar results and we start here with $e_{\mathcal{C}}^{(2)}(x^n, y^n, \varepsilon)$ for which we define

$$e_{\mathcal{C}}(\varepsilon) = \max_{x^n, y^n} e_{\mathcal{C}}^{(2)}(x^n, y^n, \varepsilon) \tag{1.15}$$

and

$$e(n, M, \varepsilon) = \min_{\mathcal{C}: |\mathcal{M}| = M} e_{\mathcal{C}}(\varepsilon). \tag{1.16}$$

An appropriate scaling

$$\alpha(D, \varepsilon) = \sup_{n, M: \frac{\log \log M}{\log n} < D} \frac{-\log e(n, M, \varepsilon)}{\log n} \tag{1.17}$$

leads to a striking result.

Theorem

$$\alpha(D, \varepsilon) = D \text{ for all } \varepsilon \in (0, 1). \tag{1.18}$$

2   Direct Coding Theorem

We use the following simple coding method. Label the members of $\binom{[n]}{\ell_n}$, the set of all ℓ_n–element subsets of $[n] = \{1, 2, \ldots, n\}$. The sender randomly selects one such subset and transmits its label and the components of x^n within this subset to the receiver. The receiver uses the joint type of y^n and x^n *within this subset* as the estimate of the joint type.

We now evaluate the performance of this method. First we count the number L of subsets where (x^n, y^n)'s local joint type is at least $\sqrt{\ell_n^{-1} \log_n^2}$ away from the true type.

For this we need the definitions

$$n(x, y) := P_{x^n y^n}(x, y)n, \tag{2.1}$$

$$\ell(x, y) := \text{local frequencies of } (x^n, y^n) \text{ in } \ell_n\text{–subset considered} \tag{2.2}$$

and

$$\vec{\ell} := \big(\ell(x, y)\big)_{(x,y) \in \mathcal{X} \times \mathcal{Y}}. \tag{2.3}$$

Clearly $\sum\limits_{x,y} \ell(x, y) = \ell_n$.

Now

$$L = \sum_{\vec{\ell} : \sum_{x,y} \left| \frac{\ell(x,y)}{\ell_n} - \frac{n(x,y)}{n} \right|^2 > \frac{\log^2 n}{\ell_n}} \prod_{x,y} \binom{n(x, y)}{\ell(x, y)}$$

and

$$L \cdot \binom{n}{\ell_n}^{-1} \leq O(n^{ab-1}) \max_{\vec{\ell} : \sum_{x,y} \left| \frac{\ell(x,y)}{\ell_n} - \frac{n(x,y)}{n} \right|^2 > \frac{\log^2 n}{\ell_n}} \frac{\sqrt{\ell_n}}{\prod\limits_{x,y} \sqrt{\ell(x, y)}}$$

$$\cdot \exp\left\{ \sum_{x,y} n(x, y) h\left(\frac{\ell(x, y)}{n(x, y)} \right) - nh\left(\frac{\ell_n}{n} \right) \right\} \tag{2.4}$$

by Stirling's formula.

This can be bounded from above by using the following auxiliary result.

Lemma. *Let positive integers* $n(x, y)$, $\ell(x, y)$, ℓ, n *satisfy*

$$\sum_{x \in \mathcal{X}, y \in \mathcal{Y}} n(x, y) = n, \quad \sum_{x \in \mathcal{X}, y \in \mathcal{Y}} \ell(x, y) = \ell, \ell(x, y) \leq n(x, y) \ \text{ for } \ x \in \mathcal{X}, y \in \mathcal{Y}.$$

Then

$$\theta \triangleq \sum_{x \in \mathcal{X}, y \in \mathcal{Y}} n(x, y) \left[h\left(\frac{\ell(x, y)}{n(x, y)} \right) - h\left(\frac{\ell}{n} \right) \right] \leq -\frac{n}{2ab\ell} \sum_{x \in \mathcal{X}, y \in \mathcal{Y}} n(x, y) \left(\frac{\ell(x, y)}{n(x, y)} - \frac{\ell}{n} \right)^2,$$

where $a = |\mathcal{X}|$, $b = |\mathcal{Y}|$.

Proof. For $\gamma(x,y) \triangleq \frac{\ell(x,y)}{n(x,y)} - \frac{\ell}{n}$ obviously $\sum_{x,y} n(x,y)\gamma(x,y) = 0$. With $C \triangleq \frac{\ell}{n}$ we can now write

$$\theta = \sum_{x,y} n(x,y)\big[h\big(C + \gamma(x,y)\big) - h(C)\big].$$

By Lagrange's interpolation formula

$$h\big(C + \gamma(x,y)\big) - h(C) = h'(C)\gamma(x,y) + \frac{h''(\xi(x,y))}{2}\,\gamma^2(x,y),$$

where $\xi(x,y)$ is between C and $C + \gamma(x,y)$.

Thus

$$\begin{aligned}
\theta &= \sum_{x,y} n(x,y)h'(C)\gamma(x,y) + \sum_{x,y} n(x,y)\frac{h''(\xi(x,y))}{2}\,\gamma^2(x,y) \\
&= \sum_{x,y} n(x,y)\frac{h''(\xi(x,y))}{2}\,\gamma^2(x,y) \\
&\leq \sum_{x,y:\gamma(x,y)\leq 0} n(x,y)\frac{h''(\xi(x,y))}{2}\,\gamma^2(x,y) \\
&= \sum_{x,y:\gamma(x,y)\leq 0} n(x,y)\left(-\frac{1}{2\xi(x,y)(1 - \xi(x,y))}\right)\gamma^2(x,y) \\
&\leq \sum_{x,y:\gamma(x,y)\leq 0} n(x,y)\left(-\frac{1}{2C}\right)\gamma^2(x,y) \\
&= -\frac{n}{2\ell} \sum_{x,y:\gamma(x,y)\leq 0} n(x,y)\gamma^2(x,y).
\end{aligned}$$

Clearly, the claimed inequality follows from the identity

$$\mu \triangleq \min_{\rho:\sum_{x,y} n(x,y)\rho(x,y)=0} \frac{\displaystyle\sum_{x,y:\rho(x,y)\leq 0} n(x,y)\rho^2(x,y)}{\displaystyle\sum_{x,y} n(x,y)\rho^2(x,y)} = \frac{1}{ab}, \tag{2.5}$$

which remains to be proved.

Obviously, the optimizing ρ has the properties:

1. $|\{(x,y) : \rho(x,y) > 0\}| = 1$
2. There exists a constant ν such that $\rho(x,b) \leq 0$ implies $\rho(a,b) = \nu$.

These two properties imply

$$\mu = \frac{(ab-1)\left(\frac{a}{ab-1}\right)^2}{a^2 + (ab-1)\left(\frac{a}{ab-1}\right)^2} = \frac{1}{ab}.$$

$\square$

We apply now the Lemma to upper bound the exponent in the exponential function and get

$$
L \cdot \binom{n}{\ell_n}^{-1} \leq O(n^{ab-1}) \max_{\vec{\ell}:\sum_{x,y}\left|\frac{\ell(x,y)}{\ell_n}-\frac{n(x,y)}{n}\right|^2 > \frac{\log^2 n}{\ell_n}} \frac{\sqrt{\ell_n}}{\prod_{x,y}\sqrt{\ell(x,y)}}
$$
$$
\cdot \exp\left\{-\frac{n}{2ab\ell_n}\sum_{x\in\mathcal{X},y\in\mathcal{Y}}n(x,y)\right\}\left(\frac{\ell(x,y)}{n(x,y)}-\frac{\ell_n}{n}\right)^2
$$

$$
\leq \max_{\vec{\ell}:\sum_{x,y}\left|\frac{\ell(x,y)}{\ell_n}-\frac{n(x,y)}{n}\right|^2 > \frac{\log^2 n}{\ell_n}} O(n^{ab})
$$
$$
\cdot \exp\left\{-\frac{\mu}{2}\frac{n}{\ell_n}\sum_{x,y}n(x,y)\right\}\left(\frac{\ell(x,y)}{n(x,y)}-\frac{\ell_n}{n}\right)^2
$$

$$
= \max_{\vec{\ell}:\sum_{x,y}\left|\frac{\ell(x,y)}{\ell_n}-\frac{n(x,y)}{n}\right|^2 > \frac{\log^2 n}{\ell_n}} O(n^{ab})
$$
$$
\cdot \exp\left\{-\frac{\mu}{2}\sum_{x,y}\frac{n\cdot\ell_n}{n(x,y)}\right\}\left(\frac{\ell(x,y)}{\ell_n}-\frac{n(x,y)}{n}\right)^2
$$

$$
\leq O(n^{ab})\exp\left\{-\frac{\mu}{2}\log^2 n\right\} \to 0 \text{ as } n \to \infty.
$$

Now the number of bits needed for sending an element of $\binom{[n]}{\ell_n}$ is $\log\binom{n}{\ell_n}$ and for sending the ℓ_n bits is ℓ_n. This amounts to a total number of $\log\binom{n}{\ell_n} + \ell_n$ bits. The accuracy achieved is $\ell_n\log^2 n$.

Therefore we get

$$
\frac{\log\delta}{\log n} = \frac{\log\ell_n}{\log n} + \frac{2\log\log n}{\log n}
$$

and

$$
\frac{\log\log M}{\log n} = \frac{\log(\ell_n\log n + \ell_n)}{\log n} = \frac{\log\ell_n}{\log n} + \frac{\log(\log n + 1)}{\log n}.
$$

If $\ell_n \gg \log n$, then

$$
\frac{\log\delta}{\log n} \approx \frac{\log\log M}{\log n}
$$

and the direct part is proved.

3   Converse of Coding Theorem (Proof in Binary Case, Exercise in General Case)

Let $\mathcal{C} = (\mathcal{E}, g)$ be a (D, α, n) code and let

$$
\mathcal{M}(x^n, y^n) = \left\{m \in \mathcal{M} : \|g(m, y^n) - P_{x^n y^n}\|_2^2 \leq \exp\{\alpha\log n + o(\log n)\}\right\}.
$$

We have
$$Q\big(\mathcal{M}(x^n, y^n)|x^n\big) > 1 - \varepsilon.$$

Select now n^β codewords independently at random according to the PD $Q(\cdot|x^n)$. Abbreviate the random code $\big(X_1^n(x^n), \dots, X_{n^\beta}^n(x^n)\big)$ as $B(x^n)$ and use $\tilde{F}(\cdot|x^n)$ to denote the uniform distribution on $B(x^n)$.

$$\Pr\left(\tilde{F}\big(\mathcal{M}(x^n, y^n)|x^n\big) < \frac{1}{2} + \varepsilon\right) \approx \sum_{k > (\frac{1}{2} - \varepsilon)n^\beta} \binom{n^\beta}{k} \varepsilon^k (1 - \beta)^{n^\beta - k}$$

$$\approx \exp\left\{-n^\beta D\left(\frac{1}{2} - \varepsilon \middle\| \varepsilon\right)\right\} \gtrsim \lambda n^\beta.$$

A y^n is called irregular with respect to x^n for a particular $B(x^n)$ or $\tilde{F}(\cdot|x^n)$ iff $\tilde{F}\big(\mathcal{M}(x^n, y^n|x^n)\big) < \frac{1}{2} + \varepsilon$.

The average number of irregular y^n is $2^{n - \lambda n^\beta}$. Therefore a choice of $B(x^n)$ exists such that the number of irregular y^n's is at most $2^{n - \lambda n^\beta}$.

According to this principle we make choices for every x^n. So we get a whole family $\big(\tilde{F}(\cdot|x^n)\big)_{x^n \in \mathcal{X}^n}$, where each member has at most $2^{n - \lambda n^\beta}$ irregular y^n's.

Now we use a constant weight error correcting code of cardinality $2^{\gamma n}$ and of minimum distance μn, where γ, μ are constants (independent of n).

Let x_1^n and x_2^n be two codewords of this code. We prove that for suitable β, $B(x_1^n) \neq B(x_2^n)$. Actually, we count the number of y^n's with

$$\left(\sum_{x,y} \big(n_{x_1^n y^n}(x, y) - n P_{x_2^n y^n}(x, y)\big)^2\right)^{\frac{1}{2}} \geq 2n^{\alpha + o(1)}.$$

For this define

$$A = \big\{t \in [n] : x_{1t} = 1 \ \text{and} \ x_{2t} = 0\big\}, B = \big\{t \in [n] : x_{1t} = 1 \ \text{and} \ x_{2t} = 1\big\},$$

$$C = \big\{t \in [n] : x_{1t} = 0 \ \text{and} \ x_{2t} = 1\big\}, \ \text{and} \ D = \big\{t \in [n] : x_{1t} = 0 \ \text{and} \ x_{2t} = 0\big\}.$$

This number of y^n's exceeds

$$2^{|B| + |D|} \sum_{|u - v| > 2n^{\frac{1}{2}\alpha + o(1)}} \binom{|A|}{u}\binom{|C|}{v} = 2^{|B| + |D|} \sum_{\ell > 2n^{\frac{1}{2}\alpha + o(1)}} \binom{|A| + |C|}{|A| - \ell} = 2^{n - \mu n^{\alpha + o(1)}}.$$

Now, if $B(x_1^n) = B(x_2^n)$, then those y^n must be irregular for at least one of x_1^n, x_2^n. Hence $2^{n - \mu n^{\alpha + o(1)}} \leq 2^{n - \lambda n^\beta}$ and thus $\alpha \geq \beta + o(1)$. Finally $M^{n^\beta} \geq 2^{rn}$ implies $M \geq 2^{rn^{1-\beta}} \geq 2^{n^{1 - \alpha - o(1)}}$. The converse is proved in the binary case.

4   Other Problems

A. The existing work on statistical inference (hypothesis testing and estimation in [4] and [3]) under communication constraints uses a "one shot" side information. It seems important to introduce and analyze interactive models.

B. Permutation invariant functions

A function F, defined on $\mathcal{X}^n \times \mathcal{Y}^n$, is called permutation invariant iff for all permutations π of the set $\{1, 2, \ldots, n\}$

$$F(x^n, y^n) = F(\pi x^n, \pi y^n), \tag{4.1}$$

where $x^n = (x_1, x_2, \ldots, x_n)$

$$\pi x^n = (x_{\pi(1)}, x_{\pi(2)}, \ldots, x_{\pi(n)}) \tag{4.2}$$

and y^n, πy^n are defined analogously.

Permutation invariant functions are actually functions of the joint empirical distribution $P_{x^n y^n}$ of the sequences x^n and y^n, where for all $x \in \mathcal{X}$, $y \in \mathcal{Y}$

$$P_{x^n y^n}(x, y) = |\{t : x_t = x, y_t = y\}| n^{-1}. \tag{4.3}$$

Examples of permutation invariant functions include, but are not limited to, sum–type functions f^n,

$$f^n(x^n, y^n) = \sum_{t=1}^{n} f(x_t, y_t), \tag{4.4}$$

such as the Hamming distance function. In identification problems, we can be interested in Boolean functions. When the problem is permutation invariant, we need to study permutation invariant Boolean functions. If we estimate the joint empirical distribution of x^n and y^n. Then $\left(P_{x^n y^n}(x, y)\right)_{x \in \mathcal{X}, y \in \mathcal{Y}}$ is a permutation invariant vector–valued function on $\mathcal{X}^n \times \mathcal{Y}^n$.

C. Approximation of continuous permutation invariant functions

Let F be a continuous function defined on the compact set $\mathcal{P}(\mathcal{X} \times \mathcal{Y})$. Define

$$\hat{F}(x^n, y^n) = F(P_{x^n y^n}). \tag{4.5}$$

If the task of the receiver is not to estimate $P_{x^n y^n}$, but to compute $\hat{F}(x^n, y^n)$, what is then the trade–off between the computation accuracy and the "communication rate" D?

This problem is closely related to the joint empirical distribution estimation problem — actually, it generalizes it.

D. Classification Problem

Let $\{\mathcal{A}_0, \mathcal{A}_1\}$ be a partition of $\mathcal{X}^n \times \mathcal{Y}^n$ and let both sets in this partition be permutation invariant. If in the model treated in this paper the task of the receiver is to determine whether or not $(x^n, y^n) \in \mathcal{A}_0$, then this is a new "classification" problem.

In case we want to determine this exactly, then we have to transmit for "most" partitions almost all bits of x^n to the receiver. We introduce now a model, which allows a much lower rate.

Let $d_1(P, P') = \|P - P'\|_1$ be the L_1–distance of P and P' in $\mathcal{P}(\mathcal{X}^n \times \mathcal{Y}^n)$. For $\mathcal{A} \subset \mathcal{X}^n \times \mathcal{Y}^n$ and $\delta > 0$ let

$$\Gamma_\delta(\mathcal{A}) = \left\{(x^n, y^n) \in \mathcal{X}^n \times \mathcal{Y}^n : d_1(P_{x^n y^n}, P_{x'^n y'^n}) \leq \delta \text{ for some } (x'^n, y'^n) \in \mathcal{A}\right\},$$

and for permutation invariant $\mathcal{A} \subset \mathcal{X}^n \times \mathcal{Y}^n$ let

$$N(\mathcal{A}) = |\{P_{x^n y^n} : (x^n, y^n) \in \mathcal{A}\}|.$$

Now, for $\varepsilon > 0$, find maximal $\delta_0, \delta_1 \geq 0$ such that

$$\frac{N\big(\Gamma_{\delta_0}(\mathcal{A}_0) \cap \mathcal{A}_1\big)}{N(\mathcal{A}_1)} \leq \varepsilon, \tag{4.6}$$

$$\frac{N\big(\Gamma_{\delta_1}(\mathcal{A}_1) \cap \mathcal{A}_0\big)}{N(\mathcal{A}_0)} \geq \varepsilon. \tag{4.7}$$

Finally, let

$$g : \mathcal{M} \times \mathcal{Y}^n \to \{0, 1\}$$

be a binary–valued function such that for all $(x^n, y^n) \in \mathcal{A}_0 \smallsetminus \Gamma_{\delta_1}(\mathcal{A}_1)$

$$Q\big(g(J, y^n) = 0 | x^n\big) \geq 1 - \varepsilon$$

and for all $(x^n, y^n) \in \mathcal{A}_1 \smallsetminus \Gamma_{\delta_0}(\mathcal{A}_0)$

$$Q\big(g(J, x^n) = 1 | x^n\big) \geq 1 - \varepsilon.$$

What is the minimum number of bits $\lceil \log M \rceil$ needed? This problem is also closely related to the joint empirical distribution estimation problem.

References

1. R. Ahlswede, Channel capacities for list codes, J. Appl. Probability, 10, 824–836, 1973.
2. R. Ahlswede, Coloring hypergraphs: A new approach to multi–user source coding, Part I, Journ. of Combinatorics, Information and System Sciences, Vol. 4, No. 1, 76–115, 1979; Part II, Journ. of Combinatorics, Information and System Sciences, Vol. 5, No. 3, 220–268, 1980.
3. R. Ahlswede and M. Burnashev, On minimax estimation in the presence of side information about remote data, Ann. of Stat., Vol. 18, No. 1, 141–171, 1990.
4. R. Ahlswede and I. Csiszár, Hypothesis testing under communication constraints, IEEE Trans. Inform. Theory, Vol. 32, No. 4, 533–543, 1986.
5. R. Ahlswede and I. Csiszár, To get a bit of information may be as hard as to get full information, IEEE Trans. Inform. Theory, Vol. 27, 398–408, 1981.
6. R. Ahlswede and G. Dueck, Identification via channels, IEEE Trans. Inform. Theory, Vol. 35, No. 1, 15–29, 1989.
7. R. Ahlswede and G. Dueck, Identification in the presence of feedback — a discovery of new capacity formulas, IEEE Trans. Inform. Theory, Vol. 35, No. 1, 30–39, 1989.
8. R. Ahlswede and J. Körner, Source coding with side information and a converse for degraded broadcast channels, IEEE Trans. Inf. Theory, Vol. 21, 629–637, 1975.
9. R. Ahlswede, E. Yang, and Z. Zhang, Identification via compressed data, IEEE Trans. Inform. Theory, Vol. 43, No. 1, 22–37, 1997.
10. R. Ahlswede and Z. Zhang, Worst case estimation of permutation invariant functions and identification via compressed data, Preprint 97–005, SFB 343 "Diskrete Strukturen in der Mathematik", Universität Bielefeld.

11. T. Berger, Rate Distortion Theory, Englewood Cliffs, NJ, Prentice–Hall, 1971.
12. I. Csiszár and J. Körner, Information Theory: Coding Theorems for Discrete Memoryless Systems, New York, Academic, 1981.
13. D. Slepian and J.K. Wolf, Noiseless coding of correlated information sources, IEEE Trans. Inform. Theory, Vol. 19, 471–480, 1973.
14. A.D. Wyner and J. Ziv, The rate–distortion function for source coding with side information at the decoder, IEEE Trans. Inform. Theory, Vol. 22, 1–10, 1976.

On Logarithmically Asymptotically Optimal Hypothesis Testing for Arbitrarily Varying Sources with Side Information

R. Ahlswede, Ella Aloyan, and E. Haroutunian*

Abstract. The asymptotic interdependence of the error probabilities exponents (reliabilities) in optimal hypotheses testing is studied for arbitrarily varying sources with state sequence known to the statistician. The case when states are not known to the decision maker was studied by Fu and Shen.

1 Introduction

On the open problems session of the Conference in Bielefeld (August 2003) Ahlswede formulated among others the problem of investigation of "Statistics for not completely specified distributions" in the spirit of his paper [1]. In this paper, in particular, coding problems are solved for arbitrarily varying sources with side information at the decoder. Ahlswede proposed to consider the problems of inference for similar statistical models. It turned out that the problem of "Hypothesis testing for arbitrarily varying sources with exponential constraint" was already solved by Fu and Shen [2]. This situation corresponds to the case, when side information at the decoder (in statistics this is the statistician, the decision maker) is absent.

The present paper is devoted to the same problem when the statistician has the possibility to make decisions after receiving the complete sequence of states of the source. This, still simple, problem may be considered as a beginning of the realization of the program proposed by Ahlswede.

This investigation is a development of results from [3]-[9] and may be continued in various directions: the cases of many hypotheses, non complete side information, sources of more general classes (Markov chains, general distributions), identification of hypotheses in the sense of [10].

2 Formulation of Results

An arbitrarily varying source is a generalized model of a discrete memoryless source, distribution of which varies independently at any time instant within a certain set. Let $\mathcal{X}$ and $\mathcal{S}$ be finite sets, $\mathcal{X}$ the source alphabet, $\mathcal{S}$ the state alphabet. $\mathcal{P}(\mathcal{S})$ is a set of all possible probability distributions P on $\mathcal{S}$. Suppose a statistician makes decisions between two conditional probability distributions

---

* Work was partially supported by INTAS grant 00738.

R. Ahlswede et al. (Eds.): Information Transfer and Combinatorics, LNCS 4123, pp. 547–552, 2006.
© Springer-Verlag Berlin Heidelberg 2006

of the source: $G_1 = \{G_1(x|s), x \in \mathcal{X}, s \in \mathcal{S}\}$, $G_2 = \{G_2(x|s), x \in \mathcal{X}, s \in \mathcal{S}\}$, thus there are two alternative hypotheses $H_1 : G = G_1$, $H_2 : G = G_2$. A sequence $\mathbf{x} = (x_1, \ldots, x_N)$, $\mathbf{x} \in \mathcal{X}^N$, $N = 1, 2, \ldots$, is emitted from the source, and sequence $\mathbf{s} = (s_1, \ldots, s_N)$ is created by the source of states. We consider the situation when the source of states is connected with the statistician who must decide which hypothesis is correct on the base of the data $\mathbf{x}$ and the state sequence $\mathbf{s}$. Every test $\varphi^{(N)}$ is a partition of the set $\mathcal{X}^N$ into two disjoint subsets $\mathcal{X}^N = \mathcal{A}_{\mathbf{s}}^{(N)} \bigcup \overline{\mathcal{A}}_{\mathbf{s}}^{(N)}$, where the set $\mathcal{A}_{\mathbf{s}}^{(N)}$ consists of all vectors $\mathbf{x}$ for which the first hypothesis is adopted using the state sequence $\mathbf{s}$.

Making decisions about these hypotheses one can commit the following errors: the hypothesis H_1 is rejected, but it is correct, the corresponding error probability is

$$\alpha_{1|2}^{(N)}(\varphi^{(N)}) = \max_{\mathbf{s} \in \mathcal{S}^N} G_1^N(\overline{\mathcal{A}}_{\mathbf{s}}^{(N)} \mid \mathbf{s}),$$

if the hypothesis H_1 is adopted while H_2 is correct, we make an error with the probability

$$\alpha_{2|1}^{(N)}(\varphi^{(N)}) = \max_{\mathbf{s} \in \mathcal{S}^N} G_2^N(\mathcal{A}_{\mathbf{s}}^{(N)} \mid \mathbf{s}).$$

Let us introduce the following error probability exponents or "reliabilities" $E_{1|2}$ and $E_{2|1}$, using logarithmical and exponential functions at the base e:

$$\varlimsup_{N \to \infty} - N^{-1} \ln \alpha_{1|2}^{(N)}(\varphi^{(N)}) = E_{1|2}, \tag{1}$$

$$\varlimsup_{N \to \infty} - N^{-1} \ln \alpha_{2|1}{}^{(N)}(\varphi^{(N)}) = E_{2|1}. \tag{2}$$

The test is called *logarithmically asymptotically optimal* (LAO) if for given $E_{1|2}$ it provides the largest value to $E_{2|1}$. The problem is to state the existence of such tests and to determine optimal dependence of the value of $E_{2|1}$ from $E_{1|2}$.

Now we collect necessary basic concepts and definitions. For $\mathbf{s}=(s_1, \ldots, s_N)$, $\mathbf{s} \in \mathcal{S}^N$, let $N(s \mid \mathbf{s})$ be the number of occurrences of $s \in \mathcal{S}$ in the vector $\mathbf{s}$. The type of $\mathbf{s}$ is the distribution

$$P_{\mathbf{s}} = \{P_{\mathbf{s}}(s), s \in \mathcal{S}\}$$

defined by

$$P_{\mathbf{s}}(s) = \frac{1}{N} N(s \mid \mathbf{s}), \qquad s \in \mathcal{S}.$$

For a pair of sequences $\mathbf{x} \in \mathcal{X}^N$ and $\mathbf{s} \in \mathcal{S}^N$, let $N(x, s|\mathbf{x}, \mathbf{s})$ be the number of occurrences of the pair $(x, s) \in \mathcal{X} \times \mathcal{S}$ in the pair of vectors $(\mathbf{x}, \mathbf{s})$. The joint type of the pair $(\mathbf{x}, \mathbf{s})$ is the distribution

$$Q_{\mathbf{x},\mathbf{s}} = \{Q_{\mathbf{x},\mathbf{s}}(x, s), x \in \mathcal{X}, s \in \mathcal{S}\}$$

defined by

$$Q_{\mathbf{x},\mathbf{s}}(x, s) = \frac{1}{N} N(x, s \mid \mathbf{x}, \mathbf{s}), \qquad x \in \mathcal{X}, s \in \mathcal{S}.$$

The conditional type of **x** for given **s** is the conditional distribution

$$Q_{\mathbf{x}|\mathbf{s}} = \{Q_{\mathbf{x}|\mathbf{s}}(x|s),\ x \in \mathcal{X},\ s \in \mathcal{S}\}$$

defined by

$$Q_{\mathbf{x}|\mathbf{s}}(x|s) = \frac{Q_{\mathbf{x},\mathbf{s}}(x, s)}{P_{\mathbf{s}}(s)}, \qquad x \in \mathcal{X}, s \in \mathcal{S}.$$

Let X and S are random variables with probability distributions $P = \{P(s), s \in \mathcal{S}\}$ and $Q = \{Q(x|s), x \in \mathcal{X}, s \in \mathcal{S}\}$. The conditional entropy of X with respect to S is:

$$H_{P,Q}(X \mid S) = -\sum_{x,s} P(s)Q(x|s) \ln Q(x|s).$$

The conditional divergence of the distribution $P \circ Q = \{P(s)Q(x|s), x \in \mathcal{X}, s \in \mathcal{S}\}$ with respect to $P \circ G_m = \{P(s)G_m(x|s), x \in \mathcal{X}, s \in \mathcal{S}\}$ is defined by

$$D(P \circ Q || P \circ G_m) = D(Q \parallel G_m | P) = \sum_{x,s} P(s)Q(x|s) \ln \frac{Q(x|s)}{G_m(x|s)}, \qquad m = 1, 2.$$

The conditional divergence of the distribution $P \circ G_2 = \{P(s)G_2(x|s), x \in \mathcal{X}, s \in \mathcal{S}\}$ with respect to $P \circ G_1 = \{P(s)G_1(x|s), x \in \mathcal{X}, s \in \mathcal{S}\}$ is defined by

$$D(P \circ G_2 || P \circ G_1) = D(G_2 \parallel G_1 | P) = \sum_{x,s} P(s)G_2(x|s) \ln \frac{G_2(x|s)}{G_1(x|s)}.$$

Similarly we define $D(G_1 \parallel G_2 | P)$.

Denote by $\mathcal{P}^N(\mathcal{S})$ the space of all types on $\mathcal{S}$ for given N, and $\mathcal{Q}^N(\mathcal{X}, \mathbf{s})$ the set of all possible conditional types on $\mathcal{X}$ for given **s**. Let $T_{P_{\mathbf{s}},Q}^{(N)}(X \mid \mathbf{s})$ be the set of vectors **x** of conditional type Q for given **s** of type $P_{\mathbf{s}}$.

It is known [3] that

$$|\mathcal{Q}^N(\mathcal{X}, \mathbf{s})| \leq (N + 1)^{|\mathcal{X}||\mathcal{S}|}, \tag{3}$$

$$(N+1)^{-|\mathcal{X}||\mathcal{S}|} \exp\{NH_{P_{\mathbf{s}},Q}(X|S)\} \leq |T_{P_{\mathbf{s}},Q}^{(N)}(X \mid \mathbf{s})| \leq \exp\{NH_{P_{\mathbf{s}},Q}(X|S)\}. \tag{4}$$

Theorem. For every given $E_{1|2}$ from $(0, \min_{P \in \mathcal{P}(\mathcal{S})} D(G_2 \parallel G_1 \mid P))$

$$E_{2|1}(E_{1|2}) = \min_{P \in \mathcal{P}(\mathcal{S})} \min_{Q:D(Q||G_1|P) \leq E_{1|2}} D(Q \parallel G_2 \mid P). \tag{5}$$

We can easily infer the following

Corollary. (Generalized lemma of Stein): When $\alpha_{1|2}^{(N)}(\varphi^{(N)}) = \varepsilon > 0$, for N large enough

$$\alpha_{2|1}^{(N)}(\alpha_{1|2}^{(N)}(\varphi^{(N)}) = \varepsilon) = \min_{P \in \mathcal{P}(\mathcal{S})} \exp\{-ND(G_1 \parallel G_2 | P)\}.$$

3 Proof of the Theorem

The proof consists of two parts. We begin with demonstration of the inequality

$$E_{2|1}(E_{1|2}) \geq \min_{P \in \mathcal{P}(\mathcal{S})} \min_{Q:D(Q\|G_1|P) \leq E_{1|2}} D(Q\|G_2|P). \tag{6}$$

For $\mathbf{x} \in T_{P_\mathbf{s},Q}^{(N)}(X|\mathbf{s})$, $\mathbf{s} \in T_{P_\mathbf{s}}^{(N)}(\mathcal{S})$, $m = 1,2$ we have,

$$G_m^N(\mathbf{x} \mid \mathbf{s}) = \prod_{n=1}^{N} G_m(x_n|s_n) = \prod_{x,s} G_m(x|s)^{N(x,s|\mathbf{x},\mathbf{s})} = \prod_{x,s} G_m(x|s)^{NP_\mathbf{s}(s)Q_{\mathbf{x}|\mathbf{s}}(x|s)} =$$

$$= \prod_{x,s} \exp\left\{NP_\mathbf{s}(s)(s)Q_{\mathbf{x}|\mathbf{s}}(x|s)\ln G_m(x|s)\right\} = \prod_{x,s} \exp\left\{N[P_\mathbf{s}(s)(s)Q_{\mathbf{x}|\mathbf{s}}(x|s)\ln G_m(x|s) - \right.$$

$$\left. - P_\mathbf{s}(s)(s)Q_{\mathbf{x}|\mathbf{s}}(x|s)\ln Q_{\mathbf{x}|\mathbf{s}}(x|s) + P_\mathbf{s}(s)(s)Q_{\mathbf{x}|\mathbf{s}}(x|s)\ln Q_{\mathbf{x}|\mathbf{s}}(x|s)]\right\} =$$

$$= \exp\left\{N\sum_{x,s}(-P_\mathbf{s}(s)(s)Q_{\mathbf{x}|\mathbf{s}}(x|s)\ln\frac{Q_{\mathbf{x}|\mathbf{s}}(x|s)}{G_m(x|s)} + P_\mathbf{s}(s)(s)Q_{\mathbf{x}|\mathbf{s}}(x|s)\ln Q_{\mathbf{x}|\mathbf{s}}(x|s))\right\} =$$

$$= \exp\left\{-N[D(Q \| G_m|P) + H_{P_\mathbf{s},Q}(X \mid S)]\right\}. \tag{7}$$

Let us show that the optimal sequence of tests $\varphi^{(N)}$ for every $\mathbf{s}$ is given by the following sets

$$\mathcal{A}_\mathbf{s}^{(N)} = \bigcup_{Q:D(Q\|G_1|P_\mathbf{s}) \leq E_{1|2}} T_{P_\mathbf{s},Q}^{(N)}(X \mid \mathbf{s}). \tag{8}$$

Using (4) and (7) we see that

$$G_m^N(T_{P_\mathbf{s},Q}^{(N)}(X|\mathbf{s}) \mid \mathbf{s}) = \mid T_{P_\mathbf{s},Q}^{(N)}(X \mid \mathbf{s}) \mid G_m^N(\mathbf{x}|\mathbf{s}) \leq \exp\{-ND(Q \| G_m|P_\mathbf{s})\}.$$

We can estimate both error probabilities

$$\alpha_{1|2}^{(N)}(\varphi^{(N)}) = \max_{\mathbf{s} \in \mathcal{S}^N} G_1^N(\overline{\mathcal{A}}_\mathbf{s}^{(N)} \mid \mathbf{s}) = \max_{\mathbf{s} \in \mathcal{S}^N} G_1^N\left(\bigcup_{Q:D(Q\|G_1|P_\mathbf{s})>E_{1|2}} T_{P_\mathbf{s},Q}^{(N)}(X \mid \mathbf{s}) \mid \mathbf{s}\right) \leq$$

$$\leq \max_{\mathbf{s} \in \mathcal{S}^N}(N+1)^{|\mathcal{X}||\mathcal{S}|} \max_{Q:D(Q\|G_1|P_\mathbf{s})>E_{1|2}} G_1^N(T_{P_\mathbf{s},Q}^{(N)}(X \mid \mathbf{s}) \mid \mathbf{s}) \leq$$

$$\leq (N+1)^{|\mathcal{X}||\mathcal{S}|} \max_{P_\mathbf{s} \in \mathcal{P}^N(\mathcal{S})} \max_{Q:D(Q\|G_1|P_\mathbf{s})>E_{1|2}} \exp\{-ND(Q\|G_1|P_\mathbf{s})\} =$$

$$= \max_{P_\mathbf{s} \in \mathcal{P}^N(\mathcal{S})} \max_{Q:D(Q\|G_1|P_\mathbf{s})>E_{1|2}} \exp\{|\mathcal{X}||\mathcal{S}|\ln(N+1) - ND(Q\|G_1|P)\} =$$

$$= \max_{P_\mathbf{s} \in \mathcal{P}^N(\mathcal{S})} \max_{Q:D(Q\|G_1|P_\mathbf{s})>E_{1|2}} \exp\{-N[D(Q\|G_1|P_\mathbf{s}) - o(1)]\} \leq$$

$$\leq \exp\{-N(E_{1|2} - o(1))\},$$

where $o(1) = N^{-1}|\mathcal{X}||\mathcal{S}|\ln(N+1) \to 0$, when $N \to \infty$. And then

$$\alpha_{2|1}^{(N)}(\varphi^{(N)}) = \max_{\mathbf{s} \in \mathcal{S}^N} G_2^N(\mathcal{A}_{\mathbf{s}}^{(N)}|\mathbf{s}) = \max_{\mathbf{s} \in \mathcal{S}^N} G_2^N(\bigcup_{Q:D(Q||G_1|P_{\mathbf{s}}) \le E_{1|2}} T_{P_{\mathbf{s}},Q}^{(N)}(X|\mathbf{s})|\mathbf{s}) =$$

$$= \max_{\mathbf{s} \in \mathcal{S}^N} \sum_{Q:D(Q||G_1|P_{\mathbf{s}}) \le E_{1|2}} G_2^N(T_{P_{\mathbf{s}},Q}^{(N)}(X|\mathbf{s})|\mathbf{s}) \le$$

$$\le (N+1)^{|\mathcal{X}||\mathcal{S}|} \max_{P_{\mathbf{s}} \in \mathcal{P}^N(\mathcal{S})} \max_{Q:D(Q||G_1|P_{\mathbf{s}}) \le E_{1|2}} \exp\{-ND(Q||G_2|P_{\mathbf{s}})\} =$$

$$= \max_{P_{\mathbf{s}} \in \mathcal{P}^N(\mathcal{S})} \max_{Q:D(Q||G_1|P_{\mathbf{s}}) \le E_{1|2}} \exp\{-N[D(Q||G_2|P_{\mathbf{s}}) - o(1)]\} =$$

$$= \exp\{-N(\min_{P_{\mathbf{s}} \in \mathcal{P}^N(\mathcal{S})} \min_{Q:D(Q||G_1|P_{\mathbf{s}}) \le E_{1|2}} D(Q||G_2|P_{\mathbf{s}}) - o(1))\}.$$

So with $N \to \infty$ we get (6)

Now we pass to the proof of the second part of the theorem. We shall prove the inequality inverse to (6). First we can show that this inverse inequality is valid for test $\varphi^{(N)}$ defined by (8). Using (4) and (7) we obtain

$$\alpha_{2|1}^{(N)}(\varphi^{(N)}) = \max_{\mathbf{s} \in \mathcal{S}^N} G_2^N(\mathcal{A}_{\mathbf{s}}^{(N)}|\mathbf{s}) = \max_{\mathbf{s} \in \mathcal{S}^N} G_2^N(\bigcup_{Q:D(Q||G_1|P_{\mathbf{s}}) \le E_{1|2}} T_{P_{\mathbf{s}},Q}^{(N)}(X|\mathbf{s})|\mathbf{s})) \ge$$

$$\ge \max_{\mathbf{s} \in \mathcal{S}^N} \max_{Q:D(Q||G_1|P_{\mathbf{s}}) \le E_{1|2}} G_2^N(T_{P_{\mathbf{s}},Q}^{(N)}(X|\mathbf{s})|\mathbf{s}) \ge$$

$$\ge \max_{P_{\mathbf{s}} \in \mathcal{P}^N(\mathcal{S})} (N+1)^{-|\mathcal{X}||\mathcal{S}|} \max_{Q:D(Q||G_1|P_{\mathbf{s}}) \le E_{1|2}} \exp\{-ND(Q||G_2|P_{\mathbf{s}})\} =$$

$$= \exp\{-N(\min_{P_{\mathbf{s}} \in \mathcal{P}^N(\mathcal{S})} \min_{Q:D(Q||G_1|P_{\mathbf{s}}) \le E_{1|2}} D(Q||G_1|P_{\mathbf{s}}) + o(1))\}.$$

So with $N \to \infty$ we get that for this test

$$E_{2|1}(E_{1|2}) \le \min_{P \in \mathcal{P}(\mathcal{S})} \min_{Q:D(Q||G_1|P) \le E_{1|2}} D(Q||G_2|P). \tag{9}$$

Then we have to be convinced that any other sequence $\tilde{\varphi}^N$ of tests defined for every $\mathbf{s} \in \mathcal{S}^N$ by the sets $\tilde{\mathcal{A}}_{\mathbf{s}}^N$ such, that

$$\alpha_{1|2}^{(N)}(\tilde{\varphi}^{(N)}) \le \exp\{-NE_{1|2}\}, \tag{10}$$

and

$$\alpha_{2|1}^{(N)}(\tilde{\varphi}^{(N)}) \le \alpha_{2|1}^{(N)}(\varphi^{(N)}),$$

in fact coincide with $\varphi^{(N)}$ defined in (8). Let us consider the sets $\tilde{\mathcal{A}}_{\mathbf{s}}^N \bigcap \mathcal{A}_{\mathbf{s}}^N$, $\mathbf{s} \in \mathcal{S}^N$. This intersection cannot be void, because in that case $\overline{\tilde{\mathcal{A}}_{\mathbf{s}}^N \bigcap \mathcal{A}_{\mathbf{s}}^N}$ will be equal to $\mathcal{X}^N = \overline{\mathcal{A}_{\mathbf{s}}^N} \bigcup \overline{\tilde{\mathcal{A}}_{\mathbf{s}}^N}$ and the probabilities $G_1^N(\overline{\mathcal{A}_{\mathbf{s}}^N}|\mathbf{s})$ and $G_1^N(\overline{\tilde{\mathcal{A}}_{\mathbf{s}}^N}|\mathbf{s})$ cannot be small simultaneously.

Now because

$$G_1^N(\overline{\mathcal{A}_{\mathbf{s}}^N \bigcap \tilde{\mathcal{A}}_{\mathbf{s}}^N}|\mathbf{s}) \le$$

$$G_1^N(\overline{\mathcal{A}_{\mathbf{s}}^N})|\mathbf{s}) + G_1^N(\overline{\tilde{\mathcal{A}}_{\mathbf{s}}^N})|\mathbf{s}) \le 2 \cdot \exp\{-NE_{1|2}\} = \exp\{-N(E_{1|2} + o(1))\},$$

from (6) we obtain

$$G_2^N(\mathcal{A}_s^N \bigcap \tilde{\mathcal{A}}_s^N|s) \leq G_2^N(\mathcal{A}_s^N|s) \leq \exp\{-N(\min_{P \in \mathcal{P}(\mathcal{S})} \min_{Q:D(Q||G_1|P) \leq E_{1|2}} D(Q||G_2|P))\}$$

and so we conclude that if we exclude from $\tilde{\mathcal{A}}_s^N$ the vectors $\mathbf{x}$ of the types $T_{P,Q}^N(X|\mathbf{s})$ with $D(Q||G_1|P) > E_{1|2}$ we do not make reliabilities of the test $\tilde{\varphi}^{(N)}$ worse. It is left to remark that when we add to $\tilde{\mathcal{A}}_s^N$ all types $T_{P,Q}^N(X|\mathbf{s})$ with $D(Q||G_1|P) \leq E_{1|2}$, that is we take $\tilde{\mathcal{A}}_s^N = \mathcal{A}_s^N$, we obtain that (6) and (9) are valid, that is the test $\varphi^{(N)}$ is optimal.

References

1. R. Ahlswede, Coloring hypergraphs: a new approach to multi-user source coding, I and II, J. Combin., Inform. System Sciences, Vol. 4, No. 1, 75-115, 1979, Vol. 5, No. 2, 220-269, 1980.

2. F.-W. Fu and S.-Y. Shen, Hypothesis testing for arbitrarily varying sources with exponential-type constraint, IEEE Trans. Inform. Theory, Vol. 44, No. 2, 892-895, 1998.

3. I. Csiszár and J. Körner, Information Theory: Coding Theorems for Discrete Memoryless Systems, Academic press, New York, 1981.

4. W. Hoeffding, Asymptotically optimal tests for multinomial distributions, Ann. Math. Stat., Vol. 36, 369-401, 1965.

5. I. Csiszár and G. Longo, On the error exponent for source coding and for testing simple statistical hypotheses, Studia Sc. Math. Hungarica, Vol. 6, 181-191, 1971.

6. G. Tusnady, On asymptotically optimal tests, Ann. Statist., Vol. 5, No. 2, 385-393, 1977.

7. G. Longo and A. Sgarro, The error exponent for the testing of simple statistical hypotheses, a combinatorial approach, J. Combin., Inform. System Sciences, Vol. 5, No 1, 58-67, 1980.

8. L. Birgé, Vitesse maximales de décroissance des erreurs et tests optimaux associés, Z. Wahrsch. verw. Gebiete, Vol. 55, 261-273, 1981.

9. E. A. Haroutunian, Logarithmically asymptotically optimal testing of multiple statistical hypotheses, Probl. Control and Inform. Theory, Vol. 19, No. 5-6, 413-421, 1990.

10. R. Ahlswede and E. Haroutunian, On logarithmically asymptotically optimal testing of hypotheses and identification: research program, examples and partial results, preprint.

On Logarithmically Asymptotically Optimal Testing of Hypotheses and Identification

R. Ahlswede and E. Haroutunian

Abstract. We introduce a new aspect of the influence of the information-theoretical methods on the statistical theory. The procedures of the probability distributions identification for $K(\geq 1)$ random objects each having one from the known set of $M(\geq 2)$ distributions are studied. N-sequences of discrete independent random variables represent results of N observations for each of K objects. On the base of such samples decisions must be made concerning probability distributions of the objects. For $N \to \infty$ the exponential decrease of the test's error probabilities is considered. The reliability matrices of logarithmically asymptotically optimal procedures are investigated for some models and formulations of the identification problems. The optimal subsets of reliabilities which values may be given beforehand and conditions guaranteeing positiveness of all the reliabilities are investigated.

"In statistical literature such a problem is referred to as one of classification or discrimination, but identification seems to be more appropriate"

Radhakrishna Rao [1].

1 Problem Statement

Let $\mathbf{X}_k = (X_{k,n}, \ n \in [N])$, $k \in [K]$, be $K(\geq 1)$ sequences of N discrete independent identically distributed random variables representing possible results of N observations, respectively, for each of K randomly functioning objects.

For $k \in [K], n \in [N], X_{k,n}$ assumes values $x_{k,n}$ in the finite set $\mathcal{X}$ of cardinality $|\mathcal{X}|$. Let $\mathcal{P}(\mathcal{X})$ be the space of all possible distributions on $\mathcal{X}$. There are $M(\geq 2)$ probability distributions $G_1, \ldots, G_M$ from $\mathcal{P}(\mathcal{X})$ in inspection, some of which are assigned to the vectors $\mathbf{X}_1, \ldots, \mathbf{X}_K$. This assignment is unknown and must be determined on the base of N–samples (results of N independent observations) $\mathbf{x}_k = (x_{k,1}, \ldots, x_{k,N})$, where $x_{k,n}$ is a result of the n-th observation of the k-th object.

When $M = K$ and all objects are different (any two objects cannot have the same distribution), there are $K!$ possible decisions. When objects are independent, there are M^K possible combinations.

Bechhofer, Kiefer, and Sobel presented investigations on sequential multiple-decision procedures in [2]. This book is concerned principally with a particular class of problems referred to as ranking problems.

R. Ahlswede et al. (Eds.): Information Transfer and Combinatorics, LNCS 4123, pp. 553–571, 2006.
© Springer-Verlag Berlin Heidelberg 2006

Chapter 10 of the book by Ahlswede and Wegener [3] is devoted to statistical identification and ranking problems.

We study models considered in [2] and [3] and variations of these models inspired by the pioneering papers by Ahlswede and Dueck [4] and by Ahlswede [5], applying the concept of optimality developed in [6]-[11] for the models with $K = 1$.

Consider the following family of error probabilities of a test

$$\alpha^{(N)}_{m_1,m_2,\dots,m_K|l_1,l_2,\dots,l_K}, \quad (m_1,m_2,\dots,m_K) \neq (l_1,l_2,\dots,l_K), \quad m_k,l_k \in [M], \quad k \in [K],$$

which are the probabilities of decisions $l_1, l_2, \dots, l_K$ when actual indices of the distributions of the objects were, respectively, $m_1, m_2, \dots, m_K$.

The probabilities to reject all K hypotheses when they are true are the following

$$\alpha^{(N)}_{m_1,m_2,\dots,m_K|m_1,m_2,\dots,m_K} = \sum_{(l_1,l_2,\dots,l_K)\neq(m_1,m_2,\dots,m_K)} \alpha^{(N)}_{m_1,m_2,\dots,m_K|l_1,l_2,\dots,l_K}.$$

We study exponential decrease of the error probabilities when $N \to \infty$ and define (using logarithms and exponents to the base e)

$$\varlimsup_{N\to\infty} -\frac{1}{N} \log \alpha^{(N)}_{m_1,m_2,\dots,m_K|l_1,l_2,\dots,l_K} = E_{m_1,m_2,\dots,m_K|l_1,l_2,\dots,l_K} \geq 0. \qquad (1)$$

These are exponents of error probabilities which we call reliabilities (in association with Shannon's reliability function [12]). We shall examine the matrix $\mathbf{E} = \{E_{m_1,m_2,\dots,m_K|l_1,l_2,\dots,l_K}\}$ and call it the reliability matrix.

Our criterion of optimality is: given M, K and values of a part of reliabilities to obtain the best (the largest) values for others. In addition it is necessary to describe the conditions under which all these reliabilities are positive. The procedure that realizes such testing is identification, which following Birgé [10], we call "logarithmically asymptotically optimal" (LAO).

Let $N(x|\mathbf{x})$ be the number of repetitions of the element $x \in \mathcal{X}$ in the vector $\mathbf{x} \in \mathcal{X}^N$, and let

$$Q = \{Q(x) = N(x|\mathbf{x})/N, \quad x \in \mathcal{X}\}$$

is the distribution, called "the empirical distribution" of the sample $\mathbf{x}$ in statistics, in information theory called "the type" [12], [13] and in algebraic literature "the composition".

Denote the space of all empirical distributions for given N by $\mathcal{P}^{(N)}(\mathcal{X})$ and by $\mathcal{T}_Q^{(N)}$ the set of all vectors of the type $Q \in \mathcal{P}^{(N)}(\mathcal{X})$.

Consider for $k \in [K]$, $m \in [M]$, divergences

$$D(Q_k\|G_m) = \sum_{x\in\mathcal{X}} Q_k(x) \log \frac{Q_k(x)}{G_m(x)},$$

and entropies

$$H(Q_k) = -\sum_{x\in\mathcal{X}} Q_k(x) \log Q_k(x).$$

We shall use the following relations for the probability of the vector $\mathbf{x}$ when G_m is the distribution of the object:

$$G_m^{(N)}(\mathbf{x}) = \prod_{n=1}^{N} G_m(x_n) = \exp\{-N[D(Q||G_m) + H(Q)]\}.$$

For $m_k \in [M]$, $k \in [K]$, when the objects are independent and G_{m_k} is the distribution of the k-th object:

$$P_{m_1,m_2,\ldots,m_K}^{(N)}(\mathbf{x}_1, \mathbf{x}_2, \ldots, \mathbf{x}_K) = \exp\{-N[\sum_{k=1}^{K} D(Q_k||G_{m_k}) + H(Q_k)]\}. \quad (2)$$

The equalities follow from the independence of N observations of K objects and from the definitions of divergences and entropies. It should be noted that the equality (2) is valid even when its left part is equal to 0, in that case for one of $\mathbf{x}_k$ the distribution Q_k is not absolutely continuous relative to G_{m_k} and $D(Q_k||G_{m_k}) = \infty$.

Our arguments will be based on the following fact: the "maximal likelihood" test accepts as the solution values $m_1, m_2, \ldots, m_k$, which maximize the probability $P_{m_1,m_2,\ldots,m_K}^{(N)}(\mathbf{x}_1, \mathbf{x}_2, \ldots, \mathbf{x}_K)$, but from (2) we see that the same solution can be obtained by minimization of the sum $\sum_{k=1}^{K} [D(Q_k||G_{m_k}) + H(Q_k)]$, that is the comparison with the help of divergence of the types of observed vectors with their hypothetical distributions may be helpful.

In the paper we consider the following models.

1. K objects are different, they have different distributions among $M \geq K$ possibilities. For simplicity we restrict ourselves to the case $K = 2, M = 2$. It is the identification problem in formulations of the books [2] and [3].

2. K objects are independent, that is some of them may have the same distributions. We consider an example for $K, M = 2$. It is surprising, but this model has not been considered earlier in the literature.

3. We investigate one object, $K = 1$, and M possible probability distributions. The question is whether the m-th distribution occurred or not. This is the problem of identification of distributions in the spirit of the paper [4].

4. Ranking, or ordering problem [5]. We have one vector of observations $\mathbf{X} = (X_1, X_2, \ldots, X_N)$ and M hypothetical distributions. The receiver wants to know whether the index of the true distribution of the object is in $\{1, 2, \ldots, r\}$ or in $\{r+1, \ldots, M\}$.

5. r-identification of distribution [5]. Again $K = 1$. One wants to identify the observed object as a member either of the subset S of $[M]$, or of its complement, with r being the number of elements in S.

Section 2 of the paper presents necessary notions and results on hypothesis testing. The models of identification for independent objects are considered in section 3 and for different objects in section 4. Section 5 is devoted to the problem of identification of an object distribution and section 6 to the problems of r-identification and ranking. Some results are illustrated by numerical examples and graphs. Many directions of further research are indicated in the course of the text and in the section 7.

2   Background

The study of interdependence of exponential rates of decrease, as the sample size N goes to the infinity, of the error probabilities $\alpha_{1|2}^{(N)}$ of the "first kind" and $\alpha_{2|1}^{(N)}$ of the "second kind" was started by the works of Hoeffding [6], Csiszár and Longo [7], Tusnády [8], Longo and Sgarro [9], Birgé [10], and for multiple hypotheses by Haroutunian [11]. Similar problems for Markov dependence of experiments were investigated by Natarajan [14], Haroutunian [15], Gutman [16] and others. As it was remarked by Blahut in his book [17], it is unfortunately confusing that the errors are denoted type I and type II, while the hypotheses are subscripted 0 and 1. The word "type" is also used in another sense to refer to the type of a measurement or the type of a vector. For this reason we do not use the names "0" and "1" for hypotheses and the name "type" for errors. Note that in [17]–[19] an application of the methods of hypothesis testing to the proper problems of information theory is developed.

It will be very interesting to combine investigation of described models with the approach initiated by the paper of Ahlswede and Csiszár [20] and developed by many authors, particularly, for the exponentially decreasing error probabilities by Han and Kobayashi [21].

In [22] Berger formulated the problem of remote statistical inference. Zhang and Berger [23] studied a model of an estimation system with compressed information. Similar problems were examined by Ahlswede and Burnashev [24] and by Han and Amari [25]. In the paper of Ahlswede, Yang and Zhang [26] identification in channels via compressed data was considered. Fu and Shen [19] studied hypothesis testing for an arbitrarily varying source.

Our further considerations will be based on the results from [11] on multiple hypotheses testing, so now we expose briefly corresponding formulations and proofs. In our terms it is the case of one object ($K = 1$) and M possible distributions (hypotheses) $G_1, \ldots, G_M$. A test $\varphi(\mathbf{x})$ on the base of N-sample $\mathbf{x} = (x_1, \ldots, x_N)$ determines the distribution.

We study error probabilities $\alpha_{m|l}^{(N)}$ for $m, l \in [M]$. Here $\alpha_{m|l}^{(N)}$ is the probability that the distribution G_l was accepted instead of true distribution G_m. For $m = l$ the probability to reject G_m when it is true, is denoted by $\alpha_{m|m}^{(N)}$ thus:

$$\alpha_{m|m}^{(N)} = \sum_{l:l \neq m} \alpha_{m|l}^{(N)}.$$

This probability is called [27] the test's "error probability of the kind m". The matrix $\{\alpha_{m|l}^{(N)}\}$ is sometimes called the "power of the test" [27].

In this paper we suppose that the list of possible hypotheses is complete. Remark that, as it was noted by Rao [1], the case, when the objects may have also some distributions different from $G_1, \ldots, G_M$, is interesting too.

Let us analyze the reliability matrix

$$
\mathbf{E} = \begin{pmatrix}
E_{1|1} & \dots & E_{1|l} & \dots & E_{1|M} \\
\dots\dots\dots\dots\dots\dots\dots\dots \\
E_{m|1} & \dots & E_{m|l} & \dots & E_{m|M} \\
\dots\dots\dots\dots\dots\dots\dots\dots \\
E_{M|1} & \dots & E_{M|l} & \dots & E_{M|M}
\end{pmatrix}
$$

with components

$$
E_{m|l} = \varlimsup_{N\to\infty} -\frac{1}{N} \log \alpha_{m|l}^{(N)}, \quad m,l \in [M].
$$

According to this definition and the definition of $\alpha_{m|l}^{(N)}$ we can derive that

$$
E_{m|m} = \min_{l:m\neq l} E_{m|l}. \tag{3}
$$

Really,

$$
E_{m|m} = \varlimsup_{N\to\infty} -\frac{1}{N} \log \sum_{l:m\neq l} \alpha_{m|l}^{(N)} =
$$

$$
= \varlimsup_{N\to\infty} -\frac{1}{N} \log \max_{l:m\neq l} \alpha_{m|l}^{(N)} + \varlimsup_{N\to\infty} -\frac{1}{N} \log \left[\left(\sum_{l:m\neq l} \alpha_{m|l}^{(N)} \right) \Big/ \max_{l:m\neq l} \alpha_{m|l}^{(N)} \right] = \min_{l:m\neq l} E_{m|l}.
$$

The last equality is a consequence of the fact that for all m and N

$$
1 \leq \left(\sum_{l:m\neq l} \alpha_{m|l}^{(N)} \right) \Big/ \max_{l:m\neq l} \alpha_{m|l}^{(N)} \leq M - 1.
$$

In the case $M = 2$, the reliability matrix is

$$
\mathbf{E} = \begin{pmatrix} E_{1|1} & E_{1|2} \\ E_{2|1} & E_{2|2} \end{pmatrix} \tag{4}
$$

and it follows from (3) that there are only two different values of elements, namely

$$
E_{1|1} = E_{1|2} \text{ and } E_{2|1} = E_{2|2}, \tag{5}
$$

so in this case the problem is to find the maximal possible value of one of them, given the value of the other.

In the case of M hypotheses for given positive and finite $E_{1|1}, E_{2|2}, \dots, E_{M-1,M-1}$ let us consider the regions of distributions

$$
\mathcal{R}_l = \{ Q : D(Q||G_l) \leq E_{l|l} \}, \quad l \in [M-1], \tag{6}
$$

$$
\mathcal{R}_M = \{ Q : D(Q||G_l) > E_{l|l}, \quad l \in [M-1] \} = \mathcal{P}(\mathcal{X}) - \bigcup_{l=1}^{M-1} \mathcal{R}_l, \tag{7}
$$

$$
\mathcal{R}_l^{(N)} = \mathcal{R}_l \bigcap \mathcal{P}^{(N)}, \quad l \in [M]. \tag{8}
$$

Let

$$E_{l|l}^* = E_{l|l}^*(E_{l|l}) = E_{l|l}, \quad l \in [M-1],$$ (9)

$$E_{m|l}^* = E_{m|l}^*(E_{l|l}) = \inf_{Q \in \mathcal{R}_l} D(Q||G_m), \quad m \in [M], \ m \neq l, \ l \in [M-1],$$ (10)

$$E_{m|M}^* = E_{m|M}^*(E_{1|1}, \ldots, E_{M-1,M-1}) = \inf_{Q \in \mathcal{R}_M} D(Q||G_m), \quad m \in [M-1],$$ (11)

$$E_{M|M}^* = E_{M|M}^*(E_{1|1}, \ldots, E_{M-1,M-1}) = \min_{l \in [M-1]} E_{M|l}^*.$$ (12)

If some distribution G_m is not absolutely continuous relative to G_l the reliability $E_{m|l}^*$ will be equal to the infinity, this means that corresponding $\alpha_{m|l}^{(N)} = 0$ for some large N.

The principal result of [11] is:

Theorem 1. If all the distributions G_m are different and all elements of the matrix $\{D(G_l||G_m)\}$, $l, m \in [M]$, are positive, but finite, two statements hold:

a) when the positive numbers $E_{1|1}, E_{2|2}, \ldots, E_{M-1,M-1}$ satisfy conditions

$$E_{1|1} < \min_{l \in [2,M]} D(G_l||G_1),$$

$$\cdots\cdots\cdots\cdots\cdots\cdots\cdots\cdots\cdots\cdots\cdots\cdots$$

$$\cdots\cdots\cdots\cdots\cdots\cdots\cdots\cdots\cdots\cdots\cdots$$ (13)

$$E_{m|m} < \min[\min_{l \in [m-1]} E_{m|l}^*(E_{l|l}), \min_{l \in [m+1,M]} D(G_l||G_m)], \quad m \in [2, M-1],$$

then there exists a LAO sequence of tests, the reliability matrix of which $E^* = \{E_{m|l}^*\}$ is defined in (9),(10),(11),(12) and all elements of it are positive;

b) even if one of conditions (13) is violated, then the reliability matrix of any such test has at least one element equal to zero (that is the corresponding error probability does not tend to zero exponentially).

The essence of the proof of Theorem 1 consists in construction of the following optimal tests sequence. Let the decision l will be taken when $\mathbf{x}$ gets into the set

$$\mathcal{B}_l^{(N)} = \bigcup_{Q \in \mathcal{R}_l^{(N)}} \mathcal{T}_Q^{(N)}, \quad l \in [M], \ N = 1, 2, \ldots.$$ (14)

The non-coincidence of the distributions G_m and the conditions (13) guarantee that the sets from (14) are not empty, they meet conditions

$$\mathcal{B}_l^{(N)} \bigcap \mathcal{B}_m^{(N)} = \varnothing, \quad l \neq m,$$

and

$$\bigcup_{l=1}^{M} \mathcal{B}_l^{(N)} = \mathcal{X}^N,$$

and so they define a sequence of tests, which proves to be LAO.

For the simplest particular case $M = 2$ elements of the reliability matrix (4) satisfy equalities (5) and for given $E_{1|1}$ from (5) and (7) we obtain the value of $E_{2|1}^* = E_{2|2}^*$:

$$E_{2|1}^*(E_{1|1}) = \inf_{Q:D(Q||G_1)\leq E_{1|1}} D(Q||G_2). \tag{15}$$

Here, according to (13), we can take $E_{1|1}$ from $(0, D(G_2||G_1))$ and $E_{2|1}^*(E_{1|1})$ will range between $D(G_1||G_2)$ and 0.

3 Identification Problem for Model with Independent Objects

We begin with study of the second model. To illustrate possibly arising developments and essential features we consider a particular case $K = 2, M = 2$. It is clear that the case with $M = 1$ is trivial. The reliability matrix is (see (1))

$$\mathbf{E} = \begin{pmatrix} E_{1,1|1,1} & E_{1,1|1,2} & E_{1,1|2,1} & E_{1,1|2,2} \\ E_{1,2|1,1} & E_{1,2|1,2} & E_{1,2|2,1} & E_{1,2|2,2} \\ E_{2,1|1,1} & E_{2,1|1,2} & E_{2,1|2,1} & E_{2,1|2,2} \\ E_{2,2|1,1} & E_{2,2|1,2} & E_{2,2|2,1} & E_{2,2|2,2} \end{pmatrix}.$$

Let us denote by $\alpha_{m_1|l_1}^{(1)}$, $\alpha_{m_2|l_2}^{(2)}$ and $E_{m_1|l_1}^{(1)}$, $E_{m_2|l_2}^{(2)}$ the error probabilities and the reliabilities as in (4) for, respectively, the first and the second objects.

Lemma. *If* $0 < E_{1|1}^{(i)} < D(G_2||G_1)$, $i = 1, 2$, *then the following equalities hold true:*

$$E_{m_1,m_2|l_1,l_2} = E_{m_1|l_1}^{(1)} + E_{m_2|l_2}^{(2)}, \ \text{if}\ m_1 \neq l_1, \ m_2 \neq l_2, \tag{16}$$

$$E_{m_1,m_2|l_1,l_2} = E_{m_i|l_i}^{(i)}, \ \text{if}\ m_{3-i} = l_{3-i}, \ m_i \neq l_i, \ i = 1,2, \tag{17}$$

Proof. From the independence of the objects it follows that

$$\alpha_{m_1,m_2|l_1,l_2}^{(N)} = \alpha_{m_1|l_1}^{(N,1)}\alpha_{m_2|l_2}^{(N,2)}, \ \text{if}\ m_1 \neq l_1, \ m_2 \neq l_2, \tag{18}$$

$$\alpha_{m_1,m_2|l_1,l_2}^{(N)} = \alpha_{m_i|l_i}^{(N,i)}(1 - \alpha_{m_{3-i}|l_{3-i}}^{(N,3-i)}), \ \text{if}\ m_{3-i} = l_{3-i}, \ m_i \neq l_i, \ i = 1,2, \tag{19}$$

According to (1), from (18) we obtain (16), from (19) and the conditions of positiveness of $E_{1|1}^{(i)}$ and $E_{2|2}^{(i)}$, $i = 1, 2$, (17) follows.

Theorem 2. *If the distributions* G_1 *and* G_2 *are different, the strictly positive elements* $E_{1,1|1,2}$, $E_{1,1|2,1}$ *of the reliability matrix* $\mathbf{E}$ *are given and bounded above:*

$$E_{1,1|1,2} < D(G_2||G_1), \ \text{and}\ E_{1,1|2,1} < D(G_2||G_1), \tag{20}$$

then the other elements of the matrix **E** *are defined as follows:*

$$E_{2,1|2,2} = E_{1,1|1,2}, \qquad E_{1,2|2,2} = E_{1,1|2,1},$$

$$E_{1,2|1,1} = E_{2,2|2,1} = \inf_{Q:\ D(Q||G_1) \le E_{1,1|1,2}} D(Q||G_2),$$

$$E_{2,1|1,1} = E_{2,2|1,2} = \inf_{Q:\ D(Q||G_1) \le E_{1,1|2,1}} D(Q||G_2), \qquad (21)$$

$$E_{2,2|1,1} = E_{1,2|1,1} + E_{2,1|1,1}, \qquad E_{2,1|1,2} = E_{2,1|1,1} + E_{1,2|2,2},$$

$$E_{1,2|2,1} = E_{1,2|1,1} + E_{1,2|2,2}, \qquad E_{1,1|2,2} = E_{1,1|1,2} + E_{1,1|2,1},$$

$$E_{m_1,m_2|m_1,m_2} = \min_{(l_1,l_2) \ne (m_1,m_2)} E_{m_1,m_2|l_1,l_2}, \quad m_1, m_2 = 1, 2.$$

If one of the inequalities (20) *is violated, then at least one element of the matrix* **E** *is equal to 0.*

Proof. The last equalities in (21) follow (as (3)) from the definition of

$$\alpha_{m_1,m_2|m_1,m_2}^{(N)} = \sum_{(l_1,l_2) \ne (m_1,m_2)} \alpha_{m_1,m_2|l_1,l_2}^{(N)}, \quad m_1, m_2 = 1, 2.$$

Let us consider the reliability matrices of each of the objects X_1 and X_2

$$\mathbf{E}^{(1)} = \begin{pmatrix} E_{1|1}^{(1)} & E_{1|2}^{(1)} \\ E_{2|1}^{(1)} & E_{2|2}^{(1)} \end{pmatrix} \qquad \text{and} \qquad \mathbf{E}^{(2)} = \begin{pmatrix} E_{1|1}^{(2)} & E_{1|2}^{(2)} \\ E_{2|1}^{(2)} & E_{2|2}^{(2)} \end{pmatrix}.$$

From (5) we know that $E_{1|1}^{(i)} = E_{1|2}^{(i)}$ and $E_{2|1}^{(i)} = E_{2|2}^{(i)}$, $i = 1, 2$. From (20) it follows that $0 < E_{1|1}^{(1)} < D(G_2||G_1)$, $0 < E_{1|1}^{(2)} < D(G_2||G_1)$. Really, if $0 < E_{1,1|1,2} < D(G_2||G_1)$, but $E_{1|1}^{(2)} \ge D(G_2||G_1)$, then from (19) and (1) we arrive to

$$\varlimsup_{N \to \infty} -\frac{1}{N} \log(1 - \alpha_{1|2}^{(N,2)}) < 0,$$

therefore index N_0 exists, such that for subsequence of $N > N_0$ we will have $1 - \alpha_{1|2}^{(N,2)} > 1$. But this is impossible because $\alpha_{1|2}^{(N,2)}$ is the probability and must be positive.

Using Lemma we can deduce that the reliability matrix **E** can be obtained from matrices $\mathbf{E}^{(1)}$ and $\mathbf{E}^{(2)}$ as follows:

$$\mathbf{E} = \begin{pmatrix} \min(E_{1|2}^{(1)}, E_{1|2}^{(2)}) & E_{1|2}^{(2)} & E_{1|2}^{(1)} & E_{1|2}^{(1)} + E_{1|2}^{(2)} \\ E_{2|1}^{(2)} & \min(E_{1|2}^{(1)}, E_{2|1}^{(2)}) & E_{1|2}^{(1)} + E_{2|1}^{(2)} & E_{1|2}^{(1)} \\ E_{2|1}^{(1)} & E_{2|1}^{(1)} + E_{1|2}^{(2)} & \min(E_{2|1}^{(1)}, E_{1|2}^{(2)}) & E_{1|2}^{(2)} \\ E_{2|1}^{(1)} + E_{2|1}^{(2)} & E_{2|1}^{(1)} & E_{2|1}^{(2)} & \min(E_{2|1}^{(1)}, E_{2|1}^{(2)}) \end{pmatrix},$$

in other words, providing, that conditions (20) are fulfilled, we find that

$$E_{1,1|1,2} = E_{1|2}^{(2)} = E_{1|1}^{(2)} \quad \text{and} \quad E_{1,1|2,1} = E_{1|2}^{(1)} = E_{1|1}^{(1)},$$

$$E_{2,1|2,2} = E_{1,1|1,2} = E_{1|2}^{(2)}, \quad E_{1,2|2,2} = E_{1,1|2,1} = E_{1|2}^{(1)},$$

$$E_{1,2|1,1} = E_{2,2|2,1} = E_{2|1}^{(2)}, \quad E_{2,1|1,1} = E_{2,2|1,2} = E_{2|1}^{(1)},$$

$$E_{2,2|1,1} = E_{2|1}^{(1)} + E_{2|1}^{(2)}, \quad E_{2,1|1,2} = E_{2|1}^{(1)} + E_{1|2}^{(2)}, \tag{22}$$

$$E_{1,2|2,1} = E_{1|2}^{(1)} + E_{2|1}^{(2)}, \quad E_{1,1|2,2} = E_{1|2}^{(1)} + E_{1|2}^{(2)},$$

$$E_{m_1,m_2|m_1,m_2} = \min\{E_{m_1|m_1}^{(1)}, E_{m_2|m_2}^{(2)}\}, \quad m_1, m_2 = 1, 2,$$

From Theorem 1 we know that if $E_{1|1}^{(i)} \in (0, D(G_2||G_1))$, $i = 1, 2$, then the tests of both objects are LAO and the elements $E_{2|1}^{(i)}$, $i = 1, 2$, can be calculated (see (15)) by

$$E_{2|1}^{(i)} = \inf_{Q:D(Q||G_1) \le E_{1|1}^{(i)}} D(Q||G_2), \quad i = 1, 2, \tag{23}$$

and if $E_{1|1}^{(i)} \ge D(G_2||G_1)$, then $E_{2|1}^{(i)} = 0$.

According to (22) and (23), we obtain, that when (20) takes place, the elements of the matrix $\mathbf{E}$ are determined by relations (21). When one of the inequalities (20) is violated, then from (23) and the first and the third lines of (22) we see, that some elements in the matrix $\mathbf{E}$ must be equal to 0 (namely, either $E_{1,2|1,1}$, or $E_{2,1|1,1}$ and others).

Now let us show that the compound test for two objects is LAO, that is it is optimal. Suppose that for given $E_{1,1|1,2}$ and $E_{1,1|2,1}$ there exists a test with matrix $\mathbf{E}'$, such that it has at least one element exceeding the respective element of the matrix $\mathbf{E}$. Comparing elements of matrices $\mathbf{E}$ and $\mathbf{E}'$ different from $E_{1,1|1,2}$ and $E_{1,1|2,1}$, from (22) we obtain that either $E_{1,2|1,1} < E'_{1,2|1,1}$, or $E_{2,1|1,1} < E'_{2,1|1,1}$, i.e. either $E_{2|1}^{(2)} < E_{2|1}^{(2)'}$, or $E_{2|1}^{(1)} < E_{2|1}^{(1)'}$. It is contradiction to the fact, that LAO tests have been used for the objects X_1 and X_2.

When it is demanded to take the same values for the reliabilities of the first and the second objects $E_{1|2}^{(1)} = E_{1|2}^{(2)} = a_1$ and, consequently, $E_{2|1}^{(1)} = E_{2|1}^{(2)} = a_2$, then the matrix $\mathbf{E}$ will take the following form

$$\mathbf{E} = \begin{pmatrix} a_1 & a_1 & a_1 & 2a_1 \\ a_2 & \min(a_1, a_2) & a_1 + a_2 & a_1 \\ a_2 & a_1 + a_2 & \min(a_1, a_2) & a_1 \\ 2a_2 & a_2 & a_2 & a_2 \end{pmatrix}.$$

4   Identification Problem for Models with Different Objects

The K objects are not independent, they have different distributions, and so the number M of the distributions is not less than K. This is the model studied in [2]. For brevity we consider the case $K = 2, M = 2$. The matrix of reliabilities will be the following:

$$\mathbf{E} = \begin{pmatrix} E_{1,2|1,2} & E_{1,2|2,1} \\ E_{2,1|1,2} & E_{2,1|2,1} \end{pmatrix}. \tag{24}$$

Since the objects are strictly dependent this matrix coincides with the reliability matrix of the first object (see (4))

$$\mathbf{E}^{(1)} = \begin{pmatrix} E_{1|1}^{(1)} & E_{1|2}^{(1)} \\ E_{2|1}^{(1)} & E_{2|2}^{(1)} \end{pmatrix},$$

because the distribution of the second object is uniquely defined by the distribution of the first one.

We can conclude that among 4 elements of the reliability matrix of two dependent objects only 2 elements are distinct, the second of which is defined by given $E_{1|1}^{(1)} = E_{1,2|1,2}$.

From symmetry it follows that the reliability matrix of the second object also may determine the matrix (24).

5   Identification of the Probability Distribution of an Object

Let we have one object, $K = 1$, and there are known $M \geq 2$ possible distributions. The question is whether r-th distribution occured, or not. There are two error probabilities for each $r \in [M]$ the probability $\alpha_{m=r|l \neq r}^{(N)}$ to accept l different from r, when r is in reality, and the probability $\alpha_{m \neq r|l=r}^{(N)}$ that r is accepted, when it is not correct.

The probability $\alpha_{m=r|l \neq r}^{(N)}$ is already known, it coincides with the probability $\alpha_{r|r}^{(N)}$ which is equal to $\sum_{l:l \neq r} \alpha_{r|l}^{(N)}$. The corresponding reliability $E_{m=r|l \neq r}$ is equal to $E_{r|r}$ which satisfies the equality (3).

We have to determine the dependence of $E_{m \neq r|l=r}$ upon given $E_{m=r|l \neq r} = E_{r|r}$, which can be assigned values satisfying conditions (13), this time we will have the conditions:

$$0 < E_{r|r} < \min_{l:l \neq r} D(G_l \| G_r), \quad r \in [M].$$

We need the probabilities of different hypotheses. Let us suppose that the hypotheses $G_1, \ldots, G_M$ have, say, probabilities $\Pr(r)$, $r \in [M]$. The only supposition we shall use is that $\Pr(r) > 0$, $r \in [M]$. We will see, that the result formulated in the following theorem does not depend on values of $\Pr(r)$, $r \in [M]$, if they all are strictly positive.

Now we can make the following reasoning for each $r \in [M]$:

$$\alpha_{m \neq r | l = r}^{(N)} = \frac{\Pr^{(N)}(m \neq r, l = r)}{\Pr(m \neq r)} = \frac{1}{\sum\limits_{m:m \neq r} \Pr(m)} \sum_{m:m \neq r} \Pr^{(N)}(m, r).$$

From here we see that for $r \in [M]$

$$E_{m \neq r | l = r} = \varlimsup_{N \to \infty} \left(-\frac{1}{N} \log \alpha_{m \neq r | l = r}^{(N)} \right) =$$

$$= \varlimsup_{N \to \infty} \frac{1}{N} \left(\log \sum_{m:m \neq r} \Pr(m) - \log \sum_{m:m \neq r} \alpha_{m | r}^{(N)} \Pr(m) \right) = \min_{m:m \neq r} E_{m | r}^*. \quad (25)$$

Using (25) by analogy with the formula (15) we conclude (with $\mathcal{R}_r$ defined as in (6) for each r including $r = M$ by the values of $E_{r|r}$ from $(0, \min\limits_{l:l \neq r} D(G_l \| G_r))$) that

$$E_{m \neq r | l = r}(E_{r|r}) = \min_{m:m \neq r} \inf_{Q \in \mathcal{R}_r} D(Q \| G_m) =$$

$$= \min_{m:m \neq r} \inf_{Q:D(Q \| G_r) \leq E_{r|r}} D(Q \| G_m), \quad r \in [M]. \quad (26)$$

We can summarize this result in

Theorem 3. For the model with different distributions, for the given sample $\mathbf{x}$ we define its type Q, and when $Q \in \mathcal{R}_r^{(N)}$ we accept the hypothesis r. Under condition that the probabilities of all M hypotheses are positive the reliability of such test $E_{m \neq r | l = r}$ for given $E_{m = r | l \neq r} = E_{r|r}$ is defined by (26).

For presentation of examples let us consider the set $\mathcal{X} = \{0, 1\}$ with only 2 elements. Let 5 probability distributions are given on $\mathcal{X}$:

$$G_1 = \{0.1, \ 0.9\}$$

$$G_2 = \{0.65, \ 0.35\}$$

$$G_3 = \{0.45, \ 0.55\}$$

$$G_4 = \{0.85, \ 0.15\}$$

$$G_5 = \{0.23, \ 0.77\}$$

On Fig. 1 the results of calculations of $E_{m \neq r | l = r}$ as function of $E_{m = r | l \neq r}$ are presented.

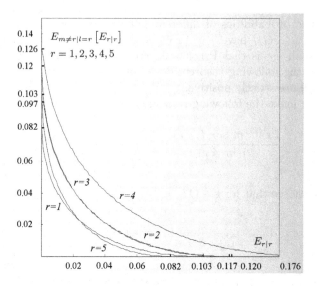

Fig. 1.

The elements of the matrix of divergences of all pairs of distributions are used for calculation of conditions (13) for this example.

$$\{D(G_m\|G_l)\}_{m\in[5]}^{l\in[5]} = \begin{pmatrix} 0 & 0.956 & 0.422 & 2.018 & 0.082 \\ 1.278 & 0 & 0.117 & 0.176 & 0.576 \\ 0.586 & 0.120 & 0 & 0.618 & 0.169 \\ 2.237 & 0.146 & 0.499 & 0 & 1.249 \\ 0.103 & 0.531 & 0.151 & 1.383 & 0 \end{pmatrix}.$$

In figures 2 and 3 the results of calculations of the same dependence are presented for 4 distributions taken from previous 5.

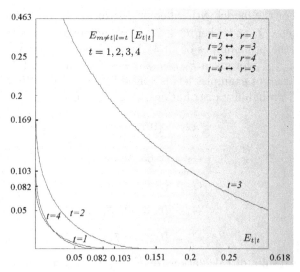

Fig. 2.

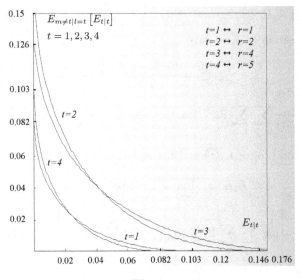

Fig. 3.

6 r-Identification and Ranking Problems

The model was introduced in [5] and named K-identification. Since in this paper the letter K is already used we speak of r-identification. Given N-sample $\mathbf{x}$ of measurements of the object the problem is to answer to the question: is the distribution of the object in the part $\mathcal{S}$ of M possible distributions or in its complement, here r is the number of elements of the set $\mathcal{S}$.

Again we can make decision on the base of the type Q of the sample $\mathbf{x}$ and suppose that before experiments all hypotheses have some positive probabilities

$$\Pr(1), \ldots, \Pr(M). \tag{27}$$

Using $(6) - (8)$ with some $E_{1,1}, \ldots, E_{M-1,M-1}$ meeting the conditions (13) when $Q \in \bigcup_{l \in \mathcal{S}} \mathcal{R}_l^{(N)}$ decision "l is in $\mathcal{S}$" follows.

The model of ranking is the particular case of the model of r-identification with $\mathcal{S} = \{1, 2, \ldots, r\}$. But conversely the r-identification problem without loss of generality may be considered as the ranking problem, to this end we can renumber the hypotheses placing the hypotheses of $\mathcal{S}$ in the r first places. Because these two models are mathematically equivalent we shall speak below only of the ranking model.

It is enough to consider the cases $r \leq \lceil M/2 \rceil$, because in the cases of larger r we can replace $\mathcal{S}$ with its complement. Remark that the case $r = 1$ was considered in section 5.

We study two error probabilities of a test: the probability $\alpha_{m \leq r|l>r}^{(N)}$ to make incorrect decision when m is not greater than r and the probability $\alpha_{m>r|l \leq r}^{(N)}$ to make error when m is greater than r. The corresponding reliabilities are

$$E_1(r) = E_{m \leq r | l > r} \text{ and } E_2(r) = E_{m > r | l \leq r}, \quad 1 \leq r \leq \lceil M/2 \rceil. \quad (28)$$

With supposition (27) we have

$$\alpha^{(N)}_{m \leq r | l > r} = \frac{\Pr^{(N)}(m \leq r, l > r)}{\Pr(m \leq r)} =$$

$$= \frac{1}{\sum_{m \leq r} \Pr(m)} \sum_{m \leq r} \sum_{l > r} \Pr^{(N)}(m, l) = \frac{1}{\sum_{m \leq r} \Pr(m)} \sum_{m \leq r} \sum_{l > r} \alpha^{(N)}_{m | l} \Pr(m). \quad (29)$$

The definition (28) of $E_1(r)$ and the equality (29) give

$$E_1(r) = \varlimsup_{N \to \infty} -\frac{1}{N} \log \alpha^{(N)}_{m \leq r | l > r} =$$

$$= \varlimsup_{N \to \infty} -\frac{1}{N} \left[\log \sum_{m \leq r} \sum_{l > r} \Pr(m) \alpha^{(N)}_{m | l} - \log \sum_{m \leq r} \Pr(m) \right] = \min_{m \leq r, l > r} E_{m | l}. \quad (30)$$

Analogously, at the same time

$$E_2(r) = \varlimsup_{N \to \infty} -\frac{1}{N} \log \alpha^{(N)}_{m > r | l \leq r} =$$

$$= \varlimsup_{N \to \infty} -\frac{1}{N} \left[\log \sum_{m > r} \sum_{l \leq r} \alpha^{(N)}_{m | l} - \log \sum_{m > r} \Pr(m) \right] = \min_{m > r, l \leq r} E_{m | l}. \quad (31)$$

For any test the value of $E_1(r)$ must satisfy the condition (compare (3) and (30))

$$E_1(r) \geq \min_{m : m \leq r} E_{m | m}. \quad (32)$$

Thus for any test meeting all inequalities from (13) for $m \leq r$ and inequality (32) the reliability $E_2(r)$ may be calculated with the equality (31). For given value of $E_1(r)$ the best $E_2(r)$ will be obtained if we use liberty in selection of the biggest values for reliabilities $E_{m | m}$, $r < m \leq M - 1$, satisfying for those m-s conditions (13). These reasonings may be illuminated by Fig.4.

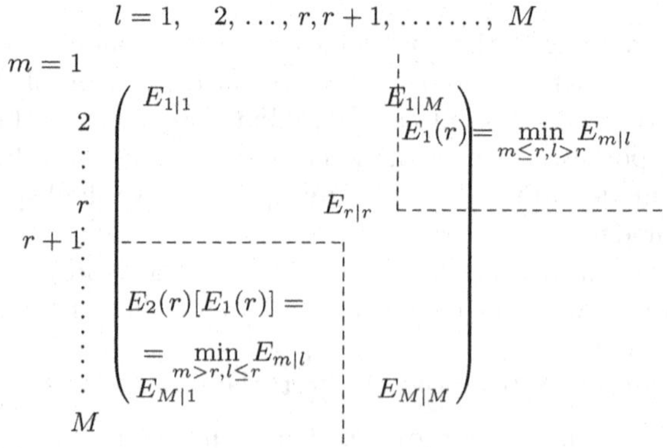

Fig. 4. Calculation of $E_2(r) [E_1(r)]$

and resumed as follows:

Theorem 4. When the probabilities of the hypotheses are positive, for given $E_1(r)$ for $m \leq r$ not exceeding the expressions on the right in (13), $E_2(r)$ may be calculated in the following way:

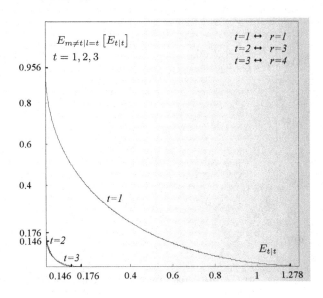

Fig. 5.

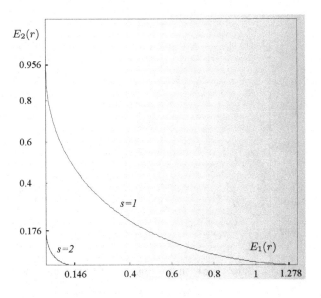

Fig. 6.

$$E_2(r)\,[E_1(r)] = \max_{\{E_{m|l},\ m,l\in[M]\}:\ \min_{m\le r,\ l>r} E^*_{m|l}=E_1(r)} \left[\min_{m>r,\ l\le r} E^*_{m|l}\right] \qquad (33)$$

with $E^*_{m|l}$ defined in (9),(10),(11),(12).

Remark. One can see from (33) that for $r = 1$ we arrive to (26) for $r = 1$.

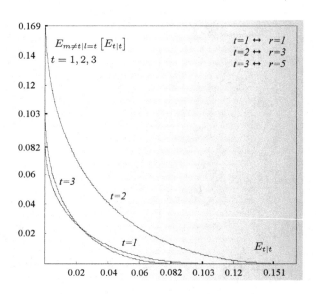

Fig. 7.

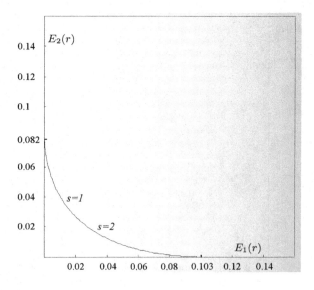

Fig. 8.

In figures 5 and 7 for 2 subsets by 3 distributions taken from 5 defined for Fig.1 the results of calculation of the dependence (26) and in figures 6 and 8 the corresponding results of the formula (33) are presented.

7 Conclusion and Extensions of Problems

The paper is a contribution to influence of the information theory methods on statistical theory. We have shown by simple examples what questions arise in different models of statistical identification.

Problems and results of the paper may be extended in several directions some of which have been already noted above.

It is necessary to examine models in which measurements are described by more general classes of random variables and processes [14]–[16], [25].

One of the directions is connected with the use of compressed data of measurements [22]–[26].

One may see perspectives in application of identification approach and methods to the authentication theory [32] and steganography [33].

Acknowledgments

The study was started during the stay of the second author in January, February of 2003 in ZiF, Bielefeld University. This author is grateful to Prof. R. Ahlswede, to Dr. L. Bäumer and to the staff of ZiF for hospitality and stimulating atmosphere.

The authors are thankful to Prof. M. Malyutov for useful discussions. The help of P. Hakobyan in formulation of Lemma and Theorem 2 in sections 3 and of S. Tonoyan and E. Aloyan in calculation of the examples and graphs of sections 5 and 6 is acknowledged with gratitude.

References

1. R.C. Rao, Linear Statistical Inference and its Applications, Wiley, New York, 1965.
2. R.E. Bechhofer, J. Kiefer, and M. Sobel, Sequential Identification and Ranking Procedures, The University of Chicago Press, Chicago, 1968.
3. R. Ahlswede and I. Wegener, Search Problems, Wiley, New York, 1987.
4. R. Ahlswede and G. Dueck, Identification via channels, IEEE Trans. Inform. Theory, Vol. 35, No. 1, 15-29, 1989.
5. R. Ahlswede, General theory of information transfer, Preprint 97–118, SFB 343 "Diskrete Strukturen in der Mathematik", Universität Bielefeld, 1997; General theory of information transfer:updated, General Theory of Information Transfer and Combinatorics, a Special Issue of Discrete Applied Mathematics, to appear.
6. W. Hoeffding, Asymptotically optimal tests for multinomial distributions, Annals. of Math. Statist., Vol. 36, 369-401, 1965.

7. I. Csiszár and G. Longo, On the error exponent for source coding and for testing simple statistical hypotheses, Studia Sc. Math. Hungarica, Vol. 6, 181-191, 1971.

8. G. Tusnady, On asymptotically optimal tests, Annals of Statist., Vol. 5, No. 2, 385-393, 1977.

9. G. Longo and A. Sgarro, The error exponent for the testing of simple statistical hypotheses, a combinatorial approach, J. of Combin., Inform. Sys. Sc., Vol. 5, No 1, 58-67, 1980.

10. L. Birgé, Vitesse maximales de décroissance des erreurs et tests optimaux associés, Z. Wahrsch. verw Gebiete, Vol. 55, 261–273, 1981.

11. E.A. Haroutunian, Logarithmically asymptotically optimal testing of multiple statistical hypotheses, Problems of Control and Inform. Theory. Vol. 19, No. 5-6, 413-421, 1990.

12. I. Csiszár I. and J. Körner, Information Theory: Coding Theorems for Discrete Memoryless Systems, Academic Press, New York, 1981.

13. I. Csiszár, Method of types, IEEE Trans. Inform. Theory, Vol. 44, No. 6, 2505-2523, 1998.

14. S. Natarajan, Large deviations, hypotheses testing, and source coding for finite Markov chains, IEEE Trans. Inform. Theory, Vol. 31, No. 3, 360-365, 1985.

15. E.A. Haroutunian, On asymptotically optimal testing of hypotheses concerning Markov chain (in Russian), Izvestia Acad. Nauk Armenian SSR. Seria Mathem. Vol. 22, no 1, 76-80, 1988.

16. M. Gutman, Asymptotically optimal classification for multiple test with empirically observed statistics, IEEE Trans. Inform. Theory, vol 35, No 2, 401-408, 1989.

17. R.E. Blahut, Principles and Practice of Information Theory, Addison-Weslay, Massachusetts, 1987.

18. R.E. Blahut, Hypotheses testing and information theory, IEEE Trans. Inform Theory, Vol. 20, No. 4, 405-417, 1974.

19. F.W. Fu and S.Y. Shen, Hypothesis testing for arbitrarily varying source with exponents type constraint, IEEE Trans. Inform. Theory, Vol. 44, No. 2, 892-895, 1998.

20. R. Ahlswede and I. Csiszár I., Hypotheses testing with communication constraints, IEEE Trans. Inform. Theory Vol. 32, No. 4, 533-542, 1986.

21. T.S. Han and K. Kobayashi, Exponential-type error probabilities for multiterminal hypothesis testing, IEEE Trans. Inform. Theory, Vol. 35, No. 1, 2-13, 1989.

22. T. Berger, Decentralized estimation and decision theory, Presented at IEEE Seven Springs Workshop on Information Theory, Mt. Kisco, NY, September 1979.

23. Z. Zhang and T. Berger, Estimation via compressed information, IEEE Trans. Inform. Theory, Vol. 34, No. 2, 198-211, 1988.

24. R. Ahlswede and M. Burnashev, On minimax estimation in the presence of side information about remote data, Annals of Statist., Vol. 18, No. 1, 141-171, 1990.

25. T.S. Han and S. Amari, Statistical inference under multiterminal data compression, IEEE Trans Inform Theory, Vol. 44, No. 6, 2300-2324, 1998.

26. R. Ahlswede, E. Yang, and Z. Zhang, Identification via compressed data, IEEE Trans. Inform. Theory, Vol. 43, No. 1, 48-70, 1997.

27. A.A. Borovkov, Mathematical Statistics (in Russian), Nauka, Novosibirsk, 1997.

28. S. Ihara, Information Theory for Continuous Systems, World Scientific, Singapore, 1993.

29. P.N. Chen, General formulas for the Neyman-Pearson type-II error exponent subject to fixed and exponential type-I error bounds, IEEE Trans. Inform. Theory, Vol. 42, No. 1, 316-323, 1996.

30. T.S. Han, Information-Spectrum Methods in Information Theory, Springer, Berlin, 2003.
31. T.S. Han, Hypothesis testing with the general source, IEEE Trans. Inform. Theory, Vol. 46, No. 7, 2415-2427, 2000.
32. U.M. Maurer, Authentication theory and hypothesis testing, IEEE Trans. Inform. Theory, Vol. 46, No. 4, 1350-1356, 2000.
33. C. Cachin, An information-theoretic model for steganography, Proc. 2nd Workshop on Information Hiding (David Ausmith, ed.), Lecture Notes in computer Science, Springer– Verlag, 1998.

Correlation Inequalities in Function Spaces

R. Ahlswede and V. Blinovsky

Abstract. We give a condition for a Borel measure on $R^{[0,1]}$ which is sufficient for the validity of an AD-type correlation inequality in the function space[1]

In [1] was proved that if $\varphi_1, \varphi_2, \varphi_3, \varphi_4$ are bounded real non negative measurable functions on the space with measure $(R^n, \mathcal{B}, \mu)$ which satisfy for all $\bar{x}, \bar{y} \in R^n$ the following inequality

$$\varphi_1(\bar{x})\varphi_2(\bar{y}) \leq \varphi_3(\bar{x} \bigvee \bar{y})\varphi_4(\bar{x} \bigwedge \bar{y}) \ a.s., \tag{1}$$

then

$$\int \varphi_1(\bar{x})\mu(d\bar{x}) \int \varphi_2(\bar{x})\mu(d\bar{x}) \leq \int \varphi_3(\bar{x})\mu(d\bar{x}) \int \varphi_4(\bar{x})\mu(d\bar{x}), \tag{2}$$

where $\mu(d\bar{x})$ is the product σ-finite measure on $\mathcal{B}$, $(\bar{x} \vee \bar{y})_i = x_i \vee y_i$, $(\bar{x} \wedge \bar{y})_i = x_i \wedge y_i$. That proof was simplified in [2] via induction on dimension n suggested in [3], [7] ,[8]. The question we consider here is how the problem can be viewed in the case of not arbitrary measure ν on R^T, when possibly $T = [0,1]$? The next theorem answers this question.

Let for arbitrary real functions $x(t), y(t)$, $t \in T = [0,1]$, $(x \bigvee y)(t) = x(t) \bigvee y(t)$ and $(x \bigwedge y)(t) = x(t) \bigwedge y(t)$. Let also $\nu_i(d\bar{x})$, $i = 1, 2, 3, 4$ be measures on the Borel sets $\mathcal{B}(\mathcal{C})$ of the linear space $\mathcal{C}$ of continuous functions from R^T with the norm $||\cdot||_\infty$, which are finite on the compact subsets of $\mathcal{C}$. Let also φ_i, $i = 1, 2, 3, 4$ be four uniformly bounded nonnegative Borel real functions from R^T.

Theorem 1. *If the following conditions are valid*

$$\varphi_1(f)\varphi_2(g) \leq \varphi_3(f \bigvee g)\varphi_4(f \bigwedge g) \tag{3}$$

$$\nu_1(A)\nu_2(B) \leq \nu_3(A \bigvee B)\nu_4(A \bigwedge B), A, B \in \mathcal{B}(\mathcal{C}), \tag{4}$$

then

$$\int \varphi_1(\bar{x})\nu_1(d\bar{x}) \int \varphi_2(\bar{x})\nu_2(d\bar{x}) \leq \int \varphi_3(\bar{x})\nu_3(d\bar{x}) \int \varphi_4(\bar{x})\nu_4(d\bar{x}). \tag{5}$$

Here $A \bigvee B = \{a \bigvee b : a \in A, b \in B\}$, $A \bigwedge B = \{a \bigwedge b : a \in A, b \in B\}$.

Condition (4) is also necessary for (5). Indeed indicator functions $I_A, I_B, I_{A \bigvee B}$, $I_{A \bigwedge B}$ satisfy (3) and substitution of them in (5) gives (4).

[1] This work is partially supported by RFFI grants No 03-01-00592 and 03-01-00098 and INTAS grant No 00-738.

R. Ahlswede et al. (Eds.): Information Transfer and Combinatorics, LNCS 4123, pp. 572–577, 2006.

We call a measure ν which satisfies the relation

$$\nu(A)\nu(B) \leq \nu(A \bigvee B)\nu(A \bigwedge B), A, B \in \mathcal{B}(\mathcal{C})$$

an FKG measure.

Proof. If

$$\int \varphi_3(\bar{x})\nu_3(d\bar{x}) \int \varphi_4(\bar{x})\nu_4(d\bar{x}) = \infty$$

then (5) follows. Next we consider that

$$\int \varphi_3(\bar{x})\nu_3(d\bar{x}) \int \varphi_4(\bar{x})\nu_4(d\bar{x}) < \infty$$

and propose at first that

$$\int \varphi_1(\bar{x})\nu_1(d\bar{x}) \int \varphi_2(\bar{x})\nu_2(d\bar{x}) < \infty. \tag{6}$$

Then $\int \varphi_i(\bar{x})\nu_i(d\bar{x}), i = 1, 2, 3, 4$ are finite Borel measures which are regular and hence there exists a compact set $\mathcal{K} \subset \mathcal{C}$, such that for given $\epsilon > 0$

$$\left| \int_{\mathcal{K}} \varphi_i(\bar{x})\nu_i(d\bar{x}) - \int \varphi_i(\bar{x})\nu_i(d\bar{x}) \right| \leq \epsilon, \ i = 1, 2, 3, 4 \tag{7}$$

and $\nu(\mathcal{K}) < \infty$. This compact set is by Ascoli's Lemma the set of equicontinuous functions $\{x_t\}$ which for some $N > 0$ satisfy the relation

$$|x_t| \leq N.$$

Without loss of generality we will consider that $\mathcal{K}$ is the set of all such functions. It is easy to see that this set is a distributive lattice. Indeed if

$$|x_i(t) - y(t)| < \epsilon, |x_i| \leq N, i = 1, 2$$

then

$$|(x_1 \bigvee x_2)(t) - y(t)| < \epsilon, \ |(x_1 \bigwedge x_2)(t) - y(t)| < \epsilon, \tag{8}$$

$$|(x_1 \bigvee x_2)(t)| < N, \ |(x_1 \bigwedge x_2)(t)| \leq N.$$

We consider the partition of the interval T into m consecutive subintervals $\Delta_i = [t_{i-1}, t_i), i = 1, 2, \ldots m - 1, t_0 = 0, \Delta_m = [t_{m-1}, 1]$ of equal length choosing m in such a way that if $t, t' \in \Delta_i$, then

$$|x_t - x_{t'}| < \delta/2. \tag{9}$$

Without loss of generality we assume that N is integer and that $\delta = L^{-1}$ for some natural L. Next we divide the interval $[-N, N]$ into $2N/\delta = 2LN$ consecutive subintervals $\Gamma_j = [s_{j-1}, s_j), \ j = 1, 2, \ldots, 2LN - 1, \Gamma_{2LN} = [s_{2LN-1}, 2LN]$ of equal length δ. At last we consider the partition of the compact $\mathcal{K}$ into the set of cylinders ($m = 2LN + 1$)

$$\pi_{t_0, t_1, \ldots, t_m}(i_0, i_1, \ldots, i_m) = \{x_t : x_{t_j} \in \Gamma_{i_j}\}, i_j = 1, 2, \ldots, 2LN.$$

Consider the finite set of rectangles

$$K(i_0, i_1, \ldots, i_m) \stackrel{\Delta}{=} K_{t_0, t_1, \ldots, t_m}(i_0, i_1, \ldots, i_m) = \{x_t : x_{t_j} \in \Gamma_{i_j}; |x_t - y_j| < \delta, t \in \Delta_j\} \subset R^T,$$

where y_j is the center of the interval Γ_j. Then from (9) it follows that

$$\mathcal{K} \subset \bigcup_{i_j} K_{t_0, t_1, \ldots, t_m}(i_0, i_1, \ldots, i_m).$$

Note also that

$$diam(K(i_0, i_1, \ldots, i_m)) = 2\delta. \tag{10}$$

Now we approximate in $L^1(R^T, \nu_i)$ functions φ_i on the compact $\mathcal{K}$ by continuous functions f_i on $\mathcal{K}$:

$$\int_{\mathcal{K}} |\varphi_i(\bar{x}) - f_i(\bar{x})| \nu_i(d\bar{x}) < \epsilon. \tag{11}$$

Using a standard procedure we can choose f_i in such a way that

$$f_i \leq \varphi_i, i = 1, 2; \ f_i \geq \varphi_i, i = 3, 4.$$

Note, that functions f_i are uniformly continuous on $\mathcal{K}$ and consequently, choosing δ sufficiently small, we can choose new functions $\xi_i, i = 1, 2, 3, 4$ on $\mathcal{K}$ such that $0 \leq f_i - \xi_i < \epsilon/\nu_i(\mathcal{K})$, $i = 1, 2$; $0 \leq \xi_i - f_i < \epsilon/\nu_i(\mathcal{K}), i = 3, 4$ and every ξ_i is constant on every set $K_{t_0, t_1, \ldots, t_m}(i_0, i_1, \ldots, i_m) \bigcap \mathcal{K}$. At last note that the family $\mathcal{A}$ of sets

$$K(i_0, i_1, \ldots, i_m) \bigcap \mathcal{K} , i_j = 1, 2, \ldots, 2LN$$

is a distributive lattice under the operations $\bigvee, \bigwedge$ on the set of indices i_j:

$$K(i_0 \bigvee i'_0, i_1 \bigvee i'_1, \ldots, i_m \bigvee i'_m) \bigcap \mathcal{K},$$

$$K(i_0 \bigwedge i'_0, i_1 \bigwedge i'_1, \ldots, i_m \bigwedge i'_m) \bigcap \mathcal{K} \in \mathcal{A}.$$

Hence we have eight families of values

$$\nu_i(i_0, i_1, \ldots, i_m) \stackrel{\Delta}{=} \nu_i(K(i_0, i_1, \ldots, i_m)), \ i = 1, 2, 3, 4,$$

$$\xi_i(i_0, i_1, \ldots, i_m) \stackrel{\Delta}{=} \xi_i(\bar{x}), \ \bar{x} \in K(i_0, i_1, \ldots, i_m)$$

and

$$\nu_1(i_0, i_1, \ldots, i_m)\xi_1(i_0, i_1, \ldots, i_m)\nu_2(j_0, j_1, \ldots, j_m)\xi_2(j_0, j_1, \ldots, j_m) \leq \tag{12}$$

$$\leq \nu_3(i_0 \bigvee j_0, i_1 \bigvee j_1, \ldots, i_m \bigvee j_m)\xi_3(i_0 \bigvee j_0, i_1 \bigvee j_2, \ldots, i_m \bigvee j_m) \times$$

$$\times \nu_4(i_0 \bigwedge j_0, i_1 \bigwedge j_1, \ldots, i_m \bigwedge j_m)\xi_4(i_0 \bigwedge j_0, i_1 \bigwedge j_1, \ldots, i_m \bigwedge j_m).$$

Hence we are in the condition (1), (2) with counting measure μ on R^{m+1}

$$\mu(A) = \sum_{i_j} \delta_{\bar{x}, (i_0, i_1, \ldots, i_m)}(A)$$

and
$$\varphi_i(i_0, i_1, \ldots, i_m) = \nu_i(i_0, i_1, \ldots, i_m)\xi_i(i_0, i_1, \ldots, i_m).$$

It follows that

$$\sum_{ij} \nu_1(i_0, i_1, \ldots, i_m)\xi_1(i_0, i_1, \ldots, i_m) \sum_{ij} \nu_2(j_0, j_1, \ldots, j_m)\xi_2(j_0, j_1, \ldots, j_m) \quad (13)$$

$$\leq \sum_{ij} \nu_3(i_0, i_1, \ldots, i_m)\xi_3(i_0, i_1, \ldots, i_m) \sum_{ij} \nu_4(i_0, i_1, \ldots, i_m)\xi_4(i_0, i_1, \ldots, i_m).$$

Because

$$\left| \sum_{ij} \nu_i(i_0, i_1, \ldots, i_m)\xi_i(i_0, i_1, \ldots, i_m) - \int \xi_i(\bar{x})\nu_i(d\bar{x}) \right| \leq 3\epsilon$$

and $\epsilon > 0$ is arbitrary from (13) it follows statement of the theorem in the case when (6) is valid. Let's consider now that (6) is not valid. Let's for example

$$\int \varphi_3(\bar{x})\nu_3(d\bar{x}) = \infty, \quad \int \varphi_4(\bar{x})\nu_4(d\bar{x}) < \infty. \quad (14)$$

Then we use the same consideration, but instead of relations (7), (11) we apply relations

$$\int_K \varphi_3(\bar{x})\nu_3(d\bar{x}) > M,$$

$$\int_K f_3(\bar{x})|\nu_3(d\bar{x}) > M - \epsilon, \ f_3 < M_1$$

correspondingly. Here M, M_1 are given constants which we will an consider arbitrary large. Repeating the proof as in the case of finite integrals we obtain that the product integrals over measures ν_3, ν_4 is arbitrary large which gives the contradiction to their finiteness. Cases other than (14) are considered similarly. This proves the theorem.

Let's show that (4) is valid if we consider the all equal for different i measures, generated by Wiener process. Actually it is not difficult to see that for the validness of (4) it is necessary and sufficient the inequality (4) to be valid for the cylinders A, B which bases are rectangles. It easily follows from the proof of Theorem.

Hence all we should do when the random process is given by its finite dimensional distributions is to check whether it has continuous modification and whether the measure on cylinders which bases are rectangles generated by the finite dimensional distributions satisfy the inequality (4).

It is easy to see that Wiener process is the case. Indeed it is well known that in finite dimensional case Gauss distribution generate the measure, satisfying (4) if $r_{i,j} \leq 0, i \neq j$, where $W = ||w_{i,j}||_{i,j=1}^n$ is the matrix inverse to the correlation matrix R [2]. In the case of Wiener process $R = ||t_i \bigvee t_j||_{i,j=1}^n, t_i > t_j, i > j$ and the inverse matrix W has nonzero elements only on the diagonal and also elements in the strip above and belong the diagonal $w_{p,p+1} = w_{p+1,p} = (t_p - t_{p+1})^{-1}$,

$p = 1, 2, \ldots, n - 1$. Hence measure ν , generated by the Wiener process ω_t satisfies (4) ($\nu_i = \nu, i = 1, 2, 3, 4$).

Also introduce one important example of the function which satisfies (3). Such function is

$$\varphi_i(\omega.) = \varphi(\omega.) = \exp\left(\int_0^1 b(t)\omega_t dt\right)$$

where $b(t)$ is some function for which the integral has sense. This functions satisfy (3) because the expression in the exponent is linear function of ω_t and the obvious relation

$$a + b = a \bigvee b + a \bigwedge b.$$

Next we introduce some corollaries of the theorem. Note that if ℓ is a nondecreasing non negative function, then functions $\varphi_1 = \varphi_3 = \ell, \varphi_2 = \varphi_4 = 1$ satisfy inequality (3). If ν_i are probability distributions then from the theorem it follows that

$$E_1(\ell) \leq E_2(\ell), \tag{15}$$

where E_i is the mathematical expectation under the probability measure ν_i. Note that the theorem is valid also if instead of $T = [0, 1]$ one consider $T = n$. In other words the theorem is valid in $n-$dimensional space. To see it is enough to make minor changes in the proof mostly concerning notations. But in this case we have the extending of the result from [3],[7],[8] to the case of the arbitrary measure, not only such that is discrete or have density. The same note is valid concerning the FKG inequality [6]. Note also that the condition that $\ell \geq 0$ is ambitious because if it is negative we can consider the truncated version of ℓ and add the positive constant and then take a limit when the level of the truncation tends to infinity. If we have probability measure which satisfies inequality (4) (all ν_i are equal) then for any pair of nondecreasing (non increasing) functions ℓ_1, ℓ_2 the following inequality is valid

$$E(\ell_1\ell_2) \geq E(\ell_1)E(\ell_2). \tag{16}$$

In the case of finite dimension it is a strengthening of the FKG inequality for the case of an arbitrary probability measure. To prove this inequality it is sufficient to use inequality (15) with $\nu_2(d\bar{x}) = \frac{\ell_2\nu(d\bar{x})}{E(\ell_2)}, \nu_1 = \nu$ and assume that ℓ_2 is non negative.

References

1. R. Ahlswede and D. Daykin, An inequality for weights of two families of sets, their unions and intersections, Z. Wahrscheinlichkeitstheorie und verw. Gebiete, 93, 183–185, 1979.
2. S. Karlin and Y. Rinott, Classes of orderings of measures and related correlation inequalities, I. Multivariate Totally Positive Distributions, Journal of Multivariate Analysis, 10, 467–498, 1980.
3. J. Kemperman, On the FKG inequality for measures in a partially ordered space, Indag.Math, 39, 313-331, 1977.

4. G. Birkhoff, Lattice Theory, Providence, RI, 1967.
5. J. Doob, Stochastic Processes, New York, Wiley, 1967.
6. C. Fortuin, P. Kasteleyn, and J. Ginibre, Correlation inequalities on some partially ordered sets, Comm. Math. Phys. 22, 89-103, 1971.
7. R. Holley, Remarks on the FKG inequalities, Comm. Math. Phys. 36, 227–231, 1974.
8. C. Preston, A generalization of the FKG inequalities, Comm. Math. Phys. 36, 233-241, 1974.

Lower Bounds for Divergence in the Central Limit Theorem

Peter Harremoës

Abstract. A method for finding asymptotic lower bounds on information divergence is developed and used to determine the rate of convergence in the Central Limit Theorem

1 Introduction

Recently Oliver Johnson and Andrew Barron [7] proved that the rate of convergence in the information theoretic Central Limit Theorem is upper bounded by $\frac{c}{n}$ under suitable conditions for some constant c. In general if $r_0 > 2$ is the smallest number such that the r'th moment do not vanish then a lower bound on total variation is $\frac{c}{n^{\frac{r_0}{2}-1}}$ for some constant c. Using Pinsker's inequality this gives a lower bound on information divergence of order $\frac{1}{n^{r_0-2}}$. In this paper more explicit lower bounds are computed. The idea is simple and follows general ideas related to the maximum entropy principle as described by Jaynes [6]. If some of the higher moments of a random variable X are known the higher moments of the centered and normalized sum of independent copies of X can be calculated. Now, maximize the entropy given these moments. This is equivalent to minimize the divergence to the normal distribution. The distribution maximizing entropy with given moment constraints can not be calculated exactly but letting n go to infinity asymptotic results are obtained.

2 Existence of Maximum Entropy Distributions

Let X be a random variable for which the moments of order $1, 2, ..., R$ exist. Wlog we will assume that $E(X) = 0$ and $Var(X) = 1$. The r'th central moment is denoted $\mu_r(X) = E(X^r)$. The *Hermite polynomials* which are orthogonal polynomials with respect to the normal distribution. The first Hermite polynomials are

$$H_0(x) = 1$$
$$H_1(x) = x$$
$$H_2(x) = x^2 - 1$$
$$H_3(x) = x^3 - 3x$$
$$H_4(x) = x^4 - 6x^2 + 3$$
$$H_5(x) = x^5 - 10x^3 + 15x$$
$$H_6(x) = x^6 - 15x^4 + 45x^2 - 15$$

R. Ahlswede et al. (Eds.): Information Transfer and Combinatorics, LNCS 4123, pp. 578–594, 2006.

$$H_7(x) = x^7 - 21x^5 + 105x^3 - 105x$$
$$H_8(x) = x^8 - 28x^6 + 210x^4 - 420x^2 + 105$$

One easily translate between moments and the Hermite moments $E(H_r(X))$. Let r_0 denote the smallest number bigger than 1 such that $E(H_r(X)) \neq 0$. Put $\gamma_0 = E(H_{r_0}(X))$.

It is well known that the normal distribution is the maximum entropy distribution for a random variable with specified first and second moment. It is also known that there exists no maximum entropy distribution if the first 3 moments are specified and the skewness is required to be non-zero [2]. In this case a little wiggle far away from 0 on the density function of the normal distribution can contribute a lot to the third moment but only contribute with a marginal decrease in the entropy. Then normal distribution is the *center of attraction* of the problem [13].

Lemma 1. *Let K be the convex set of distributions for which the first R moments are defined and satisfies the following equations and inequality*

$$E(H_r(X)) = h_r \text{ for } r < R \tag{1}$$
$$E(H_R(X)) \leq h_R.$$

If R is even then the maximum entropy distribution exists.

Proof. Let $\boldsymbol{G} \in \mathbb{R}^{R-1}$ be a vector and let $C_{\boldsymbol{G}}$ be the set of distributions satisfying the following equations

$$E(H_R(X) - h_R) \leq \sum_{n<R} G_r \cdot E(H_r(X) - h_r).$$

This inequality is equivalent to $E\left(H_R(X) - h_R - \sum_{r<R} G_r \cdot (H_r(X) - h_r)\right) \leq 0$.

We see that $C_{\boldsymbol{G}}$ is closed because $H_R(X) - h_R - \sum_{r<R} G_r(H_r(X) - h_r) \to \infty$ for $|x| \to \infty$. Therefore the intersection $K = \bigcap_{\boldsymbol{G} \in \mathbb{R}^{R-1}} C_{\boldsymbol{G}}$ is closed. Using that K is closed we get that there exists a distribution $P^* \in K$ such that the entropy is maximal. $\qquad\square$

Theorem 1. *Let C be the convex set of distributions for which the first R moments are defined and satisfies the following equations*

$$E(H_r(X)) = h_r \text{ for } r \leq R.$$

If any of the following conditions are fulfilled

- *r_0 is odd and $R = r_0 + 1$*
- *r_0 is even and $R = r_0$ and $\gamma_0 < 0$*
- *r_0 is even and $R = r_0 + 2$ and $\gamma_0 > 0$*

then the maximum entropy distribution in C exists.

Proof. Let P^* be the maximum entropy distribution in the K determined by (1).

Assume r_0 is odd and $R = r_0 + 1$. If $E_{P^*}(H_R(X)) < h_R$ then P^* maximizes entropy with respect to the conditions

$$E(H_r(X)) = h_r \text{ for } r < R.$$

which is not possible because $\gamma_0 \neq 0$ and r_0 is odd. Therefore P^* satisfy also satisfy $E(H_R(X)) = h_R$.

Assume r_0 is even and $R = r_0$ and $\gamma_0 < 0$. Assume $E_{P^*}(H_{r_0}(X)) < \gamma_0$ then P^* maximizes entropy with respect to the conditions

$$E(H_r(X)) = 0 \text{ for } 1 \leq r \leq r_0 - 1.$$

and P^* is the normal distribution $N(0,1)$. But then $E_{P^*}(H_{r_0}(X)) = 0$ and we get a contradiction.

Assume r_0 is even and $R = r_0 + 2$ and $\gamma_0 > 0$. Assume $E_{P^*}(H_R(X)) < h_R$ then P^* maximizes entropy with respect to the conditions

$$E(H_r(X)) = h_r \text{ for } r < R$$

which is not possible because $\gamma_0 > 0$ and r_0 is even. Therefore P^* satisfy also satisfy $E(H_R(X)) = h_R$. □

The most important case is when $r_0 = 3$ and skewness is non-zero. Then the theorem states that if the first 4 moments are fixed the maximum entropy distribution exists.

3   Asymptotic Lower Bounds

The method described in this section was first developed for finding lower bound for the rate of convergence in Poisson's law [5]. Let Q be a probability distribution and let X be a m-dimensional random vector such that $E_P(X) = 0$, where E_P denotes mean with respect to P. We are interested in the minimal divergence from a distribution P satisfying certain constraints of the form $E_P(X) = f(t)$ to the fixed probability distribution P. Let $I \subseteq R$ be the domain of f. We will assume that $f : I \to R^m$ is a smooth curve with $f(0) = 0$. Define

$$D_t = \inf D(P \parallel Q) \qquad (2)$$

where the infimum is taken over all P for which $E_P(X) = f(t)$. In this section we will assume that the infimum is attained, and therefore $D_0 = 0$ so if $t \curvearrowright D_t$ is smooth we also have $\frac{dD}{dt}|_{t=0} = 0$. Therefore we are interested in $\frac{d^2D}{dt^2}|_{t=0}$. To solve this problem we will use the exponential family.

Let Z be the partition function defined by

$$Z(\alpha) = E(\exp(\langle \alpha \mid X \rangle))$$

for $\alpha \in A \subseteq \mathbf{R}^m$. For $\alpha \in A$ the distribution P_α is defined by the equation

$$\frac{dP_\alpha}{dP} = \frac{\exp\left(\langle \alpha \mid X \rangle\right)}{Z(\alpha)}.$$

Then

$$E_{P_\alpha}(X) = \frac{\nabla(Z)}{Z}$$

where

$$\nabla = \begin{pmatrix} \frac{\partial}{\partial \alpha_1} \\ \frac{\partial}{\partial \alpha_2} \\ \vdots \\ \frac{\partial}{\partial \alpha_m} \end{pmatrix}.$$

We put $X(\alpha) = E_{P_\alpha}(X)$ as a function of α. Assume that

$$f(t) = E_{P_\alpha}(X).$$

Then

$$f(t) = \frac{\nabla(Z)}{Z}$$

and

$$f'(t) = \sum \frac{\partial}{\partial \alpha_i} \frac{\nabla(Z)}{Z} \cdot \frac{d\alpha_i}{dt}$$

$$= \left(\sum_i \frac{\partial}{\partial \alpha_i} \frac{\frac{\partial}{\partial \alpha_j}(Z)}{Z} \cdot \frac{d\alpha_i}{dt}\right)_j$$

$$= \left(\sum_i \frac{Z \frac{\partial^2 Z}{\partial \alpha_i \partial \alpha_j} - \frac{\partial Z}{\partial \alpha_i} \frac{\partial Z}{\partial \alpha_j}}{Z^2} \cdot \frac{d\alpha_i}{dt}\right)_j.$$

The quotient can be written as

$$\frac{Z \frac{\partial^2 Z}{\partial \alpha_i \partial \alpha_j} - \frac{\partial Z}{\partial \alpha_i} \frac{\partial Z}{\partial \alpha_j}}{Z^2} = \frac{\frac{\partial^2 Z}{\partial \alpha_i \partial \alpha_j}}{Z} - \frac{\partial Z}{\partial \alpha_i} \frac{\partial Z}{\partial \alpha_j}{Z} \frac{\partial Z}{\partial \alpha_j}$$

$$= E_{P_\alpha}(X_i \cdot X_j) - E_{P_\alpha}(X_i) \cdot E_{P_\alpha}(X_j)$$

$$= Cov_{P_\alpha}(X_i, X_j).$$

Let $\rho = \nabla \log \frac{dP_\alpha}{dP}$ denote the *score function* associated with the family P_α. Then

$$\rho = \nabla \left(\log \frac{\exp\left(\langle \alpha \mid X \rangle\right)}{Z(\alpha)}\right)$$

$$= \nabla \left(\langle \alpha \mid X \rangle - \log Z(\alpha)\right)$$

$$= X - \frac{\nabla(Z)}{Z}$$

$$= X - E_{P_\alpha}(X).$$

Therefore the Fisher information matrix I is

$$I = E_{P_\alpha} \left(\boldsymbol{\rho} \cdot \boldsymbol{\rho}^T \right)$$
$$= Cov_{P_\alpha} \left(\boldsymbol{X} \right),$$

and therefore

$$\boldsymbol{f}'(t) = I \cdot \frac{d\boldsymbol{\alpha}}{dt}.$$

For a more detailed treatment of Fisher information and families of probability measures, see [10] and [1].

Now

$$D \left(P_\alpha \parallel P \right) = \langle \boldsymbol{\alpha} \mid \boldsymbol{X} \rangle - \log \left(Z \left(\boldsymbol{\alpha} \right) \right),$$

and from now on we will consider D as a function of $\boldsymbol{\alpha}$. Then

$$\frac{dD}{dt} = \left\langle \frac{d\boldsymbol{\alpha}}{dt} \mid \boldsymbol{X} \right\rangle + \left\langle \boldsymbol{\alpha} \mid \frac{d\boldsymbol{X}}{dt} \right\rangle - \frac{\left\langle \frac{d\boldsymbol{\alpha}}{dt} \mid \nabla \left(Z \right) \right\rangle}{Z}$$
$$= \left\langle \boldsymbol{\alpha} \mid \frac{d\boldsymbol{X}}{dt} \right\rangle.$$

Therefore

$$\frac{d^2 D}{dt^2} = \left\langle \frac{d\boldsymbol{\alpha}}{dt} \mid \frac{d\boldsymbol{X}}{dt} \right\rangle + \left\langle \boldsymbol{\alpha} \mid \frac{d^2 \boldsymbol{X}}{dt^2} \right\rangle$$
$$= \left\langle I^{-1} \cdot \boldsymbol{f}'(t) \mid \boldsymbol{f}'(t) \right\rangle + \left\langle \boldsymbol{\alpha} \mid \boldsymbol{f}''(t) \right\rangle$$
$$= \left\langle I^{-1} \cdot \boldsymbol{f}'(t) \mid \boldsymbol{f}'(t) \right\rangle + \left\langle \frac{\boldsymbol{\alpha}}{t} \mid t \cdot \boldsymbol{f}''(t) \right\rangle.$$

If the last term in this equation vanish we get a lower bound on the second derivative of the divergence expressed in terms of the covariance matrix which in this case equals the fisher information matrix. Therefore it is a kind of *Cramér-Rao inequality*.

4   Skewness and Kurtosis

The quantity skewness minus excess kurtosis has been studied in several papers, see [11], [12],[4] and [8]. The quantity is well behaved in the sense that one can give bound on it for important classes of distribution. We shall see that the quantity appears in a natural way when one wants to minimize information divergence under moment constraints.

For a random variable X with finite third moment the coefficient of *skewness* τ is defined by

$$\tau = \frac{E \left((X - \mu)^3 \right)}{\sigma^3}. \tag{3}$$

where μ and σ denotes mean and spread. If X has finite forth moment the *excess kurtosis* κ is defined by

$$\kappa = \frac{E\left((X-\mu)^4\right)}{\sigma^4} - 3 \; . \tag{4}$$

Pearson proved that $\tau^2 - \kappa \leq 2$ with equality for Bernoulli random variables [11]. For unimodal distributions the upper bound is $\frac{186}{125}$ with equality for one-sided boundary-inflated uniform distributions [8]. If mean and mode coincides of the distribution is symmetric unimodal the upper bound $\frac{6}{5}$ is obtained for uniform distributions. Finally, 0 is an upper bound when the distribution is infinitely divisible, and the upper bound is attained when the distribution is normal or Poisson [12]. Using the last property squared skewness minus excess kurtosis has been proposed for testing Poisson/normal distributions versus other infinitely distributions [4].

As usual when C is a set of probability measures and Q is a probability measure we put

$$D(C \parallel Q) = \inf_{P \in C} D(P \parallel Q) \; .$$

Theorem 2. *Let X be a random variable with distribution Q which is not a Bernoulli random variable. Assume that the forth moment is finite. For $y \in [0;1]$ let C_y be the set of distributions of random variables S such that $E(S) = E(X)$ and $\frac{Var(X) - Var(S)}{Var(X)} = y$. Define*

$$g(y) = D(C_y \parallel Q) \; . \tag{5}$$

Then

$$g(0) = 0 \tag{6}$$
$$g'(0) = 0 \tag{7}$$
$$g''(0) = \frac{1}{2 - (\tau^2 - \kappa)} \; . \tag{8}$$

Proof. According to our general method we have to compute the Fisher information matrix of the random vector (X, X^2). The Fisher information is

$$\begin{pmatrix} 1 & \tau \\ \tau & \kappa + 2 \end{pmatrix}$$

and the inverse is

$$\frac{1}{2 + \kappa - \tau^2} \begin{pmatrix} \kappa + 2 & -\tau \\ -\tau & 1 \end{pmatrix} \; .$$

The derivative in zero is $(0, 1)$ and therefore the second derivative of the information divergence is $\frac{1}{2 + \kappa - \tau^2}$. $\qquad\square$

If Q is the normal distribution then the asymptotic result can be strengthened to an inequality. Let S be a random variable with $E(S) = \mu$ and $Var(S) = V_1$.

Let $N(\mu, V_2)$ denote a normal distribution with mean μ and variance $V_2 \geq V_1$. We are interested in a lower bound on $D(S \parallel N)$. The minimum of $D(S \parallel N)$ is obtained when the distribution of S is the element in the exponential family given by the densities $N_\alpha(x) = \frac{e^{\alpha(x-\mu)^2}}{Z(\alpha)} \cdot N(x)$, for which $Var(N_\alpha(x)) = V_1$. Here $Z(\alpha)$ is a constant which can be computed as $\int_{-\infty}^{\infty} e^{\alpha(x-\mu)^2} dN(\mu, V_2, x)$. All the members of the exponential family are normal distributions which simplifies the calculations considerably. We put $y = \frac{V_2 - V_1}{V_2}$.

$$D(S \parallel N(\mu, V_2)) \geq D(N(\mu, V_1) \parallel N(\mu, V_2)) \tag{9}$$

$$= D(N(0, V_1) \parallel N(0, V_2)) \tag{10}$$

$$= E_{N(0,V_1)} \left(\log \frac{dN(0, V_1)}{dN(0, V_2)} \right) \tag{11}$$

$$= E_{N(0,V_1)} \left(\log \frac{\frac{\exp\left(-\frac{x^2}{2V_1}\right)}{(2\pi V_1)^{\frac{1}{2}}}}{\frac{\exp\left(-\frac{x^2}{2V_2}\right)}{(2\pi V_2)^{\frac{1}{2}}}} \right) \tag{12}$$

$$= E_{N(0,V_1)} \left(\frac{1}{2} \log \frac{V_2}{V_1} + \frac{x^2}{2V_2} - \frac{x^2}{2V_1} \right) \tag{13}$$

$$= \frac{1}{2} \log \left(\frac{V_2}{V_1} \right) + \frac{V_1}{2V_2} - \frac{V_1}{2V_1} \tag{14}$$

$$= \frac{1}{2} (y - \log(1 + y)) . \tag{15}$$

Expanding the logarithm we get the desired lower bound

$$D(S \parallel N(\mu, V_2)) \geq \frac{1}{2} \left(y - \left(y - \frac{1}{2} y^2 \right) \right) \tag{16}$$

$$= \frac{y^2}{4}. \tag{17}$$

A similar inequality holds when the normal distribution is replaced with a Poisson distribution, but the proof is much more involved [5].

5   Rate of Convergence in Central Limit Theorem

The n'th cumulant is denoted $c_r(X)$. Let $X_1, X_2, \ldots$ be independent random variables with the same distribution as X, and put

$$U_n = \frac{\sum_{i=1}^{n} X_i}{n^{\frac{1}{2}}}.$$

Then

$$c_r\left(U_n\right) = c_r\left(\frac{\sum_{i=1}^n X_i}{n^{\frac{1}{2}}}\right)$$

$$= \frac{\sum_{i=1}^n c_r\left(X_i\right)}{n^{\frac{r}{2}}}$$

$$= \frac{c_r\left(X\right)}{n^{\frac{r}{2}-1}}\ .$$

Put $t = \frac{1}{n^{\frac{r_0}{2}-1}}$ so that $n = \left(\frac{1}{t}\right)^{\frac{2}{r_0-2}}$. Then

$$c_r\left(U_n\right) = \frac{c_r\left(X\right)}{n^{\frac{r}{2}-1}}$$

$$= c_r\left(X\right)\cdot t^{\frac{r-2}{r_0-2}}\ \text{for}\ r \geq r_0\ .$$

Put $\gamma_r = c_{r+2}\left(X\right)$ and $\gamma_0 = \gamma_{r_0-2}$ and $\delta_r = \frac{r}{r_0}$ for $r \geq r_0$ and remark that $\delta_r \geq 1$. Remark that $\gamma_0 = \frac{\mu_{r_0+2}(X)}{\sigma^{r_0+2}} = \frac{E\left(H_{r_0+2}(X)\right)}{\sigma^{r_0+2}}$. We see that the R dimensional vector $\left(c_r\left(U_n\right)\right)_r$ lies on the curve $t \curvearrowright g\left(t\right) = \left(0,1,0,0,...,0,...,\gamma_r \cdot t^{\delta_r},...\right)_r^T$. We also see that g is $C^1\left[0;\infty\right[$ and g'' is defined on $\left]0;\infty\right[$ and satisfies

$$g\left(0\right) = \left(0,1,0,...,0\right)^T$$

$$g'\left(0\right) = \left(0,0,...,0,\gamma_0,0,...,0\right)^T$$

$$t\cdot g''\left(t\right) \to \left(0,0,...,0\right)^T\ \text{for}\ t \to 0.$$

The central moments can be calculated from the cumulants by

$$\mu_r = \sum_{\lambda_1+\lambda_2+...+\lambda_k=r}\frac{1}{k!}\binom{r}{\lambda_1\ \lambda_2\ \cdots\ \lambda_k}\prod_{p=1}^k c_{\lambda_p}$$

where the summation is taken over all unordered sets $\left(\lambda_1,\lambda_2,...,\lambda_k\right)$. Therefore the curve g in the set of cumulants is mapped into a curve h in the set of central moments. The curve h is obviously $C^1\left[0;\infty\right[$ and h'' is defined on $\left]0;\infty\right[$ and satisfies

$$h\left(0\right) = \left(\mu_r\left(N\left(0,1\right)\right)\right)_r$$

$$t\cdot h''\left(t\right) \to \left(0,0,...,0\right)^T\ \text{for}\ t \to 0.$$

Then the curve h is transformed into a new curve f in the space of Hermite moments i.e. mean values of Hermite polynomials. For this curve we easily get

$$f\left(0\right) = \left(\mu_r\left(N\left(0,1\right)\right)\right)_r$$

$$t\cdot f''\left(t\right) \to \left(0,0,...,0\right)^T\ \text{for}\ t \to 0.$$

To find the derivative $f'\left(0\right)$ we can write the Hermite polynomials in terms of the cumulants as done in the appendix for $R = 8$. But to get the following general result we need an explicit formula.

Lemma 2. *For all r_0 and $R \geq r_0$ we have $\boldsymbol{f}'(0) = (0, 0, ..., 0, \gamma_0, 0, ..., 0)^T$.*

Proof. It is clear that all terms before coordinate no. r_0 are zero. Assume that $k \geq r_0$ and use the result stated in [9] that

$$E\left(H_k\left(U_n\right)\right) = k! \sum_{\substack{j_0+j_3+j_4+\cdots+j_k=n \\ 3j_3+4j_4+\cdots+kj_k=k}} \binom{n}{j_0 \; j_3 \; j_4 \; \cdots \; j_k}$$

$$\left(\frac{E\left(H_3\left(X\right)\right)}{3!n^{\frac{3}{2}}}\right)^{j_3} \left(\frac{E\left(H_4\left(X\right)\right)}{4!n^2}\right)^{j_4} \cdots \left(\frac{E\left(H_k\left(X\right)\right)}{k!n^{\frac{k}{2}}}\right)^{j_k}.$$

Using that $E\left(H_r\left(X\right)\right) = 0$ for $r < r_0$ we get

$$E\left(H_k\left(U_n\right)\right) = \frac{k!}{n^{\frac{k}{2}}} \sum_{\substack{j_0+j_{r_0}+j_{r_0+1}+\cdots+j_k=n \\ r_0 j_{r_0}+(r_0+1)j_{r_0+1}+\cdots+kj_k=k}} \binom{n}{j_0 \; j_{r_0} \; j_4 \; \cdots \; j_k}$$

$$\left(\frac{E\left(H_{r_0}\left(X\right)\right)}{r_0!}\right)^{j_{r_0}} \left(\frac{E\left(H_{r_0+1}\left(X\right)\right)}{(r_0+1)!}\right)^{r_0+1} \cdots \left(\frac{E\left(H_k\left(X\right)\right)}{k!}\right)^{j_k}.$$

Now we use that $r_0\left(j_{r_0} + j_{r_0+1} + \cdots + j_k\right) \leq r_0 j_{r_0} + (r_0+1)j_{r_0+1} + \cdots + kj_k = k$ and get

$$n - j_0 = \left(j_{r_0} + j_{r_0+1} + \cdots + j_k\right) \leq \frac{k}{r_0}.$$

Therefore

$$\binom{n}{j_0 \; j_{r_0} \; j_{r_0+1} \; \cdots \; j_k} = \frac{n_{[n-j_0]}}{j_{r_0}! \; j_{r_0+1}! \; \cdots \; j_k!}$$

$$\leq \frac{n^{\frac{k}{r_0}}}{j_{r_0}! \; j_{r_0+1}! \; \cdots \; j_k!}.$$

This gives

$$\left|E\left(H_k\left(U_n\right)\right)\right| \leq \frac{k!}{n^{\frac{k}{2}}} \sum_{\substack{j_0+j_{r_0}+j_{r_0+1}+\cdots+j_k=n \\ r_0 j_{r_0}+(r_0+1)j_{r_0+1}+\cdots+kj_k=k}} \frac{n^{\frac{k}{r_0}}}{j_{r_0}! \; j_{r_0+1}! \; \cdots \; j_k!}$$

$$\left|\frac{E\left(H_{r_0}\left(X\right)\right)}{r_0!}\right|^{j_{r_0}} \left|\frac{E\left(H_{r_0+1}\left(X\right)\right)}{(r_0+1)!}\right|^{r_0+1} \cdots \left|\frac{E\left(H_k\left(X\right)\right)}{k!}\right|^{j_k}$$

$$\leq \frac{1}{n^{\frac{k}{2}-\frac{k}{r_0}}} \sum_{r_0 j_{r_0}+(r_0+1)j_{r_0+1}+\cdots+kj_k=k} \frac{k!}{j_{r_0}! \; j_{r_0+1}! \; \cdots \; j_k!}$$

$$\left|\frac{E\left(H_{r_0}\left(X\right)\right)}{r_0!}\right|^{j_{r_0}} \left|\frac{E\left(H_{r_0+1}\left(X\right)\right)}{(r_0+1)!}\right|^{r_0+1} \cdots \left|\frac{E\left(H_k\left(X\right)\right)}{k!}\right|^{j_k}.$$

For $k = r_0$ we get $E\left(H_{r_0}\left(U_n\right)\right) = O\left(\frac{1}{n^{\frac{r_0}{2}-1}}\right)$. For $k > r_0$ we get $E\left(H_k\left(U_n\right)\right) =$
$O\left(\frac{1}{n^{\frac{k}{2}-\frac{k}{r_0}}}\right)$ and to finish the proof we just have to remark that $\frac{k}{2} - \frac{k}{r_0} \geq \frac{r_0}{2} - 1$. $\square$

This result is closely related to the results of Brown (1982), but it does not follow from his results. In the appendix detailed formulas for the Hermite moments can be found.

It is convenient to put $X_i = H_i\left(X\right)$. Then the covariance of the normal distributions is

$$COV = E\left(X_i X_j\right) = \delta_{i,j} \cdot j! \, .$$

The inverse is

$$COV^{-1} = \frac{\delta_{i,j}}{j!} \, .$$

Theorem 3. *Assume that $R \geq 2r_0$. Then the rate of convergence in the information theoretic Central Limit Theorem lower bounded by*

$$\liminf_{n\to\infty} n^{r_0-2} \cdot D\left(U_n \parallel N\left(0,1\right)\right) \geq \frac{\gamma_0^2}{2 \cdot r_0!}.$$

Proof. The conditions ensure that the maximum entropy distribution exists. We put $t = \frac{1}{n^{\frac{r_0}{2}-1}}$ and combine the previous results and get

$$\liminf_{t\to 0} \frac{D\left(U_n \parallel N\left(0,1\right)\right)}{t^2} \geq \frac{1}{2}\left(COV^{-1} \cdot f'\left(0\right) \mid f\left(0\right)\right)$$

$$= \frac{\gamma_0^2}{2 \cdot r_0!}.$$

The result is obtained by the substitution $n = n = \left(\frac{1}{t}\right)^{\frac{2}{r_0-2}}$. $\square$

If $\tilde{H}_r$ denotes the normalized Hermite polynomials $H_r / \left(r!\right)^{1/2}$ then the result can be stated as

$$\liminf_{n\to\infty} n^{r-2} \cdot D\left(U_n \parallel N\left(0,1\right)\right) \geq \frac{\left(E\left(\tilde{H}_r\left(X\right)\right)\right)^2}{2}.$$

Remark that this inequality also holds for r different from r_0. This for $r = r_0$ this can also be written as

$$\liminf_{n\to\infty} \frac{D\left(U_n \parallel N\left(0,1\right)\right)}{\left(E\left(\tilde{H}_r\left(U_n\right)\right)\right)^2} \geq \frac{1}{2}.$$

In the most important case where $r_0 = 3$ and $R = 4$ we get

$$\liminf_{t\to 0} n \cdot D\left(U_n \parallel N\left(0,1\right)\right) \geq \frac{\tau^2}{12} \, ,$$

where τ is the skewness of X_1. If $r_0 = 4$ and the excess kurtosis κ is negative we get

$$\liminf_{t\to 0} n^2 \cdot D\left(U_n \parallel N\left(0,1\right)\right) \geq \frac{\kappa^2}{48} \, . \tag{18}$$

6   Edgeworth Expansion

The sequence of maximum entropy distributions we have used is closely related
to the Edgeworth expansion as described in [3]. The first term is given by the
following theorem.

Theorem 4. *Suppose that μ_3 exists and that $|\phi|^\nu$ is integrable for some $\nu \geq 1$.
Then f_n exists for $n \geq \nu$ and as $n \to \infty$*

$$f_n(x) - \mathfrak{n}(x) - \frac{\mu_3}{6\sigma^3 n^{1/2}}\left(x^3 - 3x\right)\mathfrak{n}(x) = o\left(\frac{1}{n^{1/2}}\right)$$

uniformly in x.

Putting $t = \frac{1}{n^{1/2}}$ and $\sigma = 1$ and $\mu_3 = \gamma_0$ this can be rewritten as

$$\frac{dP_n}{dN}(x) = 1 + \frac{\gamma_0}{6}H_3(x)\cdot t + \frac{o(t)}{\mathfrak{n}(x)}.$$

In this approach the whole distribution of X is known. In general a distribution
is not determined by its moments, but with regularity conditions it is. We will
assume that X is sufficiently regular so that the distribution is determined by its
moments. Then we can put $R = \infty$ and write P^t instead of P_n. By the central
limit theorem we should put $P^o = N$ and therefore

$$\frac{dP^o}{dN}(x) = 1$$

For $t \neq 0$ we get that

$$\frac{dP^t}{dN}(x) = \frac{\exp\left(\sum_{i=1}^{\infty}\alpha_i H_i(x)\right)}{Z(\alpha)}$$

where the parameters α_i are functions of t. Then

$$\frac{\partial}{\partial t}\left(\frac{dP^t}{dN}(x)\right) = \frac{Z\cdot\left(\sum\frac{d\alpha_i}{dt}\cdot H_i(x)\exp\left(\sum_{\alpha=i}^{\infty}\alpha_i H_i(x)\right)\right) - \frac{dZ}{dt}\cdot\exp\left(\sum_{\alpha=i}^{\infty}\alpha_i H_i(x)\right)}{Z^2}.$$

For $t = 0$ we get

$$\frac{\partial}{\partial t}\left(\frac{dP^t}{dN}(x)\right)_{|t=0} = \sum\frac{d\alpha_i}{dt}_{|t=0}\cdot H_i(x)$$

$$= \frac{\gamma_0}{6}H_3(x)\ .$$

We see that the first term in the Edgeworth expansion corresponds to the asymptotic behavior of the maximum entropy distribution. If $r_0 = 4$ then a similar
expansion gives the next term in the Edgeworth expansion. To get more than
the first nontrivial term in the Edgeworth expansion would require a full power
expansion of the parameter α in terms of t.

7   An Inequality for Platykurtic Distributions

For $r_0 = 4$ the asymptotic lower bound (18) can be improved to an inequality as demonstrated in the next theorem. This is useful if one is interested in a lower bound on the divergence for a sum of random variables which are not i.i.d. Recall that a distribution is said to be *platykurtic* if the excess kurtosis is negative.

Theorem 5. *For any platykurtic random variable X with excess kurtosis κ the following inequality holds*

$$D\left(X \parallel N\left(\mu, \sigma^2\right)\right) \geq \frac{\kappa^2}{48}$$

where μ is the mean and σ^2 is the variance of X.

Proof.   Without loss of generality we may assume that $\mu = 0$ and $\sigma = 1$. Remark that

$$\kappa = E\left(X^4\right) - 3$$
$$\geq \left(E\left(X^2\right)\right)^2 - 3$$
$$= 1^2 - 3$$
$$= -2.$$

Let $H_4\left(x\right) = x^4 - 6x^2 + 3$ be the 4th Hermite polynomial. For a distribution P with

$$E\left(X^2\right) = 1$$

we have

$$E_P\left(H_4\left(X\right)\right) = \kappa. \tag{19}$$

Therefore

$$D\left(X \parallel N\left(0, 1\right)\right) \geq D\left(P_\alpha \parallel N\left(0, 1\right)\right)$$

where P_α is the distribution minimizing $D\left(P \parallel N\left(0, 1\right)\right)$ subjected to the condition (19).

Put

$$Z\left(\alpha\right) = E_N\left(\exp\left(\alpha H_4\left(X\right)\right)\right)$$

defined for $\alpha \leq 0$. For $\alpha \leq 0$ define P_α by

$$\frac{dP_\alpha}{dN} = \frac{\exp\left(\alpha H_4\left(X\right)\right)}{Z\left(\alpha\right)}.$$

Then the distribution minimizing $D\left(P \parallel N\left(0, 1\right)\right)$ is P_α for the α satisfying $E_{P_\alpha}\left(H_4\left(X\right)\right) = \kappa$. For any $\alpha \leq 0$ define κ_α by $\kappa_\alpha = E_{P_\alpha}\left(H_4\left(X\right)\right)$. Then we have to prove

$$\frac{\kappa_\alpha^2}{48} \leq D\left(P_\alpha \parallel N\left(0, 1\right)\right)$$

$$= E_{P_\alpha}\left(\log \frac{dP_\alpha}{dN}\right)$$

$$= \alpha \cdot \kappa_\alpha - \log\left(Z\left(\alpha\right)\right).$$

For $\alpha = 0$ the inequality is fulfilled so we calculate the derivative and just have to prove

$$\frac{\kappa \cdot \kappa'}{24} \geq \kappa + \alpha \cdot \kappa' - \frac{Z'}{Z} = \alpha \cdot \kappa',$$

which is equivalent to

$$24\alpha \leq \kappa = \frac{Z'}{Z}.$$

This is obvious for $\alpha = 0$ so we differentiate once more and have to prove

$$24 \geq \frac{Z \cdot Z'' - (Z')}{Z^2} = \frac{Z''}{Z} - \left(\frac{Z'}{Z}\right)^2.$$

Therefore it is sufficient to prove that $24 \geq \frac{Z''}{Z}$. Obviously

$$Z(0) = 1, \ Z'(0) = 0, \ Z''(0) \geq 0.$$

Therefore $Z \geq 1$ and we just have to prove $Z'' \leq 24$.

We have

$$Z''(\alpha) = E_N\left((H_4(X))^2 \exp(\alpha H_4(X))\right)$$

which is convex in α so if $Z'' \leq 24$ is proved for 2 values of α it is also proved for all values in between.

$$Z''(0) = E_N\left(H_4(x)^2\right) = 4! = 24.$$

The function $y \curvearrowright y^2 \exp(\alpha y)$ has derivative

$$2y \exp(\alpha y) + \alpha y^2 \exp(\alpha y) = (2 + \alpha y) y \exp(\alpha y)$$

and the 2 stationary points $(0,0)$ and $\left(-\frac{2}{\alpha}, \left(\frac{2}{\alpha}\right)^2 \exp(-2)\right)$. Now we solve the equation

$$\left(\frac{2}{\alpha}\right)^2 \exp(-2) = 24$$

and get $\alpha_0 = (\pm)\frac{e}{6^{\frac{1}{2}}} = -.150\,19$.

Now we know that $(H_4(x))^2 \exp(\alpha H_4(x)) \leq 24$ for $H_4(x) \geq 0$ and $\alpha \geq \alpha_0$. Therefore this will also hold for $\alpha = \alpha_1 = -0.18$

The equation $H_4(x) = 0$ has the solutions $\pm\left(3 \pm 6^{\frac{1}{2}}\right)^{\frac{1}{2}}$ and therefore the $H_4(x)$ is positive for

$$|x| \geq x_0 = \left(3 + 6^{\frac{1}{2}}\right)^{\frac{1}{2}} = 2.334\,4.$$

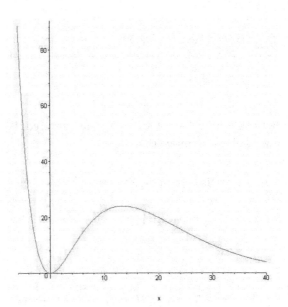

Plot of $y \curvearrowright y^2 \exp(\alpha_0 \cdot y)$.

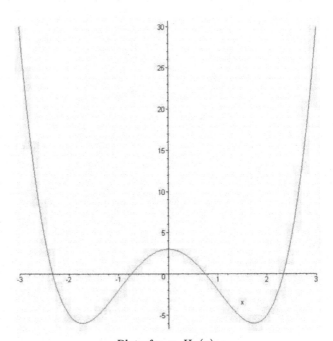

Plot of $x \curvearrowright H_4(x)$.

We have to check that

$$\frac{\int_{-x_0}^{x_0} (H_4(x))^2 \exp(\alpha_1 H_4(x)) \cdot \frac{\exp\left(-\frac{x^2}{2}\right)}{(2\pi)^{\frac{1}{2}}} \, dx}{\int_{-x_0}^{x_0} \frac{\exp\left(-\frac{x^2}{2}\right)}{(2\pi)^{\frac{1}{2}}} \, dx} \leq 24$$

which is done numerically.

We have to check that all values of $\kappa \in [-2; 0]$ are covered. For $\alpha = 0$ we get $\frac{Z'}{Z} = 0$. For $\alpha = \alpha_0$ we have to prove that

$$\frac{Z'(\alpha_0)}{Z(\alpha_0)} = \frac{\int_{-\infty}^{\infty} H_4(x) \exp(\alpha_1 H_4(x)) \cdot \frac{\exp\left(-\frac{x^2}{2}\right)}{(2\pi)^{\frac{1}{2}}} \, dx}{\int_{-\infty}^{\infty} \exp(\alpha_0 H_4(x)) \cdot \frac{\exp\left(-\frac{x^2}{2}\right)}{(2\pi)^{\frac{1}{2}}} \, dx} \leq -2$$

or equivalently

$$\int_{-\infty}^{\infty} \left(x^4 - 6x^2 + 5\right) \exp(\alpha_1 H_4(x)) \cdot \frac{\exp\left(-\frac{x^2}{2}\right)}{(2\pi)^{\frac{1}{2}}} \, dx \leq 0$$

which can easily be checked numerically since $\left(x^4 - 6x^2 + 5\right) \exp(\alpha_1 H_4(x))$ is bounded. □

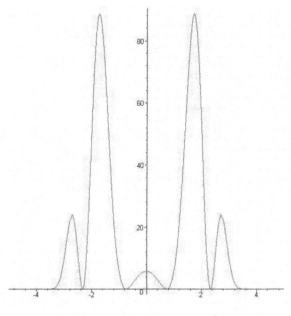

$$x \curvearrowright \left(x^4 - 6x^2 + 3\right)^2 \exp\left(\alpha \left(x^4 - 6x^2 + 3\right)\right)$$

Acknowledgement. I thank Paul Algoet for useful discussion related to Fisher information and lower bounds. I will also thank Andrew Barron for pointing my attention to the Edgeworth expansion.

References

1. S.I. Amari, Information geometry on hierarchy of probability distributions, IEEE Trans. Inform. Theory, Vol. 47, 1701–1711, 2001.
2. T.Cover and J.A. Thomas, Elements of Information Theory, Wiley, 1991.
3. W.Feller, An Introduction to Probability Theory and its Applications, Vol. 2, Wiley, New York, second edition, 1971.
4. A.K. Gupta, T.F. Móri, and G.J. Székely, Testing for Poissonity-normality vs. other infinite divisibility, Stat. Prabab. Letters, 19, 245–248, 1994.
5. P.Harremoës, Convergence to the Poisson distribution in information divergence, Technical Report 2, Mathematical department, University of Copenhagen, 2003.
6. E.T. Jaynes, Information theory and statistical mechanics, I and II, Physical Reviews, 106 and 108, 620–630 and 171–190, 1957.
7. O.Johnson and A.Barron, Fisher information inequalities and the central limit theorem, Preprint, 2001.
8. C.A.J. Klaassen, P.J. Mokveld, and B.van Es, Squared skewness minus kurtosis bounded by 186/125 for unimodal distributions, Statistical and Probability Letters, 50, 2, 131–135, 2000.
9. A.E. Kondratenko, The relation between a rate of convergence of moments of normed sums and the Chebyshev-Hermite moments, Theory Probab. Appl., 46, 2, 352–355, 2002.
10. E.Mayer-Wolf, The Cramér-Rao functional and limiting laws, Ann. of Probab., 18, 840–850, 1990.
11. K.Pearson, Mathematical contributions to the theory of evolution, xix; second supplement to a memoir on skew variation, Philos. Trans. Roy. Soc. London Ser. A 216, 429–257, 1916.
12. V.K. Rohatgi and G.J. Székely, Sharp inequalities between skewness and kurtosis, Statist. Probab. Lett., 8, 197–299, 1989.
13. F. Topsøe, Information theoretical optimization techniques, Kybernetika, 15, 1, 8 – 27, 1979.

Appendix

The moments can be calculated from the cumulants by the formulas

$$\mu_1 = 0, \ \mu_2 = c_2, \ \mu_3 = c_3, \ \mu_4 = c_4 + 3c_2^2, \ \mu_5 = c_5 + 10c_2c_3,$$
$$\mu_6 = c_6 + 15c_2c_4 + 10c_3^2 + 15c_2^3,$$
$$\mu_7 = c_7 + 21c_2c_5 + 35c_3c_4 + 105c_2^2c_3,$$
$$\mu_8 = c_8 + \tfrac{1}{2}\left(28c_2c_6 + 56c_3c_5 + 70c_4^2 + 56c_5c_3 + 28c_6c_2\right)$$
$$+ \tfrac{1}{6}\left(420c_2^2c_4 + 560c_2c_3^2 + 420c_2c_4c_2 + 560c_3c_2c_3 + 560c_3^2c_2 + 420c_4c_2^2\right) + 105c_2^4$$
$$= c_8 + 28c_2c_6 + 56c_3c_5 + 35c_4^2 + 210c_2^2c_4 + 280c_2c_3^2 + 105c_2^4,$$

where we assume that first moment is 0. If the first moment is 0 and the second is 1 we get

$$\mu_1 = 0, \ \mu_2 = 1, \ \mu_3 = c_3, \ \mu_4 = c_4 + 3, \ \mu_5 = c_5 + 10c_3, \ \mu_6 = c_6 + 15c_4 + 10c_3^2 + 15,$$
$$\mu_7 = c_7 + 21c_5 + 35c_3c_4 + 105c_3, \ \mu_8 = c_8 + 28c_6 + 56c_3c_5 + 35c_4^2 + 210c_4 + 280c_3^2 + 105 \,.$$

Then the Hermite moments can be calculated as follows.

$$E\left(H_1\left(X\right)\right) = 0$$
$$E\left(H_2\left(X\right)\right) = 1$$
$$E\left(H_3\left(X\right)\right) = c_3$$
$$E\left(H_4\left(X\right)\right) = c_4$$
$$E\left(H_5\left(X\right)\right) = c_5$$
$$E\left(H_6\left(X\right)\right) = c_6 + 10c_3^2$$
$$E\left(H_7\left(X\right)\right) = c_7 + 35c_3c_4$$
$$E\left(H_8\left(X\right)\right) = c_8 + 56c_3c_5 + 35c_4^2$$

V

Identification Entropy

R. Ahlswede

Abstract. Shannon (1948) has shown that a source $(\mathcal{U}, P, U)$ with output U satisfying Prob $(U = u) = P_u$, can be encoded in a prefix code $\mathcal{C} = \{c_u : u \in \mathcal{U}\} \subset \{0, 1\}^*$ such that for the entropy

$$H(P) = \sum_{u \in \mathcal{U}} -p_u \log p_u \leq \sum p_u ||c_u|| \leq H(P) + 1,$$

where $||c_u||$ is the length of c_u.

We use a prefix code $\mathcal{C}$ for another purpose, namely noiseless identification, that is every user who wants to know whether a u $(u \in \mathcal{U})$ of his interest is the actual source output or not can consider the RV C with $C = c_u = (c_{u_1}, \dots, c_{u||c_u||})$ and check whether $C = (C_1, C_2, \dots)$ coincides with c_u in the first, second etc. letter and stop when the first different letter occurs or when $C = c_u$. Let $L_{\mathcal{C}}(P, u)$ be the expected number of checkings, if code $\mathcal{C}$ is used.

Our discovery is an identification entropy, namely the function

$$H_I(P) = 2 \left(1 - \sum_{u \in \mathcal{U}} P_u^2 \right).$$

We prove that $L_{\mathcal{C}}(P, P) = \sum_{u \in \mathcal{U}} P_u \, L_{\mathcal{C}}(P, u) \geq H_I(P)$ and thus also that

$$L(P) = \min_{\mathcal{C}} \max_{u \in \mathcal{U}} L_{\mathcal{C}}(P, u) \geq H_I(P)$$

and related upper bounds, which demonstrate the operational significance of identification entropy in noiseless source coding similar as Shannon entropy does in noiseless data compression.

Also other averages such as $\bar{L}_{\mathcal{C}}(P) = \frac{1}{|\mathcal{U}|} \sum_{u \in \mathcal{U}} L_{\mathcal{C}}(P, u)$ are discussed in particular for Huffman codes where classically equivalent Huffman codes may now be different.

We also show that prefix codes, where the codewords correspond to the leaves in a regular binary tree, are universally good for this average.

1 Introduction

Shannon's Channel Coding Theorem for Transmission [1] is paralleled by a Channel Coding Theorem for Identification [3]. In [4] we introduced noiseless source coding for identification and suggested the study of several performance measures.

R. Ahlswede et al. (Eds.): Information Transfer and Combinatorics, LNCS 4123, pp. 595–613, 2006.
© Springer-Verlag Berlin Heidelberg 2006

Interesting observations were made already for uniform sources $P^N = (\frac{1}{N}, \ldots, \frac{1}{N})$, for which the worst case expected number of checkings $L(P^N)$ is approximately 2. Actually in [5] it is shown that $\lim_{N \to \infty} L(P^N) = 2$.

Recall that in channel coding going from transmission to identification leads from an exponentially growing number of manageable messages to double exponentially many. Now in source coding roughly speaking the range of average code lengths for data compression is the interval $[0, \infty)$ and it is $[0, 2)$ for an average expected length of optimal identification procedures. Note that no randomization has to be used here.

A discovery of the present paper is an identification entropy, namely the functional

$$H_I(P) = 2\left(1 - \sum_{u=1}^{N} P_u^2\right) \tag{1.1}$$

for the source $(\mathcal{U}, P)$, where $\mathcal{U} = \{1, 2, \ldots, N\}$ and $P = (P_1, \ldots, P_N)$ is a probability distribution.

Its operational significance in identification source coding is similar to that of classical entropy $H(P)$ in noiseless coding of data: it serves as a good lower bound.

Beyond being continuous in P it has three basic properties.

I. Concavity

For $p = (p_1, \ldots, p_N)$, $q = (q_1, \ldots, q_N)$ and $0 \leq \alpha \leq 1$

$$H_I(\alpha p + (1 - \alpha)q) \geq \alpha H_I(p) + (1 - \alpha)H_I(q).$$

This is equivalent with

$$\sum_{i=1}^{N}(\alpha p_i + (1-\alpha)q_i)^2 = \sum_{i=1}^{N}\alpha^2 p_i^2 + (1-\alpha)^2 q_i^2 + \sum_{i \neq j}\alpha(1-\alpha)p_i q_j \leq \sum_{i=1}^{N}\alpha p_i^2 + (1-\alpha)q_i^2$$

or with

$$\alpha(1 - \alpha)\sum_{i=1}^{N}p_i^2 + q_i^2 \geq \alpha(1 - \alpha)\sum_{i \neq j}p_i q_j,$$

which holds, because $\sum_{i=1}^{N}(p_i - q_i)^2 \geq 0$.

II. Symmetry

For a permutation $\Pi : \{1, 2, \ldots, N\} \to \{1, 2, \ldots, N\}$ and $\Pi P = (P_{1\Pi}, \ldots, P_{N\Pi})$

$$H_I(P) = H_I(\Pi P).$$

III. Grouping identity

For a partition $(\mathcal{U}_1, \mathcal{U}_2)$ of $\mathcal{U} = \{1, 2, \ldots, N\}$, $Q_i = \sum_{u \in \mathcal{U}_i} P_u$ and $P_u^{(i)} = \frac{P_u}{Q_i}$ for $u \in \mathcal{U}_i (i = 1, 2)$

$$H_I(P) = Q_1^2 H_I(P^{(1)}) + Q_2^2 H_I(P^{(2)}) + H_I(Q), \text{ where } Q = (Q_1, Q_2).$$

Indeed,

$$Q_1^2 2 \left(1 - \sum_{j \in \mathcal{U}_1} \frac{P_j^2}{Q_1^2} \right) + Q_2^2 2 \left(1 - \sum_{j \in \mathcal{U}_2} \frac{P_j^2}{Q_2^2} \right) + 2(1 - Q_1^2 - Q_2^2)$$

$$= 2Q_1^2 - 2 \sum_{j \in \mathcal{U}_1} P_j^2 + 2Q_2^2 - 2 \sum_{j \in \mathcal{U}_2} P_j^2 + 2 - 2Q_1^2 - 2Q_2^2$$

$$= 2 \left(1 - \sum_{j=1}^{N} P_j^2 \right).$$

Obviously, $0 \leq H_I(P)$ with equality exactly if $P_i = 1$ for some i and by concavity $H_I(P) \leq 2 \left(1 - \frac{1}{N} \right)$ with equality for the uniform distribution.

Remark. Another important property of $H_I(P)$ is Schur concavity.

2 Noiseless Identification for Sources and Basic Concept of Performance

For the source $(\mathcal{U}, P)$ let $\mathcal{C} = \{c_1, \ldots, c_N\}$ be a binary prefix code (PC) with $||c_u||$ as length of c_u. Introduce the RV U with $\text{Prob}(U = u) = P_u$ for $u \in \mathcal{U}$ and the RV C with $C = c_u = (c_{u1}, c_{u2}, \ldots, c_{u||c_u||})$ if $U = u$. We use the PC for noiseless identification, that is a user interested in u wants to know whether the source output equals u, that is, whether C equals c_u or not. He iteratively checks whether $C = (C_1, C_2, \ldots)$ coincides with c_u in the first, second etc. letter and stops when the first different letter occurs or when $C = c_u$. What is the expected number $L_{\mathcal{C}}(P, u)$ of checkings?

Related quantities are

$$L_{\mathcal{C}}(P) = \max_{1 \leq u \leq N} L_{\mathcal{C}}(P, u), \tag{2.1}$$

that is, the expected number of checkings for a person in the worst case, if code $\mathcal{C}$ is used,

$$L(P) = \min_{\mathcal{C}} L_{\mathcal{C}}(P), \tag{2.2}$$

the expected number of checkings in the worst case for a best code, and finally, if users are chosen by a RV V independent of U and defined by $\text{Prob}(V = v) = Q_v$ for $v \in \mathcal{V} = \mathcal{U}$, (see [5], Section 5) we consider

$$L_{\mathcal{C}}(P, Q) = \sum_{v \in \mathcal{U}} Q_v L_{\mathcal{C}}(P, v) \tag{2.3}$$

the average number of expected checkings, if code $\mathcal{C}$ is used, and also

$$L(P, Q) = \min_{\mathcal{C}} L_{\mathcal{C}}(P, Q) \tag{2.4}$$

the average number of expected checkings for a best code.

A natural special case is the mean number of expected checkings

$$\bar{L}_C(P) = \sum_{u=1}^{N} \frac{1}{N} L_C(P, u), \tag{2.5}$$

which equals $L_C(P, Q)$ for $Q = \left(\frac{1}{N}, \ldots, \frac{1}{N}\right)$, and

$$\bar{L}(P) = \min_{C} \bar{L}_C(P). \tag{2.6}$$

Another special case of some "intuitive appeal" is the case $Q = P$. Here we write

$$L(P, P) = \min_{C} L_C(P, P). \tag{2.7}$$

It is known that Huffman codes minimize the expected code length for PC.

This is not the case for $L(P)$ and the other quantities in identification (see Example 3 below). It was noticed already in [4], [5] that a construction of code trees balancing probabilities like in the Shannon-Fano code is often better. **In fact Theorem 3 of [5] establishes that $L(P) < 3$ for every $P = (P_1, \ldots, P_N)$!**

Still it is also interesting to see how well Huffman codes do with respect to identification, because of their classical optimality property. This can be put into the following

Problem: Determine the region of simultaneously achievable pairs $(L_C(P), \sum_u P_u \|c_u\|)$ for (classical) transmission and identification coding, where the C's are PC. In particular, what are extremal pairs? We begin here with first observations.

3   Examples for Huffman Codes

We start with the uniform distribution

$$P^N = (P_1, \ldots, P_N) = \left(\frac{1}{N}, \ldots, \frac{1}{N}\right), 2^n \le N < 2^{n+1}.$$

Then $2^{n+1} - N$ codewords have the length n and the other $2N - 2^{n+1}$ codewords have the length $n + 1$ in any Huffman code. We call the $N - 2^n$ nodes of length n of the code tree, which are extended up to the length $n + 1$ **extended nodes**.

All Huffman codes for this uniform distribution differ only by the positions of the $N - 2^n$ extended nodes in the set of 2^n nodes of length n.

The average codeword length (for data compression) does not depend on the choice of the extended nodes.

However, the choice influences the performance criteria for identification!

Clearly there are $\binom{2^n}{N-2^n}$ Huffman codes for our source.

Example 1. $N = 9$, $\mathcal{U} = \{1, 2, \ldots, 9\}$, $P_1 = \cdots = P_9 = \frac{1}{9}$.

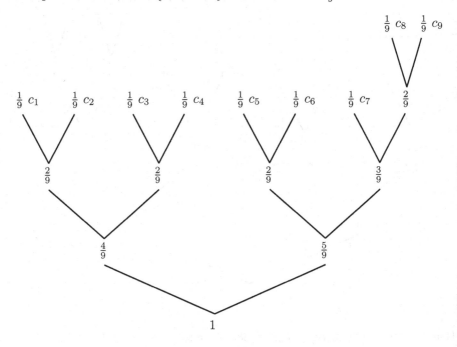

Here $L_{\mathcal{C}}(P) \approx 2.111$, $L_{\mathcal{C}}(P, P) \approx 1.815$ because

$$L_{\mathcal{C}}(P) = L_{\mathcal{C}}(c_8) = \frac{4}{9} \cdot 1 + \frac{2}{9} \cdot 2 + \frac{1}{9} \cdot 3 + \frac{2}{9} \cdot 4 = 2\frac{1}{9}$$

$$L_{\mathcal{C}}(c_9) = L_{\mathcal{C}}(c_8), L_{\mathcal{C}}(c_7) = 1\frac{8}{9}, L_{\mathcal{C}}(c_5) = L_{\mathcal{C}}(c_6) = 1\frac{7}{9},$$

$$L_{\mathcal{C}}(c_1) = L_{\mathcal{C}}(c_2) = L_{\mathcal{C}}(c_3) = L_{\mathcal{C}}(c_4) = 1\frac{6}{9}$$

and therefore

$$L_{\mathcal{C}}(P, P) = \frac{1}{9}\left[1\frac{6}{9} \cdot 4 + 1\frac{7}{9} \cdot 2 + 1\frac{8}{9} \cdot 1 + 2\frac{1}{9} \cdot 2\right] = 1\frac{22}{27} = \bar{L}_{\mathcal{C}},$$

because P is uniform and the $\binom{2^3}{9-2^3} = 8$ Huffman codes are equivalent for identification.

Remark. Notice that Shannon's data compression gives

$$H(P) + 1 = \log 9 + 1 > \sum_{u=1}^{9} P_u\|c_u\| = \frac{1}{9}3 \cdot 7 + \frac{1}{9}4 \cdot 2 = 3\frac{2}{9} \geq H(P) = \log 9.$$

Example 2. $N = 10$. There are $\binom{2^3}{10-2^3} = 28$ Huffman codes.
The 4 worst Huffman codes are maximally unbalanced.

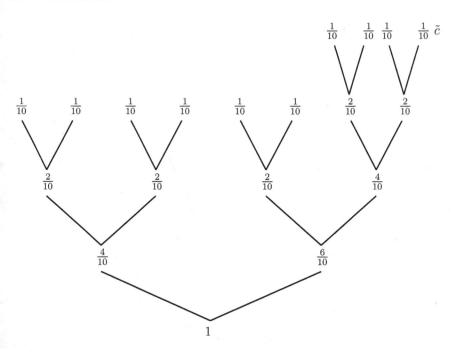

Here $L_C(P) = 2.2$ and $L_C(P, P) = 1.880$, because

$$L_C(P) = 1 + 0.6 + 0.4 + 0.2 = 2.2$$
$$L_C(P, P) = \frac{1}{10}[1.6 \cdot 4 + 1.8 \cdot 2 + 2.2 \cdot 4] = 1.880.$$

One of the 16 best Huffman codes

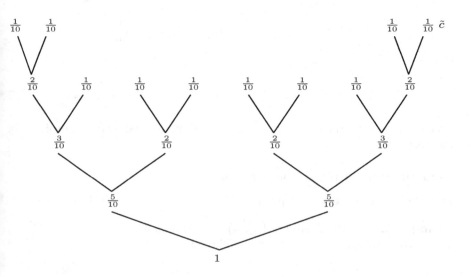

Here $L_\mathcal{C}(P) = 2.0$ and $L_\mathcal{C}(P, P) = 1.840$ because

$$L_\mathcal{C}(P) = L_\mathcal{C}(\tilde{c}) = 1 + 0.5 + 0.3 + 0.2 = 2.000$$
$$L_\mathcal{C}(P, P) = \frac{1}{5}(1.7 \cdot 2 + 1.8 \cdot 1 + 2.0 \cdot 2) = 1.840$$

Table 1. The best identification performances of Huffman codes for the uniform distribution

N	8	9	10	11	12	13	14	15
$L_\mathcal{C}(P)$	1.750	2.111	2.000	2.000	1.917	2.000	1.929	1.933
$L_\mathcal{C}(P, P)$	1.750	1.815	1.840	1.860	1.861	1.876	1.878	1.880

Actually $\lim\limits_{N \to \infty} L_\mathcal{C}(P^N) = 2$, but bad values occur for $N = 2^k + 1$ like $N = 9$ (see [5]).

One should prove that a best Huffman code for identification for the uniform distribution is best for the worst case and also for the mean.

However, for non-uniform sources generally Huffman codes are not best.

Example 3. Let $N = 4$, $P(1) = 0.49$, $P(2) = 0.25$, $P(3) = 0.25$, $P(4) = 0.01$. Then for the Huffman code $||c_1|| = 1$, $||c_2|| = 2$, $||c_3|| = ||c_4|| = 3$ and thus $L_\mathcal{C}(P) = 1 + 0.51 + 0.26 = 1.77$, $L_\mathcal{C}(P, P) = 0.49 \cdot 1 + 0.25 \cdot 1.51 + 0.26 \cdot 1.77 = 1.3277$, and $\bar{L}_\mathcal{C}(P) = \frac{1}{4}(1 + 1.51 + 2 \cdot 1.77) = 1.5125$.

However, if we use $\mathcal{C}' = \{00, 10, 11, 01\}$ for $\{1, \ldots, 4\}$ (4 is on the branch together with 1), then $L_{\mathcal{C}'}(P, u) = 1.5$ for $u = 1, 2, \ldots, 4$ and all three criteria give the same value 1.500 better than $L_\mathcal{C}(P) = 1.77$ and $\bar{L}_\mathcal{C}(P) = 1.5125$.

But notice that $L_\mathcal{C}(P, P) < L_{\mathcal{C}'}(P, P)$!

4   An Identification Code Universally Good for All P on $\mathcal{U} = \{1, 2, \ldots, N\}$

Theorem 1. Let $P = (P_1, \ldots, P_N)$ and let $k = \min\{\ell : 2^\ell \geq N\}$, then the regular binary tree of depth k defines a PC $\{c_1, \ldots, c_{2^k}\}$, where the codewords correspond to the leaves. To this code $\mathcal{C}_k$ corresponds the subcode $\mathcal{C}_N = \{c_i : c_i \in \mathcal{C}_k, 1 \leq i \leq N\}$ with

$$2\left(1 - \frac{1}{N}\right) \leq 2\left(1 - \frac{1}{2^k}\right) \leq \bar{L}_{\mathcal{C}_N}(P) \leq 2\left(2 - \frac{1}{N}\right) \tag{4.1}$$

and equality holds for $N = 2^k$ on the left sides.

Proof. By definition,

$$\bar{L}_{\mathcal{C}_N}(P) = \frac{1}{N}\sum_{u=1}^{N} L_{\mathcal{C}_N}(P, u) \tag{4.2}$$

and abbreviating $L_{C_N}(P, u)$ as $L(u)$ for $u = 1, \ldots, N$ and setting $L(u) = 0$ for $u = N + 1, \ldots, 2^k$ we calculate with $P_u \triangleq 0$ for $u = N + 1, \ldots, 2^k$

$$
\begin{aligned}
\sum_{u=1}^{2^k} L(u) =& \left[(P_1 + \cdots + P_{2^k})2^k\right] \\
& + \left[(P_1 + \cdots + P_{2^{k-1}})2^{k-1} + (P_{2^{k-1}+1} + \cdots + P_{2^k})2^{k-1}\right] \\
& + \left[(P_1 + \cdots + P_{2^{k-2}})2^{k-2} + (P_{2^{k-2}+1} + \cdots + P_{2^{k-1}})2^{k-2}\right. \\
& \quad + (P_{2^{k-1}+1} + \cdots + P_{2^{k-1}+2^{k-2}})2^{k-2} \\
& \quad \left. + (P_{2^{k-1}+2^{k-2}+1} + \cdots + P_{2^k})2^{k-2}\right] \\
& + \quad \cdots \\
& \qquad \cdot \\
& \qquad \cdot \\
& \qquad \cdot \\
& + \left[(P_1 + P_2)2 + (P_3 + P_4)2 + \cdots + (P_{2^k-1} + P_{2^k})2\right] \\
=& 2^k + 2^{k-1} + \cdots + 2 = 2(2^k - 1)
\end{aligned}
$$

and therefore

$$
\sum_{u=1}^{2^k} \frac{1}{2^k} L(u) = 2\left(1 - \frac{1}{2^k}\right). \tag{4.3}
$$

Now

$$
2\left(1 - \frac{1}{N}\right) \le 2\left(1 - \frac{1}{2^k}\right) = \sum_{u=1}^{2^k} \frac{1}{2^k} L(u) \le \sum_{u=1}^{N} \frac{1}{N} L(u) =
$$

$$
\frac{2^k}{N} \sum_{u=1}^{2^k} \frac{1}{2^k} L(u) = \frac{2^k}{N} 2\left(1 - \frac{1}{2^k}\right) \le 2\left(2 - \frac{1}{N}\right),
$$

which gives the result by (4.2). Notice that for $N = 2^k$, a power of 2, by (4.3)

$$
\bar{L}_{C_N}(P) = 2\left(1 - \frac{1}{N}\right). \tag{4.4}
$$

Remark. The upper bound in (4.1) is rough and can be improved significantly.

5   Identification Entropy $H_I(P)$ and Its Role as Lower Bound

Recall from the Introduction that

$$
H_I(P) = 2\left(1 - \sum_{u=1}^{N} P_u^2\right) \text{ for } P = (P_1 \ldots P_N). \tag{5.1}
$$

We begin with a small source

Example 4. Let $N = 3$. W.l.o.g. an optimal code $\mathcal{C}$ has the structure

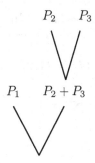

Claim.

$$\bar{L}_\mathcal{C}(P) = \sum_{u=1}^{3} \frac{1}{3} L_\mathcal{C}(P, u) \geq 2 \left(1 - \sum_{u=1}^{3} P_u^2\right) = H_I(P).$$

Proof. Set $L(u) = L_\mathcal{C}(P, u)$. $\sum_{u=1}^{3} L(u) = 3(P_1 + P_2 + P_3) + 2(P_2 + P_3)$.

This is smallest, if $P_1 \geq P_2 \geq P_3$ and thus $L(1) \leq L(2) = L(3)$. Therefore $\sum_{u=1}^{3} P_u L(u) \leq \frac{1}{3} \sum_{u=1}^{3} L(u)$. Clearly $L(1) = 1$, $L(2) = L(3) = 1 + P_2 + P_3$ and $\sum_{u=1}^{3} P_u L(u) = P_1 + P_2 + P_3 + (P_2 + P_3)^2$.

This does not change if $P_2 + P_3$ is constant. So we can assume $P = P_2 = P_3$ and $1 - 2P = P_1$ and obtain

$$\sum_{u=1}^{3} P_u L(u) = 1 + 4P^2.$$

On the other hand

$$2 \left(1 - \sum_{u=1}^{3} P_u^2\right) \leq 2 \left(1 - P_1^2 - 2 \left(\frac{P_2 + P_3}{2}\right)^2\right), \tag{5.2}$$

because $P_2^2 + P_3^2 \geq \frac{(P_2 + P_3)^2}{2}$.

Therefore it suffices to show that

$$1 + 4P^2 \geq 2\left(1 - (1 - 2P)^2 - 2P^2\right)$$
$$= 2(4P - 4P^2 - 2P^2)$$
$$= 2(4P - 6P^2) = 8P - 12P^2.$$

Or that $1 + 16P^2 - 8P = (1 - 4P)^2 \geq 0$.

We are now prepared for the first main result for $L(P, P)$.

Central in our derivations are proofs by induction based on **decomposition formulas for trees.**

Starting from the root a binary tree $\mathcal{T}$ goes via 0 to the subtree $\mathcal{T}_0$ and via 1 to the subtree $\mathcal{T}_1$ with sets of leaves $\mathcal{U}_0$ and $\mathcal{U}_1$, respectively. A code $\mathcal{C}$ for $(\mathcal{U}, P)$ can be viewed as a tree $\mathcal{T}$, where $\mathcal{U}_i$ corresponds to the set of codewords $\mathcal{C}_i$, $\mathcal{U}_0 \cup \mathcal{U}_1 = \mathcal{U}$.

The leaves are labelled so that $\mathcal{U}_0 = \{1, 2, \ldots, N_0\}$ and $\mathcal{U}_1 = \{N_0+1, \ldots, N_0+N_1\}$, $N_0 + N_1 = N$. Using probabilities

$$Q_i = \sum_{u \in \mathcal{U}_i} P_u, \qquad i = 0, 1$$

we can give the decomposition in

Lemma 1. For a code $\mathcal{C}$ for $(\mathcal{U}, P^N)$

$$L_{\mathcal{C}}((P_1, \ldots, P_N), (P_1, \ldots, P_N))$$

$$= 1 + L_{\mathcal{C}_0}\left(\left(\frac{P_1}{Q_0}, \ldots, \frac{P_{N_0}}{Q_0}\right), \left(\frac{P_1}{Q_0}, \ldots, \frac{P_{N_0}}{Q_0}\right)\right) Q_0^2$$

$$+ L_{\mathcal{C}_1}\left(\left(\frac{P_{N_0+1}}{Q_1}, \ldots, \frac{P_{N_0+N_1}}{Q_1}\right), \left(\frac{P_{N_0+1}}{Q_1}, \ldots, \frac{P_{N_0+N_1}}{Q_1}\right)\right) Q_1^2.$$

This readily yields

Theorem 2. *For every source* $(\mathcal{U}, P^N)$

$$3 > L(P^N) \geq L(P^N, P^N) \geq H_I(P^N).$$

Proof. The bound $3 > L(P^N)$ restates Theorem 3 of [5]. For $N = 2$ and any $\mathcal{C}$ $L_{\mathcal{C}}(P^2, P^2) \geq P_1 + P_2 = 1$, but

$$H_I(P^2) = 2(1 - P_1^2 - (1 - P_1)^2) = 2(2P_1 - 2P_1^2) = 4P_1(1 - P_1) \leq 1. \quad (5.3)$$

This is the induction beginning.

For the induction step use for any code $\mathcal{C}$ the decomposition formula in Lemma 1 and of course the desired inequality for N_0 and N_1 as induction hypothesis.

$$L_{\mathcal{C}}((P_1, \ldots, P_N), (P_1, \ldots, P_N))$$

$$\geq 1 + 2\left(1 - \sum_{u \in \mathcal{U}_0}\left(\frac{P_u}{Q_0}\right)^2\right) Q_0^2 + 2\left(1 - \sum_{u \in \mathcal{U}_1}\left(\frac{P_u}{Q_1}\right)^2\right) Q_1^2$$

$$\geq H_I(Q) + Q_0^2 H_I(P^{(0)}) + Q_1^2 H_I(P^{(1)}) = H_I(P^N),$$

where $Q = (Q_0, Q_1)$, $1 \geq H(Q)$, $P^{(i)} = \left(\frac{P_u}{Q_i}\right)_{u \in \mathcal{U}_i}$, and the grouping identity is used for the equality. This holds for every $\mathcal{C}$ and therefore also for $\min_{\mathcal{C}} L_{\mathcal{C}}(P^N)$. $\qquad \square$

6   On Properties of $\bar{L}(P^N)$

Clearly for $P^N = \left(\frac{1}{N}, \ldots, \frac{1}{N}\right)$ $\bar{L}(P^N) = L(P^N, P^N)$ and Theorem 2 gives therefore also the lower bound

$$\bar{L}(P^N) \geq H_I(P^N) = 2\left(1 - \frac{1}{N}\right), \tag{6.1}$$

which holds by Theorem 1 only for the Huffman code, but then for all distributions.

We shall see later in Example 6 that $H_I(P^N)$ is not a lower bound for general distributions P^N! Here we mean non-pathological cases, that is, not those where the inequality fails because $\bar{L}(P)$ (and also $L(P, P)$) is not continuous in P, but $H_I(P)$ is, like in the following case.

Example 5. Let $N = 2^k + 1$, $P(1) = 1 - \varepsilon$, $P(u) = \frac{\varepsilon}{2^k}$ for $u \neq 1$, $P^{(\varepsilon)} = \left(1 - \varepsilon, \frac{\varepsilon}{2^k}, \ldots, \frac{\varepsilon}{2^k}\right)$, then

$$\bar{L}(P^{(\varepsilon)}) = 1 + \varepsilon 2\left(1 - \frac{1}{2^k}\right) \tag{6.2}$$

and $\lim_{\varepsilon \to 0} \bar{L}(P^{(\varepsilon)}) = 1$ whereas $\lim_{\varepsilon \to 0} H_I(P^{(\varepsilon)}) = \lim_{\varepsilon \to 0}\left(2\left(1 - (1-\varepsilon)^2 - \left(\frac{\varepsilon}{2^k}\right)^2 2^k\right)\right) = 0$.

However, such a discontinuity occurs also in noiseless coding by Shannon.

The same discontinuity occurs for $L(P^{(\varepsilon)}, P^{(\varepsilon)})$.

Furthermore, for $N = 2$ $P^{(\varepsilon)} = (1 - \varepsilon, \varepsilon)$, $\bar{L}(P^{(\varepsilon)}) = 1$ $L(P^{(\varepsilon)}, P^{(\varepsilon)}) = 1$ and $H_I(P^{(\varepsilon)}) = 2(1 - \varepsilon^2 - (1 - \varepsilon)^2) = 0$ for $\varepsilon = 0$.

However, $\max_\varepsilon H_I(P^{(\varepsilon)}) = \max_\varepsilon 2(-2\varepsilon^2 + 2\varepsilon) = 1$ (for $\varepsilon = \frac{1}{2}$). Does this have any significance?

There is a second decomposition formula, which gives useful lower bounds on $\bar{L}_{\mathcal{C}}(P^N)$ for codes $\mathcal{C}$ with corresponding subcodes $\mathcal{C}_0, \mathcal{C}_1$ with uniform distributions.

Lemma 2. *For a code $\mathcal{C}$ for $(\mathcal{U}, P^N)$ and corresponding tree $\mathcal{T}$ let*

$$T_{\mathcal{T}}(P^N) = \sum_{u \in \mathcal{U}} L(u).$$

Then (in analogous notation)

$$T_{\mathcal{T}}(P^N) = N_0 + N_1 + T_{\mathcal{T}_0}(P^{(0)})Q_0 + T_{\mathcal{T}_1}(P^{(1)})Q_1.$$

However, identification entropy is not a lower bound for $\bar{L}(P^N)$. We strive now for the worst deviation by using Lemma 2 and by starting with $\mathcal{C}$, whose parts $\mathcal{C}_0, \mathcal{C}_1$ satisfy the entropy inequality.

Then inductively

$$T_T(P^N) \geq N + 2\left(1 - \sum_{u \in \mathcal{U}_0} \left(\frac{P_u}{Q_0}\right)^2\right) N_0 Q_0 + 2\left(1 - \sum_{u \in \mathcal{U}_1} \left(\frac{P_u}{Q_1}\right)^2\right) N_1 Q_1 \quad (6.3)$$

and

$$\frac{T_T(P^N)}{N} \geq 1 + \sum_{i=0}^{1} 2\left(1 - \sum_{u \in \mathcal{U}_i} \left(\frac{P_u}{Q_i}\right)^2\right) \frac{N_i Q_i}{N} \triangleq A, \text{ say.}$$

We want to show that for

$$2\left(1 - \sum_{u \in \mathcal{U}} P_u^2\right) \triangleq B, \text{ say,}$$

$$A - B \geq 0. \quad (6.4)$$

We write

$$A - B = \left[-1 + 2\sum_{i=0}^{1} \frac{N_i Q_i}{N}\right] + 2\left[\sum_{u \in \mathcal{U}} P_u^2 - \sum_{i=0}^{1} \sum_{u \in \mathcal{U}_i} \left(\frac{P_u}{Q_i}\right)^2 \frac{N_i Q_i}{N}\right]$$
$$= C + D, \text{ say.} \quad (6.5)$$

C and D are functions of P^N and the partition $(\mathcal{U}_0, \mathcal{U}_1)$, which determine the Q_i's and N_i's. The minimum of this function can be analysed without reference to codes. Therefore we write here the partitions as $(\mathcal{U}_1, \mathcal{U}_2)$, $C = C(P^N, \mathcal{U}_1, \mathcal{U}_2)$ and $D = D(P^N, \mathcal{U}_1, \mathcal{U}_2)$. We want to show that

$$\min_{P^N, (\mathcal{U}_1, \mathcal{U}_2)} C(P^N, \mathcal{U}_1, \mathcal{U}_2) + D(P^N, \mathcal{U}_1, \mathcal{U}_2) \geq 0. \quad (6.6)$$

A first idea
Recall that the proof of (5.3) used

$$2Q_0^2 + 2Q_1^2 - 1 \geq 0. \quad (6.7)$$

Now if $Q_i = \frac{N_i}{N}$ $(i = 0, 1)$, then by (6.7)

$$A - B = \left[-1 + 2\sum_{i=0}^{1} \frac{N_i^2}{N^2}\right] + 2\left[\sum_{u \in \mathcal{U}} P_u^2 - \sum_{u \in \mathcal{U}} P_u^2\right] \geq 0.$$

A goal could be now to achieve $Q_i \sim \frac{N_i}{N}$ by rearrangement not increasing $A - B$, because in case of equality $Q_i = \frac{N_i}{N}$ that does it.

This leads to a nice **problem of balancing a partition** $(\mathcal{U}_1, \mathcal{U}_2)$ of $\mathcal{U}$. More precisely for $P^N = (P_1, \ldots, P_N)$

$$\varepsilon(P^N) = \min_{\phi \neq \mathcal{U}_1 \subset \mathcal{U}} \left| \sum_{u \in \mathcal{U}_1} P_u - \frac{|\mathcal{U}_1|}{N} \right|.$$

Then clearly for an optimal $\mathcal{U}_1$

$$Q_1 = \frac{|\mathcal{U}_1|}{N} \pm \varepsilon(P^N) \quad \text{and} \quad Q_2 = \frac{N - |\mathcal{U}_1|}{N} \mp \varepsilon(P^N).$$

Furthermore, one comes to a question of some independent interest. What is

$$\max_{P^N} \varepsilon(P^N) = \max_{P^N} \min_{\phi \neq \mathcal{U}_1 \subset \mathcal{U}} \left| \sum_{u \in \mathcal{U}_1} P_u - \frac{|\mathcal{U}_1|}{N} \right|?$$

One can also go from sets $\mathcal{U}_1$ to distributions $\mathcal{R}$ on $\mathcal{U}$ and get, perhaps, a smoother problem in the spirit of game theory.

However, we follow another approach here.

A rearrangement

We have seen that for $Q_i = \frac{N_i}{N}$ $D = 0$ and $C \geq 0$ by (6.7). Also, there is "air" up to 1 in C, if $\frac{N_i}{N}$ is away from $\frac{1}{2}$. Actually, we have

$$C = -\left(\frac{N_1}{N} + \frac{N_2}{N}\right)^2 + 2\left(\frac{N_1}{N}\right)^2 + 2\left(\frac{N_2}{N}\right)^2 = \left(\frac{N_1}{N} - \frac{N_2}{N}\right)^2. \tag{6.8}$$

Now if we choose for $N = 2m$ even $N_1 = N_2 = m$, then the air is out here, $C = 0$, but it should enter the second term D in (6.5).

Let us check this case first. Label the probabilities $P_1 \geq P_2 \geq \cdots \geq P_N$ and define $\mathcal{U}_1 = \left\{1, 2, \ldots, \frac{N}{2}\right\}, \mathcal{U}_2 = \left\{\frac{N}{2} + 1, \ldots, N\right\}$. Thus obviously

$$Q_1 = \sum_{u \in \mathcal{U}_1} P_u \geq Q_2 = \sum_{u \in \mathcal{U}_2} P_u$$

and

$$D = 2\left(\sum_{u \in \mathcal{U}} P_u^2 - \sum_{i=1}^{2} \frac{1}{2Q_i} \sum_{u \in \mathcal{U}_i} P_u^2\right).$$

Write $Q = Q_1$, $1 - Q = Q_2$. We have to show

$$\sum_{u \in \mathcal{U}_1} P_u^2 \left(1 - \frac{1}{(2Q)^2}\right) \geq \sum_{u \in \mathcal{U}_2} P_u^2 \left(\frac{1}{(2Q_2)^2} - 1\right)$$

or

$$\sum_{u \in \mathcal{U}_1} P_u^2 \frac{(2Q)^2 - 1}{(2Q)^2} \geq \sum_{u \in \mathcal{U}_2} P_u^2 \left(\frac{1 - (2(1-Q))^2}{(2(1-Q))^2} \right). \tag{6.9}$$

At first we decrease the left hand side by replacing $P_1, \ldots, P_{\frac{N}{2}}$ all by $\frac{2Q}{N}$. This works because $\sum P_i^2$ is Schur-concave and $P_1 \geq \cdots \geq P_{\frac{N}{2}}$, $\frac{2Q}{N} = \frac{2(P_1 + \cdots + P_{\frac{N}{2}})}{N} \geq P_{\frac{N}{2}+1}$, because $\frac{2Q}{N} \geq P_{\frac{N}{2}} \geq P_{\frac{N}{2}+1}$. Thus it suffices to show that

$$\frac{N}{2} \left(\frac{2Q}{N} \right)^2 \frac{(2Q)^2 - 1}{(2Q)^2} \geq \sum_{u \in \mathcal{U}_2} P_u^2 \frac{1 - (2(1-Q))^2}{(2(1-Q))^2} \tag{6.10}$$

or that

$$\frac{1}{2N} \geq \sum_{u \in \mathcal{U}_2} P_u^2 \frac{1 - (2(1-Q))^2}{(2(1-Q))^2((2Q)^2 - 1)}. \tag{6.11}$$

Secondly we increase now the right hand side by replacing $P_{\frac{N}{2}+1}, \ldots, P_N$ all by their maximal possible values $\left(\frac{2Q}{N}, \frac{2Q}{N}, \ldots, \frac{2Q}{N}, q \right) = (q_1, q_2, \ldots, q_t, q_{t+1})$, where $q_i = \frac{2Q}{N}$ for $i = 1, \ldots, t$, $q_{t+1} = q$ and $t \cdot \frac{2Q}{N} + q = 1 - Q$, $t = \left\lfloor \frac{(1-Q)N}{2Q} \right\rfloor$, $q < \frac{2Q}{N}$.
Thus it suffices to show that

$$\frac{1}{2N} \geq \left(\left\lfloor \frac{(1-Q)N}{2Q} \right\rfloor \cdot \left(\frac{2Q}{N} \right)^2 + q^2 \right) \frac{1 - (2(1-Q))^2}{(2(1-Q))^2((2Q)^2 - 1)}. \tag{6.12}$$

Now we inspect the easier case $q = 0$. Thus we have $N = 2m$ and equal probabilities $P_i = \frac{1}{m+t}$ for $i = 1, \ldots, m + t = m$, say for which (6.12) goes wrong! We arrived at a very simple counterexample.

Example 6. In fact, simply for $P_M^N = \left(\frac{1}{M}, \ldots, \frac{1}{M}, 0, 0, 0 \right)$ $\lim_{N \to \infty} \bar{L}(P_M^N) = 0$, whereas

$$H_I(P_M^N) = 2 \left(1 - \frac{1}{M} \right) \text{ for } N \geq M.$$

Notice that here

$$\sup_{N,M} |\bar{L}(P_M^N) - H_I(P_M^N)| = 2. \tag{6.13}$$

This leads to the

Problem 1. Is $\sup_P |\bar{L}(P) - H_I(P)| = 2$? which is solved in the next section.

7 Upper Bounds on $\bar{L}(P^N)$

We know from Theorem 1 that

$$\bar{L}(P^{2^k}) \leq 2 \left(1 - \frac{1}{2^k} \right) \tag{7.1}$$

and come to the

Problem 2. Is $\bar{L}(P^N) \leq 2\left(1 - \frac{1}{2^k}\right)$ for $N \leq 2^k$?

This is the case, if the answer to the next question is positive.

Problem 3. Is $\bar{L}\left(\left(\frac{1}{N}, \ldots, \frac{1}{N}\right)\right)$ monotone increasing in N?

In case the inequality in Problem 2 does not hold then it should with a very small deviation. Presently we have the following result, which together with (6.13) settles Problem 1.

Theorem 3. *For* $P^N = (P_1, \ldots, P_N)$

$$\bar{L}(P^N) \leq 2\left(1 - \frac{1}{N^2}\right).$$

Proof. (The induction beginning $\bar{L}(P^2) = 1 \leq 2\left(1 - \frac{1}{4}\right)$ holds.) Define now $\mathcal{U}_1 = \left\{1, 2, \ldots, \lfloor\frac{N}{2}\rfloor\right\}$, $\mathcal{U}_2 = \left\{\lfloor\frac{N}{2}\rfloor + 1, \ldots, N\right\}$ and Q_1, Q_2 as before. Again by the decomposition formula of Lemma 2 and induction hypothesis

$$T(P^N) \leq N + 2\left(1 - \frac{1}{\lfloor\frac{N}{2}\rfloor^2}\right)Q_1\left\lfloor\frac{N}{2}\right\rfloor + 2\left(1 - \frac{1}{\lceil\frac{N}{2}\rceil^2}\right)Q_2 \cdot \left\lceil\frac{N}{2}\right\rceil$$

and

$$\bar{L}(P^N) = \frac{1}{N}T(P^N) \leq 1 + \frac{2\lfloor\frac{N}{2}\rfloor Q_1 + 2\lceil\frac{N}{2}\rceil Q_2}{N} - \frac{2}{\lfloor\frac{N}{2}\rfloor} \cdot \frac{Q_1}{N} - \frac{2Q_2}{\lceil\frac{N}{2}\rceil N} \quad (7.2)$$

Case N even: $\bar{L}(P^N) \leq 1 + Q_1 + Q_2 - \left(\frac{4}{N^2}Q_1 + \frac{4}{N^2}Q_2\right) = 2 - \frac{4}{N^2} = 2\left(1 - \frac{2}{N^2}\right) \leq 2\left(1 - \frac{1}{N^2}\right)$

Case N odd: $\bar{L}(P^N) \leq 1 + \frac{N-1}{N}Q_1 + \frac{N+1}{N}Q_2 - 4\left(\frac{Q_1}{(N-1)N} + \frac{Q_2}{(N+1)N}\right) \leq 1 + 1 + \frac{Q_2 - Q_1}{N} - \frac{4}{(N+1)N}$

Choosing the $\lceil\frac{N}{2}\rceil$ smallest probabilities in $\mathcal{U}_2$ (after proper labelling) we get for $N \geq 3$

$$\bar{L}(P^N) \leq 1 + 1 + \frac{1}{N \cdot N} - \frac{4}{(N+1)N} = 2 + \frac{1 - 3N}{(N+1)N^2} \leq 2 - \frac{2}{N^2} = 2\left(1 - \frac{1}{N^2}\right),$$

because $1 - 3N \leq -2N - 2$ for $N \geq 3$.

8 The Skeleton

Assume that all individual probabilities are powers of $\frac{1}{2}$

$$P_u = \frac{1}{2^{\ell_u}}, \quad u \in \mathcal{U}. \quad (8.1)$$

Define then $k = k(P^N) = \max_{u \in \mathcal{U}} \ell_u$.

Since $\sum\limits_{u \in \mathcal{U}} \frac{1}{2^{\ell_u}} = 1$ by Kraft's theorem there is a PC with codeword lengths

$$||c_u|| = \ell_u. \tag{8.2}$$

Notice that we can put the probability $\frac{1}{2^k}$ at all leaves in the binary regular tree and that therefore

$$L(u) = \frac{1}{2} \cdot 1 + \frac{1}{4} \cdot 2 + \frac{1}{2^3} 3 + \cdots + \frac{1}{2^t} t + \cdots + \frac{2}{2^{\ell_u}}. \tag{8.3}$$

For the calculation we use

Lemma 3. *Consider the polynomials* $G(x) = \sum\limits_{t=1}^{r} t \cdot x^t + r x^r$ *and* $f(x) = \sum\limits_{t=1}^{r} x^t$, *then*

$$G(x) = x\, f'(x) + r\, x^r = \frac{(r+1)x^{r+1}(x-1) - x^{r+2} + x}{(x-1)^2} + r\, x^r.$$

Proof. Using the summation formula for a geometric series

$$f(x) = \frac{x^{r+1} - 1}{x - 1} - 1$$

$$f'(x) = \sum_{t=1}^{r} t\, x^{t-1} = \frac{(r+1)x^r(x-1) - x^{r+1} + 1}{(x-1)^2}.$$

This gives the formula for G.
Therefore for $x = \frac{1}{2}$

$$G\left(\frac{1}{2}\right) = -(r+1)\left(\frac{1}{2}\right)^r - \left(\frac{1}{2}\right)^r + 2 + r\left(\frac{1}{2}\right)^r$$

$$= -\frac{1}{2^{r-1}} + 2$$

and since $L(u) = G\left(\frac{1}{2}\right)$ for $r = \ell_u$

$$L(u) = 2\left(1 - \frac{1}{2^{\ell_u}}\right) = 2\left(1 - \frac{1}{2^{\log \frac{1}{P_u}}}\right)$$

$$= 2(1 - P_u). \tag{8.4}$$

Therefore
$$L(P^N, P^N) \leq \sum_u P_u(2(1 - P_u)) = H_I(P^N) \tag{8.5}$$

and by Theorem 2
$$L(P^N, P^N) = H_I(P^N). \tag{8.6}$$

Theorem 4. [1] *For* $P^N = (2^{-\ell_1}, \ldots, 2^{-\ell_N})$ *with 2-powers as probabilities*

$$L(P^N, P^N) = H_I(P^N).$$

This result shows that identification entropy is a right measure for identification source coding. For Shannon's data compression we get for this source $\sum_u p_u \|c_u\| = \sum_u p_u \ell_u = -\sum_u p_u \log p_u = H(P^N)$, again an identity.

For general sources the minimal average length deviates there from $H(P^N)$, but by not more than 1.

Presently we also have to accept some deviation from the identity.

We give now a first (crude) approximation. Let

$$2^{k-1} < N \leq 2^k \tag{8.7}$$

and assume that the probabilities are sums of powers of $\frac{1}{2}$ with exponents not exceeding k

$$P_u = \sum_{j=1}^{\alpha(u)} \frac{1}{2^{\ell_{uj}}}, \ell_{u1} \leq \ell_{u2} \leq \cdots \leq \ell_{u\alpha(u)} \leq k. \tag{8.8}$$

We now use the **idea of splitting object** u into objects $u1, \ldots, u\alpha(u)$. (8.9)
Since

$$\sum_{u,j} \frac{1}{2^{\ell_{uj}}} = 1 \tag{8.10}$$

again we have a PC with codewords c_{uj} $(u \in \mathcal{U}, j = 1, \ldots, \alpha(u))$ and a regular tree of depth k with probabilities $\frac{1}{2^k}$ on all leaves.

Person u can find out whether u occurred, he can do this (and more) by finding out whether $u1$ occurred, then whether $u2$ occurred, etc. until $u\alpha(u)$. Here

$$L(us) = 2\left(1 - \frac{1}{2^{\ell_{us}}}\right) \tag{8.11}$$

and

$$\sum_{u,s} L(us)P_{us} = 2\left(1 - \sum_{u,s} \frac{1}{2^{\ell_{us}}} \cdot \frac{1}{2^{\ell_{us}}}\right) = 2\left(1 - \sum_u \left(\sum_{s=1}^{\alpha(u)} P_{us}^2\right)\right). \tag{8.12}$$

On the other hand, being interested only in the original objects this is to be compared with $H_I(P^N) = 2\left(1 - \sum_u \left(\sum_s P_{us}\right)^2\right)$, which is smaller.

---

[1] In a forthcoming paper "An interpretation of identification entropy" the author and Ning Cai show that $L_{\mathcal{C}}(P, Q)^2 \leq L_{\mathcal{C}}(P, P)L_{\mathcal{C}}(Q, Q)$ and that for a block code $\mathcal{C}$ $\min_{P \text{ on } \mathcal{U}} L_{\mathcal{C}}(P, P) = L_{\mathcal{C}}(R, R)$, where R is the uniform distribution on $\mathcal{U}$! Therefore $\bar{L}_{\mathcal{C}}(P) \leq L_{\mathcal{C}}(P, P)$ for a block code $\mathcal{C}$.

However, we get

$$\left(\sum_s P_{us}\right)^2 = \sum_s P_{us}^2 + \sum_{s\neq s'} P_{us}P_{us'} \leq 2\sum_s P_{us}^2$$

and therefore

Theorem 5

$$L(P^N, P^N) \leq 2\left(1 - \sum_u \left(\sum_{s=1}^{\alpha(u)} P_{us}^2\right)\right) \leq 2\left(1 - \frac{1}{2}\sum_u P_u^2\right). \tag{8.13}$$

For $P_u = \frac{1}{N}(u \in \mathcal{U})$ this gives the upper bound $2\left(1 - \frac{1}{2N}\right)$, which is better than the bound in Theorem 3 for uniform distributions.

Finally we derive

Corollary

$$L(P^N, P^N) \leq H_I(P^N) + \max_{1 \leq u \leq N} P_u.$$

It shows the lower bound of $L(P^n, P^N)$ by $H_I(P^N)$ and this upper bound are close.

Indeed, we can write the upper bound

$$2\left(1 - \frac{1}{2}\sum_{u=1}^N P_u^2\right) \text{ as } H_I(P^N) + \sum_{u=1}^N P_u^2$$

and for $P = \max_{1 \leq u \leq N} P_u$, let the positive integer t be such that $1 - tp = p' < p$. Then by Schur concavity of $\sum_{u=1}^N P_u^2$ we get $\sum_{u=1}^N P_u^2 \leq t \cdot p^2 + p'^2$, which does not exceed $p(tp + p') = p$.

Remark. In its form the bound is tight, because for $P^2 = (p, 1 - p)$

$$L(P^2, P^2) = 1 \text{ and } \lim_{p \to 1} H_I(P^2) + p = 1.$$

Remark. Concerning $\bar{L}(P^N)$ (see footnote) for $N = 2$ the bound $2\left(1 - \frac{1}{4}\right) = \frac{3}{2}$ is better than $H_I(P^2) + \max_u P_u$ for $P^2 = \left(\frac{2}{3}, \frac{1}{3}\right)$, where we get $2(2p_1 - 2p_1^2) + p_1 = p_1(5 - 4p_1) = \frac{2}{3}\left(5 - \frac{8}{3}\right) = \frac{14}{9} > \frac{3}{2}$.

9   Directions for Research

A. Study

$$L(P, R) \text{ for } P_1 \geq P_2 \geq \cdots \geq P_N \text{ and } R_1 \geq R_1 \geq \cdots \geq R_N.$$

B. Our results can be extended to q-ary alphabets, for which then identification entropy has the form

$$H_{I,q}(P) = \frac{q}{q-1}\left(1 - \sum_{i=1}^{N} P_i^2\right).^2$$

C. So far we have considered prefix-free codes. One also can study
   a. fix-free codes
   b. uniquely decipherable codes

D. Instead of the number of checkings one can consider other cost measures like the αth power of the number of checkings and look for corresponding entropy measures.

E. The analysis on universal coding can be refined.

F. In [5] first steps were taken towards source coding for K-identification. This should be continued with a reflection on entropy and also towards GTIT.

G. **Grand ideas:** Other data structures
   a. Identification source coding with **parallelism**: there are N identical code-trees, each person uses his own, but informs others
   b. Identification source coding with **simultaneity**: $m(m = 1, 2, \ldots, N)$ persons use simultaneously the same tree.

H. It was shown in [5] that $L(P^N) \leq 3$ for all P^N. Therefore there is a universal constant $A = \sup_{P^N} L(P^N)$. It should be estimated!

I. We know that for $\lambda \in (0,1)$ there is a subset $\mathcal{U}$ of cardinality $\exp\{f(\lambda)H(P)\}$ with probability at least λ for $f(\lambda) = (1 - \lambda)^{-1}$ and $\lim_{\lambda \to 0} f(\lambda) = 1$.

   Is there such a result for $H_I(P)$?

   It is very remarkable that in our world of source coding the classical range of entropy $[0, \infty)$ is replaced by $[0, 2)$ – singular, dual, plural – there is some appeal to this range.

References

1. C.E. Shannon, A mathematical theory of communication, Bell Syst. Techn. J. 27, 379-423, 623-656, 1948.

2. D.A. Huffman, A method for the construction of minimum redundancy codes, Proc. IRE 40, 1098-1101, 1952.

3. R. Ahlswede and G. Dueck, Identification via channels, IEEE Trans. Inf. Theory, Vol. 35, No. 1, 15-29, 1989.

4. R. Ahlswede, General theory of information transfer: updated, General Theory of Information Transfer and Combinatorics, a Special issue of Discrete Applied Mathematics.

5. R. Ahlswede, B. Balkenhol, and C. Kleinewächter, Identification for sources, this volume.

---

^{2} In the forthcoming paper mentioned in 1. the coding theoretic meanings of the two factors $\frac{q}{q-1}$ and $\left(1 - \sum_{i=1}^{N} P_i^2\right)$ are also explained.

Optimal Information Measures for Weakly Chaotic Dynamical Systems

V. Benci and S. Galatolo

Abstract. The study of weakly chaotic dynamical systems suggests that an important indicator for their classification is the quantity of information that is needed to describe their orbits. The information can be measured by the use of suitable compression algorithms. The algorithms are "optimal" for this purpose if they compress very efficiently zero entropy strings. We discuss a definition of optimality in this sense. We also show that the set of optimal algorithms is not empty, showing a concrete example. We prove that the algorithms which are optimal according to the above definition are suitable to measure the information needed to describe the orbits of the Manneville maps: in these examples the information content measured by these algorithms has the same asymptotic behavior as the algorithmic information content.

1 Introduction

The study of weakly chaotic (0-entropy) dynamical systems is a growing subject in the physical literature and in the science of complexity. There are connections with many physical phenomena: self organized criticality, anomalous diffusion process, transition to turbulence, formation of complex structures and many others.

Some conceptual instruments have been developed to understand and classify this kind of phenomena. Among them we recall the generalized entropies, notions related to fractal dimension, notions related to diffusion processes etc. To get an idea of this fast growing literature one can consult for example http://tsallis.cat.cbpf.br/biblio.htm for an updated bibliography about Tsallis generalized entropy or [3] for the diffusion entropy. The subject is still far from being understood. Many of these works are heuristic or experimental (mainly computer simulations) and few rigorous definitions and results can be found.

Conversely there are rigorous negative results ([24], [7]) about the use of generalized entropies for the construction of invariants for 0-entropy measure preserving systems.

We approach the study of weak chaos by considering the asymptotic behavior of the information needed to describe the evolution of the orbits of the system under consideration.

Roughly speaking in a positive entropy system the information relative to a generic orbit increases linearly with the time and it is proportional to the entropy of the system.

On the other hand when a system has zero entropy the information increases less than linearly with time (e.g. increases as $log(t)$, t^α with $0 < \alpha < 1, ...$) and we consider this asymptotic behavior as an indicator able to classify various kind of weakly chaotic dynamics.

R. Ahlswede et al. (Eds.): Information Transfer and Combinatorics, LNCS 4123, pp. 614–627, 2006.
© Springer-Verlag Berlin Heidelberg 2006

Conversely, when we have an experimental time series, we can measure the behavior of its information content to gain knowledge on the underlying unknown system: this is a new method in the analysis of time series.

To describe our method in a rigorous framework, we first must state precisely what we mean by information. This will be done mainly in section 2, for the moment we remark that in our framework this concept has to be well posed even for a single string, thus we need a pointwise (not average) notion of information. The most powerful concept of this kind is the AIC (Algorithmic Information Content, see the Appendix). The AIC is a flexible notion and it allows to prove many interesting and deep theorems, but unfortunately is not a computable function and some other tool needs to be considered for practical purposes. As we will see later, the use of suitable data compression algorithms allows to estimate the information content of finite strings and this is the main topic of this paper.

In order to apply these tools of information theory to dynamical systems, we translate the orbits into symbolic sequences (symbolic orbits) and we measure the information content of them.

Let (X, T) be a dynamical system where X is a compact metric space and $T : X \to X$ is continuous. Sometimes, we assume that X is equipped with a Borel invariant probability measure μ (namely a measure such that for each measurable A, $\mu(T^{-1}(A)) = \mu(A)$). In this case, in a probabilistic sense, we get a stationary system. However, the stationarity of the system is not necessary to apply our method; actually in Section 4 we will show an application to a non stationary system.

The simplest way to obtain symbolic orbits from real orbits is to consider a partition

$$\beta = \{\beta_1, ..., \beta_m\}$$

of the space. Then, given an orbit $x, T(x), T^2(x), ...$, we associate the symbolic string which lists the sets of β visited by the orbit of x.

In this way an infinite string is associated to x. Actually, we get the infinite string $\omega(x) = (\omega_i)_{i \in \mathbb{N}}$ with $\omega_i \in \{1, ..., m\}$ such that $\omega_i = j$ implies $T^i(x) \in \beta_j$. We will also consider the string $\omega^n(x) = (\omega_i)_{i \in \{1,...,n\}}$ of the first n digits of $\omega(x)$. Now, it is possible to consider the asymptotic behavior of the sequence $n \mapsto I(\omega^n)$, where $I(\omega^n)$ is the information content of ω^n, the symbolic string related to the first n steps of the orbits of x.

We remark that the association $x \mapsto \omega(x)$ depends on the choice of β. Changing the partition, the properties of the symbolic model might change as well. This is a delicate and important technical point [1] but it will be not treated in the present work.

A more refined approach to the definition of information content of orbits in dynamical systems is obtained by using open covers instead of partitions. In this

---

[1] This is the point where the construction of generalized entropies as metric invariants fails (see [24]). The very important feature of the Kolmogorov-Sinai entropy with respect to a partition is that this entropy changes continuously if we slightly modify the partition and this fact allows to avoid pathologies when taking the supremum of the entropy over all partitions.

case the open sets overlap in some points and a set of possible symbolic orbits is associated to a single orbit in X. In this case, the information content of n steps of the real orbit with respect to β is defined as the minimum information content among all the possible n steps symbolic orbits (see [13], [8]).

This approach makes the definition of information content of an orbit more involved but it allows to get rid of the choice of the cover β taking the supremum over all open covers [2] and the definition of orbit information content depends only on x and on its dynamics. Moreover the orbit information content is invariant under topological conjugation (see [25], [8] for the positive entropy case, [13] for the 0 entropy case). Thus we get new invariants for topological dynamical systems.

Moreover, using this method we recover some relation between entropy, information, initial condition sensitivity and Hausdorff dimension which are known in the positive entropy case (see [21] and [13], [11] for the zero entropy case).

Thus the main features of chaotic dynamics such as sensitivity, dimension, entropy etc. are strictly related to the orbit information content. Since the AIC is not computable it seems necessary to define a notion of information content by a lossless data compression algorithm.

The above considerations motivate the study of the various features of compression algorithms, the various possible definitions of information content and the properties which follow from them.

In the compressed string there is all the information needed to recover the original string then the length of the compressed string is an upper bound of the information content of the original one. Clearly this upper bound is sharp and it is close to the AIC if the algorithms compresses the string in an efficient way.

If a string is generated by positive entropy systems, there are many algorithms which give *optimal* results [12]; on the contrary if a string is generated by a 0-entropy system the traditional definition of optimality is not sufficient (see section 3).

In section 3, we will discuss a stronger condition of (*asymptotic optimality*, cf. Def 2) and in section 4 we will prove that this condition is sufficient to give the same results than the AIC for an an important class of dynamical systems such as the Manneville maps.

In section 5, we will show that the set of algorithms that are asymptotically optimal is not empty, producing a concrete example.

2 Information

There are different notions of information, the first one is due to Shannon. In his pioneering work, Shannon defined the quantity of information as a statistical notion using the tools of probability theory. Thus in Shannon framework, the quantity of information which is contained in a string depends on its context ([16]). For example the string *'pane'* contains a certain information when it is

---
[2] Referring to the previous footnote: in some sense the use of open covers recovers continuity.

considered as a string coming from a given language. For example this word contains a certain amount of information in English; the same string $'pane'$ contains much less Shannon information when it is considered as a string coming from the Italian language because it is much more common (in fact it means "bread"). Roughly speaking, the Shannon information of a string s is given by

$$I(s) = \log_2 \frac{1}{p(s)} \tag{1}$$

where $p(s)$ denotes the probability of s in a given context. The logarithm is taken in base two so that the information can be measured in binary digits (bits).[3]

If in a language the occurrences of the letters are independent of each other, the information carried by each letter is given by

$$I(a_i) = \log \frac{1}{p_i}.$$

where p_i is the probability of the letter a_i. Then the average information of each letter is given by

$$h = \sum_i p_i \log \frac{1}{p_i}. \tag{2}$$

Shannon called the quantity h entropy for its formal similarity with Boltzmann's entropy.

We are interested in giving a definition of quantity of information of a *single* string independent on the context, and independent on any probability measure. Of course we will require this definition to be strictly related to the Shannon entropy when we equip the space of all the strings with a suitable probability measure. Such a definition will be suitable for the analysis of experimental data, where often we have a single time series to analyze.

In our approach the intuitive meaning of *quantity of information $I(s)$* contained in s is the following one:

I(s) is the length of the smallest binary message from which you can reconstruct s.

In this framework Kolmogorov and Chaitin proposed in 1965 a definition of information content (AIC) that is defined for a single string. This definition is very flexible and allows to prove very interesting theorems, but unfortunately it is not computable; there are no algorithms to calculate the AIC of any given string.

Let us suppose to have some recursive lossless (reversible) coding procedure (for example, the data compression algorithms that are in any personal computer). Since the coded string contains all the information that is necessary to reconstruct the original string, we can consider the length of the coded string as

---

[3] From now on, we will use the symbol "log" just for the base 2 logarithm "$\log_2$" and we will denote the natural logarithm by "$\ln''$".

an approximate measure of the quantity of information that is contained in the original string.

Of course not all the coding procedures are equivalent and give the same performances, so some care is necessary in the definition of information content. For this reason we introduce various notions of *optimality* (Definitions 1 and 2) of an algorithm Z, defined by comparing its compression ratio with the empirical entropy (defined in the following section).

3   Empirical Entropy and Optimal Compression

Let us consider a finite alphabet $\mathcal{A}$ and the set of finite strings on $\mathcal{A}$, that is $\mathcal{A}^* = \bigcup_{n=1}^{\infty} \mathcal{A}^n$. Moreover we will be interested also in the set of infinite strings $\mathcal{A}^{\mathbf{N}}$; $\mathcal{A}^{\mathbf{N}}$ has a natural structure of dynamical system with the dynamics defined by the shift map

$$\sigma\left(\omega\right) = \sigma\left(\{\omega_i\}\right) = \{\omega_{i+1}\}.$$

If $\left(\mathcal{A}^{\mathbf{N}}, \sigma\right)$ is equipped with a invariant measure μ, this is called symbolic dynamical system and provides a dynamical model for an information source and the Kolmogorov entropy of the system coincides with the Shannon entropy of the information source.

Now let

$$Z : \mathcal{A}^* \to \{0,1\}^*$$

be a recursive injective function (the compression algorithm) and let us consider the information content as it is measured by Z as

$$I_Z(s) = |Z(s)| \tag{3}$$

and the relative compression ratio

$$K_Z(s) = \frac{|Z(s)|}{|s|} \tag{4}$$

where $|s|$ is the length of the string s.

In the following we will also see that choosing Z in a suitable way, it is possible to investigate interesting properties of dynamical systems.

In a zero entropy system we are interested to the asymptotic behavior of $I_Z(s^n)$ when s^n represents a n steps symbolic orbits of a dynamical system or equivalently the speed of convergence to 0 of the corresponding compression ratio $K_Z(s)$.

The empirical entropy is a quantity that can be thought to be in the middle between Shannon entropy and a pointwise definition of information content. The *empirical entropy* of a given string is a sequence of numbers $\hat{H}_l$ giving statistical measures of the average information content of the digits of the string s.

Let s be a finite string of length n. We now define $\hat{H}_l(s)$, $l \geq 1$, the lth empirical entropy of s. We first introduce the *empirical frequencies* of a word in the string s: let us consider $w \in \mathcal{A}^l$, a string on the alphabet $\mathcal{A}$ with length l;

let $s^{(m_1,m_2)} \in \mathcal{A}^{m_2-m_1}$ be the string containing the segment of s starting from the m_1-th digit up to the m_2-th digit; let

$$\delta(s^{(i+1,i+l)}, w) = \begin{cases} 1 \; if \; s^{(i+1,i+l)} = w \\ 0 \quad otherwise \end{cases} (\, 0 \leq i \leq n - l).$$

The relative frequency of w (the number of occurrences of the word w divided by the total number of l-digit sub words) in s is then

$$P(s,w) = \frac{1}{n-l+1} \sum_{i=0}^{n-l} \delta(s^{(i+1,i+l)}, w).$$

This can be interpreted as the "empirical" probability of w relative to the string s. Then the l-empirical entropy is defined by

$$\hat{H}_l(s) = -\frac{1}{l} \sum_{w \in A^l} P(s,w) \log P(s,w). \tag{5}$$

The quantity $l\hat{H}_l(s)$ is a statistical measure of the average information content of the $l-$digit long substrings of s.

As it was said before not all the coding procedures are equivalent and give the same performances, so some care is necessary in the definition of information content. For this reason we introduce the notion of *coarse optimality* of an algorithm Z, defined by comparing its compression ratio with the empirical entropy.

An algorithm Z is said to be *coarsely* optimal if its compression ratio $|Z(s)|/|s|$ is asymptotically less than or equal to $\hat{H}_k(s)$ for each k.

Next we will see how this definition is not sufficient to define a suitable notion of optimality for zero entropy strings.

Definition 1 *(Coarse Optimality). A reversible coding algorithm Z is coarsely optimal if $\forall k \in \mathbf{N}$ there is a function f_k, with $f_k(n) = o(n)$, such that for all finite strings s*

$$\frac{|Z(s)|}{|s|} \leq \hat{H}_k(s) + \frac{f_k(|s|)}{|s|}.$$

Many data compression algorithms that are used in applications are proved to be coarsely optimal.

Remark 1. *The universal coding algorithms LZ77 and LZ78 ([26],[27]) satisfy Definition 1. For the proof see [18].*

It is not difficult to prove that coarse optimal algorithms give a correct estimation of the information content in the cases of positive entropy. In this case all measures of information agree.

Theorem 1. *If (Ω, μ, σ) is a symbolic dynamical system, Z is coarsely optimal and μ is ergodic, then for μ-almost ω*

$$\lim_{n\to\infty} K_Z(\omega^n) = \limsup_{l\to\infty \; n\to\infty} \hat{H}_l(\omega^n) = \lim_{n\to\infty} \frac{AIC(\omega^n)}{n} = h_\mu(\sigma) \,.$$

Proof. The statement is substantially proved in [4], Theorem 14. In [4] the statement is given with $limsup_{n\to\infty} K_Z(\omega^n)$ instead of $\lim_{n\to\infty} K_Z(\omega^n)$. The proof of Theorem 1 follows similarly using the White's results [25] stating that under the above assumptions for almost each $\omega \in \Omega$ $limsup_{n\to\infty} \frac{AIC(\omega^n)}{n} = liminf_{n\to\infty} \frac{AIC(\omega^n)}{n}$. □

The notion of coarse optimality is not enough if we ask a coding algorithm to be able to reproduce the rate of convergence of the sequence $\hat{H}_k(s)$ as $|s| \to \infty$ for strings generated by weakly chaotic dynamical systems, for which $\lim_{|s|\to\infty} \hat{H}_k(s) = 0$. In fact in the positive entropy systems optimality implies that $\lim_{|s|\to\infty} \frac{|Z(s)|}{|s|} \leq \hat{H}_k(s)$. In the zero entropy systems, it is possible that $\hat{H}_k(s^n) \to 0$ faster than $\frac{Z(s)}{|s|}$.

For example let us consider the string 0^n1 and the $LZ78$ algorithm, then $\hat{H}_k(0^n1)$ goes like $log(n)/n$ (and also $\frac{AIC(\omega^n)}{n}$ does) while $LZ78(0^n1)/n$ goes like $\frac{n^{1/2}log(n)}{n}$ (see also [1]). Then the compression ratio of LZ78 converges to 0 much slower than the empirical entropy. This implies that coarse optimality is not sufficient to have a coding algorithm able to characterize 0-entropy strings according to the rate of convergence of their entropy to 0. For this reason, we look for an algorithm having the same asymptotic behavior as the empirical entropy. In this way even in the 0-entropy case our algorithm will give a meaningful measure of the information. The following definition (from [18]) is an approach to define optimality of a compression algorithm for the 0-entropy case.

Definition 2 (Asymptotic Optimality). *A compression algorithm Z is called asymptotically optimal with respect to $\hat{H}_k$ if*

– *it is coarsely optimal*
– *there is a function g_k with $g_k(n) = o(n)$ and $\lambda > 0$ such that $\forall s$ with $\hat{H}_k(s) \neq 0$*

$$|Z(s)| \leq \lambda|s|\hat{H}_k(s) + g_k(|Z(s)|).$$

It is not trivial to construct an asymptotically optimal algorithm. For instance the well known Lempel-Ziv compression algorithms are not asymptotically optimal. $LZ78$ is not asymptotically optimal even with respect to $\hat{H}_0$ ([18]). In [18] some examples are described of algorithms (LZ78 with RLE and LZ77) which are asymptotically optimal with respect to $\hat{H}_0$. But these examples are not asymptotically optimal for all $\hat{H}_k$ with $k \geq 1$. The asymptotic optimality of LZ77 with respect to $\hat{H}_0$ however it is enough to prove (see Section 4 , Theorem 2) that LZ77 can estimate correctly the information coming from the Manneville type maps.

The set of asymptotically optimal compression algorithms with respect to each $\hat{H}_k$ is not empty. In section 5 an example is given of a compression algorithm which is asymptotically optimal for all $\hat{H}_k$. The algorithm is similar to the Kolmogorov frequency coding algorithm. This compression algorithm is not of practical use because of its computational time cost.

To our knowledge the problem of finding a "fast" asymptotically optimal compression algorithm is still open. The Ryabko double universal code [23] seems a good candidate for solving this problem. In this direction the work in in progress.

4 Optimality with Respect to H_0 and Manneville Map

The Manneville maps $T_z : [0,1] \to [0,1]$ defined as $T_z(x) = x + x^z \ (mod \ 1)$ are dynamical systems over the unit interval.

The Manneville maps come from the theory of turbulence. They are introduced in [20] as an extremely simplified model of intermittent behavior in fluid dynamics. This maps have been studied and applied also to other areas of physics (see e.g. [2],[17],[22]). The main part of these works concentrates on the study of the case $1 \leq z \leq 2$. In the following we will consider the case $z > 2$ which gives a *weakly* chaotic dynamical system. When $z > 2$, the maps has stretched exponential sensitivity to initial conditions and information content of the orbits that increases as a power law (cf. Th. 2).

The first study of the mathematical features of complexity of the Manneville maps for $z > 2$ was done by Gaspard and Wang in [14]. Another study of the complexity of the Manneville maps was done in [13] and [5].

Next theorem states that if a compression algorithm is optimal with respect to $\hat{H}_0$ the associated information content measure is correct for the Manneville maps in the sense that it has the same asymptotic behavior as the AIC on this kind of maps.

Theorem 2. *Let us consider the dynamical system $([0,1], T_z)$ with $z > 2$. Let $\tilde{x} \in (0,1)$ be such that $T_z(\tilde{x}) = 1$. Let us consider the partition $\alpha = \{[0,\tilde{x}], (\tilde{x}, 1]\}$ of the unit interval $[0,1]$. Let Z be an universal coding which is optimal with respect to $\hat{H}_0$. If $\omega(x)$ is the symbolic orbit of x with respect to the partition α, then there are $C_1, C_2, C_3 > 0$ s.t. $\forall n$*

$$C_1 n^p \leq \int_{[0,1]} AIC(\omega^n(x))dx + C_2 \leq \int_{[0,1]} I_Z(\omega^n(x))dx \leq C_3 n^p \log_2(n) \quad (6)$$

where $p = \frac{1}{z-1}$ and the measure is the usual Lebesgue measure on the interval.

Proof. (sketch) This result is substantially proved in [4] we sketch the proof because here the statement is given in a slight more general form. The first inequality is given in [5] or [13] Proposition 33, the second inequality follows easily from Theorem 4 (Appendix). To prove the third inequality we see that

1) there are probabilistic results ([13] Section 8) showing that

$$\lim_{n \to \infty} \frac{\int_{[0,1]} nP(1, \omega^n(x))dx}{n^p} = 1$$

where $P(1, \omega)$ is the empirical probability of the digit 1 in the word ω as in the definition of empirical entropy.

2) this result and the Jensen inequality allows to calculate that

$$\lim_{n\to\infty} \frac{\int_{[0,1]} H_0(\omega_n(x))dx}{n^p log(n)} < \infty$$

and then by the 0-optimality we have an estimation of the I_Z of the string and we can conclude   □

Now we will recall some results from [18], concerning the optimality of the $LZ77$ algorithm.

Proposition 1. *The algorithm $LZ77$ is 8-optimal with respect to $\hat{H}_0$, The algorithm $LZ78 + RLE$ is 3-optimal with respect to $\hat{H}_0$.*

Since $LZ77$ and $LZ78 + RLE$ are $\hat{H}_0$ optimal then they can be applied to the Manneville map and give the same asymptotic behaviour as the AIC. In [4], there are numerical experiments with LZ77 and other algorithms illustrating this kind of phenomena.

5   Frequency Coding and Optimality

We now give a description of an asymptotically optimal coding procedure which we will call FC. This procedure is similar to the Kolmogorov Frequency Coding.

Let us consider a binary string s with length N. Let us consider the following coding procedure FC'. The procedure $FC'(s, l, k)$ (depending on two integer parameters l, k with $k < l < |s| = N$) first cuts a k digit long initial segment s_k of the string s. This initial segment will be codified in any usual way (e.g. letter by letter) at the beginning of the coded string.

The remaining part of the string is parsed in l digit long segments (words) $(w_i)_{i \leq [\frac{N-k}{l}]}$ and the remainder is a string $s_{(N-k)(mod\, l)}$ with length $(N-k)(mod\, l)$. In this way we can write

$$s = s_k w_1 ... w_{[\frac{N-k}{l}]} s_{(N-k)(mod\, l)}.$$

Also the final string $s_{(N-k)(mod\, l)}$ will be codified as you like.

This coding procedure considers each of the l-bits long word, ordered in lexicographic order, it counts how many times it appears in the sequence $(w_1, w_2, ...)$.

Thus for $1 \leq i \leq 2^l$ an integer number n_i, is associated to each of these strings.

For example, the first word $0^l = 00000...0$ and n_1 is the number of times which 0^l appears in the parsing. This number is the empirical frequency of $00000...0$ relative to the parsing $w_1, w_2...$ Then the procedure will consider the second word $0^{l-1}1 = 00000...1$ obtaining a number n_2, and so on obtaining a sequence $(n_1, ..., n_{2^l})$.

The string s will be coded by

$$FC'(s, l, k) = (N, s_k, s_{N-k(mod\, l)}, n_1, n_2, ..., n_{2^l}, W).$$

Let us explain the meaning of the number W. The string $w_1 w_2 \ldots$ is composed by $n = \sum n_i$ words from a vocabulary made of 2^l words. The number a of strings having the same empirical frequencies $(n_1, n_2, \ldots)$ as s is

$$a = \frac{n!}{\prod_i n_i!}.$$

These a strings can be ordered lexicographically and a natural number between 0 and a can be associated to each of them. The number W determines which string we choose among all these strings. With a suitable coding the number W will take not more than

$$[1 + \log a] \approx \log \left(\frac{n!}{\prod_i n_i!} \right) = \log n! - \sum_i \log(n_i!)$$

bits. A this point we can define

$$FC(s) = FC'(s, \hat{l}, \hat{k})$$

where

$$(\hat{l}, \hat{k}) = \min_{l,k} \left(|FC'(s, l, k)| \right).$$

Proposition 2. *The compression algorithm FC is asymptotically optimal with respect to $\hat{H}_k$ for all k.*

Proof. [4] We have

$$|FC'(s, l, k)| \leq \sum_{i \leq 2^l} \log n_i + \log n! - \sum_{i \leq 2^l} \log n_i! + \log N + C\,(l)$$

The first term is necessary to specify the numbers $n_1, n_2 \ldots$. $C\,(l)$ is the number of bits necessary to codify the separation symbols ",," and the parentheses in the coded string

$$(s_k, s_{N-k(mod\ l)}, n_1, n_2, \ldots, n_{2^l}, W);$$

(it depends on l but nor on N), $\log N$ is the number of bits necessary to specify N and the remaining terms are necessary to specify W.

By the Stirling formula we have the following approximation of $\ln n!$:

$$\ln n! = n \ln n - n + \frac{1}{2} \ln n + \ln(2\pi)^{\frac{1}{2}} + o(1).$$

Then

$$|FC'(s, l, k)| \leq \sum_i \log n_i + \log n! - \sum_i \log n_i! + \log N + C$$

---

[4] We remark that the first part of this proof can be made shorter by the use of results coming from the theory of types (see [10] or [15]), however we prefer to give here an elementary and self contained proof.

$$\leq \sum_i \log n_i + n \log n - \frac{n}{\ln 2} + \frac{1}{2} \log n -$$

$$- \sum_i \left(n_i \log n_i - \frac{n_i}{\ln 2} + \frac{1}{2} \log n_i \right) + \log N + C$$

$$\leq \sum_i (\frac{1}{2} \log n_i - n_i \log n_i) + n \log n + \frac{1}{2} \log n + \log N + C.$$

Now, if we denote by $p_i = n_i/n$ the "empirical probability" of the i-th word in the set $\{w_1, w_2...\}$, we have

$$|FC'(s,l,k)| \leq \sum_i \frac{1}{2} \log n_i - \sum_i np_i \log (np_i) + n \log n + \frac{1}{2} \log n + \log N + C =$$

$$= \sum_i \frac{1}{2} \log n_i - n \sum_i p_i \log n + n \sum_i p_i \log \frac{1}{p_i} + n \log n + \frac{1}{2} \log n + \log N + C =$$

$$= \frac{1}{2} \log n + \sum_i \frac{1}{2} \log n_i + n \sum_i p_i \log \frac{1}{p_i} + \log N + C =$$

$$= \frac{1}{2} \log n + \sum_i \left[\frac{1}{2} \log n_i + nh(w_i) \right] + \log N + C \qquad (7)$$

where $h(w_i) = p_i \log \frac{1}{p_i}$.

Now let us consider $l \in \mathbf{N}$. We prove that there is a $C' > 0$ such that $|FC(s)| \leq C'|s|\hat{H}_l(s) + o(|FC(s)|)$. Let us consider $k < l$ and the corresponding parsing P_k of s:

$$P_k = \{w_1 = s^{(1,k)}, w_2 = s^{(k,k+l)}...\}$$

we have that for each $k < l$ (Eq. 7)

$$|FC(s)| \leq |FC'(s,l,k)| \leq \frac{3}{2} logN + \frac{1}{2} \sum logn_i + [\frac{N-k}{l}]h^k(s) + C$$

where $h^k(s) = \sum h(w_i)$ (we recall that $N = |s|$). The number of l−length words is at most 2^l then $\forall k \sum logn_i \leq 2^l log|s|$, then

$$|FC(s)| \leq [\frac{N-k}{l}]h^k(s) + (2^l + \frac{3}{2})logN + C.$$

Now the statement follows from the remark that for some k, $h^k(s) \leq l\hat{H}_l(s)$ and that if $\hat{H}_l(s) > 0$ then $\hat{H}_l(s) \geq \frac{1}{|s|}(\log(|s|) + 1)$. The former can be proved from the following remark. Let $\omega \in A^l$, let n_ω be the number of occurrences of ω in s as in the definition of empiric entropy and let n_ω^k be the number of occurrences of the word ω in the parsing P_k (before, when k was fixed these numbers were called n_i and their dependence on k was not specified). Let also $p_\omega = \frac{n_\omega}{N-l}$ and $p_\omega^k = \frac{n_\omega^k}{[\frac{N-k}{l}]}$

be the corresponding empirical probabilities. p_ω is a convex combination of the p_ω^k, $p_\omega = \sum_k c_k p_\omega^k$ (the total empirical probability of the word w is a weighted average of the empirical probability of w in the partitions P_k). By the concavity of the entropy function we have $\sum_w p_w log p_w = l\hat{H}_l(s) \geq \sum_k c_k \, h^k(s)$ and then $h^k(s) \leq l\hat{H}_l(s)$ for some k, and this end the proof. □

References

1. F. Argenti, V. Benci, P. Cerrai, A. Cordelli, S. Galatolo, and G. Menconi, Information and dynamical systems: a concrete measurement on sporadic dynamics, Chaos, Solitons and Fractals, 13, No. 3, 461–469, 2002.
2. P. Allegrini, M. Barbi, P. Grigolini, and B.J. West, Dynamical model for DNA sequences, Phys. Rev. E, Vol. 52, no. 5, 5281-5297, 1995.
3. P. Allegrini, V. Benci, P. Grigolini, P. Hamilton, M. Ignaccolo, G. Menconi, L. Palatella, G. Raffaelli, N. Scafetta, M. Virgilio, and J. Jang, Compression and diffusion: a joint approach to detect complexity, Chaos Solitons Fractals, 15, No. 3, 517–535, 2003.
4. V. Benci, C. Bonanno, S. Galatolo, G. Menconi, and M. Virgilio, Dynamical systems and computable information , Disc. Cont. Dyn. Syst.-B, Vol. 4, No. 4, 2004.
5. C. Bonanno, S. Galatolo, The complexity of the Manneville map, work in preparation.
6. C. Bonanno C., G. Menconi, Computational information for the logistic map at the chaos threshold, Disc. Cont. Dyn. Syst.- B, 2, No. 3, 415–431, 2002.
7. F. Blume, Possible rates of entropy convergence, Ergodic Theory and Dynam. Systems, 17, No. 1, 45–70, 1997.
8. A.A. Brudno, Entropy and the complexity of the trajectories of a dynamical system, Trans. Moscow Math. Soc., 2, 127–151, 1983.
9. G.J. Chaitin, Information, Randomness and Incompleteness, Papers on Algorithmic Information Theory, World Scientific, Singapore, 1987.
10. I. Csiszár, The method of types, IEEE Trans. Inform. Theory, Vol. 44, 2505-2523, 1998.
11. S. Galatolo, Orbit complexity by computable structures, Nonlinearity, 13, 1531–1546, 2000.
12. S. Galatolo, Orbit complexity and data compression, Discrete and Continuous Dynamical Systems, 7, No. 3, 477–486, 2001.
13. S. Galatolo, Complexity, initial condition sensitivity, dimension and weak chaos in dynamical systems, Nonlinearity, 16, 1219-1238, 2003.
14. P. Gaspard and X.J. Wang, Sporadicity: between periodic and chaotic dynamical behavior, Proc. Nat. Acad. Sci. USA, 85, 4591–4595, 1988.
15. T.S. Han and K. Kobayashi, Mathematics of Information and Coding, Math. Monographs, Vol. 203, AMS, 2002.
16. A.I. Khinchin, Mathematical Foundations of Information Theory, Dover Publications, New York, 1957.
17. S. Isola, Renewal sequences and intermittency, J. Statist. Phys., 97, No. 1-2, 263–280, 1999.
18. S.R. Kosaraju and G. Manzini, Compression of low entropy strings with Lempel-Ziv algorithms, SIAM J. Comput., 29, 893–911, 2000.
19. M. Li and P. Vitanyi, An Introduction to Kolmogorov Complexity and its Applications, Springer-Verlag, 1993.

20. P. Manneville, Intermittency, self-similarity and 1/f spectrum in dissipative dynamical systems, J. Physique, 41, 1235–1243, 1980.
21. Y.B. Pesin, Dimension Theory in Dynamical Systems, Chicago Lectures in Mathematics, 1997.
22. M. Pollicott and H. Weiss, Multifractal analysis of Lyapunov exponent for continued fraction and Manneville-Pomeau transformations and applications to Diophantine approximation, Comm. Math. Phys., 207, No. 1, 145–171, 1999.
23. B. Ryabko, Twice-universal coding, Russian, Problemy Peredachi Informatsii, 20, No. 3, 24–28, 1984.
24. F. Takens and E. Verbitski, Generalized entropies: Renyi and correlation integral approach, Nonlinearity, 11, No. 4, 771–782, 1998.
25. H. White, Algorithmic complexity of points in dynamical systems, Ergodic Theory Dynam. Syst., 13, 807–830, 1993.
26. J. Ziv and A. Lempel, A universal algorithm for sequential data compression, IEEE Trans. Inform. Theory, Vol. 23, 337–342, 1977.
27. J. Ziv and A. Lempel, Compression of individual sequences via variable-rate coding, IEEE Trans. Inform. Theory, Vol. 24, 530–536, 1978.

Appendix: The Algorithmic Information Content

The most important measure for the information content is the Algorithmic Information Content (AIC). In order to define it, it is necessary to define the notion of partial recursive function. We limit ourselves to give an intuitive idea which is very close to the formal definition. We can consider a partial recursive function as a computer C which takes a program p (namely a binary string) as an input, performs some computations and gives a string $s = C(p)$, written in the given alphabet $\mathcal{A}$, as an output. The AIC of a string s is defined as the shortest binary program p which gives s as its output, namely

$$AIC(s, C) = \min\{|p| : C(p) = s\}$$

We require that our computer is a universal computing machine. Roughly speaking, a computing machine is called *universal* if it can emulate any other machine. In particular every real computer is a universal computing machine, provided that we assume that it has virtually infinite memory. For a precise definition see e.g. [19] or [9]. We have the following theorem due to Kolmogorov

Theorem 3. *If C and C' are universal computing machines then*

$$|AIC(s, C) - AIC(s, C')| \leq K(C, C')$$

where $K(C, C')$ is a constant which depends only on C and C' but not on s.

This theorem implies that the information content AIC of s with respect to C depends only on s up to a fixed constant, then its asymptotic behavior does not depend on the choice of C. For this reason from now on we will write $AIC(s)$ instead of $AIC(s, C)$. The shortest program which gives a string as its output is a sort of encoding of the string. The information which is necessary to reconstruct the string is contained in the program. We have the following result (for a proof see for example [12] Lemma 6) :

Theorem 4. *Let*

$$Z_C : \mathcal{A}^* \to \{0,1\}^*$$

be the function which associates to a string s the shortest program whose output is s itself[5] (namely, $AIC(s) = |Z_C(s)|$). If Z is any reversible recursive coding, there exists a constant M which depends only on C such that for each s

$$|Z_C(s)| \le |Z(s)| + M \tag{8}$$

The inequality (8) says that Z_C in some sense is the best possible compression procedure. Unfortunately this coding procedure cannot be performed by any algorithm (Chaitin Theorem)[6]. This is a very deep statement and, in some sense, it is equivalent to the Turing halting problem or to the Gödel incompleteness theorem. Then the AIC is not computable by any algorithm.

---

[5] If two programs of the same length produce the same string, we choose the program which comes first in lexicographic order.

[6] Actually, the Chaitin theorem states a weaker statement: an algorithm (computer program) which states that a string σ of length n can be produced by a program shorter than n, must be longer than n.

Report on Models of Write–Efficient Memories with Localized Errors and Defects

R. Ahlswede and M.S. Pinsker

Abstract. Write–efficient memories (WEM) as a model for storing and updating information were introduced by R. Ahlswede and Z. Zhang [2]. We consider now three new models of WEM with localized errors and defects, resp.

In the situation (E_+, D_-), where the encoder is informed but the decoder is not informed about the previous state of the memory we study

1. WEM codes correcting defects,
2. WEM codes detecting localized errors.

Finally, in the situation (E_+, D_+), where both, the encoder and the decoder, are informed about the previous state of the memory we study.

3. WEM codes correcting localized errors.

In all three cases we determine for binary alphabet the optimal rates under a cost constraint defined in terms of the Hamming distance.

1 Introduction

We recall first the model and a result of [2]. A write–efficient memory (WEM) is a model for storing and updating information on a rewritable medium. There is a cost $\varphi : \mathcal{X} \times \mathcal{X} \to \mathbb{R}_\infty$ assigned to changes of letters. A collection of subsets $\mathcal{C} = \{C_i : 1 \leq i \leq M\}$ of $\mathcal{X}^n$ is an (n, M, d) WEM code, if

$$C_i \cap C_j = \varnothing \quad \text{for all} \quad i \neq j \tag{1}$$

and, if

$$d_{\max} = \max_{1 \leq i,j \leq M} \max_{x^n \in C_i} \min_{y^n \in C_j} \sum_{k=1}^{n} \mathcal{C}(x_k, y_k) \leq d. \tag{2}$$

$d_{\max}$ is called the minimax correction cost with respect to the given cost function φ. The performance of a code $\mathcal{C}$ can also be measured by two parameters, namely, the maximal cost per letter $\delta_\mathcal{C} = n^{-1} d_{\max}$ and the rate of the size $R_\mathcal{C} = n^{-1} \log M$. The rate achievable with a maximal per letter cost δ is thus $R(\delta) = \sup_{\mathcal{C}:\delta_\mathcal{C} \leq \delta} R_\mathcal{C}$.

This is the most basic quantity, the storage capacity of a WEM $(\mathcal{X}^n, \varphi^n)_{n=1}^\infty$. It has been characterized in [2] for every φ.

The operational significance of a WEM code is as follows. For a set $\mathcal{M} = \{1, 2, \ldots, M\}$ of possible messages and the state $x^n = (x_1, \ldots, x_n) \in C = \bigcup_{i=1}^{M} C_i$ of the memory the encoder can store any message $i \in \mathcal{M}$ by any state y^n in C_i.

R. Ahlswede et al. (Eds.): Information Transfer and Combinatorics, LNCS 4123, pp. 628–632, 2006.

It is guaranteed that the cost of changing x^n to y^n does not exceed d, provided that the encoder knows the original state x^n. The decoder votes for message j, if $y^n \in C_j$. He does not use and therefore does not need knowledge about the previous state x^n. We are thus in a case (E_+, D_-).

In this model there is no error in encoding and decoding. Solely the cost function φ defines it. Clearly, by the forgoing explanation the encoder can update any message i stored by a state of C_i to any message j stored by a suitable state in C_j.

In this note we confine ourselves to the binary Hamming case, that is, $\mathcal{X} = \{0,1\}$ and the encoder cannot change symbols in more than $d = \delta n$, $0 \leq \delta \leq \frac{1}{2}$, positions of a state. In this case the result of [2] specialized to

$$R(\delta) = h(\delta), \tag{3}$$

where $h(\delta) = -\delta \log \delta - (1 - \delta) \log(1 - \delta)$.

WEM codes can be considered, where we take instead of $\mathcal{X}^n$ a subset with restrictions on codewords, for instance on the weight of the codewords, on the distribution of 0 and 1, etc.

Such a partition can be used in some other situations, for example WEM codes correcting defects, WEM codes detecting localized errors, i.e. the encoder has an additional information about errors.

In Section 2 we present our three models. In Section 3 – 5 we present capacity theorems for them.

2 Three Models of WEM with Errors

We study three types of WEM codes with additional properties.

1. WEM code correcting $t = \tau n$ defects (E_+, D_-)
Denote the t element subsets of $[n] = \{1, 2, \ldots, n\}$ by $\mathcal{E}_t = \binom{[n]}{t}$. Any $E \in \mathcal{E}_t$ can be the set of defect positions and any $e_E = (e_k)_{k \in E}$ can specify the defects e_k in position k. Both, E and e_E, are known to the encoder, who also knows the present state x^n whereas the decoder knows nothing.

The decoder, reading $y^n \in C_j$ votes for message j. Necessarily the C_j's are again disjoint. Moreover, for every $x^n \in C = \bigcup_{j=1}^{M} C_j$, E, and e_E there must be for every j a $y^n \in C_j$ with

$$y_k = e_k \quad \text{for} \quad k \in E \tag{4}$$

and

$$\sum_{k \in [n] \setminus E} \varphi(d_k, y_k) \leq d. \tag{5}$$

2. WEM code detecting t localized errors (E_+, D_-)
Any $E \in \mathcal{E}_t$ can be the set of positions with possible errors. At any updating E is known to the encoder. We want to be able to have a large set $\mathcal{M} = \{1, 2, \ldots, M\}$

of messages as candidates for an updating. The code $\{C_i : 1 \le i \le M\}$ has to be designed such that the decoder decides correctly, if no error occurs in the positions E, and otherwise he detects an error.

It will be shown that optimal rates are achieved with codes of the following structure:

The elements in $C = \bigcup_{i=1}^{M} C_i$ have weight $\lfloor \frac{n-t}{2} \rfloor$.

For $j \in \mathcal{M}$ and $x^n \in C$ we require that $y^n = y^n(x^n, E, j)$ satisfies $y_k = x_k$ for $k \in E$, $y^n \in C_j$, and $D(x^n, y^n) \le d$.

If for the output state v $v \in C_j$, then v is decoded into j, and if $v \notin C_j$, then necessarily $|v| > \lfloor \frac{n-t}{2} \rfloor$ and an error is detected.

Actually, we shall make sure by additional assumptions on the model that always the state sequence x^n satisfies $x_k = 0$ for $k \in E$.

This can be achieved by a.) having the same E (unknown to the decoder for all updatings) or by b.) allowing only errors which change a 1 into a 0.

In case of a detection the memory has to be cleaned by writing a 0 everywhere.

3. WEM code correcting t localized errors (E_+, D_+)

Let $E \in \mathcal{E}_t$ be the set of possible positions for errors in updating. It is known to the encoder, who also knows the present state, say x^n. He encodes a message $j \in \{1, 2, \ldots, M\}$ by a $y^n(x^n, j, E)$ satisfying $D(x^n, y^n) \le d$. For the output state v $D(y^n, v) \le t$ holds. The decoder knows x^n and bases his decoding Ψ on x^n and v. Therefore $\Psi : \mathcal{X}^n \times \mathcal{X}^n \to \mathcal{M}$ must satisfy $\Psi(x^n, v) = j$.

Of course by the distance constraint there is no loss in assuming that $y_k = x_k$ for $k \in E$. Since both, encoder and decoder, know x^n the coding problem is practically equivalent to the coding problem for transmission in the presence of localized errors, except that there is now the constraint for the codewords to have weights not exceeding d.

Remark 1. An interesting more difficult coding problem arises, if model 3 is altered such that the encoder does not remember the previous state, that is, we are in case (E_+, D_-).

3   The Capacity in Model 1

We denote by $R_{\text{def}}(\delta, \tau)$ the optimal rate, that is, the capacity of WEM codes correcting $t = \tau n$ defects at maximal cost $d = \delta n$. We need the quantity

$$R(w, \tau) = \begin{cases} (1 - \tau)h\left(\frac{w}{1-\tau}\right), & \text{if } w \le \frac{1}{2}(1 - \tau) \\ 1 - \tau, & \text{if } w \ge \frac{1}{2}(1 - \tau). \end{cases}$$

Theorem 1. For any $0 \le \tau \le 1$ and $0 \le \delta \le 1$

$$R_{\text{def}}(\delta, \tau) = R(\delta, \tau).$$

We recall the **Color Carrying Lemma.** (see [2])

The hypergraph $\mathcal{H} = (\mathcal{V}, \mathcal{E})$ carries M colors, if $M \leq \left(\ell n |\mathcal{E}| \min_{E \in \mathcal{E}} |E| \right)^{-1} \min_{E \in \mathcal{E}} |E|$.

Remark 2: Misprints in Lemma in [2] have been corrected.

Proof of Theorem 1

For $x^n \in \mathcal{X}^n$ and $x_k = e_k (k \in E)$ define

$$\mathcal{S}_d(x^n, E, e_E) = \{ y^n \in \mathcal{X}^n : y_k = e_k \text{ for } k \in E, D(y^n, x^n) \leq d \}. \tag{6}$$

Clearly, $M \leq |\mathcal{S}_d(x^n, E, e_E)|$ and $|\mathcal{S}_d(x^n, E, e_E)| = \sum_{i=0}^{d} \binom{n-t}{i} \sim \exp\{ R(\delta, \tau) n \}$.

By applying the previous lemma to the hypergraph with vertex set $\mathcal{X}^n$ and edges defined in (6) we can achieve the rate $R(\delta, \tau)$.

Remark 3: Actually, Theorem 1 is also an immediate consequence of the results of [3].

The capacity for model 2

Here $R_{\text{loc}}^{\text{detect}}(\delta, \tau)$ denotes the optimal rate (capacity) of WEM codes *detecting* $t = \tau n$, $0 \leq \tau \leq \frac{1}{2}$, localized errors with cost $d = \delta n$. The encoder knows the set E of error positions E and the previous state of the memory.

Theorem 2. *Under conditions a.) or b.)*

$$R_{\text{loc}}^{\text{detect}}(\delta, \tau) = R(\delta, \tau).$$

Remarks

4. The expression $R(\delta, \tau)$ also occurs as the capacity in correcting localized errors with constant weight δn codes (see [4]). That result also follows from the proof of Theorem 2.
5. The work of [6] does not assume restrictions on the weight of codewords.

Proof: The inequality $R_{\text{loc}}^{\text{detect}}(\delta, \tau) \leq R(\delta, \tau)$ is obvious and the opposite inequality follows again by the Color Carrying Lemma.

The capacity for model 3

We denote by $R_{\text{loc}}(\delta, \tau)$ the optimal rate (capacity) of WEM codes *correcting* $t = \tau n$, $0 < \tau \leq \frac{1}{2}$, localized errors with cost $d = \delta n$. Recall that the encoder knows the set of error positions E and both encoder and decoder, are informed about the previous state of the memory.

We define now a quantity R^L which describes the capacity in correcting $t = \tau n$ localized errors with binary constant weight $w = wn$, $0 < w \leq 1$, codes

$$R^L(w, \tau) = \begin{cases} h(w + \tau) - h(\tau), & \text{if } 0 < w \leq \frac{1}{2} \\ 1 - h(\tau), & \text{if } \frac{1}{2} - \tau \leq w \leq \frac{1}{2} \\ h(w - \tau) - h(\tau), & \text{if } \frac{1}{2} + \tau \leq w < 1 \end{cases} \tag{7}$$

(see [4]).

Combining this result with the capacity theorem of [2] for error free WEM codes one gets

Theorem 3

$$R_{\text{loc}}(\delta, \tau) = R^L(\delta, \tau).$$

Remark 6. Some facts about WEM codes in the situation (E_+, D_+) can be carried over to ordinary codes with constant weight.

Theorem 3 can be extended to the situation (E_+, D_-), when positions E of t possible errors do not change during updating, namely, we have the capacity $R^0_{\text{loc}}(\delta, \tau)$.

Theorem 4

$$R^L(w, \tau) \leq R^0_{loc}(\delta, \tau) \leq R^L(\delta, \tau), \tag{8}$$

where w is defined by the equation

$$\delta = 2w(1 - w).$$

However, if $w + \tau \geq \frac{1}{2}$, then we have equalities in (8).

References

1. R. Ahlswede and M.S. Pinsker, Report on models of write-efficient memories with localized errors and defects, Preprint 97-004 (Ergänzungsreihe), SFB 343 "Diskrete Strukturen in der Mathematik", Universität Bielefeld, 1997.
2. R. Ahlswede and Z. Zhang, Coding for write efficient memories, Inform. and Computation, Vol. 83, No. 1, 80–97, 1989.
3. R. Ahlswede, L.A. Bassalygo, and M.S. Pinsker, Binary constant weight codes correcting localized errors and defects, Probl. Inform. Trans., Vol. 30, No. 2, 10–13, 1994.
4. L.A. Bassalygo and M.S. Pinsker, Binary constant–weight codes correcting localized errors, Probl. Inform. Trans., Vol. 28, No. 4, 103–105, 1992.
5. A.V. Kusnetsov and B.S. Tsybakov, Coding a memory with defective cells, Probl. Inform. Trans., Vol. 10, No. 2, 52–60, 1974.
6. L.A. Bassalygo and M.S. Pinsker, Codes detecting localized errors, preprint.

Percolation on a k-Ary Tree

K. Kobayashi, H. Morita, and M. Hoshi

Abstract. Starting from the root, extend k branches and append k children with probability p, or terminate with probability $q = 1 - p$. Then, we have a finite k-ary tree with probability one if $0 \leq p \leq 1/k$. Moreover, we give the expectation and variance of the length of ideal codewords for representing the finite trees. Furthermore, we establish the probability of obtaining infinite tree, that is, of penetrating to infinity without termination for case $1/k \leq p \leq 1$.

1 Preliminaries

In the study of computer science and information theory, there are many occasions when we encounter combinatorial structures called trees. Most common trees appearing in this field are the rooted ordered trees. We simply denote them as trees in this paper. It would be quite important to devise efficient mechanisms to encode them for many applications such as data compression.

When we studied the pre-order coding of binary tree, we found an interesting identity [7] with respect to Catalan numbers, that is:

Theorem 1

$$\sum_{n=0}^{\infty} \frac{1}{2n+1} \binom{2n+1}{n} 2^{-(2n+1)} = 1. \tag{1}$$

The following proof provides the speed of convergence of summation to the limit one.

Proof. Let $a_n = c_{2,n} 4^{-n}$, where $c_{2,n} = \frac{1}{2n+1} \binom{2n+1}{n}$ is the Catalan number. Then we find that a_n satisfies

$$\begin{cases} (2n+4)a_{n+1} = (2n+1)a_n, n \geq 0 \\ a_0 = 1. \end{cases} \tag{2}$$

Moreover, letting $b_n = (2n+1)a_n$, we have the recurrence

$$b_{n+1} + a_{n+1} = b_n \quad \text{for } n \geq 0 \text{ with } b_0 = a_0 = 1. \tag{3}$$

By summing up (3) from $n = 0$ to N, we obtain $b_N + \sum_{n=1}^{N} a_n = b_0$. Therefore,

$$\sum_{n=0}^{N} a_n = a_0 + b_0 - b_N = 2 - \binom{2N+1}{N} 4^{-N}. \tag{4}$$

R. Ahlswede et al. (Eds.): Information Transfer and Combinatorics, LNCS 4123, pp. 633–638, 2006.
© Springer-Verlag Berlin Heidelberg 2006

From Stirling's formula, the second term of (4) can be expressed by

$$\frac{2}{\sqrt{\pi N^3}} \left(1 + O(1/N)\right). \tag{5}$$

Since (5) goes to zero as $N \to \infty$, the theorem holds.

This identity means that the pre-order coding for binary trees shows the best possible performance in the sense that its length function tightly satisfies the Kraft inequality.

On the other hand, we have shown inequalities [7] for cases of $k \geq 3$:

$$\frac{1}{2} < \sum_{n=0}^{\infty} c_{k,n} 2^{-(kn+1)} < 1, \tag{6}$$

where the $c_{k,n}$'s are the generalized Catalan numbers (see the definition (8) in the next section). The above inequalities guarantee the existence of a prefix code with the length function $kn + 1$ for k-ary trees with n internal nodes, but unfortunately denies that of a code with the length function kn. With respect this point, refer to the remark 3.

An aim of this paper is to show that the identity (1) can be generalized as in the next equation:

$$\sum_{n=0}^{\infty} \frac{1}{2n+1} \binom{2n+1}{n} p^n q^{n+1} = \begin{cases} 1 & \text{for } 0 \leq p \leq 1/2 \\ \frac{q}{p} & \text{for } 1/2 \leq p \leq 1 \end{cases}, \tag{7}$$

where $q = 1 - p$. Thus, the case $p = 1/2$ of the identity (1) corresponds to the critical point separating the conditions in the equation (7).

Remark 1. *The pre-order coding is a well known code for binary trees in the computer science. See books [3] and [6]. The natural code defined in the paper [5] of Willems, Shtarkov and Tjalkens is a truncated version of the preorder code that is used for binary trees with finite depth. We consider the whole set of (binary) trees without any depth constraint. Thus, we intend to study a code for an infinite number of combinatorial objects (various kinds of trees). Especially, Lemma 2 of their paper cannot be extended for case of non-binary alphabet without some kind of modifications.*

2   Percolation Model on k-Ary Tree

Let us consider a stochastic generation of a k-ary tree. Here, we denote a k-ary tree to be a rooted ordered tree, each internal node of which has k distinct branches, usually corresponding to k characters in an alphabet. Starting from the root, extend k branches and append k children with probability p, or terminate with probability $q = 1 - p$. Then, we have two distinct events. One is the event E_f that we ultimately obtain a finite tree, and the other one is the event

E_∞ that the coin flipping process will never be stopped, and we have an infinite tree.

From the argument by Raney[2],[4], the number $c_{k,n}$ of k-ary tree having n internal nodes is given by

$$c_{k,n} = \frac{1}{kn+1} \binom{kn+1}{n}. \tag{8}$$

Using the generalized Catalan numbers, we can express the probability of the event E_f as

$$\Pr\{E_f\} = \sum_{n=0}^{\infty} c_{k,n} p^n q^{(k-1)n+1}. \tag{9}$$

In order to evaluate the series of the above equation, let us introduce the generating function $F_{k,p}(z)$ as follows.

$$F_{k,p}(z) = \sum_{n=0}^{\infty} c_{k,n} p^n q^{(k-1)n+1} z^n. \tag{10}$$

Thus,

$$\Pr\{E_f\} = F_{k,p}(1). \tag{11}$$

With respect to this generating function, we can easily find the functional equation by the symbolic consideration.

$$F_{k,p}(z) = q + pz F_{k,p}(z)^k. \tag{12}$$

For the case $k = 2$, we can explicitly solve the functional equation as follows.

$$F_{2,p}(1) = \frac{1 - \sqrt{1-4pq}}{2p} = \frac{1 - |2p-1|}{2p} = \{\ 1 \text{ for } 0 \le p \le 1/2 \ \frac{q}{p} \text{ for } 1/2 \le p \le 1\ . \tag{13}$$

Also, for the case $k = 3$,

$$F_{3,p}(1) = \begin{cases} 1 & \text{for } 0 \le p \le 1/3 \\ \frac{\sqrt{4p-3p^2}-p}{2p} & \text{for } 1/3 \le p \le 1 \end{cases}. \tag{14}$$

In general, we have

Theorem 2. *The probability of the event E_f of having a finite k-ary tree for the extending probability p is given by*

$$\Pr\{E_f\} = F_{k,p}(1) = \begin{cases} 1 & \text{for } 0 \le p \le 1/k \\ f(p) & \text{for } 1/k \le p \le 1 \end{cases}, \tag{15}$$

where $f(p)$ is a unique real value f in the interval [0,1] satisfying the equation,

$$f^{k-1} + f^{k-2} + \cdots + f + 1 = \frac{1}{p}, \tag{16}$$

for $1/k \le p \le 1$.

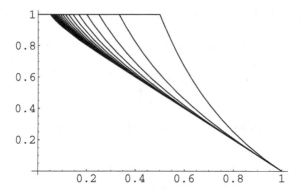

Fig. 1. Probability of getting a finite k-ary tree versus the extending probability p, (the curves correspond to the cases of $k = 2, 3, \ldots$ from the right)

Remark 2. *On pages 255–256 in a classical book[1] on probability theory, Feller described the first passage time in biased Bernoulli trial, gave a similar form of the recursion (12) for $k = 2$, and provided the same formula (13) for the ultimate winning probability. It is well known that the set of paths that start from the origin and don't touch the positive axis until the time $2n + 1$ in the discrete lattice is precisely equivalent to the set of binary trees with n internal nodes. Indeed, that is the reason of obtaining the identical form for $k = 2$, and appearing Catalan numbers in our discussion. Here, our aim is to introduce a special kind of combinatorial source, and to give a characteristic formula for specifying the length of codeword. Therefore, we consider that Theorem 2 is not a particular case of Feller's, but is an extension of his result in a different or an information theoretic context.*

Remark 3. *Previously, we showed an identity [8] with respect to the generalized Catalan numbers,*

$$\sum_{n=0}^{\infty} c_{k,n} 2^{-\{g(k)n + \log_2(k/(k-1))\}} = 1, \tag{17}$$

where $g(k) = k \log_2 k - (k - 1) \log_2(k - 1) = kh(1/k)$ and $h(p) = -p \log_2 p - (1 - p) \log_2(1 - p)$ is the binary entropy function.
    The above equation is rewritten as

$$\sum_{n=0}^{\infty} c_{k,n} \left(\frac{1}{k}\right)^n \left(\frac{k-1}{k}\right)^{(k-1)n+1} = 1. \tag{18}$$

Thus, the identity (17) corresponds to the critical case $p = 1/k$ of the equation (15).

3   Ideal Codeword Length

For case $0 \leq p \leq 1/k$, we will eventually have a finite k-ary tree with probability 1. At that time, we can consider that the tree with n internal nodes has been produced with the probability $p^n q^{(k-1)n+1}$. Here, we notice that the number of leaves (or external nodes) is $(k-1)n+1$. Thus, the ideal length of a codeword for representing the k-ary tree is $-(\log p + (k-1)\log q)n - \log q$. The expectation $\overline{L}$ of the ideal codeword length is given by

$$\overline{L} = \sum_{n=0}^{\infty} c_{k,n} p^n q^{(k-1)n+1}\{-(\log p + (k-1)\log q)n - \log q\}. \tag{19}$$

This expectation should be considered to be the *entropy* of a tree generated in our percolation model.

Therefore, we have to evaluate the sum,

$$\sum_{n=0}^{\infty} c_{k,n} p^n q^{(k-1)n+1} n = F'_{p,k}(1). \tag{20}$$

Differentiating the functional equation (12), we get

$$F'_{p,k}(1) = \frac{p}{1-kp}, \tag{21}$$

for the case $0 \le p \le 1/k$. Inserting this evaluation into the equation (19), we obtain

$$\overline{L} = -(\log p + (k-1)\log q)\frac{p}{1-kp} - \log q = \frac{h(p)}{1-kp}. \tag{22}$$

The variance $\text{var}(L)$ is calculated by

$$\begin{aligned}
\text{var}&(L) \\
&= \sum_{n=0}^{\infty} c_{k,n} p^n q^{(k-1)n+1}\{-(\log p + (k-1)\log q)n - \log q - \overline{L}\}^2 \\
&= \sum_{n=0}^{\infty} c_{k,n} p^n q^{(k-1)n+1}\{(\log p + (k-1)\log q)^2 n^2 \\
&\quad +2\log q(\log p + (k-1)\log q)n + (\log q)^2\} - \overline{L}^2. \tag{23}
\end{aligned}$$

Here, we notice from the functional equation (12) that

$$\sum_{n=0}^{\infty} c_{k,n} p^n q^{(k-1)n+1} n^2 = F'_{k,p}(1) + F''_{k,p}(1), \tag{24}$$

and

$$F''_{k,p}(1) = \frac{2 - kp - p}{(1-kp)^3} kp^2. \tag{25}$$

Substituting the equations (21),(22),(24) and (25) into (23), we have

$$\text{var}(L) = \frac{pq}{(1-kp)^3} (\log p + (k-1)\log q)^2.$$

Summarizing the previous results, we established the following theorem.

Theorem 3. *The expectation $\overline{L}$ and variance $\mathrm{var}(L)$ of the ideal length of code-words for k-ary tree generated by the extending probability $0 \le p \le 1/k$ are given by*

$$\overline{L} = \frac{h(p)}{1 - kp}, \tag{26}$$

and

$$\mathrm{var}(L) = \frac{pq}{(1 - kp)^3} \left(\log p + (k - 1) \log q \right)^2. \tag{27}$$

Acknowledgements

Authors thank to anonymous referees for their careful reviews and for pointing out a connection of Theorem 2 to a classical result by Feller which is described in the Remark 2.

References

1. W. Feller, An Introduction to Probability Theory and Its Applications, Vol. 1, Second Edition, Wiley, 1957.
2. G.N. Raney, Functional composition patterns and power series reversion, Trans. AMS, Vol. 94, 441–451, 1960.
3. D.E. Knuth, The Art of Computer Programming Vol.1: Fundamental Algorithms, Addison-Wesley, 1968.
4. R.L. Graham, D.E. Knuth, and O. Patashnik, Concrete Mathematics, Addison-Wesley, 1989.
5. F. Willems, Y. Shtarkov, and T. Tjalkens, The context-tree weighting method: basic properties, IEEE Trans. Inform. Theory, Vol. 41, No. 3, 653–664, 1995.
6. R. Sedgewick and P. Flajolet, An Introduction to the Analysis of Algorithm, Addison-Wesley, 1996.
7. K. Kobayashi and T.S. Han, On the pre-order coding for complete k-ary coding trees, Proceedings of 1996 International Symposium on Information Theory and Its Applications, 302–303, 1996.
8. K. Kobayashi, H. Morita, and M. Hoshi, Enumerative coding for k-ary trees, Proceedings of the 1997 IEEE International Symposium on Information Theory, 423, 1997.
9. K. Kobayashi, H. Morita, and M. Hoshi, Information theoretic aspects on coding of trees, Proceedings of Memorial Workshop for the 50th Anniversary of the Shannon Theory, held at Yamanashi, 43–45, 1999.
10. K. Kobayashi, H. Morita, and M.Hoshi, Coding of ordered trees, Proceedings of the 2000 IEEE International Symposium on Information Theory, 15, 2000.

On Concepts of Performance Parameters
for Channels

R. Ahlswede

Abstract. Among the mostly investigated parameters for noisy channels are code size, error probability in decoding, block length; rate, capacity, reliability function; delay, complexity of coding. There are several statements about connections between these quantities. They carry names like "coding theorem", "converse theorem" (weak, strong, ...), "direct theorem", "capacity theorem", "lower bound", "upper bound", etc. There are analogous notions for source coding.

This note has become necessary after the author noticed that Information Theory suffers from a lack of precision in terminology. Its purpose is to open a discussion about this situation with the goal to gain more clarity.

There is also some confusion concerning the scopes of analytical and combinatorial methods in probabilistic coding theory, particularly in the theory of identification. We present a covering (or approximation) lemma for hypergraphs, which especially makes strong converse proofs in this area transparent and dramatically simplifies them.

1 Channels

It is beyond our intention to consider questions of modelling, like what is a channel in reality, which parts of a communication situation constitute a channel etc. Shannon's mathematical description in terms of transmission probabilities is the basis for our discussion.

Also, in most parts of this note we speak about one–way channels, but there will be also comments on multi–way channels and compound channels.

Abstractly, let $\mathcal{I}$ be any set, whose elements are called input symbols and let $\emptyset$ be any set, whose elements are called output symbols.

An (abstract) channel $W : \mathcal{I} \to (\emptyset, \mathcal{E})$ is a set of probability distributions

$$W = \big\{ W(\cdot|i) : i \in \mathcal{I} \big\} \tag{1.1}$$

on $(\emptyset, \mathcal{E})$.

So for every input symbol i and every (measurable) $E \in \mathcal{E}$ of output symbols $W(E|i)$ specifies the probability that a symbol in E will be received, if symbol i has been sent.

The set $\mathcal{I}$ does not have to carry additional structure.

Of particular interest are channels with "time–structure", that means, symbols are words over an alphabet, say $\mathcal{X}$ for the inputs and $\mathcal{Y}$ for the outputs. Here $\mathcal{X}^n = \prod_{t=1}^{n} \mathcal{X}_t$ with $\mathcal{X}_t = \mathcal{X}$ for $t \in \mathbb{N}$ (the natural numbers) are the input

R. Ahlswede et al. (Eds.): Information Transfer and Combinatorics, LNCS 4123, pp. 639–663, 2006.
© Springer-Verlag Berlin Heidelberg 2006

words of (block)–length n and $\mathcal{Y}^n = \prod\limits_{t=1}^{n} \mathcal{Y}_t$ with $\mathcal{Y}_t = \mathcal{Y}$ for $t \in \mathbb{N}$ are the output words of length n.

Moreover, again for the purpose of this discussion we can assume that a transmitted word of length n leads to a received word of length n. So we can define a (constant block length) channel by the set of stochastic matrices

$$\mathcal{K} = \{W^n : \mathcal{X}^n \to \mathcal{Y}^n : n \in \mathbb{N}\}. \tag{1.2}$$

In most channels with time–structure there are (compatibility) relations between these matrices.

We don't have to enter these delicate issues. Instead, we present now three channel concepts, which serve as key examples in this note.

DMC: The most familiar channel is the discrete memoryless channel, defined by the transmission probabilities

$$W^n(y^n|x^n) = \prod_{t=1}^{n} W(y_t|x_t) \tag{1.3}$$

for $W : \mathcal{X} \to \mathcal{Y}$, $x^n = (x_1, \dots, x_n) \in \mathcal{X}^n$, $y^n = (y_1, \dots, y_n) \in \mathcal{Y}^n$, and $n \in \mathbb{N}$.

NDMC: The *nonstationary* discrete memoryless channel is given by a sequence $(W_t)_{t=1}^{\infty}$ of stochastic matrices $W_t : \mathcal{X} \to \mathcal{Y}$ and the rule for the transmission of words

$$W^n(y^n|x^n) = \prod_{t=1}^{n} W_t(y_t, x_t). \tag{1.4}$$

Other names are "inhomogeneous channel", "non–constant" channel.

Especially, if $\quad W_t = \begin{cases} W & \text{for } t \text{ even} \\ V & \text{for } t \text{ odd} \end{cases}$

one gets a "periodic" channel of period 2 or a "parallel" channel. (c.f. [32], [2])

ADMC: Suppose now that we have two channels $\mathcal{K}_1$ and $\mathcal{K}_2$ as defined in (1.2). Then following [3] we can associate with them an *averaged* channel

$$\mathcal{A} = \left\{ \left(\frac{1}{2}W_1^n + \frac{1}{2}W_2^n : \mathcal{X}^n \to \mathcal{Y}^n \right) : n \in \mathbb{N} \right\} \tag{1.5}$$

and when both constituents, $\mathcal{K}_1$ and $\mathcal{K}_2$ are DMC's (resp. NDMC's) we term it ADMC (resp. ANDMC).

It is a very simple channel with "strong memory", suitable for theoretical investigations. They are considered in [3] in much greater generality (any number of constituents, infinite alphabets) and have been renamed by Han and Verdu "mixed channels" in several papers (see [29]).

We shall see below that channel parameters, which have been introduced for the DMC, where their meaning is without ambiguities, have been used for general time–structured channels for which sometimes their formal or operational meaning is not clear.

NONSTATIONARITY and MEMORY, incorporated in our examples of channels, are tests for concepts measuring channel performance.

2 Three Unquestioned Concepts: The Two Most Basic, Code Size and Error Probability, Then Further Block Length

Starting with the abstract channel $W : \mathcal{I} \to (\emptyset, \mathcal{E})$ we define a *code*

$$\mathcal{C} = \big\{(u_i, D_i) : i \in I\big\} \text{ with } u_i \in \mathcal{I}, D_i \in \mathcal{E}$$

for $i \in I$ and pairwise disjoint D_i's.

$$M = |\mathcal{C}| \text{ is the code size} \tag{2.1}$$

$$e(\mathcal{C}) = \max_{i \in I} W(D_i^c | u_i) \tag{2.2}$$

is the (maximal) probability of error and

$$\overline{e}(\mathcal{C}) = \frac{1}{M} \sum_{i=1}^{M} W(D_i^c | u_i) \tag{2.3}$$

is the average probability of error.

One can study now the functions

$$M(\lambda) = \max_{\mathcal{C}} \big\{ |\mathcal{C}| : e(\mathcal{C}) \leq \lambda \big\} \text{ (resp. } \overline{M}(\lambda)) \tag{2.4}$$

and

$$\lambda(M) = \min_{\mathcal{C}} \big\{ e(\mathcal{C}) : |\mathcal{C}| = M \big\} \text{ (resp. } \overline{\lambda}(M)), \tag{2.5}$$

that is, finiteness, growth, convexity properties etc.

It is convenient to say that $\mathcal{C}$ is an (M, λ)–code, if

$$|\mathcal{C}| \geq M \text{ and } e(\mathcal{C}) \leq \lambda. \tag{2.6}$$

Now we add time–structure, that means here, we go to the channel defined in (1.2). The parameter n is called the *block length* or word length.

It is to be indicated in the previous definitions. So, if $u_i \in \mathcal{X}^n$ and $D_i \subset \mathcal{Y}^n$ then we speak about a code $\mathcal{C}(n)$ and definitions (2.4), (2.5), and (2.6) are to be modified accordingly:

$$M(n, \lambda) = \max_{\mathcal{C}(n)} \big\{ |\mathcal{C}(n)| : e\big(\mathcal{C}(n)\big) \leq \lambda \big\} \tag{2.7}$$

$$\lambda(n, M) = \min_{\mathcal{C}(n)} \big\{ e\big(\mathcal{C}(n)\big) : |\mathcal{C}(n)| = M \big\} \tag{2.8}$$

$$\mathcal{C}(n) \text{ is an } (M, n, \lambda)\text{–code, if } |\mathcal{C}(n)| \geq M, e\big(\mathcal{C}(n)\big) \leq \lambda. \tag{2.9}$$

Remark 1: One could study blocklength as function of M and λ in smooth cases, but this would be tedious for the general model $\mathcal{K}$, because monotonicity properties are lacking for $M(n, \lambda)$ and $\lambda(M, n)$.

We recall next Shannon's fundamental statement about the two most basic parameters.

3 Stochastic Inequalities: The Role of the Information Function

We consider a channel $W : \mathcal{X} \to \mathcal{Y}$ with finite alphabets. To an input distribution P, that is a PD on $\mathcal{X}$, we assign the output distribution $Q = PW$, that is a PD on $\mathcal{Y}$, and the joint distribution $\tilde{P}$ on $\mathcal{X} \times \mathcal{Y}$, where $\tilde{P}(x, y) = P(x)W(y|x)$.

Following Shannon [38] we associate with (P, W) or $\tilde{P}$ the *information function (per letter)* $I : \mathcal{X} \times \mathcal{Y} \to \mathbb{R}$, where

$$I(x, y) = \begin{cases} \log \frac{\tilde{P}(x,y)}{P(x)Q(y)} \\ 0 \end{cases} , \text{ if } \tilde{P}(x, y) = 0. \tag{3.1}$$

If X is an (input) RV with values in $\mathcal{X}$ and distribution $P_X = P$ and if Y is an (output) RV with values in $\mathcal{Y}$ and distribution $P_Y = Q$ such that the joint distribution P_{XY} equals $\tilde{P}$, *then $I(X, Y)$ is a RV*. Its distribution function will be denoted by F, so

$$F(\alpha) = \Pr\{I(X, Y) \le \alpha\} = \tilde{P}(\{(x, y) : I(x, y) \le \alpha\}). \tag{3.2}$$

We call an $(M, \overline{\lambda})$–code $\{(u_i, D_i) : 1 \le i \le M\}$ *canonical*, if $P(u_i) = \frac{1}{M}$ for $i = 1, \ldots, M$ and the decoding sets are defined by maximum likelihood decoding, which results in a (minimal) average error probability $\overline{\lambda}$.

Theorem. Shannon [38]
For a canonical $(M, \overline{\lambda})$–code and the corresponding information function there are the relations

$$\frac{1}{2} F \left(\log \frac{M}{2} \right) \le \overline{\lambda} \le F \left(\log \frac{M}{2} \right). \tag{3.3}$$

Remarks

2. Shannon carries in his formulas a blocklength n, but this is nowhere used in the arguments. The bounds hold for abstract channels (without time structure). The same comment applies to his presentation of his random coding inequality: there exists a code of length M and average probability of error

$$\overline{\lambda} \le F(\log M + \theta) + e^{-\theta}, \theta > 0.$$

3. Let us emphasize that all of Shannon's bounds involve the information function (per letter), which is highlighted also in Fano [24], where it is called mutual information. (One may argue which terminology should be used, but certainly we don't need the third "information spectrum" introduced more recently by Han!) In contrast, Fano's inequality is *not a stochastic inequality*. It works with the *average* (or expected) mutual information $I(X \wedge Y)$ (also written as $I(X; Y)$), which is a constant. Something has been given away.

4 Derived Parameters of Performance: Rates for Code Sizes, Rates for Error Probabilities, Capacity, Reliability

The concept of rate involves a renormalisation in order to put quantities into a more convenient scale, some times per unit. Exponentially growing functions are renormalized by using the logarithmic function. In Information Theory the prime example is $M(n, \lambda)$ (see 2.7). Generally speaking, with any function $f : \mathbb{N} \to \mathbb{R}_+$ (or, equivalently, any sequence $\big(f(1), f(2), f(3), \ldots\big)$ of non–negative numbers) we can associate a rate function $\mathrm{rate}(f)$, where

$$\mathrm{rate}\big(f(n)\big) = \frac{1}{n} \log f(n). \tag{4.1}$$

We also speak of the *rate at* n, when we mean

$$\mathrm{rate}_n(f) \triangleq \mathrm{rate}\big(f(n)\big) = \frac{1}{n} \log f(n). \tag{4.2}$$

This catches statements like "an increase of rate" or "rate changes".

In Information Theory f is related to the channel $\mathcal{K}$ or more specifically $f(n)$ depends on W^n. For example choose $f(n) = M(n, \lambda)$ for $n \in \mathbb{N}$, λ constant. Then $\mathrm{rate}(f)$ is a *rate function* for certain code sizes.

Now comes a *second step*: for many *stationary* systems like stationary channels (c.f. DMC) f behaves very regular and instead of dealing with a whole rate function one just wants to associate a *number* with it.

We state for the three channels introduced in Section 1 the results – not necessarily the strongest known – relevant for our discussion.

DMC: There is a constant $C = C(W)$ (actually known to equal $\max_P I(W|P)$) such that

(a) for every $\lambda \in (0,1)$ and $\delta > 0$ there exists an $n_0 = n_0(\lambda, \delta)$ such that *for all* $n \geq n_0$ there exist
$$(n, e^{(C-\delta)n}, \lambda)\text{–codes},$$

(b) for every $\lambda \in (0,1)$ and $\delta > 0$ there exists an $n_0 = n_0(\lambda, \delta)$ such that *for all* $n \geq n_0$ there does *not* exist an
$$(n, e^{(C+\delta)n}, \lambda)\text{–code}.$$

ADMC: There is a constant C (actually known to equal $\max_P \min_{i=1,2} I(W_i|P)$ [3]) such that

(a) holds

(c) for every $\delta > 0$ there *exists* a $\lambda \in (0,1)$ and an $n_0 = n_0(\lambda, \delta)$ such that *for all* $n \geq n_0$ there does *not* exist an
$$(n, e^{(C+\delta)n}, \lambda)\text{–code}.$$

NDMC: There is a sequence of numbers $\left(C(n)\right)_{n=1}^{\infty}$ (which actually can be chosen as $C(n) = \frac{1}{n}\sum_{t=1}^{n}\max_{P} I(W_t|P)$ [2]) such that

(a') for every $\lambda \in (0,1)$ and $\delta > 0$ there exists an $n_0 = n_0(\lambda,\delta)$ such that for all $n \geq n_0$ there exist

$$(n, e^{(C(n)-\delta)n}, \lambda)\text{--codes.}$$

(b') for every $\lambda \in (0,1)$ and $\delta > 0$ there exists an $n_0 = n_0(\lambda,\delta)$ such that for all $n \geq n_0$ there does *not* exist an

$$(n, e^{(C(n)+\delta)n}, \lambda)\text{--code.}$$

(This is still true for infinite output alphabets, for infinite input alphabets in general not. There the analogue of (c), say (c') is often still true, but also not always.)

Notice that with every sequence $\left(C(n)\right)_{n=1}^{\infty}$ satisfying (a') and (b') or (a') and (c') also every sequence $\left(C(n) + o(1)\right)_{n=1}^{\infty}$ does. In this sense the sequence is not unique, whereas earlier the constant C is.

The *pair* of statements ((a), (b)) has been called by Wolfowitz *Coding theorem with strong converse* and the number C has been called the *strong capacity* in [2]. For the ADMC there is no C satisfying (a) *and* (b), so this channel *does not have* a strong capacity.

The pair of statements ((a), (c)) have been called by Wolfowitz coding theorem with *weak converse* and the number C has been called in [2] the *weak capacity*. So the ADMC does have a weak capacity.

(For completeness we refer to two standard textbooks. On page 9 of Gallager [27] one reads "The converse to the coding theorem is stated and proved in varying degrees of generality in chapter 4, 7, and 8. In imprecise terms, it states that if the entropy of a discrete source, in bits per second, is greater than C, then independent of the encoding and decoding used in transmitting the source output at the destination cannot be less than some positive number which depends on the source and on C. Also, as shown in chapter 9, if R is the minimum number of binary digits per second required to reproduce a source within a given level of average distortion, and if $R > C$, then, independent of the encoding and decoding, the source output cannot be transmitted over the channel and reproduced within that given average level of distortion."

In spite of its pleasant preciseness in most cases, there seems to be no definition of the weak converse in the book by Csiszár and Körner [22].)

Now the NDMC has in general no strong and no weak capacity (see our example in Section 7)

However, if we replace the concept of capacity by that of a capacity function $\left(C(n)\right)_{n=1}^{\infty}$ then the pair ((a'), (b')) (resp. ((a'), (c'))) may be called coding theorem with strong (resp. weak) converse and accordingly one can speak about *strong (resp. weak) capacity functions*, defined modulo $o(1)$.

These concepts have been used or at least accepted – except for the author – also by Wolfowitz, Kemperman, Augustin and also Dobrushin [23], Pinsker [35]. The concept of information stability (Gelfand/Yaglom; Pinsker) defined for *sequences of numbers* and *not* – like some authors do nowadays – for a *constant only*, is in full agreement at least with the ((a), (c)) or ((a′), (c′)) concepts. Equivalent formulations are

(a′) $\inf\limits_{\lambda>0} \varliminf\limits_{n\to\infty} \left(\frac{1}{n}\log M(n,\lambda) - C(n)\right) \geq 0$

(b′) for all $\lambda \in (0,1)$ $\varlimsup\limits_{n\to\infty} \left(\frac{1}{n}\log M(n,\lambda) - C(n)\right) \leq 0$

(c′) $\inf\limits_{\lambda>0} \varlimsup\limits_{n\to\infty} \left(\frac{1}{n}\log M(n,\lambda) - C(n)\right) \leq 0.$

(For a constant C this gives (a), (b), (c).)

Remarks

4. A standard way of expressing (c) is: for rates above capacity the error probability is bounded away from 0 for *all large n*. ([25], called *"partial converse"* on page 44.)

5. There are cases (c.f. [3]), where the uniformity in λ valid in (b) or (b′) holds only for $\lambda \in (0, \lambda_1)$ with an absolute constant λ_1 – a "medium" strong converse. It also occurs in "second order" estimates of [31] with $\lambda_1 = \frac{1}{2}$.

6. There are cases where (c) (or (c′)) don't hold for constant λ's but for $\lambda = \lambda(n)$ going to 0 sufficiently fast, in one case [17] like $\frac{1}{n}$ and in another like $\frac{1}{n^4}$ [19]. In both cases $\lambda(n)$ decreases reciprocal to a polynomial and it makes sense to speak of polynomial–weak converses. The soft–converse of [12] is for $\lambda(n) = e^{o(n)}$. Any decline condition on λ_n could be considered.

7. For our key example in Section 7 ((a′), (c′) holds, but not ((a), (c)). It can be shown that for the constant $C = 0$ and any $\delta > 0$ there is a $\lambda(\delta) > 0$ such that $(n, e^{(C+\delta)n})$–codes have error probability exceeding $\lambda(\delta)$ for *infinitely many n*.

By Remark 1 this is weaker than (c) and equivalent to

$$\inf\limits_{\lambda>0} \lim\limits_{n\to\infty} \frac{1}{n}\log M(n,\lambda) = C.$$

Now comes a seemingly small twist. Why bother about "weak capacity", "strong capacity" etc. and their existence – every channel should have a capacity.

Definition: $\underline{C}$ is called the (pessimistic) capacity of a channel $\mathcal{K}$, if it is the supremum over all numbers C for which (a) holds. Since $C = 0$ satisfies (a), the number $\underline{C} = \underline{C}(\mathcal{K})$ exists. Notice that there are no requirements concerning (b) or (c) here.

To every general $\mathcal{K}$ a constant performance parameter has been assigned ! What does it do for us?

First of all the name "pessimistic" refers to the fact that another number $\overline{C} = \overline{C}(\mathcal{K})$ can be introduced, which is at least a large as $\underline{C}$.

Definition: $\overline{C}$ is called the (optimistic) capacity of a channel $\mathcal{K}$, if it is the supremum over all numbers C for which in (a) the condition "for all $n \geq n_0(\lambda, \delta)$" is replaced by "for infinitely many n" or equivalently

$$\overline{C} = \inf_{\lambda > 0} \varlimsup_{n \to \infty} \frac{1}{n} \log M(n, \lambda).$$

Here it is measured whether for every λ $R < \overline{C}$ this "rate" is occasionally, but infinitely often achievable.

(Let us briefly mention that "the reliability function" $E(R)$ is commonly defined through the values

$$\underline{E}(R) = -\varliminf_{n \to \infty} \frac{1}{n} \log \lambda(e^{Rn}, n)$$

$$\overline{E}(R) = -\varlimsup_{n \to \infty} \frac{1}{n} \log \lambda(e^{Rn}, n)$$

if they coincide. Again further differentiation could be gained by considering the sequence

$$E_n(R_n) = -\frac{1}{n} \log \lambda(e^{R_n n}, n), \ n \in \mathbb{N},$$

for sequences of rates $(R_n)_{n=1}^{\infty}$. But that shall not be pursuit here.)

In the light of old work [2] we were shocked when we learnt that these two definitions were given in [22] and that the pessimistic capacity was used throughout that book. Since the restriction there is to the DMC–situation it makes actually no difference. However, several of our Theorems had just been defined away. Recently we were even more surprised when we learned that these definitions were not new at all and have indeed been standard and deeply rooted in the community of information theorists (the pessimistic capacity $\underline{C}$ is used in [24], [42], [21] and the optimistic capacity $\overline{C}$ is used in [22] on page 223 and in [33]).

Fano [24] uses $\underline{C}$, but he at the same time emphasizes throughout the book that he deals with "constant channels".

After quick comments about the optimistic capacity concept in the next section we report on another surprise concerning $\underline{C}$.

5 A Misleading Orientation at the DMC: The Optimistic Rate Concept Seems Absurd

Apparently for the DMC the optimistic as well as the pessimistic capacities, $\overline{C}$ and $\underline{C}$, equal $C(W)$. For multi–way channels and compound channels $\{W(\cdot|\cdot, s) : s \in \mathcal{S}\}$ the optimistic view suggests a dream world.

A. Recently Cover explained that under this view for the broadcast channel $(W : \mathcal{X} \to \mathcal{Y}, V : \mathcal{X} \to \mathcal{Z})$ the rate pair $(R_{\mathcal{Y}}, R_{\mathcal{Z}}) = (C(W), C(V))$ is in the capacity region, which in fact equals $\{(R_{\mathcal{Y}}, R_{\mathcal{Z}}) : 0 \leq R_{\mathcal{Y}} \leq C(W), 0 \leq R_{\mathcal{Z}} \leq C(W)\}$.

   Just assign periodically time intervals of lengths $m_1, n_1, m_2, n_2, m_3, n_3, \ldots$ to the DMC's W and V for transmission. Just choose every interval very long in comparison to the sum of the lengths of its predecessors. Thus again and again every channel comes in its rate close, and finally arbitrary close, to its capacity. The same argument applies to the MAC, TWC etc. – so in any situation where the communicators have a choice of the channels for different time intervals.

B. The reader may quickly convince himself that $\overline{C} = \min_{s \in \mathcal{S}} C\big(W(\cdot|\cdot, s)\big) \geq \max_P \min_s I\big(W(\cdot|\cdot, s)|P\big)$ for the compound channel. For the sake of the argument choose $\mathcal{S} = \{1, 2\}$. The sender not knowing the individual channel transmits for channel $W(\cdot|\cdot, 1)$ on the m–intervals and for channel $W(\cdot|\cdot, 2)$ on the n–intervals. The receiver *can test* the channel and knows in which intervals to decode!

C. As a curious Gedankenexperiment: Is there anything one can do in this context for the AVC?

   For the semicontinuous compound channel, $|\mathcal{S}| = \infty$, the ordinary weak capacity $(((a),(c))$ hold) is unknown. We guess that optimism does not help here, because it does seem to help if there are infinitely many proper cases.

   The big issue in all problems here is of course delay. It ought to be incorporated (Space–time coding).

6   A "Paradox" for Product of Channels

Let us be given s channels $(W_j^n)_{n=1}^{\infty}, 1 \leq j \leq s$. Here $W_j^n : \mathcal{X}_j^n \to \mathcal{Y}_j^n, 1 \leq j \leq s$. The product of these channels $(W^{*n})_{n=1}^{\infty}$ is defined by

$$W^{*n} = \prod_{j=1}^{s} W_j^n : \prod_{j=1}^{s} \mathcal{X}_j^n \to \prod_{j=1}^{s} \mathcal{Y}_j^n.$$

A paper by Wyner [42] is very instructive for our discussion. We quote therefore literally the beginning of the paper (page 423) and also its Theorem with a sketch of the proof (page 425), because it is perhaps instructive for the reader to see how delicate things are even for leading experts in the field.

"In this paper we shall consider the product or parallel combination of channels, and show that (1) the *capacity of the product channel is the sum of the capacities of the component channels*, and (2) the "strong converse" holds for the product channel if it holds for each of the component channels. The result is valid for any class of channels (with or without memory, continuous or discrete) provided that the capacities exist. "Capacity" is defined here *as the supremum of those rates for which arbitrarily high reliability is achievable with block coding for sufficiently long delay.*

Let us remark here that there are two ways in which "channel capacity" is commonly defined. The first definition takes the channel capacity to be the supremum of the "information" processed by the channel, where "information" is the difference of the input "uncertainty" and the "equivocation" at the output. *The second definition, which is the one we use here, takes the channel capacity to be the maximum "error free rate".* For certain classes of channels (e.g., memoryless channels, and finite state indecomposable channels) it has been established that these two definitions are equivalent. In fact, this equivalence is the essence of the Fundamental Theorem of Information Theory. For such channels, (1) above follows directly. The second definition, however, is applicable to a broader class of channels than the first. One very important such class are time–continuous channels."

Theorem

(1) *Let C^* be the capacity of the product of s channels with capacities $C_1, C_2, \ldots, C_s$ respectively. Then*

$$C^* = \sum_{j=1}^{s} C_j. \qquad ((6.1))$$

(2) *If the strong converse holds for each of these s channels, then it holds for the product channel.*

The proof of (1) is divided into two parts. In the first (the "direct half") we will show that any $R < \sum_{j=1}^{s} C_j$ is a permissible rate. This will establish that $C^* \geq \sum_{j=1}^{s} C_j$. In the second ("weak converse") we will show that no $R > \sum_{j=1}^{s} C_j$ is a permissible rate, establishing that $C^* \leq \sum_{j=1}^{s} C_j$. The proof of (2) parallels that of the weak converse.

It will suffice to prove the theorem for the product of two channels ($s = 2$), the result for arbitrary s following immediately by induction."

Let's first remark that $C^* \geq \sum_{j=1}^{s} C_j$ for the pessimistic capacities (apparently used here) follows immediately from the fact that by taking products of codes the errors at most behave additive. By proving the reverse inequality the weak converse, statement (c) in Section 4 is *tacitly assumed* for the component channels and from there on everything is okay. The point is that this assumption does not appear as a hypothesis in the Theorem! Indeed our key example of Section 7 shows that (6.1) is in general not true. The two factor channels used in the example don't have a weak converse (or weak capacity for that matter).

The reader is reminded that having proved a weak converse for the number $\underline{C}$, the pessimistic capacity, is equivalent to having shown that the weak capacity exists.

7 The Pessimistic Capacity Definition: An Information Theoretic Perpetuum Mobile

Consider the two matrices $V^1 = \left(\begin{smallmatrix} 1 & 0 \\ 0 & 1 \end{smallmatrix}\right)$ and $V^0 = \left(\begin{smallmatrix} \frac{1}{2} & \frac{1}{2} \\ \frac{1}{2} & \frac{1}{2} \end{smallmatrix}\right)$. We know that $C(V^1) = 1$ and $C(V^0) = 0$.

Consider a NDMC $\mathcal{K}$ with $W_t \in \{V^0, V^1\}$ for $t \in \mathbb{N}$ and a NDMC $\mathcal{K}^*$ with t-th matrix W_t^* also from $\{V^0, V^1\}$ but *different* from W_t. Further consider the product channel $(\mathcal{K}, \mathcal{K}^*)$ specified by $W_1 W_1^* W_2 W_2^*$ – again a NDMC.

With the choice $(m_1, n_1, m_2, n_2, \ldots)$, where for instance $n_i \geq 2^{m_i}$, $m_{i+1} \geq 2^{n_i}$ we define channel $\mathcal{K}$ completely by requiring that $W_t = V^1$ in the m_i–length intervals and $W_t = V^0$ in the n_i–length intervals. By their growth properties we have for the pessimistic capacities $\underline{C}(\mathcal{K}) = \underline{C}(\mathcal{K}^*) = 0$. However, apparently $\underline{C}(\mathcal{K}, \mathcal{K}^*) = 1$.

8 A Way Out of the Dilemma: Capacity Functions

If $M(n, \lambda)$ fluctuates very strongly in n and therefore also $\text{rate}_n(M)$, then it does not make much sense to describe its growth by one number $\underline{C}$. At least one has to be aware of the very limited value of theorems involving that number.

For the key example in Section 7 $\underline{C}(\mathcal{K}) = \underline{C}(\mathcal{K}^*) = 0$ and on the other hand $\overline{C}(\mathcal{K}) = \overline{C}(\mathcal{K}^*) = 1$. In contrast we can choose the sequence $(c_n)_{n=1}^{\infty} = \left(\frac{1}{n} \sum_{t=1}^{n} C(W_t)\right)_{n=1}^{\infty}$ for channel $\mathcal{K}$ and $(c_n^*)_{n=1}^{\infty} = \left(\frac{1}{n} \sum_{t=1}^{n} C(W_t^*)\right)_{n=1}^{\infty}$ for channel $\mathcal{K}^*$, who are always *between* 0 and 1.

They are (even strong) capacity functions and for the product channel $\mathcal{K} \times \mathcal{K}^*$ we have the capacity function $(c_n + c_n^*)_{n=1}^{\infty}$, which equals identically 1, what it should be. Moreover thus also in general the "perpetuum mobile of information" disappears. We have been able to prove the

Theorem. *For two channels $\mathcal{K}_1$ and $\mathcal{K}_2$*

(i) *with weak capacity functions their product has the sum of those functions as weak capacity function*

(ii) *with strong capacity functions their product has the sum of those functions as strong capacity function.*

We hope that we have made clear that capacity functions in conjunction with converse proofs carry in general more information – perhaps not over, but *about channels* – than optimistic or pessimistic capacities. This applies even for channels without a weak capacity function because they can be made this way at least as large $\underline{C}$ and still satisfy (a).

Our conclusion is, that

1. when speaking about capacity formulas in non standard situations one must clearly state which definition is being used.
2. there is no "true" definition nor can definitions be justified by authority.

3. presently weak capacity functions have most arguments in their favour, also in comparison to strong capacity functions, because of their wide validity and the primary interest in direct theorems. To call channels without a strong capacity "channels without capacity" ([41]) is no more reasonable than to name an optimistic or a pessimistic capacity "the capacity".
4. we must try to help enlightening the structure of channels. For that purpose for instance $\underline{C}$ can be a useful bound on the weak capacity function, because it may be computable whereas the function isn't.
5. Similar comments are in order for other quantities in Information Theory, rates for data compression, reliability functions, complexity measures.

9 Some Concepts of Performance from Channels with Phases

In this Section we explore other capacity concepts involving the phase of the channel, which for stationary systems is not relevant, but becomes an issue otherwise. Again the NDMC $(W_t)_{t=1}^{\infty}$ serves as a genuine example. In a phase change by m we are dealing with $(W_{t+m})_{t=1}^{\infty}$. "Capacity" results for the class of channels $\{(W_{t+m})_{t=1}^{\infty} : 0 \leq m < \infty\}$ in the spirit of a compound channel, that is, for codes which are good simultaneously for all m are generally unknown. The AVC can be produced as a special case and even more so the zero–error capacity problem.

An exception is for instance the case where $(W_t)_{t=1}^{\infty}$ is almost periodic in the sense of Harald Bohr. Because these functions have a mean also $\left(C(W_t)\right)_{t=1}^{\infty}$ has a mean and it has been shown that there is a strong capacity [2].

Now we greatly simplify the situation and look only at $(W_t)_{t=1}^{\infty}$ where

$$W_t \in \left\{ \left(\begin{smallmatrix} 1 & 0 \\ 0 & 1 \end{smallmatrix}\right), \left(\begin{smallmatrix} \frac{1}{2} & \frac{1}{2} \\ \frac{1}{2} & \frac{1}{2} \end{smallmatrix}\right) \right\}$$

and thus $C(W_t) \in \{0, 1\}$. Moreover, we leave error probabilities aside and look only at $0 - 1$–sequences $(C_1, C_2, C_3, \dots)$ and the associated $C(n) = \frac{1}{n} \sum_{t=1}^{n} C_t \in [0, 1]$.

So we just play with $0 - 1$–sequences $(a_n)_{n=1}^{\infty}$ and associated Cesaro–means $A_n = \frac{1}{n} \sum_{t=1}^{n} a_t$ and $A_{m+1, m+n} = \frac{1}{n} \sum_{t=m+1}^{m+n} a_t$.

First of all there are the familiar

$$\underline{A} = \varliminf_{n \to \infty} A_n \text{ (the pessimistic mean)} \tag{9.1}$$

$$\overline{A} = \varlimsup_{n \to \infty} A_n \text{ (the optimistic mean)}. \tag{9.2}$$

We introduce now a new concept

$$\underline{\underline{A}} = \lim_{n \to \infty} \inf_{m \geq 0} A_{m+1, m+n} \text{ (the pessimistic phase independent mean)}. \tag{9.3}$$

The "inf" reflects that the system could be in any phase (*known to but not controlled by the communicators*). Next we assume that the communicators can *choose* the phase m for an intended n and define

$$\overline{\overline{A}} = \overline{\lim_{n\to\infty}} \ \sup_{m\geq 0} A_{m+1,m+n} \quad \text{(super optimistic mean)}. \tag{9.4}$$

We shall show first

Lemma

$$\underline{\lim_{n\to\infty}} \ \inf_{m\geq 0} A_{m+1,m+n} = \underline{\underline{A}} \tag{9.5}$$

$$\overline{\lim_{n\to\infty}} \ \sup_{m\geq 0} A_{m+1,m+n} = \overline{\overline{A}} \tag{9.6}$$

Proof: We prove only (9.5), the proof for (9.6) being "symmetrically" the same. We have to show that

$$\underline{\underline{A}} = \underline{\lim_{n\to\infty}} \ \inf_{m\geq 0} A_{m+1,m+n} \geq \overline{\lim_{n\to\infty}} \ \inf_{m\geq 0} A_{m+1,m+n}. \tag{9.7}$$

For every n let $m(n)$ give minimal $A_{m+1,m+n}$. The number exists because these means take at most $n+1$ different values. Let n^* be such that $A_{m(n^*)+1,m(n^*)+n^*}$ is within ε of $\underline{\underline{A}}$ and choose a much bigger N^* for which $A_{m(N^*)+1,m(N^*)+N^*}$ is within ε of the expression at the right side of (9.7) and $N^* \geq \frac{1}{\varepsilon} n^*$ holds.

Choose r such that $rn^* + 1 \leq N^* \leq (r+1)n^*$ and write

$$N^* A_{m(N^*)+1,m(N^*)+N^*} = \sum_{s=0}^{r-1} \sum_{t=m(N^*)+sn^*+1}^{m(N^*)+(s+1)n^*} a_t + \sum_{t=m(N^*)+rn^*+1}^{m(N^*)+N^*} a_t$$

$$\geq r \cdot n^* A_{m(n^*)+n^*} \geq r \cdot n^*(\underline{\underline{A}} - \varepsilon)$$

$$\geq (N^* - n^*)(\underline{\underline{A}} - \varepsilon) \geq N^*(1 - \varepsilon)(\underline{\underline{A}} - \varepsilon).$$

Finally, by changing the order of operations we get four more definitions, however, they give nothing new. In fact,

$$\inf_{m} \ \underline{\lim_{n\to\infty}} A_{m+1,m+n} = \sup_{m} \ \underline{\lim_{n\to\infty}} A_{m+1,m+n} = \underline{A} \tag{9.8}$$

$$\inf_{m} \ \overline{\lim_{n\to\infty}} A_{m+1,m+n} = \sup_{m} \ \overline{\lim_{n\to\infty}} A_{m+1,m+n} = \overline{A}, \tag{9.9}$$

because for an m_0 close to an optimal phase the first m_0 positions don't affect the asymptotic behaviour.

The list of quantities considered is not intended to be complete in any sense, but serves our illustration.

We look now at $\underline{A} \leq \underline{\underline{A}} \leq \overline{A} \leq \overline{\overline{A}}$ in four examples to see what constellations of values can occur.

We describe a $0 - 1$–sequence $(a_n)_{n=1}^{\infty}$ by the lengths of its alternating strings of 1's and 0's: $(k_1, \ell_1, k_2, \ell_2, k_3, \ldots)$

Example 1: $k_t = k$, $\ell_t = \ell$ for $t = 1, 2, \ldots$; a periodic case:

$$A = \underline{A} = \overline{A} = \overline{\overline{A}} = \frac{k}{k + \ell}.$$

Example 2: $k_t = \ell_t = t$ for $t = 1, 2, \ldots$. Use $\sum_{t=1}^{n} k_t = \sum_{t=1}^{n} \ell_t = \frac{n(n+1)}{2}$ and verify

$$0 = \underline{A} < \frac{1}{2} = \underline{A} = \overline{A} < 1 = \overline{\overline{A}}.$$

Example 3: $k_t = \sum_{s=1}^{t-1} k_s$, $\ell_t = \sum_{s=1}^{t-1} \ell_s$ for $t = 1, 2, \ldots$

$$0 = \underline{A} < \frac{1}{2} = \underline{A} < \frac{2}{3} = \overline{A} < 1 = \overline{\overline{A}}.$$

Here all four values are different.

Example 4: $k_t = \sum_{s=1}^{t-1} k_s$, $\ell_t = t$ for $t = 2, 3, \ldots, k_1 = 1$

$$0 = \underline{A} < 1 = \underline{A} = \overline{A} = \overline{\overline{A}}.$$

All four quantities say something about $(A_n)_{n=1}^{\infty}$, they all say less than the *full record*, the sequence itself (corresponding to our capacity function).

10   Some Comments on a Formula for the Pessimistic Capacity

A noticeable observation of Verdu and Han [39] is that $\underline{C}$ can be expressed for every channel $\mathcal{K}$ in terms of a stochastic limit (per letter) mutual information.

The renewed interest in such questions originated with the Theory of Identification, where converse proofs for the DMC required that output distributions of a channel, generated by an arbitrary input distribution (randomized encoding for a message), be "approximately" generated by input distributions of controllable sizes of the carriers. Already in [12] it was shown that essentially sizes of $\sim e^{Cn}$ would do and then in [30], [31] the bound was improved (strong converse) by a natural random selection approach. They termed the name "resolvability" of a channel for this size problem.

The approximation problem (like the rate distortion problem) is a "covering problem" as opposed to a "packing problem" of channel coding, but often these problems are very close to each other, actually ratewise identical for standard channels like the DMC. To establish the strong second order identification capacity for more general channels required in the approach of [30] that resolvability must equal capacity and for that the strong converse for $\mathcal{K}$ was needed.

This led them to study the ADMC [3], which according to Han [28] plaid a key role in the further development. Jacobs has first shown that there are channels with a weak converse, but without a strong converse. In his example the abstract reasoning did not give a channel capacity formula. This is reported in [32] and mentioned in [3], from where the following facts should be kept in mind.

1. The ADMC has no strong converse but a weak converse (see Section 4 for precise terminology).
2. The term weak capacity was introduced.
3. The weak capacity (and also the λ–capacity were determined for the ADMC by linking it to the familiar max min –formula for the compound channel in terms of (per letter)–mutual information.
4. It was shown that $\lim\limits_{n\to\infty} \frac{1}{n} \max_{X^n} I(X^n \wedge Y^n)$ does not describe the weak capacity in general. Compare this with Wyner's first capacity definition in Section 6.
5. It was shown that Fano's inequality, involving only the *average* mutual information $I(X^n \wedge Y^n)$, fails to give the weak converse for the ADMC.

The observation of [39] is again natural, one should use the information function of the ADMC directly rather than the max min –formula. They defined for general $\mathcal{K}$ the *sequence* of pairs

$$(\mathbf{X}, \mathbf{Y}) = (X^n, Y^n)_{n=1}^{\infty} \tag{10.1}$$

and

$$\underline{I}(\mathbf{X} \wedge \mathbf{Y}) = \sup \left\{ \alpha : \varliminf_{n\to\infty} \Pr \left\{ (x^n, y^n) : \frac{1}{n} I(x^n, y^n) \le \alpha \right\} = 0 \right\}. \tag{10.2}$$

Their general formula asserts

$$\underline{C} = \sup_{\mathbf{X}} \underline{I}(\mathbf{X} \wedge \mathbf{Y}). \tag{10.3}$$

The reader should be aware that

α.) The stochastic inequalities used for the derivation (10.3) are both (in particular also Theorem 4 of [39]) not new.

β.) Finally, there is a very important point. In order to show that a certain quantity K (for instance $\sup\limits_{\mathbf{X}} \underline{I}(\mathbf{X} \wedge \mathbf{Y})$) equals $\underline{C}$ one has to show $K \ge \underline{C}$ and then (by definition of $\underline{C}$) that $K + \delta$, any $\delta > 0$, is not a rate achievable for arbitrary small error probabilities or equivalently, that $\inf\limits_{\lambda} \varliminf\limits_{n\to\infty} \log M(n, \lambda) < K + \delta$. For this one does *not need* the *weak* converse

(b) $\inf\limits_{\lambda} \varlimsup\limits_{n\to\infty} \log M(n, \lambda) \le K$, but only

$$\inf\limits_{\lambda} \varliminf\limits_{n\to\infty} \log M(n, \lambda) \le K \tag{10.4}$$

(see also Section 4) The statement may be termed the "weak–weak converse" or the "<u>weak</u>–converse" or "occasional–converse" or whatever. Keep in mind that the fact that the weak converse does not hold for the factors led to the "information theoretic perpetuum mobile". The remark on page 1153 "Wolfowitz ... referred to the conventional capacity of Definition 1 (which is always defined) as *weak capacity*" is not only wrong, because Wolfowitz never used the term "weak capacity", it is – as we have explained – very misleading. After we have commented on the drawbacks of the pessimistic capacity, especially also for channel NDMC, we want to say that on the other hand the formula $\sup_{\mathbf{X}} \underline{I}(\mathbf{X} \wedge \mathbf{Y})$ and also its dual $\sup_{\mathbf{X}} \overline{I}(\mathbf{X} \wedge \mathbf{Y})$ are helpful in characterizing or bounding quantities of interest not only in their original context, Theory of Identification. Han has written a book [29] in which he introduces these quantities and their analogues into all major areas of Information Theory.

11   Pessimistic Capacity Functions

We think that the following concept suggests itself as one result of the discussion.

Definition: A sequence $(C_n)_{n=1}^{\infty}$ of non–negative numbers is *a* capacity sequence of $\mathcal{K}$, if

$$\inf_{\lambda>0} \varliminf_{n\to\infty} \left(\frac{1}{n} \log M(n,\lambda) - C_n \right) = 0.$$

The sequence $(\underline{C}, \underline{C}, \underline{C}, \dots)$ is a capacity sequence, so by definition there are always capacity sequences.

Replacing α by α_n in (10.2) one can characterize capacity sequences in term of sequences defined in terms of (per letter) information functions. Every channel $\mathcal{K}$ has a class of capacity sequences $\mathcal{C}(\mathcal{K})$.

It can be studied. In addition to the constant function one may look for instance at the class of functions of period m, say $\mathcal{C}(\mathcal{K}, m) \subset \mathcal{C}(\mathcal{K})$. More generally complexity measures μ for the sequences may be used and accordingly one gets say $\mathcal{C}(\mathcal{K}, \mu \leq \rho)$, a space of capacity functions of μ–complexity less than ρ.

This seems to be a big machinery, but channels $\mathcal{K}$ with no connections between W^n and $W^{n'}$ required in general constitute a *wild* class of channels. The capacity sequence space $\mathcal{C}(\mathcal{K})$ characterizes a channel in time like a capacity region for multi–way channels characterizes the possibilities for the communicators.

Its now not hard to show that for the product channel $\mathcal{K}_1 \times \mathcal{K}_2$ for any $f \in \mathcal{C}(\mathcal{K}_1 \times \mathcal{K}_2)$ there exist $f_i \in \mathcal{C}(\mathcal{K}_i); i = 1, 2,$; such that $f_1 + f_2 \geq f$. The component channels together can do what the product channel can do. This way, both, the non–stationarity and perpetuum mobile problem are taken care of.

We wonder how all this looks in the light of "quantum parallelism".

We finally quote statements by Shannon. In [37] he writes "Theorem 4, of course, is analogous to known results for the ordinary capacity C, *where the product channel has the sum of the ordinary capacities* and the sum channel has an equivalent number of letters equal to the sum of the equivalent numbers of letters for the individual channels. We conjecture, but have not been able to prove, that the equalities in Theorem 4 hold in general – not just under the conditions given". Both conjectures have been disproved (Haemers and Alon).

12   Identification

Ahlswede and Dueck, considering not the problem that the receiver wants to recover a message (*transmission problem*), but wants to decide whether or not the sent message is identical to an arbitrarily chosen one (*identification problem*), defined an $(n, N, \lambda_1, \lambda_2)$ identification (ID) code to be a collection of pairs

$$\{(P_i, \mathcal{D}_i) : i = 1, \dots, N\},$$

with probability distributions P_i on $\mathcal{X}^n$ and $\mathcal{D}_i \subset \mathcal{Y}^n$, such that the error probabilities of first resp. second kind satisfy

$$P_i W^n(\mathcal{D}_i^c) = \sum_{x^n \in \mathcal{X}^n} P_i(x^n) W^n(\mathcal{D}_i^c | x^n) \leq \lambda_1,$$

$$P_j W^n(\mathcal{D}_i) = \sum_{x^n \in \mathcal{X}^n} P_j(x^n) W^n(\mathcal{D}_i | x^n) \leq \lambda_2,$$

for all $i, j = 1, \dots, N$, $i \neq j$. Define $N(n, \lambda_1, \lambda_2)$ to be the maximal N such that a $(n, N, \lambda_1, \lambda_2)$ ID code exists.

With these definitions one has for a DMC

Theorem. (Ahlswede, Dueck [12]) *For every* $\lambda_1, \lambda_2 > 0$ *and* $\delta > 0$, *and for every sufficiently large* n

$$N(n, \lambda_1, \lambda_2) \geq \exp(\exp(n(C(W) - \delta))).$$

The next two sections are devoted to a (comparably short) proof of the following strong converse

Theorem. *Let* $\lambda_1, \lambda_2 > 0$ *such that* $\lambda_1 + \lambda_2 < 1$. *Then for every* $\delta > 0$ *and every sufficiently large* n

$$N(n, \lambda_1, \lambda_2) \leq \exp(\exp(n(C(W) + \delta))).$$

The strong converse to the coding theorem for identification via a DMC was conjectured in [12] (In case of complete feedback the strong converse was established already in [13]) and proved by Han and Verdu [31] and in a simpler way in [30]. However, even the second proof is rather complicated. The authors emphasize that they used and developed analytical methods and take the position that combinatorial techniques for instance of [6], [7] find their limitations on this kind of problem (see also Newsletter on Moscow workshop in 1994). We demonstrate now that this is not the case (see also the remarks on page XIX of [C1]).

Here we come back to the very first idea from [12], essentially to replace the distributions P_i by uniform distributions on "small" subsets of $\mathcal{X}^n$, namely with cardinality slightly above $\exp(nC(W))$.

13 A Novel Hypergraph Covering Lemma

The core of the proof is the following result about hypergraphs. Recall that a *hypergraph* is a pair $\Gamma = (\mathcal{V}, \mathcal{E})$ with a finite set $\mathcal{V}$ of vertices, and a finite set $\mathcal{E}$ of (hyper–) edges $E \subset \mathcal{V}$. We call Γ *e–uniform*, if all its edges have cardinality e. For an edge $E \in \mathcal{E}$ denote the characteristic function of $E \subset \mathcal{V}$ by 1_E.

A result from large deviation theory will be used in the sequel:

Lemma 1. *For an i.i.d. sequence $Z_1, \ldots, Z_L$ of random variables with values in $[0,1]$ with expectation $\mathbb{E}Z_i = \mu$, and $0 < \varepsilon < 1$*

$$\Pr\left\{ \frac{1}{L} \sum_{i=1}^{L} Z_i > (1+\varepsilon)\mu \right\} \leq \exp(-LD((1+\varepsilon)\mu \| \mu)),$$

$$\Pr\left\{ \frac{1}{L} \sum_{i=1}^{L} Z_i < (1-\varepsilon)\mu \right\} \leq \exp(-LD((1-\varepsilon)\mu \| \mu)),$$

where $D(\alpha \| \beta)$ is the information divergence of the binary distributions $(\alpha, 1-\alpha)$ and $(\beta, 1-\beta)$. Since

$$D((1+\varepsilon)\mu \| \mu) \geq \frac{\varepsilon^2 \mu}{2\ln 2} \ \text{for} \ |\varepsilon| \leq \frac{1}{2},$$

it follows that

$$\Pr\left\{ \frac{1}{L} \sum_{i=1}^{L} Z_i \notin [(1-\varepsilon)\mu, (1+\varepsilon)\mu] \right\} \leq 2\exp\left(-L \cdot \frac{\varepsilon^2 \mu}{2\ln 2} \right).$$

Proof: The first two inequalities are for instance a consequence of Sanov's Theorem (c.f. [21], also Lemma LD in [12]). The lower bound on D is elementary calculus.

Lemma 2. (Novel hypergraph covering, presented also in "Winter School on Coding and Information Theory, Ebeltoft, Dänemark, Dezember 1998" and in "Twin Conferences: 1. Search and Complexity and 2. Information Theory in Mathematics, Balatonelle, Ungarn, July 2000".)

Let $\Gamma = (\mathcal{V}, \mathcal{E})$ be an e–uniform hypergraph, and P a probability distribution on $\mathcal{E}$. Define the probability distribution Q on $\mathcal{V}$ by

$$Q(v) = \sum_{E \in \mathcal{E}} P(E) \frac{1}{e} 1_E(v),$$

and fix $\varepsilon, \tau > 0$. Then there exist vertices $\mathcal{V}_0 \subset \mathcal{V}$ and edges $E_1, \ldots, E_L \in \mathcal{E}$ such that with

$$\bar{Q}(v) = \frac{1}{L} \sum_{i=1}^{L} \frac{1}{e} 1_{E_i}(v)$$

the following holds:

$$Q(\mathcal{V}_0) \leq \tau,$$

$$\forall v \in \mathcal{V} \setminus \mathcal{V}_0 \ \ (1-\varepsilon)Q(v) \leq \bar{Q}(v) \leq (1+\varepsilon)Q(v),$$

$$L \leq 1 + \frac{|\mathcal{V}|}{e} \frac{2\ln 2 \log(2|\mathcal{V}|)}{\varepsilon^2 \tau}.$$

For ease of application we formulate and prove a slightly more general version of this:

Lemma 3. *Let $\Gamma = (\mathcal{V}, \mathcal{E})$ be a hypergraph, with a measure Q_E on each edge E, such that $Q_E(v) \leq \eta$ for all E, $v \in E$. For a probability distribution P on $\mathcal{E}$ define*

$$Q = \sum_{E \in \mathcal{E}} P(E)Q_E,$$

and fix $\varepsilon, \tau > 0$. Then there exist vertices $\mathcal{V}_0 \subset \mathcal{V}$ and edges $E_1, \ldots, E_L \in \mathcal{E}$ such that with

$$\bar{Q} = \frac{1}{L} \sum_{i=1}^{L} Q_{E_i}$$

the following holds:

$$Q(\mathcal{V}_0) \leq \tau,$$

$$\forall v \in \mathcal{V} \setminus \mathcal{V}_0 \ \ (1-\varepsilon)Q(v) \leq \bar{Q}(v) \leq (1+\varepsilon)Q(v),$$

$$L \leq 1 + \eta|\mathcal{V}|\frac{2\ln 2 \log(2|\mathcal{V}|)}{\varepsilon^2 \tau}.$$

Proof: Define i.i.d. random variables $Y_1, \ldots, Y_L$ with

$$\Pr\{Y_i = E\} = P(E) \text{ for } E \in \mathcal{E}.$$

For $v \in \mathcal{V}$ define $X_i = Q_{Y_i}(v)$. Clearly $\mathbb{E}X_i = Q(v)$, hence it is natural to use a large deviation estimate to prove the bounds on $\bar{Q}$. Applying Lemma 1 to the random variables $\eta^{-1}X_i$ we find

$$\Pr\left\{\frac{1}{L}\sum_{i=1}^{L} X_i \notin [(1-\varepsilon)Q(v), (1+\varepsilon)Q(v)]\right\} \leq 2\exp\left(-L \cdot \frac{\varepsilon^2 Q(v)}{2\eta ln2}\right).$$

Now we define

$$\mathcal{V}_0 = \left\{v \in \mathcal{V} : Q(v) < \frac{1}{|\mathcal{V}|}\tau\right\},$$

and observe that $Q(\mathcal{V}_0) \leq \tau$. Hence,

$$\Pr\left\{\exists v \in \mathcal{V} \setminus \mathcal{V}_0 : \frac{1}{L}\sum_{i=1}^{L} Q_{Y_i}(v) \notin [(1-\varepsilon)Q(v), (1+\varepsilon)Q(v)]\right\}$$

$$\leq 2|\mathcal{V}|\exp\left(-L \cdot \frac{\varepsilon^2 \tau}{2\eta|\mathcal{V}|ln2}\right).$$

The right hand side becomes less than 1, if

$$L > \eta|\mathcal{V}|\frac{2ln2\log(2|\mathcal{V}|)}{\varepsilon^2\tau},$$

hence there exist instances E_i of the Y_i with the desired properties.

The interpretation of this result is as follows: Q is the expectation measure of the measures Q_E, which are sampled by the Q_{E_i}. The lemma says how close the sampling average $\bar{Q}$ can be to Q. In fact, assuming $Q_E(E) = q \le 1$ for all $E \in \mathcal{E}$, one easily sees that

$$\|Q - \bar{Q}\|_1 \le 2\varepsilon + 2\tau.$$

14   Proof of Converse

Let $\{(P_i, D_i) : i = 1, \ldots, N\}$ be a $(n, N, \lambda_1, \lambda_2)$ ID code, $\lambda_1 + \lambda_2 = 1 - \lambda < 1$. Our goal is to construct a $(n, N, \lambda_1 + \lambda/3, \lambda_2 + \lambda/3)$ ID code $\{(\bar{P}_i, D_i) : i = 1, \ldots, N\}$ with KL–distributions $\bar{P}_i$ on $\mathcal{X}^n$, i.e. all the probabilities are rational with common denominator KL to be specified below.

Fix i for the moment. For a distribution T on $\mathcal{X}$ we introduce

$$T_T^n = \{x^n \in \mathcal{X}^n : \forall x \ N(x|x^n) = nT(x)\},$$

and call T *empirical distribution* if this is nonempty. There are less than $(n+1)^{|\mathcal{X}|}$ many empirical distributions.

For an empirical distribution T define

$$P_i^T(x^n) = \frac{P_i(x^n)}{P_i(T_T^n)} \text{ for } x^n \in T_T^n,$$

which is a probability distribution on T_T^n (which we extend by 0 to all of $\mathcal{X}^n$). Note:

$$P_i = \sum_{T \text{ emp. distr.}} P_i(T_T^n)P_i^T.$$

For $x^n \in T_T^n$ and

$$\alpha = \sqrt{\frac{9|\mathcal{X}||\mathcal{Y}|}{\lambda}}$$

we consider the set of *conditional typical sequences*

$$T_{W,\alpha}^n(x^n) = \{y^n \in \mathcal{Y}^n : \ldots\}.$$

It is well known that these sets are contained in the set of TW–*typical sequences* on $\mathcal{Y}^n$,

$$T_{TW,\cdot\alpha}^n = \{y^n \in \mathcal{Y}^n : \ldots\}.$$

Define now the measures Q_{x^n} by

$$Q_{x^n}(y^n) = W^n(y^n|x^n) \cdot 1_{T_{W,\alpha}^n(x^n)}(y^n).$$

By the properties of typical sequences and choice of α we have

$$\|Q_{x^n} - W(\cdot|x^n)\|_1 \leq \frac{\lambda}{9}.$$

Now with $\varepsilon = \tau = \lambda/36$ apply Lemma 3 to the hypergraph with vertex set $T_{TW,\cdot\alpha}^n$ and edges $T_{W,\alpha}^n(x^n)$, $x^n \in T_T^n$, carrying measure $W(\cdot|x^n)$, and the probability distribution P_i^T on the edge set: we get a L–distribution $\bar{P}_i^T$ with

$$\|P_i^T Q - \bar{P}_i^T Q\|_1 \leq \frac{\lambda}{9},$$

$$L \leq \exp(nI(T;W) + O(\sqrt{n})) \leq \exp(nC(W) + O(\sqrt{n})),$$

where the constants depend explicitly on α, δ, τ. By construction we get

$$\|P_i^T W^n - \bar{P}_i^T W^n\|_1 \leq \frac{\lambda}{3}.$$

In fact by the proof of the lemma we can choose $L = \exp(nC(W) + O(\sqrt{n}))$, independent of i and T.

Now chose a K–distribution R on the set of all empirical distributions such that

$$\sum_{T \text{ emp.distr.}} |P_i(T_T^n) - R(T)| \leq \frac{\lambda}{3},$$

which is possible for

$$K = \lceil 3(n+1)^{|\mathcal{X}|}/\lambda \rceil.$$

Defining

$$\bar{P}_i = \sum_{T \text{ emp.distr.}} R(T)\bar{P}_i^T$$

we can summarize

$$\frac{1}{2}\|P_i W^n - \bar{P}_i W^n\|_1 \leq \frac{\lambda}{3},$$

where $\bar{P}_i$ is a KL–distribution. Since for all $\mathcal{D} \subset \mathcal{Y}^n$

$$|P_i W^n(\mathcal{D}) - \bar{P}_i W^n(\mathcal{D})| \leq \frac{1}{2}\|P_i W^n - \bar{P}_i W^n\|_1$$

the collection $\{(\bar{P}_i, \mathcal{D}_i) : i = 1, \ldots, N\}$ is indeed a $(n, N, \lambda_1 + \lambda/3, \lambda_2 + \lambda/3)$ ID code.

The proof is concluded by two observations: because of $\lambda_1 + \lambda_2 + 2\lambda/3 < 1$ we have $\bar{P}_i \neq \bar{P}_j$ for $i \neq j$. Since the $\bar{P}_i$ however are KL–distributions, we find

$$N \leq |\mathcal{X}^n|^{KL} = \exp(n \log |\mathcal{X}| \cdot KL) \leq \exp(\exp(n(C(W) + \delta))),$$

the last if only n is large enough.

References

1. R. Ahlswede, Certain results in coding theory for compound channels, Proc. Colloquium Inf. Th. Debrecen (Hungary), 35–60, 1967.
2. R. Ahlswede, Beiträge zur Shannonschen Informationstheorie im Fall nichtstationärer Kanäle, Z. Wahrscheinlichkeitstheorie und verw. Geb. 10, 1–42, 1968. (Dipl. Thesis Nichtstationäre Kanäle, Göttingen 1963.)
3. R. Ahlswede, The weak capacity of averaged channels, Z. Wahrscheinlichkeitstheorie und verw. Geb. 11, 61–73, 1968.
4. R. Ahlswede, On two–way communication channels and a problem by Zarankiewicz, Sixth Prague Conf. on Inf. Th., Stat. Dec. Fct's and Rand. Proc., Sept. 1971, Publ. House Chechosl. Academy of Sc., 23–37, 1973.
5. R. Ahlswede, An elementary proof of the strong converse theorem for the multiple-access channel, J. Combinatorics, Information and System Sciences, Vol. 7, No. 3, 216–230, 1982.
6. R. Ahlswede, Coloring hypergraphs: A new approach to multi–user source coding I, Journ. of Combinatorics, Information and System Sciences, Vol. 4, No. 1, 76–115, 1979.
7. R. Ahlswede, Coloring hypergraphs: A new approach to multi–user source coding II, Journ. of Combinatorics, Information and System Sciences, Vol. 5, No. 3, 220–268, 1980.
8. R. Ahlswede and V. Balakirsky, Identification under random processes, Preprint 95–098, SFB 343 Diskrete Strukturen in der Mathematik, Universität Bielefeld, Problemy peredachii informatsii (special issue devoted to M.S. Pinsker), vol. 32, no. 1, 144–160, Jan.–March 1996; Problems of Information Transmission, Vol. 32, No. 1, 123–138, 1996.
9. R. Ahlswede and I. Csiszár, Common randomness in information theory and cryptography, part I: secret sharing, IEEE Trans. Information Theory, Vol. 39, No. 4, 1121–1132, 1993.
10. R. Ahlswede and I. Csiszár, Common randomness in information theory and cryptography, part II: CR capacity, Preprint 95–101, SFB 343 Diskrete Strukturen in der Mathematik, Universität Bielefeld, IEEE Trans. Inf. Theory, Vol. 44, No. 1, 55–62, 1998.
11. R. Ahlswede and G. Dueck, Every bad code has a good subcode: a local converse to the coding theorem, Z. Wahrscheinlichkeitstheorie und verw. Geb. 34, 179–182, 1976.
12. R. Ahlswede and G. Dueck, Identification via channels, IEEE Trans. Inf. Theory, Vol. 35, No. 1, 15–29, 1989.
13. R. Ahlswede and G. Dueck, Identification in the presence of feedback — a discovery of new capacity formulas, IEEE Trans. on Inf. Theory, Vol. 35, No. 1, 30–39, 1989.
14. R. Ahlswede and B. Verboven, On identification via multi–way channels with feedback, IEEE Trans. Information Theory, Vol. 37, No. 5, 1519–1526, 1991.
15. R. Ahlswede and J. Wolfowitz, The structure of capacity functions for compound channels, Proc. of the Internat. Symposium on Probability and Information Theory at McMaster University, Canada, April 1968, 12–54, 1969.
16. R. Ahlswede and Z. Zhang, New directions in the theory of identification via channels, Preprint 94–010, SFB 343 Diskrete Strukturen in der Mathematik, Universität Bielefeld, IEEE Trans. Information Theory, Vol. 41, No. 4, 1040–1050, 1995.
17. R. Ahlswede, N. Cai, and Z. Zhang, Erasure, list, and detection zero–error capacities for low noise and a relation to identification, Preprint 93–068, SFB 343 Diskrete Strukturen in der Mathematik, Universität Bielefeld, IEEE Trans. Information Theory, Vol. 42, No. 1, 55–62, 1996.

18. R. Ahlswede, P. Gács, and J. Körner, Bounds on conditional probabilities with applications in multiuser communication, Z. Wahrscheinlichkeitstheorie und verw. Geb. 34, 157–177, 1976.
19. R. Ahlswede, General theory of information transfer, Preprint 97–118, SFB 343 "Diskrete Strukturen in der Mathematik", Universität Bielefeld, 1997; General theory of information transfer:updated, General Theory of Information Transfer and Combinatorics, a Special Issue of Discrete Applied Mathematics, to appear.
20. R. Ash, Information Theory, Interscience Tracts in Pure and Applied Mathematics, No. 19, Wiley & Sons, New York, 1965.
21. T.M. Cover and J.A. Thomas, Elements of Information Theory, Wiley, Series in Telecommunications, J. Wiley & Sons, 1991.
22. I. Csiszár and J. Körner, Information Theory — Coding Theorem for Discrete Memoryless Systems, Academic, New York, 1981.
23. R.L. Dobrushin, General formulation of Shannon's main theorem of information theory, Usp. Math. Nauk., 14, 3–104, 1959. Translated in Am. Math. Soc. Trans., 33, 323–438, 1962.
24. R.M. Fano, Transmission of Information: A Statistical Theory of Communication, Wiley, New York, 1961.
25. A. Feinstein, Foundations of Information Theory, McGraw–Hill, New York, 1958.
26. R.G. Gallager, A simple derivation of the coding theorem and some applications, IEEE Trans. Inf. Theory, 3–18, 1965.
27. R.G. Gallager, Information Theory and Reliable Communication, J. Wiley and Sons, Inc., New York, 1968.
28. T.S. Han, Oral communication in 1998.
29. T.S. Han, Information – Spectrum Methods in Information Theory, April 1998 (in Japanese).
30. T.S. Han and S. Verdú, Approximation theory of output statistics, IEEE Trans. Inf. Theory, IT–39(3), 752–772, 1993.
31. T.S. Han and S. Verdú, New results in the theory of identification via channels, IEEE Trans. Inf. Theory, Vol. 39, No. 3, 752–772, 1993.
32. K. Jacobs, Almost periodic channels, Colloquium on Combinatorial Methods in Probability Theory, 118–126, Matematisk Institute, Aarhus University, August 1–10, 1962.
33. F. Jelinek, Probabilistic Information Theory, 1968.
34. H. Kesten, Some remarks on the capacity of compound channels in the semicontinuous case, Inform. and Control 4, 169–184, 1961.
35. M.S. Pinsker, Information and Stability of Random Variables and Processes, Izd. Akad. Nauk, 1960.
36. C.E. Shannon, A mathematical theory of communication, Bell System Technical Journal, Vol. 27, 379–423, 623–656, 1948.
37. C.E. Shannon, The zero error capacity of a noisy channel, IRE, Trans. Inf. Theory, Vol. 2, 8–19, 1956.
38. C.E. Shannon, Certain results in coding theory for noisy channels, Inform. and Control 1, 6–25, 1957.
39. S. Verdú and T.S. Han, A general formula for channel capacity, IEEE Trans. Inf. Theory, Vol. 40, No. 4, 1147–1157, 1994.
40. J. Wolfowitz, The coding of messages subject to chance errors, Illinois Journal of Mathematics, 1, 591–606, 1957.
41. J. Wolfowitz, Coding theorems of information theory, 3rd. edition, Ergebnisse der Mathematik und ihrer Grenzgebiete, Band 31, Springer-Verlag, Berlin-New York, 1978.
42. A.D. Wyner, The capacity of the product channel, Information and Control 9, 423–430, 1966.

Appendix: Concepts of Performance from Number Theory

We can identify the $0-1$–sequence $(a_t)_{t=1}^{\infty}$ with the set of numbers $\mathcal{A} \subset \mathbb{N}$, where

$$\mathcal{A} = \{t \in \mathbb{N} : a_t = 1\}. \tag{A.1}$$

Then the lower asymptotic density equals the pessimistic mean, so

$$\underline{d}(\mathcal{A}) = \underline{A} \tag{A.2}$$

and the upper asymptotic density equals the optimistic mean, so

$$\overline{d}(\mathcal{A}) = \overline{A}. \tag{A.3}$$

If both coincide they agree with the asymptotic density $d(\mathcal{A})$. Another well–known and frequently used concept is logarithmic density δ again with lower and upper branches

$$\underline{\delta}(\mathcal{A}) = \varliminf_{n \to \infty} \frac{1}{\log n} \sum_{\substack{a \in \mathcal{A} \\ a \le n}} \frac{1}{a} \tag{A.4}$$

$$\overline{\delta}(a) = \varlimsup_{n \to \infty} \frac{1}{\log n} \sum_{\substack{a \in \mathcal{A} \\ a \le n}} \frac{1}{a}. \tag{A.5}$$

If they are equal, then the logarithmic density $\delta(\mathcal{A}) = \underline{\delta}(\mathcal{A}) = \overline{\delta}(\mathcal{A})$ exists.

Equivalently, they can be written in the form of (lower, upper, ...) Dirichlet densities

$$\underline{\delta}(\mathcal{A}) = \varliminf_{s \to 1^+} \sum_{a \in \mathcal{A}} \frac{1}{a^s} \tag{A.6}$$

$$\overline{\delta}(\mathcal{A}) = \varlimsup_{s \to 1^+} \sum_{a \in \mathcal{A}} \frac{1}{a^s} \tag{A.7}$$

which often can be handled analytically more easily.

It is well–known that for every $\mathcal{A} \subset \mathbb{N}$

$$\underline{d}(\mathcal{A}) \le \underline{\delta}(\mathcal{A}) \le \overline{\delta}(\mathcal{A}) \le \overline{d}(\mathcal{A}). \tag{A.8}$$

Whereas the measures of the previous Section $\underset{=}{A}$ and $\overline{\overline{A}}$ are *outside* the interval $\left(\underline{d}(\mathcal{A}), \overline{d}(\mathcal{A}) \right)$ these measures are *inside*.

Operationally their meaning is not so clear except that they put more weight on the beginning of the sequence – a realistic property where time is limited.

Even though they don't seem to have an immediate information theoretical interpretation, they get one as bounds on the limit points of $(A_n)_{n=1}^{\infty}$ and also on $\underline{A}$, $\overline{A}$. For instance in a widely developed calculus on pessimistic capacities $\underline{\delta}$ helps in evaluations.

The other famous concept of density in Number Theory is

$$\sigma(\mathcal{A}) = \inf_{n \geq 1} \frac{1}{n} |\{a \in \mathcal{A} : a \leq n\}|, \tag{A.9}$$

the Schnirelmann density. It is in so far peculiar as $1 \notin \mathcal{A}$ implies already $\sigma(\mathcal{A}) = 0$.

As first application we consider a situation where the communicators have *restrictions* on transmission lengths n and on phases m, say to be members of $\mathbb{N}$ and $\mathcal{M}$. Following these rules, what are the time points at which there can be activity? One answer is the

Lemma (Schnirelmann). *Let* $0 \in \mathcal{M} \subset \mathbb{N} \cup \{0\}$ *and* $0 \in \mathbb{N} \subset \mathbb{N} \cup \{0\}$, *if* $\sigma(\mathcal{M}) + \sigma(B) \geq 1$, *then* $n \in \mathcal{M} + \mathbb{N}$ *for every* $n \in \mathbb{N}$.

But now we come closer to home.

Definition: For channel $\mathcal{K}$ we define for every $\lambda \in (0,1)$ the *Schnirelmann* λ-*capacity*

$$S(\lambda) = \sigma \left(\left\{ \frac{1}{n} \log M(n, \lambda) : n \in \mathbb{N} \right\} \right).$$

A pleasant property of σ is that $\sigma(\mathcal{A}) = \gamma$ implies

$$\frac{1}{n} |\{a \in \mathcal{A} : a \leq n\}| \geq \gamma \text{ for all } n \in \mathbb{N}. \tag{A.10}$$

Therefore $\frac{1}{n} \log M(n, \lambda) \geq S(\lambda)$ for all n. For a DMC we have for the quantity $\min_{\lambda > 0} S(\lambda) = \log M(1, 0) \leq C_{\text{zero}}(W)$.

$S(\lambda)$ lower bounds the pessimistic λ-capacity (see [15])

$$\underline{C}(\lambda) = \varliminf_{n \to \infty} \frac{1}{n} \log M(n, \lambda).$$

Remark 8: This quantity in conjunction with a weak converse has been determined (except for finitely many discontinuities in [15]) for compound channels with the average error criterion, after it was noticed in [3] that for this error concept – as opposed to the maximal error concept – there is no strong converse.

The behaviour of $\underline{C}(\lambda)$ is the same as for average errors for the case of maximal errors *and* randomisation in the encoding. Conjunction of average error criterion and randomisation lead to no improvement.

Problem: For which DMC's and for which λ do we have

$$S(\lambda) = \underline{C}(\lambda)?$$

For instance consider a BSC $\left(\begin{smallmatrix} 1-\varepsilon & \varepsilon \\ \varepsilon & 1-\varepsilon \end{smallmatrix} \right)$ and $\lambda > \varepsilon$, then $\log M(1, \lambda) = 1$. On the other hand we know that $\underline{C}(\lambda) = 1 - h(\varepsilon)$. For λ large enough it is conceivable that $\frac{1}{n} \log M(n, \lambda) \geq 1 - h(\varepsilon)$ for all $n \in \mathbb{N}$. For general channels $\mathcal{K}$ many things can happen.

Theoretically and practically it is still meaningful to investigate $S(\lambda)$ where it is smaller than $\underline{C}(\lambda)$.

Appendix: On Common Information and Related Characteristics of Correlated Information Sources

R. Ahlswede and J. Körner

Abstract. This is a literal copy of a manuscript from 1974. References have been updated. It contains a critical discussion of in those days recent concepts of "common information" and suggests also alternative definitions. (Compare pages 402–405 in the book by I. Csiszár, J. Körner "Information Theory: Coding Theorems for Discrete Memoryless Systems", Akademiai Kiado, Budapest 1981.) One of our definitions gave rise to the now well–known source coding problem for two helpers (formulated in 2.) on page 7).

More importantly, an extension of one concept to "common information with list knowledge" has recently (R. Ahlswede and V. Balakirsky "Identification under Random Processes" invited paper in honor of Mark Pinsker, Sept. 1995) turned out to play a key role in analyzing the contribution of a correlated source to the identification capacity of a channel.

Thus the old ideas have led now to concepts of operational significance and therefore are made accessible here.

1 Introduction

Let $\left\{(X_i, Y_i)\right\}_{i=1}^{\infty}$ be a sequence of pairs of random variables which are independent, identically distributed and take finitely many different values. $\{X_i\}_{i=1}^{\infty}$ and $\{Y_i\}_{i=1}^{\infty}$ are to be viewed as two correlated discrete memoryless stationary information sources (DCMSS).

In [1] a notion of "common information" was introduced for those sources. It was meant as the maximal common part of the total amount of information contained individually in each of the two sources $\{X_i\}$ and $\{Y_i\}$ and which can therefore be encoded separately by any of them without knowing the actual outcomes of the other source. It was shown in [1] that common codes of a DCMSS can use only deterministic interdependence of the sources and no further correlation can be exploited in this manner. This result was sharpened later by H.S. Witsenhausen [2]. [1]

At a first glance the results may seem unsatisfactory because the common information thus defined depends only on the zeroes of the joint pr. d. matrix and does not involve its actual values. It is therefore natural to look for other notions of common information. Motivated by the work of Gray and Wyner [3], Wyner proposed another notion of common information in [4]. He expresses the believe that he has found the right notion of common information and that the

---

[1] His result was again significantly improved in [12].

R. Ahlswede et al. (Eds.): Information Transfer and Combinatorics, LNCS 4123, pp. 664–677, 2006.

earlier one of Gács and Körner is not the right notion, because of the properties mentioned above. The quantity introduced by Wyner seems to be indeed an interesting characteristic of correlated sources. However, the present authors take the position that his notion does not reflect at all what we would mean intuitively by "common information".

In this paper some arguments are provided which substantiate this position. It is therefore natural to look again for other notions of common information. We proceed systematically and investigate several coding schemes. It will become clear in our discussion that all notions introduced heavily depend on the network used for connecting encoders and decoders. Therefore it seems to us that a question as "what is the right notion of common information of $\{X_i\}$ and $\{Y_i\}$?" is meaningless. However, we shall introduce some concepts which we believe to be natural, because they relate to some basic source coding problems.

The main aim of the present contribution is to stimulate further discussions on the subject.

A word about notation. Throughout the paper "code" shall always mean deterministic block codes and the r.v. $\tilde{X}^n$ will be said to ε-reproduce X^n if $P(\tilde{X}^n \neq X^n) < \varepsilon$. All the r.v.'s have finite ranges. The unexplained basic notation is that of Gallager [9]. For the random variables (r.v) X and Y, $H(X)$ stands for the entropy of X, $\|X\|$ denotes the cardinality of the (finite) range of X, $H(X|Y)$ is the average conditional entropy of X given Y and $I(X \wedge Y)$ denotes the mutual information of X and Y. Exp's and log's are to the base 2, $h(\varepsilon) = -\varepsilon \log \varepsilon - (1 - \varepsilon) \log(1 - \varepsilon)$, for $0 < \varepsilon < 1$.

In order to fix ideas let us first take a new look at a one–decoder scheme for $\{(X_i, Y_i)\}_{i=1}^{\infty}$ and derive some consequences of the Slepian–Wolf theorem [6]. We shall say that a triple of positive reals (R_x, R_{xy}, R_y) is an element of the rate region $\mathcal{R}_0$ iff for every $\varepsilon > 0, \delta > 0$ and sufficiently large n there exists an ε-reproduction $(\tilde{X}^n, \tilde{Y}^n)$ of (X^n, Y^n) $(X^n = X_1 \ldots X_n, Y^n = Y_1 \ldots Y_n)$ such that for some deterministic functions f_n of X^n, g_n of Y^n, t_n of (X^n, Y^n) and a "decoding function" V_n we have

(1) $(\tilde{X}^n, \tilde{Y}^n) = V_n\big(f_n(X^n), t_n(X^n, Y^n)g_n(Y^n)\big)$
(2) $\|f_n(X^n)\| \leq \exp\{n(R_x + \delta)\}$
   $\|t_n(X^n, Y^n)\| \leq \exp\{n(R_{xy} + \delta)\}$
   $\|g_n(Y^n)\| \leq \exp\{n(R_y + \delta)\}$.

Consider the quantities

(1) $A_1(X, Y) = \sup R_{xy}$
   $R_{xy} + R_x \leq H(X)$
   $R_{xy} + R_y \leq H(Y)$
   $(R_x, R_{xy}, R_y) \in \mathcal{R}_0$

and

(2) $B_1(X, Y) = \inf R_{xy}$
   $R_x + R_{xy} + R_y \leq H(X, Y)$
   $(R_x, R_{xy}, R_y) \in \mathcal{R}_0$.

It is an immediate consequence of the Slepian–Wolf theorem that $A_1(X,Y) = I(X \wedge Y)$, the mutual information, and that $B_1(X,Y) = 0$. $A_1(X,Y)$ somehow measures how much knowledge about (X,Y) is simultaneously of interest for decoding X and Y in a lossless manner. Thus we arrived at a coding theoretic interpretation of mutual information, which allows us to view this quantity as a kind of "common information" for a one–decoder network. The fact that $B_1(X,Y) = 0$ allows a simple and convincing interpretation. It means that the total entropy $H(X,Y)$ can be fully decomposed into two rates on the "sidelines", and it therefore makes sense to call $B_1(X,Y)$ the *indecomposable* entropy for a one decoder network. The two notions $A_1(X,Y)$ and $B_1(X,Y)$ are mathematically not very sophisticated; however, they help us in build up the right heuristic for two–decoder networks. Passing from the one–decoder to any two–decoder network (discussed below) the rate region *decreases* and therefore quantities defined with a "sup" decrease and those defined with an "inf" increase. It is therefore also clear that any possible reasonable notion of "common information" should lead to values *not* exceeding $A_1(X,Y) = I(X \wedge Y)$. Let us now begin with a short description of the two–decoder networks we shall deal with. Consider a DMCSS $\big\{(X_i,Y_i)\big\}_{i=1}^{\infty}$.

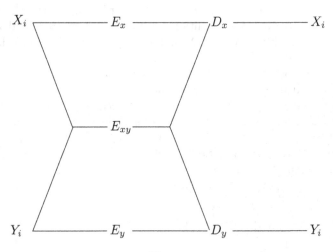

Fig. 1.

In our first network (Fig. 1) the sources $\{X_i\}_{i=1}^{\infty}$ and $\{Y_i\}_{i=1}^{\infty}$ are to be reproduced by two separate decoders, one for each of the sources. Similarly, there is one separate encoder for each of the sources, e.g. the encoder E_x can observe only $\{X_i\}_{i=1}^{\infty}$ and the block code he produces is available for the decoder D_x alone. However, there is a third encoder which allows us to exploit the correlation, since E_{xy} can observe both sources and its code is available for both individual decoders D_x and D_y. This is a modified version of a coding scheme of Gray and Wyner [3]. In their model all the three encoders can observe both sources (see Fig. 2).

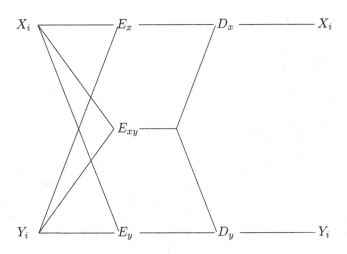

Fig. 2.

Finally, we introduce a coding scheme with four encoders (Fig. 3). The only difference between this and the coding scheme mentioned first (Fig. 1) is that the code exploiting the correlation is now supplied by two separate encoders, one for each of the sources. These codes are available for both individual decoders.

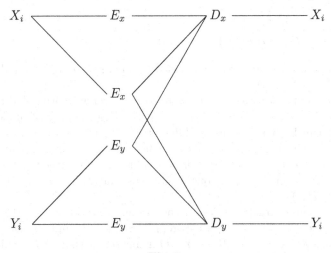

Fig. 3.

Let us denote by $\mathcal{R}_i$ the rate region of the coding problems described in figure i ($i = 1, 2, 3$). Replacing in definition (1) and (2) $\mathcal{R}_0$ by $\mathcal{R}_1$ the situation changes dramatically. Denoting an arbitrary element of $\mathcal{R}_1$ by (R_x, R_{xy}, R_y) where R_x is the rate of the code produced by E_x; R_{xy} that of E_{xy} and R_y the rate of encoder E_y, we define the quantities

(3) $A_2(X,Y) = \sup R_{xy}$
1. $(R_x, R_{xy}, R_y) \in \mathcal{R}_1$
$R_x + R_{xy} \leq H(X)$
$R_y + R_{xy} \leq H(Y)$
and

(4) $B_2(X,Y) = \inf R_{xy}$
$(R_x, R_{xy}, R_y) \in \mathcal{R}_1$
$R_x + R_y + R_{xy} \leq H(X,Y)$.

Again we refer to the first quantity defined as "common information", because it measures how much knowledge about (X,Y) is simultaneously of interest for decoding X and Y is a lossless manner. Since X and Y are decoded separately now, this quantity seems to be a natural measure. However, we prove (Corollary 1, Section 2) that $A_2(X,Y)$, which is by definition not smaller than the common information of [1], is actually equal to that quantity.

The quantity $B_2(X,Y)$ is in some sense a dual to $A_2(X,Y)$. $B_2(X,Y)$ is that minimal portion of the joint entropy $H(X,Y)$ of the DMCSS $\{(X_i, Y_i)\}_{i=1}^{\infty}$ which one has to encode by a *joint encoder* observing both sources; otherwise the coding scheme of Fig. 1 would not be optimal. In other words this entropy can not be encoded by separate encoders without a loss in the total rate, and therefore it is *indecomposable*.

Wyner [4] has earlier introduced the quantity

$C(X,Y) = \inf R_{xy}$
$(R_x, R_{xy}, R_y) \in \mathcal{R}_2$
$R_x + R_y + R_{xy} \leq H(X,Y)$.

He has independently [10] also introduced the quantity $B_2(X,Y)$ and observed that $C(X,Y) = B_2(X,Y)$.

He calls $C(X,Y)$ *the* common information. However we believe that this would be a misleading name not only because of the large variety of analogous notions which can be obtained using different coding schemes but also and more importantly because it suggests a wrong heuristic. We have explained earlier that a quantity called common information should not exceed the mutual information $I(X \wedge Y)$. However, one easily sees that $I(X \wedge Y) \leq B_2(X,Y) \leq \min\{H(X), H(Y)\}$.

A single letter characterization of the region $\mathcal{R}_2$ is known [3], [4]. We give such a characterization for $\mathcal{R}_1$ (Theorem 2, Section 2) and therefore also for the quantities $A_2(X,Y)$ and $B_2(X,Y)$. Our method is that of [5], which proves to be quite general and easily adaptable to various source coding problems. The identity $\mathcal{R}_1 = \mathcal{R}_2$ follows as a byproduct. During the preparation of this manuscript we learnt that in an independent paper and by a different method Wyner [10] also obtained Theorem 2.

In Section 3, Corollary 2, we prove the somewhat surprising fact that

$$B_2(X,Y) = I(X \wedge Y) \text{ iff } I(X \wedge Y) = A_2(X,Y).$$

The determination of the rate region $\mathcal{R}_3$ corresponding to the coding scheme of Fig. 3 is still unsolved. Stating the problem here serves three purposes:

1.) It shows the relativity of any notion of common information.

2.) The two basic coding theorems for correlated sources, that is, the Slepian–Wolf theorem and the source coding theorem in case of side information [5], [10] do not provide all the tools to deal successfully with somewhat more complex networks.

   Probably the "most canonical" network of this kind, which is intimately related to the one above, is obtained by considering a correlated source $\left\{(X_i, Y_i, Z_i)\right\}_{i=1}^{\infty}$ with three separate encoders for each source and one decoder, who wants to reproduce $\{X_i\}$ and gets side information from $\{X_i\}$ as well as from $\{Z_i\}$.

3.) Similarly to $B_2(X, Y)$ we shall introduce the quantity

   $B_2^*(X, Y) = \inf R_x^* + R_y^*$

   $R_x^* + R_x + R_y^* + R_y \leq H(X, Y)$

   $(R_x, R_y, R_x^*, R_y^*) \in \mathcal{R}_3$

   and call it the *strong indecomposable entropy* of the DMCSS $\left\{(X_i, Y_i)\right\}_{i=1}^{\infty}$.

Whereas $B_2(X, Y)$ equals $C(X, Y)$, $B_2^*(X, Y)$ seems to be a new correlation measure.

2   Auxiliary Results

This section is analogous to Section 1, Part I of [5] as far as we shall prove some convexity properties of the functions we have to deal with in the sequel. The ideas are those of Ahlswede–Körner [7], Section 4, where entropy inequalities for multiple–access channels (see [8]) were derived. Our aim is to generalize Lemmas 1 and 2 of [5].

We introduce the notation $X_1 {\rightarrow} X_2 {\rightarrow} X_3 {\rightarrow} X_4$ for saying that the r.v.'s X_1, X_2, X_3 and X_4 form a Markov chain in this order. For an arbitrary sequence $\{Z_i\}_{i \in N}$ of r.v.'s we put

$$Z^n = Z_1 Z_2 \ldots Z_n.$$

Let us be given a sequence of independent and identically distributed triples $\left\{(S_i, X_i, Y_i)\right\}_{i \in N}$. For any positive real c we put:

Definition 1. $\tau_n(c) = \left\{(R_x, R_y) : R_x \geq \frac{1}{n} H(X^n | U), R_y \geq \frac{1}{n} H(Y^n | U); \right.$
$\left. U {\rightarrow} S^n {\rightarrow} (X^n, Y^n); H(S^n | U) \geq c \right\}$
We shall write $\tau(c) = \tau_1(c)$.

This is a natural generalization of the functions $T_n(c)$ defined in [5]. We shall write $(b_1, b_2) \leq (b_1', b_2')$ iff $b_1 \leq b_1'$ and $b_2 \leq b_2'$.

Lemma 1. a) $\tau(c') \subset \tau(c)$ *for* $c \le c'$ *(monotonicity)*
b) *For any* $0 \le \alpha \le 1$ *and* $c = \alpha c_1 + (1 - \alpha)c_2$
$\alpha\tau(c_1) \oplus (1 - \alpha)\tau(c_2) \subset \tau(x)$,
where
$\alpha\tau(c_1) \oplus (1-\alpha)\tau(c_2) = \{\alpha b_1 + (1-\alpha)b_2 : b_1 \in \tau(c_1); b_2 \in \tau(c_2)\}$ *(convexity).*

Proof
a) is an immediate consequence of Definition 1. In order to prove b) we assume that $(R_x^1, R_y^1) \in \tau(c_1)$ and $(R_x^2, R_y^2) \in \tau(c_2)$, i.e. for suitable $U^{(i)}$ $(i = 1, 2)$ we have

$$H(S|U^{(i)}) \ge c_i \tag{1}$$

and $\left(H(X|U^{(i)}), H(Y|U^{(i)})\right) \le (R_x^{(i)}, R_y^{(i)})$ where $U^{(i)} \to S \to (X, Y)$. We introduce now the new quadruple of r.v.'s $\tilde{U}, \tilde{S}, \tilde{X}, \tilde{Y}$ such that

$$\Pr(\tilde{U}, \tilde{S}, \tilde{X}, \tilde{Y} = U^{(1)}, S^{(1)}, X^{(1)}, Y^{(1)}) = \alpha$$

and

$$\Pr(\tilde{U}, \tilde{S}, \tilde{X}, \tilde{Y} = U^{(2)}, S^{(2)}, X^{(2)}, Y^{(2)}) = 1 - \alpha$$

and furthermore, a r.v. I ranging over the set $\{1, 2\}$ with $\Pr(I = 1) = \alpha$ and such that $(I, \tilde{U}) \to \tilde{S} \to (\tilde{X}, \tilde{Y})$.
We have $H(\tilde{S}|\tilde{U}, I) = \alpha c_1 + (1 - \alpha)c_2 = c$. Hence

$$\left(H(\tilde{H}|\tilde{U}, I), H(\tilde{Y}|\tilde{U}, I)\right) \in \tau(c).$$

On the other hand

$$\left(H(\tilde{X}|\tilde{U}, I), H(\tilde{Y}|\tilde{U}, I)\right) = \alpha\left(H(X|U^{(1)}), H(Y|U^{(1)})\right) + (1-\alpha) \cdot \left(H(X|U^{(2)}), H(Y|U^{(2)})\right)$$

and the statement of b) follows.

Remark 1. *It follows by a usual argument (see e.g. Lemma 3 of [5]) that the set* $\tau(c)$ *remains the same if in Definition 1 we limit ourselves to r.v.'s* U *satisfying the bound*

$$\|U\| \le \|S\| + 2.$$

Lemma 2. *For all* $n \in N$ *and* $c \ge 0$

$$\tau_n(c) = \tau(c) \text{ (stationarity)}. \tag{2}$$

Proof
Let (U, S^n, X^n, Y^n) be a quadruple of r.v.'s satisfying $U \to S^n \to (X^n, Y^n)$.
We can write

$$H(X^n|U) = \sum_{i=1}^{n} H(X_i|U, X^{i-1}) \ge \sum_{i=1}^{n} H(X_i|U, X^{i-1}, S^{i-1})$$

$$= \sum_{i=1}^{n} H(X_i|U, S^{i-1}) \tag{3}$$

where the last identity follows by the fact that $U \to S^n \to (X^n, Y^n)$ and the triples (S_i, X_i, Y_i) are independent.

Similarly, one deduces that

$$H(Y^n|U) \geq \sum_{i=1}^{n} H(Y_i|U, S^{i-1}). \tag{4}$$

By the definition of $\tau(c)$ we have $\big(H(X_i|U, S^{i-1} = s^{i-1}), H(Y_i|U, S^{i-1} = s^{i-1})\big) \in \tau(c)$ for $c = H(S_i|U, S^{i-1} = s^{i-1})$ and hence by the convexity of $\tau(c)$ averaging over all the possible values of S^{i-1}, yields for the corresponding expected values

$$\big(H(X_i|U, S^{i-1}), H(Y_i|U, S^{i-1})\big) \in \tau(c_i) \tag{5}$$

where $c_i = H(S_i|U, S^{i-1})$.

This, 2, 4, and the monotonicity of $\tau(\cdot)$ yield

$$\big(H(X^n|U), H(Y^n|U)\big) \in \sum_{i=1}^{n} \tau(c_i), \tag{6}$$

where $\sum_{i=1}^{n} \tau(c_i) = \Big\{ \boldsymbol{b} : \boldsymbol{b} = \sum_{i=1}^{n} \boldsymbol{b}_i, \boldsymbol{b}_i \in \tau(c_i) \Big\}$.

From 6 and the convexity of $\tau(\cdot)$ it follows that

$$\left(\frac{1}{n} H(X^n|U), \frac{1}{n} H(Y^n|U) \right) \in \tau \left(\frac{1}{n} \sum_{i=1}^{n} c_i \right) = \tau(c^*)$$

where $c^* = \frac{1}{n} H(S^n|U)$.

Hence $\tau_n(c) \subset \tau(c)$, whereas $\tau_n(c) \supset \tau(c)$ is trivial. This completes the proof.

3   Common Information

We begin with two definitions.

Definition 2. *A triple of positive reals* (R_x, R_{xy}, R_y) *is an element of the rate region* $\mathcal{R}_1$ *iff for every* $\varepsilon > 0; \delta > 0$ *and sufficiently large* $n\big(n > n_0(\varepsilon, \delta)\big)$ *there exists an* $\varepsilon-$*reproduction* $(\tilde{X}^n, \tilde{Y}^n)$ *of* (X^n, Y^n) *satisfying the following conditions:*

There exist some deterministic functions f_n *of* X^n, g_n *of* Y^n, t_n *of* (X^n, Y^n), *and two decoding functions* V_n *and* W_n *with*

(i) $\tilde{X}^n = V_n\big(f_n(X^n), t_n(X^n, Y^n)\big)$
    $\tilde{Y}^n = W_n\big(g_n(Y^n), t_n(X^n, Y^n)\big)$

(ii) $\|f_n(X^n)\| \leq \exp\{n(R_x + \delta)\}$
    $\|t_n(X^n, Y^n)\| \leq \exp\{n(R_{xy} + \delta)\}$
    $\|g_n(Y^n)\| \leq \exp\{n(R_y + \delta)\}$.

Definition 3. $A_2(X,Y) = \sup R_{xy}$
$R_{xy} + R_x \leq H(X)$
$R_{xy} + R_y \leq H(Y)$
$(R_x, R_{xy}, R_y) \in \mathcal{R}_1$
is called the "common information" of the DMCSS $\left\{(X_i, Y_i)\right\}_{i=1}^{\infty}$.

After deriving from Theorems 1 in [5] and Lemmas 1 and 2 a single–letter description of $\mathcal{R}_1$, we shall prove that $A_2(X,Y)$ equals the common information in the sense of Gács and Körner [1]. Especially, for an X and Y having an indecomposable joint distribution (e.g.: $\forall\, x \in \mathcal{X}, y \in \mathcal{Y}\; \Pr(X = x, Y = y) > 0$) it will follow that $A_2(X,Y) = 0$.

Theorem 1. *Let* $\left\{(X_i, Y_i)\right\}_{i \in N}$ *be a discrete memoryless correlated source with finite alphabets. The rate region* $\mathcal{R}_1$ *(as defined by Definition 2.1) satisfies*

$$\mathcal{R}_1 = \left\{ \left(\frac{1}{n} H\big(X^n | t_n(X^n, Y^n)\big), \frac{1}{n} H\big(t_n(X^n, Y^n)\big) \right, \right.$$

$$\left. \frac{1}{n} H\big(Y^n | t_n(X^n, Y^n)\big) \right) n \in N; t_n : \mathcal{X}^n \times \mathcal{Y}^n + N \right\}.$$

The proof is based on the simple observation that the coding scheme of Fig. 1 can be considered as a simultaneous "source coding with side information" for the DMCSS's $\left\{(X_i^*, Y_i^*)\right\}_{i \in N}$ and $\left\{(X_i^{**}, Y_i^{**})\right\}_{i \in N}$ where (using the notation of Theorem 1 and 2 of [5])

$$X_i^* = X_i^{**} = (X_i, Y_i);\; Y_i^* = X_i;\; Y_i^{**} = Y_i$$

and where the same code has to be used for $\{X_i^*\} = \{X_i^{**}\} = \left\{(X_i, Y_i)\right\}$, serving in both cases as side information.

Now the proof of Theorem 1 in [5] literally applies and gives the assertion of the theorem.

As in [5] we shall give a single–letter description of $\mathcal{R}_1$ by rewriting our former description by means of the convexity arguments of Section 1.

Theorem 2

$$\mathcal{R}_1 = \big\{\, (R_x, R_{xy}, R_y) : R_x \geq H(X|Z), R_{xy} \geq I\big((X,Y) \wedge Z\big), R_y \geq H(Y|Z);$$
$$\|Z\| \leq \|X\| \cdot \|Y\| + 2\big\}. \tag{7}$$

Proof
We denote by $\mathcal{R}_1^*$ the set defined by the right–hand side of 7. We show first that

$$\mathcal{R}_1 \subset \mathcal{R}_1^*.$$

Suppose that for $K = t_n(X^n, Y^n)$ we have

$$R_x = \frac{1}{n} H(X^n|K), R_{xy} = \frac{1}{n} H(K) \text{ and } R_y = \frac{1}{n} H(Y^n|K).$$

We have to show that there exists a triple (X, Y, Z) such that the joint pr.d. of (X, Y) is that of the $(X_i, Y_i)'$s, $\|Z\| \leq \|X\| \cdot \|Y\| + 2$ and $R_x \geq H(X|Z)$, $R_{xy} \geq I((X, Y) \wedge Z)$ and $R_y \geq H(Y|Z)$.

It is clear that

$$n \cdot R_{xy} = H(K) \geq I(K \wedge (X^n, Y^n)) = H(X^n, Y^n) - H(X^n, Y^n|K). \tag{8}$$

The independence of the $(X_i, Y_i)'$s and 8 yield

$$\frac{1}{n}H(X^n, Y^n|K) \geq H(X, Y) - R_{xy}. \tag{9}$$

We shall apply the Lemmas of Section 1 in the following set–up: $S_i = (X_i, Y_i)$. By the definition of $\tau_n(c)$ we know that

$$\left(\frac{1}{n}H(X^n|K), \frac{1}{n}H(Y^n|K)\right) \in \tau_n\left(\frac{1}{n}H(X^n, Y^n|K)\right).$$

By Lemma 2 this gives

$$\left(\frac{1}{n}H(X^n|K), \frac{1}{n}H(Y^n|K)\right) \in \tau\left(\frac{1}{n}H(X^n, Y^n|K)\right). \tag{10}$$

Because of the monotonicity of the regions $\tau(\cdot)$ (see Lemma 1) the inequalities 9 and 10 yield

$$\left(\frac{1}{n}H(X^n|K), \frac{1}{n}H(Y^n|K)\right) \in \tau(H(X, Y) - R_{xy}). \tag{11}$$

By the definition of the region $\tau(H(X, Y) - R_{xy})$ the last relation means that there exists a triple (Z, X, Y) such that

$$R_x = \tfrac{1}{n}H(X^n|K) \geq H(X|Z), R_y = \tfrac{1}{n}H(Y^n|K) \geq H(Y|Z), \text{ and}$$
$$\|Z\| \leq \|X\| \cdot \|Y\| + 2, \tag{12}$$

whereas $H(X, Y|Z) \geq H(X, Y) - R_{xy}$. Rewriting the last inequality we get

$$R_{xy} \geq I((X, Y) \wedge Z). \tag{13}$$

Now we show that $\mathcal{R}_1^* \subset \mathcal{R}_1$ by the approximation argument of [5], Section 4. We have to prove that for every triple (Z, X, Y) with $\|Z\| \leq \|X\| \cdot \|Y\| + 2$ there exists an n and a function t_n of (X^n, Y^n) such that

$$\tfrac{1}{n}H(X^n|t_n(X^n, Y^n)) \leq H(X|Z), \tfrac{1}{n}H(Y^n|t_n(X^n, Y^n)) \leq H(Y|Z) \text{ and}$$
$$\tfrac{1}{n}H(t_n(X^n, Y^n)) \leq I((X, Y) \wedge Z).$$

It suffices to show that

$$\inf_n \inf_{\left(\frac{1}{n}H\left(X^n|t_n(X^n, Y^n)\right), \frac{1}{n}H\left(Y^n|t_n(X^n, Y^n)\right)\right) \leq (x_1, x_2)} \frac{1}{n}H\left(t_n(X^n, Y^n)\right) \leq$$
$$\inf_{(H(X|Z), H(Y|Z)) \leq (x_1, x_2)} I((X, Y) \wedge Z). \tag{14}$$

From the independence of the $(X_i, Y_i)'$s and the fact that

$$I\big((X^n, Y^n) \wedge t_n(X^n, Y^n)\big) = H\big(t_n(X^n, Y^n)\big)$$

it follows that it is enough to show for $t_n = t_n(X^n, Y^n)$

$$\sup_n \sup_{\left(\frac{1}{n}H(X^n|t_n), \frac{1}{n}H(Y^n|t_n)\right) \leq (x_1, x_2)} \frac{1}{n}H\big(X^n, Y^n | t_n(X^n, Y^n)\big) \geq$$

$$\sup_{\big(H(X|Z), H(Y|Z)\big) \leq (x_1, x_2)} H(X, Y|Z). \tag{15}$$

Now we apply the construction of [5; Section 4] to the DMCSS's $\{X_i^*, Y_i^*\}_{i \in N}$ and $\{X_i^{**}, Y_i^{**}\}_{i \in N}$ and the r.v.'s U^* and U^{**} where as in the proof of Theorem 1

$$X_i^* = X_i^{**} = (X_i, Y_i), Y_i^* = X_i; Y_i^{**} = Y_i \text{ and } U^* = U^{**} = Z.$$

Observing that the construction of [5] depends only on the joint pr. d. of (U^*, X^*, Y^*), it becomes clear that — using the notation of [5] — the choice $t_n(X^n, Y^n) \triangleq f_n(X^{*n}) = f_n(X^{**n})$ actually establishes 15.

In what follows we shall use Theorem 1 to prove a generalization of Theorem 1, p. 151 of [1]. Actually, we prove that the common information $A_2(X, Y)$ of Definition 3, which is clearly not smaller than that of [1], is equal to it. We recall from [1] the following

Definition 4. *We suppose without loss of generality that for every $x \in \mathcal{X}$ and $y \in \mathcal{Y} \Pr(X_1 = x) > 0$ and $\Pr(Y_1 = y) > 0$. We consider the stochastic matrix of the conditional probabilities $\{\Pr(X = x | Y = y)\}$ and its ergodic decomposition. Clearly, the ergodic decompositions of the matrices $\{\Pr(X = x | Y = y)\}$ and $\{\Pr(Y = y | X = x)\}$ coincide and form a partition*

$$\mathcal{X} \times \mathcal{Y} = \bigcup_j \mathcal{X}_j \times \mathcal{Y}_j$$

of $\mathcal{X} \times \mathcal{Y}$ where the $\mathcal{X}_j$'s and $\mathcal{Y}_j$'s having different subscripts are disjoint. We introduce the r.v. J such that

$$J = j \Leftrightarrow X \in \mathcal{X}_j \Leftrightarrow Y \in \mathcal{Y}_j.$$

It is clear that J is a function of both X and Y. We shall prove that the common information $A_2(X, Y)$ equals the entropy of this common function of X and Y.

Corollary 1

$$A_2(X, Y) = H(J).$$

Proof

It follows from our Theorem 2 that

$$A_2(X, Y) = \sup I\big((X, Y) \wedge Z\big)$$
$$I\big((X, Y) \wedge Z\big) + H(X|Z) \leq H(X)$$

$$I\big((X,Y)\wedge Z\big) + H(Y|Z) \le H(Y)$$
$$\|Z\| \le \|X\| \cdot \|Y\| + 2.$$

Looking at the constraint inequalities we find that from
$$H(X) \ge I\big((X,Y)\wedge Z\big) + H(X|Z) = H(X,Z) - H(Z|X,Y)$$
we get the inequality
$$H(Z|X,Y) \ge H(Z|X),$$

which gives $Z{\to}Y{\to}X$. Similarly, our other constraint gives $Z{\to}X{\to}Y$. Now we shall analyze these conditions

$$Z{\to}Y{\to}X \text{ and } Z{\to}X{\to}Y. \tag{16}$$

It follows from 16 that

$$\Pr(X{=}x, Y{=}y) > 0 \Rightarrow \Pr(Z{=}z|X{=}x, Y{=}y) = \Pr(Z{=}z|X{=}x) = \Pr(Z{=}z|Y{=}y).$$

Hence for any fixed value of Z and for every index j $\Pr(Z = z|X = \cdot, Y = \cdot)$ is

constant over $\mathcal{X}_j \times \mathcal{Y}_j$ whenever it is defined. This means that $\Pr(Z{=}z|X{=}x, Y{=}y) = \Pr(Z{=}z|J{=}j) = \sum_{\mathcal{X}_j \times \mathcal{Y}_j} \Pr(Z{=}z|X{=}\hat{x}, Y{=}\hat{y}) \cdot \Pr(X{=}\hat{x}|Y{=}\hat{y})$. The last

relation means that given any value of J the r.v. Z is conditionally independent from (X,Y), i.e. $I\big((X,Y)\wedge Z|J\big) = 0$. However since J is a function of (X,Y) we have

$$I\big((X,Y)\wedge Z\big) = I\big((X,Y,J)\wedge Z\big) = I(J\wedge Z) + I\big((X,Y)\wedge Z|J\big) \tag{17}$$

where the last equality follows by a well–known identity (see e.g. Gallager [9], formula (2.2.29) pn p. 22). Comparing the two extremities of 17 we get

$$I\big((X,Y)\wedge Z\big) \le H(J) + I\big((X,Y)\wedge Z|J\big) = H(J).$$

Taking into account that J is a deterministic function of X and a deterministic function of Y and thus it satisfies the constraints of our second definition of $A_2(X,Y)$, we conclude that $A_2(X,Y) = H(J)$.

Remark 2. *The quantity*

$$A(X,Y) = \sup R_y$$
$$R_x + R_y \le H(X)$$
$$(R_x, 0, R_y) \in \mathcal{R}_0$$

is meaningful in a one–decoder situation. It says how much information about X we can extract from Y in a "lossless manner". It is easy to see that $H(J) \le A(X,Y) \le A_2(X,Y)$ and hence that also $A(X,Y) = H(J)$.

4   Indecomposable Entropy

Definition 5. $B_2(X,Y) = \inf R_{xy}$
$(R_x, R_{xy}, R_y) \in \mathcal{R}_1$
$R_x + R_{xy} + R_y) \leq H(X,Y)$

is called the "indecomposable entropy" of the DMCSS $\{(X_i, Y_i)\}_{i \in N}$. A justification for this terminology was given in the introduction. It is clear from the
foregoing that

$$B_2(X,Y) = \inf_{H(X|Z)+H(Y|Z)+I\big((X,Y)\wedge Z\big)=H(X,Y)\|Z\|\leq\|X\|\cdot\|Y\|+2} I\big((X,Y)\wedge Z\big)$$

and

$$B_2(X,Y) \geq I(X \wedge Y) \geq A_2(X,Y).$$

Looking into the constraint on the right hand side of 5 and taking into account
that $H(X,Y|Z) + I\big((X,Y) \wedge Z\big) = (X,Y)$ always holds we conclude that the
constraint is equivalent to $H(X,Y|Z) = H(X|Z) + H(Y|Z)$. This allows us to
write

$$B_2(X,Y) = \min_{X \to Z \to Y} I\big((X,Y) \wedge Z\big)$$

$$\|Z\| \leq \|X\| \cdot \|Y\| + 2.$$

We shall prove that

Corollary 2

$$B_2(X,Y) = I(X \wedge Y) \Leftrightarrow I(X \wedge Y) = A_2(X,Y).$$

Remark 3. Since $A_2(X,Y) = H(J)$, the entropy of the ergodic class index
which is a common function of X and Y, the statement of Corollary 2 means
that $B_2(X,Y)$ equals the mutual information iff all the correlation between X and
Y is of deterministic character. Especially if X and Y have an indecomposable
joint pr.d. the corollary says that $B_2(X,Y) = I(X \wedge Y)$ implies $B_2(X,Y) = 0$.

Proof
We suppose that for a r.v. Z satisfying the constraint of minimization we have

$$I\big((X,Y) \wedge Z\big) = I(X \wedge Y).$$

Using the identity

$$H(X,Y) = I(X \wedge Y) + H(X|Y) + H(Y|X) \tag{18}$$

becomes equivalent to

$$H(X,Y|Z) = H(X|Y) + H(Y|X). \tag{19}$$

Since by our supposition $X \to Z \to Y$ we have

$$H(X,Y|Z) = H(X|Z) + H(Y|Z) = H(X|Z,Y) + H(Y|Z,X). \quad (20)$$

Comparing 19 and 20 we obtain that 18 is equivalent to the condition

$$H(X|Y) + H(Y|X) = H(X|Z,Y) + H(Y|Z,X).$$

Rewriting this we get

$$I(X \wedge Z|Y) + I(Y \wedge Z|X) = 0. \quad (21)$$

Since conditional mutual informations are non–negative, 21 is equivalent to

$$I(X \wedge Z|X) = 0 \text{ and } I(Y \wedge Z|X) = 0.$$

Hence we get that

$$X \to Y \to Z \text{ and } Z \to X \to Y.$$

Observing that this is just 16, the deduction consecutive to relation 16 in Section 2 applies and we get that $I\big((X,Y) \wedge Z\big) = H(J)$. This completes the proof.

References

1. P. Gács and J. Körner, Common information is far less than mutual information, Problems of Contr. and Inf. Th., Vol. 2, 149–162, 1973.
2. H.S. Witsenhausen, On sequences of pairs of dependent random variables, SIAM J. of Appl. Math., Vol. 28, 100–113, 1975.
3. R.M. Gray and A.D. Wyner, Source coding for a simple network, Bell System. Techn. J., Dec. 1974.
4. A.D. Wyner, The common information of two dependent random variables, IEEE Trans. Inform. Theory, Vol. IT–21, 163–179, Mar. 1975.
5. R. Ahlswede and J. Körner, Source coding with side information and a converse for degraded broadcast channels, IEEE Trans. on Inf. Th., Vol. IT–21, No. 6, 629–637, Nov., 1975.
6. D. Slepian and J.K. Wolf, Noiseless coding for correlated information sources, IEEE Trans. on Inf. Th., Vol. IT–19, 471–480, July 1973.
7. R. Ahlswede and J. Körner, On the connection between the entropies of input and output distributions of discrete memoryless channels, Proceedings of the 5th Conference on Probability Theory, Brasov 1974, Editura Academeiei Rep. Soc. Romania, Bucaresti, 13–23, 1977.
8. R. Ahlswede, Multi–way communication channels, Proc. 2nd Int. Symp. Inf. Th., Tsahkadsor, Armenian S.S.R., 1971, 23–52. Publishing House of the Hungarian Academy of Sciences, 1973.
9. R.G. Gallager, Information Theory and Reliable Communication, Wiley and Sons, New York, 1968.
10. A.D. Wyner, On source coding sith side information at the decoder, IEEE Trans. on Inf. Th., Vol. IT–21, No. 3, May 1975.
11. R. Ahlswede, P. Gács, and J. Körner, Bounds on conditional probabilities with applications in multi–user communication, Zeitschr. für Wahrscheinlichkeitstheorie und verw. Gebiete, Vol. 34, 157–177, 1976.
12. R. Ahlswede and P. Gács, Spreading of sets in product spaces and hypercontraction of the Markov operator, Ann. of Probability, Vol. 4, No. 6, 925–939, 1976.

VI

Q-Ary Ulam-Renyi Game with Constrained Lies

F. Cicalese and Christian Deppe

Abstract. The Ulam-Rényi game is a classical model for the problem
of finding an unknown number in a finite set using as few question as
possible when up to a finite number e of the answers may be lies. In
the variant, we consider in this paper, questions with q many possible
answers are allowed, q fixed and known beforehand, and lies are constrained as follows: Let $\mathcal{Q} = \{0, 1, \ldots, q - 1\}$ be the set of possible answers to a q-ary question. For each $k \in \mathcal{Q}$ when the sincere answer to the
question is k, the responder can choose a mendacious answer only from
a set $L(k) \subseteq \mathcal{Q} \setminus \{k\}$. For each $k \in \mathcal{Q}$, the set $L(k)$ is fixed before the
game starts and known to the questioner. The classical q-ary Ulam-Rényi
game, in which the responder is completely free in choosing the lies, in
our setting corresponds to the particular case $L(k) = \mathcal{Q} \setminus \{k\}$, for each
$k \in \mathcal{Q}$. The problem we consider here, is suggested by the counterpart
of the Ulam-Rényi game in the theory of error-correcting codes, where
(the counterparts of) lies are due to the noise in the channel carrying the
answers. We shall use our assumptions on noise and its effects (as represented by the constraints $L(k)$ over the possible error patterns) with
the aim of producing the most efficient search strategies. We solve the
problem by assuming some symmetry on the sets $L(k)$: specifically, we
assume that there exists a constant $d \leq q - 1$ such that $|L(k)| = d$ for
each k, and the number of indices j such that $k \in L(j)$ is equal to d. We
provide a lower bound on the number of questions needed to solve the
problem and prove that in infinitely many cases this bound is attained
by (optimal) search strategies. Moreover we prove that, in the remaining
cases, at most one question more than the lower bound is always sufficient
to successfully find the unknown number. Our results are constructive
and search strategies are actually provided. All our strategies also enjoy
the property that, among all the possible adaptive strategies, they use
the minimum amount of adaptiveness during the search process.

1 Introduction

In the q-ary Ulam-Rényi game [13,10,11] two players, classically called Paul
an Carole, first agree on fixing an integer $M \geq 0$ and a search space $U = \{0, \ldots, M - 1\}$. Then Carole thinks of a number $x_* \in U$ and Paul must find
out x_* by asking the minimum number of q-ary questions. Each q-ary question
is a list $T_0, \ldots, T_{q-1}$ of q disjoint subsets defining a partition of the set U, and
asking Carole to indicate the set T_k which contains the secret number x_*. The
parameter q is fixed beforehand. Thus a q-ary question has the form *Which set
among $T_0, T_1, \ldots, T_{q-1}$ does the secret number x_* belong to?* and the answer is
an index $k \in \mathcal{Q} = \{0, 1, \ldots, q - 1\}$, meaning that $x_* \in T_k$. It is agreed that

R. Ahlswede et al. (Eds.): Information Transfer and Combinatorics, LNCS 4123, pp. 678–694, 2006.

Carole is allowed to lie at most e times, e.g., answering $j \in \mathcal{Q}$ when actually $x_* \in T_k$, and $j \neq k$. The integer $e \geq 0$ is fixed and known to both players.

We generalize the q-ary game in the following way. We assume that if the sincere answer to a question $\mathbf{T} = \{T_0, T_1, \ldots, T_{q-1}\}$ of Paul's is k, i.e., $x_* \in T_k$, and Carole meant to lie, then she can only choose her lie in a set $L(k) \subseteq \mathcal{Q} \setminus \{k\}$, which has been fixed in advance. For each $k \in \mathcal{Q}$, the set $L(k) \subseteq \mathcal{Q} \setminus \{k\}$ of possible lies available to Carole when the correct/sincere answer is k, is fixed beforehand and known to Paul. The classical q-ary Ulam-Rényi game coincides with the case $L(k) = \mathcal{Q} \setminus \{k\}$, for each $k \in \mathcal{Q}$.

Let $\mathcal{G} : k \in \mathcal{Q} \mapsto L(k) \subseteq \mathcal{Q} \setminus \{k\}$. We may think of $\mathcal{G}$ as the set of possible noise-transitions on a channel carrying Carole's answers (see Figure 1). The q-ary Ulam-Rényi game with such a restriction on the types of Carole's lies will be called the game over the channel $\mathcal{G}$.

For each $j \in \mathcal{Q}$, let $S(j)$ be the set of integers k such that if the correct answer is k Carole can mendaciously answer j, i.e., $S(j) = \{k \in \mathcal{Q} \setminus \{j\} \mid j \in L(k)\}$. In other words $S(j)$ represents the set of possible correct answers, other than j, that Paul has to take into account when Carole answers j. In fact, an answer j could be a lie and, if so, it could have been chosen among the possible lies in $L(k)$ for some $k \in S(j)$.

We call $\mathcal{G}$ a d-regular channel iff there exists an integer $d \leq q - 1$, such that for each $k \in \mathcal{Q}$ we have $|L(k)| = |S(k)| = d$.

For any choice of the parameters q, e, M, d and for any d-regular channel $\mathcal{G}$, we are interested in determining the minimum number of questions, $N_{\mathcal{G}}^{[q]}(M, e)$ that Paul has to ask in order to infallibly guess a number $x_* \in \{0, 1, \ldots, M-1\}$, in the q-ary Ulam-Rényi game with e lies over the channel $\mathcal{G}$.

In this paper we prove that $N_{\mathcal{G}}^{[q]}(M, e)$ is independent of the particular d-regular channel and only depends on the parameter d. In fact, we exactly determine the minimum number, $N_{d,\min}^{[q]}(M, e)$, of questions which are *necessary* to infallibly guess a number $x_* \in \{0, 1, \ldots, M-1\}$, in the q-ary Ulam-Rényi game with e lies over every d-regular channel $\mathcal{G}$. Moreover, we prove that, for all sufficiently large M (the size of the search space), strategies using $N_{d,\min}^{[q]}(M, e) + 1$ questions always exist. Moreover for infinitely many values of M, strategies using exactly $N_{d,\min}^{[q]}(M, e)$ questions exist. All our strategies are implementable by procedures which use adaptiveness only once.

I. Dumitriu and J. Spencer [5] considered the the Ulam-Rényi game over an arbitrary channel (i.e., with constrained lies). Notwithstanding the wider generality of the model studied in [5], our analysis turns out to be more precise in that here the exact evaluation of the size of a shortest strategy is provided, as opposed to the asymptotic analysis of [5] where constants in the main formula evaluating the size of a shortest strategy are not taken into account.

As opposed to the ones in [5], our search strategies use the minimum possible amount of adaptiveness, among all strategies where adaptive questions are allowed. This is a very desirable property in many practical applications.

In this paper we generalize the results in [2,4] where optimal and quasi-optimal strategies (i.e., strategies whose length differs by only one from the information theoretic lower bound) are provided for all sufficiently large M and for the particular case $d = q - 1$, that is for the case when Carole can freely choose how to lie.

The problem of q-ary search with $e \geq 1$ lies was first considered by Malinowski [7] and Aigner [1] who independently evaluated the minimum number of questions Paul needs to win the game when $e = 1$, and Carole is allowed to use *free* lies, i.e., she can choose to answer j when the correct/sincere reply would be i, for all $j \neq i$. Under the hypothesis of *free* lies, the case $e = 2$ was investigated in [3] where q-ary optimal search strategies are given for the case when the search space's cardinality is q^m, for some integer $m \geq 1$, and *quasi-perfect* strategies are proved to exist in the remaining cases. Under the same hypothesis on the type of lies allowed, the case $q = 3$ was fully characterized in [8] where the author also state without proof a generalization of his result for the general case $q \geq 2$. For more variants and references on the Ulam-Rényi game we refer the interested reader to the recent survey paper [9].

2   The q-Ary Ulam-Rényi Game over a d-Regular Channel

Let Paul and Carole first fix non-negative integers $q \geq 2, d < q, M \geq 1, e \geq 0$ and a d-regular channel $\mathcal{G}$. The search space is identified with the set $U = \{0, 1, \ldots, M - 1\}$. Carole chooses a number x_* from U and Paul has to guess it by asking q-ary questions.

Typically, a q-ary *question* $\mathbf{T}$ has the form

"Which one of the sets $T_0, T_1, \ldots, T_{q-1}$ does x_* belong to ?",

where $\mathbf{T} = \{T_0, T_1, \ldots, T_{q-1}\}$ is a q-tuple of (possibly empty) pairwise disjoint subsets of U whose union is U.

Carole's answer is an integer $k \in \mathcal{Q} = \{0, 1, \ldots, q - 1\}$, telling Paul that x_* belongs to T_k.

Each answer of Carole's partitions the set of numbers which are still candidates to be the secret number x_* into three classes. If the answer to the question $\mathcal{T}$ is "k", numbers in T_k are said to *satisfy* the answer, while numbers in $\bigcup_{j \in S(k)} T_j$ *falsify* it. A number x falsifies an answer if, as a consequence of this answer, it is still to be considered as possible candidates for the unknown x_*, but, if $x = x_*$, then the last answer must be considered a lie of Carole's. The remaining numbers, i.e., numbers not in $\bigcup_{j \notin S(k)} T_j$ are *rejected* by the answer and no longer are to be considered as possible candidates for the secret number x_*. In fact, because of the rules of the game, if for some $j \notin S(k)$ the secret number had been in T_j then Carole could have not answered k. At any stage of the game, a number $y \in U$ is also rejected from consideration if it falsifies more than e answers. This accounts for the fact that Carole agreed to lie at most e times.

At any time during the game, Paul's *state* of knowledge is represented by an e-tuple $\sigma = (A_0, A_1, A_2, \ldots, A_e)$ of pairwise disjoint subsets of U, where A_i is the set of numbers falsifying exactly i answers, $i = 0, 1, 2, \ldots, e$. The *initial* state is naturally given by $(U, \varnothing, \varnothing, \ldots, \varnothing)$. A state $(A_0, A_1, A_2, \ldots, A_e)$ is *final* iff $A_0 \cup A_1 \cup A_2 \cup \cdots \cup A_e$ either has exactly one element, or is empty if Carole did not follow the rules of the game.

Suppose that Paul asks the question $T = \{T_0, T_1, \ldots, T_{q-1}\}$ when the state is $\sigma = (A_0, A_1, \ldots, A_e)$ and Carole's answer is equal to k. Let $T_k^* = \bigcup_{j \in S(k)} T_j$. Then Paul's state becomes

$$\sigma^k = (A_0 \cap T_k, \ (A_0 \cap T_k^*) \cup (A_1 \cap T_k), \ \cdots, \ (A_{e-1} \cap T_k^*) \cup (A_e \cap T_k)). \tag{1}$$

Let $\sigma = (A_0, A_1, A_2, \ldots, A_e)$ be a state. For each $i = 0, 1, 2, \ldots, e$ let $a_i = |A_i|$ be the number of elements of A_i. Then the e-tuple $(a_0, a_1, a_2, \ldots, a_e)$ is called the *type* of σ. We shall generally identify a state with its type, tacitly assuming that what holds for a given state also holds for any other state of the same type, up to renaming of the numbers.

Given a state σ, suppose questions $T_1, \ldots, T_t$ have been asked and answers $b^t = b_1, \ldots, b_t$ have been received (with $b_i \in \{0, 1, \ldots, q-1\}$). Iterated application of (1) yields a sequence of states

$$\sigma_0 = \sigma, \quad \sigma_1 = \sigma_0^{b_1}, \quad \sigma_2 = \sigma_1^{b_2}, \quad \ldots, \quad \sigma_t = \sigma_{t-1}^{b_t} = \sigma^{b^t}. \tag{2}$$

By a *strategy* S *with* n *questions* we mean a q-ary tree of depth n, where each node ν is mapped into a question T_ν, and the q edges $\eta_0, \eta_1, \ldots, \eta_{q-1}$ generated by ν are, respectively from left to right, labeled with 0, 1, $\ldots, q-1$, which represent Carole's possible answers to T_ν. Let $\eta = \eta_1, \ldots, \eta_n$ be a path in S, from the root to a leaf, with respective labels $b_1, \ldots, b_n$, generating nodes $\nu_1, \ldots, \nu_n$ and associated questions $T_{\nu_1}, \ldots, T_{\nu_n}$. Fix an arbitrary state σ. Then, according to (2), iterated application of (1) naturally transforms σ into σ^η (where the dependence on the b_j and T_j is understood). We say that strategy S is *winning* for σ iff for every path η the state σ^η is final. A strategy is said to be *nonadaptive* iff all nodes at the same depth of the tree are mapped into the same question.

We denote with $N_{\mathcal{G}}^{[q]}(M, e)$ the minimum integer n such that *there exists* a winning strategy with n questions for the q-ary Ulam-Rényi game with e errors over the (d-regular) channel $\mathcal{G}$ and a search space of cardinality M.

For every integer $q \geq 2$ and state σ of type $(a_0, a_1, \ldots, a_e)$, the *d-regular n^{th} Volume of* σ is defined by

$$V_n^{[d]}(\sigma) = \sum_{i=0}^{e} a_i \sum_{j=0}^{e-i} \binom{n}{j} d^j. \tag{3}$$

Define the *character* of σ as $\mathrm{ch}_d^{[q]}(\sigma) = \min\{n = 0, 1, 2, \ldots \mid V_n^{[d]}(\sigma) \leq q^n\}$.

Proposition 1. *Fix a state $\sigma = (A_0, A_1, \ldots, A_e)$ and a question* $\mathbf{T} = (T_0, T_1, \ldots, T_{q-1}$. *Following (1), for each* $k = 0, 1, \ldots, q-1$, *let* $\sigma^k = (A_0^k, \ldots, A_e^k)$ *be the state of knowledge of Paul after Carole has answered k to the question* $\mathbf{T}$.

(i) *For every integer $n \geq 1$ we have*

$$V_n^{[d]}(\sigma) = \sum_{k=0}^{q-1} V_{n-1}^{[d]}(\sigma^k).$$

(ii) *If σ has a winning q-ary strategy with n questions then $n \geq \mathrm{ch}_d^{[q]}(\sigma)$.*

Proof. For $i = 0, 1, \ldots, e$, let $a_i = |A_i|$ be the type of the state σ and $a_i^k = |A_i^k|$. For $i = 0, 1, \ldots, e$ and $j = 0, 1, \ldots, q-1$ let $t_i^j = |A_i \cap T_j|$. By definition we have $a_i^k = t_i^k + \sum_{j \in S(k)} t_{i-1}^j$ and $a_i = \sum_{k=0}^{q-1} t_i^k$. Thus,

$$V_n^{[d]}(\sigma) = \sum_{i=0}^{e} a_i \sum_{j=0}^{e-i} \binom{n}{j} d^j$$

$$= \sum_{i=0}^{e} \sum_{k=0}^{q-1} t_i^k \sum_{j=0}^{e-i} \binom{n}{j} d^j$$

$$= \sum_{k=0}^{q-1} \sum_{i=0}^{e} t_i^k \sum_{j=0}^{e-i} \binom{n-1}{j} d^j + \sum_{k=0}^{q-1} \sum_{i=1}^{e} \sum_{j \in S(i)} t_{i-1}^j \sum_{j=0}^{e-i} \binom{n-1}{j} d^j$$

$$= \sum_{k=0}^{q-1} \sum_{i=0}^{e} \left(t_i^k + \sum_{j \in S(i)} t_{i-1}^j \right) \sum_{j=0}^{e-i} \binom{n-1}{j} d^j$$

$$= \sum_{k=0}^{q-1} \sum_{i=0}^{e} a_i^k \sum_{j=0}^{e-i} \binom{n-1}{j} d^j = \sum_{k=0}^{q-1} V_{n-1}^{[d]}(\sigma^k).$$

The proof of (ii) follows by induction.

We define $N_{d,\min}^{[q]}(M, e) = \min\{n \mid M \sum_{j=0}^{e} \binom{n}{j} d^j \leq q^n\}$. As an immediate corollary of the above proposition we have

$$N_{\mathcal{G}}^{[q]}(M, e) \geq N_{d,\min}^{[q]}(M, e) = \mathrm{ch}_d^{[q]}(M, 0, \ldots, 0),$$

for all $q \geq 2, M \geq 1, e \geq 0, d \leq q - 1$ and for each choice of the d-regular channel $\mathcal{G}$.

By a *perfect q-ary d-regular strategy* for σ we shall mean a d-regular winning strategy for σ only requiring $\mathrm{ch}_d^{[q]}(\sigma)$ questions. Because a perfect strategy $\mathcal{S}$ uses the least possible number of questions which are also necessary in any winning strategy, $\mathcal{S}$ is *optimal*, in the sense that it cannot be superseded by a shorter strategy.

Accordingly a d-regular winning strategy for σ with $\mathrm{ch}_d^{[q]}(\sigma) + 1$ questions will be called *quasi perfect*.

The following monotonicity properties easily follow from the above definitions.

Proposition 2. *For any two states* $\sigma' = (A_0', A_1', A_2', \ldots, A_e')$ *and*

$$\sigma'' = (A_0'', A_1'', A_2'', \ldots, A_e'')$$

respectively of type $(a_0', a_1', a_2', \ldots, a_e')$ *and* $(a_0'', a_1'', a_2'', \ldots, a_e'')$, *if* $\sum_{i=0}^{j} a_i' \leq \sum_{i=0}^{j} a_i''$ *for all* $j = 0, 1, 2, \ldots, e$ *then*

(i) $\mathrm{ch}_d^{[q]}(\sigma') \leq \mathrm{ch}_d^{[q]}(\sigma'')$ *and* $V_n^{[d]}(\sigma') \leq V_n^{[d]}(\sigma'')$, *for each* $n \geq 0$
(ii) if, for some integer $n \geq 0$, *there exists a winning strategy for* σ'' *with* n *questions then there exists also a winning strategy for* σ' *with* n *questions.*

Note that $\mathrm{ch}_d^{[q]}(\sigma) = 0$ iff σ is a final state.

Let $\sigma = (A_0, A_1, A_2, \ldots, A_e)$ be a state. We say that the question

$$\mathcal{T} = \{T_0, T_1, \ldots, T_{q-1}\}$$

is *balanced for* σ iff for each $j = 0, 1, 2, \ldots, e$, we have $|A_j \cap T_i| = |A_j \cap T_{i+1}| = \frac{1}{q}|A_j|$, for $i = 0, 1, \ldots, q - 2$.

Balanced questions play a special role in Paul's strategy. In fact, if Paul asks a balanced question, Carole's answer (whether a *yes* or a *no*) has no effect on the type of the resulting state. In other words, the state resulting from a balanced question is predetermined. Moreover the information gained by Paul is maximum, since, as shown in the next lemma, a balanced question strictly decreases the character of Paul's state.

By the definition of a *balanced question* together with (3) and Proposition 1 we have the following.

Lemma 1. *Let* $\mathcal{T}$ *be a balanced question for a state* $\sigma = (A_0, A_1, A_2, \ldots, A_e)$. *Let* σ^i *be as in (1) above. Then for each* $0 \leq i < j \leq d$,

(i) $V_n^{[d]}(\sigma^i) = V_n^{[d]}(\sigma^j) = \frac{1}{q}V_{n+1}^{[d]}(\sigma)$, *for each integer* $n \geq 0$,
(ii) $\mathrm{ch}_d^{[q]}(\sigma^i) = \mathrm{ch}_d^{[q]}(\sigma^j) = \mathrm{ch}_d^{[q]}(\sigma) - 1$.

3   Encoding Strategies

We refer to [6] for background in error correcting codes. Here we shall recall few notions and fix some notations for later use. We shall assume that a d-regular q-ary channel $\mathcal{G}$ is given. We give here an example for the case $q = 6$ and $d = 3$.

Fix an integer $n > 0$ and let $x^n, y^n \in \mathcal{Q}^n = \{0, 1, \ldots, q - 1\}^n$. Recall that the Hamming distance between x^n and y^n is given by $d_H(x^n, y^n) = |\{i \mid x_i \neq y_i\}|$. The *Hamming distance* in terms of $\mathcal{G}$ (the *$\mathcal{G}$-Hamming distance*) between two symbols is defined by

$$d_{\mathcal{G}}(x, y) = \begin{cases} 0 & \text{if } x = y \\ 1 & \text{if } y \in L(x) \\ \infty & \text{otherwise.} \end{cases}$$

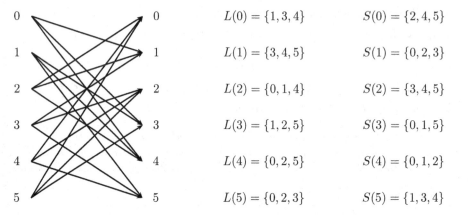

0	0	$L(0) = \{1,3,4\}$	$S(0) = \{2,4,5\}$
1	1	$L(1) = \{3,4,5\}$	$S(1) = \{0,2,3\}$
2	2	$L(2) = \{0,1,4\}$	$S(2) = \{3,4,5\}$
3	3	$L(3) = \{1,2,5\}$	$S(3) = \{0,1,5\}$
4	4	$L(4) = \{0,2,5\}$	$S(4) = \{0,1,2\}$
5	5	$L(5) = \{0,2,3\}$	$S(5) = \{1,3,4\}$

Fig. 1. Constrained lies as a Channel

The distance between two sequences is defined by $d_{\mathcal{G}}(x^n, y^n) = \sum_{i=1}^{n} d_{\mathcal{G}}(x_i, y_i)$, where $x^n, y^n \in \mathcal{Q}^n$.

Notice that for each pair of sequences $x^n, y^n \in \mathcal{Q}^n$, we have that $d_H(x^n, y^n) \leq d_{\mathcal{G}}(x^n, y^n)$, for any possible choice of the channel $\mathcal{G}$.

The *Hamming sphere* $\mathcal{B}_r^H(x^n)$ *with radius* r *and center* x^n is the set of elements of $\{0, 1, \ldots, q-1\}^n$ whose *Hamming distance* from x^n is at most r, in symbols,

$$\mathcal{B}_r^{\mathcal{G}}(x^n) = \{y^n \in \{0, 1, \ldots, q-1\}^n \mid d_H(x^n, y^n) \leq r\}.$$

The $\mathcal{G}$-*Hamming sphere* $\mathcal{B}_r^{\mathcal{G}}(x^n)$ *with radius* r *and center* x^n is the set of elements of $\{0, 1, \ldots, q-1\}^n$ whose $\mathcal{G}$-*Hamming distance* from x^n is at most r, in symbols,

$$\mathcal{B}_r^{\mathcal{G}}(x^n) = \{y^n \in \{0, 1, \ldots, q-1\}^n \mid d_{\mathcal{G}}(x^n, y^n) \leq r\}.$$

Notice that for any $x^n \in \{0, 1, \ldots, q-1\}^n$, and $r \geq 0$, we have $|\mathcal{B}_r^{\mathcal{G}}(x^n)| = \sum_{i=0}^{r} \binom{n}{i} d^i$.

Moreover, for all possible channels $\mathcal{G}$ we immediately have the relationship $\mathcal{B}_r^{\mathcal{G}}(x^n) \subseteq \mathcal{B}_r^H(x^n)$.

By a code we shall mean a q-ary code for $\mathcal{G}$ in the following sense:

Definition 1. *Let the channel* $\mathcal{G}$ *be given. A* q-*ary code* $\mathcal{C}$ *of length* n *for the channel* $\mathcal{G}$ *is a non-empty subset of* $\{0, 1, \ldots, q-1\}^n$. *Its elements are called codewords. The* minimum distance *of* $\mathcal{C}$ *is given by*

$$\delta(\mathcal{C}) = \min\left\{d \mid \mathcal{B}_{\lfloor \frac{d}{2} \rfloor}^{\mathcal{G}}(x^n) \cap \mathcal{B}_{\lfloor \frac{d}{2} \rfloor}^H(y^n) \text{ for all distinct } x^n, y^n \in \mathcal{C}\right\}$$

We say that $\mathcal{C}$ *is an* (n, M, t) *code in terms of* $\mathcal{G}$ *iff* $\mathcal{C}$ *has length* n, $|\mathcal{C}| = M$ *and* $\delta(\mathcal{C}) = t$. *Let* $\mathcal{C}_1$ *and* $\mathcal{C}_2$ *be two codes of length* n. *The* minimum distance *(in terms of* $\mathcal{G}$*) between* $\mathcal{C}_1$ *and* $\mathcal{C}_2$ *is defined by*

$$\Delta_{\mathcal{G}}(\mathcal{C}_1, \mathcal{C}_2) = \min\left\{d \mid \mathcal{B}_{\lfloor \frac{d}{2} \rfloor}^{\mathcal{G}}(x^n) \cap \mathcal{B}_{\lfloor \frac{d}{2} \rfloor}^H(y^n) \text{ for all } x^n \in \mathcal{C}_1, y^n \in \mathcal{C}_2\right\}.$$

The Hamming distance between two codes C_1 and C_2 is defined by $\Delta_H(C_1, C_2) = \min\{d_H(x^n, y^n) \mid x^n \in C_1, y^n \in C_2\}$.

Lemma 2. *Let* $n = 2, 3, \ldots$ *Then for any two integers* $1 \leq t \leq \frac{n}{2}$, *and*

$$1 \leq M \leq \frac{q^n}{\sum_{i=0}^{t-1} \binom{n}{i} q^i},$$

there exists an (n, M, t) *q-ary code* C *in terms of* $\mathcal{G}$.

The **Proof** follows directly by Gilbert's bound [6] and the relationship between the Hamming distance and the $\mathcal{G}$-*Hamming distance* between codewords.

The following lemma formally clarifies the correspondence between non-adaptive winning strategies for the Ulam-Rényi game and special families of codes. An analogous result can be found in [14].

Lemma 3. *Fix an integer* $e \geq 1$ *Let* $\sigma = (A_0, A_1, A_2, \ldots A_e)$ *be a state of type* $(a_0, a_1, a_2, \ldots, a_e)$ *and* $\mathcal{G}$ *be a d-regular channel. Let* $n \geq \mathrm{ch}_d^{[q]}(\sigma)$. *Then a d-regular non-adaptive winning strategy for* σ *with* n *questions exists if for all* $i = 0, 1, 2, \ldots, e-1$ *there are integers* $t_i \geq 2(e-i)+1$, *together with an e-tuple of q-ary codes* $\Gamma = (C_0, C_1, C_2, \ldots, C_{e-1})$, *such that each* C_i *is an* (n, a_i, t_i) *code in terms of* $\mathcal{G}$, *and* $\Delta_{\mathcal{G}}(C_i, C_j) \geq 2e - (i+j) + 1$, *(whenever* $0 \leq i < j \leq e-1$*).*

Proof. Let $\Gamma = (C_0, C_1, C_2, \ldots, C_{e-1})$ be an e-tuple of codes satisfying the hypothesis. Let

$$\mathcal{H} = \bigcup_{i=0}^{e-1} \bigcup_{x^n \in C_i} \mathcal{B}_{e-i}^{\mathcal{G}}(x^n).$$

By hypothesis, for any $i, j \in \{0, 1, \ldots, e-1\}$ and $x^n \in C_i, y^n \in C_j$ we have $d_{\mathcal{G}}(x^n, y^n) \geq 2e - (i+j) + 1$. It follows that the $\mathcal{G}$-Hamming spheres $\mathcal{B}_{e-i}^{\mathcal{G}}(x^n)$, $\mathcal{B}_{e-j}^{\mathcal{G}}(y^n)$ are pairwise disjoint. Thus

$$|\mathcal{H}| = \sum_{i=0}^{e-1} a_i \sum_{j=0}^{e-i} \binom{n}{j} d^j. \tag{4}$$

Let $\mathcal{D} = \{0, 1, \ldots, q-1\}^n \setminus \mathcal{H}$. Since $n \geq \mathrm{ch}_d^{[q]}(a_0, a_1, a_2, \ldots, a_e)$, by definition of character we have $q^n \geq \sum_{i=0}^{e} a_i \sum_{j=0}^{e-i} \binom{n}{j} d^j$. From (4) it follows that

$$|\mathcal{D}| = q^n - \sum_{i=0}^{e-1} a_i \sum_{j=0}^{e-i} \binom{n}{j} d^j \geq a_e. \tag{5}$$

Let us now fix, once and for all, a one-one function

$$f \colon A_0 \cup A_1 \cup \cdots \cup A_e \to C_0 \cup C_1 \cup \cdots \cup C_{e-1} \cup \mathcal{D},$$

that maps numbers in A_i to codewords in C_i, for $i = 0, 1, \ldots, e - 1$, and maps numbers in A_e to n-tuples in $\mathcal{D}$. The existence of the map f, is ensured by our assumptions about Γ and (5).

For each $y \in A_0 \cup A_1 \cup A_2 \cup \cdots \cup A_e$ and $j = 1, \ldots, n$ let $f(y)_j$ be the jth digit of the n-tuple $f(y) \in \{0, 1, \ldots, q - 1\}^n$. We can now exhibit the questions $T_j = (T_{j\,0}, T_{j\,1}, \ldots, T_{j\,q-1})$, $j = 1, 2, \ldots, n$ of our search strategies:

For each $j = 1, \ldots, n$ let the set $T_{j\,i} \subseteq U$ be defined by $T_{j\,i} = \{z \in \bigcup_{k=0}^{e} A_k \mid f(z)_j = i\}$. Intuitively, letting x_* denote the unknown number, T_j asks "What is the jth digit in the q-ary expansion of $f(x_*)$?"

The answers to questions $T_1, \ldots, T_n$ determine an n-tuple of digits $b^n = b_1 \cdots b_n$. We shall show that the sequence $T_1, \ldots, T_n$ yields an optimal non-adaptive winning strategy for σ. Let $\sigma_1 = \sigma^{b_1}$, $\sigma_2 = \sigma_1^{b_2}, \ldots, \sigma_n = \sigma_{n-1}^{b_n}$. Arguing by cases we shall show that $\sigma_n = (A_0^*, A_1^*, \ldots, A_e^*)$ is a final state.

By (1), for all $i = 0, 1, \ldots, e$, any $z \in A_{e-i}$ that falsifies $> i$ answers does not survive in σ_n—in the sense that $z \notin A_0^* \cup A_1^* \cup \cdots \cup A_e^*$.

Case 1. $b^n \notin \bigcup_{i=0}^{e} \bigcup_{y \in A_i} \mathcal{B}_{e-i}^{\mathcal{G}}(f(y))$.
For all $i = 0, 1, \ldots, e$, and for each $y \in A_i$ we must have $y \notin A_0^* \cup A_1^* \cup \cdots \cup A_e^*$. Indeed, the assumption $b^n \notin \mathcal{B}_{e-i}^{\mathcal{G}}(f(y))$ implies $d_{\mathcal{G}}(f(y), b^n) > e - i$, whence either y falsifies $> e - i$ of the answers to $T_1, \ldots, T_n$, or it is rejected by some answer, thus y does not survive in σ_n. We have proved that $A_0^* \cup A_1^* \cup \cdots \cup A_e^*$ is empty, and σ_n is a final state.

Case 2. $b^n \in \mathcal{B}_{e-i}^{\mathcal{G}}(f(y))$ for some $i \in \{0, 1, \ldots, e\}$ and $y \in A_i$.
Then $y \in A_0^* \cup A_1^* \cup \cdots \cup A_e^*$, because $d_{\mathcal{G}}(f(y), b^n) \le e - i$, whence y falsifies $\le e - i$ answers. Our assumptions about Γ ensure that, for all $j = 0, 1, \ldots, e$ and for all $y' \in A_j$ and $y \ne y'$, we have $b^n \notin \mathcal{B}_{e-j}^{\mathcal{G}}(f(y'))$. Thus, $d_{\mathcal{G}}(f(y'), b^n) > e - j$ and y' falsifies $> e - j$ of the answers to $T_1, \ldots, T_n$, whence y' does not survive in σ_n. This shows that for any $y' \ne y$, we have $y' \notin A_0^* \cup A_1^* \cup \cdots \cup A_e^*$. Therefore, $A_0^* \cup A_1^* \cup \cdots \cup A_e^*$ only contains the element y, and σ_n is a final state.

4 The Case $M = q^m$: Optimal Strategies with Minimum Adaptiveness

4.1 The First Batch of Questions

In this section we shall consider the case when the size of the search space is a power of q, i.e., $M = q^m$, for some integer $m \ge 1$.

By Proposition 1(ii), at least $N_{d,\min}^{[q]}(q^m, e)$ questions are *necessary* to guess the unknown number $x_* \in U = \{0, 1, \ldots, q^m - 1\}$, if up to e answers may be erroneous.[1]

---

[1] Recall that, for any two integers $e, m \ge 0$, we denote by $N_{d,\min}^{[q]}(q^m, e) = \mathrm{ch}_d^{[q]}(q^m, 0, \ldots, 0)$ the smallest integer $n \ge 0$ such that $q^n \ge q^m(\binom{n}{e}d^e + \binom{n}{e-1}d^{e-1} + \cdots + \binom{n}{2}d^2 + nd + 1)$.

As a first result of this paper we shall now prove that, for all suitably large m, $N^{[q]}_{d,\min}(q^m, e)$ questions are also *sufficient* under the following constraint: first we use a predetermined non-adaptive batch of m questions $\mathcal{D}_1, \ldots, \mathcal{D}_m$, and then, only depending on the answers, we ask the remaining $N^{[q]}_{d,\min}(q^m, e) - m$ questions in a second non-adaptive batch.

The *first batch of questions* is as follows:

For each $j = 1, 2, \ldots, m$, let $\mathcal{D}_j = (D_{j0}, D_{j1}, \ldots, D_{j\,q-1})$ denote the question "What is the j-th (q-ary) digit of x_*?" A number $y \in U$ belongs to $D_{j\,i}$ iff the jth symbol y_j of its q-ary expansion $y^m = y_1 \cdots y_m$ is equal to i.

Upon identifying Let $b_j \in \{0, 1, \ldots, q-1\}$ be the answer to question $\mathcal{D}_i$. Let $b^m = b_1 \cdots b_m$. Beginning with the initial state $\sigma = (U, \varnothing, \ldots, \varnothing)$, and repeatedly applying (1) we have that the state resulting from the answers $b_1 \cdots b_m$, is an $(e+1)$-tuple $\sigma^{b^m} = (A_0, A_1, \ldots, A_e)$, where

$$A_i = \{y \in U \mid d_{\mathcal{G}}(y^m, b^m) = i\} \qquad \text{for all } i = 0, 1, \ldots, e,$$

and we have that

$$|A_0| = 1, \quad |A_1| = md, \ldots, |A_e| = \binom{m}{e} d^e.$$

Thus σ^{b^m} has type $(1, md, \binom{m}{2}d^2, \ldots, \binom{m}{e}d^e)$. Let σ_i be the state resulting after the first i answers, beginning with $\sigma_0 = \sigma$. Since each question D_i is *balanced* for σ_{i-1}, an easy induction using Lemma 1 yields $\text{ch}^{[q]}_d(\sigma^{b^m}) = N^{[q]}_{d,\min}(q^m, e) - m$.

4.2   The Second Batch of Questions

For each m-tuple $b^m \in \{0, 1, \ldots, q-1\}^m$ of possible answers, we shall now construct a non-adaptive strategy $\mathcal{S}_{b^m}$ with $\text{ch}^{[q]}_d(1, md, \binom{m}{2}d^2, \ldots, \binom{m}{e}d^e)$ questions, which turns out to be winning for the state σ^{b^m}. Thus, let us consider the value of

$$\text{ch}^{[q]}_d\left(1, md, \binom{m}{2}d^2, \ldots, \binom{m}{e}d^e\right)$$

for $m \geq 1$.

Definition 2. *Let $e \geq 0$ and $n \geq 2e$ be arbitrary integers. The critical index $m^{[d,q]}_{n,e}$ is the largest integer $m \geq 0$ such that $\text{ch}^{[q]}_d(1, md, \binom{m}{2}d^2, \ldots, \binom{m}{e}d^e) = n$.*

Lemma 4. *Let $q \geq 2$, $e \geq 1$ and $n \geq 2e$ be arbitrary integers. Then*

$$\left\lfloor \frac{\sqrt[e]{e!}q^{\frac{n}{e}}}{d} \right\rfloor - n - e \leq m^{[d,q]}_{n,e} < \left\lfloor \frac{\sqrt[e]{e!}\,q^{\frac{n}{e}}}{d} \right\rfloor + e. \tag{6}$$

Proof. By definition, $m_{n,e}^{[d,q]} = \max\left\{ m \mid V_n^{[d]}\left(1, md, \binom{m}{2}d^2, \ldots, \binom{m}{e}d^e\right) \leq q^n \right\}$.

Setting now $m^* = \left\lfloor \frac{\sqrt[e]{e!}\,q^{\frac{n}{e}}}{d} \right\rfloor + e$, the desired right inequality is a direct consequence of the inequality $V_n^{[d]}(\sigma) > q^n$, where $\sigma = \left(1, m^*d, \binom{m^*}{2}d^2, \ldots, \binom{m^*}{e}d^e\right)$.

Indeed, we have

$$
\begin{aligned}
V_n^{[d]}(\sigma) &> V_n^{[d]}\left(0, \ldots, 0, \binom{m^*}{e}d^e\right) \\
&= \binom{m^*}{e}d^e = d^e \frac{m^*(m^*-1)\cdots(m^*-e+1)}{e!} \\
&\geq d^e \frac{\left(\frac{\sqrt[e]{e!}\,q^{\frac{n}{e}}}{d}\right)^e}{e!} = q^n.
\end{aligned}
$$

Let $\tilde{m} = \left\lfloor \frac{\sqrt[e]{e!}\,q^{\frac{n}{e}}}{d} \right\rfloor - n - e$. In order to prove the left inequality, we need to show that

$$V_{n+\tilde{m}}^{[d]}\left(q^{\tilde{m}}, 0, \ldots, 0\right) \leq q^{n+\tilde{m}},$$

thus we have to prove

$$\sum_{j=0}^{e} \binom{\tilde{m}+n}{j} d^j \leq q^n.$$

We have

$$
\begin{aligned}
\sum_{j=0}^{e} \binom{\tilde{m}+n}{j} d^j &\leq d^e \sum_{j=0}^{e} \binom{\tilde{m}+n}{j} \\
&\leq d^e \binom{\tilde{m}+n+e}{e} \\
&= d^e \frac{(\tilde{m}+n+e)(\tilde{m}+n+e-1)\cdots(\tilde{m}+n+1)}{e!} \\
&\leq d^e \frac{(\tilde{m}+n+e)^e}{e!} \\
&\leq \frac{d^e}{e!}\left(\frac{\sqrt[e]{e!}\,q^{\frac{n}{e}}}{d} - n - e + n + e\right)^e = q^n,
\end{aligned}
$$

which completes the proof.

We now prove that for all sufficiently large m there exists a second batch of $n = N_{d,\min}^{[q]}(q^m, e) - m = \mathrm{ch}_d^{[q]}(1, md, \binom{m}{2}d^2, \ldots, \binom{m}{e}d^e)$ non-adaptive questions allowing Paul to infallibly guess Carole's secret number. We first need the following lemma.[2]

---

[2] The problem of finding families of error-correcting codes with fixed reciprocal distances was also considered in [14], where the authors proved a result related to our Lemma 5 showing the existence of asymptotically optimal such families.

Lemma 5. *For any fixed integers $k = 0, 1$ and $e \geq 1$ and for all sufficiently large integers n, there exists an e-tuple of q-ary codes $\Gamma = (\mathcal{C}_0, \mathcal{C}_1, \ldots, \mathcal{C}_{e-1})$ together with integers $t_i \geq 2(e - i) + 1$ $(i = 0, 1, \ldots, e - 1)$ such that*

(i) Each $\mathcal{C}_i$ is an $(n + 2k, \binom{m_{n,e}^{[d,q]}}{i} d^i (q + 1)^k, t_i)$ code;

(ii) $\Delta_{\mathcal{G}}(\mathcal{C}_i, \mathcal{C}_j) \geq 2e - (i + j) + 1$, (whenever $0 \leq i < j \leq e - 1$.)

Proof. Let $n' = n - e^2 + 2k$. First we prove the existence of an $(n', \binom{m_{n,e}^{[d,q]}}{e-1} d^{e-1} (q + 1)^k, 2e + 1)$ code. By Lemma 4 and the trivial inequality $e! \leq \frac{(e+1)^e}{2^e}$, it follows that, for all sufficiently large n

$$\binom{m_{n,e}^{[d,q]}}{e - 1} d^{e-1} (q + 1)^k < (m_{n,e}^{[d,q]})^{e-1} d^{e-1} (q + 1)^k$$

$$< (\frac{\sqrt[e]{e!}\, q^{\frac{n}{e}}}{d} + e)^{e-1} d^{e-1} (q + 1)^k$$

$$\leq (\frac{e\, q^{\frac{n}{e}}}{d})^{e-1} d^{e-1} (q + 1)^k$$

$$\leq e^{e-1} q^{n - \frac{n}{e} + 2k}$$

$$= e^{e-1} \frac{q^{n - e^2}}{q^{\frac{n}{e} - e^2 - 2k}}$$

$$\leq \frac{q^{n - e^2 + 2k}}{\sum_{j=0}^{2e} \binom{n - e^2 + 2k}{j} q^j},$$

since $\sum_{j=0}^{2e} \binom{n - e^2 + 2k}{j} q^j$ is polynomial in n.

The existence of the desired $(n', \binom{m_{n,e}^{[d,q]}}{e-1} d^{e-1} (q + 1)^k, 2e + 1)$ code now follows from Lemma 3.2. We have proved that, for all sufficiently large n, there exists an $(n - e^2 + 2k, \binom{m_{n,e}^{[d,q]}}{e-1} d^{e-1} (q + 1)^k, 2e + 1)$ code $\mathcal{C}'$. For each $i = 0, 1, \ldots, e - 1$ let the e^2-tuple $\boldsymbol{a}_i$ be defined by

$$\boldsymbol{a}_i = \underbrace{00 \ldots 0}_{ie} \underbrace{11 \ldots 1}_{e} \underbrace{0 \ldots 0}_{e^2 - (i+1)e}.$$

Furthermore, let $\mathcal{C}_i''$ be the code obtained by appending the suffix $\boldsymbol{a}_i$ to the codewords of $\mathcal{C}'$, in symbols,

$$\mathcal{C}_i'' = \mathcal{C}' \otimes \boldsymbol{a}_i.$$

Trivially, $\mathcal{C}_i''$ is an $(n + 2k, \binom{m_{n,e}^{[d,q]}}{e-1} d^{e-1} (q+1)^k, 2e + 1)$ code for all $i = 0, 1, \ldots, e - 1$. Furthermore, we have $\Delta_{\mathcal{G}}(\mathcal{C}_i'', \mathcal{C}_j'') \geq \Delta_H(\mathcal{C}_i'', \mathcal{C}_j'') = 2e \geq 2e - (i + j) + 1$, whenever $0 \leq i < j \leq e - 1$. For each $i = 0, 1, \ldots, e - 1$, pick a subcode $\mathcal{C}_i \subseteq \mathcal{C}_i''$ with $|\mathcal{C}_i| = \binom{m_{n,e}^{[d,q]}}{i} d^i (q + 1)^k$. Then the new e-tuple of codes $\Gamma = (\mathcal{C}_0, \mathcal{C}_1, \ldots, \mathcal{C}_{e-1})$ satisfies both conditions (i) and (ii), and the proof is complete.

The following corollary implies the existence of minimum adaptiveness perfect search strategies.

Corollary 1. *Fix an integer $e \geq 0$. Then for all sufficiently large integers m and for every state σ of type $(1, md, \ldots, \binom{m}{e}d^e)$ there exists a non-adaptive winning strategy S such that the number of questions in S coincides with Berlekamp's lower bound* $\mathrm{ch}_d^{[q]}(\sigma) = N_{d,\min}^{[q]}(q^m, e) - m$.

Proof. Skipping the trivial case $e = 0$, assume $e \geq 1$. Let $n = \mathrm{ch}_d^{[q]}(\sigma)$ and $k = 0$. By definition, $n \to \infty$ as $m \to \infty$. Lemmas 5 and 3 yield a non-adaptive winning strategy with n questions for any state of type $(1, m_{n,e}^{[d,q]}d, \binom{m_{n,e}^{[d,q]}}{2}d^2,$ $\ldots, \binom{m_{n,e}^{[d,q]}}{e}d^e)$. By Definition 2, $m \leq m_{n,e}^{[d,q]}$, and a fortiori, for all sufficiently large m, a non-adaptive winning strategy with n questions exists for any state of type $(1, md, \ldots, \binom{m}{e}d^e)$. $\quad\blacksquare$

We summarize our results as follows:

Theorem 1. *Fix integers $q \geq 2, e \geq 0$, and a d-regular channel $\mathcal{G}$. Then for all sufficiently large integers m there exists a perfect winning strategy S for the Ulam-Rényi game with q-ary questions and e lies over the channel $\mathcal{G}$ and a search space of cardinality q^m, which uses adaptiveness only once. More precisely S has exactly size $N_{d,\min}^{[q]}(q^m, e)$. Therefore,*

$$N_{\mathcal{G}}^{[q]}(q^m, e) = N_{d,\min}^{[q]}(q^m, e).$$

5 Quasi-perfect Strategies Always Exist

In this section we complete our analysis by considering the case when the cardinality of the search space is not necessarily a power of q. We shall prove that for any choice of the parameter $q > 2$, $e \geq 0$, and for any d-regular channel $\mathcal{G}$ quasi-perfect strategies, always exist up to finitely many exceptional values of the search space cardinality, M. Recall that a strategy is called quasi-perfect if it uses exactly $N_{d,\min}^{[q]}(M, e) + 1$ questions.

Lemma 6. *Fix $q \geq 2, e \geq 0$. Then, for all sufficiently large n, there exists a (perfect) strategy for the state*

$$\sigma = \left((q+1)q^{m_{n,e}^{[d,q]}-1}, 0, \ldots, 0\right)$$

using $\mathrm{ch}_d^{[q]}(\sigma) = (m_{n,e}^{[d,q]} + n + 1)$ questions.

Proof. First we shall prove that $\mathrm{ch}_d^{[q]}(\sigma) > m_{n,e}^{[d,q]} + n + 1$. By definition of character it is enough to show that the $(n + m_{n,e}^{[d,q]})$-th q-ary volume of σ exceeds $q^{n+m_{n,e}^{[d,q]}}$, that is

$$V_{n+m_{n,e}^{[d,q]}}^{[d]}(\sigma) > q^{n+m_{n,e}^{[d,q]}}.$$

For $i = 0, 1, \ldots, m_{n,e}^{[d,q]} - 1$ let $\sigma_i = (a_{i\,0}, a_{i\,1}, \ldots, a_{i\,j}, \ldots, a_{i\,e})$, where

$$a_{i\,j} = (q+1)q^{m_{n,e}^{[d,q]}-1-i}\binom{i}{j}d^j,$$

For $i = 0, 1, \ldots, m_{n,e}^{[d,q]} - 2$ the state σ_{i+1} coincides with the one produced by asking a balanced question in the state σ_i. Hence by Lemma 1 we have that for all $i = 0, 1, \ldots, m_{n,e}^{[d,q]} - 2$,

$$V_{n+m_{n,e}^{[d,q]}-i}^{[d]}(\sigma_i) = q\,V_{n+m_{n,e}^{[d,q]}-i-1}^{[d]}(\sigma_{i+1}).$$

Let us now consider the state

$$\sigma_{m_{n,e}^{[d,q]}-1} = \left((q+1), (q+1)(m_{n,e}^{[d,q]}-1)d, \ldots, (q+1)\binom{m_{n,e}^{[d,q]}-1}{e}d^e\right).$$

We have

$$
\begin{aligned}
V_{n+1}^{[d]}\left(\sigma_{m_{n,e}^{[d,q]}-1}\right) &= \sum_{j=0}^{e}(q+1)\binom{m_{n,e}^{[d,q]}-1}{j}d^j\sum_{i=0}^{e-j}\binom{n+1}{i}d^i \\
&\geq \sum_{j=0}^{e}(q+1)\binom{m_{n,e}^{[d,q]}-1}{j}d^j\binom{n+1}{e-j}d^{e-j} \\
&= (q+1)d^e\binom{m_{n,e}^{[d,q]}+n}{e} \\
&\geq \frac{(q+1)d^e}{e!}(m_{n,e}^{[d,q]}+n-e)^e \\
&> \frac{(q+1)d^e}{e!}\left(\frac{\sqrt[e]{e!}q^{\frac{n}{e}}}{d}-2e-1\right)^e \\
&> \frac{q(1+\frac{1}{q})d^e}{e!}\left(\frac{1}{(1+\frac{1}{q})^{\frac{1}{e}}}\frac{\sqrt[e]{e!}q^{\frac{n}{e}}}{d}\right)^e \\
&= q^{n+1}.
\end{aligned}
$$

Thus we have the desired result

$$
\begin{aligned}
V_{n+m_{n,e}^{[d,q]}}^{[d]}(\sigma) &= V_{n+m_{n,e}^{[d,q]}}^{[d]}(\sigma_0) = q^{m_{n,e}^{[d,q]}-1}V_{n+1}^{[d]}\left(\sigma_{m_{n,e}^{[d,q]}-1}\right) \\
&> q^{m_{n,e}^{[d,q]}-1+n+1} = q^{m_{n,e}^{[d,q]}+n}.
\end{aligned}
$$

It remains to prove that there exists a q-ary winning strategy of length $n + m_{n,e}^{[d,q]} + 1$ for the state σ.

We have already proved (albeit implicitly) that there exists a strategy with $m_{n,e}^{[d,q]} - 1$ questions (the questions described above) for the state σ which leads Paul into the state $\sigma_{m_{n,e}^{[d,q]}-1}$. Therefore in order to complete the proof it is enough to show that there exists a winning strategy for the state $\sigma_{m_{n,e}^{[d,q]}-1}$ with $n+2$ questions. Indeed such a strategy is given by Lemma 5 (setting $k = 1$) and Lemma 3. Remarkably, such a strategy is a non-adaptive one. This concludes the proof.

Proposition 3. *Let $k \geq 4e^2$, then we have*

$$\sum_{j=0}^{e-1} \left(\binom{k}{j} + \binom{k+1}{j} \right) d^j \leq \sum_{j=0}^{e} \binom{k}{j} d^j.$$

Proof. If $k \geq 4e^2$ it follows

$$k^2 + 3k + 2 \geq ke^2 + 2ke + 3e.$$

This is equivalent to

$$\binom{k}{e} \geq e \binom{k+1}{e-1}.$$

It follows

$$\binom{k}{e} \geq \sum_{j=0}^{e-1} \binom{k+1}{j}.$$

With this we get

$$\sum_{j=0}^{e-1} \left(\binom{k}{j} + \binom{k+1}{j} \right) d^j \leq \sum_{j=0}^{e} \binom{k}{j} d^j.$$

Lemma 7. *Let $N_{d,\min}^{[q]}(q^m, e) \geq 4e^2$. Then we have the inequalities:*

$$N_{d,\min}^{[q]}(q^m, e) + 1 \leq N_{d,\min}^{[q]}(q^{m+1}, e) \leq N_{d,\min}^{[q]}(q^m, e) + 2.$$

Proof. Let $k = N_{d,\min}^{[q]}(q^m, e)$, then $q^m \sum_{j=0}^{e} \binom{k}{j} d^j \leq q^k$ and $q^m \sum_{j=0}^{e} \binom{k-1}{j} d^j > q^{k-1}$. It follows

$$q^{m+1} \sum_{j=0}^{e} \binom{k}{j} d^j > q^{m+1} \sum_{j=0}^{e} \binom{k-1}{j} d^j > q^k$$

and thus $N_{d,\min}^{[q]}(q^m, e) + 1 \leq N_{d,\min}^{[q]}(q^{m+1}, e)$. From $q^m \sum_{j=0}^{e} \binom{k}{j} d^j \leq q^k$ follows

$$q^{m+1} \sum_{j=0}^{e} \binom{k}{j} d^j \leq q^{k+1}. \qquad (3.1)$$

From Pascal's Identity we get $\binom{k}{j} = \binom{k+2}{j} - \binom{k}{j-1} - \binom{k+1}{j-1}$. Thus

$$q^{m+1} \sum_{j=0}^{e} \left(\binom{k+2}{j} - \binom{k}{j-1} - \binom{k+1}{j-1} \right) d^j \leq q^{k+1},$$

$$q^{m+1} \sum_{j=0}^{e} \binom{k+2}{j} d^j \leq q^{k+1} + q^{m+1} \sum_{j=0}^{e-1} \left(\binom{k}{j} + \binom{k+1}{j} \right) d^j.$$

Because of Property 3 we have:

$$q^{m+1} \sum_{j=0}^{e} \binom{k+2}{j} d^j \leq q^{k+1} + q^{m+1} \sum_{j=0}^{e} \binom{k}{j} d^j.$$

It follows by inequality 3.1

$$q^{m+1} \sum_{j=0}^{e} \binom{k+2}{j} d^j \leq q^{k+2}.$$

Thus $N^{[q]}_{d,\min}(q^{m+1}, e) \leq k + 2$.

The following theorem summarizes all it is known about shortest search strategies for the Ulam-Rényi game with e lies and q-ary questions over an arbitrary d-regular channel.

Theorem 2. *Fix integers $e \geq 0$, $q \geq 2$, and a d-regular channel $\mathcal{G}$. Then, for all sufficiently large M we have:*

$$N^{[q]}_{d,\min}(M, e) \leq N^{[q]}_{\mathcal{G}}(M, e) \leq N^{[q]}_{d,\min}(M, e) + 1.$$

Proof. Let $m = \lfloor \log_q M \rfloor$. Thus

$$N^{[q]}_{\mathcal{G}}(q^m, e) \leq N^{[q]}_{\mathcal{G}}(M, e) \leq N^{[q]}_{\mathcal{G}}(q^{m+1}, e).$$

Fix the smallest integer n such that $m \leq m^{[d,q]}_{n,e}$. By Definition 2 and Theorem 1 we have

$$N^{[q]}_{d,\min}(q^m, e) = m + n = N^{[q]}_{\mathcal{G}}(q^m, e).$$

We shall argue by cases.

Case 1. $m < m^{[d,q]}_{n,e}$. Hence $m + 1 \leq m^{[d,q]}_{n,e}$. Definition 2 and Theorem 1 yield

$$N^{[q]}_{d,\min}(q^{m+1}, e) = m + 1 + n = N^{[q]}_{\mathcal{G}}(q^{m+1}, e).$$

Thus, we have the desired result

$$N^{[q]}_{\mathcal{G}}(M, e) \leq N^{[q]}_{\mathcal{G}}(q^{m+1}, e) = N^{[q]}_{\mathcal{G}}(q^m, e) + 1 = N^{[q]}_{d,\min}(q^m, e) + 1 \leq N^{[q]}_{d,\min}(M, e) + 1.$$

Case 2. $m = m^{[d,q]}_{n,e}$. By definition we have $N^{[q]}_{d,\min}(q^{m+1}, e) \geq m + 1 + n + 1 = m + n + 2$. By Lemma 7 we have $N^{[q]}_{d,\min}(q^{m+1}, e) \leq N^{[q]}_{d,\min}(q^m, e) + 2 = m + n + 2$. Hence, $N^{[q]}_{d,\min}(q^{m+1}, e) = n + m + 2$, and by Theorem 1, we have $N^{[q]}_{\mathcal{G}}(q^{m+1}, e) = m + n + 2$.

Recall that $m = m^{[d,q]}_{n,e}$ and by Lemma 6 we have $N^{[q]}_{\mathcal{G}}((q+1)q^{m-1}, e) = m + n + 1 = N^{[q]}_{d,\min}((q+1)q^{m-1}, e)$.

Therefore, for all integers M such that $q^m \leq M \leq (q+1)q^{m-1}$, we have

$$N_{\mathcal{G}}^{[q]}(M, e) \leq m + n + 1 = N_{d,\min}^{[q]}(q^m, e) + 1 \leq N_{d,\min}^{[q]}(M, e) + 1.$$

Finally, for all integers M such that $(q+1)q^{m-1} < M < q^{m+1}$, we have

$$N_{\mathcal{G}}^{[q]}(M, e) \leq N_{\mathcal{G}}^{[q]}(q^{m+1}, e) = m + n + 2 = N_{d,\min}^{[q]}((q+1)q^{m-1}, e) + 1 \leq N_{d,\min}^{[q]}(M, e) + 1.$$

The proof is complete.

Acknowledgments

The authors are grateful to Rudolf Ahlswede for useful comments and helpful discussions.

References

1. M. Aigner, Searching with lies, J. Comb. Theory, Ser. A, 74, 43-56, 1995.
2. F. Cicalese and C. Deppe, Quasi-perfect minimally adaptive q-ary search with unreliable test, Proceedings of ISAAC, LNCS, Springer-Verlag, 527-536, 2003.
3. F. Cicalese and U. Vaccaro, Optimal strategies against a liar, Theoretical Computer Science, 230, 167-193, 2000.
4. F. Cicalese, D. Mundici, and U. Vaccaro, Least adaptive optimal search with unreliable tests, Theoretical Computer Science, Vol. 270, No. 1-2, 877-893, 2001.
5. I. Dumitriu and J. Spencer, The liar game over an arbitrary channel, preprint, 2003.
6. F.J. MacWilliams, and N.J.A. Sloane, The Theory of Error-Correcting Codes, North-Holland, Amsterdam, 1977.
7. A. Malinowski, K-ary searching with a lie, Ars Combinatoria, 37, 301-308, 1994.
8. S. Muthukrishnan, On optimal strategies for searching in presence of errors, Proc. of the 5th ACM-SIAM SODA, 680-689, 1994.
9. A. Pelc, Searching games with errors – fifty years of coping with liars, Theoret. Comput. Sci., 270, No. 1-2, 71–109, 2002.
10. A. Rényi, Napló az Információelméletről, Gondolat, Budapest, 1976. (English translation: A Diary on Information Theory, J.Wiley and Sons, New York, 1984).
11. A. Rényi, On a problem of information theory, MTA Mat. Kut. Int. Kozl. 6B, 505–516, 1961.
12. J. Spencer, Ulam's searching game with a fixed number of lies, Theoretical Comp. Sci., 95, 307-321, 1992.
13. S.M. Ulam, Adventures of a Mathematician, Scribner's, New York, 1976.
14. V.A. Zinoviev and G.L. Katsman, Universal code families, Problems of Information Transmission, 29, 2, 95-100, 1993.

Search with Noisy and Delayed Responses

R. Ahlswede and N. Cai

Abstract. It is well–known that search problems with a stochastic response matrix acting independently for the questions can be equivalently
formulated as transmission problems for a discrete memoryless channel
(DMC) with feedback.

This is explained in Chapter 3 of the book Search Problems by R.
Ahlswede and I. Wegener (Wiley 1987, translation of Suchprobleme,
Teubner 1979).

There also Ahlswede's coding scheme for the DMC and also for the
arbitrarily varying channel (AVC) achieving the capacities are described.
The latter can be viewed as a robust model for search.

In this paper we analyse this robust model with a **time delay** for
the noiseless feedback. In the terminology of search this means that the
answers are given with delay.

We determine the (asymptotically) optimal performances, that is,
find the capacities, for the cases where the **delay is constant** and
linear in the blocklength. Finally we also give the corresponding results for the DMC **with zero–error probability**.

Keywords: Search, noisy responses, liers, delay, feedback, list codes,
0–error capacity.

1 Introduction

Delay is an essential property in human interactions and especially also in engineering systems for instance those with control or communication aspects. There
have been already in the 70ties studies on delay and overflow in data compression
schemes (e.g. Wyner [22], Jelinek [19]). Recently searching with delayed answers
was considered by Cicalese, Gargano, and Vaccaro ([11], [12]) in the combinatorial model dealing with lies as considered by Renyi, Berlekamp, and Ulam (see
Deppe [16]).

It is well–known that feedback (even without delay) does not increase the
capacity of a DMC. This was first proved by Shannon [21], who also found a
formula for the zero error capacity of the DMC. This is a special case of the
result by Ahlswede [2] on the AVC for maximal probability of error, which was
completed by Ahlswede and Cai [7] omitting a convexity assumption and providing a condition for positivity of the capacity (a "trichotomy"). Here feedback
(without delay) increases the capacities. This is also true of the easier case of
average probability of error, which can be found in [4], [8].

Our first result (in Section 4) concerns the situation where the delayed feedback has a delay time upperbounded by a constant d. With a simple coding

R. Ahlswede et al. (Eds.): Information Transfer and Combinatorics, LNCS 4123, pp. 695–703, 2006.
© Springer-Verlag Berlin Heidelberg 2006

scheme for delayed feedback we show that in this case the capacities of all memoryless channels with non–delayed feedback can be achieved. Actually it is relevant here that we can do time sharing.

Next when the delay time increases linearly with the length of codes we obtain characterizations of the zero–error capacity of a DMC (Section 5) and the average–error capacity of an AVC (Section 6).

Finally we draw attention to future study of identification codes for the DMC, for which Ahlswede and Dueck [9] found the capacity in case of delayed feedback. It exceeds the capacity in the absence of feedback.

Furthermore we shall investigate the AVC for maximal probability of error, where feedback increases capacity.

In Section 2 we introduce the necessary notation and definitions. Results, on which this work is based are stated in Section 3.

2   Notation and Definitions

Let $\mathcal{K}$ be an abstract channel. Then we denote by $C_0(\mathcal{K})$, $C_{0,f}(\mathcal{K})$, and $C_{0,\ell}(\mathcal{K}, L)$ its zero–error capacity, zero–error capacity with feedback, and zero–error capacity for list codes with list size L. For a given arbitrarily varying channel $\mathcal{W}$, we denote by $C_{\mathcal{R}}(\mathcal{W})$, $C_a(\mathcal{W})$, $C_{a,f}(\mathcal{W})$, and $C_{a,\ell}(\mathcal{W}, L)$ its average–error capacities for random correlated codes (c.f. [4], [6], or [10], [14] for its definition), ordinary deterministic codes, codes with feedback and list codes with list size L.

Let $\mathcal{X}$ be our input alphabet and let $\mathcal{Y}$ be our output alphabet. We define a code with d time delayed noiseless feedback of length n, or shortly a d–feedback code of length n as a set of functions $\{f_m^{(n,d)} : m \in \mathcal{M}\}$ from $\mathcal{Y}^{n-d}$ to $\mathcal{X}^n$ such that for all $y^{n-d} = (y_1, y_2, \dots, y_{n-d}) \in \mathcal{Y}^{n-d}$,

$$f_m^{(n,d)}(y^{n-d}) = \left(f_{m,1}^{(d)}, f_{m,2}^{(d)}, \dots, f_{m,d}^{(d)}, f_{m,d+1}^{(d)}(y_1), f_{m,d+2}^{(d)}(y^2), \dots, f_{m,n}^{(d)}(y^{n-d}) \right), \tag{1}$$

where $y^i = (y_1, y_2, \dots, y_i)$ and $\mathcal{M}$ is a finite set corresponding to the set of messages. That is, for all $m \in \mathcal{M}$ $f_m^{(d)}$ is a vector valued function and its first d components are constant in $\mathcal{X}$, independent of y^{n-d} and for $t = d+1, d+2, \dots, n$, its t–th component $f_{m,t}^{(d)}$ is a function mapping y^{t-d} to $\mathcal{X}$. The information theoretical meaning is the following. At time t, the encoder sends a letter from the input alphabet $\mathcal{X}$ according to the value of the t–th component of the function $f_m^{(n,d)}$, if he wants to send the message m to the receiver (decoder), and at the same time the channel outputs a letter $y_t \in \mathcal{Y}$ according to the probabilistic rule given the channel $\mathcal{K}$ and the inputs. This output y_t arrives via a noiseless channel at the encoder at time $t + d$. Thus at time $t = 1, 2, \dots, d$, there is no feedback available, and so the encoder only can choose an input letter according to the message, which he wants to send. At time $d + 1$, the feedback starts to arrive at the encoder. At time $t = d + 1, d + 2, \dots, n$, the encoder has received the first $t - d$ outputs $y^{t-d} = (y_1, y_2, \dots, y_{t-d})$ and he may associate them with the message to choose the input letter.

We shall consider two cases. In one case the delay time d is a constant and in the other it increases linearly with the code length. It seems to us that the first is more meaningful from a practical point of view and that the second is more interesting from a mathematical point of view. The capacities are defined in the standard way and we denote them in the two cases by $C_f^{(d)}(\mathcal{K})$ and $\tilde{C}_f^{(\delta)}(\mathcal{K})$, respectively, where $\delta = \frac{d}{n}$ for the second case. Analogously the zero–error capacities in these two cases are denoted by $C_{0,f}^{(d)}$ and $\tilde{C}_{0,f}^{(\delta)}(\mathcal{K})$, respectively. These 4 capacities are non–increasing in d or δ. In particular, when $d = 0$, a d–delay feedback code is an ordinary feedback code. Similarly we define $\tilde{C}_{a,f}^{(\delta)}$ for AVC $\mathcal{W}$.

3 Known Results

In this section we report a few known results, which we will use in our proofs in Sections 5 and 6. The average–error capacity for AVC was determined by R. Ahlswede and the key tool in his proof is the following elimination technique.

Lemma 1. (Ahlswede [4]) *For an AVC $\mathcal{W}$, an integer $\gamma \in [n^2, \infty)$, any $\varepsilon, \bar{\lambda} > 0$ for sufficiently large n, there exists a random correlated code of length n assigned to a set of codes of cardinality γ with average probability of error smaller than λ and rate larger than $C_{\mathcal{R}}(\mathcal{W}) - \varepsilon$.*
Next there are results for list decoding.

Lemma 2. (Elias [17]) *Given a discrete memoryless channel W and denote by $C_{f,0}(W)$ its zero–error capacity with (non–delayed) feedback. Then for all $\varepsilon > 0$ and sufficiently large n, there exists a code with list decoding of list size $L = L(\varepsilon, W)$ (depending on ε, W but independent of n) such that the rate of code is larger than $C_{f,0}(W) - \varepsilon$.*

Lemma 3. (Ahlswede–Cai [6]) *For AVC $\mathcal{W}$, and $\varepsilon, \bar{\lambda} > 0$, there exists an $L = L(\varepsilon, \lambda, \mathcal{W})$ such that for all sufficiently large n (independent of L), there exists a code with average probability of error smaller than $\bar{\lambda}$ and rate larger than $C_{\mathcal{R}}(\mathcal{W}) - \varepsilon$.*
The previous lemma was remarkably improved.

Lemma 4. (Blinovsky–Narayan–Pinsker [10]) *For the AVC $\mathcal{W}$, there exists a constant $L = L(\mathcal{W})$ such that the average–error capacity for list codes with size of list L $C_{a,\ell}(\mathcal{W}, L) = C_{\mathcal{R}}(\mathcal{W})$.*

4 Codes with Delayed Constant Time Feedback

In this section we present a simple coding scheme to show that the constant time delay for feedback does not effect capacities or rate regions for all memoryless channels for regardless of the error concepts (zero–error, maximal–error, average error). This includes for instance arbitrarily varying channels, two way channels, multiple access channels, broadcast channels, interference channels and all other channels in the books [13] and [14], regardless whether their capacities or rate–regions are known or unknown! The coding scheme is based on the following two observations.

1. For memoryless channels the optimal rates of codes converge to capacity. That is, the rates of optimal codes are arbitrarily close to the capacity if the lengths of codes are sufficiently long. (For multi–user channels, all points in the capacity regions can be approached by codes of sufficiently long lengths.)
2. For memoryless channels output statistics at time t only depends on the inputs at time t (and the state at time t for arbitrarily varying channels).

For simplicity of notation we present the coding scheme for two terminal channels and leave its obvious extension to any multi–user channel to the reader.

Let $\{f_m^n : m \in \mathcal{M}\}$ be a code with non–delayed feedback of length n whose rate is close to the capacity. Now we construct a code $\big\{ f_{m^d}^{(nd,d)} : m^d = (m_1, m_2, \ldots, m_d)$ $\in \mathcal{M}^d \big\}$ by concatenating the code $\{f_m^n : m \in \mathcal{M}\}$ as follows. For $t = \tau n + i, 0 \leq \tau \leq d_1, 1 \leq i \leq n$ and $f_{m^d,t}^{(nd,d)}$ in (1) (i.e. the tth component of $f_{m^d}^{(nd,d)}$) we set $f_{m^d,t}^{(nd,d)} = f_{m_{\tau+1},i}$ for $f_m^n = (f_{m,1}, f_{m,2}, \ldots, f_{m,n})$.

Then the new code has the same rate as the original code with no time delayed feedback and by the observation 2 its probability of error is not larger than d times the probability of error of the original code.

Obviously the memorylessness assumption, which implies observation 2, is essential and in general the coding scheme cannot be applied to a channel with memory.

5   Zero–Error Capacity for a DMC with Linear Increasing Delay Time for Feedback

In this section we consider the case where for a $\delta \in [0, 1)$ the output of a given DMC at time t ($t \leq n - \lfloor n\delta \rfloor$) arrives at the encoder at time $t + \lfloor n\delta \rfloor$ via noiseless feedback.

Theorem 1. *For all DMC's W, and $\delta \in [0, 1)$,*

$$\tilde{C}_{f,0}^{(\delta)}(W) = (1 - \delta) C_{f,0}(W) + \delta C_0(W). \tag{2}$$

Proof

a) **Converse Part**

Let us consider zero–error codes for the following communication system and denote the capacity by $C_{f,0}^{*(\delta)}(W)$. An encoder sends messages to a decoder with a zero–error code with feedback of length n via W and for the time $t < n - \lfloor n\delta \rfloor$, the output of the channel at time t immediately arrives at the encoder via noiseless feedback and the feedback is shut down at time $n - \lfloor n\delta \rfloor$. Obviously the output of the channel at time $t \geq n - \lfloor n\delta \rfloor$ can never arrive at the encoder if the feedback delays $\lfloor n\delta \rfloor$ units of time. Therefore $\lfloor n\delta \rfloor$–feedback code may always be simulated by a code for the above communication system. Consequently

$$\tilde{C}_{f,0}^{(\delta)}(W) \leq C_{f,0}^{*(\delta)}(W). \tag{3}$$

On the other hand we may apply Shannon's well known approach in the converse proof of his zero–error coding theorem with feedback [21] and conclude that in the worst case the encoder may not reduce the messages to a list of size smaller than $M2^{-n(1-\delta)C_{0,f}(W)}$ by sending the first $n - \lfloor n\delta \rfloor$ components of the input, if the initial message set size is M. Since the feedback is shut down at time $n - \lfloor n\delta \rfloor$ and the decoder has to determine the message in the last $\lfloor n\delta \rfloor$ units of time without error, $M2^{-n(1-\delta)C_{f,0}(W)}$ must not be larger than $2^{n\delta C_0(W)}$. That is

$$C_{f,0}^{*(\delta)}(W) \leq (1 - \delta)C_{f,0}(W) + \delta C_0(W), \tag{4}$$

which together with (3) yields the converse.

b) **Direct Part**

We prove the direct part by the following coding scheme which consists of three blocks. Let ε be an arbitrarily small but positive constant.

1) Our first block of the coding scheme has length $n\left(1 - \delta - \frac{\varepsilon}{3}\right)$. By Lemma 2, there exists a constant L depending only on the channel and ε such that for sufficiently large n, there exists a zero–error code of length $n\left(1 - \delta - \frac{\varepsilon}{3}\right)$ and rate $C_{f,0}(W) - \frac{\varepsilon}{3}$ with list decoding of list size L. The encoder uses such a code in the first block. Then the decoder knows that the message falls in a list of size L after the transformation. But at this moment the encoder does not know the list and to learn the list he has to wait for the feedback.

2) The second block has length $\lfloor n\delta \rfloor$. During the time he is waiting for the feedback, the encoder may use a zero–error code of length $\lfloor n\delta \rfloor$ to sent $n\delta\left(C_0(W) - \frac{\varepsilon}{3}\right)$ bits to the decoder.

3) After $\lfloor n\delta \rfloor$ units of time, the outputs of the whole first block arrive at the encoder and so he learns the list. Now the time for the last block only leaves $n\frac{\varepsilon}{3}$ units. But it is sufficient, if n is sufficiently large, because the size of the list is a constant L. So the encoder may use a zero–error code of length $\frac{\varepsilon}{3}n$ to inform the decoder about the message in the list he sends to the decoder.

In the first two blocks the encoder sends $n\left(1 - \delta - \frac{\varepsilon}{3}\right)\left(C_{f,0}(W) - \frac{\varepsilon}{3}\right)$ bits and $n\delta\left(C_0(W) - \frac{\varepsilon}{3}\right)$ bits, respectively. So totally the rate of the code is

$$\left(1 - \delta - \frac{\varepsilon}{3}\right)\left(C_{f,0}(W) - \frac{\varepsilon}{3}\right) + \delta\left(C_0(W) - \frac{\varepsilon}{3}\right)$$
$$= (1 - \delta)C_{f,0}(W) + \delta C_0(W) - \frac{\varepsilon}{3}\left[C_{f,0}(W) + 1 - \frac{\varepsilon}{3}\right].$$

This completes the proof of the direct part.

6   Average–Error Capacity for an AVC with Linear Increasing Delayed Time Feedback

We have learnt from [8] that the average–error capacity of codes for an AVC with non–delayed feedback is equal to the capacity of random correlated codes and one cannot expect that a code with delayed feedback is better than a code with non–delayed feedback.

So for all AVC $\mathcal{W}$

$$\tilde{C}_{a,f}^{(\delta)}(\mathcal{W}) \leq C_{a,f}(\mathcal{W}) = C_{\mathcal{R}}(\mathcal{W}). \tag{5}$$

We shall show that $\tilde{C}_{a,f}^{(\delta)}$ actually is equal to $C_{\mathcal{R}}(\mathcal{W})$ and consequently the linear time delay makes no difference for feedback. That is

Theorem 2. *For all AVC,* $\delta \in [0,1)$

$$\tilde{C}_{a,f}^{(\delta)}(\mathcal{W}) = C_{\mathcal{R}}(\mathcal{W}). \tag{6}$$

In order to prove Theorem 2, it is sufficient for us to present a coding scheme asymptotically achieving $C_{\mathcal{R}}(\mathcal{W})$. Before presenting it, let us briefly review the idea in the proof of the direct part of the coding theorem for AVC [4]. R. Ahlswede first reduced the size of a domain of a random correlated code for an AVC to $O(n^2)$ by the elimination technique. Then the encoder may randomly choose a code from this domain and inform the decoder about this choice by a code of size $n^2 \lesssim 2^{no(1)}$ in the case that the channel has positive capacity. This gives a deterministic code, since we may regard the choice as a message sent to the decoder. In the case that the capacity of a channel is zero it still works if the sender and decoder have other resources to obtain common randomness e.g., the (non–delayed) feedback [8]. Along this line in our coding scheme the encoder should use an arbitrarily short block to generate randomness at the output. However the randomness does not arrive at the encoder before $\lfloor n\delta \rfloor$ units of time later. During the waiting time, the encoder may send more messages by a list code. To wait for the list via the feedback, the encoder needs another $\lfloor n\delta \rfloor$ units of time. So totally the waiting time is $2n\delta$. Consequently this naive coding scheme requires the assumption that $\delta < \frac{1}{2}$.

However, we observe that the common randomness is not necessarily to be generated by feedback, even if the capacity is zero. In fact the randomness can be sent by a (short) list code in the second block such that the encoder may use a correlated code in the next block until the outputs of the first two blocks come via feedback. Then he may use the common randomness generated by the first block and inform the decoder where the message sent in the second block locates in the list. So, totally the waiting time is around $n\delta$ units.

Proof of Theorem 2: Now we formally prove the theorem by the following coding scheme. For fixed $\varepsilon > 0$, we choose ε', $\eta_i > 0$, $i = 0, 1, 2$ such that

$$1 - \sum_{j=0}^{2} \eta_j > \delta, \tag{7}$$

where ε', η_0, η_2 are chosen arbitrarily small (depending on ε and will be specified later). Then we choose as set of messages a cartesian product

$$\mathcal{M} = \mathcal{M}_r \times \mathcal{M}_1, \tag{8}$$

with

$$|\mathcal{M}_r| = 2^{n\eta_1(C_{\mathcal{R}}(\mathcal{W})-\varepsilon')} \tag{9}$$

and

$$|\mathcal{M}_1| = 2^{n\left(1-\sum\limits_{j=0}^{2}\eta_j\right)(C_{\mathcal{R}}(\mathcal{W})-\varepsilon')}. \tag{10}$$

Then our code consists of four blocks.

1) The first block has length $n\eta_0$. We use it to build common randomness as in [8]. The difference is that in our case the common randomness is built $\lfloor n\delta \rfloor$ units of time later whereas it is built immediately in the model [8].
2) The second block has length $n\eta_1$. In this block the encoder sends message $m_r \in \mathcal{M}_r$ by a list code of constant list size, if the message, which he wants to send is $(m_r, m_1) \in \mathcal{M}_r \times \mathcal{M}_1 = \mathcal{M}$. By (9) and Lemmas 3 or 4, the code with arbitrarily small average probability of error exists.
3) The third block has length $n\left(1 - \sum\limits_{j=0}^{2}\eta_1\right)$. In this block the encoder uses a random correlated code with the domain (of the random code) $\mathcal{M}_r$ and rate $C_{\mathcal{R}}(\mathcal{W}) - \varepsilon'$. The existence of the code follows from Lemma 1. The encoder sends $m_1 \in \mathcal{M}_1$ to the decoder by the m_rth code in the domain $\mathcal{M}_r$, if he wants to send (m_r, m_1) to the decoder. He can do it by (10).
4) There are $n\eta_2$ units of time left for the last block. By (7) the outputs of the first two blocks have arrived at the encoder before the last block is started. So the common randomness generated by the first block has been built and the list of the code in the second block has come to encoding. The encoder can use a code obtained by the elimination technique of Lemma 1 to inform the decoder which message in the list he sends in the second block. Since the size of the list is a constant, η_3 can be chosen arbitrarily small.
5) The encoder knows the message m_r is in a list, but does not know which one it is at the end of the second block. Then he does nothing but waits for the end of the transmission. At end of the last block the decoder learns m_r from the last block. Then he knows which code is used in the third block and so he is able to decode m from the code in the third block. Finally he obtains (m_r, m_1).

The probability of error, clearly may be arbitrarily small and by (8) – (10) the rate is

$$\eta_1(C_{\mathcal{R}}(\mathcal{W})-\varepsilon')+\left(1 - \sum\limits_{j=0}^{2}\eta_j\right)(C_{\mathcal{R}}-\varepsilon') = [1-(\eta_0+\eta_1)](C_{\mathcal{R}}(\mathcal{W})-\varepsilon') < C_{\mathcal{R}}(\mathcal{W})-\varepsilon,$$

if we choose ε' and η_0, η_2 sufficiently small. So our proof is complete, since the converse trivially follows from (5).

Finally we note that although an algorithm of coding for feedback without delay based on [1] was studied in [20], it is a long way to find efficient algorithms of our coding schemes for AVC and zero–error codes with delayed feedback. This is so because our coding schemes contain correlated random codes and zero–error codes whose structures are unknown.

References

1. R. Ahlswede, A constructive proof of the coding theorem for discrete memoryless channels in case of complete feedback, Sixth Prague Conf. on Inf. Th., Stat. Dec. Fct's and Rand. Proc., 1–22, 1971.
2. R. Ahlswede, Channels with arbitrarily varying channel probability functions in the presence of noiseless feedback, Z. Wahrscheinlichkeitstheorie und verw. Geb. 25, 239–252, 1973.
3. R. Ahlswede, Channel capacities for list codes, J. Appl. Probability, lo, 824–836, 1973.
4. R. Ahlswede, Elimination of correlation in random codes for arbitrarily varying channels, Z. Wahrscheinlichkeitstheorie und verw. Geb. 44, 159–175, 1978.
5. R. Ahlswede and I. Wegener, Search Problems, Wiley–Interscience Series in Discrete Mathematics and Optimization, R.L. Graham, J.K. Leenstra, R.E. Tarjan, edit., English Edition of the German Edition Suchprobleme, Teubner Verlag, Stuttgart 1979, Russian Edition with Appendix by Maljutov 1981.
6. R. Ahlswede and N. Cai, Two proofs of Pinsker's conjecture concerning arbitrarily varying channels, IEEE Trans. Inform. Theory, Vol. 37, 1647–1679, 1991.
7. R. Ahlswede and N. Cai, The AVC with noiseless feedback and maximal error probability: a capacity formula with a trichotomy, in Numbers, Information and Complexity, special volume in honour of R. Ahlswede on occasion of his 60th birthday, edited by Ingo Althöfer, Ning Cai, Gunter Dueck, Levon Khachatrian, Mark S. Pinsker, Andras Sárközy, Ingo Wegener and Zhen Zhang, Kluwer Academic Publishers, Boston, Dordrecht, London, 151–176, 2000.
8. R. Ahlswede and I. Csiszár, Common randomness in information theory and cryptography II, CR. capacity, IEEE Trans. Inform. Theory, Vol. 44, 225–240, 1998.
9. R. Ahlswede and G. Dueck, Identification in the presence of feedback — a discovery of new capacity formulas, IEEE Trans. on Inf. Theory, Vol. 35, No. 1, 30–39, 1989.
10. V.M. Blinovsky, P. Narayan, and M.S. Pinsker, The capacity of arbitrarily varying channels for list decoding, Problemy Peredachii Informatsii, Vol. 31, 3–41, 1995.
11. F. Cicalese and U. Vaccaro, Coping with delay and time–outs in binary search procedures, (ISAAC 2000) D.T. Lee and Shang–Hua Teng eds, Lectures Notes in Comp. Sci., Vol. 1969, 96–107, Springer 2000.
12. F. Cicalese, L. Gargano and U. Vaccaro, On searching strategies, parallel questions and delayed answers, in. Proc. of Fun. with Algorithms (FUN01), E. Lodi, L. Pagli and N. Santoro eds., 27–42, Carleton Scientific Press, 2001.
13. T.M. Cover and J.A. Thomas, Elements of Information Theory, John Wiley, 1991.
14. I. Csiszár and J. Körner, Information Theory: Coding Theorems for Discrete Memoryless Systems, Academic, 1981.
15. I. Csiszár and P. Narayan, The capacity of the arbitrarily varying channel revised: positivity constraints, IEEE Trans. Inform. Theory, Vol. 34, 81–193, 1988.
16. C. Deppe, Solution of Ulam's search game with three lies or optimal adaptive strategies for binary three–error–correcting–codes, Dissertation, Universität Bielefeld, 1998; Discrete Math. 224, No. 1-3, 79–98, 2000.

17. P. Elias, Zero error capacity under list decoding, IEEE Trans. Inform. Theory, Vol. 34, 1070–1074, 1988.
18. T.H.E. Ericson, Exponential error bounds for random codes in the arbitrarily varying channel, IEEE Trans. Inform. Theory, Vol. 31, 42–48, 1985.
19. F. Jelinek, Buffer overflow in variable length coding of fixed rate sources, IEEE Trans. Inform. Theory, Vol. IT 14, No. 3, 490–501, 1968.
20. J.M. Ooi and G.W. Wornell, Fast iterative coding for feedback channel, ISIT 1997, 133, Ulm Germany, 1997.
21. C.E. Shannon, The zero error capacity of a noisy channel, IEEE Trans. Inform. Theory, Vol. 2, 8–19, 1956.
22. A.D. Wyner, On the probability of buffer overflow under an arbitrary bounded input–output distribution, SIAM J. Appl. Math, Vol. 27, No. 4, 544–569, 1974.

A Kraft–Type Inequality for d–Delay Binary Search Codes

R. Ahlswede and N. Cai

1 Introduction

Among the models of delayed search discussed in [1], [2], the simplest one can be formulated as the following two–player game. One player, say A, holds a secret number $m \in \mathcal{M} \triangleq \{1, 2, \ldots, M\}$ and another player, say Q, tries to learn the secret number by asking A at time i questions, like "Is $m \geq x_i$?", where x_i is a number chosen by Q. The rule is that at time $i + d$ A must answer Q's question at time i correctly and at time j Q can choose x_j according to all answers he has received. How many questions has Q at least to ask to get the secret number. Let

$$
B_d(t) = \begin{cases} 1 & \text{if } t \leq 0 \\ B_d(t-1) + B_d(t-d-1) & \text{if } t > 0. \end{cases} \tag{1}
$$

Then the main result of [1] is

Theorem AMS. (Ambainis–Bloch–Schweizer) *There exists a scheme for Q to win the game by asking t questions iff $M \leq B_d(t)$.*

We notice that the answers are determined by Q's scheme and the secret number, since A does not lie. So for a fixed scheme, for Q wining by asking t questions, each number $m \in \{1, 2, \ldots, M\} = \mathcal{M}$ gives a binary sequence of length at most t in such a way that the ith component of the sequence is zero iff the answer is "yes" if the secret number is m. Thus all successful schemes for Q define a subset in $\{0,1\}^* \triangleq \bigcup_{i=1}^{\infty} \{0,1\}$ and we shall call them d–delay binary search (d–DBS) codes. Then Theorem ABS can be restated: there exists a d–DBS code C whose codewords have at most length t iff

$$
|C| \leq B_d(t). \tag{2}
$$

For a given d–DBS code we denote by $\ell(c)$ the length of codeword c. Then $\{\ell(c) : c \in C\}$ must satisfy the Kraft inequality, because a d–DBS code has to be prefix free. However a prefix code is not necessarily a d–DBS code. The main result of the paper is a sharper Kraft–type inequality for d–DBS codes based on the work [1]. The inequality is stated and proved in the next section.

R. Ahlswede et al. (Eds.): Information Transfer and Combinatorics, LNCS 4123, pp. 704–706, 2006.
© Springer-Verlag Berlin Heidelberg 2006

2   The Inequality

Main Inequality:
For all d-DBS codes C,

$$\sum_{c \in C} B_d^{-1}(\ell(c)) \leq 1. \tag{3}$$

Lemma 1. *Let C be a d–DBS code and let L be an integer such that $\ell(c) \leq L$ for all $c \in C$, then*

$$\sum_{c \in C} B_d\big(L - \ell(c)\big) \leq B_d(L). \tag{4}$$

Proof: Originally we got the idea to prove the lemma from [1], and the result follows from Theorem ABS, and the following extension of code C. Let $|C| = M$, let S be the scheme corresponding to C on $\{0, 1, \ldots, M - 1\}$, and let

$$M^* = \sum_{c \in C} B_d\big(L - \ell(c)\big).$$

It is sufficient for us to present a successful scheme for Q to win the game by L queries if the secret number is in $\{1, 2, \ldots, M^*\}$. Let c_j be the codewords given by secret number j in scheme S and $\ell(c_j)$ be its length.

Then the scheme with L queries on $\{0, 1, \ldots, M^* - 1\}$ can be defined as follows.

1. Let $b_m = \sum_{j=0}^m B_d\big(L - \ell(c_j)\big)$ for $m = 0, 1, \ldots, M - 1$.
2. Q first simulates the scheme S. That is, Q asks "$\geq b_m$?" whenever in S "$\geq m$?" is asked, until a $j \in \{0, 1, \ldots, M - 1\}$ is found by S. In this case Q knows the "secret number" $m \in \{b_j, b_j + 1, \ldots, b_{j+1} - 1\}$. This takes $\ell(c_j)$ queries.
3. Next Q uses a scheme with $\big(L - \ell(c_j)\big)$ questions achieving $B\big(L - \ell(c_j)\big) = |\{b_j, b_j + 1, \ldots, b_{j+1} - 1\}|$ to find the "secret number" m. □

Lemma 2

$$B_d(\ell_1)B_d(\ell_2) \geq B_d(\ell_1 + \ell_2). \tag{5}$$

Proof: We proceed by induction on $\min(\ell_1, \ell_2)$ and w.l.o.g. assume $\ell_1 \leq \ell_2$.

Case $\ell_1 \leq 0$
LHS of (5) $= B_d(\ell_2) \geq B_d(\ell_2 - |\ell_1|) = B_d(\ell_1 + \ell_2)$, where "$\geq$" holds because by (1) B_d is non–decreasing.

Case $\ell_1, \ell_2 > 0$
Assume (4) holds for all $\min(\ell_1', \ell_2') < \ell_1 < \ell_2$.

LHS of (5) $= B_d(\ell_1)B_d(\ell_2)$

$$\overset{(i)}{=} \big(B_d(\ell_1 - 1) + B_d(\ell_1 - d - 1)\big)\big(B_d(\ell_2 - 1) + B_d(\ell_2 - d - 1)\big)$$
$$= B_d(\ell_1 - 1)B_d(\ell_2 - 1) + B_d(\ell_1 - 1)B_d(\ell_2 - d - 1)$$
$$+ B_d(\ell_1 - d - 1)B_d(\ell_2 - 1) + B_d(\ell_1 - d - 1)B_d(\ell_2 - d - 1)$$
$$\overset{(ii)}{\geq} B_d(\ell_1 + \ell_2 - 2) + 2B_d(\ell_1 + \ell_2 - d - 2) + B_d(\ell_1 + \ell_2 - 2d - 2)$$
$$= \big[\big(B_d(\ell_1 + \ell_2 - 1) - 1\big) + B_d\big((\ell_1 + \ell_2 - 1) - d - 1\big)\big]$$
$$\big[B_d\big((\ell_1 + \ell_2 - d - 1) - 1\big) + B_d\big((\ell_1 + \ell_2 - d - 1) - d - 1\big)\big]$$
$$\overset{(iii)}{\geq} B_d(\ell_1 + \ell_2 - 1) + B_d(\ell_1 + \ell_2 - d - 1)$$
$$\overset{(iv)}{\geq} B_d(\ell_1 + \ell_2),$$

where (i) holds by (1), (ii) holds by the induction hypothesis, and (iii) holds, because by (1) we have for all t $B_d(t) \leq B_d(t - 1) + B_d(t - d - 1)$.   □

Apply Lemma 2 to $\ell_1 = \ell(c)$ and $\ell_2 = L - \ell(c)$ for all $c \in C$, then we obtain

$$B_d\big(L - \ell(c)\big) \geq B_d^{-1}\big(\ell(c)\big)B_d(L). \tag{6}$$

Substituting (6) by (4) we get

$$\sum_{c \in C} B_d^{-1}\big(\ell(c)\big)B_d(L) \leq \sum_{c \in C} B_d\big(L - \ell(c)\big) \leq B_d(L)$$

i.e., (3).

References

1. A. Amboinis, S.A. Bloch, and D.L. Schweizer, Delayed binary search, or playing twenty questions with a procrastinator, Algorithmica, 32, 641-650, 2002.
2. F. Cicalese and V. Vaccaro, Coping with delays and time–outs in binary search proceduresm, Lectures Notes in Computer Science, Vol. 1969, Springer, 96–107, 2000.

Threshold Group Testing

P. Damaschke

Abstract. We introduce a natural generalization of the well-studied group testing problem: A test gives a positive (negative) answer if the pool contains at least u (at most l) positive elements, and an arbitrary answer if the number of positive elements is between these fixed thresholds l and u. We show that the p positive elements can be determined up to a constant number of misclassifications, bounded by the gap between the thresholds. This is in a sense the best possible result. Then we study the number of tests needed to achieve this goal if n elements are given. If the gap is zero, the complexity is, similarly to classical group testing, $O(p \log n)$ for any fixed u. For the general case we propose a two-phase strategy consisting of a DISTILL and a COMPRESS phase. We obtain some tradeoffs between classification accuracy and the number of tests.

1 Introduction

The classical version of the group testing problem is described as follows. In a set of n elements, p elements are *positive* and the other $n-p$ are *negative*. (These terms can stand for any specific property of elements. The "positive" elements are sometimes called "defective" in the group testing literature.) We denote by P the set of positive elements, hence $p = |P|$. Typically p is much smaller than n. A *group test* takes as input any set S of elements, called a *pool*. The test says YES if S contains at least one positive element (that is, $S \cap P \neq \varnothing$) and NO otherwise. The goal is to identify the set P by as few as possible tests.

Group testing is of interest in chemical and biological testing, DNA mapping, and also in several computer science applications. Many aspects of group testing have been studied in depth. One cannot even give an overview of the vast literature. Here we refer only to the book [10] and a few recent papers [7,11,12]. Many further pointers can be found there. Group testing also fits in the framework of learning Boolean functions by membership queries, as it is equivalent to the problem of learning a disjunction of p unknown Boolean variables. In [5,6] we proved a number of complexity results for learning arbitrary Boolean functions with a limited number of relevant variables.

In the present paper we study a generalization of group testing which is quite natural but has not been addressed before, to our best knowledge. Let l and u be nonnegative integers with $l < u$, called the *lower* and *upper threshold*, respectively. Suppose that a group test for pool S says YES if S contains at least u positives, and NO if at most l positives are present. If the number of positives in S is between l and u, the test can give an arbitrary answer. We suppose that l and u are *constant* and previously known. (If one is not sure about the thresholds, one can conservatively estimate l too low and u too high.)

R. Ahlswede et al. (Eds.): Information Transfer and Combinatorics, LNCS 4123, pp. 707–718, 2006.
© Springer-Verlag Berlin Heidelberg 2006

The obvious questions are: What can we figure out about P? How many tests and computations are needed? Can we do better in special cases? We refer to our problem as *threshold group testing*. We call $g := u - l - 1$ the (additive) *gap* between the thresholds. The gap is 0 iff a sharp threshold separates YES and NO, so that all answers are determined. Obviously, the classical case of group testing is $l = 0$, $u = 1$.

It should be observed that the lower sensitivity of threshold group tests compared to the classical case has some consequences that one may find puzzling first. In particular, one cannot simply test single items and thus identify the positives with n trivial tests. Instead one has to examine subsets (details follow in the technical part). However, the fact that an obvious strategy from a special case is no longer applicable is not an objection against the model itself. Anyway, our results will provide sublinear strategies.

In another generalization of group testing, the sensitivity of tests is reduced if too many negatives in S are present, i.e. if the positives are "diluted". This scenario has been studied in [4]. Algorithmically it turned out to be less interesting, in that a straightforward strategy has an asymptotically optimal number of tests. One could think of the following model that combines both aspects. Elements are samples with different concentrations of a substance to detect, and there is an upper threshold for sure detection and a lower threshold under which detection is impossible. Between the thresholds, the outcome of a test is arbitrary. These assumptions seem to be natural in some chemical test scenarios, and the problems are worth studying. This model allows verification of single "positive" elements where concentration is above the upper threshold. It is more difficult to say which other positive elements can be found efficiently (e.g. by "piggybacking" with high-concentration samples), and to what extent unknown concentrations can be estimated by thresholding. Note also that one can create tests with different amounts from different samples, i.e. presence of a sample in a pool is no longer a binary property. Perhaps combinations of techniques from [4] and the present paper are required. These questions are beyond the scope of the present article and are left for future research.

Threshold group testing as introduced here is perhaps not a suitable model for chemical testing, however one can imagine applications of a different nature that justify our model. The n elements may represent distinct "factors" that can be either present or absent: If a number of relevant factors (the positives) are present, some event (YES answer) will be observed. For example, some disorders appear only if various bad factors come together, and they will appear for sure if the bad factors exceed some limit. The relevant factors can have complex interactions and different importance, so that the outcome does not solely depend on their number, but on some monotone property of subsets. (To mention an example, there could be several types of positives, and the result is positive if positive elements of each type are present.) Now, one may wish to learn what all these risk factors are, among the n candidates. If one can create experiments where the factors can be arbitrarily and independently turned on and off (as in some knock-out experiments in cell biology), one can also use these experiments as pools in a search strategy. Thus, threshold group testing strategies may find interest in such settings, although we studied the problem mainly as a nice extension of group testing, without a concrete application in mind.

Threshold group testing is also related to another search problem, called "guessing secrets", that recently received attention due to a surprising application in Internet routing [1,9]. Here the adversary knows a set X of secrets and answers YES if all secrets are in the query set S, NO if S and X are disjoint, and arbitrarily in all other cases. Hence, this is threshold group testing in the special case $l = 0$ and $u = p$.

Overview of the paper: In Section 2 we show that the p positives can be found efficiently, subject to at most g wrongly classified items, which is inevitable in the worst case. Still, for $p \gg g$ this means a small relative error. The rest of the paper revolves round the question how many tests and computational steps are needed to achieve this best possible classification. As a first result, the computation time is bounded by $O(p^u n^{g+1})$, with the thresholds in the hidden constant. In Section 3 we discuss the case $g = 0$ and show that the asymptotic test complexity does not exceed that of classical group testing, even if no auxiliary pool with a known number of positives is available. The idea of the strategy is then enriched in Section 4 to treat the general and more realistic case $g > 0$. The main result is that the asymptotic number of tests can be made linear in p times a power of n with an arbitrarily small positive exponent (at cost of the constant factor), and computations need polynomial time. It is hard to say how far this is from optimal, but it means significant progress compared to the trivial bound of $\binom{n}{u}$ tests. Sections 5 and 6 are merely supplements and directed towards further research. We point out some possibilities to modify the algorithms and accelerate them further. Open questions are also mentioned in earlier sections when they arise.

It will become apparent that the algorithms do not need a bound on p, that is, the results hold even if the number of positives is not known in advance.

In this paper we consider only sequential test strategies. In applications with time-consuming tests it can be desirable to work in a limited number of stages where a set of tests is performed nonadaptively, i.e. without waiting for the results of each other. The outcomes of all previous stages can be used to set up the tests for the next stage. For some problems, nonadaptiveness is achievable only at cost of a higher test complexity. These tradeoffs have been investigated for various combinatorial search problems, among them classical group testing [7,10] and learning relevant variables [5,6]. Nontrivial results in this direction for threshold group testing would be interesting, too.

Notational remarks: Throughout the paper, a k-set means a set of k elements. A k-subset is a k-set being subset of some other set which is explicitly mentioned or clear from context.

We chose to present complicated bounds containing binomial coefficients in the form of powers and factorials. One might argue that resolving the binomial coefficients is not elegant, however one can easier see the dependency on the actual input parameters n and p in this way.

2 Fuzzy Identification of the Positives

First we show that we can always get close to the true set P of positives in a certain sense, using much fewer than 2^n tests (that would be needed for testing all possible pools).

Theorem 1. *If $p \geq u + g$, the seeker can identify a set P' with $|P' \setminus P| \leq g$ and $|P \setminus P'| \leq g$. For $p < u + g$, we still have the error bound $u - 1$ rather than g.*

Proof. Simply test all u-sets. Let P' be any set of alleged positives that is consistent with the answers. That means: For every pool that contained at most l members of P', the test said NO, and for every pool that contained at least u (and thus exactly u) members of P', the test said YES.

Suppose for the moment that P' has cardinality $p' \geq u$. Assume that P' has $g + 1 = u - l$ (or more) elements outside P. Add l other elements from P'. This u-set has at most l elements in P, thus it must return NO. But since it consists of u elements from P', it should also say YES, contradiction.

Similarly, suppose for the moment $p \geq u$, and assume that $g + 1$ (or more) elements of P are not in P'. Take them and add l other elements from P. This u-set is entirely in P and therefore says YES. But since it contains at most l elements from P', it should also say NO, contradiction.

For $p \geq u$ we have seen that $p' \geq p - g$. In particular, $p \geq u + g$ implies $p' \geq u$, and both directions apply. The last assertion is also obvious now. $\square$

Some lower bound on p is necessary to guarantee that the sizes of set differences are bounded by g, even if we test all 2^n subsets. For example, if $p = u - 1$ and the adversary always says NO, every $(u - 1)$-set P' is a consistent solution, even if P' and P are disjoint. But we have $u - 1 > g$ for $l > 0$. In the following we always suppose $p \geq u + g$. Our bound g on the number of misclassified elements on both sides is tight in the following sense:

Proposition 1. *For any fixed set P' with $|P' \setminus P| \leq g$ and $|P \setminus P'| \leq g$, the adversary can answer as if P' were the set of all positives.*

Proof. For pools S with $l < |S \cap P| < u$, the adversary answers arbitrarily, in particular, consistently with P'. It remains to show that the mandatory answers for all other pools S do not rule out P'. If S has at most l positives then the answer must be NO. Since S contains at most $l + g = u - 1$ elements from P', this answer is possible for P' as well. If S has at least u positives then at least $u - g = l + 1$ of them are in P'. Thus, answer YES is also allowed for P'. $\square$

We stress that this result refers only to any single solution P'. It is much more difficult to characterize the families of hypotheses P' that are consistent with some vector of answers. We leave this as an open question. However, we can say at least something about these families:

Proposition 2. *Any two solutions Q and Q' being consistent with the answers of an adversary satisfy $|Q \setminus Q'| \leq g$ and $|Q' \setminus Q| \leq g$.*

Proof. Suppose that some inequality is violated. Then, if $Q = P$, Q' cannot be a candidate. But $Q = P$ is impossible means that Q is not a candidate either.   □

Clearly, this statement is not specific to our problem. It holds similarly for any search problem where the consistent hypotheses are close to the target, with respect to some distance measure.

As an illustration we discuss what Proposition 2 means in the case $l = 0$, $u = 2$. Every candidate solution P' can be characterized by the "deviation" $\{x, x'\}$, with unique elements $x \in P \setminus P'$ and $x' \in P' \setminus P$. (One of the set differences may be empty. Then x or x' does not exist, and the deviation is a singleton set.) For deviations $\{x, x'\}$ and $\{y, y'\}$ of any two solutions, not all of x, x', y, y' can exist and be different. Hence the solution space can be represented as a star graph. Either one positive is undetected and several (maybe all) negatives are suspicious to be the only positive that is missing, or one negative is not absolved, but several alleged positives are candidates for the missing negative.

The solution spaces for $u > 2$ could be more complicated. A particularly interesting question is under what conditions a positive or negative element can be recognized as such.

Our next concern is to compute a set P' with the properties mentioned in Theorem 1 from the $\binom{n}{u}$ answers efficiently, that is, to "invert the answers". Let us call a set Q an *affirmative set* if all u-subsets of Q answered YES. Certainly, P is affirmative, as well as every subset of P. On the other hand, an affirmative set Q cannot exceed P very much: $|Q \setminus P| \leq g$, otherwise we get an obvious contradiction. This suggests the strategy to establish increasing affirmative sets, until the second condition $|P \setminus Q| \leq g$ is also satisfied. The somewhat tricky question is how to reach this situation quickly and to recognize it although P is unknown.

Theorem 2. *For $p \geq u + g$ one can find some set P' as in Theorem 1 in time* $O(\frac{u^{g+1}}{(u-1)!(g+1)!} p^u n^{g+1})$.

Proof. As our initial Q we may choose any u-set that answered YES.

Assume that we have an affirmative Q with $|P \setminus Q| > g$. There exists some $(g+1)$-set G with $Q \cap G = \varnothing$, and a set H of size at most g, such that $(Q \cup G) \setminus H$ is affirmative (and by at least one element larger than Q). To see the existence, just notice that any $G \subseteq P \setminus Q$ with $G = |g + 1|$ and $H = (Q \cup G) \setminus P$ would do. Here we use the fact that at most g members of Q are outside P.

In order to find a pair G, H with the desired property, we may try all $\binom{n}{g+1}$ candidates for G. For any fixed G we can find a suitable H, if it exists, by the following observation: H must be a hitting set for the family of u-subsets of $Q \cup G$ that said NO.

We bound the number of such sets. Since Q is affirmative, at least one element must be in G, these are $g + 1$ possibilities. Each of the other $u - 1$ elements can be in Q or be one of g elements of G. An upper bound on $|Q|$ is provided by $|Q \setminus P| \leq g$ which implies $|Q| \leq p + g$. As long as $|P \setminus Q| > g$ (true in every step except the last), this improves to p. Hence the size of our set family is less than

$\frac{g+1}{(u-1)!}(p+g)^{u-1}$. By a trivial branching algorithm, some hitting set of cardinality at most g can be found in $O(\frac{u^{g+1}}{(u-1)!}(p+g)^{u-1})$ time.

In the worst case, Q has to be augmented about p times. If, for some Q, every G fails, then at most g positives are not yet in Q, and we can output $P' := Q$.   □

There may be room for improving the time bound, mainly because the method above considers all G separately. On the other hand it seems that the number p of augmentation steps cannot be reduced substantially, since we might include up to g new negatives in Q in every step. A more ambitious question is how much computation is needed to output (some concise representation of) the complete solution space for the given answers.

We resume that, for any fixed thresholds, we can approximate P with a polynomial number of tests and in polynomial computation time, subject to a constant number of misclassified elements. But in general we will not be able to identify P exactly. The practical interpretation is: For any element in P' we have evidence that it might be a positive and should be investigated more closely, although there might be a few false positives in P'. Symmetrically, any randomly picked element not in P' is innocent with high probability, although all these elements remain suspicious, in the worst case. This might be already acceptable for applications. Also recall that a less malicious adversary (e.g. with random answers between the thresholds) which is more appropriate to model "nature" in experiments can only give more safety.

3   The Case Without Gap

In this section we study the case $g = 0$, that is, $l+1 = u$. Theorem 1 implies that we can identify P exactly. Our preliminary bound $\frac{n^u}{u!}$ on the number of queries and the complexity bound in Theorem 2 are however not very satisfactory, because the fact $g = 0$ should help. (However, we will use this preliminary result later, inside a more efficient strategy for general g.)

Suppose that we can create a pool with $u - 1$ positives. Then a simple trick (from e.g. [8]) reduces the problem to classical group testing: Take $u-1$ positives with you and add them to every pool. Then a pool says YES iff at least one further positive is present.

Still, it is interesting to ask what we can do if such an auxiliary pool is not available. For potential applications it is not clear whether one can always prepare a pool with $u - 1$ known positives right away. Instead, we first have to identify $u - 1$ positives by group tests. Another reason to address the problem is that our procedure for detecting $u - 1$ positives will also provide the idea for attacking the more complicated and more realistic case $g > 0$ in the next section.

Our approach is as follows: First test the whole set. If it says YES, split it in two halves and test them. If at least one half say YES, split it, and continue recursively. (We ignore the other half for the moment.) This way we get smaller and smaller sets that contain at least u positives. This trivial process stops if, at some point, both halves say NO.

In order to continue we need set families with a special covering property: For the moment let t, r, v be any integers with $1 \le t \le r \le v$. On a set of v elements we want a family of r-sets such that every t-set is subset of some member of this family. A lower bound on the size (number of sets of) such families is $\binom{v}{t}/\binom{r}{t}$, and for any constant t, r and large v it is known that families within a $1 + o(1)$ factor of this bound exist (the Rödl nibble, see e.g. [2]). However we cannot exploit this result here, as we need large r. Therefore we use other, rather simple covering families. More material on this kind of structure can be found e.g. in [3].

Back to a set of, say, m elements that contains at least u positives. We split our set into $u + 1$ parts of roughly equal size. Then we test the $u + 1$ pools consisting of u of these parts. Clearly, this is a covering family, i.e. we find a pool which contains at least u positives and therefore says YES. Clearly, by iterated application we can reduce m to some constant size depending on u only, using $(u + 1) \log_{(1+1/u)} m = \frac{u+1}{\ln(1+1/u)} \ln m = O(u^2 \log m)$ tests. Finally we find u positives by exhaustively testing the u-subsets. It follows:

Lemma 1. *In a set of m elements, at least u of them positives, we can identify u positives by $O(u^2 \log m)$ tests.* □

It would be interesting to close the gap to the information-theoretic lower bound $O(u \log m)$. We remark that covering families may be used right from the beginning in our procedure. Starting with binary search does not improve the worst-case bound, but it is easier as long as we get YES answers. Moreover, we may split the set in halves at random, so that u positives are in one half with high probability, as long as the actual number of positives is considerably bigger than u.

Now, take $u - 1$ of the u known positives and use them as auxiliary pool in an ordinary group testing strategy. Since $O(p \log n)$ is a query bound for group testing (see e.g. [10]), we can now formulate the final result of this section.

Theorem 3. *Threshold group testing with $g = 0$ has asymptotic query complexity no worse than $O((p + u^2) \log n)$.* □

4   Gaps Are Gulfs

Things become more complicated if there is a gap between the thresholds. We attempt to extend the idea above to the case $g > 0$. For convenience we say that we (u, a)-*cover* an m-*set* when we test all pools from a family of m/a-sets such that every u-set is contained in some pool. Here a can be any real number greater than 1. Besides the additive gap $g = u - l - 1$ between the thresholds, we also use the multiplicative gap h defined by $h := \lceil u/(l + 1) \rceil$.

In the previous section we have used $a = 1 + 1/u$, but in the following we need an a that is large compared to u. By splitting an m-set in au subsets of roughly equal size and testing any combination of u of these subsets, we obviously get an (u, a)-covering of size $\binom{au}{u} = O(e^u a^u)$, where $e = 2.718\ldots$ is Euler's number. Thus we can formulate:

Lemma 2. *Any m-set has an (u, a)-covering of size $O((ea)^u)$, independently of m. Provided that at least u positives are present, the covering can be used to find an m/a-subset with still at least u positives, by $O((ea)^u)$ tests.*   □

At least one pool in the covering gives a YES answer, however the catch is that, conversely, a YES implies only that $l + 1$ or more positives are in that pool. We cannot immediately recognize a pool with at least u positives and then continue recursively to narrow down the candidate set. This situation suggests a relaxation of the previous method.

We call a set with at least $l + 1$ positives a *heavy* set. Suppose that we have $k \leq h$ disjoint heavy m-sets that together contain at least u positives. (Note that in case $k = h$, the union has guaranteed u positives, but a smaller $k \leq h$ can be sufficient in a concrete case.) In the union of our k heavy m-sets, we find a heavy m/a-set due to Lemma 2, by $O((eka)^u)$ tests. (Just replace m with km, and a with ka.) We put this heavy subset aside and repeat the procedure with the remaining elements, as long as it generates new disjoint heavy m/a-sets. This procedure is limited to sets with $m \geq au$, since the sets in the covering family must have size at least u. We summarize the procedure in

Lemma 3. *From $k \leq h$ heavy m-sets containing a total of at least u positives, we can obtain disjoint heavy m/a-sets and a remainder with fewer than u positives. The number of tests is $O((eha)^u)$ for each of these heavy m/a-sets.*   □

This is the building block of an algorithm parameterized by a, that "distills" positive elements in heavy sets of decreasing size. We give a high-level description.

DISTILL:

At any moment, we maintain a collection of disjoint heavy sets of cardinality at most n/a^i, where i can be any non-negative integer, plus the set R of remaining elements, which is not necessarily heavy. A set with at most n/a^i but more than n/a^{i+1} elements is said to be on level $i \geq 1$. Set R is said to be on level 0. For i with $n/a^{i+1} < u \leq n/a^i$ we define the last level $(i + 1)$ where only sets of cardinality exactly u are allowed.

The initial collection consists of only one set R on level 0, containing all elements. We apply two kinds of operation:

(i) (u, a)-cover the set on level 0 and move disjoint heavy subsets of it to level 1, as long as possible.

(ii) Take the union U of h heavy sets on level i, apply the procedure from Lemma 3 to U, and move the generated disjoint heavy subsets of U to level $i + 1$. Finally move the rest of U back to R. If there are only $k < h$ sets left on level i, apply Lemma 3 to their union.

Clearly, these operations preserve disjointness of the collection, and all sets except perhaps R are heavy. The order of applying these operations is arbitrary. The process stops only when the collection has fewer than u positives on each level, since otherwise (ii) is still applicable. As the number of levels is bounded by $\log_a n$, all but $(u-1)\log_a n$ positives are eventually on the last level, where at least $l+1$ positives are in each u-set. (For the last level of u-sets, the procedure of Lemma 3 can be easily adjusted.)

For the following analysis we rewrite parameter a as $a = n^b$, where $b > 0$.

Theorem 4. DISTILL *needs* $O(\frac{e^u h^u}{l+1} h^{1/b} p n^{ub})$ *tests and packs all positives except* $\frac{u-1}{b}$ *in disjoint u-sets, such that at least $l+1$ positives are in each of these u-sets.*

Proof. It remains to analyze the number of tests. Consider any one of the, at most $\frac{p}{l+1}$, bunches of positives in the final u-sets. It has been extracted from at most h heavy sets on the previous level, by at most $O((eha)^u)$ tests due to Lemma 3. They in turn have been extracted from h^2 heavy sets, etc. Since only $\log_a n$ levels exist, all this has been done by $O(h^{\log_a n})$ applications of (i) and (ii). Hence, after $O(\frac{e^u h^u}{l+1} p a^u h^{\log n/\log a})$ tests, all positives except $(u-1)\log n/\log a$ are in disjoint u-sets. Finally set $a = n^b$. $\square$

Despite the simplicity of this analysis, the bound has a quite pleasant form: The first factor consists of constants, the dependency on p is linear, and we can arbitrarily reduce the exponent of n by choosing a small b, of course at cost of an increasing factor $h^{1/b}$ and the loss of more positives.

More seriously, DISTILL stops with a fraction of positives which is only guaranteed to be at least $1/h$ in the final u-sets. (Recall that we only know of the presence of $l+1$ positives in each u-set.) In order to increase the guaranteed density of positives beyond this limit, we invoke, in a second phase called COMPRESS, the naive algorithm from Theorem 1, but now applied to the heavy u-sets only. Since their total size is at most hp rather than n, the bounds from Theorems 1 and 2 imply immediately:

Corollary 1. *After* $O(\frac{e^u h^u}{l+1} p h^{1/b} n^{ub} + \frac{h^u}{u!} p^u)$ *tests, all positives except* $\frac{u-1}{b} + g$ *are in a set that contains at most g negatives. The amount of computation is polynomial.* $\square$

It seems possible to reduce the number of tests to $O(p\log n)$ for fixed thresholds l, u, using $O(\log n)$ levels and a potential function argument in the analysis, but we suspect that any $O(p\log n)$ algorithm would also misclassify $O(\log n)$ positives, which is of no value e.g. if $p = O(\log n)$.

5   Some Refinements

This more informal section will briefly discuss some refinements of the main result (Section 4). The aim is just to illustrate that our techniques have more potential. However, since optimality of the underlying algorithm is not known, we do not give a full treatment of these more sophisticated strategies here, and stress the ideas only.

In the case of large p, the p^u term in Corollary 1 may be prohibitive. An obvious modification of COMPRESS resolves the problem: Apply Theorem 1 to groups of t

heavy u-sets, with some parameter t. This yields a tradeoff between final density of positives and test complexity. We can create from any t heavy u-sets a new set that contains all but g of the positives contained in the t heavy sets, and at most g negatives, by $O(\frac{u^u}{u!}t^u)$ tests. The density of positives is now at least $1 - \frac{g}{t(l+1)}$. Application to all our heavy u-sets costs $\frac{u^u}{(l+1)u!}pt^{u-1}$ tests. However we lose up to $\frac{gp}{t(l+1)}$ further positives. We might also iterate the two-phase algorithm above, in order to catch more positives: The $\frac{gp}{t(l+1)}$ positives that escaped from COMPRESS can be collected in a set of size $(h-1)p$. Then the whole algorithm may be executed (with suitable parameters) on this smaller set, etc.

Next we sketch a method for reducing the exponents in the test complexity bounds from u to $g + 1$. The idea is to "shift the thresholds". As already mentioned in Section 3, if we knew already $q \leq l$ positives, we could inject them as an *auxiliary set* in every pool and reduce the threshold group testing problem for l and u to the same problem for thresholds $l - q$ and $u - q$. (Utilizing $q > l$ positives in an auxiliary set is useless, since the adversary may always say YES, so that the searcher gets no information at all.) Specifically, we propose a modification of DISTILL with improved test complexity for $l > 0$. To explain the improvement we have to refer to details of DISTILL.

First we run DISTILL in a depth-first manner, that is, we move some positives to the last level as quickly as possible. Similarly as in the proof of Theorem 4, it costs at most $h^{u+1/b}n^{ub}$ tests to produce a heavy u-set. Then we try all $\binom{u}{l}$ l-subsets as an auxiliary set. At least one of them contains exactly l positives, so that we can use it to shift the thresholds to 0 and $u - l = g + 1$, while running the original version of DISTILL on the remaining elements. The new upper threshold $g+1$ reduces the exponent of n, on the other hand we only incur another constant factor $O(\frac{u^l}{l!})$. Note that we have at least one positive in each of the final $(g+1)$-sets. Since we cannot see in advance which of the $\binom{u}{l}$ auxiliary sets has really l positives, we must find a good output among the $\binom{u}{l}$ offered results. For this purpose, we may test unions of u of the final $(g + 1)$-sets. If there is in fact one positive in every $(g+1)$-set, then all answers must be YES. Conversely, the YES answers guarantee that at least $l + 1$ positives are among the $u(g + 1)$ elements in every such pool, which means a density of positives of at least $\frac{l+1}{u(g+1)}$. Since this is a constant, the resulting set has size $O(p)$.

Finally we apply a COMPRESS phase on this set. By shifting the thresholds, the exponent of p can be reduced to $g + 1$ as well. This could work as follows. Apply the algorithm from Theorem 1 to the $O(p)$ size candidate set, but with upper threshold $g + 1$, and with help of each of the $\binom{u}{l}$ auxiliary sets. Again we have to find a good P' among the offered solutions. It suffices to rule out sets P' with large $d := |P' \setminus P|$. Then, among the surviving candidates P', a set with largest cardinality is close to P, up to constant differences (since we know that such a P' is among our candidates). Let us call a set Q *dismissive* if all $(g + 1)$-subsets of Q answer NO, together with a fixed auxiliary set. Note that $P' \setminus P$ is a dismissive set of size d, with respect to any auxiliary set, as they contain at most l positives. On the other hand, there is an auxiliary set (namely one with exactly l

positives), such that every dismissive subset of P' can have at most $d+g$ elements. Hence, by computing maximum dismissive subsets of P' from the test results, we can identify one which is close to P, with differences of constant size.

As a consequence, the positive elements can be identified subject to constantly many misclassifications by $O(pn^{(g+1)b} + p^{g+1})$ tests, with an arbitrarily small constant $b > 0$.

6 Logarithmic Distill Phase for Threshold Two

A difficulty in the DISTILL phase in Section 4 is that an uncontrollable number of positives, but fewer than u, remain in U after every application of step (ii). We simply sent them back to level 0. However if we knew more about the number of positives in a heavy set, we might proceed more carefully and avoid returns, using more clever pool sets than just covering families. This section presents a partial result in this direction: a DISTILL method for $l = 0$, $u = 2$ that needs only $O(p \log n)$ tests. Together with threshold shifting this may even lead to improvements for $g = 1$ and general u. We also conjecture that similar test sets can be created for any fixed g.

Proposition 3. *For $l = 0$ and $u = 2$, all positives but one can be collected in a set of size at most $2p$ by $O(\log n)$ tests.*

With some constant $a > 1$, we start with $(2, a)$-covering the whole set. As long as we get at least one YES, we continue recursively on heavy sets of decreasing size. If this process goes through, we eventually obtain a heavy 2-set with one or two positives by $O(\log n)$ tests. Otherwise we know that the heavy set considered last has *exactly* one positive. We repeat the procedure, always starting with all elements that are not yet in the heavy sets.

This is nothing else than depth-first DISTILL with constant parameter a. The new idea comes now: Whenever we have two heavy sets A and B, each containing exactly one positive, we can even identify one of the positives by $O(\log n)$ further tests as follows. Split A and B into sets of roughly half size, denoted A_1, A_2, B_1, B_2. Test the pools $A_i \cup B_j$ for all $i, j \in \{1, 2\}$. If the positives are w.l.o.g. in A_1 and B_1 then $A_1 \cup B_1$ answers YES, and $A_2 \cup B_2$ answers NO. If the other two pools give the same answer, i.e. both YES or both NO, the seeker concludes that the positives must be in A_1 and B_1. If they give different answers, w.l.o.g. $A_1 \cup B_2$ say YES and $A_2 \cup B_1$ says NO, then $A_1 \cup B_2$ is another candidate. In either case, A_1 surely contains a positive, and we can discard A_2. The recursive process stops when one of A and B is a singleton.

We miss at most one positive, in case that a single heavy set is left over.   □

Acknowledgments

This work was inspired by and profited a lot from discussions with Ugo Vaccaro, Ferdinando Cicalese, and Annalisa De Bonis during the author's stay at the Dipartimento di Informatica ed Applicazioni "R.M. Capocelli", Università di Salerno (Baronissi), Italy. Support from The Swedish Research Council (Vetenskapsrådet), project title "Algorithms for searching and inference in genetics", file no. 621-2002-4574, is also acknowledged. Thanks to the anonymous referees for careful reading and valuable suggestions.

References

1. N. Alon, V. Guruswami, T. Kaufman, and M. Sudan, Guessing secrets efficiently via list decoding, 13th SODA, 254-262, 2002.
2. N. Alon, J. Spencer, The Probabilistic Method, Wiley, 1992.
3. C.J. Colbourn and J.H. Dinitz (eds.), The CRC Handbook of Combinatorial Designs, CRC Press, 1996.
4. P. Damaschke, The algorithmic complexity of chemical threshold testing, 3rd CIAC, LNCS 1203, 205-216, 1997.
5. P. Damaschke, Adaptive versus nonadaptive attribute-efficient learning, Machine Learning, 41, 197-215, 2000.
6. P. Damaschke, On parallel attribute-efficient learning, J. of Computer and System Sciences, 67, 46-62, 2003.
7. A. De Bonis, L. Gasieniec, and U. Vaccaro, Generalized framework for selecuors with application in optimal group testing, 30th ICALP, LNCS, 2719, 81-96, 2003.
8. A. De Bonis and U. Vaccaro, Improved algorithms for group testing with inhibitors, Info. Proc. Letters, 66, 57-64, 1998.
9. F. Chung, R. Graham, and F.T. Leighton, Guessing secrets, Electronic J. on Combinatorics, 8, 1-25, 2001.
10. D.Z. Du and F.K. Hwang, Combinatorial Group Testing and its Applications, 2nd edition, World Scientific, 2000.
11. E.H. Hong and R.E. Ladner, Group testing for image compression, IEEE Transactions on Image Proc., 11, 901-911, 2002.
12. H.Q. Ngo and D.Z. Du, A survey on combinatorial group testing algorithms with applications to DNA library screening, DIMACS Series Discrete Math. and Theor. Computer Science, 55, AMS, 171-182, 2000.

A Fast Suffix-Sorting Algorithm

R. Ahlswede, B. Balkenhol, C. Deppe, and M. Fröhlich

1 Introduction

We present an algorithm to sort all suffixes of $x^n = (x_1, \ldots, x_n) \in \mathcal{X}^n$ lexicographically, where $\mathcal{X} = \{0, \ldots, q-1\}$. Fast and efficient sorting of a large amount of data according to its suffix structure (suffix-sorting) is a useful technology in many fields of application, front-most in the field of Data Compression where it is used e.g. for the Burrows and Wheeler Transformation (BWT for short), a block-sorting transformation ([3],[9]).

Larsson [4] describes the relationship between the BWT on one hand and suffix trees and context trees on the other hand. Then Sadakane [8] suggests a well referenced method to compute the BWT more time efficiently. Then the algorithms based on suffix trees have been improved ([6],[5],[1]).

In [3] it was observed that for an input string of size n, this transformation can be computed in $O(n)$ time and space[1] using suffix trees. While suffix trees are considered to be greedy in space – even small factors hidden in the O-notation may decide on the feasibility of an algorithm – sorting was accomplished by alternative non-linear methods: Manber and Myers [7] introduced an algorithm of $O(n \log n)$ in worst case time and $8n$ bytes of space and in [2] an algorithm based on *Quicksort* is suggested, which is fast on the average but its worst case complexity is $O(n^2 \log n)$. Most prominent in this case is the Bendson-Sedgewick Algorithm which requires $4n$ bytes and Sadakane's example of a combination of the Manber-Myers Algorithm with the Bendson-Sedgewick Algorithm with a complexity of $O(n \log n)$ worst case time using $9n$ bytes [8].

The reduction of the space requirement due to an upper bound on n seems trivial. However, it turns out that it involves a considerable amount of engineering work to achieve an improvement, while retaining an acceptable worst case time complexity. This paper proposes an algorithm, efficient in the terms described above, ideal for handling large blocks of input data. We assume that the cardinality of the alphabet (q) is smaller than the text-string (n). Our algorithm computes the suffix sorting in $O(n)$ space and $O(n^2 \log n)$ time in the worst case. It has also the property that it sorts the suffixes lexicographically according to the prefixes of length $t_2 = \log_q \lceil \frac{n}{2} \rceil$ in the worst case in linear time. After the initial sorting of length t_2, we use a Quick-sort-variant to sort the remaining part. Therefore we get the worst time $O(n^2 \log n)$. It is also possible to modify our algorithm by using Heap-sort. Then we will get a worst case time $O(n(\log n)^2)$.

---

[1] This only holds, if the space complexity of a counter or pointer is considered to be constant (e.g. 4 Bytes) and the basic operations on them (increment, comparison) are constant in time. This assumption is common in the literature and helpful for practical purposes.

R. Ahlswede et al. (Eds.): Information Transfer and Combinatorics, LNCS 4123, pp. 719–734, 2006.
© Springer-Verlag Berlin Heidelberg 2006

We use Quick-sort, because it is better in practice and has an average time of $O(n \log n)$ like Heap-sort, but with a smaller factor.

The elements of $\mathcal{X}$ are called *symbols*. We denote the symbols by their *rank* w.r.t. the order on $\mathcal{X}$. We assume that $\$ = q - 1 \in \mathcal{X}$ is a symbol not occurring in the first $n - 1$ symbols in x^n, the *sentinel* symbol.

x_i is the ith element in the sequence x^n. If $i \leq j$, then $(x_i, \ldots, x_j)$ is the factor of x^n beginning with the ith element and ending with the jth element. If $i > j$, then $(x_i, \ldots, x_j)$ is the empty sequence. A factor v of x begins at position i and ends at position j in x if $(x_i, \ldots, x_j) = v$. To conveniently refer to the factors of a sequence, we use the abbreviation x_i^j for $(x_i, \ldots, x_j)$.

2   The Initial Sorting Step

Before we tackle the problem of sorting all suffixes of a given sequence in lexicographical order we start to consider the case where we only sort the suffixes looking at the prefixes of a fixed length correctly. The simplest case is to look at all prefixes of length one, which is the case to sort all symbols occurring in the input sequence lexicographically.

2.1   Sorting of the Symbols

The sorting of the symbols of a given input sequence x^n with symbols out of a finite alphabet $\mathcal{X}$ can be done linearly in time and space complexity as follows:

We define q counters $(counter_0[0], \ldots, counter_0[q-1])$ and count for each symbol in $\{0, \ldots, q-1\}$ how often it occurs in x^n. In each step i we have to increase exactly one counter $(counter_0[x_i])$ by one. Therefore to get the frequencies of the symbols requires $O(n)$ operations. Now our alphabet is given in lexicographic order and we generate the output in the following way: First output $counter_0[0]$ many zeros, followed by $counter_0[1]$ many ones,... Obviously the generated output sequence is produced in $O(n)$ operations and the sorting is done.

2.2   Sorting a Given Prefix Length

We would like to continue the sorting of all suffixes in an iterative way by using the counting idea of the previous section. In a later step of the algorithm we need n counters. We have to take the memory already at the beginning, which allows us to use it already in the initial sorting phase. We choose t_1 such that $2^{t_1-1} < q \leq 2^{t_1}$ and t_2 such that $2^{t_1 t_2} \leq \lfloor \frac{n}{2} \rfloor < 2^{t_1(t_2+1)}$. For simplicity we assume from now on that $q = 2^{t_1}$ and $n = 2^{t_1 t_2 + 1}$.

We like to sort all suffixes such that the first t_2 symbols of each suffix are sorted lexicographically correctly.

Now we will count the number of occurrences of factors of length t_2 in our sequence x^n. We assume that $x_{n+1}, \ldots, x_{n+t_2-1} = q - 1$ and count the factors as follows. The counter$[a_1 k^{t_2-1} + a_1 k^{t_2-2} + \cdots + a_{t_2} k^0]$ counts the number of occurrences of the factor $(a_1, \ldots, a_{t_2})$. Let us define a temporary value $tmp = 2^{t_1 t_2} - 1$ and $i = n$. This is the position n of the sequence, with the factor

$(q-1, \ldots, q-1)$. Now starting at the end down to the beginning of the input sequence x^{n+t_2-1} in each step we increase $counter[tmp]$ by one, decrease i by one and we calculate:

$$tmp \rightarrow \left\lfloor \frac{tmp}{2^{t_1}} \right\rfloor + x_i 2^{t_1(t_2-1)}.$$

Notice that multiplications and divisions by powers of two can be represented by **shifts**. Let us denote

$$a >> b = \left\lfloor \frac{a}{2^b} \right\rfloor \text{ and } a << b = a2^b.$$

Furthermore notice that the $+$ operation can be replaced by a binary logical or-operation which we denote as $|$. Hence in total we need $O(n)$ operations.

By construction tmp will only take values less than $\lfloor \frac{n}{2} \rfloor = (n >> 1)$, such that we can calculate the partial sums of the entries $counter[j]$ and store them in the second half of the memory for the array $counter$

$$counter[\frac{n}{2}+j] \rightarrow \sum_{i=0}^{\frac{n}{2}-1} counter[i].$$

Obviously this calculation can also be done linearly in time:

```
i->1
counter[(n>>1)]->0
while i< (n>>1) do
    counter[(n>>1)+i] -> counter[(n>>1)+i-1] + counter[i-1]
    i-> i+1
done
```

Finally we have to write back the result of the sorting. In order to continue we introduce two further arrays of size n, one, which we denote as *pointer*, in order to describe the starting points of the suffixes, and the second one, denoted as *index*, to store the partial results of the sorting.

Again we start with $tmp = 2^{t_1 t_2} - 1$ and at position $i = n$.

```
while i>n-t_2 do
    i->i-1
    tmp->(tmp>>t_1)|x_i<<t_1(t_2-1)
    counter[tmp]->counter[tmp]-1
    index[i]->counter[tmp+(n>>1)]+counter[tmp] ;
    pointer[index[i]]->i;
done
while i>0 do
    i->i-1
    tmp->(tmp>>t_1)|x_i<<t_1(t_2-1)
    counter[tmp]->counter[tmp]-1
    index[i]->counter[tmp+(n>>1)]
    pointer[index[i]+counter[tmp]]->i
done
```

In the first loop we consider the cases where we have to take the sentinel into account (we assume that $x_{n+i} = \$$). With the starting definition of tmp the sentinel will be taken as a number greater or equal to $|\mathcal{X}| - 1$. Using the fact that it occurs only at the end of the sequence, that is with the largest entry of $count$, we can fix the position of the last t_2 entries to the starting-point of the prefix of suffixes, represented as integer tmp at that moment, plus the number of occurrences of that value tmp. In all other cases (second loop) we set $index[i]$ to the starting position of the interval of suffixes with prefix tmp.

In other words after these loops $pointer[1], \ldots, pointer[n]$ represent the starting positions of the suffixes in lexicographical order according to the prefixes of length t_2. If $index[pointer[i]] < index[pointer[j]]$ ($index[pointer[i]] > index[pointer[j]]$) then the suffix starting in $pointer[i]$ is lexicographically smaller (larger) than the suffix starting in position $pointer[j]$. If the two values are equal, then the two suffixes have a common prefix of length greater or equal to t_2.

Notice that to finish the lexicographic order in total we can continue using the two arrays $pointer$ and $index$ only, that is there is no need to look at the original input sequence to calculate the defined total order, such that the continuation is independent of the alphabet size.

3   Only Three Elements

In order to continue the sorting we first analyze how to sort and how to calculate the median of three given numbers.

3.1   Median-Position-Search of Three Elements

The median m of a triple $(n_1, n_2, n_3) \in \mathbb{N}_0^3$ is a value equal to at least one of them which is in between the two others, i.e.

$$m = n_1 \Rightarrow n_2 \leq n_1 \leq n_3 \text{ or } n_3 \leq n_1 \leq n_2,$$

$$m = n_2 \Rightarrow n_1 \leq n_2 \leq n_3 \text{ or } n_3 \leq n_2 \leq n_1,$$

$$m = n_3 \Rightarrow n_2 \leq n_3 \leq n_1 \text{ or } n_1 \leq n_3 \leq n_2.$$

Notice that we are not interested in the value itself, only in the position relative to the two others, i.e. for us there is no difference between the case $(1, 1, 1)$ and $(2, 2, 2)$. Therefore we partition the set of triples in the following way. We define 13 subsets $\mathcal{A}_1, \ldots, \mathcal{A}_{13} \subset \mathbb{N}_0^3$ in the following way: For $k \in \mathbb{N}_0$ and $l, m \in \mathbb{N}$ we define

$$\mathcal{A}_1 = \{(k, k, k)\} \quad \mathcal{A}_8 = \{(k, k + l, k + l + m)\}$$

$$\mathcal{A}_2 = \{(k, k, k + l)\} \quad \mathcal{A}_9 = \{(k, k + l + m, k + l)\}$$

$$\mathcal{A}_3 = \{(k, k + l, k)\} \quad \mathcal{A}_{10} = \{(k + l, k, k + l + m)\}$$

$$\mathcal{A}_4 = \{(k + l, k, k)\} \quad \mathcal{A}_{11} = \{(k + l, k + l + m, k)\}$$

$$\mathcal{A}_5 = \{(k, k + l, k + l)\} \quad \mathcal{A}_{12} = \{(k + l + m, k, k + l)\}$$

$$\mathcal{A}_6 = \{(k + l, k, k + l)\} \quad \mathcal{A}_{13} = \{(k + l + m, k + l, k)\}$$

$$\mathcal{A}_7 = \{(k + l, k + l, k)\}.$$

For a given triple (n_1, n_2, n_3) the median is known to us, if we know the index i with $(n_1, n_2, n_3) \in \mathcal{A}_i$. Therefore we define the following questionnaire of yes–no–questions where a question is of the following form: $a \leq b, a < b, a = b$.

```
if n_1 <= n_2 then
    if n_2 <= n_3 then m=n_2
    else if n_1 <= n_3 then m=n_3
        else m=n_1
        endif
    endif
else
    if n_3 <= n_2 then m=n_2
    else if n_1 <= n_3 then m=n_1
        else m=n_3
        endif
    endif
endif
```

Notice that we need at most three yes-no-questions and we need only two in case where the median is already in the middle.

3.2 Sorting of Three Elements

Using questions of the form mentioned in the previous section we can sort three elements using at most four questions:

```
if n_1 <= n_2 then
    if n_2 <= n_3 then
        if n_1 = n_2 then
            if n_2 = n_3 then (n_1,n_2,n_3) in A_1
            else (n_1,n_2,n_3) in A_2
            endif
        else
            if n_2 = n_3 then (n_1,n_2,n_3) in A_5
            else (n_1,n_2,n_3) in A_8
            endif
        endif
    else
        if n_1 <= n_3 then
            if n_1 = n_3 then (n_1,n_2,n_3) in A_3
            else (n_1,n_2,n_3) in A_9
            endif
```

```
     else
        if n_1 = n_2 then (n_1,n_2,n_3) in A_7
        else (n_1,n_2,n_3) in A_11
        endif
     endif
   endif
else
   if n_1 > n_3 then
     if n_2 = n_3 then (n_1,n_2,n_3) in A_4
     else if n_2 < n_3 then  (n_1,n_2,n_3) in A_12
          else  (n_1,n_2,n_3) in A_13
          endif
     endif
   else
     if n_1 = n_3 then  (n_1,n_2,n_3) in A_6
     else  (n_1,n_2,n_3) in A_10
     endif
   endif
endif
```

4   The Main Loop of the Sorting Algorithm

After the initial sorting phase we have the array *pointer*, which points to the
starting positions of the suffixes lexicographically correctly sorted according to
the prefixes of length t_2. *index* contains the partial ordering, that is if the values
are different, then the larger one is lexicographically larger than the smaller one,
if they are equal then the two suffixes have a common prefix of length greater or
equal to t_2. Finally we can calculate with the second half of the array *counter*
the positions of the intervals with common prefixes of length t_2. We use now
counter[0] to count the number of intervals where we have to continue with the
sorting, more precisely *counter*[0] points to the first free place in memory where
we can store a further interval, which is in the beginning 1 (*counter*[1] is free).

```
counter[0]->1
```

Starting the loop to get the not necessarily correctly sorted intervals *counter*[0]
is initialized with 1 because we need it in this way later and we are working on
"unsigned int".

```
i->0
while i< 2^(t_1*t_2) do
  if counter[(n>>1)+i+1]-counter[(n>>1)+i]>1 then
      counter[counter[0]+1]->counter[(n>>1)+i];
      counter[counter[0]+2]->counter[(n>>1)+i+1]-1
      while index[pointer[counter[counter[0]+2]]]>
            index[pointer[counter[counter[0]+1]]] do
            counter[counter[0]+2]->counter[counter[0]+2]-1
```

```
      done
      if counter[counter[0]+2]!=counter[counter[0]+1] then
         counter[0]->counter[0]+2
      endif
   endif
done
```

Notice that during the loop we reuse the memory in *counter* from $(n \gg 1)$ to n.

4.1   Split an Interval

We have to sort an interval from position *begin* to *end* that is *pointer*[*begin*] to *pointer*[*end*] has to be sorted but they are already of equal length *length*. We like to do the sorting by a 3 part quick-sort. The array 'smaller' contains all pointers which are smaller than the first entry (smaller defined by *index* !), the array 'equal' the pointers which are equal in the first 2*length* positions with the first one and the array 'bigger' the remaining ones. After we have split this part we have to continue with 'smaller' and 'bigger' of length *length* and with 'equal' of length $2 \cdot length$. These intervals (starting point, end point) we return to the calling function using two arrays x and y.

Given a value *val*, the index for the interval stored in *counter* at positions *counter*[*val* − 1] and *counter*[*val*], the value *length* which is the length of the common prefix already known from the previous steps (after the initial sorting it is t_2) and a flag *flag* which describes whether the intervals are stored at the beginning of counter or at the end (after the initial part at the beginning).

Now the beginning of the interval is given by *begin* = *counter*[*val* − 1] and the end position by *end* = *counter*[*val*]. Notice that the last *length* pointers of the original sequence can not occur inside this interval because they are correctly inserted in one of the previous steps due to the (virtual) sentinel symbol at the end of the input sequence. Therefore if we look at the suffixes starting at *pointer*[*begin*] and *pointer*[*end*], then we know they have by construction a common prefix of length at least *length*. But if we look at the two suffixes without the prefix of length *length*, then theses two suffixes have been sorted correctly also according to the prefix of length *length*. In other words the result of the comparison of the two pointers *pointer*[*begin*] and *pointer*[*end*] is equal to the result of the comparison of *pointer*[*begin*]+*length* and *pointer*[*end*]+*length*. We can get the result by using the values stored in the array *index*. Let us denote that *a* is lexicographic smaller than *b* with $a \prec b$ for two pointers a, b where a pointer is smaller than another one if the corresponding suffix starting at that pointer is lexicographic smaller than the other suffix. Then

$$pointer[begin] \prec pointer[end] \Leftrightarrow$$

$$pointer[begin] + length \prec pointer[end] + length.$$

Therefore if now $index[pointer[begin] + length] = index[pointer[end] + length]$ then the suffixes starting at *pointer*[*begin*] and *pointer*[*end*] have a common prefix of length at least $2 \cdot length$. Otherwise we can use the result to get the right comparison result. Notice that in this way we double the length of the comparison in each step.

Now for a given interval we like to split the interval into several parts similar to quick-sort. Therefore we take three values and calculate the median as mentioned in Section 3.1

```
n_1->index[pointer[begin]+length];
n_2->index[pointer[(begin+end)>>1]+length];
n_3->index[pointer[end]+length];


median-> (n_1 <= n_2 ?
          (n_2 <= n_3 ? n_2 : (n_1 <= n_3 ? n_3 : n_1 ) )
        : (n_3 <= n_2 ? n_2 : (n_1 <= n_3 ? n_1 : n_3 )))
```

With $currentindex = index[pointer[begin]]$ we have the value of $index[pointer[i]]$ for all $begin \le i \le end$. Now we like to split the interval into three parts, one for the pointers which are smaller than the *median* one for those which are equal and one for those which are larger. We divide the parts by changing the values of the pointers as follows:

First we need two further variables which we set to *begin* and *end* respectively.

```
s->begin
b->end
```

And we need yet another variable k for the actual position inside the interval. As long as the values of $index[pointer[k] + length] < median$ and $k \le b$ the current end of the interval we increase k by one:

```
k->begin;                /* the starting point */
while index[pointer[k]+length]<median && k<=b do
    k->k+1
done
s->k;
```

We set s to the actual value of k such that s points to the first position which is greater or equal to the *median*. In a similar way we reduce b at the end, give the first pointer which is less or equal than the *median*.

```
while index[pointer[b]+length]>median && k<=b do
    b->b-1
done
```

Remember that we have stopped the first loop in a case where

$$index[pointer[k] + length] \ge median$$

and the second one where

$$index[pointer[b] + length] \le median.$$

Now let us continue in the following way:

```
if index[pointer[k]+length]>median then
  SWAPPOINTER(k,b)
  b->b-1
```

where we denote with $SWAPPOINTER(k, b)$ the following operations:

$$tmp- > pointer[k] \quad pointer[k]- > pointer[b] \quad pointer[b]- > tmp$$

such that the two values are simply exchanged. Now we have that $index[pointer[k] + length] \leq median$ and we continue:

```
if index[pointer[k]+length]=median then
  k->k+1
  while index[pointer[k]+length]=median do
    k->k+1
  done
  else
    k->k+1 s->s+1
  endif
else
  k->k+1
  while index[pointer[k]+length]=median && k<=b do
    k->k+1
  done
endif
```

Now if $s > begin$ then the part from $begin$ to $s - 1$ stores the pointers which are smaller than the $median$ and if $b < end$ then the part from $b + 1$ to end are the pointers which are larger than the $median$. Furthermore if $s < k$ then the part from s to $k - 1$ are pointers which are equal to the $median$. Let us first continue with the case where $s = k$:

```
if s=k then
  s->end+1    /* we make the value impossible, in other   */
              /* words larger then end                     */
  while k<=b && s>end do
    if index[pointer[k]+length]<median then
      k->k+1  /* one further pointer which is smaller      */
    else
      if index[pointer[k]+length]>median then
        SWAPPOINTER(k,b);
        b->b-1 /* add to bigger interval                   */
      else
        s->k   /* s is getting a value <= end and          */
               /* the loop stops.                          */
        k->k+1 /* they are equal                           */
```

```
        endif
      endif
    done
  endif
```

Now we have found at least one pointer which is equal to the median. We have to continue similarly as before but if $index[pointer[k] + length] < median$ then we have to exchange in addition the pointers in positions k and s and we have to increase also s. Furthermore the only stop situation for the loop occurs if $k > b$.

```
while k<=b do
  if index[pointer[k]+length]<median then
    SWAPPOINTER(k,s);
    k->k+1
    s->s+1
  else
    if index[pointer[k]+length]>median then
      SWAPPOINTER(k,b);
      b->b-1 /* add to bigger */
    else
      k->k+1 /* they are equal */
    endif
  endif
done
```

Now we have the three parts

$$begin, \ldots, s - 1, \text{ the pointers which are smaller}$$

$$s, \ldots, b, \text{ the pointers which are equal}$$

$$b + 1, \ldots, end, \text{ the pointers which are larger.}$$

If $s - 1 < begin$ or $b + 1 > end$ then the corresponding intervals are empty. In order to use these parts in the future, we have to update the values of index for the current pointers. Notice that equal to the *median* means that they have a common prefix of a length at least $2 \cdot length$.

For the first interval (if it exists) nothing has to be done, because the values of *index* are already at the starting point of the interval. The new starting point of the second part is

```
currentindex->currentindex+s-begin
```

Of course the second part contains at least one pointer by construction (the pointer which is used to calculate the median has a common prefix to itself !).

```
if s>begin && s<=b then
  k->s
  while k<=b do
```

```
    index[pointer[k]]->currentindex;
    k->k+1
  done
endif
```

Finally we have to calculate the starting point of the last interval (if it exists)

```
currentindex->currentindex+b+1-s;
```

```
if b+1<=end then
  k->b+1
  while k<=end do
    index[pointer[k]]->currentindex
    k->k+1
  done
endif
```

Now we have to continue with our sorting algorithm on the constructed intervals. But before we start to consider the interval from s to b of length $2 \cdot length$ we like to finish all intervals of length $length$ in order to double the compared lengths of the prefixes again. For that reason we store that interval at the opposite end of the array $counter$ on which we are working at the moment. After the initial part we are working at the beginning to store our intervals, such that we store the interval from s to b at the end. After we have finished all intervals which we have to compare of length $length$ we start to work at the end with the intervals sorted correctly of length $2 \cdot length$ and store all intervals we produce of length $4 \cdot length$ at the beginning. Notice that the total number of intervals we have to store is always less than n such that if we need more space at the end it is free at the beginning of the array $counter$ and vice versa. To add these intervals we define a function $INSTOCOUNTER(FROM, TO, FLAG)$ where $FROM,TO$ are the boundaries of the interval which we have to add and $FLAG$ describes where. If we are working at the end of $counter$ we use $counter[n]$ similarly to $counter[0]$ for the beginning part. To delete one interval at the end we have to increase $counter[n]$ such that we need two different rules to add an interval at the end:

```
INSTOCOUNTER(FROM,TO,FLAG) {
  switch(FLAG) {
  case 0: {
    counter[counter[0]]->(FROM);
    counter[0]->counter[0]+1
    counter[counter[0]]->(TO);
    counter[0]->counter[0]+1
    break;
  }
  case 1: {
    counter[n]->counter[n]-1
    counter[counter[n]]_>(TO);
```

```
counter[n]->counter[n]-1
counter[counter[n]]->(FROM);
break;
  }
default: { /* case 2 */
  counter[counter[n]]->(FROM);
  counter[counter[n]+1]->(TO);
  counter[n]->counter[n]-2;
  break;
  }
  }
}
```

Now the insertion of the intervals using the function $INSTOCOUNTER$ can be done as follows:

```
if s-begin>1 then
  INSTOCOUNTER(begin,s-1,2-(flag<<1))
endif
if b-s>0 then
  INSTOCOUNTER(s,b,flag)
endif
if end-b>1 then
  INSTOCOUNTER(b+1,end,2-(flag<<1))
endif
```

4.2   Calling the Sorting Procedure

To conclude the description of the whole algorithm it remains to describe the step between the initial sorting phase and the calling of the procedure to split a given interval.

We are starting in a situation where we have given the three arrays *counter*, *pointer* and *index* and we know, that if we use the values stored in *index* as rule for the comparison of two pointers then the result is correct according to the first t_2 symbols (from the initial sorting part).

As mentioned earlier we like to use the array *counter* from both sides. At the beginning we use a variable *length* which describes the length of the common prefix correctly sorted. This variable is initialized with t_2 from the initial sorting phase. In order to double the length in each loop we have to use the information stored in *index* to sort all suffixes according to the first $2 \cdot length$ symbols correctly. After that we use the information to double the length again and so on. *counter*[0] is already used to describe the first free position in memory at the beginning of *counter*. Analogously we use *counter*[n] in order to do the same procedure at the end. Therefore we have to store at the same time intervals sorted with prefixes of length *length* and of length $2 \cdot length$. If there is no further one of length *length* we start to sort them of length $2 \cdot length$ and produce new ones of length $4 \cdot length$. Notice that the total number of intervals can not be more then $n \gg 1$ such that to store them with starting and ending point we need at most n values in the memory. Furthermore out of the initial sorting part some pointers at the end of the input sequence (exactly t_2 many) are already correctly sorted such that the memory requirement is strictly less than $n - 2$ (we make an initial sorting at least of length 2). For typical files we need only something like $n \gg 2$ entries in memory, but in worst case $n - t_2$ is needed as we can see by the following example:

Take a deBruijn sequence of length 2^{n-1} copy the sequence and concatenate the two. The property of a deBruijn sequence is, that if we are looking at a linear shift-register of length $n - 1$ then these sequences have maximal period, or more precisely every binary sequence of length $n - 1$ occurs exactly once. Now if we have a length of $t_2 = n - 1$ then each prefix occurs in the constructed sequence exactly twice and hence we have n intervals from which only $n - 1$ are getting correctly sorted at the initial phase.

Now at the beginning we have no interval to sort of length $2 \cdot length$:

```
counter[n]->n;
```

We are starting the main loop.

```
/* as long as there is something to compare            */
while(counter[0]>1) do
    /* starting with the beginning part (at the end)    */
```

We call this loop twice because first we like to sort every interval of length *length* correctly, after that we continue at the end of *counter* and sort the intervals of length $2 \cdot length$. If there are further intervals of length $4 \cdot length$ then we can find them at the beginning of *counter*.

```
/* as long as we have something to compare of length "length" */
    while counter[0]>1 do
        counter[0]->counter[0]-2

        switch(counter[counter[0]+1]-counter[counter[0]]) {
                /* +1 is the number of elements !            */
```

Notice that using the procedure of Section 4.1 the calculation of the median is only efficient if we have enough elements to sort. Therefore in case where we have intervals of a small length we sort directly:

With only two entries we need in the worst case two questions in order to sort them

```
        case 1: {  /* only two entries                     */
            m1->index[pointer[counter[counter[0]]]+k]
                    /* a shortcut to store them in order not    */
            m2->index[pointer[counter[counter[0]+1]]+k]
                    /* to calculate them twice                  */
            if m1=m2 then
```

The two values are equal, that means the two suffixes are equal of length "2*lengthänd therefore we add it at the end of counter.

```
INSTOCOUNTER(counter[counter[0]],
             counter[counter[0]+1],1)
  else
```

They are different so that we can compare them

```
if m1<m2 then
```

The beginning value of the interval is smaller than the end, therefore we do not have to exchange the order and we can update the index.

```
index[pointer[counter[counter[0]+1]]]->
       index[pointer[counter[counter[0]+1]]]+1;
  else
```

We have to swap them and to update the index of the beginning pointer.

```
SWAPPOINTER(counter[counter[0]],
            counter[counter[0]+1])
index[counter[counter[0]]]->
       index[counter[counter[0]]]+1
   endif
  endif
  break;
} /* end of case interval of length 2              */
```

An interval with three elements we can sort as described in Section 3.2. We call a function *sort3* which needs as parameters the array *counter*, the position (*counter*[*counter*[0]]) in counter to get the boundaries for the interval to sort, the arrays *pointer* and *index*, a *flag* which describes how to insert a new interval to continue with, the *length* of the already compared prefixes and finally the length *n* (necessary to insert a new interval using the function *INSTOCOUNTER*).

```
case 2: { /* interval of length 3                  */
    sort3(counter,counter[counter[0]],pointer,
          index,1,length,n)
```

Either everything is sorted or we are getting an interval back which starts with the same first $2 \cdot length$ symbols and that is we have to add them to the end of counter.

```
    break;
} /* end of interval of length 3                   */
```

In all other cases we call the function described in Section 4.1 which we denote as *splitcount*.

```
default: { /* the general case                        */
     splitcount(counter,counter[0]+1,pointer,index,
               1,length,n);
  break;
} /* end of the general case                          */
} /* end of the switch                                */
```

Now we can stop the loop for sorting intervals with length *length* and look at the intervals of length $2 \cdot length$.

```
done /* inner loop: counter[0]>1                       */
length->(length<<1)
```

The length llengthïs finished, that is we can continue with "2*lengthïn order not to copy the end to the beginning and continue the main loop we repeat the whole procedure with exchanging the role of the beginning of the array counter and the end of it. Of course $counter[0] = 1$, in other words at the beginning there is no interval of $4 \cdot length$ which we have to compare. Now we have to start the loop at the end:

```
while counter[n]<n do
  switch(counter[counter[n]+1]-counter[counter[n]]) {
     /* +1 is the number of elements !                 */
     case 1: { /* only two elements                     */
        /* two shortcuts                                */
        m1=index[pointer[counter[counter[n]]]+length];
        m2=index[pointer[counter[counter[n]+1]]+length];
        if m1=m2 then
```

The two values are equal and we have to add a new interval at the beginning of the array *counter* using the function *INSTOCOUNTER*.

```
           INSTOCOUNTER(counter[counter[n]],
                        counter[counter[n]+1],0);
        else /* we can compare them                     */
           if m1<m2 then
              index[pointer[counter[counter[n]+1]]]++;
           else
              SWAPPOINTER(counter[counter[n]],
                          counter[counter[n]+1]);
              index[counter[counter[n]]]++;
           endif
        endif
        break;
     }  /* end of the case with only two elements.      */
```

As before we also consider a separate case with only three elements using the function *sort3* as before but with $flag = 0$.

```
case 2: {
    sort3(counter,counter[counter[n]],pointer,
                               index,0,length,n);
    break;
} /* and of case 2.                                  */
```

Again in all others cases we use the function *splitcount*.

```
default: {
    splitcount(counter,counter[n]+1,pointer,
                               index,0,length,n);
    break;
}
} /* and of the switch                                */
  /* continue with the next interval.                 */
counter[n]->counter[n]+2
done /* end of the loop counter[n]<n                   */
length->(length<<1) /* double again and return to the  */
/* first loop: counter[0]>1.                           */
done
```

References

1. B. Balkenhol and S. Kurtz, Space efficient linear time computation of the Burrows and Wheeler transformation, Number, Information and Complexity, Special volume in honour of R. Ahlswede on occasion of his 60th birthday, editors I. Althöfer, N. Cai, G. Dueck, L. Khachatrian, M. Pinsker, A. Sárközy, I. Wegener, and Z. Zhang, Kluwer Acad. Publ., Boston, Dordrecht, London, 375–384, 1999.
2. J. Bentley and R. Sedgewick, Fast algorithm for sorting and searching strings, Proceedings of the ACM–SIAM Symposium on Discrete Algorithms, 360–369, 1997.
3. M. Burrows and D.J. Wheeler, A block–sorting lossless data compression algorithm, Technical report, Digital Systems Research Center, 1994.
4. N.J. Larsson, The context trees of block sorting compression, Proceedings of the IEEE Data Compression Conference, Snowbird, Utah, March 30 – April 1, IEEE Computer Society Press, 189–198, 1998.
5. N.J. Larsson, Structures of string matching and data compression, PhD thesis, Dept. of Computer Science, Lund University, 1999.
6. N.J. Larsson and K. Sadakane, Faster suffix–sorting, Technical Report LU–CS–TR: 99–214, LUNDFD6/(NFCS–3140)/1–20/(1999), Dept. of Computer Science, Lund University, 1999.
7. U. Manber and E.W. Myers, Suffix arrays: A new method for on–line string searches, SIAM Journal on Computing, 22, 5, 935–948, 1993.
8. K. Sadakane, A fast algorithm for making suffix arrays and for Burrows–Wheeler transformation, Proceedings of the IEEE Data Compression Conference, Snowbird, Utah, March 30 – April 1, IEEE Computer Society Press, 129–138, 1998.
9. M. Schindler, A fast block–sorting algorithm for lossless data compression, Proceedings of the Conference on Data Compression, 469, 1997.

Monotonicity Checking

M. Kyureghyan

Abstract. In our thesis we cosidered the complexity of the monotonicity checking problem: given a finite poset and an unknown real-valued function on it find out whether this function is monotone. Two decision models were considered: the comparison model, where the queries are usual comparisons, and the linear model, where the queries are comparisons of linear combinations of the input. This is a report on our results.

1   Introduction

The monotonicity checking problem is: Given a finite poset P and an unknown real-valued function f on it, find out whether this function is order preserving on P, that is, whether $f(x) \leq f(y)$ for any $x < y$ in P. We consider the worst-case complexity of the monotonicity checking problem. That is the number of tests an optimal monotonicity checking algorithm performs in the worst case.

The first model of monotonicity checking considered here is the comparison model where we can compare the values of f at any two elements of the poset P. The naive way of monotonicity checking in this model is comparing the values of the function at any two cover pairs of the poset ((x, y) is a cover pair in P if $x < y$ ant there is no z in P with $x < z < y$). But it turns out that for some posets one can do better comparing the values also at some incomparable elements. The linear model is a generalization of the comparison model, here the queries are comparisons of real linear combinations of the values of the input function. In this model monotonicity checking problem is equivalent to the polyhedral membership problem on a certain class of polyhedra.

A class of probabilistic algorithms monotonicity testing algorithms was considered in [3]. In [7] the computational complexity of monotonicity checking algorithms was studied.

In Section 2 we briefly describe the main results that we have for the comparison model. In Section 3 we derive a general lower bound on complexity of monotonicity checking in the linear model. We apply this bound to get a lower bound on the complexity of finding simultaneously the minimum and the maximum in a sequence of real numbers.

2   The Comparison Model

Let P be an n-element poset. Let $C(P)$ be the complexity of monotonicity checking on poset P in the comparison model.

If P is not connected, that means its cover graph is not connected, then it is not difficult to show that $C(P)$ is equal to the sum of monotonicity checking

R. Ahlswede et al. (Eds.): Information Transfer and Combinatorics, LNCS 4123, pp. 735–739, 2006.
© Springer-Verlag Berlin Heidelberg 2006

complexities on the subposets that we get by inducing the order relation in P on the connected components of the cover graph. Hence, we can restrict our attention to connected posets. One can show that

Proposition 1. *Let P be an n-element connected poset. Then $C(P) \geq n - 1$.*

As an example of an element poset with $C(P) = n - 1$ one can take the linear order.

To get an upper bound on $C(P)$ note that one can perform monotonicity checking by sorting the values of the function. As sorting of a string of n numbers requires $O(n \log n)$ comparisons we have $C(P) = O(n \log n)$.

An interesting toping of research is the investigation of properties of monotonicity checking. It turns out, for example, that: *There are posets with strictly less complexity of monotonicity checking than some of its subposets.*

Another topic of particular interest is the complexity of monotonicity checking of the Boolean lattice B_n. J. Kahn conjectured that the order of $C(B_n)$ is greater than 2^n. We have shown the following bounds.

Theorem 1. *For the complexity $C(B_n)$ of monotonicity checking on the Boolean lattice of order n, for $n \geq 6$,*

$$\frac{9}{7}2^n - \frac{1}{7}2^{r_1+1} - 1 \leq C(B_n) \leq 11 \ q \ 2^{n-3} + 2^{n-r_2} \ C_{r_2} \tag{1}$$

holds, where where $n = 3k + r_1$ with $0 \leq r_1 \leq 2$ and $n = 4q + r_2$ with $r_2 \in \{0, 2, 3, 5\}$, and $C_0 = 0$, $C_2 = 3$, $C_3 = 9$, $C_5 = 59$.

3   The Linear Model

The *Polyhedral Membership Problem* (PMP) is: Given a polyhedron $\mathcal{P} \subset \mathbb{R}^n$ and an unknown point $x \in \mathbb{R}^n$ decide whether $x \in \mathcal{P}$ using linear comparisons.

Let P be a poset defined on a ground set $\{x_1, \ldots, x_n\}$. The monotonicity checking problem on P can be reformulated as a PMP: Given an n-tuple $(f(x_1) \ldots f(x_n)) \in \mathbb{R}^n$, find out whether $f(x_i) - f(x_j) \geq 0$ for all $x_i > x_j$ in P. The polyhedron

$$\left\{ (f(x_1), \ldots, f(x_n)) \in \mathbb{R}^n : f : P \to \mathbb{R}, f \text{ is monotone on } P \right\}$$

is called the *monotone polyhedron of the poset P*. Thus the linear complexity of monotonicity checking on a poset P coincides with the complexity of PMP on the monotone polyhedron of P.

In [8] the authors have found a lower bound on the complexity of PMP on polyhedron $\mathcal{P}$ in terms of the facial structure of $\mathcal{P}$. Later this lower bound was improved in [5], the following theorem was proven.

Theorem 2. *([5]) For the complexity $D(\mathcal{P})$ of PMP on a polyhedron $\mathcal{P}$*

$$2^{D(\mathcal{P})} \binom{D(\mathcal{P})}{n-s} \geq |\mathcal{F}_s(\mathcal{P})| 2^{n-s} \tag{2}$$

holds, for each $0 \leq s \leq n$.

To apply the bound from the theorem above to monotonicity checking complexity we need to estimate the number of faces of a monotone polyhedron. One can describe precisely the entire facial structure of $\mathcal{M}(P)$ the monotone polyhedron of a poset P, using the order structure of P. We call a partition $\{P_1, \ldots, P_s\}$ of P *connected partition* if each P_i, $1 \leqslant i \leqslant s$, is connected in the cover graph of the poset P. Define a binary relation $\prec$ on a partition $\{P_1, \ldots, P_s\} = P$ of P by setting

$$P_i \prec P_j \text{ if } x < y \text{ for some } x \in P_i, \ y \in P_j.$$

We call a partition *compatible* if the transitive closure of the relation "$\prec$"is a partial order (is antisymmetric).

Theorem 3. *([2], [6]) [1] There is a one–to–one correspondence between the set of faces of monotone polyhedron $\mathcal{M}(P)$ of a poset P and the set of connected and compatible partitions of P. Moreover, every connected and compatible partition of P into s parts corresponds to an s–dimensional face.*

Let $C_l(P)$ denote the complexity of monotonicity checking on a poset P in the linear model. From Theorem 2 and Theorem 3 we get the following theorem.

Theorem 4. *Let P be an n-element poset. Then we have*

$$2^{C_l(P))} \binom{C_l(P)}{n-s} \geqslant |\mathcal{J}_s(P)| 2^{n-s} \tag{3}$$

for each $0 \leqslant s \leqslant n$, where $\mathcal{J}_s(P)$ is the number of connected and compatible partition of P into s parts.

Let a sequence of n real numbers be given. And the minimum and the maximum of this sequence are to be determined simultaneously. The complexity of this problem when the queries are usual comparisons is $\lceil \frac{3n}{2} \rceil - 2$, (see [1], ch. 4). In [9] it was asked about the linear complexity of this problem. We apply the Theorem 4 to get a lower bound on the complexity of this problem.

Theorem 5. *To find simultaneously the minimum and the maximum in a sequence of n real numbers one needs at least $1.23n$ linear comparisons, for n large enough.*

Proof. Let $a_1, \ldots, a_n$ be real numbers. We establish the lower bound considering a seemingly easier problem: let us show that at least $1.23n$ linear queries are needed to check the hypothesis that a_1 is a minimum and a_n is a maximum of this sequence. This is equivalent to showing that the monotonicity checking

---

[1] In these papers not monotone polyhedra of posets but order polytopes were considered. The order polytope of a poset P is defined as

$$\Big\{ (f(x_1), \ldots, f(x_n)) \ : \ f : P \to [0,1], f \text{ is monotone on } P \Big\}.$$

However their proof can be easily modified to get the statement for monotone polyhedra.

complexity of the n-element poset P with Hasse diagram is not less than $1.23n$, for n large enough.

The number of connected and compatible partitions of poset P into $i+2$ parts is $\binom{n-2}{i}2^{n-2-i}$. Indeed, to get a partition of P into $i+2$ parts, choose i elements of rank one from P each as a one-element part of the partition, and the rest of these elements take in the same part of the partition with the minimum or the maximum element. This is a connected and compatible partition into i parts. It is easy to see that in this way we construct all connected and compatible partitions of poset P.

According to the Theorem 4 we get

$$2^{C_l(P)}\binom{C_l(P)}{k} \geq \binom{n-2}{k}2^{2k} \qquad (4)$$

for any $0 \leq k \leq n-2$.

Since the linear model is not slower than the comparison model we have $n-1 \leq C_l(P) \leq 1.5n$. Let $C_l(P) = c(n)n$. We want to show that $c(n)$ is greater than 1.23, for n large enough. Assume, to the opposite that for some n we have $c(n) \leq 1.23$. Then

$$2^{1.23n}\binom{\lceil 1.23n \rceil}{k} \geq 2^{C_l(P)}\binom{C_l(P)}{k}. \qquad (5)$$

Taking $i = \lceil \frac{n}{10} \rceil$ in (4) we get from (5)

$$2^{1.23n}\binom{\lceil 1.23n \rceil}{\lceil \frac{9n}{10} \rceil} \geq \binom{n-2}{\lceil \frac{9n}{10} \rceil}2^{\frac{9n}{5}}. \qquad (6)$$

Using the estimate $\binom{n}{i} = \exp(nh(i/n) + O(\log n))$, for $0 \leq i \leq n$, where $h(p) = -p\ln p - (1-p)\ln(1-p)$ is the entropy function, we get from (6)

$$2^{1.23n}e^{1.23nh(\frac{9}{12.3})\theta_1(1)+o(n)} \geq 2^{\frac{9n}{5}}e^{nh(\frac{9}{10})\theta_2(1)+o(n)}, \qquad (7)$$

where $\theta_1(1) \to 1, \theta_2(1) \to 1$ as $n \to \infty$. For n large enough this would mean

$$1.23 + (1.23h(\frac{9}{12.3}) - h(\frac{9}{10}))\log_2 e \geq \frac{9}{5} \qquad (8)$$

however this is not true as can be checked by tables for the entropy or direct computer calculation. $\qquad \square$

References

1. M. Aigner, Combinatorial Search, Wiley-Teubner Series in Computer Science, Stuttgart, 1988.
2. L. Geissinger, A polytope associated to a finite ordered set, preprint.
3. O. Goldreich, S. Goldwasser, E. Lehman, D. Ron, and A. Samorodnitsky, Testing monotonicity, Combinatorica, 301-337, 2000.

4. M. Kyureghyan, Monotonicity checking, PhD Thesis, Universität Bielefeld, Bielefeld, 2004.
5. J. Moravek and P. Pudlak, New lower bound for the polyhedral membership problem with an application to linear programming, Mathematical Foundations of Computer Science, (Prague, 1984), Lecture Notes in Comput. Sci., 176, Springer, Berlin, 416–424, 1984.
6. R. Stanley, Two poset polytopes, Discrete Comput. Geom., 1, 9-23, 1986.
7. A. Voronenko, On the complexity of recognizing monotonicity, Mathematical problems in cybernetics, No. 8 (Russian), Mat. Vopr. Kibern., 8, 301–303, 1999.
8. A. Yao and R. Rivest, On the Polyhedral Decision Problem, SIAM J. Comput., 9, 343-347, 1980.
9. A. Yao, On the complexity of comparison problems using linear decision trees, Proc. IEEE 16th Annual Symposium on foundations of Computer Science, 85-89, 1975.

Algorithmic Motion Planning: The Randomized Approach

S. Carpin

Abstract. Algorithms based on randomized sampling proved to be the only viable algorithmic tool for quickly solving motion planning problems involving many degrees of freedom. Information on the configuration space is acquired by generating samples and finding simple paths among them. Paths and samples are stored in a suitable data structure. According to this paradigm, in the recent years a wide number of algorithmic techniques have been proposed and some approaches are now widely used. This survey reviews the main algorithms, outlining their advantages and drawbacks, as well as the knowledge recently acquired in the field.

1 Introduction

Robot motion planning algorithms are currently being used in many fields beyond robotics, like structural studies in biology, computer graphics, computer assisted surgery, and many others (see [52] for an excellent review). In particular, their application to bioinformatics, for problems like protein folding or ligand binding, appears to be one of the most important, promising and challenging directions. The successful application of these algorithms relies on the availability of an abstract formulation that can be used to model many different scenarios, as well as on an algorithmic machinery capable to solve difficult instances in a reasonable time. The planning of the robot motion is performed as a search in a suitable space, called the *configuration space*. Since the early days of research in algorithmic motion planning it was recognized that even the basic version of this problem is PSPACE-complete [16], [66], and the best deterministic algorithm known so far is exponential in the dimension of the configuration space [15]. At the same time, real world problems ask for algorithms able to solve problems with tenths of degrees of freedom, and the demand is continuously growing. Around the mid nineties a new approach was introduced, and this boosted the research in the field, as well as the practical use of these algorithms. This approach is based on the use of randomization. Randomized sampling is used to acquire information over the problem instance being solved without the necessity of a systematic deterministic processing of the data. This information is stored in a data structure which captures the connectivity of the configuration space. Such a data structure is usually composed by nodes, i.e. points in the configuration space, and links, i.e. simple valid paths connecting nodes. Complicated paths are then obtained by moving from sample to sample through the simple paths. For sake of completeness it has to be mentioned that another randomized motion

R. Ahlswede et al. (Eds.): Information Transfer and Combinatorics, LNCS 4123, pp. 740–768, 2006.
© Springer-Verlag Berlin Heidelberg 2006

planner previously introduced (RPP, randomized path planner, illustrated in [8])
will not be considered here, as the underlying idea is different. In fact the RPP
performs a gradient descent over a potential field and performs random motions
when it gets stuck in a local minima different from the goal. Then, no sampling
is involved, differently from all the algorithms later described in this survey.

The sampling based approach has some appealing properties. Differently from
many previously developed planners, it works for virtually all robots, without
imposing any constraint on the dimension of their configuration space. It is
then immediately applicable whenever a configuration state space can be used
to model a real world problem. It is this aspect that allows their application in
many different fields. For example, randomized robot motion planners have been
used for multi-robot systems [81], closed chain systems [23], [33], [84], deformable
objects [9], [49], and protein folding or ligand docking [4], [10], [75], [76], [77], [78].
Randomized motion planning algorithms are also well suited for practical parallel
implementations, as illustrated in [3], [17], [19] and [35], thus allowing further
performance gains. Finally, the implementation of these algorithms is usually
quite simple.

The price to pay is completeness. The algorithms described in this survey
obtain probabilistic completeness rather than deterministic completeness. This
means that if a solution exists, the probability to find it converges to 1 when the
processing time approaches infinity. So, when a randomized algorithm fails to
find a solution, it could be the case that a solution does not exist at all, or that
in the alloted execution time the sampling process has not been able to get the
information needed to solve the search. A compromise somehow considered in
this context is the so called *resolution completeness*, as illustrated for example
in [20] and [74]. An algorithm is resolution complete if when it fails to find
a solution we know that either this does not exist, or it requires a sampling
resolution below a certain fixed and known threshold.

The randomized framework has many degrees of freedom. For example, different sampling techniques can be used, or different data structures, or different
resampling strategies, and so on. This motivates the great number of randomized
algorithms proposed up to now. Some algorithms and techniques proved to be
very efficient and are now part of the standard literature. In addition to computational aspects, also the theoretical foundations of this approach have been
addressed, so that a certain understanding about the power and the limitations
of the probabilistic framework has been obtained. On the other hand it has to be
acknowledged that many aspects are still not completely clear, and deeper investigation is needed. This survey focuses on some of the most common algorithms
and gives an overall perspective on the most widely used ideas in the field. For
every algorithm pseudo code is provided, as well as practical indications. It also
provides a significant amount of references to the most significant papers, so that
the reader can quickly get to the original sources if needed. Basic convergence
ideas and results are illustrated but proofs are not provided, and the interested
reader is referred to the cited papers.

The survey is organized as follows. Section 2 provides the formal statement of the robot motion planning problem in the context of randomized algorithms. All the algorithms will be formulated in the framework of this abstract formulation, thus being not tied to the robotics scenario. By following this approach, readers interested in applying them in other fields will find their task easier. Algorithms based on the so called *probabilistic roadmaps* are discussed in section 3. The basic formulation, as well as improvements, are illustrated and discussed. The section also provides some of the fundamental theoretical results concerning probabilistic convergence. Following the same approach, section 4 reviews algorithms whose underlying data structure is a tree rather than a graph. Finally, conclusions are offered in section 5.

2   Problem Statement

This section provides a brief formulation of the problem under consideration. Extensive treatment of the basic computational aspects can be found in [32], [39], [51], [70], [71] and [73]. The robot motion planning (RMP) problem is dealt with by using the configuration state space approach introduced by Lozàno-Perez in [61]. Every robot is associated with a set of degrees of freedom which specify its placement in the workspace. The combination of the degrees of freedom assumes values in a space called the *configuration space*, usually indicated as $\mathcal{C}$. The configuration space is partitioned into two subsets, the space of free configurations, $\mathcal{C}_{free}$, and the space of obstacle configurations, $\mathcal{C}_{obs}$. The space of free configurations is the subset of licit configurations. The space of obstacle configurations is its complement, i.e. $\mathcal{C}_{obs} = \mathcal{C} \setminus \mathcal{C}_{free}$. In the case of a robotic system, a configuration belongs to $\mathcal{C}_{obs}$ if it corresponds to a robot placement where the robot collides with obstacles, or with itself, or if it places the robot in a position not satisfying its physical or operative constraints. In a different scenario, like for example in bioinformatics, the physical systems will be a protein or a ligand rather than a robot. In this case the configuration still specifies the placement of the chemical compound in the space, but its validity is determined by the associated energetic level. Given two points $x_{start} \in \mathcal{C}_{free}$ and $x_{goal} \in \mathcal{C}_{free}$, the RMP problem requires to compute a continuous function $f : [0, 1] \rightarrow \mathcal{C}_{free}$ such that $f(0) = x_{start}$ and $f(1) = x_{goal}$. This function could even not exist, for example if $\mathcal{C}_{free}$ is disconnected and x_{start} and x_{goal} belong to different connected components. An ideal RMP algorithm should be able to determine this situation, and to stop the computation as soon as it is possible to establish that a solution cannot be found. As we will see, randomized algorithms fail to accomplish this goal, so it is usually necessary to bound the computation time to avoid infinite loops while trying to solve unsolvable problem instances. In the randomized motion planning framework, the availability of a collision checker function is assumed. Given a configuration, the collision checker determines if it belongs to $\mathcal{C}_{free}$ or to $\mathcal{C}_{obs}$. Formally, it can be specified as follows:

$$\forall c \in \mathcal{C} \quad Check(c) = \begin{cases} 0 \text{ if } & c \in \mathcal{C}_{obs} \\ 1 \text{ if } & c \in \mathcal{C}_{free} \end{cases} \quad (1)$$

The availability of the collision checker is fundamental for randomized motion planners, as they exploit it to validate samples and acquire information about $\mathcal{C}$. From a practical point of view it has to be outlined that there is a reasonable choice of algorithms to be used as collision checkers like [14], [29], [31], [50], [60], [63], and some efficient implementations are freely available to the scientific community.

A strictly related problem is the *kinodynamic motion planning problem*. In the kinodynamic framework not only kinematic constraints are considered, i.e. constraints on the position, but also dynamic constraints have to be satisfied. Dynamic constraints involve the time derivatives of the configuration space. For example, there can be speed or acceleration bounds. Kinodynamic constraints are constraints involving both kinematic and dynamic. For instance, a bound on the maximal speed (dynamic constraint) could be not absolute but rather related to the position (kinematic constraint) of the robot in its environment. This could allow higher speeds in wide areas and force slow motions when the robot is near to an obstacle. In the kinodynamic literature, the term configuration space is usually replaced by the term *state space*, and is often indicated as X rather than $\mathcal{C}$. As a consequence, the space of free states is indicated as X_{free} and its complement is indicated as X_{obs}. In what follows we will adopt this terminology and notation. A point in the state space is defined as $(x, \dot{x})$, i.e. it stores not only the configuration, but also its time derivative. An instance of the kinodynamic motion planning problem requires to find a trajectory p, i.e. a time parameterized path in the time interval $[0, T]$, from a start point $(x_s, \dot{x}_s) \in X_{free}$ to a goal point $(x_g, \dot{x}_g) \in X_{free}$ or a goal region $X_{goal} \subset X_{free}$. The goal region is often introduced as the kinematic constraints could limit the set of reachable configurations and then a single point could be not reachable. The trajectory should not violate the kinodynamic constraints, i.e.

$$\forall t \in [0, T] \quad p(t) \in X_{free}.$$

The exact solution of the kinodynamic problem is known to be NP-hard, while approximated dynamic programming based solutions have been illustrated in [25], [26] and [27].

3 Graphs Based Motion Planners

While it has to be acknowledged that some of the basic ideas can be first found in the early paper [30], the first widely used randomized algorithm for RMP was introduced in [43], which extends the early versions found in [40], [65] and [5]. The algorithm operates in two stages, the first called *learning stage* and the second called *query stage*. Given an instance of the RMP problem, in the learning stage the algorithm samples the configuration space and builds an undirected graph $G = (V, E)$ which captures the information gathered. The graph is called *probabilistic roadmap* (PRM). In the query stage, the PRM is used to solve specific RMP problem instances which are then reduced to a graph search over the roadmap. The availability of the following elements is assumed:

- a subroutine *Check* which computes the function described in equation 1
- a subroutine *Distance* which computes a distance between two configurations, i.e. it computes a function

$$D : \mathcal{C} \times \mathcal{C} \to \mathbb{R}^+ \cup \{0\}. \tag{2}$$

- a fast and possibly incomplete deterministic planner P. Different implementations can opt for different planners.

3.1   The Learning Stage

The learning stage aims to build a graph which captures the connectivity of the subset $\mathcal{C}_{free}$. Two different approaches are commonly used. The first one is illustrated in Algorithm 1.

Algorithm 1. Learning stage

1: $V \leftarrow \varnothing \quad E \leftarrow \varnothing$
2: **loop**
3:    Generate a random configuration $c \in \mathcal{C}_{free}$
4:    $V \leftarrow V \cup \{c\}$
5:    $V_n \leftarrow \{v \in V \quad | \quad Distance(c, v) < M\}$
6:    **for all** vertices $v \in V_n$ in order of increasing $Distance(c, v)$ **do**
7:       **if** c and v can be connected by P and they do not lie in the same connected component **then**
8:          $E \leftarrow E \cup \{(c, v)\}$

Samples randomly generated over $\mathcal{C}$ are accepted if they belong to $\mathcal{C}_{free}$, and discarded otherwise. The collision checker is used to take this decision. When a sample is accepted, it becomes a vertex of the PRM graph (lines 2 and 3). After a vertex is added, the algorithm checks if it is possible to add edges between the inserted vertex and vertices already in the graph. A subset of neighboring vertices is selected and the planner P is run to determine if the new vertex can be connected to them. Two vertices are neighbors if their distance D is less than a fixed threshold M. This choice is made for sake of efficiency, as it is unlikely that a couple of far apart vertices will be connected by the planner P, and on the other hand we wish to minimize the number of calls to the simple planner. Another often used efficiency driven choice is to limit the size of the neighbors set to a maximum size, say K. This heuristic is also chosen to limit the number of calls to the deterministic planner. If the planner succeeds in finding a free path between them, and if the nodes do not belong to the same connected component of the graph, an edge connecting the two vertices is added to the graph. The second approach consists in generating all the samples first, and then verifying their connectivity later on, instead of interleaving samples generation connection attempts.

For what concerns the planner P, different techniques have been evaluated [2] and [28]. In the vast majority of implementations the deterministic planner simply connects the two points with a straight segment and verifies whether it lies in $\mathcal{C}_{free}$ or not. The validation is performed by selecting a set of intermediate points along the segment and by calling the collision checker on each of them. The segment is declared to be valid if all the intermediate points lie in $\mathcal{C}_{free}$. It is evident that this approach is inherently error prone, as just a discretization is used to determine the validity of the entire segment. Recently however a new algorithm which computes exact collision checking has been introduced [72], and its use appears appealing in the context of randomized RMP, as it can be used to perform exact validation of the edges. In other cases the local planner P is a potential field based planner like [44]. If the resulting graph contains more than one connected component (see Figure 1), the learning stage is often followed by a roadmap improving substep.

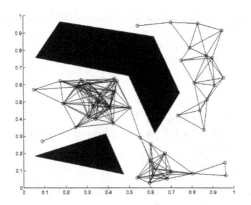

Fig. 1. An example of a Probabilistic Roadmap built over $\mathcal{C} = [0,1]^2$. Black regions indicate $\mathcal{C}_{obs}$. In this case the PRM consists of two connected components and a roadmap improving stage could succeed in merging them.

The improving is performed by selecting some vertices assumed to lie in difficult regions and then trying to add more samples in their neighborhood. Different criteria are used to decide whether a vertex lies in difficult region or not. A quite common technique consists in associating a weight to each vertex, the weight being determined by the number of edges originating at that vertex. Then, vertices are selected randomly, and the probability that they are selected is proportional to the inverse of their weight. This technique aims to add edges to nodes poorly connected to the roadmap, thus improving its quality. It is however worth noting that if the configuration space is indeed disconnected, this step could even be not necessary, meaning that it could not improve the overall roadmap quality at all. On the other hand, due to the probabilistic nature, it is impossible to determine if the improving step is indeed needed or not.

3.2   The Query Stage

In the query stage two configurations x_{start} and x_{goal} are given, and the algorithm is required to produce a path between them (see Algorithm 2). The algorithm tries to connect x_{start} and x_{goal} to vertices in the graph G. This trial is done by using the same technique used to insert edges in the roadmap, i.e. vertices are probed by increasing distance order. Vertices farther than a fixed threshold are again not considered. If it is not possible to connect both x_{start} and x_{goal} to the roadmap G, the algorithm reports failure. Otherwise, let us assume that x_{start} has been connected to a vertex v_s and that x_{goal} has been connected to a vertex v_g. Then, a graph search is performed to verify whether v_s and v_g belong to the same connected component of G. If the search succeeds, then a path is returned, by taking the sequence of segments associated with the edges connecting v_s and v_g, plus the segments connecting x_{start} with v_s and x_{goal} with v_g.

Algorithm 2. Query stage

$V_s \leftarrow \{v \in V \mid Distance(x_{start}, v) < M\}$
if the planner P can find a path between x_{goal} and a vertex in V_s then
    Let $v_s \in V_s$ be that vertex
else
    return Failure
$V_g \leftarrow \{v \in V \mid Distance(x_{goal}, v) < M\}$
if the planner P can find a path between x_{goal} and a vertex in V_g then
    Let $v_g \in V_g$ be that vertex
else
    return Failure
if a path $e_1, e_2, \ldots, e_n$ between v_s and v_g is found then
    return the overall solution path $(x_{start}, v_s), e_1, e_2, \ldots, e_n, (v_g, x_{goal})$
else
    return Failure

3.3   Considerations on the PRM Algorithm

The PRM algorithm has been implemented an tested under many different conditions and a significant amount of knowledge has been outlined. This motivated the development of further research. First, it is evident that the learning stage will take much more time than then query stage. This mainly because some of the operations used during construction time, like collision checking are computationally challenging and are to be performed many times. For this reason the learning stage is worth only if many queries will benefit from the created roadmap. This is not the case when the user is interested in the *single shot* instance, i.e. just one instance of the problem has to be solved. Similarly, if the robot is required to move in a dynamic environment, the PRM will be no more valid as obstacles move, thus vanishing the time spent to build it. Second, the original PRM method used uniform sampling over $\mathcal{C}$ while generating samples.

This choice is reasonable because the PRM being built should be later used to solve multiple queries, thus allowing problem instances where x_{start} and x_{goal} are placed in arbitrary regions of $\mathcal{C}_{free}$. But this sampling strategy has its own disadvantages. The main one is its low probability to place samples in narrow regions. The difficulty of discovering narrow passages in $\mathcal{C}$ is one of the motivations which led to the development of many refinements we will illustrate later.

3.4 Theoretical Analysis of the PRM Algorithm

In this section we briefly sketch the theoretical properties of the motion planners based on the probabilistic roadmaps approach. As the subject is pretty broad and a detailed treatment would require a significant amount of space, we provide just the main concepts and we refer the interested reader to the cited bibliography. The PRM algorithm is probabilistic complete, meaning that if enough time is alloted to the learning stage, this will eventually end up creating a roadmap that will determine the solution of every solvable RMP problem instance. While this can somehow be intuitive, as the uniform sampling process over $\mathcal{C}_{free}$ will eventually cover it completely, from a practical point of view it would be precious to know how many nodes should be in the roadmap in order to get a desired probability of success. This is of course related to shape of $\mathcal{C}_{free}$, and to the placement of x_{start} and x_{goal} therein. We here give some basic results found in the literature. In [41] basic speculations concerning the original version of the PRM algorithm are formulated. The work does not address the vast range of possible heuristics that can be employed to improve its performance but rather deals with the algorithm formerly illustrated. Even under this simplified assumption, significant indications can be drawn. There are three parameters playing an important role in the overall planning performance. Let us suppose that there exists a path p connecting x_{start} and x_{goal}. The first relevant parameter is L, the length of p. The second parameter is ε which is the Euclidean distance of p from $\mathcal{C}_{obs}$, and the third parameter is N, the number of vertices in the graph (i.e. $N = |V|$). The first result proved by the authors is the following.

Theorem 1. *Let* $p : [0, L] \to \mathcal{C}_{free}$ *be a path connecting* x_{start} *and* x_{goal} *and let* ε *be its distance from* $\mathcal{C}_{obs}$. *Let* d *be the dimension of the configuration space. Then the probability that the PRM will fail to connect* x_{start} *and* x_{goal} *is at most*

$$\frac{2L}{\varepsilon}(1 - \alpha\omega_d)^N \qquad (3)$$

where $\alpha = \varepsilon^d/(2^d|\mathcal{C}_{free}|)$ *and* ω_d *is the volume of the unit ball in the* $d-$*dimensional space.*

Theorem 1 illustrates some common aspects arising while dealing with the probabilistic convergence of randomized algorithms for motion planning. The result is quite intuitive. First, the bound claims that the probability of failure decreases while increasing the number of samples in the roadmap. The second point is that the failure probability increases with the increase of length. This is some-how expected. Long paths require more information, i.e. more samples, to be caught. Also the dependence on ε, is intuitive. This takes into account the problems arising when narrow passages are present in $\mathcal{C}_{free}$, as more samples will be needed to discover them. Theorem 1 considers just the minimum distance between the path p and $\mathcal{C}_{free}$, thus the bound is somehow too defensive and tends to overestimate failure probability. The authors then provide a second bound which considers a mean distance between the path and the obstacle space.

Theorem 2. *Let p be a path of length L connecting x_{start} and x_{goal}, i.e. p : $[0, L] \rightarrow \mathcal{C}_{free}$. Let $\varepsilon(t)$ be the distance between the path and $\mathcal{C}_{obs}$ at instant t. Then, the failure probability is bounded by*

$$6 \int_0^L \frac{(1 - \frac{\alpha_d}{2^d}\omega_d\varepsilon^d(t))^N}{\varepsilon(t)} dt \tag{4}$$

where $\alpha_d = 2^{-d}|\mathcal{C}_{free}|$ and ω_d is the volume of the unit ball in the $d-$dimensional space.

Both bounds depend either on ε or $\varepsilon(t)$, and also on α. This outlines that this bound is far from being trivial to compute as the exact computation of these parameters is not easy. They depend on the path p and on the shape of $\mathcal{C}_{free}$. Thus it is not immediate to get an answer to the following question: given this instance of the RMP problem, how many vertices should the algorithm generate in order to achieve a given bound on the failure probability? They nevertheless indicate that the algorithm indeed converges, and which parameters play an important role in this convergence schema.

A different set of analysis is presented in [7], [35], [42] and [48]. In particular [48] provides an alternative formulation of the problem in terms of measure theory over probability spaces and this appears to be a promising direction for further developments in the theoretical analysis.

3.5   Lazy PRMs

By profiling the execution of the basic PRM algorithm previously illustrated, one would notice that most of the time is spent to compute the *Check* function, i.e. to execute the collisions detection algorithm. It has however to be observed that in the single shot framework the majority of the edges of the roadmap will not be used, as they will not lie on the final path connecting x_{start} and x_{goal}. Starting from this point, the *lazy PRM* algorithm introduced in [11] gains a considerable speedup by postponing edge validation (and then collision detection) until it is really needed. The algorithm works as follows. It first builds an initial roadmap by placing N_{init} random samples uniformly distributed over the configuration space. As in the basic PRM algorithm, these will be the vertices of the roadmap. An edge is added between two vertices if they are closer than a certain distance, measured according to a generic metric function, like 2. No collision detection is performed while generating these edges. To answer a query, the shortest path

Algorithm 3. Lazy PRM

1: **Routine ValidPath(**$Path, G(V, E)$**)**
2: **INPUT** $Path$: path to validate
3: **INPUT/OUTPUT** $G = (V, E)$: roadmap
4: **RETURN**: *true* if the path is valid, *false* otherwise
5: **for all** vertices $v_i \in Path$ **do**
6:   **if** $v_i \notin C_{free}$ **then**
7:     Remove vertex v_i from V
8:     Remove all edges originating at v_i from E
9:     **return** *false*
10: **for all** edges $(v_i, v_j) \in Path$ **do**
11:   **if** $(v_i, v_j) \notin C_{free}$ **then**
12:     Remove (v_i, v_j) from E
13:     **return** *false*
14: **return** *true*

between x_{start} and x_{goal} is sought, by using the A^* algorithm [67]. Only if a path is found, then the *path validation* stage is activated (see Algorithm 3). This substep first checks all the vertices along the path to verify if they are valid or not. If not, they are removed from the roadmap, as well as all the edges originating at these vertices, and the path validation stage is stopped. If all vertices lie in C_{free}, then edges lying along that path are checked for validity. As in the PRM algorithm the validation is done by discretizing the segment and by calling the collision checker to validate each intermediate point. If all the edges lie in C_{free} the problem is solved. If an edge is determined to be not valid, it is removed from the roadmap and the path validation stage terminates. If the path found by A^* is not valid, then A^* is run again, and if a new path is produced it is again tested for validity. Every time A^* is executed again, it will start from an updated roadmap $G(V, E)$ missing vertices or edges which caused previously determined paths to be not valid. When A^* fails to find a path, a node enhancement stage is run, to improve the quality of the roadmap, and then the A^* search and path validation cycle is repeated.

The authors illustrate a quantitative comparison between the classic PRM algorithm and the lazy PRM implementation over a real world industrial robotic scenario, and a significant speedup is outlined. Lazy edge validation techniques are now widely used while implementing randomized planners ([68], [69]). A similar technique, called *fuzzy roadmap*, has also been used while applying PRM like algorithms to manipulation problems ([64]). In this case the authors postpone edges' validation, but they perform samples validation immediately. The technique appears to be competitive in certain configuration spaces.

A further extension of the ideas introduced with Lazy and Fuzzy PRMs was introduced in [79] and [80]. The authors move along the same lines, i.e. they postpone the validation of nodes or edges, but this is not done following a fixed pattern. Instead the methods they developed incorporate those techniques and then allow their planning system to be biased towards one or the other.

3.6   The Problem of Narrow Passages

The original PRM approach relies on uniform sampling over $\mathcal{C}$. This has the advantage of being easy to implement and to guarantee that eventually the entire $\mathcal{C}_{free}$ will be covered. But this has also its own drawbacks. Most notably, uniform sampling poorly deals with narrow passages. In fact, if we indicate with $\mu(S)$ the measure of the set S, by using random sampling over $\mathcal{C}$ the probability of placing a sample inside S is $\mu(S)/\mu(\mathcal{C})$. This clearly indicates the inability of the algorithm to quickly place samples in these small volume regions. Along the years a number of techniques have been developed and are being developed for overcoming this difficulty. In the following, we describe some of them.

Gaussian sampling. One way to address the narrow passages problem, is to change the underlying probability distribution. In [12] a Gaussian sampling is used. The authors start from the consideration that in the PRM algorithm most of the time is spent while adding nodes to the roadmap. This fraction is far greater than the time spent to generate samples. Thus they speculate on the possibility of spending more time to generate "good" samples, so that less points need to be added to the roadmap. The idea is to have many samples in cluttered regions and just a few in large open areas. This is achieved by the algorithm illustrated in Algorithm 4.

Algorithm 4. The Gaussian sampling strategy

1: **loop**
2:    $c_1 \leftarrow$ random configuration
3:    $dist \leftarrow$ distance chosen according to a Gaussian distribution
4:    $c_2 \leftarrow$ random configuration at distance $dist$ from c_1
5:    **if** $c_1 \in \mathcal{C}_{free}$ **and** $c_2 \notin \mathcal{C}_{free}$ **then**
6:        add c_1 to the graph
7:    **else if** $c_2 \in \mathcal{C}_{free}$ **and** $c_1 \notin \mathcal{C}_{free}$ **then**
8:        add c_2 to the graph
9:    **else**
10:        discard c_1 and c_2

It can be observed that two samples are generated but just one is added to the graph. Experimental results illustrated by the authors give evidence that this technique increases the overall roadmap quality , by having more samples in narrow areas and less in wide areas. This sampling technique is inspired by image processing based techniques (like *blurring*), and the mathematical details are the following. A Gaussian probability distribution in a d-dimensional configuration space is defined as

$$\phi(c, \sigma) = \frac{1}{(2\pi\sigma^2)^{d/2}} e^{-\frac{c^T c}{2\sigma^2}} \tag{5}$$

Fig. 2. Left figure shows a configuration space subset of R^2 where black regions indicate $\mathcal{C}_{obs}$. Right figure shows the desired sampling density obtained by the Gaussian sampling strategy (the darker the region, the more the sampling probability).

The function $Obs(c)$ is defined to assume the value 0 if $c \in \mathcal{C}_{free}$ and 1 otherwise. According to the image perception literature ([67]), the following function *blurs* obstacles

$$f(c, \sigma) = \int Obs(y)\phi(c - y, \sigma)dy \tag{6}$$

Finally, the function g is defined as

$$g(c, \sigma) = \max(0, f(c, \sigma) - Obs(c)). \tag{7}$$

Then, by using this technique it is possible to draw samples having $g(c, \sigma)$ as probability distribution. Note that g is null inside obstacles, so that this guarantees that samples are generated just in $\mathcal{C}_{free}$ (see Figure 2). While appealing in itself, this sampling distribution relies on a good choice of the σ parameter, which indicates how close should samples be to the obstacles. It is easy to see that different values of σ could be more suitable in different areas of $\mathcal{C}$. This problem, although while using a different RMP algorithm, is addressed in [18].

Obstacle based PRM. The problem of narrow passages is tackled also in [1]. The authors address the problem in the context of *Obstacle Based Probabilistic Roadmaps* (OBPRM from now on). OBPRM [5] are a variant of the PRM algorithm where nodes generation is performed with the goal of placing samples near to, or in contact with the obstacles. The underlying idea is that in this way it is possible to correctly operate even in the presence of cluttered environments, as narrow passages are the result of facing obstacles. It is then appropriate to say that OBPRM address this problem since its roots. In addition to this, it is outlined that not only the sampling process plays an important role to discover narrow regions, but also roadmap connection can be carried out using different approaches that influence the overall performance. The authors illustrate then a set of different approaches both for nodes generation and roadmap connection. Nodes generation is performed by using three different methods. The first one generates configurations in contact with obstacles (see Algorithm 5).

Algorithm 5. Algorithm for creating samples on the boundary of the obstacle O_j

1: GENERATE_CONTACT_CONFIGURATION
2: determine a point p inside the obstacle O_j
3: let M be a set of directions emanating from p
4: **for all** $m \in M$ **do**
5:     use binary search to determine a point lying on the boundary of O_j along the direction m

The second generates samples in free space, but near to the boundary of C_{free}. These points can be obtained by a slight modification of Algorithm 5 so that free space points rather than contact points are generated. The third one aims to create *shells* of configurations around obstacles, so that paths in those difficult regions can be quickly found. Shells are obtained by retaining some of the valid samples generated while looking for contact and free configurations in the previous steps. During the connection stage three different local planners are used. The first one is the usual straight line planner used in the simplest PRM implementation. The second one is the so called rotate–at–s, where s is a number between 0 and 1 [2]. While seeking a path between the configurations c_1 and c_2, this planner tries to translate the robot from c_1 to an intermediate configuration along the line connecting c_1 and c_2. Then it rotates the robot and it tries to translate it to the final c_2 configuration. The value s is the fraction along the straight line where the rotation is performed. The last planner is an A^*-like planner. During roadmap connection, three different stages are carried out. The first one, called *Simple Connection* utilizes the simplest planner (straight line), which is called many times to try many cheap connections among samples belonging to the same obstacle. The second stage, called *Connecting Components*, tries to create connections between disjointed roadmap components. The third stage, called *Growing Components*, has the same goal, but while trying to create connections, it can also generate new samples if needed. This is done by enhancing the map adding nodes near to small components and by keeping valid samples lying in segments not entirely accepted. In the second and third stages, while trying to connect two nodes the local planners are used in this order: straight line, rotate–at–$1/2$, rotate–at–0, rotate–at–1, A^*. Thus, by following an increasing planning cost strategy, expensive local planners are used just when cheaper and simpler planners fail to succeed.

The authors report extensive simulation results illustrating that significant speedups can be obtained by using the combination of techniques they propose and also provide experimentally driven recommendations about the techniques to utilize. The use of many different sampling and connecting techniques however has its own drawback, as many parameters should be fixed and it can then be non trivial to determine a suitable combined tuning. An additional drawback is that the analysis of the probability distribution of the samples is hard to determine.

Medial axis probabilistic roadmaps. Always in the attempt to overcome the problem of detecting narrow passages, a slightly different approach that uses sampling on the medial axis of $\mathcal{C}_{free}$ was proposed in [59], [82] and [83]. In this framework, called MAPRM, medial axis PRM, samples are not generated on the surface of the obstacles, but rather over $\mathcal{C}$ and are then *retracted* into the medial axis of $\mathcal{C}_{free}$. The medial axis of the configuration space are the points in $\mathcal{C}_{free}$ with maximal distance from $\mathcal{C}_{obs}$. Formally, for $x \in \mathcal{C}_{free}$, we define $B_{\mathcal{C}_{free}}(x)$ to be the largest closed ball centered in x and completely lying in $\mathcal{C}_{free}$. The medial axis of the space of free configurations is defined as the set of points whose associated $B_{\mathcal{C}_{free}}$ are maximal with respect to inclusion, i.e.:

$$MA(\mathcal{C}_{free}) = \{x \in \mathcal{C}_{free} | \nexists y \in \mathcal{C}_{free} \quad \text{with} \quad B_{\mathcal{C}_{free}}(x) \subsetneq B_{\mathcal{C}_{free}}(y)\}.$$

Classical properties of medial axis ensures that the network or medial axis associated with the configuration space captures the connectivity of the space itself. Using this *reduced* representation one is ensured that no significant information is lost. The strength of the algorithm relies on the ability to efficiently push or pull a sample to a medial axis. The authors report algorithms for dealing both with two and three dimensional workspaces, while here (see Algorithm 6) we sketch the simple algorithm for retracting a sample into the medial axis of a two dimensional environment. Once a sample c over $\mathcal{C}$ is generated, the nearest point lying on the boundary of $\mathcal{C}_{free}$ is determined (here it is indicated as n). Then, if c lies in $\mathcal{C}_{free}$, it is retracted to the medial axis, otherwise the point to move is n. The line to move along in order to reach the medial axis is determined by c and n, while the direction depends whether c is in $\mathcal{C}_{free}$ or not. The reader

---

Algorithm 6. Medial axis retraction algorithm

1: Let $c \in \mathcal{C}$
2: Among the points in $\partial \mathcal{C}_{free}$ determine the nearest to c, and call it n
3: **if** $c \in \mathcal{C}_{free}$ **then**
4:     $\boldsymbol{d} \leftarrow \boldsymbol{nc}$
5:     $s \leftarrow c$
6: **else**
7:     $\boldsymbol{d} \leftarrow \boldsymbol{cn}$
8:     $s \leftarrow n$
9: Move s along the direction $\boldsymbol{d}$. Stop moving s when n is not the unique nearest point of $\partial \mathcal{C}_{free}$ to s

---

should note that the sample added is not c itself, but s after it has been moved along the direction $\boldsymbol{nc}$ or $\boldsymbol{cn}$. The point stops to move when at least two points in the boundary are at the same distance. Thus, s is indeed in the medial axis. In order to efficiently perform the computation, the retraction step (line 9) is performed by using a bisection technique. The overall PRM algorithm generates samples uniformly over $\mathcal{C}$ and then retracts them over the medial axis. Then a graph is built by connecting those samples, and this constitutes the probabilistic roadmap. The rational of this approach is the following: to discover narrow passages, it is no more necessary to generate samples into the narrow passages themselves, but rather to generate samples that once retracted end up into the medial axis associated with the narrow passages, thus increasing the probability of discovering them. This is achieved by sampling over the entire $\mathcal{C}$ rather than over just $\mathcal{C}_{free}$.

Planning in dilated spaces. Along the same lines of OBPRM, some authors [36] pushed the idea of sampling on obstacle surfaces even further, by allowing the generation of samples lying outside $\mathcal{C}_{free}$. The planning is divided into two stages. First a roadmap is created in a *dilated* configuration space. This means that if a sample lies inside $\mathcal{C}_{obs}$, but its distance from $\mathcal{C}_{free}$ is smaller than a certain threshold δ, it is retained rather than discarded. This step is performed using the classical PRM algorithm where the space $\mathcal{C}_{free}$ is substituted by its dilatation $\mathcal{C}_{free}^{dil}$

$$\mathcal{C}_{free}^{dil} = \mathcal{C}_{free} \cup \{c \in \mathcal{C}_{obs} \mid Distance(c, \mathcal{C}_{free}) < \delta\}.$$

In the second stage, samples inside $\mathcal{C}_{obs}$ are pulled into $\mathcal{C}_{free}$. This is done by resampling in their neighborhood. Next, edges have to be created, and also in this case resampling could be needed to push edges into $\mathcal{C}_{free}$. Algorithm 7 illustrates the complete algorithm. In the Figure, $U_v(v)$ is the resampling region associated with vertex v, while $U_e(v_1, v_2)$ is the resampling region asso-

---

Algorithm 7. Algorithms for creating a valid roadmap starting from a roadmap created in then dilated configuration space

---

1: Generate a Roadmap $R' = (V', E')$ in the dilated space $\mathcal{C}_{free}^{dil}$.
2: $V \leftarrow \varnothing$ $E \leftarrow \varnothing$
3: **for all** $v' \in V'$ **do**
4:    **if** $v' \in \mathcal{C}_{free}$ **then**
5:        $V \leftarrow V \cup \{v'\}$
6:        $p(v') = v'$
7:    **else**
8:        pick up to k samples in $U_v(v')$ and add to V the first one lying in $\mathcal{C}_{free}$ (if any)
9:    let $p(v')$ be the vertex added to V (if any)
10: **for all** $(v_1, v_2) \in E'$ **do**
11:    **if** $(p(v_1), p(v_2)) \in \mathcal{C}_{free}$ **then**
12:        $E \leftarrow E \cup \{(p(v_1), p(v_2))\}$
13:    **else**
14:        Resample in $U_e(p(v_1), p(v_2))$. Let R be this sample set
15:        **if** by using samples in R a path connecting $p(v_1)$ and $p(v_2)$ is found **then**
16:            add the samples and the edges to V and E respectively

---

ciated with the edge (v_1, v_2). By using the dilated configuration space, narrow passages are easier to detect, as they are widened. The choice of the value of δ is very important. Taking it too small would not give too many advantages over the basic PRM algorithm, but taking a too big value of δ has its disadvantages too, as entire obstacles could then disappear. The authors illustrate the encouraging results of their simulation and offer some hints about practical implementation. The resampling region $U_v(v)$ is a sphere centered in v while $U_e(v_1, v_2)$ is a square.

Moreover the authors found that by using a series of decreasingly dilated spaces better results can be obtained.

4   Tree Based Motion Planners

Samples and edges can also be organized in a tree data structure rather than in a graph. With this approach efficient planners have been implemented. The first aspect is that they are well suited for addressing the *single-shot* motion planning problem. Moreover, while growing a tree it is possible to utilize the motion equations of the robot, thus obtaining paths complying with kinodynamic constraints. This section will illustrate mainly two classes of algorithms, the so called RRT based planners and the planners based on the concept of expansive spaces. For sake of completeness it has to be pointed out that a similar approach was presented in [62].

4.1   Rapidly Exploring Random Trees

Rapidly Exploring Random Trees (RRT) are a recently introduced class of RMP algorithm that can be used both for systems involving kinodynamic constraints or not [22], [47], [53], [55], [56], [57], [58]. RRT proved to be suitable for being used in very different real world applications, as illustrated in [13] and [46]. In addition to the *Check* and *Distance* routines used in the PRM framework, the RRT algorithm assumes the availability of the following elements:

- a set U of inputs to be applied to the system
- an incremental simulator, i.e. a procedure that given a state $x(t) \in X$ and an input $u \in U$, produces the state $x(t + \Delta t)$, provided that the input u has been applied over the given time interval.

Then, by including system's equations into the incremental simulator, the planner is able to directly produce paths satisfying the kinodynamic constraints. If no such constraints are given, i.e. the planner is required to produce a path for a holonomic robot, no incremental simulation takes place and simple interpolation is performed, as every motion is allowed.

The basic version of the algorithm is composed by the algorithms given in Algorithm 8. In the provided pseudocode the methods $T.init$, $T.add\_vertex$ and $T.add\_edge$ update the tree being built. In particular $T.init$ creates a tree consisting of a single node which is its root (x_{init} in our case). The method $T.add\_vertex$ adds a node to the tree and the method $T.add\_edge$ establish a parent–child relationship between nodes in the tree. To be precise, it defines that the second parameter is a child node of the first one. In addition, this method also associates which input is needed in order to move the robot from the parent state to the child state. This information is passed as last parameter.

The algorithm starts by building a tree rooted at the starting point x_{start}. Samples are then randomly generated over $\mathcal{C}_{free}$. If the incremental simulator is included, the *NEW_STATE* subroutine (called in line 3 of the *EXTEND* substep)

Algorithm 8. Algorithms for the construction of a RRT

1: BUILD_RRT(x_{init})
2: T.init(x_{init})
3: **for** $k = 1$ to K **do**
4:     x_{rand} ← RANDOM_STATE()
5:     EXTEND(T,x_{rand})
6: **return** T

1: EXTEND(T, x)
2: x_{near} ← NEAREST_NEIGHBOR(x, T)
3: **if** NEW_STATE($x, x_{near}, x_{new}, u_{new}$) **then**
4:     T.add_vertex(x_{new})
5:     T.add_edge($x_{near}, x_{new}, u_{new}$)
6:     **if** $x_{new} = x$ **then**
7:         **return** *Reached*
8:     **else**
9:         **return** *Advanced*
10:     **return** *Trapped*

---

chooses an input u, either randomly or by determining the one which will give a new state as closed as possible to the new random state, and determines a new state to be added to the tree.

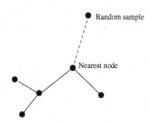

Fig. 3. Extension of a RRT. Starting from a random sample x_{rand} generated over the state space X, the nearest RRT node is found (x_{near}), and a new node is created as its child. The new node is placed along the segment connecting x_{near} and x_{rand} if the system is holonomic, otherwise it is generated by applying the incremental simulator to x_{near}.

The routine returns *Reached* if the new sample can be reached, *Advanced* if it cannot be reached but a new state has been added, or *Trapped* if no new state has been produced. If the simulator is not needed, *NEW_STATE* simply tries to place the new state at a fixed distance from x_{near} along the segment connecting it with x (see Figure 3).

The rationale behind the RRT algorithm is the following (see Figure 4). Let x be a node in the tree T and let $V(x)$ be its associated Voronoi region, i.e. the set of states nearer to x than to any other state in T. Then, by uniformly sampling over the state space, it is more likely to place samples into a big Voronoi region rather than into a small one. If a sample is placed into $V(x)$, then x will be chosen to be extended. Thus the tree is biased to grow towards unexplored regions.

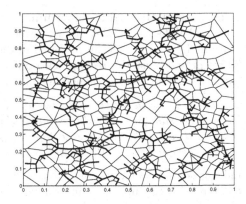

Fig. 4. A RRT built over an obstacle free $[0,1]^2$ state space and the associated Voronoi regions

The schema depicted in Algorithm 8 simply builds a RRT which explores the state space starting from the given x_{start} point. If a couple of points is given in a *single shot* framework, a significant speedup is obtained by growing two trees, one from x_{start} and the other from x_{goal}. The *RRT-Connect* algorithm [47], designed for problems not involving kinodynamic constraints, exploits this technique, as well as a greedy tree extension to cut down planning time even more. Algorithm 9 illustrates this improved version. An even more aggressive behavior can be obtained by using always the *CONNECT* routine instead of alternating it with *EXTEND*. This version of the planner is often indicated as *RRTConCon*. In this way, as soon as a promising direction is discovered, the tree is expanded in that direction as much as it is possible thus decreasing exploration time. In the provided pseudocode the function *PATH* is used to extract a path connecting x_{start} and x_{goal} once the trees T_a and T_b have been connected.

Algorithm 9. The RRT-Connect algorithm

1: CONNECT(T, q)
2: **repeat**
3:     $S \leftarrow$ EXTEND(T,x_{rand})
4: **until** NOT $S = Advanced$
5: **return** S
1: RRT_CONNECT(x_{start}, x_{goal})
2: T_a.init(x_{start})
3: T_b.init(x_{goal})
4: **for** $k = 1$ to K **do**
5:     $x \leftarrow$ RANDOM_STATE()
6:     **if not** (EXTEND(T_a,x) = *Trapped*) **then**
7:         **if** CONNECT(T_b,x_{new}) = *Reached* **then**
8:             **return** PATH(T_a,T_b)
9:     swap trees T_a and T_b
10: **return** *Failure*

Probabilistic completeness of the RRT algorithm. The RRT algorithm has been proved to be probabilistic complete under rather mild hypothesis. As for the PRM algorithm, detailed proofs can be found in the aforementioned references, while here we provide just the results. Two different theorems are valid, one for holonomic systems and one for nonholonomic systems.

Theorem 3. *Let x_{init} and x_{goal} lie in the same connected component of a non-convex, bounded, open, n-dimensional connected component of an n-dimensional state space. The probability that a RRT constructed from x_{init} will find a path to x_{goal} approaches one as the number of RRT vertices approaches infinity.*

A similar theorem holds for nonholonomic systems. In what follows it is assumed that in the *NEW_STATE* routine the input u is uniformly randomly chosen over the set of available inputs U.

Theorem 4. *In the same hypothesis of Theorem 3, let further assume that $|U|$ is finite, Δt is constant, and no two RRT vertices lie in a specified $\varepsilon > 0$, according to the used metric distance. Let also assume that there exists a sequence $u_1, \ldots, u_n$ of inputs that when applied to x_{init} will lead the system to the state x_{goal}. Then the probability that a RRT initialized at x_{start} will contain a vertex in the X_{goal} region approaches 1 as the number of vertices approaches infinity.*

While valuable in itself, probabilistic convergence only ensures convergence to the solution when the number of vertices, and then the computation time, approaches infinity. As with PRM, it would be highly useful to have a rate of convergence of the planner, in order to be able to predict the expected time. Again, while some results have been obtained, they are expressed in terms of environment specific quantities not easy to determine. The following theorems apply to single RRT and assume that an instance of the RMP problem is given in terms of a starting point x_{goal} and of a goal region X_{goal}, i.e. the robot is required to reach a region rather than a point. Both theorems rely on the following definition.

Definition 1. *A sequence of subsets $\mathcal{A} = \{A_1, A_2, \ldots, A_k\}$ of the state space X is an attraction sequence if $A_0 = \{x_{init}\}$, $A_k = X_{goal}$, and for each A_i there exists a basin $B_i \subseteq X$ such that:*

1. for all $x \in A_{i-1}$, $y \in A_i$, and $z \in X \setminus B_i$, the metric distance yields $Distance(x, y) < Distance(x, z)$
2. for all $x \in B_i$, there exists a number m such that the sequence of inputs $\{u_1, u_2, \ldots, u_m\}$ selected by the EXTEND algorithm will bring the state into $A_i \subseteq B_i$

Given a set S, we indicate with $\mu(S)$ its measure.

Theorem 5. *Let assume that an attraction sequence $\mathcal{A} = \{A_1, A_2, \ldots, A_k\}$ of length k exists, and let*

$$p = min_i\{\mu(A_i)/\mu(X_{free})\}.$$

Then, the expected number of iterations required to connect x_{start} and X_{goal} is not greater than k/p.

Theorem 6. *If an attraction sequence of length k exists and $\delta \in (0,1]$, then the probability that the RRT finds a path after n iterations is at least $1 - \exp(-np\delta^2/2)$, where $\delta = 1 - k/(np)$, and p is defined as in Theorem 5.*

Both theorems 5 and 6 suffer from the dependence on k, the length of the attraction sequence. While given a solvable instance of the RMP problem one can assume the existence of such sequence, the number of elements in it depends heavily on the shape of the environment and is far from being easy to compute.

Considerations on the RRT algorithm. The *EXTEND* substep of the RRT algorithm starts by determining the nearest node to the last generated sample. This has at least two implications from a practical point of view. First, the search has to be performed over the whole set of nodes generated so far. If one does not adopt a suitable data structure, but rather scans the whole sequence, this will yield a quadratic dependence. This problem has been addressed in [6], where the problem of efficient neighbor search is tackled and a solution is proposed, yielding an overall $n \log n$ complexity. It has however to be pointed out that the techniques utilized are quite involved, and could take some effort to be implemented. The second important issue related to neighbor search is the metric used. The choice of a good metric is a fundamental problem, as the use of an inappropriate one could lead the tree to grow towards the wrong direction. Ideally, the value returned by the *Distance* function should reflect the *cost to go*, while moving from one state to another. It is evident that this cost depends on the underlying system model and that the metric should then be strictly related to it. This problem is well discussed in [21]. Recently, the possibility of obtaining resolution completeness rather then probabilistic completeness in the RRT framework has been developed in [20]. Although important from a practical point of view, it is not addressed here as it does not fall into the probabilistic scenario. Finally it has to be mentioned that an implementation of most RRT based algorithms (and also some PRM based) is available in the freely available Motion Strategy Library software [54], so that one can test different versions in order to find out which one better fits its needs.

4.2   Planning in Expansive Spaces

An approach similar to RRT has been proposed in [34], [37], [38], and [45]. As the RRT planner, this planner efficiently builds a tree data structure. We first illustrate the basic version which does not deal with kinodynamic constraints. Given an instance of the motion planning problem, the algorithm starts building two trees, one rooted at the start point x_{start}, and the other rooted at the goal point x_{goal}. As for the PRM algorithm, we assume that a tree T consists of a couple of sets $T = (V, E)$, where V is the set of tree nodes and E is the set of edges between nodes. The two trees are iteratively expanded by using the same algorithm, and the process terminates if it is possible to find a free path connecting the two trees. Algorithm 10 illustrates this iterative approach. The algorithm terminates when either a solution is found or the maximum number of iterations is reached.

Algorithm 10. Expansive planner

1: Let T_1 be a tree rooted at x_{start} and with no other nodes
2: Let T_2 be a tree rooted at x_{goal} and with no other nodes
3: **for** MAX_ITERATIONS times **do**
4:    EXPANSION(T_1)
5:    EXPANSION(T_2)
6:    **if** CONNECT(T_1, T_2) **then**
7:       return the PATH connecting x_{start} and x_{goal}
8: return FAILURE

Algorithm 11 illustrates how a tree $T = (V, E)$ can be expanded. In what follows, let $B_d(s)$ be the ball of radius d centered in s. At each step the algorithm associates a *weight* with every node in V. The weight of the node $s \in V$ is the number of sampled nodes of V lying in $B_d(s)$, i.e.

$$w(x) = |V \cap B_d(s)|.$$

The goal of the weight function is to avoid oversampling in regions already explored and to rather bias the expansion towards unexplored areas of the configuration space. In this respect both RRT and the expansive planner aim to the same goal, the only difference being in the technique used to identified poorly explored zones.

Algorithm 11. Expansion algorithm

1: EXPANSION(T)
2: Pick a sample s from V with probability proportional to $1/w(s)$
3: Let K be a set of N samples lying in $B_d(s)$
4: **for all** $k \in K$ **do**
5:    compute $w(k)$ and retain k with probability proportional to $1/w(k)$
6:    **if** k is retained and $k \in \mathcal{C}_{free}$ and the segment $(s, k) \in \mathcal{C}_{free}$ **then**
7:       $V \leftarrow V \cup \{s\}$
8:       $E \leftarrow E \cup \{(s, k)\}$

Finally, Algorithm 12 illustrates how connection between trees is verified. To limit the number of useless trials, the algorithm ignores node couples too far apart. Again, this because it is assumed that it unlikely to find a free segment connecting far–away samples. The algorithm assumes that a path can be found if a couple of nodes is close enough. In that case the segment is stored so that it can be later used to produce the path connecting x_{start} with x_{goal}.

   The above algorithm can be adapted in order to deal with kinodynamic constraints. In this case, a single tree is built, but the samples space is not X but rather

$$\mathcal{CT} = \mathcal{C}_{free} \times [0, +\infty]$$

which is called *space×time*. The subset of free valid configurations of $\mathcal{CT}$ is indicated as $\mathcal{CT}_{free}$. Along the same lines of the RRT algorithm, the availability of

---

Algorithm 12. Connection

1: CONNECT(T_1, T_2)
2: **for all** $x \in V_1$ **do**
3:   **for all** $y \in V_2$ **do**
4:     **if** $Distance(x, y) <$ Threshold **then**
5:       **if** $(x, y) \in C_{free}$ **then**
6:         store (x, y)
7:         return TRUE
8: return FALSE

---

an incremental simulator and of a set of inputs U_l is assumed. In this context it is assumed that inputs in U_l are piecewise constant functions with at most l pieces. Algorithm 13 illustrates how it is possible to generate a trajectory complying with the kinodynamic constraints. Again, a problem instance is formulated in terms of a start point and of a goal region, which the authors call *ENDGAME* and will be here indicated as *EG*. In Algorithm 13 the function *INTEGRATE* computes how the system evolves when the input u is applied while the system is in state s. In other words, it implements the incremental simulator previously introduced.

Algorithm 13. Randomized kinodynamic motion planner

1: let T be a tree whose root is $(x_{start}, 0)$
2: **for** at most MAX_ITERATIONS times **do**
3:   Pick a sample s from V with probability $1/w(s)$
4:   Pick an input u from U_l uniformly at random
5:   $s' =$INTEGRATE(s, u)
6:   **if** $s' \in CT$ **then**
7:     $V \leftarrow V \cup \{s'\}$
8:     $E \leftarrow E \cup \{(s, s')\}$
9:   **if** $s' \in EG$ **then**
10:     Terminate with success

Extensive results over simulations and real robots are illustrated in [37]. The trials involved both nonholonomic robots and systems performing in dynamic environments, i.e. with moving obstacles. Detailed results provide evidence that the devised algorithms lead to real time compliant systems.

Probabilistic convergence. The former algorithms have been formulated in a framework based on the *expansiveness* concept introduced in [38]. We here report the generalized results illustrated in [37] which concerns the kinodynamic motion planner. Given (s, t) and $(s', t') \in CT_{free}$ we say that (s', t') is reachable from (s, t) if there exists a control function that leads to an admissible trajectory from (s, t) to (s', t'). If such a trajectory can be obtained by applying just the inputs of the U_l set, then we say that (s', t') is l-reachable from (s, t). According to these definitions, it is possible to define the set of points *reachable* and *l-reachable* from from a point $p = (s, t)$. We indicate the first set as $R(p)$ and the second as $R_l(p)$. Then, given a subset of $S \subset \mathcal{CT}_{free}$, it is possible to define its reachable sets:

$$R(S) = \bigcup_{p \in S} R(p)$$

$$R_l(S) = \bigcup_{p \in S} R_l(p)$$

Definition 2. *Let* $\beta \in [0, 1)$ *be a constant and Let* $S \subset \mathcal{CT}$. *The* lookout *of the set* S *is*

$$Lookout_\beta(S) = \{p \in S \mid \mu(R_l(p)) \setminus S) \geq \beta\mu(R(S) \setminus S)\}$$

Definition 3. *Let* α, β *be constants in* $[0, 1]$. *For any* $p \in \mathcal{CT}_{free}$, $R(p)$ *is* (α, β)-*expansive if for every connected subset* $S \subseteq R(p)$,

$$\mu(Lookout_\beta(S)) \geq \alpha\mu(S).$$

$\mathcal{CT}_{free}$ *is* (α, β)-expansive *if for every* $p \in \mathcal{CT}_{free}$, $R(p)$ *is* (α, β)-*expansive.*

The following theorem proves that the algorithm will reach the *ENDGAME* region *EG*, i.e. will succeed in finding a solution, with high probability.

Theorem 7. *Let* $\mathcal{X}$ *be the reachability of the start point* (s, t) *and let* $g = \mu(EG \cap \mathcal{X})$ *be strictly positive. Let* $\mathcal{X}$ *be* (α, β)-*expansive. Let* $\gamma \in (0, 1]$ *be a constant. Let* T *be a tree rooted at* (s, t) *with* r *nodes. The probability that* T *has a node in* EG *is at least* $1 - \gamma$ *if*

$$r \geq \frac{k}{\alpha} \ln \frac{2k}{\gamma} + \frac{2}{g} \ln \frac{2}{\gamma},$$

where $k = (1/\beta) \ln(2/g)$.

A set of similar results and definitions holds for the basic planner that does not deal with kinodynamic constraints. It is again evident that probabilistic convergence can be proved, but convergence rate is difficult to measure in terms of the problem instance to be solved.

5   Conclusions

We outlined the major algorithms developed in the last years in the field of randomized motion planning. When dealing with practical implementations a number of issues have to be taken into consideration. First, it is necessary to have algorithms for computing the *Check* function and the *Distance* function. Collision detection is a challenging problem in itself, but fortunately there exist very efficient algorithms whose implementations are freely available to the scientific community. These algorithms often assume that the description of the objects is given in terms of meshes of triangles. This is a very favorable hypothesis, as many CAD systems export this type of representation for solid objects. However different algorithms exhibit different performances in various operating scenarios and a preliminary evaluation is needed in order to select the one better fitting the needs of the problem to be solved. The choice of the collision detector is extremely important, as most of the time spent by planners is devoted to collision checking, either for validating samples or edges connecting samples. For what concerns the computation of the distance function, many more alternatives are possible. Luckily, this is much easier to implement, so it is viable to try different definitions and than rely on the one giving better experimental results. Common choices are the L_1, L_2 or L_∞ norms over the configuration space $\mathcal{C}$. While computing distances between configurations, it is also usual to assign different weights to the degrees of freedom, or to normalize them to a given common interval. This is often the case when both translational and rotational joints are present, as rotating and translating can have a different impact on the geometry of the system. If the problem involves kinodynamic constraints, there is also the need for the incremental simulator. In this case no general rules or software libraries can be used, as this strongly depends on the underlying system model. In this case, as anticipated, the distance function, should reflect the effort needed to drive the system between two points in the state space. Ad hoc norms should then be designed and implemented.

The choice of the planning algorithm to use is driven by many different factors. If the problem involves kinodynamic constraints than the choice is on one of the two algorithms illustrated in section 4. Up to now no analytic comparison is possible and also no fair experimental comparison has been performed. On the other hand both algorithms proved to be suitable for being used in real world applications, they address the same class of problems, and they require the same components. It is then somehow difficult to give general indications on the one which could better fit the needs or could be easier to implement. If the problem to be solved involves just kinematic constraints, then a wider choice of algorithms is available. In the single shot scenario, tree based algorithms and lazy PRM perform better, while in a situation where many queries should be answered, also the use of the basic PRM algorithm appears appropriate. If the operating environment stems a configuration space with narrow passages, then one of the outlined refined PRM algorithms is the choice. It is somehow evident that in general no algorithm is better, but rather the environment influences the performance. The opportunity of having more planners to be used in different regions is addressed in [24], but up to now no criteria have been proved. It is indeed not easy to efficiently determine during the computation which strategy is better.

Another important issue is the sampling and resampling strategy. The vast majority of the proposed planners use uniform sampling over either $\mathcal{C}$ or a suitable subset. This leads to easy implementation, but has the outlined drawbacks. The use of a different sampling schema, like Gaussian sampling, could overcome these limitations, but up to now very few planners have been developed according to this paradigm. For what concerns resampling, associating a weight to each vertex and then choosing vertices to resample with a probability proportional to the inverse of the weight is easy to implement. The easiest weight to compute is the number of edges outgoing from a vertex. Also the weight suggested in Algorithm 11 is easy to compute, but could take more time.

This paper illustrated the most widely used algorithmic techniques, namely graph based and tree based. The field is however continuously growing, and more and more refinements are being proposed, so that an exhaustive listing is doomed to early obsolescence. Instead, some of the algorithms which proved to have more influence in the related scientific literature have been illustrated. The user needing to implement them should have had concrete indications about their strength and limitations, and will not find too much difficulties in adapting them to its specific needs.

References

1. N.M. Amato, O. Burchand Bayazit, L.K. Dale, H. Jones, and D. Vallejo, Obprm: an obstacle-based prm for 3d workspaces, In P.K. Agarwal, L.E. Kavraki, and M.T. Mason, editors, Robotics : The Algorithmic Perspective, The Third Workshop on the Algorithmic Foundations of Robotics, 156–168, A.K. Peters, 1998.
2. N.M. Amato, O.B. Bayazit, L.K. Dale, and C. Jone, Choosing good distances metrics and local planners for probabilistic roadmap methods, IEEE Transactions on Robotics and Automation, 16, 4, 442–447, 2000.
3. N.M. Amato and L.K. Dale, Probabilistic roadmaps are embarrassingly parallel, Proceedings of the IEEE International Conference on Robotics and Automation, 688–694, 1999.
4. N.M. Amato, K.A. Dill, and G.Song, Using motion planning to map protein folding landscapes and analyze folding kinetics of known native structures, Proceedings the 6th International Conference on Computational Molecular Biology (RECOMB), 2–11, 2002.
5. N.M. Amato and Y. Wu, A randomized roadmap method for path and manipulation planning, Proceedings of the IEEE International Conference on Robotics and Automation, 113–120, 1996.
6. A. Atramentov and S.M. LaValle, Efficient nearest neighbor searching for motion planning, Proceedings of the IEEE Conference on Robotics and Automation, 632–637, Washington, May 2002.
7. J. Barraquand, L. Kavraki, J.C. Latombe, T. Li, R. Motwani, and P. Raghavan, A random sampling scheme for robot path planning, International Journal of Robotics Research, 15, 6, 759–774, 1997.
8. J. Barraquand and J.C. Latombe, Robot motion planning: a distributed representation approach, The International Journal of Robotics Research, 10, 6, 628–649, 1991.
9. O. Burchan Bayazit, Jyh-Ming Lien, and N.M. Amato, Probabilistic roadmap motion planning for deformable objects, Proceedings of the IEEE International Conference on Robotics and Automation, 2126–2133, 2002.
10. O. Burchan Bayazit, G. Song, and N.M. Amato, Ligand binding with obprm and user input, Proceedings of the IEEE International Conference on Robotics and Automation, 2001.

11. R. Bohlin and L.E. Kavraki, Path planning using lazy prm, Proceedings of the IEEE International Conference on Robotics and Automation, 1469–1474, Seoul, May 2000.
12. V. Boor, M.H. Overmars, and A.F. van der Stappen, The gaussian sampling theory for probabilistic roadmap planners, Proceedings of the IEEE International Conference on Robotics and Automation, 1018–1023, Detroit, May 1999.
13. J. Bruce and M. Veloso, Real-time randomized path planning for robot navigation, Proceedings of the IEEE/RSJ Intl. Conference on Intelligent Robots and Systems, 2383–2388, 2002.
14. S. Cameron, Enhancing gjk: computing minimum distance and penetration between convex polyhedra, Proceedings of the IEEE Conference on Robotics and Automation, 3112–3117, 1997.
15. J. Canny, The Complexity of Robot Motion Planning, MIT Press, Cambridge (MA), 1988.
16. J. Canny, Some algebraic and geometric computations in pspace, Proceedings of the IEEE Symposium on foundations of computer science, 460–467, 1988.
17. S. Carpin and E. Pagello, On parallel rrts for multi-robot systems, Proceedings of the 8th conference of the Italian Association for Artificial Intelligence, 834–841, 2002.
18. S. Carpin and G. Pillonetto, Robot motion planning using adaptive random walks, Proceedings of the IEEE International Conference on Robotics and Automation, 3809–3814, 2003.
19. D. Challou, D. Boley, M. Gini, V. Kumar, and C. Olson, Parallel search algorithms for robot motion planning, In K. Gupta and A.P. del Pobil, editors, Practical Motion Planning, 115–132, John Wiley & Sons, 1998.
20. P. Cheng and S. M. LaValle, Resolution complete rapidly-exploring random trees, Proceedings of the IEEE International Conference on Robotics and Automation, 267–272, 2002.
21. P. Cheng and S.M LaValle, Reducing metric sensitivity in randomized trajectory design, Procceedings of the IEEE/RSJ Conference on Intelligent Robots and Systems, 2001.
22. P. Cheng, Z. Shen, and S. M. LaValle, RRT-based trajectory design for autonomous automobiles and spacecraft, Archives of Control Sciences, 11, 3-4, 167–194, 2001.
23. J. Cortes, T. Simeon, and J.P. Laumond, A random loop generator for planning the motions of closed kinematics chains using prm methods, Proceedings of the IEEE International Conference on Robotics and Automation, 2141–2146, 2002.
24. L.K. Dale and N.M. Amato, Probabilistic roadmaps - putting it all together, Proceedings of the IEEE International Conference on Robotics and Automation, 1940–1947, Seoul, May 2001.
25. B. Donald and P. Xavier, Provably good approximation algorithms for optimal kinodynamic planning for cartesian robots and open chain manipulators, Algorithmica, 14, 6, 480–530, 1995.
26. B. Donald and P. Xavier. Provably good approximation algorithms for optimal kinodynamic planning: robots with decoupled dynamic bounds, Algorithmica, 14, 6, 443–479, 1995.
27. B. Donald, P. Xavier, J. Canny, and J. Reif, Kinodynamic motion planning, Journal of the ACM, 40, 5, 1048–1066, November 1993.
28. R. Geraerters and M.H. Overmars, A comparative study of probabilistic roadmap planners, In J.D. Boissonat, J. Burdick, K. Goldberg, and S. Hutchinson, editors, Workshop on Algorithmic Foundations of Robotics, Advanced Robotics Series, Springer, 2002.

29. E.G. Gilbert, D.W. Johnson, and S.S. Keerthi, A fast procedure for computing the distance between complex objects in three-dimensional space, IEEE Journal of Robotics and Automation, 4, 2, 193–203, 1988.

30. B. Glavina, Solving findpath by combination of goal-directed and randomized search, Proceedings of the IEEE Conference on Robotics and Automation, 1718–1723, 1990.

31. S. Gottschalk, M. C. Lin, and D. Manocha, Obb-tree: a hierarchical structure for rapid interference detection, Proc. of ACM Siggraph'96, 1996.

32. D. Halperin, L. Kavraki, and J.C. Latombe, Robotics, In J.E. Goodman and J. O'Rourke, editors, CRC Handbook of Discrete and Computational Geometry, chapter 41, 755–778, CRC Press, 1997.

33. L. Han and N. Amato, A kinematics-based probabilistic roadmap method for closed chain systems, In B. Donald, K. Lynch, and D. Rus, editors, Algorithmic and Computational Robotics - New Directions, 233–246, A.K. Peters, 2000.

34. D. Hsu, Randomized single-query motion planning in expansive spaces, PhD thesis, Department of Computer Sciece, Stanford University, 2000.

35. D. Hsu, L.E. Kavraki, J.C. Latombe, and R. Motwani, Capturing the connectivity of high-dimensional geometric spaces by parallelizable random sampling techniques, In P.M. Pardalos and S. Rajasekaran, editors, Advances in Randomized Parallel Computing, 159–182, Kluwer Academic Publishers, Boston, MA, 1999.

36. D. Hsu, L.E. Kavraki, J.C. Latombe, R. Motwani, and S. Sorkin, On finding narrow passages with probabilistic roadmap planners, In P.K. Agarwal, L.E. Kavraki, and M.T. Mason, editors, Robotics : The Algorithmic Perspective, The Third Workshop on the Algorithmic Foundations of Robotics, 142–153, A.K. Peters, 1998.

37. D. Hsu, R. Kindel, J.C. Latombe, and S. Rock, Randomized kinodynamic motion planning with moving obstacles, In B. Donald, K. Lynch, and D. Rus, editors, Algorithmic and Computational Robotics - New Directions, A.K. Peters Ltd, 2000.

38. D. Hsu, J.C. Latombe, and R. Motwani, Path planning and expansive configuration spaces, International Journal of Computational Geometry and Applications, 9, 495–512, 1999.

39. Y.K. Hwang and N. Ahuja, Gross motion planning - a survey, ACM Computational Surveys, 24, 3, 219–290, 1992.

40. L. Kavraki and J.C. Latombe, Randomized preprocessing of configuration space for fast path planning, Proceedings of the IEEE Conference on Robotics and Automation, 2138–2145, 1994.

41. L.E. Kavraki, M.N. Kolountzakis, and J.C. Latombe, Analysis of probabilistic roadmaps for path planning, IEEE Transactions on Robotics and Automation, 14, 1, 166–171, 1998.

42. L.E. Kavraki, J.C. Latombe, R. Motwani, and P. Raghavan, Randomized query processing in robot path planning, Journal of Computer and Systems Science, 57, 1, 50–60, 1998.

43. L.E. Kavraki, P. Švestka, J.C. Latombe, and M.H. Overmars, Probabilistic roadmaps for path planning in high-dimensional configuration spaces, IEEE Transactions on Robotics and Automation, 12, 4, 566–580, 1996.

44. O. Khatib, Real-time obstacle avoidance for manipulators and mobile robots, International Journal of Robotic Research, 5, 1, 90–98, 1986.

45. R. Kindel, D. Hsu, J.C. Latombe, and S. Rock, Kinodynamic motion planning admnist moving obstacles, Proceedings of the IEEE International Conference on Robotics and Automation, 537–543, 2000.

46. J. Kufner, K. Nishiwaki, S. Kagami, M. Inaba, and H. Inoue, Motion planning for humanoid robots under obstacle and dynamic balance constraints, Proceedings of the IEEE International Conference on Robotics and Automation, 692–698, 2001.

47. J.J. Kufner and S.M. LaValle, Rrt-connect: an efficient approach to single-query path planning, Proceedings of the IEEE Conference on Robotics and Automation, 995–1001, San Francisco, April 2001.

48. A. Ladd and L.E. Kavraki, Generalizing the analysis of prm, Proceedings of the IEEE International Conference on Robotics and Automation, 2120–2125, 2002.

49. F. Lamiraux and L.E. Kavraki, Planning paths for elastic objects under manipulation constraints, International Journal of Robotics Research, 20, 3, 188–208, 2001.

50. E. Larsen, S. Gottschalk, M.C. Lin, and D. Manocha, Fast distance queries with rectangular swept sphere volumes, Proceedings of the IEEE International Conference on Robotics and Automation, 3719–3726, 2000.

51. J.C. Latombe, Robot Motion Planning, Kluver Academic Publishers, 1990.

52. J.C. Latombe, Motion planning: a journey of robots, molecules, digital actors, and other artifacts, The International Journal of Robotics Research - Special Issue on Robotics at the Millennium, 18, 11, 1119–1128, 1999.

53. S. M. LaValle, From dynamic programming to RRTs: algorithmic design of feasible trajectories, In A. Bicchi, H. I. Christensen, and D. Prattichizzo, editors, Control Problems in Robotics, 19–37, Springer-Verlag, Berlin, 2002.

54. S.M. LaValle, Msl - the motion strategy library software, version 2.0, http://msl.cs.uiuc.edu.

55. S.M. LaValle, Rapidly-exploring random trees, Technical Report TR 98-11, Computer Science Department Iowa State University, October, 1998.

56. S.M. LaValle and J. J. Kuffner, Randomized kinodynamic planning, Proceedings of the IEEE International Conference on Robotics and Automation, 473–479, 1999.

57. S.M. LaValle and J.J. Kufner, Randomized kinodynamic planning, International Journal of Robotics Research, 20, 5, 378–400, 2001.

58. S.M. LaValle and J.J. Kufner, Rapidly-exploring random trees: progress and prospects, In B. Donald, K. Lynch, and D. Rus, editors, Algorithmic and Computational Robotics, New Directions, 45–59, A.K. Peters, 2001.

59. J.-M. Lien, S. L. Thomas, and N.M. Amato, A general framework for sampling on the medial axis of the free space, Proceedings of the IEEE International Conference on Robotics and Automation, 2003.

60. M.C. Lin and J.F. Canny, A fast algorithm for incremental distance calculation, Proceedings of the IEEE Conference on Robotics and Automation, 1008–1014, 1991.

61. T. Lozano-Pérez, Spatial planning: a configuration space approach, IEEE Transactions on Computers, C-32, 2, 108–120, 1983.

62. E. Mazer, J.M. Ahuactzin, and P. Bessiére, The ariadne's clew algorithm, Journal of Artificial Intelligence Research, 9, 295–316, 1998.

63. B. Mirtich, V-clip: fast and robust polyhedral collision detection, ACM Transaction on Graphics, 17, 3, 177–203, 1998.

64. C. Nielsen and L.E. Kavraki, A two-level fuzzy prm for manipulation planning, Proceedings of The IEEE/RSJ International Conference on Intelligent Robots and Systems, 1716–1722, 2000.

65. M. Overmars and P. Švestka, A probabilistic learning approach to motion planning, In K. Goldberg, D. Halperin, J.C. Latombe, and R. Wilson, editors, Algorithmics Foundations of Robotics. A K Peters Ltd., 1995.

66. J.H. Reif, Complexity of the mover's problem and generalization, Proceedings of the 20th IEEE Symposium on Foundations of Computer Science, 421–427, 1979.
67. S. Russel, P. Norwig, Artificial Intelligence - A modern approach, Prentice Hall International, 1995.
68. G. Sánchez and J.C. Latombe, A single-query bi-directional probabilistic roadmap planner with lazy collision checking, International Journal of Robotics Reserach, 21, 1, 5–26, 2002.
69. G. Sánchez-Ante, Path planning using a single-query bi-directional lazy collision checking planner, In C.A. Coello Coello, editor, MICAI 2002, number 2313 in LNAI, 41–50. Springer-Verlag, 2002.
70. J.T. Schwartz and M. Sharir, Algorithmic motion planning in robotics, In Handbook of Theoretical Computer Science, volume 1, chapter 8, 392–430, Elsevier, 1990.
71. J.T. Schwartz, M. Sharir, and J. Hopcroft, editors, Planning, Geometry, and Complexity of Robot Motion, Ablex Publishing, 1987.
72. F. Schwarzer, M. Saha, and J.C. Latombe, Exact collision checking of robot paths, In J.D. Boissonat, J. Burdick, K. Goldberg, and S. Hutchinson, editors, Workshop on Algorithmic Foundations of Robotics, Advanced Robotics Series, Springer, 2002.
73. M. Sharir, Algorithmic motion planning, IEEE Computer, 22, 3, 9–20, 1989.
74. T. Siméon, S. Leroy, and J.P. Laumond, Path coordination for multiple mobile robots: a resolution-complete algorithm, IEEE Transactions on Robotics and Automation, 18, 1, 42–49, 2002.
75. A.P. Singh, J.C. Latombe, and D.L. Brutlag, A motion planning approach to flexible ligand binding, Proceedings of the Int. Conf. on Intelligent Systems for Molecular Biology, 252–261, 1999.
76. G. Song and N.M. Amato, A motion planning approach to folding: from paper craft to protein folding, Proceedings of the IEEE International Conference on Robotics and Automation, 948–953, 2001.
77. G. Song and N.M. Amato, Using motion planning to study protein folding, Proceedings the 5th International Conference on Computational Molecular Biology (RECOMB), 287–296, 2001.
78. G. Song and N.M. Amato, A motion-planning approach to folding: from paper craft to protein folding, IEEE Transactions on Robotics and Automation, 20, 1, 60–71, 2004.
79. G. Song, S. Miller, and N.M. Amato, Customizing prm roadmaps at query time, Proceedings of the IEEE International Conference on Robotics and Automation, 1500–1505, Seoul, May 2001.
80. G. Song, S. Thomas, and N. M. Amato, A general framework for prm motion planning, Proceedings of the IEEE International Conference on Robotics and Automation, 4445–4450, 2003.
81. P. Švestka and M.H. Overmars, Coordinated path planning for multiple robots, Robotics and Autonomous Systems, 23, 3, 125–152, 1998.
82. S.A. Wilmarth, N.M. Amato, and P.F. Stiller, Motion planning for a rigid body using random networks on the medial axis of the free space, Proceedings of the 15th Annual ACM Symposium on Computational Geometry, 173–180, 1999.
83. S.A. Wilmarth, N.M. Amato, and P.F. Stiller, A probabilistic roadmap planner with sampling on the medial axis of the free space, Proceedings of the IEEE International Conference on Robotics and Automation, 1024–1031, 1999.
84. J.H. Yakey, S.M. LaValle, and L.E. Kavraki, Randomized path planning for linkages with closed kinematic chains, IEEE Transactions on Robotics and Automation, 17, 6, 951–958, 2001.

VII

Information Theoretic Models in Language Evolution

R. Ahlswede, E. Arikan, L. Bäumer, and C. Deppe*

Abstract. We study a model for language evolution which was introduced by Nowak and Krakauer ([12]). We analyze discrete distance spaces and prove a conjecture of Nowak for all metrics with a positive semidefinite associated matrix. This natural class of metrics includes all metrics studied by different authors in this connection. In particular it includes all ultra-metric spaces.

Furthermore, the role of feedback is explored and multi-user scenarios are studied. In all models we give lower and upper bounds for the fitness.

1 Introduction

The human language is used to store and transmit information. Therefore there is a significant interest in the mathematical models of language development. These models aim to explain how natural selection can lead to the gradual emergence of human language. Nowak and coworkers ([12], [13]) created a mathematical model, in which they introduced the *fitness of a language* as a measure for the communicative performance of a signalling system. In this model the signals can be misinterpreted with certain probabilities. In this case it was shown that the performance of such systems is intrinsically limited, meaning that the fitness can not be increased over a certain threshold by adding more and more signals to the repertoire of the communicators. This limitation can be overcome by concatenating signals or phonemes to form words, which increases significantly the fitness.

In the model the signals are elements of a given distance space. The fitness of the distance space is then defined as the supremum of the fitness values taken over all languages. In [13] and [5] the fitnesses of different metric spaces were investigated. Nowak conjectures that the fitness of a product-space is equal to the product of the fitnesses of the individual spaces. In the following we will refer to this conjecture as *product conjecture*.

In this paper we analyze discrete distance spaces. We prove the product conjecture for this model under assumptions which are sufficiently general so that the result includes all the models of metric spaces considered before in [12], [13] and [5].

We also show in this model that Hamming codes asymptotically achieve the maximal possible fitness.

---

* Supported in part by INTAS-00-738.

R. Ahlswede et al. (Eds.): Information Transfer and Combinatorics, LNCS 4123, pp. 769–787, 2006.
© Springer-Verlag Berlin Heidelberg 2006

This model for simple signalling systems and their fitness suggests the investigations of other classical information theoretical problems in this context. We will start this direction of research by considering feedback problems and transmission problems for multiway channels. In the feedback model that we introduced we show that feedback-fitness can be bigger than the fitness without feedback.

In [14] a relation between Shannon's noisy coding theorem and the fitness of a language is shown. They show that Shannon's error probability is inversely proportional to the fitness function.

2   Definitions, Notations and Known Results

We consider a special case of a model which was introduced in [13]. In this model a group of individuals can communicate about a given number of objects. We denote this set of objects by

$$\mathcal{O} = \{o_1, \ldots, o_N\}.$$

These are objects of the environment, other individuals, concepts or actions. Each object is mapped to a sequence of signals by the function

$$r : \mathcal{O} \to \mathcal{X}^n.$$

We represent each signal-sequence by a sequence of length n, where $\mathcal{X}$ is the set of all possible signals in the language. We call a signal-sequence, which describes an object, a word of the language. It is possible, that several objects are mapped to the same word. We assume that we have a distance function

$$d : \mathcal{X} \times \mathcal{X} \to \mathbb{R}_+$$

and $(\mathcal{X}, d)$ forms a distance space. We always write $\mathcal{X}$ for the distance space, if it is clear which distance function we use. If d satisfies, in addition, the following triangle inequality: $d(x, z) \leq d(x, y) + d(y, z)$ for all $x, y, z \in \mathcal{X}$ and $d(x, y) = 0$ holds only for $x = y$, then $(\mathcal{X}, d)$ is called a metric space.

We denote by x_t for $1 \leq t \leq n$ the t-th letter of a word x^n, thus $x^n = (x_1, \ldots, x_n)$. The distance between two words is defined by $d^n(x^n, y^n) = \sum_{t=1}^n d(x_t, y_t)$, where $x^n, y^n \in \mathcal{X}^n$.

As in [13] we define the similarity of two words by $s : \mathcal{X}^n \times \mathcal{X}^n \to \mathbb{R}_+$, where

$$s(x^n, y^n) = exp(-d^n(x^n, y^n)).$$

We call a family

$$\mathcal{L} = \{x^n(i) : i = 1, \ldots, N\}$$

with $x^n(i) = r(o_i)$ a language for N objects in $\mathcal{X}^n$. Note that in this way it is allowed to use the same word in order to describe different objects.

The probability of understanding y^n when x^n was signalled is given by

$$p(x^n, y^n) = \frac{s(x^n, y^n)}{\sum_{i=1}^N s(x^n, x^n(i))}.$$

We assume that successful communication is of benefit to speaker and listener. Thus for each correct transmitted word for the i-th object both get a payoff a_i, which defines the value of this object. We assume here that $a_i = 1$ for all i.

With this restriction we define the fitness of a language $\mathcal{L}$ of length N in $\mathcal{X}^n$ by

$$F(\mathcal{L}, \mathcal{X}^n) = \sum_{i=1}^{N} p(x^n(i), x^n(i)).$$

The fitness of the distance space $\mathcal{X}^n$ is then defined as the maximal possible value of the fitness of all languages in $\mathcal{X}^n$. Thus

$$F(\mathcal{X}^n) = \sup\{F(\mathcal{L}, \mathcal{X}^n) : \mathcal{L} \text{ language in } \mathcal{X}^n\}.$$

If we restrict the languages to be for a fixed number N of objects we define correspondingly:

$$F(\mathcal{X}^n, N) = \sup\{F(\mathcal{L}, \mathcal{X}^n) : \mathcal{L} \text{ language in } \mathcal{X}^n \text{ for } N \text{ objects}\}.$$

The next statement shows how the fitness values behave if we form languages of product type.

Let $\mathcal{L}$ be a language in the space $\mathcal{X}$ then the product language $\mathcal{L}^n$ is defined as the n-fold Cartesian product of $\mathcal{L}$, i.e., $\mathcal{L}^n = \times_{k=1}^n \mathcal{L}_k$, with $\mathcal{L}_k = \mathcal{L}$ for all k and the elements of the family $\mathcal{L}^n$ consist of all possible concatenations of n words from $\mathcal{L}$.

Proposition 1. *Let $\mathcal{L}$ be a language in the space $\mathcal{X}$. Then*

$$F(\mathcal{L}^n, \mathcal{X}^n) = F(\mathcal{L}, \mathcal{X})^n$$

and therefore

$$F(\mathcal{X}^n) \geq F(\mathcal{X})^n.$$

In [13] the authors considered three models for $\mathcal{X}$.

1. $\mathcal{X} = [0, a] \subset \mathbb{R}$ and $d(x, y) = |x - y|$,
2. $\mathcal{X} = [0, 1) \subset \mathbb{R}$ and $d(x, y) = \min\{|x - y|, 1 - |x - y|\}$,
3. $\mathcal{X} = \{0, d\}$ and $d(x, y) = \begin{cases} 0, \text{ if x=y} \\ d, \text{ else} \end{cases}$

For the model 1 they obtained the

Theorem (NKD, [5])

1. $F([0, a]) = 1 + \frac{a}{2}$.
2. $F([0, a] \times [0, b]) = F([0, a]) F([0, b])$.
3. $F([0, a]^n) = \left(1 + \frac{a}{2}\right)^n$.

Motivated by some experiments and this result Nowak formulated the following

Conjecture 1 (Product conjecture). *Let $(\mathcal{X}, d)$ be a metric space, then*

$$F(\mathcal{X}^n) = (F(\mathcal{X}))^n.$$

3   The Product Conjecture

Let $(\mathcal{X}, d)$ be a finite distance space. For a language $\mathcal{L}$ with N words (of length 1, that is letters) from $\mathcal{X}$ we introduce a language vector $\lambda = (\lambda_x)_{x \in \mathcal{X}}$, with

$$\lambda_x = \frac{Number \ of \ occurences \ of \ the \ word \ x}{N},$$

so that λ is a probability distribution (PD) on $\mathcal{X}$. With these definitions we can denote

$$F(\mathcal{L}, \mathcal{X}) = F(\mathcal{X}, \lambda) = \sum_x \frac{\lambda_x}{\sum_y \lambda_y e^{-d(x,y)}}.$$

For the fitness of the space $\mathcal{X}$ we can write

$$F(\mathcal{X}) = \max_\lambda F(\mathcal{X}, \lambda).$$

For a PD λ on $\mathcal{X}$, let λ^n denote the product-form distribution on $\mathcal{X}^n$ with marginals λ.

Property 1 now takes the form $F(\mathcal{X}^n, \lambda^n) = F(\mathcal{X}, \lambda)^n$ and $F(\mathcal{X}^n) \geq F(\mathcal{X})^n$. The product conjecture states that equality holds here for any metric space.

Supposition. In the following we shall assume, unless stated otherwise, that

(i) the diameter $D(\mathcal{X})$ of the set $\mathcal{X}$, defined as the maximum of $d(x, y)$ over all pairs (x, y) in $\mathcal{X}$, is finite, and
(ii) the matrix $[e^{-d(x,y)}]_{x,y \in \mathcal{X}}$ is positive semi-definite (psd.),

that is a self-adjoint square matrix with $A = A^T$ (Hermitian matrix) and all of whose eigenvalues are nonnegative. In our case all matrices are Hermitian because they are symmetric. We shall prove the product conjecture for such spaces. Recall that $d^n(x^n, y^n) = \sum_{t=1}^n d(x_t, y_t)$ is of sum-type.

We note that if $[e^{-d(x,y)}]_{x,y \in \mathcal{X}}$ is psd., then $[e^{-d^n(x^n, y^n)}]_{x^n, y^n \in \mathcal{X}^n}$ is psd. This follows from the fact that $[e^{-d^n(x^n, y^n)}]$ is the nth tensor power of $[e^{-d(x,y)}]$.

3.1   A Lower Bound on $F(\mathcal{X})$

Since $F(\mathcal{X}) \geq F(\mathcal{X}, \lambda)$ for all PDs λ on $\mathcal{X}$, we obtain a lower bound on $F(\mathcal{X})$ for any choice of λ. Let λ^* be a PD that achieves the minimum in

$$\min_\lambda \sum_x \sum_y \lambda_x \lambda_y e^{-d(x,y)}. \tag{1}$$

Since by assumption the matrix $[e^{-d(x,y)}]$ is psd., the necessary and sufficient conditions for λ^* to achieve this minimum are given by the Karush-Kuhn-Tucker conditions, namely,

$$\sum_y \lambda_y^* e^{-d(x,y)} \geq c, \ \ \text{for all } x \text{ with equality if } \lambda_x^* > 0, \tag{2}$$

where c is a constant whose value can be found by multiplying the two sides of the inequality by λ_x^* and summing over x,

$$c = \sum_x \lambda_x^* \sum_y \lambda_y^* e^{-d(x,y)}. \tag{3}$$

It turns out that the parameter $R_0(\mathcal{X})$ defined by

$$R_0(\mathcal{X}) = -\log c \tag{4}$$

plays a crucial role here. In terms of this parameter, we notice that

$$F(\mathcal{X}, \lambda^*) = \sum_x \lambda_x^* \frac{1}{e^{-R_0(\mathcal{X})}} = e^{R_0(\mathcal{X})}. \tag{5}$$

This gives us the following lower bound.

Proposition 2. *Under our Supposition for a space $\mathcal{X}$,*

$$F(\mathcal{X}) \geq e^{R_0(\mathcal{X})}, \tag{6}$$

where

$$R_0(\mathcal{X}) = -\log \min_\lambda \sum_x \sum_y \lambda_x \lambda_y e^{-d(x,y)}. \tag{7}$$

As an example we note that for $\mathcal{X}$ as Hamming space, $\mathcal{X} = \{0,1\}$ with

$$d(x,y) = \begin{cases} 0 \text{ , if } x{=}y \\ 1 \text{ , else} \end{cases},$$

$R_0(\mathcal{X}) = \log[2/(1 + e^{-1})]$ and the lower bound is

$$F(\mathcal{X}) \geq \frac{2}{1 + e^{-1}}. \tag{8}$$

For use in the next section we note that $R_0(\mathcal{X}^n) = nR_0(\mathcal{X})$. This follows by observing that the optimality conditions (2) written for the space $\mathcal{X}^n$ are satisfied by a product-form distribution with marginals equal to λ^*. (We note the similarity of this result to the "parallel channel theorem" in [7], Chapter 5).

3.2   An Upper Bound

The following upper bound combined with the above lower bound establishes the product conjecture.

Proposition 3. *For all $n \geq 1$*

$$F(\mathcal{X}^n) \leq e^{nR_0(\mathcal{X})+o(n)}.$$

Before proving this proposition, let us show that the product conjecture follows as a consequence.

Theorem 1. *For spaces satisfying our Supposition, the fitness function is given by* $F(\mathcal{X}^n) = e^{nR_0(\mathcal{X})}$.

Proof: Suppose to the contrary that for some m, $F(\mathcal{X}^m) \geq e^{mR_0(\mathcal{X})+\epsilon}$ for some $\epsilon > 0$. Then, by the fact that $F((\mathcal{X}^m)^k) \geq (F(\mathcal{X}^m))^k$, we have $F(\mathcal{X}^{mk}) \geq e^{km(R_0(\mathcal{X})+\epsilon/m)}$. Since ϵ/m is not an $o(m)$ term, this contradicts Proposition 2. Hence, we must conclude that for all $m \geq 1$, $F(\mathcal{X}^m) \leq e^{mR_0(\mathcal{X})}$. Since the reverse inequality $F(\mathcal{X}^m) \geq e^{mR_0}$ has already been established, the conclusion follows. □

Proof of Proposition 2: Fix $n \geq 1$ arbitrarily. Let λ be any PD on $\mathcal{X}^n$. Let S be the support set of λ. For each $x \in S$, define

$$A_x = \sum_y \lambda_y e^{-d(x,y)}$$

Note that for all $x \in S$

$$e^{-nD} \leq A_x \leq 1$$

where $D = D(\mathcal{X})$ is the diameter of $\mathcal{X}$ which is finite by assumption. Fix $\delta > 0$ arbitrarily and put $K = \lceil nD/\delta \rceil$. For $k = 1, \ldots, K$ define

$$S_k = \{x \in S : e^{-k\delta} < A_x \leq e^{-(k-1)\delta}\}$$

Note that these sets form a partition of S. So, we may write and justify afterwards

$$F(\mathcal{X}^n, \lambda) = \sum_{k=1}^K \sum_{x \in S_k} \lambda_x \frac{1}{A_x} \tag{9}$$

$$= \sum_k \lambda(S_k) \sum_{x \in S_k} \frac{\lambda_x}{\lambda(S_k)} \frac{1}{A_x} \tag{10}$$

$$\leq \sum_k \lambda(S_k) \frac{e^\delta}{\sum_{x \in S_k} \frac{\lambda_x}{\lambda(S_k)} A_x} \tag{11}$$

$$= \sum_k \lambda(S_k) \frac{e^\delta}{\sum_{x \in S_k} \frac{\lambda_x}{\lambda(S_k)} \sum_{y \in S} \lambda_y e^{-d(x,y)}} \tag{12}$$

$$\leq \sum_k \lambda(S_k) \frac{e^\delta}{\sum_{x \in S_k} \frac{\lambda_x}{\lambda(S_k)} \sum_{y \in S_k} \lambda_y e^{-d(x,y)}} \tag{13}$$

$$= \sum_k \frac{e^\delta}{\sum_{x \in S_k} \frac{\lambda_x}{\lambda(S_k)} \sum_{y \in S_k} \frac{\lambda_y}{\lambda(S_k)} e^{-d(x,y)}} \tag{14}$$

$$\leq \sum_k \frac{e^\delta}{e^{-nR_0(\mathcal{X})}} \tag{15}$$

$$= K e^\delta e^{nR_0(\mathcal{X})} \tag{16}$$

In (10) we have used $\lambda(S_k) = \sum_{x \in S_k} \lambda_x$. Inequality (11) follows by the following argument. For shorthand put $p_x = \lambda_x/\lambda(S_x)$ and recall that, for all $x \in S_k$,

$e^{-k\delta} < A_x \le e^{-(k-1)\delta}$. Then,

$$\sum_{x \in S_k} p_x \frac{1}{A_x} \le \sum_{x \in S_k} p_x \frac{1}{e^{-k\delta}} \qquad (17)$$

$$= \frac{1}{\sum_{x \in S_k} p_x e^{-k\delta}} \qquad (18)$$

$$\le \frac{e^\delta}{\sum_{x \in S_k} p_x A_x} \qquad (19)$$

In line (15), we used the assumption (ii) that the distance matrix is psd., hence $R_0(\mathcal{X}^n) = nR_0(\mathcal{X})$. The remaining inequalities are self-explanatory. Now, we may choose $\delta = \sqrt{n}$, say, then $K \approx \sqrt{n}$, and we have

$$F(\mathcal{X}^n, \lambda) \le e^{nR_0(\mathcal{X})+o(n)}.$$

Since the upper bound holds uniformly for all PDs λ, the fitness of the space is also upper-bounded by $e^{nR_0(\mathcal{X})+o(n)}$. This completes the proof. □

Remark 1

1. *It does not follow from the above results that $F(\mathcal{X}, \lambda)$ is a concave function of λ.*

2. *The proof can possibly be extended to any distance space with a bounded distance function but generalization to arbitrary distance spaces is not at all obvious.*

3. *The assumption about the positive semidefiniteness of the distance matrix appears to be essential. The Hamming metric, the metrics $|x-y|$ and $(x-y)^2$ defined on real spaces satisfy this constraint, as we show in the next section.*

3.3   A Connection Between Fitness and Parameters of Communication Channels

It is noteworthy that the Nowak fitness has an interesting relationship to pairwise error probabilities in noisy channels. Given a discrete memoryless channel W : $\mathcal{A} \to \mathcal{B}$, the Bhattacharyya distance (B-distance) between two input letters $a, a' \in \mathcal{A}$ is defined as

$$d_B(a, a') = -\log \sum_{b \in \mathcal{B}} \sqrt{W(b|a)W(b|a')}.$$

The cutoff rate parameter of the channel is defined as

$$R_0(W) = -\log \min_\lambda \left[\sum_{a \in \mathcal{A}} \sum_{a' \in \mathcal{A}} \lambda_a \lambda_{a'} e^{-d(a,a')} \right],$$

where the minimum is over all PDs $\lambda = \{\lambda_a : a \in \mathcal{A}\}$.

To illustrate the connection between fitness and channel coding, let $\mathcal{X} = \{0,1\}$ with d the Hamming metric. The Hamming distance $d(x,y)$ for any $x,y \in \mathcal{X}$ equals the B-distance $d_B(x,y)$ of a binary symmetric channel $W : \mathcal{X} \to \mathcal{X}$ with crossover probability ϵ chosen so that $d_B(0,1) = 1$, i.e., $\sqrt{4\epsilon(1-\epsilon)} = e^{-1}$. For W chosen this way, the cutoff rate of the BSC equals $R_0(W) = \log[2/(1+e^{-1})]$. Thus, the $R_0(\mathcal{X})$ that appears as the exponent in the fitness growth rate for space $(\mathcal{X}, d)$ can be identified as the cutoff rate $R_0(W)$ of the associated BSC W.

This type of association between the metrics considered by Nowak et al. and B-distances of DMC's can be established in certain other cases as well. E.g., the metric $|x-y|$ is the B-distance for an exponential noise channel $W : X \to X+N$, where $X \geq 0$ is the channel input and $X + N$ is the channel output with N equal to an independent exponentially distributed random variable with intensity $\mu = 2$ (mean $1/2$). Likewise, the metric $(x - y)^2$ can be interpreted as the B-distance for a Gaussian noise channel. Whenever a distance d can be associated with the B-distance of a channel, the matrix $[e^{-d(x,y)}]$ is a Gramm matrix and hence psd. Thus, the product conjecture holds for such distances on finite spaces.

This association between the fitness model and noisy communication channels is significant in that it explains the confoundability of phonemes as the result of the phonemes being sent across a noisy channel. This association also helps interpret Nowak's formula in terms of well-studied concepts in information theory, such as pairwise error probabilities and average list sizes in list-decoding.

3.4   Embedding of Distance Spaces

Let (X, d) and (X', d') be two distance spaces. Then (X, d) is said isometrically embeddable into (X', d') if there exists a mapping Φ (the isometric embedding) from X to X' such that $d(x,y) = d'(\Phi(x), \Phi(y))$ for all $x,y \in X$. For any $p \geq 1$, the vector space $\mathcal{R}^m$ can be endowed with the l_p-norm defined by $||x||_p = (\sum_{k=1}^m |x_k|^p)^{\frac{1}{p}}$ for $x \in \mathcal{R}^m$. The associated metric is denoted by d_{l_p}. The metric space (R^m, d_{l_p}) is abbreviated as l_p^m. A distance space is said to be l_p-embeddable, if (X, d) is isometrically embeddable into the space l_p^m for some integer $m \geq 1$. We call a distance space psd., if the corresponding matrix $[e^{-d(x,y)}]$ is psd.

Lemma 1. *If a distance space is psd., then all distance subspaces are also psd. Furthermore all distance spaces, which can be isometrically embedded in a subspace of a psd. distance space are psd.*

Proof: If $[e^{-d(x,y)}]$ is psd., then for all non-zero vectors x in $\mathcal{R}^n$ we have

$$x^T [e^{-d(x,y)}] x \geq 0.$$

This property remains if we delete a finite number of columns and rows of $[e^{-d(x,y)}]$. Therefore the remaining space is still psd.   □

With the help of this lemma it is possible for us to establish the product conjecture for an arbitrary distance space whenever it is possible to embed it in a larger distance space which is psd. The following theorems of Vestfried and Fichet are very useful.

Theorem 1 (V, [16]). *Any separable ultrametric space is l_2-embeddable.*

Theorem 2 (F, [6]). *Any metric space with 4 points is l_1-embeddable.*

We describe now in a proposition situations where this technique applies. Recall that in an ultra-metric space for any three points a, b, c holds $d(a, b) \leq \max(d(a, c), d(c, b))$.

Proposition 4. *1. All ultra-metric spaces are psd.*
2. All finite metric spaces with up to 4 elements are psd.
3. There exist some metric spaces with 5 elements which are not psd.
4. For every distance space there exists a scaling, such that the space becomes psd.

Proof: 1. follows from the theorem of Vestfried and Lemma 1.
2. follows from the theorem of Fichet. To show 3. consider the following metric space on five points: Let for $i \neq j$ $d(i, j) = a$ if $i, j \in \{1, 2, 3\}$ and $d(i, j) = \frac{a}{2}$ otherwise. Then if $0 < a < 7.07 \cdot 10^{-6}$ the corresponding matrix is not psd. 4. follows, because the matrix $[e^{-\alpha d(x,y)}]$ converges for $\alpha \to \infty$ to the identity matrix. $\square$

4   A Hamming Code Is a Good Language

In the previous section we have shown that the product conjecture is true in particular for the Hamming model. The optimal fitness is attained at $\lambda = (\frac{1}{2^n}, \ldots, \frac{1}{2^n})$. But this means, that one has to use all possible words in the language to achieve the optimal fitness. In general the memory of the individuals is restricted. For this reason we look for languages, which use only a fraction of all possible words, but have large fitness.

We consider simple and perfect codes: The Hamming codes ([8]). A q-ary block-code of length n is a map c from a finite set $\mathcal{O}$ to $\{0, 1, \ldots, q-1\}^n$. $c(o)$ with $o \in \mathcal{O}$ is called a codeword and $\mathcal{C} = \{c(o) : o \in \mathcal{O}\}$ is called the code. Thus we can view each code as a language. There exists a lot of work about codes (see [11]). A special class of codes are the t-error correcting block-codes. These codes have the property that for two different codewords the Hamming-distance is larger than $2t+1$. For a block-code of length n the weight-distribution $(A_0, A_1, \ldots, A_n)$ and the distance distribution $(B_0, \ldots, B_n)$ are defined. A_i denotes the number of codewords of weight i and B_i is the number of ordered pairs of codewords (u, v) such that $d(u, v) = i$ divided by the number of messages. We summarize the properties of the single-error-correcting Hamming-codes.

Proposition 5. *1. Hamming codes exist for the lengths $2^k - 1$.*
2. Their number of codewords is $N = 2^{2^k - 1 - k}$. The minimal distance is 3.
3. The weight distribution is the same for each word.

In [9] and [10] it is shown that the weight distribution is very easy to calculate. Let $(A_0, A_1, \ldots, A_n)$ be the distance-distribution of the Hamming-code $\mathcal{C}$, then we define the Hamming weight enumerator by

$$W_{\mathcal{C}}(x) = \sum_{i=0}^{n} A_i x^i.$$

Theorem 3 (McW, [9],[10]). *Let* $(A_0, A_1, \ldots, A_n)$ *be the distance-distribution of the Hamming-code* $\mathcal{C}$, *then the Hamming weight enumerator of this code is given by*

$$W(x) = \frac{1}{n+1} \left((1+x)^n + n(1-x)(1-x^2)^{\frac{n-1}{2}} \right).$$

With $F_H(n)$ we denote the fitness of a Hamming Code of length n.

Theorem 2. *The fitness of the Hamming code approaches asymptotically the optimal fitness. Not only* $\lim_{n \to \infty} \frac{1}{n} F_H(n) = \lim_{n \to \infty} \frac{1}{n} F(\mathcal{X}^n)$ *and* $\lim_{n \to \infty} \frac{F_H(n)}{F(\mathcal{X}^n)} = 1$, *but even the stronger condition*

$$\lim_{n \to \infty} F_H(n) - F(\mathcal{X}^n) = 0$$

holds.

Proof

The fitness of the Hamming code can be expressed using the weight enumerator W.

$$F_H(n) = \frac{2^{2^k - 1 - k}}{W(exp(-1))} = \frac{2^{n - \log_2(n+1)}}{W(exp(-1))}.$$

We now show that the difference $F(\mathcal{X}^n) - F_H(n)$ goes to zero.

$$F(\mathcal{X}^n) - F_H(n) = \left(\frac{2}{1 + exp(-1)} \right)^n - \frac{2^{n - \log_2(n+1)}}{W(exp(-1))}$$

$$= \left(\frac{2}{1 + e^{-1}} \right)^n - \frac{2^{n - \log_2(n+1)}}{\frac{1}{n+1}(1+e^{-1})^n + n(1-e^{-1})(1-e^{-2})^{\frac{n-1}{2}}}$$

$$= \left(\frac{2}{1 + e^{-1}} \right)^n - \frac{2^n}{(1+e^{-1})^n + n(n+1)(1-e^{-1})(1-e^{-2})^{\frac{n-1}{2}}}$$

$$= \frac{2^n n(n+1)(1-e^{-1})(1-e^{-2})^{\frac{n-1}{2}}}{(1+e^{-1})^{2n} + (1+e^{-1})^n n(n+1)(1-e^{-1})(1-e^{-2})^{\frac{n-1}{2}}}$$

$$\leq \frac{2^n n(n+1)(1-e^{-1})(1-e^{-2})^{\frac{n-1}{2}}}{(1+e^{-1})^{2n}},$$

$$= \frac{\left(2\sqrt{(1-e^{-2})}\right)^n n(n+1)(1-e^{-1})}{((1+e^{-1})^2)^n \sqrt{(1-e^{-2})}}.$$

The last term goes to zero if n goes to infinity, because

$$\frac{2\sqrt{1-e^{-2}}}{(1+e^{-1})^2} < 1,$$

$\left(\frac{2\sqrt{1-e^{-2}}}{(1+e^{-1})^2} < 0.995\right)$. Since the difference is always positive the proof is complete.   □

Next we show that *ratewise* the fitness of the Hamming space is attained if we choose the middle level as a language.

Suppose that n is even and let the language $\mathcal{L}$ consist of all words x^n with exactly $\frac{n}{2}$ ones, i.e. $w(x^n) = \frac{n}{2}$. If we fix any word from this language then there are $\binom{\frac{n}{2}}{j}^2$ words in $\mathcal{L}$ at a distance of $2j$, $(j = 0, \ldots, \frac{n}{2})$. Therefore the fitness of $\mathcal{L}$ is

$$F(\mathcal{L}, \mathcal{X}^n) = \frac{\binom{n}{\frac{n}{2}}}{\sum_{j=0}^{\frac{n}{2}} \binom{\frac{n}{2}}{j}^2 e^{-2j}}.$$

Let $j^*(n)$ denote the j for which the summand in the denominator is maximal and let $\tau^*(n) = \frac{j^*(n)}{n}$. Then we can estimate the rate of the fitness of $\mathcal{L}$ as follows. Let $\epsilon > 0$.

$$\frac{1}{n} \log F(\mathcal{L}, \mathcal{X}^n) = \frac{1}{n} \log \binom{n}{\frac{n}{2}} - \frac{1}{n} \log \sum_{j=0}^{\frac{n}{2}} \binom{\frac{n}{2}}{j}^2 e^{-2j}$$

$$\geq \frac{1}{n} \log \binom{n}{\frac{n}{2}} - \frac{1}{n} \log \left((\frac{n}{2} + 1) \binom{\frac{n}{2}}{j^*(n)}^2 e^{-2j^*(n)} \right)$$

$$= \frac{1}{n} \log \binom{n}{\frac{n}{2}} - \frac{1}{n} \log \left(\frac{n}{2} + 1 \right) - \frac{1}{\frac{n}{2}} \log \binom{\frac{n}{2}}{2\tau^*(n) \cdot \frac{n}{2}} + 2\tau^*(n) \log(e),$$

which we can bound further for sufficiently large n by

$$\geq 1 - 0 + \min_{\tau} \{ -h(2\tau) + 2\tau \log(e) \} - \epsilon, \tag{20}$$

where h is the binary entropy function, $h(\tau) = -\tau \log \tau - (1 - \tau) \log(1 - \tau)$.

We can find the minimum of the convex function $-h(2\tau) + 2\tau \log(e)$ by looking at the root of the first derivative. The first derivative is $2\log(2\tau) - 2\log(1 - 2\tau) + 2\log(e)$, which is zero for $\tau = \frac{1}{2(1+e)}$. Substituting this in (20) we can conclude that for sufficiently large n

$$\frac{1}{n} \log F(\mathcal{L}, \mathcal{X}^n) \geq 1 - \log(1 + e^{-1}) - \epsilon.$$

The opposite inequality $\frac{1}{n} \log F(\mathcal{L}, \mathcal{X}^n) \leq 1 - \log(1 + e^{-1})$ is also true because we know from Theorem 1 that for the Hamming space $F(\mathcal{X}^n) = \left(\frac{2}{1+e^{-1}} \right)^n$. Therefore we can summarize our result in the following theorem.

Theorem 3. *Let $\mathcal{L}$ be the language in the Hamming space $\mathcal{X}^n$ that consists of all words of weight $\frac{n}{2}$. Then the fitness of the language $\mathcal{L}$ is ratewise optimal, i.e.,*

$$\lim_{n \to \infty} \frac{1}{n} \log F(\mathcal{L}, \mathcal{X}^n) - \frac{1}{n} \log F(\mathcal{X}^n) = 0.$$

Theorem 4. *Let c be a fixed integer and $\mathcal{L}$ be the language in the Hamming space $\mathcal{X}^n$ that consists of all words of weight $\frac{n}{2}$ with $\lceil \frac{c}{2} \rceil$ fixed position with 0's and $\lfloor \frac{c}{2} \rfloor$ fixed positions with 1's. Then the fitness of the language $\mathcal{L}$ is also ratewise optimal, i.e.,*

$$\lim_{n \to \infty} \frac{1}{n} \log F(\mathcal{L}, \mathcal{X}^n) - \frac{1}{n} \log F(\mathcal{X}^n) = 0.$$

Proof: We assume that n and c are even. Following the same idea as in Theorem 3 we get for $\epsilon > 0$.

$$\frac{1}{n} \log F(\mathcal{L}, \mathcal{X}^n) = \frac{1}{n} \log \left(\frac{1}{2^c} \binom{n}{\frac{n}{2}} \right) - \frac{1}{n} \log \sum_{j=0}^{\frac{n-c}{2}} \binom{\frac{n-c}{2}}{j}^2 e^{-2j}$$

$$\geq \frac{1}{n} \log \left(\frac{1}{2^c} \binom{n}{\frac{n}{2}} \right) - \frac{1}{n} \log \left((\frac{n-c}{2} + 1) \binom{\frac{n-c}{2}}{j^*(n)}^2 e^{-2j^*(n)} \right)$$

$$= \frac{1}{n} \log \left(\frac{1}{2^c} \binom{n}{\frac{n}{2}} \right) - \frac{1}{n} \log \left(\frac{n-c}{2} + 1 \right) - \frac{1}{\frac{n}{2}} \log \left(\frac{\frac{n-c}{2}}{2\tau^*(n) \cdot \frac{n}{2}} \right) + 2\tau^*(n) \log(e),$$

which we can bound further for sufficiently large n by

$$\geq 1 - 0 + \min_{\tau} \left\{ -h(2\tau) + 2\tau \log(e) \right\} - \epsilon. \tag{21}$$

□

5   A Language with Noiseless Feedback

In this section we consider a language with noiseless feedback. The channel model is well known in Information Theory ([3], [2]). It can be described in our language model as follows. Individual A signalled a letter (word of length 1) and is informed which letter individual B understood (because of some reaction of B). Individual A has a special strategy for each object. After n repetitions of this procedure B notices some object with a certain probability. We denote the set of objects like before by $\mathcal{O} = \{o_1, \ldots, o_N\}$.

The functions

$$st_j(o_i, y^{j-1})$$

for $j = 1, \ldots, n$ define the next signal given by the speaker if he wants to speak about object i and the listener understands $y^n \in \{0,1\}^n$. Thus

$$st_j : \mathcal{O} \times \{0,1\}^{j-1} \to \{0,1\}.$$

We define the set of error vectors by

$$\mathcal{E} = \{0,1\}^n.$$

This is the set of all possible error vectors. Let $e^n = (e^n(1), \ldots, e^n(n)) \in \mathcal{E}$, if $e^n(t) = 1$ then an error happened at the t-th position, otherwise $e^n(t) = 0$.

We set $0^n = (0, \ldots, 0)$ a vector of length n. The error vector and the strategy determine what the the speaker says. Thus we have a function

$$st : \mathcal{O} \times \mathcal{E} \to \{0, 1\}^n$$

where $st(o_i, e^n)$ is defined by

$$(st_1(o_i), st_2(o_i, st_1(o_1) + e_1 = y^1), \ldots, st_n(o_i, st_{n-1}(o_{n-1}, y^{n-2}) + e_{n-1} = y^{n-1})).$$

We define the feedback-language as $\mathcal{L}^{st} = (st(o_t, 0^n))_{t=1}^N$. We need a distance-function to define the fitness in this case. We define the similarity for two words as follows. $s(x^n, y^n) = e^{-t}$, where

$$t = \begin{cases} \min\{w(e^n) : st(o_i, e^n) \oplus e^n = y^n\} & \text{if } \exists y^n : st(o_i, e^n) \oplus e^n = y^n \\ 0 & \text{otherwise} \end{cases}.$$

The feedback fitness of a strategy is defined as

$$F^f(st, \mathcal{X}^n) = \sum_{t=1}^{|\mathcal{L}^{st}|} \sum_{e^n : st(o_t, e^n) \in \mathcal{L}} \frac{1}{\sum_{e^n : st(o_t, e^n) \in \mathcal{L}} s(st(o_t, 0^n), st(o_t, e^n))}$$

and the fitness is defined as the maximal possible value of the fitness of all strategies in $\mathcal{X}^n$. Thus

$$F^f(\mathcal{X}^n) = \sup_{st}\{F^f(st, \mathcal{X}^n)\}.$$

This is a generalization of the model without feedback. If the speaker just ignores the feedback, we get the same model like before. We write F^f for all fitness definitions, if we consider the fitness with feedback.

Proposition 6

$$F(\mathcal{X}^n) = F(\mathcal{X}^n, \mathcal{X}^n) = F^f(\mathcal{X}^n, \mathcal{X}^n).$$

Proof: The property holds because, if we use all possible words of a language, all similarities between the words occur in the fitness formula in the summands just in another order. $\quad \Box$

Now we will give an example where the feedback-fitness is bigger than the usual fitness. For the case $n = 3$ we know that $F(\{0, 1\}^3) = \left(\frac{2}{1 + exp(-1)}\right)^3$. We will show now that the fitness can be increased with feedback. We give an example for a feedback-language with seven objects and a bigger fitness.

Example: Strategy f: Map the i-th object to the binary representation of i. If a 1 is understood as a 0 start saying 0.

	o_1	o_2	o_3	o_4	o_5	o_6	o_7
$t = 0$	001	010	011	100	101	110	111
	000	000	000	000	000	000	000
$t = 1$	011	110	111	101	100	111	110
	101	011	010	110	111	100	100
	100	001	001	001	001	001	001
$t = 2$	010	111	110	010	010	010	010
	111	100	100	111	110	101	101
$t = 3$	110	101	101	011	011	011	011

Obviously $\mathcal{L}^{st} = \{0,1\}^n \backslash 0^n$.

It holds $F^f(\mathcal{L}^*) = 3, 19 > F(\mathcal{X}^n, \mathcal{X}^n)$. Our strategy can be generalized and gives a lower bound for the feedback fitness.

Proposition 7

$$F^f(\mathcal{X}^n) \geq \frac{2^n - 1}{\left(\sum_{j=0}^n \binom{n}{j} e^{-j}\right) - e^{-1}}.$$

Proof: Use the generalization of the strategy in the example and the result follows. □

It is also possible to give a trivial upper bound.

Proposition 8

$$F^f(\mathcal{X}^n, N) \leq \frac{N}{1 + (N-1)e^{-n}}.$$

Proof: The smallest possible similarity between two different words is e^{-n}. Thus we assume that all similarities of all possible words are as small as possible and get the upper bound for the fitness. □

6 List-Language

In a "list-language", we divide the words of a language $\mathcal{L}$ into lists (subfamilies). For example words about food, words about danger e.t.c.. The goal of the listener is just to find out about which list the speaker speaks. To simplify the situation we assume that all words of the language $\mathcal{L}$ belong to exactly one list, all lists are of the same size l and we look only at languages with $l|N$, ($N = |\mathcal{L}|$). In general, if $|\mathcal{L}| = l \cdot k + r$ with $r < l$, we have r lists of size $l + 1$ and $k - r$ lists of size l, i.e., here we assume that $r = 0$ and call such a language an l-list-language.

We denote the lists by $\mathcal{L}_i$ for $i = 1, \ldots, k$. We set $L(x^n) = \mathcal{L}_i$, if the word x^n belongs to the list $\mathcal{L}_i$.

In a list-language the individuals get some profit, if the listener understands the list of the speaker. Therefore we define

$$F^l(\mathcal{L}, \mathcal{X}^n) = \sum_{x^n \in \mathcal{L}} \sum_{y^n \in L(x^n)} p(x^n, y^n).$$

Then naturally the question of the best l-list-language arises:

$$F^l(\mathcal{X}^n) = \sup\{F^l(\mathcal{L}, \mathcal{X}^n) : \mathcal{L} \text{ is } l - \text{list} - \text{language in } \mathcal{X}^n\}.$$

Next we calculate the fitness of list-languages in a special case, namely that of *constant similarity*. Let $C > 0$ be a constant and let d be the following metric on $\mathcal{X}$

$$d(x, y) = \begin{cases} 0 & \text{, if } x=y \\ C & \text{, else} \end{cases}.$$

In this case the following proposition holds.

Proposition 9. $F^l(\mathcal{L}, \mathcal{X}) \leq F(\mathcal{L}, \mathcal{X}) + l - 1$

Proof

$$F^l(\mathcal{L}, \mathcal{X}) = \sum_{x \in \mathcal{L}} \sum_{y \in L(x)} \frac{exp(-d(x, y))}{\sum_{z \in \mathcal{L}} exp(-d(x, z))}$$

$$= F(\mathcal{L}, \mathcal{X}) + \sum_{x \in \mathcal{L}} \sum_{y \in L(x), y \neq x} \frac{exp(-d(x, y))}{\sum_{z \in \mathcal{L}} exp(-d(x, z))}$$

$$= F(\mathcal{L}, \mathcal{X}) + \frac{Nexp(-C)(l - 1)}{1 + (N - 1)exp(-C)} \leq F(\mathcal{L}, \mathcal{X}) + l - 1.$$

$\square$

7   Multi-access-Language

In this section we will consider the following situation. Two individuals speak simultaneously. There is some interference and one individual wants to understand both. We look at two models. In the first model the speakers use the same language, in the second model they use different languages. Such models are well known in Information Theory. They were introduced in [1].

7.1   Model I

In this model $\mathcal{X} = \{0, 1\}$ and $\mathcal{Y} = \{0, 1, 2\}$. The individuals can only speak words which contain the signals 0 and 1. The listener understands 0 if both use the signal 0. He understands 1, if one individual uses the signal 0 and the other the signal 1 and he understands 2 if both use the signal 1. The listener understands some word in $\mathcal{Y}^n$. We search now for a language with the biggest multi-access-fitness. This model is known in information theory as the binary adder channel.

We set $d((x^n, y^n), (v^n, w^n)) = d_H(x^n + y^n, v^n + w^n)$, where $x^n + y^n = (x_1 + y_1, \ldots, x_n + y_n)$ and define the fitness of a multi-access-adder-language as

$$F_A(\mathcal{L}, \mathcal{X}^n) = \sum_{i=1}^{N} \sum_{j=1}^{N} p((x(i)^n, y(j)^n), (x(i)^n, y(j)^n)).$$

The probability and the similarity are defined as before. We will consider an example for $n = 2$. The language contains all elements of $\{0,1\}$ exactly once. Thus we get the following table:

	$r(o_1) = 00$	$r(o_2) = 01$	$r(o_3) = 10$	$r(o_4) = 11$
$r(o_1) = 00$	00	01	10	11
$r(o_2) = 01$	01	02	11	12
$r(o_3) = 10$	10	11	20	21
$r(o_4) = 11$	11	12	21	22

Now we have for example $d((01,01),(10,01)) = d(02,11) = 2$. The fitness of this language is $F_A(\mathcal{L}, \mathcal{X}^n) \approx 2.83$.

Proposition 10
$$F_A(\mathcal{X}^n) \leq F(\{0,\ldots,2|\mathcal{X}|\}).$$

We know consider a generalization of this model. The speaker uses two different languages over the same distance space. We search for two languages which have the biggest common multi-access-fitness.

$$F_A(\mathcal{L}, \mathcal{M}, \mathcal{X}^n) = \sum_{i=1}^{m} \sum_{j=1}^{k} p((x(i)^n, y(j)^n), (x(i)^n, y(j)^n)).$$

For example let $\mathcal{L} = (00, 01, 10, 11)$ and $\mathcal{M} = (00, 11)$. Then we get the following table.

	00 01 10 11
00	00 01 10 11
11	11 12 21 22

The fitness of this language is $F_A(\mathcal{L}, \mathcal{M}) = 2,71$.

7.2   Model II

In this model $\mathcal{X} = \mathcal{Y} = \{0,1\}$ and $d((x^n, y^n), (x^n, y^n)) = d_H(x^n \oplus y^n, x^n \oplus y^n)$, where $\oplus$ is the sum modulo $|\mathcal{X}| = 2$ in all components. All other definition are the same. Let us look at our example:

	00 01 10 11
00	00 01 10 11
01	01 00 11 10
10	10 11 00 01
11	11 10 01 00

All words are contained four times in the table. Thus this language attains the maximum, because the product conjecture holds. This can be generalized.

Theorem 5. *The optimal fitness for the adder model II is attained, if the language consists of all possible codewords.*

Another configuration with the same fitness as the previous example:

$$
\begin{array}{c|c}
 & 00\ 01 \\
\hline
00 & 00\ 01 \\
10 & 10\ 11
\end{array}
$$

If we allow two different languages for the two speakers, we find more configurations, which attain the optimal fitness.

Let us look at our example:

$$
\begin{array}{c|c}
 & 00\ 01\ 10\ 11 \\
\hline
00 & 00\ 01\ 10\ 11
\end{array}
$$

8 Broadcast to Two Different Languages

In this section we will consider the following situation. We have two individuals with two different languages $\mathcal{L} = (x(1), \dots, x(N))$ and $\mathcal{M} = (y(1), \dots, y(N))$ on the same distance space $(\mathcal{X}, d)$, such that $x(i)$ describes the same object as $y(i)$ for all $i = 1, \dots, N$. Our goal is to find a good language for a third individual, which wants to communicate with both of them simultaneously. In Information Theory this kind of models were introduced in [4].

We define the fitness between two languages as

$$
F(\mathcal{L}, \mathcal{M}) = \sum_{i=1}^{N} \frac{exp(-d(x(i), y(i)))}{\sum_{j=1}^{N} exp(-d(x(i), y(j)))}.
$$

There exists also examples in human language, where both people can speak in their own language and understand each other. An example is a conversation between a Swede and a Dane, who both speak in their language.

We define the fitness of a broadcast-language $\mathcal{N}$ as

$$
F_B\left(\mathcal{N}, (\mathcal{L}, \mathcal{M}), \mathcal{X}^n\right) = \frac{1}{2}\left(F\left(\mathcal{N}, \mathcal{L}\right) + F\left(\mathcal{N}, \mathcal{M}\right)\right).
$$

Proposition 11

$$
F_B\left(\mathcal{N}, (\mathcal{L}, \mathcal{M}), \mathcal{X}^n\right) \geq \frac{1}{2}\max\{F(\mathcal{L}, \mathcal{X}^n) + 1, F(\mathcal{M}, \mathcal{X}^n) + 1\}.
$$

9 Language Without Multiplicity

In all previous sections we allowed multiplicity of words. That means the individuals were allowed to use one word for more than one object. We will show that in the case without multiplicity there are examples, where the fitness of a product space is bigger than the product of the fitnesses of the single spaces.

Again we consider the set of objects

$$
\mathcal{O} = \{o_1, \dots, o_N\}
$$

and now each object is mapped to a sequence of signals by the injective function

$$r : \mathcal{O} \to \mathcal{X}^n.$$

We call the languages of this type *injective* and denote the corresponding fitness values by F_{in}.

We consider the metric space $(\mathcal{M} = \{a, b, c\}, d)$, where the distance is defined as follows:

d	a	b	c
a	0	0.01	3
b	0.01	0	3
c	3	3	0

In this case holds:

$$F_{in}(\mathcal{M}) = F_{in}(\{a, c\}, \mathcal{M}) = \frac{2}{1 + e^{-3}} > F_{in}(\{a, b, c\}, \mathcal{M}),$$

but for the product we have:

$$F_{in}(\mathcal{M}^2) = F_{in}(\{aa, ac, cb, cc\}, \mathcal{M}^2) > F_{in}(\{aa, ac, ca, cc\}, \mathcal{M}^2).$$

Thus $F_{in}(\mathcal{M})^2 < F_{in}(\mathcal{M}^2)$. This means the product conjecture does not hold for injective languages. The reason for this behavior is, that the distance between a and b is very small and the optimal fitness does not consist of all possible letters. In the product space we can use the unused letter to improve the fitness. This counterexample does not work in the original problem, because in the case of such a finite metric space it is always better to choose all elements with a certain multiplicity.

Acknowledgment. The authors would like to thank V. Blinovsky and E. Telatar for discussions on these problems and P. Harremoes for drawing their attention to the counter-example in the case without multiplicity.

References

1. R. Ahlswede, Multi-way communication channels, Proceedings of 2nd International Symposium on Information Theory, Thakadsor, Armenian SSR, Sept. 1971, Akademiai Kiado, Budapest, 23-52, 1973.
2. R. Ahlswede, I. Wegener, Suchprobleme, Teubner, 1979, English translation: Wiley, 1987, Russian translation: MIR, 1982.
3. E.R. Berlekamp, Block coding for the binary symmetric channel with noiseless, delayless feedback in H.B.Mann, Error Correcting Codes, Wiley, 61-85, 1968.
4. T.M. Cover, Broadcast channels, IEEE Trans. Inform. Theory, Vol. 18, 2-14, 1972.
5. A. Dress, The information storage capacity of a metric space, preprint.
6. B. Fichet, Dimensionality problems in l_1-norm representations, in Classification and Dissimilarity Analysis, Lecture Notes in Statistics, Vol. 93, 201-224, Springer-Verlag, Berlin, 1994.

7. R.G. Gallager, Information Theory and Reliable Communication, New York, Wiley, 1968.

8. R.V. Hamming, Error detecting and error correcting codes, Bell Sys. Tech. Journal, 29, 147-160, 1950.

9. F.J. MacWilliams, Combinatorial problems of elementary group theory, Ph.D. Thesis, Harvard University, 1962.

10. F.J. MacWilliams, A theorem on the distribution of weights in a systematic code, Bell syst. Tech. J., Vol. 42, 79-94, 1963.

11. F.J. MacWilliams and N.J.A. Sloane, The Theory of Error-Correcting Codes, Elsevier Science Publishers B.V., 1977.

12. M.A. Nowak and D.C. Krakauer, The evolution of language, PNAS 96, 14, 8028-8033, 1999.

13. M.A. Nowak, D.C. Krakauer, and A. Dress, An error limit for the evolution of language, Proceedings of the Royal Society Biological Sciences Series B, 266, 1433, 2131-2136, 1999.

14. J.B. Plotkin and M.A. Nowak, Language evolution and information theory, J. theor. Biol. 205, 147-159, 2000.

15. C.E. Shannon, The zero-error capacity of a noisy channel, IRE Trans. Inform. Theory E, 8–19, 1956.

16. A.F. Timan and I.A. Vestfried, Any seperable ultrametric space is isometrically embeddable in l_2, Funk. Anal. Pri. 17, 1, 85-86, 1983.

Zipf's Law, Hyperbolic Distributions and Entropy Loss

P. Harremoës and F. Topsoe*

Abstract. Zipf's law – or Estoup-Zipf's law – is an empirical fact of computational linguistics which relates rank and frequency of words in natural languages. The law suggests modelling by distributions of "hyperbolic type". We present a satisfactory general definition and an information theoretical characterization of the resulting *hyperbolic distributions*. When applied to linguistics this leads to a property of stability and flexibility, explaining that a language can develop towards higher and higher expressive powers without changing its basic structure.

Keywords: Zipf's law, hyperbolic distributions, entropy loss.

1 Zipf's Law

Consider word usage in a comprehensive section of a language such as a novel, a collection of newspaper texts or some other material, in the following referred to as "the text". The text will contain a number of distinct words, each occurring with a certain frequency. The words may be characterized by their *rank*. The most frequent word in the text has rank 1, the second most frequent word has rank 2 and so on.

In 1916 the French stenographer J.B. Estoup noted that rank (r) and frequency (F) in a French text were related by a "hyperbolic" law which states that $r \cdot F$ is approximately constant, cf. [1]. This observation became well known after studies by the American linguist George Kingsley Zipf (1902–1950). He collected his findings in the monograph "Human Behavior and the Principle of Least Effort" from 1949, cf. [6]. Zipf could confirm that the hyperbolic rank-frequency relationship appeared to be a general empirical law, valid for any comprehensive text and with a surprisingly high accuracy. Because of Zipf's careful studies, the law is now known as *Zipf's law*.

In [6] Zipf argues that in the development of a language, a certain *vocabulary balance* will eventually be reached as a result of two opposing forces, the force of *unification* and the force of *diversification*. The first force tends to reduce the vocabulary and corresponds to a principle of least effort seen from the point of view of the speaker, whereas the second force has the opposite effect and is connected with the auditors wish to associate meaning to speech. Though Zipf does not transform these ideas into a mathematical model, we note his basic

* Peter Harremoës is supported by a post-doc stipend from the Villum Kann-Rasmussen Foundation and both authors are supported by the Danish Natural Science Research Council and by INTAS (project 00-738).

R. Ahlswede et al. (Eds.): Information Transfer and Combinatorics, LNCS 4123, pp. 788–792, 2006.
© Springer-Verlag Berlin Heidelberg 2006

consideration as a two-person game, however without a precise definition of the cost-functions involved.

Zipf's study relied on very thorough empirical investigations. He used James Joyce's *Ulysses* with its 260.430 running words as his primary example. Ulysses contains 29.899 different words. The hyperbolic rank-frequency relationship is illustrated by plotting the points (r, F_r); $r \leq 29.899$ on doubly logarithmic paper with F_r the number of occurrences in the text of the word with rank r. The result is quite striking and clearly reveals the closeness to an exact hyperbolic law $r \cdot F_r = C$. Some of the frequencies found by Zipf are listed in Table 1.

Table 1. Rank-frequency in Ulysses (adapted after [6])

r	F_r	$r \cdot F_r$
10	2.653	26.530
20	1.311	26.220
100	265	26.500
500	50	25.000
2000	12	24.000
5000	5	25.000
10000	2	20.000
20000	1	20.000
29899	1	29.899

If we model the rank-frequency relation by a probability distribution we are led to a *harmonic distribution*, which we shall here take to mean a distribution over a section of the natural numbers, here $\{1, 2, \ldots, 29.899\}$, for which the i'th point probability is proportional to $\frac{1}{i}$. According to Zipf, cf. notes to Chapter two in [6], the choice of Ulysses was made as it was expected that a harmonic distribution would *not* be found in a large and artistically sophisticated text as this [1].

The positive findings have led to the general acknowledgement of Zipf's law as an empirical fact[2]. However, there is of course something dubious about this. Clearly, in the above example, 29.899 is no sacred number. The phenomenon is a limiting phenomenon — a phenomenon of vocabulary balance in Zipf's words — and, given the time, James Joyce would surely have used more words or be forced to introduce new words in order to increase his expressive power. This points to a need for models based on probability distributions over the entire set $\mathbb{N}$ of natural numbers. A key goal of the research reported on here is to define precisely a class of distributions, called *hyperbolic distributions*[3], which serves this purpose.

---

[1] Our theoretical findings later point to the expectation that sophisticated texts as Ulysses (with a high bit rate) will follow Zipf's law more closely than other texts.

[2] Linguists today have some reservations about the law and seek more precise relationships and associated models. This search is facilitated by modern computer technology. The reader may want to visit http://www.ucl.ac.uk/english-usage/ in this connection.

[3] The literature dealing with Zipf's law does operate with a notion of hyperbolic distributions, but, typically, these are not precisely defined and also incorporate what we called harmonic distributions above, hence allowing distributions with finite support.

Shannon used Zipf's law to estimate the entropy of English words in his well–known study [5] from 1951. Other studies include an interesting paper from 1961 by B. Mandelbrot who essentially argues that a purely random mechanism will generate a text obeying Zipf's law, cf. [3]. As put by Schroeder, cf. [4], "a monkey hitting typewriter keys *at random* will also produce a "language" obeying Zipf's law".

Apparently then, Zipf's considerations with two opposing forces and a move towards vocabulary balance cannot be put on a sound mathematical footing. Some comments are in order. Firstly, other routes to Zipf's law than via the type-writing monkey are of course possible on purely logical grounds and here Zipf's game-theoretic oriented reflections appear sound. Also note that Mandelbrot in his paper [3] operates with game-theoretic elements via coding considerations. We believe that such considerations contain the key to a better understanding, cf. the section to follow.

Though the route to Zipf's law from the point of view of linguistic develop-ment is of course interesting, we shall not be much concerned with it but rather accept the end result in whichever way it is arrived at and try to characterize in information-theoretic terms the distributions that occur.

2   Hyperbolic Distributions

In a condensed form we shall now give the definitions and results needed for the theoretical part of the manuscript. Further details can be found in [2].

We shall only define hyperbolic distributions over $\mathbb{N}$ and only consider dis-tributions P for which the point probabilities are ordered $(p_1 \geq p_2 \geq \cdots)$ and positive. Clearly, for all i, $p_i \leq \frac{1}{i}$. The condition we shall look at goes in the other direction. Precisely, P is said to be *hyperbolic* if, given any $a > 1$, $p_i \geq i^{-a}$ for infinitely many i.

Any distribution with infinite entropy $H(P)$ is hyperbolic. Clearly, when we use such distributions for our linguistic modelling, this will lead to a high expres-sive power. It is surprising that the same effect can be achieved with distributions of finite entropy. Therefore, for the present study, hyperbolic distributions with finite entropy have our main interest. It is easy to give examples of such distrib-utions: For $i \geq 2$, take p_i proportional to $i^{-1}(\log i)^{-c}$ for some $c > 2$. Also note that any convex combination of distributions with ordered point probabilities, which assigns positive weight to at least one hyperbolic distribution, is again hyperbolic. These distributions are thus plentiful and yet, as we shall explain, have very special properties.

The special properties are connected with the *Code Length Game*, pertaining to any *model* $\mathcal{P} \subseteq M_+^1(\mathbb{N})$, the set of distributions over $\mathbb{N}$. By $K(\mathbb{N})$ we denote the set of (idealized) *codes* over $\mathbb{N}$, i.e. the set of $\kappa : \mathbb{N} \to [0; \infty]$ for which $\sum_1^\infty \exp(-\kappa_i) = 1$. The Code Length Game for $\mathcal{P}$ is a two–person zero–sum game. In this game, Player I chooses $P \in \mathcal{P}$ and Player II chooses $\kappa \in K(\mathbb{N})$. The game is defined by taking the average code length $\langle \kappa, P \rangle$ as cost function, seen from the point of view of Player II.

We put $H_{\max}(\mathcal{P}) = \sup\{H(P)|P \in \mathcal{P}\}$. It turns out that the game is in equilibrium with a finite value if and only if $H_{\max}(co(\mathcal{P})) = H_{\max}(\mathcal{P}) < \infty$. If so, the value of the game is $H_{\max}(\mathcal{P})$ and there exists a distribution P^*, the $H_{\max}$-*attractor*, such that $P_n \to P^*$ (say, in total variation) for every sequence $(P_n)_{n \geq 1} \subseteq \mathcal{P}$ for which $H(P_n) \to H_{\max}(\mathcal{P})$. Normally, one expects that $H(P^*) = H_{\max}(\mathcal{P})$. However, cases with *entropy loss*, $H(P^*) < H_{\max}(\mathcal{P})$, are possible. This is where the hyperbolic distributions come in.

Theorem 1. *Assume that $P^* \in M^1_+(\mathbb{N})$ is of finite entropy and has ordered point probabilities. Then a necessary and sufficient condition that P^* can occur as $H_{\max}$-attractor in a model with entropy loss is that P^* is hyperbolic. If this condition is fulfilled then, for every h with $H(P^*) \leq h < \infty$, there exists a model $\mathcal{P} = \mathcal{P}_h$ with P^* as $H_{\max}$-attractor and $H_{\max}(\mathcal{P}_h) = h$. In fact, $\mathcal{P}_h = \{P|\langle \kappa^*, P \rangle \leq h\}$ is the largest such model. Here, κ^* denotes the code adapted to P^*, i.e. $\kappa^*_i = -\ln p^*_i \, ; i \geq 1$.*

3   Hyperbolic Distributions and Zipf's Law

Put negatively, hyperbolic distributions are connected with entropy loss. However, we find it more appropriate to view these distributions as, firstly, distributions expressing the basic underlying structure of a model (they are $H_{\max}$-attractors) and, secondly, as guarantors of stability. In the context of computational linguistics this translates into a potential to enrich the language to higher and higher expressive powers without changing the basic structure of the language.

Consider an ideal language where the frequencies of words are described by a hyperbolic distribution P^* with finite entropy. Small children use the few words they know with relative frequencies very different from the probabilities given by P^*. They only form simple sentences, and at this stage the number of bits per word will be small, i.e. the entropy of the child's distribution is small. The parents talk to their children at a lower bit rate than they normally use, but with a higher bit rate than their children. Thereby new words and grammatical structures will be presented to the child. At a certain stage the child will be able to communicate at a reasonably high rate (about $H(P^*)$). Now the child knows all the basic words and structures of the language. The child is able to increase its bit rate still further. Bit rates higher than $H(P^*)$ are from now on obtained by the introduction of specialized words, which occur seldom in the language as a whole. This can continue during the rest of the life. Therefore one is able to express even complicated ideas without changing the basic structure of the language, indeed there is no limit, theoretically, to the bit rate at which one can communicate without change of basic structure.

One may speculate that modelling based on hyperbolic laws lies behind the phenomenon that "we can talk without thinking". We just start talking using basic structure of the language and then from time to time stick in more informative words and phrases in order to give our talk more semantic content, and in doing so, we use more infrequent words and structures, thus not violating basic principles – hence still speaking recognizably Danish, English or what the case may be.

Another consideration: If Alice, who we consider to be an expert, wants to get a message across to Bob and if Alice knows the level of Bob (layman or expert), Alice can choose the appropriate entropy level, h, and use that level, still maintaining basic structural elements of the language. Speaking to the layman, Alice will get the message across, albeit at a lower bit rate, by choosing h sufficiently small, and if Alice addresses another expert, she can choose a much higher level h and increase the bit rate considerably. The considerations here point to an acceptance of the maximal models of Theorem 1 as natural models to consider.

We believe that the interpretation of Zipf's law in the light of Theorem 1 is fundamental. Naturally, it raises a number of questions. More qualitative considerations are desirable, the dynamic modelling should be considered, the fact that the hyperbolic distributions are multiple parameter distributions poses certain problems which are connected with the apparent fundamental difficulty — perhaps impossibility – of estimating statistically the entropy of models as those considered. Basically these questions seem to offer a fruitful new area of research which will also be of relevance for other fields than computational linguistics, in particular perhaps for branches of biology and physics.

Our approach leads to assertions which can be tested empirically. Thus language development should evolve in the direction of Zipf's law. There exist primitive computer models of language development and some preliminary investigations [4] provides supportive evidence. Further work is evidently needed. In view of the difficulties involved in the study of long term effects in the development of natural languages, it appears that experiments based on computer models is the most realistic way forward.

Acknowledgements

Thanks go to Joshua Plotkin, Thomas Mikosch and Dorota Glowacka for inspiration and useful references, and to Jim Hurford for useful discussions.

References

1. J.B. Estoup, Gammes Sténographique, Paris, 1916.
2. P. Harremöes and F. Topsøe, Maximum entropy fundamentals, Entropy, Vol. 3, 191–226, 2001.
3. B.B. Mandelbrot, On the theory of word frequencies and on related Markovian models of discourse, in R. Jacobsen (ed.), Structures of Language and its Mathematical Aspects, New York, American Mathematical Society, 1961.
4. M. Schroeder, Fractals, Chaos, Power Laws, New York, W.H. Freeman, 1991.
5. C.E. Shannon, Prediction and entropy of printed english, Bell Syst. Tech. J., Vol. 30, 50–64, 1951.
6. G.K. Zipf, Human Behavior and the Principle of Least Effort, Addison-Wesley, Cambridge, 1949.

---

[4] Private communication with J. Hurford, University of Edinburgh.

Bridging Lossy and Lossless Compression by Motif Pattern Discovery

A. Apostolico*,**, M. Comin, and L. Parida

Abstract. We present data compression techniques hinged on the notion of a *motif*, interpreted here as a string of intermittently solid and
wild characters that recurs more or less frequently in an input sequence
or family of sequences. This notion arises originally in the analysis of
sequences, particularly biomolecules, due to its multiple implications
in the understanding of biological structure and function, and it has
been the subject of various characterizations and study. Correspondingly, motif discovery techniques and tools have been devised. This task
is made hard by the circumstance that the number of motifs identifiable in general in a sequence can be exponential in the size of that
sequence. A significant gain in the direction of reducing the number
of motifs is achieved through the introduction of *irredundant* motifs,
which in intuitive terms are motifs of which the structure and list of
occurrences cannot be inferred by a combination of other motifs' occurrences. Although suboptimal, the available procedures for the extraction of some such motifs are not prohibitively expensive. Here we
show that irredundant motifs can be usefully exploited in lossy compression methods based on textual substitution and suitable for signals
as well as text. Actually, once the motifs in our lossy encodings are
disambiguated into corresponding lossless codebooks, they still prove
capable of yielding savings over popular methods in use. Preliminary
experiments with these fungible strategies at the crossroads of lossless
and lossy data compression show performances that improve over popular methods (i.e. GZip) by more than 20% in lossy and 10% in lossless
implementations.

Keywords: Pattern discovery, pattern matching, motif, lossy and lossless data compression, off-line textual substitution, grammar based
codes, grammatical inference.

* Extended abstracts related to this work were presented at the 2003 and 2004 IEEE
Conference on Data Compression, Snowbird, Utah, March 2003 and March 2004,
and included in those *Proceedings*. Part of this work was performed while visiting
the ZiF Research Center of the University of Bielefeld within the framework of
Project "General Theory of Information Transfer and Combinatorics".
** Work performed in part while on leave at IBM T.J. Watson Center. Supported in
part by an IBM Faculty Partnership Award, by NATO Grant PST.CLG.977017,
by the Italian Ministry of University and Research under the National Projects
"Bioinformatics and Genomics Research" and FIRB RBNE01KNFP, and by the
Research Program of the University of Padova.

R. Ahlswede et al. (Eds.): Information Transfer and Combinatorics, LNCS 4123, pp. 793–813, 2006.
© Springer-Verlag Berlin Heidelberg 2006

1 Introduction

Data compression methods are partitioned traditionally into lossy and lossless. Typically, lossy compression is applied to images and more in general to signals susceptible to some degeneracy without lethal consequence. On the other hand, lossless compression is used in situations where fidelity is of the essence, which applies to high quality documents and perhaps most notably to text files. Lossy methods rest mostly on transform techniques whereby, for instance, cuts are applied in the frequency, rather than in the time domain of a signal. By contrast, lossless textual substitution methods are applied to the input in native form, and exploit its redundancy in terms of more or less repetitive segments or patterns.

When textual substitution is applied to digital documents such as fax, image or audio signal data, one could afford some loss of information in exchange for savings in time or space. In fact, even natural language can easily sustain some degrees of indeterminacy where it is left for the reader to fill in the gaps. The two versions below of the opening passage from the Book1 of the Calgary Corpus, for instance, are equally understandable by an average reader and yet when applied to the entire book the first variant requires 163,837 less bytes than the second one, out of 764,772.

DESCRIPTION OF FARMER OAK – AN INCIDENT When Farmer Oak smile., the corners .f his mouth spread till the. were within an unimportant distance .f his ears, his eye. were reduced to chinks, and ...erging wrinklered round them, extending upon ... countenance li.e the rays in a rudimentary sketch of the rising sun. His Christian name was Gabriel, and on working days he was a young man of sound judgment, easy motions, proper dress, and ...eral good character. On Sundays, he was a man of misty views rather given to postponing, and .ampered by his best clothes and umbrella : upon ... whole, one who felt himself to occupy morally that ... middle space of Laodicean neutrality which ... between the Communion people of the parish and the drunken section, – that ... he went to church, but yawned privately by the t.ime the cong.egation reached the Nicene creed,- and thought of what there would be for dinner when he meant to be listening to the sermon.

DESCRIPTION OF FARMER OAK – AN INCIDENT When Farmer Oak smiled, the corners of his mouth spread till they were within an unimportant distance of his ears, his eyes were reduced to chinks, and diverging wrinkles appeared round them, extending upon his countenance like the rays in a rudi-mentary sketch of the rising sun. His Christian name was Gabriel, and on working days he was a young man of sound judgment, easy motions, proper dress, and general good character. On Sundays he was a man of misty views, rather given to postponing, and hampered by his best clothes and umbrella : upon the whole, one who felt himself to occupy morally that vast middle space of Laodicean neutrality which lay between the Communion people of the parish and the drunken section, – that is, he went to church, but yawned privately by the time the congregation reached the Nicene creed,- and thought of what there would be for dinner when he meant to be listening to the sermon.

In practice, the development of optimal lossless textual substitution methods is made hard by the circumstance that the majority of the schemes are

NP-hard [27]. Obviously, this situation cannot improve with lossy ones. As an approximation, heuristic off-line methods of textual substitution can be based on greedy iterative selection as follows (see e.g., [2,6,10]). At each iteration, a substring w of the text x is identified such that encoding a maximal set of non-overlapping instances of w in x yields the highest possible contraction of x; this process is repeated on the contracted text string, until substrings capable of producing contractions can no longer be found. This may be regarded as inferring a "straight line" grammar [15,16,19] by repeatedly finding the production or rule that, upon replacing each occurrence of the "definition" by the corresponding "nonterminal", maximizes the reduction in size of the current text string representation. Recent implementations of such greedy off-line strategies [6] compare favorably with other current methods, particularly as applied to ensembles of otherwise hardly compressible inputs such as biosequences. They also appear to be the most promising ones in terms of the achievable approximation to optimum descriptor sizes [19].

Off-line methods can be particularly advantageous in applications such as mass production of CD-ROMs, backup archiving, and any other scenario where extra time or parallel implementation may warrant the additional effort imposed by the encoding (see, e.g., [14]).

The idea of trading some amount of errors in reconstruction in exchange for increased compression is ingrained in Rate Distortion Theory [11,12], and has been recently revived in a number of papers, mostly dealing with the design and analysis of lossy extensions of Lempel-Ziv on-line schemata. We refer to, e.g., [17,18,21], and references therein. In this paper, we follow an approach based on the notion of a *motif*, a kind of redundancy emerged particularly in molecular biology and genomic studies. In loose terms, a motif consists of a string of intermittently solid and wild characters, and appearing more or less frequently in an input sequence. Because motifs seem to be implicated in many manipulations of biological as well as more general sequences, techniques for their discovery are of broad interest. We refer to the quoted literature for a more comprehensive discussion. In a nutshell, the role of motifs in our constructions is to capture the auto-correlation in the data by global pattern discovery. The combinatorial structure of our motifs is engineered to minimize redundancy in the "codebook". The presence of a controlled number of don't care characters enhances the compression achievable in the subsequent stage of off-line greedy textual substitution.

In general, the motif discovery and use is made particularly difficult by the fact that the number of candidate motifs in a sequence grows exponentially with the length of that string. Fortunately, a significant reduction in the basis of candidate motifs is possible in some cases. In the context of our textual substitution schemes, for instance, it comes natural to impose that the motif chosen at each iteration satisfies certain maximality conditions that prevent forfeiting information gratuitously. To begin with, once a motif is chosen it seems reasonable to exploit the set of its occurrences to the fullest, compatibly with self-overlaps. Likewise, it seems reasonable to exclude from consideration motifs that could be enriched in terms of solid characters without prejudice in the corresponding set of occurrences.

Recently, a class of motifs called "irredundant" has been identified along these lines that grows linearly with input size [7,8,9]. We examine here the application of such motifs to various scenarios of lossy and lossless compression. As it turns out, significant savings can be obtained with this approach.

This paper is organized as follows. In the next section, we recall some basic definitions and properties, and the combinatorial facts subtending to our construction. Section 3 is devoted to the description of our method and the section that follows lists preliminary experiments. Conclusions and plans of future work close the paper.

2   Notions and Properties

Let $s = s_1 s_2 ... s_n$ be a *string* of length $|s| = n$ over an alphabet Σ. We use suf_i to denote the suffix $s_i s_{i+1} ... s_n$ of s and $s[i]$ for the i-th symbol. A character from Σ, say σ, is called a *solid* character and '.' is called a "don't care" character.

Definition 1. *($\sigma_1 \prec, =, \preceq \sigma_2$) If σ_1 is a don't care character then $\sigma_1 \prec \sigma_2$. If both σ_1 and σ_2 are identical characters in Σ, then $\sigma_1 = \sigma_2$. If either $\sigma_1 \prec \sigma_2$ or $\sigma_1 = \sigma_2$ holds, then $\sigma_1 \preceq \sigma_2$.*

Definition 2. *(p occurs at l, Cover) A string, p, on $\Sigma \cup$ '.', occurs at position l in s if $p[j] \preceq s[l + j - 1]$ holds for $1 \leq j \leq |p|$. String p is said to cover the interval $[l, l + |p| - 1]$ on s.*

A *motif* is any element of Σ or any string on $\Sigma \cdot (\Sigma \cup \{.\})^* \cdot \Sigma$.

Definition 3. *(k-Motif m, Location list $\mathcal{L}_m$) Given a string s on alphabet Σ and a positive integer k, $k \leq |s|$, a string m on $\Sigma \cup$ '.' is a motif with location list $\mathcal{L}_m = (l_1, l_2, \ldots, l_q)$, if all of the following hold: (1) $m[1], m[|m|] \in \Sigma$, (2) $q \geq k$, and (3) there does not exist a location l, $l \neq l_i$, $1 \leq i \leq q$ such that m occurs at l on s (the location list is of maximal size).*

The first condition ensures that the first and last characters of the motif are solid characters; if don't care characters are allowed at the ends, the motifs can be made arbitrarily long in size without conveying any extra information. The third condition ensures that any two distinct location lists must correspond to distinct motifs.

Using the definition of motifs, the different 2-motifs are as follows: $m_1 = ab$ with $\mathcal{L}_{m_1} = \{1, 5\}$, $m_2 = bc$ with $\mathcal{L}_{m_2} = \{2, 6\}$, $m_3 = cd$ with $\mathcal{L}_{m_3} = \{3, 7\}$, $m_4 = abc$ with $\mathcal{L}_{m_4} = \{1, 5\}$, $m_5 = bcd$ with $\mathcal{L}_{m_5} = \{2, 6\}$ and $m_6 = abcd$ with $\mathcal{L}_{m_6} = \{1, 5\}$.

Notice that $\mathcal{L}_{m_1} = \mathcal{L}_{m_4} = \mathcal{L}_{m_6}$ and $\mathcal{L}_{m_2} = \mathcal{L}_{m_5}$. Using the notation $\mathcal{L} + i = \{x + i | x \in \mathcal{L}\}$, $\mathcal{L}_{m_5} = \mathcal{L}_{m_6} + 1$ and $\mathcal{L}_{m_3} = \mathcal{L}_{m_6} + 2$ hold. We call the motif m_6 *maximal* as $|m_6| > |m_1|, |m_4|$ and $|m_5| > |m_2|$. Motifs m_1, m_2, m_3, m_4 and m_5 are non-maximal motifs.

We give the definition of maximality below. In intuitive terms, a motif m is *maximal* if we cannot make it more specific or longer while retaining the list $\mathcal{L}_m$ of its occurrences in s.

Definition 4. *($m_1 \preceq m_2$) Given two motifs m_1 and m_2 with $|m_1| \leq |m_2|$, $m_1 \preceq m_2$ holds if $m_1[j] \preceq m_2[j+d]$, with $d \geq 0$ and $1 \leq j \leq |m_1|$.*

We also say in this case that m_1 is a *sub-motif* of m_2, and that m_2 *implies* or *extends* or *covers* m_1. If, moreover, the first characters of m_1 and m_2 match then m_1 is also called a *prefix* of m_2. For example, let $m_1 = ab..e$, $m_2 = ak..e$ and $m_3 = abc.e.g$. Then $m_1 \preceq m_3$, and $m_2 \not\preceq m_3$. The following lemma is straightforward to verify.

Lemma 1. *If $m_1 \preceq m_2$ then $\exists\, d \mid \mathcal{L}_{m_1} \supseteq \mathcal{L}_{m_2} + d$, and if $m_1 \preceq m_2$, $m_2 \preceq m_3$, then $m_1 \preceq m_3$.*

Definition 5. *(Maximal Motif) Let m_1, m_2, ..., m_k be the motifs in a string s. A motif m_i is maximal in composition if and only if there exists no m_l, $l \neq i$ with $\mathcal{L}_{m_i} = \mathcal{L}_{m_l}$ and $m_i \preceq m_l$. A motif m_i, maximal in composition, is also maximal in length if and only if there exists no motif m_j, $j \neq i$, such that m_i is a sub-motif of m_j and $|\mathcal{L}_{m_i}| = |\mathcal{L}_{m_j}|$. A maximal motif is a motif that is maximal both in composition and in length.*

Requiring maximality in composition and length limits the number of motifs that may be usefully extracted and accounted for in a string. However, the notion of maximality alone does not suffice to bound the number of such motifs. It can be shown that there are strings that have an unusually large number of maximal motifs without conveying extra information about the input.

A maximal motif m is *irredundant* if m and the list $\mathcal{L}_m$ of its occurrences cannot be deduced by the union of a number of lists of other maximal motifs. Conversely, we call a motif m *redundant* if m (and its location list $\mathcal{L}_m$) can be deduced from the other motifs *without* knowing the input string s. More formally:

Definition 6. *(Redundant motif) A maximal motif m, with location list $\mathcal{L}_m$, is redundant if there exist maximal sub-motifs m_i, $1 \leq i \leq p$, such that $\mathcal{L}_m = \mathcal{L}_{m_1} \cup \mathcal{L}_{m_2} \ldots \cup \mathcal{L}_{m_p}$, (i.e., every occurrence of m on s is already implied by one of the motifs $m_1, m_2, \ldots, m_p$).*

Definition 7. *(Irredundant motif) A maximal motif that is not redundant is called an irredundant motif.*

We use $\mathcal{B}_i$ to denote the set of irredundant motifs in suf_i. Set $\mathcal{B}_i$ is called the *basis* for the motifs of suf_i. Thus, in particular, the basis $\mathcal{B}$ of s coincides with $\mathcal{B}_1$.

Definition 8. *(Basis) Given a sequence s on an alphabet Σ, let $\mathcal{M}$ be the set of all maximal motifs on s. A set of maximal motifs $\mathcal{B}$ is called a basis of $\mathcal{M}$ iff the following hold: (1) for each $m \in \mathcal{B}$, m is irredundant with respect to $\mathcal{B} - \{m\}$, and, (2) let $\mathbf{G}(\mathcal{X})$ be the set of all the redundant maximal motifs generated by the set of motifs $\mathcal{X}$, then $\mathcal{M} = \mathbf{G}(\mathcal{B})$.*

In general, $|\mathcal{M}| = \Omega(2^n)$. The natural attempt now is to obtain as small a basis as possible. Before getting to that, we examine some basic types of maximality.

Lemma 2. *Let m be a maximal motif with no don't care and $|\mathcal{L}_m| = 1$, then $m = s$.*

Proof. Any motif with those properties can be completed into s, by the notion of maximality.

Lemma 3. *Let m be a maximal motif with at least one don't care, then $|\mathcal{L}_m| \geq 2$.*

Proof. Under the hypothesis, it must be $|m| > 1$. The rest is a straightforward consequence of the notion of maximality.

Lemmas 2 and 3 tell us that, other than the string s itself and the characters of the alphabet, the only maximal motifs of interest have more than one occurrence. Solid motifs, i.e., motifs that do not contain any don't care symbol, enjoy a number of nice features that make it pedagogically expedient to consider them separately. Let the equivalence relation $\equiv_s$ be defined on a string s by setting $y \equiv_s w$ if $\mathcal{L}_y = \mathcal{L}_w$. Recall that the *index* of an equivalence relation is the number of equivalence classes in it. The following well known fact from [13] shows that the number of maximal motifs with no don't care is linear in the length of the text string. It descends from the observation that for any two substrings y and w of s, if $\mathcal{L}_w \cap \mathcal{L}_y$ is not empty then y is a prefix of w or vice versa.

Fact 1. The index k of the equivalence relation $\equiv_x$ obeys $k \leq 2n$.

When it comes to motifs with at least one don't care, it is desirable to obtain as small a basis as possible. Towards this, let x and y be two strings with $m = |x| \leq |y| = n$. The *consensus* of x and y is the string $z_1 z_2 ... z_m$ on $\Sigma \cup \text{`.`}$ defined as: $z_i = x_i$ if $x_i = y_i$ and $z_i = \text{`.`}$ otherwise ($i = 1, 2, ..., m$). Deleting all leading and trailing don't care symbols from z yields a (possibly empty) motif that is called the *meet* of x and y. The following general property [7] (cf. proof given in the Appendix) shows that the irredundant 2-motifs are to be found among the pairwise meets of all suffixes of s.

Theorem 1. *Every irredundant 2-motif in s is the meet of two suffixes of s.*

An immediate consequence of Theorem 1 is a linear bound for the cardinality of our set of irredundant 2-motifs: by maximality, these motifs are just some of the $n - 1$ meets of s with its own suffixes. Thus

Theorem 2. *The number of irredundant 2-motifs in a string x of n characters is $O(n)$.*

With its underlying convolutory structure, Theorem 1 suggests a number of immediate ways for the extraction of irredundant motifs from strings and arrays, using available pattern matching with or without FFT. We refer to [1] for a nice discussion of these alternatives. Specific "incremental" algorithms are also available [7] that find all irredundant 2-motifs in time $O(n^3)$. The paradigm explored there is that of iterated updates of the set of base motifs $\mathcal{B}_i$ in a string under consecutive unit symbol extensions of the string itself. Such an algorithm is thus incremental and single-pass, which may lend itself naturally to applications of the kind considered here. The construction used for our experiments must take into account additional parameters related to the density of solid characters, the maximum motif length and minimum allowed number of occurrences. This algorithm is described next.

3 The Pattern Discovery Algorithm

The algorithm follows a standard approach to association discovery: it begins by computing elementary patterns of high *quorum* and then successively extends motifs one solid character at a time until this growth must stop. In general, one drawback of this approach is that the number of patterns at each step grows very rapidly. In our case, the patterns being grown are chosen among $O(n^2)$ substrings of pairwise suffix meets, so that no more than $O(n^3)$ candidates are considered overall. Trimming takes place at each stage, based on the quorum, to keep the overall number of growing patterns bounded. Thus our basis can be detected in polynomial time.

The algorithm makes recurrent use of a routine that solves the following

Set Union Problem, SUP(n,q). Given n sets $S_1, S_2 \ldots, S_n$ on q elements each, find all the sets S_i such that $S_i = S_{i_1} \cup S_{i_2} \cup \ldots \cup S_{i_p}$, $i \neq i_j$, $1 \leq j \leq p$. We present an algorithm in Appendix 6 to solve this problem in time $O(n^2 q)$.

Input Parameters. The input parameters are: (1) the string s of length n, (2) the quorum k, which is the minimum number of times a pattern must appear, and (3) the maximum number D of consecutive '.' characters allowed in a motif. For convenience in exposition, the notion of a motif is relaxed to include singletons consisting of just one character.

For the rest of the algorithm we will let $m_1.^d m_2$ denote the string obtained by concatenating the elements m_1 followed by d '.' characters followed by the element m_2. Also, recall that $\mathcal{L}_m = \{i | m \text{ occurs at } i \text{ on } s\}$.

Computing the Basis

The algorithm proceeds in the following steps. M is the set of motifs being constructed.

1. $M = M' \leftarrow \{m' = \sigma \in \Sigma \text{ and } \mathcal{L}_{m'} \geq k\}$

2. (a) Let $\sigma_i.^d m$, with $0 \leq d \leq D$, denote the left extension of the motif m along a meet. For each motif $m' \in M'$, use meets to compute all of its possible left extensions and store them in the set M''.
   For every $m'' \in M''$, if $|\mathcal{L}_{m''}| < k$ then $M'' \leftarrow M'' - \{m''\}$

(b) Remove all redundant motifs.
For each $m_i \in M$, with $m_i \preceq m_j''$ for some $m_j'' \in M''$,
if $\exists\, m_{i_1}'', m_{i_2}'', \ldots m_{i_p}'' \in M''$, $p \geq 1$ such that
$$m_i \preceq m_{i_j}'' \text{ and}$$
$$\mathcal{L}_{m_i} = \mathcal{L}_{m_{i_1}''+f_1} \cup \mathcal{L}_{m_{i_2}''+f_2} \ldots \cup \mathcal{L}_{m_{i_p}''+f_p} \text{ then}$$
$$M \leftarrow M - \{m_i\}.$$
The above is solved using an instance of the SUP() problem.
(c) Update the basis and M'.
$$M \leftarrow M \cup M''; \quad M' \leftarrow M''$$
3. The previous step is repeated until no changes occur to M.

The algorithm works by iterated extensions of motifs previously in M, where at each iteration a motif is extended (to the left) by zero or more don't cares followed by precisely one solid character. Thus, at the end of the i-th iteration of Step 2, the set M contains motifs with at most $i+1$ solid characters. As there can be no more that n solid characters in a meet, the number of iterations is bounded by n. Since motifs come from meets at all times, and at most one new motif is considered at one iteration for every ending position on a meet, we have that M' and M'' are each bounded by $O(n^2)$, whereas the elements in M are $O(n^3)$ at all times (in fact, much fewer in practice). At each iteration we have $O(n^2)$ extensions to perform. By solving the $\mathbf{SUP}(n^3, n)$, the algorithm must now try and cover each motif in M by using the new $O(n^2)$ ones in M''. Step 2-b ensures that no motif currently in the set M can be deduced with its location list from the union of other discovered motifs. In other words, the elements of M are irredundant *relative to* M itself, in the sense that no member of M can be inferred from the others. The following claim gives a sharper characterization of the set M.

Theorem 3. *Let $M^{(i)}$ be the set generated by the pattern discovery algorithm at step $i, 0 \leq i \leq n$. Then $M^{(i)}$ contains every k-motif m such that:*

1. *m is a substring of the meet of two suffixes, with at most $i+1$ solid characters and density D.*
2. *m is irredundant relative to the elements of $M^{(i)}$.*
   Moreover,
3. *for every k-motif with these properties not in $M^{(i)}$, there are motifs in $M^{(i)}$ capable of generating it.*
4. *$M^{(i)}$ is a minimum cardinality set with such properties.*

Proof. The claim holds trivially prior to the first iteration. In fact, by initialization $M = M^{(0)} = \{m = \sigma$ is a substring of a meet and m has at least k occurrences $\}$. Clearly, the elements of $M^{(0)}$ are mutually irredundant since they correspond each to a distinct character and hence there is no way to express one of them using the others. $M^{(0)}$ is also exhaustive of such motifs, so that no character of quorum k is left out. At the same time, $M^{(0)}$ is the smallest set capable of generating itself. The first time Step 2 is executed this generates all distinct motifs in the form $\sigma_1.{}^d\sigma_2$ that are substrings of meets of quorum k.

These motifs are stored in M''. As they are all distinct, they cannot express each other, and the only possibility is for them to obliterate single characters. The latter are now in M', which coincides with M. Through Step 2, a single-character motif $m = \sigma$ is eliminated precisely when it can be synthesized by two-character motifs that either begin or end by σ. As all and only such singletons are eliminated, this propagates all claimed properties. Assuming now the claim true up to step $i - 1 \geq 2$, consider the i-th execution of Step 2. We note that, in an application of Step 2-b, any motif in the set which is used to eliminate motif m must coincide with m on each and every solid character of m. Therefore, no one of the newly introduced motifs with exactly $i + 1$ solid characters can be expressed and discarded using *different* motifs with $i + 1$ solid characters, or (even worse) motifs with less than i solid characters. Consequently, no such motif can be discarded by this execution of Step 2-b. Also, no collection of motifs formed solely by members of $M^{(h)}$ with $h < i$ can be used to discard other motifs, since, by the operation of the algorithm, any such action would have to have already taken place at some prior iteration. Finally, no mixed collection formed by some new motifs in M'' and motifs currently in M can obliterate motifs in M. In fact, such an action cannot involve only suffixes of the members of M'', or it would have had to be performed at an earlier iteration. Then, it must involve prefixes of motifs in M''. But any such prefix must have been already represented in M, by the third invariant condition.

In conclusion, the only thing that can happen is that motifs currently in M are obliterated by motifs with $i + 1$ solid characters, currently in M''. At the beginning of step i, all and only the qualifying k-motifs with i characters are by hypothesis either present directly or generated by motifs in $M^{(i-1)}$, which represents also the smallest possible base for the collection of these motifs. The algorithm extends all the motifs in $M^{(i-1)}$ along a meet of two suffixes, hence all candidate extensions are considered. Now, the net worth of Step 2 is in the balance between the number of newly introduced motifs with $i + 1$ solid characters and the number of motifs with i solid characters that get discarded. Assume for a contradiction that a base $\hat{M}$ exists at the outset which is smaller than $M^{(i-1)}$. Clearly, this savings cannot come from a reduction in the number of motifs with $i + 1$ solid characters, since eliminating any one of them would leave out a qualifying motif and play havoc with the notion of a base. Hence, such a reduction must come from the elimination of some extra motifs in $M^{(i-1)}$. But we have argued that all and only the motifs in $M^{(i-1)}$ that can be eliminated are in fact eliminated by the algorithm. Thus $M^{(i)}$ must have minimum cardinality. □

Theorem 4. *The set M at the outset is unique.*

Proof. Let h be the first iteration such that from $M^{(h-1)}$ we can produce two sets, $M^{(h)}$ and $\bar{M}^{(h)}$, such that $M^{(h)} \neq \bar{M}^{(h)}$ but $|M^{(h)}| = |\bar{M}^{(h)}|$. Clearly, the members of M'' that come from extensions of $M^{(h-1)}$ must belong both to $M^{(h)}$ and $\bar{M}^{(h)}$. Hence the two sets must differ by way of an alternate selection of the members of $M^{(h-1)}$ that are eliminated on behalf of the motifs in M''. But it is clear that any motif that could be, but is not eliminated by M'' will fail to comply with the second clause of the preceding theorem. Hence no option exists in the choice of motifs to be eliminated.

4   Implementation and Experiments

Each phase of our *steepest descent* paradigm alternates the selection of the pattern to be used in compression with the actual substitution and encoding. The sequence representation at the outset is finally pipelined into some of the popular encoders and the best one among the overall scores thus achieved is retained. By its nature, such a process makes it impossible to base the selection of the best motif at each stage on the actual compression that will be conveyed by this motif in the end. The decision performed in choosing the pattern must be based on an estimate, that also incorporates the peculiarities of the scheme or rewriting rule used for encoding. In practice, we estimate at $\log i$ the number of bits needed to encode the integer i (we refer to, e.g., [5] for reasons that legitimate this choice). In one scheme (hereafter, $Code_1$) [6], we eliminate all occurrences of m, and record in succession m, its length, and the total number of its occurrences followed by the actual list of such occurrences. Letting $|m|$ denote the length of m, f_m the number of occurrences of m in the text string, $|\Sigma|$ the cardinality of the alphabet and n the size of the input string, the compression brought about by m is estimated by subtracting from the $f_m|m|\log|\Sigma|$ bits originally encumbered by this motif on s, the expression $|m|\log|\Sigma| + \log|m| + f_m \log n + \log f_m$ charged by encoding, thereby obtaining:

$$G(m) = \tag{1}$$
$$(f_m - 1)|m|\log|\Sigma| - \log|m| - f_m \log n - \log f_m.$$

This is accompanied by a fidelity loss $L(m)$ represented by the total number of don't cares introduced by the motif, expressed as a fraction of the original length. If d such gaps were introduced, this would be:

$$L(m) \quad = \quad \frac{f_m d \log|\Sigma|}{f_m|m|\log|\Sigma|} \quad = \quad \frac{d}{|m|}. \tag{2}$$

Other encodings are possible (see, e.g., [6]). In one scheme (hereafter, $Code_2$), for example, every occurrence of the chosen pattern m is substituted by a pointer to a common dictionary copy, and we need to add one bit to distinguish original characters from pointers. The space originally occupied by m on the text is in this case $(\log|\Sigma| + 1)f_m|m|$, from which we subtract $|m|\log|\Sigma| + \log|m| + \log|f_m| + f_m(\log D + 1)$, where D is the size of the dictionary, in itself a parameter to be either fixed a priori or estimated.

The tables and figures below were obtained from preliminary experiments. The major burden in computations is posed by the iterated updates of the motif occurrence lists, that must follow the selection of the best candidate at each stage.

This requires maintaining motifs with their occurrences in a doubly linked list as in Fig. 1: following each motif selection, the positions of the text covered by its occurrences are scanned horizontally. Next, proceeding vertically from each such position, the occurrences of other motifs are removed from their respective lists.

To keep time manageable, most of the experiments were based on a small number of iterations, typically in the range 250-3,000. For Book1, for instance, more than $30k$ motifs could be extracted. Each one of these would convey some compression if used, but time constraints allowed only less than 10% to be implemented. In the pattern discovery stage, a maximum length for motifs was enforced at about 40 to 50 characters, and a threshold of 5 or 6 was put on the overall number of don't care allowed in a single motif, irrespective of its length. The collection of these measures made it possible to test the method on a broad variety of inputs. By the same token, the resulting scores represent quite a conservative estimate of its potential.

The tables summarize scores related to various inputs under various acceptances of loss. Table 1 refers to 8-bit grey-level images as a function of the don't care density allowed (last column). The next table, Table 2 shows results on black and white pictures. These are similar except in this case the loss of one bit translates into that of 1 byte. By their nature, binary or dithered images such as in facsimile transmission seem to be among the most promising applications of our method. At the same time, it has already been reported that "directional" lossy textual substitution methods can compete successfully even with chrominance oriented methods like JPEG [1]. In view of the results in [6], off-line lossy variants of the kind presented here should perform just as well and probably better. Table 3 shows results for musical records sampled at 8 bits. For this family of inputs, the motif extraction phase alone seems to present independent interest in applications of contents-based retrieval.

Tables 5, 6, and 7 cover inputs from the Calgary Corpus and some yeast families. DNA sequences represent interesting inputs for compression, in part because of the duality between compression and structural inference or classification, in part due to the well know resiliency of bio-sequences towards compression (see, e.g., [6] and references therein). Particularly for text (we stress that lossy compression of bio-sequences is a viable classification tool), lossy compression may be not very meaningful without some kind of reconstruction. As suggested at the beginning of the paper, this might be left to the user in some cases. Otherwise, Table 4 list results obtained by exact completions of the motifs involved in implementation of all of our lossy schemata. It suggests that the bi-lateral context offered by motifs lends itself to better *predictors* than the traditional ones based on the left context alone. In any case, the iteration of motif extraction at several consecutive levels of hierarchy unveils structural properties and descriptors akin to unconventional grammars.

We use Figure 2 to display encodings corresponding to the images from Table 1. The single most relevant parameter here is represented by the density of don't care, which is reported in the last column of the table and also evidenced by the black dots injected in figures at the last column. As mentioned, the maximum length of motifs extracted had to be limited by practical considerations. Even so, it was found that images rarely produce motifs longer than a few tens of characters. More severe consequences of these practical restrictions came from the need to limit the number of motifs actually deployed in compression, which was kept at those with at least 5 to 10 occurrences, corresponding to a quite limited dictionary of 1,000 to 2,000 entries. Interpolation was carried out by averaging from the two solid characters adjacent to each gap. The corresponding discrepancies from the original pixel values reach into 16% in terms of % *number* of inexact pixels, but was found to be only a few percentage points if the variation in *value* of those pixels was measured instead as a percentage of the affected pixels (next to last column of Table 8, and entirely negligible (a fraction of a percent, see last column in Table 8) when averaged over all pixels. This is demonstrated in the reconstructed figures, that show little perceptible change.

As mentioned, our main interest was testing the breadth of applicability of the method rather that bringing it to the limit on any particular class of inputs. This is the scope of future work. In the experiments reported here, the number of iterations (hence, motifs selected or vocabulary size) was in the range of 250 to 1,000 and slightly higher (3,000) for the Calgary Corpus. The length of motifs was limited to few tens of characters and their minimum number of occurrences to 20 or higher. Typically, motifs in the tens of thousands were excluded from consideration on these grounds, which would have been provably capable of contributing savings.

Table 1. Lossy compression of gray-scale images (1 pixel = 1 byte)

file	file len	GZip len [%compr]	$Codec_2$ [%compr]	$Codec_1$ [%compr]	%Diff gzip	%loss	'.'/ char
bridge	66336	$61657_{[7.05]}$	$60987_{[8.06]}$	$57655_{[13.08]}$	6.49	0.42	1/4
			$60987_{[8.06]}$	$50656_{[23.63]}$	17.84	14.29	1/3
camera	66336	$48750_{[26.51]}$	$47842_{[27.88]}$	$46192_{[30.36]}$	5.25	0.74	1/6
			$48044_{[27.57]}$	$45882_{[30.83]}$	5.88	2.17	1/5
			$47316_{[28.67]}$	$43096_{[35.03]}$	11.60	9.09	1/4
lena	262944	$234543_{[12.10]}$	$226844_{[13.73]}$	$210786_{[19.83]}$	10.13	4.17	1/4
			$186359_{[29.13]}$	$175126_{[33.39]}$	25.33	20.00	1/3
peppers	262944	$232334_{[11.64]}$	$218175_{[17.03]}$	$199605_{[23.85]}$	14.09	6.25	1/4
			$180783_{[31.25]}$	$173561_{[33.99]}$	25.30	20.00	1/3

5 LZW Encoding

Ziv and Lempel designed a class of compression methods based on the idea of back-reference: while the text file is scanned, substrings or *phrases* are identified and stored in a *dictionary*, and whenever, later in the process, a phrase or concatenation of phrases is encountered again, this is compactly encoded by suitable pointers or indices [20,30,31]. In view of Theorem 1 and of the good performance

Fig. 1. Compression and reconstruction of images. The original is on the first column, next to its reconstruction by interpolation of two closest solid pixels. Black dots used in the figures of the last column are used to display the distribution of the don't care characters. Compression of "Bridge" at 1/4 and 1/3 (shown here) '.'/char densities yields savings of 6.49% and 17.84% respectively. Correspondingly. 0,31% and 12,50% of the pixels differ from original after reconstruction. The lossy compression of Camera at 1/4 '.'/char density saves 11.60% over GZip. Only 6.67% of pixels differ from the original after reconstruction. Gains by "Lena" at 1/4 and 1/3 (shown) '.'/char density are respectively of 10,13% and 25,33%, while interpolation leaves resp. 3,85% and 10,13% differences from original. For "Peppers" (last row), the gains at 1/4 and 1/3 (shown) '.'/char densities were respectively 14,09% (5,56% the corresponding difference) and 25,30% (16,67% diff).

Table 2. Lossy compression of binary images

file	file len	GZip len [%compr]	$Codec_2$ [%compr]	$Codec_1$ [%compr]	%Diff GZip	%loss	'.'/char
ccitt7	513229	$109612_{[78.64]}$	$98076_{[80.89]}$	$91399_{[82.19]}$	16.62	16.67	1/5
			$93055_{[81.87]}$	$90873_{[82.29]}$	17.10	16.67	1/4
			$92658_{[81.95]}$	$85391_{[83.36]}$	22.10	25.00	1/3
test4	279213	$58736_{[78.96]}$	$57995_{[79.23]}$	$54651_{[80.42]}$	6.95	0,91	1/4
			$57714_{[79.32]}$	$54402_{[80.51]}$	7.38	1.27	1/3

Table 3. Lossy compression of music (1 sample = 1 byte)

file	file len	GZip len [%compr]	$Codec_2$ [%compr]	$Codec_1$ [%compr]	%Diff GZip	%loss	'.'/char
crowd	128900	$103834_{[19.44]}$	$92283_{[28.41]}$	$86340_{[33.01]}$	16.85	16.67	1/3
eclipse	196834	$171846_{[12.96]}$	$148880_{[24.36]}$	$139308_{[29.22]}$	18,93	9.09	1/4
			$114709_{[41.72]}$	$111058_{[43.57]}$	35.37	25.00	1/3

Table 4. Lossy vs. lossless performance

file	file dim	GZip [%compr]	$Codec_1$ [%compr]	%loss	'.'/char	Lossless [%compr]	%Diff GZip
bridge	66336	$61657_{[7.05]}$	$50656_{[23.63]}$	14.29	1/3	$59344_{[10.54]}$	3.75
camera	66336	$48750_{[26.51]}$	$43096_{[35.03]}$	9,09	1/4	$45756_{[31.02]}$	6.14
lena	262944	$234543_{[12.10]}$	$175126_{[33.39]}$	20.00	1/3	$199635_{[24.07]}$	14.88
peppers	262944	$232334_{[11.64]}$	$199605_{[23.85]}$	6.25	1/4	$211497_{[19.56]}$	8.97
			$173561_{[33.99]}$	20.00	1/3	$195744_{[25.55]}$	15.75
ccitt7	513229	$109612_{[78.64]}$	$90873_{[82.29]}$	16.67	1/4	$97757_{[80.09]}$	10.82
			$85391_{[83.36]}$	25.00	1/3	$89305_{[82.59]}$	18.53
test4	279213	$58736_{[78.96]}$	$54402_{[80.51]}$	1.27	1/3	$54875_{[80.34]}$	6.57
crowd	128900	$103834_{[19.44]}$	$86340_{[33.01]}$	16.67	1/3	$96903_{[24.82]}$	6.68
eclipse	196834	$171846_{[12.96]}$	$139308_{[29.22]}$	9.09	1/4	$159206_{[19.11]}$	7.36
			$111058_{[43.57]}$	25.00	1/3	$151584_{[22.98]}$	11.97

of motif based off-line compression [8], it is natural to inquire into the structure of ZL and ZLW parses which would use these patterns in lieu of exact strings. Possible schemes along these lines include, e.g., adaptations of those in [26], or more radical schemes in which the innovative add-on inherent to ZLW phrase growth is represented not by one symbol alone, but rather by that symbol plus the longest match with the substring that follows some previous occurrence of the phrase. In other words, the task of vocabulary build-up is assigned to the growth of (candidate), perhaps irredundant, 2-motifs.

Table 5. Lossless compression of Calgary Corpus

file	file len	GZip [%compr]	$Codec_1$ [%compr]	%loss	'.'/char	Lossless [%compr]	%Diff GZip
bib	111261	$35063_{[68.49]}$	$36325_{[67.35]}$	3,70	1/3	$37491_{[66.30]}$	6.92
book1	768771	$313376_{[60.01]}$	$245856_{[68.01]}$	12.50	1/3	$277180_{[63.95]}$	11.55
book2	610856	$206687_{[66.16]}$	$197199_{[67.72]}$	4,35	1/4	$202713_{[66.81]}$	1.92
geo	102400	$68493_{[33.11]}$	$40027_{[60.91]}$	16.67	1/4	$63662_{[37.83]}$	7.05
news	377109	$144840_{[61.59]}$	$144541_{[61.67]}$	0.42	1/3	$144644_{[61.64]}$	0.14
obj1	21504	$10323_{[51.99]}$	$8386_{[61.00]}$	16.67	2/5	$9221_{[57.12]}$	10.68
obj2	246814	$81631_{[66.93]}$	$71123_{[71.18]}$	20.00	1/2	$83035_{[66.36]}$	-1.72
paper1	53161	$18577_{[65.06]}$	$19924_{[62.52]}$	1.75	1/3	$20174_{[62.05]}$	-8.60
paper2	82199	$29753_{[63.80]}$	$29920_{[63.60]}$	0.76	1/2	$30219_{[63.24]}$	-1.57
pic	513216	$56422_{[89.01]}$	$52229_{[89.82]}$	0.56	1/3	$52401_{[89.79]}$	7.13
progc	39611	$13275_{[66.49]}$	$13840_{[65.06]}$	1.32	1/2	$14140_{[64.30]}$	-6.52
progl	71646	$16273_{[77.29]}$	$17249_{[75.92]}$	0.58	1/3	$17355_{[75.78]}$	-6.65
progp	49379	$11246_{[77.23]}$	$12285_{[75.12]}$	0.64	1/3	$12427_{[74.83]}$	-10.50

Table 6. Lossless compression of sequences from DNA yeast families

file	file len	GZip [%compr]	$Codec_1$ [%compr]	%loss	'.'/char	Lossless [%compr]	%Diff GZip
Spor EarlyII	25008	$8008_{[67.98]}$	$6990_{[72.05]}$	0.45	1/3	$7052_{[71.80]}$	11.94
Spor EarlyI	31039	$9862_{[68.23]}$	$8845_{[71.50]}$	0.36	1/3	$8914_{[71.28]}$	9.61
Helden CGN	32871	$10379_{[68.43]}$	$8582_{[73.89]}$	1.33	1/3	$8828_{[73.14]}$	14.94
Spor Middle	54325	$16395_{[69.82]}$	$14839_{[72.68]}$	0.36	1/4	$14924_{[72.53]}$	8.97
Helden All	112507	$33829_{[69.93]}$	$29471_{[73.81]}$	1.56	1/4	$29862_{[73.46]}$	11.73
Spor All	222453	$68136_{[69.37]}$	$56323_{[74.68]}$	1.61	1/3	$57155_{[74.31]}$	16.12
All Up 400k	399615	$115023_{[71.22]}$	$93336_{[76.64]}$	14.29	1/3	$106909_{[73.25]}$	7.05

Of the existing versions of the method, we recapture below the parse known as Ziv-Lempel-Welch, which is incarnated in the compress of UNIX. For the encoding, the dictionary is initialized with all the characters of the alphabet. At the generic iteration, we have just read a segment s of the portion of the text still to be encoded. With σ the symbol following this occurrence of s, we now proceed as follows: If $s\sigma$ is in the dictionary we read the next symbol, and repeat with segment $s\sigma$ instead of s. If, on the other hand, $s\sigma$ is not in the dictionary, then we append the dictionary index of s to the output file, and add $s\sigma$ to the dictionary; then reset s to σ and resume processing from the text symbol following σ. Once s is initialized to be the first symbol of the source text, "s belongs to the dictionary" is established as an invariant in the above loop. Note that the resulting set of phrases or codewords obeys the *prefix closure* property, in the sense that if a codeword is in the set, then so is also every one of its prefixes.

LZW is easily implemented in linear time using a tree data structure as the substrate [30,31], and it requires space linear in the number of phrases at the

Table 7. Synopsis of compression rates for sequences in the yeast DNA by various lossless methods. The figure in parenthesis is the percentage gain of $Codec_1$ versus other methods.

File	File Len	Huffman Pack [%diff]	LZ-78 Compress [%diff]	LZ-77 GZip [%diff]	BWT BZip [%diff]	$Codec_1$ Lossless
Spor EarlyII	25008	$7996_{[13.4]}$	$7875_{[11.7]}$	$8008_{[13.6]}$	$7300_{[3.5]}$	7052
Spor EarlyI	31039	$9937_{[11.5]}$	$9646_{[8.2]}$	$9862_{[10.6]}$	$9045_{[1.5]}$	8914
Helden CGN	32871	$10590_{[20.0]}$	$10223_{[15.8]}$	$10379_{[17.6]}$	$9530_{[8.0]}$	8828
Spor Middle	54325	$17295_{[15.9]}$	$16395_{[9.9]}$	$16395_{[9.9]}$	$15490_{[3.8]}$	14924
Helden All	112507	$36172_{[21.1]}$	$33440_{[12.0]}$	$33829_{[13.3]}$	$31793_{[6.5]}$	29862
Spor All	222453	$70755_{[23.8]}$	$63939_{[11.9]}$	$68136_{[19.2]}$	$61674_{[7.9]}$	57155
All Up 400k	399615	$121700_{[13.8]}$	$115029_{[7.6]}$	$115023_{[7.6]}$	$112363_{[5.1]}$	106909

Table 8. Compression, fidelity and loss in reconstruction of grey scale images

File	File len	GZip len [%compr]	$Codec_1$ [%compr]	Diff % GZip	%Loss	'.'/ car	% Loss recon pix	% Loss all pix
bridge	66336	$61657_{[7.05]}$	$57655_{[13.08]}$	6.49	0.42	1/4	5.67	0.02
			$50656_{[23.63]}$	17.84	14.29	1/3	7.69	0.90
camera	66336	$48750_{[26.51]}$	$43090_{[35.03]}$	11.60	9.09	1/4	0.78	0.05
lena	262944	$234543_{[12.10]}$	$210786_{[19.83]}$	10.13	4.17	1/4	7.26	0.27
			$175126_{[33.39]}$	25.33	20.00	1/3	5.11	0.81
peppers	262944	$232334_{[11.64]}$	$199605_{[23.85]}$	14.09	6.25	1/4	1.53	0.08
			$173561_{[33.99]}$	25.30	20.00	1/3	3.29	0.52

Table 9. Lossy/Lossless compression of gray-scale images using LZW-like encoding

File	File len	GZip len	LZW-like lossy	% Diff GZip	% Loss	LZW-like lossless	% Diff GZip	'.' / car
bridge	66.336	61.657	38.562	37.46	0.29	38.715	37.21	1/4
			38.366	37.78	5.35	42.288	31.41	1/3
camera	66.336	48.750	34.321	29.60	0.00	34.321	29.60	1/6
			34.321	29.60	0.06	34.321	29.60	1/5
			32.887	32.54	6.16	35.179	27.84	1/4
lena	262.944	234.543	120.308	48.71	1.36	123.278	47.44	1/4
			123.182	47.48	7.32	135.306	42.31	1/3
peppers	262.944	232.334	117.958	49.23	1.75	121.398	47.75	1/4
			119.257	48.67	4.45	129.012	44.47	1/3

outset. Another remarkable property of LZW is that the encoding and decoding are perfectly symmetrical, in particular, the dictionary is recovered while the decompression process runs (except for a special case that is easily taken care of).

We test the power of ZLW encoding on the motifs produced in greedy off-line schemata such as above. Despite the apparent superiority of such greedy off-line

Table 10. Lossy/Lossless compression of the Calgary Corpus by LZW-like encoding

File	File len	GZip len	LZW-like lossy	% Diff GZip	% Loss	LZW-like lossless	% Diff GZip	'.' / car
bib	111.261	35.063	26.679	23.91	14.11	34.174	2.54	1/4
			26.679	23.91	14.11	34.174	2.54	1/3
geo	102.400	68.493	30.951	54.81	19.43	57.098	16.64	1/4
			33.334	51.33	20.63	58.038	15.26	1/3
news	377.109	144.840	104.807	27.64	11.42	128.429	11.33	1/4
			106.483	26.48	19.95	153.243	-5.80	1/3
obj1	21.504	10.323	8.447	18.17	8.38	9.642	6.60	1/4
			7.409	28.23	18.03	9.849	4.59	1/3
			6.736	34.75	21.61	8.521	17.46	2/5
obj2	246.814	81.631	56.810	30.41	12.75	67.857	16.87	1/3
			53.094	34.96	19.88	67.117	17.78	1/2
paper1	53.161	18.577	16.047	13.62	13.63	19.411	-4.49	1/3
			15.626	15.89	18.22	19.198	-3.34	1/2
paper2	82.199	29.753	23.736	20.22	15.19	28.743	3.39	1/3
			22.519	24.31	36.44	35.390	-18.95	1/2
pic	513.216	56.422	36.491	35.32	0.56	36.599	35.13	1/4
progc	39.611	13.275	11.576	12.80	6.88	12.812	3.49	1/4
			11.381	14.27	8.69	12.854	3.17	1/3
			11.010	17.06	24.86	15.246	14.85	1/2
progl	71.646	16.273	14.828	8.88	2.72	15.601	4.13	1/4
			14.748	9.37	6.99	16.149	0.76	1/3
			14.676	9.81	7.70	16.261	0.07	2/5
progp	49.379	11.246	10.287	8.53	3.83	10.879	3.26	1/4
			10.265	8.72	7.43	11.328	-0.73	1/3
			10.416	7.38	7.12	11.569	-2.87	2/5

approaches in capturing long range repetitions, one drawback is in the encoding of references, which are bi-directional and thus inherently more expensive than those in ZLW. Our exercise consists thus of using the motifs selected in the greedy off-line to set up an initial vocabulary of motif phrases, but then encode these and their outgrowth while we carry out a parse of the source string similar to that of ZLW. Assuming that we have already selected the motifs to be used, this adaptation of ZLW to motifs requires to address primarily the following problems:

1. We need to modify the parsing in such a way that for every chosen motif, every one of its occurrences is used in the parsing.
2. We need to modify the dictionary in order to accommodate motifs in addition to strings. This is to be done while retaining *prefix closure*.

To ensure that a motif is correctly detected and deployed in the parsing, it has to be stored in the dictionary before its first occurrence is detected. This requires building a small dictionary that needs to be sent over to the decoder together

Table 11. Lossy/Lossless compression of sequences from DNA yeast families by LZW-like encoding

File	File len	GZip len	LZW-like lossy	% Diff GZip	% Loss	LZW-like lossless	% Diff GZip	'.' / car
Spor EarlyII	25.008	8.008	6.137	23.36	16.93	7.430	7.22	1/4
			6.163	23.04	12.01	7.052	11.94	1/3
Spor EarlyI	31.039	9.862	7.494	24.01	14.66	8.865	10.11	1/4
			7.494	24.01	14.66	8.865	10.11	1/3
Helden CGN	32.871	10.379	7.728	25.54	15.36	9.330	10.11	1/3
Spor Middle	54.325	16.395	11.565	29.46	0.38	11.672	28.81	1/4
			11.555	29.52	0.43	11.703	28.62	1/3
Helden All	112.507	33.829	25.873	23.52	18.83	32.029	5.32	1/4
			26.010	23.11	18.86	32.182	4.87	1/3
Spor All	222.453	68.136	48.035	29.50	18.98	60.042	11.88	1/4
			47.896	29.71	19.00	59.955	12.01	1/3
All Up 400k	399.615	115.023	90.120	21.65	18.62	110.659	3.79	1/3

with the encoded string. In order to enforce the prefix closure, all prefixes of a motif are inserted in the dictionary together with that motif.

With the dictionary in place, the parse phase of the algorithm proceeds in much the same way as in the original scheme, with the proviso that once a motif is chosen, then all of its occurrences are to be deployed. For this, the algorithm looks at each stage for the longest substring in the tree that does not interfere with the next motif occurrence already allocated from previous stages. The motif chosen in this way is then concatenated with the symbol following it and the result is inserted in the dictionary. In order to avoid the insertion of undesired don't cares in text regions not encoded by motifs, that symbol is treated as mismatching all other characters at this stage of the search.

Decoding is easier. The recovery follows closely the standard ZLW, except for initialization of the dictionary. The only difference is thus that now the decoder receives, as part of the encoding, also an initial dictionary containing all motifs utilized, which are used to initialize the tree.

The tables below summarize results obtained on gray-scale images (Table 9, 1 pixel = 1 byte), the Calgary Corpus (Table 10), and genetic data (Table 11). For each case, the compression is reported first for lossy encoding with various don't care densities, then also for the respective lossless completions.

6   Conclusion

Irredundant motifs seem to provide an excellent repertoire of codewords for grammar based compression and syntactic inference of documents of various kinds. Various completion strategies and possible extensions (e.g., to nested descriptors) and generalizations (notably, to higher dimensions) suggest that the notions explored here can develop in a versatile arsenal of data compression methods capable of bridging lossless and lossy textual substitution in a way that is both aesthetically pleasant and practically advantageous. Algorithms for efficient motif extraction as well as for their efficient deployment in compression are highly desirable from this perspective. In particular, algorithms for computing the statistics for maximal sets of *non-overlapping* occurrences for each motif should be set up for use in gain estimations, along the lines of the constructions given in [10] for solid motifs. Progress in these directions seems not out of reach.

Acknowledgments

The authors are indebted to R. Ahlswede for the warm hospitality and congenial atmosphere provided at the ZiF Meetings in Bielefeld. D. Epstein and, independently, S. Lonardi brought reference [19] to the attention of one of the authors. The anonymous referees of [8,9] gave useful comments on those preliminary versions of parts of this paper.

References

1. M.J. Atallah, Y. Genin, and W. Szpankowski, Pattern matching image compression: algorithmic and empirical results, IEEE Transactions on PAMI, Vol. 21, No. 7, 614–629, 1999.
2. A. Apostolico, On the efficiency of syntactic feature extraction and approximate schemes for data compression, Proceedings of the 5th International Conference on Pattern Recognition., 982–984, Miami, Florida, 1980.
3. A. Apostolico and M.J. Atallah, Compact recognizers of episode sequences, Information and Computation, 174, 180-192, 2002.
4. A. Apostolico and Z. Galil (Eds.), Pattern Matching Algorithms, Oxford University Press, New York, 1997.
5. A. Apostolico and A. Fraenkel, Robust transmission of unbounded strings using Fibonacci representations, IEEE Trans. Inform. Theory, Vol. 33, No. 2, 238–245, 1987.
6. A. Apostolico and S. Lonardi, Off-line compression by greedy textual substitution, Proceedings of the IEEE, Vol. 88, No. 11, 1733–1744, 2000.
7. A. Apostolico and L. Parida, Incremental paradigms of motif discovery, Journal of Computational Biology, Vol. 11, No. 1, 15-25, 2004.
8. A. Apostolico and L. Parida, Compression and the wheel of fortune, Proceedings of IEEE DCC Data Compression Conference, Computer Society Press, 143–152, 2003.
9. A. Apostolico, M. Comin, and L. Parida, Motifs in Ziv-Lempel-Welch clef, Proceedings of IEEE DCC Data Compression Conference, 72–81, Computer Society Press, 72–81, 2004.
10. A. Apostolico and F.P. Preparata, Data structures and algorithms for the string statistics problem, Algorithmica, 15, 481–494, 1996.
11. T. Berger, Rate Distortion Theory: A Mathematical Basis for Data Compression, Prentice Hall, Englewood Cliffs, N.J., 1971.
12. T. Berger and J.D. Gibson, Lossy source coding, IEEE Trans. Inform. Theory, Vol. 44, No. 6, 2693-2723, 1998.

13. A. Blumer, J. Blumer, A. Ehrenfeucht, D. Haussler, M.T. Chen, and J. Seiferas, The smallest automaton recognizing the subwords of a text, Theoretical Computer Science, 40, 31-55, 1985.
14. S. DeAgostino and J.A. Storer, On-line versus off-line computation in dynamic text compression, Inform. Process. Lett., Vol. 59, No. 3, 169–174, 1996.
15. K.S. Fu and T.L. Booth, Grammatical inference: introduction and survey – part I, IEEE Transactions on Systems, Man and Cybernetics, 5, 95–111, 1975.
16. K.S. Fu and T.L. Booth, Grammatical inference: introduction and survey – part II, IEEE Transactions on Systems, Man and Cybernetics, 5, 112–127, 1975.
17. J.C. Kieffer, A survey of the theory of source coding with a fidelity criterion, IEEE Trans. Inform. Theory, Vol. 39, No. 5, 1473–1490, 1993.
18. I. Kontoyiannis, An implementable lossy version of Lempel-Ziv algorithm –part 1: optimality for memoryless sources, IEEE Trans. Inform. Theory, Vol. 45, 2293–2305, 1999.
19. E. Lehman and A. Shelat, Approximation algorithms for grammar based compression, Proceedings of the eleventh ACM-SIAM Symposium on Discrete Algorithms (SODA 2002), 205-212, 2002.
20. A. Lempel and J. Ziv, On the complexity of finite sequences, IEEE Trans. Inform. Theory, Vol. 22, 75–81, 1976.
21. T. Luczak and W. Szpankowski, A suboptimal lossy data compression algorithm based on approximate pattern matching, IEEE Trans. Inform. Theory, Vol. 43, No. 5, 1439–1451, 1997.
22. C. Neville-Manning, I.H. Witten, and D. Maulsby, Compression by induction of hierarchical grammars, In DCC: Data Compression Conference, IEEE Computer Society TCC, 244–253, 1994.
23. L. Parida, I. Rigoutsos, and D. Platt, An output-sensitive flexible pattern discovery algorithm, Combinatorial Pattern Matching (CPM 2001), (A. Amir and G. Landau, Eds.), LNCS, Vol. 2089, 131–142, 2001.
24. L. Parida, I. Rigoutsos, A. Floratos, D. Platt, and Y. Gao, Pattern discovery on character sets and real-valued data: linear bound on irredundant motifs and polynomial time algorithms, Proceedings of the eleventh ACM-SIAM Symposium on Discrete Algorithms (SODA 2000), January 2000, 297–308, 2000.
25. I. Rigoutsos, A. Floratos, L. Parida, Y. Gao, and D. Platt, The emergence of pattern discovery techniques in computational biology, Journal of Metabolic Engineering, Vol. 2, No. 3, 159-177, 2000.
26. I. Sadeh, On approximate string matching, Proceedings of DCC 1993, IEEE Computer Society Press, 148–157, 1993.
27. J.A. Storer, Data Compression: Methods and Theory, Computer Science Press, 1988.
28. J. Wang, B. Shapiro, and D. Shasha (Eds.), Pattern Discovery in Biomolecular Data: Tools, Techniques, and Applications, Oxford University Press, 1999.
29. M. Waterman, Introduction to Computational Biology, Chapman and Hall, 1995.
30. J. Ziv and A. Lempel, A universal algorithm for sequential data compression, IEEE Trans. Inform. Theory, Vol. 23, No. 3, 337-343, 1977.
31. J. Ziv and A. Lempel, Compression of individual sequences via variable-rate coding, IEEE Trans. Inform. Theory, Vol. 24, No. 5, 530-536, 1978.

Appendix

Proof of Theorem 1
Let m be a 2-motif in $\mathcal{B}$, and $\mathcal{L}_m = (l_1, l_2, \ldots, l_p)$ be its occurrence list. The claim is true for $p = 2$. Indeed, let $i = l_1$ and $j = l_2$, and consider the meet m' of suf_i and suf_j. By the maximality in composition of m, we have that $m' \preceq m$. On the other hand, for any motif $\hat{m}$ with occurrences at i and j it must be $\hat{m} \preceq m'$, whence, in particular, $m \preceq m'$. Thus, $m = m'$. Assume now $p \geq 3$ and that there is no pair of indices i and j in $\mathcal{L}_m$ such that m is the meet of suf_i and suf_j. Again, for any choice of i and j in $\mathcal{L}_m$, we must have that $m \preceq m'$, where m' denotes as before the meet of suf_i and suf_j. Therefore, we have that $m \preceq m'$ but $m \neq m'$ for all choices of i and j. Assume now one such choice is made. By the maximality of m, it cannot be that m' is the meet of all suffixes with beginning in $\mathcal{L}_m$. Therefore, there must be at least one index k such that m' differs either from the meet of suf_k and suf_i or from the meet of suf_k and suf_j, or from both. Let, to fix the ideas, m'' be this second meet. Since $m \preceq m''$ and $m \preceq m'$ then $\mathcal{L}_{m'}$ and $\mathcal{L}_{m''}$ are sublists of $\mathcal{L}_m$, by Lemma 1. In other words, $\mathcal{L}_m$ can be decomposed into two or more lists of maximal motifs such that their union implies m and its occurrences. But this contradicts the assumption that m is irredundant. □

The Set Union Problem, SUP(n, q). Given n sets $S_1, S_2 \ldots, S_n$ on q elements each, find all the sets S_i such that $S_i = S_{i_1} \cup S_{i_2} \cup \ldots \cup S_{i_p}$, $i \neq i_j$, $1 \leq j \leq p$.

This is a very straightforward algorithm (this contributes an additive term to the overall complexity of the pattern detection algorithm): For each set S_i, we first obtain the sets S_j $j \neq i, j = 1 \ldots n$ such that $S_j \subset S_i$. This can be done in $O(nq)$ time (for each i). Next, we check if $\cup_j S_j = S_i$. Again this can be done in $O(nq)$ time. Hence the total time taken is $O(n^2 q)$.

Reverse–Complement Similarity Codes

A. D'yachkov*, D. Torney, P. Vilenkin, and S. White

Abstract. In this paper, we discuss a general notion of *similarity function* between two sequences which is based on their common subsequences. This notion arises in some applications of molecular biology [14]. We introduce the concept of *similarity codes* and study the logarithmic asymptotics for the size of optimal codes. Our mathematical results announced in [13] correspond to the *longest common subsequence* (LCS) similarity function [2] which leads to a special subclass of these codes called *reverse-complement* (RC) similarity codes. RC codes for *additive* similarity functions have been studied in previous papers [9,10,11,12].

Keywords: sequences, subsequences, similarity, DNA sequences, codes, code distance, rate of codes, insertion-deletion codes.

1 Introduction

This paper in organized as follows. In Section 2, we define similarity functions, similarity codes and a reverse-complement operation for codewords. In Section 3, we consider some types of similarity functions. In Section 4, we describe some biological applications that give motivation for our study. Finally, in Section 5 we obtain random coding bounds on the rate of reverse-complement similarity codes. We restrict ourselves to only one special similarity function, namely, the length of a longest common subsequence of given sequences.

2 Notations and Definitions

In the whole paper, the symbol $\triangleq$ denotes definitional equalities.

2.1 Similarity Functions and Similarity Codes

Consider a finite alphabet A. For a positive integer n denote by A^n the set of all sequences (codewords) of length n: $\mathsf{A}^n \triangleq \big\{ \mathbf{x} = (x_1, \ldots, x_n), x_i \in \mathsf{A} \big\}$. A subset $\mathcal{C} \subset \mathsf{A}^n$ is called a *(fixed-length) code* of *length* n. We will use symbol $\mathsf{A}^\star$ to denote the set of all finite codewords, i.e. the union of all A^n over all $n \geq 1$.

Definition 1. *A symmetric function* $\mathcal{S}^{(n)}$ *defined on the set* $\mathsf{A}^n \times \mathsf{A}^n$ *is called a (fixed-length) similarity function if* $0 \leq \mathcal{S}^{(n)}(\mathbf{x}, \mathbf{y}) \leq \mathcal{S}^{(n)}(\mathbf{x}, \mathbf{x})$ *for any* $\mathbf{x}, \mathbf{y} \in \mathsf{A}^n$. *We call a similarity family a set of fixed-length similarity functions* $\mathcal{S}' = \{\mathcal{S}^{(n)}, n \geq 1\}$.

* The work of Arkadii D'yachkov and Pavel Vilenkin was supported by the Russian Foundation of Basic Research, Grant 01-01-00495, and INTAS-00-738.

R. Ahlswede et al. (Eds.): Information Transfer and Combinatorics, LNCS 4123, pp. 814–830, 2006.

For a code $C \subset \mathsf{A}^n$ and a given similarity function $S^{(n)}$ consider the following similarity parameters:

- *the minimum self-similarity:* $S_{\mathrm{s}}^{(n)}(C) \triangleq \min\limits_{\mathbf{x} \in C} S^{(n)}(\mathbf{x}, \mathbf{x});$
- *the maximum cross-similarity:* $S_{\mathrm{c}}^{(n)}(C) \triangleq \max\limits_{\substack{\mathbf{x}, \mathbf{y} \in C \\ \mathbf{x} \neq \mathbf{y}}} S^{(n)}(\mathbf{x}, \mathbf{y});$
- *the similarity threshold:* $S_{\mathrm{t}}^{(n)}(C) \triangleq S_{\mathrm{s}}^{(n)}(C) - S_{\mathrm{c}}^{(n)}(C).$

A code $C \subset \mathsf{A}^n$ is called a similarity code if the value $S_{\mathrm{s}}^{(n)}(C)$ is large enough and the value $S_{\mathrm{c}}^{(n)}(C)$ is small enough.

Definition 2. *Let S and Δ be nonnegative numbers. A code $C \subset \mathsf{A}^n$ is called a similarity code with threshold Δ (St-code) if $S_{\mathrm{t}}^{(n)}(C) \geq \Delta$. A code $C \subset \mathsf{A}^n$ is called a similarity code with parameters (S, Δ) (Sp-code) if $S_{\mathrm{s}}^{(n)}(C) \geq S + \Delta$ and $S_{\mathrm{c}}^{(n)}(C) \leq S$.*

2.2   Rates of Similarity Codes

Let a similarity family $S' = \{S^{(n)}, n \geq 1\}$ be fixed. For a given n denote by $\mathcal{M}_{\mathrm{t}}^{(n)}$ and $\mathcal{M}_{\mathrm{p}}^{(n)}$ the maximum sizes of St-code and Sp-code with given parameters:

$$\mathcal{M}_{\mathrm{t}}^{(n)}(\Delta) \triangleq \max\left\{|C| : C \subset \mathsf{A}^n, \ S_{\mathrm{t}}^{(n)}(C) \geq \Delta\right\}, \quad \Delta \geq 0,$$

$$\mathcal{M}_{\mathrm{p}}^{(n)}(S, \Delta) \triangleq \max\left\{|C| : C \subset \mathsf{A}^n, \ S_{\mathrm{s}}^{(n)}(C) \geq S + \Delta, \ S_{\mathrm{c}}^{(n)}(C) \leq S\right\}, \quad S \geq 0, \Delta \geq 0.$$

We would like to study the logarithmic asymptotics of these values described by the following *rate functions*:

$$\mathcal{R}_{\mathrm{t}}(d) \triangleq \varlimsup_{n \to \infty} \frac{\log \mathcal{M}_{\mathrm{t}}^{(n)}(dn)}{n}, \quad d \geq 0,$$

$$\mathcal{R}_{\mathrm{p}}(s, d) \triangleq \varlimsup_{n \to \infty} \frac{\log \mathcal{M}_{\mathrm{p}}^{(n)}(sn, dn)}{n}, \quad s \geq 0, d \geq 0.$$

The following relation between these functions is obvious.

Proposition 1. $\mathcal{R}_{\mathrm{t}}(d) = \max\limits_{s \geq 0} \mathcal{R}_{\mathrm{p}}(s, d).$

2.3   Pseudodistance Functions

Let $S^{(n)}$ be a similarity function on A^n. Consider the following function

$$\mathcal{D}^{(n)}(\mathbf{x}, \mathbf{y}) \triangleq \frac{S^{(n)}(\mathbf{x}, \mathbf{x}) + S^{(n)}(\mathbf{y}, \mathbf{y})}{2} - S^{(n)}(\mathbf{x}, \mathbf{y}), \quad \mathbf{x}, \mathbf{y} \in \mathsf{A}^n. \tag{1}$$

One can see that $\mathcal{D}^{(n)}(\mathbf{x}, \mathbf{y}) = \mathcal{D}^{(n)}(\mathbf{y}, \mathbf{x}) \geq 0$ and $\mathcal{D}^{(n)}(\mathbf{x}, \mathbf{x}) = 0$ for any $\mathbf{x}, \mathbf{y} \in \mathsf{A}^n$. The function $\mathcal{D}^{(n)}$ is called a *pseudodistance function* corresponding to the similarity function $\mathcal{S}^{(n)}$. On the other hand, let $\mathcal{D}^{(n)}$ be a function with the properties mentioned above and let $\mathcal{W}$ be a positive function on A^n such that $\mathcal{D}(\mathbf{x}, \mathbf{y}) \leq \mathcal{W}(\mathbf{x})$ for any $\mathbf{x}, \mathbf{y} \in \mathsf{A}^n$. Then there exists a similarity function on A^n having the form

$$\mathcal{S}^{(n)}(\mathbf{x}, \mathbf{y}) \triangleq \frac{\mathcal{W}(\mathbf{x}) + \mathcal{W}(\mathbf{y})}{2} - \mathcal{D}^{(n)}(\mathbf{x}, \mathbf{y}).$$

Let $\mathcal{S}^{(n)}$ be a similarity function, $\mathcal{D}^{(n)}$ be the corresponding pseudodistance function and $\mathcal{C}$ be a similarity code with threshold Δ. Then $\mathcal{D}^{(n)}(\mathbf{x}, \mathbf{y}) \geq \Delta$ for any $\mathbf{x}, \mathbf{y} \in \mathsf{A}^n$, $\mathbf{x} \neq \mathbf{y}$. If $\mathcal{D}^{(n)}$ is a distance function, then $\mathcal{C}$ is a code with distance Δ. This relation between similarity codes and well-known distance codes is useful for the problem of obtaining bounds on the rates of similarity codes.

For a function $\mathcal{D}^{(n)}$ becomes a real distance we need the triangle inequality and also need that $\mathcal{D}^{(n)}(\mathbf{x}, \mathbf{y}) > 0$ for any $\mathbf{x} \neq \mathbf{y}$. The second property will be held if for example $\mathcal{S}^{(n)}(\mathbf{x}, \mathbf{y}) < \mathcal{S}^{(n)}(\mathbf{x}, \mathbf{x})$ for any $\mathbf{x} \neq \mathbf{y}$. And the triangle inequality for $\mathcal{D}^{(n)}$ is equivalent the following relation

$$\mathcal{S}^{(n)}(\mathbf{x}, \mathbf{z}) + \mathcal{S}^{(n)}(\mathbf{y}, \mathbf{z}) \leq \mathcal{S}^{(n)}(\mathbf{x}, \mathbf{y}) + \mathcal{S}^{(n)}(\mathbf{z}, \mathbf{z}) \text{ for any } \mathbf{x}, \mathbf{y}, \mathbf{z} \in \mathsf{A}^n.$$

2.4   Reverse-Complement Codes

Now we introduce the additional *reverse-complement condition* which is important for the most significant biological application of similarity codes. Assume that a *complement operation* on A is fixed, i.e. for any element $a \in \mathsf{A}$ a complement element $\bar{a} \in \mathsf{A}$ is specified. We assume that $\bar{\bar{a}} = a$ for any element $a \in \mathsf{A}$.

For a sequence $\mathbf{x} = (x_1, x_2, \ldots, x_n) \in \mathsf{A}^n$ define a *reverse-complement sequence* $\overleftarrow{\mathbf{x}} \in \mathsf{A}$ that has the form

$$\overleftarrow{\mathbf{x}} \triangleq (\overline{x_n}, \ldots, \overline{x_2}, \overline{x_1}). \tag{2}$$

Definition 3. *A code $\mathcal{C} \subset \mathsf{A}^n$ is called a reverse-complement code (RC-code) if for any $\mathbf{x} \in \mathcal{C}$ the corresponding reverse-complement sequence $\overleftarrow{\mathbf{x}} \in \mathcal{C}$. If in addition $\overleftarrow{\mathbf{x}} \neq \mathbf{x}$ for any $\mathbf{x} \in \mathcal{C}$, then we call a code $\mathcal{C}$ the dual reverse-complement code (RCd-code). Any RCd-code is composed of pairs of distinct reverse-complement codewords.*

The notion of RC-codes generalizes the notion of reversible codes [1] that corresponds to a trivial complement operation $\bar{a} = a$.

3   Some Types of Similarity Families

3.1   Subsequence Similarity Function

One of the natural (especially for biological applications) approach for measuring similarity between two sequences is based on common subsequences. Denote by

$CS(\mathbf{x}, \mathbf{y})$, $\mathbf{x}, \mathbf{y} \in \mathsf{A}^\star$, the set of all sequences $\mathbf{z} \in \mathsf{A}^\star$ such that $\mathbf{z}$ is a subsequence of both $\mathbf{x}$ and $\mathbf{y}$. Let a subset $ACS(\mathbf{x}, \mathbf{y}) \subset CS(\mathbf{x}, \mathbf{y})$ be defined for any $\mathbf{x}, \mathbf{y} \in \mathsf{A}^\star$ such that $ACS(\mathbf{x}, \mathbf{y}) = ACS(\mathbf{y}, \mathbf{x})$ and $\mathbf{x} \in ACS(\mathbf{x}, \mathbf{x})$ for any $\mathbf{x} \in \mathsf{A}^\star$. We call $ACS(\mathbf{x}, \mathbf{y})$ a set of *admissible common subsequences* of $\mathbf{x}$ and $\mathbf{y}$. The functional ACS itself will be referred to as the *rule* that singles out admissible common subsequences among all common subsequences. Now define the following function:

$$\mathcal{S}(\mathbf{x}, \mathbf{y}) \triangleq \max_{\mathbf{z} \in ACS(\mathbf{x}, \mathbf{y})} \mathcal{W}(\mathbf{z}), \quad \mathbf{x}, \mathbf{y} \in \mathsf{A}^\star. \tag{3}$$

We call it a *subsequence similarity function* based on the weight function $\mathcal{W}$ and a rule ACS. One can see that such function satisfies all terms of definition 1.

If we restrict ourselves to the fixed-length sequences, then we obtain a fixed-length similarity function. Taking all possible lengths we get a similarity family. Now we consider some examples of similarity families obtained by this method.

3.2 LCS Similarity Function

Let $ACS(\mathbf{x}, \mathbf{y}) = CS(\mathbf{x}, \mathbf{y})$. Put $\mathcal{W}(\mathbf{z})$ equals the length of the sequence $\mathbf{z}$. Then equality (3) defines the well-known measure of similarity between two sequences — the length of the longest common subsequence of $\mathbf{x}$ and $\mathbf{y}$. It is usually denoted by $LCS(\mathbf{x}, \mathbf{y})$.

Consider the corresponding fixed-length similarity function $\mathcal{S}^{(n)}$ on A^n. In this case the pseudodistance function (1) coincides with the well-known *Levenstein distance*, i.e. the minimum number of insertion–deletion transformations that transform one sequence to another.

LCS similarity function can be generalized in several ways. We can assign weights to different symbols from A. Then the weight function $\mathcal{W}$ can be defined as the component-wise sum of alphabetic weights:

$$\mathcal{W}(\mathbf{z}) \triangleq \sum_i w(z_i). \tag{4}$$

In the case $w(z) \equiv 1$ this approach gives the LCS similarity function.

Another way of generalization is to consider non-trivial rules $ACS(\mathbf{x}, \mathbf{y})$. We would like to mention the paper [15] where the authors consider the following rule: a sequence $\mathbf{z} \in CS(\mathbf{x}, \mathbf{y})$ is called admissible if any consecutive elements z_i and z_{i+1} which are separated in $\mathbf{x}$ are also separated in $\mathbf{y}$ and vice versa. This rule has a good biological motivation (see below).

3.3 Additive Similarity Functions

A number of fixed-length similarity functions can be defined taking the component-wise sum of alphabetic similarities. Let $\mathcal{S}$ be a similarity function on the alphabet A. For an integer $n \geq 1$ put

$$\mathcal{S}^{(n)}(\mathbf{x}, \mathbf{y}) \triangleq \sum_{i=1}^{n} \mathcal{S}(x_i, y_i), \quad \mathbf{x}, \mathbf{y} \in \mathsf{A}^n.$$

In this case $\mathcal{S}^{(n)}$ is called an *additive similarity function* corresponding to the *alphabetic similarity function* $\mathcal{S}$. Obviously, in this case the corresponding pseudodistance function $\mathcal{D}^{(n)}$ also has the form of the component-wise sum of alphabetic pseudodistances.

Consider a special case of additive similarity functions. Namely, let w be an arbitrary positive function on the alphabet A. We call w the *alphabetic weight function*. Define the following alphabetic similarity function $\mathcal{S}(x, y) \triangleq$

$$
\begin{cases}
0, & if\, x \neq y, \\
w(x), & if\, x = y, \\
x, y \in \text{A}
\end{cases}
$$

. We call $\mathcal{S}$ the *alphabetic weight similarity function* and the corresponding additive similarity function $\mathcal{S}^{(n)}$ the *weight additive similarity function*. One can see that the corresponding function $\mathcal{D}^{(n)}$ is a real distance function. It has the form of component-wise sum of the following alphabetic distance function

$$
\mathcal{D}(x, y) = \frac{w(x) + w(y)}{2}, \quad x, y \in \text{A}, \quad x \neq y \quad \text{and} \quad \mathcal{D}(x, x) = 0. \tag{5}
$$

As an example consider the weight function $w(x) \equiv 1$. In this case $\mathcal{D}^{(n)}(\mathbf{x}, \mathbf{y})$ is the Hamming distance between vectors $\mathbf{x}$ and $\mathbf{y}$, and $\mathcal{S}^{(n)}(\mathbf{x}, \mathbf{y})$ is the *Hamming similarity* between $\mathbf{x}$ and $\mathbf{y}$, i.e., the number of coordinates i such that $x_i = y_i$. If $\mathcal{C} \subset \text{A}^n$ is a similarity code with threshold Δ, then $\mathcal{C}$ is also a code with Hamming distance Δ.

Note that weight additive similarity functions can be considered as the special case of subsequence similarity functions. For this we need to take the additive weight function (4) and define the following rule: a sequence $\mathbf{z} \in \mathrm{CS}(\mathbf{x}, \mathbf{y})$ is admissible if any element z_i has the same indices in $\mathbf{x}$ and $\mathbf{y}$.

For a biologically motivated case of quaternary alphabet and a special weight function such codes were considered in [9] without the reverse-complement condition. The general case was studied in [11]. The results are based on the mentioned relationship between similarity codes and distance codes. Some of the methods which are used for Hamming distance [5] can be generalized to the weight distance (5).

4   Biological Motivation

4.1   DNA Sequences and Their Properties

A *DNA molecule* is a sequence which is composed of consecutive *bases*. There exist 4 types of bases: *adenine* (A), *citosine* (C), *guanine* (G), and *thymine* ($\mathcal{T}$). A DNA molecule is directed, i.e., one endpoint of it can be considered as the *beginning* and another one as the *end*. Due to this fact, the one-to-one correspondence exists between DNA molecules and codewords over the quaternary alphabet A $\triangleq \{\text{A}, \text{C}, \text{G}, \mathcal{T}\}$. We will use the term *DNA sequence* to indicate both a DNA molecule and a codeword that corresponds to it.

An important property of DNA sequences that plays the fundamental role for many practical purposes is that a pair of oppositely oriented molecules can form a *duplex* which is based on hydrogen bonds between some base pairs. Namely, a pair "A—$\mathcal{T}$" forms *two* bonds, a pair "C—G" forms *three* bonds, and any other pair is called a *mismatch* because it does not form any bond. This leads to the following natural complement operation on the alphabet A:

$$\overline{\text{A}} \triangleq \mathcal{T}, \quad \overline{\text{C}} \triangleq \text{G}, \quad \overline{\text{G}} \triangleq \text{C}, \quad \overline{\mathcal{T}} \triangleq \text{A}. \tag{6}$$

Thus, hydrogen bonds occur only in pairs that are formed by complement bases.

Fig. 1 shows an example of a duplex. Using the notations of molecular biology we denote the beginning and the end of each molecule by the symbols $5'$ and $3'$, respectively. In the given example two pairs "A—$\mathcal{T}$" and three pairs "C—G" form $2 \cdot 2 + 3 \cdot 3 = 13$ hydrogen bonds. Other 4 pairs of bases are mismatches and form no hydrogen bonds. A process of forming a duplex from single strands is referred to as *DNA hybridization*.

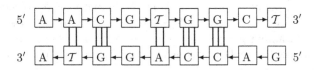

Fig. 1. A duplex formed by the DNA sequences $\mathbf{x} = \text{AACG}\mathcal{T}\text{GGC}\mathcal{T}$ and $\mathbf{y} = \text{GACCAGG}\mathcal{T}\text{A}$

An important characteristic of a duplex is its *energy*. It can be considered as a measure of a duplex stability. It is connected with an important physics characteristic called the *melting temperature* of a duplex. If a temperature of the environment grows higher than this level, then all bonds collapse and the duplex splits back to the single strands. In other words, the more stable a duplex is, the greater temperature is necessary to melt it.

Let $\mathbf{x}, \mathbf{y} \in \text{A}^*$ be two DNA sequences. Let $\mathcal{E}(\mathbf{x}, \mathbf{y})$ be the energy of the duplex formed by $\mathbf{x}$ and $\mathbf{y}$. Regardless of the exact definition of this function (which will be discussed later) several properties follow from its physical meaning:

(E1) It is obvious that $\mathcal{E}(\mathbf{x}, \mathbf{y}) = \mathcal{E}(\mathbf{y}, \mathbf{x})$.

(E2) Let a sequence $\mathbf{x}$ be fixed. We wish to maximize the value $\mathcal{E}(\mathbf{x}, \mathbf{y})$ over all $\mathbf{y} \in \text{A}^*$, i.e., find a sequence $\mathbf{y}$ forming the most stable duplex with $\mathbf{x}$. One can easily understand that the most stable duplex appears if and only if there are no mismatches in it, i.e., it contains only complement base pairs. One can see that this holds iff $\mathbf{y}$ is the reverse-complement (2) sequence to $\mathbf{x}$ based on the complement operation (6). Thus, we obtain

$$\max_{\mathbf{y}} \mathcal{E}(\mathbf{x}, \mathbf{y}) = \mathcal{E}(\mathbf{x}, \overleftarrow{\mathbf{x}}). \tag{7}$$

(E3) We assume that $\mathcal{E}(\mathbf{x}, \mathbf{y}) = \mathcal{E}(\overleftarrow{\mathbf{x}}, \overleftarrow{\mathbf{y}})$. This assumption is based on the fact that the duplexes that are formed by the given DNA sequences have the same complement pairs located in the same order. Only the mismatches between them may differ, see Fig. 2. We assume that such differences do not affect the value of stability.

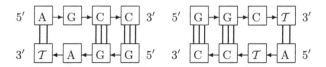

Fig. 2. Duplexes formed by the DNA sequences AGCC and GGA$\mathcal{T}$ with their reverse-complements GGC$\mathcal{T}$ and A$\mathcal{T}$CC

4.2   Using DNA Sequences as Tags

Properties of DNA sequences described above yield many interesting applications. One of the popular application is DNA computing. We consider the technique of using DNA sequences as *tags* for molecular objects. Assume that we have a pool containing p types of some molecular objects. Each object of the i-th type is marked by a DNA sequence $\mathbf{x}_i$, $i = 1, \ldots, p$. These sequences are referred to as *capture tags*.

Our aim is to separate the objects. To do this we consider a set of other p DNA sequences $\{\mathbf{y}_1, \ldots, \mathbf{y}_p\}$ which are called *address tags*. Then a solid support is taken. It is divided into p separate zones. Many copies of an address tag $\mathbf{y}_j$ are immobilized onto the corresponding j-th zone that physically segregates them. Then the support is placed into the pool (Fig. 3).

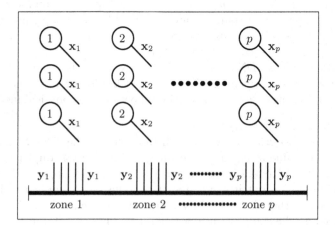

Fig. 3. A pool with capture tags $\mathbf{x}_i$ and address tags $\mathbf{y}_i$

Each pair of DNA sequences in a pool may form a duplex (except immobilized address tags). In particular, any capture tag $\mathbf{x}_i$ may form a duplex with an address tag $\mathbf{y}_j$. In this case the corresponding object of the i-th type finds itself settled on the j-th zone of the support. Since there are many copies of each object and many copies of each address tag, one can finally find any type of object settled on j-th zone for any $j = 1, \ldots, p$.

For the simplicity assume that the energy function $\mathcal{E}$ expresses exactly the melting temperature of a duplex. Assume that for an index $j \in \{1, 2, \ldots, p\}$ a certain temperature range separates the large value $\mathcal{E}(\mathbf{x}_j, \mathbf{y}_j)$ from the small values $\mathcal{E}(\mathbf{x}_i, \mathbf{y}_j)$ for $i \neq j$. This means that there exists a temperature range at which all duplexes on the j-th zone melt except those which are formed by $\mathbf{x}_j$ and $\mathbf{y}_j$. Finally, only the objects of the j-th type will be settled on the corresponding zone and that separates them from the other types, see Fig. 4. Whenever this condition holds for all values j, we are able to separate all types of objects.

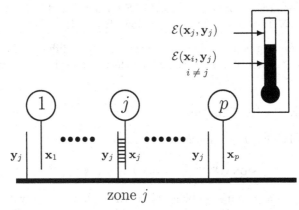

Fig. 4. Separation of the j-th objects

Consider two simple conditions which should be imposed on the DNA sequences that are used in the process. Obviously, the tags of the same type should be different, i.e., $\mathbf{x}_i \neq \mathbf{x}_j$ and $\mathbf{y}_i \neq \mathbf{y}_j$ for any $i \neq j$. In addition, we require that not only $\mathcal{E}(\mathbf{x}_i, \mathbf{y}_j)$ must be small enough for $i \neq j$ but also $\mathcal{E}(\mathbf{x}_i, \mathbf{x}_j)$ for any i and j. This requirement is based on the fact that a capture tag may also form a duplex with another capture tag. If this occur, then two molecular objects will be attached to themselves. If such duplexes are stable enough, then this may affect the experiment and the separating procedure. In particular, this yields $\mathbf{x}_i \neq \mathbf{y}_j$ for any i and j. Thus, all $2p$ tags should be mutually different.

4.3   Statements of the Problems

Let p be a fixed positive integer. The model which was introduced above leads to the following problems:

Problem π_1: for given numbers S and $\Delta \geq 0$ construct $2p$ different DNA tags $\mathbf{x}_1, \ldots, \mathbf{x}_p, \mathbf{y}_1, \ldots, \mathbf{y}_p \in \mathsf{A}^\star$ such that $\mathcal{E}(\mathbf{x}_i, \mathbf{y}_i) \geq S + \Delta$ for any i, $\mathcal{E}(\mathbf{x}_i, \mathbf{y}_j) \leq S$ for any $i \neq j$, and $\mathcal{E}(\mathbf{x}_i, \mathbf{x}_j) \leq S$ for any i and j.

Problem π_2: for a given number $\Delta \geq 0$ construct $2p$ different tags for which there exists a number S such that these tags solve problem π_1 for the pair (S, Δ).

One can see that in problem π_1 a pair (S, Δ) specifies the exact temperature range (starting at the point S and having the length Δ) which is used for separating. In problem π_2 a number Δ specifies only the size of this range but not the exact location of it.

We will not consider these general problems but use a certain simplification. Namely, we put $\mathbf{y}_i \triangleq \overleftarrow{\mathbf{x}}_i$ for all i. This choice is motivated by the equality (7) and the requirement that tags $\mathbf{x}_i$ and $\mathbf{y}_i$ should form the stable duplex for any i. This reduction leads to the following problems.

Problem π_1': for given numbers S and $\Delta \geq 0$ construct p different tags $\mathbf{x}_1, \ldots,$ $\mathbf{x}_p$ such that: (a) $\mathbf{x}_i \neq \overleftarrow{\mathbf{x}}_j$ for any i and j; (b) $\mathcal{E}(\mathbf{x}_i, \overleftarrow{\mathbf{x}}_i) \geq S + \Delta$ for any i; (c) $\mathcal{E}(\mathbf{x}_i, \overleftarrow{\mathbf{x}}_j) \leq S$ for any $i \neq j$; (d) $\mathcal{E}(\mathbf{x}_i, \mathbf{x}_j) \leq S$ for any i and j; (e) $\mathcal{E}(\overleftarrow{\mathbf{x}}_i, \overleftarrow{\mathbf{x}}_j) \leq S$ for any i and j. Note that condition (e) follows from (d) and the property (E3).

Problem π_2': for a given number $\Delta \geq 0$ construct p different tags $\mathbf{x}_1, \ldots, \mathbf{x}_p$ such that there exists a number S for which these tags solve problem π_1' for the pair (S, Δ).

4.4   DNA Similarity Function

Consider the following *DNA similarity function*

$$\mathcal{S}_{\mathrm{DNA}}(\mathbf{x}, \mathbf{y}) \triangleq \mathcal{E}(\mathbf{x}, \overleftarrow{\mathbf{y}}), \quad \mathbf{x}, \mathbf{y} \in \mathsf{A}^\star. \tag{8}$$

Using (8) we can formulate problem π_1' in the following equivalent form: for given numbers $S \geq 0$ and $\Delta \geq 0$ construct $2p$ mutually different DNA sequences $\mathbf{x}_1, \ldots, \mathbf{x}_p, \overleftarrow{\mathbf{x}}_1, \ldots, \overleftarrow{\mathbf{x}}_p$ such that:

(a) $\mathcal{S}_{\mathrm{DNA}}(\mathbf{x}_i, \mathbf{x}_i) \geq S + \Delta$ for any i; (b) $\mathcal{S}_{\mathrm{DNA}}(\mathbf{x}_i, \mathbf{x}_j) \leq S$ for any $i \neq j$.

One can see that these conditions coincide with the definition of similarity RCd-code with parameters (S, Δ) (see definitions 2 and 3). Similarly, problem π_2' leads to similarity RCd-code with threshold Δ.

Let us discuss possible ways for measuring duplex energy $\mathcal{E}(\mathbf{x}, \mathbf{y})$. This problem does not seem to be easy and the authors do not know a satisfactory answer. To start with, any pair of DNA sequences may form different duplexes. Figures 1 and 2 show the simplest configuration in which DNA sequences of the same length are located exactly opposite each other. But it is known that many other configurations are also possible. We can single out four configurations which form the basis for all possible locations of molecules in a duplex. We call them *normal location*, *overhang*, *loop*, and *hairpin*. They are shown on Fig. 5. Note that a hairpin is based on a single DNA sequence. Since these configurations appear in different combinations, a pair of long DNA sequences may form large number of duplexes.

Even if we choose a specific duplex configuration, it is not obvious how to measure its energy value. One possible method is to calculate the number of

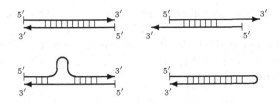

Fig. 5. Basic configurations of DNA duplexes. Top left: normal; top right: overhang; bottom left: loop; bottom right: hairpin.

hydrogen bonds. Thus, a duplex shown on figures 1 and 2 have energy values 13 and 8, respectively. This model leads to additive similarity functions. But it is too simple and rough model. For example, it is known that gaps between sequences bonded bases powerfully decrease the stability of a duplex.

A better approach is the *nearest-neighbor interactions method* which is presented in [6]. In this method, the stability function has the form of the sum which is taken not over single positions of the DNA sequences but over pairs of them. For example, the stability of the duplex shown on Fig. 1 should be calculated as the sum

$$e(\mathrm{AA}/\mathrm{A}\mathcal{T}) + e(\mathrm{AC}/\mathcal{T}\mathrm{G}) + e(\mathrm{CG}/\mathrm{GG}) + \cdots$$

The values of the function e are chosen by experimental methods.

Unfortunately, in [6] authors consider only the stability of *DNA double helixes*, i.e., duplexes with normal configuration and no mismatches (in other words, duplexes formed by pairs of DNA sequences $(\mathbf{x}, \overleftarrow{\mathbf{x}})$).

A survey of different methods of calculating the stability function can be found in [7]. We should only note that more precise methods can be found much more difficult for mathematical analysis.

Thus, we can only work with more or less precise model describing the chemical process under consideration. We can consider the general subsequence method considered in section 3.1. A rule ACS used in this method singles out physically realizable duplex configurations and a weight function $\mathcal{W}$ is used to measure the stability of a given configuration. In the rest of the paper we will consider the simplest model in which weight function is equal to the length of the sequence and all subsequences are considered as physically realizable. In the paper [15] the authors consider a special rule (see comment in section 3.2) which is based on the idea that if there is a gap in one molecule, then there should be a corresponding gap in another.

5   Bounds on the Rate of LCS Similarity Codes

5.1   Definitions

In the present section we study the LCS similarity function defined in section 3.2. We will use symbol q to denote the size of the alphabet $|\mathsf{A}|$. Denote by $\mathrm{LCS}(\mathbf{x}, \mathbf{y})$ the length of a longest common subsequence $\mathbf{z} \in \mathrm{CS}(\mathbf{x}, \mathbf{y})$, $\mathbf{x}, \mathbf{y} \in \mathsf{A}^{\star}$.

The function $\mathrm{LCS}(\mathbf{x}, \mathbf{y})$ is the similarity measure that will be considered in the rest of the paper.

Obviously, $\mathrm{LCS}(\mathbf{x}, \mathbf{x})$ equals the length of the sequence $\mathbf{x}$ itself. This property simplifies definition 2 because $\mathrm{LCS}(\mathbf{x}, \mathbf{x}) = n$ for all $\mathbf{x} \in \mathsf{A}^n$. Thus, we will study codes defined as follows.

Definition 4. *A fixed-length code $\mathcal{C} \subset \mathsf{A}^n$ is called an LCS code with similarity S if $\mathrm{LCS}(\mathbf{x}, \mathbf{y}) \leq S$ for any $\mathbf{x}, \mathbf{y} \in \mathcal{C}$, $\mathbf{x} \neq \mathbf{y}$. One can see that this condition is equivalent to the following one: the Levenstein distance between any two different codewords is not less than $n - S$. Although these codes are well known and have been studied before, the lower bound on the rate which is proved in this section seems to be new.*

Definition 5. *Let a complement operation on the alphabet A be fixed (see section 2.4). We call a set $\mathcal{C} \subset \mathsf{A}^n$ a reverse-complement LCS (RC-LCS) code with similarity S if $\mathcal{C}$ satisfied definition 4 and $\overleftarrow{\mathbf{x}} \in \mathcal{C}$ for any $\mathbf{x} \in \mathcal{C}$.*

Definition 6. *A set $\mathcal{C} \subset \mathsf{A}^n$ will be referred to as a dual reverse-complement LCS (dRC-LCS) code with similarity S if $\mathcal{C}$ satisfied definition 5 and $\overleftarrow{\mathbf{x}} \neq \mathbf{x}$ for any $\mathbf{x} \in \mathcal{C}$.*

Denote by $\mathcal{M}_{\mathrm{LCS}}^{(n)}(S)$, $\mathcal{M}_{\mathrm{RC-LCS}}^{(n)}(S)$, $\mathcal{M}_{\mathrm{dRC-LCS}}^{(n)}(S)$ the maximum sizes of codes defined above. Consider the corresponding similarity rate functions $\mathcal{R}_{\mathrm{LCS}}(s)$, $\mathcal{R}_{\mathrm{RC-LCS}}(s)$, $\mathcal{R}_{\mathrm{dRC-LCS}}(s)$ defined as follows:

$$\mathcal{R}_{\star}(s) \triangleq \varlimsup_{n \to \infty} \frac{\log \mathcal{M}_{\star}^{(n)}(sn)}{n}, \quad s \geq 0. \tag{9}$$

We will obtain a lower bound on these functions. Denote the pseudodistance function (1) for LCS similarity function by $\mathcal{D}_{\mathrm{id}}^{(n)}$. It has the form $\mathcal{D}_{\mathrm{id}}^{(n)}(\mathbf{x}, \mathbf{y}) = n - \mathrm{LCS}(\mathbf{x}, \mathbf{y})$, $\mathbf{x}, \mathbf{y} \in \mathsf{A}^n$, and coincides with the Levenstein distance [2,?], i.e., the minimum number of insertion-deletion transformations that transform $\mathbf{x}$ to $\mathbf{y}$. An LCS code with similarity S is equivalent to a code with insertion-deletion distance $D = n - S$. Thus, using notations of coding theory we will consider the distance rate functions $\mathcal{R}_{\mathrm{id}}(d) = \mathcal{R}_{\mathrm{LCS}}(1 - d)$, $\mathcal{R}_{\mathrm{RC-id}}(d) = \mathcal{R}_{\mathrm{RC-LCS}}(1 - d)$, and $\mathcal{R}_{\mathrm{dRC-id}}(d) = \mathcal{R}_{\mathrm{dRC-LCS}}(1 - d)$, $0 \leq d \leq 1$.

5.2 Additional Statements

Let us consider two propositions which will be used in the proof of lower bound on the rate functions for LCS similarity codes.

The first statement is considered with the size of spheres for the Levenstein metric. For the case $q = 2$ the statement was proved in [3]. Although the proof for the general case is the same, we perform it below for the completeness of the paper.

Proposition 2. *[3]Let n and m be integers, $0 \leq m \leq n$. For an arbitrary sequence $\mathbf{y} \in \mathsf{A}^m$ denote by $\mathbf{B}(\mathbf{y}, n) \subset \mathsf{A}^n$ the set of all sequences $\mathbf{x} \in \mathsf{A}^n$ that include $\mathbf{y}$ as a subsequence, i.e., that can be obtained from $\mathbf{y}$ by $n - m$ insertions.*

Then for the fixed n and m the size of $\mathbf{B}(\mathbf{y}, n)$ does not depend on $\mathbf{y} \in \mathsf{A}^m$ and has the form

$$|\mathbf{B}(\mathbf{y}, n)| = \sum_{k=0}^{n-m} \binom{n}{k}(q-1)^k \triangleq B(m, n). \tag{10}$$

Proof. First we prove that for the given integer m the size of a set $\mathbf{B}(\mathbf{y}, n)$ does not depend on $\mathbf{y} \in \mathsf{A}^m$ and we can denote it by $B(m, n)$. We use induction over m. For $m = 0$ and $m = 1$ the statement is trivial. Assume that it is proved for all integers less than $m \geq 2$. Take a vector $\mathbf{y} = (y_1, \ldots, y_m) \in \mathsf{A}^m$. Consider the subsequence $\widetilde{\mathbf{y}} \triangleq (y_2, \ldots, y_m) \in \mathsf{A}^{m-1}$. Divide the set $\mathbf{B}(\mathbf{y}, n)$ into the sum of mutually disjoint sets $\mathbf{B}_k(\mathbf{y}, n)$, $k = 1, \ldots, n - m + 1$, where the set $\mathbf{B}_k(\mathbf{y}, n)$ is composed of vectors $\mathbf{x} \in \mathbf{B}(\mathbf{y}, n)$ such that $x_i \neq y_1$ for $i = 1, \ldots, k - 1$, and $x_k = y_1$. Obviously, any such vector $\mathbf{x}$ belongs to the set $\mathbf{B}(\mathbf{y}, n)$ if and only if the sequence $(x_{k+1}, \ldots, x_n) \in \mathsf{A}^{n-k}$ contains a subsequence $\widetilde{\mathbf{y}}$. Thus,

$$|\mathbf{B}_k(\mathbf{y}, n)| = (q-1)^{k-1}|\mathbf{B}(\widetilde{\mathbf{y}}, n - k)| = (q-1)^{k-1}B(m-1, n-k),$$

where we used the induction hypothesis. Thus, the size $|\mathbf{B}_k(\mathbf{y}, n)|$ is the same for all vectors $\mathbf{y}$. This proves that the size $|\mathbf{B}(\mathbf{y}, n)|$ also does not depend on $\mathbf{y}$.

To complete the proof, take a vector $\mathbf{y} = (0, \ldots, 0)$. The equality (10) for it is trivial.

Proposition 2 is proved.

For a random coding bound we will also need the logarithmic asymptotics of the function $B(m, n)$.

Proposition 3. *Let a number μ be fixed, $0 \leq \mu \leq 1$. Then $\beta_q(\mu) \triangleq \lim_{n \to \infty}$*
$$\frac{\log_q B(\mu n, n)}{n} = \begin{cases} 1, & 0 \leq \mu \leq \frac{1}{q}, \\ h_q(\mu) + (1 - \mu) \log_q(q-1), & \frac{1}{q} \leq \mu \leq 1, \end{cases} \text{ where the function}$$
$h_q(\mu) = -\mu \log_q \mu - (1 - \mu) \log_q(1 - \mu)$ is the entropy function.

The proof of proposition 3 is a simple analysis of the logarithmic asymptotics. It is omitted here.

Proposition 4. *Let $\mathbf{x} \in \mathsf{A}^n$ be an arbitrary sequence. Put $m \triangleq \mathrm{LCS}(\mathbf{x}, \overleftarrow{\mathbf{x}})$. Then for the pair $(\mathbf{x}, \overleftarrow{\mathbf{x}})$ there exist a longest common subsequence $\mathbf{y} \in \mathrm{CS}(\mathbf{x}, \overleftarrow{\mathbf{x}})$, $\mathbf{y} \in \mathsf{A}^m$, such that $\mathbf{y} = \overleftarrow{\mathbf{y}}$.*

Proof. Introduce some additional notations. Put $[n] \triangleq \{1, 2, \ldots, n\}$. Consider the set of *index pairs* $[n]^2 \triangleq [n] \times [n]$. Notations which are given below are motivated by representation of a pair $(i, j) \in [n]^2$ in the form of a segment as shown at Fig. 6. We say that a pair $(i_1, j_1) \in [n]^2$ *does not intersect* a pair (i_2, j_2) if either $i_1 < i_2$ and $j_1 < j_2$, or $i_1 > i_2$ and $j_1 > j_2$. Otherwise we say that (i_1, j_1) *intersects* (i_2, j_2). Moreover, we say that (i_1, j_1) intersects (i_2, j_2) *from the left* if $i_1 \leq i_2$ and $j_1 \geq j_2$. On the other hand, if $i_1 \geq i_2$ and $j_1 \leq j_2$, then we say that (i_1, j_1) intersects (i_2, j_2) *from the right*.

We will call an index pair $(i, j) \in [n]^2$ *admissible* for a sequence pair $(\mathbf{x}, \mathbf{y}) \in \mathsf{A}^n \times \mathsf{A}^n$ if $x_i = y_j$. One can see that a common subsequence of a sequence

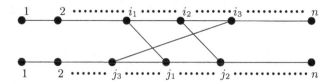

Fig. 6. The pairs (i_1, j_1) and (i_2, j_2) do not intersect; the pair (i_3, j_3) intersects (i_1, j_1) and (i_2, j_2) from the right

pair is uniquely specified by a set of mutually nonintersecting admissible index pairs.

Consider the following reverse-complement operation for index pairs: $\overleftarrow{(i,j)} \triangleq (n+1-j, n+1-i)$. If an index pair (i,j) is admissible for a sequence pair $(\mathbf{x}, \mathbf{y})$, then the pair $\overleftarrow{(i,j)}$ is admissible for $(\overleftarrow{\mathbf{y}}, \overleftarrow{\mathbf{x}})$. In particular, if (i,j) is admissible for $(\mathbf{x}, \overleftarrow{\mathbf{x}})$, then $\overleftarrow{(i,j)}$ is also admissible for $(\mathbf{x}, \overleftarrow{\mathbf{x}})$.

Take a sequence pair $(\mathbf{x}, \overleftarrow{\mathbf{x}})$. Let $m = \mathrm{LCS}(\mathbf{x}, \overleftarrow{\mathbf{x}})$ and $\mathbf{y} \in A^m$ be a common subsequence

$$\mathbf{y} = (x_{i_1}, \ldots, x_{i_m}) = (\overleftarrow{x}_{j_1}, \ldots, \overleftarrow{x}_{j_m}),$$

where $\overleftarrow{\mathbf{x}} = (\overleftarrow{x}_1, \ldots, \overleftarrow{x}_n)$, $\overleftarrow{x}_i = \overline{x}_{n+1-i}$. Consider the corresponding system of mutually nonintersecting admissible pairs:

$$\sigma = \{(i_1, j_1), \ldots, (i_m, j_m)\} \subset [n]^2.$$

We will divide the whole set σ into three mutually disjoint subsets: σ_c, σ_r and σ_ℓ. The first subset σ_c is composed of pairs $(i,j) \in \sigma$ such that $\overleftarrow{(i,j)} \in \sigma$. We call them *correct pairs*. One can see that if $\sigma = \sigma_c$, then $\mathbf{y} = \overleftarrow{\mathbf{y}}$.

If $\overleftarrow{(i,j)} \notin \sigma$, then there exists a pair $(i', j') \in \sigma$ that intersects with $\overleftarrow{(i,j)}$ because otherwise we could include the admissible pair $\overleftarrow{(i,j)}$ into σ and obtain a common subsequence of length $m+1 > \mathrm{LCS}(\mathbf{x}, \overleftarrow{\mathbf{x}})$ what is impossible. Moreover, one can see that if $\overleftarrow{(i,j)}$ intersects several pairs from the set σ, then it intersects all of them identically, i.e., either from the left or from the right. This can be easily proved by contradiction using the fact that pairs from σ are mutually nonintersecting. Thus, if a pair $(i,j) \in \sigma$ is not included into σ_c, then we put it into σ_ℓ or σ_r depending on how $\overleftarrow{(i,j)}$ intersects other pairs from σ.

Assume that $(i,j) \in \sigma \backslash \sigma_c$ and $\overleftarrow{(i,j)}$ intersects $(i', j') \in \sigma$. Then either $(i,j) \in \sigma_r$ and $(i', j') \in \sigma_\ell$ or $(i,j) \in \sigma_\ell$ and $(i', j') \in \sigma_r$. This follows from the fact that if $\overleftarrow{(i,j)}$ intersects (i', j') from the right, then $\overleftarrow{(i', j')}$ intersects (i,j) from the left and vise versa.

One can see that if we remove the set σ_ℓ from σ, then we can add to σ all pairs $\overleftarrow{(i,j)}$ such that $(i,j) \in \sigma_r$. All pairs in the new set are admissible and mutually nonintersecting. Thus we obtain a new common subsequence $\mathbf{y}_r$ having the length $m - |\sigma_\ell| + |\sigma_r|$. Similarly, removing the set σ_ℓ we can obtain a common subsequence $\mathbf{y}_\ell$ having the length $m - |\sigma_r| + |\sigma_\ell|$. Since both these values can not exceed m we obtain that $|\sigma_\ell| = |\sigma_r|$ and both lengths are equal to m. Moreover, one can see that both new sets are composed of correct pairs and, thus, both new subsequences $\mathbf{y}_r$ and $\mathbf{y}_\ell$ are self-reverse-complement: $\overleftarrow{\mathbf{y}_r} = \mathbf{y}_r$ and $\overleftarrow{\mathbf{y}_\ell} = \mathbf{y}_\ell$.

Proposition 4 is proved.

5.3   Lower Bound on the Rate of LCS Similarity Codes

Theorem 1. *Denote by $\widehat{d}_q$ the unique root d, $0 < d < \frac{q-1}{q}$, of the equation*

$$1 + d(1 - 2\log_q(q-1)) - 2h_q(d) = 0,$$

where $h_q(d)$ is the entropy function. Then all distance rate functions $\mathcal{R}_{\mathrm{id}}(d)$, $\mathcal{R}_{\mathrm{RC-id}}(d)$, and $\mathcal{R}_{\mathrm{dRC-id}}(d)$ satisfy the same inequality $\mathcal{R}_\star(d) > 0$ for $0 \leq d < \widehat{d}_q$ and

$$\mathcal{R}_\star(d) \geq 1 + d(1 - 2\log_q(q-1)) - 2h_q(d), \quad 0 \leq d \leq \widehat{d}_q. \tag{11}$$

Proof. We will use the random coding method. Fix integers $M > 0$ and $n > 0$ and consider a family composed of M sequences $\mathcal{C} = \{\mathbf{x}_1, \ldots, \mathbf{x}_M\} \subset \mathsf{A}^n$. Each sequence $\mathbf{x}_i$ is choose randomly according to the uniform distribution on the set A^n independently from other sequences.

Let a number $S \geq 0$ be fixed. Consider the following three conditions on a sequence $\mathbf{x}_i \in \mathcal{C}$:

(a) $\mathrm{LCS}(\mathbf{x}_i, \mathbf{x}_j) \leq S$ for all $j \neq i$; (b) $\mathrm{LCS}(\mathbf{x}_i, \overleftarrow{\mathbf{x}_i}) \leq S$; (c) $\mathbf{x}_i \neq \overleftarrow{\mathbf{x}_i}$.

If a sequence $\mathbf{x}_i \in \mathcal{C}$ satisfies condition (a), then we call it *LCS-good within $\mathcal{C}$*. If $\mathbf{x}_i$ satisfies both conditions (a) and (b), then we call it *RC-LCS-good within $\mathcal{C}$*. Finally, if $\mathbf{x}_i$ satisfies all conditions (a)–(c), then we call it *dRC-LCS-good within $\mathcal{C}$*. Otherwise we will call a sequence $\mathbf{x}_i$ $(\star)$-*bad within $\mathcal{C}$*.

Note that condition (a) is *mutual*, i.e. it corresponds to a pair of sequences within a family, while conditions (b) and (c) correspond to a single sequence. It will be shown that due to this property the random coding bound essentially depends only on the first condition. This yields that the bound is the same for all three types of codes under consideration.

To obtain a random coding bound we need to study the probabilities that conditions (a)–(c) do not hold for a randomly chosen sequence.

Consider the set $K(n, S) \triangleq \{(\mathbf{x}, \mathbf{y}) \in \mathsf{A}^n \times \mathsf{A}^n : \mathrm{LCS}(\mathbf{x}, \mathbf{y}) > S\}$. If $\mathrm{LCS}(\mathbf{x}, \mathbf{y}) > S$, then there exists a common subsequence $\mathbf{z} \in \mathrm{CS}(\mathbf{x}, \mathbf{y})$ having length S. Thus, the following upper bound is true

$$|K(n, S)| \leq q^S \left(B(S, n)\right)^2.$$

Let sequences $\mathbf{x}$ and $\mathbf{y}$ are chosen independently according to the uniform distribution from A^n. Then consider the following probability

$$P_a(S, n) \triangleq \Pr\{\mathrm{LCS}(\mathbf{x}, \mathbf{y}) > S\} = \frac{|K(n, S)|}{q^{2n}} \leq \frac{(B(S, n))^2}{q^{2n-S}}.$$

From (3) we obtain

$$\varlimsup_{n \to \infty} \frac{\log_q P_a(sn, n)}{n} \le \pi_q(s) \triangleq 2\beta_q(s) + s - 2. \tag{12}$$

Let $\pi_q(s) < 0$ for a number s. The we use the following well-known method. For a random family $\mathcal{C}$ described above and an index i we have

$$\Pr\{\mathbf{x}_i \text{ is LCS-bad within } \mathcal{C}\} \le \sum_{j \ne i} \Pr\{\text{LCS}(\mathbf{x}_i, \mathbf{x}_j) > S\} = (M-1)P_a(S, n).$$
$$\tag{13}$$

Let $S = sn$, $n \to \infty$. Then $P_a(sn, n) < Cq^{n\pi(s)}$ for some constant C and large enough values of n. Take $M \triangleq q^{-n\pi(s)}/(2C)$. Then the mathematical expectation of the number of LCS-bad codewords in $\mathcal{C}$ is less than $M/2$. Thus, there exists a family $\mathcal{C}$ contains more than $M/2$ LCS-good sequences. Note that all LCS-good sequences are mutually different if $S < n$. Removing all bad sequences from this family we obtain an LCS similarity code of size $M \triangleq q^{-n\pi(s)}/(4C)$. Thus, the following bound on the similarity rate function holds

$$\mathcal{R}_{\text{LCS}}(s) \ge -\pi(s). \tag{14}$$

Finally note that the function $\pi(s) = s$ for $0 \le s \le \frac{1}{q}$, and for larger values of s the function $\pi(s)$ decreases to the value $\pi(1) = -1$. Thus, at the interval $\frac{1}{q} \le s \le 1$ these exists a unique root $\widehat{s}_q$ of the equation $\pi(s) = 0$, and the bound (14) holds for the values $\widehat{s}_q \le s \le 1$. Finally, considering the new variable $d = 1 - s$ we obtain the statement for the distance rate function $\mathcal{R}_{\text{id}}(d)$.

Now consider RC-LCS codes. For a random sequence $\mathbf{x} \in A^n$ consider the following probability

$$P_b(S, n) \triangleq \Pr\{\text{LCS}(\mathbf{x}, \overleftarrow{\mathbf{x}}) > S\}.$$

From proposition 4 it follows that if $\text{LCS}(\mathbf{x}, \overleftarrow{\mathbf{x}}) = S' > S$, then there exists a sequence $\mathbf{z} \in \text{CS}(\mathbf{x}, \overleftarrow{\mathbf{x}})$ having length S' and such that $\mathbf{z} = \overleftarrow{\mathbf{z}}$. A self-reverse complement sequence is specified by half of its length. Thus, we obtain the following bound

$$P_b'(S', n) \triangleq \Pr\{\text{LCS}(\mathbf{x}, \overleftarrow{\mathbf{x}}) = S'\} \le \frac{q^{\lceil S'/2 \rceil} B(S', n)}{q^n}.$$

The logarithmic asymptotic of this value has the form (see (12)):

$$\varlimsup_{n \to \infty} \frac{\log_q P_b'(s'n, n)}{n} \le \beta_q(s') + \frac{s'}{2} - 1 = \pi_q(s').$$

Again let s be a number such that $\pi(s) < 0$. Then this function decreases and $\pi(s') < \pi(s)$ for $s' > s$. Thus, $P_b(sn, n) = \Pr\{\text{LCS}(\mathbf{x}, \overleftarrow{\mathbf{x}}) > sn\} \to 0$ as $n \to \infty$.

Based on the random family $\mathcal{C} = \{\mathbf{x}_1, \ldots, \mathbf{x}_M\}$ consider a reversed-complement family

$$\mathcal{C}' = \{\mathbf{x}_1, \ldots, \mathbf{x}_M \overleftarrow{\mathbf{x}}_1, \ldots, \overleftarrow{\mathbf{x}}_M\}.$$

Note that if a pair $(\mathbf{x}, \mathbf{y})$ has the uniform distribution over $\mathsf{A}^n \times \mathsf{A}^n$, then the same is true for the pairs $(\mathbf{x}, \overleftarrow{\mathbf{y}})$ and $(\overleftarrow{\mathbf{x}}, \overleftarrow{\mathbf{y}})$. Thus, the following upper bound holds (see (13)):

$$\Pr\left\{\mathbf{x}_i \text{ is RC-LCS-bad within } \mathcal{C}'\right\} \le (2M - 1)P_a(S, n) + P_b(S, n).$$

After this we perform the same reasoning as before and obtain the same bound for RC-LCS codes.

For dRC-LCS code we must consider the additional (c) condition for good codewords. One can see that the following inequality is true:

$$P_c(n) \triangleq \Pr\left\{\mathbf{x} = \overleftarrow{\mathbf{x}}\right\} \le q^{-\lfloor n/2 \rfloor} \to 0, \quad n \to \infty.$$

The random coding method for dRC-LCS codes will be based on the following inequality

$$\Pr\left\{\mathbf{x}_i \text{ is dRC-LCS-bad within } \mathcal{C}'\right\} \le (2M - 1)P_a(S, n) + P_b(S, n) + P_c(n),$$

where the last two terms are asymptotically small. After this we use the same method as before.

Theorem 1 is proved.

We do not have a special upper bound on the distance rate functions under consideration. One can see that the Hamming distance of a code is not less than the Levenstein distance. Thus, any upper bound on the rate of codes with Hamming distance [4,?] can also serve an upper bound for the rate of LCS similarity codes as well as RC-LCS and dRC-LCS codes. Although this bound is definitely rough, the authors do not know better results.

We also do not know the critical point, i.e. the value $d_0(q)$ such that $\mathcal{R}_{\mathrm{id}}(d) > 0$ for $0 \le d < d_0(q)$ and $\mathcal{R}_{\mathrm{id}}(d) = 0$ for $d \ge d_0(q)$. Known upper and lower bounds yield the inequality $\widehat{d}_q \le d_0(q) \le \frac{q-1}{q}$. For some numerical values of q these bounds have the form

q	$\widehat{d}_q$	$\frac{q-1}{q}$
2	0.133404	0.5
3	0.213527	0.6666
4	0.270294	0.75
5	0.313882	0.8
6	0.349016	0.8333
7	0.378281	0.8571
8	0.403244	0.875
9	0.424926	0.8888
10	0.444028	0.9
95	0.763101	0.9895
96	0.764152	0.9896
97	0.76519	0.9897
98	0.766212	0.9898
99	0.767221	0.9899
100	0.768216	0.99

Probably, the critical point problem is connected with the problem of the expected LCS length for independent random sequences. This problem is carefully studied in [8]. The method which is considered there allows to obtain a better lower bound than stated in Theorem 1 for LCS-codes. This better bound yields the following inequalities for the critical point: $d_0(2) \geq 0.16237$, $d_0(3) \geq 0.23419$, $d_0(4) \geq 0.29176$. But we do not know how to generalize that method for reverse-complement codes.

References

1. J.L. Massey, Reversible codes, Information and Control, 7, 369-380, 1964.
2. V.I. Levenshtein, Binary codes capable of correcting deletions, insertions, and reversals, J. Soviet Phys.—Doklady, 10, 707–710, 1966.
3. V.I. Levenshtein, Elements of coding theory (in Russian), in: Discrete Mathematics and Mathematical Problems of Cybernetics, Moscow, Nauka, 207–305, 1974.
4. R.J. McEliece, E.R. Rodemich, H.Jr. Rumsey, and L.R. Welch, New upper bounds on the rate of a code via the Delsarte—MacWilliams inequalities, IEEE Trans. Inform. Theory, Vol. 23, No. 2, 157–166, 1977.
5. F.J. MacWilliams and N.J.A. Sloane, The Theory of Error-Correcting Codes, Amsterdam, The Netherlands, North Holland, 1977.
6. K.J. Breslauer, R. Frank, H. Blöcker, and L.A. Marky, Predicting DNA duplex stability from the base sequence, Proc. Nat. Acad. Sci. USA (Biochemistry), 83, 3746–3750, 1986.
7. M.S. Waterman (ed.), Mathematical Methods for DNA Sequences, CRC Press, Inc. Boca Raton, Florida, 1989.
8. V. Dancik, Expected length of longest common subsequences, Ph.D. dissertation, University of Warwick, 1994.
9. A.G. D'yachkov and D.C. Torney, On similarity codes, IEEE Trans. Inform. Theory, Vol. 46, No. 4, 1558–1564, 2000.
10. A.G. D'yachkov, D.C. Torney, P.A. Vilenkin, and P.S. White, Reverse–complement similarity codes for DNA sequences, Proc. of ISIT–2000, Sorrento, Italy, July 2000.
11. P.A. Vilenkin, Some asymptotic problems of combinatorial coding theory and information theory (in Russian), Ph.D. dissertation, Moscow State University, 2000.
12. V.V. Rykov, A.J. Macula, C.M.Korzelius, D.C. Engelhart, D.C. Torney, and P.S. White, DNA sequences constructed on the basis of quaternary cyclic codes, Proc. of 4-th World Multiconference on Systemics, Cybernetics and Informatics, Orlando, Florida, USA, July 2000.
13. A.G. D'yachkov, D.C. Torney, P.A. Vilenkin, and P.S.White, On a class of codes for the insertion-deletion metric, Proc. of ISIT–2002, Lausanne, Switzerland, July 2002.
14. A.G. D'yachkov, P.L. Erdos, A.J. Macula, V.V. Rykov, D.C. Torney, C.-S. Tung, P.A. Vilenkin, and P.S. White, Exordium for DNA codes, Journal of Combinatorial Optimization, Vol. 7, No. 4, 2003.
15. A.G. D'yachkov, A.J. Macula, W.K. Pogozelski, T.E. Renz, V.V. Rykov, and D.C. Torney, A weighted insertion–deletion stacked pair thermodynamic metric for DNA codes, The Tenth International Meeting on DNA Computing. Milano-Bicocca, Italy, 2004.

On Some Applications of Information Indices in Chemical Graph Theory*

E.V. Konstantinova

Abstract. Information theory has been used in various branches of science. During recent years it is applied extensively in chemical graph theory for describing chemical structures and for providing good correlations between physico–chemical and structural properties by means of information indices. The application of information indices to the problem of characterizing molecular structures is presented in the paper. The information indices based on the distance in a graph are considered with respect to their correlating ability and discriminating power.

Keywords and Phrases: information theory, Shannon formula, information indices, molecular graphs, correlating ability, discriminating power.
Mathematics Subject Classification 2000: 94A15, 94A17, 94C15.

1 Introduction

We briefly review here selected topics in chemical graph theory, in particular, being concerned with the use of information theory to characterizing molecular structure. Chemical graph theory is interested in the nature of chemical structure [1]. All structural formulae of chemical compounds are molecular graphs where vertices represent atoms and edges represent chemical bonds. Figure 1 gives the schematic representation of the derivation of a molecular graph from an alkane molecule. The molecular graph is the hydrogen–suppressed one. That is the commonly used representation in chemical graph theory because hydrogen atoms are small and so add very little to the overall size of the molecule. Using the molecular graph one can obtain, for example, the carbon–number index (the number of carbon atoms in the hydrocarbon molecule) which is known since 1844 as one of the first topological indices used in chemistry to characterize molecular structures. Basically the topological index expresses in numerical form the 2–dimensional topology of the chemical species it presents. Topological indices are designed by transforming a molecular graph into a number. Topological indices possess the remarkable ability of being able to correlate and predict a very wide spectrum of properties for a vast range of molecular species. The carbon–number index is well–known to provide an effective measure of the molecular volume: for the members of homologous series, the molecular volume is known to be directly proportional to the carbon–number index [2].

The construction and investigation of topological indices which could uniquely characterize molecular topology is one of the main directions of chemical graph theory. Among the most important such trends in chemical graph theory are:

---

* This work was partially supported by the RFBR grant 06-01-00694.

R. Ahlswede et al. (Eds.): Information Transfer and Combinatorics, LNCS 4123, pp. 831–852, 2006.
© Springer-Verlag Berlin Heidelberg 2006

1 – unique representation of compounds;

2 – isomer discrimination (by developing indices with high discriminating power);

3 – structure–property relationships and structure–activity correlations; (An universally accepted paradigm is that similar molecules have similar properties. In other words, structures that differ little in the mathematical invariant properties will differ little also in their physical, chemical and biological properties. The above immediately reveals a strategy to attack the problems of structure–property relationship. Rather than directly trying to relate a property to structure, we may instead investigate different mathematical properties (invariants) of a structure and then follow with property–property correlations in which relatively simpler mathematical properties are used to express more complicated or convoluted physico–chemical and biological properties.)

4 – design of compounds of desired properties; (The development of a single drug can take anything from 8–20 years and involve a cost of some $ 100 million. The current success rate in the testing of possible new drugs is only around one in 10.000. In drug design, one can synthesize a

large number of derivatives from a "lead" structure. It is unusual that one to test 200.000 or more chemicals to discover a molecule that is marketable. In many cases, one might be interested to know the property of a molecule not yet synthesized. Then the only solution is to estimate properties using theoretical parameters which can be calculated for any arbitrary chemical structure, real or hypothetical.)

5 – enumeration and construction of compounds of certain classes;

We concentrate our attention on the second and third trends. Moreover, we consider all the aspects with respect to the information theory application.

It is well-known that application of some ideas from one scientific field to another one often gives a new view on the problems. Information, one of the most general ideas in contemporary science, should be expected to penetrate in various branches of science. Indeed, applications of information theory to molecular graphs have produced results which are important in chemistry.

The science of information theory has grown mainly out of the pioneering studies of Shannon [3], Ashby [4], Brillouin [5], and Kolmogorov [6]. There is more than one version of information theory. In Shannon's statistical information theory, information is measured as reduced uncertainty of the system. Ashby [4] describes information as a measure of variety. In the algorithmic theory of Kolmogorov, the quantity of information is defined as the minimal length of a program which allows a one–to–one transformation of an object (set) into another.

Applying information theory to material structures like atoms, molecules, crystals, etc., as well as to different mathematical structures like sets, groups, graphs, etc., the interpretation given by Mowshovitz [7–10] in 1968 is more appropriate.

Let a given system I having n elements be regarded according to a certain equivalence relation, into k equivalence classes with cardinalities n_i. Considering

Alkane Molecule

$\downarrow$

$$C_3H_8$$

Chemical formula

$\downarrow$

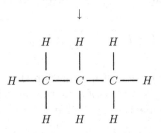

Structural formula

$\downarrow$

Molecular hydrogen–suppressed graph

$\downarrow$

$$n = 3$$

Topological index

Fig. 1. Schematic representation of the derivation of the molecular graph and the carbon–number index from an alkane molecule

all the n elements partitioned into k classes, we can define the probability p_i, $i = 1, \ldots, k$, for a randomly selected element of this system to be found in the i-th class. Therefore, a finite probability scheme may be associated with the following structure:

$$\begin{pmatrix} 1 & 2 & 3 & \ldots & k \\ n_1 & n_2 & n_3 & \ldots & n_k \\ p_1 & p_2 & p_3 & \ldots & p_k \end{pmatrix}$$

where $n = \sum_{i=1}^{k} n_i$, $p_i = n_i/n$ and $\sum_{i=1}^{k} p_i = 1$.

The information content of a system I with n elements is defined by the relation [5]

$$I = nlog_2 n - \sum_{i=1}^{k} n_i \log_2 n_i \qquad (1)$$

The logarithm is taken to base 2 for measuring the information contents in bits. Another information measure is the mean information content of one element of the system I defined by means of the total information content or by the Shannon relation [3] (which is also called the binary entropy of a finite probability scheme):

$$\overline{I} = I/n = - \sum_{i=1}^{k} p_i \log_2 p_i, \qquad (2)$$

where $p_i = n_i/n$. The application of information theory to different systems or structures is based on the possibility of constructing a finite probability scheme for every system. One can mention here, that the criterium for partitioning the elements of a given system is not unique. The number of information measures is equal to the number of ways in which a set of n elements may be partitioned into different subsets, that is, the number of Young diagrams for a given n. It is always possible to select for any system several information measures, each of them closely connected with certain properties of the system. They reflect the essence of the idea of information, given by Ashby [4] as a measure of the variety in a given system. This idea was used in graph theory and in chemical graph theory for characterizing graphs as well as molecular graphs and molecular structures.

At first information theory was applied to graphs in 1955 by Rashevsky [11], who defined the so–called topological information content of the graph I_{top}. His definition is based on the partitioning of the vertices of a given graph into classes of equivalent vertices having the same valences. Trucco [12,13] in 1956 made this definition more precise on the basis of an automorphism group of the graphs. In the latter case, two vertices are considered equivalent if they belong to the same orbit of the automorphism group, i.e., if they can interchange preserving the adjacency of the graph. Later [14] the topological information was used by Rashevsky in studying the possibility of self-generation of the life on earth. As for chemical structures, information theory has been successfully applied in the study of various molecular properties [15–17], in the field of molecular dynamics [18,19] and quantum chemistry [20–23], in the description of the electronic structure of atoms [24], in the interpretation of the Pauli principle and Hund rule [25].

A molecular topology determines a large number of molecular properties. It was found in the last years that some biological activities of molecules, and even carcinogenecity, are closely related to a molecular topology. Thus, it is of a pertinent interest for chemistry (as well as for other natural sciences) to have some quantitative measure reflecting the essential features of a given topological structure. As it was mentioned above such measures are usually called topological indices in chemical graph theory. A lot of such indices have been suggested in the last 50 years [26–40]. They have usually correlated more or less satisfactorily with the molecular properties but could not discriminate well between structural isomers, often providing the same index for different isomers. The first topological index reflecting the topological structure of a molecular graph was proposed by Harry Wiener in 1947 [26]. The Wiener number W was defined as the sum of all edges between all pairs of carbon atoms in hydrocarbons. It gives a good correlation with the thermodynamic properties of saturated hydrocarbon molecules but doesn't discriminate well among structural isomers.

Bonchev and Trinajstić [41] applied information theory to the problem of characterizing molecular structures and molecular topology [42–46] by means of information indices [47] which are just the quantitative measures of a given topological structure. The advantage of such kind of indices is in that they may be used directly as simple numerical descriptors in a comparison with physical, chemical or biological parameters of molecules in quantitative structure–property relationships and in structure–activity relationships [48–50]. It can also be noted that information indices normally have greater discriminating power for isomers than the respective topological indices. The reasons for this are that information indices are not restricted to integral values as topological indices frequently are and information indices are formed from a summation of different magnitudes which is usually greater in number than that for the topological indices.

We present here some results concerning information indices applications to characterizing molecular structures. The paper is organized in the following way. First of all, the information indices based on the distance in a graph are reviewed. Then the numerical results of discriminating tests of indices on structural isomers and graphs are presented. At last, the correlating ability of information indices is demonstrated on the several classes of organic and organometallic compound.

2   Information Indices Based on the Distance in a Graph

One can start looking for possible information indices among the graph invariants. Information indices are constructed for various matrices (adjacency matrix, incidence matrix, distance matrix, layer matrix) and also for some topological indices such as the Wiener number.

In 1977 Bonchev and Trinajstić [41] introduced an information on distances to explain the molecular branching that is the critical parameter determining the relative magnitude of various molecular thermodynamic properties. Firstly they used the information indices defined by Rashevsky for graphs. However, these indices are not suitable for describing branching properties of graphs since they cannot reflect the essence of branching. This may be exemplified by considering trees with five vertices presented in Figure 2. The five vertices are partitioned in different orbits in the above three graphs: T_1 $(2,2,1)$, T_2 $(2,1,1,1)$, T_3 $(4,1)$. Using Eq.(1), the following values for the information content in bits are obtained: $I_{T_1} = 7.61, I_{T_2} = 9.61, I_{T_3} = 3.61$. One can see that this index cannot

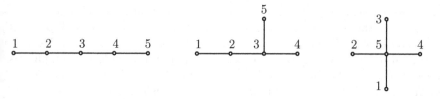

$$T_1 \; : \; \{1,5\}, \; \{2,4\}, \; \{3\} \qquad T_2 \; : \; \{4,5\}, \; \{1\}, \; \{2\}, \; \{3\} \qquad T_3 \; : \; \{1,2,3,4\}, \; \{5\}$$

Fig. 2. Trees with five vertices and their orbits

reproduce the obvious fact that the branching increases from a chain, through a branched tree, to a star.

So another approach to find an appropriate information measure of branching was used. One of the graph invariants is the distance matrix. Let G be a connected graph with the set of vertices $V(G)$, $n = |V(G)|$. The distance $d(u,v)$ between vertices u and v in a graph G is the length of the shortest path that connects these vertices. The distance matrix $D =\parallel d_{ij} \parallel, i,j = 1,\ldots,n$, contains the distances $d_{ij} = d(i,j)$ between the different pairs of connected vertices. Branching is connected with the distance matrix in an obvious way, since with increasing branching the distances in the graph become smaller. This can easily be seen from the distance matrices of the trees T_1, T_2, T_3 presented in Figure 2:

$$D(T_1) = \begin{vmatrix} 0\,1\,2\,3\,4 \\ 1\,0\,1\,2\,3 \\ 2\,1\,0\,1\,2 \\ 3\,2\,1\,0\,1 \\ 4\,3\,2\,1\,0 \end{vmatrix} \qquad D(T_2) = \begin{vmatrix} 0\,1\,2\,3\,3 \\ 1\,0\,1\,2\,2 \\ 2\,1\,0\,1\,1 \\ 3\,2\,1\,0\,2 \\ 3\,2\,1\,2\,0 \end{vmatrix} \qquad D(T_3) = \begin{vmatrix} 0\,2\,2\,2\,1 \\ 2\,0\,2\,2\,1 \\ 2\,2\,0\,2\,1 \\ 2\,2\,2\,0\,1 \\ 1\,1\,1\,1\,0 \end{vmatrix}$$

Wiener [26] first made use of the connection between the distance matrix and branching defining the topological index

$$W = \frac{1}{2} \sum_{i,j=1}^{n} d_{ij} \tag{3}$$

However, the Wiener number often has the same value for different graphs. For reducing degeneracies, Bonchev and Trinajstić introduced an information on distances I_D in a graph, considering all the matrix elements of distance matrix d_{ij} as elements of a finite probability scheme associated with the graph in question. Let the distance of a value i appears $2n_i$ times in the distance matrix, where $1 \leq i \leq d(G)$ and $d(G) = \max_{i,j \in V(G)} d(i,j)$ is the diameter of a graph. Then n^2 matrix elements d_{ij} are partitioned into $d(G) + 1$ groups, and $d(G) + 1$ group contains n zeros which are the diagonal matrix elements. With each one of these $d(G) + 1$ groups can be associated a certain probability for a randomly chosen distance d_{ij} to be in the i–th group:

$$\begin{pmatrix} 0 & 1 & 2 & \cdots & d(G) \\ n & 2n_1 & 2n_2 & \cdots & 2n_{d(G)} \\ \frac{1}{n} & p_1 & p_2 & \cdots & p_{d(G)} \end{pmatrix}$$

where $p_i = 2n_i/n^2$ and $p_0 = n/n^2 = 1/n$.

The information on distances of a given graph will then, according to Eqs.(1),(2) be

$$I = n^2 \, \log_2 n^2 - n \, \log_2 n - \sum_{i=1}^{d(G)} 2n_i \, \log_2 2n_i \qquad (4)$$

$$\bar{I} = -\frac{1}{n} \, \log_2 \frac{1}{n} - \sum_{i=1}^{d(G)} \frac{2n_i}{n^2} \, \log_2 \frac{2n_i}{n^2} \qquad (5)$$

Since D is a symmetric matrix, one can consider, for simplicity of discussion, only the upper triangular submatrix that does preserve all properties of the information measure. In that case, the following expressions for the mean and total information on distances are obtained:

$$I_D^E = \frac{n(n-1)}{2} \, \log_2 \frac{n(n-1)}{2} - \sum_{i=1}^{d(G)} n_i \, \log_2 n_i \qquad (6)$$

$$\bar{I}_D^E = -\sum_{i=1}^{d(G)} \frac{2n_i}{n(n-1)} \, \log_2 \frac{2n_i}{n(n-1)}, \qquad (7)$$

where $\frac{n(n-1)}{2}$ is the total number of upper off–diagonal elements in the distance matrix D. These information indices correspond to the information on the distribution of distances in the graph according to their equality or nonequality and depend on the partitioning of the total number of distances into classes.

Using Eq.(7), one can obtain the following information on distances in the three graphs with five vertices presented in Figure 2:

$$T_1 : \begin{array}{|cccc|c} 1 & 2 & 3 & 4 & i \\ 4 & 3 & 2 & 1 & n_i \\ \frac{4}{10} & \frac{3}{10} & \frac{2}{10} & \frac{1}{10} & p_i \end{array} \qquad T_2 : \begin{array}{|ccc|c} 1 & 2 & 3 & i \\ 4 & 4 & 2 & n_i \\ \frac{4}{10} & \frac{4}{10} & \frac{2}{10} & p_i \end{array} \qquad T_3 : \begin{array}{|cc|c} 1 & 2 & i \\ 4 & 6 & n \\ \frac{4}{10} & \frac{6}{10} & p \end{array}$$

$$\bar{I}_D^E(T_1) = 1.85 \qquad \bar{I}_D^E(T_2) = 1.52 \qquad \bar{I}_D^E(T_3) = 0.97$$

One can see that $\bar{I}_D^E$ as well as $I_D^E = \frac{n(n-1)}{2} \, \bar{I}_D^E$ reproduces the branching properties of trees T_1, T_2, T_3 decreasing regularity with increased branching.

Moreover, Bonchev and Trinajstić have shown that I_D^E is a rather sensitive measure of branching having different values for all trees with $n = 4, 5, 6, 7, 8$ (the total number of trees is 45). The number of all possible distributions of d_{ij}, i.e., number of different I_D^E, increases rapidly with the increase in the number of vertices in the graph. This makes I_D^E an appropriate quantity for distinguishing structural isomers.

However, there is another possible information measure which can be defined on the basis of distances in the graph. Bonchev and Trinajstić introduced the information index I_D^W as the information on the realized distances in a given graph which depends on the partitioning of the total distance. It is an information on the partitioning the Wiener number (which is the total distance of the graph) into groups of distances of the same absolute values. Since the Wiener number is given by formula $W = \sum_{i=1}^{d(G)} i\, n_i$ and following Eqs.(1) and (2), we obtain

$$I_D^W = W \log_2 W - \sum_{i=1}^{d(G)} i\, n_i \log_2 i \tag{8}$$

$$\overline{I}_D^W = -\sum_{i=1}^{d(G)} n_i \frac{i}{W} \log_2 \frac{i}{W} \tag{9}$$

For the three five–vertices trees presented in Figure 2, the following values are obtained:

$$T_1 : \quad I_D^W = 62.93, \quad \overline{I}_D^W = 3.15,$$
$$T_2 : \quad I_D^W = 57.55, \quad \overline{I}_D^W = 3.20,$$
$$T_3 : \quad I_D^W = 52.00, \quad \overline{I}_D^W = 3.25$$

It is easy to see that I_D^W decreases with branching. It is a more sensitive quantity than the Wiener number since it can distinguish two graphs having the same Wiener number but different i and n_i. It was checked that $\overline{I}_D^W$ increases regularity with branching at lower values of n, but at higher ones some irregularity occur and it cannot be used as a good measure of branching. As for I_D^W, it is a sensitive measure of branching having different values for all trees with $n = 4, 5, 6, 7, 8$ (the total number of trees is 45). Figure 3 presents the pair of trees having the same value of the Wiener number and the different values of I_D^E and I_D^W.

In [41] the values of both information measures I_D^E and I_D^W were inspected in comparison with several topological indices such as the Wiener number, the greatest eigenvalue of the characteristic polynomial χ_1, the sum of the polynomial coefficients (or Hosoya index) [28], the information on polynomial coefficients I_{pc}, and Randić connectivity index χ_R [29]. The inspection of these values indicates the great sensitivity of the two information indices I_D^E, I_D^W to all structural details of the tree graphs. There are no two graphs among the 45 graphs examined which have the same information on the graph distances. All the other listed indices are not so specific, and they often have the same value for different graphs. The same results were obtained for information indices $\bar{I}_D^E$ and $\bar{I}_D^W$. Their values were tested on the set of 45 tree graphs. There are no two graphs having the same values of these indices.

Thus the information measures introduced on the basis of distance matrix appear to be very appropriate indices for discrimination of graphs. The number of different values I_D^E for the graphs having the same number of vertices is limited by the number of all possible distributions $n(n-1)/2$ graph edges into k different groups. Since the number increases rapidly with increasing values of n, one may expect the information on graph distances to have a good ability of differentiation between structural isomers even for very large systems. It was one of the main result obtained by Bonchev and Trinajstić.

It was confirmed later by Konstantinova, Paleev and Diudea [51–53] that the information approach allows to design very sensitive information indices based on the distance in a graph. The information distance index of vertex i was introduced in [51] and defined as follows:

$$H_D(i) = -\sum_{j=1}^{n} \frac{d_{ij}}{d(i)} \, \log_2 \frac{d_{ij}}{d(i)}, \tag{10}$$

where $d(i) = \sum_{j=1}^{n} d_{ij}$ is the distance of a vertex i. Then the information distance index of graph vertices takes the form

$$H_D^n = \sum_{i=1}^{n} H_D(i) \tag{11}$$

The same approach was applied to the layer matrix $\lambda =\parallel \lambda_{ij} \parallel, i = 1, \ldots, n, j = 1, \ldots, d(G)$, where λ_{ij} is equal to the number of vertices located at a distance j from vertex i. The information layer index of graph vertices is defined by formula

$$H_\lambda^n = \sum_{i=1}^{n} H_\lambda(i) = -\sum_{i=1}^{n} \sum_{j=0}^{e(i)} \frac{\lambda_{ij}}{n} \log_2 \frac{\lambda_{ij}}{n}, \tag{12}$$

where $e(i) = \max_{v \in V(G)} d(i, v)$ is the vertex eccentricity. It will be shown later that indices H_D^n and H_λ^n have a great discriminating power among structural isomers.

One more information index based on the distance matrix was considered by Skorobogatov et. al [54] in structure–activity correlations.

The information index H_2 is defined by the relation

$$H_2 = -\sum_{i=1}^{k} \frac{d(i)k_i}{2W} \log_2 \frac{d(i)k_i}{2W}, \tag{13}$$

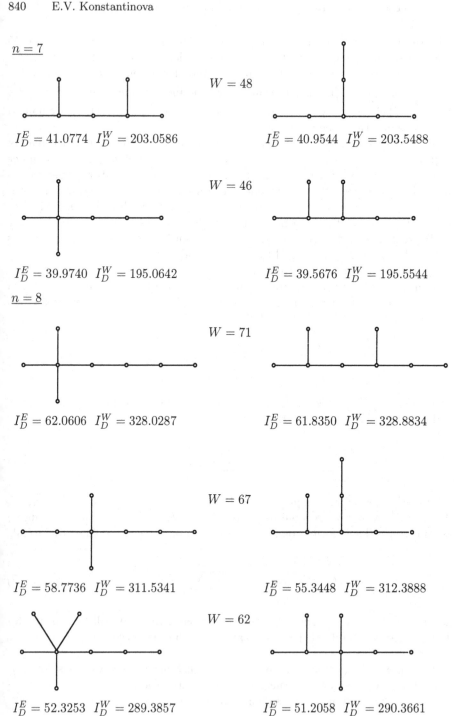

<u>$n = 7$</u>

$W = 48$

$I_D^E = 41.0774$ $I_D^W = 203.0586$ $I_D^E = 40.9544$ $I_D^W = 203.5488$

$W = 46$

$I_D^E = 39.9740$ $I_D^W = 195.0642$ $I_D^E = 39.5676$ $I_D^W = 195.5544$

<u>$n = 8$</u>

$W = 71$

$I_D^E = 62.0606$ $I_D^W = 328.0287$ $I_D^E = 61.8350$ $I_D^W = 328.8834$

$W = 67$

$I_D^E = 58.7736$ $I_D^W = 311.5341$ $I_D^E = 55.3448$ $I_D^W = 312.3888$

$W = 62$

$I_D^E = 52.3253$ $I_D^W = 289.3857$ $I_D^E = 51.2058$ $I_D^W = 290.3661$

Fig. 3. The pair of trees having the same value of the Wiener number and the different values of I_D^E and I_D^W

where $k_i, i = 1, \ldots, k$, is the number of vertices having the distance $d(i)$. This index gives the linear correlations with information mass–spectrum indices on the several classes of organic and organometallic compounds [55–58].

D'yachkov, Konstantinova and Vilenkin [59] consider the entropy H_D, the marginal entropy H_D^i and the information I_D based on the distance matrix as follows

$$H_D \equiv -\sum_{i=1}^{n}\sum_{j=1}^{n} \frac{d_{ij}}{2W} \log_2 \frac{d_{ij}}{2W} = 1 + log_2 W - \frac{1}{W}\sum_{i=1}^{d(G)} n_i\, i \log_2 i \quad (14)$$

$$H_D^i \equiv -\sum_{i=1}^{n} \frac{d(i)}{2W} \log_2 \frac{d(i)}{2W} = 1 + log_2 W - \frac{1}{2W}\sum_{i=1}^{n} d(i) \log_2 d(i) \quad (15)$$

$$I_D \equiv 2H_D^i - H_D = 1 + log_2 W - \frac{1}{W}\left(\sum_{i=1}^{n} d(i) \log_2 d(i) - \sum_{i=1}^{d(G)} n_i\, i \log_2 i\right) \quad (16)$$

where n_i is the number of vertex pairs being at a distance i from each other and $W = \sum_{i=1}^{d(G)} i\, n_i$.

Let l_i be the number of matrix elements equal to i. The entropy H_λ, the marginal entropies H_λ^j and H_λ^i and the information I_λ are defined by

$$H_\lambda \equiv -\sum_{i=1}^{n}\sum_{j=1}^{d(G)} \frac{\lambda_{ij}}{n(n-1)} \log_2 \frac{\lambda_{ij}}{n(n-1)} = \log_2 n(n-1) - \frac{1}{n(n-1)}\sum_{i=1}^{max} l_i\, i \log_2 i$$
$$(17)$$

$$H_\lambda^j \equiv -\sum_{j=1}^{d(G)} \frac{2n_j}{n(n-1)} \log_2 \frac{2n_j}{n(n-1)} = \log_2 n(n-1) - \frac{1}{n(n-1)}\sum_{j=1}^{d(G)} 2n_j \log_2 2n_j$$
$$(18)$$

$$H_\lambda^i \equiv -n\frac{n}{n(n-1)} \log_2 \frac{n}{n(n-1)} = \log_2 n \quad (19)$$

$$I_\lambda \equiv H_\lambda^i + H_\lambda^i - H_\lambda = \log_2 n - \frac{1}{n(n-1)}\left(\sum_{j=1}^{d(G)} 2n_j \log_2 2n_j - \sum_{i=1}^{max} l_i\, i \log_2 i\right) \quad (20)$$

The information indices $\overline{I}_D^W, H_D$ and H_λ^j are based on the vector n_i and the constants n and W that leads to their correlations. In particular, since $\overline{I}_D^W = log_2 W - \frac{1}{W}\sum_{i=1}^{d(G)} n_i\, i \log_2 i$ then following Eq.(14) we immediately obtain $H_D = \overline{I}_D^W + 1$. So it is enough to study the only index among them. The index $\overline{I}_D^W$ was investigated in discriminating tests as the most known one [60].

3   Discriminating Tests

The discriminating power [45,61] is one of the basic characteristics of a topological index I and corresponds to a measure of its ability to distinguish among the nonisomorphic graphs (the structural isomers) by distinct numerical values of index I. The theoretical evaluation of index sensitivity S on a fixed set M of nonisomorphic graphs can be achieved by the formula

$$S = \frac{N - N_I}{N}, \qquad (21)$$

where $N = |M|$ is the number of graphs in a set M and N_I is the number of degeneracies of an index I within set M. According to the definition, $S = 1$ means that among the elements of the set considered, no two nonisomorphic graphs have the same value of the index I.

Bonchev and Trinajstić [41] investigated the discriminating power of information and topological indices between 45 alkane trees. Basak et. al [62], have continued these investigations on the set of 45 alkane trees as well as on the set of 19 monocyclic graphs. Razinger, Chretien and Dubois [61] explicitly pointed out the fact that the discriminating power of the Wiener number is very low in alkane series. The first discriminating tests among the polycyclic graphs were done by Konstantinova and Paleev [51] on the set of 1020 subgraphs of the regular hexagonal and square lattices. Later Konstantinova [52] has tested information and topological indices for 2562 subgraphs of the regular hexagonal lattice. Graphs of this class represent the molecular structures of unbranched cata-condensed benzenoid hydrocarbons. The discriminating powers of topological and information indices as well as the Wiener polynomial derivatives were studied by Konstantinova and Diudea [53] on 3006 subgraphs of the regular hexagonal lattice and on the set of 347 cycle–containing graphs with ten vertices and three to eight–membered cycle.

An exhaustive analysis of 13 information and topological indices based on the distance in a graph was performed by Konstantinova and Vidyuk [60] on $1\,443\,032$ polycyclic graphs and $3\,473\,141$ trees. The information indices $I_D, H_D^i,$ $I_\lambda, H_\lambda, H_D^n, H_\lambda^n, H_2, \overline{I}_D^W$ presented in section 2 and the topological indices such as the Wiener number, the Schultz number, the Balaban number and the Randić number were examined in the discriminating tests. The formulae for topological indices are given below.

The Schultz molecular topological index [63] is defined by

$$MTI = \sum_{v \in V(G)} deg(v) \cdot d(v) + \sum_{v \in V(G)} deg(v)^2, \qquad (22)$$

where $deg(v)$ is the vertex degree. This index has found interesting applications in chemistry [38]. Its discriminating power was investigated by Dobrynin [64] for cata–condensed benzenoid graphs.

The average distance sum connectivity was introduced by Balaban [33] and defined by

$$J = \frac{m}{m-n+2} \sum_{u,v \in V(G)} (d(u) \cdot d(v))^{-\frac{1}{2}}, \tag{23}$$

where m is the number of edges in a graph G.

The Randić index χ [29] is based on the molecular connectivity and is achieved by formula

$$\chi = \sum_{u,v \in V(G)} (deg(u) \cdot deg(v))^{-\frac{1}{2}} \tag{24}$$

The numerical results of discriminating tests for the indices under consideration are given below on the sets of polycyclic graphs and trees.

3.1 Polycyclic Graphs

The polycyclic graphs embedded to the regular hexagonal, square and trigonal lattices are tested. The hexagonal graphs correspond to the structural formulae of planar polycyclic aromatic hydrocarbons [65,66]. The values of 12 information and topological indices were calculated for 849 285 hexagonal, 298 382 square and 295 365 triangular graphs. The calculation accuracy for all indices is 10^{-13}. The discriminating powers of indices were obtained in accordance with Eq.(21) and the results are given in Table 1, where N is the number of graphs in the respective class. Table 1 shows the discriminating power of indices on the sets of N hexagonal, square and triangular graphs:

Table 1.

N	I_D	H_D^i	I_λ	H_λ	H_D^n	H_2	H_λ^n
849 285	0.999	0.999	0.997	0.993	0.999	0.999	0.997
298 382	0.997	0.995	0.954	0.811	0.998	0.994	0.906
295 365	0.984	0.982	0.844	0.466	0.992	0.981	0.585

N	$\overline{I}_D^W$	J	χ	MTI	W
849 285	0.659	0.998	0.0001	0.004	0.0006
298 382	0.133	0.993	0.005	0.002	0.0003
295 365	0.021	0.986	0.407	0.0008	0.0001

The data show that the information indices give much more discriminating power. The indices H_D^n, I_D, H_D^i, H_2 have the best result ($S = 0.999$) for hexagonal graphs. All topological indices, exception of J, could not discriminate between these graphs. The same situation is observed for square and triangular graphs. The degeneracy is high for W, MTI, and very low for H_D^n, I_D. The information index $\overline{I}_D^W$ discriminates not bad among hexagonal graphs but it doesn't discriminate among square and triangular graphs. The opposite situation is observed for the Randić index χ. Its discriminating power is the lowest one on hexagonal graphs and the highest one on triangular graphs.

3.2 Trees

Similar results were obtained on the set of trees. A tree is a connected acyclic graph. The discriminating power of indices was calculated on the set of $3\,490\,528$ trees up to 21 vertices. The obtained data are given in Table 2. The highest sensitivity corresponds to the information index H_D^n. There are no degeneracies of this index, i.e. $S = 1$, on the set of trees up to 17 vertices (the number of trees is $N = 81\,134$). Two trees with 18 vertices and two trees having 19 vertices give the same values of H_D^n. There are only 14 degeneracies on the set of trees on 20 vertices, $N = 823\,065$, and only 12 degeneracies on the set of trees on 21 vertices, $N = 2\,144\,505$. This index could be used for characterizing molecular structures of alkane isomers. Topological indices show a very low discriminating power. Table 2 shows the discriminating power of indices on the set of N trees up to 21 vertices:

Table 2.

N	I_D	H_D^i	I_λ	H_λ	H_D^n	H_2	H_λ^n	$\overline{I}_D^W$	J
$3\,490\,528$	0.998	0.912	0.985	0.321	0.999	0.907	0.428	0.683	0.907

N	χ	MTI	W
$3\,490\,528$	0.017	0.00004	0.00002

So in this section the discriminating power of information indices were considered in comparison with the topological indices. The data obtained on the sets of nonisomorphic graphs and trees indicate that in common the information indices have greater discriminating power than the topological ones. Another basic characteristic of a topological index is its correlating ability with a molecular property. The information approach to the correlations between structure and reactive capability of molecules is considered in the next section.

4 Correlating Ability of Information Indices

One of the key problems of modern chemistry is to find a relation between a structure and a reactive capability of a molecule. The reactive capability of a molecule can be characterized by it's mass–spectrum which contains the information on the ways of a molecule fragmentation and displays the "behavior" of some molecular fragments which can be interpreted as subgraphs of a molecular graph.

Let us define the information index of the chemical mass–spectrum using the Shannon relation. From the information theory point of view, the mass–spectrum is the distribution of probabilities $p_i = \frac{A_i}{A}, i = 1, \ldots, k$, of the ions formation, where A_i is the mass–spectrum amplitude of the i-th ion,

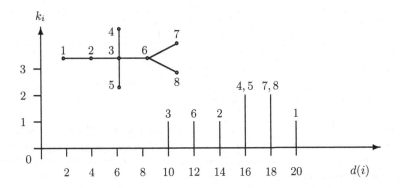

Fig. 4. The distance vertex spectrum

$A = \sum_{i=1}^{k} A_i$, and k is the number of peaks in the mass–spectrum. The amplitude information index HA is defined by

$$HA = -\sum_{i=1}^{k} \frac{A_i}{A} \cdot log_2 \frac{A_i}{A} \qquad (25)$$

On the other hand, a molecular graph that represents a structural formula of a molecule could be used for defining specific structural features of a molecule by means of information indices based on the distance in a molecular graph. As it was mentioned in introduction the topological index is designed by transforming a molecular graph into a number and it expresses in a numerical form the topology of the chemical species it presents. Moreover, it was shown by Skorobogatov et. al [54] that some information indices have a "chemical" spectral interpretation. Let us consider the information index H_2 that is based on the vertex distance $d(i)$ and the number k_i of vertices having the distance $d(i)$, and let define the pairs $(d(i), k_i)$ as the points in Euclidean plane. Then the distance vertex spectrum can be pictured on the plane by the lines $\{(d(i), k_i), (d(i), 0)\}, i = 1, \ldots, k$. Figure 4 shows the distance vertex spectrum for the tree. As one can see, there is even the visual correspondence between the chemical mass–spectra and the topological spectra of molecular graphs.

One more topological spectrum called the autometricity vertex spectrum was defined on the basis of a layer matrix. It could happen that some rows of this matrix are the same. It means that the corresponding vertices belong to one and the same class of autometricity. By this way the vertex set is divided into the autometricity classes. The autometricity vertex spectrum is defined by the autometricity classes and the number n_i^λ of vertices in the i–th class of autometricity.

Figure 5 shows the autometricity vertex spectrum for the same tree. Its canonical layer matrix is the following one:

$$\lambda^* = \begin{vmatrix} 1\,2\,3\,1 \\ 1\,1\,3\,2 \\ 3\,3\,1 \\ 2\,3\,2 \\ 1\,3\,3 \\ 4\,3 \end{vmatrix} \quad \begin{matrix} 7,8 \\ 1 \\ 6 \\ 2 \\ 4,5 \\ 3 \end{matrix}$$

The rows are ordered with respect to their lengths and then the rows are ordered lexicographically. The numeration of autometricity classes corresponds to the numeration of rows in the canonical layer matrix. Let us notice, that there is a finite probability scheme on the vertex set with respect to the autometricity ratio and one can define the information index $H_a = -\sum_{i=1}^{k} \frac{n_i^\lambda}{n} \log_2 \frac{n_i^\lambda}{n}$.

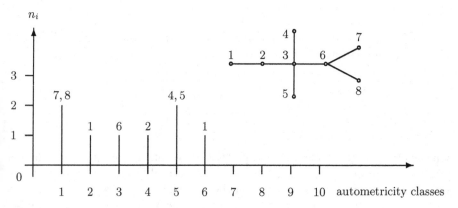

Fig. 5. The autometricity vertex spectrum

The information indices H_2, H_a and HA based on the topological as well as the chemical spectra provide the presentation of a molecular graph and a chemical structure in terms of the same quantitative scale because their values are expressed in information bits. These indices are suitable ones for finding structure–activity correlations.

At first this approach was applied to the class of ferrocene derivatives C_pFeC_5 H_4R, where R is a substituent [54]. The linear correlations between the information indices H_2 and HA, and H_a and HA were found. It was shown that the initial set of molecular structures is divided into three subsets by the linear regression. In the cases considered the correlation ratio ranges from 0.94 to 0.975. The example of the autometricity vertex spectrum for a molecular structure of this class is given in Figure 6.

Figure 7 shows the correlations between the information indices H_a and HA on the set of ferrocene derivatives. The correlation ratio r and the number of structures n for each subset are presented.

Similar results were obtained for information indices H_2 and HA. The correlations between them were found as follows:

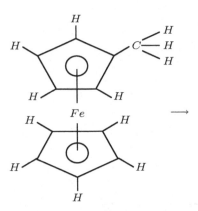

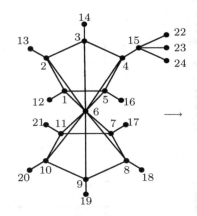

a) the structure formula
of methylferrocene

b) the molecular graph
of methylferrocene

$$\lambda = \begin{vmatrix} 1 & 3 & 9 & 7 & 3 \\ 1 & 3 & 3 & 9 & 7 \\ 4 & 9 & 7 & 3 \\ 4 & 3 & 9 & 7 \\ 1 & 3 & 9 & 10 \\ 10 & 10 & 3 \\ 4 & 12 & 7 \\ 4 & 9 & 10 \end{vmatrix}$$

$$\begin{aligned} 12, 13, 17, 18, 19, 20, 21 &= X_1 \\ 22, 23, 24, &= X_2 \\ 1, \; 2, \; 7, \; 8, \; 9, \; 10, 11 &= X_3 \\ 15 &= X_4 \\ 14, 16 &= X_5 \\ 6 &= X_6 \\ 4 &= X_7 \\ 3, \; 5 &= X_8 \end{aligned}$$

c) the canonical layer matrix of the methylferrocene molecular graph

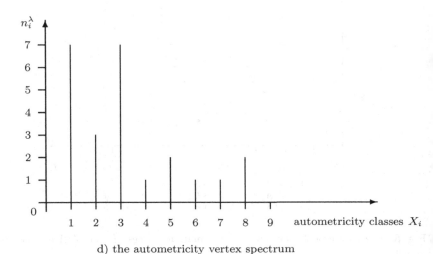

d) the autometricity vertex spectrum

Fig. 6. The example of the autometricity vertex spectrum for the molecular graph of methylferrocene

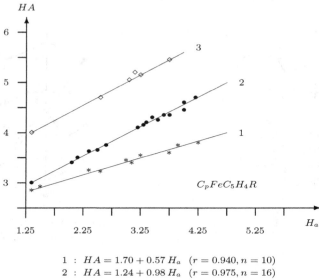

$$1 : HA = 1.70 + 0.57\,H_a \quad (r = 0.940, n = 10)$$
$$2 : HA = 1.24 + 0.98\,H_a \quad (r = 0.975, n = 16)$$
$$3 : HA = 3.91 + 0.41\,H_a \quad (r = 0.940, n = 6)$$

Fig. 7. The correlations between the information indices H_a and HA on the set of ferrocene derivatives

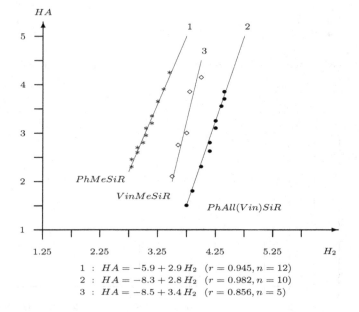

$$1 : HA = -5.9 + 2.9\,H_2 \quad (r = 0.945, n = 12)$$
$$2 : HA = -8.3 + 2.8\,H_2 \quad (r = 0.982, n = 10)$$
$$3 : HA = -8.5 + 3.4\,H_2 \quad (r = 0.856, n = 5)$$

Fig. 8. The correlations between the information indices H_2 and HA on the set of arylsilanes

$$1 \; : \; HA = 1.09 + 0.76\,H_2 \quad (r = 0.950, n = 9)$$
$$2 \; : \; HA = 1.30 + 0.99\,H_2 \quad (r = 0.975, n = 17)$$
$$3 \; : \; HA = 3.96 + 0.40\,H_2 \quad (r = 0.950, n = 6)$$

Later this approach was used for finding correlation on several classes (~ 20) of organic and organometallic compounds by Prof.Yu.S.Nekrasov, etc. [55–58]. In particular, they have found the correlations on the set of arylsilanes. There are three subsets of arylsilanes and each of them has specific structure peculiarities. Figure 8 shows the correlations between the information indices H_2 and HA on the set of arylsilanes. Line 1 corresponds to the set of phenylmethylsilanes, line 2 corresponds to the set of phenylallyl– and phenylvinylsilanes and line 3 corresponds to the set of vinylmethylsilanes. As one can see from the picture the correlation ratio for all cases ranges from 0.856 to 0.982.

5   Conclusion

Information theory is a useful tool in developing the new invariants for characterizing molecular structures. The only limit to the design of invariants is the imagination and resourcefulness of investigators. The situation is similar to a search for a system of codes and a design of codes for chemical structures in particular. There are graph theoretical invariants considered in the mathematical literature that have not yet been tested for possible chemical applications. Such may or may not be of interest in chemistry, but without testing we will not know. If some existing mathematical invariants are shown to correlate with some of the known molecular properties, the findings may be of considerable interest for chemistry - not only because such invariants may offer a novel predictive tool in structure–property studies, or an alternative route to a certain molecular property, but they may give additional insight into structure–property relationships. Finally, such results may show novel mathematical concepts of use in chemistry.

Acknowledgement

The author is grateful to the referees for their remarks and comments. Moreover, the author gratefully thanks Prof. Vladimir Levenshtein for drawing her in the information theory business. The author also cordially thanks Prof. Rudolf Ahlswede for hospitality.

References

1. N. Trinajstić, Chemical Graph Theory, 2nd revised ed; CRC Press, Boca Raton, FL, 1992.
2. D.H. Rouvray, The limits of applicability of topological indices, J. Mol. Struc. (Theochem), 185, 187–201, 1989.
3. C. Shannon and W. Weaver, Mathematical Theory of Communications, University of Illinois, Urbana, 1949.

4. W.R. Ashby, An Introduction to Cybernetics, Wiley, New York, 1956.
5. L. Brillouin, Science and Information Theory, Academic, New York, 1956.
6. A.N. Kolmogorov, On logic basis of information theory, Probl. Peredachi Inf., 5, 3–7, 1996.
7. A. Mowshovitz, The information content of digraphs and infinite graphs, Bull. Math. Biophys., 30, 225–240, 1968.
8. A. Mowshovitz, An index of the relative complexity of a graph, Bull. Math. Biophys., 30, 175–204, 1968.
9. A. Mowshovitz, Graphs with prescribed information content, Bull. Math. Biophys., 30, 387–414, 1968.
10. A. Mowshovitz, Entropy measures and graphical structure, Bull. Math. Biophys., 30, 533–546, 1968.
11. N. Rashevsky, Life, information theory and topology, Bull. Math. Biophys., 17, 229–235, 1955.
12. E. Trucco, A note of the information content of graphs, Bull. Math. Biophys., 17, 129–135, 1956.
13. E. Trucco, On the informational content of graphs–compound symbols, different states for each point, Bull. Math. Biophys., 18, 237–245, 1956.
14. N. Rashevsky, Life, information theory, probability and physics, Bull. Math. Biophys. , 22, 351–364, 1960.
15. H. Morovitz, Some order–disorder considerations in living systems, Bull. Math. Biophys. , 17, 81–86, 1955.
16. M. Valentinuzzi and M.E. Valentinuzzi, Information content of chemical structures and some possible biological applications, Bull. Math. Biophys. , 25, 11–27, 1963.
17. D. Bonchev, D. Kamenski, and V. Kamenska, Symmetry and information content of chemical structures, Bull. Math. Biophys. , 38, 119–133, 1976.
18. R.B. Bernstein and R.D. Levine, Entropy and chem. change I: characterization of product (and reactant) energy distributions in reactive molecular collisions: information and entropy deficiency, J. Chem. Phys., 57, 434–449, 1972.
19. A. Ben-Shaul, R.D. Levine, and R.B. Bernstein, Entropy and chem. change II: analysis of product energy distributions: temperature and entropy deficiency, J. Chem. Phys., 57, 5427–5447, 1972.
20. C. Aslangul, R. Constanciel, R. Daudel, and P. Kottis, Aspects of the localizability of electrons in atoms and molecules: Loge theory and related methods, Adv. Quantum Chem., 6, 93–141, 1972.
21. R. Daudel, R.F. Bader, M.E. Stephens, and D.S. Borett, The electron pair in chemistry, Can. J. Chem., 52, 1310–1320, 1974.
22. C. Aslangul, R. Constanciel, R. Daudel, L. Esnault, and E. Ludena, The Loge theory as a starting point for variational calculations, I. General formalism, Int. J. Quantum Chem., 8, 499–522, 1974.
23. F. Fratev, V. Enchev, O.E. Polansky, and D. Bonchev, A theoretical–information study on the electron delocalization (aromaticity) of annulenes with and without bond alternation, THEOCHEM, 88, 105–118, 1982.
24. D. Bonchev, Information indices for atoms and molecules, MATCH, 7, 65–113, 1979.
25. D. Bonchev, Information theory interpretation of the Pauli principle and Hund rule, Intern. J. Quantum Chem., 19, 673–679, 1981.
26. H. Wiener, Structural determination of paraffin boiling points, J. Am. Chem. Soc., 69, 17–20, 1947.
27. H. Wiener, Vapor pressure–temperature relationships among the branched paraffin hydrocarbons, J. Chem. Phys., 52, 425–430, 1948.

28. H. Hosoya, Topological index, a newly proposed quantity characterizing the topological nature of structural isomers of hydrocarbons, Bull. Chem. Soc. Jpn., 44, 2332–2339, 1971.

29. M. Randić, On characterization of molecular branching, J. Am. Chem. Soc., 69, 6609–6615, 1975.

30. R.C. Entringer, D.E. Jackson, and D.A. Snyder, Distance in graphs, Czechoslovak Math.J., 2, 283–297, 1976.

31. J.K. Doyle and J.E. Graver, Mean distance in graphs, Discrete Math., 17, 147–154, 1977.

32. L.B.Kier and L.H.Hall, Derivation and significance of valence molecular connectivity, J. Pharm. Sci., 70, 583–589, 1981.

33. A.T. Balaban, Topological indices based on topological distances in molecular graphs, Pure Appl. Chem., 55, 199–206, 1983.

34. D.H. Rouvray, Should we have designs on topological indices? chemical applications of topology and graph theory. Studies in Physical and Theoretical Chemistry, King R.B. (ed.), Elsevier, Amsterdam, 28, 159–177, 1983.

35. M. Randić, On molecular identification numbers, J. Chem. Inf. Comput. Sci., 24, 164–175, 1984.

36. R.B. King and D.H. Rouvray (eds.), Graph Theory and Topology in Chemistry, Elsevier, Amsterdam, 1987.

37. M. Randić, Generalized molecular descriptors, J. Math. Chem., 7, 155–168, 1991.

38. S. Nikolić, N. Trinajstić, and Z .Mihalić, The Wiener index: developments and applications, Croat. Chem. Acta, 68, 105–129, 1995.

39. R.C. Entringer, Distance in graphs: trees, J. Combin. Math. Combin. Comput., 24, 65–84, 1997.

40. M.V. Diudea and I. Gutman, Wiener-type topological indices, Croat. Chem. Acta., 71, 21–51, 1998.

41. D. Bonchev and N. Trinajstić, Information theory, distance matrix, and molecular branching, J. Chem. Phys., 38, 4517–4533, 1977.

42. D. Bonchev and N. Trinajstić, On topological characterization of molecular branching, Int. J. Quantum Chem., S12, 293–303, 1978.

43. D. Bonchev, J.V. Knop, and N. Trinajstić, Mathematical models of branching, MATCH, 6, 21–47, 1979.

44. D. Bonchev, O. Mekenyan, and N. Trinajstić, Topological characterization of cyclic structure, Int. J. Quantum Chem., 17, 845–893, 1980.

45. D. Bonchev, O. Mekenyan, and N. Trinajstić, Isomer discrimination by topological information approach, J. Comput. Chem., 2, 127–148, 1981.

46. D. Bonchev and N. Trinajstić, Chemical information theory, structural aspects, Intern. J. Quantum Chem. Symp., 16, 463-480, 1982.

47. D. Bonchev, Information-Theoretic Indices for Characterization of Chemical Structures, Research Studies Press, Chichester, 1983.

48. L.B. Kier and L.H. Hall, Molecular Connectivity in Chemistry and Drug Research, Academic Press, New York, 1976.

49. L.B. Kier and L.H. Hall, Molecular Connectivity in Structure–Activity Analysis, Research Studies Press, Letchworth, 1986.

50. A.T. Balaban, A. Chirac, I. Motoc, and Z. Simon, Steric Fit in Quantitative Structure–Activity Relationships, Lecture Notes in Chemistry, N.15, Springer, Berlin, 1980.

51. E.V. Konstantinova and A.A. Paleev, Sensitivity of topological indices of polycyclic graphs (Russian), Vychisl. Sistemy, 136, 38–48, 1990.

52. E.V. Konstantinova, The discrimination ability of some topological and information distance indices for graphs of unbranched hexagonal systems, J. Chem. Inf. Comput. Sci., 36, 54–57, 1996.

53. E.V. Konstantinova and M.V. Diudea, The Wiener polynomial derivatives and other topological indices in chemical research, Croat. Chem. Acta, 73, 383–403, 2000.

54. V.A. Skorobogatov, E.V. Konstantinova, Yu.S. Nekrasov, Yu.N. Sukharev, and E.E. Tepfer, On the correlation between the molecular information topological and mass-spectra indices of organometallic compounds, MATCH, 26, 215–228, 1991.

55. Yu.N. Sukharev, Yu.S. Nekrasov, N.S. Molgacheva, and E.E. Tepfer, Computer processing and interpretation of mass–spectral information, Part IX - Generalized characteristics of mass–spectra, Org. Mass Spectrom.,28, 1555–1561, 1993.

56. Yu.S. Nekrasov, E.E. Tepfer, and Yu.N. Sukharev, On the relationship between the mass–spactral and structural indices of arylsilanes, Russ. Chem. Bull., 42, 343–346, 1993.

57. Yu.S. Nekrasov, Yu.N. Sukharev, N.S. Molgacheva, and E.E. Tepfer, Generalized characteristics of mass–spectra of aromatic compounds and their correlation with the constants of substituents, Russ. Chem. Bull., 42, 1986–1990, 1993.

58. Yu.S. Nekrasov, Yu.N. Sukharev, E.E. Tepfer, and N.S. Molgacheva, Establishment of correlations between the structure and reactivity of molecules in the gas phase based on information theory, Russ. Chem. Bull., 45, 2542–2546, 1996

59. A.G. D'yachkov, E.V. Konstantinova, and P.A. Vilenkin, On entropy and information of trees, in progress.

60. E.V. Konstantinova and M.V. Vidyuk, Discriminating tests of information and topological indices, animals and trees, J Chem Inf Comput Sci., Vol. 43, No. 6, 1860-1871, 2003.

61. M. Razinger, J.R. Chretien, and J.K. Dubois, Structural selectivity of topological indices in alkane series, J. Chem. Inf. Comput. Sci., 25, 23–27, 1985.

62. C. Raychaudhary, S.K. Ray, J.J. Ghosh, A.B. Roy, and S.C. Basak, Discrimination of isomeric structures using information theoretic topological indices, J. Comput. Chem., 5, 581–588, 1984.

63. I. Gutman, Selected properties of the Schultz molecular index, J. Chem. Inf. Comput. Sci., 34, 1087–1089, 1994.

64. A.A. Dobrynin, Discriminating power of the Schultz index for cata–condensed benzenoid graphs, MATCH, 38, 19–32, 1998.

65. I. Gutman and S.J. Cyvin, Introduction to the Theory of Benzenoid Hydrocarbons, Springer–Verlag, Berlin, 1989.

66. I.Gutman and S.J.Cyvin (eds.), Advances in the Theory of Benzenoid Hydrocarbons, Springer–Verlag, Berlin, 1990.

Largest Graphs of Diameter 2 and Maximum Degree 6

S.G. Molodtsov

Abstract. The results of computer generation of the largest graphs of diameter 2 and maximum degree 6 are presented. The order of such graphs is equal 32. There are exactly 6 graphs of diameter 2 and maximum degree 6 on 32 vertices including one vertex-transitive graph.

1 Introduction

We consider simple (without loops and multiple edges), finite, and undirected graphs. The *order* of the graph is the number of its vertices. The *degree* of a vertex is the number of incident edges to that vertex. The *distance* between two vertices is the length of the shortest path between them. The *diameter* of a graph is the maximum distance between any two vertices. Denote by *(d,k)-graph* the graph of diameter k and maximum vertex degree d.

Degree/Diameter Problem. *To find the largest possible number of vertices $n(d, k)$ of (d, k)-graphs for various d and k.*

The state of general problem is presented in [1,2]. We focus only on graphs of diameter 2. A well-known bound on the largest order of $(d, 2)$-graphs is *Moore bound*: $n(d, 2) \leq d^2 + 1$. This bound is attainable only for $d = 1, 2, 3, 7$ and, possibly, 57 [3]. Excluding $d = 57$, the corresponding graphs K_2, C_2, the Petersen graph and the Hoffman–Singleton graph are unique. The existing of a $(57, 2)$-graph on $n = 57^2 + 1 = 3250$ vertices is not known. For the remaining d, it is shown that $n(d, 2) \leq d^2 - 1$ [4]. It is known that this bound is attainable only for $d = 4, 5$ [5]. The largest known order of (6,2)-graphs was 32 [1,2]. A theoretic possible bound for these graphs is $n = 6^2 - 1 = 35$.

Our purpose is to construct all the nonisomorphic largest graphs of diameter 2 and maximum vertex degree 6.

2 Computer Graph Generation of Graphs of Diameter 2

An efficient algorithm and the computer program GENM for the generation of all nonisomorphic graphs from a given set of vertices have been developed [6]. This algorithm is based on the enumerative variant of the branch-and-bound method [7] and constructs all canonical adjacency matrices in decreasing order. An adjacency matrix of a graph G is called *canonical* if it is maximum adjacency matrix of G.

R. Ahlswede et al. (Eds.): Information Transfer and Combinatorics, LNCS 4123, pp. 853–857, 2006.
© Springer-Verlag Berlin Heidelberg 2006

The algorithm of graph generation is a stepwise procedure. On each step a partially filed matrix is constructed. To generate graphs of diameter 2 efficiently, a partially filed matrix is tested on possibility of completion up to an adjacency matrix of a graph of diameter 2. For this purpose, the following criteria were used.

Let V be the set of vertices of the $(d, 2)$-graph G. Let H be a spanning subgraph of G. Denote by $A_H(v)$ and $B_H(v)$ the subsets of vertices at distance 1 and 2 from the vertex $v \in V$ in the subgraph H, respectively. Then $C_H(v) = V \backslash (v \cup A_H(v) \cup B_H(v))$ is the subset of vertices at distance at least 3 from the vertex v in H. Denote by $\deg_G(v)$ the degree of a vertex v in G. Then $d_H(v) = \deg_G(v) - \deg_H(v)$ is called the *free valency* of a vertex v in H. Let $I_H \subset V$ be the subset of vertices such that $d_H(i) = 0$ for every $i \in I_H$.

Criterion of admissibility. *The sum of free valences of all vertices from $A_H(i)$ is not less than the number of vertices in $C_H(i)$ for any $i \in I_H$:*

$$\sum_{j \in A_H(i)} d_H(j) \geq |C_H(i)|.$$

Criterion of forcing. *If the sum of free valences of all vertices from $A_H(i)$ is equal to the number of vertices in $C_H(i)$ for some $i \in I_H$,*

$$\sum_{j \in A_H(i)} d_H(j) = |C_H(i)|,$$

then all edges (u, v), where $u, v \in A_H(i)$, and all edges (u, v), where $u \in A_H(i), v \in B_H(i)$ of G presence in the subgraph H.

The proof of the both criteria follows from the fact that each vertex from $C_H(i)$ should be adjacent at least with one vertex from $A_H(i)$ in the graph G. Therefore, the criterion of admissibility is used to cut off the current branch during the generation of $(d, 2)$-graphs. The criterion of forcing helps to fill in some elements of the partially filed matrix on an each step.

3   Results

A graph G is called a *regular* graph of degree d if all vertices of G have the same degree d. The following statement shows that first of all one should search the largest $(d, 2)$-graphs among regular graphs of degree d.

Statement. *If G is a $(d, 2)$-graph on n vertices and $n > d(d - 1) + 1$ then G is a regular graph of degree d.*

Proof. Let $\deg(v)$ be a degree of a vertex v in graph G. Consider an vertex $v \in V$. Let A and B be the subsets of vertices at distance 1 and 2 from the vertex v, respectively. Since A, B, and $\{v\}$ partition V, $n = |A| + |B| + 1$.

Obviously, $|A| = \deg(v)$. Note that each vertex $a \in A$ can be adjacent at most with $\deg(a) - 1$ vertices of B. Since G is the (d,2)-graph, each vertex of B must be adjacent with some vertex of A and $\deg(a) \leq d$ for any $a \in A$. Thus

$$|B| \leq \sum_{a \in A} (\deg(a) - 1) \leq \sum_{a \in A} (d - 1) = (d - 1) \cdot |A| = d \cdot \deg(v) - \deg(v)$$

Hence, $n = |A| + |B| + 1 \leq d \cdot \deg(v) + 1$. Since $n > d(d - 1) + 1$, we have $d \cdot \deg(v) + 1 > d(d - 1) + 1$ and $\deg(v) > d - 1$. Therefore, $\deg(v) = d$ for any $v \in V$.   □

It follows that all (6,2)-graphs on more than 31 vertices are regular graph of degree 6. We have searched the (6,2)-graphs starting on 35 vertices. Such graphs on 35, 34, and 33 vertices are not found. Exactly 6 graphs of diameter 2 and degree 6 on 32 vertices are constructed. Their lists of adjacent and the order of the automorphism group are shown below.

$| \operatorname{Aut}(G_1) |= 6$
1–2,3,4,5,6,7; 2–3,8,9,10,11; 3–12,13,14,15; 4–8,16,17,18,19; 5–9,20,21,22,23;
6–10,24,25,26,27; 7–28,29,30,31,32; 8–20,21,24,28; 9–16,25,26,29; 10–17,18,22,30;
11–19,23,27,31,32; 12–16,17,23,25,28; 13–16,22,24,31,32; 14–18,20,22,27,29;
15–19,21,24,26,30; 16–27,30; 17–21,29,31; 18–23,26,32; 19–22,25,29; 20–25,30,31;
21–27,32; 22–28; 23–24,30; 24–29; 25–32; 26–28,31; 27–28

$| \operatorname{Aut}(G_2) |= 2$
1–2,3,4,5,6,7; 2–3,8,9,10,11; 3–12,13,14,15; 4–8,16,17,18,19; 5–9,20,21,22,23;
6–10,24,25,26,27; 7–28,29,30,31,32; 8–20,21,24,28; 9–16,25,26,29; 10–17,22,30,31;
11–18,19,23,27,32; 12–16,17,23,25,28; 13–16,22,24,30,32; 14–18,20,22,27,29;
15–19,21,24,26,31; 16–27,31; 17–21,29,32; 18–23,26,30; 19–22,25,29; 20–25,31,32;
21–27,30; 22–28; 23–24,31; 24–29; 25–30; 26–28,32; 27–28

$| \operatorname{Aut}(G_3) |= 144$
1–2,3,4,5,6,7; 2–3,8,9,10,11; 3–12,13,14,15; 4–16,17,18,19,20; 5–16,21,22,23,24;
6–17,21,25,26,27;7–28,29,30,31,32;8–16,17,21,28,29;9–18,19,22,25,30;10–20,22,26,27,31;
11–20,23,24,25,32; 12–16,26,27,30,32; 13–17,23,24,30,31; 14–18,19,21,31,32;
15–20,22,25,28,29; 16–25,31; 17–22,32; 18–23,26,28; 19–24,27,29; 20–21,30;
21–30; 22–32; 23–27,29; 24–26,28; 25–31; 26–29; 27–28

$| \operatorname{Aut}(G_4) |= 64$
1–2,3,4,5,6,7; 2–8,9,10,11,12; 3–8,13,14,15,16; 4–9,17,18,19,20; 5–10,21,22,23,24;
6–11,25,26,27,28; 7–12,29,30,31,32; 8–17,21,25,29; 9–13,22,26,30; 10–14,18,27,31;
11–15,19,23,32; 12–16,20,24,28; 13–17,23,28,31; 14–19,20,26,32; 15–18,22,24,30;
16–19,22,27,29; 17–24,27,32; 18–25,28,29; 19–21,31; 20–23,25,30; 21–26,28,30;
22–25,32; 23–27,29; 24–26,31; 25–31; 26–29; 27–30; 28–32

$| \operatorname{Aut}(G_5) |= 48$
1–2,3,4,5,6,7; 2–8,9,10,11,12; 3–8,13,14,15,16; 4–9,17,18,19,20; 5–10,21,22,23,24;
6–11,25,26,27,28; 7–12,29,30,31,32; 8–17,21,25,29; 9–13,22,26,30; 10–14,18,27,31;
11–15,19,23,32; 12–16,20,24,28; 13–17,23,28,31; 14–19,20,26,32; 15–18,24,25,30;
16–19,22,27,29; 17–24,27,32; 18–22,28,29; 19–21,31; 20–23,25,30; 21–26,28,30;
22–25,32; 23–27,29; 24–26,31; 25–31; 26–29; 27–30; 28–32

$| \operatorname{Aut}(G_6) |= 1920$

1–2,3,4,5,6,7; 2–8,9,10,11,12; 3–8,13,14,15,16; 4–9,17,18,19,20; 5–10,21,22,23,24; 6–11,25,26,27,28; 7–12,29,30,31,32; 8–17,21,25,29; 9–13,22,26,30; 10–14,18,27,31; 11–15,19,23,32; 12–16,20,24,28; 13–17,23,28,31; 14–20,21,26,32; 15–18,24,25,30; 16–19,22,27,29; 17–24,27,32; 18–22,28,29; 19–21,26,31; 20–23,25,30; 21–28,30; 22–25,32; 23–27,29; 24–26,31; 25–31; 26–29; 27–30; 28–32

The total time of generation of all (6,2)-graphs on 35, 34, 33, and 32 vertices is approximately 400 hours on an 1.8 GHz Celeron. Note that the last (6,2)-graph of the lists is a vertex-transitive graph.

Table 1 contains the order and the number N of known largest $(d, 2)$-graphs and vertex-transitive $(d, 2)$-graphs for $3 \leq d \leq 7$. Here $n_t(d, 2)$ is the largest order of vertex-transitive $(d, 2)$-graphs.

Table 1. The known largest order of $(d, 2)$-graphs

d	$n(d, 2)$	N	$n_t(d, 2)$	N
3	10	1	10	1
4	15	1	13	1
5	24	1	20	1
6	32	6	32	1
7	50	1	50	1

The presence of the vertex-transitive graph among the largest (6,2)-graphs is according to the fact that the largest known $(d, 2)$-graphs for some d have been found among vertex-transitive graphs [8,9].

Acknowledgment

This research was started when the author was taking part in the seminar of the ZiF Research Group "General Theory of Information Transfer and Combinatorics". The author thanks the coordinator of the seminar Prof. R. Ahlswede and the directorate of ZiF for organization of the very useful scientific cooperation, and Prof. V. I. Levenshtein for attracting author's attention to degree/diameter problem.

References

1. M.J. Dinneen, New results for the degree/diameter problem, Networks, 24, 359–367, 1994.
2. The degree/diameter problem for graphs, http://maite71.upc.es/grup_de_grafs/ grafs/taula_delta_d.html
3. A.J. Hoffman and R.R. Singleton, On Moore graphs with diameters 2 and 3, IBM J. Res. Develop., 4, 497–504, 1960.
4. P. Erdös, S. Fajtlowicz, and A.J. Hoffman, Maximum degree in graphs of diameter 2, Networks, 10, 87–90, 1980.

5. B. Elspas, Topological constrains on interconnected-limited logic, Proc. 5th Ann Symp. Switching Circuit Theory and Logic Design 133–147, 1964.

6. S.G. Molodtsov, Computer-aided generation of molecular graphs, MATCH, 30, 213–224, 1994.

7. I.A. Faradjev, Constructive enumeration of combinatorial objects, Problemes Combinatoires et Theorie des Graphes, Colloque Internationale CNRS 260, 131–135, 1978.

8. L. Cambell, G.E. Carlsson, M.J. Dinneen, V. Faber, M.R. Fellows, M.A. Langston, J.W. Moore, A.P. Mullhaupt, and H.B. Sexton, Small diameter symmetric networks from linear groups, IEEE Trans. Comput., 41, 218–220, 1992.

9. B.D. McKay, M. Miller, and J. Širáň, A note on large graphs of diameter two and given maximum degree, J. Combin. Theory Ser. B, 74, 110–118, 1998.

VIII
An Outside Opinion

R. Ahlswede

In order to get an outside opinion about Network Coding we cite here the Network Coding Homepage (www.networkcoding.info) of a leading expert, Ralf Koetter.

Welcome to the Network Coding Coding Home Page. This site is meant to provide a service to the community by summarizing the main developments in network coding. Our hope is that this site can serve as a repository and resource for researchers and scientists in the field.

1 Network Coding Example

Like many fundamental concepts, network coding is based on a simple basic idea which was first stated in its beautiful simplicity in the the seminal paper by R. Ahlswede, N. Cai, S.-Y. R. Li, and R. W. Yeung, "Network Information Flow", (IEEE Transactions on Information Theory, IT-46, pp. 1204-1216, 2000). The core notion of network coding is to allow and encourage mixing of data at

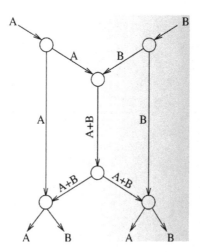

intermediate network nodes. A receiver sees these data packets and deduces from them the messages that were originally intended for the data sink. In contrast to traditional ways to operate a network that try to avoid collisions of data streams as much as possible, this elegant principle implies a plethora of surprising results.

R. Ahlswede et al. (Eds.): Information Transfer and Combinatorics, LNCS 4123, pp. 858–860, 2006.
© Springer-Verlag Berlin Heidelberg 2006

One of the most exciting opportunities of the approach is the use of random mixing of data streams, thus freeing up the symmetrizing properties of random coding arguments in the analysis of networks. Not only is network coding a fresh and sharp tool that has the potential to open up stagnant fundamental areas of research, but due to its cross-cutting nature it naturally suggests a unified treatment of previously segmented areas. A striking example of the type of unification that network coding makes possible is the recently found elegant and complete characterization of the capacity of multicast networks, which was possible only through the joint treatment of coding and routing.

The principle of network coding is easiest explained with an example (from Ahlswede et al.) that kicked off the field of network coding. In this example two sources having access to bits A and B at a rate of one bit per unit time have to communicate these bits to two sinks so that both sinks receive both bits per unit time. All links have a capacity of one bit per unit time. The network problem can be satisfied with the transmissions outlined in the example but cannot be satisfied with only forwarding of bits at intermediate packet nodes.

2   Network Coding Bibliography of Kötter with More Than 100 Papers Since the Start of the Subject in the Year 2000 with the Paper of Ahlswede et al. Just Mentioned

We just highlight here 3 contributions. (There are other important contributions!) The main result of Ahlswede et al., a Min–Max–Theorem, saying that a string of bits can be sent **simultaneously** from one source to sinks $1, 2, \ldots, k$ at a rate determined by $\min_{1 \le i \le n} F_i$, where F_i is the standard Max–Flow from the source to sink i.

Significant improvements were made in

S.-Y. R. Li, R. W. Yeung, and N. Cai, Linear network coding, IEEE Transactions on Information Theory, Vol. 49, Issue 2, 371–381, 2003,

where $\min_{1 \le i \le n} F_i$ can be achieved with linear coding (the Best Paper Award winner of the IEEE Information Theory Society for the year 2005).

P. Sanders, S. Egner, L. Tolhuizen, Polynomial Time Algorithms for Network Information Flow, Proc. of the 15th annual ACM Symposium on Parallel Algorithms and Architectures, 286 - 294, 2003,

where upon our suggestion during this ZIF project a polynomial coding algorithm was given. The results were then merged with those of other authors in

S. Jaggi, P. Sanders, P. A. Chou, M. Effros, S. Egner, K. Jain, and L. Tolhuizen, Polynomial time algorithms for multicast network code construction, IEEE Transactions on Information Theory, Vol. 51, Issue 6, 1973 – 1982, 2005.

Y. Wu, K. Jain, S.-Y. Kung, A unification of Edmond's Graph Theorem and Ahlswede et al's Network Coding Theorem, in Proc. 42nd Annual Allerton Conference on Communication, Control, and Computing, Sept 29 - Oct 1, 2004.
http://www.princeton.edu/ yunnanwu/papers/WuJK2005.pdf

where an important connection indicated in the title was made.

Recently a **first survey Theory of Network Coding in tutorial presentation** was given by R. W. Yeung, S.-Y. R. Li, N. Cai, and Z. Zhang; submitted to Foundations and Trends in Communications and Information Theory.
http://iest2.ie.cuhk.edu.hk/ whyeung/post/netcode/main.pdf

Problems in Network Coding and Error Correcting Codes Appended by a Draft Version of S. Riis "Utilising Public Information in Network Coding"

S. Riis and R. Ahlswede

1 Introduction

In most of todays information networks messages are send in packets of information that is not modified or mixed with the content of other packets during transmission. This holds on macro level (e.g. the internet, wireless communications) as well as on micro level (e.g. communication within processors, communication between a processor and external devises).

Today messages in wireless communication are sent in a manner where each active communication channel carries exactly one "conversation". This approach can be improved considerably by a cleverly designed but sometimes rather complicated channel sharing scheme (network coding). The approach is very new and is still in its pioneering phase. Worldwide only a handful of papers in network coding were published year 2001 or before, 8 papers in 2002, 23 papers in 2003 and over 25 papers already in the first half of 2004; (according to the database developed by R. Koetters). The first conference on Network Coding and applications is scheduled for Trento, Italy April 2005. Research into network coding is growing fast, and Microsoft, IBM and other companies have research teams who are researching this new field. A few American universities (Princeton, MIT, Caltech and Berkeley) have also established research groups in network coding.

The holy grail in network coding is to plan and organize (in an automated fashion) network flow (that is allowed to utilise network coding) in a feasible manner. With a few recent exceptions [5] most current research does not yet address this difficult problem.

The main contribution of this paper is to provide new links between Network Coding and combinatorics. In this paper we will elaborate on some remarks in [8], [9]. We will show that the task of designing efficient strategies for information network flow (network coding) is closely linked to designing error correcting codes. This link is surprising since it appears even in networks where transmission mistakes never happen! Recall that traditionally error correction, is mainly used to reconstruct messages that have been scrambled due to unknown (random) errors. Thus error correcting codes can be used to solve network flow problems even in a setting where errors are assumed to be insignificant or irrelevant.

Our paper is the first paper that use error correcting codes when channels are assumed to be error-free. The idea of linking Network Coding and Error Correcting Codes when channels are not error-free was already presented in [4].

R. Ahlswede et al. (Eds.): Information Transfer and Combinatorics, LNCS 4123, pp. 861–897, 2006.
© Springer-Verlag Berlin Heidelberg 2006

In this paper Cai and Yeung obtained network generalizations of the Hamming bound, the Gilbert-Varshamov bound, as well as the singleton bound for classical error-correcting codes.

2   The Basic Idea and Its Link to Work by Euler

The aim of the section is to illustrate some of the basic ideas in network coding. To illustrate the richness of these ideas we will show that solving the flow problem for certain simple networks, mathematically is equivalent to a problem that puzzled Euler and was first solved fully almost 200 years later! First consider the network in figure 1.

The task is to send the message x from the upper left node, to the lower right node labelled $r : x$ (indicating that the node is required to receive x) as well as to send the message y from the upper right node, to the lower left node labelled $r : y$. Suppose the messages belong to a finite alphabet $A = \{1, 2, \ldots, n\}$. If the two messages are sent as in ordinary routing (as used on the world wide web or in an ordinary wireless network) there is a dead lock along the middle channel where message x and message y will clash. If instead we send the message $s_{x,y} = S(x, y) \in A$ through the middle channel, it is not hard to show that the problem is solvable if and only if the matrix $(s_{i,j})_{i,j \in A}$ forms a latin square (recall that a latin square of order n is an $n \times n$ matrix with entries $1, 2, \ldots, n$ appearing exactly once in each row and in each column). We can now link this observation to work by Euler! Consider the extension of the previous flow problem in figure 2.

Now the task is to send the message x and the message y to each of the five nodes at the bottom. To do this each of the matrices $\{s_{x,y}\}$ and $\{t_{x,y}\}$ must, according to the previous observation, be latin squares. However, the latin squares must also be orthogonal i.e. if we are given the value $s \in A$ of the entry $s_{x,y}$ and the value $t \in A$ of the entry $t_{x,y}$, the values of x and y must be uniquely determined. Thus, we notice that:

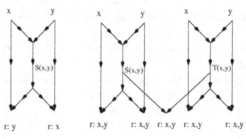

Fig. 1.   Fig. 2.

Proposition 1. *There is a one-to-one correspondence between solutions to the flow problem in figure 2 with alphabet A and pairs of orthogonal latin squares of order |A|.*

The problem of deciding when there exist such two orthogonal latin squares has an interesting history. Euler knew (c.1780) that there was no orthogonal

Latin square of order 2 and he knew constructions when n is odd or divisible by 4. Based on much experimentation, Euler conjectured that orthogonal Latin squares did not exist for orders of the form $4k + 2, k = 0, 1, 2, \ldots$. In 1901, Gaston Tarry proved (by exhaustive enumeration of the possible cases) that there are no pairs of orthogonal Latin squares of order 6 - adding evidence to Euler's conjecture. However, in 1959, Parker, Bose and Shrikhande were able to construct two orthogonal latin squares of order 10 and provided a construction for the remaining even values of n that are not divisible by 4 (of course, excepting $n = 2$ and $n = 6$). From this it follows:

Proposition 2 ((corollary to the solution to Euler's question)). *The flow problem in figure 2 has a solution if and only if the underlying alphabet does not have 2 or 6 elements.*

The flow problem in figure 2 might be considered somewhat 'unnatural' however the link to orthogonal latin squares is also valid for very natural families of networks. The multicast problem $N_{2,4,2}$ defined below has for example recently been shown to be essentially equivalent to Eulers question [6].

3   Network Coding and Its Links to Error Correcting Codes

The task of constructing orthogonal latin squares can be seen as a special case of constructing error correcting codes. There is, for example, a one-to-one correspondence between orthogonal latin squares of order $|A|$ and $(4, |A|2, 3)$ $|A|$-ary error correcting codes.[1]

Next consider the flow problem in figure 3.

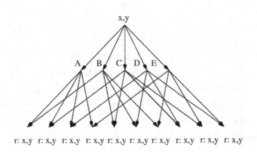

Fig. 3.

Assume each channel in this multi-cast network has the capacity to carry one message (pr. unit time). Assume that the task is to send two messages

---

[1] Recall that a (n, c, d) r-ary error correcting code C consists of c words of length n over an alphabet containing r letters. The number d is the minimal hamming distance between distinct words $w, w' \in C$.

$x, y \in A$ from the top nodes to each of the 10 bottom nodes. It can be shown that this flow problem has a solution over the alphabet A if and only if there exist an $(5, |A|^2, 4)$ $|A|$-ary error correcting code. It has been shown that there exit such codes if and only if $|A| \notin \{2, 3, 6\}$. The flow-problem in figure 3 can be generalized. Consider a network $N_{k,r,s}$ such that it consists of k messages $x_1, x_2, \ldots, x_k \in A$, that are transmitted from a source node. The source node is connected to a layer containing r nodes, and for each s element subset of r (there are $\binom{r}{s} = \frac{r!}{(r-s)!r!}$ such) we have a terminal node. The task is to insure that each message $x_1, x_2, \ldots, x_k \in A$ can be reconstructed in each of the terminal nodes. Notice the previous network flow problem is $N_{2,5,2}$. In general it can be shown [9], [8]:

Proposition 3. *The flow problem $N_{k,r,s}$ has a solution if and only if there exists an $(r, |A|k, r - s + 1)$ $|A|$-ary error correcting code.*[2]

Essentially, there is a one-to-one correspondence between solutions to the network flow problem $N_{2,4,2}$ and $(4, 4, 3)$ binary error correcting codes, i.e. orthogonal latin squares. Thus despite of the fact that the flow problem in figure 2 has a topology very different from the $N_{2,4,2}$ problem, the two problems essentially have the same solutions!

Next, consider the famous Nordstrom-Robinson code: This code is now known to be the unique binary code of length 16, minimal distance 6 containing 256 words. The point about this code is that it is non-linear, and is the only $(16, 256, 6)$ binary code. Again we can apply the proposition to show that the multi-cast problem $N_{8,16,11}$ has no linear solution over the field F_2, while it has a non-linear solution. Are phenomena like this just rare isolated incidences or much more widespread?

4   The Classical Theory for Error Correcting Needs Extensions

The previous sections indicate how it is possible to recast and translate network flow problems into the theory of error correcting codes (thus, using standard results in coding theory, it is possible to translate network flow problems into questions about finite geometries). Another approach is outlined in [7].

In [9], [8] the first example with only non-linear solutions was constructed. Unlike other examples this construction seems to go beyond standard results in error correcting codes. The construction is based on the network in figure 4. The network N in figure 4 has the curious property (like $N_{8,16,11}$) that the maximal through-put can only be achieved if non-linear flows are allowed (i.e non-linear boolean functions are needed in any solution). Furthermore it turns out that any code optimizing the vertical flows has to be a "minimal distance code" [9], [8]. This phenomena is interesting since a minimal distance code from a traditional perspective is very bad (as it essentially has the worst possible error correcting

---

[2] The fact that known bounds on maximum distance separable codes can be applied to bound the required alphabet-size was shown in [10].

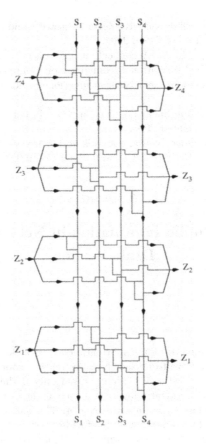

Fig. 4.

capability). This example is one of a collection of examples that suggests that the classical theory of error correcting codes needs to be extended and developed in order to serve as a basis for network coding. See also [3], [1], [2] more results pointing in this direction.

References

1. R. Ahlswede, Remarks on Shannon's secrecy systems, Probl. of Control and Inf. Theory, 11, 4, 301308, 1982.
2. R. Ahlswede and G. Dueck, Bad codes are good ciphers, Probl. of Control and Inf. Theory, 11, 5, 337351, 1982.
3. R Ahlswede and L.H Khachatrian, The diametric theorem in hamming spaces optimal anticodes, Advances in Applied Mathematics, 20, 429 449, 1996.
4. N. Cai and R.W. Yeung, Network coding and error correction, in ITW 2002 Bangalore, 119122, 2002.
5. S. Deb, C. Choute, M. Medard, and R. Koetter, Data harvesting: A random coding approach to rapid dissemination and efficient storage of data, in INFOCOM, submitted.

6. R. Dougherty, C. Freiling, and K. Zeger, Linearity and solvability in multicast networks, in Proceeding of CISS, 2004.
7. C. Fragouli and E. Soljanin, A connection between network coding and convolutional codes, in IEEE International Conference on Communications, 2004.
8. S. Riis, Linear versus non-linear boolean functions in network flow, in Proceeding of CISS 2004.
9. S Riis, Linear versus non-linear boolean functions in network flow (draft version), Technical report, November 2003.
10. M. Tavory, A. Feder, and D. Ron, Bounds on linear codes for network multicast, Technical Report 33, Electronic Colloquium on Computational Complexity, 2003.

Appendix

Utilising Public Information in Network Coding
Draft Version

S. Riis

Abstract. We show that an information network flow problem N in which n messages have to be sent to n destination nodes has a solution (that might utilize Network Coding) if and only if the directed graph G_N (that appears by identifying each output node with its corresponding input node) has *guessing number* $\geq n$. The *guessing number* of a (directed) graph G is a new concept defined in terms of a simple cooperative game. We generalize this result so it applies to general information flow networks.

We notice that the theoretical advantage of Network Coding is as high as one could have possibly hoped for: for each $n \in N$ we define a network flow problem N_n with n input nodes and n output nodes for which the optimal through-put using Network Coding is n times as large as what can be obtained by vector routing or any other technique that does not allow interference (between messages) . In the paper we obtain a characterisation of the set of solutions for each flow problem N_n.

1 Network Coding

1.1 A. The Wave Approach to Information Network Flow

In recent years a new area called Network Coding has evolved. Like many fundamental concepts, Network Coding is based on a simple mathematical model of network flow and communication first explicitly stated in its simplicity in [3]. Recently, ideas related to Network Coding have been proposed in a number of distinct areas of Computer Science (e.g. broadcasting in wireless networks [25,?,23], data security [4], distributed network storage [6,?] and wireless sensor networks

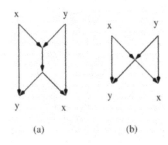

Fig. 1.

[15]). Network Coding has also a broad interface with various Mathematical disciplines (error correcting codes [18,5,10], circuit complexity [17], information theory [11], algebra [13,12] and graph theory).

The basic idea underlying Network Coding has been explained in numerous papers e.g. [13,3,17,7].The idea can be illustrated by considering the "butterfly" network in figure 1a.

The task is to send the message x from the upper left corner to the lower right corner and to send the message y from the upper right corner to the lower left corner. The messages $x, y \in A$ are selected from some finite alphabet A. Assume each information channel can carry at most one message at a time. If the messages x and y are sent simultaneously there is a bottleneck in the middle information channel. On the other hand if we, for example, send $x \oplus y \in A$ through the middle channel, the messages x and y can be recovered at 'output' nodes at the bottom of the network.

The network in figure 1a can be represented as the network in figure 1b. In this representation (which we will call the 'circuit representation') each node in the network computes a function $f : A \times A \rightarrow A$ of its inputs, and sends the function value along $each$ outgoing edge. Historically, it is interesting to note that in this slightly different (but mathematically equivalent) form, the idea behind Network Coding (i.e. the power of using non-trivial boolean functions rather than "pushing bit") was acknowledged already in the 70s (though never emphasized or highlighted) in research papers in Circuit Complexity (see e.g. [22,20,16,21,2]). It is also worth mentioning that in Complexity Theory many lower bounds are proved under the assumption that the algorithm is $conservative$ or can be treated as such. Conservative means that the input elements of the algorithm are atomic unchangeable elements that can be compared or copied but can not be used to synthesize new elements during the course of the algorithm. From a perspective of Circuit Complexity, Network Coding is an interesting theory of information flows since it corresponds to unrestricted models of computation.

Information flow in networks falls naturally within a number of distinct paradigms. Information flow can, for example, be treated in a fashion similar to traffic of cars in a road system. In this view each message is treated as **a packet** (e.g. a car) with a certain destination. Messages (cars!) cannot be copied, or divided. This way of treating messages is almost universally adopted in today's

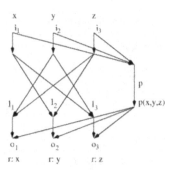

Fig. 2.

information networks (e.g. wireless communication, communication on the web, communication within processors or communication between processors and external devices). Another, less used possibility, is to treat messages in some sense as **a liquid** that can be divided and sent along different routes before they reach their destination. This approach (like, for example, in vector routing [9]) allows messages to be spread out and distributed over large parts of the network. Another and more radical approach is to treat messages as **"waves"**. Recall that the signals carrying the messages are digital (discrete) and thus certainly do not behave like waves. It is, however, possible to transmit and handle the digital (discrete) signals in a fashion where the messages (not the bits carrying the messages) behave like waves subject to interference and super position. More specifically, assume A is a (finite) alphabet of distinct (wave) signals that can be sent through a channel. The superposition of (wave) signals $w_1, w_2 \in A$ creates a new (wave) signal $w = w_1 \oplus w_2 \in A$. Thus mathematically, in the wave picture the set A of wave signals forms a (finite) commutative group with neutral element $0 \in A$ representing the zero-signal.

The network in figure 2 illustrates the point that in specific network topologies there can be quite a large advantage of treating messages as waves. The task of the network is to send messages x, y and z from the source (input) nodes i_1, i_2 and i_3 to the three output nodes o_1, o_2 and o_3. The receiver (output) node o_1 requires x, node o_2 requires y and node o_3 requires z. We assume that channels are one way and that the messages are only sent downwards in the figure. All crossings in the figure are 'bridges' and it is, for example, only possible to move from i_1 to o_1 by moving through channel p.

If messages are treated as packets (cars) like in traditional routing, or if messages are treated as a liquid, there is no point in sending information through l_1, l_2 or l_3. All messages x, y and z must pass through the channel labelled with p (for 'public'). This clearly creates a bottleneck in channel p if we assume that only one message can pass at a time.

If, however, messages are treated as waves we can send $p(x, y, z) := x \oplus y \oplus z$, the superposition of the messages x, y and z, through channel p. And we can send superpositions $l_1 := -(y \oplus z)$, $l_2 := -(x \oplus z)$ and $l_3 := -(x \oplus y)$ through the nodes with these labels. Node o_1 can take the superposition of l_1 and $p(x, y, z)$

and then reconstruct the message x $= -(y \oplus z) \oplus (x \oplus y \oplus z)$. Similarly, node o_2 (or o_3) can take the superposition of l_2 (or l_3) and $p(x, y, z)$ and then reconstruct the message $y = -(x \oplus z) \oplus (x \oplus y \oplus z)$ (or $z = -(x \oplus y) \oplus (x \oplus y \oplus z)$). This shows that the wave approach allows us to eliminate the bottleneck in channel p in figure 2. Notice also that the wave approach increases the overall network performance (of the network in figure 1) by a factor 3. [3]

In general the advantage of the wave approach (compared to any approach that does not allow interference) can be as large as one could have possibly hoped for. We will later notice that there exists information flow networks (with n source nodes and n receiver nodes) for which the optimal throughput is n times larger using the wave approach. Actually, there are even networks where the success rate for each active channel using the wave approach are close (as close as we wish) to n times the success rate for each active channel in a routing solution. The wave approach usually requires more information channels to be involved than traditional routing (or other methods that do not allow interference). Yet, by allowing interference, the total network performance divided by number of active information channels can for some network topologies be close to n times higher than any approach that is unable to utilise interference.

Network Coding allows messages to be sent within the wave paradigm. In fact superpositioning of signals (described above) represents an important type of Network Coding we will refer to as *Linear Network Coding* (see also [14]). Although Linear Network Coding represents a very important subclass of network coding, in general network coding involves methods that go beyond linear network coding. Certain network problems have no linear solutions, but require the application of non-linear boolean functions [17,7]. Non-Linear Network Coding has no obvious physical analogue. Rather general Network Coding represents a paradigm of information flow based on a mathematical model where 'everything goes' and where there are no a priory restrictions on how information is treated. Thus in Network Coding packets might be copied, opened and mixed. And sets of packets might be subject to highly complex non-linear boolean transformations.

2   Coherence: Utilising Apparently Useless Information

2.1   A. Guessing Game with Dice

While I was researching various flow problems related to Circuit Complexity it became clear that a key problem is to characterize and formalism what pieces of information are "useful" and what pieces of information are genuinely "useless" . It became clear that this distinction can be very deceptive. A piece of information that is useless in one context, can sometime be very valuable in a slightly different context [17].

---

[3] Notice that this increase of a factor 3 comes at a certain expense. In the routing approach only 7 channels are active (namely, $(i_1, p), (i_2, p), (i_3, p), (p, o_1), (p, o_2), (p, o_3)$ and channel p), while in the Network Coding solution all 19 channels are active. The success rate $\frac{3}{19}$ for each active channel is higher in the Network Coding solution than in the ordinary solution $\frac{1}{7}$.

To illustrate the problem, consider the following situation [19]: Assume that n players each has a fair s-sided dice (each dice has its sides labelled as $1, 2, \ldots, s$). Imagine that all players (simultaneously) throws their dice in such a manner that no player knows the value of their own dice.

1. What is the probability that each of the n players is able to guess correctly the value of their own dice?
2. Assume that each player knows the values of all other dice, but has no information about the value of their own dice. What is the probability that each of the n players correctly guesses the value of their own dice? (**Hint:** *The probability is NOT $(\frac{1}{s})^n$- The players can do much better than uncoordinated guessing!!*)
3. Assume the ith player receives a value $v_i = v_i(x_1, x_2, \ldots, x_{i-1}, x_{i+1}, \ldots, x_n) \in \{1, 2, \ldots, s\}$ that is allowed to depend on all dice values except the i'th player's own dice. What is the probability that each of the n players correctly manages to guess the value of their own dice?

In question 1 the probability that each player is right is $\frac{1}{s}$ and thus with probability $(\frac{1}{s})^n$ all n players successfully manage to guess correctly their own dice' value simultaneously. Maybe somewhat surprisingly in question 2, the answer depends on the 'protocol' adopted by the players! An optimal protocol appears, for example, if the players agree in advance to assume that the sum of all n dice' values is divisible by s. *This protocol ensures that all players simultaneously 'guess' the value of their own dice with probability $\frac{1}{s}$.*

Question 3, can be answered using a minor modification of the protocol just discussed. Let v_i be defined as the sum $x_1 \oplus x_2 \oplus \ldots, \oplus x_{i-1} \oplus x_{i+1} \oplus \ldots \oplus x_n$ modulo s. Each player then 'guesses' that $x_i = -v_i$ modulo s. Again, the probability that all n players simultaneously guess the correct value of their own dice is $\frac{1}{s}$.

2.2   B. Playing the Guessing Game on a Graph

We will now define a generalisation of the dice guessing game that is (surprisingly?) directly related to a certain type (the so called multiple-unicast type) of information flow problems.

Definition

The Guessing Game (G, s) is a cooperative game defined as follows: Assume that we are given a directed graph $G = (V, E)$ on a vertex set $V = \{1, 2, \ldots, n\}$ representing n players. Each player $v \in \{1, 2, \ldots, n\}$ sends the value of their dice $\in \{1, 2, \ldots, s\}$ to each player $w \in \{1, 2, \ldots, n\}$ with $(v, w) \in E$. Or in other words each node w receives dice' values from a set $A_w := \{v \in V : (v, w) \in E\}$. Each player has to guess the value of their own die. We want to calculate (assuming the players in advance have agreed on an optimal protocol) the probability that all the players (nodes) simultaneously guess their dice values. Question 2 (in Section 1 (A)) corresponded to the case where G is the complete graph on n nodes.

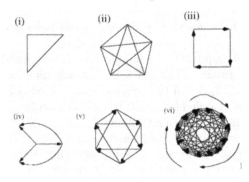

Fig. 3.

Definition

A (cooperative) guessing strategy for the Guessing Game (G, s) is a set of functions $f_\omega : \{1, 2, \ldots, s\}^{A_\omega} \to \{1, 2, \ldots, s\}$ with $\omega \in \{1, 2, \ldots, n\}$. Notice that each player (node) ω is assigned exactly one function f_ω.

In figure 3, we consider six simple examples:

In (i) and (ii) corresponds to the dice guessing game we already considered (with 3 and 5 players). The players have a guessing strategy that succeeds with probability $\frac{1}{s}$. In the guessing game based on (iii) (or in general the cyclic graph on n points) an optimal protocol appears if each node 'guesses' that its own dice value is the same as the value as it receives. This strategy succeeds if each of the four dice has the same value i.e. with probability $(\frac{1}{s})^3$ (or in general $(\frac{1}{s})^{n-1}$). Though this probability is low, it is s times higher than if the players just make uncoordinated random guesses.

In (iv) the graph contains no cycles so the players cannot do any better than just guessing i.e. the players can achieve probability at most $(\frac{1}{s})^4$.

In (v) it can be shown that there are a number of distinct guessing strategies that guarantee the players' success with probability $(\frac{1}{s})^4$ (one, optimal strategy appears by dividing the graph into two disjoint cycles (triangles)).

Finally, in (vi) we consider a graph with 12 nodes (one for each hour on a clock) and edges from (i, j) if the 'time' from i to j is at most 5 hours. We will show (and this will be fairly simple given the general methods we develop) that the players in the Guessing Game (G, s) have an optimal guessing strategy that ensures that the players with probability $(\frac{1}{s})^7$ (i.e. with a factor s^5 better than pure uncoordinated guessing) all simultanously guess the value of their own dice.

Definition

A graph $G = (V, E)$ has for $s \in N$ *guessing number* $k = k(G, s)$ if the players in the Guessing Game (G, s) can choose a protocol that guarantees success with probability $(\frac{1}{s})^{|V|-k}$.

Thus the guessing number of a directed graph is a measure of how much better than pure guessing the players can perform. If the players can achieve a factor s^k better than pure random uncoordinated guessing, the graph has guessing number $k = k(G, s)$. Notice that a directed graph has a guessing number for each $s = 2, 3, 4, \ldots$.

For many directed graphs (though not all) the guessing number is independent of s. The directed graphs in figure 3 have guessing numbers $4, 2, 1, 0, 2$ and 5 (independently of $s \geq 2$). From the definition there is no reason to believe that the guessing number of a directed graph is in general an integer. Yet remarkably many graphs have integer guessing numbers. Later we will show that there exist directed graphs for which the guessing number $k = k(G, s)$ (for alphabet of size $s \in N$) of a graph is not an integer. We will show that there exist graphs where the guessing number $k(G, s)$ even fails to be an integer for each $s \in \{2, 3, 4, \ldots, \}$.

Observation(A)

In the Guessing Game (G, s) the graph G allows the players to do better than pure uncoordinated guessing if and only if G contains a cycle.

Observation(B)

A graph $G = (V, E)$ contains a cycle if and only if its guessing number is ≥ 1. If a graph contains k disjoint cycles its guessing number $\geq k$ (for each $s \geq 2$). A graph is reflexive if and only it has guessing number $|V|$. Assume that the set of nodes V in the graph G can be divided in r disjoint subsets $V_1, V_2, \ldots, V_r$ of nodes such that the restriction of G to each subset V_j is a clique. Then the graph G has guessing number $\geq |V| - r$ (for each $s \geq 2$).

From Observation (A), we notice the curious fact that, *the players have a "good" strategy that ensures that they all succeed with higher probability than uncoordinated random guessing if and only if the players have a "bad" strategy that insures they never succeed.*

Sometimes it is convenient to focus on certain more limited guessing strategies.

Definition Let B be a class of functions $f : A^d \to A$ for $d = 1, 2, 3, \ldots$. An important class appears if we let A denote a fixed algebraic structure (e.g. a group, a ring or a vector space) of $s = |A|$ elements, and let the class $B = LIN$ consist of all homomorphisms (linear maps) $A^d \to A$ for $d = 1, 2, 3, \ldots$. If all the functions f_w belong to the class B we say the players have chosen a guessing strategy in B. If $B = LIN$ we say that the players use a linear guessing strategy.

Definition A graph $G = (V, E)$ has *guessing number* $k = k_B(G, s)$ with respect to the functions in B if the players in the Guessing Game (G, s) have a protocol with all guessing functions in B that guarantee success with probability $(\frac{1}{s})^{|V|-k}$. We say G has (special) linear guessing number $k_{lin} = k_{lin}(G, s)$ if the players have a linear guessing strategy that guarantee success with probability $\geq (\frac{1}{s})^{|V|-k}$.

3   Network Coding and Guessing Games

In this section we show that mathematically there is a very close link between Network Coding and the guessing games we just defined. We will show that each information flow problem is equivalent to a problem about directed graphs.

The translation between information networks and directed graphs is most clean if we represent information networks such that we place all computations (Network Codings) in the nodes of the network. As already indicated, we refer to this representation as the Circuit representation. This representation is slightly more economical (usually save a few nodes) than the standard representation in Network Coding. The representation is more in line with circuit complexity, where the task of the network in general is a computational task. Formally, each source node is associated with a variable. Each node computes a function of incoming edges signals. Each outgoing edge from a node transmits the same signal (function value of node). Each receiver node is required to produce a specific input variable.

In general given an information flow problem N (in the Circuit representation) we obtain a directed graph G_N by identifying each source node with the corresponding receiver node.

In figure 4 we see a few examples of simple information networks together with their corresponding directed graphs.

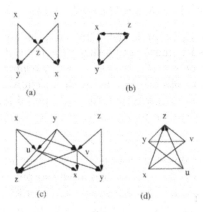

Fig. 4.

The information network N in figure 4a (or figure 1b) is the usual 'butterfly' network (presented in circuit represention). If we identify the input node (source node) x with the output node (receiver node) x, and identify input node (source node) y with the output node (receiver node) y, we get the graph in figure 4b.

The information network in figure 4c does not have any obvious symmetries, but when input and output nodes are identified we get the directed graph in figure 4d that clearly contains a number of symmetries. The translation shows that nodes x and u (as well as y and v) are equivalent points. The guessing number of the graph in (b) as well as the graph in (d) can be shown to have the value 2.

In general we let $C_{multiple-unicast}$ (the class of multiple-unicast directed information networks) consist of information networks N for which for some $n \in N$, n messages $m_1, m_2, \ldots, m_n \in A$ (selected from some alphabet A) has to be sent from input (source) nodes $i_1, i_2, \ldots, i_n$ to output nodes $o_1, o_2, \ldots, o_n$. Somewhat formally, to each source node i_j is associated a variable x_j and each node w (except the source nodes) are assigned a function symbol f_w representing a function f_w that is mapping all incoming signals $a_1, a_2, \ldots, a_{k_w}$ to an element $a = f(a_1, a_2, \ldots, a_{k_w}) \in A$. Each outgoing edge from a node transmits the same signal (the function value a of the node). Each receiver node is required to produce a specific input variable.

For an information network $N \in C_{multiple-unicast}$ we associate a directed graph G_N that appears by identifying each source (input) node i_j in N with its corresponding receiver (output) node o_j. If N has n input nodes, n output nodes and m inner nodes ($2n + m$ nodes in total) the graph G_N has $n + m$ nodes.

We are now ready to state the surprising link that shows that each information flow problem is equivalent to a problem about directed graphs.

Theorem 1

An information Network flow problem $N \in C_{multiple-unicast}$ with n input/output nodes has a solution over alphabet A with $|A| = s$ elements if and only if the graph G_N has guessing number $\geq n$.

The main point of the theorem is that it replaces the flow problem - a problem that mathematically speaking involves slightly complicated concepts like *set of source nodes, set of receiver nodes* as well as set *of requirements (demands)* that specifies the destination of each input - with an equivalent problem that can be expressed in pure graph theoretic terms (no special input or output nodes). Actually we show the theorem in a slightly stronger form:

Theorem 2

The solutions (over alphabet A with $|A| = s$) of an information network flow problem $N \in C_{multiple-unicast}$ with n input/output nodes are in one-to-one correspondence with the guessing strategies (over alphabet A with $|A| = s$) that ensure that the players in the guessing game played on G_N have success with probability $(\frac{1}{s})^{|G_N|-n}$ (where $|G_N|$ is the number of nodes in G_N).

The following simple observation highlights (in a quite geometric fashion) the difference between Network coding and traditional routing:

Observation(C)

An information flow network $N \in C$ has through put k using ordinary routing (i.e. pushing each message along a unique path) if and only the graph G_N contains k disjoint cycles.

Consider the three information flow problems in figure 5(i-iii). They are in circuit representation (i.e. all functions are placed in their nodes and each outgoing edge from a node transmits the same function value). The three information networks in 5(i)-(iii) are non-isomorphic and are clearly distinct. However if we identify the sauce nodes and the receiver nodes in each of the networks we get the *same* directed graph in figure 5 (iv).

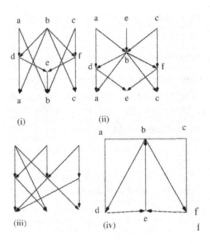

Fig. 5.

According to Theorem 2 there is a one-to-one correspondence between so-lutions of each of the three information networks 5(i)-5(iii) and the successful strategies in the Guessing Game (G, s). Thus, the set of solutions to each of the three information networks 5(i)-5(iii) are in a natural one-to-one correspon-dence. Before we prove Theorem 1 and Theorem 2, let us have a closer look at the networks in figure 5. A (cooperative) strategy for the players in the guessing game with the directed graph in figure 5 (iv) consists of 6 functions $g_1, g_2, \ldots, g_6$ such that:

$$a^{guess} = g_1(b, d)$$
$$b^{guess} = g_2(a, c, e)$$
$$c^{guess} = g_3(b, f)$$
$$d^{guess} = g_4(a, b)$$
$$e^{guess} = g_5(d, f)$$
$$f^{guess} = g_6(b, c)$$

For all players to guess their own message correctly we must have $a^{guess} = a$ i.e. we must have $a = g_1(b, d)$. Thus assuming that we work under the conditional sit-uation with $a^{guess} = a$, we can substitute a with $g_1(b, d)$ leading to the equations:

$$b^{guess} = g_2(g_1(b, d), c, e)$$
$$c^{guess} = g_3(b, f)$$
$$d^{guess} = g_4(g_1(b, d), b)$$
$$e^{guess} = g_5(d, f)$$
$$f^{guess} = g_6(b, c)$$

Now pick any equation of the form $x^{guess} = h$ where x does not appear in the expression h. We might for example assume $c = g_3(b, f)$ (i.e. the $c^{guess} = c$). Substituting c with $g_3(b, f)$ in the equations we get:

$$b^{guess} = g_2(g_1(b,d), g_3(b,f), e)$$
$$d^{guess} = g_4(g_1(b,d), b)$$
$$e^{guess} = g_5(d,f)$$
$$f^{guess} = g_6(b, g_3(b,f))$$

This system of equations contains still one equation of the form $x^{guess} = h$ where x does not appear in the expression h. Let $e = g_5(d,f)$ (assuming $e^{guess} = g_5(d,f)$) and substitute this into the equations we get:

$$b^{guess} = g_2(g_1(b,d), g_3(b,f), g_5(d,f))$$
$$d^{guess} = g_4(g_1(b,d), b)$$
$$f^{guess} = g_6(b, g_3(b,f))$$

For any fixed choice of functions g_1, g_2, g_3, g_4, g_5 and g_6 let $0 \le p \le 1$ denote the probability that a random choice of b, d and f satisfies the equations:

$$b = g_2(g_1(b,d), g_3(b,f), g(d,f))$$
$$d = g_4(g_1(b,d), b)$$
$$f = g_6(b, g_3(b,f))$$

It is not hard to show that the probability that $a^{guess} = a$, $c^{guess} = c$ and $e^{guess} = e$ with probability $(\frac{1}{s})^3$ (for a more general argument see the proof of Theorem 2). Thus the conditional probability that the remaining players all guess correctly their own dice value is p and the probability that all players are correct is $p(\frac{1}{s})^3$. Thus - in agreement with Theorem 1 - the guessing number of the graph in figure 4 (iv) is 3 if and only if there exist functions $g_1, g_2, \ldots, g_6$ such that the equations hold for all b, d and f (i.e. hold with probability 1).

As it happens we can solve the equations by turning the alphabet A into a commutative group $(A, \oplus)$ and the by letting $g_1(b,d) = b \oplus d$, $g_2(\alpha, \beta, \gamma) = \alpha \ominus \gamma$, $g_3(b,f) = b \oplus f$, $g_4(\alpha, \beta) = \alpha \ominus \beta$, $g_5(d,f) = d$ and $g_6(\alpha, \beta) = \ominus \alpha \oplus \beta$. Thus the players have a (cooperative) guessing strategy that ensures that all players simultaneously are able to guess their own message correctly with probability $(\frac{1}{s})^3$. One strategy is given by:

$$a^{guess} = b \oplus d$$
$$b^{guess} = a \ominus e$$
$$c^{guess} = b \oplus f$$
$$d^{guess} = a \ominus b$$
$$e^{guess} = d$$
$$f^{guess} = c \ominus b$$

Figure 6 (i)-(iii) shows how this strategy naturally corresponds to network codings in the three information flow problems in figure 5(i)-(iii). Figure 6 (iv) shows the strategy as a guessing strategy.

4   Proof of Theorems

Before we prove Theorem 1 and Theorem 2 we need a few formal definitions of information networks. As already pointed out, the translation between information networks and directed graphs is most clean if we represent information networks such that we place all computations (Network Codings) in the nodes of the network. An information flow network N (in circuit representation) is an acyclic directed graph with all source nodes (input nodes) having in-degree 0 and all receiver nodes (output nodes) having outdegree 0. Each source node is associated with a variable from a set Γ_{var} of variables. In the receiver node there is assigned a demand i.e. variable from Γ_{var}. In each node w that is not a source, there is assigned a function symbol f_w. The function symbols in the network are all distinct.

Messages are assumed to belong to an alphabet A. Sometimes we assume A has additional structure (e.g. a group, a ring or a vector space). Each outgoing edge from a node transmits the same signal (function value of node).

An actual information flow is given by letting each function symbol f represent an actual function $\tilde{f} : A^d \to A$ where d is the number of incoming edges to the node that is associated the function symbol f. The information flow is a solution, if the functions compose such that each demand always is met.

We let $C_{multiple-unicast}$ denote the class of information networks N for which for some $n \in N$, n messages $m_1, m_2, \ldots, m_n \in A$ (selected from some alphabet A) have to be sent from nodes $i_1, i_2, \ldots, i_n$ to output nodes $o_1, o_2, \ldots, o_n$.

Let $C_{multiple-unicast}$ be an information network in this model. We define the graph G_N by identifying node i_1 with o_1, node i_2 with o_2, ... and node i_j with o_j in general for $j = 1, 2, \ldots, n$.

Theorem 1 follows directly from Theorem 2. So to prove the theorems it suffices to prove Theorem 2.

Proof of Theorem 2: Let N be an information network with input (source) nodes $i_1, i_2, \ldots, i_n$, output (receiver) nodes $o_1, o_2, \ldots, o_n$ and inner nodes n_1, $n_2, \ldots, n_m$. The network N is acyclic so we can assume that we have ordered the nodes as $i_1 < i_2 < \ldots < i_n < n_1 < n_2 < \ldots < n_m < o_1 < o_2 < \ldots < o_n$ such that any edge (i, j) in N has $i < j$ in the ordering. Any selection of coding functions (whether they form a solution or not) can then be written as:

$$z_1 = f_1(x_1, x_2, \ldots, x_n)$$
$$z_2 = f_2(x_1, x_2, \ldots, x_n, z_1)$$
$$z_3 = f_3(x_1, x_2, \ldots, x_n, z_1, z_2)$$
$$\ldots\ldots\ldots\ldots$$
$$z_m = f_m(x_1, x_2, \ldots, x_n, z_1, z_2, \ldots, z_{m-1})$$
$$x_1^o = g_1(x_1, x_2, \ldots, x_n, z_1, z_2, \ldots, z_m)$$
$$x_2^o = g_2(x_1, x_2, \ldots, x_n, z_1, z_2, \ldots, z_m)$$
$$\ldots\ldots\ldots\ldots$$
$$x_n^o = g_n(x_1, x_2, \ldots, x_n, z_1, z_2, \ldots, z_m)$$

where x_j is the variable denoting the value assigned to the input node i_j, z_j is the variable denoting the value computed by the inner node n_j and x_j^o is the variable denoting the output value computed by the node o_j for $j = 1, 2, \ldots, n$ for a given choice of values of $x_1, x_2, \ldots, x_n \in \{1, 2, \ldots, s\}$.

Next consider the corresponding graph G_N we get by identifying nodes i_r and o_r for $r = 1, 2, \ldots, n$. We consider the guessing strategy given by the functions above i.e. the strategy given by:

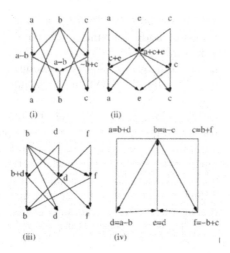

Fig. 6.

$$z_1^{guess} = f_1(x_1^{real}, x_2^{real}, \ldots, x_n^{real})$$
$$z_2^{guess} = f_2(x_1^{real}, x_2^{real}, \ldots, x_n^{real}, z_1^{real})$$
$$z_3^{guess} = f_3(x_1^{real}, x_2^{real}, \ldots, x_n^{real}, z_1^{real}, z_2^{real})$$

$$\ldots\ldots\ldots$$

$$z_m^{guess} = f_m(x_1^{real}, x_2^{real}, \ldots, x_n^{real}, z_1^{real}, z_2^{real}, \ldots, z_{m-1}^{real})$$
$$x_1^{guess} = g_1(x_1^{real}, x_2^{real}, \ldots, x_n^{real}, z_1^{real}, z_2^{real}, \ldots, z_m^{real})$$
$$x_2^{guess} = g_2(x_1^{real}, x_2^{real}, \ldots, x_n^{real}, z_1^{real}, z_2^{real}, \ldots, z_m^{real})$$

$$\ldots\ldots\ldots$$

$$x_n^{guess} = g_n(x_1^{real}, x_2^{real}, \ldots, x_n^{real}, z_1^{real}, z_2^{real}, \ldots, z_m^{real})$$

Conversely each guessing strategy for G_N can be written on this form and can thus be viewed as an attempt to solve the information flow problem N. To prove Theorem 2 we show that the guessing strategy succeeds with probability $(\frac{1}{s})^m$ if and only if the corresponding information flow functions solves the information Network problem. This boils down to showing that the probability that all inner nodes guess their own dice values correctly is $(\frac{1}{s})^m$. To see this assume we have shown this. Then the probability all players guess correctly is at most as large as the probability all players corresponding to inner nodes $n_1, n_2, \ldots, n_m$ guess correctly. Thus all the players guess simultaneously their own die values correctly with probability $\leq (\frac{1}{s})^m$. Equality holds if and only if the conditional probability x_j^{guess}, $j = 1, 2, \ldots, n$ takes the correct value with probability 1. This happens if and only if the functions $f_1, f_2, \ldots, f_m, g_1, \ldots, g_n$ form a solution for the information flow problem. So to complete the proof of Theorem 2 it suffices to show:

Lemma 3. For any set of functions $f_1, \ldots, f_m$ and $g_1, \ldots, g_n$ the probability that players $n_1, n_2, \ldots, n_m$ (i.e. players in nodes corresponding to inner nodes in the information Network) guess their own die values correctly is $(\frac{1}{s})^m$ (i.e. independent of the chosen guessing functions).

Proof: We are asking for the probability $z_j^{guess} = z_j^{real}$ for $j = 1, 2, \ldots, m$ where

$$z_1^{guess} = f_1(x_1^{real}, x_2^{real}, \ldots, x_n^{real})$$
$$z_2^{guess} = f_2(x_1^{real}, x_2^{real}, \ldots, x_n^{real}, z_1^{real})$$
$$z_3^{guess} = f_3(x_1^{real}, x_2^{real}, \ldots, x_n^{real}, z_1^{real}, z_2^{real})$$

$$\ldots\ldots\ldots\ldots$$

$$z_m^{guess} = f_m(x_1^{real}, x_2^{real}, \ldots, x_n^{real}, z_1^{real}, z_2^{real}, \ldots, z_{m-1}^{real})$$

The number of choices of $x_1^{real}, x_2^{real}, \ldots, x_n^{real}$ and $z_1^{real}, z_2^{real}, \ldots, z_m^{real}$ is s^{n+m}. We want to count the number of "successful" choices for which $z_j^{guess} = z_j^{real}$ for $j = 1, 2, \ldots, m$. That is the number of choices for which:

$$z_1^{real} = f_1(x_1^{real}, x_2^{real}, \ldots, x_n^{real})$$
$$z_2^{real} = f_2(x_1^{real}, x_2^{real}, \ldots, x_n^{real}, z_1^{real})$$
$$z_3^{real} = f_3(x_1^{real}, x_2^{real}, \ldots, x_n^{real}, z_1^{real}, z_2^{real})$$

$$\ldots\ldots\ldots\ldots$$

$$z_m^{real} = f_m(x_1^{real}, x_2^{real}, \ldots, x_n^{real}, z_1^{real}, z_2^{real}, \ldots, z_{m-1}^{real})$$

But for each choice of $x_1^{real}, x_2^{real}, \ldots, x_n^{real}$ there is exactly one choice of $z_1^{real}, z_2^{real}, \ldots, z_m^{real}$. Thus the number of successful choices is s^n. The probability is $\frac{\text{number of successful choices}}{\text{number of choices}} = \frac{s^n}{s^{n+m}} = \frac{1}{s^m}$.   ♣

4.1   A. Standard Representation

There are a few slightly different ways to represent flows in information networks. In the previous section we considered the Circuit representation. We call the standard (and "correct") way of representing information flows in Network Coding the **Standard representation**. If we use the standard representation we get slightly different versions of Theorem 1 and Theorem 2. The actual theorems can be stated in the same way (verbatim)! The Theorems are modified to fit the standard representation in the way the graph G_N is defined.

An information network N is a directed acyclic multi-graph. Each source node has indegree 0, while each receiver node has outdegree 0. Associated with each source node is a variable from a set Γ_{var} of variables. Each outgoing edge is associated with a distinct function symbol with an argument for each incoming edge. Each receiver node has a list of demands which is a subset of variables from Γ_{var}. In the receiver node there is assigned a function symbol for each demand. All function symbols are distinct.

Messages are assumed to belong to an alphabet A. An actual flow (using Network Coding) is given by letting each function symbol f represents an actual function $\tilde{f} : A^d \to A$ where d is the number of incoming edges to the node that is associated with the function symbol f. The flow (that might utilise Network Coding) is a solution if the functions compose such that each demand is given by the functional expression of the involved terms.

We let $C_{multiple-unicast}$ denote the class of information networks N for which for some $n \in N$, n messages $m_1, m_2, \ldots, m_n \in A$ (selected from some alphabet A) has to be send from nodes $i_1, i_2, \ldots, i_n$ to output nodes $o_1, o_2, \ldots, o_n$.

We convert a given information network $N \in C_{multiple-unicast}$ to a directed graph G_N as follows:

Step 1: For each variable or function symbol assigned to an edge or a node we introduce a node in the new graph G_N.

Step 2: We identify nodes i_1 with o_1, i_2 with o_2, ... and i_j with o_j in general for $j = 1, 2, \ldots, n$.

With this translation of N to G_N Theorem 1 and Theorem 2 remain valid (verbatim).

5   General Results for Information Networks

Theorem 1 and Theorem 2 only apply for information networks $N \in C_{multiple-unicast}$. In this section we generalize the results so they essentially cover all (!) instantaneous information networks.

Let N be an information network and let A be an (finite) alphabet with s elements. For a selection of fixed network functions $\bar{f}$ we define the networks N's *global success rate* $p(N, s, \bar{f})$ of a specific network coding flow as the probability that *all* outputs produce the required outputs if all inputs are selected random-ly with independent probability distribution. The *maximal global success rate* $p(N, s)$ of the information flow network N (over alphabet of size s) is defined as the supremum of all global success rates $p(N, s, \bar{f})$ that can be achieved by any choice of coding functions $\bar{f}$. Since the set of functions $\bar{f}$ (for a fixed finite alphabet A) is finite $p(N, s)$ is the maximal global success rate of $p(N, s, \bar{f})$.

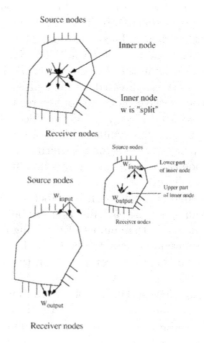

Fig. 7.

Assume that N is an information network over an alphabet A with s elements. Assume that N has n source nodes (input nodes) and that r of these are required by at least one receiver node (output node). Usually, $n = r$ since in most information networks each source message is required by at least one receiver node.

Definition

We define the *source transmission bandwidth* $k = k(N, s)$ of the information network N (over alphabet of size s) as $k(N, s) = \log_s(p(N, s)) + r$.

The notion is motivated by the Theorem 4 below, and can be viewed as a generalisation of the guessing number of a graph.

Notice, that a network has source transmission bandwidth k if all output nodes simultaneously can calculate their required messages with probability s^k higher that what can be achieved by the "channel free" network. An information network N that sends n distinct source messages (each message is required at one or more receiver nodes), has source transmission bandwidth $k(N, s) = n$ if and only if it has is solvable (in the sense of network coding) over an alphabet of size s.

For each directed graph $G = (V, E)$ we want to define an information flow problem $N_G = (W, F)$ with $|V|$ source nodes (input nodes) and $|V|$ receiver nodes (output nodes). Expressed slightly informally, we define N_G by splitting each node $w \in V$ into two nodes w_{input} and w_{output} (thus the vertex set W consists of two copies of V). For each edge $(w, v) \in E$ we add an edge $(w_{input}, v_{output}) \in F$. Let $N_G = (W, F)$ denote the flow problem that appears through this transformation where each output node v_{output} requires the message assigned to v_{input}. Notice that the information network N_G usually is very far from being solvable since most source (input) nodes have no path to its corresponding receiver (output) node.

Observation

Let G be a directed graph. Then $N_G \in C_{multiple-unicast}$ and G has guessing number $k = k(G, s)$ if and only if N_G has source transmission bandwidth $k = k(N_G, s)$.

For each $p \in [0, 1]$ there is a one-to-one correspondence between guessing strategies $\bar{f}$ in the Guessing Game (G, s) that achieve success with probability p and information flows $\bar{f}$ in N_G that have global success rate p.

The Observation is too trivial to deserve to be called a theorem. It is however quite interesting since it shows that the notion of source transmission bandwidth generalizes the guessing number of a directed graph.

We now introduce a simple move we call "split". Given an information network $N = (V, E)$ (with E being a multiset) the move "split" can be applied to any inner node $w \in V$ in N (a node is an inner node if it is not a source or a receiver node). The move "split" copy the inner node w into two nodes w_{input} and w_{output}. In other words the move convert the vertex set V to the set $V' = V \cup \{w_{input}, w_{output}\} \setminus \{w\}$ containing all point in V but with two copies (w_{input} and

w_{output}) of w. For each outgoing edge $(w, u) \in E$ from w we introduce an edge $(w_{input}, u) \in E'$ (with the same multiplicity as (w, u)). For each incoming edge $(u, w) \in V$ we introduce an edge $(u, w_{input}) \in E'$ (with the same multiplicity as (w, u)).

The information network $N' = (V', E')$ has as source (input) nodes all source (input) nodes in V together with $\{w_{input}\}$. The set of receiver (output) nodes consists of the receiver (output) nodes in V together with $\{w_{output}\}$. We associate a new variable z to the node w_{input} and node w_{output} demands (requires) z. All other nodes keep their demands.

In figure 7, we see how the split move can be applied. We say that the information network N' appear from the information network N by a "reverse split move", if N appears from N' using a split move.

The split move always result in an information network that have no solution (since there is no path from the source node w_{input} to the receiver node w_{output}).

The next Theorem can be considered as a generalization of Theorem 1 and Theorem 2.

Theorem 4

Let N and N' be two information networks that appear from each other by a sequence of split and inverse split moves (in any order). The network N and N' has the same source bandwidth (i.e. $k(N, s) = k(N', s)$)

More specifically let N be an information flow network, let A be an alphabet with $|A| = s$ letters and assume $\bar{f}$ is a selection of coding functions over this alphabet. Assume N has source messages $x_1, x_2, \ldots, x_r$ (they might be transmitted from more than one input edge). Assume that the coding functions have global success rate $p = p(N, s, \bar{f}) \in [0, 1]$. Let N' be any information network that appears from N by application of the split move. Then N' with the coding functions $\bar{f}$) has global success rate $p(N', s, \bar{f}) = \frac{p}{s}$.

In general if N has global success rate p (over alphabet A) any network N' that appears from N by application of r split moves and t reverse split moves (in any order) has global success rate $p \times s^{t-r}$.

Proof: The first part follows from the more detailed second part since each application of the split rule increase the value of r by one and each application of the inverse split rule decrease the value of r by one.

Assume that the information network $N = (V, E)$ has global success rate $p = p(N, s, \bar{f}) \in [0, 1]$ with respect to the coding functions $\bar{f}$. Let $w \in V$ be any inner node in N. Replace (split) w into two nodes w_{input} and w_{output} as already explained. The incoming coding function to node w_{output} is the same function as the inner coding function to node w in the network N. Each outgoing coding function of w_{input} is the same as each outgoing function for node w. The network N' has got a new input node. Let us calculate the probability $p(N', s, \bar{f})$ that all output nodes produce the correct outputs. The probability that node w_{output} produce the correct output is exactly $\frac{1}{s}$. Assume now $w_{output} = w_{input}$. The conditional probability (i.e. the probability given $z_{output} = z_{input}$) that all output nodes in the network N produce the correct output is $p = p(N, s, \bar{f})$.

But, then the probability all output nodes in N' produce the correct output is exactly $\frac{p}{s}$.

The second part of the theorem follows from the first part. Assume $\bar{f}$ is a selection of coding functions such that $p(N, s, \bar{f}) = p(N, s)$ (the alphabet is finite so there are only finitely many functions $\bar{f}$ and thus there exists functions that achieve the maximum value $p(N, s)$). We already showed that $p(N', s, \bar{f}) = \frac{p(N, s)}{s}$. We claim that $p(N', s) = p(N', s, \bar{f})$. Assume that $p(N', s, \bar{g}) > p(N', s, \bar{f})$. But then $p(N, s, \bar{g}) = s \times p(N', s, \bar{g}) > s \times p(N', s, \bar{f}) = p(N, s, \bar{f})$ which contradicts the assumption that $\bar{f}$ was an optimal coding function for the information network N (over alphabet of size s).   ♣

6   Utilising Public Information

6.1   A. Another Game

Consider the directed graph in figure 8(i) (introduced in 4 (iv)). Each node has to derive their own message. This is, of course, impossible and we know that the best the players can hope for (if they use a suitable coordinated guessing strategy) is that they are all correct on s^3 distinct inputs (out of the s^6 different input). If the players have access to s^3 public messages and these are carefully chosen, it is possible for the players (through a cooperative strategy) to ensure that each player can derive his/her own message.

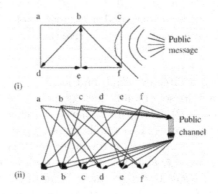

Fig. 8.

If, for example, the values $a \ominus b \ominus d \in A, c \ominus b \ominus f \in A$ as well as $e \ominus d \in A$ are common knowledge (broadcast through public messages) each node can derive its own message (since $a = (a \ominus b \ominus d) \oplus b \oplus d, b = (e \ominus d) \ominus (a \ominus b \ominus d) \oplus (a \ominus e), c = (c \ominus b \ominus f) \oplus b \oplus f, d = (a \ominus b) \oplus (a \ominus b \ominus d), e = (e \ominus d) \oplus d$ and $f = (c \ominus b) \ominus (c \ominus b \ominus f))$.

Another equivalent way of stating this is to consider the bipartite flow problem in figure 8 (ii), with public channel of bandwidth 3. Notice that figure 8 (i) and figure 8 (ii) are different representations of the problems that are mathematically equivalent.

Are the solutions (public messages $a \ominus b \ominus d \in A, c \ominus b \ominus f \in A$ as well as $e \ominus d \in A$) in figure 8 (i) and figure 8 (ii) optimal? Is it possible to send fewer than s^3 message through the public channel (and still have all players being able to deduce their own message)? From the analysis of the guessing game in figure 5 (iv) we know that the probability that the players in nodes a, c and e guess their own messages is independent (for any guessing strategy) and thus nodes a, c and e guess correctly their own message with probability $(\frac{1}{s})^3$. We claim that if node a, c and e in general are able to derive their own message they must have access to at least s^3 distinct messages in the public channel. To see this assume that it were possible for the players in figure 8 (i) to deduce their own messages from a public channel that sends $< s^3$. The players could then all agree to guess *if* the public channel is broadcasting a specific message m they agreed on in advance. Since there are less than s^3 public messages there is a message m that is broadcast with probability $> (\frac{1}{s})^3$). This contradicts the fact that the players (especially the players in nodes a, b and c) cannot do better than $(\frac{1}{s})^3$. Thus the solutions in figure 8 (i) (and in figure 8 (ii)) are optimal.

Let $G = (V, E)$ be a directed graph. Assume like before that each node is being assigned a message x randomly chosen from a fixed finite alphabet A containing $s = |A|$ elements. Like in the guessing game each node transmits its message (dice value) along all outgoing edges. In other words each node j knows the messages (dice values) of exactly all nodes i with $(i, j) \in E$.

The task of the players is to deduce their own message. This is of course impossible (unless the graph is reflexive) since in general the players have no direct access to their own message (dice values). The task of the players is to cooperate and agree on a protocol and a behaviour of a public channel that ensure that all players are always able to derive their own messages.

Definition

Let $G = (V, E)$ be a directed graph and let A denote an alphabet with s letters. Let P be a finite set of public messages. Consider the following Public Channel Game (G, A, P). The game is played as follows. Each node $j \in V$ is assigned to a message $x_j \in A$. A public message $p = p(x_1, x_2, \ldots, x_n) \in P$ (given by a function $p : A^n \to P$) is broadcast to all nodes. Each node j has access to the message $p \in P$ as well as x_i for each i with $(i, j) \in E$. In the game each player j needs to deduce the content of their own message x_j.

Each player (node) $v \in \{1, 2, \ldots, n\}$ sends its message to each person $w \in \{1, 2, \ldots, n\}$ with $(v, w) \in E$. Or in other words each node w receives messages from a set $A_w := \{v \in V : (v, w) \in E\}$. The task is to design the function $p(x_1, x_2, \ldots, x_n)$ such that each player always (i.e. for any choice of $x_1, x_2, \ldots, x_n \in S$) can deduce their own message. If this is possible, we say that the Public Channel Game (G, A, P) has a solution.

Definition

A directed graph $G = (V, E)$ has *(general) linear guessing number* $k = k_s$ if the Public Channel Game (G, A, P) has solution for some A with $|A| = s$ and with $P = s^{|V|-k}$.

In the case of figure 3(iii) each player would be able to calculate his own dice value if, for example, $x_1 \oplus x_4, x_2 \oplus x_4$ and $x_3 \oplus x_4$ modulo s were known public information. [To see this, notice that node 1 receives x_4 from which it can calculate $x_1 = (x_1 \oplus x_4) \ominus x_4$, node $i = 2, 3$ receives x_{i-1} from which it can calculate $x_i = (x_i \oplus x_4) \ominus (x_{i-1} \oplus x_4) \oplus x_{i-1}$. Finally, node 4 receives x_3 from which it can calculate $x_4 = (x_3 \oplus x_4) \ominus x_3$].

For any information network N we can apply the split move until all inner nodes have been spilt. In this case N becomes a bipartite graph B_N with no inner nodes. Notice that B_N is uniquely determined by N.

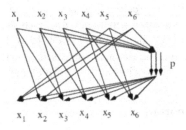

Fig. 9.

This example suggests that it is always possible to replace the guessing part of guessing game, and instead let all players have access to a suitable public channel of information. We will show (Corollary 10) that this is possible for linear solutions (also sometimes called matrix linear) for the guessing game, but it is never possible if only non-linear solutions exists. Notice, that this analysis is only meaningful when the alphabet (i.e. the dice values) can be organized as a vector space U (of dimension d) over a finite field F (with a number q of elements being a prime power). The number $|U|$ of element of U is given by $s := q^d$.

Theorem 5

Assume that the alphabet U is a vector space of dimension d over a finite field F with q elements (i.e. q is a prime power). Then the following statements are equivalent:

(1) The players have a linear guessing strategy in the Guessing Game (G, U) that succeeds with probability $(\frac{1}{q^d})^k$

(2) G has a special linear guessing number $k = k_{lin}(G, q^d)$.

(3) The Public Channel Game (G, U, U^k) has a solution (possible non-linear).

(4) The Bipartite information flow problem B_G associated to G has a solution (over U and possible non-linear) that uses a public channel P of bandwidth k.

(5) The Bipartite information flow problem associated to G has a linear solution (over U) that uses a public channel of bandwidth k.

(6) The Public Channel Game (G, U, U^k) has a linear solution.

From this we get:

Theorem 6
Assume that the alphabet U is a finite dimensional vector space over a finite field F. The nodes in a directed graph G can calculate their messages (selected from U) if they have access to a public channel of bandwidth $\leq k$ if and only if the (special) linear guessing number of G is $\geq |V| - k$.

Theorem 6 explain the terminology of the *(special) linear guessing number*. In the case where the alphabet is a vectorspace the linear guessing number (in sense of linear maps) agree with the *(general) linear guessing number*. The two notions of linear guessing numbers agree when they are both defined. The general linear guessing number is, however, defined for all $s \in \{2, 3, 4, \ldots, \}$, while the special linear guessing number only is defined when s is a prime power (since a finite dimensional vector space always has a number of elements being a prime power).

6.2   B. Proof of Theorem 5

First notice that (1) and (2) are equivalent (by definition). We claim:

Lemma 7
(1) implies (3):

Proof: We are given a graph $G = (V, E)$ and we consider the Guessing Game (G, U, U^k), for U being a vector space of dimension d over a field F with q elements (q being a prime power). The number k is given by (1). We assume that the players have a linear guessing strategy, i.e. a strategy where all functions $f_w : U^{r_w} \to U$ are linear (i.e. given by a $r_w d \otimes d$ matrix with entries in F). Furthermore we assume this linear guessing strategy makes it possible for the players to guess correctly all their own dice values with probability $(\frac{1}{q^d})^k$.

Consider $\tilde{U} := U^{|V|}$, the linear subspace of vectors $(v_1, v_2, \ldots, v_{|V|}) \in U^{|V|}$ with $v_j \in U$ for $j = 1, 2, \ldots, |V|$. Let $W \subseteq \tilde{U}$ denote the linear subspace of dice values for which the players all successfully guess their own dice value (while using the linear guessing strategy for which we assume that it exists). Since the strategy is successful with probability $(\frac{1}{q})^{dk}$ and since the number of points in $\tilde{U}$ is $q^{d|V|}$ the number of points in W is $q^{d|V|-dk}$. Since W is a linear subspace with $q^{d|V|-kd}$ points, its dimension $d|V| - dk$ must be an integer (thus dk must be an integer while k might not be an integer).

For each vector $u \in \tilde{U}$ we consider the linear "side" space $u + W$. Let $u_1, u_2, \ldots, u_l$ denote a maximal family of vectors with $W, u_1 + W, u_2 + W, \ldots, u_l + W$ all being disjoint. It follows that $l = q^{dk} - 1$, i.e. that there are q^{dk} disjoint side spaces of W and that $W \cup_j (u_j + W) = U$.

We can now convert this into a solution to the Public Channel Game (G, U, U^k). We do this by broadcasting a public message as follows: Assume each node in V has been assigned a value from U. The information of all dice values are contained in a vector $u \in \tilde{U}$. There exists exactly one index $j \in \{1, 2, \ldots, l\}$ such that $u \in$

$u_j + W$. Broadcast the index $j \in \{0, 1, 2, \ldots, q^{dk} - 1\}$ by selecting a bijection from $\{0, 1, \ldots, q^{dk} - 1\}$ to U. Now each node can calculate its own message by correcting their guess (they could have played the Guessing Game) by the suitable projection of u_j.

This shows that the Public Channel Game (G, U, U^k) has a solution (possible non-linear) with the public message being selected from the set of public messages U^k. ♣

In this construction, the public channel broadcasts different messages for each index $j \in \{1, 2, \ldots, l\}$. In general this map is not linear. We will show that any non-linear strategy can be turned into a linear strategy.

Lemma 8

(4) implies (5)

Before we prove this implication we make a few general observations and definitions. Assume the flow problem has a solution with the public channel broadcasting

$$p_1(x_1, x_2, \ldots, x_n), \ldots, p_w(x_1, x_2, \ldots, x_n).$$

Since $p_j : A^n \to A$ and A is a field, each function p_j can be expressed as a polynomial $p_j \in A[x_1, x_2, \ldots, x_n]$. Each output node o_j receives $p_1, p_2, \ldots, p_w \in A$ as well as $x_{j_1}, x_{j_2}, \ldots, x_{j_v} \in A$. The task of output node o_j is to calculate $x_j \in A$. For any $q \in A[x_1, x_2, \ldots, x_n]$ let $L(q) \in A[x_1, x_2, \ldots, x_n]$ denote the sum of all monomials (with the original coefficients) of q that only contains one variable (e.g. x_j, x^3, or x_j^7). In other words $L(q)$ consists of q where the constant term, as well as all monomials containing more than one variable, have been removed. If for example $q = 5x_1x_3 - 7x_1x_2 + 3x_1 - 5x_2 + 1$, then $L(q) = 3x_1 - 5x_2$.

We first consider the special case where the alphabet U is a one dimensional vector space (i.e a finite field) rather than a general finite (dimensional) vector space.

Lemma 9

A bipartite information flow problem B has a solution with public information given by polynomials $p_1, p_2, \ldots, p_w \in A[x_1, x_2, \ldots, x_n]$ then B has a solution with public information given by linear expressions $l_1, l_2, \ldots, l_w \in A[x_1, x_2, \ldots, x_n]$.

Remark: In general non-linear flows cannot be eliminated from information networks. In a general network a non-linear solution might for example involve that two nodes send messages $(x + y)$ and $(y + z)$ to a node r where their product $(x + y)(y + z) = xy + xz + yz + y^2 = xy + xz + yz + y$ is being calculated. Removing mixed monomials would lead to $L(x + y) = x + y$ and $L(y + z) = y + z$ to be sent to node r where $L((x + y)(y + z)) = y^2$ must be calculated. Since it is not possible to derive y^2 (or y) from $x + y$ and $y + z$ the process of removing monomials with mixed variables fails in general. The networks in [17] and [7] show that certain flow problems only have non-linear solutions. For such networks any attempt of removing non-linear terms (not just using local linearisation) will fail. The point of the lemma is that the network B together with any public channel is structured in such a fashion that allows us to remove mixed terms and then replace the resulting function with linear functions. Information networks in which only two messages are transmitted provide another case where linearisation is always possible [8].

Proof of Lemma 9: We apply the operator L that removes all monomials with two or more distinct variables. The public information then becomes $L(p_1), L(p_2), \ldots, L(p_w)$. These functions can be realized (since there are no restrictions on the public channel and all functions $A^n \to A^w$ can be calculated). Using the same argument we can remove all mixed terms and insure that each output node o_j receives a function of its inputs (the input from input nodes as well as from the public channel). This completes the proof of Lemma 9.   ♣

In general, when A is a vector space the operator L removes all monomials that contain variables associated to distinct inputs.

Thus it is easy to prove Theorem 5. We have shown (1) → (3) (Lemma 7) , as well as (4) → (5) (Lemma 8).

The implications (5) → (6) → (1) as well as (3) ↔ (4) are all almost trivial and are left as easy exercises for the reader. This completes the proof of Theorem 5. Theorem 6 follows as an easy corollary.

7   Some Corollaries

In general the Guessing Game (G, s) might only have non-linear optimal guessing strategies. When this happens G has linear guessing number k_{lin} that is strictly smaller than G's guessing number k. We have the following characterisation:

Corollary 10

Let $G = (V, E)$ be a graph and let U be a finite vector space. The linear guessing number k_{lin} of G over U is smaller or equal to the guessing number k of G. Equality holds if and only if the Public Channel Game $(G, U, U^{|V|-k})$ is solvable.

We have seen that the problem of solving information network flow problems (of class C) can be restated to that of calculating the guessing number of a graph. The linear guessing number of a graph is an important concept:

Corollary 11

The information flow problem $N \in C$ with n input/output nodes has a linear solution (i.e. a solution within the "wave paradigm") over an alphabet of size s if and only if G_N has its linear guessing number $k(G, s) \geq n$ (which happens if and only if $k(G, s) = n$).

8   Algebraic Solution to the Case Where $|A| = 2$

Consider a directed graph $G = (V, E)$. In this section we show that the linear guessing number (for an alphabet of size 2) has an algebraic definition. Assume

G has no edges $(v, v) \in E$ for $v \in V$. We say $G' = (V, E')$ is a subgraph of G if $E' \subseteq E$. The reflexive closure $ref(G)$ of $G' = (V, E')$ is the graph that appear if we add all the edges (v, v) to the edge set E'.

Theorem 12

Let G be a directed graph. Assume that the alphabet A contains two elements. Then G has linear guessing number $k = k_{lin,2}$ (for alphabet of size 2) if and only if there exists a subgraph G' of G such that the rank of the incident matrix for $ref(G') = k$.

Proof: We assume $A = \{0, 1\}$. Assume G has linear guessing number k. According the Theorem 5 G has linear guessing number k if and only if the Public Channel Game $(G, 2)$ has a linear solution S with a public channel of bandwidth k. We say an edge $(v_1, v_2) \in E$ in G is active (with respect to the solution S) if the message in v_1 affects the guessing function in v_2. Let $E' \subseteq E$ consists of all active edges in G. Let $G' = (V, E')$ be the subgraph of G that consists of all active edges in G.

Consider a node $w \in V$ such that $(v_1, w), (v_2, w), \ldots, (v_d, w)$ are all active incoming edges. The (linear) signal being send to node w is $s = m_{v_1} + m_{v_2} + \ldots + m_{v_d}$ i.e. the sum of all incoming signals, as well as the signals that are send from the public channel. Next we assume that the rank of $ref(G')$ is k for some $G' \subseteq G$. Let $l_1(x_1, x_2, \ldots, x_n), l_2(x_1, x_2, \ldots, x_n), \ldots, l_k(x_1, x_2, \ldots, x_n)$ denote the k linearly independent rows of $ref(G')$. Send these signals as public messages. Let w be an arbitrary node. The node receives a signal $m_{v_1} + m_{v_2} + \ldots + m_{v_r}$ from the channels in G'. The node w needs to derive m_w so it suffice to show that the node w can derive $m_{v_1} + m_{v_2} + \ldots + m_{v_d} + x_w$ from the public messages. But, the row $m_{v_1} + m_{v_2} + \ldots + m_{v_d} + m_w$ appears in $ref(G')$ and thus it belong to the span of the k vectors $l_1(x_1, x_2, \ldots, x_n), l_2(x_1, x_2, \ldots, x_n), \ldots, l_k(x_1, x_2, \ldots, x_n)$ that are send through the public channel.   ♣

For a graph G let $Ref(G)$ denote the reflexive closure of G. Let $rank(G)$ denote the rank over the field $\{0, 1\}$ of the incident matrix of G.

Theorem 13

Assume the alphabet $A = \{0, 1\}$ only contains two elements. Let G be a graph. Then the Public Channel Game $(G, \{0, 1\}, \{0, 1\}^k)$ has a solution if and only if

$$k \geq min_{G' \subseteq G} rank(Ref(G'))$$

9   More Games

Suppose $N \in C_{multiple-unicasts}$ is an information network where some nodes have indegree > 2. For each node n with indegree $d > 2$ we can replace the incoming d edges with a tree with d leaves and a root in n (see figure 11).

Theoretically this replacement restricts the power of the information network since not all functions $f : A^d \to A$ can be written as a composition of d functions

$g_j : A^2 \rightarrow A$, with $j = 1, 2, \ldots, d$. Let S_d denote the class of d-ary functions $f : A^d \rightarrow A$ that can be written as a composition of d, 2-ary functions.

Given a directed graph $G = (V, E)$ and assume that each node with indegree d can only compute functions that belong to S_d. How does this affect the guessing number of the graph? How does it affect the set of solutions?

The network in figure 2 corresponds to a type of games that can be described as follows:

- Public Channel Game Variant(G, s): As before let $G = (V, E)$ be a graph on a vertex set $V = \{1, 2, \ldots, n\}$ of persons. The game is played by n players. Each player is assigned a message selected from some alphabet $\{1, 2, \ldots, s\}$. Each person $w \in \{1, 2, \ldots, n\}$ receive *the function value* (a value in $\{1, 2, \ldots, s\}$) from the set $A_w = \{v \in V : (v, w) \in E\} \subseteq V$. Each player also have access to a public information channel p. How many messages should the public channel p be able to broadcast for all players to be able to deduce there own message? Problem 3 (in section I(A)) corresponded to the case where G is the complete graph on n nodes.

As we already pointed out there exists graphs G for which the dice guessing game only can achieve maximal probability, if the players uses non-linear functions.

We will show (and this will follow as a corollary of Theorem 16) that:

Theorem 14

Assume that the public information is given by a function $p : A^n \rightarrow A$. Then the Public Channel Game Variant (K_n) played on the complete graph K_n has a solution if and only if there exists a commutative group $(A, \oplus)$ structure on the alphabet A and there exists n permutations $\pi_1, \pi_2, \ldots, \pi_n \in S_A$ of elements in A such that the public channel broadcast

$$p(x_1, x_2, \ldots, x_n) = \pi_1 x_1 \oplus \pi_2 x_2 \oplus \ldots \oplus \pi_n x_n$$

Roughly, Theorem 14 states that the set of solutions consists of all the "obvious" solutions (where $p(x_1, x_2, \ldots .x_n) = x_1 \oplus x_2 \oplus \ldots \oplus x_n$ for a commutative group), together with all "encryptions" $\pi : A \rightarrow A$ of these.

10   On the Power of Network Coding

In this section we show that the advantage of (linear) Network Coding compared to any method that does not allows "interference" is as high as one could possible have hoped for. Consider the information networks N in figure 10. The networks corresponds to the Guessing Game on the complete graph K_n.

Theorem 15

For each n there is a network N with n input nodes and n output nodes such that the through-put is n times higher than any method that does not allow interference.

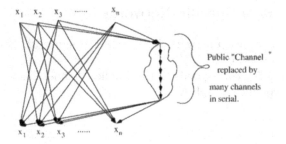

Fig. 10.

For any $n \in N$ and for any $\epsilon > 0$ there exists a network $N(n, \epsilon)$ such that the through-put divided by the number of active channel using Network Coding, is $n - \epsilon$ times as high as the maximal through-put divided by the number of active channels using methods that does not allow interference.

If each inner node is required to have in-degree (and out-degree) ≥ 2 the result remains valid.

Proof: For each $n \geq 2$ (and each $\epsilon > 0$ we base the construction on the network in figure 10. Assume that the public channel consists of m channels in serial. In any "solution" (operating at rate $\frac{1}{n}$) that does not allow mixture of datapackets all messages must go through these m channels. Thus the number of active channels is $m+2$. In the Network Coding solution (operating at rate 1) all $n(n-1)+(m+2)$ channels are active. We can choose m such that $n \times (\frac{(m+2)}{n(n-1)+(m+2)}) > n - \epsilon$. For this m the through-put divided by the number of active channel (in the Network Coding solution) is $n - \epsilon$ times as high as the maximal through-put divided by the number of active channels using methods that does not allow interference.

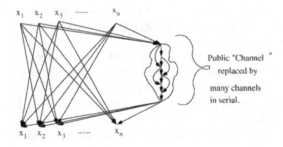

Fig. 11.

The serial construction in this proof might be considered unacceptable. It might be argued that the cost of using the serial channels ought to count as 1 rather than m. To overcome this criticism we can modify the serial channels as indicated in figure 11 and select m such that each path through the public channel still must involve $\geq m$ active channels (m chosen as before).   ♣

11   Analysis of Specific Networks

Consider the information network N_n sketched in figure 12. The network $N(3)$ is displayed in in figure 2.

The networks N_n corresponds to the Public Channel Game Variant (K_n, s) played on the complete graph K_n.

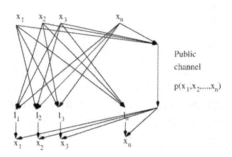

Public

channel

$p(x_1, x_2, \ldots, x_n)$

Fig. 12.

Consider, again the information network $N(3)$ in figure 2. The three output nodes receive the messages $l_1(x_2, x_3) \in A, l_2(x_1, x_3) \in A$ and $l_3(x_1, x_2) \in A$. Besides this, each output node has access to public message $p = p(x_1, x_2, x_3) \in A$. We notice that a solution to the flow problem associated with N_3 consists of six functions $l_1, l_2, l_3, r_1, r_2, r_3 : A \times A \to A$ as well as one function $p : A \times A \times A \to A$ such that $x_1 = r_1(p(x_1, x_2, x_3), l_1(x_2, x_3)), x_2 = r_2(p(x_1, x_2, x_3), l_2(x_1, x_3))$ and $x_3 = r_3(p(x_1, x_2, x_3), l_3(x_1, x_2))$.

The solution we already considered can be achieved (within the framework of linear Network Coding) as follows: Let $(A, \oplus)$ be an Abelian group, let $p(x_1, x_2, x_3) := x_1 \oplus x_2 \oplus x_3$, let $l_i(x, y) := x \oplus y$ for $i = 1, 2, 3$ and let $r_i(x, y) := x \ominus y$ for $i = 1, 2, 3$. We leave to the reader to check that this defines a solution to the flow problem associated with the network N_3.

Actually, for each Abelian group $(A, \oplus)$ and for any three permutations $\pi_1, \pi_2, \pi_3 : A \to A$ the network has a solution with $p(x_1, x_2, x_3) := \pi_1 x_1 \oplus \pi_2 x_2 \oplus \pi_3 x_3$, $l_1(x_2, x_3) := \pi_2 x_2 \oplus \pi_3 x_3, l_2(x_1, x_3) := \pi_1 x_1 \oplus \pi_3 x_3$ and $l_3(x_1, x_2) := \pi_1 x_1 \oplus \pi_2 x_2$. We will show that all solutions are essentially of this form. More generally let N_n denote the network:

The network N_n has n input nodes. These transmit messages $x_1, x_2, \ldots, x_n \in A$. The messages $x_1, x_2, \ldots, x_n$ are independent so we assume that the network cannot exploit hidden coherence in the data. The network N_n has n internal nodes $l_1, l_2, \ldots, l_n$. The node l_j is connected to each input node *except* the node that transmits message x_j. The network has n output nodes that are required to receive the messages $x_1, x_2, \ldots, x_{n-1}$ and x_n (one message for each output node). The node required to receive x_j is connected to l_j as well as to the public channel p. The public channel broadcasts one message $p = p(x_1, x_2, \ldots, x_n) \in A$ to all output nodes. First we notice that:

Observation

The network N_n has a solution over any (finite) alphabet A. Using routing only one message can be transmitted at a time. Thus the through-put using Network coding is n-times as large as the through-put using any type of routing method that does not allow interference. This is optimal since any network problem with n input nodes that is solvable using network coding can be solved using routing if the bandwidth is increased by a factor n.

The next Theorem gives a complete classification of the set of solutions (all utilising Network coding) to the network N_n.

Theorem 16

Consider the network flow problem N_n over a finite alphabet A. Assume $n \geq 3$. Let $p : A^n \to A$ be any function. The network flow problem N_n has a solution with public information p if and only if for some group composition $\oplus$ on A that makes $(A, \oplus)$ an Abelian group, there exist n permutations $\pi_1, \pi_2, \ldots, \pi_n : A \to A$ such that $p(x_1, x_2, \ldots, x_n) = \oplus_{j=1}^n \pi_j x_j$.

Proof: In general if Theorem 16 have been shown for N_r for some $r \geq 3$ the theorem is also valid for each N_s with $s \geq r$. Thus to prove the theorem it suffices to show that the theorem is valid for N_3.

Let $p : A^3 \to A$ be defined by $p(x_1, x_2, x_3)$. Assume that the network has a solution when the public signal is given by p. The function $p : A^3 \to A$ must be 'latin' (i.e. $f_{a,b}(z) := p(a, b, z)$, $g_{a,c}(y) := p(a, y, c)$ and $h_{b,c}(x) := p(x, b, c)$ for each $a, b, c \in A$ define bijections $f_{a,b}, g_{a,c}, h_{b,c} : A \to A$). Notice that p defines a latin cube of order $|A|$. The functions $l_1.l_2, l_3 : A^2 \to A$ are also forced to be latin i.e. they define three latin squares each of order $|A|$. In order to proceed we need to prove a number of lemmas.

Lemma 17

Denote one element in A by 1. The network N_3 has a solution for some functions $l_1, l_2, l_3 : A^2 \to A$ if and only if the network N_3 has a solution when $l_1(x_2, x_3) := p(1, x_2, x_3)$, $l_2(x_1, x_3) := p(x_1, 1, x_3)$ and $l_3(x_1, x_2) = p(x_1, x_2, 1)$.

Proof of Lemma 17: We introduce a new and interesting type of argument that might be useful when reasoning about 'latin' network flow in general. For each output node we draw a triangle with a coding function assigned to each corner. The triangle corresponding to the output node that required output x_1 has assigned $p(x_1, x_2, x_3), l_1(x_2, x_3)$ and x_1 to its corners. If p and l_1 are functions that produce a solution to the network flow problem, $x_1 \in A$ can uniquely be calculated from $p(x_1, x_2, x_3) \in A$ and $l_1(x_2, x_3) \in A$ (i.e. there exists a (latin) function $f : A^2 \to A$ such that $x_1 = f(p(x_1, x_2, x_3), l_1(x_2, x_3)))$. Notice, that any coding function assigned to one of the corners can be calculated uniquely from the two other functions. More specifically $l_1(x_2, x_3) \in A$ is uniquely determined by $x_1 \in A$ and $p(x_1, x_2, x_3) \in A$. And the value $p(x_1, x_2, x_3)$ is uniquely determined by x_1 and $l_1(x_2, x_3)$. We say that a triangle with a coding function assigned to each corner is 'latin' if each of the three coding functions can be calculated from the two other functions. For any solution of the network flow problem N_3 each of the following three triangles are latin:

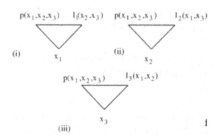

Fig. 13.

Letting $x_1 = 1$ in triangle (i) we notice that $p(1, x_2, x_3)$ can be calculated from $l_1(x_2, x_3)$ and conversely we notice that $l_1(x_2, x_3)$ can be calculated from $p(1, x_2, x_3)$. Thus we can replace the function $l_1(x_2, x_3)$ with the function $l_1(x_2, x_3) = p(1, x_2, x_3)$. Similarly, by letting $x_2 = 1$ in triangle (ii) and letting $x_3 = 1$ in triangle (iii) we obtain a solution with $l_2(x_1, x_3) = p(x_1, 1, x_3)$ and $l_3(x_1, x_2) = p(x_1, x_2, 1)$. This completes the proof.   ♣

Lemma 18

Assume that there is a solution to the flow problem N_3 with public information given by $p : A^3 \rightarrow A$. Then the latin function $p(x_1, x_2, x_3)$ determines (uniquely) two latin functions (i.e two latin squares) $l : A^2 \rightarrow A$ (l stands for 'left') and $r : A^2 \rightarrow A$ (r stands for 'right') defined by the two equations:

$$- p(1, l(x_1, x_2), x_3) = p(x_1, x_2, x_3)$$
$$- p(x_1, r(x_2, x_3), 1) = p(x_1, x_2, x_3)$$

Proof of Lemma 18: Certainly (since p is latin), there exist uniquely defined functions $l', r' : A^3 \rightarrow A$ such that $p(1, l'(x_1, x_2, x_3), x_3) = p(x_1, x_2, x_3)$ and $p(x_1, r'(x_1, x_2, x_3), 1) = p(x_1, x_2, x_3)$. To show Lemma 18 it suffices to show that l' is independent of x_3 and that r' is independent of x_1. Consider the two latin triangles:

Fig. 14.

In each triangle (iv) and (v) each coding function is uniquely determined by the two other coding functions in the triangle. Thus there exists f, g : $A^2 \to A$ such that $p(x_1, x_2, x_3) = f(p(x_1, x_2, 1), x_3)$ and such that $p(x_1, x_2, x_3) = g(x_1, p(1, x_2, x_3))$. Let $l(x_1, x_2) = l'(x_1, x_2, 1)$ and let $r(x_2, x_3) = r'(1, x_2, x_3)$ and notice that $p(x_1, x_2, 1) = p(1, l(x_1, x_2), 1)$ and $p(1, x_2, x_3) = p(1, r(x_2, x_3), 1)$. But then $p(x_1, x_2, x_3) = f(p(x_1, x_2, 1), x_3) = f(p(1, l(x_1.x_2), 1), x_3) = p(1, l(x_1, x_2), x_3)$ and $p(x_1, x_2, x_3) = g(x_1, p(1, x_2, x_3)) = g(x_1, p(1, r(x_2, x_3), 1) = p(x_1, r(x_2, x_3), 1)$. Thus l and r satisfies the same equations that uniquely determined l' and r' and thus $l'(x_1, x_2, x_3) = l(x_1, x_2)$ and $r'(x_1, x_2, x_3) = r(x_2, x_3)$. This completes the proof.   ♣

Lemma 19

Assume that $p : A^3 \to A$ has a solution and that $p(x_1, x_2, x_3) = p(1, l(x_1, x_2), x_3)$ and assume that $p(x_1, x_2, x_3) = p(x_1, r(x_2, x_3), 1)$. Then the functions $l, r : A^2 \to A$ satisfy the equation $r(l(x_1, x_2), x_3) = l(x_1, r(x_2, x_3))$.

Proof: Since p is latin and $p(x_1, x_2, x_3) = p(1, r(l(x_1, x_2), x_3), 1) = p(1, l(x_1, r(x_2, x_3)), 1)$.   ♣

The next three lemma are straight forward to prove.

Lemma 20

Assume $p(x_1, x_2, x_3)$ allows a solution and that $l(x_1, x_2)$ and $r(x_2, x_3)$ are defined such that $p(1, l(x_1, x_2), x_3) = p(x_1, x_2, x_3)$ and $p(x_1, r(x_2, x_3), 1) = p(x_1, x_2, x_3)$. Then for each pair $\pi_1, \pi_3 : A \to A$ of permutations $p'(x_1, x_2, x_3) = p(\pi_1 x_1, x_2, \pi_3 x_3)$ allows a solution and $l'(x_1, x_2) = l(\pi_1 x_1, x_2)$ and $r'(x_2, x_3) = r(x_2, \pi_3 x_3)$ satisfies the equations

$$p'(1, l'(x_1, x_2), x_3) = p'(x_1, x_2, x_3) \text{ and } p'(x_1, r'(x_2, x_3), 1) = p'(x_1, x_2, x_3).$$

Lemma 21

There exists permutations $\pi_1, \pi_3 : A \to A$ such that $l(\pi_1 x_1, 1) = x_1$ and such that $r(1, \pi_3 x_3) = x_3$.

Lemma 22

If $p(x_1, x_2, x_3)$ is a solution, there is another solution $p'(x_1, x_2, x_3) = p(\pi_1 x_1, x_2, \pi_3 x_3)$ such that the two functions $l'(x_1, x_2)$ and $r'(x_2, x_3)$ that satisfy the equations $p'(1, l'(x_1, x_2), x_3) = p'(x_1, x_2, x_3)$, $p'(x_1, r'(x_2, x_3), 1) = p'(x_1, x_2, x_3)$ as well as $l'(x_1, 1) = x_1$ and $r'(1, x_3) = x_3$.

Without loss of generality (possibly after having replaced x_1 and x_3 by $\pi_1 x_1$ and $\pi_3 x_3$) we can assume that we are given a latin function $p(x_1, x_2, x_3)$ and two latin functions $l(x_1, x_2)$ and $r(x_2, x_3)$ that satisfies $l(x_1, 1) = x_1$, $r(1, x_3) = x_3$, and have $l(x_1, r(x_2, x_3)) = r(l(x_1, x_2), x_3)$ for all $x_1, x_2, x_3 \in A$. But, then $r(x_1, x_3) = r(l(x_1, 1), x_3) = l(x_1, r(1, x_3)) = l(x_1, x_3)$ and thus $l = r$. But then l is transitive i.e. $l(x_1, l(x_2, x_3)) = l(l(x_1, x_2), x_3)$. Furthermore since $l(x, 1) = x$ and $l(1, x) = r(l, x) = x$ we notice that l defines a group operation on A. Thus we have shown that for any function $p(x_1, x_2, x_3)$ that allows a solution to the network flow problem N_3, there exist permutations $\pi_1, \pi_3 : A \to A$ such that if we let $p'(x_1, x_2, x_3) = p(\pi_1 x_1, x_2, \pi_3 x_3)$ then there is a group structure $*$ on A such that $p'(x_1, x_2, x_3) = p'(1, x_1 * x_2 * x_3, 1)$ for all x_1, x_2, x_3. But then there is a permutation $\pi : A \to A$ such that if we let $p''(x_1, x_2, x_3) = \pi(p'(x_1, x_2, x_3))$ then $p''(1, b, 1) = b$ for all $b \in A$. Notice, that $p''(x_1, x_2, x_3) = \pi(p'(x_1, x_2, x_3)) = \pi(p'(1, x_1 * x_2 * x_3, 1)) = p''(1, x_1 * x_2 * x_3, 1) = x_1 * x_2 * x_3$. This shows:

Lemma 23

Let $p : A^3 \to A$ be the public information in the network N_3. Then, if there is a solution to the network flow problem N_3, there exists a group composition $*$ on A such that 'essentially' $p(x_1, x_2, x_3) = x_1 * x_2 * x_3$ (modulo the application of suitable permutations to x_1, x_3 and p (or x_2)).

Lemma 24

Let $(A, *)$ be a group and let $p(x_1, x_2, x_3) = x_1 * x_2 * x_3$. Then the flow problem N_3 has a solution if and only if $(A, *)$ is a commutative group.

Proof: Assume that $p(x_1, x_2, x_3) = x_1 * x_2 * x_3$ (or just $x_1 x_2 x_3$ for short) allows a solution. Then we have the following 'derivation' from latin triangle with coding functions $p(a, b, c) = abc$, $p(a, 1, c) = ac$ and b.

Fig. 15.

Figure 15, represents the fact that b can be uniquely determined from abc and ac. But, then given $c^{-1}bc$ and ac we can calculate $abc = (ac)c^{-1}bc$ and thus we can determine b. Now ac can take any value (depending on a) and thus this equation is useless in calculating b. This shows that b is uniquely determined from $c^{-1}bc$. The expression $c^{-1}bc$ must be independent of c and thus $c^{-1}bc = 1^{-1}b1 = b$. But, then $bc = cb$ for all $a, b, c \in A$ which shows that the group $(A, *)$ must be a commutative group. The converse is rather obvious, since if $(A, *)$ is an ablean group and $p(x_1, x_2, x_3) = x_1 x_2 x_3$, we get a solution by letting $l_1(x_1, x_2) = x_1 x_2$, $l_2(x_1, x_3) = x_1 x_3$ and $l_3(x_1, x_2) = x_1 x_2$. This completes the proof of Lemma 19 which in turn clearly implies the theorem for N_3. This in turn easily implies the validity of Theorem 2 for general N_n with $n \geq 3$.   ♣

References

1. S. Medard, R.Koetter, M. Acedanski, and S. Deb, How good is random linear coding based distribued network storage? to appear.
2. A. Aggarwal and J.S. Vitter, The input/output complexity of sorting and related problems, Communications of the ACM, 31, 9, 1116–1126, 1988.
3. R Ahlswede, N Cai,S.Y.R. Li, and R.Yeung, Network information flow, IEEE Trans. Inf. Theory, Vo. 46, No. 4, 1204-1216, 2000.

4. K.R. Bhattad and K. Narayanan, Weakly secure network coding, to appear.

5. N. Cai and R.W. Yeung, Network coding and error correction, ITW Bangalore, 119–122, 2002.

6. S. Deb, C. Choute, M. Medard, and R. Koetter, Data harvesting: A random coding approach to rapid dissemination and efficient storage of data, INFOCOM, 2005, submitted.

7. R. Dougherty, C. Freiling, and K. Zeger, Insufficiency of linear coding in network information flow, Technical report, February 2004.

8. R. Dougherty, C. Freiling, and K. Zeger, Linearity and solvability in multicast networks, Proceeding of CISS, 2004.

9. R. Dougherty, C. Freiling, and K. Zeger, Network routing capacity, IEEE/ACM TRANSACTIONS ON NETWORKING, October 2004, submitted.

10. C. Fragouli and E. Soljanin, A connection between network coding and convolutional codes, IEEE International Conference on Communications, 2004.

11. T. Ho, M. Medard, and R. Koetter, An information theoretic view of network management, Proceeding of the 2003 IEEE Infocom, 2003.

12. R. Koetter and M. Medard. An algebraic approach to network coding, Proceedings of the 2001 IEEE International Symposium on Information Theory, 2001.

13. R. Koetter and M. Medard, Beyond routing: an algebraic approach to network coding, Proceedings of the 2002 IEEE Infocom, 2002.

14. S.-Y.R. Li, R.W.Yeung, and N. Cai, Linear network codes, IEEE Trans. Inform. Theory, Vol. 49 ,371-381, 2003.

15. K. Rabaey, J. Petrovic, and D. Ramchandran, Overcomming untuned radios in wireless networks with network coding, to appear.

16. L.G. Pippenger and N. Valiant, Shifting graphs and their applications, JACM, 23, 423–432, 1976.

17. S. Riis, Linear versus non-linear boolean functions in network flow, Proceeding of CISS, 2004.

18. S. Riis and R. Ahlswede, Problems in network coding and error correcting codes, NetCod 2005.

19. M. Thorup and S. Riis, Personal communication, 1997.

20. L. Valiant, Graph-theoretic arguments in low-level complexity LNCS, Springer Verlag, No. 53, 162-176, 1997.

21. L. Valiant, On non-linear lower bounds in computational complexity, Proc. 7th ACM Symp. on Theory of Computing, 45–53, 1975.

22. L. Valiant, Why is boolean circuit complexity theory difficult? In M.S. Patterson, editor, Springer Lecture Series, 84–94, 1992.

23. C. Boudec J.Y. Widmer, and J. Fragouli, Low-complexity energy-efficient broadcasting in wireless ad-hoc networks using network coding, to appear.

24. Y. Wu, P.A. Chou, and S.Y. Kung, Information exchange in wireless networks with network coding and physical-layer broadcast, Technical Report MSR-TR-2004-78, Microsoft Technical Report, Aug. 2004.

25. R. Yeung and Z. Zhang, Distributed source coding for satellite communications, IEEE Trans. Inform. Theory, Vol. 45, 1111–1120, 1999.

IX

On the Thinnest Coverings of Spheres and Ellipsoids with Balls in Hamming and Euclidean Spaces

I. Dumer*, M.S. Pinsker**, and V.V. Prelov***

Abstract. In this paper, we present some new results on the thinnest coverings that can be obtained in Hamming or Euclidean spaces if spheres and ellipsoids are covered with balls of some radius ε. In particular, we tighten the bounds currently known for the ε-entropy of Hamming spheres of an arbitrary radius r. New bounds for the ε-entropy of Hamming balls are also derived. If both parameters ε and r are linear in dimension n, then the upper bounds exceed the lower ones by an additive term of order $\log n$. We also present the uniform bounds valid for all values of ε and r.

In the second part of the paper, new sufficient conditions are obtained, which allow one to verify the validity of the asymptotic formula for the size of an ellipsoid in a Hamming space. Finally, we survey recent results concerning coverings of ellipsoids in Hamming and Euclidean spaces.

1 Introduction

Let $\mathbb{E}^n$ be the Hamming space of binary vectors $x = (x_1, \ldots, x_n)$ of length n. Given an integer ε, let $B^n(y, \varepsilon)$ be the ball of radius ε centered at the point $y \in \mathbb{E}^n$, i.e.,

$$B^n(y, \varepsilon) := \{x \in \mathbb{E}^n \mid d(x, y) \le \varepsilon\},$$

where $d(x, y)$ is the Hamming distance between x and y, that is the number of coordinate positions in which $x_i \ne y_i$. We say that a subset $M_\varepsilon(A) \subseteq \mathbb{E}^n$ forms an ε-covering of a set $A \subseteq \mathbb{E}^n$ if A belongs to the union of the balls $B^n(y, \varepsilon)$ centered at points $y \in M_\varepsilon(A)$, i.e.,

$$A \subseteq \bigcup_{y \in M_\varepsilon(A)} B^n(y, \varepsilon).$$

The ε-*entropy* [1] $H_\varepsilon(A)$ of a set A is the logarithm of the size of its minimal ε-covering:

$$H_\varepsilon(A) := \log \min |M_\varepsilon(A)|,$$

---

* Supported in part by NSF grants CCF-0622242 and CCF-0635339.

** Supported in part by the Russian Foundation for Basic Research (project no. 03–01–00592) and INTAS (project 00-738).

*** Supported in part by ZIF (project "General Theory of Information Transfer and Combinatorics"), the Russian Foundation for Basic Research (project no. 03-01-00592) and INTAS (project 00-738).

R. Ahlswede et al. (Eds.): Information Transfer and Combinatorics, LNCS 4123, pp. 898–925, 2006.
© Springer-Verlag Berlin Heidelberg 2006

where the minimum is taken over all ε-coverings $M_\varepsilon(A)$ and log denotes logarithm base 2.

Given a vector $v = (v_1, \ldots, v_n)$, $v_i \in [0, \infty)$, $i = 1, \ldots, n$, the ellipsoid E_v^n is defined by the equality

$$E_v^n := \left\{ x \in \mathbb{E}^n \mid \sum_{i=1}^n v_i x_i \le 1 \right\}, \tag{1}$$

where all operations are performed over real numbers. Note that the inequality $\sum_{i=1}^n v_i x_i \le 1$ is equivalent to the inequality $\sum_{i=1}^n v_i x_i^2 \le 1$. This fact explains the notion of ellipsoid for E_v^n by analogy with that in the Euclidean space.

Below we study the asymptotic behavior (as $n \to \infty$) of the ε-entropy of an arbitrary ellipsoid E_v^n and, in particular, the ε-entropy of a ball $B^n(r) := B^n(0, r)$ and a sphere

$$S^n(r) := \{ x \in \mathbb{E}^n \mid d(x, 0) = r \},$$

where r is an integer.

The following notation will be used throughout the paper. Consider the binary entropy function

$$h(t) := -t \log t - (1 - t) \log(1 - t), \quad 0 \le t \le 1.$$

Then, given any integer n and real-valued vectors $P = P(n) = (p_1, \ldots, p_n)$ and $Q = Q(n) = (q_1, \ldots, q_n)$ such that $0 \le p_i \le 1$, $0 \le q_i \le 1$, $i = 1, \ldots, n$, define the function

$$h(P, Q) := \sum_{i=1}^n h(p_i, q_i) = h(P) - h(Q),$$

where

$$h(p_i, q_i) := h(p_i) - h(q_i), \quad h(P) := \sum_{i=1}^n h(p_i), \quad h(Q) := \sum_{i=1}^n h(q_i).$$

Here and below, we sometimes omit an argument n or a superscript n in our notation of vectors, ellipsoids, balls, and spheres.

In this paper, we also consider an arbitrary ellipsoid $E_a = E_a^n$ defined in the n-dimensional Euclidean space $\mathbb{R}^n$ as

$$E_a^n := \left\{ x = (x_1, \ldots, x_n) \in \mathbb{R}^n \mid \sum_{i=1}^n \frac{x_i^2}{a_i^2} \le 1, \right\} \tag{2}$$

where $a = (a_1, \ldots, a_n)$ is a real-valued vector with n positive coordinates. We will use the same notation $B^n(y, \varepsilon)$ and $S^n(r)$ for balls and spheres in the Euclidean space $\mathbb{R}^n$ as the one used in the Hamming space $\mathbb{E}^n$. The definitions of $B^n(y, \varepsilon)$, $S^n(r)$ and the ε-entropy $H_\varepsilon(A)$ of a bounded set $A \subset \mathbb{R}^n$ in the

Euclidean space are absolutely similar to those in the Hamming space. The only difference arises in the definition of the function $d(x, y)$: now $d(x, y)$ is the
Euclidean distance between real-valued vectors x and y. Also, in the Euclidean space, the radii r, ε, and many other parameters can take arbitrary positive
values instead of integers used in the Hamming space. In both spaces we are
interested in the asymptotic ε-entropy of any ellipsoid E_v^n or E_a^n

Note that ellipsoids in the Hamming spaces arise in various problems related to
combinatorics, decoding, and data compression. For example, ellipsoids in the
form of (1) can be considered as Boolean threshold functions. Also, ellipsoids
emerge in maximum likelihood decoding for binary memoryless channels. On
the other hand, the problem of coverings of spheres and ellipsoids in Euclidean
spaces often arises in vector quantizers.

In the next section, we better the bounds recently obtained for the ε-entropy
of spheres in the Hamming space. We also extend these results for balls and
derive new uniform bounds. In Section 3, we shortly survey a few results already
known for the size and the ε-entropy of ellipsoids in the Hamming space. There
we also derive some new sufficient conditions for the validity of the asymptotic
formula for the size of an ellipsoid. Section 4 is devoted to recent results obtained
for the ε-entropy of ellipsoids in Euclidean spaces.

2   Covering of Spheres and Balls in Hamming Spaces

Below we assume in this section that n, r, and ε are some positive integers such
that the ratios
$$\rho := r/n, \quad \sigma := \varepsilon/n$$
satisfy condition
$$0 < \sigma < \rho \le 1/2.$$

Firstly, note that the whole space $\mathbb{E}^n$ can be considered as a special case of a ball
$B^n(r)$ when $r = n$. It is shown (see, e.g., [2,3]) that the normalized ε-entropy
$\overline{H}_\varepsilon(\mathbb{E}^n) := H_\varepsilon(\mathbb{E}^n)/n$ satisfies equality

$$\overline{H}_\varepsilon(\mathbb{E}^n) = 1 - h(\sigma) + O\left(\frac{\log n}{n}\right), \quad n \to \infty,$$

for any fixed σ. Non-uniform lower and upper bounds on the ε-entropy of spheres
in the Hamming space are obtained in [4]. It is proven there that for any fixed $\rho <
1/2$ and σ there exist some constants $c(\rho, \sigma)$ and $C(\rho, \sigma)$, which are independent
of n but can depend on ρ and σ such that the normalized ε-entropy $\overline{H}_\varepsilon(S(r))$
satisfies the following inequalities:

$$h(\rho) - h(\sigma) + \frac{c(\rho, \sigma)}{n} \le \overline{H}_\varepsilon(S(r)) \le h(\rho) - h(\sigma) + \frac{3 \log n}{2n} + \frac{C(\rho, \sigma)}{n}.$$

The following theorem improves this bound and also gives the *uniform* (with
respect to parameters ρ and σ) lower and upper bounds on the ε-entropy of

spheres in the Hamming space. Such bounds are of independent interest, and can be used, for example, in deriving the upper bounds for the ε-entropy of ellipsoids.

Theorem 1

1. For all n, $r \leq n/2$, and $\varepsilon < r$, the normalized ε-entropy of an n-dimensional sphere $S(r)$ satisfies the following lower and upper bounds:

$$h(\rho) - h(\sigma) - \frac{\log(2n)}{2n} \leq \overline{H}_\varepsilon(S(r)) \leq h(\rho) - h(\sigma) + \frac{3\log n}{2n} + \frac{c}{n}, \qquad (3)$$

where c is an absolute constant. Moreover, the lower bound in (3) can be improved if σ or both σ and ρ are constants independent of n. Namely,

2. If σ is a constant, then there exists a constant $c_1(\sigma)$ such that

$$\overline{H}_\varepsilon(S(r)) \geq h(\rho) - h(\sigma) + \frac{c_1(\sigma)}{n}. \qquad (4)$$

3. If both ρ and σ are constants, then there exists a constant $c_2(\rho, \sigma)$ such that

$$\overline{H}_\varepsilon(S(r)) \geq h(\rho) - h(\sigma) + \frac{\log n}{2n} + \frac{c_2(\rho, \sigma)}{n}. \qquad (5)$$

Proof. This theorem will be proven using Proposition 1 given in the sequel. However, the lower bounds in (3) and (4) are almost trivial. Indeed, we first use the following well-known inequalities (see, e.g., (10.16) and (10.20) in [5]):

$$(8n\rho(1-\rho))^{-1/2} 2^{nh(\rho)} \leq |S(r)| \leq (2\pi n\rho(1-\rho))^{-1/2} 2^{nh(\rho)} \qquad (6)$$

and

$$|B(\varepsilon)| \leq 2^{nh(\sigma)}. \qquad (7)$$

Applying (6) and (7) we obtain the lower (packing) bound

$$\overline{H}_\varepsilon(S(r)) \geq \frac{1}{n} \log \frac{|S(r)|}{|B(\varepsilon)|} \geq h(\rho) - h(\sigma) -$$

$$\frac{\log 8n\rho(1-\rho)}{2n} \geq h(\rho) - h(\sigma) - \frac{\log(2n)}{2n},$$

which gives the left inequality in (3). Similarly, note that for $\sigma < 1/2$, we have

$$\frac{|B(\varepsilon)|}{|S(\varepsilon)|} \leq \sum_{i=0}^{\varepsilon} \left(\frac{\sigma}{1-\sigma}\right)^i \leq \frac{1-\sigma}{1-2\sigma}, \qquad (8)$$

which follows from the inequality

$$|S(\varepsilon - 1)|/|S(\varepsilon)| \leq \sigma/(1-\sigma).$$

Therefore, using (6) and (8), we obtain inequality (4):

$$\overline{H}_\varepsilon(S(r)) \geq \frac{1}{n}\log\frac{|S(r)|}{|B(\varepsilon)|} = \frac{1}{n}\log\frac{|S(r)|}{|S(\varepsilon)|} - \frac{1}{n}\log\frac{|B(\varepsilon)|}{|S(\varepsilon)|}$$

$$\geq h(\rho) - h(\sigma) - \frac{\log(2n)}{2n} + \frac{\log(2\pi n\sigma(1-\sigma))}{2n} - \frac{1}{n}\log\frac{1-\sigma}{1-2\sigma}.$$

To proceed with the other bounds of Theorem 1, we shall extensively use the following notation. Given any integer $w \in [r - \varepsilon, r + \varepsilon]$, we consider a point $x \in S(w)$ and the subset

$$A^r(x, \varepsilon) := B(x, \varepsilon) \cap S(r).$$

Given any $y \in S(r)$, we also consider the set

$$A^w(y, \varepsilon) := B(y, \varepsilon) \cap S(w).$$

Now we can prove that $\overline{H}_\varepsilon(S(r))$ can be tightly related to the function

$$\Omega(r, \varepsilon) := \max_w |A^r(x, \varepsilon)|, \quad x \in S(w). \tag{9}$$

Namely, we shall prove that the ε-entropy of the sphere $S(r)$ satisfies the bounds

$$\frac{1}{n}\log\frac{|S(r)|}{\Omega(r,\varepsilon)} \leq \overline{H}_\varepsilon(S(r)) \leq \frac{1}{n}\log\frac{|S(r)|}{\Omega(r,\varepsilon)} + \frac{\log(n\ln 2)}{n} + \frac{\log 2h(\rho)}{n}. \tag{10}$$

1. The lower bound in (10) is based on a standard packing argument. Here we use the fact that each ball $B(x, \varepsilon)$ can cover at most $\Omega(r, \varepsilon)$ points on $S(r)$.

2. The upper bound is obtained using random coverings. Given a fixed integer N, we choose the radius w, for which equality (9) holds. Then we perform N trials choosing the centers x of an ε-covering independently and uniformly on $S(w)$. Then any point $y \in S(r)$ is covered in one trial with the same probability

$$\gamma = \frac{|A^w(y, \varepsilon)|}{|S(w)|}. \tag{11}$$

Now we employ a straightforward argument (used by Bassalygo in 1965; see also Lemma 6 in [4]) that states that for any $x \in S(w)$ and any $y \in S(r)$, the two sets $A^r(x, \varepsilon)$ and $A^w(y, \varepsilon)$ cover the same fraction of spheres $S(r)$ and $S(w)$, respectively. Therefore,

$$\gamma = \frac{|A^r(x, \varepsilon)|}{|S(r)|}. \tag{12}$$

Now note that any $y \in S(r)$ is not covered in N trials with probability $P_N = (1 - \gamma)^N$. Choosing

$$N = \frac{1 + \ln|S(r)|}{\gamma}, \tag{13}$$

we obtain $P_N \leq (e|S(r)|)^{-1}$. Therefore the whole sphere $S(r)$ is covered in N trials with a probability $\overline{P}_N(S(r)) \geq 1 - P_N \cdot |S(r)| \geq 1 - e^{-1}$. In this case, there exists an ε-covering of size N. Using (12), we obtain the upper bound

$$\overline{H}_\varepsilon(S(r)) \leq \frac{\log N}{n} = \frac{1}{n} \log \frac{|S(r)|}{\Omega(r,\varepsilon)} + \frac{1}{n} \log(1 + \ln|S(r)|). \tag{14}$$

Now the upper bound in (10) follows from (6).

Thus, we see that to prove Theorem 1 we need to derive rather tight lower and upper bounds on the quantity $\Omega(r,\varepsilon)$. These bounds - derived in the sequel in Proposition 1 - conclude the proof of Theorem 1. In fact, slightly more precise calculations show that the universal constant c belongs to the interval $(0,2)$. □

Proposition 1

1. For all n, $r \leq n/2$, and $\varepsilon < r$, the following lower and upper bounds hold:

$$nh(\sigma) - \log 8n \leq \log \Omega(r,\varepsilon) \leq nh(\sigma). \tag{15}$$

Moreover,

2. If σ is a constant, then there exists a constant $c(\sigma)$ such that

$$\log \Omega(r,\varepsilon) \leq nh(\sigma) - (\log n)/2 + c(\sigma).$$

3. If ρ and σ are constants, then there exists a constant $c(\rho,\sigma)$ such that

$$\log \Omega(r,\varepsilon) \leq nh(\sigma) - \log n + c(\rho,\sigma).$$

Proposition 1 will be proven in the Appendix.

In the following theorem, we present the corresponding non-uniform and uniform lower and upper bounds for the ε-entropy of balls in the Hamming space.

Theorem 2

1. For all n, $r \leq n/2$, and $\varepsilon < r$, the normalized ε-entropy of an n-dimensional ball $B(r)$ satisfies the following lower and upper bounds:

$$h(\rho) - h(\sigma) - \frac{\log(2n)}{2n} \leq \overline{H}_\varepsilon(B(r)) \leq h(\rho) - h(\sigma) + \frac{2\log n}{n} + \frac{C}{n}, \tag{16}$$

where C is an absolute constant. Moreover,

2. If σ is a constant, then there exists a constant $C_1(\sigma)$ such that

$$\overline{H}_\varepsilon(B(r)) \geq h(\rho) - h(\sigma) + \frac{C_1(\sigma)}{n}. \tag{17}$$

3. If both $\rho < 1/2$ and σ are constants, then there exists a constant $C_2(\rho,\sigma)$ such that

$$h(\rho) - h(\sigma) + \frac{\log n}{2n} + \frac{C_2(\rho,\sigma)}{n} \leq \overline{H}_\varepsilon(B(r)) \leq h(\rho) - h(\sigma) + \frac{3\log n}{2n} + \frac{C}{n}. \tag{18}$$

Proof. Note first that the lower bounds in (16) - (18) immediately follow from the corresponding lower bounds in (3) - (5) since $S(r) \subset B(r)$ for all r.[1]

Before starting to prove the upper bounds in (16)–(18), note that for all r and ε we have

$$H_\varepsilon(B(r)) \leq H_\varepsilon(S(r)) + \log n \qquad (19)$$

since $B(r) = \bigcup_{k=0}^{r} S(k)$ and $H_\varepsilon(S(k)) \leq H_\varepsilon(S(r))$ for all $k = 0, \ldots, r$. Therefore, to derive some upper bounds for $H_\varepsilon(B(r))$ we can use (19) together with the upper bounds in (3) and (4). By doing this, we obtain the upper bounds which are worse than those in (16)–(18). The latter means that the trivial inequality (19) does not allow us to prove the upper bounds in (16)–(18).

In order to prove these upper bounds, let us use the following modification of the random choice method described in the proof of Theorem 1. Namely, let an integer N be fixed. Then we choose N points $x^{(1)}, \ldots, x^{(N)}$ of an ε-covering for the ball $B(r)$ randomly according to the following rule. In the i-th trial, $i = 1, \ldots, N$, independently of all other trials, we choose an integer u, $\varepsilon + 1 \leq u \leq r$, at random with probability

$$p(u) = \frac{|S(u)|}{|B(r)| - |B(\varepsilon)|}, \qquad \varepsilon + 1 \leq u \leq r.$$

Given u, the i-th point $x^{(i)}$ of the ε-covering is chosen randomly on the sphere $S(u')$, where

$$u' := \left\lfloor \frac{u - \varepsilon}{1 - 2\sigma} \right\rfloor.$$

Similarly to the proof of Theorem 1 (see also formula (89) in the Appendix), here we also use the fact that our choice of u' maximizes the covering area $A^u(x, \varepsilon)$ generated by the ball $B(x, \varepsilon)$ with any center $x \in S(u')$, so that

$$|A^u(x, \varepsilon)| = \Omega(u, \varepsilon).$$

Now we estimate the probability γ_y that any fixed point $y \in S(u)$ is covered by a ball $B(x^{(i)}, \varepsilon)$. First, note that similarly to equalities (11) and (12), for any $x \in S(u')$ and $y \in S(u)$ we have the equality

$$\frac{\left|A^{u'}(y, \varepsilon)\right|}{|S(u')|} = \frac{|A^u(x, \varepsilon)|}{|S(u)|}.$$

Then we use inequality (15) for $\Omega(u, \varepsilon)$ and see that any point $y \in S(u)$ is covered by a ball $B(x^{(i)}, \varepsilon)$ with probability

$$\gamma_y \geq \frac{|S(u)|}{|B(r)| - |B(\varepsilon)|} \cdot \frac{\Omega(u, \varepsilon)}{|S(u)|}$$

$$\geq \gamma_r := \frac{2^{n h(\sigma) - \log 8n}}{|B(r)| - |B(\varepsilon)|}. \qquad (20)$$

---

[1] Note that the size of the covering is always bounded from below by $|B(r)|/|B(\varepsilon)|$. This fact allows one to remove the term $-(\log n)/2n$ from the lower bound (16). We omit the corresponding calculations.

It is important that for any fixed point $y \in B(r)$, the probability γ_y of its covering in one trial is bounded from below in (20) by a quantity γ_r that does not depend on y. The rest of the proof of the theorem almost coincides with that of Theorem 1 and therefore we omit it here. The main difference arises from the fact that now we use a bigger set $B(r)$ in (20) instead of $S(r)$ employed in (12). Note also that now we take

$$N = \frac{1}{\gamma_r}(1 + \ln|B(r)|),$$

instead of (13), but this change does not affect the asymptotics of $\overline{H}_\varepsilon(B(r))$. Finally, we apply the inequalities

$$|B(r)| \le \frac{1-\rho}{1-2\rho}|S(r)|$$

(cf. (8)) in the proof of (18) and

$$|B(r)| \le 2^{nh(\rho)}$$

in the proof of (16) and (17). This completes the proof of Theorem 2.   □

The main conclusion we derive from Theorems 1 and 2 is that for both spheres and balls with fixed relative radii ρ and ε, the corresponding upper and lower bounds differ only by an additive term of order $\log n$. Note that this term is related to our random-covering algorithm used for upper bounds. Thus, any further tightening of the above bounds is only possible if there exist constructive coverings that surpass their randomly chosen counterparts.

3   Covering of Ellipsoids in Hamming Spaces

3.1   The Size of Ellipsoids

Consider an ellipsoid $E_v = E_v^n$ defined in (1). The first interesting question is the problem of finding the number of different ellipsoids in $\mathbb{E}^n$. It was proven long ago [6] that this number is upper bounded by 2^{n^2}. The problem had been addressed in many publications for over 140 years until Zuev proved [7] that the number of different ellipsoids is lower bounded by $2^{n^2(1-10/\ln n)}$ thus solving the problem up to the exponential order of 2^{n^2}.

Another important problem is to derive the size of any ellipsoid E_v. This quantity $|E_v|$ is necessary, for example, to write out the packing (Hamming) bound for the ε-entropy of E_v. The main term of the asymptotics of $\log|E_v|$ was found by Pinsker [8]. To state his result, let us consider the vector $P^* = P^*(n) = (p_1^*, \dots, p_n^*)$ with components

$$p_i^* = p_i^*(n) := \left(1 + 2^{\lambda v_i}\right)^{-1}, \quad i = 1, \dots, n, \tag{21}$$

where parameter $\lambda = \lambda(n)$ is defined by the equalities

$$
\begin{cases}
\sum_{i=1}^{n} v_i \left(1 + 2^{\lambda v_i}\right)^{-1} = 1 & if \quad \frac{1}{2}\sum_{i=1}^{n} v_i > 1, \\
\lambda = 0 & if \quad \frac{1}{2}\sum_{i=1}^{n} v_i \leq 1.
\end{cases}
\tag{22}
$$

Let

$$
\mathcal{H}_n := h(P^*(n)). \tag{23}
$$

Then the following statement holds.

Theorem 3. *[8]. If*

$$
\lim_{n\to\infty} \frac{\mathcal{H}_n}{\log n} = \infty, \tag{24}
$$

then

$$
\mathcal{H}_n(1 + o(1)) \leq \log|E_v| \leq \mathcal{H}_n, \quad n \to \infty, \tag{25}
$$

and, in particular,

$$
\log|E_v| = \mathcal{H}_n(1 + o(1)), \quad n \to \infty. \tag{26}
$$

The proof of the upper bound in (25) given in [8]) (see Lemma 1 there) is rather simple and short and therefore for reader's convenience we reproduce it here. Indeed, given the uniform distribution $\Pr(x) = 2^{-n}$, $x \in \mathbb{E}^n$, we have

$$
\log|E_v| = \log \Pr(E_v) + n = \log \Pr\left\{\sum_{i=1}^{n} v_i X_i \leq 1\right\} + n, \tag{27}
$$

where X_i, $i = 1, \ldots, n$, are independent binary random variables taking values $0, 1$ with probabilities $\Pr(0) = \Pr(1) = 1/2$. It is clear that if $\sum_{i=1}^{n} v_i/2 > 1$, then

$$
\Pr\left\{\sum_{i=1}^{n} v_i X_i \leq 1\right\} = \Pr\left\{2^{-\lambda\sum_{i=1}^{n} v_i X_j} \geq 2^{-\lambda}\right\} \leq 2^{\lambda}\mathbf{E}\,2^{-\lambda\sum_{i=1}^{n} v_i X_i}
$$

$$
= 2^{\lambda}\prod_{i=1}^{n}\mathbf{E}\,2^{-\lambda v_i X_i} = 2^{\lambda}\prod_{i=1}^{n}\frac{1 + 2^{-\lambda v_i}}{2}. \tag{28}
$$

Therefore, formulas (27) and (28) imply

$$
\log|E_v| \leq \lambda + \sum_{i=1}^{n}\log\left(1 + 2^{-\lambda v_i}\right)
$$

$$
= \sum_{i=1}^{n}\left[\lambda v_i\left(1 + 2^{\lambda v_i}\right)^{-1} + \log\left(1 + 2^{-\lambda v_i}\right)\right]
$$

$$
= \sum_{i=1}^{n} h(p_i^*) = h(P^*) = \mathcal{H}_n.
$$

For the case $\sum_{i=1}^{n} v_i/2 \leq 1$, we have $\lambda = 0$ and $p_i^* = 1/2$, $i = 1, \ldots, n$, and therefore the right-hand side inequality in (25) is trivially fulfilled.

The proof of the lower bound in (25) is more involved. The idea is to construct a sequence of sets $A_{n,u} \subseteq E_v$, $n = 1, 2, \ldots$ that are the direct products of balls of a special radius u in the corresponding subspaces and then to evaluate the asymptotic behavior of the size of $A_{n,u}$. For details of the proof we refer to [8].

Without loss of generality, below we will assume that the coefficients v_i, $i = 1, \ldots, n$, of an ellipsoid E_v form a non-increasing sequence, i.e.,

$$v_1 \geq v_2 \geq \ldots \geq v_n. \tag{29}$$

This can always be attained by renumbering the elements of the sequence $\{v_i\}$. Further, we will consider two different cases:

(i) the whole sequence $\{v_i\}_{i=1}^{\infty}$ is given, i.e., v_i does not depend on n for any fixed $i = 1, 2, \ldots$;

(ii) "scheme of series": each $v_i = v_i(n)$ can depend on n but for any given n the elements $v_i(n)$, $i = 1, \ldots, n$, satisfy condition (29).

The following statement, for the case (i), gives a simple necessary and sufficient condition on coefficients $\{v_i\}$ under which the main condition (24) of Theorem 3 is fulfilled.

Proposition 2. *Given a non-increasing sequence $\{v_i\}_{i=1}^{\infty}$, condition (24) holds if and only if*

$$\lim_{n \to \infty} v_n = 0. \tag{30}$$

Corollary 1. *Given a non-increasing sequence $\{v_i\}_{i=1}^{\infty}$, Theorem 3 can be reformulated as follows: if $\lim_{n \to \infty} v_n = 0$, then inequalities (25) and equality (26) are valid.*

Proof. It can easily be seen that

$$\mathcal{H}_n = \max_P h(P), \tag{31}$$

where $\mathcal{H}_n$ is defined in (23) and the maximum in (31) is taken over all vectors $P = (p_1, \ldots, p_n)$ such that

$$0 \leq p_1, \ldots, p_n \leq 1/2, \quad \sum_{i=1}^{n} v_i p_i \leq 1.$$

We will use equality (31) below.

1. Assume first that $\lim_{n \to \infty} v_n \neq 0$, i.e., there exists a positive constant v such that $v_n \geq v > 0$ for all n. Then we have

$$\mathcal{H}_n = \max_P h(P) \leq \max_{P'} h(P'), \tag{32}$$

where the second maximum in (32) is taken over all vectors $P' = (p'_1, \ldots, p'_n)$ such that

$$0 \le p'_1, \ldots, p'_n \le 1/2, \quad \sum_{i=1}^{n} v p'_i \le 1.$$

It is clear that

$$\max_{P'} h(P') = nh\left(\frac{1}{nv}\right) = \frac{1}{v}(\log n)(1 + o(1)), \quad n \to \infty. \tag{33}$$

Relations (32) and (33) show that condition (24) is not satisfied if $\lim_{n \to \infty} v_n \ne 0$.

2. Assume now that $\lim_{n \to \infty} v_n = 0$. Let $k = k(n) = \alpha n$ where α, $0 < \alpha < 1$, is a fixed constant.[2] Then we clearly have

$$\mathcal{H}_n \ge \max_{\widehat{P}(k)} \sum_{i=n-k+1}^{n} h(\widehat{p}_i(k)) = \begin{cases} kh(1/(kv_{n-k+1})) & \text{if} \quad kv_{n-k+1} \ge 2, \\ k & \text{if} \quad kv_{n-k+1} \le 2, \end{cases} \tag{34}$$

where the maximum in (34) is taken over all k-dimensional vectors $\widehat{P}(k) = (\widehat{p}_{n-k+1}(k), \ldots, \widehat{p}_n(k))$ such that

$$0 \le \widehat{p}_{n-k+1}(k), \ldots, \widehat{p}_n(k) \le 1/2, \quad \sum_{i=n-k+1}^{n} v_{n-k+1}\widehat{p}_i(k) \le 1.$$

Let a rather large positive constant $A > 2$ be fixed. Consider first a subset $\mathbb{N}_1$ of integers n for which

$$kv_{n-k+1} \le A, \quad n \in \mathbb{N}_1.$$

For such $n \in \mathbb{N}_1$, inequality (34) shows that

$$\mathcal{H}_n \ge kh(1/A) = \alpha nh(1/A), \tag{35}$$

and (24) holds. On the other hand, for $n \in \mathbb{N}_2$ for which

$$kv_{n-k+1} > A, \quad n \in \mathbb{N}_2,$$

it follows from (34) that

$$\mathcal{H}_n \ge \frac{\log(kv_{n-k+1})}{v_{n-k+1}}. \tag{36}$$

Therefore, if

$$v_{n-k+1} \le \frac{1}{\log k}, \quad n \in \mathbb{N}_2,$$

then we have

$$\mathcal{H}_n \ge (\log A \cdot \log k), \quad n \in \mathbb{N}_2. \tag{37}$$

---

[2] More precisely, we should write $k = \lceil \alpha n \rceil$ instead of $k = \alpha n$ but this difference does not affect further asymptotic relations.

At the same time if

$$v_{n-k+1} \geq \frac{1}{k^c}, \quad n \in \mathbb{N}_2,$$

where c, $0 < c < 1$, is a constant, then we obtain from (36) that

$$\mathcal{H}_n \geq \frac{\log k - \log(1/v_{n-k+1})}{v_{n-k+1}} \geq \frac{(1-c)\log k}{v_{n-k+1}}, \quad n \in \mathbb{N}_2. \tag{38}$$

Hence, inequalities (37) and (38) imply that

$$\mathcal{H}_n \geq \min\left\{\log A, \frac{1-c}{v_{n-k+1}}\right\}(\log k), \quad n \in \mathbb{N}_2. \tag{39}$$

Finally, taking $A \to \infty$ rather slowly, we conclude from (35) and (39) that

$$\lim_{n\to\infty} \frac{\mathcal{H}_n}{\log n} = \infty.$$

Proposition 2 is proved. $\qquad\square$

Consider now the general case - "scheme of series" - where each $v_i = v_i(n)$ may depend on n but for any given n the sequence is non-increasing. For such a situation the following statement holds.

Proposition 3. *Assume that for any given n the sequence $v_i = v_i(n)$, $i = 1, \ldots, n$, does not increase. Then*

(a) *If $\lim\limits_{n\to\infty} v_n \neq 0$, then condition (24) is not fulfilled.*

(b) *If there exists a sequence $k = k(n)$ such that*

$$\begin{cases} v_{n-k+1} \to 0 & \text{as} \quad n \to \infty, \\ \liminf\limits_{n\to\infty} \frac{\log k}{\log n} > 0, \end{cases}$$

then condition (24) does hold.

Proof. The proof of part (a) does not differ from that of Proposition 2. To prove the second statement, note that in the proof of the direct part of Proposition 2 we have only used the facts that $\lim\limits_{n\to\infty} (\log k/\log n) > 0$ for subsequence $k = k(n)$ and that $v_{n-k+1} \to 0$ as $n \to \infty$ (which in turn follows from condition (30) used in Proposition 2). Now the two latter conditions are introduced in the formulation of Proposition 3. Therefore, taking into account this observation, we can claim that Proposition 3 holds. $\qquad\square$

Remark 2. It easily follows from inequality (34) (which also holds for the scheme of series) that if there exists a sequence $k = k(n)$ such that

$$\sum_{i=n-k+1}^{n} v_i \leq 2 \quad \text{and} \quad \frac{k}{\log n} \to \infty \quad \text{as} \quad n \to \infty$$

or

$$\sum_{i=n-k+1}^{n} v_i > 2 \quad \text{and} \quad \frac{kh\left(1/\sum_{i=n-k+1}^{n} v_i\right)}{\log n} \to \infty \quad \text{as} \quad n \to \infty,$$

then condition (24) is fulfilled. In particular, this holds if

$$\sum_{i=1}^{n} v_i \leq 2$$

or

$$\sum_{i=1}^{n} v_i > 2 \quad \text{and} \quad \frac{kh\left(1/\sum_{i=1}^{n} v_i\right)}{\log n} \to \infty \quad \text{as} \quad n \to \infty.$$

Remark 3. It should be mentioned that using some results known from the theory of limit theorems and, in particular, the theory of large deviations for the sums of independent but not-identically distributed random variables, the asymptotics of $\log|E_v|$ can be expressed in a different form. Moreover, in certain special cases, the next terms of the asymptotics of $\log|E_v|$ can also be found [9].

3.2   The ε-Entropy of Ellipsoids

Below in this subsection, we will always assume, as was already mention earlier, that the coefficients $v_i = v_i(n)$, $i = 1, \ldots, n$, do not increase for any given n and, moreover, that $\varepsilon = \varepsilon(n)$ is an integer such that $1 \leq \varepsilon < n/2$. Otherwise (i.e., if $\varepsilon \geq n/2$) the equality

$$H_\varepsilon(E_v^n) = O(\log n), \quad n \to \infty,$$

holds for arbitrary ellipsoids E_v^n since $E_v^n \subseteq \mathbb{E}^n$ and it is obvious that $H_\varepsilon(\mathbb{E}^n) = O(\log n)$ (cf. also (2)).

Lower bound
Using relation (26) for the size of ellipsoid E_v and inequality (7) for the size of the ball $B(\varepsilon)$, we can easily write out the packing (Hamming) bound for the ε-entropy $H_\varepsilon(E_v^n)$:

$$H_\varepsilon(E_v^n) \geq \log^+ \frac{|E_v|}{|B(\varepsilon)|} \geq [\mathcal{H}_n - nh(\varepsilon/n)]^+ (1 + o(1)), \quad n \to \infty, \qquad (40)$$

if condition (24) holds. Here and throughout the rest of the paper we use notation $u^+ := \max\{u, 0\}$. It is clear that, in general, this lower bound is not asymptotically tight, and we will discuss this fact later.

Let

$$\widehat{\mathcal{H}}_k = h(\widehat{P}^*(k)), \quad k = 1, \ldots, n, \qquad (41)$$

where $\widehat{P}^*(k) = (\widehat{p}^*_{n-k+1}, \ldots, \widehat{p}^*_n)$ is defined similar to $P^*(n)$ (cf. (21) and (22)):

$$\widehat{p}^*_i = \widehat{p}^*_i(k) := \left(1 + 2^{\widehat{\lambda}v_i}\right)^{-1}, \quad i = n - k + 1, \ldots, n, \tag{42}$$

and parameter $\widehat{\lambda} = \widehat{\lambda}(k)$ is defined by the equalities

$$\begin{cases} \sum_{i=n-k+1}^n v_i \left(1 + 2^{\widehat{\lambda}v_i}\right)^{-1} = 1 & if \quad \frac{1}{2}\sum_{i=n-k+1}^n v_i > 1, \\ \widehat{\lambda} = 0 & if \quad \frac{1}{2}\sum_{i=n-k+1}^n v_i \le 1. \end{cases} \tag{43}$$

In particular, comparing relations (21)–(23) and (41)–(43), we observe that

$$\widehat{\lambda}(n) = \lambda(n), \quad \widehat{P}^*(n)P^*(n), \quad \widehat{\mathcal{H}}_n = \mathcal{H}_n.$$

Now, define the quantity

$$R_n(\varepsilon) := \max_{\{k:\, 2\varepsilon < k \le n\}} \left[\widehat{\mathcal{H}}_k - kh(\varepsilon/k)\right]. \tag{44}$$

Then the following *generalized packing (Hamming) bound* holds ([4,10]).

Theorem 4. *Let condition (24) be satisfied. Then*

$$H_\varepsilon(E_v^n) \ge R_n^+(\varepsilon) + o(\mathcal{H}_n), \quad n \to \infty. \tag{45}$$

In particular, if [3]

$$R_n(\varepsilon) \asymp \mathcal{H}_n, \quad n \to \infty, \tag{46}$$

then

$$H_\varepsilon(E_v^n) \ge R_n(\varepsilon)(1 + o(1)), \quad n \to \infty. \tag{47}$$

Proof. The proof of inequality (45) is very simple. Indeed, consider the projection of ellipsoid E_v^n into the subspace $\widehat{\mathbb{E}}^k$ spanned over the last k coordinates. Then we obtain the sub-ellipsoid

$$\widehat{E}_v^k := \left\{ (x_{n-k+1}, \ldots, x_n) \mid \sum_{i=n-k+1}^n v_i x_i \le 1 \right\}.$$

Also, after such a projection the ball $B^n(\varepsilon)$ becomes $B^k(\varepsilon)$. Therefore,

$$H_\varepsilon(E_v^n) \ge \max_{2\varepsilon < k \le n} H_\varepsilon(\widehat{E}_v^k) \ge \max_{2\varepsilon < k \le n} \log^+ \frac{|\widehat{E}_v^k|}{|B^k(\varepsilon)|}. \tag{48}$$

(Note that we maximize over all k, $2\varepsilon < k \le n$ because for $k \le 2\varepsilon$ we have only trivial equality $\log^+\left(|\widehat{E}_v^k|/|B^k(\varepsilon)|\right) = O(\log k)$ since $|\widehat{E}_v^k| \le 2^k$.)

---

[3] Below the notation $R_n \asymp \mathcal{H}_n$, $n \to \infty$, means that there exist some constants c_1 and c_2 such that the inequalities $0 < c_1 \le R_n/\mathcal{H}_n \le c_2 < \infty$ hold for all sufficiently large n.

Assume that the maximum in (44) is achieved with $k_0 = k_0(n)$. If $\widehat{\mathcal{H}}_{k_0} / \log k_0 \to \infty$ as $n \to \infty$, then Theorem 3 and inequality (48) show that

$$H_\varepsilon(E_v^n) \geq \left[\widehat{\mathcal{H}}_{k_0} - k_0 h(\varepsilon/k_0)\right]^+ + o(\widehat{\mathcal{H}}_{k_0})$$

from which (45) immediately follows. On the other hand, if $\widehat{\mathcal{H}}_{k_0} / \log k_0 \leq C < \infty$ for all $k_0 = k_0(n)$, then

$$R_n^+(\varepsilon) < \widehat{\mathcal{H}}_{k_0} = o(\mathcal{H}_n)$$

and therefore inequality (45) is reduced to the trivial inequality

$$H_\varepsilon(E_v^n) \geq o(\mathcal{H}_n), \quad n \to \infty$$

which is always correct.   □

As a remark to the lower bound (47), note that condition (46) holds if, for example, there exists a constant β, $0 < \beta < 1$, such that $h(\varepsilon/n) \leq \beta(\mathcal{H}_n/n)$ for all sufficiently large n.

Upper bound

Let us introduce the quantity

$$K_n(\varepsilon) := \max_P \min_{\mathcal{E}} \sum_{i=1}^n h^+(p_i, e_i) = \max_P \min_{\mathcal{E}} \sum_{i=1}^n [h(p_i) - h(e_i)]^+, \tag{49}$$

where $\max \min$ is taken over all vectors $P = (p_1, \ldots, p_n)$ and $\mathcal{E} = (e_1, \ldots, e_n)$ such that

$$\begin{cases} \sum_{i=1}^n v_i p_i \leq 1, & 0 \leq p_i \leq 1/2, \quad i = 1, \ldots, n, \\ \sum_{i=1}^n e_i \leq \varepsilon, & 0 \leq e_i \leq 1/2, \quad i = 1, \ldots, n. \end{cases} \tag{50}$$

Theorem 5. *[4].   The inequality*

$$H_\varepsilon(E_v^n) \leq K_n(\varepsilon) + O(\sqrt{n \log n}), \quad n \to \infty, \tag{51}$$

holds for any $\varepsilon \geq 1$.

Sketch of the proof. Here, we present only a sketch of the proof of the upper bound. The full proof can be found in [4]. Also, now we will give a more detailed proof of an inequality which is not obvious but has been omitted there.

Recall first that we continue to assume that coefficients v_i, $i = 1, \ldots, n$, do not increase (see (46)). It immediately follows from this assumption that

$$H_\varepsilon(E_v^n) \leq H_\varepsilon(\widetilde{E}_v^n), \tag{52}$$

where the ellipsoid $\widetilde{E}_v^n$ is defined by the equality

$$\widetilde{E}_v^n = \left\{ x \in {}^n \mid v_s \sum_{i=1}^s x_i + v_{2s} \sum_{i=s+1}^{2s} x_i + \ldots + v_{ts} \sum_{i=(t-1)s+1}^{ts} x_i \leq 1 \right\}.$$

Here we assume without loss of generality that s and $t = n/s$ are some integers which will be chosen later. We obviously have

$$\widetilde{E}_v^n = \bigcup_{q_1,\dots,q_k} \prod_{i=1}^{t} S^s(sq_i), \tag{53}$$

where the union in (53) is taken over all numbers $q_1, \dots, q_t$ such that $0 \le q_i \le 1$, $q_i s$ are integers, $i = 1, \dots, t$, and $\sum_{i=1}^{t} sq_i v_{is} \le 1$.

Taking $t = \sqrt{n/\log n}$ and using the uniform upper bound (3) for the ε-entropy of spheres $S^s(sq_i)$, it is not difficult to derive from (53) that

$$H_\varepsilon(\widetilde{E}_v^n) \le \max_Q \min_L \sum_{i=1}^{n} [h(q_i) - h(l_i)]^+ + O\left(\sqrt{n \log n}\right), \quad n \to \infty, \tag{54}$$

where $\max\min$ in (54) is taken over all vectors $Q = (q_1, \dots, q_n)$ and $L = (l_1 \dots, l_n)$ such that

$$\begin{cases} v_s \sum_{i=1}^{s} q_i + v_{2s} \sum_{i=s+1}^{2s} q_i + \dots + v_{ts} \sum_{i=(t-1)s+1}^{ts} q_i \le 1, & \text{for } 0 \le q_i \le 1/2, \quad i = 1, \dots, n, \\ \sum_{i=1}^{n} l_i \le \varepsilon, & \text{for } 0 \le l_i \le 1/2, \quad i = 1, \dots, n. \end{cases} \tag{55}$$

For any given vectors Q and L satisfying (55), let $P' = (p_1', \dots, p_n')$ and $\mathcal{E}' = (e_1', \dots, e_n')$ be the vectors whose components are given by the equalities

$$p_i' = \begin{cases} 0 & if \quad i \in [1, s], \\ q_{i-s} & if \quad i \in [s+1, n], \end{cases} \tag{56}$$

and

$$e_i' = \begin{cases} 0 & if \quad i \in [1, s], \\ l_{i-s} & if \quad i \in [s+1, n]. \end{cases} \tag{57}$$

It is clear that for any such vectors Q, L, P' and $\mathcal{E}'$ the following relations hold as $n \to \infty$:

$$\sum_{i=1}^{n} [h(q_i) - h(l_i)]^+ - \sum_{i=1}^{n} [h(p_i') - h(e_i')]^+ = \sum_{i=(t-1)s+1}^{ts} [h(q_i) - h(l_i)]^+ \le s = O\left(\sqrt{n \log n}\right). \tag{58}$$

Finally, we claim that (52), (54) and (58) imply

$$H_\varepsilon(E_v^n) \le H_\varepsilon(\widetilde{E}_v^n) \le \max_P \min_{\mathcal{E}} \sum_{i=1}^{n} [h(p_i) - h(e_i)]^+ + O\left(\sqrt{n \log n}\right), \quad n \to \infty, \tag{59}$$

where $\max\min$ in (59) is taken over all vectors P and $\mathcal{E}$ satisfying conditions (50). Inequality (59) is equivalent to the desired upper bound (51). Below we explain in more detail the proof of (59) because, as we already mentioned above, it has been omitted in [4].

To do this, assume now that vectors Q and L considered above are optimal, i.e., max min in (54) is achieved on this pair of vectors. Therefore, to prove (59) it is sufficient to show that

$$\min_{\mathcal{E}} \sum_{i=1}^{n} [h(p_i') - h(e_i)]^+ - \sum_{i=1}^{n} [h(q_i) - h(l_i)]^+ = O\left(\sqrt{n \log n}\right), \quad n \to \infty, \quad (60)$$

where minimum is taken over all vectors $\mathcal{E}$ satisfying the second-line conditions in (50). Taking into account inequality (58), we see that (60) holds if

$$\sum_{i=1}^{n} [h(\widehat{e}_i) - h(e_i')]^+ = O\left(\sqrt{n \log n}\right), \quad n \to \infty, \quad (61)$$

where $\widehat{\mathcal{E}} = (\widehat{e}_1, \ldots, \widehat{e}_n)$ is a vector for which the first sum in (60) achieves its minimum.

It can easily be seen that the components of $\widehat{\mathcal{E}}$ satisfy the equality

$$\widehat{e}_i = \min\{p_i', \nu\}, \quad i = 1, \ldots, n, \quad (62)$$

where parameter ν is chosen in such a way that

$$\min\{p_i', \nu\} = \varepsilon. \quad (63)$$

Similarly, $L = (l_1, \ldots, l_n)$ satisfies the equality

$$l_i = \min\{q_i, \mu\}, \quad i = 1, \ldots, n, \quad (64)$$

where the parameter ν is defined by the equation

$$\min\{q_i, \mu\} = \varepsilon. \quad (65)$$

It easily follows from (62)–(65) and relations (56) and (57) that

$$\sum_{i=1}^{n} [h(\widehat{e}_i) - h(e_i')]^+ \leq (n - m) [h(\mu + \mu s/(n - m)) - h(\mu)], \quad (66)$$

where the integer m can be found from the relations

$$q_m \leq \mu < q_{m+1}.$$

It is obvious that

$$(n - m) [h(\mu + \mu s/(n - m)) - h(\mu)] \leq c \cdot s = O\left(\sqrt{n \log n}\right), \quad n \to \infty. \quad (67)$$

Finally, we observe that inequalities (67) and (66) imply inequality (61) and therefore the desired inequality (60) is proved. $\qquad \square$

Optimization problem

Comparing the lower and upper bounds in (45) (or (47)) and (51) we observe that they have different forms. Therefore the first problem is to find a relationship between the quantities $R_n(\varepsilon)$ and $K_n(\varepsilon)$ defined in (44) and (49), respectively. In other words, we need to find $\max\min$ of the right-hand side of (49). To our surprise, this problem turned out to be rather involved; in particular, no explicit expressions were found for it. The corresponding results obtained to date are formulated below.

Note first that the following proposition holds.

Proposition 4. *[10]. The inequality $K_n(\varepsilon) > 0$ holds if and only if*

$$\sum_{i=n-2\varepsilon+1}^{n} v_i/2 < 1. \tag{68}$$

The proof of this proposition is rather simple and can be found in [10]. Below we will always assume that condition (68) is fulfilled (note that in [10], we did not introduce this obvious assumption in the formulations of the statements).

The next theorem shows that $K_n(\varepsilon)$ can differ from $R_n(\varepsilon)$ by at most $1/2$.

Theorem 6. *[10]. The following inequalities hold:*

$$R_n(\varepsilon) \le K_n(\varepsilon) < R_n(\varepsilon) + 1/2 \tag{69}$$

provided that assumption (68) is fulfilled.

Thus, this theorem together with lower and upper bounds (47) and (51) leads to the following consequence.

Corollary 2. *If condition (46) is satisfied and*

$$\frac{\mathcal{H}_n}{\sqrt{n \log n}} \to \infty, \quad n \to \infty,$$

then

$$H_\varepsilon(E_v^n) = R_n(\varepsilon)(1 + o(1)), \quad n \to \infty.$$

Note that the left inequality in (69) is rather trivial and easily follows from the definitions of $R_n(\varepsilon)$ and $K_n(\varepsilon)$. But at the same time, the right inequality in (69) is proven in the main theorem of paper [10] using rather convoluted arguments. This theorem gives a solution to the optimization problem arising in the definition of quantity $K_n(\varepsilon)$. To formulate this theorem, we need to introduce some new definitions. Namely, for any integer $k \in [2\varepsilon, n]$ we use a real parameter $t \in [0, \varepsilon/(k+1)]$ and define the quantities r_t and ε_t by the equalities

$$r_t = 1 - v_{n-k}t, \quad \varepsilon_t = \varepsilon - t.$$

Then we consider two k-dimensional vectors

$$\widehat{P}^*(k,t) = (\widehat{p}_{n-k+1}^*(k,t), \ldots, \widehat{p}_n^*(k,t))$$

and

$$\widehat{\mathcal{E}}^*(k,t) = (\widehat{e}^*_{n-k+1}(k,t), \dots, \widehat{e}^*_n(k,t))$$

with components

$$\widehat{p}^*_i(k,t) := \left(1 + 2^{\widehat{\lambda}v_i}\right)^{-1}, \quad i = n-k+1, \dots, n,$$

$$\widehat{e}^*_i(k,t) := \varepsilon_t/k, \quad i = n-k+1, \dots, n,$$

where the parameter $\widehat{\lambda} = \widehat{\lambda}(k,t)$ is defined by the equalities

$$\begin{cases} \sum_{i=n-k+1}^n v_i \left(1 + 2^{\widehat{\lambda}v_i}\right)^{-1} = r_t, & \text{if } \frac{1}{2}\sum_{i=n-k+1}^n v_i > r_t, \\ \widehat{\lambda} = 0, & \text{if } \frac{1}{2}\sum_{i=n-k+1}^n v_i \le r_t. \end{cases}$$

For the degenerate case where $r_t \le 0$, we take $\widehat{p}_i(k,t) \equiv 0$ for all $i = n-k+1, \dots, n$. The main theorem from [10] can be formulated as follows.

Theorem 7. *[10].   Quantity $K_n(\varepsilon)$ satisfies the equality*

$$K_n(\varepsilon) = \max_{k,t}\left[h(\widehat{P}^*(k,t)) - h(\widehat{\mathcal{E}}^*(k,t))\right] = \max_{k,t}\left[\sum_{i=n-k+1}^n h(\widehat{p}^*_i(k,t)) - kh(\varepsilon_t/k)\right],$$

(70)

where the maximum is taken over all k and t such that $k \in [2\varepsilon, n]$ and $0 \le t < \varepsilon/(k+1)$. Moreover, there exists a pair (k,t) such that the $\max\min$ on the right-hand side of (49) is obtained on the pair of vectors $(\overline{P}(k,t), \overline{\mathcal{E}}(k,t))$ where

$$\overline{P}(k,t) := (\mathcal{T}(k,t), \widehat{P}^*(k,t)), \quad \overline{\mathcal{E}}(k,t) := (\mathcal{T}(k,t), \widehat{\mathcal{E}}^*(k,t)), \quad \mathcal{T}(k,t) := (\overline{0}_{n-k-1}, t).$$

Thus, in general we do not know the optimal pair (k,t), i.e., the pair on which the right-hand side of (70) achieves its maximum. But we can claim that the function $h(\widehat{P}^*(k,t)) - h(\widehat{\mathcal{E}}^*(k,t))$ can have at most one local maximum on the half open interval $t \in [0, \varepsilon/(k+1))$ for any given k. The proof of this statement can be found in [10]. On the other hand, the next proposition gives a sufficient condition under which the maximum of the function above is achieved on the pair $(n, 0)$.

Proposition 5. *[10].   If*

$$\widehat{p}^*_{n-k}(k, \varepsilon/(k+1)) \ge \varepsilon/k$$

(71)

for all $k = 2\varepsilon, \dots, n-1$, then

$$K_n(\varepsilon) = R_n(\varepsilon) = \mathcal{H}_n - nh(\varepsilon/n).$$

Remark 4. In particular, this proposition gives a simple sufficient condition under which the generalized lower packing bound for the ε-entropy of ellipsoids coincides with the classical one. In [11], it is shown that for a special class ellipsoids E_v^n whose coefficients v_i, $i = 1, \ldots, n$, can take only two possible values, sufficient condition (71) can be improved. Namely, it is proved in [11] that there exists a threshold value ε_0 (for which some formula is given) such that $K_n(\varepsilon) = \mathcal{H}_n - nh(\varepsilon/n)$ if $\varepsilon < \varepsilon_0$ and $K_n(\varepsilon) > \mathcal{H}_n - nh(\varepsilon/n)$ if $\varepsilon > \varepsilon_0$. Some explicit asymptotic expressions for the ε-entropy of such ellipsoids have also been obtained in the above paper.

4 Covering of Ellipsoids in Euclidean Spaces

In this section, we consider the problem of finding the asymptotic size of an optimal ε-covering (or ε-entropy) of arbitrary ellipsoids $E_a = E_a^n$. The latter is represented in the form (2) in the n-dimensional Euclidean space $\mathbb{R}^n$. The problem of asymptotically optimal ε-coverings has been studied for long for balls B_r^n of an arbitrary radius r, which includes the important special case of the whole space $\mathbb{R}^n$. For the ball $B^n(r)$, various bounds on the minimum covering size are obtained in [12]. By changing the scale in $\mathbb{R}^n$, one can always replace any ε-covering of $B^n(r)$ using the *unit balls* to cover a ball $B^n(r/\varepsilon)$. Therefore, the problem of coverings with balls of radius ε is often replaced in the Euclidean spaces by using the unit balls instead. One particularly important result is obtained by Rogers [12] who proved that for $n \geq 9$, the thinnest unit covering has size

$$|M_1(B^n(r))| \leq \begin{cases} Cn(\log n)r^n, & \text{if } r \geq n, \\ Cn^{5/2}r^n, & \text{if } r < n, \end{cases}$$

where C is an absolute constant (see also [13]). A related question is the covering problem of the surface of a unit sphere in $\mathbb{R}^n$ with caps of a given half angle θ. In [14], it was shown by using a random coding argument that the minimum number of caps of half angle θ required to cover the unit Euclidean n-sphere $S^n(1)$ is $\exp[-n \log \sin \theta + o(n)]$ as $n \to \infty$.

In case $\mathbb{R}^n$ or a set A of infinite volume $V(A) = \infty$, any ε-covering has also infinite size. Therefore, for a given sequence $M_\varepsilon = M_\varepsilon(\mathbb{R}^n)$ of ε-coverings for the space $\mathbb{R}^n$, it is customary to consider the *lower density* for such a sequence which is defined as the asymptotic infimum of the average number of balls covering a point in $B^n(r)$:

$$\delta(M_\varepsilon(\mathbb{R}^n)) := \lim_{r \to \infty} \inf \frac{\sum\limits_{y \in M_\varepsilon} V(B^n(y, \varepsilon) \cap B^n(r))}{V(B^n(r))}.$$

The main problem is to define the minimum density $\delta(M_\varepsilon(\mathbb{R}^n))$ taken over all ε-coverings. Here we refer to the monograph by Rogers [15], book [16], and a survey [17], where one can also find an extensive bibliography on this subject. Some results on optimal coverings of other sets can be found in [18] and [19]. Below we briefly describe some new results obtained in the recent paper [20] for optimal coverings of ellipsoids in Euclidean spaces.

4.1 Lower Bound

Given an ellipsoid E_a^n, define the quantity $\mathcal{R} = \mathcal{R}_n(\varepsilon)$ as

$$\mathcal{R}_n(\varepsilon) := \sum_{i=1}^n \log^+ (a_i/\varepsilon) = \sum_{i \,:\, a_i > \varepsilon} \log(a_i/\varepsilon). \qquad (72)$$

We begin with a lower bound on the ε-entropy of an ellipsoid E_a^n, which holds for all dimensions n and vectors a.

Theorem 8. *[18] (Generalized packing bound). For any ellipsoid E_a^n, its ε-entropy satisfies inequality*

$$H_\varepsilon(E_a^n) \geq \mathcal{R}_n(\varepsilon). \qquad (73)$$

Proof. The proof of this theorem is very simple and similar to that of Theorem 4 in the Hamming space. Indeed, consider the projection of an ellipsoid E_a^n into the subspace $\mathbb{R}^k$ spanned over the last k coordinates. Then we obtain the sub-ellipsoid

$$\widehat{E}_a^k : \left\{ (x_{n-k+1}, \ldots, x_n) \mid \frac{x_{n-k+1}^2}{a_{n-k+1}^2} + \cdots + \frac{x_n^2}{a_n^2} \leq 1 \right\}.$$

By dividing the volume of $\widehat{E}_a^k$ by the volume of $B^k(\varepsilon)$, we define the minimum number of covering balls of radius ε in $\mathbb{R}^k$. Thus, we obtain the bound

$$H_\varepsilon(E_a^n) \geq \max_{1 \leq k \leq n} \log \frac{V(\widehat{E}_a^k)}{V(B^k(\varepsilon))}.$$

By applying the formulas known for volumes of balls and ellipsoids, we conclude from the previous formula that

$$H_\varepsilon(E_a^n) \geq \max_{1 \leq k \leq n} \sum_{i=n-k+1}^n \log(a_i/\varepsilon) = \sum_{i=1}^n \log^+ (a_i/\varepsilon),$$

where the last equality simply reflects the fact that $\log(a_i/\varepsilon) \leq 0$, if $(a_i/\varepsilon) \leq 1$.
□

4.2 Asymptotic Upper Bound

Below, we will assume that $n \geq 2$, for the case $n = 1$ we obviously have $H_\varepsilon(E_a^1) = \log\lceil a/\varepsilon \rceil$. Also we will assume that parameters n, a and ε vary in such a way that $\mathcal{R}_n(\varepsilon) \to \infty$. For example, n and ε can be fixed while components a_i grow. Therefore our asymptotic setting $\mathcal{R}_n(\varepsilon) \to \infty$ (or, briefly $\mathcal{R} \to \infty$) will also serve as a *limiting* condition for all other conditions in the theorem below.

Theorem 9. *[18]. The ε-entropy of an ellipsoid E_a^n satisfies inequality*

$$H_\varepsilon(E_a^n) \le \mathcal{R}_n(\varepsilon)(1 + o(1)), \quad \mathcal{R} \to \infty, \tag{74}$$

provided that

$$\log\left[\max_i\{a_i/\varepsilon)\}\right] = o\left(\frac{\mathcal{R}^2}{m\log n}\right), \quad \mathcal{R} \to \infty, \tag{75}$$

where $\mathcal{R}_n(\varepsilon)$ is defined in (72) and m is the number of half-axes a_i of length greater than ε:

$$m = |\{\, i : a_i > \varepsilon\}|.$$

From the lower and upper bounds in (73) and (74) we immediately obtain the following obvious corollary.

Corollary 3. *The asymptotic equality*

$$H_\varepsilon(E_a^n) = \mathcal{R}_n(\varepsilon)(1 + o(1)), \quad \mathcal{R} \to \infty,$$

holds provided condition (75) is fulfilled.

Remark 5. Note that condition (75) and Theorem 9 always hold if

$$\lim_{\mathcal{R} \to \infty} \frac{\mathcal{R}}{m\log n} = \infty,$$

due to the fact that $\log\left[\max_i\{a_i/\varepsilon\}\right] \le \mathcal{R}$ by definition (72). In particular, Theorem 9 holds if a_n or some other coefficients grow for fixed n and ε, in which case $\mathcal{R} \to \infty$.

Theorem 9 is proved in [20]. The main idea of the proof is rather close to that of the upper bound in the Hamming space. Namely, here we also divide the interval of integers $[1, n]$ into a number of subintervals and then cover an ellipsoid E_a^n with a finite number of subsets each of which is a *direct product of the balls* (of lesser dimensions). Note however that - as opposed to the Hamming spaces - we need to employ a quantization procedure (by continuity of the space $\mathbb{R}^n$), which later leads to some restrictions on the parameters that describe Euclidean ellipsoids. Then, by using a slight generalization of the above Rogers result on the upper bound on the size of the minimum covering of balls, we obtain a general upper bound on the ε-entropy $H_\varepsilon(E_a^n)$, which depends on the quantization parameters and our partitioning of the interval $[1, n]$. Finally, we estimate the asymptotic bound by optimizing both the quantization and partitioning parameters.

References

1. A.N. Kolmogorov and V.M. Tikhomirov, ε-entropy and ε-capacity, Uspekhi Mat. Nauk, Vol. 14, 3–86, 1959.
2. P. Delsart and P. Piret, Do most binary linear codes achieve the Goblick bound on the covering radius? , IEEE Trans. Inf. Theory, Vol. 32, No. 6, 826–828, 1986.

3. G. Cohen, I. Honkala, S. Litsyn, and A. Lobstein, Covering Codes, Amsterdam, Elsevier, 1997.
4. I.I. Dumer, M.S. Pinsker, and V.V. Prelov, Epsilon-entropy of an ellipsoid in a Hamming space, Problemi Peredachi Informatsii, Vol. 38, No. 1, 3–18, 2002.
5. F.J. MacWilliams and N.J.A. Sloane, The Theory of Error-Correcting Codes, Amsterdam - New-York - Oxford, North-Holland Publishing Company, 1977.
6. L. Schlafli, Gesammelte mathematische Abhandlungen. Band 1, Basel, Verlag Birkhauser, 1850 (in German).
7. Yu.A. Zuev, Combinatorial-probabilistic and geometrical methods in threshold logic, Discrete Math., Vol. 3, No. 2, 47–57, 1991.
8. M.S. Pinsker, Entropy of an ellipsoid in a Hamming space, Problemi Peredachi Informatsii, Vol. 36, No. 4, 47–52, 2000.
9. V.V. Prelov, On the entropy of ellipsoids in the Hamming space, Proc. Seventh Int. Workshop, Algebraic and Combinatorial Coding Theory , Bansko, Bulgaria., 269–272, 2000.
10. I.I. Dumer, M.S. Pinsker, and V.V. Prelov, An optimization problem concerning the computation of the epsilon-entropy of an ellipsoid in a Hamming space, Problemi Peredachi Informatsii, Vol. 38, No. 2, 3–18, 2002.
11. V.V Prelov and E.C. van der Meulen, On the epsilon-entropy for a class of ellipsoids in Hamming space, Problemi Peredachi Informatsii, Vol. 38, No. 2, 19–32, 2002.
12. C.A. Rogers, Covering a sphere with spheres, Mathematika, Vol. 10, 157–164, 1963.
13. D.L. Donoho, Counting bits with Kolmogorov and Shannon, Technical report, Stanford University, No. 2000-38, 1–28, 2000.
14. A.D. Wyner, Random packings and coverings of the unit N-sphere, Bell Systems Technical Journal, Vol. 46, 2111–2118, 1967.
15. C.A. Rogers, Packing and Covering, Cambridge, Cambridge Univ. Press, 1964.
16. J.H. Conway and N.J.A. Sloane, Sphere Packings, Lattices and Groups, New York, Springer-Verlag, 1988.
17. G.F. T'oth, New results in the theory of packing and covering, in Convexity and its Applications, P.M. Gruber and J.M. Wills, Ed., Birkhauser Verlag, Basel, 318–359, 1983.
18. G. Pisier, The Volume of Convex Bodies and Banach Space Geometry, Cambridge Tracts in Mathematics, 94. Cambridge; New York, Cambridge University Press, 1989.
19. B. Carl and I. Stephani, Entropy, Compactness and the Approximation of Operators, Cambridge Tracts in Mathematics, 98. Cambridge, New York, Cambridge University Press, 1990.
20. I. Dumer, M.S. Pinsker, and V.V. Prelov, On coverings of ellipsoids in Euclidean spaces, IEEE Trans. Inf. Theory, Vol. 50, No. 10, pp. 2348–2356, 2004.

Appendix

Proof of Proposition 1. Obviously, $\Omega(r, \varepsilon) \leq 2^{nh(\sigma)}$. To tighten the bounds on $\Omega(r, \varepsilon)$, we also consider parameters

$$\theta := r - w, \quad s := \lfloor (\varepsilon - r + w)/2 \rfloor, \quad t := \max\{0, -\theta\}, \qquad (76)$$

in which case

$$\theta + 2s = \begin{cases} \varepsilon, & \text{if } w = (r - \varepsilon) \pmod 2, \\ \varepsilon - 1, \text{if } w \neq (r - \varepsilon) \pmod 2. \end{cases} \qquad (77)$$

Next, define the subset

$$a^r(x, \varepsilon) := S(x, \theta + 2s) \cap S(r).$$

Our proof of Proposition 1 includes a few steps, which are done in the following lemmas.

Lemma 1. *For any $x \in S(w)$,*

$$|a^r(x, \varepsilon)| = \binom{w}{s}\binom{n-w}{\theta+s}. \tag{78}$$

$$1 \le \frac{|A^r(x, \varepsilon)|}{|a^r(x, \varepsilon)|} \le \frac{1-\rho}{1-\rho-\sigma}. \tag{79}$$

Proof. Given any $x \in S(w)$, any point $y_x \in A^r(x, \varepsilon)$ can be obtained from x if and only if we replace i ones and $i+\theta$ zeros in x, where $i \ge t$ and $2i + \theta \le \varepsilon$. Thus, we obtain $a^r(x, \varepsilon)$ by taking $i = s$, and $A^r(x, \varepsilon)$ is obtained by taking all $i = t, ..., s = \lfloor (\varepsilon - \theta)/2 \rfloor$. This proves (78) and also gives the size

$$|A^r(x, \varepsilon)| = \sum_{i=t}^{s} \binom{w}{i}\binom{n-w}{\theta+i}.$$

Note that $\varepsilon < r \le n - r$. Therefore for any $i = t, ..., s$:

$$\frac{i}{w} \le \frac{s}{w} \le \frac{(\varepsilon-\theta)/2}{r-\theta} \le \frac{\sigma}{\rho+\sigma}; \tag{80}$$

$$\frac{\theta+i}{n-w} \le \frac{\theta+s}{n-w} \le \frac{(\varepsilon+\theta)/2}{n-r+\theta} \le \frac{\sigma}{1-\rho+\sigma}. \tag{81}$$

Note also that bounds (80) and (81) are also limited by $1/2$. Therefore (80) and (81) give

$$\sum_{i=t}^{s} \binom{w}{i}\binom{n-w}{\theta+i} \le \binom{w}{s}\sum_{i=t}^{s}\binom{n-w}{\theta+i}. \tag{82}$$

The latter sum in (82) is upper-bounded by the size of the ball $B(\theta+s)$ in E^{n-w}. This is estimated by inequalities (8) and (81). Here we replace parameter σ in (8) by $(\theta+s)/(n-w)$ and obtain inequality (79). $\qquad\square$

Inequality (79) also shows that the sets $A^r(x, \varepsilon)$ and $a^r(x, \varepsilon)$ differ in size only by a multiplicative constant, if σ is bounded away from $1/2$. Let

$$a^r(w, \varepsilon) := \binom{w}{s}\binom{n-w}{\theta+s}, \quad w \in [r-\varepsilon, r+\varepsilon],$$

be the size of a set $a^r(x, \varepsilon)$. Our goal is to derive the bounds on

$$\Phi(r, \varepsilon) := \max_{w} \log a^r(w, \varepsilon).$$

According to (77), $a^r(x, \varepsilon)$ belongs either to the sphere $S(x, \varepsilon-1)$ or to the bigger sphere $S(x, \varepsilon)$. Therefore for all integers n, $r \le n/2$, and $\varepsilon < r$,

$$\Phi(r, \varepsilon) \le nh(\sigma) - \log(2\pi\sigma(1-\sigma)n)/2 \le nh(\sigma) - \log(\sigma n)/2. \tag{83}$$

Below we will reduce the residual term $-\log(\sigma n)/2$ to $-\log(\sigma n)$. To do so, we use the following two lemmas.

Lemma 2. $\Phi(r, \varepsilon)$ *is achieved only if*

$$w = (r - \varepsilon) \pmod 2, \tag{84}$$

$$s = (\varepsilon + w - r)/2, \tag{85}$$

in which case

$$\Phi(r, \varepsilon) = \max_w \log \binom{w}{s} \binom{n-w}{\varepsilon - s}. \tag{86}$$

Proof. According to (76) and (77), for any $w \neq (r - \varepsilon) \pmod 2$, the smaller weight $w - 1$ keeps the same parameter s but increases θ by 1. Thus,

$$\frac{a^r(w-1, \varepsilon)}{a^r(w, \varepsilon)} = \binom{w-1}{s} \binom{n-w+1}{\varepsilon - s} \Big/ \left\{ \binom{w}{s} \binom{n-w}{\varepsilon - s - 1} \right\}$$

$$= \frac{w-s}{w} \cdot \frac{n-w+1}{\varepsilon - s} \geq 1,$$

due to inequalities

$$\frac{\varepsilon - s}{n - w + 1} < 1/2 < 1 - \frac{s}{w},$$

which in turn follow from (80) and (81). □

Later, we will consider the case $s = 0$ separately. For all other cases, we rewrite (86) using (6), (80), and (81). This gives the upper bounds

$$\Phi(r, \varepsilon) \leq \max_w \{g(w) - \log(\pi^2 s(\varepsilon - s))/2\} \tag{87}$$

$$\leq \max_w \{g(w) - (\log 4 \min\{s, \varepsilon - s\}\varepsilon)/2\},$$

where

$$g(w) \stackrel{\text{def}}{=} wh\left(\frac{s}{w}\right) + (n-w)h\left(\frac{\varepsilon - s}{n - w}\right)$$

depends only on w for given n, r, and ε. To tighten bound (83), we use the following straightforward statement.

Lemma 3. *For any $x \in (0, 1/2]$ and any $\lambda \in [-x, x]$,*

$$h(x + \lambda) \leq \begin{cases} h(x) + \lambda \log \frac{1-x}{x} - \frac{\lambda^2 \log e}{2x(1-x)}, & \text{if } \lambda \leq 0, \\ h(x) + \lambda \log \frac{1-x}{x}, & \text{if } \lambda \geq 0. \end{cases} \tag{88}$$

Proof. For $\lambda \leq 0$, we use the first three terms in Taylor's expansion series of $h(x)$:

$$h(x + \lambda) = h(x) + \lambda h'(x) + \lambda^2 h''(y)/2, \quad y \in [x + \lambda, x].$$

Given $\lambda \leq 0$, we see that $h''(y) = -\log e/(y(1-y))$ achieves its maximum on the interval $[x + \lambda, x]$ if $y = x$. This gives the upper inequality (88). For $\lambda \geq 0$, we use that $h''(y) < 0$ on the whole interval $[0, 1]$. □

Remark. Since $h''(y) \leq -4 \log e$ for all y, the term $-2\lambda^2 \log e$ can be added in (88), when $\lambda \geq 0$. However, we use bound (88), since further improvements do not change its asymptotics.

Below in our bound (87), we use real parameters w and $s = s(w)$ defined in (85). We also use parameters

$$v := \frac{\rho - \sigma}{1 - 2\sigma} n, \tag{89}$$

$$\lambda_1 := \frac{s}{w} - \sigma, \quad \lambda_2 := \frac{\varepsilon - s}{n - w} - \sigma. \tag{90}$$

In this case the number $s(v)$ satisfies equalities

$$\frac{s}{v} = \frac{\varepsilon - s}{n - v} = \sigma.$$

From (90) we also see that

$$w\lambda_1 = -(n - w)\lambda_2. \tag{91}$$

Below we show that $\Phi(r, \varepsilon)$ can in essence be upper-bounded by taking $w = v$, given the following rather loose condition on $\rho := r/n$ and $\sigma := \varepsilon/n$.

Lemma 4. *Let n, $r \leq n/2$, and $\varepsilon < r$ satisfy condition*

$$n\sigma(\rho - \sigma)(1 - 2\sigma) \geq \log n \tag{92}$$

for all n large enough. Then for sufficiently large n,

$$\Phi(r, \varepsilon) \leq nh(\sigma) - \log(n\sigma) - \frac{1}{2} \log \frac{\rho - \sigma}{1 - 2\sigma}. \tag{93}$$

Proof. Note that λ_1 and λ_2 have different signs according to (91). Below we consider any weight $w \leq v$, in which case $\lambda_1 \leq 0$ and $\lambda_2 \geq 0$ (the opposite case $w > v$ can be treated in a similar way). Then we can use the following inequalities:

$$h\left(\frac{s}{w}\right) \leq h(\sigma) + \lambda_1 \log \frac{1 - \sigma}{\sigma} - \frac{\lambda_1^2 \log e}{2\sigma(1 - \sigma)}; \tag{94}$$

$$h\left(\frac{\varepsilon - s}{n - w}\right) \leq h(\sigma) + \lambda_2 \log \frac{1 - \sigma}{\sigma}. \tag{95}$$

Now we combine inequalities (94) and (95) with equality (91), thus excluding all first-order terms:

$$g(w) \leq nh(\sigma) - \frac{w\lambda_1^2 \log e}{2\sigma(1 - \sigma)}. \tag{96}$$

Now consider any $w \in (r - \varepsilon, v]$. (Recall that $w = r - \varepsilon$ corresponds to the boundary case $s = 0$ and will be considered later). Let $c_* = 1 - 2\sigma$. Also, let $c = w/v$, where $c \in [c_* + 1/v, 1]$. We rewrite (85) and (90) as

$$s(w) = v(c - c_*)/2,$$
$$\lambda_1 = -c_*(1 - c)/2c.$$

In this case, we can combine (87) and (96) as

$$\Phi(r,\varepsilon) \leq \Phi_c(r,\varepsilon),$$

where

$$\Phi_c(r,\varepsilon) := nh(\sigma) - \frac{\log(4\sigma n)}{2} - \frac{c_*^2(1-c)^2}{8c\sigma(1-\sigma)}v\log e - \frac{\log v(c-c_*)}{2}$$

is a function of c. Direct verification shows that its derivative

$$\frac{\partial\Phi_c(r,\varepsilon)}{\partial c} = \frac{c_*^2 v\log e}{8\sigma(1-\sigma)}(c^{-2}-1) - \frac{\log e}{2(c-c_*)}$$

is positive for $n \to \infty$ on some interval $c \in [c_* + 1/v, c']$, where $c' \to 1$ provided that condition (92) holds. Thus, we take $w = v$ and $s = s(v)$ and obtain the upper bound:

$$\Phi(r,\varepsilon) \leq \Phi_1(r,\varepsilon) \leq g(w) - (\log 4s(v)\varepsilon)/2,$$

which in turn gives (93).

Finally, consider the boundary case $s = 0$. Then

$$a^r(r-\varepsilon,\varepsilon) = \binom{n-r+\varepsilon}{\varepsilon} \leq \binom{n}{\varepsilon} \cdot \left(1 - \frac{r-\varepsilon}{n}\right)^\varepsilon,$$

which gives the estimate

$$\log a^r(r-\varepsilon,\varepsilon) \leq nh(\sigma) - n\sigma(\rho-\sigma).$$

Now we can directly verify that this estimate also satisfies our bound (93) given condition (92), since $n\sigma(\rho-\sigma) > \log n$. □

Our next goal is to obtain a lower bound on $a^r(w,\varepsilon)$. In fact, we show that the lower bound derived below differs from (93) by an additive term of order $(\log n)/2$.

Lemma 5. *For all n, $r \leq n/2$, and $\varepsilon < r$,*

$$\Phi(r,\varepsilon) \geq \begin{cases} nh(\sigma) - \frac{1}{2}\log(8n\sqrt{2}) + \frac{1}{2}\log\sigma, & \text{if } \frac{\sigma(\rho-\sigma)}{1-2\sigma}n < 1, \\ nh(\sigma) - \log(8n\sqrt{2}) - \frac{1}{2}\log\frac{\rho-\sigma}{1-2\sigma}, & \text{otherwise.} \end{cases} \tag{97}$$

Thus, for all ρ and σ,

$$\Phi(r,\varepsilon) \geq nh(\sigma) - \log 8n. \tag{98}$$

Proof. We choose two different values of w and s to satisfy conditions (84). Namely, let $w_1 := \lfloor v \rfloor = v - \tau$, where $\tau \in [0,1)$ and $w_2 := w_1 + 1$. Let $s_1 := s(w_1)$ and $s_2 := s(w_2)$ be defined by (85). Obviously, one of the numbers $w_i, i = 1, 2$, satisfies conditions (84) and therefore can be used in conjunction with (86). Our goal is to show that both numbers w_i allow us to obtain the quantity $a^r(w_i,\varepsilon)$ that satisfies (97).

Note that by definition of v in (89)

$$s_1 = \frac{\varepsilon + w_1 - r}{2} = \frac{\varepsilon + v - r}{2} - \frac{\tau}{2} = \sigma v - \frac{\tau}{2},$$

in which case we have inequalities:

$$\frac{s_1}{w_1} < \sigma, \quad \frac{s_1 + 1}{w_1} > \sigma, \quad \frac{\varepsilon - s_1}{n - w_1} > \sigma.$$

Now we estimate $a^r(w_1, \varepsilon)$ using inequality

$$\binom{w_1}{s_1} = \binom{w_1}{s_1 + 1} \cdot \left(\frac{s_1 + 1}{w_1 - s_1} \right) > \sigma \binom{w_1}{s_1 + 1},$$

and the lower bound from (6):

$$a^r(w_1, \varepsilon) \geq \binom{w_1}{s_1 + 1} \cdot \binom{n - w_1}{\varepsilon - s_1} \cdot \sigma \tag{99}$$

$$\geq ((s_1 + 1)(\varepsilon - s_1))^{-1/2} \cdot 2^{nh(\sigma)} \cdot \sigma/8$$

$$\geq (\max(2\sigma v, 2) \cdot \varepsilon)^{-1/2} \cdot 2^{nh(\sigma)} \cdot \sigma/8.$$

Here we employed trivial inequality

$$s_1 + 1 < \sigma v + 1 \leq \max(2\sigma v, 2).$$

Next, we estimate $a^r(w_2, \varepsilon)$ in a similar way. We use the quantity

$$s_2 = \frac{\varepsilon + v - r}{2} + \frac{1 - \tau}{2} = \sigma v + \frac{1 - \tau}{2},$$

which satisfies inequalities

$$\frac{s_2}{w_2} > \sigma, \quad \frac{\varepsilon - s_2}{n - w_2} < \sigma, \quad \frac{\varepsilon - s_2 + 1}{n - w_2} > \sigma,$$

and gives the same estimates

$$\binom{n - w_2}{\varepsilon - s_2} > \binom{n - w_2}{\varepsilon - s_2 + 1} \cdot \left(\frac{\varepsilon - s_2 + 1}{n - w_2} \right) > \sigma \binom{n - w_2}{\varepsilon - s_2 + 1},$$

$$a^r(w_2, \varepsilon) \geq \binom{w_2}{s_2} \cdot \binom{n - w_2}{\varepsilon - s_2 + 1} \cdot \sigma$$

$$\geq (s_2(\varepsilon - s_2 + 1))^{-1/2} \cdot 2^{nh(\sigma)} \cdot \sigma/8$$

$$\geq (\max(2\sigma v, 2) \cdot \varepsilon)^{-1/2} \cdot 2^{nh(\sigma)} \cdot \sigma/8.$$

Finally, we substitute v from (89) in (99) and obtain (97) for both pairs. To replace (97) with (98), we use the fact that $\sigma \in [1/n, 1/2)$ and $\rho - \sigma \leq (1 - 2\sigma)/2$. □

Now we see that Proposition 1 follows from bounds (79), (83), (93), and (97). □

Appendix: On Set Coverings in Cartesian Product Spaces

R. Ahlswede

Abstract. Consider $(X, \mathcal{E})$, where X is a finite set and $\mathcal{E}$ is a system of subsets whose union equals X. For every natural number $n \in \mathbb{N}$ define the cartesian products $X_n = \prod_1^n X$ and $\mathcal{E}_n = \prod_1^n \mathcal{E}$. The following problem is investigated: how many sets of $\mathcal{E}_n$ are needed to cover X_n? Let this number be denoted by $c(n)$. It is proved that for all $n \in \mathbb{N}$

$$\exp\{C \cdot n\} \leq c(n) \leq \exp\{Cn + \log n + \log \log |X|\} + 1.$$

A formula for C is given. The result generalizes to the case where X and $\mathcal{E}$ are not necessarily finite and also to the case of non–identical factors in the product. As applications one obtains estimates on the minimal size of an externally stable set in cartesian product graphs and also estimates on the minimal number of cliques needed to cover such graphs.

1 A Covering Theorem

Let X be a non–empty set with finitely many elements and let $\mathcal{E}$ be a set of non–empty subsets of X with the property $\bigcup_{E \in \mathcal{E}} E = X$. (We do not introduce an index set for $\mathcal{E}$ in order to keep the notations simple). For $n \in \mathbb{N}$, the set of natural numbers, we define the cartesian product spaces $X_n = \prod_1^n X$ and $\mathcal{E}_n = \prod_1^n \mathcal{E}$. The elements of $\mathcal{E}_n$ can be viewed as subsets of X_n.

We say that $\mathcal{E}'_n \subset \mathcal{E}_n$ covers X_n or is a *covering* of X_n, if $X_n = \bigcup_{E_n \in \mathcal{E}'_n} E_n$. We are interested in obtaining bounds on the numbers $c(n)$ defined by

$$c(n) = \min_{\mathcal{E}'_n \text{ covers } X_n} |\mathcal{E}'_n|, \quad n \in \mathbb{N}. \tag{1}$$

Clearly, $c(n_1 + n_2) \leq c(n_1) \cdot c(n_2)$ for $n_1, n_2 \in \mathbb{N}$. Example 1 below shows that equality does not hold in general. Denote by Q the set of all probability distributions on the finite set $\mathcal{E}$, denote by $1_E(\cdot)$ the indicator function of a set E, and define K by

$$K = \max_{q \in Q} \min_{x \in X} \sum_{E \in \mathcal{E}} 1_E(x) q_E. \tag{2}$$

Theorem 1. With $C = \log K^{-1}$ the following estimates hold:

a) $c(n) \geq \exp\{C \cdot n\}$, $n \in \mathbb{N}$.
b) $c(n) \leq \exp\{C \cdot n + \log n + \log \log |X|\} + 1$, $n \in \mathbb{N}$.
c) $\lim_{n \to \infty} \frac{1}{n} \log c(n) = C$.

R. Ahlswede et al. (Eds.): Information Transfer and Combinatorics, LNCS 4123, pp. 926–937, 2006.
© Springer-Verlag Berlin Heidelberg 2006

Proof. c) is a consequence of a) and b). In order to show a) let us assume that $\mathcal{E}_{n+1}^*$ covers X_{n+1} and that $|\mathcal{E}_{n+1}^*| = c(n + 1)$.

Write an element E_{n+1} of $\mathcal{E}_{n+1}^*$ as $E^1 E^2 \ldots E^{n+1}$ and denote by $_x X_{n+1}$ the set of all those elements of X_{n+1} which have x as their first component. Finally, define a probability distribution q^* on $\mathcal{E}$ by

$$q_E^* = \left|\{E_{n+1} \mid E_{n+1} \in \mathcal{E}_{n+1}^*, E^1 = E\}\right| c^{-1}(n + 1) \text{ for } E \in \mathcal{E}. \tag{3}$$

In order to cover the set $_x X_{n+1}$ we need at least $c(n)$ elements of $\mathcal{E}_{n+1}^*$. This and the definition of q^* yield

$$c(n + 1) \sum_{E \in \mathcal{E}} 1_E(x) q_E^* \geq c(n). \tag{4}$$

Since 1 holds for all $x \in X$ we obtain

$$c(n + 1) \min_{x \in X} \sum_{E \in \mathcal{E}} 1_E(x) q_E^* \geq c(n) \tag{5}$$

and therefore also

$$c(n + 1) \max_{q \in Q} \min_{x \in X} \sum_{E \in \mathcal{E}} 1_E(x) q_E \geq c(n). \tag{6}$$

Inequality a) is an immediate consequence of 1.

We prove now b). Let r be an element of Q for which the maximum in 1 is assumed. Denote by r_n the probability distribution on $\mathcal{E}_n$, which is defined by

$$r_n(E_n) = \prod_{t=1}^{n} r_E t, \ E_n = E^1 E^2 \ldots E^n \in \mathcal{E}_n. \tag{7}$$

Let N be a number to be specified later. Select now N elements $E_n^{(1)}, \ldots, E_n^{(N)}$ of $\mathcal{E}_n$ *independently* of each other according to the random experiment $(\mathcal{E}_n, r_n)$. If every $x_n \in X_n$ is covered by $\{E_n^{(1)}, \ldots, E_n^{(N)}\}$ with positive probability then there exists a covering of X_n with N sets. Let $x_n = (x^1, \ldots, x^n)$ be any element of X_n. Define $\mathcal{E}(x_n)$ by

$$\mathcal{E}(x_n) = \{E_n \mid E_n \in \mathcal{E}_n, x_n \in E_n\}. \tag{8}$$

Clearly, $\mathcal{E}(x_n) = \prod_1^n \{E \mid E \in \mathcal{E}, x^t \in E\}$ and therefore

$$r_n\big(\mathcal{E}(x_n)\big) = \prod_{t=1}^{n} \left(\sum_E 1_E(x^t) r_E \right). \tag{9}$$

Recalling the definitions for r and K we see that $\sum_E 1_E(x^t)r_E \geq K$ and that

$$r_n\big(\mathcal{E}(x_n)\big) \geq K^n. \tag{10}$$

This implies that x_n is *not* contained in anyone of the N selected sets with a probability smaller than $(1 - K^n)^N$ and therefore X_n is not covered by those sets with a probability smaller than $|X|^n(1 - K^n)^N$. Thus there exist coverings of cardinality N for all N satisfying

$$|X|^n(1 - K^n)^N < 1. \tag{11}$$

Since $(1 - K^n)^N \leq \exp\{-K^n N\}$ one can choose any N satisfying

$\exp\{-K^n N\} \leq \exp\{-\log|X|n\}$ or (equivalently) $N \geq \exp\{\log K^{-1} \cdot n + \log n + \log\log|X|\}$.

The proof is complete.

Probabilistic arguments like the one used here have been applied frequently in solving combinatorial problems, especially in the work of Erdös and Renyi. The cleverness of the proofs lies in the choice of the probability distribution assigned to the combinatorial structures. The present product distribution has been used for the first time by Shannon [2] in his proof of the coding theorem of Information Theory. For the packing problem defined in section 5 the present approach will not yield asymptotically optimal results.

Example 1

$$X = \{0,1,2,3,4\}, \ \mathcal{E} = \big\{\{x, x+1\} \mid x \in X\big\}.$$

The addition is understood   mod 5. Clearly, $c(1) = 3$. We list the elements of X_2 as follows:

00	22	02	20	43	14	34	41
01	23	03	30	44	24	40	42
10	32	12	21	04			
11	33	13	31				

The elements in every column are contained in a set which is an element of $\mathcal{E}_2$. Therefore $c(2) \leq 8 < c(1)^2$. Since in the present case $K^{-1} = \frac{5}{2}$ and since $c(2) \geq c(1)K^{-1} = \frac{15}{2} > 7$ we obtain that actually $c(2) = 8$. Moreover, since $\lim_{n \to \infty} \frac{1}{n}\log c(n) = \log \frac{5}{2}$ there exists infinitely many n with $c(2n) < c^2(n)$.

2   Generalizations of the Covering Theorem

Let $(X^t, \mathcal{E}^t)_{t=1}^\infty$ be a sequence of pairs, where X^t is an arbitrary non–empty set and $\mathcal{E}^t$ is an arbitrary system of non–empty subsets of X^t. For every $n \in \mathbb{N}$ set $X_n = \prod_{t=1}^n X^t$, $\mathcal{E}_n = \prod_{t=1}^n \mathcal{E}^t$, and define $c(n)$ again as the smallest cardinality of a covering of X_n. Define Q^t, $t \in \mathbb{N}$, as the set of all probability distributions on $\mathcal{E}^t$ which are concentrated on a *finite* subset of $\mathcal{E}^t$. Finally, set

$$K^t = \sup_{q^t \in Q^t} \inf_{x^t \in X^t} \sum_{E^t \in \mathcal{E}^t} 1_{E^t}(x^t) q^t_{E^t} \quad \text{and} \quad C^t = \log(K^t)^{-1} \quad \text{for} \quad t \in \mathbb{N}.$$

A. The case of identical factors

Let us assume that $(X^t, \mathcal{E}^t) = (X, \mathcal{E})$ for $t \in \mathbb{N}$. This implies that also $Q^t = Q$, $K^t = K$, and $C^t = C$ for $t \in \mathbb{N}$.

Corollary 1.
a) $c(n) \geq \exp\{C \cdot n\}$, $n \in \mathbb{N}$
b) For every $\delta > 0$ there exists an n_δ such that $c(n) \leq \exp\{C \cdot n + \delta n\}$ for $n \geq n_\delta$.
c) $\lim_{n \to \infty} \frac{1}{n} \log c(n) = C$.

Proof. If $c(1) = \infty$, then also $c(n) = \infty$ and a) is obviously true. b) holds in this case, because $K = 0$. If $c(1) < \infty$, then also $c(n) < \infty$. Replacing "max" by "sup" and "min" by "inf" the proof for a) of the theorem carries over verbally to the present situation. We prove now b). Choose r^* such that

$$\left| \log K^{-1} - \log \left(\inf_{x \in X} \sum_{E \in \mathcal{E}} 1_E(x) r_E^* \right)^{-1} \right| < \frac{\delta}{2}. \tag{12}$$

Let $\mathcal{E}^*$ be the finite support of r^*. We define an equivalence relation on X by

$$x \sim x' \text{ iff } \{E | E \in \mathcal{E}^*, x \in E\} = \{E | E \in \mathcal{E}^*, x' \in E\}. \tag{13}$$

Thus we obtain at most $2^{|\mathcal{E}^*|}$ many equivalence classes. Denote the set of equivalence classes by $\overline{X}$ and let $\overline{E}$ be the subset of $\overline{X}$ obtained from E by replacing it's elements by their equivalence classes. Write $\overline{\mathcal{E}} = \{\overline{E} | E \in \mathcal{E}\}$, $\overline{X}_n = \prod_1^n \overline{X}$, and $\overline{\mathcal{E}}_n = \prod_1^n \overline{\mathcal{E}}$. A covering of $\overline{X}_n$ induces a covering of X_n with the same cardinality. If follows from the theorem and from 2 that

$$c(n) \leq \exp\left\{ Cn + \frac{\delta}{2}n + \log n + \log\log 2^{|\mathcal{E}^*|} \right\} + 1. \tag{14}$$

This implies b). c) is again a consequence of a) and b).

B. Non–identical factors

Corollary 2. *Assume that* $\max_t |\mathcal{E}^t| \leq a < \infty$. *Then for all* $n \in \mathbb{N}$:
a) $c(n) \geq \exp\left\{ \sum_{t=1}^n C^t \right\}$
b) $c(n) \leq \exp\left\{ \sum_{t=1}^n C^t + \log n + \log\log 2^a \right\} + 1$.

Proof. Introducing equivalence relations in every X^t as before in X we see that it suffices to consider the case $\max_t |X^t| \leq 2^a$. a) is proved as in case of identical factors. We show now b). Since $\mathcal{E}^t$ is finite there exists an r^t for which C^t is assumed. Replace the definition of r_n given in 1 by

$$r_n(E_n) = \prod_{t=1}^r r^t(E^t) \quad \text{for all} \quad E_n = E^1 \dots E^n \in \mathcal{E}_n. \tag{15}$$

By the argument which led to 1 we obtain now

$$|2^a|^n \left(1 - \prod_{t=1}^n K^t\right)^N < 1 \tag{16}$$

and therefore b).

The condition on the $\mathcal{E}^t$'s can actually be weakened to the following uniformity condition:

For every $\delta > 0$ there exists an m_δ and r^t's with supports of cardinality smaller then m_δ such that

$$\left|\log(K^t)^{-1} - \log\left(\inf_{x^t \in X^t} \sum_{E^t \in \mathcal{E}^t} 1_{E^t}(x^t) r_{E^t}^t\right)^{-1}\right| \le \delta. \tag{17}$$

The upper bound on $c(n)$ which one then obtains is of course only of a sharpness as the one in b) of corollary 1.

Remark 1. *One can assign weights to the elements of $\mathcal{E}_n$ and than ask for coverings with minimal total weight. It may be of some interest to elaborate conditions on the weight function under which the covering theorem still holds. The weight function will of course enter the definition of K.*

3   Hypergraphs: Duality

In this and later sections we consider only finite sets and products of finite sets, even though the results obtained can easily be generalized along the lines of section 2 to the infinite case. Thus we have the benefit of notational simplicity.

Let $X = \{x(i)|i = 1, \ldots, a\}$ be a non–empty finite set and let $\mathcal{E} = (E(j)|j = 1, \ldots, b)$ be a *family* of subsets of X. The pair $H = (X, \mathcal{E})$ is called a hypergraph (see [3]), if

$$\bigcup_{j=1}^b E(j) = X \quad \text{and} \quad E(j) \ne \varnothing \quad \text{for} \quad j = 1, \ldots, b. \tag{18}$$

The $x(i)$'s are called vertices and the $E(j)$'s are called edges. A hypergraph is called simple, if $\mathcal{E}$ is a set of subsets of X. For the problems studied in this paper we can limit ourselves without loss of generality to simple hypergraphs and we shall refer to them shortly as hypergraphs. A hypergraph is a graph (without isolated vertices), if $|E(j)| \le 2$ for $j = 1, \ldots, b$. Interpreting $E(1), \ldots, E(b)$ as points $e(1), \ldots, e(b)$ and $x(1), \ldots, x(a)$ as sets $X(1), \ldots, X(a)$, where

$$X(j) = \{e(i)|i \le a, x(j) \in E(i)\} \tag{19}$$

one obtains the dual hypergraph $H^* = (E^*, \mathcal{X}^*)$. A hypergraph is characterized by it's incidence matrix A. The incidence matrix of H^* is the conjugate of A. Let

$H^t = (X^t, \mathcal{E}^t)$, $t \in \mathbb{N}$, be hypergraphs. For $n \in \mathbb{N}$ we define cartesian product hypergraphs $H_n = \prod_{t=1}^{n} H^t$ by

$$H_n = (X_n, \mathcal{E}_n). \tag{20}$$

The covering theorem can be interpreted as a statement about edge coverings of cartesian product hypergraphs. We are looking now for the dual statement. One easily verifies that

$$H_n^* = (E_n^*, \mathcal{X}_n^*) = \prod_{t=1}^{n} (H^t)^*. \tag{21}$$

This means that the dual of the product hypergraph is the product of the dual hypergraphs. A set $T \subset X$ is called a transversal (or support) in $H = (X, \mathcal{E})$ if

$$T \cap E \neq \varnothing \quad \text{for all} \quad E \in \mathcal{E}. \tag{22}$$

Denote the smallest cardinality of transversals in H_n (resp. H_n^*) by $t(n)$ (resp. $t^*(n)$). A transversal in H_n is a covering in H_n^*, and vice versa. Denoting the smallest cardinality of coverings in H_n^* by $c^*(n)$ we thus have

$$t(n) = c^*(n), t^*(n) = c(n), n \in \mathbb{N}. \tag{23}$$

Let now P be the set of all probability distributions on X and define K^* by

$$K^* = \max_{p \in P} \min_{E \in \mathcal{E}} \sum_{x \in X} 1_E(x) p_x. \tag{24}$$

K^* plays the same role for H_n^* as K does for H_n. The covering theorem implies

Corollary 3. *With $C^* = \log K^{*-1}$ the following estimates hold for $n \in \mathbb{N}$:*

a) $t(n) = c(n) \geq \exp\{C^* \cdot n\}$
b) $t(n) \leq \exp\{C^* \cdot n + \log + \log\log |\mathcal{E}|\} + 1$.

Of course the dual results to Corollaries 1, 2 also hold. There is generally no simple relationship between K and K^*. By choosing $\mathcal{E}$ as $\{\{x\} \mid x \in X\} \cup \{X\}$ we obtain $c(n) = 1, t(n) = |X|^n$, and therefore $K^* < K$ in this case. $K > K^*$ occurs for the dual problem. It may be interesting (and not too hard) to characterize hypergraphs for which $K^* = K$. We show now that K (resp. K^*) can be expressed as a function of P (resp. Q).

Lemma 1.
a) $K = \max_{q \in Q} \min_{x \in X} \sum_{E \in \mathcal{E}} 1_E(x) q_E = \min_{p \in P} \max_{E \in \mathcal{E}} \sum_{x \in X} 1_E(x) p_x = \overline{K}$,
b) $K^* = \max_{p \in P} \min_{E \in \mathcal{E}} \sum_{x \in X} 1_E(x) p_x = \min_{q \in Q} \max_{x \in X} \sum_{E \in \mathcal{E}} 1_E(x) q_E$.

Proof. We have to show a) only since b) follows by dualization. P and Q are convex and compact in the supremum norm topology. The function $f(p,q) = \sum_{x \in X} \sum_{E \in \mathcal{E}} 1_E(x) p_x q_E$ is linear and continuous in both variables p and q. Therefore von Neumann's Minimax Theorem ([4]) is applicable and yields

$$\max_q \min_p \sum_x \sum_E 1_E(x) p_x q_E = \min_p \max_q \sum_x \sum_E 1_E(x) p_x q_E = M, \text{ say.} \quad (25)$$

Write K as $\max_q \min_{\delta_{x_0}} \sum_x \sum_E 1_E(x) \delta(x, x_0) q_E$, where δ_{x_0} is the probability distribution concentrated on x_0 and $\delta(\cdot, \cdot)$ is Kronecker's symbol. We see that $K \geq M$ and similarly that $M \geq \overline{K}$. For all p and q we have

$$\max_E \sum_x 1_E(x) p_x \geq \sum_E q_E \sum_x 1_E(x) p_x = \sum_x p_x \sum_E 1_E(x) q_E \geq \min_x \sum_E 1_E(x) q_E.$$
$$(26)$$

This implies $\overline{K} \geq K$ and thus $\overline{K} = K$. In studying infinite hypergraphs one could make use of more general Minimax Theorems, which have been proved by Kakutani, Wald, Nikaido, and others.

4   Applications to Graphs

Let $G = (X, U)$ be a non–oriented graph without multiple edges. Define Γx by

$$\Gamma x = \{ y | y \in X, (x, y) \in U \}, x \in X. \quad (27)$$

Γx is the set of vertices connected with x by an edge. The graph G is completely described by X and Γ and we therefore also write $G = (X, \Gamma)$. Given a sequence of graphs $(G^t)_{t=1}^\infty$ then we define for every $n \in \mathbb{N}$ the cartesian product graphs $G_n = (X_n, \Gamma_n) = \prod_{t=1}^n G^t$ by

$$X_n = \prod_{t=1}^n X^t, \Gamma_n x_n = \prod_{t=1}^n \Gamma^t x^t \quad (28)$$

for all $x_n = (x^1, \ldots, x^n) \in X_n$. (This product has also been called the cardinal product in the literature). Two vertices $x_n = (x^1, \ldots, x^n)$ and $y_n = (y^1, \ldots, y^n)$ of G_n are connected by an edge if and only if they are connected component–wise. In the sequel we shall show that the covering theorem leads to estimates for some fundamental graphic parameters in case of product graphs.

A. The coefficient of external stability
Given a graph $G = (X, \Gamma)$, a set $S, S \subset X$, is said to be externally stable if

$$\Gamma x \cap S \neq \varnothing \quad \text{for all } x \in S^c \quad (29)$$

or (equivalently) if

$$\bigcup_{x \in S} (\Gamma x \cup \{x\}) = X. \quad (30)$$

The coefficient of external stability $s(G)$ of a graph G is defined by

$$s(G) = \min_{S \text{ ext. stable}} |S|. \tag{31}$$

Finally, denote by $Q(X, \Gamma)$ the set of all probability distributions on $\{\Gamma y | y \in X\}$.

Corollary 4. *Let* $G = (X, \Gamma)$ *be a finite graph with all loops included, that is* $x \in \Gamma x$ *for all* $x \in X$. *With* $\overline{C} = \log \left(\max_{q \in Q(X,\Gamma)} \min_{x \in X} \sum_{y \in X} 1_{\Gamma y}(x) q_{\Gamma y} \right)^{-1}$ *and* $s(n) = s\left(\prod_1^n G \right)$ *the following estimates hold for* $n \in \mathbb{N}$:

a) $s(n) \geq \exp\{\overline{C}n\}$
b) $s(n) \leq \exp\{\overline{C}n + \log n + \log \log |X|\} + 1$.

Proof. Since $x \in \Gamma x$ by assumption we also have that $x_n \in \Gamma_n x_n = \prod_{t=1}^n \Gamma x^t$. According to 4 $S_n \subset X_n$ is externally stable if and only if $\bigcup_{x_n \in S_n} \Gamma_n x_n = X_n$. Consider the hypergraph $H = (X, \mathcal{E})$, where $\mathcal{E} = \{\Gamma x | x \in X\}$, and it's product $H_n = (X_n, \mathcal{E}_n)$. An externally stable set S_n corresponds to a covering of X_n by edges of H_n, and vice versa. The corollary follows therefore from the covering theorem.

B. Clique coverings
We recall that a clique in G is simply a complete subgraph of G. A clique is maximal if it is not properly contained in another clique.

Lemma 2. *Given* $G_n = \prod_1^n G$, *where* G *is a graph with an edge set containing all loops. The maximal cliques* M_n *in* G_n *are exactly those cliques which can be written as* $M_n = \prod_{t=1}^n M^t$, *where the* M^t's *are maximal cliques in* G.

Proof. Products of maximal cliques are a maximal clique in the product graph. It remains to show the converse. Define
$$B^t = \left\{ x^t \mid \exists y_n = (y^1, \ldots, y^t, \ldots, y^n) \in M_n \text{ with } y^t = x^t \right\}; t = 1, 2, \ldots, n.$$
The B^t's are cliques and therefore $B_n = \prod_{t=1}^n B^t$ is a clique in G_n containing M_n. Since M_n is maximal we have that $M_n = B_n$ and also that the B^t's are maximal. The system of cliques $\left\{ M_n^{(i)} \mid i = 1, \ldots, m \right\}$ covers G_n if $\bigcup_{i=1}^m M_n^{(i)} = X_n$. We denote by $m(n)$ the smallest number of cliques needed to cover G_n. Define $\mathcal{M}$ as the set of all maximal cliques in G and define $Q(\mathcal{M})$ as set of all probability distributions on $\mathcal{M}$.

Corollary 5. *Let* G *be a finite graph with all loops in the edge set.* *With* $L = \log \left(\max_{q \in Q(\mathcal{M})} \min_{x \in X} \sum_{M \in \mathcal{M}} 1_M(x) q_M \right)^{-1}$ *the following estimates hold for* $n \in \mathbb{N}$:
a) $m(n) \geq \exp\{Ln\}$
b) $m(n) \leq \exp\{Ln + \log n + \log \log |X|\} + 1$.

Proof. It follows from Lemma 2 that clique coverings for G_n are simply edge coverings of the hypergraph $H_n = \prod_1^n H$, where $H = (X, \mathcal{M})$. The corollary is a consequence of the covering theorem.

Remark 2. *A clique covering of G_n can be interpreted as a colouring of the dual graph G_n^c. This graph can be written as $\prod_1^n G^c$, where the product is to be understood as follows: two vertices $x_n = (x^1, \ldots, x^n)$, $y_n = (y^1, \ldots, y^n) \in X_n$ are joined by an edge if for at least one t, $1 \leq t \leq n$, x^t and y^t are joined. Thus the corollary 5 gives estimates for minimal colorings of *-product graphs. The result of the present section generalize of course to the case of non–identical factor and also to the so called strong product.*

5   A Packing Problem and It's Equivalence to a Problem by Shannon

A. The problem
Instead of asking how many edges are needed to cover the set of all vertices of the hypergraph $H_n = (X_n, \mathcal{E}_n)$ one may ask how many non–intersecting edges can one pack into X_n. Formally, $\mathcal{E}_n' \subset \mathcal{E}_n$ is called a packing in H_n if $E_n \cap E_n' = \varnothing$ for all $E_{n'} E_n' \in \mathcal{E}_n'$. Define the maximal packing number $\pi(n)$ by

$$\pi(n) = \max_{\mathcal{E}_n' \text{ is packing in } H_n} |\mathcal{E}_n'|, \ n \in \mathbb{N}. \tag{32}$$

Using the argument which led to 1 one obtains

$$\pi(n+1) \leq \pi(n) \left(\min_{q \in Q} \max_n \sum_{E \in \mathcal{E}} 1_E(x) q_E \right)^{-1}. \tag{33}$$

The inequality goes in the other direction and the roles of "max" and "min" are exchanged, because we are dealing with packings rather than with coverings. We know from Lemma 1 that $\min_{q \in Q} \max_x \sum_{E \in \mathcal{E}} 1_E(x) q_E = K^*$. Since obviously $\pi(n) \leq t(n)$ inequality 5 becomes trivial. Equality does not hold in general.

Example 2
$X = \{0, 1, 2\}$, $E(j) = \{j, j+1\}$ for $j = 0, 1, 2$. The addition is understood mod 3. In this case $K^* = \frac{2}{3}$ and therefore $t(n) \geq \left(\frac{3}{2}\right)^n$. However, $\pi(n) = 1$ for all $n \in \mathbb{N}$.

B. The dual problem. Independent sets of vertices
The packing problem for the dual hypergraph means the following for the original hypergraph: How many vertices can we select from X such that no two of them are contained in an edge? We are simply asking for the largest cardinality of a strongly independent set of vertices. We recall that $I \subset X$ is called a strongly independent set if and only if

$$|I \cap E| \leq 1 \quad \text{for all} \ E \in \mathcal{E}. \tag{34}$$

$W \subset X$ is called a weakly independent set if and only if

$$|W \cap E| < |E| \quad \text{for all} \ E \in \mathcal{E}. \tag{35}$$

One easily verifies that a strongly independent set is also weakly independent provided that $|E| \geq 2$ for all $E \in \mathcal{E}$. (Loops are excluded.) If $H = H(G)$ is the hypergraph of a graph G without loops, then the two concepts are the same. A weakly independent set for $H(G)$ is simply an internally stable set for $G = (X, \Gamma)$, and conversely. $V \subset X$ is said to be an internally stable set of G if $V \cap \Gamma V = \varnothing$. This implies that no element of V has a loop. We would like to call a set $J \subset X$ with no 2 vertices joined by an edge a Shannon stable or briefly S–stable set of a graph, because this concept has been used by Shannon in [1] and because the difference between the two notions of stability seems not to have been emphasized enough in the literature even though it is significant for product graphs. In an S–stable set elements with loops are permitted. An internally stable set is S–stable. The converse is not necessarily true. $T \subset X$ is a transversal in G if every edge has at least one vertex in T. The complement of an internally stable set in G is a transversal in G, and vice versa. The same relationship holds for weakly independent sets and transversals in hypergraphs. Let $v(G_n)$ be the coefficient of internal stability of G_n, that is, the largest cardinality which can be obtained by an internally stable set in G_n and let $t(G_n)$ be the smallest cardinality for a transversal in G_n. We have

$$t(G_n) = |X|^n - v(G_n), n \in \mathbb{N}. \tag{36}$$

Denoting by $w(G_n)$ the largest cardinality of a weakly independent set in H_n and writing $t(H_n) = t(n)$ we also obtain

$$t(H_n) = |X|^n - w(H_n), n \in \mathbb{N}. \tag{37}$$

Our estimates for $t(H_n)$ (see section 3) can be translated into estimates for $w(H_n)$. However, those hypergraph results have no implications for $t(G_n)$ and $v(G_n)$. This is due to the fact that $H(G_n) \neq \prod_1^n H(G)$ in general. Actually, $v(G_n)$ is not a very interesting function of n. If $G = (X, U)$ is such that U contains all loops then also G_n contains all loops and $v(G_n) = 0$ for all $n \in \mathbb{N}$. If there exists an element $x \in X$ without a loop, then $_xX_{n-1}$ is internally stable in G_n and therefore $\lim_{n \to \infty} \frac{1}{n} \log v(G_n) = \log |X|$. This is also true in this case for $j(G_n)$, the largest cardinality of an S–stable set in G_n. Similarly one can show that $w(H_n) \equiv 0$ if H contains all loops and $\lim_{n \to \infty} \frac{1}{n} \log w(H_n) = \log |X|$ otherwise. In summarizing our discussion we can say that the following problems are unsolved:

1. 1.) The transversal–problem for graphs not containing all loops in the edge set $\big(t(G_n)\big)$.
2. 2.) The S–stability–problem for graphs with all loops in the edge set $\big(j(G_n)\big)$.
3. 3.) The strong independence–problem for hypergraphs $\big(i(H_n)\big)$.
4. 4.) The packing problem for hypergraphs $\big(\pi(n)\big)$.

A solution of 3.) for all hypergraphs is equivalent to a solution of 4.) for all hypergraphs, because the problems are dual to each other. Moreover, we notice that 2.) is a special case of 3.). Suppose that G is a graph with all loops in the edge set and that $H(G)$ is the hypergraph associated with G, then an S–stable set in G_n is a strongly independent set in $H(G)_n$, and conversely. We show that 4.) is a special case of 2.) and therefore that all three problems are equivalent. Let $H = (X, \mathcal{E})$ be a hypergraph. Define a graph $G(H)$ as follows:

Choose $\mathcal{E}$ as set of vertices and join $E, E' \in \mathcal{E}$ by an edge if and only if $E \cap E' \neq \varnothing$. $G(H)$ is a graph with all loops in the edge set and the packings of H_n are in one to one correspondence to the S–stable sets of $G(H)_n$.

C. Shannon's zero error capacity

Problem 2.) is due to Shannon [1]. It is a graph theoretic formulation of the information theoretic problem of determining the maximal number of messages which can be transmitted over a memoryless noisy channel with error probability zero. $\lim_{n \to \infty} \frac{1}{n} \log j(G_n)$ was called in [1] the zero error capacity C_o, say. Using our standard argument (see 1 and 5 one can show that for $G = (X, U)$, where U contains all loops,

$$
j(G_{n+1}) \leq j(G_n) \left(\min_{p \in P} \max_{E \in U} \sum_{x \in X} 1_E(x) p_x \right)^{-1}. \tag{38}
$$

This implies that

$$
C_o \leq \log \left(\min_{p \in P} \max_{E \in U} \sum_{x \in X} 1_E(x) p_x \right)^{-1}. \tag{39}
$$

It has been shown in [6] that for bipartite graphs $j(G_n) = \left(j(G) \right)^n$ for all $n \in \mathbb{N}$ and hence that $C_o = \log j(G)$ in this case. The proof uses the marriage theorem. The simplest non–bipartite graph for which C_o is unknown is the pentagon graph. It was shown in [1] that in this case

$$
\frac{1}{2} \log 5 \leq C_o \leq \log \frac{5}{2}. \tag{40}
$$

The lower bound is an immediate consequence of the equation $j(G_2) = 5$. The upper bound follows also from 5. No improvement has been made until now on any of those bounds. We have been able to prove that

$$
j(G_3) = 10, \ j(G_4) = 25, \ j(G_5) = 50, \ \text{and} \ j(G_6) = 125. \tag{41}
$$

We conjecture that

$$
j(G_{2n}) = 5^n, \ j(G_{2n+1}) = 2.5^n \ \text{for all} \ n \in \mathbb{N}, \tag{42}
$$

but so far we have no proof for $n > 6$. 5 would imply $C_o = \frac{1}{2} \log 5$. The result announced in 5 and results which go beyond this (including colouring problems) will appear elsewhere. We would like to mention that we came to the covering problem by trying to understand the results of [5] from a purely combinatorial point of view. Those results can be understood as statements about "packings with small overlapping and an additional weight assignment". It seems to us that the methods of [5] allow refinements which may be helpful for the *construction* of minimal coverings. We expect that the covering theorem has applications in Approximation Theory, in particular for problems involving ε–entropy (see [7]).

It might also be of some interest to compare our estimates with known results (see [8]) on coverings with convex sets in higher dimensional spaces.

References

1. C.E. Shannon, The zero–error capacity of a noisy channel, IRE Trans. Inform. Theory, IT-2, 8–19, 1956.
2. C.E. Shannon, Certain results in Coding Theory for noisy channels, Inform. and Control, 1, 6–25, 1957.
3. C. Berge, Graphes et Hypergraphes, Monographies universitaires de mathématiques. Dunod Paris, 1970.
4. J. von Neumaun, O. Morgenstern, Theory of Games and Economic Behaviour, Princeton, Princeton University Press, 1944.
5. R. Ahlswede, Channel capacities for list codes, J. Appl. Probability, 10, 824–836, 1973.
6. A.G. Markosian, The number of internal stability in a cartesian product graph and it's application to information theory (In Russian), Presented at the Second International Symposium on Information Theory, Sept., Tsahkadsor, Armenian SSR, 2–8, 1971.
7. A. N. Kolmogoroff, W.M. Tichomirow, ε–Entropie and ε–Kapazität von Mengen in Funktionalräumen (Translation from the Russian), Arbeiten zur Informationstheorie III, Mathematische Forschungsberichte, VEB Deutscher Verlag der Wissenschaften, Berlin 1960..
8. C.A. Rogers, Packing and Covering, Cambridge Tracts in Mathematics and Mathematical Physics, University Press, Cambridge, 1964.

Testing Sets for 1-Perfect Code

S.V. Avgustinovich and A. Yu. Vasil'eva

This paper continues the research of [1,2]. In [1] it was shown that a 1-perfect code is uniquely determined by its vertices at the middle levels of hypercube and in [2] the concerned formula was obtained. Now we prove that the vertices at the r-th level, $r \leq (n-1)/2$, of such a code of length n uniquely determine all code vertices at the lower levels.

1. Main result. We denote n-dimensional vector space over $GF(2)$ by E^n and call it a hypercube. We consider Hamming metric in E^n, i.e. the distance $\rho(\mathbf{x}, \mathbf{y})$ between vertices $\mathbf{x}$ and $\mathbf{y}$ of hypercube is equal to the number of positions in which the vertices differ. Hamming weight $wt(\mathbf{x})$ of vertex $\mathbf{x}$ is equal to the number of nonzero positions of $\mathbf{x}$. Denote by $S_r(\mathbf{x})$ $(B_r(\mathbf{x}))$ a sphere (a ball) of radius r with the center $\mathbf{x}$. The sphere $S_r(\mathbf{0}) = W_r$ centered in the all-zero vertex $\mathbf{0}$ is called a r-th level of hypercube.

A ϑ-*centered function* $f : E^n \to \mathbf{R}$ is a function such that the sum of its values in a ball of the radius 1 is equal to ϑ. A *perfect binary single-error-correcting code* C (briefly a *1-perfect code*) of length n is a subset of E^n such that the set $\{B_1(\mathbf{x}) \mid \mathbf{x} \in C\}$ is a partition of E^n. The characteristic function of a 1-perfect code is 1-centered.

Let Φ be a family of real functions over E^n and $A, B \subseteq E^n$. We call A a B-*testing set* for the family Φ if for any $f, g \in \Phi$ the following condition holds: if $f(x) = g(x)$ for any $x \in A$ then $f(x) = g(x)$. in other words if one knows the values of an arbitrary function $f \in \Phi$ over A then one can find all values of f over the set B. Let Φ^ϑ be the family of all ϑ-centered functions over E^n. The main result and an easy consequence are

Theorem 1. *Let ϑ be real, $\mathbf{v} \in E^n$ and $r \leq (n-1)/2$. Then the $S_r(\mathbf{v})$ is the $B_r(\mathbf{v})$-testing set for the family Φ^ϑ.*

The next Theorem can be easily derived from the previous Theorem.

Theorem 2. *Let $\mathbf{v} \in E^n$ and $r \leq (n-1)/2$. The set of vertices of a 1-perfect code of weight less than r is uniquely determined by the set of all code vertices of weight r.*

2. Local spectra. A k-dimensional *face* γ of the hypercube is the set of all vertices with fixed $n - k$ coordinates. Fix a vertex $\mathbf{x} \in \gamma$. The local spectrum of a function $f : E^n \to \mathbf{R}$ in the face γ with respect to the vertex $\mathbf{x}$ is the $(k+1)$-dimensional vector

$$l(\mathbf{x}) = (l_0(\mathbf{x}), l_1(\mathbf{x}), \ldots, l_k(\mathbf{x})), \quad \text{where } l_i(\mathbf{x}) = \sum_{\mathbf{y} \in \gamma, \ \rho(\mathbf{x}, \mathbf{y})=i} f(\mathbf{y}).$$

R. Ahlswede et al. (Eds.): Information Transfer and Combinatorics, LNCS 4123, pp. 938–940, 2006.
© Springer-Verlag Berlin Heidelberg 2006

We can consider the partial ordering in E^n: $\mathbf{x} \preceq \mathbf{y}$ if $x_i \leq y_i$, $i = 1, \ldots, n$, where $\mathbf{x} = (x_1, \ldots, x_n)$ and $\mathbf{y} = (y_1, \ldots, y_n)$. For a vertex $\mathbf{x}$ of hypercube we denote

$$\gamma_{\mathbf{x}} = \{\mathbf{y} \in E^n \mid \mathbf{y} \preceq \mathbf{x}\}, \gamma_{\mathbf{x}}^{\perp} = \{\mathbf{y} \in E^n \mid \mathbf{y} \succeq \mathbf{x}\}.$$

Pares of faces like this are called orthogonal. Note that $\dim \gamma_{\mathbf{x}} = wt(\mathbf{x})$ and $\dim \gamma_{\mathbf{x}}^{\perp} = n - wt(\mathbf{x})$. Denote by $l(\mathbf{x}) = (l_0(\mathbf{x}), \ldots, l_{wt(\mathbf{x})}(\mathbf{x}))$ the local spectrum of a function f in the face $\gamma_{\mathbf{x}}$ with respect to $\mathbf{x}$ and by $l^{\perp}(\mathbf{x}) = (l_0^{\perp}(\mathbf{x}), \ldots, l_{n-wt(\mathbf{x})}^{\perp}(\mathbf{x}))$ the local spectrum of a function f in the face $\gamma_{\mathbf{x}}^{\perp}$ with respect to the vertex $\mathbf{x}$.

The local spectra of a perfect code in two orthogonal faces with respect to their common vertex were proved [5] to be in the tight interdependence. More generally, it holds

Theorem 3. *Let f be a 0-centered function, $\mathbf{x} \in W_i$, $i \leq (n-1)/2$. Then for any j, $0 \leq j \leq n - i$, it holds*

$$l_j^{\perp}(\mathbf{x}) = \sum_{q+r=j} (-1)^q p_r^i l_q(\mathbf{x}), \quad where \quad p_{2t}^i = -p_{2t+1}^i = (-1)^t \binom{(n-1)/2 - i}{t}.$$

The proof of this theorem repeats almost literally the proof of its special case concerned perfect codes [5]. Now we are ready to prove Theorem 1 in the the case $\vartheta = 0$.

3. Proof of Theorem 1. Let $\vartheta = 0$ and f be an arbitrary 0-centered function over E^n. Without loss of generality we consider $\mathbf{v} = \mathbf{0}$. We will prove that the r-th level of hypercube is the D_r-testing set for 0-centered functions, where $D_r = W_0 \bigcup W_1 \bigcup \ldots \bigcup W_r$. We will show by induction on i that r-th level W_r of hypercube is W_i-testing set for 0-centered functions.

The base of induction: $i = 0$. Analogously with the case of 1-perfect codes [3,4] it can be obtained that

$$\sum_{\mathbf{a} \in E^n, \; wt(\mathbf{a})=r} f(\mathbf{a}) = p_r^0 f(\mathbf{0})$$

Thus $f(\mathbf{0})$ is uniquely determined by the values of the function f over W_r.

The step of induction. Suppose that W_r is proved to be the D_{i-1}-testing set for 0-centered functions. Let $\mathbf{x} \in W_i$, $i = 1, \ldots, r - 1$. From Theorem 3 we have:

$$l_{r-i}^{\perp}(\mathbf{x}) = \sum_{q=0}^{r-i} (-1)^q p_{r-i-q}^i l_q(\mathbf{x}), \; f(\mathbf{x}) = l_0(\mathbf{x}) = \frac{1}{p_{r-i}} \left(l_{r-i}^{\perp}(\mathbf{x}) - \sum_{q=1}^{r-i} (-1)^q p_{r-i-q}^i l_q(\mathbf{x}) \right).$$

Here p_{r-i} is nonzero and $l_{r-i}^{\perp}(\mathbf{x}) = \sum_{\mathbf{a} \succeq \mathbf{x}, \; wt(\mathbf{a})=r} f(\mathbf{a})$, $l_q(\mathbf{x}) = \sum_{\mathbf{y} \preceq \mathbf{x}, \; wt(\mathbf{y})=i-q} f(\mathbf{y})$.

By induction supposition the last sum is uniquely determined by the values of f over the r-th level. Hence $f(\mathbf{x})$ is also uniquely determined by the values of f over the r-th level and W_r is the D_i-testing set. Theorem 1 is proved.

References

1. S.V. Avgustinovich, On a property of perfect binary codes, Discrete Analysis and Operation Research (in Russian), Vol. 2, No. 1, 4-6, 1995.
2. S.V. Avgustinovich and A. Y. Vasil'eva, Reconstruction of centered functions by its values on two middle levels of hypercube, Discrete Analysis and Operation Research (in Russian), Vol. 10, No. 2, 3-16, 2003.
3. S.P. Lloyd, Binary block coding, Bell Syst. Techn. J., Vol. 36. No. 2, 517-535, 1957.
4. H.S. Shapiro and D.L. Slotnick, On the mathematical theory of error correcting codes, IBM J. Res. Develop. 3., No. 1, 25-34, 1959.
5. A.Y. Vasil'eva, Local spectra of perfect binary codes, Discrete Analysis and Operation Research (in Russian), Vol. 6, No. 1, 16-25, 1999.

On Partitions of a Rectangle into Rectangles with Restricted Number of Cross Sections

R. Ahlswede and A.A. Yudin*

Abstract. Consider partitions of a rectangle into rectangles with restricted number of cross sections. The problem has an information theoretic flavor (in the spirit of [3]): richness of two dimensional pictures under 1–dimensional restrictions of from local information to global information.

We present two approaches to upper bound the cardinality of such partitions: a combinatorial recursion and harmonic analysis.

1 The Problem

Given a partition of a rectangle into rectangles, such that any line parallel to a side of the initial rectangle intersects at most n rectangles of the partition. The task is to estimate the maximal number of rectangles of the partition. We make now our concept of partition precise and consider a more general problem.

Definition 1. We say that a family of rectangles $\mathcal{S}$ gives a *partition* of the rectangle F, if

$$\bigcup_{A \in \mathcal{S}} A = F \qquad (1.1)$$

and for $\forall A, B \in \mathcal{S}$, $A \neq B$ the intersection of A and B contains at most boundary points.

Definition 2. We say that the partition $\mathcal{S}_F$ has the property (m, n), if any line parallel to the *base* of the rectangle F intersects at most m rectangles of the partition, and perpendicular to the base — at most n rectangles.

The problem is to find

$$f(m, n) = \sup \operatorname{card}(\mathcal{S}_F) = \sup |\mathcal{S}_F|, \qquad (1.2)$$

where the supremum is taken over all partitions $\mathcal{S}_F$ with property (m, n).

It is evident that $f(m, n) = f(n, m)$. In case $m = n$ one readily sees that

$$f(n, n) \geq 2 \, f(n-1, n-1) + 2 \geq 2^n \text{ for } n \geq 2. \qquad (1.3)$$

It seems natural to conjecture that the first inequality is actually an equality. By successively improving our methods we derive here a decreasing sequence of upper bounds on $f(m, n)$ and finally come close to the conjectured bound.

---

* Recently we learnt that a paper on such problems appeared already in 1993: D.J. Kleitman "Partitioning a rectangle into many sub-rectangles so that a line can meet only a few", DIMACS Series in Discrete Mathematics and Theoretical Computer Science, Volume 9, 95-107, 1993.

R. Ahlswede et al. (Eds.): Information Transfer and Combinatorics, LNCS 4123, pp. 941–954, 2006.
© Springer-Verlag Berlin Heidelberg 2006

2   A Crude Estimate

Let us look at

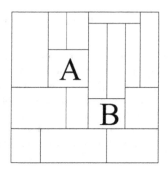

Fig. 1.

We associate with the partition $\mathcal{S}_F$ a graph $G_{\mathcal{S}}$. Every $A \in \mathcal{S}_F$ corresponds to a vertex of $G_{\mathcal{S}}$. For all $A, B \in \mathcal{S}_F$, $A \neq B$, there is an edge that connects them, iff A and B have common boundary, different from a point.

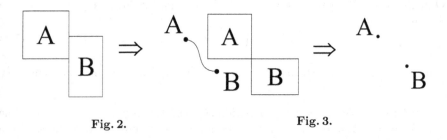

Fig. 2.   Fig. 3.

Now we state some properties of the graph $G_{\mathcal{S}}$.

1) We establish an upper bound for degrees of the vertices of this graph.

A vertex A is incident to as many edges as the number of rectangles which have a common boundary with the rectangle A. It is evident that along any horizontal side there are at most m of such rectangles, and along any vertical side there are at most n of them, and thus around the rectangle A there are at most $2(m + n)$ rectangles.

We have, therefore that for every vertex $v \in G_{\mathcal{S}}$

$$\deg v \leq 2(m + n). \tag{2.1}$$

2) Now we find an upper bound for the *diameter* of the graph $G_{\mathcal{S}}$.

Given 2 vertices A and B (see Figure 4), let the vertex X correspond to the rectangle that contains the point of intersection of the lines ℓ_1 and ℓ_2, $A \in \ell_1$ and $\ell_1 \| \alpha\delta$, $B \in \ell_2$ and $\ell_2 \| \gamma\delta$.

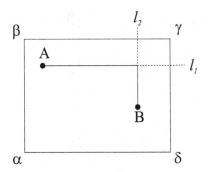

Fig. 4.

By definition, ℓ_1 intersects at most m rectangles of the partition, and ℓ_2 intersects at most n rectangles, hence the vertices A and X are connected by at most m edges, and X and B by at most n edges.

Therefore the diameter $d(G_\mathcal{S}) \leq m + n$. Thus we have

1) $\forall v \in G_\mathcal{S}$ $\deg v \leq 2(m+n)$
2) $d(G_\mathcal{S}) \leq m + n$.

By inequality (1) on page 171 of [1]

$$f(m,n) \leq 1 + 2(m+n)\frac{(2(m+n)-1)^{m+n}-1}{2(m+n)-2} \leq \big(2(m+n)\big)^{m+n+1}. \quad (2.2)$$

3   A Sharper Estimate

Choose on the side $\alpha\beta$ a rectangle P with the biggest height h.

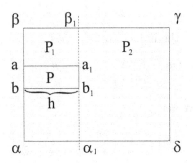

Fig. 5.

Then partition the rectangle $\alpha\beta\gamma\delta$ by line $\alpha_1\beta_1$ in two rectangles P_1 on the left side and P_2 on the right side of $\alpha_1\beta_1$.

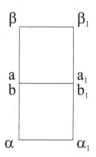

Fig. 6.

Adjust the rectangle P_1 by throwing away the rectangle P, and joining together the resulting two ones in the following way:

We get a new rectangle P_1^*.

Now we find the parameters (m', n') of the rectangles P_1^* and P_2.

For P_1^* it is evident that $m' \leq m$ and $n' \leq n-1$. For P_2 it is evident that $m' \leq m-1$, $n' \leq n$.

Thus we get

$$f(m,n) \leq 1 + f(m, n-1) + f(m-1, n). \tag{3.1}$$

By adding 1 to both sides of this inequality and denoting $f(m,n)+1$ by $g(m,n)$, we get

$$g(m,n) \leq g(m, n-1) + g(m-1, n). \tag{3.2}$$

By induction on $m+n$ we prove that

$$g(m,n) \leq \binom{m+n}{n}. \tag{3.3}$$

Indeed,

$$g(n,1) = g(1,n) = n+1 \leq \binom{1+n}{n}$$

and finally by (3.2) and the induction hypothesis

$$g(m,n) \leq \binom{m+n-1}{n-1} + \binom{m+n-1}{n} = \binom{m+n}{n}.$$

In the case $m = n$ we get

$$f(n,n) \leq \binom{2n}{n} - 1 \sim c\frac{4^n}{\sqrt{n}}. \tag{3.4}$$

4   A Still Sharper Estimate

The beginning is the same as in Section 3. Let us examine the estimate there more carefully.

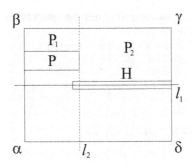

Fig. 7.

Let a horizontal line ℓ_1 intersect m rectangles of the rectangle P_1. Then there exists a rectangle H of the partition from the side $\gamma\delta$ reaching the line ℓ_2. The parameters of rectangles P_1 and P_2, after the reconstruction satisfy

$$m'(P_1^*) \leq m \text{ and } n'(P_1^*) \leq n - 1$$

$$m'(P_2^*) \leq m - s \text{ and } n'(P_2^*) \leq n - 1.$$

So in this case the following inequality holds

$$f(m,n) \leq f(m, n-1) + f(m-1, n-1) + 1.$$

Again, by adding 1 to both sides of this inequality and denoting $f(m,n) + 1$ by $g(m,n)$, we obtain

$$g(m,n) \leq g(m, n-1) + g(m-1, n-1). \tag{4.1}$$

Thus if there exists a line ℓ_1, which intersects m rectangles of the rectangle P_1, then the inequality (4.1) holds.

Assume now that there is no such a line ℓ_1, then the parameters m' and n' for P_1^* and P_2^* are

$$m'(P_1^*) \leq m - 1 \text{ and } n'(P_1^*) \leq n - 1,$$

$$m'(P_2^*) \leq m - 1 \text{ and } n'(P_2^*) \leq n.$$

Thus the following inequality holds

$$g(m,n) \leq g(m-1, n-1) + g(m-1, n). \tag{4.2}$$

Combining (4.1) and (4.2) we get the following inequalities

$$g(m,n) \leq \begin{cases} g(m-1, n-1) + g(m-1, n) \\ g(m, n-1) + g(m-1, n-1) \end{cases} \tag{4.3}$$

We search for an estimate of the form

$$g(m,n) \leq c \, \lambda^{(m+n)}, \tag{4.4}$$

where the constants c and λ are not known yet. We will determine them later.

We get the estimate by induction. First let us determine λ. We have

$$
\begin{aligned}
g(m,n) &\leq \max\{g(m-1,n-1)+g(m-1,n), g(m,n-1)+g(m-1,n-1)\} \\
&\leq \max\{c\lambda^{m+n-2}+c\lambda^{m+n-1}, c\lambda^{m+n-1}+c\lambda^{m+n-2}\} \\
&\leq c(\lambda^{m+n-2}+\lambda^{m+n-1}) = c\lambda^{m+n}\left(\frac{1}{\lambda^2}+\frac{1}{\lambda}\right).
\end{aligned}
$$

To be able to perform induction we need

$$
\frac{1}{\lambda^2}+\frac{1}{\lambda} \leq 1,
$$

from which after calculations we find that we can set $\lambda_0 = \frac{\sqrt{5}+1}{2}$.

The constant c in (4.4) is determined from the initial step of the induction:

$$
g(1,n) = n+1 \leq c\left(\frac{\sqrt{5}+1}{2}\right)^{n+1}, \quad \text{and thus } c = \sup_n \frac{n+1}{\left(\frac{\sqrt{5}+1}{2}\right)^{n^2}} \leq 1.
$$

Then we get

$$
g(m,n) \leq \left(\frac{\sqrt{5}+1}{2}\right)^{m+n}. \tag{4.5}
$$

In the case $m = n$ this implies

$$
f(n,n) \leq \left(\frac{\sqrt{5}+1}{2}\right)^{2n} - 1 < (1,62)^{2n}, \tag{4.6}
$$

which is of course better than $\frac{4^n}{\sqrt{n}}$ in (3.4), but much worse than the expected 2^n.

5   The Best Estimate Found so far by the Method of Recursion

First we give some concepts.

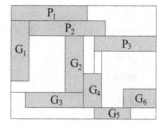

Fig. 8.

Again, given a rectangle F and a partition S of it, we call a sequence of rectangles $P_1, P_2, \ldots, P_k$ ($P_i \neq P_j, i \neq j$) a chain from the side $(\alpha\beta)$ to the side $(\gamma\delta)$ if any line $\ell \| (\alpha\beta)$ which intersects F, intersects at least one of $P_1, \ldots, P_k$.

Example: P_1, P_2, P_3 is a chain, $G_1, G_2, G_3, G_4, G_5, G_6$ is a chain.
We call a chain minimal if the number of rectangles in it is minimal. A minimal chain is extremal, if the sum of the lengths along the side $(\alpha\delta)$ of the rectangles of this chain is maximal (that is among all minimal chains we take the chain of maximum length).
It is straightforward to see that extremal chains exist.

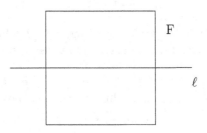

Fig. 9.

Indeed, as ℓ intersects no more than m rectangles of S, which evidently form a chain, there are chains. Since the cardinality of S is a finite number, there is a chain with minimal number of elements, that is a minimal chain. Now the existence of extremal chains is evident, as the number of all chains, and so of minimal ones, is finite.

We classify now a partition S of the rectangle F by the cardinality of its horizontal and vertical minimal chains.

We begin the consideration with the chains of cardinality ≥ 2. Note that if there is a chain of cardinality one, then the partition S has the form

Fig. 10.

and this case is trivial to handle.
First we prove the following result.

Theorem 1. *Let a partition S with parameters m, n of a rectangle F have an extremal chain of cardinality ≥ 5. Then*

$$|S| \leq 2f(m-3, n-1) + f(m-2, n) + 2.$$

Proof

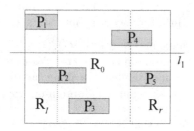

Fig. 11.

Without loss of generality let the rectangles P_1, P_2, P_3, P_4 and P_5 form an extremal chain. Partition the rectangle F into three rectangles R_ℓ, R_0, R_r (as in Figure 11).

We estimate the parameters m', n' for the rectangles R_ℓ, R_0, R_r. As after the adjustment the rectangle P_1 will be thrown away, we have

$$n' \text{ for } R_\ell \le n - 1. \tag{5.1}$$

Now notice that P_3, P_4, P_5 cannot intersect R_ℓ. Indeed, assume that $P_3 \cap R_\ell \ne \varnothing$. Then P_1, P_3, P_4, P_5 form a chain of cardinality 4, which is a contradiction, since the chain P_1, P_2, P_3, P_4, P_5 is extremal. Similarly $P_4 \cap R_\ell = R_\ell \cap P_5 = \varnothing$.

Now we show that any line ℓ_1 intersects at most $m - 3$ rectangles of the partition $\mathcal{S}$, which belong to rectangle R_ℓ. Assume to the opposite that ℓ_1 intersects not less than $m - 2$ rectangles of R_ℓ. Denote by Q_1 the right most one of them. As there are no more than m rectangles lying on ℓ_1, there are at most 2 of them that do not intersect R_ℓ. Denote them by Q_2 and Q_3. Again we come to a contradiction, since P_1, Q_1, Q_2, Q_3 form a chain, while P_1, P_2, P_3, P_4, P_5 is extremal. Thus we showed that

$$m' \text{ for } R_\ell \le m - 3. \tag{5.2}$$

Similarly for R_r

$$m' \le m - 3 \text{ and } n' \le n - 1. \tag{5.3}$$

For R_0 evidently

$$m' \le m - 2 \text{ and } n' \le n. \tag{5.4}$$

Combining (5.1), (5.2), (5.3) and (5.4) we get the claim of Theorem 1.

Corollary. *If any partition $\mathcal{S}$ with parameters m and n has an extremal chain of cardinality ≥ 5, then*

$$f(m, n) \le 2^{\frac{m+n}{2}}.$$

Indeed, if the condition is fulfilled, then by Theorem 1

$$g(m, n) \le 2g(m - 3, n - 1) + g(m - 2, n),$$

where $g(m, n) = f(m, n) + 1$. Assuming, that the following estimate

$$g(m, n) \le c \cdot \lambda^{\frac{m+n}{2}}$$

is correct for proper c and λ, we get that

$$g(m,n) \leq 2c\lambda^{\frac{m+n}{2}} \cdot \lambda^{-2} + c\lambda^{\frac{m+n}{2}} \cdot \lambda^{-1} = c\lambda^{\frac{m+n}{2}} \left(\frac{2}{\lambda^2} + \frac{1}{\lambda} \right) = c2^{\frac{m+n}{2}}, \text{ for } \lambda = 2.$$

As $f(1,n) = n$, the constant c can be found from the condition

$$1 \leq n \leq c \cdot 2^{\frac{n+1}{2}} \text{ for all } n \in \mathbb{N}.$$

It is evident that we can take $c = 1$.

From now on only partitions with extremal chains of cardinality ≤ 4 will be considered.

First we consider partitions $\mathcal{S}$ with chains of length 2.

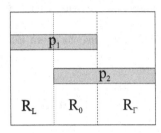

Fig. 12.

Now we find all possible collections of parameter values for rectangles R_ℓ, R_0, R_r.

N	R_ℓ m', n'	R_r m', n'	R_0 m', n'	Inequality	Solutions $\leq \sqrt{2}$
1	$m-1, n-3$	$m-1, n-3$	$m, n-2$	$\frac{2}{\lambda^4} + \frac{1}{\lambda^2} \leq 1$	yes
2	$m-1, n-2$	$m-1, n-3$	$m, n-3$	$\frac{2}{\lambda^3} + \frac{1}{\lambda^4} \leq 1$	yes
3	$m-1, n-3$	$m-1, n-2$	$m-1, n-2$	$\frac{2}{\lambda^3} + \frac{1}{\lambda^4} \leq 1$	yes
4	$m-1, n-2$	$m-1, n-2$	$m-2, n-2$	$\frac{2}{\lambda^3} + \frac{1}{\lambda^4} \leq 1$	yes
5	$m-1, n-3$	$m-1, n-2$	$m, n-3$	$\frac{2}{\lambda^3} + \frac{1}{\lambda^4} \leq 1$	yes
6	$m-1, n-2$	$m-1, n-2$	$m-1, n-3$	$\frac{2}{\lambda^3} + \frac{1}{\lambda^4} \leq 1$	yes
7	$m-1, n-3$	$m-1, n-2$	$m-1, n-3$	$\frac{2}{\lambda^3} + \frac{1}{\lambda^4} \leq 1$	yes
8	$m-1, n-2$	$m-1, n-1$	$m-2, n-3$	$\frac{1}{\lambda^2} + \frac{1}{\lambda^3} + \frac{1}{\lambda^5} \leq 1$	no
9	$m-1, n-2$	$m-1, n-3$	$m, n-3$	$\frac{2}{\lambda^3} + \frac{1}{\lambda^4} \leq 1$	yes
10	$m-1, n-1$	$m-1, n-3$	$m-1, n-3$	$\frac{1}{\lambda^2} + \frac{2}{\lambda^4} \leq 1$	yes
11	$m-1, n-2$	$m-1, n-2$	$m-1, n-3$	$\frac{2}{\lambda^3} + \frac{1}{\lambda^4} \leq 1$	yes
12	$m-1, n-1$	$m-1, n-2$	$m-2, n-3$	$\frac{1}{\lambda^2} + \frac{1}{\lambda^3} + \frac{1}{\lambda^5} \leq 1$	no
13	$m-1, n-2$	$m-1, n-2$	$m, n-4$	$\frac{2}{\lambda^3} + \frac{1}{\lambda^4} \leq 1$	yes
14	$m-1, n-1$	$m-1, n-2$	$m-1, n-4$	$\frac{1}{\lambda^2} + \frac{1}{\lambda^3} + \frac{1}{\lambda^5} \leq 1$	no
15	$m-1, n-2$	$m-1, n-1$	$m-1, n-4$	$\frac{1}{\lambda^3} + \frac{1}{\lambda^2} + \frac{1}{\lambda^5} \leq 1$	no
16	$m-1, n-1$	$m-1, n-1$	$m-2, n-4$	$\frac{2}{\lambda^2} + \frac{1}{\lambda^6} \leq 1$	no

Here a few words about the table. The general procedure of finding an upper bound for $f(m,n)$ is the same as earlier: for each row of the values we get a recursion for $g(m,n) = f(m,n) + 1$. Then we search for an upper bound of the form $g(m,n) \leq c \cdot \lambda^{m+n}$ with constants c and λ to be specified. The inequalities in the 5-th column of the table guarantee that we can perform induction to prove the upper bound. So, their solutions can be specified as appropriate values for λ. The last column shows that in all cases (except those with entry "no", which are discussed below) one obtains the upper bound by setting $\lambda = \sqrt{2}$. The value of c can be taken to be equal 1.

As it can be seen from the table we get only two inequalities that give solutions $> \sqrt{2}$, namely

$$\frac{1}{\lambda^2} + \frac{1}{\lambda^3} + \frac{1}{\lambda^5} \leq 1 \text{ and } \frac{2}{\lambda^2} + \frac{1}{\lambda^6} \leq 1.$$

But in these cases we can put $\lambda = 1.5$.

Now consider the case, when the partition S has an extremal chain of length ≥ 3.

As there are just a few cases, we discuss them all. The following 4 cases are possible.

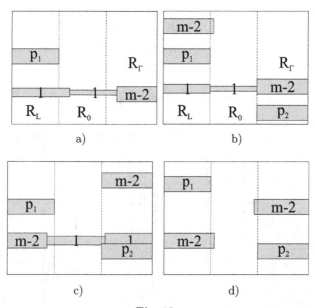

a)   b)

c)   d)

Fig. 13.

In Figures a, b, c, d a rectangle with number x inside means that a line intersecting it would intersect exactly x rectangles of partition S.

According to the four cases we obtain respectively:

a) $g(m,n) \leq g(m-1,n-1) + g(m-1,n-1) + g(m-2,n-2)$
b) $g(m,n) \leq g(m-2,n-1) + g(m-1,n-1) + g(m-2,n-1)$
c) $g(m,n) \leq g(m-1,n-1) + g(m-2,n-1) + g(m-2,n-1)$
d) $g(m,n) \leq g(m-2,n-1) + g(m-2,n-1) + g(m-2,n-1)$

The calculations show that the worst case is a). We get the inequality $\frac{2}{\lambda^2} + \frac{1}{\lambda^4} \leq$ 1 with solution $\lambda = \sqrt{\sqrt{2}+1}$.

Thus we proved

Theorem 2. *For all* $m, n \geq 1$ *we have* $f(m,n) \leq (\sqrt{2}+1)^{\frac{m+n}{2}}$. *Thus* $f(n,n) \leq$ $(2.414)^n$.

Remark 1. As it follows from Theorem 1, it is enough to examine partitions with extremal chains of cardinality ≤ 4

2. So far we studied configurations only for chains in one dimension.

One can also consider pairs of chains. In that more general setting, the study of the following cases (see Figure 14) would yield sharper upper bounds for $f(m,n)$.

$\gamma(\ell_2) \setminus \gamma(\ell_1)$	2	3	4
2	(2,2)	(2,3)	(2,4)
3		(3,3)	(3,4)
4			(4,4)

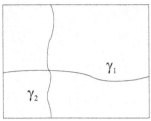

Let γ_1 and γ_2 be chains of cardinalities $\ell(\gamma_1)$ and $\ell(\gamma_2)$ resp.

Fig. 14.

Such an analysis, even for the case (2,2), is rather complicated.

6   Harmonic Analysis

Given a partition S of a rectangle F.

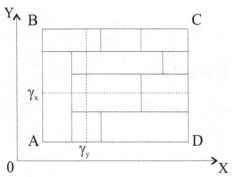

Fig. 15.

We identify the opposite sides of the rectangle, so that F becomes a torus and consider our problem on the torus F.

We have a partition S of the torus into rectangles such that any coordinate circle on F intersects S in the following way:

Each γ_x intersects at most m rectangles and each γ_y at most n rectangles of the partition.

The problem is to estimate the number of rectangles in partition S, that is $|S|$.

Without loss of generality we may assume that $ABCD$ is a square with sides of length 1.

Let $L_y(\varepsilon)$ be a strip of width ε on the torus that is parallel to axis OX,

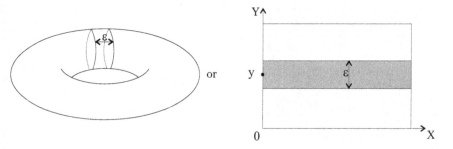

Fig. 16.

where y is the OY coordinate of the center of that strip. Let

$$\Phi_\varepsilon(y) = \sum_{P \in S \text{ and } P \cap L_y(\varepsilon) \neq \varnothing} 1,$$

that is $\Phi_\varepsilon(y)$ is equal to the number of rectangles of the partition S, which have nonempty intersection with $L_y(\varepsilon)$.

Write the Fourier series for $\Phi_\varepsilon(y)$:

$$\Phi_\varepsilon(y) = c_0 + \sum_{m : m \neq 0} c_m \, \ell(m, y).$$

By the identity of Parseval we get

$$\int_0^1 \Phi_\varepsilon^2(y) dy = c_0^2 + \sum_{m : m \neq 0} c_m^2 \text{ and thus } \int_0^1 \Phi_\varepsilon^2(y) dy \geq c_0^2.$$

Now find c_0:

$$c_0 = \int_0^1 \left(\sum_{P \in S \text{ and } P \cap L_y(\varepsilon) \neq \varnothing} 1 \right) dy = \sum_{P \in S} \int_{\{y : P \cap L_y(\varepsilon) \neq \varnothing\}} dy = \sum_{P \in S} (\mu(P) + 2\varepsilon) = \sum_{P \in S} \mu(P) + 2\varepsilon |S|,$$

where $\mu(P)$ is the length of the side of P parallel to axis OY. Let us now look at the integral $\int_0^1 \Phi_\varepsilon(y)dy$ from another point of view. $\Phi_\varepsilon(y)$ is equal to the number of rectangles of the partition, which intersect the boundary of the strip $L_y(\varepsilon)$, evidently if there are no rectangles that are completely contained in $L_y(\varepsilon)$.

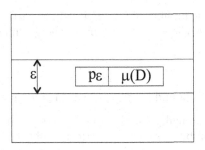

Fig. 17.

As the partition $\mathcal{S}$ has the property (m, n), we can estimate $\Phi_\varepsilon(y) \leq 2m$, if we don't take into consideration the rectangles, for which $\mu(P) < \varepsilon$.

By restricting ourselves to the set of rectangles $\mathcal{S}(\varepsilon) = \{P \in \mathcal{S} : \mu(P) > \varepsilon\}$, we obtain

$$\left(\sum_{P \in \mathcal{S}(\varepsilon)} \mu(P) + 2\varepsilon|\mathcal{S}(\varepsilon)| \right) \cdot c_0 \leq c_0^2 \leq \int_0^1 \Phi_\varepsilon^2(y)dy \leq \max_y \Phi_\varepsilon(y) \int_0^1 \Phi_\varepsilon(y)dy \leq 2m \cdot c_0$$

and hence

$$\sum_{P \in \mathcal{S}(\varepsilon)} \mu(P) + 2\varepsilon|\mathcal{S}(\varepsilon)| \leq 2m.$$

Therefore

$$2\varepsilon|\mathcal{S}(\varepsilon)| \leq 2m - \sum_{P \in \mathcal{S}(\varepsilon)} \mu(P) \text{ implying } |\mathcal{S}(\varepsilon)| \leq \frac{m}{\varepsilon}.$$

Thus we proved that the number of rectangles of the partition $\mathcal{S}$ with height $\geq \varepsilon$ does not exceed $\frac{m}{\varepsilon}$. Dealing similarly with the other dimension, we get that the number of rectangles of the partition $\mathcal{S}$ with length $\geq \varepsilon$ does not exceed $\frac{n}{\varepsilon}$. In summary

Theorem 3. *The number of rectangles of partition $\mathcal{S}$ with exception of those, both sides of which have length less than ε, does not exceed*

$$\frac{m}{\varepsilon} + \frac{n}{\varepsilon} = \frac{m+n}{\varepsilon}.$$

Corollary. *Given a square K_n on checked paper with the length of side 2^n, which is partitioned into rectangles (that consist of entire checks), so that any* *line parallel to a side of the square intersects at most n rectangles, then the number of rectangles of the partition is $\leq 2n \cdot 2^n$.*

Proof: Apply to K_n homothety with coefficient 2^{-n}, and take in the previous claim

$$\varepsilon = 2^{-n}.$$

Then we get that the number of rectangles of the partition which have at least one dimension not less than $\varepsilon = 2^{-n}$ (and in the given case they all do satisfy this condition) is

$$f(n,n) \leq \frac{n+n}{2^{-n}} = 2n \cdot 2^N.$$

Recall that the lower bound for $f(n,n)$, which is conjectured to be tight, is 2^n.

Remark: The main difficulty for the analytic approach to this problem is the tight interlacing of combinatorial and metrical — old structures of the partition.

It seems, that if there are "small" rectangles in the structure of the partition, then the partition can be adjusted so that these small rectangles are removed, while preserving the combinatorial structure. But we don't know yet how to do it.

References

1. B. Bollobás, Extremal Graph Theory, Academic Press, London, 1978.
2. J. Bourgain, Harmonic analysis and combinatorics: how much may they contribute to each other? Published in Mathematics: Frontiers and Perspectives, V. Arnold, M. Atiyah, P. Lax, and B. Mazur, Editors, 13–32, 2000.
3. R. Ahlswede, Coloring hypergraphs: A new approach to multi–user source coding, Part I, Journ. of Combinatorics, Information and System Sciences, Vol. 4, No. 1, 76–115, 1979; Part II, Vol. 5, No. 3, 220–268, 1980.

On Attractive and Friendly Sets in Sequence Spaces

R. Ahlswede and L. Khachatrian

Abstract. To a large extent the present work is far from being conclusive, instead, new directions of research in combinatorial extremal theory are started. Also questions concerning generalizations are immediately noticeable.

The incentive came from problems in several fields such as Algebra, Geometry, Probability, Information and Complexity Theory. Like several basic combinatorial problems they may play a role in other fields. For scenarios of interplay we refer also to [9].

1 Introduction: New Problems and Results

A. A New Isoperimetric Problem: Boundaries with Intensity k, a Counterexample to Keane's Conjecture

For $\mathcal{X} = \{0,1\}$ let d be the Hamming distance in $\mathcal{X}^n = \prod_1^n \mathcal{X}$ and let the pair $\mathcal{H}^n = (\mathcal{X}^n, d)$ be the Hamming space.

$$S(x^n) = \left\{ y^n \in \mathcal{X}^n : d(x^n, y^n) = 1 \right\} \tag{1.1}$$

is the sphere of radius 1 with center $x^n \in \mathcal{X}^n$.

For any set $V \subset \mathcal{X}^n$ we define

$$\Gamma(V) = \left\{ y^n \in \mathcal{X}^n : d(x^n, y^n) \leq 1 \text{ for some } x^n \in V \right\} \tag{1.2}$$

and

$$B(V) = \Gamma(V) \smallsetminus V \tag{1.3}$$

as the (outer) boundary of V. Harper [4] considered

$$b(n, N) = \min_{V \subset \mathcal{X}^n, |V| = N} |B(V)| \tag{1.4}$$

and established his well–known Isoperimetric Theorem (in graphic language also called "Vertex Isoperimetric Theorem").

Notice that the points in $B(V)$ have distance 1 with at least one point of V. Our generalisation to a boundary of intensity k is

$$B_k(V) = \left\{ y^n \in \mathcal{X}^n \smallsetminus V, |S(y^n) \cap V| \geq k \right\}. \tag{1.5}$$

Obviously $B_1(V) = B(V)$.

R. Ahlswede et al. (Eds.): Information Transfer and Combinatorics, LNCS 4123, pp. 955–970, 2006.
© Springer-Verlag Berlin Heidelberg 2006

Problem 1: Determine

$$b_k(n, N) \triangleq \min_{V \subset \mathcal{X}^n : |V| = N} |B_k(V)| \tag{1.6}$$

and the structure of optimal V.

We have not yet solved it, but we introduced related problems 1–4 and solved problem 2 "ratewise" and problem 3, 4 exactly.

However, we made progress on a problem related to problem 1. For an upset $\mathcal{U}$ we consider $B_k(\mathcal{X}^n \setminus \mathcal{U}) = \mathcal{U}_k \triangleq \{x \in \mathcal{U} : |S(x) \cap (\mathcal{X}^n \setminus \mathcal{U})| \geq k\}$.

M. Keane [Oral communication] defined in the eighties the function

$$f(n, k) \triangleq \max\{|\mathcal{U}_k| : \mathcal{U} \text{ upset in } \mathcal{X}^n\} 2^{-n}$$

and conjectured that

$$\lim_{k \to \infty} \left(\sup_n f(k, n) \right) = 0.$$

Observation: For $k \geq \frac{n}{2}$

$$f(k, n) = \binom{n}{k} 2^{-n}.$$

Indeed, by the AZ–identity [5] for any $\mathcal{A} \subset 2^{[n]}$

$$\sum_{X \subset [n]} \frac{W_{\mathcal{A}}(X)}{|X| \binom{n}{|X|}} \equiv 1,$$

where

$$W_{\mathcal{A}}(X) = \left| \bigcap_{X \supset A \in \mathcal{A}} A \right|,$$

and therefore

$$\sum_{|X| \geq k} W_{\mathcal{U}}(X) \leq \max_{\ell \geq k \geq \frac{n}{2}} \ell \binom{n}{\ell} = k \binom{n}{k}$$

or

$$2^n f(k, n) \leq \frac{1}{k} \sum_{|X| \geq k} W_{\mathcal{U}}(X) \leq \binom{n}{k}.$$

This looks reassuring, however, Keane's conjecture is false.

Example 1

$$[n] = \Omega_1 \dot\cup \Omega_2 \cup \cdots \cup \Omega_m, |\Omega_i| = x, m = \frac{n}{x}.$$

$$\mathcal{A} = \{A \in 2^{[n]} : \exists i, \text{ s.t. } \Omega_i \subset A\} = \mathcal{U}\{\Omega_1, \ldots, \Omega_{\frac{n}{x}}\},$$

$$\mathcal{U}_x = \{B \in \mathcal{A} : \exists i, \text{ s.t. } \Omega_i \subset B \text{ and } \Omega_j \not\subset B \text{ for all } j \neq i\},$$

and

$$|\mathcal{U}_x| = \frac{n}{x} \cdot (2^x - 1)^{\frac{n}{x} - 1} = g(x, n), \text{ say.}$$

$\max_n g(x, n)$ is assumed at $n = x \cdot 2^x - \frac{x}{1 + x \log 2}$.

Hence, we may assume $n = n(x) = x \cdot (2^x - 1)$. For this n we consider

$$\frac{g(x, n)}{2^n} = \frac{(2^x - 1)^{2^x - 1}}{2^{x(2^x - 1)}} = \left(\frac{2^x - 1}{2^x}\right)^{2^x - 1} = \frac{1}{\left(1 + \frac{1}{2^x - 1}\right)^{2^x - 1}}$$

and therefore $\lim\limits_{x \to \infty} \frac{g(x,n)}{2^n} = \lim\limits_{x \to \infty} \left(1 + \frac{1}{2^x - 1}\right)^{-(2^x - 1)} = e^{-1}$.   □

B. The Smallest Rich World Problem

Consider

$$I_k(V) = \{y^n \in V : |S(y^n) \cap V| \geq k\} \tag{1.7}$$

and

Problem 2: Determine

$$r_k(n) = \min_{\substack{\phi \neq V = I_k(V) \\ V \subset \mathcal{X}^n}} |V| \left(= \min_{V \subset \mathcal{X}^n, \; \phi \neq V, \; |I_k(V)| = |V|} |V| \right). \tag{1.8}$$

We call an optimal V k–friendly set (or k–best world). We report now the much more general Problem 7 in [6], which was solved "ratewise" in [7].

For $\varphi : \mathcal{X} \times \mathcal{X} \to \mathbb{R}$, $\mathcal{X}$ a finite set, define

$$\alpha = \min_{x, y \in \mathcal{X}} \varphi(x, y), \quad \beta = \max_{x, y \in \mathcal{X}} \varphi(x, y), \tag{1.9}$$

and the sum–type function $\varphi_n : \mathcal{X}^n \times \mathcal{X}^n \to \mathbb{R}$, where $\varphi_n(x^n, y^n) = \sum\limits_{t=1}^{n} \varphi(x_t, y_t)$ for $x^n = (x_1, \ldots, x_2)$ and $y^n = (y_1, \ldots, y_2)$.

Now for any closed interval $L \subset [\alpha, \beta]$, any positive real number ρ, and any positive integer n call a set $S \subset \mathcal{X}^n$ with the property

$$\left| \left\{ y^n \in S : \frac{1}{n} \varphi_n(x^n, y^n) \in L \right\} \right| \geq 2^{n\rho} \text{ for all } x^n \in S \tag{1.10}$$

(n, L, ρ)–good and denote by $N(n, L, \rho)$ the smallest cardinality of (n, L, ρ)–good sets.

For the set $\{y^n \in S : \frac{1}{n} \varphi_n(x^n, y^n) \in L\}$ we also write $B(x^n, L, S)$. In the case $\alpha = 0$ and $L = [0, \beta]$ it is the intersection of S with a ball with center x^n and φ–radius β.

Inequality (1.10) says that every point in S has $2^{n\rho}$ points in S in its neighbourhood. In this sense S is a "rich world". The definition of $N(n, L, \rho)$ catches the goal to make the "world small".

One readily can show that $\lim\limits_{n \to \infty} \frac{1}{n} \log N(n, L, \rho)$ exists, because $N(n_1 + n_2, L, \rho) \geq N(n_1, L, \rho) \cdot N(n_2, L, \rho)$. We denote the limit by $\sigma(L, \rho)$. Its characterisation requires a few concepts.

Let (U, X, Y) be a triple of RV's with values in $\mathcal{U} \times \mathcal{X} \times \mathcal{X}$. We say that (X, Y) is a matching through U, if for conditional entropies

$$H(X|U) = H(Y|U) \text{ and } H(Y|XU) = H(X|YU). \tag{1.11}$$

Finally, we set
$$\mathcal{Q}(\mathsf{L}, \rho) = \big\{(X, U) : \text{for some } Y (X, Y) \text{ is matched through } U, \mathbb{E}_{\varphi}(X, Y) \in \mathsf{L},$$
and $H(Y|XU) \geq \rho\big\}. \tag{1.12}$

Theorem AC. [7]
$$\sigma(L, \rho) = \min_{(X,U) \in Q(L,\rho)} H(X|U).$$

Actually, we can bound the cardinality of $\mathcal{U}$ by $|\mathcal{X}|^2 + 4$.
Furthermore, we can limit the distributions P_{XY} to those with equal marginals.

Remark: Problem 2 relates to the case $\mathcal{X} = \{0, 1\}$, $\mathsf{L} = [k, n]$ and φ_n as Hamming distance.

C. k–Attractive Sets
In another direction we consider

$$\tilde{V}_k = \{y^n \in \mathcal{X}^n : |\, S(y^n) \cap V| \geq k\}. \tag{1.13}$$

Clearly,
$$\tilde{V}_k = B_k(V) \,\dot{\cup}\, I_k(V). \tag{1.14}$$

Analogous to Problem 1 is

Problem 3: Determine

$$a_k(n) \triangleq \min_{\substack{\phi \neq V : |V| \leq |\tilde{V}_k| \\ V \subset \mathcal{X}^n}} |V| \tag{1.15}$$

and the structure of solutions.
An optimal V is called k–attractive set. We call a V with $|V| \leq |\tilde{V}_k|$ a k–admissible set.

Example 2: Define for $k > 1$

$$\mathcal{X}_+^k = \big\{x^k = (x_1, \dots, x_k) \in \mathcal{X}^k : \sum_{t=1}^{k} x_t \equiv 0 \mod 2\big\}$$

and
$$\mathcal{X}_-^k = \mathcal{X}^k \smallsetminus \mathcal{X}_+^k.$$

Notice that for $V = \mathcal{X}_+^k * (0, 0, \dots, 0) \subset \mathcal{X}^n$ we have $\tilde{V}_k = \mathcal{X}_-^k * (0, 0, \dots, 0)$ and that V is k–admissible.
Moreover, let us write

$$Z_+^k = \mathcal{X}_+^k * (0, 0, \dots, 0),\ Z_-^k = \mathcal{X}_-^k * (0, 0, \dots, 0) \subset \mathcal{X}^n \tag{1.16}$$

and observe that for any $\pi \in \Sigma_n$, the symmetric group acting on $\{1, 2, \ldots, n\}$, also $\pi \, Z_+^k$ is k–admissible.

Furthermore, for any $y^n \in GF(2)^n$ also $Z_+^k + y^n$ is k–admissible. Quite surprisingly, these sets lead to all k–attractive sets.

Theorem 1. *For $k \geq 2$ the k–attractive subsets of $\mathcal{X}^n$ are of the form $(\pi \, Z_+^k) + y^n$. In particular for a k–attractive subset V of $\mathcal{X}^n$*

$$|V| = |\tilde{V}_k| = 2^{k-1}.$$

For 1–attractive sets

$$|V| = 1, |\tilde{V}_1| = n.$$

D. k–Pairs

We prove Theorem 1 by deriving it from the solution of the somewhat more general Problem 4:

For $A, B \subset \mathcal{X}^n$ we call (A, B) a k–pair, if

$$\sum_{x^n \in B} |A \cap S(x^n)| \geq k|B|. \tag{1.17}$$

It is *admissible*, if $|A| \leq |B|$, and it is *optimal*, if

$$|A| = \min_{(A', B') \text{admissible } k\text{–pair}} |A'|.$$

Determine all optimal k–pairs.

Theorem 2. *For $k \geq 2$ the optimal k–pairs are of the form*

$$(\pi \, Z_+^k + y^n, \pi \, Z_-^k + y^n).$$

The implication of Theorem 1 is readily established.

Clearly, if V is k–admissible, then $(V, \tilde{V}_k)$ is an admissible k–pair. The class of $(V, \tilde{V}_k)$ corresponding to attractive V's constitute a subclass of the class of optimal k–pairs. By Theorem 2 and Example 1 these classes are actually equal.

E. Results for Lopsided Sets in Combinatorial Language

Lopsided sets where introduced in [10] in the study of convex sets. There are several equivalent definitions [11]. We use here the terminology of set theory.

Let $\mathcal{S} \subset 2^{[n]}$, $\mathcal{S}^* = 2^{[n]} \setminus \mathcal{S}$. The set $\mathcal{S}$ is called lopsided, if for all $A \in 2^{[n]}$ either there exists a $B \subset [n] \setminus A$ such that for all $C \subset A \quad B \cup C \in \mathcal{C}$ or there exists a $B' \subset A$ such that for all $C' \subset [n] \setminus A \quad B' \cup C' \in \mathcal{S}^*$.

When Levon told me that Andreas Dress asked him to prove Theorem 3 below, which is considered basic for lopsided sets, I told him that he could and should do it in one afternoon, because this might be helpful for his career.

He followed the suggestion, but his simple proof earned him no benefits.

Theorem 3. *For a set $C \subset 2^{[n]}$ define*

$$L(C) \triangleq \{Y \subset [n] : \{Y \cap C : C \in C\} = 2^Y\}.$$

Then

$$|C| \le |L(C)|. \tag{1.18}$$

Obviously $L(C)$ is a downset for every $C \subset 2^{[n]}$ and if C is a downset, then $L(C) = C$. So for downsets there is equality in (1.18).

Recall now the standard push-down operation:
given $C \subset 2^{[n]}$ and $x \in [n]$ for $A \in C$

$$T(A, x) = \begin{cases} A \smallsetminus \{x\}, & \text{if } x \in A \text{ and } A \smallsetminus \{x\} \notin C \\ A & \text{otherwise} \end{cases} \tag{1.19}$$

$$T(C, x) = \{T(A, x) : A \in C\}. \tag{1.20}$$

Now readily verify

$$|T(C, x)| = |C| \text{ and } L(T(C, x)) \subset L(C). \tag{1.21}$$

After finitely many, say m, push–down operations $T(C, x_1), T((C, x_1), x_2), \ldots$ we get a downset D and by (1.21) $|D| = |C|$ and $D = L(D) \subset L(C)$, which proves Theorem 3.

We are going now for equality characterization in (1.18). We already now equality for downsets.

Symmetrically, if C is an upset, then $L(C) = \bar{C} = \{[n] \smallsetminus C : C \in C\}$ and again there is equality.

Defining push–up operation U analogously to the push–down operation T, then by symmetry

$$|U(C, x)| = |C| \text{ and } L(U(C, x)) \subset L(C). \tag{1.22}$$

We say now that a set $A \subset 2^{[n]}$ is accessible for $C \subset 2^{[n]}$, if starting from C one can obtain A by consecutively applying finitely many push–down and push–up operations. (For example $T(C, x_1), U(T(C, x_1), x_2), T(U(T(C, x_1), x_2), x_3), \ldots$.

By (1.21), (1.22) $|A| = |C|$, if A is an accessible set for C.

Theorem 4. *Let $D_1, \ldots, D_m \subset 2^{[n]}$ be the downsets accessible for a set $C \subset 2^{[n]}$, then*

$$\bigcup_{i=1}^{m} D_i = L(C). \tag{1.23}$$

In particular $|L(C)| = |C|$ if and only if there exists a unique downset accessible for C.

This result holds, because by (1.21), (1.22) any downset D accessible for C satisfies $D \subset L(C)$ and thus $\bigcup_{i=1}^{m} D_i \subset L(C)$. Further, for any $S \in L(C)$ we get from

C by down pushing in all $x \in \bar{S}$ downset $\mathcal{D}_S$ with $S \in \mathcal{D}_S$ and thus equality in (1.23).

Moreover, if there is exactly one $\mathcal{D}_i$, then $|L(C)| = |\mathcal{D}_i| = |C|$ and, conversely the equation $|L(C)| = |C|$ implies that $m = 1$, because the union of two or more sets of cardinality $|C|$ each, would imply $\left| \bigcup_{i=1}^{m} \mathcal{D}_i \right| > |C|$.

However, in order to understand the structure of C' with equality in (1.18) it is important to notice that **accessibility is not commutative**: if $\mathcal{D}$ is accessable for C, then C needs not be accessable for $\mathcal{D}$.

Example 3: Let $n = 3$, $C = \{\phi, \{1\}, \{2,3\}, \{3\}\}$, then $L(C) = \{\phi, \{1\}, \{2\}, \{3\}\}$ and $|L(C)| = |C| = 4$. However, C is not accessable for downsets and upsets as can be checked.

2   Main Auxiliary Old and New Results for the Proof of Theorem 2

We make essentially use of Harper's Edge Isoperimetric Theorem ([1], correct proofs in [2] – [3]). For $C \subset \mathcal{X}^n$ define

$$\emptyset(C) = \left\{ (x^n, y^n) : x^n \in C, y^n \in \mathcal{X}^n \setminus C, d(x^n, y^n) = 1 \right\}. \tag{2.1}$$

For a number M define

$$\omega(M) = \min_{C \subset \mathcal{X}^n, |C| = M} |\emptyset(C)|. \tag{2.2}$$

Theorem (Edge Isoperimetry). *The minimum in (2.2) is assumed for a generalized cylinder.*

We recall the definition of a generalized cylinder.

Every positive integer M can uniquely be written in a binary expansion

$$M = 2^{n_1} + 2^{n_2} + \cdots + 2^{n_s}, n_1 > n_2 > \cdots > n_s \geq 0. \tag{2.3}$$

Let us use the picture in (2.4) for the set $\mathcal{X}^m$, then we can present

$$Z(M) = \mathcal{X}^{n_1} \times \{0\}^{n-n_1}$$
$$\dot{\cup}\, \mathcal{X}^{n_2} \times \{0\}^{n-n_2} \times \{1\} \times \{0\}^{n-n_1-1}$$
$$\vdots$$

as union of the disjoint sets

n_1				000 ... 0
n_2			000100 ... 0	
n_3		00100100 ... 0		
n_4	0010100100 ... 0			

$$\vdots$$

(2.4)

Instead of minimizing the number $\text{out}(C) = |\varnothing(C)|$ of outgoing "edges" we can equivalently maximize the number $\text{int}(C) = |I(C)|$ of internal "edges", because all vertices have degree n and thus

$$\text{out}(C) + \text{int}(C) = n|C|.$$

We refer then to the dual form of this optimisation problem.
Define

$$i(M) = \max_{C:|C|=M} \text{int}(C). \tag{2.5}$$

For the proof of our Uniqueness Theorem below we need

Lemma 1. *For $M < 2^k$ necessarily*

$$i(M) < k \cdot \frac{M}{2}.$$

Proof: We can calculate $i(M)$ from Theorem H_1 using the representation (2.4).
Indeed

$$i(M) = n_1 \, 2^{n_1-1} + n_2 \, 2^{n_2-1} + n_3 \, 2^{n_3-1} + \cdots + n_s \, 2^{n_s-1}$$
$$+ \, 2^{n_2} + 2 \cdot 2^{n_3} + \cdots + (n_s - 1)2^{n_s}. \tag{2.6}$$

For $M < 2^k$ necessarily

$$k \geq n_1 + 1 \geq n_2 + 2 \geq n_3 + 3 \geq \ldots \qquad . \tag{2.7}$$

We have to show that $k \, M > 2 \, i(M)$ or that

$$k(2^{n_1} + 2^{n_2} + \cdots + 2^{n_s}) > n_1 \, 2^{n_1} + n_2 \, 2^{n_2} + \cdots + n_s \, 2^{n_s}$$
$$+ \, 1 \cdot 2 \cdot 2^{n_2} + 2 \cdot 2 \cdot 2^{n_3} + 3 \cdot 2 \cdot 2^{n_4} + \ldots$$

or that

$$(k-n_1) \cdot 2^{n_1} + (k-n_2)2^{n_2} + \cdots + (k-n_s)2^{n_s} > 1 \cdot (2 \cdot 2^{n_2}) + 2(2 \cdot 2^{n_3}) + \cdots + . \tag{2.8}$$

Now, by (2.7) $(k - n_i) \geq i$ and the RHS in (2.8) does not exceed the LHS even if we ignore the term $(k - n_s)2^{n_s}$, which is positive. Thus (2.8) holds and the Lemma is proved.

Uniqueness Theorem. *Generalized cylinders are up to permutations $\pi \in \Sigma_n$ and additions with x^n in $GF(2)^n$ the only solutions in Theorem H_2 (at least for $M = 2^k$).*

Proof for the case $M = 2^k$: Let $C \subset \mathcal{X}^n$ be optimal. Then

$$\text{int}(C) = i(M) = k \cdot 2^{k-1}. \tag{2.9}$$

Consider

$$C_1 = \big\{(x_2,\ldots,x_n) : (1,x_2,\ldots,x_n) \in C\big\}, C_0 = \big\{(x_2,\ldots,x_n):(0,x_2,\ldots,x_n)\in C\big\}. \tag{2.10}$$

We proceed by induction on n.

We have to show that either $C_1 = \varnothing$ or $C_0 = \varnothing$ or that $C_1 = C_0$.

Case $|C_1| = |C_0|$:
Consider

$$\mathrm{int}(C) = \mathrm{int}(C_1) + \mathrm{int}(C_0) + |C_1 \cap C_0| \le 2i(2^{k-1}) + |C_1 \cap C_0|$$
$$\le 2(k-1)2^{k-2} + |C_1 \cap C_0| = (k-1)2^{k-1} + |C_1 \cap C_0| = k \cdot 2^{k-1}. \tag{2.11}$$

Since $\mathrm{int}(C) = k \cdot 2^{k-1}$, necessarily $|C_1 \cap C_0| = 2^{k-1}$ and thus $C_1 = C_0$.

Case $|C_1| > |C_0|$:
Here $|C_1| = 2^{k-1} + 2^{\ell_2} + \ldots, \ell_2 > 0$, and $|C_0| = 2^k - |C_1|$.

Subcase $2^k > |C_1| = 2^{k-1} + 2^{k-2} + \ldots$:
Here $|C_0| \le 2^{k-3} + \cdots < 2^{k-2}$.
By Lemma 1 $\mathrm{int}(C_1) < k\frac{|C_1|}{2}$ and $\mathrm{int}(C_0) \le (k-2)\frac{|C_0|}{2}$.
Therefore by (2.11)

$$\mathrm{int}(C) < k\frac{|C_1|}{2} + (k-2)\frac{|C_0|}{2} + |C_0| = k\frac{|C|}{2} = k \cdot 2^{k-1},$$

which contradicts (2.9).

Subcase $2^k > |C_1| = 2^{k-1} + 0 + \ldots$:
Here $|C_0| = 2^{k-2} + \cdots < 2^{k-1}$.
The estimate of $\mathrm{int}(C_1)$ is more tricky. We use the representation (2.4) for C_1 and obtain

$$\mathrm{int}(C_1) \le (k-1)2^{k-2} + (|C_1| - 2^{k-1}) + i(|C_1| - 2^{k-1}). \tag{2.12}$$

Since by Lemma 1 $i(|C_1| - 2^{k-1}) < (k-2)\frac{|C_1|-2^{k-1}}{2}$ and $\mathrm{int}(C_0) < (k-1)\frac{|C_0|}{2}$, we have

$$\mathrm{int}(C) < (k-1)\left(2^{k-2} + \frac{|C_0|}{2}\right) + k\frac{|C_1| - 2^{k-1}}{2} = k\frac{|C_0| + |C_1|}{2} - \left[\frac{|C_0|}{2} - (k-1)2^{k-2} + k\frac{2^{k-1}}{2}\right]$$

in contradiction to $\mathrm{int}(C) = k \cdot 2^{k-1}$, because the term in brackets is positive.

3   Further Auxiliary Results

We use again the dual form and apply it to $C = A \cup B$.

Lemma 2. *For an admissible k–pair (A, B) we have*

$$|A \cup B| \geq 2^k. \tag{3.1}$$

Proof: Let $e(A, B)$ be the number of edges between A and B. Then by our assumptions

$$\text{int}(C) \geq e(A, B) \geq k|B|.$$

Since also $|B| \geq |A|$, therefore also

$$\text{int}(C) \geq k\frac{|C|}{2}. \tag{3.2}$$

We assume now that (3.1) does not hold and derive a contradiction. If now $M = |C| < 2^k$, then by (3.2)

$$i(M) \geq \text{int}(C) \geq k\frac{M}{2}. \tag{3.3}$$

Next we use this result to derive a lower bound on $|A|$.

Lemma 3. *For an admissible k–pair (A, B) necessarily*

$$|A| \geq 2^{k-1}.$$

Proof: For an admissible k–pair (A, B) label the elements in B as $b_1 \ldots b_r$ such that

$$|A \cap S(b_i)| \leq |A \cap S(b_{i+1})| \text{ for } i = 1, \ldots, r - 1$$

and define for $j = |B| - |A|$

$$B' = B - \{b_1, \ldots, b_j\}.$$

Notice that (A, B') is an admissible k–pair with $|A| = |B'|$ and thus $|A| \geq \frac{|A \cup B'|}{2}$. Since by Lemma 2 $|A \cup B'| \geq 2^k$, the result follows.

Lemma 4. *For an optimal k–pair (A, B) necessarily*
  (a) $|A| = |B| = 2^{k-1}$
  (b) $A \cap B = \varnothing$.

Proof: We know from Lemma 3 that (Z_+^k, Z_-^k) is a minimal k–pair and thus $|A| = 2^{k-1}$.
  Suppose that $|B| = 2^{k-1} + b$, $b > 0$.
  Then by the procedure described in the proof of Lemma 2 we can get a minimal k–pair (A, B') with $|B'| = 2^{k-1} + 1$.

By assumption

$$\text{int}(A \cup B') \geq \sum_{b \in B'} |A \cap S(b)| \geq k|B'| \geq k \cdot 2^{k-1} + k. \tag{3.4}$$

However, since $|A \cup B'| \leq 2^k + 1$ we have

$$\text{int}(A \cup B') \leq i(2^k + 1) = k \cdot 2^{k-1} + 1. \tag{3.5}$$

Now (3.5) contradicts (3.4) for $k \geq 2$. (b) follows with Lemma 2.

Remark: For $k = 1$ $\left(\{00 \ldots 0\}, \{x^n : \sum_{t=1}^{n} x_t = 1\}\right)$ is an optimal 1–pair and (a) does not hold!

4   Proof of Theorem 2

We know already that for an optimal k–pair (A, B) necessarily $A \cap B = \varnothing$, $|A| = |B| = 2^{k-1}$. Since also $\text{int}(A \cup B) \geq k|B| = k2^{k-1}$ and by Theorem H_1 $\text{int}(A \cup B) \leq i(2^k) = k \cdot 2^{k-1}$, we conclude that

$$\text{int}(A \overset{.}{\cup} B) = k \cdot 2^{k-1}. \tag{4.1}$$

By the Uniqueness Theorem $A \cup B$ is a cylinder Z and w.l.o.g. $Z = \boxed{k} 00 \ldots 0$. Since every element in Z has degree k and $|B|k = 2^{k-1} \cdot k$, necessarily

$$|A \cap S(b)| = k \text{ for all } b \in B. \tag{4.2}$$

Now (4.1) and (4.2) imply

$$e(A, B) = \text{int}(A \cup B), \tag{4.3}$$

that is, all edges are between the sets (or no edges are in A or in B). Clearly

$$|B \cap S(a)| = k \text{ for all } a \in A. \tag{4.4}$$

Finally, we can assume w.l.o.g. that $00 \ldots 0 \in B$. Then no singleton can be in B and by (4.2) the first k of the n singletons must be in A.

By the degree condition (4.4) then all doubletons with 1's in $\{1, 2, \ldots, k\}$ must be in B, next all tripletons with 1's in $\{1, 2, \ldots, k\}$ must be in A, etc. This completes the proof.

Problem 4: How do the results generalize from $\mathcal{X} = \{0, 1\}$ to $\mathcal{X} = \{0, \ldots, \alpha - 1\}$?

5   Further Observations

On large boundaries of intensity k.

We adapt the *convention*: $x = x^n$.

Recall the definition of $\tilde{V}_k$ in (1.13). For $1 \leq k \leq n$ and $0 \leq N \leq 2^n$ we define now

$$\gamma(n, N, k) = \max_{|V| = N} |\tilde{V}_k|. \tag{5.1}$$

Fact I: γ is not decreasing in N. Unlike Theorem 1 it is here very difficult to obtain exact results. We discuss therefore some special cases of the function γ.

Case $k = 1$: For not too large N it is clearly optimal to choose V as a 1–error correcting code, that is,

$$\min_{x,y \in V} d(x,y) \geq 3. \tag{5.2}$$

Here $|\tilde{V}| = n|V|$ and this is optimal.

Case $k \geq 2$: The situation is now quite different, because the points in V should be not too far apart

Fact II: The points in $A(V) = \{x \in V : d(V \setminus \{x\}, x) \geq 2\}$ don't contribute to $\tilde{V}_k$ for $k \geq 2$. We can therefore assume that $A(V) = \varnothing$.

Fact III: Let us associate with V the graph $G_2 = G_2(V) = (V, \mathcal{E})$, where

$$\mathcal{E} = \{\{x,y\} : x,y \in V \text{ and } d(x,y) = 2\}.$$

We can assume that $G_2(V)$ is *connected*, because $x \in \mathcal{X}$ can have distance 1 only with vertices in the same connected component.

Case $k = 2$: By induction on n one gets

Lemma 5. *If $G_2(V)$ is connected and $|V| \geq 2$, then $|V| \leq |\tilde{V}_2|$.*
    We derive now a lower bound on $\alpha(n, N, k)$ for smaller k.

Lemma 6. *For $k \leq \log N + 1$*

$$\alpha(n, N, k) = \Omega\left(\frac{N \cdot \log N}{k - 1}\right).$$

Proof: Being concerned only about the **order of growth** we make the simplifying assumption

$$k - 1 | \log N.$$

Recall the definition of $\mathcal{X}_+^k$ in Section 3 and choose $C = \frac{\log N}{k-1}$ of its copies $\mathcal{X}_{+\ell}^k$, $1 \leq \ell \leq L$, and define $W^k = \mathcal{X}_{+1}^k \times \cdots \times \mathcal{X}_{+L}^k \times \{0\} \times \cdots \times \{0\} \subset \mathcal{X}^n$, where the factor $\{0\}$ occurs exactly $n - L \cdot k$ times.
    Clearly, $|W^k| = 2^{(k-1)L} = N$ and for $V = W^k$ we have

$$\tilde{V}_k = \{x \oplus e_j : x \in W^k, 1 \leq j \leq kL\},$$

where e_j has a 1 in the j-th position and 0 otherwise.
    Hence $|\tilde{V}_k| = |W^k|\frac{kL}{k} = NL$, because $\mathcal{X}_{+\ell}^k + e_i = \mathcal{X}_{+\ell}^k + \ell$, for $k(\ell - 1) + 1 \leq i, j \leq k\ell$.

On a dual form of the vertex isoperimetric theorem in the Hamming space $(\mathcal{X}^n, d)$.

(A, B) with $A, B \subset \mathcal{X}^n$ is an (n, d)–pair, if

$$d_H(a, b) \geq d \text{ for } a \in A, b \in B.$$

It was shown in [AK] that

$$\max\{|A||B| : (A, B) \text{ is } (n, d)\text{–pair}\}$$

is assumed for $(A, B) = \{a_t^n \mathcal{X}^n : w(a^n) \leq \ell\}, \{b^n \in \mathcal{X}^n : w(b^n) \geq d + \ell\}$ with a suitable ℓ.

Here we show that $\ell = \left\lceil \frac{n-d}{2} \right\rceil$.

Lemma 7. *For every* $n \in \mathbb{N}$, $0 \leq d \leq n$, *and* $0 \leq \ell \leq n$ *the values* $f_{n,d}(\ell) =$

$$\left(\sum_{i=0}^{\ell} \binom{n}{i} \right) \left(\sum_{i=d+\ell}^{n} \binom{n}{i} \right) \text{ satisfy}$$

$$f_{n,d}\left(\left\lceil \frac{n-d}{2} \right\rceil \right) \geq f_{n,d}(\ell).$$

Proof: By symmetry it suffices to show that $f_{n,d}(\ell) \leq f_{n,d}(\ell+1)$, if $n \geq 2\ell+1+d$, that is,

$$\left(\sum_{i=0}^{\ell} \binom{n}{i} \right) \left(\sum_{i=d+\ell}^{n} \binom{n}{i} \right) \leq \left(\sum_{i=0}^{\ell} \binom{n}{i} \right) \left(\sum_{i=d+\ell+1}^{n} \binom{n}{i} \right)$$

iff

$$\left(\sum_{i=0}^{\ell} \binom{n}{i} \right) \binom{n}{d+\ell} \leq \binom{n}{\ell+1} \left(\sum_{i=d+\ell+1}^{n} \binom{n}{i} \right). \tag{5.3}$$

We prove inequality (5.3), by induction on n for all ℓ, d with $n \geq 2\ell + 1 + d$.

For $n = 2$, that is, $\ell = 0$, (5.3) obviously holds. Therefore we assume (5.3) to be true for $n' < n$. We consider first the cases $\ell = 0$, and $n = 2\ell + 1 + d$.

a) $d = 1$:

$$\left(\sum_{i=0}^{\ell} \binom{n}{i} \right) \binom{n}{\ell+1} \leq \binom{n}{\ell+1} \left(\sum_{i=\ell+2}^{n} \binom{n}{i} \right).$$

b) $\ell = 0$:

$$\binom{n}{0}\binom{n}{d} = \binom{n}{d} \leq \binom{n}{1} \left(\sum_{i=d+1}^{n} \binom{n}{i} \right) = n\binom{n}{d+1} + \sum_{i=d+2}^{n} \binom{n}{i}.$$

Since $\binom{n}{d} \leq n\binom{n}{d+1}$ $(*)$ holds for $\ell = 0$.

c) $n = 2\ell + 1 + d$ implies

$$\left(\sum_{i=0}^{\ell}\binom{n}{i}\right)\binom{n}{d+\ell} = \left(\sum_{i=0}^{\ell}\binom{n}{i}\right)\binom{n}{\ell+1} = \left(\sum_{i=d+\ell+1}^{n}\binom{n}{i}\right)\binom{n}{\ell+1}.$$

Let now $\ell \geq 1$, $d \geq 2$, $n \geq 2\ell + d + 2$:

$$\left(\sum_{i=0}^{\ell}\binom{n}{i}\right)\binom{n}{d+\ell} = \left(\sum_{i=0}^{\ell}\binom{n-1}{i} + \sum_{i=0}^{\ell-1}\binom{n-1}{i}\right)\left(\binom{n-1}{d+\ell} + \binom{n-1}{d+\ell-1}\right)$$

$$= \left(\sum_{i=0}^{\ell}\binom{n-1}{i}\right)\binom{n-1}{d+\ell} + \left(\sum_{i=0}^{\ell}\binom{n-1}{i}\right)\binom{n-1}{d+\ell-1} + \left(\sum_{i=0}^{\ell-1}\binom{n-1}{i}\right)\binom{n-1}{d+\ell}$$

$$+ \left(\sum_{i=0}^{\ell-1}\binom{n-1}{i}\right)\binom{n-1}{d+\ell-1}$$

$$\leq \binom{n-1}{\ell+1}\left(\sum_{i=d+\ell+1}^{n-1}\binom{n-1}{i}\right) + \binom{n-1}{\ell+1}\left(\sum_{i=d+\ell}^{n-1}\binom{n-1}{i}\right) + \binom{n-1}{\ell}\left(\sum_{i=d+\ell+1}^{n-1}\binom{n-1}{i}\right)$$

$$+ \binom{n-1}{\ell}\sum_{i=d+\ell}^{n-1}\binom{n-1}{i}$$

$$= \left(\binom{n-1}{\ell+1} + \binom{n-1}{\ell}\right)\left(\sum_{i=d+\ell+1}^{n-1}\binom{n-1}{i} + \sum_{i=d+\ell}^{n-1}\binom{n-1}{i}\right) = \binom{n}{\ell+1}\left(\sum_{i=d+\ell+1}^{n}\binom{n-1}{i}\right).$$

Remark: The result must be known and also have a simpler proof!

6   Concluding Conjectures

On k–pairs

We consider

$$\alpha(n, N, k) = \max\{|B| : A, B \subset \mathcal{X}^n, |A| \leq N, (A, B) \text{ is } k\text{–pair}\}.$$

Conjectures

The following constructions give the "asymptotic" value of $\alpha(n, N, k)$:

a.) If $N = \binom{m}{k-1}$ and $2(k-1) < m \leq n$, then $A = \binom{[m]}{k-1} \times \binom{[n-m]}{0}$, $B = \binom{[m]}{k} \times \binom{[n-m]}{0}$

$\alpha(n, N, k) = N \cdot \frac{m-k+1}{k}$ $m = 2k = n$?

b.) If $N = \binom{k}{j} + \binom{k}{j-2}$ with $2 \leq j \leq \frac{k}{2}$, then $\alpha(n, N, k) \sim N\frac{j}{k-j+1}$.

Use $B = \binom{[k]}{j-1} \times \binom{[n-k]}{0}$, $A = \left(\binom{[k]}{j} \cup \binom{[k]}{j-2}\right) \times \binom{[n-k]}{0}$.

On edge isoperimetry

For $A \subset \mathcal{X}^n$ define $E_A = \{(x,y) : x,y \in A,\ x \oplus y = e_i$ for some $i\}$, where $e_i = (0,0,\ldots,1,0,\ldots,0)$, with "1" in component i. Then

$$f_2(N) = \max_{\substack{A \subset \mathcal{X}^n \\ |A|=N}} |E_A| f_2(2^k) = k \cdot 2^k \text{ by Harper and } f_2(N) \le N \log N. \quad (6.1)$$

Define now $E_r(A) = \left\{(x_1, x_2, \ldots, x_r) : x_i \in A,\ \bigoplus_{j=1}^{r} x_j = e_i \text{ for some } i\right\}$ and

$$f_r(N) = \max_{\substack{A \subset \mathcal{X}^n \\ |A|=N}} |E_r(A)|.$$

We can write $E_r(A) = \bigcup_{x \in A} B_x$, where

$$B_x = \left\{(x, x_2, \ldots, x_r) : \bigoplus_{j=2}^{r} x_j = e_i \oplus x\right\} \text{ and}$$

$$|B_x| = \left| \left\{(x_2, \ldots, x_r) : \bigoplus_{j=2}^{r} x_j = e_i\right\} \right| \le f_{r-1}.$$

Consequently $f_r(2^k) = (2^k)^{r-1} \cdot k$ (Harper's cylinder).

Conjecture

$$f_r(N) \le N f_{r-1}(N) \le \cdots \le N^{r-2} f_2(N) \le N^{r-1} \log N.$$

References

1. L.H. Harper, Optimal assignment of numbers to vertices, J. Soc. Industr. Appl. Math., 12, 1, 385–393, 1964.

2. H.H. Lindsey, Assignment of numbers to vertices, Amer. Math. Monthly 71, 508–516, 1964.

3. A.J. Bernstein, Maximal connected arrays on the n–cube, SIAM J. Appl. Math. 15, 6, 1485–1489, 1967.

4. L.H. Harper, Optimal numberings and isoperimetric problems on graphs, J. Combin. Theory 1, 385–393, 1966.

5. R. Ahlswede and Z. Zhang, An identity in combinatorial extremal theory, Adv. in Math., Vol. 80, No. 2, 137–151, 1990.

6. R. Ahlswede and Z. Zhang, On multi–user write–efficient memories, IEEE Trans. Inform. Theory, Vol. 40, No. 3, 674–686, 1994.

7. R. Ahlswede and N. Cai, Models of multi–user write–efficient memories and general diametric theorems, Preprint 93–019, SFB 343 "Diskrete Strukturen in der Mathematik", Universität Bielefeld, Information and Computation, Vol. 135, No. 1, 37–67, 1997.

8. R. Ahlswede and G. Katona, Contributions to the geometry of Hamming spaces, Discrete Mathematics 17, 1–22, 1977.

9. R. Ahlswede, Advances on extremal problems in number theory and combinatorics, European Congress of Mathematics, Barcelona 2000, Vol. I, 147–175, Carles Casacuberta, Rosa Maria Miró–Roig, Joan Verdera, Sebastiá Xambó–Descamps, edit., Progress in Mathematics, Vol. 201, Birkhäuser Verlag, Basel–Boston–Berlin, 2001.

10. J. Lawrence, Lopsided sets and orthant–intersection of convex sets, Pacific J. Math. 104, 155–173, 1983.

11. H.J. Bandelt, V. Chepoi, A. Dress, and J. Koolen, Theory of lopsided set systems, Preprint.

Remarks on an Edge Isoperimetric Problem

C. Bey

Abstract. Among all collections of a given number of k-element subsets of an n-element groundset find a collection which maximizes the number of pairs of subsets which intersect in $k - 1$ elements.

This problem was solved for $k = 2$ by Ahlswede and Katona, and is open for $k > 2$.

We survey some linear algebra approaches which yield to estimations for the maximum number of pairs, and we present another short proof of the Ahlswede-Katona result.

1 Introduction

Let $G = (\mathcal{V}, \mathcal{E})$ be a simple graph. Given a subset $\mathcal{M}$ of the vertex set $\mathcal{V}$, we put $B_G(\mathcal{M}) := \{\{u, v\} \in \mathcal{E} : u \in \mathcal{M}, v \notin \mathcal{M}\}$, the *edge-boundary* of $\mathcal{M}$, and $I_G(\mathcal{M}) := \{\{u, v\} \in \mathcal{E} : u, v \in \mathcal{M}\}$, the set of *inner edges* spanned by $\mathcal{M}$. Two edge isoperimetric problems (EIP's) for G are the determination of the numbers $B_G(m) := \min\{|B_G(\mathcal{M})| : \mathcal{M} \subseteq \mathcal{V}, |\mathcal{M}| = m\}$ and $I_G(m) := \max\{|I_G(\mathcal{M})| : \mathcal{M} \subseteq \mathcal{V}, |\mathcal{M}| = m\}$. If G is regular with degree d, then both problems are equivalent since $2I_G(\mathcal{M}) + B_G(\mathcal{M}) = d|\mathcal{M}|$ holds for every $\mathcal{M} \subseteq \mathcal{V}$. We refer to [6] for a survey on edge isoperimetric problems on graphs.

Here we consider the Johnson graph $G = J(n, k)$. Let $[n]$ denote the set $\{1, 2, \ldots, n\}$ and $\mathcal{V}_k^n$ the set of all k-element subsets of $[n]$. The graph $J(n, k)$ has vertex set $\mathcal{V}_k^n$, and edge set $\{\{A, B\} : |A \cap B| = k - 1\}$. Thus, looking at incidence vectors, $J(n, k)$ is the graph whose vertices are the $\{0, 1\}$-sequences of length n and weight k, and whose edges are those pairs of sequences with Hamming distance 2. The Johnson graph $J(n, k)$ is an adjacency relation of the Johnson scheme, which is the natural setting for studying constant weight codes (cf. [10]).

Note that $J(n, k)$ is regular with degree $k(n - k)$.

We write $B_{n,k}(m)$ for $B_{J(n,k)}(m)$ and $B(\mathcal{M})$ resp. $I(\mathcal{M})$ for $B_{J(n,k)}(\mathcal{M})$ resp. $I_{G(n,k)}(\mathcal{M})$.

The EIP of the graph $J(n, 2)$ was solved by Ahlswede and Katona in [3]. Solutions for the case $k = 2$ also appeared in [1,7] and [16]. In order to state the result, we need the following definition.

Let $m = \binom{d}{2} + t$ with $0 \le t < d$. The *quasi-complete* graph C_n^m on n vertices is obtained from the complete graph on d vertices by adding a vertex of degree t and $n - 1 - d$ isolated vertices. The *quasi-star* S_n^m is the complement of the graph $C_n^{\binom{n}{2}-m}$.

Theorem 1 ([3]). *For every $0 \le m \le \binom{n}{2}$ the minimum boundary $B_{n,2}(m)$ is attained for the quasi-complete graph C_n^m or for the quasi-star S_n^m.*

R. Ahlswede et al. (Eds.): Information Transfer and Combinatorics, LNCS 4123, pp. 971–978, 2006.

The EIP for $J(n, k)$ with $k > 2$ is open. The papers [12] and [2] study continuous versions of the EIP, whose solutions yield the numbers $B_{n,k}(m)$ for certain values of m. A natural conjecture on the structure of optimal solutions for the EIP is also disproved in [2].

This note contains a survey on some estimations for the EIP based on eigenvalues, as well as a short and new proof for Theorem 1.

2   Estimations Via Eigenvalues

Given a set $\mathcal{M} \subseteq \mathcal{V}_k^n$ of vertices of $J(n, k)$ and a set $P \subseteq [n]$ we denote by

$$d(P) = d_{\mathcal{M}}(P) = |\{M \subseteq \mathcal{M} : P \subseteq M\}|$$

the *degree* of P in $\mathcal{M}$. If $p \in [n]$ we write also $d(p)$ for $d(\{p\})$.

Obviously,

$$\sum_{P \in \mathcal{V}_{k-1}^n} d(P)^2 = 2|I(\mathcal{M})| + k|\mathcal{M}| = k(n - k + 1)|\mathcal{M}| - |B(\mathcal{M})|. \tag{1}$$

Thus, determining the maximum sum of squares of degrees $d_{\mathcal{M}}(P)$, $P \in \mathcal{V}_{k-1}^n$, over all subsets $\mathcal{M} \subseteq \mathcal{V}_k^n$ of a given size is equivalent to the EIP.

In [9], de Caen proves the following inequality for the case $k = 2$: Given $\mathcal{M} \subseteq \mathcal{V}_2^n$,

$$\sum_{P \in \mathcal{V}_1^n} d(P)^2 \le \frac{2}{n-1}|\mathcal{M}|^2 + (n-2)|\mathcal{M}|. \tag{2}$$

Of course, this inequality can be checked using Theorem 1, but the calculations are somewhat involved. De Caen's inequality was generalized to arbitrary k in [5]: For every $0 \le p \le n$ and $\mathcal{M} \subseteq \mathcal{V}_k^n$ we have

$$\sum_{P \in \mathcal{V}_p^n} d(P)^2 \le \frac{\binom{k}{p}\binom{k-1}{p}}{\binom{n-1}{p}}|\mathcal{M}|^2 + \binom{k-1}{p-1}\binom{n-p-1}{k-p}|\mathcal{M}|. \tag{3}$$

Equality holds in (3) for $0 < p < k < n$ if and only if $\mathcal{M} = \varnothing$ or $\mathcal{M} = \mathcal{V}_k^n$ or $\mathcal{M}$ is a star or a complement of a star, or $n = k + 1$ and $\mathcal{M} \subseteq \mathcal{V}_k^{k+1}$ is arbitrary. Here a *star* is the set of all k-element subsets which contain a fixed element from $[n]$, and the complement of $\mathcal{M}$ is $\mathcal{V}_k^n \setminus \mathcal{M}$.

The proofs of the above estimations in [9,5] utilize a positive semidefinite matrix in the Bose-Mesner-algebra of the Johnson scheme (cf. [10]). This matrix is essentially the p-element versus k-element subsets incidence matrix multiplied with it transposed. A more transparent proof, using this matrix, can be given using the following theorem. Recall that a semiregular graph is a bipartite graph such that the degrees of the vertices are constant on each bipartition.

Theorem 2. *Let $G = (\mathcal{V}, \mathcal{E})$ be a connected semiregular graph with bipartition $\mathcal{V} = \mathcal{V}_0 \cup \mathcal{V}_1$ and degrees d_0, d_1. Let μ_2 be the second largest eigenvalue of the adjacency matrix of G. For $\mathcal{M} \subseteq \mathcal{V}_1$ and $v \in \mathcal{V}_0$ put*

$$d(v) = |\{v_1 \in \mathcal{M} \ : \{v, v_1\} \in \mathcal{E}\}| \,.$$

Then

$$\sum_{v \in \mathcal{V}_0} d(v)^2 \le \left(\frac{d_0 d_1 - (\mu_2)^2}{|\mathcal{V}_1|} \right) |\mathcal{M}|^2 + (\mu_2)^2 |\mathcal{M}| \,.$$

Proof: Let A be the adjacency matrix of G. We have

$$A = \begin{pmatrix} 0 & W \\ W^\top & 0 \end{pmatrix},$$

where W is the $|\mathcal{V}_0| \times |\mathcal{V}_1|$-matrix describing adjacency between the bipartitions V_0 and V_1. It is known that the nonzero eigenvalues of A are exactly the (positive and negative) square roots of the nonzero eigenvalues of $W^\top W$, with equal multiplicities correspondingly. We denote the eigenvalues of the latter matrix by $\mu_1^2, \mu_2^2, \ldots$, with $\mu_1 > \mu_2 > \cdots \ge 0$. Since G is biregular and connected, the largest eigenvalue is $\mu_1^2 = d_0 d_1$, and the corresponding eigenspace is one-dimensional and generated by the all one vector which we denote by j. In particular, μ_2 is indeed the second largest eigenvalue.

Now let φ be the characteristic row vector of $\mathcal{M}$ (of length V_1), and $\varphi = \varphi_1 + \varphi_2 + \ldots$ be the orthogonal decomposition of φ according to the eigenspaces of $W^\top W$. Note that $|\mathcal{M}| = \varphi^\top j = \varphi_1^\top j$, thus φ_1 is the constant vector with entry $|\mathcal{M}|/|\mathcal{V}_1|$, and $\varphi_1^\top \varphi_1 = |\mathcal{M}|^2/|\mathcal{V}_1|$. Now we have

$$\sum_{v \in \mathcal{V}_0} d(v)^2 = (W\varphi)^\top W\varphi = \mu_1^2 \varphi_1^\top \varphi_1 + \sum_{i \ge 2} \mu_i^2 \varphi_i^\top \varphi_i$$

$$\le d_0 d_1 \varphi_1^\top \varphi_1 + \mu_2^2 (|\mathcal{M}| - \varphi_1^\top \varphi_1) = \left(\frac{d_0 d_1 - \mu_2^2}{|\mathcal{V}_1|} \right) |\mathcal{M}|^2 + \mu_2^2 |\mathcal{M}| \,.$$

Inequality (3) now follows from Theorem 2. The computation of the corresponding eigenvalue μ_2 (cf. [5]) uses the known eigenvalues of the Johnson scheme (cf. [10]).

Using (1) and (3) with $p = k - 1$ we get the following estimation for the EIP of the graph $J(n, k)$:

$$B_{n,k}(m) \ge \frac{n}{\binom{n}{k}} \, m \left(\binom{n}{k} - m \right) . \tag{4}$$

Equality holds for $0 < m < \binom{n}{k}$ if and only if $m = \binom{n-1}{k-1}$ or $m = \binom{n-1}{k}$ or $n = k + 1$.

This recalls a general edge isoperimetric inequality using Laplace eigenvalues due to Alon and Milman:

Theorem 3 ([4]). *Let* $G = (\mathcal{V}, \mathcal{E})$ *be a simple graph, and* λ_2 *be the second smallest Laplace eigenvalue of* G. *Then, for every* $0 \leq m \leq |\mathcal{V}|$,

$$B_G(m) \geq \frac{\lambda_2}{|\mathcal{V}|} \, m \, (|\mathcal{V}| - m) \, .$$

Recall that the Laplace eigenvalues of a graph G are the eigenvalues of the difference of the degree matrix and the adjacency matrix of G, where the degree matrix is the diagonal matrix having the degrees of G on its diagonal. For regular graphs the Laplace eigenvalues are the differences of the degree and the adjacency eigenvalues.

The second smallest Laplace eigenvalue of the Johnson graph $J(n, k)$ is n. This is again an easy computation using the eigenvalues of the Johnson scheme. Thus, inequality (4) and hence also (3) follow from Theorem 3. In fact, it is also easy to deduce Theorem 2 from Theorem 3 (at least if the bipartite graph in Theorem 2 is strongly regular when viewed from the bipartition V_1).

We continue with an estimation for the EIP in the case $k = 2$. Our Theorem 2 can be considered as a bipartite version of the following result.

Theorem 4 ([13]). *Let* $G = (\mathcal{V}, \mathcal{M})$ *be a graph and* μ_1 *be the largest eigenvalue of the adjacency matrix of* G. *For* $v \in \mathcal{V}$ *let* $d(v) = d_{\mathcal{M}}(v)$ *be the degree of* v. *Then*

$$\sum_{v \in \mathcal{V}} d(v)^2 \leq (\mu_1)^2 |\mathcal{V}| \, .$$

If G *is connected then equality holds if and only if* G *is regular or semiregular.*

In order to apply this theorem one needs bounds on the largest eigenvalue of a graph. Many such bounds have been established, and we refer to [8] for a survey.

For example, a result by Schwenk [19] and Hong [14] says that the largest eigenvalue μ_1 of a connected graph $G = (\mathcal{V}, \mathcal{M})$ satisfies

$$\mu_1 \leq \sqrt{2|\mathcal{M}| - |\mathcal{V}| + 1} \, .$$

This bound together with Theorem 4 however yields a weaker estimation for the $k = 2$ case of the EIP than inequality (2). A best possible upper bound on the largest eigenvalue (called index) of a graph in terms of the number of edges was obtained by Rowlinson [17]:

Theorem 5 ([17]). *Among all graphs with* n *vertices and* m *edges exactly the quasi-complete graph* C_n^m *has largest index.*

Note that this result together with the Theorem 4 comes close to the optimum of the EIP of $J(n, 2)$ in half of all cases (but does not yield to sharp estimations for all quasi-complete graphs due to the equality characterization in Theorem 4). Indeed, the two previous theorems yield a better estimation for the EIP of $J(n, 2)$ than inequality (2). It seems worthwhile to study analogues of Theorem 5 for hypergraphs in order to improve the estimation (4).

3   A Short Proof for the Case $k = 2$

Recall the majorization (or dominance) order for sequences: If $d = (d_1, \ldots, d_n)$ and $e = (e_1, \ldots, e_n)$ are real vectors we say that d is majorized by e if for all $j = 1, \ldots, n$ the sum of the largest j entries of d is not larger than the corresponding sum for e, and if equality holds for $j = n$. Equivalently, d is majorized by e if d can be obtained from e by a sequence of alterations each of which replaces two entries d_i, d_j with $d_i < d_j$ by $d_i + x$, $d_j - x$ with $0 \leq 2x \leq d_j - d_i$.

We look for graphs with a fixed number of vertices and a fixed number of edges which maximize the sum of squares of degrees. For this it is sufficient to look among graphs having a degree sequence which is not majorized by any other degree sequence. Indeed, this follows by elementary means or by noting that the sum of squares of degrees is a symmetric and convex function of the degrees. A characterization of such degree sequences is known, and gives in fact a characterization of all degree sequences of graphs:

Given an integer vector $d = (d_1, \ldots, d_n)$ with nonincreasing entries (i.e. a partition of $d_1 + \cdots + d_n$), the number $r := \max\{i : d_i \geq i\}$ is called the *rank* of d. Further, the sequence $d' = (d'_1, \ldots, d'_{n-1}, \ldots)$ with $d'_i = |\{j : d_j \geq i\}|$ is called the *conjugate* sequence (or partition).

Theorem 6 (Ruch, Gutman [18]). *Let $d = (d_1, \ldots, d_n)$ be a rank r sequence of nonincreasing nonnegative integers. Then d is a degree sequence of a graph with n vertices and m edges if and only if $2m = d_1 + \cdots + d_n$ and*

$$d_1 + \cdots + d_i + i \leq d'_1 + \cdots + d'_i \text{ holds for all } 1 \leq i \leq r. \tag{5}$$

Degree sequences of graphs which satisfy equality in (5) for all $i = 0, \ldots, r$ are called *threshold sequences*, the corresponding graphs are called *threshold graphs*. These graphs were introduced in a different manner in [11] and have many characterizations. For example, threshold graphs are exactly those graphs whose degree sequences are not realized by any other nonisomorphic graph. Also, the threshold sequences and their permutations are exactly the extreme points of the convex hull of all degree sequences of graphs on a fixed number of vertices. We refer to [15] for more details.

Theorem 6 says that a partition $d = (d_1, \ldots, d_n)$ is the degree sequence of a graph on n vertices if and only if d is majorized by a threshold sequence.

Our proof of the Ahlswede-Katona theorem will proceed by some easy operations on the diagrams of a threshold sequences. Recall that the (Young) diagram of a sequence $(d_1, \ldots, d_n)$ of nonincreasing integers is the array of $d_1 + \cdots + d_n$ boxes having n left-justified rows with row i containing d_i boxes, $i = 1 \ldots, n$. The *square* of a rank r partition d is the square consisting of the r^2 boxes in the left upper corner of the diagram of d. We will identify sequences of nonincreasing integers with their diagrams. Figure 1 shows the degree sequences of a quasi-star and a quasi-complete graph. Both types of graphs are threshold graphs. Note also that a rank r diagram of a quasi-star has $n - 1$ boxes in each of its first $r - 1$ rows, and a rank r diagram of a quasi-complete graph has at most $r + 1$ columns.

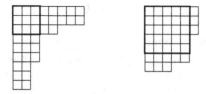

Fig. 1. Diagrams of S_9^{18} and C_9^{18}

Fig. 2. Moving a box

A simple calculation shows that $\sum_i d_i'^2 = \sum_i d_i^2 + \sum_i d_i$ holds for all threshold sequences $(d_1, \ldots, d_n)$. Maximizing $\sum_i d_i^2$ over all threshold sequences with fixed $\sum_i d_i$ is thus equivalent to maximizing $\sum_i d_i'^2$. Let us place for each threshold sequence d the weight $i + j - 1$ in the box of row i and column j of the diagram of d. Then our maximization is equivalent to finding a threshold sequence which has largest sum of weights among all threshold sequences with a fixed number $2m = d_1 + \cdots + d_n$ of boxes.

Let us now turn to the proof of Theorem 1. We proceed by induction on the number m of edges. Suppose that G is an optimal threshold graph with m edges and nonincreasing degree sequence $(d_1, \ldots, d_n)$. We may assume that $d_1 = n - 1$. Indeed, in the other case the complement $\overline{G}$ of G has a vertex of degree $n - 1$, and is optimal among all graphs with $\binom{n}{2} - m$ edges (since the edge boundaries of G and its complement are equal). But if we can show that $\overline{G}$ can be taken to be a quasi-star or a quasi-complete graph, then G can be so too. The graph obtained from G by removing a vertex of degree $d_1 = n - 1$ is an optimal graph with $m - (n - 1)$ edges. By induction, we can assume that it is either a quasi-star or a quasi-complete graph. In the first case G itself is a quasi-star and we are done. Thus assume that the second case occurs. We will show how to transfer the diagram d of G's degree sequence into one of the diagrams of the quasi-star or the quasi-complete graph while maintaining the sum of weights. Since d is threshold, the boxes to the right of d's square have a mirror counterpart below the square. In the following, when we perform operations with the boxes lying to right of the square of a threshold diagram, we always assume that the same (transposed) operations are done with the boxes below the square, such that the resulting diagram will be again a threshold one.

Let r be the rank of d, and let $1 + s$ be the number of boxes in the $(r + 1)$-th column of d. We may assume that $r \geq 3$ and $n - 1 - r \geq 2$, since otherwise d is the diagram of a quasi-star resp. quasi-complete graph. We consider two cases.

First, let $r \geq n - 1 - r$. If $1 + s \geq n - 1 - r$, then moving the last box of the first row to the first empty place in the $(r + 1)$-th column (and doing the transposed operation below the square) will yield a threshold sequence with a larger sum of weights, contradicting the optimality of G (see Figure 2).

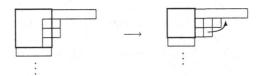

Fig. 3. Flipping and Moving a box

Thus $1 + s < n - 1 - r$. But then we flip the hook consisting of the $s + 1$ boxes in the $(r + 1)$-th column and the last $n - 1 - r$ boxes in the first row to obtain a threshold diagram d' with $n - 1 - r$ boxes in the $(r + 1)$-th column and $r + s + 1$ boxes in the first row, while maintaining the sum of weights. If $s \geq 1$, then, as above, moving the last box from the first row of d' to the $(r + 1)$-th column yields a contradiction. Thus we have $s = 0$. But then d' is the diagram of a quasi-complete graph, and we are done with the case $r \geq n - 1 - r$. Let $r < n - 1 - r$. Then we flip the last s boxes of the $(r + 1)$-th column and the second up to r-th box of the r-th column to obtain a threshold sequence d'' with $2r - 2$ boxes in the second row and $r - 1 + s$ boxes in the third row, while maintaining the sum of weights (see Figure 3).

If $0 < s < r - 1$, then moving in the diagram d'' the last box of the third row to the first empty place of the second row increases the sum of weights, contradicting the optimality of G. If $s = r - 1 \ (\geq 2)$, then moving the last two boxes of the third row to the first two empty places of the second row will increase the sum weights. Thus we have again $s = 0$. If $r \geq 4$, then d'' has rank $r - 1$, and moving the box in row and column $r - 1$ to the first empty place of the second row will increase the sum of weights. Thus we have $r = 3$. But then d'' is the diagram of a quasi-star, and we are also done with the case $r < n - 1 - r$.

References

1. R. Aharoni, A problem in rearrangements of $(0, 1)$-matrices. Discrete Math. 30, 191–201, 1980.

2. R. Ahlswede and N. Cai, On edge-isoperimetric theorems for uniform hypergraphs, Preprint 93-018, Diskrete Strukturen in der Mathematik SFB 343, Univ. Bielefeld, 1993; this volume.

3. R. Ahlswede and G.O.H. Katona, Graphs with maximal number of adjacent pairs of edges, Acta Math. Acad. Sci. Hungar. 32, 1-2, 97–120, 1978.

4. N. Alon and V.D. Milman, λ_1-isoperimetric inequalities for graphs, and supercon-centrators, J. Combin. Theory Ser. B 38, 73–88, 1985.

5. C. Bey, An upper bound on the sum of squares of degrees in a hypergraph, Discrete Math. 269, 259–263, 2003.

6. S.L. Bezrukov, Edge isoperimetric problems on graphs, Graph theory and combinatorial biology (Balatonlelle, 1996), János Bolyai Math. Soc., Budapest, 157–197, 1999.

7. R.A. Brualdi, and E.S. Solheid, Some extremal problems concerning the square of a(0, 1)-matrix, Linear Mult. Alg. 22, 57–73, 1987.

8. D. Cvetkovic and P. Rowlinson, The largest eigenvalue of a graph: A survey, Linear Mult. Alg. 28, 3–33, 1990.

9. D. de Caen, An upper bound on the sum of squares of degrees in a graph, Discrete Math. 185, 245–248, 1998.

10. P. Delsarte, An Algebraic Approach to the Association Schemes of Coding Theory, Philips Res. Rep. Suppl. 10, 1973.

11. P.L. Hammer, and V. Chvátal, Aggregation of inequalities in integer programming, Ann. Discrete Math. 1, 145–162, 1977.

12. L.H. Harper, On a problem of Kleitman and West, Discrete Math. 93, 2-3, 169–182, 1991.

13. M. Hofmeister, Spectral radius and degree sequence, Math. Nachr. 139, 37–44, 1988.

14. Y. Hong, A bound on the spectral radius of graphs, Linear Algebra Appl. 108, 135–139, 1988.

15. N.V.R. Mahadev and U.N. Peled, Threshold graphs and related topics, Ann. Discrete Math. 56, 1995.

16. D. Olpp, A conjecture of Goodman and the multiplicities of graphs. Austral. J. Comb. 14, 267–282, 1996.

17. P. Rowlinson, On the maximal index of graphs with a prescribed number of edges, Linear Algebra Appl. 110, 43–53, 1988.

18. E. Ruch and I. Gutman, The branching extent of graphs, J. Combinatorics, Inform. Syst. Sci. 4, 285–295, 1979.

19. A. Schwenk, New derivations of spectral bounds for the chromatic number, Graph Theory Newsletter 5, 77, 1975.

Appendix: On Edge–Isoperimetric Theorems for Uniform Hypergraphs

R. Ahlswede and N. Cai

1 Introduction

Denote by $\Omega = \{1, \ldots, n\}$ an n–element set. For all $A, B \in \binom{\Omega}{k}$, the k–element subsets of Ω, define the relation $\sim$ as follows:
$A \sim B$ iff A and B have a common shadow, i.e. there is a $C \in \binom{\Omega}{k-1}$ with $C \subset A$ and $C \subset B$. For fixed integer α, our goal is to find a family $\mathcal{A}$ of k–subsets with size α, having as many as possible $\sim$ –relations for all pairs of its elements. For $k = 2$ this was achieved by Ahlswede and Katona [2] many years ago. However,

it is surprisingly difficult for $k \geq 3$, in particular there is no complete solution even for $k = 3$. Perhaps, the reason is the complicated behaviour for "bad α" so that the most natural and reasonable conjecture, which will be described in the last section and was mentioned already in [2], is false. Actually, our problem can

also be viewed as a kind of isoperimetric problem in the sense of Bollobás and Leader ([4], see also [6]). They gave two versions. Partition the vertex set V of a graph $G = (V, E)$ into 2 parts A and A^c such that for fixed α $|A| = \alpha$ and

I. The subgraph induced by A has maximal number of edges
   or
II. The number of edges connecting vertices from A and A^c is as small as possible.

When G is regular, the two versions are equivalent. In our case we define $G = (V, E)$ by $V = \binom{\Omega}{k}$ and $E = \big\{\{A, B\} \subset V \colon A \neq B \text{ and } A \sim B\big\}$. Thus the original problem is an edge–isoperimetric problem for a certain regular graph. In order

to solve our problem, in Section 2 we reduce it to another kind of problem, which we call "sum of ranks problem": For a lattice with a rank function find a downset of given size with maximal sum of the ranks of its elements. Similar questions were studied in [3], [6], and [8]. In Section 3, we go over to a continuous version

of the problem and solve it for $k = 3$ and "good α". Some of the auxiliary results and ideas there extend also to general k. A related but much simpler

result concerning a moment problem is presented in Section 4.

2 From Edge–Isoperimetric to Sum of Ranks Problem

In this section we reduce the edge–isoperimetric problem to the sum of ranks problem. Denote by $\mathcal{L}(n, k) = (S_{n,k}, \leq)$ the lattice defined by

R. Ahlswede et al. (Eds.): Information Transfer and Combinatorics, LNCS 4123, pp. 979–1005, 2006.

$$S_{n,k} = \left\{ (x_1, \ldots, x_k) : 1 \le x_1 < x_2 \cdots < x_k \le n, x_i \in \mathbb{Z}^+ \right\}$$

and $(x_1, \ldots, x_k) \le (x'_1, \ldots, x'_k) \Leftrightarrow x_i \le x'_i (1 \le i \le k)$. For $x^k \in S_{n,k}$, the rank of x^k is defined as $|x^k| = \sum_{i=1}^k x_i$ and for $W \subset S_{n,k}$, let $||W|| = \sum_{x^k \in W} |x^k|$. In addition we let $A = \{x_1, \ldots, x_k\} \in \binom{\Omega}{k}$, with elements labelled in increasing order, correspond to $x^k = \Phi(A) \triangleq (x_1, \ldots, x_k) \in S_{n,k}$, and, similarly, $\mathcal{A} \subset \binom{\Omega}{k}$ to $\Phi(\mathcal{A}) = \{\Phi(A) : A \in \mathcal{A}\}$. Moreover, for $\mathcal{A} \subset \binom{\Omega}{k}$ we introduce

$$\mathcal{P}(\mathcal{A}) = \left\{ (A, B) \in \mathcal{A}^2 : A \sim B \right\}.$$

Using for $A \in \mathcal{A}$ and $1 \le i < j \le n$ the following "pushing to the left" or so–called switching operator $O_{i,j}$, which is frequently employed in combinatorial extremal theory:

$$O_{i,j}(A) = \begin{matrix} (A \smallsetminus \{j\}) \cup \{i\} \text{ if } (A \smallsetminus \{j\}) \cup \{i\} \notin \mathcal{A}, \ j \in A, \text{ and } i \notin A \\ A \qquad\qquad\qquad\qquad\qquad \text{otherwise,} \end{matrix}$$

one can prove, by standard arguments, that for fixed α an $\mathcal{A} \subset \binom{\Omega}{k}$ with $|\mathcal{A}| = \alpha$, which maximizes $|\mathcal{P}(\mathcal{A})|$, can be assumed to be within a family of subsets, which are invariant under the pushing to left operator. It is also easy to see that such subsets correspond to a downset in $\mathcal{L}(n, k)$.

Lemma 1. For $\alpha \in \mathbb{Z}^+$ $\max_{|\mathcal{A}|=\alpha} |\mathcal{P}(\mathcal{A})|$ is assumed by an $\mathcal{A} \subset \binom{\Omega}{k}$ s.t. $\Phi(\mathcal{A})$ is a downset in $\mathcal{L}(n, k)$.

Now we are ready to show the first of our main results.

Theorem 1. For fixed $\alpha \in \mathbb{Z}^+$, maximizing $|\mathcal{P}(\mathcal{A})|$ for $\mathcal{A} \subset \binom{\Omega}{k}$, $|\mathcal{A}| = \alpha$, is equivalent to finding a downset W in $\mathcal{L}(n, k)$ with $|W| = \alpha$ and maximal $||W||$.

Proof. Assume that $\mathcal{A} \subset \binom{\Omega}{k}$, $W = \Phi(\mathcal{A})$ is a downset in $\mathcal{L}(n, k)$, and $|\mathcal{A}| = \alpha$.

For every $x^k \in W$ there are exactly

$$(x_{i+1} - x_i - 1)\binom{k-i}{k-1-i} = (x_{i+1} - x_i - 1)(k - i) \tag{1.1}$$

y^k's with $y^k \le x^k$, whose first i components coincide with those of x^k and the $(i+1)$-st components differ, and for which A and B have a common shadow if $x^k = \Phi(A)$ and $y^k = \Phi(B)$. (Here $x_0 \triangleq 0$.) By (1.1), for $x^k = \Phi(A)$ fixed, there is a total of

$$\sum_{i=0}^{k-1}(x_{i+1} - x_i - 1)(k - i) = \sum_{i=1}^k (k - i + 1)x_i - \sum_{i=0}^{k-1}(k - i)x_i - \sum_{i=0}^{k-1}(k - i)$$

$$= \sum_{i=1}^k x_i - \binom{k+1}{2} = |x^k| - \binom{k+1}{2} \tag{1.2}$$

B's with $\Phi(B) = y^k \leq x^k$, $B \sim A$, and with $\Phi(B) \in \mathcal{A}$, because $\Phi(\mathcal{A})$ is a downset. Consequently

$$|\ \mathcal{P}(\mathcal{A})\ | = 2 \sum_{x^k \in W} |x^k| - 2\binom{k+1}{2}|\mathcal{A}| = 2||W|| - 2\alpha\binom{k+1}{2}. \qquad (1.3)$$

Thus our theorem follows from Lemma 1 and (1.3).

From now on we study our problem in the "sum–rank" version.

3   From the Discrete to a Continuous Model

A natural idea to solve a discrete problem for "good parameters" is to study the related continuous problem. Every $z^k \in \mathbb{Z}^k$ we let correspond to a cube $C(z^k) \triangleq \{x^k : \lceil x_i \rceil = z_i\}$ in $\mathbb{R}^k$. This mapping sends our $S_{U,k}$ for $U \in \mathbb{Z}^+$ to $\overset{\sim}{\rightarrow} S_{U,k} \triangleq \{x^k : 0 < x_1 < x_2 \cdots < x_k \leq U, \lceil x_i \rceil \neq \lceil x_j \rceil, \text{ if } i \neq j\}$. Thus, keeping the partial order "$\leq$", we can "embed" our $\mathcal{L}(U,k)$ into a "continuous lattice" $\overset{\sim}{\rightarrow} \mathcal{L}(U,k) = (\overset{\sim}{\rightarrow} S_{U,k}, \leq)$. Moreover, the image $\overset{\sim}{\rightarrow} W \triangleq \Phi(W)$ of a downset W in $\mathcal{L}(U,k)$ is a downset in $\overset{\sim}{\rightarrow} \mathcal{L}(U,k)$, with (finite) integer–components for maximal points. Let μ be the Lebesgue measure on $\mathbb{R}^{k'}$, and let $k' \leq k$ be specified by the context. For $W \subset \mathbb{R}^k$, define

$$||W|| = \int_W |x^k| d\mu, \quad \text{where} \quad |x^k| = \sum_j x_j. \qquad (3.1)$$

Let $\mathcal{D}$ be the set of downsets in $\overset{\sim}{\rightarrow} \mathcal{L}(U,k)$ with finitely many maximal points. Since it is of no consequence if we add or substract a set of measure zero, we will frequently exchange "$<$" (or "$>$") and "$\leq$" (or "$\geq$") in the sequel. It is enough in our problem for "good α" to consider $\max_{\mu(\overset{\sim}{\rightarrow} W) = \alpha, \overset{\sim}{\rightarrow} W \in \mathcal{D}} ||W||$ in $\overset{\sim}{\rightarrow} \mathcal{L}(U,k)$, and the following lemma is the desired bridge.

Lemma 2. *Suppose that $\overset{\sim}{\rightarrow} W \in \mathcal{D}$ has only maximal points with integer components, and so for a $W \subset \mathcal{L}(U,k)$ $\overset{\sim}{\rightarrow} W = \Phi(W)$.*

Then

$$|| \overset{\sim}{\rightarrow} W|| = ||W|| - \frac{k}{2}\alpha, \quad \text{where} \quad \alpha = \mu(\overset{\sim}{\rightarrow} W). \qquad (3.2)$$

Proof.

$$|| \overset{\sim}{\rightarrow} W|| = \sum_{z^k \in W} ||C(z^k)|| = \sum_{z^k \in W} \int_{C(z^k)} |x^k| \mu(dx^k)$$
$$= \sum_{z^k \in W} \int_{z_k-1}^{z_k} dx_k \cdots \int_{z_1-1}^{z_1} dx_1 \sum_{j=1}^k x_j$$
$$= \sum_{z^k \in W} \sum_{i=1}^k \int_{z_i-1}^{z_i} x_i dx_i = \sum_{z^k \in W} \sum_{i=1}^k \frac{1}{2}(2z_i - 1) \qquad (3.3)$$

and (3.2) follows, because $|W| = \mu(\overset{\sim}{\rightarrow} W)$. We say that $W \in \mathcal{D}$ can be reduced to $W' \in \mathcal{D}$, if $\mu(W') = \mu(W)$ and $||W'|| \geq ||W||$.

4 Cones and Trapezoids

Next we define cones and trapezoids, which will play important role in our problem. A cone in $\overset{\sim}{\to} S_{U,k}$ is a set

$$K_k(u) = \{x^k \in R^k : 0 < x_1 < \cdots < x_k \leq u \text{ and } \lceil x_i \rceil \neq \lceil x_j \rceil \text{ for } i \neq j\}, \text{ with } u \leq U. \tag{4.1}$$

Clearly, $\overset{\sim}{\to} S_{U,k}$ is a cone itself. It can be denoted by $K_k(U)$. A trapezoid $R_k(v, u)$ in $K_k(U)$ is a downset below $(v, u \ldots u)$, where $0 < v \leq u \leq U$, i.e.

$$R_k(v, u) \triangleq \{x^k \in \overset{\sim}{\to} S_{U,k} : x_1 \leq v, x_k \leq u\} \tag{4.2}$$

and therefore $K_k(u) = R_k(u, u)$. Moreover, for $W \subset K_k(u)$ set

$$\overline{W}^{(u)} \triangleq K_k(u) \smallsetminus W \tag{4.3}$$

and

$$\hat{W}^{(u)} \triangleq \{(\lfloor u \rfloor, \ldots, \lfloor u \rfloor) - x^k : x^k \in \overline{W}^{(u)}\}. \tag{4.4}$$

For integral u one can easily verify that

$$W = \hat{V}^{(u)} \text{ for } V = \hat{W}^{(u)} \tag{4.5}$$

and

$$R_k(v, u) = \hat{K}_k^{(u)}(u - v). \tag{4.6}$$

Lemma 3. *For $W \in \mathcal{D}$ and $W \subset K_k(u)$, $u \leq U$,*

$$||W|| = ||K_k(u)|| - k\lfloor u \rfloor \mu(\hat{W}^{(u)}) + ||\hat{W}^{(u)}||. \tag{4.7}$$

Proof. According to the definitions of "$^\wedge(u)$" and "$|| \ ||$",

$$\begin{aligned}
||W|| &= \int_W |x^k| \mu(dx^k) = \int_{K_k(u) \smallsetminus \overline{W}^{(u)}} |x^k| \mu(dx^k) \\
&= ||K_k(u)|| - \int_{\overline{W}^{(u)}} |x^k| \mu(dx^k) \\
&= ||K_k(u)|| - \int_{\hat{W}^{(u)}} \sum_{j=1}^k (\lfloor u \rfloor - x_j) \mu(dx^k) \\
&= ||K_k(u)|| - k\lfloor u \rfloor \mu(\hat{W}^{(u)}) + ||\hat{W}^{(u)}||.
\end{aligned}$$

Notice that for $u \notin \mathbb{Z}^+$ $\hat{W}^{(u)}$ is not in $\mathcal{L}(u, k)$.

Corollary 1. *For $u \in \mathbb{Z}^+$*

$$||K_k(u)|| = \frac{ku}{2}\mu(K_k(u)). \tag{4.8}$$

Proof. One can verify (4.8) by standard techniques in calculus for evaluating integrals, however, Lemma 3 provides a very elegant and simple way.

By (4.7) for $W \subset K_k(u)$

$$||W|| - ||\hat{W}^{(u)}|| = ||K_k(u)|| - ku \, \mu(\hat{W}^{(u)}) \tag{4.9}$$

and by (4.5) and (4.7) one can exchange the roles of W and $\hat{W}$. Therefore we have

$$||\hat{W}^{(u)}|| - ||W|| = ||K_k(u)|| - ku \, \mu(W). \tag{4.10}$$

"Adding (4.9) and (4.10)" and using the fact $\mu(K_k(u)) = \mu(W) + \mu(\hat{W}^{(u)})$, we obtain (4.8). Next we establish a connection between $||K_k(u)||$ and $\mu(K_k(u))$ for not necessarily integral u. It can elegantly be expressed in terms of densities. We define the density of $W \subset \mathbb{R}^{k'}$ ($k' \leq k$ defined by context) as

$$d_{k'}(W) = \frac{||W||}{\mu(W)} \quad \text{and set} \quad d = d_k. \tag{4.11}$$

Then Corollary 1 takes the form

$$d(K_k(u)) = \frac{k}{2} u, \ u \in \mathbb{Z}^+. \tag{4.12}$$

We extend this formula to general u.

Lemma 4. *For $u \leq U$ not necessarily integers, denote by $\theta \triangleq \{u\} = u - \lfloor u \rfloor$ the fractional part of u. Then*

(i) $\mu(K_k(u)) = \binom{\lfloor u \rfloor}{k} + \theta \binom{\lfloor u \rfloor}{k-1}$,

(ii) $||K_k(u)|| = \frac{ku}{2} \mu(K_k(u)) + \frac{k-1}{2} \theta(1 - \theta) \binom{\lfloor u \rfloor}{k-1}$

and therefore

(iii) $d(K_k(u)) = \frac{ku}{2} + \dfrac{\frac{k-1}{2}\theta(1-\theta)}{\frac{1}{k}(\lfloor u \rfloor + 1 - k) + (k-1)\theta}$.

Proof. By its definition

$$K_k(u) = K_k(\lfloor u \rfloor) \cup \{x^k : \lfloor u \rfloor < x_k \leq u \text{ and } (x_1, \dots, x_{k-1}) \in K_{k-1}(\lfloor u \rfloor)\}$$
$$\triangleq K_k(\lfloor u \rfloor) \cup J \text{ (say)}. \tag{4.13}$$

On the other hand, according to the correspondence Φ between the discrete and the continuous models,

$$\mu(K_k(\lfloor u \rfloor)) = \binom{\lfloor u \rfloor}{k}, \mu(K_{k-1}(\lfloor u \rfloor)) = \binom{\lfloor u \rfloor}{k-1}. \tag{4.14}$$

Therefore $\mu(J) = \theta \binom{\lfloor u \rfloor}{k-1}$ and consequently (i) holds. Now

$$||K_k(u)|| = ||K_k(\lfloor u \rfloor)|| + ||J||. \tag{4.15}$$

By Corollary 1 and (4.14)

$$||K_k(\lfloor u \rfloor)|| = \frac{k\lfloor u \rfloor}{2}\binom{\lfloor u \rfloor}{k}. \tag{4.16}$$

Furthermore, by (4.8) for $k-1$ and by (4.14)

$$\begin{aligned}||J|| &= \mu\big(K_{k-1}(\lfloor u \rfloor)\int_{\lfloor u \rfloor}^{u} x_k\, dx_k + \int_{\lfloor u \rfloor}^{u} dx_k ||K_{k-1}(\lfloor u \rfloor)|| \\ &= (\lfloor u \rfloor + \tfrac{\theta}{2})\theta\binom{\lfloor u \rfloor}{k-1} + \theta\tfrac{k-1}{2}\lfloor u \rfloor\binom{\lfloor u \rfloor}{k-1}.\end{aligned} \tag{4.17}$$

Combination of these three identities gives

$$||K_k(u)|| = \frac{k\lfloor u \rfloor}{2}\binom{\lfloor u \rfloor}{k} + \left(\lfloor u \rfloor + \frac{\theta}{2} + \frac{k-1}{2}\lfloor u \rfloor\right)\theta\binom{\lfloor u \rfloor}{k-1}$$

and thus

$$||K_k(u)|| = \frac{k\lfloor u \rfloor}{2}\binom{\lfloor u \rfloor}{k} + \left(\frac{k+1}{2}\lfloor u \rfloor + \frac{\theta}{2}\right)\theta\binom{\lfloor u \rfloor}{k-1}. \tag{4.18}$$

This and (i) imply

$$\begin{aligned}||K_k(u)|| - \tfrac{ku}{2}\mu(K_k(u)) &= -\tfrac{k\theta}{2}\binom{\lfloor u \rfloor}{k} + \left(\tfrac{\lfloor u \rfloor}{2} - \tfrac{k-1}{2}\theta\right)\theta\binom{\lfloor u \rfloor}{k-1} \\ &= -\tfrac{k\theta}{2}\binom{\lfloor u \rfloor}{k} + \tfrac{\lfloor u \rfloor}{2}\theta\binom{\lfloor u \rfloor}{k-1} - \tfrac{k-1}{2}\theta^2\binom{\lfloor u \rfloor}{k-1} \\ &= -\tfrac{\theta\lfloor u \rfloor}{2}\binom{\lfloor u \rfloor - 1}{k-1} + \tfrac{\lfloor u \rfloor}{2}\theta\binom{\lfloor u \rfloor}{k-1} - \tfrac{k-1}{2}\theta^2\binom{\lfloor u \rfloor}{k-1} \\ &= \tfrac{\lfloor u \rfloor}{2}\theta\binom{\lfloor u \rfloor - 1}{k-2} - \tfrac{k-1}{2}\theta^2\binom{\lfloor u \rfloor}{k-1} = \tfrac{k-1}{2}\theta\binom{\lfloor u \rfloor}{k-1} - \tfrac{k-1}{2}\theta^2\binom{\lfloor u \rfloor}{k-1},\end{aligned}$$

and therefore (ii).

Remark 1 (to Lemma 4).
Actually, we can derive a somewhat more general result along the same lines.
Let $J_k(u, u') \triangleq \{(x_1, \ldots, x_k) \mid u < x_1 < \cdots < x_k \leq u' \text{ and } \lceil x_i \rceil \neq \lceil x_j \rceil, \text{ for } i \neq j\}$, $u < u' \in \mathbb{R}$, $\theta \triangleq \lceil u \rceil - u$ and $\theta' = u' - \lfloor u' \rfloor \triangleq \{u'\}$, then

$$\mu\big(J_k(u, u')\big) = \binom{\lfloor u' \rfloor - \lceil u \rceil}{k} + \binom{\lfloor u' \rfloor - \lceil u \rceil}{k-1}(\theta + \theta') + \theta\theta'\binom{\lfloor u' \rfloor - \lceil u \rceil}{k-2} \tag{4.19}$$

and

$$||J_k(u, u')|| - k(u + u') = \frac{k-1}{2}[(\theta' - \theta)[1 - (\theta + \theta')]]\binom{\lfloor u' \rfloor - \lceil u \rceil}{k-1} - \frac{\theta\theta'}{2}(\theta' - \theta)\binom{\lfloor u' \rfloor - \lceil u \rceil}{k-2}. \tag{4.20}$$

This can be seen as follows.
By shifting the origin, we can assume w.l.o.g., that $u = -\theta$, $\theta \in [0, 1)$, i.e. $\lfloor u \rfloor = 0$. Then

$$\begin{aligned}J_k(u, u') = &K_k(\lfloor u' \rfloor) \cup \big(\{x_1 : -\theta < x_1 \leq 0\} \times \{(x_2, \ldots, x_k) : (x_2, \ldots, x_k) \in K_{k-1}(\lfloor u' \rfloor)\} \\ &\cup \big(\{(x_1, \ldots, x_{k-1}) : (x_1, \ldots, x_{k-1}) \in K_{k-1}(\lfloor u' \rfloor)\} \times \{x_k : \lfloor u' \rfloor < x_k \leq u'\}\big) \\ &\cup \big(\{x_1 : -\theta < x_1 \leq 0\} \times \{(x_2, \ldots, x_{k-1}) \in K_{k-2}(\lfloor u' \rfloor)\} \times \{x_k : \lfloor u' \rfloor < x_k \leq u'\}\big)\end{aligned}$$

and by the same argument as the one used in the proof of Lemma 4 we obtain (4.19) and (4.20).

5   The Cases $k = 2, 3$

Using the same idea as in the proof of Theorem 1 in [2] simple calculations lead to two alternatives.

Lemma 5. *For* $k = 2$, $U \in \mathbb{Z}^+$ *and* $W \in \mathcal{D}$ *consider*

$$m_1(W) \triangleq \max\{x : (x, y) \in W \text{ for some } y\}. \tag{5.1}$$

Then

(i) W *can be reduced to a trapezoid, if* $m_1(W) \leq \frac{U}{2}$

   and

(ii) W *can be reduced to a cone, if* $m_1(W) \geq \frac{U}{2}$.

Now we turn our attention to $k = 3$ and drop all subscripts k (for example write $K(U)$ instead of $K_3(U)$ and so on).

For $W \subset K(U)$ we call the 2–dimensional set

$$S_u(W) \triangleq \{(x, y) : (x, y, u) \in W \text{ and } (x, y, u + \varepsilon) \notin W \text{ for all } \varepsilon > 0\} \tag{5.3}$$

a Z–surface of W at u.

We call this surface *regular*, when for some $(x, y) \in S_u(W)$ and some $\varepsilon > 0$ $(x, y, u + \varepsilon) \in K(U)$. Therefore $S_u(W)$ is irregular iff $u = U$. The $Y-$ and X–surfaces are defined analogously. We present now the basic idea of "moving

top layers from lower density to higher density".

Observe first that the condition $\mu\big(R(v, u)\big) = \alpha$ (for fixed α) forces v to depend continuously on u, say

$$v = V_\alpha(u). \tag{5.4}$$

There are again two alternatives.

Lemma 6. *For* $k = 3$, $u \leq U$, *and* $U \in \mathbb{Z}^+$ *any trapezoid* $R(v, u)$ *can be reduced to a cone or the trapezoid* $R\big(V_\alpha(U), U\big)$.

Proof. Fix α and $U \in \mathbb{Z}^+$. Then $\|R\big(V_\alpha(u), u\big)\|$ is a continuous function in u, which achieves a maximal value. So, if the lemma is not true, then there are a $U \in \mathbb{Z}^+$, an α, and a u_0 with $v_0 \triangleq V_\alpha(u_0) < u_0 < U$ and $R(v_0, u_0)$ achieves the maximal value. $R(v_0, u_0)$ has one regular Z–surface and one regular X-surface, namely

$$\begin{aligned} S_1 \quad &\triangleq \big\{(x, y) : 0 < x < y \leq \lceil u_0 \rceil - 1, x \leq v_0 \text{ and } \lceil x \rceil \neq \lceil y \rceil\big\} \\ \text{and } S_2 \quad &\triangleq \big\{(y, z) : \lceil v_0 \rceil < y < z \leq u_0 \text{ and } \lceil y \rceil \neq \lceil z \rceil\big\}. \end{aligned} \tag{5.6}$$

(c.f. Figure 1)

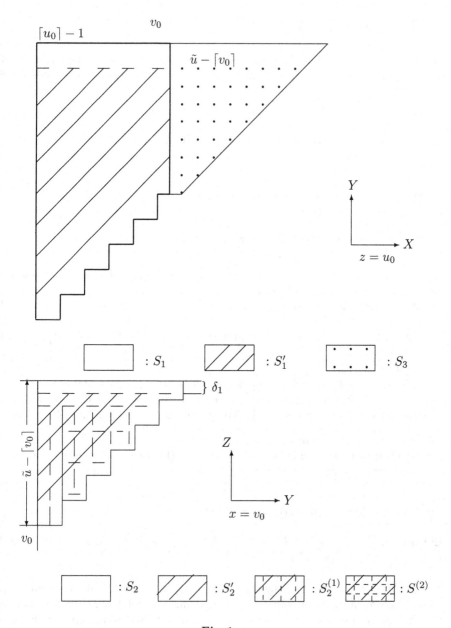

Fig. 1.

Case 1: $d(S_1) + u_0 < d(S_2) + v_0.$ (5.7)

Choose $\delta_1, \delta_2 > 0$ and define

$$
\begin{aligned}
D_1 &= S_1 \times \{z : u_0 - \delta_1 < z \le u_0\} \\
\text{and } D_2 &= \{x : v_0 < x \le v_0 + \delta_2\} \times S_2'.
\end{aligned}
$$ (5.9)

They satisfy

$$\mu(D_1) = \mu(D_2), \tag{5.10}$$

$$\delta_1 \leq u_0 - \left(\lceil u_0 \rceil - 1\right), \delta_2 \leq \left(\lfloor v_0 \rfloor + 1\right) - v_0, \tag{5.11}$$

and

$$d(S_1) + u_0 < d(S_2'') + v_0 \leq d(S_2') + v_0, \tag{5.12}$$

where

$$S_2'' \triangleq S_2 \smallsetminus \left\{(y, z) : u_0 - \delta_1 < z \leq u_0\right\} \tag{5.13}$$

and

$$S_2' \triangleq \begin{cases} S_2'' \smallsetminus \left\{(y, z) : v_0 < y \leq v_0 + 1\right\} & \text{if } v_0 \in \mathbb{Z}^+ \\ S_2'' & \text{otherwise.} \end{cases} \tag{5.14}$$

The second inequality in (5.12) follows from Lemma 4 and our choice is possible by (5.7). Then

$$R' \triangleq \left(R(v_0, u_0) \smallsetminus D_1\right) \cup D_2 \in \mathcal{D} \tag{5.15}$$

is a trapezoid with measure α.
However by (5.9) - (5.14),

$$\begin{aligned}
\|R'\| - \|R(v_0, u_0)\| &= \|D_2\| - \|D_1\| \\
&= \left[\mu(S_2') \int_{v_0}^{v_0 + \delta_2} x\, dx + \delta_2 \|S_2'\|\right] - \left[\|S_1\|\delta_1 + \mu(S_1) \int_{u_0 - \delta_1}^{u_0} z\, dz\right] \\
&= \left[(\mu(S_2')\delta_2)\left(v_0 + \tfrac{\delta_2}{2}\right) + \left(\delta_2 \mu(S_2')\right)d(S_2')\right] - \left[(\mu(S_1)\delta_1)d(S_1) + (\mu(S_1)\delta_1)\left(u_0 - \tfrac{\delta_1}{2}\right)\right] \\
&= \mu(D_2)\left[v_0 + \tfrac{\delta_2}{2} + d(S_2')\right] - \mu(D_1)\left[d(S_1) + u_0 - \tfrac{\delta_1}{2}\right] \\
&= \mu(D_1)\left[\left(d(S_2') + v_0\right) - \left(d(S_1) + u_0\right) + \tfrac{\delta_1 + \delta_2}{2}\right] > 0,
\end{aligned}$$

a contradiction. Here the fourth equality follows from $\mu(S_2')\delta_2 = \mu(D_2)$ and $\mu(S_1)\delta_1 = \mu(D_2)$ (by (5.9)), the fifth equality follows from (5.10) and the inequality follows from (5.12).

Case 2: $d(S_1) + u_0 > d(S_2) + v_0$. One can come to a contradiction just like in case 1.

Case 3: $d(S_1) + u_0 = d(S_2) + v_0$. $\tag{5.16}$

S_2 is a "shifted cone". One can calculate $d(S_2)$ and conclude with (5.16)

$$\lceil u_0 \rceil - 2 > v_0. \tag{5.17}$$

Consequently the following two surfaces are not empty:

$$\begin{aligned}
S_1' &\triangleq \left\{(x, y) : 0 < x < y \leq \lceil u_0 \rceil - 2, x \leq v_0 \text{ and } \lceil x \rceil \neq \lceil y \rceil\right\} \\
\text{and } S_2^{(1)} &\triangleq \left\{(y, z) : \lceil v_0 \rceil < y < z \leq u_0 - 1 \text{ and } \lceil y \rceil \neq \lceil z \rceil\right\} \\
&= S_2 \smallsetminus \left\{(y, z) : u_0 - 1 < z \leq u_0\right\}.
\end{aligned} \tag{5.19}$$

(See Figure 1) Assume first that

$$\mu(S_1') \geq \mu(S_2^{(1)}). \tag{5.20}$$

Let

$$
\begin{aligned}
D_1 &\triangleq \left\{ (x,y,z) \in R(v_0,u_0) : u_0 - 1 < z \leq u_0 \right\} \\
&= S_1 \times \left\{ z : \lceil u_0 \rceil - 1 < z \leq u_0 \right\} \cup S_1' \times \left\{ z : u_0 - 1 < z \leq \lceil u_0 \rceil - 1 \right\} \\
&\triangleq D_1' \cup D_1'',
\end{aligned}
$$

$$
\begin{aligned}
D_2 &\triangleq \left\{ (x,y,z) \in S_U : v_0 < x \leq x_0, z \leq u_0 - 1 \right\} \\
&= \left\{ x : v_0 < x \leq \lceil v_0 \rceil \right\} \times S_2^{(1)} \cup \left[\bigcup_{i \geq 2} \left(\left\{ x : \lceil v_0 \rceil + i - 1 < x \leq v^{(i)} \right\} \times S_2^{(i)} \right) \right],
\end{aligned}
\tag{5.22}
$$

where

$$S_2^{(i)} = S_2^{(i-1)} \setminus \left\{ (y,z) : \lceil v_0 \rceil + 2 - i < x \leq \lceil v_0 \rceil + 3 - i \right\},$$

the last $v^{(i)}$ equals x_0, for the other i's $v^{(i)} = \lceil v_0 \rceil + i$, and finally x_0 is specified by

$$\mu(D_1) = \mu(D_2), \text{ if such an } x_0 \text{ exists.}$$

Otherwise continue with Case 4. Introduce now

$$R' = \big(R(v_0,u_0) \setminus D_1 \big) \cup D_2.$$

R' is a trapezoid with measure α. Now we have, with justifications given afterwards,

$$
\begin{aligned}
\|D_1\| &= \left[\mu(S_1)\left(u_0 - \tfrac{u_0 - \lceil u_0 \rceil + 1}{2}\right)(u_0 - \lceil u_0 \rceil + 1) + \|S_1\|(u_0 - \lceil u_0 \rceil + 1) \right] \\
&\quad + \left[\mu(S_1')\left(\lceil u_0 \rceil - \tfrac{\lceil u_0 \rceil - u_0}{2} - 1\right)(\lceil u_0 \rceil - u_0) + \|S_1'\|(\lceil u_0 \rceil - u_0) \right] \\
&= \mu(D_1')\left(d(S_1) + u_0 - \tfrac{u_0 - \lceil u_0 \rceil + 1}{2}\right) + \mu(D_1'')\left[d(S_1') + \lceil u_0 \rceil - \tfrac{\lceil u_0 \rceil - u_0}{2} - 1\right] \\
&= \left[\mu(D_1')d(S_1) + \mu(D_1'')d(S') \right] + (u_0 - 1)(\mu(D_1') + \mu(D_1'')) \\
&\quad + \tfrac{1}{2}\mu(D_1')(u_0 - \lceil u_0 \rceil + 1) + \tfrac{1}{2}(\lceil u_0 \rceil - u_0)(2\mu(D_1') + \mu(D_1'')) \\
&< \mu(D_1)(d(S_1) + u_0 - 1) + \tfrac{1}{2}(u_0 - \lceil u_0 \rceil + 1)\mu(D_1') + \tfrac{1}{2}(\lceil u_0 \rceil - u_0)(2\mu(D_1') + \mu(D_1'')) \\
&= \mu(D_1)(d(S_1) + u_0 - 1) + \tfrac{1}{2}\left[\tfrac{\mu^2(D_1')}{\mu(S_1)} + 2\tfrac{\mu(D_1')\mu(D_1'')}{\mu(S_1')} + \tfrac{\mu(D_1'')^2}{\mu(S_1')} \right] \\
&< \left(d(S_1) + u_0 - 1 + \tfrac{\mu(D_1)}{2\mu(S_1')} \right)\mu(D_1).
\end{aligned}
\tag{5.23}
$$

Here the second and the fourth equality are obtained by

$$\mu(D_1') = \mu(S_1)\big(u_0 - \lceil u_0 \rceil + 1\big) \text{ and } \mu(D_1'') = \mu(S_1')\big(\lceil u_0 \rceil - u_0\big).$$

The first inequality follows from $d(S_1) > d(S_1')$ and $\mu(D_1) = \mu(D_1') + \mu(D_1'')$ and the second one follows from $\mu(S_1) > \mu(S_1')$. Similarly, since $d(S_2^{(1)}) < d(S_1^{(i)})$ and $\mu(S_2^{(1)}) > d(S_2^{(i)})$ for $i \geq 2$

$$\|D_2\| > \left(d(S_2^{(1)}) + v_0 + \frac{\mu(D_2)}{2\mu(S_2^{(1)})} \right)\mu(D_2). \tag{5.24}$$

Finally, as S_2 and $S_2^{(1)}$ are shifted cones, by (iii) in Lemma 4, (5.6), (5.16), and (5.19)

$$d(S_2^{(1)}) + v_0 > d(S_2) - 1 + v_0 = d(S_1) + u_0 - 1. \tag{5.25}$$

So a contradiction $||R'|| - ||R(v_0, u_0)|| = ||D_2|| - ||D_1|| > 0$ follows from (5.19), (5.23), and (5.25). Therefore (5.20) must be false, i.e.

$$\mu(S_1') < \mu(S_2^{(1)}). \tag{5.26}$$

Let now $\overset{\sim}{\rightarrow} u \triangleq \lceil u_0 \rceil - 2$, $S_3 \triangleq K(\overset{\sim}{\rightarrow} u) \smallsetminus S_1'$ (c.f. Figure 1), $\xi = 1 - \{v_0\}$, and $\eta = u_0 - (\lceil u_0 \rceil - 1)$, then by (5.26)

$$\mu(S_3) - \mu(S_1') > \mu(S_3) - \mu(S_2^{(1)}) = (\overset{\sim}{\rightarrow} u - \lceil v_0 \rceil)(\xi - \eta), \tag{5.27}$$

and by (i) in Lemma 4

$$\mu(S_3) = \frac{1}{2}[(\overset{\sim}{\rightarrow} u - \lceil v_0 \rceil)^2 - (\overset{\sim}{\rightarrow} u - \lceil v_0 \rceil) + 2\xi(\overset{\sim}{\rightarrow} u - \lceil v_0 \rceil)] = \frac{\overset{\sim}{\rightarrow} u - \lceil v_0 \rceil}{2}(\overset{\sim}{\rightarrow} u - \lceil v_0 \rceil \quad 1 + 2\xi \ . \tag{5.28}$$

However, by their definitions

$$\mu(S_1') + \mu(S_3) = \mu\big(K(\overset{\sim}{\rightarrow} u)\big) = \frac{1}{2}(\overset{\sim}{\rightarrow} u^2 - \overset{\sim}{\rightarrow} u). \tag{5.29}$$

Adding (5.27) to (5.29) we obtain

$$\mu(S_3) > \frac{1}{4}(\overset{\sim}{\rightarrow} u - 1)\overset{\sim}{\rightarrow} u + \frac{1}{2}(\overset{\sim}{\rightarrow} u - \lceil v_0 \rceil)(\xi - \eta). \tag{5.30}$$

(5.28) and (5.30) imply

$$(\overset{\sim}{\rightarrow} u - \lceil v_0 \rceil)(\overset{\sim}{\rightarrow} u - \lceil v_0 \rceil - 1 + \xi + \eta) > \frac{\overset{\sim}{\rightarrow} u}{2}(\overset{\sim}{\rightarrow} u - 1). \tag{5.31}$$

Simplifying (5.31), we obtain

$$(\overset{\sim}{\rightarrow} u - \lceil v_0 \rceil)^2 > \frac{\overset{\sim}{\rightarrow} u^2}{2} + \frac{\overset{\sim}{\rightarrow} u}{2} - \lceil v_0 \rceil - (\xi + \eta)(\overset{\sim}{\rightarrow} u - \lceil v_0 \rceil) > \frac{\overset{\sim}{\rightarrow} u^2}{2} - \frac{3}{2}\overset{\sim}{\rightarrow} u + \lceil v_0 \rceil$$

$$(\text{as } \overset{\sim}{\rightarrow} u \geq \lceil v_0 \rceil, \text{ see } (5.17) \text{ and as } \xi + \eta \leq 2)$$

$$= \frac{1}{2}\left(\overset{\sim}{\rightarrow} u - \frac{3}{2}\right)^2 - \frac{9}{8} + \lceil v_0 \rceil, \text{ i.e.}$$

$$\overset{\sim}{\rightarrow} u - \lceil v_0 \rceil > \frac{\sqrt{2}}{2}\overset{\sim}{\rightarrow} u - \frac{3\sqrt{2}}{4}, \text{ or}$$

$$\lceil v_0 \rceil < \left(1 - \frac{\sqrt{2}}{2}\right)\overset{\sim}{\rightarrow} u + \frac{3\sqrt{2}}{4} = \left(1 - \frac{\sqrt{2}}{2}\right)\overline{u} - 1 + \frac{5\sqrt{2}}{4}, \tag{5.32}$$

where $\overline{u} \triangleq \lceil u_0 \rceil - 1 = \overset{\sim}{\rightarrow} u + 1$. On the other hand, by (iii) in Lemma 4 and (5.16) with $\eta' = \{u_0\}$

$$d(S_1) = d(S_2) + v_0 - u_0 \leq \left(u_0 + \lceil v_0 \rceil + \frac{\eta'(1 - \eta')}{\overline{u} - \lceil v_0 \rceil - 1}\right) + v_0 - u_0$$

$$= v_0 + \lceil v_0 \rceil + \frac{\eta'(1 - \eta')}{\overline{u} - \lceil v_0 \rceil - 1}. \tag{5.33}$$

Consider that S_1 is the union of a rectangle and a 2–dimensional cone (a triangle).

$$\|S_1\| = \tfrac{1}{2}\left(\lceil v_0\rceil^2 - \lceil v_0\rceil\right)\lceil v_0\rceil + v_0\left(\bar{u} - \lceil v_0\rceil\right)\left(\lceil v_0\rceil + \tfrac{v_0+\bar{u}-\lceil v_0\rceil}{2}\right) \tag{5.34}$$
$$= \tfrac{1}{2}\left[\lceil v_0\rceil^2\left(\lceil v_0\rceil - 1\right) + v_0\left(\bar{u} - \lceil v_0\rceil\right)\left(v_0 + \lceil v_0\rceil + \bar{u}\right)\right],$$

and

$$\mu(S_1) = \frac{1}{2}\left(\lceil v_0\rceil^2 - \lceil v_0\rceil\right) + v_0\left(\bar{u} - \lceil v_0\rceil\right). \tag{5.35}$$

(5.33) - (5.35) imply

$$\left(v_0 + \lceil v_0\rceil + \tfrac{\eta'(1-\eta')}{\bar{u}-\lceil v_0\rceil-1}\right)\left(\tfrac{1}{2}\left(\lceil v_0\rceil^2 - \lceil v_0\rceil\right) + v_0\left(\bar{u} - \lceil v_0\rceil\right)\right)$$
$$\geq \tfrac{1}{2}\left[\lceil v_0\rceil^2\left(\lceil v_0\rceil - 1\right) + v_0\left(\bar{u} - \lceil v_0\rceil\right)\left(v_0 + \lceil v_0\rceil + \bar{u}\right)\right], \text{ i.e.}$$

$$\lceil v_0\rceil\left(\lceil v_0\rceil - 1\right)\tfrac{\eta'(1-\eta')}{\bar{u}-\lceil v_0\rceil-1} \geq v_0\left(\bar{u} - \lceil v_0\rceil\right)\left(\bar{u} - v_0 - \lceil v_0\rceil - \tfrac{2\eta'(1-\eta')}{\bar{u}-\lceil v_0\rceil-1}\right) - v_0\left(\lceil v_0\rceil^2 - \lceil v_0\rceil\right)$$
$$= v_0\left(\bar{u}^2 - 3\lceil v_0\rceil\bar{u} + \lceil v_0\rceil^2\right) + v_0\lceil v_0\rceil + v_0\left(\bar{u} - \lceil v_0\rceil\right)\left[\left(\lceil v_0\rceil - v_0\right) - \tfrac{2\eta'(1-\eta')}{\bar{u}-\lceil v_0\rceil-1}\right]$$
$$\geq \left(\lceil v_0\rceil - 1\right)\left[\left(\bar{u}^2 - 3\bar{u}\lceil v_0\rceil + \lceil v_0\rceil^2\right) + \lceil v_0\rceil - \left(\bar{u} - \lceil v_0\rceil\right)\tfrac{2\eta'(1-\eta')}{\bar{u}-\lceil v_0\rceil-1}\right],$$

i.e.

$$\bar{u}^2 - 3\bar{u}\lceil v_0\rceil + \lceil v_0\rceil^2 \leq \left(2\bar{u} - \lceil v_0\rceil\right)\tfrac{\eta'(1-\eta')}{\bar{u}-\lceil v_0\rceil-1} - \lceil v_0\rceil \tag{5.36}$$
$$\leq \tfrac{1}{4}\tfrac{2\bar{u}-\lceil v_0\rceil}{\bar{u}-\lceil v_0\rceil-1} - \lceil v_0\rceil.$$

Comparing (5.32) and (5.36), one can conclude

$$\left[\left(1 - \tfrac{\sqrt{2}}{2}\right) + \tfrac{5\sqrt{2}-4}{4\bar{u}}\right]^2 - 3\left[\left(1 - \tfrac{\sqrt{2}}{2}\right) + \tfrac{5\sqrt{2}-4}{4\bar{u}}\right] + 1$$
$$< \tfrac{1}{\bar{u}} \cdot \tfrac{1}{2\sqrt{2\bar{u}}-5\sqrt{2}} - \tfrac{1}{\bar{u}}\left[\left(1 - \tfrac{\sqrt{2}}{2}\right) + \tfrac{\sqrt{2}-4}{4\bar{u}}\right]$$
$$= \tfrac{1}{\bar{u}}\left(\tfrac{1}{2\sqrt{2\bar{u}}-5\sqrt{2}} - \tfrac{5\sqrt{2}-4}{4\bar{u}}\right) - \tfrac{1}{\bar{u}}\left(1 - \tfrac{\sqrt{2}}{2}\right), \text{ or}$$

$$\left(1 - \tfrac{\sqrt{2}}{2}\right)^2 - 3\left(1 - \tfrac{\sqrt{2}}{2}\right) + 1 < \tag{5.37}$$
$$\tfrac{1}{4\bar{u}}\left(3\sqrt{2} + 2\right) + \tfrac{1}{\bar{u}}\left(\tfrac{1}{2\sqrt{2\bar{u}}-5\sqrt{2}} - \tfrac{4(5\sqrt{2}-4)+(5\sqrt{2}-4)}{16\bar{u}}\right).$$

One can check that (5.37) does not hold unless $\bar{u} < 8$, or $\lceil u_0\rceil \leq 8$. However,

it is not difficult to check that (5.16) and (5.26) cannot hold simultaneously for $4 < u \leq 8$. Finally using the condition $U \notin \mathbb{Z}^+$ it follows that $U \geq 4$. One can also check the lemma for $3 < u \leq 4$.

Case 4

If an x_0 with $\mu(D_1) = \mu(D_2)$ does not exist, i.e. D_1 is too big to find a D_2 with the same measure, we choose a proper h, $0 < h < 1$, such that for

$$D_1 \triangleq \left\{(x, y, z) \in R(v_0, u_0) : u_0 - h < z \leq u_0\right\} \text{ and}$$
$$D_2 \triangleq \left\{(x, y, z) \in S_U : v_0 < x < y \leq u_0 - h\right\}, \quad \mu(D_1) = \mu(D_2).$$

D_2 is a shifted cone. By the arguments leading to Lemma 4, (c.f. (4.18), (4.19) in Remark to Lemma 4) we get for its density

$$d(D_2) \geq 3\lceil v_0 \rceil + \tfrac{3}{2}\left[u_0 - h - \lceil v_0 \rceil - \left(1 - \{v_0\}\right)\right] - \frac{\{v_0\}\left(1 - \{v_0\}\right)}{\left|u_0 - h - \lceil v_0 \rceil\right|^{+} + 2\left(1 - \{v_0\}\right)}$$

$$= \tfrac{3}{2}(u_0 + v_0 - h) - \frac{\{v_0\}\left(1 - \{v_0\}\right)}{\left|u_0 - h - \lceil v_0 \rceil\right|^{+} + 2\left(1 - \{v_0\}\right)} \, .$$

However, by (5.16) and Lemma 4

$$d(D_1) = d(S_1) + u_0 - \frac{h}{2} = d(S_2) + v_0 - \frac{h}{2} \leq v_0 + \lceil v_0 \rceil + u_0 - \frac{h}{2} + \frac{1}{4}.$$

Then

$$d(D_2) - d(D_1) \geq \tfrac{u_0}{2} + \tfrac{v_0}{2} - \lceil v_0 \rceil - h - \tfrac{1}{4} - \frac{\{v_0\}\left(1 - \{v_0\}\right)}{\left|u_0 - h - \lceil v_0 \rceil\right|^{+} + 2\left(1 - \{v_0\}\right)}$$

$$> \tfrac{1}{2}\left(u_0 - \lceil v_0 \rceil\right) - h - \tfrac{3}{4}.$$

Thus by (5.16), for $u_0 > 8$

$$d(D_2) > d(D_1).$$

For $\lceil u_o \rceil \leq 8$ we check it directly.

Remark 2. *For $m \in \mathbb{Z}^{+}$ denote by $\mathcal{D}_m$ the set of downsets of $\overset{\sim}{\to} \mathcal{L}(U)\big(\overset{\triangle}{=}\overset{\sim}{\to}$ $\mathcal{L}(U,3)\big)$ with m maximal points. We can show that $\max_{\mu(W)=\alpha, W \in \mathcal{D}_m} \|W\|$ can be achieved, as well.*

More precisely, define a metric on the set $\left\{(x^i, y^i, z^i)_{i=1}^k : (x^i, y^i, z^i) \in \mathbb{R}^3\right\}$ as the sum of Euclidean (or L_1-) metrics of the k components points. Then for fixed $\mu(W) = \alpha$, $W \in \mathcal{D}_m$, $\|W\|$ is a continuous function of its maximal points.

6   On Regular Surfaces

Lemma 7. *Every $W \in \mathcal{D}$ can be reduced to a $W' \in \mathcal{D}$, which has of each of the regular $X-, Y-$ and $Z-$ surfaces at most one (for $U \in Z^{+}$).*

Proof. Suppose there exists a W that canot be reduced to such kind of W'. W.l.o.g. by Remark 1 we assume W achieves $\underset{m' \leq m}{\to} \max_{\mu(W)=\alpha, W \in \mathcal{D}_{m'}} \|W\|$, (recalling $\mathcal{D} = \bigcup_{m=1}^{\infty} \mathcal{D}_m$ by its definition).

Case 1: Suppose W has at least 2 regular $z-$surfaces, say S_i at i, for $i = 1, 2$, and

$$d(S_1) + u_1 \leq d(S_2) + u_2. \tag{6.1}$$

Using the same method as in the proof of Lemma 6, Case 1, one can obtain a contradiction. Furthermore, we can see that W has 2 regular X–surfaces iff $\hat{W}^{(u)}$ has 2 regular Z–surfaces. Since W and $\hat{W}^{(u)}$ must achieve the maximal value simultaneously, we are left with **Case 2:** W has at least 2 regular Y–surfaces S_1 at v_1 and S_2 at v_2 with

$$d(S_1) + v_1 \leq d(S_2) + v_2 \tag{6.2}$$

and of each of the regular $Z-$ and $X-$ surfaces at most one. Let $S_2' = S_2$, if $v_2 \notin Z$, and otherwise let $S_2' = S_2 \setminus \{(x,z) \mid v_1 < z \leq v_1 + 1\}$. Since W has no 2 regular Z–surfaces nor X–surfaces, S_2 is rectangular, consequently $d(S_2') > d(S_2)$. Thus we can use S_2' to replace S_2 and play the same game as before to arrive at a contradiction.

7   Main Result in Continuous Model, $k = 3$

Theorem 2. *For $U \in \mathbb{Z}^+$ and fixed α every $W \in \mathcal{D}$ with $\mu(W) = \alpha$ can be reduced to a cone or the trapezoid $R(V_\alpha(U), U)$.*

Proof. Assume the theorem is not true. Then by Remark 1 and Lemma 6 there exists a $W \in \mathcal{D}$ with m maximal points achieving maximal value of $\|W\|$ over $\bigcup_{m' \leq m} \mathcal{D}_{m'}$, which is neither a cone nor a trapezoid. Moreover, by Lemma 7 we can assume that W has at most one regular $X-$, at most one regular $Y-$, and at most one regular $Z-$ surface.

Case 1: W has only one (regular or irregular) Z–surface at $u \leq U$. Then W has one or two maximal points, whose third components must be u. **Subcase 1.1:** W has one maximal point, say $P = (w, v, u)$. Because $v = \lceil u \rceil - 1$ implies W is a trapezoid, we assume $w < v \leq \lceil u \rceil - 1$. Thus, W has one Z–surface S_1 and one Y–surface, which are shown in Figure 2 (a).
We are going to use the same idea as before. However, it is not enough to exchange the layers. Instead of it we will exchange cylinders. (a) Suppose $w \geq u - \lceil v \rceil$.
We choose $0 < h_1 < u - \lceil v \rceil$ and define $S_2 \triangleq \{(y, z) : v < y < z \leq u - h_1$ and $\lceil y \rceil \neq \lceil z \rceil\}$, $D_1 = S_1 \times \{z : u - h_1 < z \leq u\}$, $D_2 \triangleq \{x : 0 < x \leq w\} \times S_2$, and $W' = (W \setminus D_1) \cup D_2$ such that

$$\mu(D_1) = \mu(D_2). \tag{7.1}$$

Then $W' \in \mathcal{D}$ and furthermore, if we denote $\{v\}$ by θ and use the arguments of the proof of Lemma 4 (see Remark to Lemma 4), then we obtain

$$d(S_2) - (v + u - h_1) = \frac{(\theta' - \overline{\theta})[1 - (\theta' + \overline{\theta})] - \overline{\theta}\theta'(\theta' - \overline{\theta})(\lfloor u - h_1 \rfloor - \lceil v \rceil)^{-1}}{(\lfloor u - h_1 \rfloor - \lceil v \rceil - 1) + 2(\theta' + \overline{\theta}) + 2\overline{\theta}\theta'(\lfloor u - h_1 \rfloor - \lceil v \rceil)^{-1}} \triangleq \eta_1, \tag{7.2}$$

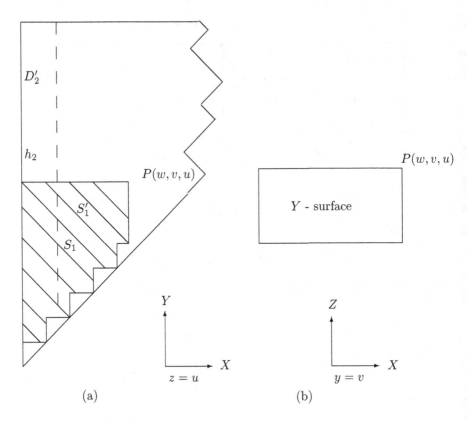

Fig. 2 (a).

where $\theta' \triangleq \{u - h_1\}$ and $\overline{\theta} = 1 - \theta = \lceil v \rceil - v$, if $u - h_1 - \lceil v \rceil > 1$. By Lemma 4

and Corollary 2,

$$d(S_1) - v \leq \frac{\theta(1 - \theta)}{\lceil v \rceil - 1 + 2\theta} \triangleq \eta_2. \tag{7.3}$$

Consequently

$$d(S_2) - \big(d(S_1) + u\big) \geq -h_1 + \eta_1 - \eta_2. \tag{7.4}$$

Therefore, by simple calculation

$$\|W'\| - \|W\| = \|D_2\| - \|D_1\|$$
$$= \mu(D_2)\big(d(S_2) + \tfrac{w}{2}\big) - \mu(D_1)\big(d(S_1) + u - \tfrac{h_1}{2}\big) \tag{7.5}$$
$$= \mu(D_2)\big[d(S_2) - \big(d(S_1) + u\big) + \tfrac{w}{2} + \tfrac{h_1}{2}\big] \geq \mu(D_2)\big[\tfrac{w}{2} - \tfrac{h_1}{2} + \eta_1 - \eta_2\big].$$

By (7.2),

$$\eta_1 \geq -\frac{\overline{\theta}(1 - \overline{\theta})}{\lfloor u - h_1 \rfloor - \lceil v \rceil - 1 + 2\overline{\theta}} = \frac{-\theta(1 - \theta)}{\lfloor u - h_1 \rfloor - \lceil v \rceil - 1 + 2(1 - \theta)}. \tag{7.6}$$

$$P(w, v, u)$$

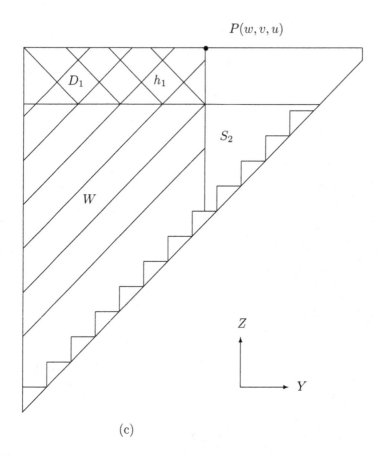

(c)

Fig. 2 (b).

Thus, (7.3) and (7.6) imply

$$\eta_1 - \eta_2 \geq -\frac{1}{2}. \tag{7.7}$$

However, when $h_1 \leq u - \lceil v \rceil - 1$, (7.5) and (7.2) imply the contradiction

$$||W'|| > ||W||. \tag{7.8}$$

When $u - \lceil v \rceil - 1 \leq h_1 < u - \lceil v \rceil$, S_2 becomes a rectangle (c.f. Figure 3) and $d(S_2) = v + u - h_1 + \frac{\bar{\theta}}{2} - \frac{u - \lceil v \rceil - h_1}{2}$. Then use

$$\eta_1 = \frac{1 - \theta}{2} - \frac{u - \lceil v \rceil - h_1}{2}, \tag{7.9}$$

and (7.8) holds again. (b) If $w < u - \lceil v \rceil$, then we choose $0 < h_2 < w$ and let $S_1' = S_1 \setminus \{(x, y) : 0 < x \leq h_2\}$, $S_2' = \{(y, z) : v \leq y < z < u, \lceil y \rceil \neq \lceil z \rceil\}$, $D_1' \triangleq S_1' \times \{z : \lceil v \rceil < z \leq u\}$, and $D_2' = S_2 \times \{x : 0 < x \leq h_2\}$ with

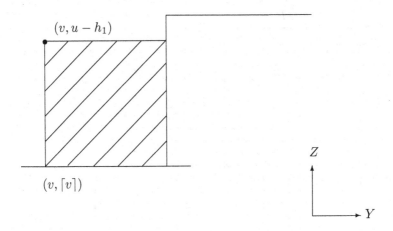

Fig. 3.

$\mu(D_1') = \mu(D_2')$. Considering $(W \setminus D_1') \cup D_2'$ in a similar way we arrive at a contradiction. (c.f. Figure 2 (a)) **Subcase 1.2:** W has 2 maximal points.

According to our assumption on regular surfaces the Z–surface S_1 of W must be as in Figure 4.

Then we follow the same reasoning as in the previous subcase in the shadow part (i.e. exchange cylinders in the shadow part $\{(x, y, z) \in S_U \mid x \leq v_0\}$, where v_0 is the smaller first component in the 2 maximal points) and obtain a contradiction.

Case 2: W has 2 Z–surfaces. Since W and $\hat{W}$ always simultaneously achieve their maximum, we can assume $\hat{W}$ has 2 Z–surfaces too, because otherwise we can use $\hat{W}$, which has been studied in Case 1 already, instead of W. However, $\hat{W}$ has 2 Z–surfaces iff W has one regular X–surface, and

$$\{(0, y, z) \in S_U\} \setminus W \neq \varnothing. \qquad (7.10)$$

Thus we can assume W has one regular X–surface and (7.10) holds.

Then by our assumption W has 2 maximal points, say $P_1 = (w_1, v_1, U)$ and $P_2 = (w_2, v_2, u)$ and $v_1 < \lceil U \rceil - 1$. **Subcase 2.1:** $\lceil v_1 \rceil \geq \lfloor u \rfloor$. Then $w_1 < w_2$,

because P_2 is maximal. Recalling that in our proof under subcase 1.1 we only exchange the points (x, y, z) with $x \leq w$, and $y \geq \lceil v \rceil$, in the present case we can use the plane $x = w_1$ to cut S_U into 2 parts and repeat the same reasoning as in subcase 1.1 to obtain a contradiction in the part $x \geq w_1$.

Moreover, for this kind of W's, $\hat{W}^{(U)}$ has 2 maximal points, $\hat{P}_1 = (\hat{w}_1, \hat{v}_1, U)$ and $\hat{P}_2 = (\hat{w}_2, \hat{v}_2, \hat{u})$ with $\hat{w}_1 = U - \lceil v_1 \rceil$, $\hat{v}_1 = U - v_1$, $\hat{w}_2 = U - u$, $\hat{v}_2 = U - \lceil w_1 \rceil$, $\hat{u} = U - w_1$, i.e. $\hat{w}_1 = \lceil \hat{v}_1 \rceil - 1$, $\hat{v}_2 = \lceil \hat{u} \rceil - 1$ and $\hat{w}_2 \geq \hat{w}_1$. Therefore, the following subcase 2.2 can be cancelled from our list. **Subcase 2.2:** $w_1 = \lceil v_1 \rceil - 1$,

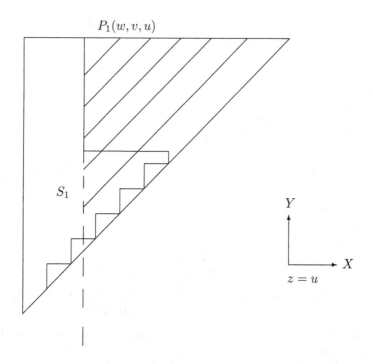

Fig. 4.

$v_2 = \lceil u \rceil - 1$, and $w_2 \geq w_1$. **Subcase 2.3:** $w_1 = \lceil v_1 \rceil - 1$, $v_2 = \lceil u \rceil - 1$, $w_2 < w_1$,

and $\lceil v_1 \rceil < u$. In this subcase, there are one regular Z–surface and one regular Y–surface passing P_1.

Denote by $S_1 = \big\{(x,y) : y \leq v_1 \lceil x \rceil \neq \lceil y \rceil \big\}$ the irregular Z–surface, by S_2 the regular X–surface at w_2, a shifted cone, and by $S_3 = \big\{(y,z) : \lceil y \rceil \neq \lceil z \rceil$, $(0,y,z) \in S_U \smallsetminus W \big\}$ as in Figure 5.

Then $\overset{\sim}{\rightarrow} W \triangleq W \cap \big\{(x,y,z) : y > v_1 \big\}$ is a cylinder with base S_2. Therefore we can assume

$$v_2 - v_1 = \lceil u \rceil - 1 - v_1 > U - u, \tag{7.11}$$

because otherwise, by Lemma 5, we can replace $\overset{\sim}{\rightarrow} W$ by a cylinder with the same size 2–dimensional trapezoid base and the same height, and then reduce W to a downset with 2 regular Y–surfaces. If $d(S_1) + U < d(S_3)$, then we can repeat our reasoning as before and arrive at a contradiction. So we only need to consider

$$d(S_1) + U \geq d(S_3), \tag{7.12}$$

which, in fact, is also impossible. By Lemma 4

$$d(S_1) = v_1 + \frac{\theta(1 - \theta)}{\mid \lfloor v_1 \rfloor - 1 \mid^+ + 2\theta} \triangleq v_1 + \eta. \tag{7.13}$$

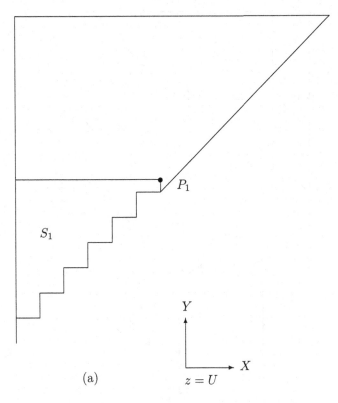

(a)

$z = U$

Fig. 5 (a).

Partitioning S_3 into a rectangle S_3' and a (2–dimensional) cone S_3', we obtain

$$||S_3|| = \frac{1}{2}(\lceil u \rceil - 1 + v_1 + U + u)\mu(S_3') + (U + \lceil u \rceil - 1)\mu(S_3''), \qquad (7.14)$$

$$\mu(S_3') = (\lceil u \rceil - 1 - v_1)(U - u), \mu(S_3'') = \binom{U - (\lceil u \rceil - 1)}{2}, \qquad (7.15)$$

and

$$\mu(S_3) = \mu(S_3') + \mu(S_3''). \qquad (7.16)$$

(see Figure 5 (c).) Thus, it follows from (7.12) – (7.16) that

$$\frac{1}{2}[U - u - (\lceil u \rceil - 1) + v_1](\lceil u \rceil - 1 - v_1)(U - u) - (\lceil u \rceil - 1 - v_1)\binom{U - (\lceil u \rceil - 1)}{2} + \eta\,\mu(S_3) \geq 0.$$
$$(7.17)$$

(7.11) and (7.17) imply

$$\eta\,\mu(S_3) > (\lceil u \rceil - 1 - v_1)\binom{U - (\lceil u \rceil - 1)}{2}. \qquad (7.18)$$

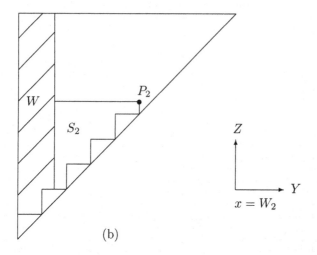

(b)

Fig. 5 (b).

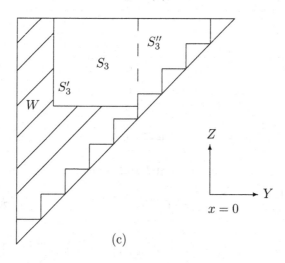

(c)

Fig. 5 (c).

However, by (7.15) and (7.16)

$$\frac{\mu(S_3)}{(\lceil u \rceil - 1 - v_1)\binom{U-(\lceil u \rceil - 1)}{2}} = \frac{U-u}{\binom{U-(\lceil u \rceil - 1)}{2}} + \frac{1}{\lceil u \rceil - 1 - v_1} \leq 4, \text{ if } U - (\lceil u \rceil - 1) \geq 2.$$

(7.19)

On the other hand, by the definition of η, $\eta \leq \frac{1}{4}$, which contradicts (7.18) and (7.19). When $U - \lceil u \rceil - 1 \leq 1$, we can directly derive a contradiction.

Thus we are left with the case $w_1 < \lceil v_1 \rceil - 1$ (and $\lceil v_1 \rceil < u$), i.e. both of the regular $X-$ and Y-surfaces pass through P_1, or in other words neither of the surfaces passes through P_2 unless P_2 shares one of them with P_1. In fact, all of the following 3 subcases are not new to us.

Subcase 2.4: There is no regular surface passing through P_2, i.e. $P_2 = (\lceil u \rceil - 2, \lceil u \rceil - 1, u)$. Then the top part of W, namely, $W_t \triangleq W \cap \{(x, y, z) : z > u\}$ is a cylinder with a 2 dimensional trapezoid $R_2(w_1, v_1)$ (its irregular Z–surface) as base. By similar reasoning with Lemma 5 as after (7.11) we can assume $v_1 = \lfloor u \rfloor$, which has been treated in the subcase 2.1.

Subcase 2.5: P_1 and P_2 share a regular X–surface, i.e. $w_1 = w_2$ and $v_2 = \lceil u \rceil - 1$. Then $\hat{W}^{(U)}$ falls into subcase 2.4.

Subcase 2.6: P_1 and P_2 share a regular Z–surface, i.e. $v_1 = v_2$, and $w_2 = \lceil v_2 \rceil - 1$. Then $\hat{W}^{(U)}$ falls into subcase 2.3.

8   A Last Auxiliary Result

Lemma 8. *For* $U \in \mathbb{Z}^+, U \geq 6$, $\alpha = \binom{U}{3} - \binom{m}{3} < \frac{1}{2}\binom{U}{3}$ *and* $m \in \mathbb{Z}^+$

$$\|R(V_\alpha(U), U)\| > \|K(u)\|, \quad if \ \mu(K(u)) = \alpha = \mu(R_\alpha(U), U). \tag{8.1}$$

Proof. At first let us restrict ourselves to $U \geq 12$. We know from (i) in Lemma 4 that

$$6\mu(K(u)) = 6\binom{\lfloor u \rfloor}{3} + 6\theta\binom{\lfloor u \rfloor}{2} = (u-1)^3 - \left\{ \left[3\left(\theta - \frac{1}{2}\right)^2 + \frac{1}{4} \right] \lfloor u \rfloor - (1-\theta)^3 \right\}. \tag{8.2}$$

Therefore,

$$d(K(u)) \geq \frac{3}{2}u > \frac{3}{2}\left[6\mu(K(u))^{\frac{1}{3}} + 1 \right] = \frac{3}{2}\left[(6\alpha)^{\frac{1}{3}} + 1 \right]. \tag{8.3}$$

On the other hand for $\eta > 0$, by (8.2)

$$[u - (1 + \eta)]^3 = (u-1)^3 - 3\eta(u-1)^2 + 3\eta^2 - \eta^3$$
$$= 6\mu(K(u)) - \left[3\eta(u-1)^2 - 3\eta^2(u-1) + \eta^3 - \left[3\left(\theta - \frac{1}{2}\right)^2 + \frac{1}{4} \right] \lfloor u \rfloor + (1-\theta)^3 + \eta^3 \right]$$
$$= 6\mu(K(u)) - \left[3\eta\lfloor u \rfloor^2 - 3\left(2\eta\overline{\theta} + \eta^2 - \overline{\theta}(1-\overline{\theta}) + \frac{1}{3}\right)\lfloor u \rfloor + (\overline{\theta} + \eta)^3 \right]$$
$$\leq 6\mu(K(u)) - 3\lfloor u \rfloor\left[\lfloor u \rfloor\eta - (2\eta + \eta^2 + \frac{1}{3}) \right],$$
$$\text{where} \quad \overline{\theta} \triangleq 1 - \theta. \tag{8.4}$$

Let

$$\eta = \xi - \frac{2\theta\overline{\theta}}{(\lfloor u \rfloor - 2) + 6\theta} > 0 \tag{8.5}$$

and η, ξ will be defined later. Then by (8.4) and (8.5),

$$d\big(K(u)\big) \leq \frac{3}{2}\left[6\mu\big(K(u)\big)^{\frac{1}{3}} + (1+\xi)\right]. \tag{8.6}$$

when

$$\lfloor u \rfloor \geq \frac{2\eta + \eta^2 + \frac{1}{3}}{\eta}, \tag{8.7}$$

Choose $\xi_1 = 0.12$ and $\xi_2 = 0.035$, to estimate $d\big(K(u)\big)$ and $d\big(K(U)\big)$, resp. By our assumption $u \geq 7\frac{20}{29}$, if $U = 12$, and $u > 8$, if $U \geq 13$. Then one can verify (8.6) with (8.5) for u, ξ_1 (or U, ξ_2). So, by (8.7)

$$d\big(K(u)\big) \leq \frac{3}{2}\left[\big(6\mu(u)\big)^{\frac{1}{3}} + 1 + \xi_1\right] = \frac{3}{2}\left(\big(6\alpha\big)^{\frac{1}{3}} + 1 + \xi_1\right) \tag{8.8}$$

$$d\big(K(U)\big) \leq \frac{3}{2}\left[\big(6\mu\big(K(U)\big)\big)^{\frac{1}{3}} + 1 + \xi_2\right]. \tag{8.9}$$

Setting $\alpha = \lambda\mu\big(K(U)\big)$, by Lemmas 3 and 4, (8.3), and (8.8), we obtain

$$\|R\big(V_\alpha(U), U\big)\| - \|K(u)\| = \frac{3}{2}U\mu\big(K(u)\big) - 3\big(\mu\big(K(U)\big) - \alpha\big) + \|\hat{R}^{(U)}\big(V_\alpha(U), U\big)\| - \|K(u)\|$$
$$\geq \frac{3}{2}\Big\{\left[\big(6\mu\big(K(U)\big)\big)^{\frac{1}{3}} + 1 + \xi_2\right]\big(2\alpha - \mu\big(K(U)\big)\big) + \left[\big(6[\mu\big(K(U)\big) - \alpha]\big)^{\frac{1}{3}} + 1\right]$$
$$\cdot\big(\mu\big(K(U)\big) - \alpha\big) - \left[\big(6\alpha\big)^{\frac{1}{3}} + 1 + \xi_1\right]\alpha\Big\} = \frac{3}{2}\sqrt[3]{6}\mu\big(K(u)\big)f(\lambda), \text{ where}$$
$$f(\lambda) = 2\lambda - 1 + (1-\lambda)^{\frac{1}{3}} - \lambda^{\frac{1}{3}} - \frac{\xi_2 + (\xi_1 - 2\xi_2)\lambda}{\big(6\mu\big(K(U)\big)\big)^{\frac{1}{3}}},$$
$$\tag{8.10}$$

is concave in λ. Let $\varepsilon_1 = \dfrac{2.7}{\big(6\mu\big(K(U)\big)\big)^{\frac{1}{3}}}$, $\varepsilon_2 = \dfrac{2.68/2^{\frac{5}{3}}}{\big(6\mu\big(K(U)\big)\big)^{\frac{1}{3}}}$ and $M \in \mathbb{Z}^+$ be specified by

$$\binom{M}{3} \leq \frac{1}{2}\binom{U}{3} < \binom{M+1}{3}. \tag{8.11}$$

Then

$$\varepsilon_1 < \frac{3}{U} = \frac{\binom{U-1}{2}}{\binom{U}{3}} = \frac{\binom{U}{3} - \binom{U-1}{3}}{\mu\big(J(U)\big)}, \tag{8.12}$$

and as $\dfrac{\left[2\binom{M+1}{2}\right]^3}{\left[6\binom{M}{3}\right]^2} = \dfrac{M(M-1)}{(M+1)}$, by (8.11) and $M > 9$ (when $U > 12$),

$$\frac{1}{2}\frac{\binom{M}{2}}{\binom{U}{3}} = \frac{1}{4}\left[\frac{M(M-1)}{(M+1)^2}\right]^{\frac{1}{3}}\frac{\big(6\binom{M+1}{3}\big)^{\frac{1}{3}}}{\mu\big(K(U)\big)} > \frac{3}{2}\left(\frac{1}{2}\right)^{\frac{2}{3}}\left[\frac{M(M-1)}{(M+1)}\right]^{\frac{1}{3}}\frac{1}{\left[6\mu\big(K(U)\big)\right]^{\frac{1}{3}}}$$
$$\geq \frac{3}{2^{\frac{5}{3}}}(0.72)^{\frac{1}{3}}\frac{1}{\left[6\mu\big(K(U)\big)\right]^{\frac{1}{3}}} = \frac{2.68884\ldots}{2^{\frac{5}{3}}}\frac{1}{\left[6\mu\big(K(U)\big)\right]^{\frac{1}{3}}} > \varepsilon_2. \tag{8.13}$$

However, with Taylor's expansion,

$$f(\varepsilon_1) \geq 2\varepsilon_1 - \frac{4}{3}\varepsilon_1 + \frac{4}{9}\varepsilon_1^2 - \varepsilon_1^{\frac{4}{3}} - \frac{\xi_2 + (\xi_1 - 2\xi_2)\varepsilon_1}{\big(6\mu\big(K(U)\big)\big)^{\frac{1}{3}}}$$
$$= \frac{1}{\left[6\mu\big(K(U)\big)\right]^{\frac{1}{3}}}\left(\frac{2}{3} \times 2.7 - 2.7 \times \varepsilon_1^{\frac{1}{3}} - \xi_2\right) \tag{8.14}$$
$$+ \frac{\varepsilon_1}{\left[6\mu\big(K(U)\big)\right]^{\frac{1}{3}}}\left[\frac{4}{9} \times 2.7 - (\xi_1 - 2\xi_2)\right] > 0.$$

Moreover, set $g(x) = (1+x)^{\frac{4}{3}} - (1-x)^{\frac{4}{3}}$. Then

$$g(0) = g''(0) = 0, g'(0) = \frac{8}{3} \quad \text{and} \quad g''(x) > -0.6254,$$

when $0 \le x \le 2\varepsilon_2 < 0.1551$. Thus, by the definition of ε_2 and Taylor's expansion again

$$
\begin{aligned}
f\left(\tfrac{1}{2} - \varepsilon_2\right) &= -2\varepsilon_2 + \left(\tfrac{1}{2}\right)^{\frac{4}{3}} g(2\varepsilon_2) - \frac{\xi_1\left(\frac{1}{2}-\varepsilon_2\right)+2\xi_2\varepsilon_2}{\left[6\mu\big(K(U)\big)\right]^{\frac{1}{3}}} \\
&\ge -(2\varepsilon_2) + \left(\tfrac{1}{2}\right)^{\frac{4}{3}} \tfrac{8}{3}(2\varepsilon_2) - 0.6254(2\varepsilon_2)^3 - \frac{\xi_1\left(\frac{1}{2}-\varepsilon_2\right)+2\xi_2\varepsilon_2}{\left[6\mu\big(K(U)\big)\right]^{\frac{1}{3}}} \qquad (8.15) \\
&= 2\varepsilon_2\left[-1 + \tfrac{2^{\frac{5}{3}}}{3} - 0.6254(2\varepsilon_2)^2 - \tfrac{2^{\frac{2}{3}}}{2.68}\left[\tfrac{1}{2}\xi_1(1-2\varepsilon_2) + \xi_2(2\varepsilon_2)\right]\right] \\
&\ge 2\varepsilon_2[-1 + 1.05826\cdots - 0.0150\cdots - 0.0332\ldots] > 0.
\end{aligned}
$$

(8.14), (8.15) and the convexity of f imply $f(\lambda) > 0$, when $\lambda \in \left[\varepsilon_1, \tfrac{1}{2} - \varepsilon_2\right]$, or, in other words, if $U \ge 12$ and $\varepsilon_1\mu\big(K(U)\big) \le \alpha \le \left(\tfrac{1}{2} - \varepsilon_2\right)\mu\big(K(U)\big)$, then $\|R\big(V_\alpha(U), U\big)\| > \|K(U)\|$. On the other hand (8.12) and the assumption on α together imply $\alpha > \varepsilon_1\mu\big(K(U)\big)$. Moreover it follows from the assumption on α, (8.11) and (8.13), that $\alpha \le \left(\tfrac{1}{2} - \varepsilon_2\right)\mu\big(K(U)\big)$, unless

$$\alpha = \binom{U}{3} - \binom{M+1}{3} \quad \text{and} \quad \binom{M}{3} \le \alpha \le \binom{M+1}{3}, \qquad (8.16)$$

where M is defined by (8.11).

However (8.16) implies $\hat{R}^{(U)}\big(V_\alpha(U), U\big) = K(M+1)$ and $u \in [M, M+1]$. Therefore

$$\hat{R}^{(U)}\big(V_\alpha(U), U\big) \smallsetminus K(u) = \left\{(x,y,z) : u < z \le M+1, 0 < x < y < M, \lceil x \rceil \ne \lceil y \rceil\right\} \triangleq \Delta, \quad \text{say.} \qquad (8.17)$$

This and Lemma 4 imply

$$d(\Delta) = M + \frac{M+1+u}{2} \ge 2M + \frac{1}{2}. \qquad (8.18)$$

Moreover, one can easily check in our case (i.e. $U \ge 12$) that $M \ge \frac{3}{4}U$, which together with (8.18) means that

$$d(\Delta) > \frac{3}{2}U. \qquad (8.19)$$

This and Lemmas 3, 4 imply

$$
\begin{aligned}
\|R\big(V_\alpha(U),U\big)\| - \|K(U)\| &= \tfrac{3}{2}U\mu\big(K(U)\big) - \tfrac{3}{2}U\big(\mu\big(K(U)\big) - \alpha\big) \\
&\quad + \left(\|\hat{R}^{(U)}\big(V_\alpha(U),U\big)\| - \|K(u)\|\right) = \tfrac{3}{2}U\left[\alpha - \big(\mu\big(K(U)\big) - \alpha\big)\right] + \|\Delta\| \qquad (1) \\
&= \tfrac{3}{2}U\big(\mu\big(K(u) - \hat{R}^{(U)}\big(V_\alpha(U),U\big)\big) + \|\Delta\| = \big(d(\Delta) - \tfrac{3}{2}U\big)\mu(\Delta) > 0.
\end{aligned}
$$

i.e. so far, we have shown (8.1) for $U \geq 12$. Finally, we check (8.1) directly for $U = 6, 7, \ldots, 11$.

Remark 3. *For $U < 6$, there is no room for $\alpha = \binom{U}{3} - \binom{M}{2} < \frac{1}{2}\binom{U}{3}$.*

9 Main Result for $k = 3$ and Good α

Now let us return to our main problem in the discrete model. Denote by $R^*(v, u)$ the downset of $(v, u - 1, u)$ $(v, u \in \mathbb{Z}^+)$ in $\mathcal{L}(U, 3)$ and by $K^*(u)$ the downset of $(u - 2, u - 1, u)$ $(u \in \mathbb{Z}^+)$ in $\mathcal{L}(U, 3)$. Then Lemmas 2,3, and 8 and Theorems 1 and 2 together imply immediately this solution.

Theorem 3. *Let $U \in \mathbb{Z}^+$, $U \geq 6$, then*

(i) *For $\alpha = \binom{U}{3} - \binom{m}{3} \leq \frac{\binom{U}{3}}{2}$ for some $m \in \mathbb{Z}^+$,*
$\max_{|\mathcal{A}|=\alpha} \mathcal{P}(\mathcal{A})$ *is achieved by $R^*(U - m, U)$.*

(ii) *For $\alpha = \binom{m}{3} \geq \frac{\binom{U}{3}}{2}$ for some $m \notin \mathbb{Z}^+$*
$\max_{|\mathcal{A}|=\alpha} \mathcal{P}(\mathcal{A})$ *is achieved by $K^*(m)$.*

10 A False Natural Conjecture for $k = 3$ and General α; There Is "Almost" No "Order" at All

We conclude our paper by taking a look at general α. Both, the result for $k = 2$ in [2] and our result for $k = 3$ and good α suggest that the following conjecture is reasonable, namely, that for $k = 3$ and α with

$$\binom{U}{3} - \binom{a+1}{3} < \alpha < \binom{U}{3} - \binom{a}{3} \leq N(\alpha) < \frac{\binom{U}{3}}{2}, \qquad (10.1)$$

where $a \in \mathbb{Z}^+$ and $N(\alpha)$ is a function depending only on α, if U is big enough, the following configuration W is optimal for maximizing $\mathcal{P}(\mathcal{A})$:

(i) take the $\binom{U}{3} - \binom{a+1}{3}$ points (x, y, z) with $x \leq U - (a + 1)$ in $S_{U,3}$

(ii) add the $\alpha - \left[\binom{U}{3} - \binom{a+1}{3}\right]$ points $(U - a, y, z)$ where (y, z) are points of a quasi–star or a quasi–complete graph in the sense of [2] according to the value of $\alpha - \left[\binom{U}{3} - \binom{a+1}{3}\right]$.

However, this conjecture, which has been made by several authors, is false.

Example 1: For $\alpha_0 \triangleq \left[\binom{U}{3} - \binom{U-2}{3}\right] - (U - 2) - (U - 3) = \binom{U}{3} - \binom{U-2}{3} - 2U + 5$ (when U is big enough), the W described above is $S_1 \setminus (S_2 \cup S_3)$ where $S_1 \triangleq \{(x, y, z) \in S_{U,3} : x = 1, 2\}$.

$$S_2 \triangleq \{(2, 3, U), (2, 4, U), \ldots, (2, U - 2, U), (2, U - 1, U)\},$$

and S_3 is listed in (10.2) below.

Now let us consider the configuration W' with $W' \triangleq S_1 \smallsetminus (S_2 \cup S_3')$, where S_3' is also listed in (10.2).

$$S_3 : (2,3,U-1), (2,4,U-1), \ldots, (2,U-2,U-1), (2,U-3,U-2), (2,U-4,U-2)$$
$$S_3' : (1,2,U), (1,3,U), (1,4,U), \ldots, (1,U-2,U), (1,U-1,U). \tag{10.2}$$

Thus, $\|S_3\| > \|S_3'\|$ when $U > 10$ and therefore $\|W\| < \|W'\|$. This example tells us that a solution for general α, even when $k = 3$, is much more challenging. Actually, if we pay a little bit more attention to it, we will find a deeper result just at our hands. People working on these kinds of problems usually wish to find "an order", more precisely a nested optimal sequence such as

$$W_1 \subset W_2 \subset W_3 \subset \cdots$$

where W_i is optimal for size i. It is not surprising that in many cases, obviously including our problem, there is no order at all. In these cases, and in particular for our case, we define M_k as the maximal integer s.t. the optimal nested chain with length M_k i.e. the optimal nested chain

$$W_1 \subset W_2 \subset W_3 \subset \cdots \subset W_{M_k} \tag{10.3}$$

exists. Considering our problem we only need to study the α-s with $\alpha \leq \frac{1}{2}\binom{U}{3}$, because we can take "complements". Therefore we wish M_k to be close to $\frac{1}{2}\binom{U}{3}$. In fact in [2], it was shown that $M_2 \geq \frac{1}{2}\binom{U}{2} - \frac{U}{2}$, and that therefore M_2 is asymptotically equal to $\frac{1}{2}\binom{U}{3}$ (i.e. $\frac{\frac{1}{2}\binom{U}{2}-M_2}{\binom{U}{2}} \to 0$).

However, it is surprising that there is a jump between M_2 and M_3, because M_3 is asymptotically close to zero as can be seen from the following result.

Theorem 4.

$$M_3 < \binom{U}{3} - \binom{U-2}{3} \triangleq \alpha_2 \ \ for \ \ U > U_0. \tag{10.4}$$

Proof. Assume the result is false. Then there is a nested optimal chain $W_1 \subset W_2 \subset \cdots \subset W_{\alpha_2}$.

Let α_0, W and W' be defined as in Example 1 and set $\alpha_1 \triangleq \binom{U}{3} - \binom{U-1}{3}$. Then (when U is big enough) $\alpha_1 < \alpha_0 < \alpha_2$ and therefore $W_{\alpha_1} \subset W_{\alpha_0} \subset W_{\alpha_2}$. First of all, we draw attention to the fact that in the proofs in Section 3, we actually have already proved that the optimal configurations in Theorem 3 are unique (except if $\alpha = \frac{1}{2}\binom{U}{3}$.) Therefore, $W_{\alpha_1} = R^*(1,U)$ and $W_{\alpha_2} = R^*(2,U)$ or

$$(1, U-1, U) \in W_{\alpha_1} \ \ \text{and} \ \ (2, U-1, U) \in W_{\alpha_2} \tag{10.5}$$

and so

$$(1, U - 1, U) \in W_{\alpha_0}. \qquad (10.6)$$

Consequently,

$$W_{\alpha_0} \neq W'. \qquad (10.7)$$

Moreover, there exists an $(x_0, y_0, z_0) \in W_{\alpha_0}$ with $x_0 \geq 3$, because otherwise by Theorems 2 and 3 in [2] $||W_{\alpha_0}|| = ||W||$, which would contradict Example 1 (here W and W' are defined as in Example 1). However, $(x_0, y_0, z_0) \notin R^*(2, U) = W_{\alpha_2} \supset W_{\alpha_0}$, a contradiction.

11 A Related Topic: The Maximal Moments for the Family of Measurable Symmetric Downsets

Next let us drop the condition $\lceil x \rceil \neq \lceil y \rceil$, $\lceil y \rceil \neq \lceil z \rceil$ used in the definition of $S_{U,3}$ in previous sections, i.e. consider the lattice $\alpha'(U, 3) \triangleq (S'_{U,3}, \leq)$, $S'_{U,3} \triangleq \{(x, y, z) \in R^3 : 0 \leq x \leq y \leq z\}$. The problem becomes more smooth and therefore much simpler. To see this, we mention here two observations.

(a) To guarantee the formula analogous to (4.8), we don't have to require $u \in \mathbb{Z}^+$.

(b) One can simply derive a lemma analogous to Lemma 6, by standard methods in calculus (such as to take right derivatives and so on).

In fact, in a similar but much simpler way we can prove the following result.

Theorem 5. *For $U \in R$ let $I_U = [0, U]^3 \subset \mathbb{R}^3$ and let $\mathcal{F}_\alpha$ be the family of the Lebesgue measurable subsets S of I_U, satisfying*

(i) For every $S \in \mathcal{F}_\alpha$ $\mu(S) = \alpha$.
(ii) For every permutation π on $\{1, 2, 3\}$ and every $S \in \mathcal{F}_\alpha$ $(x_1, x_2, x_3) \in S$ implies $x_{\pi(1)}, x_{\pi(2)}, x_{\pi(3)} \in S$.
(iii) For every $S \in \mathcal{F}_\alpha$, $(x, y, z) \in S$ and $(x', y', z') \leq (x, y, z)$. Also $(x', y', z') \in S$.

Then $\max_{S \in \mathcal{F}_\alpha} ||S||$, where $||S|| = \int_S (x + y + z) dx\, dy\, dz$, is achieved by a set $S^* \in \mathcal{F}_\alpha$ of the form

$$S^* = \begin{cases} \{(x, y, z) : \min\{x, y, z\} \leq v\} & \text{for some } v = v(\alpha), \text{ if } \alpha \leq \frac{U^3}{2} \\ \{(x, y, z) : 0 \leq x, y, z \leq u\} & \text{for some } u = u(\alpha), \text{ if } \alpha \geq \frac{U^3}{2}. \end{cases}$$

References

1. R. Ahlswede and N. Cai, On edge-isoperimetric theorems for uniform hypergraphs, Preprint 93-018, SFB 343 "Diskrete Strukturen in der Mathematik", Universität Bielefeld, 1993.
2. R. Ahlswede and G. Katona, Graphs with maximal number of adjacent pairs of edges, Acta. Math. Sci. Hungaricae Tomus 32, 1–2, 97–120, 1978.
3. R. Ahlswede and G. Katona, Contributions to the geometry of Hamming spaces, Discrete Math. 17, 1–22, 1977.
4. B. Bollobás and I. Leader, Edge–isoperimetric inequalities in the grid, Combinatorica 11, 4, 299–314, 1991.
5. L.H. Harper, Optimal assignments of numbers to vertices, SIAM J. Appl. Math. 12, 131–135, 1964.
6. R. Ahlswede, Simple hypergraphs with maximal number of adjacent pairs of edges, J. Combinatorial Theory, Series B, Vol. 28, No. 2, 164–167, 1980.
7. S.L. Bezrukov and V.P. Boronin, Extremal ideals of the lattice of multisets with respect to symmetric functionals (in Russian), Diskretnaya Matematika 2, No. 1, 50–58, 1990.
8. R. Ahlswede and I. Althöfer, The asymptotic behaviour of diameters in the average, Preprint 91–099, SFB 343, Diskrete Strukturen in der Mathematik, J. Combinatorial Theory B., Vol. 61, No. 2, 167–177, 1994.
9. R. Ahlswede and N. Cai, On partitioning and packing products with rectangles, Combin. Probab. Comput. 3, no. 4, 429–434, 1994.

Appendix: Solution of Burnashev's Problem and a Sharpening of the Erdős/Ko/Rado Theorem

R. Ahlswede

Motivated by a coding problem for Gaussian channels, Burnashev came to the following *Geometric Problem* (which he stated at the Information Theory Meeting in Oberwolfach, Germany, April 1982). For every $\delta > 0$, does there exist a constant $\lambda(\delta) > 0$ such that the following is true: "Every finite set $\{x_1, \ldots, x_N\}$ in a Hilbert space H has a subset $\{x_{i_1}, \ldots, x_{i_M}\}$, $M \geq \lambda(\delta)N$, without 'bad' triangles. (A triangle is *bad*, if one side is longer than $1 + \delta$ and the two others are shorter ($\leq$) than 1)"?

This is the case for Euclidean spaces. (A good exercise before the further reading!) We show that this is *not* so for infinite–dimensional Hilbert spaces. The proof is based on a sharpening of the famous Erdős–Ko–Rado Theorem and was given at the same meeting.

The publication of this note from 1982 was originally planned in a forthcoming book on Combinatorics by G. Katona. Since the completion of this book is still unclear and on the other hand the method of generated sets of [1] and the method of pushing and pulling of [2] are now available there is realistic hope that this direction of work with its open problems can now be continued. Therefore it should be made known and the late publication is justified.

The solution was found by a funny chance event: Burnashev pronounced the name "Hilbert" in the Russian way like "Gilbert", which gave us the inspiration to view the problem in a sequence space.

Let h be the Hamming distance. Define

$$G_k^n \triangleq \left(\left\{ (a_1, a_2, a_3, \ldots) : a_t \in \left\{ 0, \frac{1}{\sqrt{2}} \right\}, 1 \leq t \leq n;\ a_t = 0, t > n;\ \sum_{t=1}^{n} a_t = \frac{k}{\sqrt{2}} \right\}, h \right),$$

Obviously, for $1 \leq k \leq n$ $\quad G_k^n \subset H = \ell_2$ and for $a^n, b^n \in G_k^n$

$$h(a^n, b^n) \leq 2 \Leftrightarrow \|a^n - b^n\|_2 \leq 1. \tag{1}$$

We call $X \subset G_k^n$ **good**, if it contains no bad triangle. It suffices to show that for some k

$$g_k(n) \triangleq \max_{X \subset G_k^n,\ \text{good}} |X| = o\left(\binom{n}{k} \right). \tag{2}$$

Using the representation of subsets of an n–set as $(0 - 1)$–incidence vectors the determination of $q_2(n)$ leads to an extremal problem of independent interest, whose solution provides in all but one case an amazing sharpening of the well-known Erdős/Ko/Rado Theorem.

This says that for any family $\mathcal{B} \subset \mathcal{P}_\ell(\{1, \ldots, n\})$ of all ℓ–element subsets of an n–set with the

R. Ahlswede et al. (Eds.): Information Transfer and Combinatorics, LNCS 4123, pp. 1006–1009, 2006.

Intersection Property: $B \cap B' \neq \varnothing$ $\forall B, B' \in \mathcal{B}$
necessarily

$$|\mathcal{B}| \leq \binom{n-1}{\ell-1}, \text{ if } n \geq 2\ell. \tag{3}$$

Our result is the

Theorem. *Let* $n \geq 2\ell$, $\ell \geq 2$. *For any* $\mathcal{A} \subset \mathcal{P}_\ell\big((1, 2, \ldots, n)\big)$ *with the*

Triangle Property: $\forall A, B, C \in \mathcal{A} : A \cap B \neq \varnothing, B \cap C \neq \varnothing \Rightarrow A \cap C \neq \varnothing$
we have

$$|\mathcal{A}| \leq \begin{cases} n & \text{if } \ell = 2 \text{ and } n \equiv 0 \mod 3 \\ \binom{n-1}{\ell-1} & \text{otherwise.} \end{cases} \tag{4}$$

Moreover, this bound is best possible.

Proof: The Triangle Property implies that $\mathcal{A}$ can be partitioned into families $\mathcal{A}(1), \ldots, \mathcal{A}(T)$ such that

(a) The families $\mathcal{A}(t)$, $1 \leq t \leq T$, have the Intersection Property.
(b) The sets $A(t) \triangleq \cup\{A : A \in \mathcal{A}(t)\}$, $1 \leq t \leq T$, are disjoint.
(c) The numbers $\alpha_t \triangleq |A(t)|$ satisfy $\ell \leq \alpha_t \leq n$ for $1 \leq t \leq T$.

This and (3) imply

$$|\mathcal{A}| = \sum_{1 \leq t \leq T} |\mathcal{A}(t)| \leq \sum_{t : \alpha_t < 2\ell} \binom{\alpha_t}{\ell} + \sum_{t : \alpha_t \geq 2\ell} \binom{\alpha_t - 1}{\ell - 1}. \tag{5}$$

Case 1: The second sum equals 0
By Pascal's identity for $q \geq p \geq \ell$

$$\binom{q}{\ell} + \binom{p}{\ell} = \binom{q}{\ell} + \binom{p-1}{\ell} + \binom{p-1}{\ell-1} \leq \binom{q}{\ell} + \binom{p-1}{\ell} + \binom{q}{\ell-1} = \binom{q+1}{\ell} + \binom{p-1}{\ell}$$

and therefore with (b)

$$|\mathcal{A}| \leq \sum_{t : \alpha_t < 2\ell} \binom{\alpha_t}{\ell} \leq \binom{2\ell-1}{\ell} \frac{n}{\lfloor 2\ell-1 \rfloor} + \binom{(2\ell-1)x}{\ell}, \text{ where } x = \frac{n}{2\ell-1} - \frac{n}{\lfloor 2\ell-1 \rfloor}. \tag{1}$$

Case 2: The second sum does not equal 0
We show first that for $2\ell > \gamma \geq \ell$, $\beta \geq \ell$, $\gamma + \beta \leq n$

$$\binom{\gamma}{\ell} + \binom{\beta-1}{\ell-1} \leq \binom{\gamma+\beta-1}{\ell-1}. \tag{7}$$

Clearly,

$$\ell(\gamma+\beta-1) \cdots (\gamma+\beta-\ell+1) \geq \ell(\beta-1) \cdots (\beta-\ell+1) + \ell\gamma^{\ell-1}$$
$$\geq \ell(\beta-1) \cdots (\beta-\ell+1) + \gamma(\gamma-1) \cdots (\gamma-\ell+2)(\gamma-\ell+1),$$

since $\gamma - \ell + 1 \leq \ell$, and thus (7) follows.

Using (7) we can shift terms from the first sum to the second sum in (5) and obtain finally an upper bound of the form

$$\sum_{i \in I} \binom{\rho_i - 1}{\ell - 1}; \quad \sum \rho_i \leq n, \rho_i \geq 2\ell,$$

which is obviously smaller than $\binom{n-1}{\ell-1}$.

Thus we have

$$|\mathcal{A}| \leq \max\left(\binom{n-1}{\ell-1}, \binom{2\ell-1}{\ell}\frac{n}{\lfloor 2\ell-1 \rfloor} + \binom{(2\ell-1)x}{\ell}\right) \tag{8}$$

where $x = \frac{n}{2\ell-1} - \frac{n}{\lfloor 2\ell-1 \rfloor}$.

For $\ell = 2$ thus $|\mathcal{A}| \leq \begin{cases} n-1 & \text{if } n \not\equiv 0 \mod 3 \\ n & \text{if } n \equiv 0 \mod 3 \end{cases}$.

In all other cases it suffices to show

$$\binom{n-1}{\ell-1} \geq \binom{2\ell-1}{\ell-1}\frac{n}{\lfloor 2\ell-1 \rfloor} + \binom{(2\ell-1)x}{\ell}.$$

In case $n = 2\ell$ this is true, because

$$\frac{2\ell}{\lfloor 2\ell-1 \rfloor} = 1 \text{ and } \binom{(2\ell-1)x}{\ell} = \binom{1}{\ell} = 0.$$

In case $n \geq 2\ell + 1$ we have

$$\binom{n-1}{\ell-1} \geq \binom{2\ell-1}{\ell-1}\frac{n}{2\ell-1}, \text{ because } (n-1)(n-\ell+1) = n(n-\ell)+\ell-1 \geq n(\ell+1) \tag{9}$$

and thus $(2\ell-1)(n-1)(n-2)\cdots\cdots(n-\ell+1) \geq (2\ell-1)(2\ell-2)\cdots\cdots(\ell+1)n$.

Since for $0 \leq x \leq 1$ $x\binom{2\ell-1}{\ell} \geq \binom{(2\ell-1)x}{\ell}$ (9) yields the result.

Translation of the result for G_2^n yields

$$g_2(n) = \begin{cases} n & \text{for } n \equiv 0 \mod 3 \\ n-1 & \text{otherwise} \end{cases}.$$

Thus we have as

Corollary. (Negative answer to Burnashev's Question for *every* $\delta > 0$)

$$\lim_{n \to \infty} g_2(n) \cdot |G_2^n|^{-1} = \lim_{n \to \infty} \frac{2}{n-1} = 0.$$

Problems

1. Let $M(N)$ be the guaranteed cardinality of a largest good subset of an N–set in H. We have just shown that $M(N) \leq 0(\sqrt{N})$.
   What is the exact asymptotic growth of $M(N)$?
2. What is the best choice of $\lambda(\delta)$ for the n–dimensional Euclidean space?
3. Generalize the Theorem to families of sets with the property:

$$|A \cap B| \geq d, |B \cap C| \geq d \Rightarrow |A \cap C| \geq d.$$

References

1. R. Ahlswede and L.H. Khachatrian, The complete intersection theorem for systems of finite sets, Preprint 95–066, SFB 343 "Diskrete Strukturen in der Mathematik", European J. Combinatorics, 18, 125–136, 1997.
2. R. Ahlswede and L.H. Khachatrian, A pushing–pulling method: new proofs of intersection theorems, Preprint 97–043, SFB 343 "Diskrete Strukturen in der Mathematik", Universität Bielefeld, Combinatorica 19(1), 1–15, 1999.

Realization of Intensity Modulated Radiation Fields Using Multileaf Collimators

T. Kalinowski

Abstract. In the treatment of cancer using high energetic radiation the problem arises how to irradiate the tumor without damaging the healthy tissue in the immediate vicinity. In order to do this as efficiently as possible intensity modulated radiation therapy (IMRT) is used. A modern way to modulate the homogeneous radiation field delivered by an external accelerator is to use a multileaf collimator in the static or in the dynamic mode. In this paper several aspects of the construction of optimal treatment plans are discussed and some algorithms for this task are described.

1 Introduction

In cancer treatment high energetic radiation is used to destroy the tumor. To achieve this goal the irradiation process must be planned in such a way that the tumor (target volume) receives a sufficiently high dose while the organs close to it (organs at risk) are not damaged. In clinical practice the radiation is delivered by a linear accelerator which is part of a gantry that can be rotated about the treatment couch (see Figure 1).

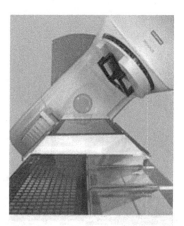

Fig. 1. A linear accelerator with a treatment couch

The first step in the treatment planning after the target volume and the organs at risk have been localized is to discretize the radiation beam head into bixels and the irradiated volume into voxels. Then a set of gantry angles from which

R. Ahlswede et al. (Eds.): Information Transfer and Combinatorics, LNCS 4123, pp. 1010–1055, 2006.

Fig. 2. The leaf pairs of a multileaf collimator

radiation is released has to be determined. In order to increase the efficiency of the treatment it is often desirable to modulate the intensity profile of the radiation beam. So for each gantry angle an intensity function is prescribed, i.e. an amount of radiation released at each bixel. Finally, we have to find a way to realize this modulation. Here we consider only the last step of this planning process. That is as our starting point we take an intensity function for a fixed irradiation angle. We assume that the radiation head is a rectangle and choose a partition into equidistant cells as discretization. Then the intensity function can be described as a nonnegative matrix whose entries are the desired doses at the corresponding bixels. A modern approach to the modulation of homogeneous fields is the usage of a multileaf collimator (MLC). A multileaf collimator consists of one pair of metal leaves for each row of the intensity matrix (see Figure 2). These leaves can be inserted between the beam head and the patient in order to protect parts of the irradiated area. So differently shaped homogeneous fields are generated and by superimposing a number of these the given modulated intensity can be realized. There are two essentially different ways to generate intensity modulated fields with multileaf collimators: in the static mode (stop–and–shoot) the beam is switched off when the leaves are moving while in the dynamic mode the beam is switched on during the whole treatment and the modulation is achieved by varying the speed of the leaf motion. Two important criteria for the quality of a treatment plan are the total irradiation time and the total treatment time. The total irradiation time should be small since there is always a small amount of radiation transmitted through the leaves, and if the used model ignores this leaf transmission the error increases with the total irradiation time. A small total treatment time is desirable for efficiency reasons. In the dynamic mode the two criteria coincide. So here the problem is to determine a velocity function for each leaf such that the given intensity is realized in the shortest possible time. In the static mode the whole treatment consists of the irradiation and the intervals in between when the leaves are moved. Thus we have two parameters which influence the total treatment time: the irradiation time and the number of homogeneous fields that are needed. How these parameters have to be weighted depends on the used technology: the longer the time intervals between the different fields are, the more important becomes the reduction of the number of fields. The lengths of these time intervals is influenced by the leaf velocity and by the so called verification and record overhead, which is the time necessary to check the correct positions of the leaves. In a more realistic model one should also take the shapes of the fields into account, because clearly the necessary leaf travel time between two fields depends on the shapes of these fields (see [18,2]). The dynamic mode has the advantage of a smaller total treatment time, but the static mode involves no leaf movement with radiation on and so the verification of the correct realization of the treatment plan is easier which makes the method less sensitive to malfunctions of the technology.

There are additional machine–dependent restrictions which have to be considered when determining the leaf positions:

Interleaf collision constraint (ICC): In some widely used MLC's it is forbidden that opposite leaves of adjacent rows overlap, because otherwise these leaves collide. So leaf positions as illustrated in Figure 3 are not allowed.

Fig. 3. Leaf position that is excluded by the ICC. The shading indicates the area that is covered by the leaves

Tongue and groove constraint: In order to reduce leakage radiation between adjacent leaves the commercially available MLC's use a tongue–and–groove (or similar) design (see Figure 4) with the effect that there is a small overlap of the regions that are covered by adjacent leaves.

Consider two bixels x and y that are adjacent along a column and two homogeneous fields, where in the first field x is irradiated and y is covered and in the second field y is irradiated and x is covered. Then in the composition of these fields along the border of x and y there is a narrow strip (the overlap of the regions that are covered by the leaves in the rows of x and y, respectively) that receives no radiation at all. Figure 5 illustrates this for the intensity map $\left(\begin{smallmatrix} 2 & 3 \\ 3 & 4 \end{smallmatrix}\right)$.

To avoid this effect one may require that two bixels that are adjacent along a column are irradiated simultaneously for the time the lower of the two doses is delivered. Then the border region receives this lower dose. If this is the case for all the relevant pairs of adjacent bixels the treatment plan is said to satisfy the tongue and groove constraint.

In this paper we collect some of the known algorithms for the intensity modulation of radiation beams with multileaf collimators in a unified notation.

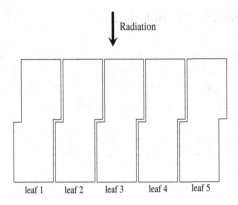

Fig. 4. The principle of the tongue–and–groove design. The picture shows a cut through the leaf bank perpendicular to the direction of leaf motion.

2 Static Methods

This chapter is organized as follows. In the first section some notation is introduced and we describe a linear programming formulation of the total irradiation time minimization (taken from [7]). In the second section we prove that the minimization of the number of homogeneous fields that are needed is an NP–complete problem. The remaining sections are devoted to the discussion of some concrete algorithms.

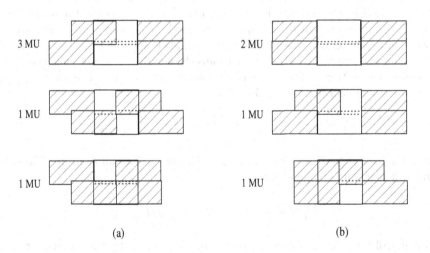

Fig. 5. Two realizations of the same intensity map. (a) The overlap of bixels $(1,1)$ und $(2,1)$ receives no radiation because of the tongue and groove effect. (b) The overlaps of bixels that are adjacent along a column receive the smaller one of the doses delivered to the overlapping bixels.

2.1   Notation and LP–Formulation

Throughout we use the notation $[n] := \{1, 2, \ldots, n\}$ for positive integers n. The given intensity function can be considered as a nonnegative integer matrix $A = (a_{i,j})_{\substack{1 \le i \le m \\ 1 \le j \le n}}$. A segment is a matrix that corresponds to a leaf position of an MLC. This is made precise in the following definition.

Definition 1. *A segment is an $m \times n$–matrix $S = (s_{i,j})$, such that there exist integers l_i, r_i $(i \in [m])$ with the following properties:*

$$1 \le l_i \le r_i + 1 \le n + 1 \qquad\qquad (i \in [m]), \qquad\qquad (1)$$

$$s_{i,j} = \begin{cases} 1 \; \textit{if } l_i \le j \le r_i \\ 0 \; \textit{otherwise} \end{cases} \qquad (i \in [m], \; j \in [n]). \qquad (2)$$

So $l_i - 1$ and $r_i + 1$ have to be interpreted as the positions of the i–th left and right leaf, respectively. A *segmentation* of A is a representation of A as a sum of segments, i.e.

$$A = \sum_{i=1}^{k} u_i S_i$$

with segments S_i and positive numbers u_i $(i = 1, 2, \ldots, k)$. Every segmentation corresponds in the obvious way to a treatment plan realizing the given intensity matrix A. Our goal is to minimize the total number of monitor units (TNMU) and the number of segments (NS), which in the segmentation correspond to $\sum_{i=1}^{k} u_i$ and k, respectively. First of all, observe that in general it is not possible to minimize both of these parameters simultaneously. For the segmentation problem with ICC this was shown by an example in [9]. Here we give an example that is independent of the ICC, that means the simultaneous minimization is not possible, no matter if the ICC is taken into account or not. The matrix $\left(\begin{smallmatrix} 2 & 6 & 3 \\ 4 & 5 & 6 \end{smallmatrix}\right)$ has a segmentation with 6 monitor units

$$\left(\begin{smallmatrix} 2 & 6 & 3 \\ 4 & 5 & 6 \end{smallmatrix}\right) = 3 \left(\begin{smallmatrix} 0 & 1 & 1 \\ 1 & 1 & 1 \end{smallmatrix}\right) + 1 \left(\begin{smallmatrix} 1 & 1 & 0 \\ 1 & 1 & 1 \end{smallmatrix}\right) + 1 \left(\begin{smallmatrix} 1 & 1 & 0 \\ 0 & 1 & 1 \end{smallmatrix}\right) + 1 \left(\begin{smallmatrix} 0 & 1 & 0 \\ 0 & 0 & 1 \end{smallmatrix}\right),$$

and this cannot be done with 3 segments. However, if we allow to use 7 monitor units, 3 segments are sufficient:

$$\left(\begin{smallmatrix} 2 & 6 & 3 \\ 4 & 5 & 6 \end{smallmatrix}\right) = 4 \left(\begin{smallmatrix} 0 & 1 & 0 \\ 1 & 1 & 1 \end{smallmatrix}\right) + 2 \left(\begin{smallmatrix} 1 & 1 & 1 \\ 0 & 0 & 1 \end{smallmatrix}\right) + 1 \left(\begin{smallmatrix} 0 & 0 & 1 \\ 0 & 1 & 0 \end{smallmatrix}\right).$$

But it will be an easy consequence of Lemma 1 below, that for a single row A, i.e. in the case $m = 1$, both parameters can be minimized simultaneously. By $\mathcal{F}$ we denote the subsets of $V := [m] \times [n]$ that correspond to segments, that is

$$\mathcal{F} = \{T \subseteq V : \text{There exists a segment } S \text{ with } ((i, j) \in T \iff s_{i,j} = 1)\}.$$

Now an LP–relaxation of the TNMU–minimization problem is given by:

$$(P) \begin{cases} \displaystyle\sum_{T \in \mathcal{F}} f(T) \ \to \ \min & \text{subject to} \\[2ex] f(T) \geq 0 & \forall T \in \mathcal{F}, \\[2ex] \displaystyle\sum_{T \in \mathcal{F}:(i,j) \in T} f(T) = a_{i,j} \ \forall (i,j) \in V. \end{cases}$$

In order to show that a certain algorithm is optimal with respect to the TNMU one can use the dual of this program:

$$(D) \begin{cases} \displaystyle\sum_{(i,j) \in V} a_{i,j} g(i,j) \ \to \ \max \text{ subject to} \\[2ex] \displaystyle\sum_{(i,j) \in T} g(i,j) \leq 1 & \forall T \in \mathcal{F}. \end{cases}$$

Following [7] one can define the functions g_s $(1 \leq s \leq m)$ by

$$g_s(i,j) = \begin{cases} 1 \text{ if } i = s, \ a_{i,j} \geq a_{i,j-1} \text{ and } a_{i,j+1} < a_{i,j} \\ -1 \text{ if } i = s, \ a_{i,j} < a_{i,j-1} \text{ and } a_{i,j+1} \geq a_{i,j} \\ 0 \text{ otherwise,} \end{cases}$$

where we put $a_{i,0} = a_{i,n+1} = 0$ for all i. It is easy to see that the g_s are feasible for (D) and that

$$\sum_{(i,j) \in V} a_{i,j} g_s(i,j) = \sum_{j=1}^{n} \max\{0, a_{s,j} - a_{s,j-1}\}.$$

Thus

$$\max_{1 \leq i \leq m} \sum_{j=1}^{n} \max\{0, a_{i,j} - a_{i,j-1}\}$$

is a lower bound for the TNMU of a segmentation, and in order to show the optimality of a given algorithm it is sufficient to show that it realizes this bound.

In order to include the ICC into our model we have to add the following conditions to the definition of a segment:

$$(\text{ICC}) \quad l_i \leq r_{i+1} + 1, \qquad r_i \geq l_{i+1} - 1 \qquad (i \in [m-1]). \tag{3}$$

2.2 NS–Minimization is NP–Complete

According to [1] R.E. Burkard showed that the NS–minimization is NP–complete for $m \geq 2$. Here we describe a formulation of the $(m = 1)$–case which was found independently by the author and the authors of [1], and yields the NP–completeness in this case. The intensity map is an n–dimensional row vector

$\mathbf{a} = (a_1, a_2, \ldots, a_n)$ with nonnegative integer entries, and a segment is an n–dimensional $(0, 1)$–vector $\mathbf{s} = \mathbf{s}(l, r)$ with

$$s_i(l, r) = \begin{cases} 1 \text{ if } l \leq i \leq r \\ 0 \text{ otherwise,} \end{cases}$$

for some integers l and r. Now the decision version of the NS–minimization problem is the following: given a row vector $\mathbf{a} = (a_1, \ldots, a_n)$ and an integer N, is there a segmentation of $\mathbf{a}$ with at most N segments? In order to prove the NP–completeness of this problem we give a network flow formulation of the segmentation problem. We define the digraph $\Gamma = (V, E)$, where

$$V = [n + 1],$$
$$E = \{(i, j) : 1 \leq i < j \leq n + 1\}$$

Now a flow $y_{l,r+1} > 0$ on an arc $(l, r+1) \in E$ can be associated with the vector $y_{l,r+1}\mathbf{s}(l, r)$. Then a segmentation of $\mathbf{a}$ corresponds to a flow on Γ, such that the net flow at vertex i is $-d_i$, where $d_i := a_i - a_{i-1}$ for $i = 1, 2, \ldots, n$ and $d_{n+1} := -a_n$, i.e.

$$\sum_{j=1}^{i-1} y_{j,i} - \sum_{j=i+1}^{n+1} y_{i,j} = -d_i \qquad (i \in [n + 1]).$$

In order to count the segments in the considered segmentation we introduce the $(0, 1)$–variables $x_{i,j}$ for $1 \leq i < j \leq n+1$, where $x_{l,r+1} = 1$ iff the segment $\mathbf{s}(l, r)$ has nonzero coefficient. So we can write the problem of finding a segmentation with minimal number of segments as the following fixed charge network flow problem:

$$\sum_{1 \leq i < j \leq n+1} x_{i,j} \to \min \qquad \text{subject to} \qquad (4)$$

$$y_{i,j} \leq L x_{i,j} \qquad (1 \leq i < j \leq n + 1) \qquad (5)$$

$$\sum_{j=1}^{i-1} y_{j,i} - \sum_{j=i+1}^{n+1} y_{i,j} = -d_i \qquad (i \in [n + 1]) \qquad (6)$$

$$x_{i,j} \in \{0, 1\}, \ y_{i,j} \in \mathcal{R}_+ \qquad (1 \leq i < j \leq n + 1), \qquad (7)$$

where L is an upper bound for the coefficients in the segmentation, e.g. the maximum entry of A.

Lemma 1. *There is an optimal solution to* (4)–(7) *with* $y_{i,j} = x_{i,j} = 0$ *for all* (i, j) *with* $d_i \leq 0$ *or* $d_j \geq 0$.

Proof. Let $(\mathbf{x}, \mathbf{y})$ be an optimal solution and assume there is positive flow $y_{i,j}$ on some arc (i, j) with $d_i \leq 0$ or $d_j \geq 0$. Let

$$\phi(\mathbf{x}, \mathbf{y}) = |\{(i, j) \in E : y_{i,j} > 0 \text{ and } d_i \leq 0 \text{ or } d_j \geq 0\}|.$$

We construct another optimal solution $(\mathbf{x}', \mathbf{y}')$ with $\phi(\mathbf{x}', \mathbf{y}') < \phi(\mathbf{x}, \mathbf{y})$. Repeating this step if necessary, we finally obtain a solution $(\mathbf{x}'', \mathbf{y}'')$ with $\phi(\mathbf{x}'', \mathbf{y}'') = 0$, and thus $(\mathbf{x}'', \mathbf{y}'')$ is the required solution. Let $(i_1, i_2, \ldots, i_t)$ be a path with the following properties:

1. $y_{i_k, i_{k+1}} > 0$ and $(d_{i_k} \leq 0$ or $d_{i_{k+1}} \geq 0)$ for $1 \leq k \leq t - 1$.
2. For $i < i_1$, $y_{i,i_1} > 0$ implies $(d_i > 0$ and $d_{i_1} < 0)$.
3. For $i > i_t$, $y_{i_t,i} > 0$ implies $(d_{i_t} > 0$ and $d_i < 0)$.

Such a path with $t \geq 2$ exists by assumption.

Case 1: $d_{i_1} > 0$ and $d_{i_t} < 0$.
Let $\alpha = \min\{y_{i_k, i_{k+1}} : 1 \leq k \leq t - 1\}$, and put

$$y'_{i_k, i_{k+1}} = y_{i_k, i_{k+1}} - \alpha \quad (1 \leq k \leq t - 1),$$

$$x'_{i_k, i_{k+1}} = \begin{cases} 1 \text{ if } y'_{i_k, i_{k+1}} > 0, \\ 0 \text{ if } y'_{i_k, i_{k+1}} = 0, \end{cases}$$

$$x'_{i_1, i_t} = 1,$$

$$y'_{i_1, i_t} = y_{i_1, i_t} + \alpha$$

and $x'_{i,j} = x_{i,j}$, $y'_{i,j} = y_{i,j}$ for all the remaining (i, j). Obviously, the transition from $(\mathbf{x}, \mathbf{y})$ to $(\mathbf{x}', \mathbf{y}')$ preserves the net flows at the vertices, hence $(\mathbf{x}', \mathbf{y}')$ is a feasible solution. Now for at least one $k \in [t - 1]$, $x_{i_k, i_{k+1}} = 1$ and $x'_{i_k, i_{k+1}} = 0$ and since the only x–component which might change from 0 to 1 is x_{i_1, i_t}, we obtain

$$\sum_{1 \leq i < j \leq n+1} x'_{i,j} \leq \sum_{1 \leq i < j \leq n+1} x_{i,j},$$

hence $(\mathbf{x}', \mathbf{y}')$ is also optimal. Finally, for a $k \in [t - 1]$ with $y_{i_k, i_{k+1}} = \alpha$, $y'_{i_k, i_{k+1}} = 0$. And since (i_1, i_t) is the only arc with increasing flow and does not contribute to ϕ,

$$\phi(\mathbf{x}', \mathbf{y}') < \phi(\mathbf{x}, \mathbf{y}).$$

Case 2: $d_{i_1} > 0$ and $d_{i_t} \geq 0$.
Since the net flow in i_t is nonpositive there is some $i_{t+1} > i_t$ with $y_{i_t, i_{t+1}} > 0$ and by condition 3, $d_{i_{t+1}} < 0$. Now we make the same construction as in Case 1 with the path $(i_1, \ldots, i_t, i_{t+1})$.

Case 3: $d_{i_1} \leq 0$ and $d_{i_t} < 0$.
Since the net flow on i_1 is nonnegative there is some $i_0 < i_1$ with $y_{i_0, i_1} > 0$ and by condition 2, $d_{i_0} > 0$. Now we make the same construction as in Case 1 with the path $(i_0, i_1, \ldots, i_t)$.

Case 4: $d_{i_1} \leq 0$ and $d_{i_t} \geq 0$.
As in the Cases 2 and 3, there are $i_0 < i_1$ and $i_{t+1} > i_t$ with $y_{i_0, i_1} > 0$, $y_{i_t, i_{t+1}} > 0$, $d_{i_0} > 0$ and $d_{i_{t+1}} < 0$, and we can make the same construction with the path $(i_0, i_1, \ldots, i_{t+1})$.

So we can restrict our search to the arc set

$$E_0 = \{(i,j) \ : \ 1 \le i < j \le n+1, \ d_i > 0, \ d_j < 0\}$$

and thus we have reduced the problem to a fixed charge transportation problem with sources $S = \{i \ : \ d_i > 0\}$ and sinks $T = \{j \ : \ d_j < 0\}$.

Example 1. *The segmentation*

$$\begin{pmatrix} 2\,4\,1\,3\,1\,4 \end{pmatrix} = 2\begin{pmatrix} 1\,1\,0\,0\,0\,0 \end{pmatrix} + \begin{pmatrix} 0\,1\,0\,0\,0\,0 \end{pmatrix} + \begin{pmatrix} 0\,1\,1\,1\,1\,1 \end{pmatrix} + 2\begin{pmatrix} 0\,0\,0\,1\,0\,0 \end{pmatrix}$$
$$+ \, 3\begin{pmatrix} 0\,0\,0\,0\,0\,1 \end{pmatrix} \tag{8}$$

corresponds to the flow in Fig. 6.

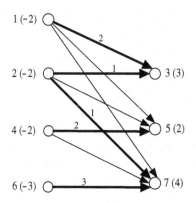

Fig. 6. The flow corresponding to the segmentation (8). The numbers in parentheses are the net flows at the vertices.

Remark 1. *Observe that in an optimal flow satisfying the conditions of Lemma 1 the sum of the flows over all arcs, i.e. the TNMU of the corresponding segmentation, equals the sum of the net flows at the sinks. Clearly, this is a lower bound for the TNMU, hence the corresponding segmentation is also optimal with respect to the TNMU. So for $m = 1$, in contrast to the general case, the TNMU and the NS can be minimized simultaneously.*

Using this transportation formulation we can now prove the NP–completeness.

Theorem 1. *The NS–minimization problem is NP–complete.*

Proof. The problem is obviously in NP, and to show the NP–hardness we reduce the $0-1$–knapsack problem: given positive integers $c_1, \ldots, c_{n-1}, K$, is there a subset $I \subseteq \{1, \ldots, n-1\}$ with $\sum_{i \in I} c_i = K$? We put

$$a_i = \sum_{j=1}^{i} c_j \quad (i = 1, 2, \ldots, n-1) \text{ and } a_n = K,$$

and claim that the answer to the $0-1-$knapsack problem $(c_1, c_2, \ldots, c_{n-1}, K)$ is yes iff the answer to the NS-minimization problem $(a_1, a_2, \ldots, a_n, n-1)$ is yes. We distinguish 3 cases.

Case 1: $K > \sum\limits_{i=1}^{n-1} c_i = a_{n-1}$.

The answer to the knapsack problem $(c_1, c_2, \ldots, c_{n-1}, K)$ is no, and in the transportation problem corresponding to the segmentation we have n sources and 1 sink, so we need n edges with nonzero flow and hence the answer to the NS–minimization problem $(a_1, \ldots, a_n, n-1)$ is also no.

Case 2: $K = \sum\limits_{i=1}^{n-1} c_i = a_{n-1}$.

The answer to the knapsack problem $(c_1, c_2, \ldots, c_{n-1}, K)$ is yes, and in the transportation problem corresponding to the segmentation we have $n-1$ sources and 1 sink, so $n-1$ edges with nonzero flow are sufficient and the answer to the NS–minimization problem $(a_1, \ldots, a_n, n-1)$ is also yes.

Case 3: $K < \sum\limits_{i=1}^{n-1} c_i = a_{n-1}$.

In the transportation problem we have $n-1$ sources with supplies $c_1, c_2, \ldots, c_{n-1}$ and 2 sinks with demands K and $a_{n-1} - K$. At every source there must be at least one outgoing arc with nonzero flow. So altogether there are at least $n-1$ edges with nonzero flow and $n-1$ arcs are sufficient iff at every source there is exactly one outgoing arc with nonzero flow. But this is equivalent to the existence of a subset $I \subset \{1, 2, \ldots, n-1\}$ with $\sum\limits_{i \in I} c_i = K$.

Due to this result it is reasonable to look for a good approximative algorithm for the NS–minimization.

2.3   The Algorithm of Galvin, Chen and Smith

In [8] the authors propose a heuristic algorithm which aims at finding a segmentation with a small NS. As several of the algorithms below it works according to the following general strategy: depending on the given matrix A a coefficient $u > 0$ and a number of segments $S_1, S_2, \ldots, S_t$ are determined such that $A' = A - u(S_1 + S_2 + \cdots + S_t)$ is still nonnegative, and then the algorithm is iterated with A' instead of A. It is clear that this always yields a segmentation of A. Since in a number of algorithms the maximal entry of A is a parameter, it is convenient to give it a name. So let L denote the maximal entry of the considered matrix A for the rest of this thesis. The algorithm from [8] works as follows

1. In a preliminary step eliminate the background intensity, that is put $A := A - uJ$, where u is the smallest entry of A and J is the $m \times n$ all–one matrix.
2. Let u be the smallest integer such that $\frac{1}{2}u(u+1) \geq L$.
3. Mark all the entries of A which are greater or equal to u, i.e. which can be irradiated with u MU.

4. Determine a sequence of segments whose sum is a $(0,1)$−matrix S which has a 1 at position (i,j) iff the entry (i,j) is marked.

5. Put $A := A - uS$, $u := u - 1$ and continue with step 3.

If we do not consider interleaf collision constraints the rows can be treated independently and step 4 can be realized as follows. In each row i, we find the maximal intervals of entries which are greater than or equal to u. These intervals can be described by their left and right boundaries, that is by numbers $l_{i,1}, \ldots, l_{i,t(i)}$ and $r_{i,1}, \ldots, r_{i,t(i)}$, such that

$$1 \le l_{i,1}, \quad r_{i,t(i)} \le n,$$

$$l_{i,k} \le r_{i,k} \qquad\qquad (1 \le k \le t(i)),$$

$$r_{i,k} < l_{i,k+1} - 1 \qquad\qquad (1 \le k \le t(i) - 1),$$

$$a_{i,j} \begin{cases} \ge u \text{ if } l_{i,k} \le j \le r_{i,k} \text{ for some } k, \\ < u \text{ otherwise.} \end{cases}$$

With the additional convention that $t(i) = 0$ for rows without entries greater or equal to u and $l_{i,k} = n + 1$, $r_{i,k} = n$ for $k > t(i)$, the whole procedure is summarized in Algorithm 1.

Algorithm 1. Galvin, Chen and Smith

$u := \min\{a_{i,j} \;:\; 1 \le i \le m, 1 \le j \le n\}$
$A := A - uJ$
$u := \min\left\{k \;:\; \frac{1}{2}k(k+1) \ge L\right\}$
while $A \ne 0$ **do**
    for $i = 1$ to m **do**
        determine $l_{i,1}, l_{i,2}, \ldots, l_{i,t(i)}, r_{i,1}, r_{i,2}, \ldots, r_{i,t(i)}$
    $t := \max\limits_{1 \le i \le m} t(i)$
    for $1 \le k \le t$ let S_k be the segment determined by the $l_{i,k}, r_{i,k}$
    $S := \sum\limits_{k=1}^{t} S_k$
    $A := A - uS;\ u := u - 1$

Example 2. *We will illustrate some of the described algorithms by construction of a segmentation for the benchmark matrix (from [4,14])*

$$A = \begin{pmatrix} 4 & 5 & 0 & 1 & 4 & 5 \\ 2 & 4 & 1 & 3 & 1 & 4 \\ 2 & 3 & 2 & 1 & 2 & 4 \\ 5 & 3 & 3 & 2 & 5 & 3 \end{pmatrix}.$$

For $u = 3$ we obtain $S = \begin{pmatrix} 1 & 1 & 0 & 0 & 1 & 1 \\ 0 & 1 & 0 & 1 & 0 & 1 \\ 0 & 1 & 0 & 0 & 0 & 1 \\ 1 & 1 & 1 & 0 & 1 & 1 \end{pmatrix}$ *with residual matrix* $\begin{pmatrix} 1 & 2 & 0 & 1 & 1 & 2 \\ 2 & 1 & 1 & 0 & 1 & 1 \\ 2 & 0 & 2 & 1 & 2 & 1 \\ 2 & 0 & 0 & 2 & 2 & 0 \end{pmatrix}.$

For $u = 2$ we obtain $S = \begin{pmatrix} 0 & 1 & 0 & 0 & 0 & 1 \\ 1 & 0 & 0 & 0 & 0 & 0 \\ 1 & 0 & 1 & 0 & 1 & 0 \\ 1 & 0 & 0 & 1 & 1 & 0 \end{pmatrix}$ *with residual matrix* $\begin{pmatrix} 1 & 0 & 0 & 1 & 1 & 0 \\ 0 & 1 & 1 & 0 & 1 & 1 \\ 0 & 0 & 0 & 1 & 0 & 1 \\ 0 & 0 & 0 & 0 & 0 & 0 \end{pmatrix}.$

So the total segmentation is

$$
\begin{pmatrix} 4&5&0&1&4&5 \\ 2&4&1&3&1&4 \\ 2&3&2&1&2&4 \\ 5&3&3&2&5&3 \end{pmatrix} = 3 \begin{pmatrix} 1&1&0&0&0&0 \\ 0&1&0&0&0&0 \\ 0&1&0&0&0&0 \\ 1&1&1&0&0&0 \end{pmatrix} + 3 \begin{pmatrix} 0&0&0&0&1&1 \\ 0&0&0&1&0&0 \\ 0&0&0&0&0&1 \\ 0&0&0&0&1&1 \end{pmatrix} + 3 \begin{pmatrix} 0&0&0&0&0&0 \\ 0&0&0&0&0&1 \\ 0&0&0&0&0&0 \\ 0&0&0&0&0&0 \end{pmatrix}
$$

$$
+ 2 \begin{pmatrix} 0&1&0&0&0&0 \\ 1&0&0&0&0&0 \\ 1&0&0&0&0&0 \\ 1&0&0&0&0&0 \end{pmatrix} + 2 \begin{pmatrix} 0&0&0&0&0&1 \\ 0&0&0&0&0&0 \\ 0&0&1&0&0&0 \\ 0&0&0&1&1&0 \end{pmatrix} + 2 \begin{pmatrix} 0&0&0&0&0&0 \\ 0&0&0&0&0&0 \\ 0&0&0&0&1&0 \\ 0&0&0&0&0&0 \end{pmatrix} + 1 \begin{pmatrix} 1&0&0&0&0&0 \\ 0&1&1&0&0&0 \\ 0&0&0&1&0&0 \\ 0&0&0&0&0&0 \end{pmatrix}
$$

$$
+ 1 \begin{pmatrix} 0&0&0&1&1&0 \\ 0&0&0&0&1&1 \\ 0&0&0&0&0&1 \\ 0&0&0&0&0&0 \end{pmatrix}.
$$

2.4   The Algorithm of Bortfeld *et al.*

The first segmentation algorithm which is optimal with respect to the TNMU was introduced in [3]. Again we neglect additional constraints like ICC, and so the rows can be treated independently. Let $\mathbf{a} = (a_1, a_2, \ldots, a_n)$ be a row of A. In addition we put $a_0 = a_{n+1} = 0$ and $L = \max\limits_{1 \leq i \leq n} a_i$. Now, for $1 \leq k \leq L$, we determine the index sets

$$
P_k = \{i \in [n] \ : \ a_{i-1} < k \leq a_i\}, \quad Q_k = \{i \in [n] \ : \ a_i \geq k > a_{i+1}\},
$$

and put $P = \bigcup_k P_k$, $Q = \bigcup_k Q_k$ where the unions have to be understood in the multiset sense. Observe that, for each k, $|P_k| = |Q_k|$, and that

$$
c := \sum_{k=1}^{L} |P_k| = \sum_{i=1}^{n} \max\{0, a_i - a_{i-1}\}.
$$

If $P = (p_1, p_2, \ldots, p_c)$ and $Q = (q_1, q_2, \ldots, q_c)$ are ordered such that $q_i \geq p_i$ for all i, then we can write $\mathbf{a}$ as a sum of c segments $\mathbf{b}^{(1)}, \ldots, \mathbf{b}^{(c)}$ defined by

$$
b_j^{(i)} = \begin{cases} 1 \text{ if } p_i \leq j \leq q_i, \\ 0 \text{ otherwise.} \end{cases}
$$

In [3] two variants of the segmentation algorithm are deduced from this. For the sweep technique P and Q are ordered independently by magnitude, i.e.

$$
p_1 \leq p_2 \leq \ldots \leq p_c, \quad q_1 \leq q_2 \leq \ldots \leq q_c.
$$

For the close–in technique the P_k and the Q_k are ordered by magnitude, each element of a P_k is paired with the corresponding element of Q_k and the resulting pairs (p, q) are ordered by the magnitude of the first component.

Combining the segmentations of the single rows one can produce segmentations for general intensity matrices.

Example 3. *For the second row of* $A = \begin{pmatrix} 4 & 5 & 0 & 1 & 4 & 5 \\ 2 & 4 & 1 & 3 & 1 & 4 \\ 2 & 3 & 2 & 1 & 2 & 4 \\ 5 & 3 & 3 & 2 & 5 & 3 \end{pmatrix}$ *we obtain*

$$P_1 = \{1\}, \; P_2 = \{1,4,6\}, \; P_3 = \{2,4,6\}, \; P_4 = \{2,6\},$$
$$Q_1 = \{6\}, \; Q_2 = Q_3 = \{2,4,6\}, \; Q_4 = \{2,6\}$$

and the sequence of pairs (p,q) *using the sweep technique is*

$$(1,2),(1,2),(2,2),(2,4),(4,4),(4,6),(6,6),(6,6),(6,6),$$

while the close–in technique yields

$$(1,6),(1,2),(2,2),(2,2),(4,4),(4,4),(6,6),(6,6),(6,6).$$

The corresponding segmentations of the whole matrix are

$$\begin{pmatrix} 4 & 5 & 0 & 1 & 4 & 5 \\ 2 & 4 & 1 & 3 & 1 & 4 \\ 2 & 3 & 2 & 1 & 2 & 4 \\ 5 & 3 & 3 & 2 & 5 & 3 \end{pmatrix} = 1 \begin{pmatrix} 1 & 1 & 0 & 0 & 0 & 0 \\ 1 & 1 & 0 & 0 & 0 & 0 \\ 1 & 1 & 0 & 0 & 0 & 0 \\ 1 & 0 & 0 & 0 & 0 & 0 \end{pmatrix} + 1 \begin{pmatrix} 1 & 1 & 0 & 0 & 0 & 0 \\ 1 & 1 & 0 & 0 & 0 & 0 \\ 1 & 1 & 1 & 0 & 0 & 0 \\ 1 & 0 & 0 & 0 & 0 & 0 \end{pmatrix} + 1 \begin{pmatrix} 1 & 1 & 0 & 0 & 0 & 0 \\ 0 & 1 & 0 & 0 & 0 & 0 \\ 0 & 1 & 1 & 1 & 1 & 1 \\ 1 & 1 & 1 & 0 & 0 & 0 \end{pmatrix}$$

$$+ 1 \begin{pmatrix} 1 & 1 & 0 & 0 & 0 & 0 \\ 0 & 1 & 1 & 1 & 0 & 0 \\ 0 & 0 & 0 & 0 & 1 & 1 \\ 1 & 1 & 1 & 1 & 1 & 0 \end{pmatrix} + 1 \begin{pmatrix} 0 & 1 & 0 & 0 & 0 & 0 \\ 0 & 0 & 0 & 1 & 0 & 0 \\ 0 & 0 & 0 & 0 & 0 & 1 \\ 1 & 1 & 1 & 1 & 1 & 0 \end{pmatrix} + 1 \begin{pmatrix} 0 & 0 & 0 & 1 & 1 & 1 \\ 0 & 0 & 0 & 1 & 1 & 1 \\ 0 & 0 & 0 & 0 & 0 & 1 \\ 0 & 0 & 0 & 0 & 1 & 1 \end{pmatrix} + 2 \begin{pmatrix} 0 & 0 & 0 & 0 & 1 & 1 \\ 0 & 0 & 0 & 0 & 0 & 1 \\ 0 & 0 & 0 & 0 & 0 & 0 \\ 0 & 0 & 0 & 0 & 1 & 1 \end{pmatrix}$$

$$+ 1 \begin{pmatrix} 0 & 0 & 0 & 0 & 1 & 1 \\ 0 & 0 & 0 & 0 & 0 & 1 \\ 0 & 0 & 0 & 0 & 0 & 0 \\ 0 & 0 & 0 & 0 & 0 & 0 \end{pmatrix} + 1 \begin{pmatrix} 0 & 0 & 0 & 0 & 0 & 1 \\ 0 & 0 & 0 & 0 & 0 & 1 \\ 0 & 0 & 0 & 0 & 0 & 0 \\ 0 & 0 & 0 & 0 & 0 & 0 \end{pmatrix} \quad and$$

$$\begin{pmatrix} 4 & 5 & 0 & 1 & 4 & 5 \\ 2 & 4 & 1 & 3 & 1 & 4 \\ 2 & 3 & 2 & 1 & 2 & 4 \\ 5 & 3 & 3 & 2 & 5 & 3 \end{pmatrix} = 1 \begin{pmatrix} 1 & 1 & 0 & 0 & 0 & 0 \\ 1 & 1 & 1 & 1 & 1 & 1 \\ 1 & 1 & 1 & 1 & 1 & 1 \\ 1 & 1 & 1 & 1 & 1 & 1 \end{pmatrix} + 1 \begin{pmatrix} 1 & 1 & 0 & 0 & 0 & 0 \\ 1 & 1 & 0 & 0 & 0 & 0 \\ 1 & 1 & 1 & 0 & 0 & 0 \\ 1 & 1 & 1 & 1 & 1 & 1 \end{pmatrix} + 1 \begin{pmatrix} 1 & 1 & 0 & 0 & 0 & 0 \\ 0 & 1 & 0 & 0 & 0 & 0 \\ 0 & 1 & 0 & 0 & 0 & 0 \\ 1 & 1 & 1 & 0 & 0 & 0 \end{pmatrix}$$

$$+ 1 \begin{pmatrix} 1 & 1 & 0 & 0 & 0 & 0 \\ 0 & 1 & 0 & 0 & 0 & 0 \\ 0 & 0 & 0 & 0 & 1 & 1 \\ 1 & 0 & 0 & 0 & 0 & 0 \end{pmatrix} + 1 \begin{pmatrix} 0 & 1 & 0 & 0 & 0 & 0 \\ 0 & 0 & 0 & 1 & 0 & 0 \\ 0 & 0 & 0 & 0 & 0 & 1 \\ 1 & 0 & 0 & 0 & 0 & 0 \end{pmatrix} + 1 \begin{pmatrix} 0 & 0 & 0 & 1 & 1 & 1 \\ 0 & 0 & 0 & 1 & 0 & 0 \\ 0 & 0 & 0 & 0 & 0 & 1 \\ 0 & 0 & 0 & 0 & 1 & 1 \end{pmatrix} + 2 \begin{pmatrix} 0 & 0 & 0 & 0 & 1 & 1 \\ 0 & 0 & 0 & 0 & 0 & 1 \\ 0 & 0 & 0 & 0 & 0 & 0 \\ 0 & 0 & 0 & 0 & 1 & 0 \end{pmatrix}$$

$$+ 1 \begin{pmatrix} 0 & 0 & 0 & 0 & 1 & 1 \\ 0 & 0 & 0 & 0 & 0 & 1 \\ 0 & 0 & 0 & 0 & 0 & 0 \\ 0 & 0 & 0 & 0 & 0 & 0 \end{pmatrix} + 1 \begin{pmatrix} 0 & 0 & 0 & 0 & 0 & 1 \\ 0 & 0 & 0 & 0 & 0 & 1 \\ 0 & 0 & 0 & 0 & 0 & 0 \\ 0 & 0 & 0 & 0 & 0 & 0 \end{pmatrix}.$$

2.5 The Algorithm of Engel

Engel proposes an algorithm which is optimal with respect to the TNMU and almost optimal with respect to the NS. The theoretical result underlying that algorithm is

Theorem 2 ([7]). *The minimal TNMU of a segmentation of* A *equals*

$$c(A) := \max_{1 \le i \le m} c_i(A), \; where \tag{9}$$

$$c_i(A) := \sum_{j=1}^{n} \max\{0, a_{i,j} - a_{i,j-1}\}. \tag{10}$$

(Recall that $a_{i,0} = a_{i,n+1} = 0$ *for all* i.)

Using this terminology the reason for the optimality of the algorithm of Bortfeld *et al.* can be summarized as follows: if A' is the residual matrix after the first step, then by construction

$$c_i(A') = c_i(A) - 1, \tag{11}$$

for all i with $c_i(A) > 0$, in particular $c(A') = c(A) - 1$, and thus after $c(A)$ steps A is reduced to the zero matrix. The drawback of this method is that *a priori* all the segments have coefficient 1, and thus the NS is rather large. Obviously, if the algorithm yields the same segment S in u different steps, these can be combined to obtain one segment with coefficient u. In view of (11) this amounts to the search for a pair (u, S) of a positive integer u and a segment S such that $A - uS$ is still nonnegative and

$$c_i(A - uS) = c_i(A) - u,$$

for all i with $c_i(A) > 0$. But this condition is unnecessary strong: we only need

$$c(A - uS) = c(A) - u, \tag{12}$$

i.e.

$$c_i(A - uS) \leq c(A) - u \tag{13}$$

for all i. For the choice of the coefficient u it is a suggestive strategy to take the maximal u for which there exists a segment S such that $A - uS$ is nonnegative and (12) is true. Let u_{max} be this maximal possible value for u. According to [7], u_{max} can be determined as follows. We put

$$d_{i,j} = a_{i,j} - a_{i,j-1} \qquad (1 \leq i \leq m, \ 1 \leq j \leq n+1)$$

and consider some segment S, given by $l_1, \ldots, l_m$ and $r_1, \ldots, r_m$. One can prove (see [7]) that it is no restriction to assume that, for all i, either $l_i = r_i + 1$ or $(d_{i,l_i} > 0$ and $d_{i,r_i+1} < 0)$, and that under these assumptions $c_i(A - uS) \leq c(A) - u$ is equivalent to $u \leq v_i(l_i, r_i)$, where

$$v_i(l, r) = \begin{cases} g_i(A) & \text{if } l = r+1, \\ g_i(A) + \min\{d_{i,l}, -d_{i,r+1}\} & \text{if } l \leq r \text{ and } g_i(A) \leq |d_{i,l} + d_{i,r+1}|, \\ (d_{i,l} - d_{i,r+1} + g_i(A))/2 & \text{if } l \leq r \text{ and } g_i(A) > |d_{i,l} + d_{i,r+1}|, \end{cases}$$

with $g_i(A) := c(A) - c_i(A)$. For convenience we denote the set of pairs (l, r) to which we restrict our search in row i by $\mathcal{I}_i$, that is we put

$$\mathcal{I}_i := \{(l, r) \ : \ 1 \leq l \leq r + 1 \leq n + 1$$
$$\text{and either } l = r + 1 \text{ or } (d_{i,l} > 0 \text{ and } d_{i,r+1} < 0)\}.$$

Clearly the nonnegativity of $A - uS$ is equivalent to $u \leq w_i(l_i, r_i)$ for all i, where

$$w_i(l, r) = \begin{cases} \infty & \text{if } l = r+1, \\ \min_{l \leq j \leq r} a_{i,j} & \text{if } l \leq r. \end{cases}$$

Now we put, for $1 \leq i \leq m$ and $(l, r) \in \mathcal{I}_i$,

$$\hat{u}_i(l, r) = \min\{v_i(l, r), w_i(l, r)\},$$

and for $1 \leq i \leq m$,

$$\tilde{u}_i = \max_{(l,r) \in \mathcal{I}_i} \hat{u}_i(l, r). \tag{14}$$

Then

$$u_{max} = \min_{1 \leq i \leq m} \tilde{u}_i. \tag{15}$$

In order to construct a segment S such that, for $u = u_{max}$, $A - uS$ is nonnegative and (12) is true, we just have to find, for every $i \in [m]$, a pair $(l_i, r_i) \in \mathcal{I}_i$ with

$$\hat{u}_i(l_i, r_i) \geq u_{max}.$$

A trivial way of doing this is to take a pair (l_i, r_i) where the maximum in (14) is attained, i.e. with $\hat{u}_i(l_i, r_i) = \tilde{u}_i$. These (l_i, r_i) can be computed simultaneously with the calculation of u_{max} and this method yields $mn + n - 1$ as an upper bound for the NS of the segmentation (see [7]). But there are better constructions for S after the determination of u_{max}. We describe a construction of S which, on randomly generated test matrices, yields slightly better results than the one given in [7]. We put

$$q(A) = |\{(i, j) \in [m] \times [n] \ : \ d_{i,j} \neq 0\}|, \tag{16}$$

and choose a segment S so that $q(A - uS)$ is minimized. To make this precise, for $1 \leq i \leq m$ and $(l, r) \in \mathcal{I}_i$, we put

$$p_i(l, r) = \begin{cases} 2 \text{ if } d_{i,l} = -d_{i,r+1} = u_{max}, \\ 1 \text{ if } d_{i,l} = u_{max} \neq -d_{i,r+1} \text{ or } d_{i,l} \neq u_{max} = -d_{i,r+1}, \\ 0 \text{ if } l = r + 1 \text{ or } (d_{i,l} \neq u_{max} \text{ and } -d_{i,r+1} \neq u_{max}). \end{cases}$$

Now for (l_i, r_i) we choose among the pairs $(l, r) \in \mathcal{I}_i$ with $\hat{u}_i(l, r) \geq u_{max}$ one with maximal value of $p_i(l, r)$, and if there are several of these we choose one with maximal value of $r - l$.

Example 4. *For the benchmark matrix the algorithm yields*

$$\begin{pmatrix} 4 & 5 & 0 & 1 & 4 & 5 \\ 2 & 4 & 1 & 3 & 1 & 4 \\ 2 & 3 & 2 & 1 & 2 & 4 \\ 5 & 3 & 3 & 2 & 5 & 3 \end{pmatrix} = 4 \begin{pmatrix} 1 & 1 & 0 & 0 & 0 & 0 \\ 0 & 0 & 0 & 0 & 0 & 1 \\ 0 & 0 & 0 & 0 & 0 & 1 \\ 1 & 0 & 0 & 0 & 0 & 0 \end{pmatrix} + 2 \begin{pmatrix} 0 & 0 & 0 & 0 & 1 & 1 \\ 0 & 0 & 0 & 1 & 0 & 0 \\ 1 & 1 & 1 & 0 & 0 & 0 \\ 0 & 1 & 1 & 1 & 1 & 0 \end{pmatrix} + 1 \begin{pmatrix} 0 & 0 & 0 & 1 & 1 & 1 \\ 1 & 1 & 1 & 1 & 0 & 0 \\ 0 & 0 & 0 & 1 & 1 & 0 \\ 1 & 1 & 1 & 0 & 0 & 0 \end{pmatrix}$$

$$+ 1 \begin{pmatrix} 0 & 0 & 0 & 0 & 1 & 1 \\ 1 & 1 & 0 & 0 & 0 & 0 \\ 0 & 1 & 0 & 0 & 0 & 0 \\ 0 & 0 & 0 & 0 & 1 & 1 \end{pmatrix} + 1 \begin{pmatrix} 0 & 1 & 0 & 0 & 0 & 0 \\ 0 & 1 & 0 & 0 & 0 & 0 \\ 0 & 0 & 0 & 0 & 1 & 0 \\ 0 & 0 & 0 & 0 & 1 & 1 \end{pmatrix} + 1 \begin{pmatrix} 0 & 0 & 0 & 0 & 0 & 1 \\ 0 & 1 & 0 & 0 & 0 & 0 \\ 0 & 0 & 0 & 0 & 0 & 0 \\ 0 & 0 & 0 & 0 & 1 & 1 \end{pmatrix}.$$

2.6   The Algorithm of Kalinowski

In [11] the approach of [7] is generalized to include the ICC. For this purpose we reformulate Theorem 2 as follows: let $\overrightarrow{G_0} = (V, E_0)$ be a digraph with

$$V = [m] \times [n+1] \cup \{s, t\},$$
$$E_0 = E_1 \cup E_2 \quad \text{where}$$
$$E_1 = \{(s, (i,1)) \; : \; i \in [m]\} \cup \{((i, n+1), t) \; : \; i \in [m]\},$$
$$E_2 = \{((i,j), (i, j+1)) \; : \; i \in [m], j \in [n-1]\},$$

and define a weight function δ on $\overrightarrow{G_0}$ (depending on A) by

$$\begin{aligned}
\delta(s, (i,1)) &= a_{i,1} & i \in [m], \\
\delta((i, n+1), t) &= 0 & i \in [m], \\
\delta((i,j), (i, j+1)) &= \max\{0, d_{i,j+1}\} & i \in [m], j \in [n].
\end{aligned}$$

An equivalent formulation of Theorem 2 is

Theorem 2' 1. *The minimal TNMU of a segmentation of A (without ICC) equals the maximal weight of an (s,t)–path in $\overrightarrow{G_0}$ with respect to A.*

The vertices $(i, n+1)$ $(i \in [m])$ are not necessary here, since the arcs $((i,n), (i, n+1))$ have weight 0 anyway. But we have added them to avoid case distinctions below. In order to model the ICC in the graph we have to add some additional arcs. We define the digraph $\overrightarrow{G} = (V, E)$ with $E = E_0 \cup E_3 \cup E_4$, where

$$\begin{aligned}
E_3 &= \{((i,j), (i+1, j)) \; : \; 1 \le i \le m-1, \; 1 \le j \le n-1\}, \\
E_4 &= \{((i,j), (i-1, j)) \; : \; 2 \le i \le m, \; 1 \le j \le n-1\},
\end{aligned}$$

and we extend δ to E by

$$\begin{aligned}
\delta((i,j), (i+1, j)) &= -a_{i,j} & 1 \le i \le m-1, \; 1 \le j \le n-1, \\
\delta((i,j), (i-1, j)) &= -a_{i,j} & 2 \le i \le m, \; 1 \le j \le n-1.
\end{aligned}$$

In Figure 7 the construction is illustrated for the matrix

$$A = \begin{pmatrix} 4 & 5 & 0 & 1 & 4 & 5 \\ 2 & 4 & 1 & 3 & 1 & 4 \\ 2 & 3 & 2 & 1 & 2 & 4 \\ 5 & 3 & 3 & 2 & 5 & 3 \end{pmatrix}.$$

The main result of [11] is

Theorem 3. *The minimal TNMU of a segmentation of A with ICC equals the maximal weight of an (s,t)–path in $\overrightarrow{G}$ with respect to A.*

We denote this maximal weight by $c(A)$:

$$c(A) = \max\{\delta(P) \; : \; P \text{ is an } (s,t) - \text{path in } \overrightarrow{G}\}.$$

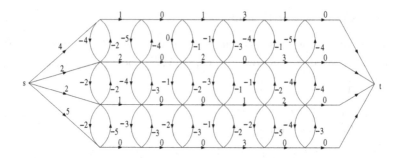

Fig. 7. The weighted digraph corresponding to the benchmark matrix

The proof of the theorem consists of two parts. First with an (s,t)–path P in $\overrightarrow{G}$ we associate a function $g_P : [m] \times [n] \to \{0, 1, -1\}$ such that g is dually feasible for the TNMU–minimization, and for some (s,t)–path P with $\delta(P) = c(A)$ we have

$$\sum_{(i,j)\in[m]\times[n]} a_{i,j}g(i,j) = \delta(P).$$

From this by duality we conclude that the TNMU of a segmentation is greater or equal to $c(A)$. The function g_P that does the job is

$$g_P(i,j) = \begin{cases} 1 \text{ if } \{(i,j),(i,j+1),(i,j+2)\} \subset P, \ d_{i,j} \geq 0, \ d_{i,j+1} < 0, \\ 1 \text{ if } \{(i,j),(i,j+1)\} \subset P, \ (i,j+2) \notin P, \ d_{i,j} \geq 0, \\ -1 \text{ if } \{(i,j),(i,j+1),(i,j+2)\} \subset P, \ d_{i,j} < 0, \ d_{i,j+1} \geq 0, \\ -1 \text{ if } (i,j) \in P, \ (i,j+1) \notin P, \\ -1 \text{ if } \{(i-1,j),(i,j),(i+1,j)\} \subset P, \\ -1 \text{ if } (i,j) \notin P, \ (i,j+1) \in P, \ d_{i,j+1} \geq 0, \\ 0 \text{ otherwise.} \end{cases}$$

The second part of the proof is the construction of a segmentation of A with TNMU $c(A)$. For this we put $A_0 = A$, and in the i-th step we construct a segment $S = S_i$ such that $c(A_{i-1} - S) = c(A_{i-1}) - 1$, and put $A_i = A_{i-1} - S$. So for $k = c(A)$ after k steps we obtain $c(A_k) = 0$ and this implies $A_k = 0$. By construction

$$A = \sum_{i=1}^{k} S_i$$

is the required segmentation. In order to describe the construction of S for a fixed A we put

$$\alpha_1(i,j) = \max\{\delta(P) \ : \ P \text{ is an } (s,(i,j)) - \text{path in } \overrightarrow{G}\},$$
$$\alpha_2(i,j) = \max\{\delta(P) \ : \ P \text{ is an } ((i,j),t) - \text{path in } \overrightarrow{G}\},$$
$$\alpha(i,j) = \alpha_1(i,j) + \alpha_2(i,j),$$

and define two subsets of $[m] \times [n]$,

$$V_1 = \{(i,j) \in [m] \times [n] : d_{i,j} \geq 0, \ d_{i,j+1} < 0\},$$
$$V_2 = \{(i,j) \in V_1 : \alpha(i,j) = c(A), \ \alpha_1(i,j) = a_{i,j}\}.$$

Now the segment S (described by the l_i, r_i ($i \in [m]$)) can be constructed according to Algorithm 2. The proofs that the g_P have the claimed properties, and that the algorithm yields the required results are quite technical and we omit them here (see [11] for the details).

In order to reduce the NS one can proceed analogous to the algorithm of Engel. In [12] is described a backtracking algorithm, that determines a pair (u, S) of an integer u and a segment S such that $A - uS$ is nonnegative, $c(A - uS) = c(A) - u$ and u is maximal under the condition that a segment with these properties exists.

Example 5. *For the benchmark matrix the algorithm from [12] yields*

$$\begin{pmatrix} 4 & 5 & 0 & 1 & 4 & 5 \\ 2 & 4 & 1 & 3 & 1 & 4 \\ 2 & 3 & 2 & 1 & 2 & 4 \\ 5 & 3 & 3 & 2 & 5 & 3 \end{pmatrix} = 3 \begin{pmatrix} 0 & 0 & 0 & 0 & 1 & 1 \\ 0 & 0 & 0 & 0 & 0 & 1 \\ 0 & 0 & 0 & 0 & 0 & 1 \\ 0 & 0 & 0 & 0 & 1 & 1 \end{pmatrix} + 3 \begin{pmatrix} 1 & 1 & 0 & 0 & 0 & 0 \\ 0 & 1 & 0 & 0 & 0 & 0 \\ 0 & 0 & 0 & 0 & 0 & 0 \\ 1 & 0 & 0 & 0 & 0 & 0 \end{pmatrix} + 1 \begin{pmatrix} 0 & 0 & 0 & 0 & 0 & 1 \\ 0 & 0 & 0 & 1 & 1 & 1 \\ 0 & 0 & 0 & 0 & 1 & 0 \\ 0 & 1 & 1 & 1 & 1 & 0 \end{pmatrix}$$

$$+ 1 \begin{pmatrix} 0 & 0 & 0 & 1 & 1 & 1 \\ 0 & 0 & 0 & 1 & 0 & 0 \\ 0 & 1 & 1 & 0 & 0 & 0 \\ 1 & 1 & 1 & 0 & 0 & 0 \end{pmatrix} + 1 \begin{pmatrix} 0 & 1 & 0 & 0 & 0 & 0 \\ 1 & 0 & 0 & 0 & 0 & 0 \\ 1 & 1 & 0 & 0 & 0 & 0 \\ 1 & 1 & 1 & 1 & 1 & 0 \end{pmatrix} + 1 \begin{pmatrix} 1 & 1 & 0 & 0 & 0 & 0 \\ 1 & 1 & 1 & 1 & 0 & 0 \\ 1 & 1 & 1 & 1 & 1 & 1 \\ 0 & 0 & 0 & 0 & 0 & 0 \end{pmatrix}.$$

Algorithm 2. Segment $S(A, V_2)$

> **for** $(i,j) \in V_2$ **do**
> $\quad l_i := \max\{j' \leq j : a_{i,j'} = 0\} + 1$
> $\quad r_i := j$
> **for** $i = 1$ to $i_1 - 1$ **do**
> 5: $\quad l_i := l_{i_1}; r_i := l_i - 1$
> **for** $i = i_t + 1$ to m **do**
> $\quad l_i := l_{i_t}; r_i := l_i - 1$
> **for** $k = 1$ to $t - 1$ **do**
> $\quad$ **if** $j_k > j_{k+1}$ **then**
> 10: $\quad i := i_k$
> $\quad\quad$ **while** $i < i_{k+1}$ and $l_i > r_{i_{k+1}} + 1$ **do**
> $\quad\quad\quad i := i + 1$
> $\quad\quad\quad r_i := l_{i-1} - 1$
> $\quad\quad\quad l_i := \max\{j \leq r_i : a_{ij} = 0\} + 1$
> 15: $\quad\quad$ **for** $i' = i + 1$ to $i_{k+1} - 1$ **do**
> $\quad\quad\quad r_{i'} := r_{i_{k+1}}; l_{i'} := r_{i'} + 1$
> $\quad$ **else**
> $\quad\quad i := i_{k+1}$
> $\quad\quad$ **while** $i > i_k$ and $l_i > r_{i_k} + 1$ **do**
> 20: $\quad\quad\quad i := i - 1$
> $\quad\quad\quad r_i := l_{i+1} - 1$
> $\quad\quad\quad l_i := \max\{j \leq r_i : a_{ij} = 0\} + 1$
> $\quad\quad$ **for** $i' = i_k + 1$ to $i - 1$ **do**
> $\quad\quad\quad r_{i'} := r_{i_k}; l_{i'} := r_{i'} + 1$

2.7   The Algorithm of Xia and Verhey

In [24] another heuristic method for the construction of a segmentation with small NS is proposed. Here again the general principle is to determine a coefficient u and a segment S and to continue with $A - uS$. The coefficient is chosen to be a power of 2 which is close to half of the maximal entry of A, precisely

$$u = 2^{\lceil \log L \rceil - 1},$$

where the base of the logarithm is 2. The next step towards the algorithm is the observation that in the two–column case every $(0, 1)$–matrix is a segment. So for a two–column matrix A the segment corresponding to the coefficient u may be defined by

$$s_{i,j} = \begin{cases} 1 \text{ if } a_{i,j} \geq u, \\ 0 \text{ otherwise.} \end{cases}$$

In the whole segmentation process every power of 2 between 1 and $2^{\lceil \log L \rceil - 1}$ appears at most once as a coefficient, and thus the NS is at most $\lceil \log L \rceil$. The straightforward generalization of this method to an n–column matrix A is to divide A into two–column submatrices, and apply the algorithm to these submatrices. (If n is odd one has to add a dummy $(n + 1)$–th column with all entries equal to 0.) This yields $\lceil \frac{n}{2} \rceil \lceil \log L \rceil$ as an upper bound for the NS. Actually, this bound can be replaced by

$$\sum_{k=1}^{\lceil \frac{n}{2} \rceil} \lceil \log L_k \rceil,$$

where L_k is the maximal entry of the submatrix which consists of the columns $2k - 1$ and $2k$. Obviously, it is not very efficient to treat the two–column submatrices independently, because it may be possible to combine some segments for different two–column submatrices to obtain a single segment for the whole matrix. The authors of [24] propose two ways of doing this. The sliding window technique determines the coefficient always according to the leftmost nonzero two–column submatrix, say columns j and $j + 1$. Then the leaves are set to obtain the largest possible extension of a leaf setting for columns j and $j + 1$. The reducing level technique determines the coefficient according to the maximal entry of the whole matrix A and sets the leaves such that the irradiated area, i.e. the number of 1's in the segment S, is maximal.

Example 6. *The segmentation of the benchmark matrix using the sliding window technique is*

$$\begin{pmatrix} 4 & 5 & 0 & 1 & 4 & 5 \\ 2 & 4 & 1 & 3 & 1 & 4 \\ 2 & 3 & 2 & 1 & 2 & 4 \\ 5 & 3 & 3 & 2 & 5 & 3 \end{pmatrix} = 4 \begin{pmatrix} 1 & 1 & 0 & 0 & 0 & 0 \\ 0 & 1 & 0 & 0 & 0 & 0 \\ 0 & 0 & 0 & 0 & 0 & 1 \\ 1 & 0 & 0 & 0 & 0 & 0 \end{pmatrix} + 2 \begin{pmatrix} 0 & 0 & 0 & 0 & 0 & 0 \\ 1 & 0 & 0 & 0 & 0 & 0 \\ 1 & 1 & 1 & 0 & 0 & 0 \\ 0 & 1 & 1 & 1 & 1 & 1 \end{pmatrix} + 1 \begin{pmatrix} 0 & 1 & 0 & 0 & 0 & 0 \\ 0 & 0 & 1 & 1 & 1 & 1 \\ 0 & 1 & 0 & 0 & 0 & 0 \\ 1 & 1 & 1 & 0 & 0 & 0 \end{pmatrix}$$

$$+ 2 \begin{pmatrix} 0 & 0 & 0 & 0 & 1 & 1 \\ 0 & 0 & 0 & 1 & 0 & 0 \\ 0 & 0 & 0 & 0 & 1 & 0 \\ 0 & 0 & 0 & 0 & 1 & 0 \end{pmatrix} + 1 \begin{pmatrix} 0 & 0 & 0 & 1 & 1 & 1 \\ 0 & 0 & 0 & 0 & 0 & 1 \\ 0 & 0 & 0 & 1 & 0 & 0 \\ 0 & 0 & 0 & 0 & 1 & 1 \end{pmatrix} + 1 \begin{pmatrix} 0 & 0 & 0 & 0 & 1 & 1 \\ 0 & 0 & 0 & 0 & 0 & 1 \\ 0 & 0 & 0 & 0 & 0 & 0 \\ 0 & 0 & 0 & 0 & 0 & 0 \end{pmatrix} + 1 \begin{pmatrix} 0 & 0 & 0 & 0 & 0 & 1 \\ 0 & 0 & 0 & 0 & 0 & 1 \\ 0 & 0 & 0 & 0 & 0 & 0 \\ 0 & 0 & 0 & 0 & 0 & 0 \end{pmatrix},$$

and with the reducing level technique we obtain

$$
\begin{pmatrix} 4 & 5 & 0 & 1 & 4 & 5 \\ 2 & 4 & 1 & 3 & 1 & 4 \\ 2 & 3 & 2 & 1 & 2 & 4 \\ 5 & 3 & 3 & 2 & 5 & 3 \end{pmatrix} = 4 \begin{pmatrix} 1 & 1 & 0 & 0 & 0 & 0 \\ 0 & 1 & 0 & 0 & 0 & 0 \\ 0 & 0 & 0 & 0 & 0 & 1 \\ 1 & 0 & 0 & 0 & 0 & 0 \end{pmatrix} + 4 \begin{pmatrix} 0 & 0 & 0 & 0 & 1 & 1 \\ 0 & 0 & 0 & 0 & 0 & 1 \\ 0 & 0 & 0 & 0 & 0 & 0 \\ 0 & 0 & 0 & 0 & 1 & 0 \end{pmatrix} + 2 \begin{pmatrix} 0 & 0 & 0 & 0 & 0 & 0 \\ 0 & 0 & 0 & 0 & 0 & 0 \\ 1 & 1 & 1 & 0 & 0 & 0 \\ 0 & 1 & 1 & 1 & 0 & 0 \end{pmatrix}
$$

$$
+ 2 \begin{pmatrix} 0 & 0 & 0 & 0 & 0 & 0 \\ 1 & 0 & 0 & 0 & 0 & 0 \\ 0 & 0 & 0 & 0 & 1 & 0 \\ 0 & 0 & 0 & 0 & 0 & 1 \end{pmatrix} + 1 \begin{pmatrix} 0 & 1 & 0 & 0 & 0 & 0 \\ 0 & 0 & 1 & 1 & 1 & 0 \\ 0 & 1 & 0 & 0 & 0 & 0 \\ 1 & 1 & 1 & 0 & 0 & 0 \end{pmatrix} + 1 \begin{pmatrix} 0 & 0 & 0 & 1 & 0 & 0 \\ 0 & 0 & 0 & 0 & 0 & 0 \\ 0 & 0 & 0 & 1 & 0 & 0 \\ 0 & 0 & 0 & 0 & 1 & 1 \end{pmatrix} + 1 \begin{pmatrix} 0 & 0 & 0 & 0 & 0 & 1 \\ 0 & 0 & 0 & 0 & 0 & 0 \\ 0 & 0 & 0 & 0 & 0 & 0 \\ 0 & 0 & 0 & 0 & 0 & 0 \end{pmatrix}.
$$

In [17] four variations of the Xia–Verhey–algorithm are compared to the algorithm of Galvin, Chen and Smith and the algorithm of Bortfeld *et al.* The three alternative versions of the Xia–Verhey–algorithm differ in the choice of the coefficient u. In the first one it is $u = \lceil \frac{L}{2} \rceil$, in the second one it is the nearest integer to the average of the nonzero entries of A, and in the third one it is the median of the nonzero entries of A. The essential result of the comparison is that none of these variants is most efficient in all cases (neither for random test matrices nor for clinical examples), but the original version of Xia and Verhey yields on average the smallest NS. The NS can be reduced by a factor of 2 compared to Bortfeld's algorithm at the cost of an increase of the TNMU by about 50%.

2.8   The Algorithm of Siochi

In [18] a segmentation algorithm is described which is the basis of the Siemens IMFAST algorithm, as implemented in the commercial IMRT planning system CORVUS. This algorithm minimizes a more realistic measure for the total treatment time which takes into account both the irradiation time and the leaf travel time. For the segmentation $A = \sum_{t=1}^{k} u_t S_t$ we put

$$
\tau = \sum_{t=1}^{k} \frac{u_t}{D} + \sum_{t=2}^{k} \max\{T_{VR}, \delta_t\}, \tag{17}
$$

where D is the dose rate (in MU/min), T_{VR} is the verification and record (V&R) overhead and

$$
\delta_t = \max_{1 \le i \le m} \max \left\{ \frac{|l_i^{(t)} - l_i^{(t-1)}|}{v}, \frac{|r_i^{(t)} - r_i^{(t-1)}|}{v} \right\}
$$

is the leaf travel time between segments $t - 1$ and t. Here v is the leaf speed, and $l_i^{(t-1)}, r_i^{(t-1)}$ ($i \in [m]$) and $l_i^{(t)}, r_i^{(t)}$ ($i \in [m]$) are the parameters of the segments $t - 1$ and t, respectively. The second sum starts at $t = 2$ since it is assumed that the leaves are set to the first position before the treatment starts. The motivation for taking the maximum in the second sum in (17) instead of the sum of the two values is that we can already start the V&R–cycle in rows where the leaves have already stopped while in others they are still moving, and according to [18] the V&R of the last leaf pair is negligible compared to that of all the others combined. Now the proposed algorithm is a combination of two

parts called *extraction* and *rod pushing*. The extraction part is closely related
to the algorithm of Galvin, Chen and Smith, but formulated in a way that
allows to include ICC and tongue and groove constraints. The rod pushing part
is essentially a reformulation of the algorithm of Bortfeld *et al.* in a geometric
setting, but also adjustable to additional constraints. First we describe the basic
algorithm without additional constraints, and after that we show how the two
parts have to be modified to include the constraints.

The basic algorithm

Rod pushing: The matrix A is visualized as a rectangular $m \times n$–array of
rods, where the rod at position (i,j) consists of $a_{i,j}$ cubes. In the beginning
all the rods stand on a plane π. Fig. 8 illustrates this for the matrix

$$\begin{pmatrix} 1 & 4 & 2 \\ 3 & 2 & 1 \\ 1 & 2 & 2 \end{pmatrix}.$$

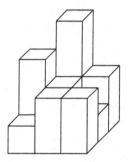

Fig. 8. Visualization of an intensity matrix as an array of rods

Now we push some of the rods up in order to achieve a situation where, for
all $h > 0$, the positions of the cubes at height h (above π) can be used to
describe a segment. The position of any rod (i,j) is uniquely determined by
the height of its lowest cube (the base of the rod), which we call $b(i,j)$. That
is, the rod (i,j) occupies the cubes

$$(i,j,b(i,j)), (i,j,b(i,j)+1), \ldots, (i,j,t(i,j)),$$

where $t(i,j) := b(i,j) + a_{i,j} - 1$ is the height of the highest cube (the top)
of the rod. Now the rod pushing procedure can be described as follows:

for $i = 1$ to m **do**
 $b(i,1) := 1, t(i,1) = a_{i,1}$
 for $j = 2$ to n **do**
  if $a_{i,j} > a_{i,j-1}$ **then**
   $b(i,j) := b(i,j-1); t(i,j) = b(i,j) + a_{i,j} - 1$
  else
   $t(i,j) := t(i,j-1); b(i,j) = t(i,j) - a_{i,j} + 1$

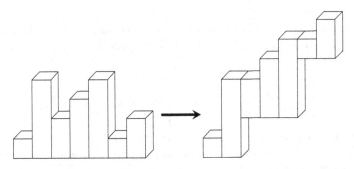

Fig. 9. The rod pushing process for one row

Fig. 9 illustrates the rod pushing process corresponding to the segmentation

$$\begin{pmatrix} 1\,4\,2\,3\,4\,1\,2 \end{pmatrix} = \begin{pmatrix} 1\,1\,0\,0\,0\,0\,0 \end{pmatrix} + \begin{pmatrix} 0\,1\,0\,0\,0\,0\,0 \end{pmatrix} + \begin{pmatrix} 0\,1\,1\,1\,1\,0\,0 \end{pmatrix}$$
$$+ \begin{pmatrix} 0\,1\,1\,1\,1\,0\,0 \end{pmatrix} + \begin{pmatrix} 0\,0\,0\,1\,1\,0\,0 \end{pmatrix} + \begin{pmatrix} 0\,0\,0\,0\,1\,1\,1 \end{pmatrix} + \begin{pmatrix} 0\,0\,0\,0\,0\,0\,1 \end{pmatrix}$$

By construction, for every h, the cubes at height h describe a segment, the sum of these segments is A and the maximal height of a cube in row i is $\sum_{j=1}^{n} \max\{0, a_{i,j} - a_{i,j-1}\}$. So by Theorem 2 the result is optimal with respect to the TNMU, and one can check that the same segmentation is obtained by the algorithm of Bortfeld $et\ al.$ using the sweep technique.

Extraction: This step consists of the determination of a sequence of coefficients $u_1, u_2, \ldots, u_{k_0}$ and corresponding segments $S_1, \ldots, S_{k_0}$ such that the residual matrix

$$A' = A - \sum_{i=1}^{k} u_i S_i$$

is nonnegative. The optimization algorithm does an exhaustive search on a certain set of pairs $(k_0, \mathbf{u})$ where k_0 is a positive integer and $\mathbf{u}$ is a k_0–tuple $\mathbf{u} = (u_1, \ldots, u_{k_0})$ of positive integers (to be defined below). For each of these pairs, a sequence of segments $(S_1, \ldots, S_{k_0})$ is determined as follows.

$A_0 := A$
for $i = 1$ to k_0 do
   determine a segment S_i with respect to the matrix A_{i-1} and the coefficient u_i as in the algorithm of Galvin $et\ al.$
   $A_i := A_{i-1} - u_i S_i$

Then the rod pushing procedure is applied to A_{k_0} and the pairs $(k_0, \mathbf{u})$ are evaluated according to the total treatment time τ for the segmentation that results from the combination of the two parts. Finally, the $(u_1, \ldots, u_{k_0})$ yielding the smallest τ is chosen for the segmentation together with the corresponding $(S_1, \ldots, S_{k_0})$ and the subsequent rod pushing segments. The search space is restricted by the following conditions, which have been found to be strong enough to make the search computationally feasible but weak enough to give good solutions ([18]).

1. $u_1 \geq u_2 \geq \ldots \geq u_{k_0}$.
2. $u_1 \leq \max\{\lfloor L/2 \rfloor, \hat{u}\}$, where $\hat{u}$ is the extract value yielding the best result if we extract only one segment before applying the rod pushing, i.e. if $k_0 = 1$.
3.

$$\sum_{i=1}^{k_0} u_i \leq \max_{1 \leq i \leq m} \sum_{j=1}^{n} \max\{0, a_{i,j} - a_{i,j-1}\}.$$

Example 7. *To illustrate the algorithm we assume that the size of the cells in our benchmark matrix is* $1\,cm \times 1\,cm$, *the leaf speed* $v = 1\,cm/sec$, *the verification and record overhead* $T_{VR} = 2\,sec$ *and the dose rate* $D = 60\,MU/min$. *The best solution if only one segment is extracted is obtained with* $\hat{u} = 3$ *and so the extraction sequences* $(u_1, \ldots, u_{k_0})$ *with* $3 \geq u_1 \geq u_2 \geq \ldots \geq u_{k_0} > 0$ *and* $\sum_{i=1}^{k_0} u_i \leq 10$ *have to be checked. The result is that the segmentation with only one extraction is already the optimal one, namely*

$$\begin{pmatrix} 4 & 5 & 0 & 1 & 4 & 5 \\ 2 & 4 & 1 & 3 & 1 & 4 \\ 2 & 3 & 2 & 1 & 2 & 4 \\ 5 & 3 & 3 & 2 & 5 & 3 \end{pmatrix} = 3 \begin{pmatrix} 1 & 1 & 0 & 0 & 0 & 0 \\ 0 & 1 & 0 & 0 & 0 & 0 \\ 0 & 1 & 0 & 0 & 0 & 0 \\ 1 & 1 & 1 & 0 & 0 & 0 \end{pmatrix} + \begin{pmatrix} 1 & 1 & 0 & 0 & 0 & 0 \\ 1 & 0 & 0 & 0 & 0 & 0 \\ 1 & 0 & 0 & 0 & 0 & 0 \\ 1 & 0 & 0 & 0 & 0 & 0 \end{pmatrix} + \begin{pmatrix} 0 & 1 & 0 & 0 & 0 & 0 \\ 1 & 1 & 1 & 1 & 0 & 0 \\ 1 & 0 & 0 & 0 & 0 & 0 \\ 1 & 0 & 0 & 0 & 0 & 0 \end{pmatrix}$$

$$+ \begin{pmatrix} 0 & 0 & 0 & 1 & 1 & 1 \\ 0 & 0 & 0 & 1 & 0 & 0 \\ 0 & 0 & 1 & 0 & 0 & 0 \\ 0 & 0 & 0 & 1 & 1 & 0 \end{pmatrix} + \begin{pmatrix} 0 & 0 & 0 & 0 & 1 & 1 \\ 0 & 0 & 0 & 1 & 1 & 1 \\ 0 & 0 & 1 & 1 & 1 & 1 \\ 0 & 0 & 0 & 1 & 1 & 0 \end{pmatrix} + \begin{pmatrix} 0 & 0 & 0 & 0 & 1 & 1 \\ 0 & 0 & 0 & 0 & 0 & 1 \\ 0 & 0 & 0 & 0 & 1 & 1 \\ 0 & 0 & 0 & 0 & 1 & 1 \end{pmatrix} + \begin{pmatrix} 0 & 0 & 0 & 0 & 1 & 1 \\ 0 & 0 & 0 & 0 & 0 & 1 \\ 0 & 0 & 0 & 0 & 0 & 1 \\ 0 & 0 & 0 & 0 & 1 & 1 \end{pmatrix}$$

$$+ \begin{pmatrix} 0 & 0 & 0 & 0 & 0 & 1 \\ 0 & 0 & 0 & 0 & 0 & 1 \\ 0 & 0 & 0 & 0 & 0 & 1 \\ 0 & 0 & 0 & 0 & 1 & 1 \end{pmatrix}$$

where we have

$$\tau = (10 + 2 + 3 + 4 + 3 + 2 + 2 + 2)\,sec = 28\,sec.$$

Interleaf collision constraint: The ICC forbids the overlapping of opposite leaves in adjacent rows, that is we must have

$$l_i \leq r_{i+1} + 1 \text{ and } r_i \geq l_{i+1} - 1 \quad (1 \leq i \leq m-1).$$

Extraction: In the extraction step with coefficient u we have to find, in each row i, an interval of entries greater or equal to u. If we fix two adjacent columns j and $j+1$ and require that every nonempty of the intervals intersects at least one of these columns, then the ICC is automatically satisfied, since then $l_i \leq j+1$ and $r_i \geq j$ for all i with $l_i \leq r_i$, and for the zero–rows of the segment it is obvious how to choose (l_i, r_i) with $l_i = r_i + 1$ in order to satisfy the ICC. Now we can do this for all possible pairs of adjacent columns j, $j+1$ and finally choose the segment with the largest irradiated area.

Rod pushing: A violation of the ICC can only occur, if for some (i,j) we have $b(i,j) > t(i+1,j) + 1$ or $t(i,j) < b(i+1,j) - 1$ (see Fig. 10, where the segment corresponding to height 3 is $\left(\begin{smallmatrix} 1 & 0 & 0 \\ 0 & 0 & 1 \end{smallmatrix}\right)$ and thus violates the ICC).

If this is the case we may push up the rod $(i+1,j)$ (resp. (i,j)) (see Algorithm 3 for the details).

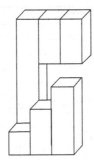

Fig. 10. The segment corresponding to height 3 violates the ICC

Tongue and groove effect: We have to make sure that it does not occur that in one segment the bixel (i, j) is exposed and the bixel $(i \pm 1, j)$ is covered, while in some later step (i, j) is covered and $(i \pm 1, j)$ is exposed.

Extraction: A sufficient condition to avoid the tongue and groove effect between different extract matrices is that, for every extract matrix $S^{(t)} = \left(s_{i,j}^{(t)} \right)$ with coefficient u_t, we have

$$a_{i,j} - u_t \geq a_{i+1,j} \quad \text{if } s_{i,j}^{(t)} = 1 \text{ and } s_{i+1,j}^{(t)} = 0 \tag{18}$$

$$a_{i,j} - u_t \geq a_{i-1,j} \quad \text{if } s_{i,j}^{(t)} = 1 \text{ and } s_{i-1,j}^{(t)} = 0. \tag{19}$$

This implies that if in some later step t' the bixel $(i \pm 1, j)$ is exposed (i.e. $s_{i\pm1,j}^{(t')} = 1$) then it is also possible to expose bixel (i, j) and so the tongue and groove underdosage is avoided. In order to achieve the validity of (18) and (19) one proceeds as follows

construct a segment $S = (s_{i,j})$ as above
repeat
  for (i, j) with $s_{i,j} = 1$ and (18) or (19) is violated **do**
  $s_{i,j} := 0$
  change entries from 1 to 0 so that a segment satisfying the ICC results
until no entry has to be changed

Rod pushing: In the rod pushing process the tongue and groove effect can be avoided using a modification of the basic method similar to Algorithm 3. Instead of the corrections in lines 14 to 19 and 22 to 27 of this algorithm one has to use

$$\left. \begin{array}{l} t(i+1, j) := t(i, j) \\ b(i+1, j) := t(i+1, j) - a_{i+1,j} + 1 \end{array} \right\} \begin{array}{l} \text{if } a_{i,j} < a_{i+1,j} \\ \text{and } t(i, j) > t(i+1, j), \end{array}$$

$$\left. \begin{array}{l} b(i, j) := b(i+1, j) \\ t(i, j) := b(i, j) + a_{i,j} - 1 \end{array} \right\} \begin{array}{l} \text{if } a_{i,j} < a_{i+1,j} \\ \text{and } b(i, j) < b(i+1, j), \end{array}$$

$$\left. \begin{array}{l} t(i, j) := t(i+1, j) \\ b(i, j) := t(i, j) - a_{i,j} + 1 \end{array} \right\} \begin{array}{l} \text{if } a_{i,j} \geq a_{i+1,j} \\ \text{and } t(i, j) < t(i+1, j), \end{array}$$

Algorithm 3. Rod pushing with ICC

---

    for $i = 1$ to m **do**
        $b(i, 1) = 1$; $t(i, 1) = a_{i,1}$
    for $j = 2$ to n **do**
        for $i = 1$ to m **do**
5:        **if** $a_{i,j} > a_{i,j-1}$ **then**
            $b(i, j) := b(i, j - 1)$; $t(i, j) = b(i, j) + a_{i,j} - 1$
        else
            $t(i, j) := t(i, j - 1)$; $b(i, j) = t(i, j) - a_{i,j} + 1$
        Choose $i_0 \in [m]$ with $t(i_0, j) \geq t(i, j)$ for all i
10:     **for** $i = i_0 - 1$ downto 1 **do**
        if $t(i, j) < b(i + 1, j) - 1$ **then**
            $t(i, j) := b(i + 1, j) - 1$; $b(i, j) := t(i, j) - a_{i,j} + 1$
        if $b(i, j) > t(i + 1, j) + 1$ **then**
            $t(i + 1, j) := b(i, j) - 1$; $b(i + 1, j) := t(i + 1, j) - a_{i+1,j} + 1$
15:     **for** $i = i_0 + 1$ to m **do**
        if $t(i, j) < b(i - 1, j) - 1$ **then**
            $t(i, j) := b(i - 1, j) - 1$; $b(i, j) := t(i, j) - a_{i,j} + 1$
        if $b(i, j) > t(i - 1, j) + 1$ **then**
            $t(i - 1, j) := b(i, j) - 1$; $b(i - 1, j) := t(i - 1, j) - a_{i+1,j} + 1$

$$\left. \begin{array}{l} b(i + 1, j) := b(i, j) \\ t(i + 1, j) := b(i + 1, j) + a_{i+1,j} - 1 \end{array} \right\} \quad \begin{array}{l} \text{if } a_{i,j} \geq a_{i+1,j} \\ \text{and } b(i, j) > b(i + 1, j). \end{array}$$

These corrections make sure that, for two rods that are adjacent along a column, the shorter one has its base above or at the same level as the longer one and has its top below or at the same level as the longer one. Thus the resulting segments also satisfy the ICC.

2.9   The Algorithm of Kamath *et al.*

Another segmentation algorithm is described in [13]. Here the authors consider more general constraints which are motivated by the design of some MLCs:

Minimum separation constraint (MSC): The distance between the left and the right leaf in every row can not be smaller than a minimum distance $\delta_0 \geq 0$. In our terminology this means

$$r_i - l_i \geq \delta_0 - 1 \qquad (i \in [m]).$$

Leaf interdigitation constraint (LIC): The distance between opposite leaves in adjacent rows is at least δ_1 for some $\delta_1 \geq 0$, i.e.

$$r_{i+1} - l_i \geq \delta_1 - 1, \; r_i - l_{i+1} \geq \delta_1 - 1 \qquad (i \in [m - 1]).$$

For $\delta_1 = 0$ this is just the ICC.

The proposed algorithm constructs segmentations in which the leaves always move from left to right. So the segmentation can be described by

$$I_L^{(i)}(1) \le I_L^{(i)}(2) \le \dots \le I_L^{(i)}(n), \ I_R^{(i)}(1) \le I_R^{(i)}(2) \le \dots \le I_R^{(i)}(n) \qquad (i \in [m]),$$

where $I_L^{(i)}(j)$ and $I_R^{(i)}(j)$ denote the numbers of monitor units that have been delivered when the left and the right leaf, respectively, in row i passes column j. These numbers can be translated into segments as follows. Let

$$S^{(t)} = \left(s_{i,j}^{(t)} \right)$$

denote the segment corresponding to the leaf position when the t-th monitor unit is delivered. Then

$$s_{i,j}^{(t)} = \begin{cases} 1 \text{ if } I_R^{(i)} < t \le I_L^{(i)}(j) \\ 0 \text{ otherwise.} \end{cases}$$

The condition that must be satisfied in order to generate the matrix A is

$$I_L^{(i)}(j) - I_R^{(i)}(j) = a_{i,j} \qquad (i \in [m], \ j \in [n]).$$

First neglecting the leaf interdigitation constraint a segmentation is build up from segmentations of the single rows as described in Algorithm 4. Observe that

---

Algorithm 4. Basic segmentation

---

for $i = 1$ to m do
$\quad I_L^{(i)}(1) = a_{i,1}; \ I_R^{(i)} = 0$
$\quad$ for $j = 2$ to n do
$\quad\quad I_L^{(i)}(j) = I_L^{(i)}(j-1) + \max\{0, a_{i,j} - a_{i,j-1}\}$
$\quad\quad I_R^{(i)}(j) = I_R^{(i)}(j-1) + \max\{0, a_{i,j-1} - a_{i,j}\}$

---

this is another formulation of the rod pushing part of Siochi's algorithm: $I_L^{(i)}(j)$ and $I_R^{(i)}(j) + 1$ correspond to $t(i,j)$ and $b(i,j)$, respectively. The essential result on Algorithm 4 is

Theorem 4.

1. *Algorithm 4 is optimal with respect to the TNMU even when bidirectional leaf movement is permitted. (Theorem 3 in [13])*
2. *If there exists a segmentation of A satisfying the MSC then the segmentation constructed using Algorithm 4 satisfies the MSC. (Theorem 5 in [13])*

In order to construct a segmentation satisfying the LIC it is proposed to modify the $I_L^{(i)}(j)$, $I_R^{(i)}(j)$ obtained by Algorithm 4 until the LIC is satisfied. If, as a result of Algorithm 4, $I_R^{(i)}(j) > 0$ for some $i \in [m]$, $j \le \delta_1$ there is no segmentation satisfying the LIC. So w.l.o.g. we may assume $I_R^{(i)}(j) = 0$ for all $i \in [m]$, $1 \leq j \leq \delta_m$. An LIC–violation occurs iff $I_L^{(k)}(j - \delta_1) < I_R^{(i)}(j)$ for some $i \in [m]$, $j \in [n]$, $k \in \{i+1, i-1\}$. Among all violations we determine one with minimal j and eliminate it by putting

$$I_L^{(k)}(j - \delta_1) := I_R^{(i)}(j),$$

and modifying the $I_L^{(k)}(j')$ $(j' > j - \delta_1)$ appropriately (see Algorithm 5 for the details). The main result on Algorithm 5 is

Theorem 5 (Theorem 6 in [13]).

1. *Algorithm 5 terminates.*
2. *If Algorithm 5 terminates with a violation of the MSC, then there is no segmentation satisfying MSC and LIC.*
3. *Otherwise the algorithm yields a segmentation satisfying MSC and LIC and having minimal TNMU under these conditions.*

Algorithm 5. : Elimination of LIC violations

---

   while The MSC is satisfied and the LIC is violated **do**

      $j_0 := \min\{j \in [n] : \exists i \in [m] \text{ with } I_L^{(k)}(j-\delta_1) < I_R^{(i)}(j) \text{ for some } k \in \{i+1, i-1\}\}$

      choose i and $k \in \{i+1, i-1\}$ with $I_L^{(k)}(j_0 - \delta_1) < I_R^{(i)}(j_0)$

      $I_L^{(k)}(j_0 - \delta_1) := I_R^{(i)}(j_0)$

5:    $I_R^{(k)}(j_0 - \delta_1) := I_L^{(k)}(j_0 - \delta_1) - a_{k, j_0 - \delta_1}$

      for $j = j_0 - \delta_1 + 1$ to n **do**

         $I_L^{(k)}(j) := \max\left\{I_L^{(k)}(j), I_L^{(k)}(j-1) + \max\{0, a_{i,j} - a_{i,j-1}\}\right\}$

         $I_R^{(k)}(j) := I_L^{(k)}(j) - a_{k,j}$

If $\delta_1 = 0$ lines 4–9 of Algorithm 5 can be replaced by

$\Delta := I_R^{(i)}(j_0) - I_L^{(k)}(j_0)$

for $j = j_0$ to n **do**

   $I_L^{(k)}(j) := I_L^{(k)}(j) + \Delta$

   $I_R^{(k)}(j) := I_R^{(k)}(j) + \Delta$

In this case the algorithm coincides with the ICC–version of Siochi's rod pushing method, no MSC–violation can occur and thus always a TNMU–optimal segmentation is obtained.

2.10 The Algorithm of Boland, Hamacher and Lenzen

In [2] is given a network flow formulation of the TNMU–minimization which also includes the ICC. The set of segments is identified with the set of paths from D to D' in the layered digraph $G = (V, E)$, constructed as follows. The vertices in the i–th layer correspond to the possible pairs (l_i, r_i) $(1 \leq i \leq m)$, and two additional vertices D and D' are added:

$$V = \{(i, l, r) : i = 1, \ldots, m; \ l = 1, \ldots, n+1; \ r = l-1, \ldots, n+1\} \cup \{D, D'\}.$$

Between two vertices (i, l, r) and $(i+1, l', r')$ there is an arc if the corresponding leaf positions are consistent with the ICC, i.e. if $l' \le r+1$ and $r' \ge l-1$. In addition E contains all arcs from D to the first layer, all arcs from the last layer m to D' and the arc (D', D), so

$$E = E_+(D) \cup E_-(D') \cup \bigcup_{i=1}^{m-1} E(i) \cup \{(D', D)\}, \text{ where}$$

$$E_+(D) = \{(D, (1, l, r)) \; : \; (1, l, r) \in V\},$$
$$E_-(D') = \{((m, l, r), D') \; : \; (m, l, r) \in V\},$$
$$E(i) = \{((i, l, r), (i+1, l', r')) : l' \le r+1, \; r' \ge l-1\}.$$

There is a bijection between the possible leaf positions and the cycles in G. This is illustrated in Fig. 11 which shows two cycles in G for $m = 4$, $n = 2$, corresponding to the segments

$$\begin{pmatrix} 1 & 0 \\ 0 & 1 \\ 1 & 1 \\ 1 & 0 \end{pmatrix} \text{ (straight lines)} \quad \text{and} \quad \begin{pmatrix} 0 & 1 \\ 1 & 1 \\ 1 & 0 \\ 0 & 1 \end{pmatrix} \text{ (dotted lines).}$$

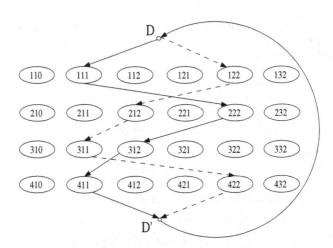

Fig. 11. The vertices of G for $m = 4$, $n = 2$ and two cycles

With a segment S, given by $(l_1, r_1), (l_2, r_2), \ldots, (l_m, r_m)$, we associate a unit flow on the cycle

$$D, (1, l_1, r_1), (2, l_2, r_2), \ldots, (m, l_m, r_m), D', D.$$

Then any positive combination of segments defines a circulation $\phi : E \to \mathcal{R}_+$ on G. For instance,

$$3 \begin{pmatrix} 1 & 0 \\ 0 & 1 \\ 1 & 1 \\ 1 & 0 \end{pmatrix} + 2 \begin{pmatrix} 0 & 1 \\ 1 & 1 \\ 1 & 0 \\ 0 & 1 \end{pmatrix} = \begin{pmatrix} 3 & 2 \\ 2 & 5 \\ 5 & 3 \\ 3 & 2 \end{pmatrix}$$

corresponds to 3 units of flow on $(D, (1,1,1), (2,2,2), (3,1,2), (4,1,1), D')$, 2 units of flow on $(D, (1,2,2), (2,1,2), (3,1,1), (4,2,2), D')$ and 5 units of flow on (D', D). The amount of radiation that is released at bixel (i, j) equals the sum of the flows going through the vertices (i, l, r) with $l \le j \le r$, hence the conditions that must be satisfied by the circulation in order to correspond to a segmentation of A are

$$\sum_{l=1}^{j} \sum_{r=j}^{n} \sum_{l'=1}^{r+1} \sum_{r'=\max\{l,l'\}-1}^{n} \phi((i,l,r),(i+1,l',r')) = a_{i,j}, \tag{20}$$

for $1 \le i \le m-1$, $1 \le j \le n$, and

$$\sum_{l=1}^{j} \sum_{r=j}^{n} \phi((m,l,r), D') = a_{m,j}, \tag{21}$$

for $1 \le j \le n$. Since all the flow must go through the arc (D', D), the TN-MU of the segmentation corresponding to ϕ equals $\phi(D', D)$. Thus the TNMU–minimization problem can be solved by finding a circulation satisfying conditions (20) and (21) and having minimal cost with respect to the cost function $\alpha : E \to \mathcal{R}_+$,

$$\alpha(e) = \begin{cases} 1 \text{ if } e = (D, D'), \\ 0 \text{ otherwise.} \end{cases}$$

The graph G can be expanded to a graph $\hat{G} = (\hat{V}, \hat{E})$ so that, instead of the constraints (20) and (21), the structure of $\hat{G}$ together with a capacity function on $\hat{E}$ forces the circulation to represent a segmentation of A.

$$\hat{V} = \{(i,l,r)^1, (i,l,r)^2 \ : \ 1 \le i \le m, \ 1 \le l \le r+1 \le n+1\}$$
$$\cup \{(i,j) \ : \ 1 \le i \le m, \ 0 \le j \le n\} \cup \{D, D'\}.$$

The arcs set of $\hat{G}$ is $\hat{E} = \hat{E}^{\text{old}} \cup \hat{E}^1 \cup \hat{E}^2$, where

$$\hat{E}^{\text{old}} = \{((i,l,r)^2, (i+1,l',r')^1) \ : \ ((i,l,r),(i+1,l',r')) \in E\}$$
$$\cup \{(D, (1,l,r)^1) \ : \ (1,l,r)^1 \in \hat{V}\}$$
$$\cup \{((m,l,r)^2, D') \ : \ (m,l,r)^2 \in \hat{V}\}$$
$$\cup \{(D', D)\},$$
$$\hat{E}^1 = \{((i,l,r)^1, (i,l-1)) \ : \ (i,l,r)^1 \in \hat{V}\}$$
$$\cup \{((i,r),(i,l,r)^2) \ : \ (i,l,r)^2 \in \hat{V}\},$$
$$\hat{E}^2 = \{((i,j-1),(i,j)) \ : \ i \in [m], j \in [n]\}.$$

Now a segment with parameters l_i, r_i $(i \in [m])$ corresponds to the cycle

$$D, (1,l_1,r_1)^1, (1,l_1-1), (1,l_1), \ldots, (1,r_1), (1,l_1,r_1)^2,$$
$$(2,l_2,r_2)^1, (2,l_2-1), (2,l_2), \ldots, (2,r_2), (2,l_2,r_2)^2,$$
$$\ldots$$
$$(m,l_m,r_m), (m,l_m-1), (m,l_m), \ldots, (m,r_m), (m,l_m,r_m)^2, D', D$$

Figure 12 shows the cycles in $\hat{G}$ corresponding to the cycles in Figure 11.

Now the flow on the arc $((i, j-1), (i, j))$ equals the amount of radiation released at bixel (i, j) in the corresponding segmentation, and introducing lower and upper capacities $\underline{u}$ and $\overline{u}$ on the arcs of $\hat{G}$ by

$$\underline{u}(e) = \begin{cases} 0 & \text{if } e \in \hat{E}^{\text{old}} \cup \hat{E}^1 \\ a_{i,j} & \text{if } e = ((i, j-1), (i, j)) \in \hat{E}^2 \end{cases} \tag{22}$$

$$\overline{u}(e) = \begin{cases} \infty & \text{if } e \in \hat{E}^{\text{old}} \cup \hat{E}^1 \\ a_{i,j} & \text{if } e = ((i, j-1), (i, j)) \in \hat{E}^2 \end{cases} \tag{23}$$

Now in order to obtain another reformulation of the TNMU–minimization problem one just has to make sure that the flow on the edge $((i, l, r)^1, (i, l-1))$ equals the flow on the edge $((i, r), (i, l, r)^2)$, since both of these correspond to the amount of radiation that is released while $l_i = l$ and $r_i = r$.

Theorem 6 ([2]). *The TNMU–minimization problem is equivalent to the network flow problem*

$$\phi(D', D) \rightarrow \min$$

subject to ϕ a circulation in $\hat{G} = (\hat{V}, \hat{E})$ with lower and upper capacities $\underline{u}$ and $\overline{u}$, defined by (22) and (23), and satisfying, for all $(i, l, r)^{1,2} \in \hat{V}$,

$$\phi((i, l, r)^1, (i, l-1)) = \phi((i, r), (i, l, r)^2). \tag{24}$$

This formulation is quite close to a pure Min–Cost–Network–Flow problem. But the standard algorithms for this problem type have to be adjusted in order to include the side constraint (24). Doing this one obtains a polynomial time algorithm for the TNMU–minimization with ICC (see [2] and [15]).

Example 8. *A segmentation of the benchmark matrix with ICC that is optimal with respect to the TNMU, and thus corresponds to an optimal flow on the appropriate $\hat{G}$ is the one given in Example 5.*

2.11   The Algorithm of Baatar and Hamacher

Another TNMU–optimal segmentation algorithm was proposed in [1]. For this a digraph $G = (V, E)$ is constructed as follows: The vertex set consists of $2m$ layers $L_1, R_1, L_2, R_2, \ldots, L_m, R_m$ and two additional vertices D and D'. Here, for $i = 1, 2, \ldots m$,

$$L_i = \{(i, 1, 1), (i, 1, 2), \ldots, (i, 1, n+1)\}, \quad R_i = \{(i, 2, 0), (i, 2, 1), \ldots, (i, 2, n)\}$$

Now arcs between L_i and R_i correspond to possible leaf positions in row i, that is

$$E_1 = \{((i, 1, l), (i, 2, r)) \ : \ 1 \leq i \leq m, \ 1 \leq l \leq r+1 \leq n+1\},$$

and arcs between R_i and L_{i+1} correspond to leaf positions satisfying the ICC, that is

$$E_2 = \{((i, 2, r), (i+1, 1, l)) \ : \ 1 \leq i \leq m-1, \ 1 \leq l \leq r+1 \leq n+1\},$$

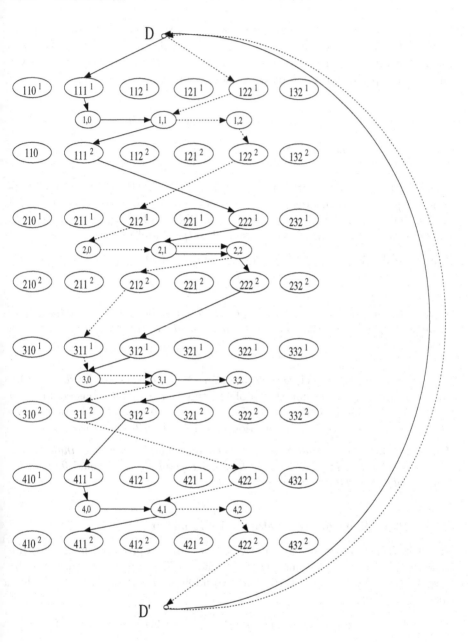

Fig. 12. The vertices of $\hat{G}$ for $m = 4$, $n = 2$ and two cycles

and finally all the arcs between D and L_1 and between R_m and D' are added, that is

$$E = E_1 \cup E_2 \cup E_3 \cup E_4 \cup \{((m,2,r), D') \ : \ 0 \leq r \leq n\} \cup \{(D', D)\},$$

where

$$E_3 = \{(D, (1,1,l)), \; : \; 1 \le l \le n+1\} \text{ and } E_4 = \{((m,2,r), D') \; : \; 0 \le r \le n\}.$$

Figure 13 shows G for $m = 3$ and $n = 4$.

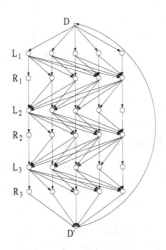

Fig. 13. The digraph G for $m = 3$ and $n = 4$

As in the description of Hamacher's algorithm we associate to a segment with parameters $l_1, r_1, l_2, r_2, \ldots, l_m, r_m$ a unit flow on the cycle

$$D, (1,1,l_1), (1,2,r_1), (2,1,l_2), (2,2,r_2), \ldots, (m,1,l_m,1), (m,2,r_m), D', D.$$

So every segmentation corresponds to a circulation on G (but not conversely, since the ICC between the left leaf of row i and the right leaf of row $i+1$ is not reflected in the structure of the digraph). In the circulation corresponding to a segmentation the total flow going through vertex $(i,1,l)$ equals the number of monitor units for which the left leaf is positioned at $j-1$, i.e. for which $l_i = j$, and similarly, the total flow going through $(i,2,r)$ equals the number of monitor units for which $r_i = j$. Let ϕ be a circulation on G, and denote by $\phi(v)$ the total flow going through $v \in V$. In [1] it is shown that in order to find a circulation corresponding to a segmentation with minimal TNMU, we may assume that, for all $i \in [m]$,

$$\phi(i,1,1) = a_{i,1}, \; \phi(i,2,n) = a_{i,n} \text{ and } \phi(i,2,0) = \phi(i,1,n+1) = 0.$$

Fixing these values, necessary and sufficient conditions for the $\phi(v)$ to correspond to a segmentation of A are (see [1])

$$\phi(i,1,j) - l^*_{i,j} = \phi(i,2,j) - r^*_{i,j} \geq 0 \qquad (i \in [m], j \in [n]), \qquad (25)$$

$$\sum_{j=1}^{k} \phi(i,1,j) \geq \sum_{j=1}^{k-1} \phi(i+1,2,j) \qquad (1 \leq i \leq m-1, 1 \leq k \leq n), \qquad (26)$$

$$\sum_{j=1}^{k} \phi(i,1,j) \leq \sum_{j=1}^{k-1} \phi(i-1,2,j) \qquad (2 \leq i \leq m, 1 \leq k \leq n), \qquad (27)$$

$$\phi(i,1,l) \in \mathcal{Z} \qquad (1 \leq i \leq m,\ 2 \leq l \leq n), \qquad (28)$$

$$\phi(i,2,r) \in \mathcal{Z} \qquad (1 \leq i \leq m,\ 1 \leq r \leq n-1). \qquad (29)$$

where

$$l^*_{i,j} = \max\{0, a_{i,j} - a_{i,j-1}\} \quad \text{and} \quad r^*_{i,j} = \max\{0, a_{i,j} - a_{i,j+1}\}.$$

So the task to determine $\phi(v)$ corresponding to a segmentation with minimal TNMU leads to the mixed integer program

$$\left. \begin{array}{rl} T \rightarrow \min & \text{subject to} \\ T = a_{i,1} + \sum_{j=2}^{n} \phi(i,1,j) & \text{and } (25)-(29). \end{array} \right\} \qquad (30)$$

The constraint matrix of this problem is totally unimodular, as is shown in [1] using the theorem of Ghouila–Houri, and so in order to determine the $\phi(v)$ it is sufficient to solve the LP–relaxation of (30). When the $\phi(v)$ are determined a segmentation with minimal TNMU can be constructed by iteratively extracting unit flows along cycles in G taking in each layer the leftmost vertex with positive throughput. In [1] it is also proposed to reduce the NS by a greedy strategy: an integer program is solved to determine the maximal u such that there is a cycle in G along which flow u can be extracted and the residual network still satisfies (25)–(29).

2.12 The Algorithm of Langer, Thai and Papiez

In [14] the authors give a mixed integer linear program formulation of the segmentation problem which can be used to find a segmentation with minimal NS among those with minimal TNMU. The model also allows to include additional constraints. In order to describe segmentations binary variables $l^{(t)}_{i,j}$ and $r^{(t)}_{i,j}$ are introduced for $i \in [m]$, $j \in [n]$ and $t \in [T]$, where T is an upper bound for the TNMU, for instance

$$T = \sum_{(i,j) \in [m] \times [n]} a_{i,j}.$$

$l^{(t)}_{i,j}$ takes value 1 if bixel (i,j) is covered by the left leaf while the t–th MU is delivered, but takes value 0 otherwise. Similarly, $r^{(t)}_{i,j}$ takes value 1 if bixel (i,j) is covered by the right leaf. Then the segment delivering the t–th MU is given by

$$s_{i,j}^{(t)} = 1 - l_{i,j}^{(t)} - r_{i,j}^{(t)}. \tag{31}$$

Observe that this equation, together with $s_{i,j}^{(t)}, l_{i,j}^{(t)}, r_{i,j}^{(t)} \in \{0, 1\}$, also implies that the opposite leaves in row i do not overlap, i.e. that for no (i, j, t) both $l_{i,j}^{(t)}$ and $r_{i,j}^{(t)}$ equal 1. The geometric properties of the leaves are modelled by the following constraints:

$$l_{i,j+1}^{(t)} \le l_{i,j}^{(t)} \qquad (i \in [m], \ j \in [n-1], \ t \in [T]), \tag{32}$$

$$r_{i,j}^{(t)} \le r_{i,j+1}^{(t)} \qquad (i \in [m], \ j \in [n-1], \ t \in [T]). \tag{33}$$

Furthermore, for a segmentation of A we obtain the constraints

$$\sum_{t=1}^{T} s_{i,j}^{(t)} = a_{i,j} \qquad (i \in [m], j \in [n]). \tag{34}$$

The MUs can be counted by introducing new binary variables $z^{(t)}$ ($t \in [T]$), where $z^{(t)}$ takes value 1 iff $s_{i,j}^{(t)} = 1$ for at least one pair (i, j), formally

$$\sum_{(i,j) \in [m] \times [n]} s_{i,j}^{(t)} \le mn z^{(t)} \qquad (t \in [T]). \tag{35}$$

Now the TNMU–minimization problem can be formulated as

$$\sum_{t=1}^{T} z^{(t)} \to \min \quad \text{subject to (31)–(35).} \tag{36}$$

Let T_0 denote the optimal value of the objective function, i.e. the minimal TN-MU. Observe that the determination of T_0 as the solution of (36) can be replaced by the calculation of the minimal TNMU according to Theorem 2. The next step is to find, among all the segmentations with T_0 MU, one with minimal NS. For this new binary variables $g^{(t)}$ ($t \in [T_0 - 1]$) are introduced, where $g^{(t)}$ takes value 1 if $s_{i,j}^{(t)} \ne s_{i,j}^{(t+1)}$ for some $(i, j) \in [m] \times [n]$. The global variable $g^{(t)}$ is described by local binary variables

$$\sigma_{i,j}^{(t)} = \alpha_{i,j}^{(t)} + \beta_{i,j}^{(t)}, \tag{37}$$

where $\alpha_{i,j}^{(t)}$ and $\beta_{i,j}^{(t)}$ are binary variables satisfying

$$-\alpha_{i,j}^{(t)} \le s_{i,j}^{(t+1)} - s_{i,j}^{(t)} \le \beta_{i,j}^{(t)}. \tag{38}$$

$g^{(t)}$ can take the value 0 only if all the $\sigma_{i,j}^{(t)}$ are zero, and this yields

$$\sum_{(i,j) \in [m] \times [n]} \sigma_{i,j}^{(t)} \le mn g^{(t)} \qquad (t \in [T_0]). \tag{39}$$

So the NS–minimization (with minimal TNMU) is

$$\sum_{t=1}^{T_0} g^{(t)} \rightarrow \text{ min subject to } (31)\text{--}(35),(37)\text{--}(39). \tag{40}$$

The authors of [14] suggest to solve this program by standard branch and bound techniques as implemented in commercial packages as CPLEX. Special restrictions can be included by adding constraints to the program. So the ICC corresponds to

$$l_{i,j}^{(t)} + r_{i+1,j}^{(t)} \leq 1 \quad (i \in [m-1],\ j \in [n],\ t \in [T]), \tag{41}$$

$$r_{i,j}^{(t)} + l_{i+1,j}^{(t)} \leq 1 \quad (i \in [m-1],\ j \in [n],\ t \in [T]). \tag{42}$$

The method for the segmentation problem with ICC described in [14] is to increase the number T' of monitor units step by step, starting with $T' = T_0$, and in each step try to find a feasible solution with T' monitor units. This procedure can be shortened by determining the minimal TNMU for a segmentation with ICC according to Theorem 3 and fixing T' at this value.

The tongue and groove condition is described by

$$-1 \leq s_{i+1,j}^{(t)} - s_{i,j}^{(t)} + s_{i,j}^{(t')} - s_{i+1,j}^{(t')} \leq 1$$
$$(i \in [m-1],\ i \in [n],\ 1 \leq t < t' \leq T). \tag{43}$$

The drawback of this method is that it requires to solve integer programs with a huge number of variables, and so it seems to be applicable only to very small problems.

Example 9. *For the benchmark matrix the algorithm yields the same result as Engel's algorithm.*

2.13   The Algorithm of Dai and Zhu

The algorithm proposed in [6] searches for a segmentation with a small NS. Again this is done by choosing a segment S and a coefficient u such that $A' = A - uS$ is nonnegative and continuing with A'. The criterion for the choice of u and S is the complexity of the residual matrix A', where the complexity of a matrix A is the number of segments necessary for a segmentation of A using some other algorithm. Obviously, the result of this method depends on the algorithm that is used to measure the complexity.

Recall that L denotes the maximum entry of A. For $u \in [L]$ we determine in each row i maximal intervals of entries greater than or equal to u. As in the section on the algorithm of Galvin *et al.* these can be described by numbers

$l_{i,1,u}, \ldots, l_{i,t(i,u),u}$ and $r_{i,1,u}, \ldots, r_{i,t(i,u),u}$, such that

$$1 \leq l_{i,1,u}, \; r_{i,t(i,u),u} \leq n$$

$$l_{i,k,u} \leq r_{i,k,u} \qquad\qquad (1 \leq k \leq t(i,u))$$

$$r_{i,k,u} < l_{i,k+1,u} - 1 \qquad\qquad (1 \leq k \leq t(i,u) - 1)$$

$$a_{i,j} \begin{cases} \geq u \text{ if } l_{i,k,u} \leq j \leq r_{i,k,u} \text{ for some } k, \\ < u \text{ otherwise.} \end{cases}$$

Now for all $u \in [L]$ the complexities of $A - uS$ are computed for all the $\prod_{i=1}^{m} t(i, u)$ possible segments, and among all the tested pairs (u, S) one with minimal complexity of $A - uS$ is chosen. If there are several pairs with minimal complexity of the residual matrix, we choose one with maximal irradiated area, i.e. with maximal number of 1's in S. The ICC can easily be included into the algorithm by excluding segments that violate the ICC from the complexity checking. The obvious drawback of this algorithm is the time complexity. The number of segments that have to be checked grows exponentially with the size of the matrix, and so the method becomes infeasible for moderate problem sizes.

Example 10. *We consider segmentation without ICC and use the algorithm of Bortfeld et al. with the sweep technique for the calculation of the complexity. Then our benchmark matrix has complexity 9, and the first extracted matrix is*

$$3 \begin{pmatrix} 1 & 1 & 0 & 0 & 0 & 0 \\ 0 & 1 & 0 & 0 & 0 & 0 \\ 0 & 0 & 0 & 0 & 0 & 1 \\ 1 & 1 & 1 & 0 & 0 & 0 \end{pmatrix},$$

where the residual matrix $\begin{pmatrix} 1 & 2 & 0 & 1 & 4 & 5 \\ 2 & 1 & 1 & 3 & 1 & 4 \\ 2 & 3 & 2 & 1 & 2 & 1 \\ 2 & 0 & 0 & 2 & 5 & 3 \end{pmatrix}$ *has complexity 6. Continuing we obtain the segmentation*

$$\begin{pmatrix} 4 & 5 & 0 & 1 & 4 & 5 \\ 2 & 4 & 1 & 3 & 1 & 4 \\ 2 & 3 & 2 & 1 & 2 & 4 \\ 5 & 3 & 3 & 2 & 5 & 3 \end{pmatrix} = 3 \begin{pmatrix} 1 & 1 & 0 & 0 & 0 & 0 \\ 0 & 1 & 0 & 0 & 0 & 0 \\ 0 & 0 & 0 & 0 & 0 & 1 \\ 1 & 1 & 1 & 0 & 0 & 0 \end{pmatrix} + 3 \begin{pmatrix} 0 & 0 & 0 & 0 & 1 & 1 \\ 0 & 0 & 0 & 0 & 0 & 1 \\ 0 & 0 & 0 & 0 & 0 & 0 \\ 0 & 0 & 0 & 0 & 1 & 1 \end{pmatrix} + 1 \begin{pmatrix} 0 & 0 & 0 & 1 & 1 & 1 \\ 1 & 1 & 1 & 1 & 1 & 1 \\ 1 & 1 & 1 & 1 & 1 & 1 \\ 0 & 0 & 0 & 1 & 1 & 0 \end{pmatrix}$$

$$+ 1 \begin{pmatrix} 1 & 1 & 0 & 0 & 0 & 0 \\ 1 & 0 & 0 & 0 & 0 & 0 \\ 1 & 1 & 1 & 0 & 0 & 0 \\ 0 & 0 & 0 & 1 & 1 & 0 \end{pmatrix} + 1 \begin{pmatrix} 0 & 1 & 0 & 0 & 0 & 0 \\ 0 & 0 & 0 & 1 & 0 & 0 \\ 0 & 1 & 0 & 0 & 0 & 0 \\ 1 & 0 & 0 & 0 & 0 & 0 \end{pmatrix} + 1 \begin{pmatrix} 0 & 0 & 0 & 0 & 0 & 1 \\ 0 & 0 & 0 & 1 & 0 & 0 \\ 0 & 0 & 0 & 0 & 1 & 0 \\ 1 & 0 & 0 & 0 & 0 & 0 \end{pmatrix}.$$

As indicated by the example, the algorithm yields quite good results compared to other algorithms, in particular it is essentially more NS–efficient than the algorithm used for the calculation of the complexity. This is confirmed in [6] by a number of numerical experiments.

2.14   Reduction of Leaf Motion

After the segments and their coefficients have been determined by some algorithm we still have the freedom to choose the order in which the corresponding homogeneous fields are delivered. In order to reduce the total treatment time it is suggestive to choose an order which minimizes the leaf travel time between consecutive segments. The minimization of the overall leaf travel time is equivalent to the search for a Hamiltonian path of minimal weight on the complete graph which has the segments as vertices and a weight function μ on the edges, defined as follows: for two segments S and S', given by l_i, r_i $(i \in [m])$ and l_i', r_i' $(i \in [m])$, respectively, we put

$$\mu(S, S') = \max_{1 \leq i \leq m} \max\{|l_i - l_i'|, |r_i - r_i'|\}.$$

Clearly, $\mu(S, S') = \mu(S', S)$, $\mu(S, S') \geq 0$ with equality iff $l_i = l_i'$ and $r_i = r_i'$ for all $i \in [m]$ and

$$\mu(S, S'') = \max_{1 \leq i \leq m} \max\{|l_i - l_i''|, |r_i - r_i''|\}$$
$$\leq \max_{1 \leq i \leq m} \max\{|l_i - l_i'| + |l_i' - l_i''|, |r_i - r_i'| + |r_i' - r_i''|\}$$
$$\leq \mu(S, S') + \mu(S', S'').$$

Thus μ is a metric and there are good approximations for a minimal Hamiltonian path ([10]). When the number NS is not relatively small, as is the case for practical problems, it is even possible to solve the Hamiltonian path problem exactly.

Example 11. *Using the version of Kalinowski's algorithm that heuristically reduces the NS we obtain the segmentation*

$$
\begin{pmatrix}
16 & 10 & 0 & 5 & 4 & 12 \\
1 & 13 & 16 & 6 & 2 & 14 \\
6 & 6 & 3 & 3 & 15 & 2 \\
8 & 15 & 3 & 0 & 13 & 11 \\
3 & 16 & 13 & 6 & 9 & 3 \\
7 & 16 & 10 & 11 & 14 & 6
\end{pmatrix}
= 8
\begin{pmatrix}
0 & 0 & 0 & 0 & 0 & 1 \\
0 & 0 & 0 & 0 & 0 & 1 \\
0 & 0 & 0 & 0 & 1 & 0 \\
0 & 0 & 0 & 0 & 1 & 1 \\
0 & 0 & 0 & 0 & 1 & 0 \\
0 & 1 & 1 & 1 & 1 & 0
\end{pmatrix}
+ 8
\begin{pmatrix}
1 & 1 & 0 & 0 & 0 & 0 \\
0 & 1 & 1 & 0 & 0 & 0 \\
0 & 0 & 0 & 0 & 0 & 0 \\
1 & 1 & 0 & 0 & 0 & 0 \\
0 & 1 & 1 & 0 & 0 & 0 \\
0 & 1 & 0 & 0 & 0 & 0
\end{pmatrix}
+ 4
\begin{pmatrix}
1 & 0 & 0 & 0 & 0 & 0 \\
0 & 1 & 1 & 1 & 0 & 0 \\
1 & 1 & 0 & 0 & 0 & 0 \\
0 & 1 & 0 & 0 & 0 & 0 \\
0 & 1 & 1 & 1 & 0 & 0 \\
1 & 0 & 0 & 0 & 0 & 0
\end{pmatrix}
$$

$$
+ 3
\begin{pmatrix}
0 & 0 & 0 & 1 & 1 & 1 \\
0 & 0 & 0 & 0 & 0 & 1 \\
0 & 0 & 0 & 0 & 1 & 0 \\
0 & 0 & 0 & 0 & 1 & 1 \\
0 & 0 & 0 & 0 & 0 & 1 \\
0 & 0 & 0 & 0 & 1 & 1
\end{pmatrix}
+ 2
\begin{pmatrix}
1 & 1 & 0 & 0 & 0 & 0 \\
0 & 0 & 1 & 0 & 0 & 0 \\
1 & 1 & 1 & 1 & 1 & 0 \\
0 & 1 & 1 & 0 & 0 & 0 \\
1 & 1 & 0 & 0 & 0 & 0 \\
0 & 0 & 1 & 1 & 1 & 1
\end{pmatrix}
+
\begin{pmatrix}
0 & 0 & 0 & 1 & 0 & 0 \\
0 & 0 & 1 & 1 & 1 & 1 \\
0 & 0 & 0 & 0 & 1 & 0 \\
0 & 0 & 0 & 0 & 1 & 0 \\
0 & 0 & 0 & 1 & 0 & 0 \\
0 & 0 & 0 & 1 & 1 & 1
\end{pmatrix}
+
\begin{pmatrix}
0 & 0 & 0 & 1 & 1 & 1 \\
0 & 0 & 0 & 0 & 0 & 1 \\
0 & 0 & 1 & 1 & 1 & 1 \\
0 & 0 & 0 & 0 & 1 & 0 \\
0 & 1 & 1 & 1 & 1 & 0 \\
1 & 0 & 0 & 0 & 0 & 0
\end{pmatrix}
$$

$$
+
\begin{pmatrix}
1 & 0 & 0 & 0 & 0 & 0 \\
1 & 1 & 1 & 1 & 1 & 1 \\
0 & 0 & 0 & 0 & 0 & 0 \\
0 & 0 & 0 & 0 & 0 & 0 \\
1 & 1 & 0 & 0 & 0 & 0 \\
1 & 0 & 0 & 0 & 0 & 0
\end{pmatrix}
+
\begin{pmatrix}
1 & 0 & 0 & 0 & 0 & 0 \\
0 & 0 & 0 & 0 & 0 & 0 \\
0 & 0 & 0 & 0 & 0 & 0 \\
0 & 1 & 1 & 0 & 0 & 0 \\
0 & 0 & 0 & 0 & 0 & 0 \\
1 & 0 & 0 & 0 & 0 & 0
\end{pmatrix}.
$$

If we deliver the segments in this order the length of the corresponding Hamiltonian path is $5 + 1 + 4 + 5 + 4 + 3 + 5 + 1 = 28$. *Using a minimum spanning tree approximation for the Hamiltonian path we obtain the delivery order* $1, 4, 6, 7, 5, 2, 3, 8, 9$ *with a length of* $3 + 3 + 3 + 3 + 1 + 1 + 1 + 1 = 16$.

First numerical results for the reduction of leaf motion when the algorithms of Engel and Kalinowski are used are shown in 1.

For algorithms using a sweep technique, such that the leaves move always in one direction there is nothing to do, since the leaf motion is automatically minimized.

Lemma 2. *Let* $A = \sum_{t=1}^{k} u_t S^{(t)}$ *be a segmentation obtained by some algorithm using a sweep–technique, that is if* $l_i^{(t)}$ *and* $r_i^{(t)}$ *are the parameters of* $S^{(t)}$ $(t \in [k])$

Table 1. Reduction of leaf motion by a minimum spanning tree approximation of the minimal Hamiltonian path for the algorithms of Kalinowski ([12]) and Engel ([7]). P_{old} is the Hamiltonian path corresponding to the order in which the segments are constructed by the algorithm and P_{new} is the approximation of a minimal Hamiltonian path. The results are averaged over 1000 15×15−matrices with random entries from $\{0, 1, \ldots, 16\}$.

	Engel		Kalinowski	
L	$\mu(P_{\text{old}})$	$\mu(P_{\text{new}})$	$\mu(P_{\text{old}})$	$\mu(P_{\text{new}})$
3	112.5	106.5	64.7	49.3
4	126.1	119.1	81.5	59.9
5	136.7	128.2	128.3	85.7
6	145.9	136.4	141.5	93.3
7	152.0	141.9	152.1	99.4
8	157.4	146.7	163.4	105.4
9	163.2	151.2	172.7	111.7
10	166.6	154.3	179.7	116.5
11	170.5	158.3	187.5	121.2
12	174.2	161.0	193.4	124.4
13	178.1	165.1	199.8	128.4
14	180.6	167.0	206.5	131.6
15	183.1	169.5	211.0	134.8
16	185.4	171.9	217.4	138.5

then we have, for all $i \in [m]$,

$$l_i^{(1)} \leq l_i^{(2)} \leq \ldots \leq l_i^{(k)} \quad \text{and} \quad r_i^{(1)} \leq r_i^{(2)} \leq \ldots \leq r_i^{(k)}.$$

Then $(S^{(1)}, S^{(2)}, \ldots, S^{(k)})$ is a Hamiltonian path of minimal weight in the complete graph with vertex set $\{S^{(1)}, S^{(2)}, \ldots, S^{(k)}\}$ and weight function μ.

Proof. Let π be an arbitrary permutation of $[k]$. We have to show that

$$\sum_{t=1}^{k-1} \mu\left(S^{(\pi(t))}, S^{(\pi(t+1))}\right) \geq \sum_{t=1}^{k-1} \mu\left(S^{(t)}, S^{(t+1)}\right).$$

The crucial observation is that, for $1 \leq t \leq t'' \leq t' \leq k$, we have

$$\mu\left(S^{(t)}, S^{(t')}\right) \geq \mu\left(S^{(t)}, S^{(t'')}\right),$$

which follows directly from the definition of μ. If $\pi(t) < \pi(t+1)$ for all $t \in [k-1]$ there is nothing to do. Otherwise put

$$t_0 = \min\{t \ : \ \pi(t) > \pi(t+1)\},$$

$$t_1 = \begin{cases} k & \text{if } \pi(t_0) = k, \\ \min\{t \ : \ \pi(t+1) > \pi(t_0)\} & \text{otherwise,} \end{cases}$$

$$t_2 = \min\{t \ : \ \pi(t) > \pi(t_1)\}.$$

Table 2. Test results without ICC. The columns labeled Gal, Bor, X–V and Eng correspond to the algorithm of Galvin et al. [8], the algorithm of Bortfeld et al. [3], the algorithm of Xia and Verhey [24] and the algorithm of Engel [7], respectively.

	TNMU				NS			
L	Gal	Bor	X–V	Eng	Gal	Bor	X–V	Eng
3	17.0	14.0	16.6	14.0	11.3	14.0	11.1	9.7
4	31.3	17.9	22.4	17.9	15.4	17.9	14.1	10.9
5	32.7	21.8	25.0	21.8	16.2	21.8	15.1	11.7
6	39.5	25.6	37.7	25.6	17.7	25.6	17.9	12.5
7	50.9	29.5	38.8	29.5	20.1	29.5	16.2	13.1
8	60.5	33.3	46.3	33.3	21.3	33.3	20.2	13.7
9	60.3	37.1	51.0	37.1	22.1	37.1	20.2	14.2
10	71.0	40.9	53.9	40.9	23.1	40.9	20.5	14.7
11	83.5	44.8	55.7	44.8	25.1	44.8	21.6	15.1
12	84.5	48.6	81.1	48.6	25.7	48.6	21.8	15.5
13	98.2	52.4	83.3	52.4	26.5	52.4	22.4	15.8
14	108.7	56.2	83.5	56.2	27.2	56.2	22.8	16.2
15	128.2	60.1	83.5	60.1	27.9	60.1	23.5	16.5
16	93.6	63.8	93.6	63.8	29.4	63.8	23.9	16.8

Now we may replace π by the permutation π' given by

$$\pi(1), \ldots, \pi(t_2 - 1), \pi(t_1), \pi(t_1 - 1), \ldots, \pi(t_2), \pi(t_1 + 1), \ldots, \pi(k).$$

To see this, assume first $t_2 > 1$ and $t_1 < k$. Then $\pi(t_2 - 1) < \pi(t_1) < \pi(t_2)$ and $\pi(t_1) < \pi(t_2) < \pi(t_1 + 1)$, hence

$$\sum_{t=1}^{k-1} \mu\left(S^{(\pi'(t))}, S^{(\pi'(t+1))}\right) = \sum_{t=1}^{k-1} \mu\left(S^{(\pi(t))}, S^{(\pi(t+1))}\right) - \mu\left(S^{(\pi(t_2-1))}, S^{(\pi(t_2))}\right)$$

$$- \mu\left(S^{(\pi(t_1))}, S^{(\pi(t_1+1))}\right) + \mu\left(S^{(\pi(t_2-1))}, S^{(\pi(t_1))}\right) + \mu\left(S^{(\pi(t_2))}, S^{(\pi(t_1+1))}\right)$$

$$\leq \sum_{t=1}^{k-1} \mu\left(S^{(\pi(t))}, S^{(\pi(t+1))}\right).$$

Similarly,

$$\sum_{t=1}^{k-1} \mu\left(S^{(\pi'(t))}, S^{(\pi'(t+1))}\right) \leq \sum_{t=1}^{k-1} \mu\left(S^{(\pi(t))}, S^{(\pi(t+1))}\right)$$

if $t_2 = 1$ or $t_1 = k$. Repeating this replacement if necessary, we obtain the permutation $1, 2, \ldots, k$, and the lemma is proved.

2.15   Numerical Results

In this subsection the performance of some of the algorithms is compared based on the segmentation of 15×15–matrices. As in [24] for every algorithm we

construct segmentations of 1000 matrices with random entries from $\{0, 1, \ldots, L\}$
($L = 3, 4, \ldots, 16$) and determine the average TNMU and the average NS. We
used the results from [24] for the algorithm of Galvin et al., the algorithm of
Bortfeld et al. and the algorithm of Xia and Verhey, and we implemented the
algorithm of Engel, the algorithm of Kamath and the algorithm of Kalinowski
in C++. The results are shown in Tables 2 and 3.

Table 3. Test results with ICC. The columns labeled Gal, Bor, X–V, Kam and Kal
correspond to the algorithm Galvin et al. [8], the algorithm of Bortfeld et al. [3], the
algorithm of Xia and Verhey [24], the algorithm of Kamath [13] and the algorithm of
Kalinowski [12], respectively.

	TNMU					NS				
L	Gal	Bor	X–V	Kam	Kal	Gal	Bor	X–V	Kam	Kal
3	19.7	17.7	19.5	15.4	15.4	13.4	17.7	13.3	15.4	12.6
4	40.5	22.8	29.6	19.5	19.5	20.4	22.8	18.6	19.5	14.5
5	40.1	27.9	30.9	23.6	23.6	20.4	27.9	19.0	23.6	16.0
6	44.2	32.8	46.8	27.6	27.6	21.5	32.8	20.3	27.6	17.2
7	67.1	37.9	45.6	31.7	31.7	27.1	37.9	20.0	31.7	18.2
8	72.3	42.8	63.4	35.7	35.7	28.2	42.8	24.3	35.7	19.1
9	72.3	47.8	67.1	39.8	39.8	28.3	47.8	24.3	39.8	19.9
10	76.5	52.6	68.6	43.8	43.8	28.9	52.6	25.7	43.8	20.7
11	81.4	57.6	68.6	47.7	47.7	30.9	57.6	25.7	47.7	21.3
12	106.8	62.4	101.1	51.8	51.8	34.8	62.4	27.0	51.8	21.9
13	101.1	67.3	100.6	55.7	55.7	35.5	67.3	26.9	55.7	22.5
14	112.7	72.2	100.0	59.8	59.8	35.6	72.2	26.9	59.8	23.0
15	116.0	77.1	98.0	63.8	63.8	35.9	77.1	26.7	63.8	23.5
16	154.5	82.0	124.9	67.7	67.7	41.7	82.0	30.0	67.7	24.0

With respect to the computation time all of the considered algorithms are
acceptable: On an 1.3GHz PC the computation of the whole column for the
algorithm of Kalinowski took 40 minutes, and for all the other algorithms the
whole column can be computed in a few minutes.

3   Dynamic Methods

Another approach to the generation of intensity modulated irradiation fields is
to use an MLC in the dynamic mode. That means the beam is always switched
on and the modulation is realized by varying the speed of the leaves. In the
literature two different variants can be found according to the starting positions
of the leaves. For the sweep technique both leaves start at the left end of the
field and move always to the right, while for the close–in technique the leaves
start at opposite ends of the field and move towards each other. Obviously, with
a single run of the close–in technique only profiles with a single maximum can be
generated, and profiles where the gradient is too small are also excluded due to

the finiteness of the maximal leaf velocity. In contrast, the sweep technique can be used to generate arbitrary profiles, and several authors have derived equations for the leaf velocities realizing a given profile and minimizing the total irradiation time ([5,16,20,21,4,19]). In the first section we sketch the basic principle that is common to all of these approaches and in the second section we describe how the tongue and groove effect can be avoided using a method introduced in [22].

3.1 The Basic Principle

The leaf trajectories are determined independently for each row, and thus we only describe the realization of a single row profile. Let $a_0, a_1, \ldots, a_n$ be the required doses at the equidistant points $x_0, x_1, \ldots, x_n$, where x_0 and x_n are the coordinates of the left and the right end of the field, respectively. Denote by $t_L(x)$ $(t_R(x))$ the time when the left (right) leaf passes the point with coordinate x. Then the dose delivered at x is proportional to $t_L(x) - t_R(x)$ and by scaling the time appropriately we may assume

$$a_j = t_L(x_j) - t_R(x_j).$$

Denote the maximal leaf velocity by $\hat{v}$ and the velocities of the left and the right leaf by $v_L(x)$ and $v_R(x)$, respectively. Suppose $t_L(x_j)$ and $t_R(x_j)$ are already known. Then in order to minimize the time that is needed to generate the profile over the interval $[x_j, x_{j+1}]$ we put, for $x_j \leq x < x_{j+1}$,

$$v_R(x) = \hat{v} \text{ if } a_{j+1} \geq a_j \quad \text{and} \quad v_L(x) = \hat{v} \text{ if } a_{j+1} < a_j. \tag{44}$$

First assume $a_{j+1} \geq a_j$. Then

$$a_{j+1} = t_L(x_{j+1}) - t_R(x_{j+1}) = t_L(x_{j+1}) - \left(t_R(x_j) + \frac{\Delta x}{\hat{v}} \right),$$

where $\Delta x = x_{j+1} - x_j$. We interpolate the profile between x_j and x_{j+1} linearly, so $v_L(x)$ is constant for $x_j \leq x < x_{j+1}$, and we obtain

$$v_L(x) = \frac{\Delta x}{t_L(x_{j+1}) - t_L(x_j)} = \frac{\hat{v}}{1 + (a_{j+1} - a_j)\frac{\hat{v}}{\Delta x}}, \tag{45}$$

and analogously, if $a_{j+1} < a_j$,

$$v_R(x) = \frac{\Delta x}{t_R(x_{j+1}) - t_R(x_j)} = \frac{\hat{v}}{1 - (a_{j+1} - a_i)\frac{\hat{v}}{\Delta x}}. \tag{46}$$

The generation of the whole profile is complete when the left leaf reaches x_n. The time it takes for the left leaf to cross the interval $[x_j, x_{j+1}]$ is

$$\frac{\Delta x}{\hat{v}} \text{ if } a_{j+1} \leq a_j \quad \text{and} \quad \frac{\Delta x}{\hat{v}} + (a_{j+1} - a_j) \text{ if } a_{j+1} \geq a_i.$$

Thus the total irradiation time is

$$t_L(x_n) = \frac{x_n - x_0}{\hat{v}} + \sum_{j=0}^{n-1} \max\{a_{j+1} - a_j, 0\}.$$

To see that this is optimal under the condition that both leaves start at x_0, observe that, for $0 \le j \le n - 1$,

$$a_{j+1} = t_L(x_{j+1}) - t_R(x_{j+1}) \text{ and } a_j = t_L(x_j) - t_R(x_j),$$

thus, if $a_{j+1} > a_j$,

$$t_L(x_{j+1}) - t_L(x_j) = t_R(x_{j+1}) - t_R(x_j) + a_{j+1} - a_j \ge \frac{\Delta x}{\hat{v}} + a_{j+1} - a_j.$$

Clearly the total irradiation time for a multiple row intensity map is just the maximum of the irradiation time over the rows.

This method can be refined in several ways. So [19] and [20] include a compensation for the transmission through the leaves, and [21] takes into account the finite acceleration of the leaves.

3.2   The Tongue and Groove Effect

As in the static mode the tongue and groove design of the MLCs causes under-dosage in the border region between adjacent rows, as illustrated in Figure 14.

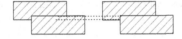

Fig. 14. Suppose the method from the previous section yields the same constant ve-locity for all the depicted leaves. Then the strip between the dotted lines receives only half of the dose that is required in both rows.

In [22] there is proposed a method to avoid this effect in the sense that after the correction the border region always receives the lower of the two relevant doses. The procedure is very similar to the tongue and groove correction of Siochi's rod pushing algorithm.

Synchronization of two rows: Consider two adjacent rows $(a_{i,0}, \dots, a_{i,n})$ and $(a_{i+1,0}, \dots, a_{i+1,n})$ and denote by $t_L^{(k)}(x)$, $t_R^{(k)}(x)$ ($k \in \{i, i+1\}$) the times when the left (resp. right) leaf of row k passes x. We determine inductively leaf velocities $v_L^{(k)}$ and $v_R^{(k)}$ on the intervals $[x_j, x_{j+1}]$ such that the given profile is generated without tongue and groove underdosage. Suppose the leaf motion up to the point x_j is already determined. First we compute the velocities and the corresponding $t_L^{(k)}(x_{j+1})$, $t_R^{(k)}(x_{j+1})$ according to (44)–(45). Tongue and groove underdosage occurs iff

$$t_R^{(i)}(x_{j+1}) > t_R^{(i+1)}(x_{j+1}) \quad \text{and} \quad t_L^{(i)}(x_{j+1}) > t_L^{(i+1)}(x_{j+1}), \qquad (47)$$

or the same with the roles of i and $i + 1$ interchanged. We call the pair of rows synchronized if

$$t_R^{(i)}(x_{j+1}) = t_R^{(i+1)}(x_{j+1}) \quad \text{or} \quad t_L^{(i)}(x_{j+1}) = t_L^{(i+1)}(x_{j+1}).$$

Then in order to avoid the tongue and groove effect it is sufficient to change the velocities in such a way that the rows are synchronized. By symmetry we may assume that (47) holds. Then we just have to slow down both leaves in row $i + 1$. Precisely, if $a_{i,j+1} \leq a_{i+1,j+1}$, we put

$$t_L^{(i+1)}(x_{j+1}) := t_L^{(i)}(x_{j+1}),$$
$$t_R^{(i+1)}(x_{j+1}) := t_L^{(i+1)}(x_{j+1}) - a_{i+1,j+1},$$

and if $a_{i,j+1} > a_{i+1,j+1}$ we put

$$t_R^{(i+1)}(x_{j+1}) := t_R^{(i)}(x_{j+1}),$$
$$t_L^{(i+1)}(x_{j+1}) := t_R^{(i+1)}(x_{j+1}) + a_{i+1,j+1}.$$

This is illustrated in Figure 15.

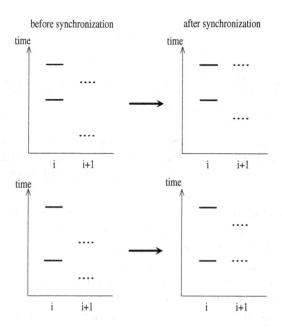

Fig. 15. The synchronization for two rows. The straight lines stand for $t_L^{(i)}(x_{j+1})$ and $t_R^{(i)}(x_{j+1})$, the dotted lines stand for $t_L^{(i+1)}(x_{j+1})$ and $t_R^{(i+1)}(x_{j+1})$.

Finally the new velocities for $x_j \leq x < x_{j+1}$ are computed:

$$v_L^{(i+1)}(x) = \frac{\Delta x}{t_L^{(i+1)}(x_{j+1}) - t_L^{(i+1)}(x_j)},$$

$$v_R^{(i+1)}(x) = \frac{\Delta x}{t_R^{(i+1)}(x_{j+1}) - t_R^{(i+1)}(x_j)}.$$

Synchronization of more than two rows: For general treatment plans the synchronization of leaf trajectories is based on the iterated synchronization of two rows. To correct the leaf trajectories between the points x_j and x_{j+1} first a row i_0 with slowest left leaf is determined, i.e. with $t_L^{(i_0)}(x_{j+1}) \geq t_L^{(i)}(x_{j+1})$ for all $i \in [m]$. Now the whole synchronization is described in Algorithm 6. The algorithm terminates since in every step some leaves are slowed down, but never a left leaf arrives later at x_{j+1} than the one in row i_0, and so in the worst case finally $t_L^{(i)}(x_{j+1}) = t_L^{(i_0)}(x_{j+1})$ for all i.

Algorithm 6. : Synchronization of leaf motion

---

repeat
   finished:=true
   for $i = i_0 - 1$ downto 1 **do**
     if rows i and $i + 1$ are not synchronized **then**
       finished:=false
       synchronize rows i and $i + 1$
   for $i = i_0$ to $m - 1$ **do**
     if rows i and $i + 1$ are not synchronized **then**
       finished:=false
       synchronize rows i and $i + 1$
until finished

In [23] the authors argue that it might not be necessary to fully synchronize the leaf motion. This is because in the overlap region that is covered by both leaves the depth of each leaf is only half of the full leaf–depth and taking account of the difference between the transmission through the full depth and the half depth they derive a criterion for 'partial synchronization' which assures that the overlap region receives at least the lower of the two relevant doses.

4 Summary and Discussion

To realize intensity modulated radiation fields using a multileaf collimator in the static mode it is necessary to determine a sequence of leaf positions and corresponding irradiation times such that the superposition of the homogeneous fields yields the required modulated intensity. This amounts to the problem of representing a nonnegative integer matrix as a positive integer combination of certain $(0,1)$–matrices, so called segments. In order to optimize the treatment this segmentation has to be chosen in such a way that the total number of monitor units and the number of segments are small. Ignoring machine–dependent constraints the construction of a segmentation with minimal number of monitor units can be done in polynomial time, for instance by the algorithm of Bortfeld et al. [3] or the algorithm of Engel [7]. In contrast, the minimization of the number of segments is NP–complete already for a single row, and thus probably one has to be satisfied with an approximative algorithm for this problem. Our variant of Engel's algorithm seems to be very good in this respect, but there remains the problem to find a theoretical bound for the quality of the approximation. Other algorithms use heuristic principles to reduce the number of segments, but are no longer optimal with respect to the monitor units.

For the segmentation problem with interleaf collision constraint the algorithm of Kamath et al. [13], the algorithm of Baatar and Hamacher [1] and the algorithm of Kalinowski [12] minimize the number of monitor units in polynomial time. In addition, [1] and [12] propose greedy heuristics for the reduction of the number of segments, but these algorithms have the drawback that the computation time grows rapidly with the problem size.

The algorithm of Langer et al. formulates the segmentation problem as a mixed integer program and finds solutions that are optimal in first instance with respect to the monitor units and in second instance with respect to the segments. Also machine–dependent constraints are easily included. However, as for the algorithm of Dai and Zhu, due to computational complexity the method is not applicable for problem sizes that arise in practice.

One difficulty that comes up when the tongue and groove constraint is taken into account is that the local strategy of most of the published algorithms is no longer applicable since the different segments are not independent of each other. The first question which has to be addressed in this context is if it is necessary to avoid the tongue and groove effect totally, or if it is sufficient to reduce it to some extend. In the second case some quantitative measure for this acceptable underdosage has to be developed, and a corresponding objective function for the minimization has to be constructed.

When the multileaf collimator is used in the dynamic mode the leaf velocities have to be chosen so that the required intensity is generated. It is possible to determine optimal leaf velocities for an unidirectional sweep of the leaves across the field. Also the tongue and groove underdosage can be totally avoided by synchronization of the leaf motion.

References

1. D. Baatar and H.W. Hamacher, New LP model for multileaf collimators in radiation therapy, contribution to the conference ORP3, University of Kaiserslautern, 2003.
2. N. Boland, H.W. Hamacher, and F. Lenzen, Minimizing beam-on time in cancer radiation treatment using multileaf collimators, Networks, 43, 226, 2004.
3. T.R. Bortfeld, D.L. Kahler, T.J. Waldron, and A.L. Boyer, X–ray field compensation with multileaf collimators, Int. J. Radiat. Oncol. Biol. Phys., 28, 723–730, 1994.

4. A.L. Boyer and C.Y. Yu. Intensity–modulated radiation therapy with dynamic multileaf collimators, Semin. Radiat. Oncol., 9, 48–59, 1999.

5. D.J. Convery and M.E. Rosenbloom, The generation of intensity–modulated fields for conformal radiotherapy by dynamic collimation, Phys. Med. Biol., 37, 6, 1359–1374, 1992.

6. J. Dai and Y. Zhu, Minimizing the number of segments in a delivery sequence for intensity–modulated radiation therapy with a multileaf collimator, Med. Phys., 28, 2113–2120, 2001.

7. K. Engel, A new algorithm for optimal multileaf collimator field segmentation, Preprint 03/5, Fachbereich Mathematik, Uni Rostock, 2003.

8. J.M. Galvin, X.G. Chen, and R.M. Smith, Combining multileaf fields to modulate fluence distributions, Int. J. Radiat. Oncol. Biol. Phys., 27, 697–705, 1993.

9. M. Haufschild and U.M. Korn, Mit Mathematik gegen Krebs – Optimale Einstellung eines Gerätes in der Strahlentherapie, contribution to Jugend–forscht, 2003.

10. D. Jungnickel, Graphen, Netzwerke und Algorithmen, BI–Wissenschaftsverlag, Mannheim, 1994.

11. T. Kalinowski, An algorithm for optimal multileaf collimator field segmentation with interleaf collision constraint, Preprint 03/2, Fachbereich Mathematik, Uni Rostock, 2003.

12. T. Kalinowski, An algorithm for optimal multileaf collimator field segmentation with interleaf collision constraint 2, Preprint 03/8, Fachbereich Mathematik, Uni Rostock, 2003.

13. S. Kamath, S. Sahni, J. Li, J. Palta, and S. Ranka, Leaf sequencing algorithms for segmented multileaf collimation, Phys. Med. Biol., 48, 307–324, 2003.

14. M. Langer, V. Thai, and L. Papiez, Improved leaf sequencing reduces segments of monitor units needed to deliver IMRT using multileaf collimators, Med. Phys., 28, 2450–2458, 2001.

15. F. Lenzen, An integer programming approach to the multileaf collimator problem, Master's thesis, University of Kaiserslautern, Dept. of Mathematics, 2000.

16. L. Ma, A.L. Boyer, L. Xing, and C.M. Ma, An optimized leaf setting algorithm for beam intensity modulation using dynamic multileaf collimators, Phys. Med. Biol., 43, 1629–1643, 1998.

17. W. Que, Comparison of algorithms for multileaf collimator field segmentation, Med. Phys., 26, 2390–2396, 1999.

18. R.A.C. Siochi, Minimizing static intensity modulation delivery time using an intensity solid paradigm, Int. J. Radiat. Oncol. Biol. Phys., 43, 671–680, 1999.

19. S.V. Spirou and C.S. Chui, Generation of arbitrary intensity profiles by dynamic jaws or multileaf collimators, Med. Phys., 21, 1031–1041, 1994.

20. J. Stein, T. Bortfeld, B. Dörschel, and W. Schlegel, Dynamic X–ray compensation for conformal radiotherapy by means of multi–leaf collimation, Radiother. Oncol., 32, 163–173, 1994.

21. R. Svensson, P. Källman, and A. Brahme, An analytical solution for the dynamic control of multileaf collimators, Phys. Med. Biol., 39, 37–61, 1994.

22. J.P.C. van Santvoort and B.J.M. Heijmen, Dynamic multileaf collimation without 'tongue-and-groove' underdosage effects, Phys. Med. Biol., 41, 2091–2105, 1996.

23. S. Webb, T. Bortfeld, J. Stein, and D. Convery, The effect of stair–step leaf transmission on the 'tongue–and–groove problem' in dynamic radiotherapy with a multileaf collimator, Phys. Med. Biol., 42, 595–602, 1996.

24. P. Xia and L. Verhey, Multileaf collimator leaf–sequencing algorithm for intensity modulated beams with multiple static segments, Med. Phys., 25, 1424–1434, 1998.

Sparse Asymmetric Connectors in Communication Networks⋆

R. Ahlswede and H. Aydinian

Abstract. An (n, N, d)–connector is an acyclic digraph with n inputs and N outputs in which for any injective mapping of input vertices into output vertices there exist n vertex disjoint paths of length d joining each input to its corresponding output. We consider the problem of construction of sparse $(n, N, 2)$–connectors (depth 2 connectors) when $n \ll N$. The probabilistic argument in [1] shows the existence of $(n, N, 2)$–connectors of size (number of edges) $O(N)$ if $n \leq N^{1/2-\varepsilon}$, $\varepsilon > 0$. However, the known explicit constructions with $n \leq \sqrt{N}$ in [6],[1],[2] are of size $O(N\sqrt{n})$. Here we present a simple combinatorial construction for $(n, N, 2)$–connectors of size $O(N \log_2 n)$. We also consider depth 2 fault–tolerant connectors under arc or node failures.

Keywords: connector, rearrangeable network, fault–tolerant connector.

1 Introduction

An (n, N)–network is a directed acyclic graph with n distinguished vertices called inputs and N other distinguished vertices called outputs. All other vertices are called links. The size of the network is the number of edges, and the depth is the length of the longest path from an input to an output.

An (n, N, d)–connector (also called a rearrangeable network) is a network of depth d $(n \leq N)$, such that for every injective mapping of input vertices into output vertices there exist n vertex disjoint paths joining each input to its corresponding output. Connectors are useful architectures for parallel machines. Their study started from pioneering works [13], [14], [8], [3], in connection with practical problems arisen in designing of switching networks for telephone traffic.

Symmetric connectors, i.e. (n, n, d)–connectors are well studied. Pippenger and Yao [11] obtained lower and upper bounds for the size of an (n, n, d)–connector: $\Omega(n^{1+1/d})$ and $O(n^{1+1/d}(\log n)^{1/d})$, respectively. The best known explicit construction for odd depth $2i + 1$ is $O(n^{1+1/(i+1)})$, due to Pippenger [12]. Hwang and Richards [7] gave an explicit construction for depth 2 connectors of size $O(n^{5/3})$. For asymmetric connectors Oruc [10] gave constructions for depth $\Omega(\log_2 N + \log_2 n)$ of size $O((N + n) \log_2 n)$ and for depth $\Omega(\log_2 N + \log_2^2 n)$ of size $O(N + n \log_2 n)$. Hwang and Richards [7] gave an explicit construction for $(n, N, 2)$–connectors of size $(1+o(1))N\sqrt{n}$ if $n \leq \sqrt{N}$. Baltz, Jäger and Srivastav [1], [2] showed by a probabilistic argument the existence of $(n, N, 2)$–connectors

⋆ Supported by DFG-Schwerpunkt Nr.1126 "Algorithmik großer und komplexer Netzwerke".

of size $O(N)$, if $n \leq N^{1/2-\varepsilon}$, $\varepsilon > 0$. They also extended the result of Hwang and Richards and improved it by a constant factor.

We are interested in the case when $n \ll N$. Such connectors with $d = 2$ are of particular interest (refering to [1,2]) in the design of sparse electronic switches. A challenging problem is to construct linear–sized $(n, N, 2)$–connectors. In this paper we show how to construct $(n, N, 2)$–connectors of size $O(N \log_2 n)$.

Let us consider the following standard approach to build connectors of depth 2. Suppose the vertex set $V = \mathcal{I} \cup \mathcal{L} \cup \mathcal{O}$ of a graph $G = (V, E)$ is partitioned into input vertices $\mathcal{I}$ with $|\mathcal{I}| = n$, link vertices $\mathcal{L}$ with $|\mathcal{L}| = L$ and output vertices $\mathcal{O}$ with $|\mathcal{O}| = N$. Then one can easily see that G is an $(n, N, 2)$–connector if the following two conditions are satisfied.

C1: $\mathcal{I}$ and $\mathcal{L}$ are completely connected, that is inputs and links form a complete bipartite graph.

C2: $|\Gamma(S)| \geq |S|$ holds for every $S \subset \mathcal{O}$ with $|S| \leq n$, i.e. the Hall's condition (shortly H–condition) is fulfilled for every n-set of output vertices $\mathcal{O}$.

In this paper we consider connectors satisfying conditions C1,C2. In Section 3 we give a simple explicit construction for such connectors of size $(1+o(1))N \log_2 n$ for all $n < N^{1/\sqrt{\log_2 N}}$.

In Section 4 we consider reliable or fault-tolerant connectors. An $(n, N, 2)$–connector is called t-edge fault-tolerant if in spite of any t or less edge failures it still remains an $(n, N, 2)$–connector. Correspondingly, it is called t-vertex fault-tolerant if t or less vertex failures are admissible in the connector.

We show that our construction can be used to obtain t-edge/vertex fault–tolerant connectors.

2   Preliminary

We need some notation and definitions from extremal set theory. Let $[n] := \{1, \ldots, n\}$. Given $n, k \in \mathbb{N}$ we denote

$$2^{[n]} = \left\{A : A \subset [n]\right\}, \qquad \binom{[n]}{k} = \left\{A \in 2^{[n]} : |A| = k\right\}.$$

For a family $\mathcal{A} \subset \binom{[n]}{k}$ we use the notation $\|\mathcal{A}\| := |\cup_{A \in \mathcal{A}} A|$. The shadow of $\mathcal{A} \subset \binom{[n]}{k}$ is defined by

$$\partial(\mathcal{A}) = \left\{F \subset \binom{[n]}{k-1} : \exists A \in \mathcal{A} : F \subset A\right\}.$$

Define also the colex order for the k-sets of $\mathbb{N}$ denoted by $\binom{\mathbb{N}}{k}$: For $A, B \in \binom{\mathbb{N}}{k}$

$$A \prec B \quad \Leftrightarrow \quad \max\left((A \setminus B) \cup (B \setminus A)\right) \in B.$$

Let $\mathcal{C}(k, m)$ denote the first m members of $\binom{\mathbb{N}}{k}$ in the colex ordering.

The following fact is well known (see [4] or [5]). Every positive integer m can be uniquely represented in the following form called the k–cascade representation of m:

$$m = \binom{a_k}{k} + \binom{a_{k-1}}{k-1} + \cdots + \binom{a_t}{t} \tag{2.1}$$

with $a_k > a_{k-1} > \cdots > a_t \geq t \geq 1$.

Recall now the well–known Kruskal–Katona Theorem (see [4],[5]).

Theorem KK. *Let $\mathcal{A}$ be a family of k–sets with $|\mathcal{A}| = m$ and let m have the cascade representation (2.1).*
Then we have

$$|\partial(\mathcal{A})| \geq |\partial(\mathcal{C}(k,m))|, \tag{2.2}$$

or equivalently

$$|\partial(\mathcal{A})| \geq \binom{a_k}{k-1} + \cdots + \binom{a_t}{t-1}.$$

For our purposes we need the following

Corollary KK (see [5]). *Let $\mathcal{A} \subset \binom{\mathbb{N}}{k}$ with*

$$|\mathcal{A}| \leq \binom{2k-1}{k} + \binom{2k-3}{k-1} + \cdots + \binom{3}{2} + \binom{1}{1}. \tag{2.3}$$

Then $|\partial(\mathcal{A})| \geq |\mathcal{A}|$ and the bound (2.3) is best possible.

3   The Construction

Given $n < N$, let k be the smallest integer such that

$$n \leq \binom{2k-1}{k} + \binom{2k-3}{k-1} + \cdots + \binom{3}{2} + \binom{1}{1}. \tag{3.1}$$

We take the k–cascade representation of

$$N = \binom{a_k}{k} + \binom{a_{k-1}}{k-1} + \cdots + \binom{a_t}{t} \tag{3.2}$$

and put

$$L = \binom{a_k}{k-1} + \binom{a_{k-1}}{k-2} + \cdots + \binom{a_t}{t-1}. \tag{3.3}$$

Consider now the set $\mathcal{C}(k, N)$. It is known (see [4],[5]) that

$$\partial(\mathcal{C}(k,N)) = \mathcal{C}(k-1, |\partial(\mathcal{C}(k,N)|)),$$

moreover, $|\partial(\mathcal{C}(k,N))| = L$.

Our construction now is straightforward. The vertex set of our graph $G = (V,E)$ is $V = \mathcal{I} \cup \mathcal{L} \cup \mathcal{O}$ with $|\mathcal{I}| = n$, $|\mathcal{L}| = L$, $|\mathcal{O}| = N$.

C1. The inputs $\mathcal{I}$ and links $\mathcal{L}$ form a complete bipartite graph. We identify the outputs with $\mathcal{C}(k, N)$, and the links with $\partial(\mathcal{C}(k, N))$. Thus the vertices of $\mathcal{O}$ and $\mathcal{L}$ correspond respectively to the initial N k-sets and L $(k-1)$-sets of $\mathbb{N}$ in the colex ordering. The edges between links and outputs are defined in a natural way, by inclusion. Clearly, each vertex of $\mathcal{O}$ has k neighbors (k subsets from $\binom{\mathbb{N}}{k-1}$).

C2. Corollary KK implies that $|\Gamma(S)| > |S|$ for every subset $S \subset \mathcal{O}$ with $|S| \le n$. Thus we have the following

Proposition. *The described construction (called ∂–construction) gives an (n, N)– connector of size*

$$|E| = nL + Nk. \tag{3.4}$$

Let us estimate $|E|$ in terms of parameters n and N. By (3.1) we have

$$\Theta(\log_2 n) = k \le \log_2 n. \tag{3.5}$$

By (3.2) and (3.3) we have

$$N = \Theta\left(\binom{a_k}{k}\right), \qquad L = O\left(\frac{k}{a_k} N\right), \tag{3.6}$$

where $a_k \ge \frac{k}{e} \binom{a_k}{k}^{1/k}$ and hence

$$a_k = \Omega\left(\log_2 n N^{1/\log_2 n}\right). \tag{3.7}$$

Finally, (3.4),(3.5),(3.6) and (3.7) give

$$|E| = nL + Nk = N \log_2 n \left(O\left(\frac{n}{a_k}\right) + 1\right).$$

If now $n = O\left(\log_2 n N^{1/\log_2 n}\right)$, then

$$|E| = O(N \log_2 n).$$

In particular, we have the following

Theorem 1. *For all N and $n = O\left(N^{1/\sqrt{\log_2 N}}\right)$ there are explicitly constructible $(n, N, 2)$–connectors of size $(1 + o(1))N \log_2 n$.*

4 Fault–Tolerant Connectors

In this section we show that by ∂–construction we can also obtain t–fault– tolerant connectors.

Theorem 2. *Let $G = (V, E)$ be an (n, N)–connector given by ∂–construction, and let k be the degree of output vertices defined from $\binom{2k-3}{k} < n \le \binom{2k-1}{k}$. Then G is a $(k-1)$–edge fault–tolerant connector.*

To prove the theorem we need the following

Lemma 1. *Let $\mathcal{A} \subset \binom{\mathbb{N}}{k}$. Then*

$$|\partial \mathcal{A}| - |\mathcal{A}| \ge k - 1 \tag{4.1}$$

if (a) $|\mathcal{A}| \le \binom{2k-1}{k} - k + 1$, or (b) $\binom{2k-1}{k} - k + 1 < |\mathcal{A}| \le \binom{2k-1}{k}$ and $\|\mathcal{A}\| \ne 2k - 1$.

Proof. Case (a) can be easily derived from Theorem KK by simple calculations.
Case (b): The cascade representation of $|\mathcal{A}|$ for this case is

$$|\mathcal{A}| = \binom{2k-2}{k} + \binom{2k-3}{k-1} + \cdots + \binom{a_1}{1},$$

where $1 \leq a_1 \leq k-1$, or

$$|\mathcal{A}| = \binom{2k-1}{k}.$$

Hence by Theorem KK

$$|\partial(\mathcal{A})| \geq \binom{2k-2}{k-1} + \binom{2k-3}{k-2} + \cdots + \binom{k}{1} + \binom{a_1}{0} = \binom{2k-1}{k}.$$

Note that $\|\mathcal{A}\| > 2k - 1$ (since $\|\mathcal{A}\| \neq 2k - 1$). We use now the result in
Moers [9] which generalizes the Kruskal–Katona Theorem. In our special case,
$\|\mathcal{A}\| > 2k - 1$, it says that the minimum of $|\partial\mathcal{A}|$ is attained for $\|\mathcal{A}\| = 2k$.
Moreover, an optimal family achieving the minimum is

$$\mathcal{A}^* = \mathcal{C}(k, |\mathcal{A}| - 1) \cup \big(\{1, \ldots, k-1, 2k\}\big).$$

Observe now that

$$|\partial(\mathcal{A}^*)| = \binom{2k-1}{k} + k - 1.$$

$\square$

Proof of Theorem 2
Let $G = (V, E)$ be an $(n, N, 2)$–connector given by ∂–construction. Given an
injective mapping $\phi : \mathcal{I} \to \mathcal{O}$, let $\mathcal{M} = \phi(\mathcal{I}) \subset \mathcal{O}$. Define also $\mathcal{L}' = \Gamma(\mathcal{M})$.
Let now G' be the subgraph of G induced by the vertices $\mathcal{I} \cup \mathcal{L}' \cup \mathcal{M}$. Let also
E_1 and E_2 be the edge sets of G' induced by $\mathcal{I} \cup \mathcal{L}'$ and $\mathcal{L}' \cup \mathcal{M}$, respectively.
Suppose now t ($t \leq k - 1$) edge–failures have occurred in G'. Since $\mathcal{I} \cup \mathcal{L}'$ is
a complete bipartite graph, we note that the "worst" case is that all faults are
from E_2. Thus it is sufficient to show that after deletion of t edges in E_2 the
resulting graph satisfies the H–condition.
Suppose now $S \subseteq \mathcal{M}$. Remind that $n = |\mathcal{I}| = |\mathcal{M}| \leq \binom{2k-1}{k}$. Let $\mathcal{A} \subset \binom{N}{k}$
be the family identified with S. If $\mathcal{A}$ satisfies one of the conditions of Lemma 1,
then (4.1) means that $|\Gamma(S)| - |S| \geq k - 1$. This clearly implies that the removal
of any $k - 1$ link–vertices (and hence any $k - 1$ edges) from G' results in a graph,
where the output vertices satisfy the H–condition. Thus it remains to consider
the case $|\mathcal{A}| > \binom{2k-1}{k} - k + 1$ with $\|\mathcal{A}\| = 2k - 1$.
This case is easier to examine in terms of the incidence matrix H for $\mathcal{A}$ and
$\partial(\mathcal{A})$ (equivalently, the adjacency matrix of the corresponding bipartite graph
induced by $\Gamma(S) \cup S$). Let the columns of H be labeled by elements of $\mathcal{A}$ and
the rows by elements of $\partial(\mathcal{A})$.
Without loss of generality we may assume that $\mathcal{A} \subset \binom{[2k-1]}{k-1}$. Suppose first
that $\mathcal{A} = \binom{[2k-1]}{k}$ and consider the corresponding incidence matrix H of size
$\binom{2k-1}{k} \times \binom{2k-1}{k}$.

Each column and each row of H contains exactly k ones. Delete now any $k - i - 1$ $(1 \leq i \leq k - 1)$ columns in H.

Lemma 2.

(i) Every row in the resulting matrix H' contains at least $i + 1$ ones.
(ii) Every two rows of H' contain together at least $k + i$ ones.

Proof. (i) is obvious. (ii) follows from the fact that any two rows have at most one 1 in common and hence two rows in H' have at least $2k - (k - i) = k + i$ ones. Thus the incidence matrix H' of every family $\mathcal{A}$ with $|\mathcal{A}| = \binom{2k-1}{k} - k + 1 + i$ has the properties (i) and (ii). $\qquad \Box$

Exchange now any $t \leq k - 1$ ones in H' into zeros (which corresponds to deletion of t edges in E_2), obtaining a new matrix H^*. In view of (i) and (ii) H^* contains at most one all–zero row. Thus the sum of all columns of H^* contains at most one zero coordinate if $i < k - 1$ and no zero coordinate if $i = k - 1$. This implies that for the corresponding bipartite graph G^* (with t deleted edges) $|\Gamma(S)| \geq |S|$ holds for every set of output vertices S with $\binom{2k-1}{k} - k + 1 < |S| \leq \binom{2k-1}{k}$.
This completes the proof of Theorem 2. $\qquad \Box$

Clearly, if G is a t-vertex fault–tolerant (n, N)–connector, then it is also a t-edge fault–tolerant (n, N)–connector. Thus in general we can speak about t–fault–tolerant connectors under edge–vertex failures.

The following statement directly follows from the proof of Theorem 2.

Corollary 1. *The ∂–construction gives a $(k-1)$–fault–tolerant (n, N)–connector if*

$$\binom{2k - 3}{k - 1} < n \leq \binom{2k - 1}{k} - k + 1.$$

5   A Direction of Further Research

Our main tool are results on shadows of subfamilies of the family of subsets of a finite set, which is a poset by the inclusion relation. It is conceivable that shadow properties of other posets give better results.

References

1. A. Baltz, G. Jäger, and A. Srivastav, Elementary constructions of sparse asymmetric connectors, Proceedings of the 23rd Conference on Foundations of Software Technology and Theoretical Computer Science (FSTTCS), Mumbai, India, LNCS 2914, 13–22, 2003.
2. A. Baltz, G. Jäger, and A. Srivastav, Constructions of sparse asymmetric connectors with number theoretic methods, Networks, Vol. 45, No. 3, 119-124, 2005.
3. V.E. Beneš, Optimal rearrangeable multistage connecting networks, Bell System Tech. J. 43, 1641-1656, 1964.
4. B. Bollobas, Combinatorics, Cambridge University Press, 1986.

5. K. Engel, Sperner Theory, Cambridge University Press, 1997.
6. C. Clos, A study of non–blocking switching networks, Bell System Tech. J. 32, 406–424, 1953.
7. P. Feldman, J. Friedman, and N. Pippenger, Wide-sense nonblocking networks, SIAM J. Discr. Math. 1, 158 -173, 1988.
8. F.K. Hwang and G.W. Richards, A two–stage rearrangeable broadcast switching network, IEEE Trans. on Communications 33, 1025–1035, 1985.
9. M. Moers, A generalization of a theorem of Kruskal, Graphs and Combinatorics 1, 167–183, 1985.
10. A.Y. Oruc, A study of permutation networks: some generalizations and tradeoffs, J. of Parall. and Distr. Comput., 359–366, 1994.
11. N. Pippenger and A.C. Yao, On rearrangeable networks with limited depth, SIAM J. Algebraic Discrete Methods 3, 411–417, 1982.
12. N. Pippenger, On rearrangeable and nonblocking switching networks, J. Comput. System Sci. 17, 145–162, 1987.
13. C.E. Shannon, Memory requirements in a telephone exchange, Bell System Tech. J. 29, 343–349, 1950.
14. D. Slepian, Two theorems on a particular crossbar switching network, unpublished manuscript, 1952.

X

Finding $C_{\mathrm{NRI}}(\mathcal{W})$, the Identification Capacity of the AVC $\mathcal{W}$, if Randomization in the Encoding Is Excluded

R. Ahlswede

For a DMC W it is not hard to show that $C_{\mathrm{NRI}}(W)$ equals the logarithm of the number of different row-vectors in W (see [A145]). Also for AVC $\mathcal{W}_0$ with 0-1-matrices only as transmission matrices $C_{\mathrm{NRI}}(\mathcal{W}_0)$ equals the capacity $C(\mathcal{W}_0)$ for transmission, which in turn equals Shannon's zero error capacity $C_0(\overline{W})$, where $\overline{W} = \frac{1}{|\mathcal{W}_0|} \sum_{W \in \mathcal{W}_0} W$ (see [A7]).

But for general $\mathcal{W}$ $C_{\mathrm{NRI}}(\mathcal{W})$ can be larger than $C(\mathcal{W})$. For example by Theorem 5 of [A145] for

$$\mathcal{W} = \left\{ \begin{pmatrix} 1 & 0 \\ 0 & 1 \end{pmatrix}, \begin{pmatrix} 1-\delta & \delta \\ \delta & 1-\delta \end{pmatrix} \right\}, \ \delta \in \left(0, \frac{1}{2}\right)$$

$$C_{\mathrm{NRI}}(\mathcal{W}) > C(\mathcal{W}) = 1 - h(\delta),$$

where the identity is a very special case of the capacity theorem of [A6]. Computer results by B. Balkenhol suggest that $C_{\mathrm{NRI}}(\mathcal{W}) < 1$. The heart of the matter is the following coding problem.

We denote by $\mathcal{B}(u^n, d) \subset \{0,1\}^n$ the Hamming ball with radius d. For numbers $1 < \beta < \delta < \frac{1}{2}$ and $\lambda \in (0,1)$ find a subset $A \subset \{0,1\}^n$ as large as possible such that for all $x^n \in A$

$$\left| \mathcal{B}(x^n, n\delta) \cap \left[\bigcup_{y^n \in A \setminus \{x^n\}} B(y^n, n\beta) \right] \right| \leq \lambda |B(x^n, n\delta)|.$$

As special cases we get

a) $\delta = \beta$: Here the problem is the essence of the Coding Theorem for the BSC.

b) $\lambda = 0$: Then the problem reduces to a generalization of the problem of maximal error correcting codes (where in addition $\delta = \beta$).

R. Ahlswede et al. (Eds.): Information Transfer and Combinatorics, LNCS 4123, p. 1063, 2006.
© Springer-Verlag Berlin Heidelberg 2006

Intersection Graphs of Rectangles and Segments

R. Ahlswede and I. Karapetyan

Let F be a finite family of sets and $G(F)$ be the intersection graph of F (the vertices of $G(F)$ are the sets of family F and the edges of $G(F)$ correspond to intersecting pairs of sets). The transversal number $\tau(F)$ is the minimum number of points meeting all sets of F. The independent (stability) number $\alpha(F)$ is the maximum number of pairwise disjoint sets in F. The clique number $\omega(F)$ is the maximum number of pairwise intersecting sets in F. The coloring number $q(F)$ is the minimum number of classes in a partition of F into pairwise disjoint sets.

The following problem was raised by Gýarfás and Lehel [1]. Suppose that some rectangles with sides parallel to coordinate axes are given in the plane. Is there a constant c such that

$$\tau(F) \leq c\alpha(F)?$$

Gýarfás and Lehel showed [1]

$$\left\lfloor \frac{3}{2}\alpha(F) \right\rfloor \leq \tau(F) \leq \alpha^2(F).$$

J. Beck [2] improved the upper bound to $c\alpha \log^2 \alpha$, where $\alpha = \alpha(F)$. Gy. Károlyi [2] improved it to

$$c\alpha \log \alpha.$$

Fon-Der-Flaass and Kostochka [3] improved the lower bound to

$$\left\lfloor \frac{5}{3}\alpha \right\rfloor.$$

Statement 1. Let $F = \{R_1, R_2, \ldots, R_n\}$ be a family of rectangles in the plane. Then $\tau(F) \leq 2(k+1)\alpha(F)$, where $k = \max\limits_{1 \leq i \leq n} \left\lceil \frac{\ell_i(R_i)}{\omega_i(R_i)} \right\rceil$ and $\ell_i(R_i)$ is the length of R_i and $\omega_i(R_i)$ is the width of R_i.

Corollary. It F is a family of squares in the plane, then

$$\tau(F) \leq 4\alpha(F).$$

Statement 1*. If F is a family of unit squares, then $\tau(F) \leq 2\alpha(F)$.

Statement 2. Let $F = \{R_1, R_2, \ldots, R_n\}$ consist of only k noncongruent rectangles $(k \leq n)$ then

$$\tau(F) \leq 4k\alpha(F),$$

For a family F of rectangles in the plane Asplund and Grünbaum in [4] proved that $q(F) \leq 4\omega^2(F) - 3\omega(F)$ and if $G(F)$ is triangle-free then $q(F) \leq 6$. Burling [5] showed that $q(F)$ cannot be bounded by any function of $\omega(F)$ for

R. Ahlswede et al. (Eds.): Information Transfer and Combinatorics, LNCS 4123, pp. 1064–1065, 2006.
© Springer-Verlag Berlin Heidelberg 2006

three-dimensional boxes. Akiyama, Hosono and Urabe [6] proved, if a family F of unit squares and $G(F)$ is triangle-free, then $q(F) \leq 3$ and they stated the following

Conjecture. $q(F) \leq \omega(F) + 1$ for a family F of unit squares in the plane. This conjecture is evidently not correct.

Counterexample. Replace each vertex of a pentagon (C_5) by a k-clique. Obviously the defined graph is intersection graph of unit squares in the plane, with clique number $2k$, while the coloring number is $3k$.

Statement 3. If F is a family of squares in the plane, then $q(F) \leq 4\omega(F) - 3$ and if $G(F)$ is triangle-free then $q(F) \leq 3$.

We note that statement 3 was independently proved by Perepelitsa [7].

Problem (Asplund, Grünbaum [4]). Suppose we have m concurrent straight lines L_i in a plane. Given m and ω, does there exist a family of segments F (in the plane) each of them parallel to one of the lines L_i, such that

$$q(F) = m\omega(F)?$$

In [4] two examples with $m = 2, \omega = 3, q = 5$, and $m = 2, \omega = 4, q = 6$ are given.

Statement 4. For $m = 2$ and every k there exists a family of segments (parallel to the concurrent lines L_1, L_2) with $\omega(F) = k$, such that

$$q(F) = 2k \quad \text{if} \quad k \quad \text{is even}$$

and

$$q(F) = 2k - 1 \quad \text{if} \quad k \quad \text{is odd}.$$

Statement 5. If F is a family of segments parallel to the concurrent lines L_1, L_2 in the plane and $G(F)$ has no induced odd cycles $(C_{2k+1}, k \geq 2)$ then $G(F)$ is a perfect graph.

Recall that a graph G is said to be perfect if $q(G') = \omega(G')$ for every induced subgraph G' of G.

References

1. A. Gýarfás and J. Lehel, Covering and coloring problems for relatives of intervals, Discrete Math. 55, 167-180, 1985.
2. G. Károlyi, On point covers of parallel rectangles, Periodica Mathematica Hungarica, Vol. 23, 2, 105-107, 1991.
3. D.G. Fon-Der-Flaass and A.V. Kostochka, Covering boxes by points, Discrete Math. 120, 269-275, 1993.
4. E. Asplund and B. Grünbaum, On a coloring problem, Math. Scand. 8, 181-188, 1960.
5. J. Burling, On coloring problems of families of prototypes, Ph.D. Thesis, University of Colorado Boulder, CO, 1965.
6. J. Akiyama, K. Hosono and H. Urabe, Some combinatorial problems, Discrete Math. 116, 291-298, 1993.
7. I. Perepelitsa, The estimates of the chromatic number of the intersection graphs of figures on the plane (personal communication).

Cutoff Rate Enhancement

E. Arikan

1 Cutoff Rate

The cutoff rate of a discrete memoryless channel (DMC) $W : \mathcal{X} \to \mathcal{Y}$ is defined as

$$R_0(W) = \max_Q - \log \sum_{y \in \mathcal{Y}} \left[\sum_{x \in \mathcal{X}} Q(x)\sqrt{W(y|x)} \right]^2 \tag{1}$$

where the maximum is over all probability distributions on $\mathcal{X}$. This parameter serves as a figure of merit for coding applications. There is also a decoding algorithm known as 'sequential decoding' that can readily achieve rates up to R_0.

2 Parallel Channels Theorem

Gallager [1, p.149] shows that if W is the parallel combination of two independent DMC's W_1 and W_2, then $R_0(W) = R_0(W_1) + R_0(W_2)$. A precise definition of parallel combination is as follows. If $W_1 : \mathcal{X}_1 \to \mathcal{Y}_1$ and $W_2 : \mathcal{X}_2 \to \mathcal{Y}_2$, then $W : \mathcal{X}_1 \times \mathcal{X}_2 \to \mathcal{Y}_1 \times \mathcal{Y}_2$ and

$$W(y_1 y_2 | x_1 x_2) = W_1(y_1 | x_1) W_2(y_2 | x_2). \tag{2}$$

In particular, if we consider the channel W^n that consists of n parallel uses of the same DMC W, this theorem gives

$$R_0(W^n) = n R_0(W) \tag{3}$$

3 Massey's Example

If the parallel channels are not independent, the above theorem no longer applies. The following example due to Massey [2] illustrates this. Suppose W is the parallel combination of two binary erasure channels (BEC) W_1 and W_2 that are completely correlated in the sense that erasures on the two channels always occur simultaneously. Then, a simple calculation shows that

$$R_0(W) = \log \frac{4}{1 + 3\epsilon} \tag{4}$$

$$R_0(W_1) = R_0(W_2) = \log \frac{2}{1 + \epsilon} \tag{5}$$

and

$$R_0(W) < R_0(W_1) + R_0(W_2) \tag{6}$$

for all $0 < \epsilon < 1$, where ϵ is the erasure probability.

R. Ahlswede et al. (Eds.): Information Transfer and Combinatorics, LNCS 4123, pp. 1066–1068, 2006.
© Springer-Verlag Berlin Heidelberg 2006

This inequality is obtained under the assumption that completely independent encoding and decoding are carried out on the subchannels. The decoders for W_1 and W_2 do not know each other's code, and do not in fact observe each other's channel output. This result is counter-intuitive in many ways and should make one think hard about the meaning of cutoff rate.

We may note that the capacities of these channels are $C(W) = 2(1 - \epsilon)$, $C(W_1) = C(W_2) = (1 - \epsilon)$ bits, and so $C(W) = C(W_1) + C(W_2)$. Certainly, the capacity $C(W)$ represents the ultimate achievable rate by the channel W and it cannot be increased by splitting or otherwise. What is unexpected is that the capacity is not decreased by this operation and in fact the cutoff rate is increased!

4   Ahlswede, Balkenhol, Cai Example

Ahlswede et al. [3] considered the case where W is the parallel combination of two completely correlated binary symmetric channels (BSC) W_1 and W_2. Thus, errors on W_1 and W_2 occur always simultaneously. The capacities are given by (for $\epsilon \neq 1/2$)

$$C(W) = 1 + (1 - \mathcal{H}(\epsilon)) \tag{7}$$

$$C(W_1) = C(W_2) = 1 - \mathcal{H}(\epsilon) \tag{8}$$

where

$$\mathcal{H}(\epsilon) = -\epsilon \log(\epsilon) - (1 - \epsilon) \log(1 - \epsilon),$$

and ϵ is the probability of error. In this case, the capacity is decreased if one ignores the correlation between the two subchannels. The sum capacity under independent encoding and decoding equals $C(W_1) + C(W_2)$, which is less than $C(W)$ for all $0 < \epsilon < 1/2$.

However, Ahlswede et al. proposed a scheme that employs independent encoding on the two subchannels W_1 and W_2 and a two-stage decoding algorithm, where the decoder for W_1 passes its estimate of the error locations to the decoder for W_2 This scheme achieves rates up to $C(W_1)$ on channel W_1, and rates up to 1 bit on W_2, for a sum rate that equals the capacity $C(W)$ of the original channel.

It appears that Massey's example is an example of a very specific type correlation between subchannels where decoder cooperation has no payoff. In general, we will be interested in decoding strategies where the decoders help each other by passing their final decisions.

If we look at the cutoff rates for this example, we obtain that

$$R_0(W) = 1 + \log \frac{2}{1 + \sqrt{4\epsilon(1 - \epsilon}} \tag{9}$$

$$R_0(W_1) = R_0(W_2) = \log \frac{2}{1 + \sqrt{4\epsilon(1 - \epsilon}} \tag{10}$$

If one employs the two-stage procedure, with the decoder for W_1 passing its estimates of error locations to the decoder for W_2, then the sum cutoff rate equals the cutoff rate $R_0(W)$ of the original channel

$$R_0(W_1) + R_0(W_2|W_1) = R_0(W) \tag{11}$$

where we introduced the notation $R_0(W_2|W_1)$ to denote the cutoff rate under the condition the two-stage decoding rule.

On the other hand if decoders work independently, the sum cutoff rate is smaller than that of the original channel

$$R_0(W) > R_0(W_1) + R_0(W_2) \tag{12}$$

for all $0 < \epsilon < 1/2$.

Although, there is no improvement (and no loss) in the cutoff rate under the two-stage decoding scheme, there may be complexity advantages by splitting the decoder into two parts as it avoids decoding a large code and instead decodes two smaller ones in succesion.

5   Open Problem

The above examples raise several questions.

- Given a channel W, is it always possible to split it into two subchannels W_1 and W_2 such that under two-stage decoding $R_0(W) < R_0(W_1) + R_0(W_2|W_1)$? What is the largest achievable gain?
- Identify the class of channels for which independent encoding and decoding improves the cutoff rate, i.e., $R_0(W) < R_0(W_1) + R_0(W_2)$.
- Consider splitting the nth extension W^n of W in the same way. What is the largest achievable gain in terms of the sum cutoff rate?
- Another direction is to consider the above problems under L-way splitting, where W is split into L subchannels $W_1, \ldots, W_L$, independent encoding is applied on each subchannel, and the decoding is done in L stages where every decoder observes the entire output of channel W plus the estimates of decoders that precede it.
- The above problems may be studied also under the assumption that the inputs to the subchannels are correlated in some way. What is the sum cutoff rate in this case?

References

1. R.G. Gallger, Information Theory and Reliable Communication, Wiley, 1968.
2. J.L. Massey, Capacity, cutoff rate, and coding for a direct-detection optical channel, IEEE Trans. Comm., Vol. 29, No. 11, 1615-1621, 1981.
3. R. Ahlswede, B. Balkenhol, and N. Cai, Parallel error correcting codes, IEEE Trans. Inform. Theory, Vol. 48, No. 4, 959-962, 2002.

Some Problems in Organic Coding Theory

S. Artmann

1 Introduction

Organic coding theory describes, analyses, and simulates the structure and function of organic codes (OC), viz., of codes used in living systems as sets of arbitrary rules of encoding and decoding between two independent subsystems [3]. A very well-known example of OC is the genetic code (GC), a degenerate quaternary code of length 3 whose codewords (mRNA triplets) encode amino acids, which are component parts of the primary structure of proteins, and the beginning and end of encoding mRNA sequences. From a semiotic point of view, GC is of great interest because it breaks free from a pure symbolic way of encoding which can be characterized as resulting in codes wherein the mutual Kolomogorov complexity of the encoding and the encoded structure is nearly equal to the Kolomogorov complexity of each one of these structures [2]. GC is analysed thoroughly since fifty years. But organic coding theory is still in its infancy. Many problems of formal, empirical, and methodological nature are open. In the following, I present three of them: the empirical problem of the evolutionary function of OCs (2), the methodological problem of the arbitrariness of OCs (3), and the formal problem of selecting an adequate model of OCs (4).

2 Why Organic Codes?

A long time it was believed that GC is a unique phenomenon in nature. Today the picture has changed radically: There are many more OCs in living systems, e.g., mRNA splicing codes, membrane adhesion codes, signal transduction codes between the intercellular and intracellular messenger systems, pattern codes in ontogeny, and so on. From a first systematical survey of OCs known today, it was conjectured that every macro-evolutionary step in the history of life consists in establishing a new OC [3]: without splicing codes no eukaryotes, without adhesion and transduction codes no multicellular organisms, and without pattern codes no animals. This conjecture must be subjected to further empirical research on the number, structure, function, origin and evolution of OCs.

3 Why These Organic Codes?

If the concept of OC is ontologically useful in biology, that is, if this concept is necessary to define an important class of biological objects, then biological research on OC is threatened to be trapped in an epistemological deadlock. One of the fundamental characteristics of OCs is their arbitrariness: The encoding

R. Ahlswede et al. (Eds.): Information Transfer and Combinatorics, LNCS 4123, pp. 1069–1072, 2006.
© Springer-Verlag Berlin Heidelberg 2006

resp. decoding rules between the encoding and the encoded structure of an OC are not motivated by some extra-codical affinities between both structures. This well-known property of the signs of human language [14] was described in the case of GC as its origin in a "frozen accident" [6] whose singular occurence is not deducible from physico-chemical laws but can only be explained by the narration of some contingent boundary conditions. If this is true then the only message that, from an evolutionary point of view, the existence of GC can deliver us is that it is more advantageous for self-reproduction to have any GC than to have no GC. Its internal structure passes on pure natural chance as the only source of evolutionary innovation. More generally, the concept of OC is, in this case, an index of an epistemological limitation that biological research will never overcome because the structure of OC can only be accepted as "gratuit" [11], i.e., given for free. Since this is not a satisfying situation for biology, the "frozen accident"-theorem of the arbirariness of OC must be subjected to further research which should combine empirical and formal aspects of the evolution of OCs [10].

4   How to Model?

Formal models of OC that can help to show a way out of the methodological problem should fulfil the following methodological requirements. Syntactically, the model has to allow an adequate and clear representation of the encoding and the encoded structure in a uniform conceptual framework so that none of the ontological dualisms can appear that haunt the theory of human language (e.g., between the material and the mental aspect of a code). Semantically, the model must be able to describe its semantics by its own means so that the difference between the object-level and the meta-level of the model is not reflected in any kind of epistemological dualism. Pragmatically, the model has to simulate the modelled object so that it can stimulate empirical research with virtual experiments. Petri Nets (PNs), a species of bipartite graphs, fulfil all three requirements [12,4]. PNs allow a formal and graphic representation of concurrent processes in one topological structure. The semantics of a PN is representable as a PN. A PN simulates the processes that can happen in its framework by modelling the dynamics of tokens marking the component parts of a PN. Therefore it is not surprising that PNs are used for the simulation of biological systems, especially of metabolic pathways [5,7,8]. The most simplest PN is defined as follows [Bra80]: A directed net is a triple $N = (S, T; F)$ so that

1. $S \cap T = \varnothing$
2. $S \cup T \neq \varnothing$
3. $F \subseteq (S \times T) \cup (T \times S)$
4. $dom(F) \cup codom(F) = S \cup T$.

S (the set of places) and T (the set of transitions) represent the two kinds of nodes of a PN, and F (the flow relation) is a subset of all such arcs between S- and T-nodes that connect a S-node to a T-node or a T- to a S-node.

We propose that for the modelling of OCs interfaces are the most important PN structures. They are defined (for simple nets wherein two nodes that are connected in the same way to the same sets of nodes are identified) as follows [16]: $N = (S, T; F)$ is a simple net, and $B = (SB, TB; FB)$ is a subnet of N. $Q \subseteq SB \cup TB$ is a set of distinguished nodes of B where $Z = (SZ, TZ; FZ)$ with $SZ = S - (SB - Q)$, $TZ = T - (TB - Q)$, and $FZ = F \cap (SZ \times TZ \cup TZ \times SZ)$ is also a subnet of N. Q is called the interface of B and Z, and Z is called the environment of B with respect to Q (in N), iff 1) there exists no path from an element of B to an element of Z not containing an element of Q, and 2) $F \cap (Q \times Q) = \varnothing$ Such interfaces can serve in PNs as representations of arbitrary encoding and decoding rules between two subnets that represent the encoding and the encoded structures connected by a code. As far as I can see, the potential of interfaces to model OCs was not realized until now. A first step towards the use of PN interfaces would be a systematic investigation into the behaviour of different interface structures in different types of environment. This could be done by using PNs with interface structures in Artificial Life experiments on virtual ecosystems wherein PNs compete against each other as virtual organisms. The use of PNs can help to solve an open problem in Artificial Life: Simulations of the evolution of ecosystems like Tierra [13], Echo [9], and Avida [1] surely show interesting evolutionary dynamics but they seem to be rather too limited in comparison to the requirements of open-ended evolutions of complexity [15]. My proposal is that one of the main reasons of this limitation is the neglect of OCs in these programs because each OC opens up vast new areas where natural and virtual evolution can search for unforeseen solutions of adaptive problems.

Acknowledgment

I thank Andr Skusa (Bielefeld) and Peter Dittrich (Jena) for stimulating discussions.

References

1. C. Adami, Introduction to Artificial Life, New York 1998
2. S. Artmann, Using semiotics in artificial life, to be published in: Joachim Schult (ed.), Proceedings of Workshop "Biosemiotics. Practical Applications and Consequences for the Sciences" (Ernst-Haeckel-Haus Jena, May 2002)
3. M. Barbieri, The Organic Codes. An Introduction to Semantic Biology, Cambridge 2003
4. W. Brauer (ed.), Net Theory and Applications: Proceedings of the Advanced Course on General Net Theory of Processes and Systems, Hamburg, 1979 (Lecture Notes in Computer Science, Vol. 84), Berlin, Heidelberg, New York 1980
5. M. Chen and R. Hofestdt, Quantitative Petri net model of gene regulated metabolic networks in the cell, in: In Silico Biology, 3, 0029, 2003.
6. F.H.C. Crick, The origin of the genetic code, Journal of Molecular Biology, 38, 367–379, 1968.

7. H.J. Genrich, R. Küffner, and K. Voss, Executable petri net models for the analysis of metabolic pathways, in: K. Jensen (ed.): DAIMI PB, No. 547: Workshop Proceedings Practical Use of High-level Petri Nets, University of Aarhus, Department of Computer Science, 1–14, 2000.

8. P.J.E. Goss and J. Peccoud, Quantitative modeling of stochastic systems in molecular biology by using stochsatic petri nets, Proceedings of The National Academy of Sciences USA, 6750–6755, 1995.

9. P.T. Hraber, T. Jones, and S. Forrest, The ecology of Echo, Artificial Life, 3, 165–190, 1997.

10. S. Kauffman, The Origins of Order, Oxford, 1993.

11. J. Monod, Le Hasard et la Ncessit, Paris, 1970.

12. C.A. Petri, Kommunikation mit Automaten, Bonn, 1962.

13. T.S. Ray, An approach to the synthesis of life, in: C. Langton et al. (eds.), Artificial Life II, Santa Fe Institute Studies in the Sciences of Complexity, vol. XI, Redwood City, 371–408, 1991.

14. F. de Saussure, Cours de Linguistique Gnrale, Geneva, 1916.

15. R. M. Smith and M.A. Bedau, Is echo a complex adaptive system?, Evolutionary Computation, 8, 419–442, 2000.

16. K. Voss, Interface as a basic concept for systems specification and verification, in: K. Voss, H.J. Genrich, and G. Rozenberg (eds.), Concurrency and Nets - Advances in Petri Nets, Berlin, Heidelberg, New York, 585–604, 1987.

Generalized Anticodes in Hamming Spaces

H. Aydinian

Let $\mathcal{H}(n) = \{0,1\}^n$ denote the binary Hamming space with the Hamming distance d_H. The Hamming weight is denoted by wt_H. Given integers $l \geq 1$, $1 \leq \delta < n$, let $\mathcal{A} \subset \mathcal{H}(n)$ satisfy the Condition (D): for every subset $A \subset \mathcal{A}$ with $|A| = l+1$ there exist two distinct points $a, b \in A$ with $d_H(a,b) \leq \delta$.

Problem. Determine or estimate

$$\mathcal{D}(n, \delta, l) := \max |\mathcal{A}|.$$

Remark 1. The same problem can be considered for other metric spaces. Among them the q–ary Hamming space $\mathcal{H}_q(n) = \{0, 1, \ldots, q-1\}^n$ and the Johnson space $\mathcal{J}(n, k) = \{x \in \mathcal{H}(n) : \mathrm{wt}_H(x) = k\}$ with Johnson distance $d_J = \frac{1}{2}d_H$ are most important from coding point of view.

Comment. For $l = 1$ we have the Isodiametric Problem: Given n, δ, find the largest size of a set of points $\mathcal{A} \subset \mathcal{H}(n)$ with diameter δ ($\delta := \max_{a,b \in \mathcal{A}} d_H(a,b)$). This problem was solved by Kleitman [1] who showed that Hamming balls of diameter δ have optimal size. Let $B_e(x)$ be the Hamming ball of radius e centered at x, that is $B_e(x) := \{a \in \mathcal{H}(n) : d_H(a,x) \leq e\}$. Let also $b_e := |B_e(x)|$.

Theorem (Kleitman [1]). For $0 < \delta < n$

$$\mathcal{D}(n, \delta, 1) = \begin{cases} b_e, & \text{if } \delta = 2e \\ b_e + \binom{n-1}{e}, & \text{if } \delta = 2e+1 \end{cases}$$

Remark 2. The corresponding problems for the Johnson space and Hamming space $\mathcal{H}_q(n)$ were solved by Ahlswede and Khachatrian in [2] and [3], respectively.

The problem is open for $l = 2$.
Consider the case when $\delta = 2e$ is even. Observe then that two disjoint Hamming balls $B_e(x)$ and $B_e(y)$ of radius $e \leq \frac{1}{2}n$ satisfy Condition (D), i.e. for every three points $a_1, a_2, a_3 \in B_e(x) \cup B_e(y)$ at least two are at distance $\leq \delta$. Thus we have

$$\mathcal{D}(n, 2e, 2) \geq 2b_e.$$

Can we do it better?

Let us consider the following construction of a competitor set, which in some cases is "better" than two balls.

Construction. For an integer n and $0 \leq r \leq e \leq (n-1)/3$ define

$$S_r := \{(a,b) \ : \ a \in \mathcal{H}(n - 3r - 1), \ \mathrm{wt}_H(a) \leq e - r, b \in \mathcal{H}(3r + 1)\},$$

R. Ahlswede et al. (Eds.): Information Transfer and Combinatorics, LNCS 4123, pp. 1073–1074, 2006.
© Springer-Verlag Berlin Heidelberg 2006

Observe that S_r satisfies Condition (D) with $\delta = 2e$. Thus, for $1 \le e \le (n-1)/3$, we have

$$\mathcal{D}(n, 2e, 2) \ge \max_{0 \le r \le e} |S_r|.$$

Examples

1. $n = 15$, $\delta = 8$.

   $2b_4 = 2 \sum_{i=0}^{4} \binom{15}{i} = 2 \cdot 1941.$

   $S_4 = 2^{13} = 2 \cdot 4096 > 2b_4.$

2. $n = 19$, $\delta = 8$.

   $2b_4 = 2 \sum_{i=0}^{4} \binom{19}{i} = 2 \cdot 5036.$

   $S_3 = 10 \cdot 2^{10} = 2 \cdot 5120 > 2b_4.$

Remark 3. As an application we note that the function $\mathcal{D}(n, \delta, l + 1)$ can be used to obtain better upper bounds for $(n, \delta + 1)$–codes $\mathcal{C} \subset \mathcal{H}(n)$ correcting $e = \lfloor \delta/2 \rfloor$ errors (see [4]). In fact we have

$$|\mathcal{C}| \le \frac{2^n \cdot l}{\mathcal{D}(n, \delta, l)}.$$

References

1. D.J. Kleitman, On a combinatorial conjecture of Erdős, J. Combin. Theory, 1, 209–214, 1996.
2. R. Ahlswede and L.H. Khachatrian, The complete intersection theorem for systems of finite sets, European J. Combin., 18, 125–136, 1997.
3. R. Ahlswede and L.H. Khachatrian, The diametric theorem in Hamming spaces – optimal anticodes, Advances in Applied Mathematics, 20, 429–449, 1997.
4. R. Ahlswede, H. Aydinian and L.H. Khachatrian, Perfect codes and related concepts, Designs, Codes and Cryptography, 22, 221–237, 2001.

Two Problems from Coding Theory

V. Blinovsky

We suggest to prove the two following conjectures about the properties of some special functions.

First consider the following polynome in $\lambda \in [0, (q-1)/q]$, $q = 2, 3, \ldots$

$$f_L^q(\lambda) = \sum_{j_i:\, \sum_{i=1}^{q} j_i = L+1} \left(1 - \frac{\max\{j_1, \ldots, j_q\}}{L+1}\right) \binom{L+1}{j_1, \ldots, j_q} \left(\frac{\lambda}{q-1}\right)^{L+1-j_q} (1-\lambda)^{j_q}. \tag{1}$$

It arise in the problem of obtaining the upper bound for the rate of multiple pa cking of q−ary Hamming space. The problem is to prove that this function is $\bigcap$−convex.

In the binary case $q = 2$ it can be shown that

$$f_L^2(\lambda) = \sum_{i=1}^{\ell} \frac{\binom{2i-2}{i-1}}{i} (\lambda(1-\lambda))^i,$$

where $\ell = \left\lfloor \frac{L}{2} \right\rfloor$ and

$$(f^2)_L'' = -\ell \binom{2\ell}{\ell} (\lambda(1-\lambda))^{\ell-1}.$$

Hence in binary case this conjecture is valid. Problem is to prove the convexity of $f_L^q(\lambda)$ for the arbitrary q.

Next problem arise when I obtain the upper bound for the reliability function for list-of−L decoding in binary symmetric channel. It is necessary to prove the inequality

$$\varphi_L^q(\xi) = \sum_{i=1}^{\ell} (-1)^{i+q} \binom{L+1}{i} (a_{i,q} + a_{i,q-1}) \ln \frac{\cosh\left(\frac{L+1}{2} - i\right)\xi}{\cosh \frac{L+1}{2}\xi} \leq 0, \tag{2}$$

where $q = 1, 2, \ldots, \ell$; $\xi \in R^1$, and

$$a_{i,q} = \frac{(L+1)\binom{L-i-q-1}{L-2q-1} + i\binom{L-i-q-1}{L-2q-2}}{\binom{2q}{q}}.$$

We conjecture that function $\varphi_L^q(\xi)$ is $\bigcap$−convex (from which inequality (2) easily follows).

R. Ahlswede et al. (Eds.): Information Transfer and Combinatorics, LNCS 4123, p. 1075, 2006.
© Springer-Verlag Berlin Heidelberg 2006

Private Capacity of Broadcast Channels

N. Cai

The broadcast channel was introduced by T. M. Cover in 1972 [7]. In its simplified version, it has one sender (or encoder) E and two users (or decoders) D_l, $l = 1$, 2. The sender E is required to send the messages m_1 and m_2 uniformly chosen from the message sets $\mathcal{M}_1$ and $\mathcal{M}_2$ respectively to D_1 and D_2 correctly with probability close to one. That is, the sender encodes the message (m_1, m_2) to an input sequence x^n over a finite input alphabet $\mathcal{X}$ and sends it to the two users via two noisy channels W^n and V^n, respectively. The first (second) user D_1 (D_2) decodes the output sequence y^n over the finite output alphabet $\mathcal{Y}$ of the channel W^n (the output sequence z^n over the finite output alphabet $\mathcal{Z}$ of the channel V^n) to the first message $\hat{m}_1$ (the second message $\hat{m}_2$). In general, the capacity regions for this kind of channels are still unknown. Their determination is probably one of the hardest open problems in Multi-user Shannon Theory.

The wiretap channel was introduced by A. D. Wyner in 1975 [12] and generalized by I. Csiszàr and J. Körner in 1978 [6], which can be considered as applying a "broadcast channel" for a different purpose. Namely, a wiretap channel has the same statistical properties as a broadcast channel and the difference is that one of the receivers, say D_2, now is assumed to be an illegal user, or an eavesdropper. There is only one message from $\mathcal{M}(= \mathcal{M}_1)$ to be transmitted (i. e., $\mathcal{M}_2$ does not exist at all). The requirement for the wiretap channel is that the legal receiver D_1 should be able to recover the message from $\mathcal{M}$ correctly with high probability whereas the eavesdropper D_2 should obtain no significant knowledge about the message. To protect the security of the data randomization at encoder is allowed. The capacity regions for wiretap channels were determined ([12] and [6]).

However, in the real life the answer to the question who is legal user often depends on the sources of message. For example, each customer is only legal for the statement of his own account but illegal for the others when a bank distributes the statements of balances of accounts to its customers. Similar situations are for letters, emails,..... This motivates us to study a communication system, in which a sender sends different messages to different users (receivers) and each user is only legal for his own message but illegal for the others. The following model was introduced by N. Cai and K. Y. Lam [5] in 2000. Two messages M_1 and M_2 uniformly distributed on two finite sets $\mathcal{M}_1$ and $\mathcal{M}_2$ respectively, are encoded to an input sequence x^n and x^n is sent to the two users via two noisy channels W^n and V^n, respectively. Like in broadcast channel, we require that the first (second) user D_1 (D_2) decodes the output sequence y^n of the channel W^n (the output sequence z^n of the channel V^n) to the first message $\hat{m}_1$ (the second message $\hat{m}_2$) such that the decoding error $Pr\{\hat{m}_1 \neq m_1\}$ ($Pr\{\hat{m}_1 \neq m_1\}$) is arbitrarily small when the length n of the code increases. In addition we want the information of the other message obtained by each user to vanish i.e., $\frac{1}{n}I(M_2; Y^n) \longrightarrow 0$ and

R. Ahlswede et al. (Eds.): Information Transfer and Combinatorics, LNCS 4123, pp. 1076–1078, 2006.
© Springer-Verlag Berlin Heidelberg 2006

$\frac{1}{n}I(M_1; Z^n) \longrightarrow 0$, where Y^n and Z^n are the output random variables of W^n and V^n generated by the pair (M_1, M_2) of messages. The capacity region of this communication system is called private capacity region of broadcast channel.

A broadcast channel is called deterministic if there exists a pair of functions $\phi : \mathcal{X} \longrightarrow \mathcal{Y}$ and $\psi : \mathcal{X} \longrightarrow \mathcal{Z}$ such that $W(y|x) = 1$ iff $y = \phi(x)$ and $V(z|x) = 1$ iff $z = \psi(x)$. Its ("non-private") capacity region was determined by M. S. Pinsker [11] in 1978. It was shown in [5] that the coding theorem for private capacity of deterministic broadcast channel without randomization at encoder is equivalent to the following combinatorial problem:

for a given $0 - 1$ matrix A (over a proper field) find an all 1 submatrix of its nth Kronecker power with maximum size.

The problem for Yao's lower bound [13] to two-way communication complexity (e.g., [2]) has been studied by several authors and in general it is still open. The private capacity region of deterministic broadcast channel was determined by N. Cai and K. Y. Lam [5]. The main technique in the proof of the direct part is coloring (or binning) on rows and columns of the $0 - 1$ matrix.

In general the coding problem to broadcast channel with private capacity is still widely open. This provides us research area in multi-user Information Theory.

Problem 1 is to find private capacity region of general broadcast channels or a class of broadcast channels if it would be too hard for general channels.

Problem 2 is to find good inner bound and/or outer bound to the private capacity. In particular, can one obtain an inner bound by applying coloring (binning) technique to Marton's inner bound for (ordinary) capacity region of broadcast channels [9]?

An interesting direction in Information Theory started by J. Massey [10] 1981, is splitting a single channel to two correlated channels to reduce the coding complexity (more results are found in E. Arikan, [4]), which contains rich problems for research. As an example J. Massey presented a channel with an input alphabet $\{00, 01, 10, 11\}$, such that with probability ϵ an input symbol $x \in \{00, 01, 10, 11\}$ is erased (as ?) and otherwise is correctly transmitted to the output. He showed that one may reduce the coding complexity by splitting the channel to two correlated binary eraser channels without loosing the total capacity. We notice that one can apply two codes to the two correlated binary eraser channels separately such that the receiver for each channel can correctly decode the message to him but has no knowledge about the message to the other receiver. This provides us the next problem.

Problem 3 is to study how to split a single channel to two correlated channels for protection of privacy or for reducing the coding complexity and protecting privacy simultaneously.

The two channels obtained by splitting are more similar to an interference channel (introduced and the non-single-letter capacity region determined by R. Ahlswede [1], 1973 ; the best inner bound by T. S. Han and K. Kobayashi [8],

1981; capacity region in the case of noiseless transmission for one receiver, by R. Ahlswede and N. Cai [3]), whose single-letter capacity region in general is still unknown.

Problem 4 is to study the private capacity region for the interference channel and other multi-user channels.

References

1. R. Ahlswede, Multi-way communication channels, Proceeding 2nd international Symp. on Inform. Theory, Tsahkadsor, Armenia, USSR, 1971, Akadémiai Kiadó, Budapest, pp.23-52, 1973.
2. R. Ahlswede and N. Cai, On communication complexity of vector-valued functions, IEEE Trans. on Inform. Theory, vol. IT-40, no. 6, 2062-2067, 1994.
3. R. Ahlswede and N. Cai, Codes with the identifiable parent property and the multi-access channel, Part I, preprint, 2003.
4. E. Arikan, Creating correlated channels for coding gains, lecture in ZIF Universität Bielefeld, 2003.
5. N. Cai and K. Y. Lam, How to broadcast privacy: Secret coding for deterministic broadcast channels, in Numbers, Information, and Complexity (Festschrift for Rudolf Ahlswede), eds: I. Althöfer, N. Cai, G. Dueck, L. Khachatrian, M. Pinsker, A. Sarkozy, I. Wegener, and Z. Zhang, Kluwer, 353-368, 2000.
6. I. Csiszár and J. Körner, Broadcast channels with confidential messages, IEEE Trans. on Inform. Theory, vol. IT-24, pp. 339-348, 1978.
7. T. M. Cover, Broadcast channels, IEEE Trans. on Inform. Theory, vol. IT-18, pp 2-14, 1972.
8. H. S. Han and K. Kobayashi, A new achievable rate region for interference channel, IEEE Trans. on Inform. Theory, vol. IT-27, pp. 49-60, 1981.
9. K. Marton, A coding theorem for the discrete memoryless broadcast channels, IEEE Trans. on Inform. Theory, vol. IT-25, pp. 306-311, 1979.
10. J. L. Massey, Capacity, cutoff rate, and coding for direct-detection optical channel, IEEE Trans. Com., v. 27, pp. 1615-1621, 1981.
11. M. S. Pinsker, Capacity region of noiseless broadcast channels, Prob. Inform. Trans., vol. 14, pp.28-32, 1978.
12. A. D. Wyner, The wire-tap Channels, Bell System Tech. J., vol. 54, pp. 1355-1387.
13. A. Yao, Some complexity questions related to distributive computing, in proc. 11th Annu. ACM Symp. Theory Comput., pp. 209-213, 1979.

A Short Survey on Upper and Lower Bounds for Multidimensional Zero Sums

C. Elsholtz

After giving some background on sums of residue classes we explained the following problem on multidimensional zero sums which is well known in combinatorial number theory:

Let $f(n,d)$ denote the least integer such that any choice of $f(n,d)$ elements in $\mathbb{Z}_n^d$ contains a subset of size n whose sum is zero. Harborth [12] proved that $(n-1)2^d + 1 \leq f(n,d) \leq (n-1)n^d + 1$. The lower bound follows from the example in which there are $n-1$ copies of each of the 2^d vectors with entries 0 or 1. The upper bound follows since any set of $(n-1)n^d + 1$ vectors must contain, by the pigeonhole principle, n vectors which are equivalent modulo n.

If d is fixed, the upper bound was improved considerably by Alon and Dubiner [2] to $c_d\, n$. Erdős, Ginzburg, and Ziv [6] proved that $f(n,1) = 2n-1$ and Kemnitz conjectured that $f(n,2) = 4n - 3$. There are partial results due to Kemnitz [14], as well as Gao [8], [9], [10], [11], Rónyai [16] and Thangadurai [17].

For example, Rónyai [16] proved that for primes p one has $f(p,2) \leq 4p - 2$, which implies that $f(n,2) \leq 4.1\,n$. Gao [11] extended this to powers of primes: $f(p^a, 2) \leq 4p^a - 2$.

If n is fixed but d is increasing not very much is known.

$$f(2^a,d) = (2^a - 1)2^d + 1 \qquad \text{see Harborth [12],}$$
$$f(3,3) = 19 \qquad \text{see Harborth [12], Brenner [3],}$$
$$f(3,4) = 41 \qquad \text{see Brown and Buhler [4], Brenner [3], Kemnitz [14],}$$
$$91 \leq f(3,5) \leq 121 \qquad \text{see Kemnitz [13],}$$
$$f(3,18) \geq 300 \times 2^{12} \qquad \text{see Frankl, Graham, Rödl [7],}$$
$$f(3,d) \geq 2.179^d \text{ for } d \geq d' \qquad \text{see Frankl, Graham, Rödl [7],}$$
$$f(n,d) \geq (1.125)^{\lfloor \frac{d}{3} \rfloor} (n-1)2^d + 1 \quad \text{for odd } n, \text{ see Elsholtz [5],}$$
$$f(n,d) = o(n^d) \qquad \text{if } n \text{ is fixed and } d \text{ goes to infinity, see Alon and Dubiner,}$$
$$f(3,d) = O(\tfrac{3^d}{d}) \qquad \text{see Meshulam [15] .}$$

The proof of our lower bound was the first nontrivial lower bound for any value of $f(n,d)$ with $n > 3$. It is based on a set of 9 vectors considered by Harborth [12] tp prove that $f(3,3) = 19$, but we consider these in $\mathbb{Z}_n^3$ (not only in $\mathbb{Z}_3^3$).

$$\begin{pmatrix}2\\1\\2\end{pmatrix}, \begin{pmatrix}0\\0\\0\end{pmatrix}, \begin{pmatrix}0\\0\\1\end{pmatrix}, \begin{pmatrix}0\\1\\0\end{pmatrix}, \begin{pmatrix}0\\1\\1\end{pmatrix}, \begin{pmatrix}1\\0\\0\end{pmatrix}, \begin{pmatrix}1\\0\\1\end{pmatrix}, \begin{pmatrix}1\\1\\2\end{pmatrix}, \begin{pmatrix}1\\2\\2\end{pmatrix}.$$

R. Ahlswede et al. (Eds.): Information Transfer and Combinatorics, LNCS 4123, pp. 1079–1080, 2006.
© Springer-Verlag Berlin Heidelberg 2006

We proved:
Let $n \geq 3$ be odd. If any n vectors taken from a multiset of the above 9 vectors add to $0 \in \mathbb{Z}_n^3$, then necessarily one has taken n times the very same vector. Therefore, taking $n - 1$ copies of each vector only avoids zero sums of length n.

Remarks
It seems conceivable that starting off from other examples for small fixed n and d, one might be able to improve the lower bound. But it is not at all obvious that this will work for any particular value of $f(n, d)$. The proof in Elsholtz [5] is in principle elementary, and might be accessible to generalization, possibly with the help of computers. Albeit, an exhaustive search to determine values of $f(n, d)$ seems out of reach even for moderate sized n and d.

I hope that this survey of existing results convinces the reader that there is still a huge gap between the known upper and lower bounds.

References

1. N. Alon and M. Dubiner, Zero-sum sets of prescribed size, Combinatorics, Paul Erdős is eighty, Vol. 1, 33–50, Bolyai Soc. Math. Stud., János Bolyai Math. Soc., Budapest, 1993.
2. N. Alon and M. Dubiner, A lattice point problem and additive number theory. Combinatorica 15, 301–309, 1995.
3. J.L. Brenner, Problem 6298, Amer. Math. Monthly, 89, 279–280, 1982.
4. T.C. Brown, J.P. Buhler, A density version of a geometric Ramsey theorem, J. Combin. Theory Ser. A 32, 20–34, 1982.
5. C. Elsholtz, Lower bounds for multidimensional zero sums, Combinatorica 24, No. 3, 351 - 358, 2004.
6. P. Erdős, A. Ginzburg, and A. Ziv, Theorem in the additive number theory, Bull. Res. Council Israel 10F, 41–43, 1961.
7. P. Frankl, R.L. Graham, and V. Rödl, On subsets of abelian groups with no three term arithmetic progression, J. Combin. Theory Ser. A 45, 157–161, 1987.
8. W. Gao, On zero-sum subsequences of restricted size, J. Number Theory, 61, 97–102, 1996.
9. W. Gao, A combinatorial problem on finite abelian groups, J. Number Theory, 58, 100–103, 1996.
10. W. Gao, Two zero-sum problems and multiple properties, J. Number Theory, 81, 254-265, 2000.
11. W. Gao, Note on a zero-sum problem. J. Combin. Theory Ser. A 95, 387–389, 2001.
12. H. Harborth, Ein Extremalproblem für Gitterpunkte. J. Reine Angew. Math. 262/263, 356–360, 1973.
13. A. Kemnitz, Extremalprobleme für Gitterpunkte. Ph.D. Thesis, Technische Universität Braunschweig, 1982.
14. A. Kemnitz, On a lattice point problem. Ars Combin. 16, B, 151–160, 1983.
15. R. Meshulam, On subsets of finite abelian groups with no 3-term arithmetic progressions. J. Combin. Theory Ser. A 71, 168–172, 1995.
16. L. Rónyai, On a conjecture of Kemnitz, Combinatorica 20, 569–573, 2000.
17. R. Thangadurai, On a conjecture of Kemnitz, C. R. Math. Acad. Sci. Soc. R. Can. 23, no. 2, 39–45, 2001.

Binary Linear Codes That Are Optimal for Error Correction

T. Kløve

When a binary linear $[n, k]$ code C is used for error correction on the binary symmetric channel with bit error probability p, and the decoding algorithm is maximum likelihood decoding, the probability of correct decoding $P_{\mathrm{cd}}(C, p)$ is given by

$$P_{\mathrm{cd}}(C, p) = \sum_{i=0}^{n} \alpha_i \, p^i (1 - p)^{n-i},$$

where α_i is the number of correctable errors (coset leaders) of weight i.

An $[n, k]$ code C is **optimal for** p where $0 < p < 1/2$, if

$$P_{\mathrm{cd}}(C, p) \geq P_{\mathrm{cd}}(D, p)$$

for all $[n, k]$ codes D.

General problem: For given n, k, and p find an $[n, k]$ code that is optimal for p.

Slepian [3] and Fontaine and Peterson [1] found optimal codes for all p for some combinations of small n and k.

To present some conjectures, we find it convenient to describe the codes using *modular representation*. The modular representation of a $k \times n$ generator matrix G is the sequence $(x_0, x_1, \ldots, x_{2^k-1})$ where let x_j is the number of times the binary representation of j appears as a column in G. This representation determines G, and hence the code generated by G, up to equivalence (column permutations).

It is easy to see that for an optimal code we have $x_0 = 0$. Therefore, we leave this out and say that $\mathbf{x} = (x_1, \ldots, x_{2^k-1})$ represents the code generated by G. We also say that $\mathbf{x}$ is optimal when the corresponding code is optimal.

Based on some limited computations, we put forward some conjectures and problems.

Conjecture, dimension 2. For all $n \geq 2$ and all $p \in (0, 1/2)$ the following table gives an optimal $[n, 2]$ code:

n	representing vector
$3m + 2$	$(1, 1, 0) + m(1, 1, 1)$
$3m + 3$	$(2, 1, 0) + m(1, 1, 1)$
$3m + 4$	$(2, 1, 1) + m(1, 1, 1)$

R. Ahlswede et al. (Eds.): Information Transfer and Combinatorics, LNCS 4123, pp. 1081–1083, 2006.
© Springer-Verlag Berlin Heidelberg 2006

Conjecture, dimension 3. For all $n \geq 3$ and all $p \in (0, 1/2)$ the following table gives an optimal $[n, 3]$ code:

n	representing vector	range
$7m + 3$	$(1, 1, 0, 1, 0, 0, 0) + m(1, 1, 1, 1, 1, 1, 1)$	all p
$7m + 4$	$(1, 1, 0, 1, 0, 1, 0) + m(1, 1, 1, 1, 1, 1, 1)$	all p
$7m + 5$	$(1, 1, 0, 1, 0, 1, 1) + m(1, 1, 1, 1, 1, 1, 1)$	all p
$7m + 6$	$(1, 1, 0, 1, 1, 1, 1) + m(1, 1, 1, 1, 1, 1, 1)$	all p
$7m + 7$	$(2, 1, 0, 1, 1, 1, 1) + m(1, 1, 1, 1, 1, 1, 1)$	all p
$7m + 8$	$(2, 1, 1, 1, 1, 1, 1) + m(1, 1, 1, 1, 1, 1, 1)$	$p \leq p_n$
	$(2, 2, 0, 2, 1, 1, 0) + m(1, 1, 1, 1, 1, 1, 1)$	$p \geq p_n$
$7m + 9$	$(2, 2, 1, 1, 1, 1, 1) + m(1, 1, 1, 1, 1, 1, 1)$	all p

Remark. There seems to be no proof by simple induction on m since $\mathbf{x} + 1$ may be non-optimal even if $\mathbf{x}$ is optimal. For example, (for $k = 3$): $(1, 1, 0, 1, 0, 0, 1)$ is optimal for all p, but $(2, 2, 1, 2, 1, 1, 2)$ is not optimal for any p.

The values of p_{7m+8} have been computed for $m \leq 15$, and are given by the following table.

n	p_n	n	p_n
8	0.5	15	0.3069390306
22	0.3496295543	29	0.3020650080
36	0.2994471641	43	0.2936566721
50	0.2934421679	57	0.2932733749
64	0.2945611733	71	0.2959098813
78	0.2977592280	85	0.2995913456
92	0.3016067869	99	0.3035678969
106	0.3055858254	113	0.3075366414

We see that the value of p_{7m+8} is approximately 0.3 for $m \leq 15$.

Questions: Does $\lim_{m \to \infty} p_{7m+8}$ exist?
What is the limit (if it exists)?

Conjecture, general k. For any given k there exists a **finite** set

$$\{\mathbf{x}_1, \mathbf{x}_2, \ldots, \mathbf{x}_r\} \subset \{0, 1, 2, \ldots\}^{2^k - 1}$$

with the property that for any $n \geq k$ and any $p \in (0, 1/2)$ there exist a $j \in \{1, 2, \ldots, r\}$ and an $m \geq 0$ such that $\mathbf{x}_j + m\mathbf{1}$ represents an optimal $[n, k]$ code for p.

Problem. Find such sets for particular $k \geq 4$.

Related problems

We briefly mentions some related problems.

1. Consider the same problem for non-binary linear codes.

2. Consider the same problem (with maximum likelihood decoding) for non-linear codes. Do they sometimes do better than linear codes of the same length and size?

3. Similar problems can be stated for *error detection*. The probability of undetected error for a binary $[n, k]$ code C with weight distribution $A_0, A_1, \ldots, A_n$ is given by

$$P_{\text{ue}}(C, p) = \sum_{i=1}^{n} A_i \, p^i (1 - p)^{n-i}.$$

The code C is optimal for error detection for a given p, if

$$P_{\text{ue}}(C, p) \leq P_{\text{ue}}(D, p)$$

for all $[n, k]$ codes D. We note that a code may be optimal for error correction without being optimal for error detection and vice versa. It has been shown that for any $k \leq 4$ and for any n there exists a code that is optimal for all p (see [2, pp.85–91]). For $k = 2$ and $k = 3$ the following tables give optimal codes for error detection for all p (for the table for $k = 4$, see [2]).

n	representing vector
$3m + 2$	$(1, 1, 0) + m(1, 1, 1)$
$3m + 3$	$(1, 1, 1) + m(1, 1, 1)$
$3m + 4$	$(2, 1, 1) + m(1, 1, 1)$

n	representing vector
$7m + 3$	$(1, 1, 0, 1, 0, 0, 0) + m(1, 1, 1, 1, 1, 1, 1)$
$7m + 4$	$(1, 1, 0, 1, 0, 0, 1) + m(1, 1, 1, 1, 1, 1, 1)$
$7m + 5$	$(1, 1, 1, 1, 1, 0, 0) + m(1, 1, 1, 1, 1, 1, 1)$
$7m + 6$	$(1, 1, 1, 1, 1, 1, 0) + m(1, 1, 1, 1, 1, 1, 1)$
$7m + 7$	$(1, 1, 1, 1, 1, 1, 1) + m(1, 1, 1, 1, 1, 1, 1)$
$7m + 8$	$(2, 1, 1, 1, 1, 1, 1) + m(1, 1, 1, 1, 1, 1, 1)$
$7m + 9$	$(2, 2, 1, 1, 1, 1, 1) + m(1, 1, 1, 1, 1, 1, 1)$

For $k \geq 5$ nothing is known.

References

1. A.B. Fontaine and W.W. Peterson, Group code equivalence and optimum codes, IEEE Trans. Inform. Theory, Vol. 5, Special supplement, 60-70, 1959.
2. T. Kløve and V. Korzhik, Error Detecting Codes: General Theory and Their Application in Feedback Communication Systems, Boston, Kluwer Acad. Press, 1995.
3. D. Slepian, A class of binary signaling alphabets, Bell Syst. Tech. J., Vol. 35, 203–234, 1956.

Capacity Problem of Trapdoor Channel

K. Kobayashi

Abstract. The capacity problem of the trapdoor channel is one of famous long-standing problems. The trapdoor channel considered by
Blackwell[1] is a typical example of channel with memory. Ash[2] expressed the channel by using two channel matrices with four states,
while the expression does not necessarily make the problem tractable.
We provided an interesting recurrence giving the explicit expression of
the conditional distributions of the channel output sequences given input
sequences of length n [5]. In order to determine the capacity of this special channel, we have to understand the fractal structure buried in the
channel matrices between input and output sequences of long length.

1 Preliminaries

The actions of the trapdoor channel are described as follows. The input alphabet $\mathcal{X}$ and the output alphabet $\mathcal{Y}$ are binary. The channel has two trapdoors.
Initially, there is a symbol $s \in \{0,1\}$ called the *initial* symbol on one of trapdoors, and no symbol on another trapdoor. Just after the first input symbol
x_1 moves onto the empty trapdoor, one of trapdoors opens equiprobably, and
the symbol on the opened door falls to become an output symbol $y_1 \in \{s, x_1\}$.
After the door is closed, we are back to the same situation as at the initial instant, but there is a symbol s or x_1 on the non-empty door depending on the
output $y_1 = x_1$ or $y_1 = s$, respectively. This process is repeated until an output sequence $y_1 y_2 \ldots y_n$ has emitted from the channel for the input sequence
$x_1 x_2 \ldots x_n$.

Without loss of generality, we can assume that the initial symbol is 0 as far
as our concern is on the capacity problem.

2 Recursion of Channel Matrices

Let $P_{n|s}(\mathbf{x}, \mathbf{y})$ be the conditional probability of output sequence $\mathbf{y}$ by n equiprob-
able trapdoor actions for the input sequence $\mathbf{x}$ with the initial symbol s. Then,
we can show that the conditional probability matrices obey the following recur-
sions:

$$P_{n+1|0} = \begin{bmatrix} P_{n|0} & 0 \\ \frac{1}{2}P_{n|1} & \frac{1}{2}P_{n|0} \end{bmatrix}, \tag{1}$$

and

$$P_{n+1|1} = \begin{bmatrix} \frac{1}{2}P_{n|1} & \frac{1}{2}P_{n|0} \\ 0 & P_{n|1} \end{bmatrix}, \tag{2}$$

R. Ahlswede et al. (Eds.): Information Transfer and Combinatorics, LNCS 4123, pp. 1084–1087, 2006.

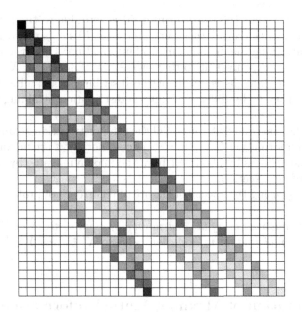

Fig. 1.

where the initial matrices are defined as

$$P_{0|0} = P_{0|1} = [1].\tag{3}$$

Fig. 1 is the density plot of the channel matrix $P_{5|0}$ where dark color corresponds to high probability. You can easily recognize the fractal structure contained in this channel.

As far as we study the capacity problem, we can further restrict to study the following half channel,

$$W_{n+1} = \left[\frac{1}{2}P_{n|1} \quad \frac{1}{2}P_{n|0}\right],\tag{4}$$

because the difference between the capacity of $P_{n|0}$ and that of W_n is at most one. Therefore, we will concentrate our discussions on the maximization of mutual information for the lower half channel W_n by its input distribution.

Remark 1: By investigating the form of the channel matrices (1) and (2), we can easily deduce that the zero error capacity of the trapdoor channel is one half. This result was already obtained by Ahlswede and Kaspi ([3]).

3   Problem to Be Solved

By the chain rule, we can decompose the mutual information of input and output sequences, $X^{n+1} = X^n X_{n+1}$ and $Y^{n+1} = Y^n Y_{n+1}$ of length $n + 1$ as follows.

$$I(X^{n+1}; Y^{n+1}) = I(X^n; Y^n) + I(Y_{n+1}; X^n|Y^n)$$
$$+ I(X_{n+1}; Y^n|X^n) + I(X_{n+1}; Y_{n+1}|X^nY^n).\tag{5}$$

Here we can show that the third term of (5) is zero due to the form of (4) and the recursion (1). Moreover, the fourth term of (5) is also zero for the input distribution attaining the maximal mutual information. Thus, the maximization of $I(X^{n+1}; Y^{n+1})$ is equivalent to that of $I(X^n; Y^n) + I(Y_{n+1}; X^n | Y^n)$.

Thus, the marginal distribution of the distribution attaining the maximal $I(X^{n+1}; Y^{n+1})$ might be different from the distribution attaining the maximal $I(X^n; Y^n)$.

Actually, the marginal distribution of the total distribution attaining the maximum mutual information $I(X^{n+1}; Y^{n+1})$ does "not" give the maximum value of $I(X^n; Y^n)$ and reduces by about 0.1 from the maximum value by the numerical calculations of next section.

Therefore, by using the special structure of recurrence of probability matrices we need

"*a systematic construction of the distribution of input sequences of length $(N+1)$ giving the maximal mutual information between X^{n+1} and Y^{n+1} from that of length N.*"

4   Computation of Maximal Mutual Information

We calculated[6] the maximum values of the mutual information $I(X^n; Y^n)$ for the channel W_n by finding the optimum input distribution of p_{X^n} for small n up to 10. Fig. 2 shows the maximum values of $I(X^n; Y^n)$ for $n = 1, \ldots, 10$. Fig. 3 and Fig. 4 show the first order and the second order difference of the data of Fig. 2, respectively.

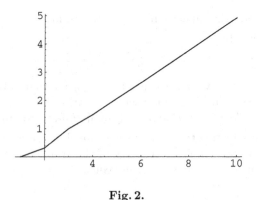

Fig. 2.

Note that as the second order difference is attenuated to zero in damped oscillation (Fig. 4), the increment of $I(X^n; Y^n)$ tends to a constant value around $0.572 \cdots$ (Fig. 3). Here we emphasize that the asymptotic increment of $I(X^n; Y^n)$ is the capacity to be determined. In [5], we argued that the capacity of trapdoor channel satisfy $C > 0.54$.

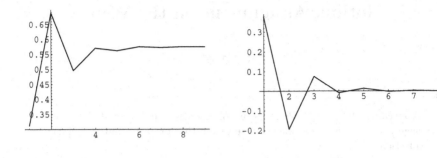

Fig. 3.   Fig. 4.

References

1. D. Blackwell, Information Theory, in Modern Mathematics for the Engineer, Second Series, E.F.Beckenbach, Ed., McGraw-Hill, 1961.
2. R.B. Ash, Information Theory, Interscience Publishers, 1965 and Dover Publications, Inc., 1990.
3. R. Ahlswede and A. Kaspi, Optimal coding strategies for certain permuting channels, Trans. Inf. Theory, Vol. 33, No. 3, 310-314, 1987.
4. R. Ahlswede, J.P. Ye, and Z. Zhang, Creating order in sequence spaces with simple machines, Information and Computation, Vol. 89, No. 1, 47–94, 1990.
5. K. Kobayashi, H. Morita, and M. Hoshi, An input/output recursion for the trapdoor channel, Proceedings of IEEE International Symp. on Inform. Theory, 423, 2002.
6. K. Kobayashi, H. Morita, and M.Hoshi, Some considerations on the trapdoor channel, Proceedings of 3rd Asian-European Workshop on Inform. Theory, 9–10, 2003.

Hotlink Assignment on the Web

E.S. Laber

Abstract. Here, we explain the problem of assigning hotlinks to web pages. We indicate the known results for this problem and the open questions.

1 Introduction

Due the expansion of the Internet at unprecedented rates, continuing efforts are being made in order to improve its performance. An important approach is improving the design of web sites [3,7]. A web site can be viewed as a directed graph where nodes represent web pages and arcs represent hyperlinks. In this case, the node that corresponds to the home page is a source node. Hence, when a user searches for an information i in a web site, it traverses a directed path in the corresponding graph, from the source node to the node that contains i. We assume that the user always knows which link leads to the desired information. In this context, we define hotlinks as additional hyperlinks added to web pages in order to reduce the number of accessed pages per search [7].

Since a "nice" web page cannot contain much information, the number of hotlinks inserted in each page should be limited. This scenario motivates the problem of inserting at most one hotlink in each web page, so as to minimize the number of accesses required to locate an information. Two goals are considered: minimizing the maximum number of accesses and minimizing the average number of accesses.

Here, we consider the version of the problem in which the given web site is represented by a rooted directed tree T, where only the leaves contain information to be searched by the user. We assume that the user always follows a hotlink (u, v) from node u when searching for a leaf in the subtree rooted by v. Due to this assumption, the insertion of a hyperlink (u, v) must be followed by the deletion of any other arc that ends in v. As a result, the graph obtained after inserting a hotlink in a tree is also a tree, as shown in Figure 1. In this figure, we represent a tree T on the left, where the small triangle represents a subtree T' of T. At the right side, we represent the tree obtained from T by inserting a hotlink from the root of T to the root of T'.

1.1 Problem Definition

An instance of the hotlink assignment problem is defined by a directed tree $T = (V, E)$, rooted at a node $r \in V$. We say that a node v is a descendant of another node u in T when the only path in T that connects r to v contains u. In this case, we also have that u is ancestor of v. A node u is a proper

R. Ahlswede et al. (Eds.): Information Transfer and Combinatorics, LNCS 4123, pp. 1088–1092, 2006.

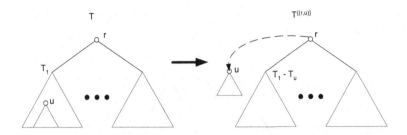

Fig. 1. The insertion of a hotlink in a tree leads to another tree

descendant (ancestor) of v if u is a descendant (ancestor) of v and $u \neq v$. We use $T_u = (V_u, E_u)$ to denote the subgraph of T induced by the descendants of u, that is, $V_u = \{v \in V \mid v$ is descendant of $u\}$, and $E_u = \{(v, w) \in E \mid v, w \in V_u\}$. Furthermore, we use L (L_u) to denote the subset of V (V_u) that contains all leaves of T (T_u).

Given $T = (V, E)$, a solution to the WCHS problem is a *hotlink assignment*, defined as a set $A \subset V \times V$. A hotlink assignment A is *feasible* iff it satisfies the following conditions:

(i) for every arc $(u, v) \in A$, v is descendant of u in T;

(ii) there are no pair of arcs $(u, a), (v, b) \in A$ that simultaneously satisfy the following conditions: u is a proper ancestor of v, v is a proper ancestor of a and a is ancestor of b;

(iii) for every node $u \in V$, there is at most one arc $(u, v) \in A$.

Figure 2 is used to illustrate the motivation behind condition (ii). It represents a hotlink assignment that does not satisfy this condition. In this figure, each triangle represents a subtree and the root node of each subtree is indicated by a circle. Hotlinks are represented by dashed arrows. Recall that we assume that the user always follows a hotlink (u, a) from node u when searching for a leaf in T_a. Since b is a node of T_a, we conclude that the hotlink (v, b) will never be followed by the user. In this case, condition (ii) prevents the addition of such a hotlink.

As we have mentioned before, the addition of hotlinks to a tree produces another tree that we denote by *improved tree*.

Definition 1. *Given* $T = (V, E)$, *and a feasible hotlink assignment* A, *the improved tree obtained from* T *through* A *is defined as* $T^A = (V, (E - X) \cup A)$, *where* $X = \{(u, v) \in E \mid (y, v) \in A$ *for some* $y \in V\}$.

In the definition above, X is the set of arcs in E whose heads receive some hotlink from A. Given an improved tree T^A rooted at r, we use $d_A(u)$ to denote the distance between r and u in T^A, that is, number of arcs in the path from r to u in T^A.

Below we consider two possible goals for the hotlink assignment problem.

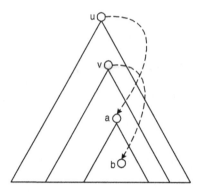

Fig. 2. A hotlink assignment that violates condition ii

The Worst Case Hotlink Search Problem (WCHS). Define

$$h(T^A) = \max_{l \in L} d_A(l)$$

Observe that some internal nodes of T may become leaves in T^A. By definition, these nodes do not belong to L. Thus, $h(T^A)$ may not be the height of T^A

In this context, we define an optimal hotlink assignment to T as a feasible hotlink assignment A^* such that

$$h(T^{A^*}) = \min_{A} h(T^A),$$

where A is minimized over all possible feasible assignments. The objective of the WCHS problem is to find an optimal hotlink assignment for T.

The Average Case Hotlink Search Problem (WCHS). Let p be a given probability distribution over the nodes in L. For every $l \in L$, we use $p(l)$ to denote the probability of l being accessed. Define

$$c(T^A) = \sum_{l \in L} p(l) d_A(l)$$

In this context, we define an optimal hotlink assignment to T as a feasible hotlink assignment A^* such that

$$c(T^{A^*}) = \min_{A} c(T^A),$$

where A is minimized over all possible feasible assignments. The objective of the ACHS problem is to find an optimal hotlink assignment for T.

2   Known Results

The WCHS Problem: it was considered by Pessoa et. al. in [8], where the following results are presented

1. an exact algorithm that runs in $O(n(nm)^{2.882})$ time requiring $O(n(nm)^{1.441})$ space, where n and m are the number of nodes (internal and external) and the number of leaves in T, respectively.
2. a $(14/3)$-approximate algorithm for the same problem, that runs in a $O(n \log m)$ time and requires a linear space.
3. A proof that $h(T^{A^*}) \leq 1.441 \log n + 1$

The exact algorithm is based on dynamic programming. However, it is worth mentioning that the straightforward application of this technique leads to an exponential algorithm. In order to obtain a polynomial algorithm, it is first proved that there is an optimal hotlink assignment A^* such that the height of T^{A^*} is $O(\log n)$. Then, a special decomposition technique is used to devise an optimal algorithm with an exponential running time on the height of T^{A^*}.

The ACHS Problem: For a complete discussion of this problem and some variants, including practical aspects, we refer to [6]. Bose et al. [2] show that

$$\frac{E(p)}{\log(d+1)} \leq c(T^{A^*}),$$

where $E(p)$ is the entropy [1] of the access probability distribution p and d is the maximum node outdegree in the tree. In [5], Kranakis et al. present an $O(n^2)$ time algorithm that produces a feasible assignment A such that

$$c(T^A) \leq \frac{E(p)}{\log(d+1) - (d/(d+1)) \log d} + \frac{d+1}{d}.$$

By using the same techniques employed in [8] to devise a polynomial time algorithm for WCHS, it is possible to obtain a polynomial time algorithm for the version of ACHS where the height of the input tree T is fixed.

We remark that no algorithm with constant factor approximation is known for the ACSH problem even for trees with uniform distribution probability.

2.1 Some Variants

In [4], Fuhrmann et al. considered the version of ACHS where multiple hotlinks can be assigned from a single node. For k-regular complete trees, they proved upper bounds for general distributions and lower bounds for uniform distributions.

Bose et al. [2] show that the ACHS problem is $\mathcal{NP}$-complete when the input graph, instead of a tree, is a DAG (Directed Acyclic Graph). The NP-Completness holds even for an uniform distribution.

3 Open Questions

The main open question related to the hotlink assignment problems described here is if there exists a polynomial time algorithm for the ACHS problem. The results in [8] imply that one can solve this problem by giving a a logarithmic upper bound on the **height** of T^{A^*}.

Some other interesting problems are

1. Find a polynomial time algorithm with constant approximation ratio for the ACSH.
2. Improve the current known bounds for the WCHP: $\log n \leq h(T^{A^*}) \leq 1.441 \log n + 1$.

We point out that a better upper bound on $h(T^{A^*})$ automatically implies on a polynomial algorithm for WCHP with lower time complexity [8].

References

1. N. Abramson, Information Theory and Coding, McGraw Hill, 1963.
2. P. Bose, E. Kranakis, D. Krizanc, M.V. Martin, J. Czyzowicz, A. Pelc, and L. Gasieniec, Strategies for hotlink assignments, International Symposium on Algorithms and Computation, 23–34, 2000.
3. M.C. Drott, Using web server logs to improve site design, Proceedings of ACM Conference of Computer Documentation, 43–50, 1998.
4. S. Fuhrmann, S.O. Krumke, and H.-C. Wirth, Multiple hotlink assignment, Proceedings of the Twenty-Seventh International Workshop on Graph- Theoretic Concepts in Computer Science, 2001.
5. E. Kranakis, D. Krizanc, and S. Shende, Approximate hotlink assignment, Lecture Notes in Computer Science, 2223, 756–767, 2001.
6. M.V. Martin, Enhancing hyperlink structure for improving web performance, PhD. Thesis, Carleton University, 2002.
7. M. Perkowitz and O. Etzioni, Towards adaptive web sites: conceptual framework and case study, Computer Networks, Amsterdam, Netherlands, Vol. 31, No. 11–16, 1245–1258, 1999.
8. A.A. Pessoa, E.S. Laber, and C. Souza, On the worst case search in trees with hotlinks, Technical Report MCC 23/03, Departmento de Informatica, PUC-Rio, 2003.

The Rigidity of Hamming Spaces

V.S. Lebedev

1 Notations and Definitions

We call a set B a base of a metric space L if every point of L is uniquely determined by its distances to the points of B.

The minimal possible number of points of a base is called the rigidity of the metric space and is denoted by $r(L)$.

n-dimensional q-ary Hamming space is denoted by $H_{n,q}$.

The rigidity of the binary Hamming space was introduced in 1963 by P.Erdos and A.Renyi (see [4]) for solving the following weighings problem: what is the minimal number $M(n)$ of weighings on an spring scale to determine all false coins in a set of n coins (predetermined case). It is easy to see that the rigidity of the binary Hamming space differs not more than on one from $M(n)$.

B.Lindstrom [5] and later D.G.Cantor and W.H. Mills [2] proved that

$$M(n) = \frac{2n}{log_2 n}(1 + o(1))$$

2 The Rigidity of q-ary Hamming Spaces

Consider the case when q is fixed and n goes to infinity.

Theorem 1. (G.Kabatianskii). Asymptotic bounds:

$$\frac{2n}{log_q n}(1 + o(1)) \le r(H_{n,q}) \le \frac{c(q)n}{log_q n}(1 + o(1)),$$

where $c(q) = 2log_q(1 + (q - 1)q)$.

The upper bound followed from random coding method of constructing a base of $H_{n,q}$. We suppose that for the considered problem random choice does not give the final answer.

Theorem 2. (G.Kabatianskii and V.Lebedev). For $q = 3$ and $q = 4$ we have

$$r(H_{n,q}) = \frac{2n}{log_q n}(1 + o(1)).$$

For $q = 3$ and $q = 4$ we give explicit constructions. For $q > 4$ such constructions are not known.

Open problem. Find $r(H_{n,q})$ for $q > 4$.

We conjecture that $r(H_{n,q}) = \frac{2n}{log_q n}(1 + o(1))$ for all q.

R. Ahlswede et al. (Eds.): Information Transfer and Combinatorics, LNCS 4123, pp. 1093–1094, 2006.
© Springer-Verlag Berlin Heidelberg 2006

References

1. M. Aigner, Combinatorial Search, 1988.
2. D.G. Cantor and W.H. Mills, Determination of a subset from certain combinatorial properties, Can. J. Math., 18, 42-48, 1966.
3. V. Chvatal, Mastermind, Combinatorica, 3, 325-329, 1983.
4. P. Erdos and A. Renyi, On two problems of information theory, Publ. Math. Inst. Hung. Acad. Sci., 8, 241-254, 1963.
5. B. Lindstrom, On a combinatory detection problem 1, Publ. Math. Inst. Hung. Acad. Sci., 9, 195-207, 1964.
6. B. Lindstrom, On a combinatorial problem in number theory, Can. Math. Bull., 8, 477-490, 1965.

A Conjecture in Finite Fields

U. Leck

1 The Conjecture

Let $q \equiv 1 \pmod 4$ be a prime power. For a primitive element α of $\mathbb{F} := \mathrm{GF}(q)$ and an element $0 \neq t \in \mathbb{F}$ we define a graph $G = G(\alpha, t)$ on the vertex set $\mathbb{F}$ by $E(G) = E_1 \cup E_2$, where

$$E_1 = \left\{ \{\alpha^{2i}, \alpha^{2i+1}\} \mid i = 0, 1, \ldots, (q-3)/2 \right\},$$
$$E_2 = \left\{ \{\alpha^{2i+1} + t, \alpha^{2i+2} + t\} \mid i = 0, 1, \ldots, (q-3)/2 \right\}.$$

It is easy to verify that $E_1 \cap E_2 = \varnothing$. Hence, all vertices of G have degree two, except 0 and t which have degree one. In other words, one component of G is a path connecting 0 and t, and all other components are cycles (of even length). Moreover, if $0 \neq t, t' \in \mathbb{F}$, then $G(\alpha, t)$ and $G(\alpha, t')$ are isomorphic [4], justifying the notation $G(\alpha)$.

Conjecture 1 ([4]). *For every prime power $q \equiv 1 \pmod 4$ there is a primitive element α of $\mathrm{GF}(q)$ such that $G(\alpha)$ has just one component (i.e. is a path of length $q - 1$).*

The above conjecture has been confirmed by computer for all primes powers $q \equiv 1 \pmod 4$ with $q < 2000$.

2 Background

Let G be like above, and consider the collection $\mathcal{G} = \{G + x \mid x \in \mathbb{F}\}$ of graphs, where $G + x$ is the graph on the vertex set $\mathbb{F}$ an with edge set $\{\{u + x, v + x\} \mid \{u, v\} \in E(G)\}$. It turns out that every $\{u, v\} \subset \mathbb{F}$ occurs as an edge in exactly two graphs from $\mathcal{G}$ and that any two members of $\mathcal{G}$ share exactly one edge. In other words, $\mathcal{G}$ is an *orthogonal double cover* (ODC) of the complete graph K_q (on $\mathbb{F}$) by G, and this ODC is generated by the additive group of $\mathrm{GF}(q)$.

ODCs arose first in the study of Armstrong representations of minimum size for key and functional dependencies in the relational database model [1], the paper [2] is a comprehensive survey on ODCs.

Let P_n denote the path with n vertices.

Conjecture 2 ([3]). *Let $T \neq P_4$ be a tree on $n \geq 3$ vertices. Then there exists an ODC of K_n by T.*

In general, this conjecture is far from being settled, some results have been obtained for special classes of trees (see [2]).

R. Ahlswede et al. (Eds.): Information Transfer and Combinatorics, LNCS 4123, pp. 1095–1096, 2006.
© Springer-Verlag Berlin Heidelberg 2006

Perhaps the most appealing special case is $T = P_n$, i.e. we are interested in ODCs of complete graphs by Hamiltonian paths. If the graph G constructed above has just one component, then the corresponding collection $\mathcal{G}$ is an ODC of K_q by P_q for the corresponding prime power q. Moreover, $\mathcal{G}$ has an additional property: It is *two-colorable*. This means, we can two-color $\binom{\mathbb{F}}{2}$ such that in each path from $\mathcal{G}$ incident edges receive different colors. Such two-colorable ODCs are particularly valuable because of the following multiplication theorem.

Theorem 1 ([6]). *Let $n \geq 3$ be an integer, and let $q \geq 5$ be a prime power. If there are an ODC of K_n by P_n and a two-colorable ODC of K_q by P_q, then there is an ODC of K_{qn} by P_{qn}.*

Two-colorable ODCs of K_m by P_m have been constructed for $m \in \{2^3, 2^4, 2^5\}$ in [8], for all m of the form $4a^2 + 1$, $8a^2 + 2$, $2a^2 + 2a + 1$, or $4a^2 + 4a + 2$ in [5], and for all m of the form $(2a + 1)^2$ or $2(2a^2 + 1)^2$ in [7]. Together with the above theorem this gives an infinite class of ODCs of complete graphs by Hamiltonian paths.

References

1. J. Demetrovics, Z. Füredi, and G.O.H. Katona, Minimum matrix representations of closure operations, Discrete Appl. Math., 11, 115-128, 1985.
2. H.-D.O.F. Gronau, M. Grüttmüller, S. Hartmann, U. Leck, and V. Leck, On orthogonal double covers of graphs, Designs Codes Cryptogr., 27, 49-93, 2002.
3. H.-D.O.F. Gronau, R.C. Mullin, and A. Rosa, Orthogonal double covers of complete graphs by trees, Graphs Combin., 13, 251-262, 1997.
4. S. Hartmann, U. Leck, and V. Leck, A conjecture on orthogonal double covers by paths, Congr. Numer., 140, 187-193, 1999..
5. S. Hartmann, U. Leck, and V. Leck, More orthogonal double covers of complete graphs by hamiltonian paths, Preprint 6/02, Universität Rostock, Fachbereich Mathematik, 2002, submitted.
6. J.D. Horton and G.M. Nonay, Self-orthogonal Hamilton path decompositions, Discrete Math., 97, 251-264, 1991.
7. U. Leck, A class of 2-colorable orthogonal double covers of complete graphs by hamiltonian paths, Graphs Combin., 18, 155-167, 2002.
8. V. Leck, On orthogonal double covers by Hamilton paths, Congr. Numer., 135, 153-157, 1998.

Multiparty Computations in Non-private Environments

M. Liśkiewicz

1 Backgrounds

Private multi-party computations is an intensively studied subject of modern cryptography. In general, private computation can be defined as follows: Consider a set of players, where each player knows an individual secret. The goal is to compute a function depending on these secrets such that after the computation none of the players knows anything about the secrets of others that cannot be derived from his own input and the result of the function. To compute the function, the players exchange messages with each other using secure links. For a formal definition of cryptographically secure privacy see [8] and for privacy in information theoretic sense see [5,2].

It is well known that any Boolean function can privately be computed on any 2-connected network of links. However, many networks which are used in praxis, are not 2-connected. In [3] we have showd that there are many Boolean functions, even simple ones like parity, disjunction, and conjunction, that cannot privately be computed if the underlying network is 1-connected but not 2-connected. Hence, an interesting problem arises how to compute functions in such non-private environments in such a way that each of them learns as little as possible about the inputs of the other players.

Communication with secrecy constraints has been studied for two-party model in [7]. This model corresponds to the simplest setting where a network consists of two 2-connected components that have only one bridge node in common. Leakage of information in the information-theoretical sense for two parties has been considered in [1,6].

2 Our Contribution

We consider the computation of Boolean functions $f : \{0,1\}^n \to \{0,1\}$ on a network of n players $P_1, \ldots, P_n$ equipped with random tapes (the random bits are private). Initially, each player knows a single bit of the input x. To compute the function the players can send messages to other players via point-to-point communication using secure links where the link topology is given by an undirected graph G. When the computation stops, all players should know the value $f(x)$. The goal is to compute $f(x)$ such that every P_i learns as little as possible about $x_1, \ldots, x_{i-1}, x_{i+1}, \ldots, x_n$. In the following we define a measure for the leakage of information for a protocol computing f. We argue that in the case when no assumption about the probability distribution for the input $x_1, \ldots, x_n$ is made, this is a suitable measure.

R. Ahlswede et al. (Eds.): Information Transfer and Combinatorics, LNCS 4123, pp. 1097–1099, 2006.

Let $\mathcal{A}$ denote a protocol computing function f and let $c_1, c_2, c_3, \ldots$ denote a fixed enumeration of all communication strings seen by any player during the execution of $\mathcal{A}$. By C_i we denote a random variable of the communication string seen by player P_i, and by R_i a random string provided to P_i.

Definition 1 ([4]). *Let C_i be a random variable of the communication string seen by player P_i while executing $\mathcal{A}$. Then for $a, b \in \{0, 1\}$ and for every random string R_i provided to P_i, define the* information source *of P_i on a, b, and R_i as*

$$\mathcal{S}_{\mathcal{A}}(i, a, b, R_i) = \{(\mu_x(c_1), \mu_x(c_2), \ldots) \mid x \in \{0, 1\}^n \wedge x_i = a \wedge f(x) = b\}$$

where $\mu_x(c_k) = \Pr[C_i = c_k | R_i, x]$ and the probability is taken over the random variables $R_1, \ldots, R_{i-1}, R_{i+1}, \ldots, R_n$ of all other players.

Basically $\mathcal{S}_{\mathcal{A}}(i, a, b, R_i)$ is the set of all different probability distributions on the communication strings observed by P_i when the input x of the players varies over all possible bit strings with $x_i = a$ and $f(x) = b$. We also define $s_{\mathcal{A}}(i, a, b, R_i) = |\mathcal{S}_{\mathcal{A}}(i, a, b, R_i)|$, where $|A|$ denotes the cardinality of the set A, and $s_{\mathcal{A}}(i, a, b) = \max_{R_i} s_{\mathcal{A}}(i, a, b, R_i)$ for a given protocol $\mathcal{A}$. Note that for protocol $\mathcal{A}$ the properties:

- $s_{\mathcal{A}}(i, a, b) = 1$ for every $a, b \in \{0, 1\}$, and
- $\mathcal{A}$ is private with respect to player P_i (for a formal definition see e.g. [3]),

are equivalent. Finally, for a network $G = (V, E)$ with $|V| = n$, let

$$s_G(i, a, b) = \min_{\mathcal{A}} s_{\mathcal{A}}(i, a, b).$$

In [4] we show that it is sufficient to consider only bridge players (i.e. players that correspond to bridge nodes in G) when considering protocols with minimum loss of information. We proved that for any protocol $\mathcal{A}$ on an 2-edge-connected G there exists a protocol $\mathcal{A}'$ on G computing the same function as $\mathcal{A}$ such that $\mathcal{A}'$ is private with respect to every internal player and the loss of $\mathcal{A}'$ to each bridge player P_q is at most the loss of $\mathcal{A}$ to this bridge player (i.e. $s_{\mathcal{A}'}(q, a, b) \leq s_{\mathcal{A}}(q, a, b)$ for every a, b). Below we show that the players can easily distinguish the distributions.

Let P_q be a bridge player of G, $a, b \in \{0, 1\}$, and R_q be the random string provided to P_q. Let

$$X = \{x \in \{0, 1\}^n \mid x_i = a \wedge f(x) = b\}$$

and for any communication string c_i let

$$\psi(c_i) = \{x \in X \mid \mu_x(c_i) > 0\}.$$

Obviously, for every c_i that can be observed by P_q on some input $x \in X$, P_q can deduce that $x \in \psi(c_i)$. If $s_{\mathcal{A}}(q, a, b) = s_G(q, a, b) = 1$, then we have either $\psi(c_i) = X$ or $\psi(c_i) = \varnothing$. Thus P_q does not learn anything in this case.

Theorem 1 ([4]). *If $s_G(q, a, b) > 1$, then for any protocol $\mathcal{A}$ and every communication string c_i that can be observed by P_q on input $x \in X$, $\psi(c_i)$ is a non-trivial subset of X, i.e. $\varnothing \neq \psi(c_i) \subsetneq X$, and there exist at least $s_G(q, a, b)$ different such sets.*

Let μ and μ' be two probability distributions over the same set of elementary events. The *fidelity* measures the similarity of μ and μ' and is defined by $F(\mu, \mu') = \sum_c \sqrt{\mu(c) \cdot \mu'(c)}$.

Theorem 2 ([4]). *If $\mathcal{A}$ is an optimal protocol for player P_q on a and b, i.e. $s_{\mathcal{A}}(q, a, b) = s_G(q, a, b)$, then for every random string R_q and all probability distributions $\mu \neq \mu'$ in $\mathcal{S}_{\mathcal{A}}(q, a, b, R_q)$ we have $F(\mu, \mu') = 0$.*

3   Open Problems

In case, when the bridge player can communicate with each 2-connected component only once, the size of the information source while communicating in one order can be exponentially larger than the size obtained by communication in another order. This is true, even if we restrict ourselves to symmetric functions.

Problem 1 ([4]). *Is it possible to minimize the loss of more than one bridge players simultaneously for general functions?*

There are several ways to quantify leakage of information for multiparty protocols. In this paper we have discussed *the information source*. However one can investigate also some other measures, i.g. in an information-theoretic setting.

Problem 2. *What is the answer to Problem 1 if one defines the loss of information by a protocol based on the mutual information?*

References

1. R. Bar-Yehuda, B. Chor, E. Kushilevitz, and A. Orlitsky, Privacy, additional information, and communication, IEEE Trans. Inform. Theory, 39, 1930–1943, 1993.
2. M. Ben-Or, S. Goldwasser, and A. Wigderson, Completeness theorems for noncryptographic fault-tolerant distributed computation, 20th STOC, 1–10, 1988.
3. M. Bläser, A. Jakoby, M. Liśkiewicz, and B. Siebert, Private computation — k-connected versus 1-connected networks, 22nd CRYPTO, LNCS 2442, 194–209. Springer, 2002.
4. M. Bläser, A. Jakoby, M. Liśkiewicz, and B. Siebert, Privacy in non-private environments, SIIM TR A-03-10, Unverity Lübeck, 2003.
5. D. Chaum, C. Crépeau, and I. Damgård, Multiparty unconditionally secure protocols, 20th STOC, 11–19, 1988.
6. Eyal Kushilevitz, Privacy and communication complexity, SIAM J. Discrete Math., 5(2), 273–284, 1992.
7. A. Orlitsky and A. El Gamal, Communication with secrecy constraints, 16th STOC, 217–224, 1984.
8. A. C.-C. Yao, Protocols for secure computations, 23rd FOCS, 160–164, 1982.

Some Mathematical Problems Related to Quantum Hypothesis Testing

Hiroshi Nagaoka

Abstract. We present two open problems related to the asymptotics of quantum hypothesis testing, together with some discussions about their mutual relation, the classical counterparts and a possible conjecture.

1 Preliminaries

Let A be a Hermitian matrix, $\{a_i\} \subset \mathbb{R}$ be its eigenvalues and $\{\pi_i\}$ be the orthogonal projection matrices for the corresponding eigenspaces, so that we have

$$\pi_i = \pi_i^* = \pi_i^2, \quad \pi_i \pi_j = 0 \quad \text{if} \quad i \neq j,$$

$$\sum_i \pi_i = I, \quad \text{and} \quad A = \sum_i a_i \pi_i.$$

We define [1]

$$\{A > 0\} := \sum_{i:\, a_i > 0} \pi_i,$$

which is the orthogonal projection matrix for the linear subspace spanned by the eigenvectors corresponding to the positive eigenvalues. As the notation suggests, $\{A > 0\}$ has an analogy with the notion of "the event that a random variable A takes a positive value" in the usual probability theory. We can similarly define $\{A \geq 0\}$, $\{A < 0\}$, $\{A > B\} = \{A - B > 0\}$, etc.

Let ρ and σ be arbitrary density matrices of the same size; i.e., ρ and σ are positive-semidefinite matrices with trace 1. Define

$$g_n(a) := \text{Tr}\left[\rho^{\otimes n}\left\{\rho^{\otimes n} - e^{na}\sigma^{\otimes n} > 0\right\}\right]$$

for $a \in \mathbb{R}$ and $n \in \mathbb{N}$, where $^{\otimes n}$ denotes the nth tensor power with respect to the usual tensor product of matrices

$$\begin{bmatrix} a_{11} & \cdots & a_{1k} \\ \vdots & & \vdots \\ a_{k1} & \cdots & a_{kk} \end{bmatrix} \otimes B = \begin{bmatrix} a_{11}B & \cdots & a_{1k}B \\ \vdots & & \vdots \\ a_{k1}B & \cdots & a_{kk}B \end{bmatrix}.$$

It is then shown [1] that $g_n : \mathbb{R} \to [0, 1]$ is monotonically nonincreasing and that

$$\lim_{n \to \infty} g_n(a) = \begin{cases} 1 \text{ if } a < D(\rho \,\|\, \sigma) \\ 0 \text{ if } a > D(\rho \,\|\, \sigma) \end{cases},$$

where $D(\rho \,\|\, \sigma)$ is the quantum relative entropy defined by

$$D(\rho \,\|\, \sigma) := \text{Tr}\left[\rho(\log \rho - \log \sigma)\right].$$

R. Ahlswede et al. (Eds.): Information Transfer and Combinatorics, LNCS 4123, pp. 1100–1103, 2006.

2   The First Problem

Problem I   *Determine the following convergence rates:*

$$\eta(a) := -\lim_{n\to\infty} \frac{1}{n} \log\left(1 - g_n(a)\right) \quad for \quad a < D(\rho \,\|\, \sigma),$$

and

$$\eta^c(a) := -\lim_{n\to\infty} \frac{1}{n} \log g_n(a) \quad for \quad a > D(\rho \,\|\, \sigma).$$

3   A Setting of Quantum Hypothesis Testing

Given two $k \times k$ density matrices ρ, σ and a $k^n \times k^n$ Hermitian matrix T_n satisfying $0 \le T_n \le I$, let

$$\alpha_n[T_n] := 1 - \text{Tr}\left[\rho^{\otimes n} T_n\right] \quad and \quad \beta_n[T_n] := \text{Tr}\left[\sigma^{\otimes n} T_n\right].$$

Considering the hypothesis testing for the two hypotheses $H_0 : \rho^{\otimes n}$ and $H_1 : \sigma^{\otimes n}$ on the *true* state, and regarding T_n as the POVM (positive-operator valued measure) $(T_n(H_0), T_n(H_1)) = (T_n, I - T_n)$ which represents a test for the hypotheses, $\alpha_n[T_n]$ and $\beta_n[T_n]$ are interpreted as the error probabilities of the 1st kind and 2nd kind, respectively;

$$\alpha_n[T_n] = \text{Prob}\left\{H_1 \text{ is accepted} \,|\, H_0 \text{ is true}\right\},$$
$$\beta_n[T_n] = \text{Prob}\left\{H_0 \text{ is accepted} \,|\, H_1 \text{ is true}\right\}.$$

Let

$$\beta_n^*(\varepsilon) := \min\left\{\beta_n[T_n] \,|\, T_n : \alpha_n[T_n] \le \varepsilon\right\}.$$

Then we have the following *quantum Stein's lemma* [2,3]:

$$-\lim_{n\to\infty} \frac{1}{n} \log \beta_n^*(\varepsilon) = D(\rho \,\|\, \sigma) \quad for \quad 0 < \forall\varepsilon < 1.$$

This result can be restated as

$$\lim_{n\to\infty} \beta_n^*\left(e^{-nr}\right) = \begin{cases} 0 \text{ if } r < D(\sigma \,\|\, \rho) \\ 1 \text{ if } r > D(\sigma \,\|\, \rho) \end{cases}.$$

4   The Second Problem

Problem II   *Determine the following convergence rates:*

$$R(r) := -\lim_{n\to\infty} \frac{1}{n} \log \beta_n^*\left(e^{-nr}\right) \quad for \quad r < D(\sigma \,\|\, \rho)$$

and

$$R^c(r) := -\lim_{n\to\infty} \frac{1}{n} \log\left(1 - \beta_n^*\left(e^{-nr}\right)\right) \quad for \quad r > D(\sigma \,\|\, \rho).$$

5   Relation Between the Two Problems

Assuming the existence of the limits in the definitions of $\eta(a)$ and

$$\zeta(a) := -\lim_{n \to \infty} \frac{1}{n} \log h_n(a),$$

$$\zeta^c(a) := -\lim_{n \to \infty} \frac{1}{n} \log(1 - h_n(a)),$$

where

$$h_n(a) := \operatorname{Tr}\left[\sigma^{\otimes n}\left\{\rho^{\otimes n} - e^{na}\sigma^{\otimes n} > 0\right\}\right],$$

we can prove [1] that

$$R(r) = \sup\left\{\zeta(a) \,|\, a \in \mathbb{R} \text{ s.t. } \eta(a) \geq r\right\}$$
$$= \inf\left\{a + \eta(a) \,|\, a \in \mathbb{R} \text{ s.t. } \eta(a) < r\right\}$$

and

$$R^c(r) = \sup_{a \in \mathbb{R}} \min\left\{\zeta^c(a), r + a\right\}$$
$$= \inf_{a \in \mathbb{R}} \max\left\{\zeta^c(a), r + a\right\}.$$

6   The Classical (or Commutative) Case

Assume $\rho\sigma = \sigma\rho$. Then we have

$$\rho = U \begin{bmatrix} p(1) & & 0 \\ & \ddots & \\ 0 & & p(k) \end{bmatrix} U^* \quad \text{and} \quad \sigma = U \begin{bmatrix} q(1) & & 0 \\ & \ddots & \\ 0 & & q(k) \end{bmatrix} U^*$$

for some unitary matrix U and some probability distributions $p = (p(1) \ldots, p(k))$ and $q = (q(1) \ldots, q(k))$, and

$$g_n(a) = \operatorname{Prob}\left\{p^n(X^n) - e^{na}q^n(X^n) > 0\right\}$$

$$= \operatorname{Prob}\left\{\frac{1}{n}\sum_{t=1}^{n} \log \frac{p(X_t)}{q(X_t)} > a\right\},$$

$$h_n(a) = \operatorname{Prob}\left\{p^n(Y^n) - e^{na}q^n(Y^n) > 0\right\}$$

$$= \operatorname{Prob}\left\{\frac{1}{n}\sum_{t=1}^{n} \log \frac{p(Y_t)}{q(Y_t)} > a\right\},$$

where p^n and q^n are the nth i.i.d. extensions of p and q, and $X^n = (X_1, \ldots, X_n)$ and $Y^n = (Y_1, \ldots, Y_n)$ are i.i.d. random variables obeying p^n and q^n, respectively. In this case, the quantum relative entropy turns out to be the Kullback-Leibler divergence;

$$D(\rho \,\|\, \sigma) = D(p \,\|\, q) = \sum_{x=1}^{n} p(x) \log \frac{p(x)}{q(x)}.$$

Furthermore, letting

$$\psi(\theta) := \log \sum_{x=1}^{k} p(x)^{1+\theta} q(x)^{-\theta},$$

$$\varphi(a) := \max_{\theta \in \mathbb{R}} \left(\theta a - \psi(\theta) \right),$$

we obtain from Cramér's theorem for large deviations that

$$\eta(a) = \varphi(a) \quad \text{for} \quad a \leq D(p \parallel q),$$

$$\eta^{c}(a) = \varphi(a) \quad \text{for} \quad a \geq D(p \parallel q),$$

$$\zeta(a) = a + \varphi(a) \quad \text{for} \quad a \geq -D(q \parallel p),$$

$$\zeta^{c}(a) = a + \varphi(a) \quad \text{for} \quad a \leq -D(q \parallel p).$$

Applying these relations to those in the previous section, we can represent $R(r)$ and $R^{c}(r)$ in terms of $\psi(\theta)$ or $\varphi(a)$, which are nothing but the theorem of Hoeffding and that of Han-Kobayashi [4], respectively.

7   An Optimistic Conjecture

A possible conjecture is:

The classical results are extended to the general quantum case by simply replacing $\sum_x p(x)^{1+\theta} q(x)^{-\theta}$ in the definition of $\psi(\theta)$ with $\text{Tr} \left[\rho^{1+\theta} \sigma^{-\theta} \right]$.

References

1. H. Nagaoka and M. Hayashi, An information-spectrum approach to classical and quantum hypothesis testing, e-print, quant-ph/0206185, 2002.
2. F. Hiai and D. Petz, The proper formula for relative entropy and its asymptotics in quantum probability, Commun. Math. Phys., Vol. 143, 99–114, 1991.
3. T. Ogawa and H. Nagaoka, Strong converse and Stein's lemma in quantum hypothesis testing, IEEE Trans. Inform. Theory, Vol. 46, 2428-2433, 2000.
4. T.S. Han and K. Kobayashi, The strong converse theorem for hypothesis testing, IEEE Trans. Inform. Theory, Vol. 35, 178–180, 1989.

Designs and Perfect Codes

F.I. Solov'eva

A *Steiner triple system of order* n (briefly $STS(n)$) is a family of 3-element blocks (subsets or triples) of the set $N = \{1, 2, \ldots, n\}$ such that each not ordered pair of elements of N appears in exactly one block.

Two STS-s of order n are called *isomorphic* if there exists a permutation on the set N which transforms them into one another. It is well known that $STS(n)$ exists if and only if $n \equiv 1$ or $3 \pmod 6$.

The number $N(n)$ of nonisomorphic Steiner triple systems STS-s of order n satisfies to the following bounds

$$(e^{-5}n)^{\frac{n^2}{6}} \le N(n) \le (e^{-1/2}n)^{\frac{n^2}{6}}.$$

The lower bound was proved by Egorychev in 1980 using the result on permanents of double stochastic matrices, see [1,2], the upper bound is straightforward. It is well known that for $n = 15$ there are 80 nonisomorphic Steiner triple systems of order 15.

Problem 1. Improve these bounds on the number of nonisomorphic Steiner triple systems of order $n > 15$.

A code C is *perfect* if for any vector x from n-dimensional metric space over the Galois field $GF(2)$ with the Hamming metric there exists exactly one vector $y \in C$ such that $d(x, y) \le 1$.

Problem 2. Can every Steiner triple system of order n, $STS(n)$, $n = 2^m - 1$, $m > 4$, be embedded into some perfect binary code of length n?

Remind that a *Steiner quadruple system* SQS(n) of order n is a collection of 4-element subsets of N, such that each not ordered 3-element subset of N is contained in exactly one block.

Problem 3. Can every $STS(n)$, $n \equiv 1$ or $3 \pmod 6$ be extended to some SQS(n+1).

The result is true for $n = 15$.

Problem 4. Can every $SQS(n)$ be embedded into some perfect binary extended code of length $n > 16$?

Avgustinovich, 1995, see [3], proved that any perfect code of length n is uniquely determined by its codewords of weight $(n - 1)/2$.

It gives the following upper bound on the number of different perfect binary codes of length n:

$$N_n \le 2^{2^{n - \frac{3}{2}\log n + \log\log(en)}}.$$

R. Ahlswede et al. (Eds.): Information Transfer and Combinatorics, LNCS 4123, pp. 1104–1105, 2006.
© Springer-Verlag Berlin Heidelberg 2006

The best lower bound on the number of different perfect binary codes of length n given by Krotov in 2000 [4] is

$$N_n \geq 2^{2^{\frac{n+1}{2} - \log_2(n+1)}} \cdot 3^{2^{\frac{n-3}{4}}} \cdot 2^{2^{\frac{n+5}{4} - \log_2(n+1)}}.$$

Problem 5. Improve the bounds on the number of different perfect binary codes.

Remark. The investigation of the problem 5 can be started with the length 15. It is known [5] that there exist Steiner quadruple system of order 16 that can not be embedded into an extended perfect code of length 16. Analogous result was recently done for Steiner triple systems of order 15, see [6].

References

1. G.P. Egorychev, Solution of the van der Waerden problem for permanents, Preprint IFSO-13 M Akad. Nauk SSSR Sibirsk. Otdel., Inst. Fiz., Krasnoyarsk, 1980.
2. G.P. Egorychev, The solution of the van der Waerden problem for permanents, Adv. in Math., 42, 3, 299–305, 1981.
3. S.V. Avgustinovich, On a property of perfect binary codes, Discrete Analysis and Operation Research, 2, 1, 4–6, 1995.
4. S.D. Krotov, Lower bounds on the number of m-quasigroups of order 4 and the number of perfect binary codes, Discrete Analysis and Operation Research 1, 7, 47–53, 2000.
5. F.Hergert, Algebraische methoden für nichtlineare codes, Dissertation, Technische Hochschule Darmstadt, 1985.
6. P.R.J. Ostergard, O.Pottonen, There exist Steiner triple systems of order 15 that do not occur in a perfect binary one-error-correcting code, submitted.

Special Issue of Discrete Applied Mathematics
"General Theory of Information Transfer and Combinatorics" List D

Survey on Topics in Coding Theory

[1] R. Ahlswede, General theory of information transfer: updated.

Surveys on Topics in Coding Theory

[2] R. Dodunekova, M. Dodunekov, and E. Nikolova, A survey on proper codes.
[3] S. Györi, Coding for a multiple access OR channel: a survey.
[4] F.I. Solov'eva, On perfect binary codes.

Papers

[5] R. Ahlswede, Ratewise optimal nonsequential search strategies under constraints on the tests.
[6] R. Ahlswede and H. Aydinian, On diagnosibility of large multiprocessor networks.
[7] R. Ahlswede, B. Balkenhol, C. Deppe, H. Mashurian, and T. Partner, T-shift synchronization codes.
[8] R. Ahlswede and V. Blinovsky, Multiple packing in sum-type metric spaces.
[9] R. Ahlswede, J. Cassaigne, and A. Sárközy, On the correlation of binary sequences.
[10] R. Ahlswede, F. Cicalese, and C. Deppe, Searching with lies under error transition cost constraints.
[11] A. Apostolico and R. Giancarlo, Periodicity and repetitions in parametrized strings.
[12] C. Bey, The edge-diametric theorem in Hamming spaces.
[13] M. Dutour, M. Deza, and M. Shtogrin Filling of a given boundary by p-gons and related problems.
[14] Z. Füredi and M. Ruszinkó, Large Convex Cones in Hypercubes.
[15] A. Kostochka and G. Yu, Minimum degree conditions for h-linked graphs.
[16] G. Kyureghyan, Minimal Polynomials of the modified de Bruijn sequences.
[17] M. Kyureghyan, Recursive Constructions of N-Polynomials over $GF(2^s)$.
[18] V. Levenshtein, E. Konstantinova, E. Konstantinov, and S. Molodtsov, Reconstruction of a graph from 2-vicinities of its vertices.
[19] U. Tamm, Size of downsets in the pushing order and a problem of Berlekamp.

R. Ahlswede et al. (Eds.): Information Transfer and Combinatorics, LNCS 4123, pp. 1106–1108, 2006.
© Springer-Verlag Berlin Heidelberg 2006

Levon H. Khachatrian

(1954–2002)

(by Harout Aydinian with Rudolf Ahlswede)
His sudden death came as a shock to all those who had the privilege of knowing him well, to those who admired him for his human qualities and his contributions in Number Theory and Combinatorics. Levon Khachatrian died in Bielefeld, Germany, on January 30, 2002 of heart attack.

He was born on January 5, 1954 in Yerevan, Armenia. Khachatrian received his M.S. degree from the State University of Yerevan in 1976 and the degree Candidate of Science (an equivalent of Ph.D.) from the Computing Center of the Soviet Academy of Sciences, Moscow, Russia, in 1983. He served two years 1976–1978 at the Research Institute of Mathematical Machines, Yerevan, as an engineer. During the years 1978–1981 Khachatrian was a postgraduate student at the Institute of Problems of Information Transmission, Moscow, where he wrote his doctoral thesis Ärrays with some peoperties of regularityünder the direction of Victor Zinoviev. Then he joined the Institute of Problems of Infromatics and Automation, Armenian Academy of Sciences, Yerevan, as a resarch scientist. He wrote a series of papers on pseudorandom arrays and sequences, fault-tolerant coding and networks. Khachatrian received the position of a Leading Research Fellowship of the Institute in 1990 and continued to work there until midyear 1991.

In 1991 Khachatrian was invited to to visit the University of Bielefeld, as a guest of a Research Project (SFB 343, Diskrete Strukturen in der Mathematik). Since then he remained at the Bielefeld University for the rest of his life. In Bielefeld Khachatrian began a very fruitful collaboration with R. Ahlswede, which continued for more than ten years. As a result, two famous problems of Erdős were completely settled during the years 1993–1995. The first one in Number theory, the Coprimality problem raised in 1962, and the second, in Extremal Set Theory, the Intersection Problem raised in the seminal paper by Erdős, Ko and Rado, written already in 1938 and published in 1961. Khachatrian together with R. Ahlswede, proved the long-standing conjecture of Frankl, which, in particular, implies the famous "$4m$–conjecture" of Erdős, Ko and Rado (1938), for which Erdős awarded 500$.

The powerful methods developed for solving these problems and their extensions have been successfully applied to obtain many other significant results in Combinatorial Number Theory and Extremal Combinatorics. Among them, the proof of a conjecture of Erdős and Graham on sets of integers with pairwise common divisors (1996), Density inequalities for sets of multiples (1995), Results on primitive and cross-primitive sequences (1997), Solution of the Diametric Theorem for Hamming spaces (1998) (all with R.Ahlswede) etc. A survey of these results can be found in [R. Ahlswede, Advances in Extremal Problems in Number Theory and Combinatorics, European Congress of Mathematics, Barcelona 2000, vol.1, 147–175].

During the next years Khachatrian continued his intensive work in Number Theory and Combinatorics. A series of remarkable papers on divisibility (primitivity) of sequences of integers have been written with R. Ahlswede and A. Sárközy. Extremal problems under dimension constraints was another subject introduced and studied by Khachatrian (with R. Ahlswede and H. Aydinian) during the last years.

Being with the University of Bielefeld, Khachatrian gave courses of lectures for graduate students, (in Combinatorics, Number Theory, Coding Theory, Graph Theory, Complexity Theory, Cryptography, etc.). His lectures were distinguished by lucidity and simplicity of presentation. He avoided unnecessary formalism, however, always giving rigorous proofs. Many complicated and involved proofs became attractive and friendly in his presentation. Often he found unexpected simple proofs, for known results, during his preparations of the lectures.

****

Levon was a kind and modest man, and a wonderful friend. The mathematical community lost a brilliant mathematician.

He will be missed by his friends and colleagues.

Bibliography of Publications by Rudolf Ahlswede

1967

[1] Certain results in coding theory for compound channels, Proc. Colloquium Inf. Th. Debrecen (Hungary), 35–60.

1968

[2] Beiträge zur Shannonschen Informationstheorie im Fall nichtstationärer Kanäle, Z. Wahrscheinlichkeitstheorie und verw. Geb. 10, 1–42.

[3] The weak capacity of averaged channels, Z. Wahrscheinlichkeitstheorie und verw. Geb. 11, 61–73.

1969

[4] Correlated decoding for channels with arbitrarily varying channel probability functions, (with J. Wolfowitz), Information and Control 14, 457–473.

[5] The structure of capacity functions for compound channels, (with J. Wolfowitz), Proc. of the Internat. Symposium on Probability and Information Theory at McMaster University, Canada, April 1968, 12–54.

1970

[6] The capacity of a channel with arbitrarily varying channel probability functions and binary output alphabet, (with J. Wolfowitz), Z. Wahrscheinlichkeitstheorie und verw. Geb. 15, 186–194.

[7] A note on the existence of the weak capacity for channels with arbitrarily varying channel probability functions and its relation to Shannon's zero error capacity, Ann. Math. Stat., Vol. 41, No. 3, 1027–33.

1971

[8] Channels without synchronization, (with J. Wolfowitz), Advances in Applied Probability, Vol. 3, 383–403.

[9] Group codes do not achieve Shannon's channel capacity for general discrete channels, Ann. Math. Stat., Vol. 42, No. 1, 224–240.

[10] Bounds on algebraic code capacities for noisy channels I, (with J. Gemma), Information and Control, Vol. 19, No. 2, 124–145.

[11] Bounds on algebraic code capacities for noisy channels II, (with J. Gemma), Information and Control, Vol. 19, No. 2, 146–158.

1973

[12] Multi–way communication channels, Proceedings of 2nd International Symposium on Information Theory, Thakadsor, Armenian SSR, Sept. 1971, Akademiai Kiado, Budapest, 23–52.

[13] On two–way communication channels and a problem by Zarankiewicz, Sixth Prague Conf. on Inf. Th., Stat. Dec. Fct's and Rand. Proc., Sept. 1971, Publ. House Chechosl. Academy of Sc., 23–37.

[14] A constructive proof of the coding theorem for discrete memoryless channels in case of complete feedback, Sixth Prague Conf. on Inf. Th., Stat. Dec. Fct's and Rand. Proc., Sept. 1971, Publ. House Czechosl. Academy of Sc., 1–22.

[15] The capacity of a channel with arbitrarily varying additive Gaussian channel probability functions, Sixth Prague Conf. on Inf. Th., Stat. Dec. Fct's and Rand. Proc., Sept. 1971, Publ. House Czechosl. Academy of Sc., 39–50.

R. Ahlswede et al. (Eds.): Information Transfer and Combinatorics, LNCS 4123, pp. 1109–1124, 2006.
© Springer-Verlag Berlin Heidelberg 2006

[16] Channels with arbitrarily varying channel probability functions in the presence of noiseless feedback, Z. Wahrscheinlichkeitstheorie und verw. Geb. 25, 239–252.

[17] Channel capacities for list codes, J. Appl. Probability, lo, 824–836.

1974

[18] The capacity region of a channel with two senders and two receivers, Ann. Probability, Vol. 2, No. 5, 805–814.

[19] On common information and related characteristics of correlated information sources, (with J. Körner), presented at the 7th Prague Conf. on Inf. Th., Stat. Dec. Fct's and Rand. Proc., included in "Information Theory" by I. Csiszár and J. Körner, Acad. Press, 1981. Recently included in General Theory of Information Transfer and Combinatorics, Report on a Research Project at the ZIF (Center of interdisciplinary studies) in Bielefeld Oct. 1, 2001 – August 31, 2004, edited by R. Ahlswede with the assistance of L. Bäumer and N. Cai.

1975

[20] Approximation of continuous functions in p–adic analysis, (with R. Bojanic), J. Approximation Theory, Vol. 15, No. 3, 190–205.

[21] Source coding with side information and a converse for degraded broadcast channels, (with J. Körner), IEEE Trans. Inf. Theory, Vol. IT–21, 629–637.

[22] Two contributions to information theory, (with P. Gács), Colloquia Mathematica Societatis János Bolyai, 16. Topics in Information Theory, I. Csiszár and P. Elias Edit., Keszthely, Hungaria, 1975, 17–40.

1976

[23] Bounds on conditional probabilities with applications in multiuser communication, (with P. Gács and J. Körner), Z. Wahrscheinlichkeitstheorie und verw. Geb. 34, 157–177.

[24] Every bad code has a good subcode: a local converse to the coding theorem, (with G. Dueck), Z. Wahrscheinlichkeitstheorie und verw. Geb. 34, 179–182.

[25] Spreading of sets in product spaces and hypercontraction of the Markov operator, (with P. Gács), Ann. Prob., Vol. 4, No. 6, 925–939.

1977

[26] On the connection between the entropies of input and output distributions of discrete memoryless channels, (with J. Körner), Proceedings of the 5th Conference on Probability Theory, Brasov 1974, Editura Academeiei Rep. Soc. Romania, Bucaresti 1977, 13–23.

[27] Contributions to the geometry of Hamming spaces, (with G. Katona), Discrete Mathematics 17, 1–22.

[28] The number of values of combinatorial functions, (with D.E. Daykin), Bull. London Math. Soc., 11, 49–51.

1978

[29] Elimination of correlation in random codes for arbitrarily varying channels, Z. Wahrscheinlichkeitstheorie und verw. Geb. 44, 159–175.

[30] An inequality for the weights of two families of sets, their unions and intersections, (with D.E. Daykin), Z. Wahrscheinlichkeitstheorie und verw. Geb. 43, 183–185.

[31] Graphs with maximal number of adjacent pairs of edges, (with G. Katona), Acta Math. Acad. Sc. Hung. 32, 97–120.

1979

[32] Suchprobleme, (with I. Wegener), Teubner Verlag, Stuttgart, Russian Edition with Appendix by Maljutov 1981 (Book).

[33] Inequalities for a pair of maps $S \times S \to S$ with S a finite set, (with D.E. Daykin), Math. Zeitschrift 165, 267–289.

[34] Integral inequalities for increasing functions, (with D.E. Daykin), Math. Proc. Comb. Phil. Soc., 86, 391–394.

[35] Coloring hypergraphs: A new approach to multi–user source coding I, Journ. of Combinatorics, Information and System Sciences, Vol. 4, No. 1, 76–115.

1980

[36] Coloring hypergraphs: A new approach to multi–user source coding II, Journ. of Combinatorics, Information and System Sciences, Vol. 5, No. 3, 220–268.

[37] Simple hypergraphs with maximal number of adjacent pairs of edges, J. Comb. Theory, Ser. B, Vol. 28, No. 2, 164–167.

[38] A method of coding and its application to arbitrarily varying channels, J. Combinatorics, Information and System Sciences, Vol. 5, No. 1, 10–35.

1981

[39] To get a bit of information may be as hard as to get full information, (with I. Csiszár), IEEE Trans. Inf. Theory, IT–27, 398–408.

[40]] Solution of Burnashev's problem and a sharpening of Erdős–Ko–Rado, Siam Review, to appear in a book by G. Katona. Recently included in General Theory of Information Transfer and Combinatorics, Report on a Research Project at the ZIF (Center of interdisciplinary studies) in Bielefeld Oct. 1, 2001 – August 31, 2004, edited by R. Ahlswede with the assistance of L. Bäumer and N. Cai.

1982

[41] Remarks on Shannon's secrecy systems, Probl. of Control and Inf. Theory, Vol. 11, No. 4, 301–318.

[42] Bad Codes are good ciphers, (with G. Dueck), Probl. of Control and Inf. Theory, Vol. 11, No. 5, 337–351.

[43] Good codes can be produced by a few permutations, (with G. Dueck), IEEE Trans. Inf. Theory, IT–28, No. 3, 430–443.

[44] An elementary proof of the strong converse theorem for the multiple–access channel, J. Combinatorics, Information and System Sciences, Vol. 7, No. 3, 216–230.

[45] Jacob Wolfowitz (1910–1981), IEEE Trans. Inf. Theory, Vol. IT–28, No. 5, 687–690.

1983

[46] Note on an extremal problem arising for unreliable networks in parallel computing, (with K.U. Koschnick), Discrete Mathematics 47, 137–152.

[47] On source coding with side information via a multiple–access channel and related problems in multi–user information theory, (with T.S. Han), IEEE Trans. Inf. Theory, Vol. IT–29, No. 3, 396–412.

1984

[48] A two family extremal problem in Hamming space, (with A. El Gamal and K.F. Pang), Discrete mathematics 49, 1–5.

[49] Improvements of Winograd's Result on Computation in the Presence of Noise, IEEE Trans. Inf. Theory, Vol. IT–30, No. 6, 872–877.

1985

[50] The rate–distortion region for multiple descriptions without excess rate, IEEE Trans. Inf. Theory, Vol. IT–31, No. 6, 721–726.

1986

[51] Hypothesis testing under communication constraints, (with I. Csiszár), IEEE Trans. Inf. Theory, Vol. IT–32, No. 4, 533–543.

[52] On multiple description and team guessing, IEEE Trans. Inf. Theory, Vol. IT–32, No. 4, 543–549.

[53] Arbitrarily varying channels with states sequence known to the sender, invited paper at a Statistical Research Conference dedicated to the memory of Jack Kiefer and Jacob Wolfowitz, held at Cornell University, July 1983, IEEE Trans. Inf. Theory, Vol. IT–32, No. 5, 621–629.

1987

[54] Optimal coding strategies for certain permuting channels, (with A. Kaspi), IEEE Trans. Inf. Theory, Vol. IT–33, No. 3, 310–314.

[55] Search Problems, (with I. Wegener), English Edition of with Supplement of recent Literature, R.L. Graham, J.K. Leenstra, R.E. Tarjan (Ed.), Wiley–Interscience Series in Discrete Mathematics and Optimization.

[56] Inequalities for code pairs, (with M. Moers), European J. of Combinatorics 9, 175–181.

[57] Eight problems in information theory
— a complexity problem
— codes as orbits
Contributions to "Open Problems in Communication and Computation", T.M. Cover and B. Gopinath (Ed.), Springer–Verlag.

[58] On code pairs with specified Hamming distances, Colloquia Mathematica Societatis János Bolyai 52, Combinatorics, Eger (Hungary), 9–47.

1989

[59] Identification via channels, (with G. Dueck), IEEE Trans. Inf. Theory, Vol. 35, No. 1, 15–29.

[60] Identification in the presence of feedback — a discovery of new capacity formulas, (with G. Dueck), IEEE Trans. Inf. Theory, Vol. 35, No. 1, 30–39.

[61] Contributions to a theory of ordering for sequence spaces, (with Z. Zhang), Problems of Control and Information Theory, Vol. 18, No. 4, 197–221.

1990

[62] A general 4–words inequality with consequences for 2–way communication complexity, (with N. Cai and Z. Zhang), Advances in Applied Mathematics, Vol. 10, 75–94.

[63] Coding for write–efficient memory, (with Z. Zhang), Information and Computation, Vol. 83, No. 1, 80–97.

[64] Creating order in sequence spaces with simple machines, (with Jian–ping Ye and Z. Zhang), Information and Computation, Vol. 89, No. 1, 47–94.

[65] An identity in combinatorial extremal theory, (with Z. Zhang), Adv. in Math., Vol. 80, No. 2, 137–151.

[66] On minimax estimation in the presence of side information about remote data, (with M.V. Burnashev), Ann. of Stat., Vol. 18, No. 1, 141–171.

[67] Extremal properties of rate–distortion functions, IEEE Trans. Inf. Theory, Vol. 36, No. 1, 166–171.

[68] A recursive bound for the number of complete K–subgraphs of a graph, (with N. Cai and Z. Zhang), "Topics in graph theory and combinatorics" in honour of G. Ringel on the occasion of his 70th birthday, R. Bodendiek, R. Henn (Eds), 37–39.

[69] On cloud–antichains and related configurations, (with Z. Zhang), Discrete Mathematics 85, 225–245.

1991

[70] Reusable memories in the light of the old AV– and new OV–channel theory, (with G. Simonyi), IEEE Trans. Inf. Theory, Vol. 37, No. 4, 1143–1150.

[71] On identification via multi–way channels with feedback, (with B. Verboven), IEEE Trans. Inf. Theory, Vol. 37, No. 5, 1519–1526.

[72] Two proofs of Pinsker's conjecture concerning AV channels, (with N. Cai), IEEE Trans. Inf. Theory, Vol. 37, No. 6, 1647–1649.

1992

[73] Diametric theorems in sequence spaces, (with N. Cai and Z. Zhang), Combinatorica, Vol. 12, No. 1, 1–17.

[74] On set coverings in Cartesian product spaces, *Ergänzungsreihe* des SFB 343 "Diskrete Strukturen in der Mathematik", Universität Bielefeld, Nr. 92–005. Recently included in General Theory of Information Transfer and Combinatorics, Report on a Research Project at the ZIF (Center of interdisciplinary studies) in Bielefeld Oct. 1, 2001 – August 31, 2004, edited by R. Ahlswede with the assistance of L. Bäumer and N. Cai.

[75] Rich colorings with local constraints, (with N. Cai and Z. Zhang), Preprint 89–011, SFB 343 "Diskrete Strukturen in der Mathematik", Universität Bielefeld, J. Combinatorics, Information & System Sciences, Vol. 17, Nos. 3-4, 203–216.

1993

[76] Asymptotically dense nonbinary codes correcting a constant number of localized errors, (with L.A. Bassalygo and M.S. Pinsker), Proc. III International workshop "Algebraic and Combinatorial Coding Theory", June 22–28, 1992, Tyrnovo, Bulgaria, Comptes rendus de l' Académie bulgare des Sciences, Tome 46, No. 1, 35–37.

[77] The maximal error capacity of AV channels for constant list sizes, IEEE Trans. Inf. Theory, Vol. 39, No. 4, 1416–1417.

[78] Nonbinary codes correcting localized errors, (with L.A. Bassalygo and M.S. Pinsker), IEEE Trans. Inf. Theory, Vol. 39, No. 4, 1413–1416.

[79] Common randomness in information theory and cryptography, Part I: Secret sharing, (with I. Csiszár), IEEE Trans. Inf. Theory, Vol. 39, No. 4, 1121–1132.

[80] A generalization of the AZ identity, (with N. Cai), Combinatorica 13 (3), 241–247.

[81] On partitioning the n–cube into sets with mutual distance 1, (with S.L. Bezrukov, A. Blokhuis, K. Metsch, and G.E. Moorhouse), Applied Math. Lett., Vol. 6, No. 4, 17–19.

[82] Communication complexity in lattices, (with N. Cai and U. Tamm), Applied Math. Lett., Vol. 6, No. 6, 53–58.

[83] Rank formulas for certain products of matrices, (with N. Cai), Preprint 92–014, SFB 343 "Diskrete Strukturen in der Mathematik", Universität Bielefeld, Applicable Algebra in Engineering, Communication and Computing, 2, 1–9.

[84] On extremal set partitions in Cartesian product spaces, (with N. Cai), Preprint 92–034, SFB 343 "Diskrete Strukturen in der Mathematik", Universität Bielefeld, Combinatorics, Probability & Computing 2, 211–220.

1994

[85] Note on the optimal structure of recovering set pairs in lattices: the sandglass conjecture, (with G. Simonyi), Preprint 91–082, SFB 343 "Diskrete Strukturen in der Mathematik", Universität Bielefeld, Discrete Math., 128, 389–394.

[86] On extremal sets without coprimes, (with L.H. Khachatrian), Preprint 93–026, SFB 343 "Diskrete Strukturen in der Mathematik", Universität Bielefeld, Acta Arithmetica, LXVI 1, 89–99.

[87] The maximal length of cloud–antichains, (with L.H. Khachatrian), Preprint 91–116, SFB 343 "Diskrete Strukturen in der Mathematik", Universität Bielefeld, Discrete Mathematics, Vol. 131, 9–15.

[88] The asymptotic behaviour of diameters in the average, (with I. Althöfer), Preprint 91–099, SFB 343 "Diskrete Strukturen in der Mathematik", Universität Bielefeld, J. Combinatorial Theory, Series B, Vol. 61, No. 2, 167–177.

[89] 2–way communication complexity of sum–type functions for one processor to be informed, (with N. Cai), Preprint 91–053, SFB 343 "Diskrete Strukturen in der Mathematik", Universität Bielefeld, Problemy Peredachi Informatsii, Vol. 30, No. 1, 3–12.

[90] Messy broadcasting in networks, (with H.S. Haroutunian and L.H. Khachatrian), Preprint 93–075, SFB 343 "Diskrete Strukturen in der Mathematik", Universität Bielefeld, Special volume in honour of J.L. Massey on occasion of his 60th birthday. Communications and Cryptography (Two sides of one tapestry), R.E. Blahut, D.J. Costello, U. Maurer, T. Mittelholzer (Ed.), Kluwer Acad. Publ., 13–24.

[91] Binary constant weight codes correcting localized errors and defects, (with L.A. Bassalygo and M.S. Pinsker), Preprint 93–025, SFB 343 "Diskrete Strukturen in der Mathematik", Universität Bielefeld, Probl. Peredachi Informatsii, Vol. 30, No. 2, 10–13 (In Russian); Probl. of Inf. Transmission, 102–104.

[92] On sets of words with pairwise common letter in different positions, (with N. Cai), Preprint 91–050, SFB 343 "Diskrete Strukturen in der Mathematik", Universität Bielefeld, Proc. Colloquium on Extremal Problems for Finite Sets, Visograd, Bolyai Siciety Math. Studies, 3, Hungary, 25–38.

[93] On multi–user write–efficient memories, (with Z. Zhang), IEEE Trans. Inf. Theory, Vol. 40, No. 3, 674–686.

[94] On communication complexity of vector–valued functions, (with N. Cai), Preprint 91–041, SFB 343 "Diskrete Strukturen in der Mathematik", Universität Bielefeld, IEEE Trans. Inf. Theory, Vol. 40, No. 6, 2062–2067.

[95] On partitioning and packing products with rectangles, (with N. Cai), Preprint 93–008, SFB 343 "Diskrete Strukturen in der Mathematik", Universität Bielefeld, Combinatorics, Probability & Computing 3, 429–434.

[96] A new direction in extremal theory for graphs, (with N. Cai and Z. Zhang), J. Combinatorics, Information & System Sciences, Vol. 19, No. 3–4, 269–280.

[97] Asymptotically optimal binary codes of polynomial complexity correcting localized errors, (with L.A. Bassalygo and M.S. Pinsker), Preprint 94–055, SFB 343 "Diskrete Strukturen in der Mathematik", Universität Bielefeld, Proc. IV International workshop on Algebraic and Combinatorial Coding Theory, Novgorod, Russia, 1–3.

1995

[98] Localized random and arbitrary errors in the light of AV channel theory, (with L.A. Bassalygo and M.S. Pinsker), Preprint 93–036, SFB 343 "Diskrete Strukturen in der Mathematik", Universität Bielefeld, IEEE Trans. Inf. Theory, Vol. 41, No. 1, 14–25.

[99] Edge isoperimetric theorems for integer point arrays, (with S.L. Bezrukov), Preprint 94–067, SFB 343 "Diskrete Strukturen in der Mathematik", Universität Bielefeld, Applied Math. Letters, Vol. 8, No. 2, 75–80.

[100] New directions in the theory of identification via channels, (with Z. Zhang), Preprint 94–010, SFB 343 "Diskrete Strukturen in der Mathematik, Universität Bielefeld, IEEE Trans. Inf. Theory, Vol. 41, No. 4, 1040–1050.

[101] Towards characterising equality in correlation inequalities, (with L.H. Khachatrian), Preprint 93–027, SFB 343 "Diskrete Strukturen in der Mathematik", Universität Bielefeld, European J. of Combinatorics 16, 315–328.

[102] Maximal sets of numbers not containing $k + 1$ pairwise coprime integers, (with L.H. Khachatrian), Preprint 94–080, SFB 343 "Diskrete Strukturen in der Mathematik", Universität Bielefeld, Acta Arithmetica LXX II, 1, 77–100.

[103] Density inequalities for sets of multiples, (with L.H. Khachatrian), Preprint 93–049, SFB 343 "Diskrete Strukturen in der Mathematik", Universität Bielefeld, J. of Number Theory, Vol. 55, No. 2., 170–180.

[104] A splitting property of maximal antichains, (with P.L. Erdős and N. Graham), Preprint 94–048, SFB 343 "Diskrete Strukturen in der Mathematik", Universität Bielefeld, Combinatorica 15 (4), 475–480.

1996

[105] Sets of integers and quasi–integers with pairwise common divisor, (with L.H. Khachatrian), Preprint 95–036, SFB 343 "Diskrete Strukturen in der Mathematik", Universität Bielefeld, Acta Arithmetica, LXXIV.2, 141–153.

[106] A counterexample to Aharoni's "Strongly maximal matching" conjecture, (with L.H. Khachatrian), included in "Report on work in progress in combinatorial extremal theory", Ergänzungsreihe des SFB 343 "Diskrete Strukturen in der Mathematik", Universität Bielefeld, Nr. 95–004, Discrete Mathematics 149, 289.

[107] Erasure, list, and detection zero–error capacities for low noise and a relation to identification, (with N. Cai and Z. Zhang), Preprint 93–068, SFB 343 "Diskrete Strukturen in der Mathematik", Universität Bielefeld, IEEE Trans. Inf. Theory, Vol. 42, No. 1, 55–62.

[108] Optimal pairs of incomparable clouds in multisets, (with L.H. Khachatrian), Preprint 93–043, SFB 343 "Diskrete Strukturen in der Mathematik", Universität Bielefeld, Graphs and Combinatorics 12, 97–137.

[109] Sets of integers with pairwise common divisor and a factor from a specified set of primes, (with L.H. Khachatrian), Preprint 95–059, SFB 343 "Diskrete Strukturen in der Mathematik", Universität Bielefeld, Acta Arithmetica LXX V 3, 259–276.

[110] Cross–disjoint pairs of clouds in the interval lattice, (with N. Cai), Preprint 93–038, SFB 343 "Diskrete Strukturen in der Mathematik", Universität Bielefeld, The Mathematics of Paul Erdős, Vol. I; R.L. Graham and J. Nesetril, ed., Algorithms and Combinatorics B, Springer Verlag, Berlin/Heidelberg/New York, 155–164.

[111] Identification under random processes, (with V. Balakirsky), Preprint 95–098, SFB 343 "Diskrete Strukturen in der Mathematik", Universität Bielefeld, Problemy peredachii informatsii (special issue devoted to M.S. Pinsker), vol. 32, no. 1, 144–160, Jan.–March 1996; Problems of Information Transmission, Vol. 32, No. 1, 123–138.

[112] Report on work in progress in combinatorial extremal theory: Shadows, AZ–identity, matching. *Ergänzungsreihe* des SFB 343 "Diskrete Strukturen in der Mathematik", Universität Bielefeld, Nr. 95–004.

[113] Fault–tolerant minimum broadcast networks, (with L. Gargano, H.S. Haroutunian, and L.H. Khachatrian), Preprint 94–032, SFB 343 "Diskrete Strukturen in der Mathematik", Universität Bielefeld, Networks, Vol. 27, No. 4, 1293–1307.

[114] The complete nontrivial–intersection theorem for systems of finite sets, (with L.H. Khachatrian), Preprint 95–102, SFB 343 "Diskrete Strukturen in der Mathematik", Universität Bielefeld, J. Combinatorial Theory, Series A, 121–138.

[115] Incomparability and intersection properties of Boolean interval lattices and chain posets, (with N. Cai), Preprint 93–037, SFB 343 "Diskrete Strukturen in der Mathematik", Universität Bielefeld, European J. of Combinatorics 17, 677–687.

[116] Classical results on primitive and recent results on cross–primitive sequences, (with L.H. Khachatrian), Preprint 93–042, SFB 343 "Diskrete Strukturen in der Mathematik", Universität Bielefeld, The Mathematics of P. Erdős, Vol. I; R.L. Graham and J. Nesetril, ed., Algorithms and Combinatorics B, Springer Verlag, Berlin/Heidelberg/ New York, 104–116.

[117] Intersecting Systems, (with N. Alon, P.L. Erdős, M. Ruszinko, L.A. Székely), Preprint Supplement 01–003, SFB 343 "Diskrete Strukturen in der Mathematik", Universität Bielefeld, Combinatorics, Probability and Computing 6, 127–137.

[118] Some properties of fix–free codes, (with B. Balkenhol and L.H. Khachatrian), Proceedings First INTAS International Seminar on Coding Theory and Combinatorics 1996, Thahkadzor, Armenia, 20–33, 6–11 October 1996.

[119] Higher level extremal problems, (with N. Cai and Z. Zhang), Preprint 92–031, SFB 343 "Diskrete Strukturen in der Mathematik", Universität Bielefeld, Comb. Inf. & Syst. Sc., Vol. 21, No. 3–4, 185–210.

1997

[120] On interactive communication, (with N. Cai and Z. Zhang), Preprint 93–066, SFB 343 "Diskrete Strukturen in der Mathematik", Universität Bielefeld, IEEE Trans. Inf. Theory, Vol. 43, No. 1, 22–37.

[121] Identification via compressed data, (with E. Yang and Z. Zhang), Preprint 95–007, SFB 343 "Diskrete Strukturen in der Mathematik", Universität Bielefeld, IEEE Trans. Inf. Theory, Vol. 43, No. 1, 48–70.

[122] The complete intersection theorem for systems of finite sets, (with L.H. Khachatrian), Preprint 95–066, SFB 343 "Diskrete Strukturen in der Mathematik", European J. Combinatorics, 18, 125–136.

[123] Universal coding of integers and unbounded search trees, (with T.S. Han and K. Kobayashi), Preprint 95–001, SFB 343 "Diskrete Strukturen in der Mathematik", Universität Bielefeld, Trans. Inf. Theory, Vol. 43, No. 2, 669–682.

[124] Number theoretic correlation inequalities for Dirichlet densities, (with L.H. Khachatrian), Preprint 93–060, SFB 343 "Diskrete Strukturen in der Mathematik", Universität Bielefeld, J. Number Theory, Vol. 63, No. 1, 34–46.

[125] General edge–isoperimetric inequalities, Part 1: Information theoretical methods, (with Ning Cai), Preprint 94–090, SFB 343 "Diskrete Strukturen in der Mathematik", Universität Bielefeld, European J. of Combinatorics 18, 355–372.

[126] General edge–isoperimetric inequalities, Part 2: A local–global principle for lexicographical solutions, (with Ning Cai), Preprint 94–090, SFB 343 "Diskrete Strukturen in der Mathematik", Universität Bielefeld, European J. of Combinatorics 18, 479–489.

[127] Models of multi–user write–efficient memories and general diametric theorems, (with N. Cai), Preprint 93–019, SFB 343 "Diskrete Strukturen in der Mathematik", Universität Bielefeld, Information and Computation, Vol. 135, No. 1, 37–67.

[128] Shadows and isoperimetry under the sequence–subsequence relation, (with N. Cai), Preprint 95-045, SFB 343 "Diskrete Strukturen in der Mathematik", Universität Bielefeld, Combinatorica 17 (1), 11–29.

[129] Counterexample to the Frankl/Pach conjecture for uniform, dense families, (with L.H. Khachatrian), Preprint 95–114, SFB 343 "Diskrete Strukturen in der Mathematik", Universität Bielefeld, Combinatorica 17 (2), 299–301.

[130] Correlated sources help the transmission over AVC, (with N. Cai), Preprint 95–106, SFB 343 "Diskrete Strukturen in der Mathematik", Universität Bielefeld, IEEE Trans. Inf. Theory, Vol. 135, No. 1, 37–67.

1998

[131] Common randomness in Information Theory and Cryptography, Part II: CR capacity, (with I. Csiszár), Preprint 95–101, SFB 343 "Diskrete Strukturen in der Mathematik", Universität Bielefeld, IEEE Trans. Inf. Theory, Vol. 44, No. 1, 55–62.

[132] The diametric theorem in Hamming spaces — optimal anticodes, (with L.H. Khachatrian) Preprint 96–013, SFB 343 "Diskrete Strukturen in der Mathematik", Universität Bielefeld, Proceedings First INTAS International Seminar on Coding Theory and Combinatorics 1996, Thahkadzor, Armenia, 1–19, 6–11 October 1996; Advances in Applied Mathematics 20, 429–449.

[133] Information and Control: Matching channels, (with N. Cai), Preprint 95–035, SFB 343 "Diskrete Strukturen in der Mathematik", Universität Bielefeld, IEEE Trans. Inf. Theory, Vol. 44, No. 2, 542–563.

[134] Zero–error capacity for models with memory and the enlightened dictator channel, (with N. Cai and Z. Zhang), IEEE Trans. Inf. Theory, Vol. 44, No. 3, 1250–1252.

[135] Code pairs with specified parity of the Hamming distances, (with Z. Zhang), Preprint 96–058, SFB 343 "Diskrete Strukturen in der Mathematik", Universität Bielefeld, Discrete Mathematics 188, 1–11.

[136] Isoperimetric theorems in the binary sequences of finite lengths, (with Ning Cai), Applied Math. Letters, Vol. 11, No. 5, 121–126.

[137] The intersection theorem for direct products, (with H. Aydinian and L.H. Khachatrian), Preprint 97–051, SFB 343 "Diskrete Strukturen in der Mathematik", Universität Bielefeld, European Journal of Combinatorics 19, 649–661.

1999

[138] Construction of uniquely decodable codes for the two–user binary adder channel, (with V.B. Balakirsky), Preprint 97–016, SFB 343 "Diskrete Strukturen in der Mathematik", IEEE Trans. Inf. Theory, Vol 45, No. 1, 326–330.

[139] Arbitrarily varying multiple–access channels, Part I. Ericson's symmetrizability is adequate, Gubner's conjecture is true, (with N. Cai), Preprint 96–068, SFB 343 "Diskrete Strukturen in der Mathematik", Universität Bielefeld, IEEE Trans. Inf. Theory, Vol. 45, No. 2, 742–749.

[140] Arbitrarily varying multiple–access channels, Part II. Correlated sender's side information, correlated messages, and ambiguous transmission, (with N. Cai), Preprint 97–006, SFB 343 "Diskrete Strukturen in der Mathematik", Universität Bielefeld, IEEE Trans. Inf. Theory, Vol. 45, No. 2, 749–756.

[141] A pushing–pulling method: new proofs of intersection theorems, (with L.H. Khachatrian), Preprint 97–043, SFB 343 "Diskrete Strukturen in der Mathematik", Universität Bielefeld, Combinatorica 19(1), 1–15.

[142] A counterexample in rate–distortion theory for correlated sources, (with Ning Cai), Preprint 97–034, SFB 343 "Diskrete Strukturen in der Mathematik", Universität Bielefeld, Applied Math. Letters, Vol. 12, No. 7, 1–3.

[143] On maximal shadows of members in left–compressed sets, (with Zhen Zhang), Preprint 97-026, SFB 343 "Diskrete Strukturen in der Mathematik", Universität Bielefeld, Proceedings of the Rostock Conference, Discrete Applied Math. 95, 3–9.

[144] A counterexample to Kleitman's conjecture concerning an edge–isoperimetric problem, (with Ning Cai), Preprint 00–121, SFB 343 "Diskrete Strukturen in der Mathematik", Universität Bielefeld, Combinatorics, Probability and Computing 8, 301–305.

[145] Identification without randomization, (with Ning Cai), Preprint 98–075, SFB 343 "Diskrete Strukturen in der Mathematik", Universität Bielefeld, IEEE Trans. Inf. Theory, Vol. 45, No. 7, 2636–2642.

[146] On the quotient sequence of sequences of integers, (with L.H. Khachatrian and A. Sárközy), Preprint 98–068, SFB 343 "Diskrete Strukturen in der Mathematik", Universität Bielefeld, Acta Arithmetica, XCI.2, 117–132.

[147] On the counting function for primitive sets of integers, (with L.H. Khachatrian and A. Sárközy), Preprint 98–077, SFB 343 "Diskrete Strukturen in der Mathematik", Universität Bielefeld, J. Number Theory 79, 330–344.

[148] On the Hamming bound for nonbinary localized–error–correcting codes, (with L.A. Bassalygo and M.S. Pinsker), Preprint 99–077, SFB 343, Diskrete Strukturen in der Mathematik, Universität Bielefeld, Problemy Per. Informatsii, Vol. 35, No. 2, 29–37, Probl. of Inf. Transmission, Vol. 35, No. 2, 117–124.

[149] Asymptotical isoperimetric problem, (with Z. Zhang), Proceedings 1999 IEEE ITW, Krüger National Park, South Africa, June 20–25, 85–87.

[150] Nonstandard coding method for nonbinary codes correcting localized errors, (with L. Bassalygo and M. Pinsker), Proceedings 1999 IEEE ITW, Krüger National Park, South Africa, June 20–25, 78–79.

2000

[151] On prefix–free and suffix–free sequences of integers, (with L.H. Khachatrian and A. Sárközy), Preprint 99–050, SFB 343 "Diskrete Strukturen in der Mathematik", Universität Bielefeld, Numbers, Information and Complexity, Special volume in honour of R. Ahlswede on occasion of his 60th birthday, editors I. Althöfer, N. Cai, G. Dueck, L.H. Khachatrian, M. Pinsker, A. Sárközy, I. Wegener, and Z. Zhang, Kluwer Acad. Publ., Boston, Dordrecht, London, 1–16.

[152] Splitting properties in partially ordered sets and set systems, (with L.H. Khachatrian), Preprint 94–071, SFB 343 "Diskrete Strukturen in der Mathematik", Universität Bielefeld, Numbers, Information and Complexity, Special volume in honour of R. Ahlswede on occasion of his 60th birthday, editors I. Althöfer, N. Cai, G. Dueck, L.H. Khachatrian, M. Pinsker, A. Sárközy, I. Wegener, and Z. Zhang, Kluwer Acad. Publ., Boston, Dordrecht, London, 29–44.

[153] The AVC with noiseless feedback and maximal error probability: A capacity formula with a trichotomy, (with N. Cai), Preprint 96–064, SFB 343 "Diskrete Strukturen in der Mathematik", Universität Bielefeld, Numbers, Information and Complexity, Special volume in honour of R. Ahlswede on occasion of his 60th birthday, editors I. Althöfer, N. Cai, G. Dueck, L.H. Khachatrian, M. Pinsker, A. Sárközy, I. Wegener, and Z. Zhang, Kluwer Acad. Publ., Boston, Dordrecht, London, 151–176.

[154] A diametric theorem for edges, (with L.H. Khachatrian), Preprint 97–100, SFB 343 "Diskrete Strukturen in der Mathematik", Universität Bielefeld, J. Comb. Theory, Series A 92, 1–16.

[155] Network information flow, (with Ning Cai, S.Y. Robert Li, and Raymond W. Yeung), Preprint 98–033, SFB 343 "Diskrete Strukturen in der Mathematik", Universität Bielefeld, IEEE Trans. Inf. Theory, Vol. 46, No. 4, 1204–1216.

2001

[156] On perfect codes and related concepts, (with H. Aydinian and L.H. Khachatrian), Preprint 98–080, SFB 343 "Diskrete Strukturen in der Mathematik", Universität Bielefeld, Designs, Codes and Cryptography, 22, 221–237.

[157] Quantum data processing, (with Peter Löber), Preprint 99–087, SFB 343 "Diskrete Strukturen in der Mathematik", Universität Bielefeld, IEEE Trans. Inf. Theory, Vol. 47, No. 1, 474–478.

[158] On primitive sets of squarefree integers, (with L.H. Khachatrian and A. Sárközy), Preprint 99–093, SFB 343 "Diskrete Strukturen in der Mathematik", Universität Bielefeld, Periodica Mathematica Hungarica Vol. 42 (1–2), 99–115.

[159] Advances on extremal problems in number theory and combinatorics, European Congress of Mathematics, Barcelona 2000, Vol. I, 147–175, Carles Casacuberta, Rosa Maria Miró–Roig, Joan Verdera, Sebastiá Xambó–Descamps (Eds.), Progress in Mathematics, Vol. 201, Birkhäuser Verlag, Basel–Boston–Berlin.

[160] An isoperimetric theorem for sequences generated by feedback and feedback–codes for unequal error protection, (with N. Cai and C. Deppe), Preprint 00–115, SFB 343 "Diskrete Strukturen in der Mathematik", Universität Bielefeld, Problemy Peredachi Informatsii, No. 4, 63–70, 2001, Translation Problems of Information Transmission, Vol. 37, No. 4, 332–338.

2002

[161] Strong converse for identification via quantum channels, (with A. Winter), Preprint 00–122, SFB 343 "Diskrete Strukturen in der Mathematik", Universität Bielefeld, IEEE Trans. Inf. Theory, Vol. 48, No. 3, 569–579.

[162] Parallel error correcting codes, (with B. Balkenhol and N. Cai), Preprint 00–114, SFB 343 "Diskrete Strukturen in der Mathematik", Universität Bielefeld, IEEE Trans. Inf. Theory, Vol. 48, No. 4, 959–962.

[163] Semi–noisy deterministic multiple–access channels: coding theorems for list codes and codes with feedback, (with N. Cai), Preprint 00–022, SFB 343 "Diskrete Strukturen in der Mathematik", Universität Bielefeld, IEEE Trans. Inf. Theory, Vol. 48, No. 8, 2953–2962.

[164] The t–intersection problem in the truncated Boolean lattice, (with C. Bey, K. Engel, and L.H. Khachatrian), Preprint 99–104, SFB 343 "Diskrete Strukturen in der Mathematik", Universität Bielefeld, European Journal of Combinatorics 23, 471–487.

[165] Undirectional error control codes and related combinatorial problems, (with H. Aydinian and L.H. Khachatrian), in Proceedings of Eight International workshop on Algebraic and Combinatorial Coding Theory, 8–14 September, Tsarskoe Selo, Russia, 6–9.

2003

[166] Forbidden (0,1)–vectors in hyperplanes of $\mathbb{R}^n$: The restricted case, (with H. Aydinian and L.H. Khachatrian), Designs, Codes and Cryptography, 29, 17–28.

[167] Cone dependence — a basic combinatorial concept, (with L.H. Khachatrian), Preprint 00–117, SFB 343 "Diskrete Strukturen in der Mathematik", Universität Bielefeld, Designs, Codes and Cryptography, 29, 29–40.

[168] More about shifting techniques, (with H. Aydinian and L.H. Khachatrian), Preprint 00–127, SFB 343 "Diskrete Strukturen in der Mathematik", Universität Bielefeld, European Journal of Combinatorics 24, 551–556.

[169] On lossless quantum data compression and quantum variable–length codes, (with Ning Cai), Chapter 6 in "Quantum Information Processing", Gerd Leuchs, Thomas Beth (Eds.), Wiley–VCH Verlag, Weinheim, Germany, 66–78.

[170] Maximum number of constant weight vertices of the unit n–cube contained in a k–dimensional subspace, (with H. Aydinian and L.H. Khachatrian), Preprint 99–118, SFB 343 "Diskrete Strukturen in der Mathematik", Universität Bielefeld, Combinatorica, Vol. 23 (1), 5–22.

[171] A complexity measure for families of binary sequences (with L.H. Khachatrian, C. Mauduit, and A. Sárközy), Preprint Supplement 01–007, SFB 343 "Diskrete Strukturen in der Mathematik", Universität Bielefeld, Periodica Mathematica Hungarica, Vol. 46 (2), 107–118.

[172] Extremal problems under dimension constraints, (with H. Aydinian and L.H. Khachatrian), Preprint 00–116, SFB 343 "Diskrete Strukturen in der Mathematik", Universität Bielefeld, Discrete Mathematics, Special issue: EuroComb'01 – Edited by J. Nesetril, M. Noy and O. Serra, Vol. 273, No. 1–3, 9–21.

[173] Maximal antichains under dimension constraints, (with H. Aydinian and L.H. Khachatrian), Discrete Mathematics, Special issue: EuroComb'01 – Edited by J. Nesetril, M. Noy and O. Serra, Vol. 273, No. 1–3, 23–29.

[174] Large deviations in quantum information theory, (with V. Blinovsky), Probl. of Inf. Transmission, Vol. 39, Issue 4, 373–379.

2004

[175] On Bohman's conjecture related to a sum packing problem of Erdős, (with H. Aydinian and L.H. Khachatrian), Proceedings of the American Mathematical Society, Vol. 132, No. 5, 1257–1265.

[176] On shadows of intersecting families, (with H. Aydinian and L.H. Khachatrian), Combinatorica 24 (4), 555–566.

[177] On lossless quantum data compression with a classical helper, (with Ning Cai), IEEE Trans. Inf. Theory, Vol. 50, No. 6,

[178] On the density of primitive sets, (with L.H. Khachatrian and A. Sárközy), Preprint 00–112, SFB 343 "Diskrete Strukturen in der Mathematik", Universität Bielefeld, J. Number Theory 109, 319-361.

2005

[179] Katona's Intersection Theorem: Four Proofs, (with L.H. Khachatrian), Combinatorica 25 (1), 105-110.

[180] Forbidden (0,1)–vectors in Hyperplanes of $\mathbb{R}^n$: The unrestricted case, Designs, Codes and Cryptography 37, 151-167.

2006

[181] Intersection theorems under dimension constraints part I: the restricted case and part II: the unrestricted case, (with H. Aydinian and L.H. Khachatrian), J. Comb. Theory, Series A 113, 483-519.

[182] Search with noisy and delayed responses, (with N. Cai), General Theory of Information Transfer and Combinatorics, Report on a Research Project at the ZIF (Center of interdisciplinary studies) in Bielefeld Oct. 1, 2001 – August 31, 2004, edited by R. Ahlswede with the assistance of L. Bäumer and N. Cai.

[183] Watermarking identification codes with related topics in common randomness, (with N. Cai), General Theory of Information Transfer and Combinatorics, Report on a Research Project at the ZIF (Center of interdisciplinary studies) in Bielefeld Oct. 1, 2001 – August 31, 2004, edited by R. Ahlswede with the assistance of L. Bäumer and N. Cai.

[184] Large families of pseudorandom sequences of k symbols and their complexity, Part I, (with C. Mauduit and A. Sárközy), General Theory of Information Transfer and Combinatorics, Report on a Research Project at the ZIF (Center of interdisciplinary studies) in Bielefeld Oct. 1, 2001 – August 31, 2004, edited by R. Ahlswede with the assistance of L. Bäumer and N. Cai.

[185] Large families of pseudorandom sequences of k symbols and their complexity, Part II, (with C. Mauduit and A. Sárközy), General Theory of Information Transfer and Combinatorics, Report on a Research Project at the ZIF (Center of interdisciplinary studies) in Bielefeld Oct. 1, 2001 – August 31, 2004, edited by R. Ahlswede with the assistance of L. Bäumer and N. Cai.

[186] A Kraft–type inequality for d–delay binary search codes, (with N. Cai), General Theory of Information Transfer and Combinatorics, Report on a Research Project at the ZIF (Center of interdisciplinary studies) in Bielefeld Oct. 1, 2001 – August 31, 2004, edited by R. Ahlswede with the assistance of L. Bäumer and N. Cai.

[187] Sparse asymmetric connectors in communication networks, (with H. Aydinian) General Theory of Information Transfer and Combinatorics, Report on a Research Project at the ZIF (Center of interdisciplinary studies) in Bielefeld Oct. 1, 2001 – August 31, 2004, edited by R. Ahlswede with the assistance of L. Bäumer and N. Cai.

[188] A strong converse theorem for quantum multiple access channels, (with N. Cai), General Theory of Information Transfer and Combinatorics, Report on a Research Project at the ZIF (Center of interdisciplinary studies) in Bielefeld Oct. 1, 2001 – August 31, 2004, edited by R. Ahlswede with the assistance of L. Bäumer and N. Cai.

[189] Codes with the identifiable parent property and the multiple–access channel, (with N. Cai) General Theory of Information Transfer and Combinatorics, Report on a Research Project at the ZIF (Center of interdisciplinary studies) in Bielefeld Oct. 1, 2001 – August 31, 2004, edited by R. Ahlswede with the assistance of L. Bäumer and N. Cai.

[190] Estimating with randomized encoding the joint empirical distribution in a correlated source, (with Zhen Zhang), General Theory of Information Transfer and Combinatorics, Report on a Research Project at the ZIF (Center of interdisciplinary studies) in Bielefeld Oct. 1, 2001 – August 31, 2004, edited by R. Ahlswede with the assistance of L. Bäumer and N. Cai, (Preliminary version: Worst case estimation of permutation invariant functions and identification via compressed data, (with Zhen Zhang), Preprint 97–005, SFB 343 "Diskrete Strukturen in der Mathematik", Universität Bielefeld).

[191] On attractive and friendly sets in sequence spaces, (with L.H. Khachatrian), General Theory of Information Transfer and Combinatorics, Report on a Research Project at the ZIF (Center of interdisciplinary studies) in Bielefeld Oct. 1, 2001 – August 31, 2004, edited by R. Ahlswede with the assistance of L. Bäumer and N. Cai.

[192] Information theoretic models in language evolution, (with E. Arikan, L. Bäumer and C. Deppe), General Theory of Information Transfer and Combinatorics, Report on a Research Project at the ZIF (Center of interdisciplinary studies) in Bielefeld Oct. 1, 2001 – August 31, 2004, edited by R. Ahlswede with the assistance of L. Bäumer and N. Cai. Accepted as Language Evolution and Information Theory in ISIT, Chicago June 27 - July 2, 2004.

[193] A fast suffix–sorting algorithm, (with B. Balkenhol, C. Deppe, and M. Fröhlich), General Theory of Information Transfer and Combinatorics, Report on a Research Project at the ZIF (Center of interdisciplinary studies) in Bielefeld Oct. 1, 2001 – August 31, 2004, edited by R. Ahlswede with the assistance of L. Bäumer and N. Cai.

[194] On partitions of a rectangle into rectangles with restricted number of cross sections, (with A. Yudin), General Theory of Information Transfer and Combinatorics, Report on a Research Project at the ZIF (Center of interdisciplinary studies) in Bielefeld Oct. 1, 2001 – August 31, 2004, edited by R. Ahlswede with the assistance of L. Bäumer and N. Cai.

[195] On concepts of performance parameters for channels, General Theory of Information Transfer and Combinatorics, Report on a Research Project at the ZIF (Center of interdisciplinary studies) in Bielefeld Oct. 1, 2001 – August 31, 2004, edited by R. Ahlswede with the assistance of L. Bäumer and N. Cai, (Original version: Concepts of performance parameters for channels, Preprint 00–126, SFB 343 "Diskrete Strukturen in der Mathematik", Universität Bielefeld).

[196] Report on models of write–efficient memories with localized errors and defects, (with M.S. Pinsker), General Theory of Information Transfer and Combinatorics, Report on a Research Project at the ZIF (Center of interdisciplinary studies) in Bielefeld Oct. 1, 2001 – August 31, 2004, edited by R. Ahlswede with the assistance of L. Bäumer and N. Cai, (Original version: Report on models of write–efficient memories with localized errors and defects, (with M.S. Pinsker), Preprint 97–004, Ergänzungsreihe, SFB 343 "Diskrete Strukturen in der Mathematik", Universität Bielefeld).

[197] Correlation inequalities in function spaces, (with V. Blinovsky), General Theory of Information Transfer and Combinatorics, Report on a Research Project at the ZIF (Center of interdisciplinary studies) in Bielefeld Oct. 1, 2001 – August 31, 2004, edited by R. Ahlswede with the assistance of L. Bäumer and N. Cai.

[198] Solution of Burnashev's Problem and a sharpening of the Erdős/Ko/Rado Theorem, General Theory of Information Transfer and Combinatorics, Report on a Research Project at the ZIF (Center of interdisciplinary studies) in Bielefeld Oct. 1, 2001 – August 31, 2004, edited by R. Ahlswede with the assistance of L. Bäumer and N. Cai.

[199] Transmission, identification and common randomness capacities for wire-tape channels with secure feedback from the decoder, (with Ning Cai), General Theory of Information Transfer and Combinatorics, Report on a Research Project at the ZIF (Center of interdisciplinary studies) in Bielefeld Oct. 1, 2001 – August 31, 2004, edited by R. Ahlswede with the assistance of L. Bäumer and N. Cai.

[200] On set coverings in Cartesian product spaces, General Theory of Information Transfer and Combinatorics, Report on a Research Project at the ZIF (Center of interdisciplinary studies) in Bielefeld Oct. 1, 2001 – August 31, 2004, edited by R. Ahlswede with the assistance of L. Bäumer and N. Cai.

[201] Identification for sources, (with B. Balkenhol and C. Kleinewächter), General Theory of Information Transfer and Combinatorics, Report on a Research Project at the ZIF (Center of interdisciplinary studies) in Bielefeld Oct. 1, 2001 – August 31, 2004, edited by R. Ahlswede with the assistance of L. Bäumer and N. Cai, (Original version: Identification for sources, (with B. Balkenhol and C. Kleinewächter), Preprint 00–120, SFB 343 "Diskrete Strukturen in der Mathematik", Universität Bielefeld).

[202] Secrecy Systems for Identification Via Channels with Additive-Like Instantaneous Block Encipherers, (with Ning Cai and Zhaozhi Zhang), General Theory of Information Transfer and Combinatorics, Report on a Research Project at the ZIF (Center of interdisciplinary studies) in Bielefeld Oct. 1, 2001 – August 31, 2004, edited by R. Ahlswede with the assistance of L. Bäumer and N. Cai.

[203] Identification Entropy, General Theory of Information Transfer and Combinatorics, Report on a Research Project at the ZIF (Center of interdisciplinary studies) in Bielefeld Oct. 1, 2001 – August 31, 2004, edited by R. Ahlswede with the assistance of L. Bäumer and N. Cai.

[204] On Logarithmically Asymptotically Optimal Hypothesis Testing for Arbitrarily Varying Sources with Side Information, (with Evgueni Haroutunian and Ella Aloyan), General Theory of Information Transfer and Combinatorics, Report on a Research Project at the ZIF (Center of interdisciplinary studies) in Bielefeld Oct. 1, 2001 – August 31, 2004, edited by R. Ahlswede with the assistance of L. Bäumer and N. Cai.

[205] On Logarithmically Asymptotically Optimal Testing of Hypothesis and Identification, (with Evgueni Haroutunian), General Theory of Information Transfer and Combinatorics, Report on a Research Project at the ZIF (Center of interdisciplinary studies) in Bielefeld Oct. 1, 2001 – August 31, 2004, edited by R. Ahlswede with the assistance of L. Bäumer and N. Cai.

[206] Problems in Network coding and error correcting codes, (with S. Riis), General Theory of Information Transfer and Combinatorics, Report on a Research Project at the ZIF (Center of interdisciplinary studies) in Bielefeld Oct. 1, 2001 – August 31, 2004, edited by R. Ahlswede with the assistance of L. Bäumer and N. Cai.

[207] On edge-isoperimetric theorems for uniform hypergraphs, (with N. Cai), General Theory of Information Transfer and Combinatorics, Report on a Research Project at the ZIF (Center of interdisciplinary studies) in Bielefeld Oct. 1, 2001 – August 31, 2004, edited by R. Ahlswede with the assistance of L. Bäumer and N. Cai.

to appear

[208] General theory of information transfer: updated, General Theory of Information Transfer and Combinatorics, a Special issue of Discrete Applied Mathematics, (Original version: General theory of information transfer, Preprint 97–118, SFB 343 "Diskrete Strukturen in der Mathematik", Universität Bielefeld).

[209] Searching with lies under error transition cost constraints, (with F. Cicalese and C. Deppe), General Theory of Information Transfer and Combinatorics, a Special issue of Discrete Applied Mathematics.

[210] On diagnosability of large multiprocessor networks, (with H. Aydinian), General Theory of Information Transfer and Combinatorics, a Special issue of Discrete Applied Mathematics.

[211] Rate–wise optimal non–sequential search strategies under a cardinality constraint on the tests, General Theory of Information Transfer and Combinatorics, a Special issue of Discrete Applied Mathematics.

[212] On the correlation of binary sequences, (with J. Cassaigne and A. Sárközy), General Theory of Information Transfer and Combinatorics, a Special issue of Discrete Applied Mathematics.

[213] Multiple packing in sum-type metric spaces, (with V. Blinovsky), General Theory of Information Transfer and Combinatorics, a Special issue of Discrete Applied Mathematics.

[214] On q–ary codes correcting all unidirectional errors of a limited magnitude, (with H. Aydinian, L.H. Khachatrian, L.M. Tolhuizen), Special issue dedicated to the memory of Varshamov, Abstract included in Proceedings of the International workshop on Algebraic and Combinatorial Coding Theory (ACCT), Kranevo, Bulgaria, June 19 - 25, 2004.

[215] About the number of step functions with restrictions, (with V. Blinovsky), Probability Theory and Applications.

[216] Construction of asymmetric connectors of depth two, (with H. Aydinian) Special issue of J. Combinatorial Theory, Series A in memory of Jacobus H. van Lint.

[217] An interpretation of identification entropy, (with N. Cai), IEEE Trans. Inf. Theory.

[218] Maximal sets of integers not containing $k + 1$ pairwise coprimes and having divisors from a specified set of primes, (with V. Blinovsky), Special Issue of J. Combinatorial Theory, Series A, in memory of Jacobus H. van Lint.

submitted

[219] T–Shift synchronization Codes", (with B. Balkenhol, C. Deppe, H. Mashurian and T. Partner), General Theory of Information Transfer and Combinatorics, a Special issue of Discrete Applied Mathematics.

[220] On switching for hypergraphs, (with V. Blinovsky), European J. of Combinatorics.

[221] The final form of Tao's inequality relating conditional expectation and conditional mutual information, IEEE Trans. Inf. Theory.

[222] Another diametric theorem in Hamming spaces: optimal group anticodes, Proc. Information Theory Workshop, Punta del Este, Uruguay, March 13-17, 2006.

[223] Maximal sets of ideals without coprimes, (with V. Blinovsky), J. Number Theory.

[224] Capacity of quantum arbitrarily varying channels, (with V. Blinovsky), IEEE Trans. Inf. Theory.

[225] Error control codes for parallel asymmetric channels, (with H. Aydinian), IEEE Trans. Inf. Theory.

[226] Nonbinary error correcting codes with noiseless feedback, localized errors or both, (with C. Deppe and V. Lebedev), Annals of European Academy of Sciences.

Author Index

Lecture Notes in Computer Science

For information about Vols. 1–4238

please contact your bookseller or Springer

Vol. 4279: N. Kobayashi (Ed.), Programming Languages and Systems. XI, 423 pages. 2006.

Vol. 4278: R. Meersman, Z. Tari, P. Herrero (Eds.), On the Move to Meaningful Internet Systems 2006: OTM 2006 Workshops, Part II. XLV, 1004 pages. 2006.

Vol. 4277: R. Meersman, Z. Tari, P. Herrero (Eds.), On the Move to Meaningful Internet Systems 2006: OTM 2006 Workshops, Part I. XLV, 1009 pages. 2006.

Vol. 4276: R. Meersman, Z. Tari (Eds.), On the Move to Meaningful Internet Systems 2006: CoopIS, DOA, GADA, and ODBASE, Part II. XXXII, 752 pages. 2006.

Vol. 4275: R. Meersman, Z. Tari (Eds.), On the Move to Meaningful Internet Systems 2006: CoopIS, DOA, GADA, and ODBASE, Part I. XXXI, 1115 pages. 2006.

Vol. 4274: Q. Huo, B. Ma, E.-S. Chng, H. Li (Eds.), Chinese Spoken Language Processing. XXIV, 805 pages. 2006. (Sublibrary LNAI).

Vol. 4273: I. Cruz, S. Decker, D. Allemang, C. Preist, D. Schwabe, P. Mika, M. Uschold, L. Aroyo (Eds.), The Semantic Web - ISWC 2006. XXIV, 1001 pages. 2006.

Vol. 4272: P. Havinga, M. Lijding, N. Meratnia, M. Wegdam (Eds.), Smart Sensing and Context. XI, 267 pages. 2006.

Vol. 4271: F.V. Fomin (Ed.), Graph-Theoretic Concepts in Computer Science. XIII, 358 pages. 2006.

Vol. 4270: H. Zha, Z. Pan, H. Thwaites, A.C. Addison, M. Forte (Eds.), Interactive Technologies and Sociotechnical Systems. XVI, 547 pages. 2006.

Vol. 4269: R. State, S. van der Meer, D. O'Sullivan, T. Pfeifer (Eds.), Large Scale Management of Distributed Systems. XIII, 282 pages. 2006.

Vol. 4268: G. Parr, D. Malone, M. Ó Foghlú (Eds.), Autonomic Principles of IP Operations and Management. XIII, 237 pages. 2006.

Vol. 4267: A. Helmy, B. Jennings, L. Murphy, T. Pfeifer (Eds.), Autonomic Management of Mobile Multimedia Services. XIII, 257 pages. 2006.

Vol. 4266: H. Yoshiura, K. Sakurai, K. Rannenberg, Y. Murayama, S. Kawamura (Eds.), Advances in Information and Computer Security. XIII, 438 pages. 2006.

Vol. 4265: L. Todorovski, N. Lavrač, K.P. Jantke (Eds.), Discovery Science. XIV, 384 pages. 2006. (Sublibrary LNAI).

Vol. 4264: J.L. Balcázar, P.M. Long, F. Stephan (Eds.), Algorithmic Learning Theory. XIII, 393 pages. 2006. (Sublibrary LNAI).

Vol. 4263: A. Levi, E. Savaş, H. Yenigün, S. Balcısoy, Y. Saygın (Eds.), Computer and Information Sciences – ISCIS 2006. XXIII, 1084 pages. 2006.

Vol. 4262: K. Havelund, M. Núñez, B. Wolff, G. Roşu (Eds.), Formal Approaches to Software Testing and Runtime Verification. VIII, 255 pages. 2006.

Vol. 4261: Y. Zhuang, S. Yang, Y. Rui, Q. He (Eds.), Advances in Multimedia Information Processing - PCM 2006. XXII, 1040 pages. 2006.

Vol. 4260: Z. Liu, J. He (Eds.), Formal Methods and Software Engineering. XII, 778 pages. 2006.

Vol. 4259: S. Greco, Y. Hata, S. Hirano, M. Inuiguchi, S. Miyamoto, H.S. Nguyen, R. Słowiński (Eds.), Rough Sets and Current Trends in Computing. XXII, 951 pages. 2006. (Sublibrary LNAI).

Vol. 4257: I. Richardson, P. Runeson, R. Messnarz (Eds.), Software Process Improvement. XI, 219 pages. 2006.

Vol. 4256: L. Feng, G. Wang, C. Zeng, R. Huang (Eds.), Web Information Systems – WISE 2006 Workshops. XIV, 320 pages. 2006.

Vol. 4255: K. Aberer, Z. Peng, E.A. Rundensteiner, Y. Zhang, X. Li (Eds.), Web Information Systems – WISE 2006. XIV, 563 pages. 2006.

Vol. 4254: T. Grust, H. Höpfner, A. Illarramendi, S. Jablonski, M. Mesiti, S. Müller, P.-L. Patranjan, K.-U. Sattler, M. Spiliopoulou, J. Wijsen (Eds.), Current Trends in Database Technology – EDBT 2006. XXXI, 932 pages. 2006.

Vol. 4253: B. Gabrys, R.J. Howlett, L.C. Jain (Eds.), Knowledge-Based Intelligent Information and Engineering Systems, Part III. XXXII, 1301 pages. 2006. (Sublibrary LNAI).

Vol. 4252: B. Gabrys, R.J. Howlett, L.C. Jain (Eds.), Knowledge-Based Intelligent Information and Engineering Systems, Part II. XXXIII, 1335 pages. 2006. (Sublibrary LNAI).

Vol. 4251: B. Gabrys, R.J. Howlett, L.C. Jain (Eds.), Knowledge-Based Intelligent Information and Engineering Systems, Part I. LXVI, 1297 pages. 2006. (Sublibrary LNAI).

Vol. 4250: H.J. van den Herik, S.-C. Hsu, T.-s. Hsu, H.H.L.M. Donkers (Eds.), Advances in Computer Games. XIV, 273 pages. 2006.

Vol. 4249: L. Goubin, M. Matsui (Eds.), Cryptographic Hardware and Embedded Systems - CHES 2006. XII, 462 pages. 2006.

Vol. 4248: S. Staab, V. Svátek (Eds.), Managing Knowledge in a World of Networks. XIV, 400 pages. 2006. (Sublibrary LNAI).

Vol. 4247: T.-D. Wang, X. Li, S.-H. Chen, X. Wang, H. Abbass, H. Iba, G. Chen, X. Yao (Eds.), Simulated Evolution and Learning. XXI, 940 pages. 2006.

Vol. 4246: M. Hermann, A. Voronkov (Eds.), Logic for Programming, Artificial Intelligence, and Reasoning. XIII, 588 pages. 2006. (Sublibrary LNAI).

Vol. 4245: A. Kuba, L.G. Nyúl, K. Palágyi (Eds.), Discrete Geometry for Computer Imagery. XIII, 688 pages. 2006.

Vol. 4244: S. Spaccapietra (Ed.), Journal on Data Semantics VII. XI, 267 pages. 2006.

Vol. 4243: T. Yakhno, E.J. Neuhold (Eds.), Advances in Information Systems. XIII, 420 pages. 2006.

Vol. 4242: A. Rashid, M. Aksit (Eds.), Transactions on Aspect-Oriented Software Development II. IX, 289 pages. 2006.

Vol. 4241: R.R. Beichel, M. Sonka (Eds.), Computer Vision Approaches to Medical Image Analysis. XI, 262 pages. 2006.

Vol. 4239: H.Y. Youn, M. Kim, H. Morikawa (Eds.), Ubiquitous Computing Systems. XVI, 548 pages. 2006.